Honoring America

For Americans, the flag has always had a special meaning. It is a symbol of our nation's freedom and democracy.

Flag Etiquette

Over the years, Americans have developed rules and customs concerning the use and display of the flag. One of the most important things every American should remember is to treat the flag with respect.

- The flag should be raised and lowered by hand and displayed only from sunrise to sunset. On special occasions, the flag may be displayed at night, but it should be illuminated.
- The flag may be displayed on all days, weather permitting, particularly on national and state holidays and on historic and special occasions.
- No flag may be flown above the American flag or to the right of it at the same height.
- The flag should never touch the ground or floor beneath it.

- The flag may be flown at half-staff by order of the president, usually to mourn the death of a public official.
- The flag may be flown upside down only to signal distress.
- The flag should never be carried flat or horizontally, but always carried aloft and free.
- When the flag becomes old and tattered, it should be destroyed by burning. According to an approved custom, the Union (stars on blue field) is first cut from the flag; then the two pieces, which no longer form a flag, are burned.

★ ★ ★ ★ ★ ★ ★ ★

The American's Creed

I believe in the United States of America as a Government of the people, by the people, for the people, whose just powers are derived from the consent of the governed; a democracy in a republic; a sovereign Nation of many sovereign States; a perfect union, one and inseparable; established upon those principles of freedom, equality, justice, and humanity for which American patriots sacrificed their lives and fortunes.

I therefore believe it is my duty to my Country to love it; to support its Constitution; to obey its laws; to respect its flag, and to defend it against all enemies.

The Pledge of Allegiance

I pledge allegiance to the Flag of the United States of America and to the Republic for which it stands, one Nation under God, indivisible, with liberty and justice for all.

The Star-Spangled Banner

O! say, can you see, by the dawn's early light,
What so proudly we hail'd at the twilight's last gleaming?
Whose broad stripes and bright stars, thro' the perilous fight,
O'er the ramparts we watched were so gallantly streaming?
And the rockets' red glare, the bombs bursting in air,
Gave proof thro' the night, that our flag was still there.
O! say, does that Star-Spangled Banner yet wave
O'er the land of the free and the home of the brave?

On the shore, dimly seen thro' the mist of the deep,
Where the foe's haughty host in dread silence reposes,
What is that which the breeze, o'er the towering steep,
As it fitfully blows, half conceals, half discloses?
Now it catches the gleam of the morning's first beam,
In full glory reflected now shines on the stream.
'Tis the Star-Spangled Banner. O long may it wave
O'er the land of the free and the home of the brave.

And where is that band who so vauntingly swore,
That the havoc of war and the battle's confusion
A home and a country should leave us no more?
Their blood has wash'd out their foul footstep's pollution.
No refuge could save the hireling and slave
From the terror of flight or the gloom of the grave,
And the Star-Spangled Banner in triumph doth wave
O'er the land of the free and the home of the brave.

O thus be it e'er when free men shall stand
Between their lov'd home and war's desolation,
Blest with vict'ry and peace, may the Heav'n-rescued land
Praise the pow'r that hath made and preserv'd us a nation.
Then conquer we must, when our cause it is just,
And this be our motto, "In God is our Trust."
And the Star-Spangled Banner in triumph shall wave
O'er the land of the free and the home of the brave.

Glencoe

World Geography

Glencoe

World Geography

Senior Author

Richard G. Boehm, Ph.D.

New York, New York Columbus, Ohio Chicago, Illinois Peoria, Illinois Woodland Hills, California

About the Authors

Senior Author
RICHARD G. BOEHM

Richard G. Boehm, Ph.D., was one of seven authors of *Geography for Life*, national standards in geography, prepared under Goals 2000: Educate America Act. He was also one of the authors of the *Guidelines for Geographic Education*, in which the five themes of geography were first articulated. In 1990, Dr. Boehm was designated "Distinguished Geography Educator" by the National Geographic Society. In 1991 he received the George J. Miller award from the National Council for Geographic Education (NCGE) for distinguished service to geographic education. He was President of the NCGE and has twice won the *Journal of Geography* award for best article. He has received the NCGE's "Distinguished Teaching Achievement" award and presently holds the Jesse H. Jones Distinguished Chair in Geographic Education at Southwest Texas State University in San Marcos, Texas.

NATIONAL GEOGRAPHIC SOCIETY

The National Geographic Society, founded in 1888 for the increase and diffusion of geographic knowledge, is the world's largest nonprofit scientific and educational organization. Since its earliest days, the Society has used sophisticated communication technologies, from color photography to holography, to convey geographic knowledge to a worldwide membership. The School Publishing Division supports the Society's mission by developing innovative educational programs—ranging from traditional print materials to multimedia programs including CD-ROMs, videos, and software.

Contributing Writer

Bob Haddad, ethnomusicologist, developed the World Music features in the Student and Teacher Wraparound editions of *Glencoe World Geography.* He also chose the music selections for Glencoe's *World Music: A Cultural Legacy* program and authored the accompanying teacher guide. Mr. Haddad is founder of Music of the World and president of Owl's Head Music.

Glencoe/McGraw-Hill

A Division of The ***McGraw·Hill*** *Companies*

Send all inquiries to:
Glencoe/McGraw-Hill
8787 Orion Place
Columbus, Ohio 43240-4027

ISBN 0-02-664173-9 (Student Edition) ISBN 0-07-824931-7 (Teacher Wraparound Edition)

Printed in the United States of America

2 3 4 5 6 7 8 9 071/043 06 05 04 03 02

Consultants

General Content Consultant
KAY E. WELLER, Ph.D.
Assistant Professor of Geography
University of Northern Iowa
Cedar Falls, Iowa

Geography Consultant
SARI J. BENNETT, Ph.D.
Director of Geographic Education
University of Maryland, Baltimore County
Baltimore, Maryland

Cultural Consultant
JOSEPH P. STOLTMAN, Ph.D.
Professor of Geography
Western Michigan University
Kalamazoo, Michigan

Religion Consultants
CHARLES H. LONG, Ph.D.
Professor of Religion (Retired)
University of California
Santa Barbara, California

BLUMA ZUCKERBROT-FINKELSTEIN
Instructor, University of Memphis
Memphis, Tennessee

SHABBIR MANSURI
Founding Director
SUSAN L. DOUGLASS
Affiliated Scholar
Council on Islamic Education
Fountain Valley, California

Introduction to Geography
BURRELL E. MONTZ, Ph.D.
Professor of Geography and Environmental Studies
Binghamton University
Binghamton, New York

United States and Canada
JOSEPH D. ENEDY, Ph.D.
Professor of Geography
James Madison University
Harrisonburg, Virginia

Latin America
CYRUS B. DAWSEY, Ph.D.
Professor of Geography
Auburn University
Auburn, Alabama

Europe
FIONA M. DAVIDSON, Ph.D.
Associate Professor of Geography
University of Arkansas
Fayetteville, Arkansas

DOUGLAS C. WILMS, Ph.D.
Professor Emeritus of Geography
East Carolina University
Greenville, North Carolina

Russia
GARY J. HAUSLADEN, Ph.D.
Associate Professor of Geography
University of Nevada
Reno, Nevada

OLGA MEDVEDKOV, Ph.D.
Professor of Geography
Wittenberg University
Springfield, Ohio

North Africa, Southwest Asia, and Central Asia
NORMAN BETTIS, Ph.D.
Professor of Education
Illinois State University
Normal, Illinois

MOUNIR A. FARAH, Ph.D.
Associate Director of Middle Eastern Studies
University of Arkansas
Fayetteville, Arkansas

JOSEPH HOBBS, Ph.D.
Professor of Geography
University of Missouri
Columbia, Missouri

Africa South of the Sahara
EZEKIAL KALIPENI, Ph.D.
Associate Professor of Geography
University of Illinois
Urbana, Illinois

GARTH A. MYERS, Ph.D.
Assistant Professor of African & African-American Studies, and Geography
University of Kansas
Lawrence, Kansas

South Asia
WILLIAM R. STRONG, Ph.D.
Professor of Geography
University of North Alabama
Florence, Alabama

East Asia
KENJI K. OSHIRO, Ph.D.
Professor of Urban Affairs and Geography
Wright State University
Dayton, Ohio

SUSAN M. WALCOTT, Ph.D.
Assistant Professor of Anthropology and Geography
Georgia State University
Atlanta, Georgia

Southeast Asia
JON D. GOSS, Ph.D.
Associate Professor of Geography
University of Hawaii
Honolulu, Hawaii

RALPH LENZ, Ph.D.
Professor of Geography
Wittenberg University
Springfield, Ohio

Australia, Oceania, and Antarctica
ALYSON L. GREINER, Ph.D.
Assistant Professor of Geography
Oklahoma State University
Stillwater, Oklahoma

LAURIE MOLINA, Ph.D.
Program Director, Geography Education and Technology
Florida State University
Tallahassee, Florida

Teacher Reviewers

ROBERT M. ASHLEY
Red Bud High School
Red Bud, Illinois

NATE COLLINS
Buffalo Gap High School
Swoope, Virginia

PAUL T. GRAY, JR.
Russellville High School
Russellville, Arkansas

LINDA L. HAMMON
Charles W. Akins High School
Austin, Texas

MARK C. JONES
St. Luke's School
New Canaan, Connecticut

DANIEL J. LANGEN
Oak Hills High School
Cincinnati, Ohio

DONNA MERLAU
Onondaga Central Junior-Senior High School
Nedrow, New York

PAUL RANTA
North Garland High School
Garland, Texas

BETTY SCOOPMIRE
J.H. High School
Greenville, North Carolina

CHERIE VELA
ABC Secondary School
Cerritos, California

FRED WALK
Normal Community High School
Normal, Illinois

AMY ZORN
Anna High School
Anna, Ohio

Contents

Unit 6

Unit 7

Features

GLOBAL CONNECTION

GEOGRAPHY AND HISTORY

NATIONAL GEOGRAPHIC

Viewpoint

CASE STUDY on the Environment

SKILLBUILDER

MAP & GRAPH SKILLBUILDER

CRITICAL THINKING SKILLBUILDER

TECHNOLOGY SKILLBUILDER

STUDY & WRITING SKILLBUILDER

Geography Lab Activities

WORLD CULTURE

Maps

Reference Atlas

Geography Skills Handbook

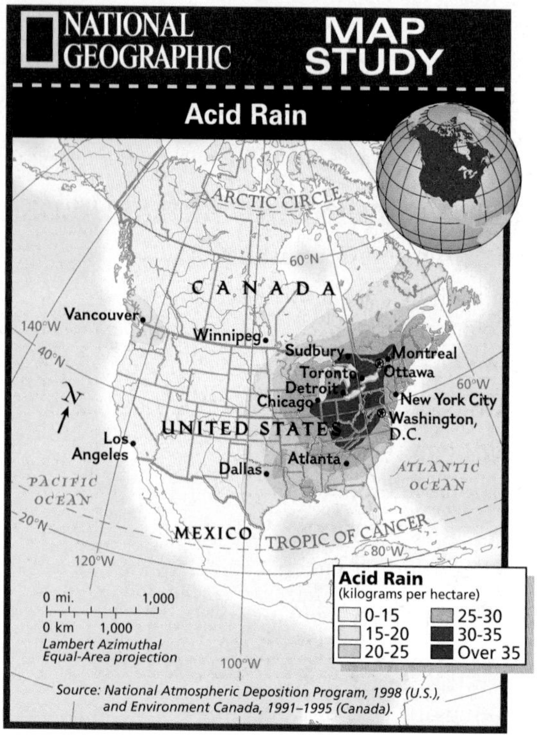

NATIONAL GEOGRAPHIC MAP STUDY

Acid Rain

Source: National Atmospheric Deposition Program, 1998 (U.S.), and Environment Canada, 1991–1995 (Canada).

UNIT 1

The World

UNIT 2

The United States and Canada

UNIT 3

Latin America

UNIT 4

Europe

UNIT 5

Russia

UNIT 9
East Asia

UNIT 10
Southeast Asia

UNIT 11
Australia, Oceania, and Antarctica

Graphs, Charts, and Diagrams

NATIONAL GEOGRAPHIC DIAGRAM STUDY

Inside the Earth

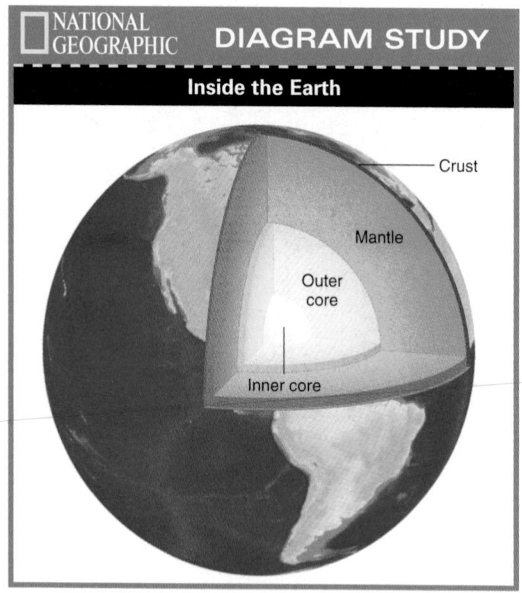

NATIONAL GEOGRAPHIC GRAPH STUDY

World Trade

$billion GDP
$billion Trade
40,000
35,000
30,000
25,000
20,000
15,000
10,000
5000
$bn 0
7000
6000
5000
4000
3000
2000
1000
0 $bn
World GDP
World Trade
1980
1990
2000

Gross domestic product (GDP) is the value of goods and services produced within a country in a year.

Sources: World Trade Organization; IMF World Economics Outlook, 2001

NATIONAL GEOGRAPHIC **CHART STUDY**

Jobs in Geography

Geography Field	Description	Applications/Careers
Physical Geographer	Studies Earth's features and the geographic forces shaping them	Forecasting weather, tracing causes and effects of pollution, conserving wilderness areas
Human Geographer	Analyzes human aspects of culture—population, language, ethnicity, religion, government	Developing cultural policies for international organizations, such as the United Nations
Economic Geographer	Examines human economic activities and their relationship to the environment	Urban planning, focusing on the location of industries or transportation routes
Regional Geographer	Studies geographic features of a particular place or region	Assisting government and business in making decisions related to a region
Environmental Specialist	Focuses on all aspects of the environment	Advising government and business on ways of protecting the environment
Geographic Educator	Teaches about geography	Teaching geography at all educational levels; serving as consultant to business and government

Primary Sources

NG = National Geographic

Geography Skills for Life

If you think that geography means memorizing a list of states and their capitals, think again. Geography is a broad and ever-changing subject. It includes the study of Earth's physical features, as well as the countless and fascinating ways that humans, animals, and plants interact with the world around them.

To understand how our world is connected, some geographers have broken down the study of geography into five themes. The Five Themes of Geography are (1) location, (2) place, (3) human-environment interaction, (4) movement, and (5) regions.

Most recently, geographers have begun to look at geography in a different way. Geography educators have created a set of eighteen learning standards called *Geography for Life.* Each of these eighteen standards is organized into six essential elements, which are explained for you below. Being aware of these elements will help you sort out what you are learning about geography.

1 The World in Spatial Terms

Each time you tell a classmate how to get to your home or give directions to a school visitor, you use geography. You are thinking about places in terms of their location in space. Knowing how to read maps, give directions, and create your own mental maps of spaces around you will help you throughout life as you travel down new paths and into unfamiliar locations.

◄ Street signs on Monhegan Island, Maine, U.S.

2 Places and Regions

Are you a Texan? How about a city dweller, a beachcomber, or a Midwesterner? Places and regions exert such a powerful force in our lives that many people define themselves in terms of a specific place or region. Places and regions impact how people live, their career choices, culture, language, and even their view of the world. Learning more about places and regions will build your understanding and respect for the similarities and differences among the world's people.

▼ Woman weaving decorative cloth, Antigua, Guatemala

▲ Satellite photo of tropical storm Irene, south of Cuba

3 Physical Systems

When a skateboarder checks the weather forecast before heading outside for the nearest hill, she uses geographic information. Earth's physical processes, including climate, erosion, and earthquakes, shape the pattern of Earth's surface. Understanding the past and future effects of Earth's forces may someday guide your choice of where and how to live.

▼ Curbside recycling in Palm Springs, California, U.S.

4 Human Systems

Do you use a cell phone or a pager? Communications systems like these rely on geographic information to send and receive signals. How humans settle Earth, use resources, and build the transportation, communications, and economic systems that keep life going are part of the geography of modern society. As you become a contributing member of the workforce, you will help to shape this network of economic interdependence on Earth's surface.

▲ Passenger train in Leipzig, Germany

5 Environment and Society

Separating recyclables from the rest of your trash requires that you apply geographic perspectives in everyday life. As a responsible citizen of Earth, you will continually be called upon to weigh human needs against the limitations of the physical world. You will use geographic perspectives to make informed decisions about how best to protect Earth's physical environment.

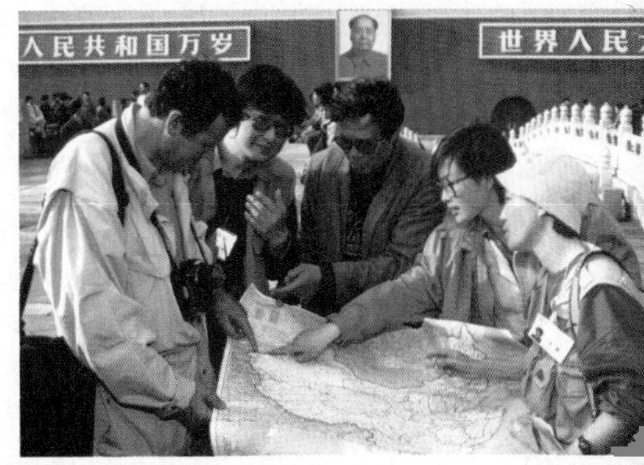

▲ Tourists in Tiananmen Square, Beijing, China

6 The Uses of Geography

In school you practice core geography skills as you study the history of Earth, read maps, and interpret historical events with respect to location. Your future use of geography may be less direct, but far more important. Applying geographic principles and perspectives will help you build a better future for the world's people and safeguard the best interests of our planet.

Reading for Information

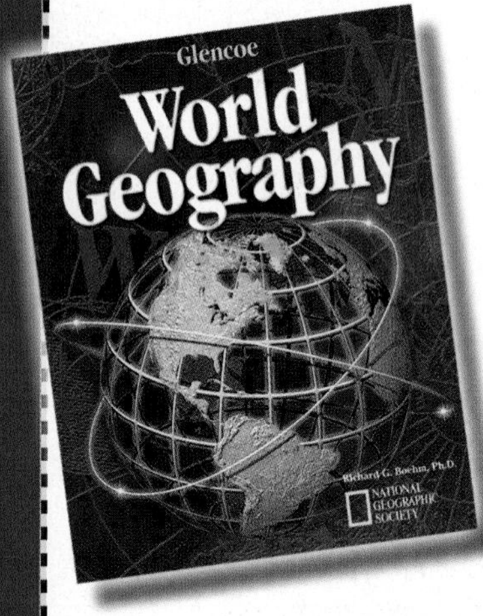

Think about your textbook as a tool that helps you learn more about the world around you. It is an example of nonfiction writing—it describes real-life events, people, ideas, and places. Here is a menu of reading strategies that will help you become a better textbook reader. As you come to passages in your textbook that you don't understand, refer to these reading strategies for help.

Before You Read

Set a Purpose

- Why are you reading the textbook?
- How does the subject relate to your life?
- How might you be able to use what you learn in your own life?

Preview

- Read the chapter title to find what the topic will be.
- Read the subtitles to see what you will learn about the topic.
- Skim the photos, charts, graphs, or maps. How do they support the topic?
- Look for vocabulary words that are boldfaced. How are they defined?

Draw From Your Own Background

- What have you read or heard about concerning new information on the topic?
- How is the new information different from what you already know?
- How will the information that you already know help you understand the new information?

As You Read

Question

- What is the main idea?
- How do the photos, charts, graphs, and maps support the main idea?

Connect

- Think about people, places, and events in your own life. Are there any similarities with those in your textbook?
- Can you relate the textbook information to other areas of your life?

Predict

- Predict events or outcomes by using clues and information that you already know.
- Change your predictions as you read and gather new information.

Visualize

- Pay careful attention to details and descriptions.
- Create graphic organizers to show relationships that you find in the information.

After You Read

Summarize

- Describe the main idea and how the details support it.
- Use your own words to explain what you have read.

Assess

- What was the main idea?
- Did the text clearly support the main idea?
- Did you learn anything new from the material?
- Can you use this new information in other school subjects or at home?
- What other sources could you use to find more information about the topic?

Look for Clues As You Read

- **Comparison-and-Contrast Sentences:**

  Look for clue words and phrases that signal comparison, such as *similarly, just as, both, in common, also,* and *too.*

  Look for clue words and phrases that signal contrast, such as *on the other hand, in contrast to, however, different, instead of, rather than, but,* and *unlike.*

- **Cause-and-Effect Sentences:**

  Look for clue words and phrases such as *because, as a result, therefore, that is why, since, so, for this reason,* and *consequently.*

- **Chronological Sentences:**

  Look for clue words and phrases such as *after, before, first, next, last, during, finally, earlier, later, since,* and *then.*

REFERENCE ATLAS

NATIONAL GEOGRAPHIC

ATLAS KEY

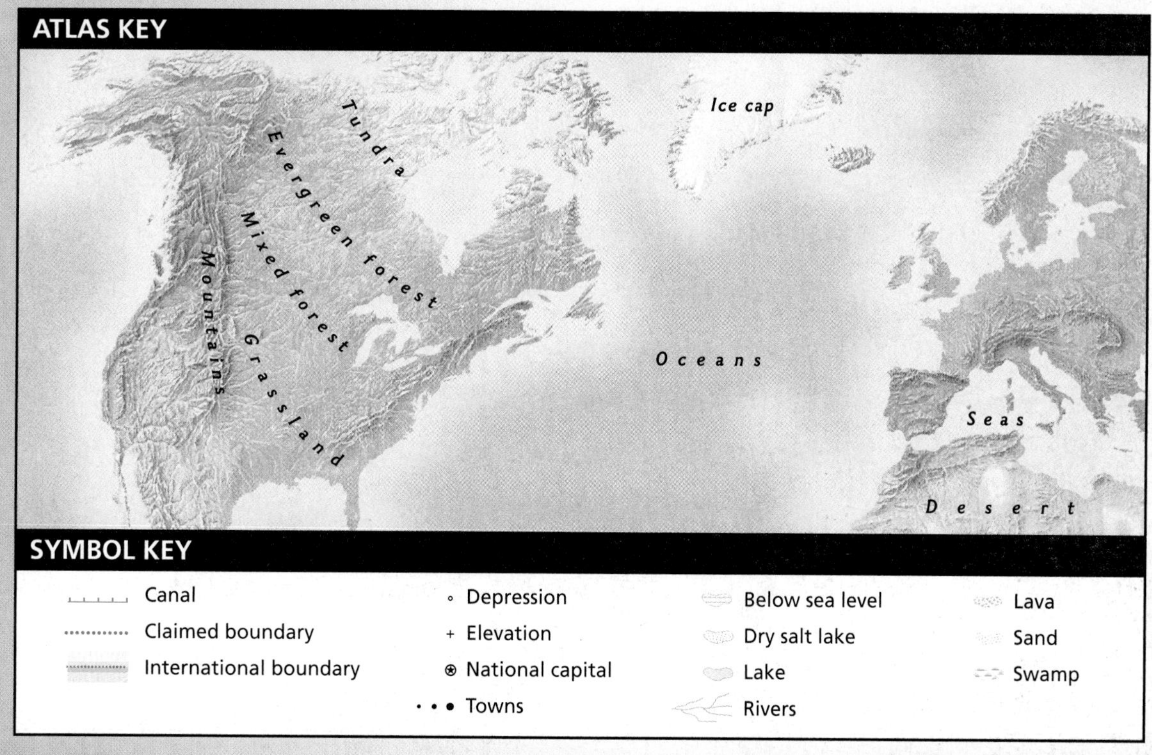

SYMBOL KEY

- Canal
- Claimed boundary
- International boundary
- Depression
- Elevation
- National capital
- Towns
- Below sea level
- Dry salt lake
- Lake
- Rivers
- Lava
- Sand
- Swamp

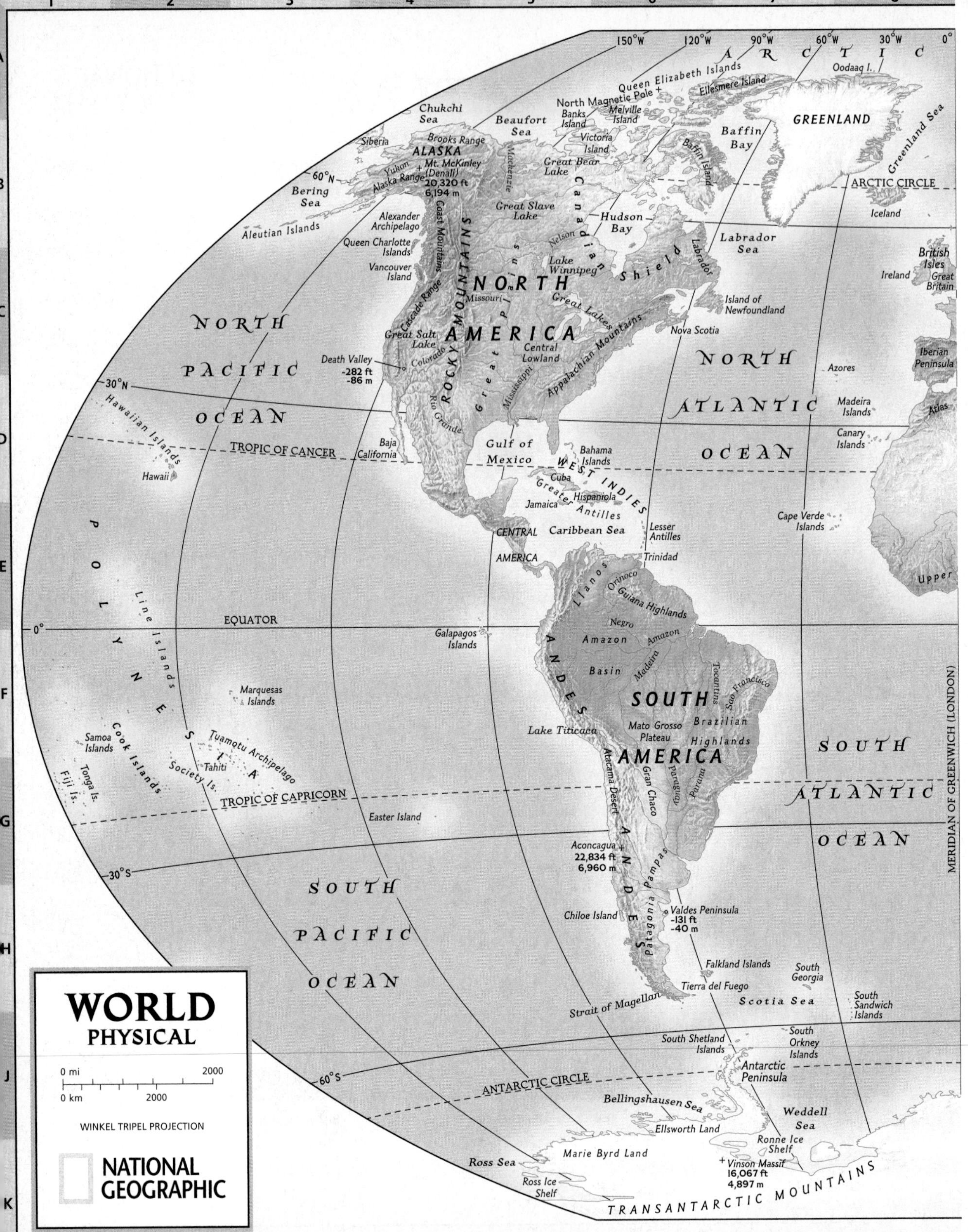
WORLD
PHYSICAL
WINKEL TRIPEL PROJECTION
0 mi 2000
0 km 2000
NATIONAL GEOGRAPHIC
NORTH AMERICA
SOUTH AMERICA
NORTH PACIFIC OCEAN
SOUTH PACIFIC OCEAN
NORTH ATLANTIC OCEAN
SOUTH ATLANTIC OCEAN
ARCTIC
GREENLAND
ALASKA
POLYNESIA
TROPIC OF CANCER
TROPIC OF CAPRICORN
EQUATOR
ARCTIC CIRCLE
ANTARCTIC CIRCLE
MERIDIAN OF GREENWICH (LONDON)
Mt. McKinley (Denali) 20,320 ft 6,194 m
Death Valley -282 ft -86 m
Aconcagua 22,834 ft 6,960 m
Valdes Peninsula -131 ft -40 m
Vinson Massif 16,067 ft 4,897 m
ROCKY MOUNTAINS
ANDES
TRANSANTARCTIC MOUNTAINS
Hudson Bay
Gulf of Mexico
Caribbean Sea
WEST INDIES
CENTRAL AMERICA
Amazon Basin
Brazilian Highlands
Canadian Shield
Great Plains
Great Lakes
Bering Sea
Labrador Sea
Scotia Sea
Weddell Sea
Bellingshausen Sea
Ross Sea

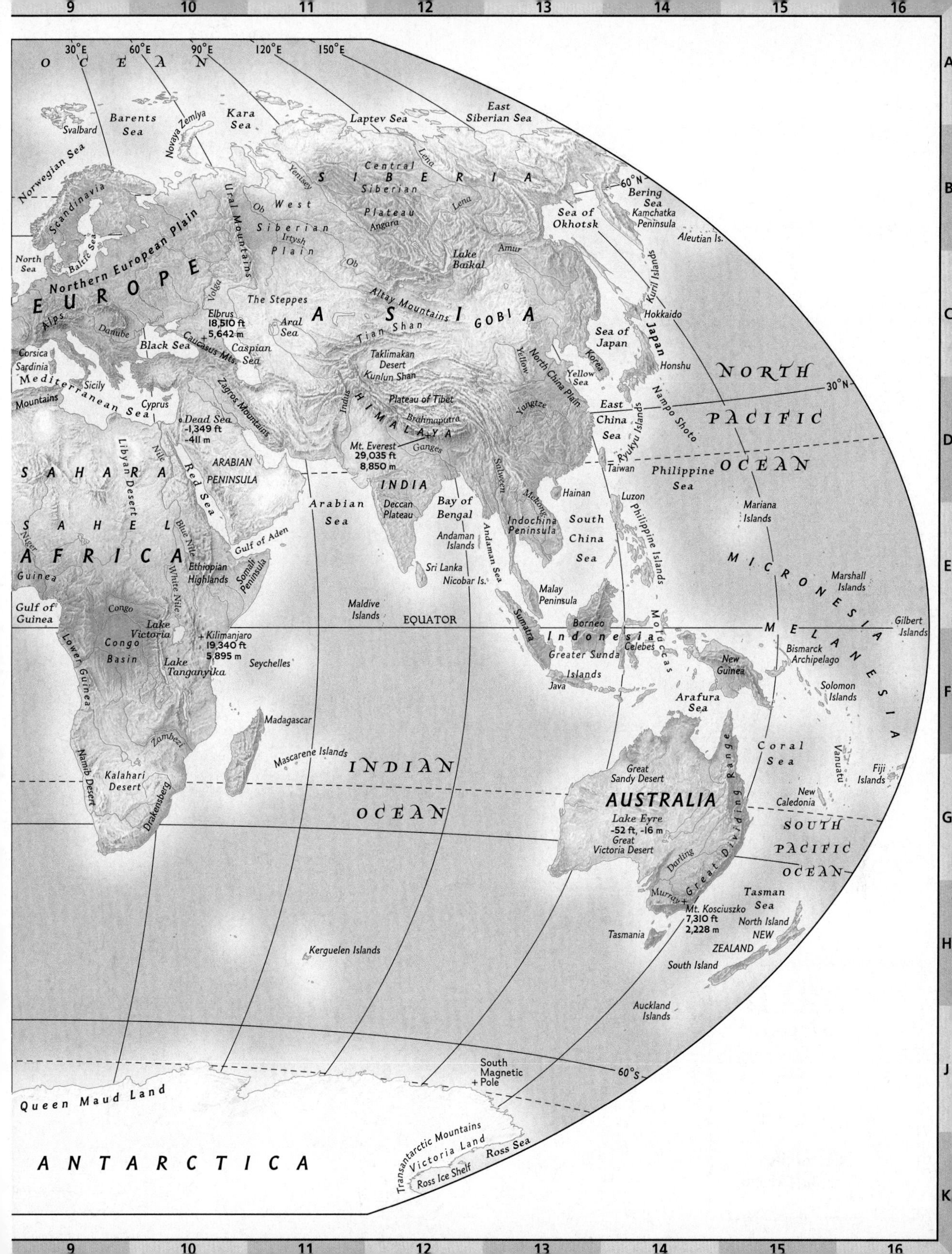

9
10
11
12
13
14
15
16
A
B
C
D
E
F
G
H
J
K
30°E
60°E
90°E
120°E
150°E
O C E A N
Svalbard
Barents Sea
Novaya Zemlya
Kara Sea
Laptev Sea
East Siberian Sea
Norwegian Sea
Lena
Yenisey
Central Siberian Plateau
S I B E R I A
60°N
Bering Sea
Kamchatka Peninsula
Aleutian Is.
Sea of Okhotsk
Scandinavia
Ural Mountains
Ob
West Siberian Plain
Irtysh
Angara
Lena
Lake Baikal
Amur
North Sea
Baltic Sea
Northern European Plain
EUROPE
Volga
Ob
Kuril Islands
The Steppes
Altay Mountains
Hokkaido
Alps
Elbrus 18,510 ft 5,642 m
Aral Sea
A S I A
GOBI A
Danube
Tian Shan
Sea of Japan
Japan
Black Sea
Caucasus Mts.
Caspian Sea
Taklimakan Desert
Yellow
North China Plain
Korea
Honshu
Corsica
Sardinia
Mediterranean Sea
Sicily
Kunlun Shan
Yellow Sea
NORTH
Mountains
Cyprus
Zagros Mountains
Indus
HIMALAYA
Plateau of Tibet
Brahmaputra
Yangtze
Nampo Shoto
30°N
Dead Sea -1,349 ft -411 m
East China Sea
Ryukyu Islands
PACIFIC
Mt. Everest 29,035 ft 8,850 m
Ganges
Nile
Libyan Desert
SAHARA
Red Sea
ARABIAN PENINSULA
Taiwan
Philippine Sea
OCEAN
INDIA
Salween
Hainan
Arabian Sea
Deccan Plateau
Bay of Bengal
Mekong
Luzon
Mariana Islands
SAHEL
Blue Nile
Indochina Peninsula
South China Sea
Philippine Islands
Niger
AFRICA
Gulf of Aden
Andaman Islands
Andaman Sea
MICRONESIA
White Nile
Ethiopian Highlands
Somali Peninsula
Sri Lanka
Nicobar Is.
Marshall Islands
Guinea
Malay Peninsula
Gulf of Guinea
Congo
Maldive Islands
EQUATOR
Borneo
Moluccas
Gilbert Islands
Lake Victoria
Kilimanjaro 19,340 ft 5,895 m
Sumatra
Indonesia
Celebes
MELANESIA
Congo Basin
Greater Sunda Islands
Bismarck Archipelago
Lake Tanganyika
Seychelles
New Guinea
Lower Guinea
Java
Solomon Islands
Arafura Sea
Madagascar
Zambezi
Mascarene Islands
INDIAN OCEAN
Great Dividing Range
Coral Sea
Vanuatu
Great Sandy Desert
Fiji Islands
Namib Desert
Kalahari Desert
Drakensberg
AUSTRALIA
New Caledonia
Lake Eyre -52 ft, -16 m
SOUTH PACIFIC OCEAN
Great Victoria Desert
Darling
Murray
Tasman Sea
Mt. Kosciuszko 7,310 ft 2,228 m
North Island
Tasmania
NEW ZEALAND
Kerguelen Islands
South Island
Auckland Islands
South Magnetic Pole
60°S
Queen Maud Land
Transantarctic Mountains
Victoria Land
Ross Sea
Ross Ice Shelf
ANTARCTICA

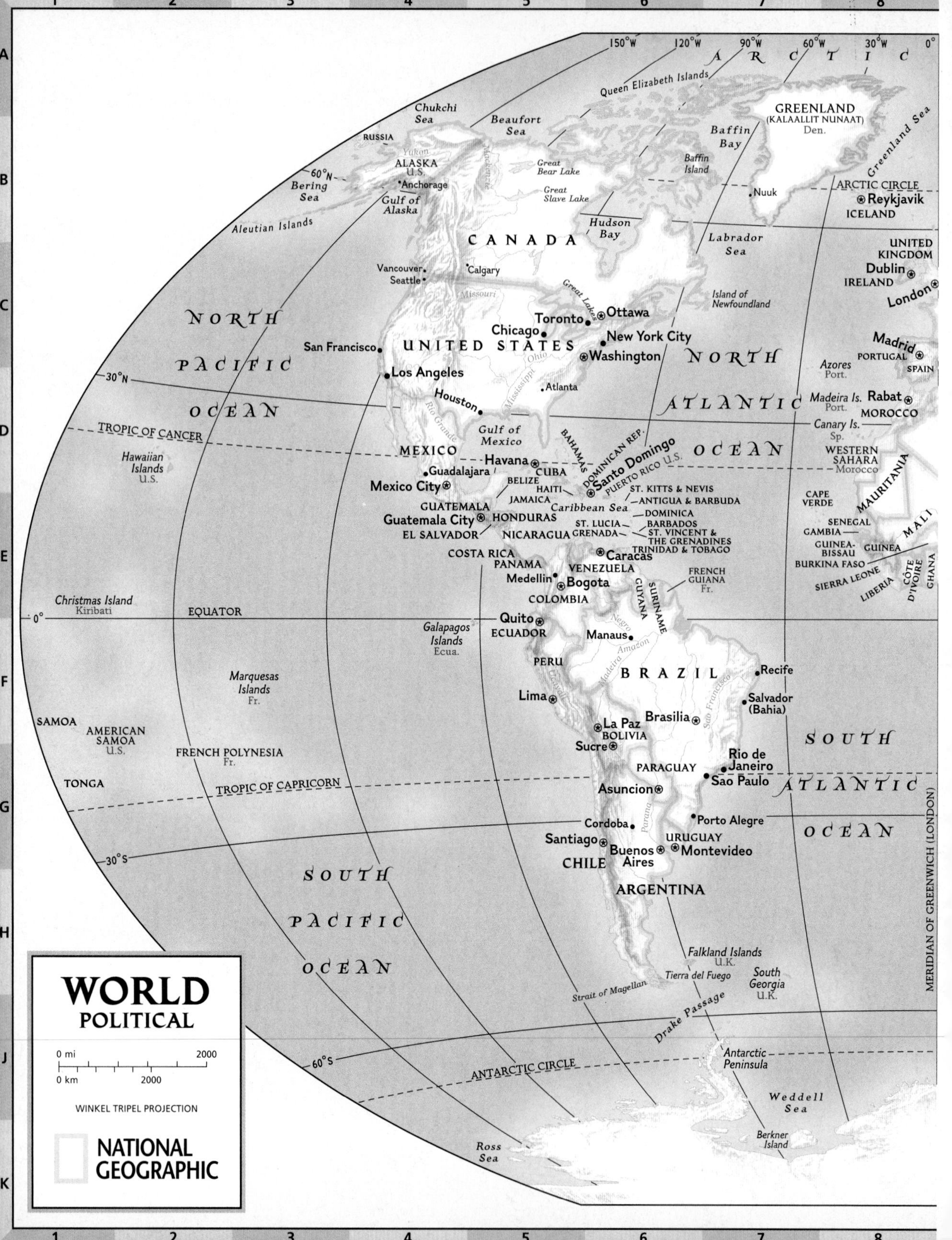

WORLD
POLITICAL
0 mi
2000
0 km
2000
WINKEL TRIPEL PROJECTION
NATIONAL GEOGRAPHIC
ARCTIC
NORTH PACIFIC OCEAN
SOUTH PACIFIC OCEAN
NORTH ATLANTIC OCEAN
SOUTH ATLANTIC OCEAN
150°W
120°W
90°W
60°W
30°W
0°
60°N
30°N
0°
30°S
60°S
ARCTIC CIRCLE
TROPIC OF CANCER
EQUATOR
TROPIC OF CAPRICORN
ANTARCTIC CIRCLE
MERIDIAN OF GREENWICH (LONDON)
Queen Elizabeth Islands
GREENLAND
(KALAALLIT NUNAAT)
Den.
Greenland Sea
Chukchi Sea
Beaufort Sea
RUSSIA
Yukon
ALASKA
U.S.
Anchorage
Bering Sea
Gulf of Alaska
Aleutian Islands
Mackenzie
Great Bear Lake
Great Slave Lake
Baffin Bay
Baffin Island
Nuuk
Reykjavik
ICELAND
Hudson Bay
Labrador Sea
CANADA
UNITED KINGDOM
Dublin
IRELAND
London
Vancouver
Seattle
Calgary
Missouri
Great Lakes
Island of Newfoundland
Ottawa
Toronto
Chicago
New York City
Washington
San Francisco
UNITED STATES
Ohio
Madrid
PORTUGAL
SPAIN
Azores
Port.
Los Angeles
Mississippi
Atlanta
Madeira Is.
Port.
Rabat
MOROCCO
Houston
Rio Grande
Gulf of Mexico
Canary Is.
Sp.
BAHAMAS
DOMINICAN REP.
Hawaiian Islands
U.S.
MEXICO
Havana
CUBA
Santo Domingo
PUERTO RICO U.S.
WESTERN SAHARA
Morocco
Guadalajara
BELIZE
Mexico City
HAITI
JAMAICA
ST. KITTS & NEVIS
ANTIGUA & BARBUDA
MAURITANIA
CAPE VERDE
GUATEMALA
Caribbean Sea
Guatemala City
HONDURAS
DOMINICA
BARBADOS
ST. LUCIA
GRENADA
ST. VINCENT & THE GRENADINES
TRINIDAD & TOBAGO
SENEGAL
GAMBIA
MALI
EL SALVADOR
NICARAGUA
GUINEA-BISSAU
GUINEA
COSTA RICA
PANAMA
Caracas
VENEZUELA
BURKINA FASO
SIERRA LEONE
LIBERIA
CÔTE D'IVOIRE
GHANA
Medellin
Bogota
COLOMBIA
GUYANA
SURINAME
FRENCH GUIANA
Fr.
Christmas Island
Kiribati
Quito
ECUADOR
Galapagos Islands
Ecua.
Negro
Manaus
Amazon
Madeira
PERU
BRAZIL
Recife
Marquesas Islands
Fr.
Ucayali
São Francisco
Lima
Salvador
(Bahia)
Brasilia
SAMOA
AMERICAN SAMOA
U.S.
La Paz
BOLIVIA
FRENCH POLYNESIA
Fr.
Sucre
Rio de Janeiro
PARAGUAY
Sao Paulo
TONGA
Asuncion
Cordoba
Parana
Porto Alegre
Santiago
URUGUAY
Buenos Aires
Montevideo
CHILE
ARGENTINA
Falkland Islands
U.K.
South Georgia
U.K.
Tierra del Fuego
Strait of Magellan
Drake Passage
Antarctic Peninsula
Weddell Sea
Berkner Island
Ross Sea

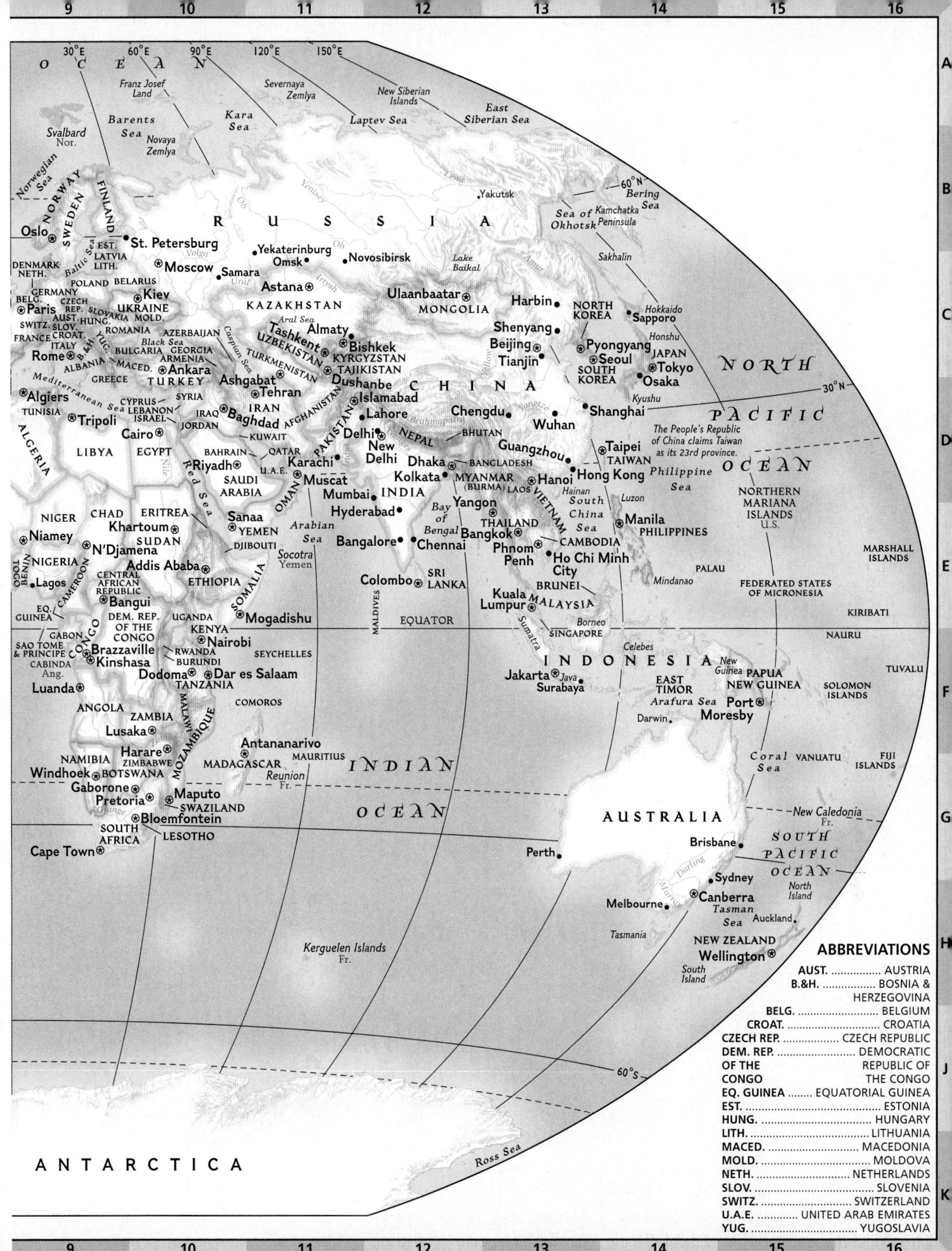

9
10
11
12
13
14
15
16
A
B
C
D
E
F
G
H
J
K
30°E
60°E
90°E
120°E
150°E
O C E A N
Franz Josef Land
Severnaya Zemlya
New Siberian Islands
East Siberian Sea
Laptev Sea
Kara Sea
Barents Sea
Svalbard Nor.
Novaya Zemlya
Norwegian Sea
NORWAY
SWEDEN
FINLAND
Lena
Yenisey
Ob
Yakutsk
60°N
Bering Sea
Sea of Okhotsk
Kamchatka Peninsula
R U S S I A
Oslo
St. Petersburg
EST.
LATVIA
LITH.
Baltic Sea
Volga
Yekaterinburg
Omsk
Novosibirsk
Lake Baikal
Amur
Sakhalin
DENMARK
NETH.
Moscow
Samara
Ural
Irtysh
POLAND
BELARUS
GERMANY
BELG.
CZECH REP.
Kiev
UKRAINE
Astana
Ulaanbaatar
MONGOLIA
Harbin
NORTH KOREA
Hokkaido
Sapporo
Paris
SLOVAKIA
MOLD.
AUST.
HUNG.
SWITZ.
SLOV.
CROAT.
ROMANIA
KAZAKHSTAN
Aral Sea
FRANCE
ITALY
B.&H.
YUG.
AZERBAIJAN
Black Sea
BULGARIA
GEORGIA
ARMENIA
Caspian Sea
Tashkent
UZBEKISTAN
Almaty
Bishkek
KYRGYZSTAN
TURKMENISTAN
TAJIKISTAN
Shenyang
Beijing
Tianjin
Yellow
Pyongyang
Seoul
SOUTH KOREA
Honshu
JAPAN
Tokyo
Osaka
N O R T H
Rome
ALBANIA
MACED.
GREECE
Ankara
TURKEY
Ashgabat
Dushanbe
C H I N A
Mediterranean Sea
Algiers
TUNISIA
CYPRUS
SYRIA
LEBANON
ISRAEL
Tehran
IRAN
AFGHANISTAN
Islamabad
Lahore
Chengdu
Yangtze
Wuhan
Shanghai
Kyushu
30°N
P A C I F I C
Tripoli
IRAQ
Baghdad
JORDAN
Cairo
PAKISTAN
Brahmaputra
NEPAL
BHUTAN
Delhi
New Delhi
Ganges
Indus
Guangzhou
Taipei
TAIWAN
The People's Republic of China claims Taiwan as its 23rd province.
ALGERIA
LIBYA
EGYPT
Nile
KUWAIT
BAHRAIN
QATAR
Karachi
Dhaka
BANGLADESH
O C E A N
Riyadh
Red Sea
U.A.E.
SAUDI ARABIA
Muscat
OMAN
Kolkata
MYANMAR (BURMA)
Hanoi
LAOS
VIETNAM
Hong Kong
Philippine Sea
Hainan
NORTHERN MARIANA ISLANDS U.S.
Mumbai
INDIA
Hyderabad
Bay of Bengal
Yangon
South China Sea
Luzon
NIGER
CHAD
ERITREA
Khartoum
Sanaa
YEMEN
Arabian Sea
THAILAND
Bangkok
Manila
PHILIPPINES
Niamey
SUDAN
DJIBOUTI
Bangalore
Chennai
CAMBODIA
Phnom Penh
Ho Chi Minh City
MARSHALL ISLANDS
TOGO
BENIN
NIGERIA
N'Djamena
Addis Ababa
Socotra Yemen
CAMEROON
Lagos
CENTRAL AFRICAN REPUBLIC
ETHIOPIA
SOMALIA
Colombo
SRI LANKA
BRUNEI
PALAU
Mindanao
FEDERATED STATES OF MICRONESIA
EQ. GUINEA
Bangui
DEM. REP. OF THE CONGO
UGANDA
KENYA
Mogadishu
MALDIVES
Kuala Lumpur
MALAYSIA
Borneo
KIRIBATI
EQUATOR
GABON
SAO TOME & PRINCIPE
CONGO
Brazzaville
Kinshasa
CABINDA Ang.
Nairobi
RWANDA
BURUNDI
SEYCHELLES
Sumatra
SINGAPORE
Celebes
NAURU
I N D O N E S I A
New Guinea
PAPUA NEW GUINEA
Luanda
Dodoma
Dar es Salaam
TANZANIA
Jakarta
Java
Surabaya
EAST TIMOR
Arafura Sea
Port Moresby
TUVALU
SOLOMON ISLANDS
ANGOLA
COMOROS
MALAWI
ZAMBIA
Darwin
Lusaka
MOZAMBIQUE
Antananarivo
NAMIBIA
Harare
ZIMBABWE
MAURITIUS
MADAGASCAR
Reunion Fr.
I N D I A N
Coral Sea
VANUATU
FIJI ISLANDS
Windhoek
BOTSWANA
Gaborone
Pretoria
Maputo
SWAZILAND
O C E A N
Orange
Bloemfontein
AUSTRALIA
New Caledonia Fr.
SOUTH AFRICA
LESOTHO
S O U T H
P A C I F I C
O C E A N
Cape Town
Perth
Brisbane
Darling
Sydney
Murray
Canberra
North Island
Melbourne
Tasman Sea
Auckland
Tasmania
Kerguelen Islands Fr.
NEW ZEALAND
Wellington
South Island
60°S
Ross Sea
ANTARCTICA
ABBREVIATIONS
AUST. .... AUSTRIA
B.&H. .... BOSNIA & HERZEGOVINA
BELG. .... BELGIUM
CROAT. .... CROATIA
CZECH REP. .... CZECH REPUBLIC
DEM. REP. OF THE CONGO .... DEMOCRATIC REPUBLIC OF THE CONGO
EQ. GUINEA .... EQUATORIAL GUINEA
EST. .... ESTONIA
HUNG. .... HUNGARY
LITH. .... LITHUANIA
MACED. .... MACEDONIA
MOLD. .... MOLDOVA
NETH. .... NETHERLANDS
SLOV. .... SLOVENIA
SWITZ. .... SWITZERLAND
U.A.E. .... UNITED ARAB EMIRATES
YUG. .... YUGOSLAVIA

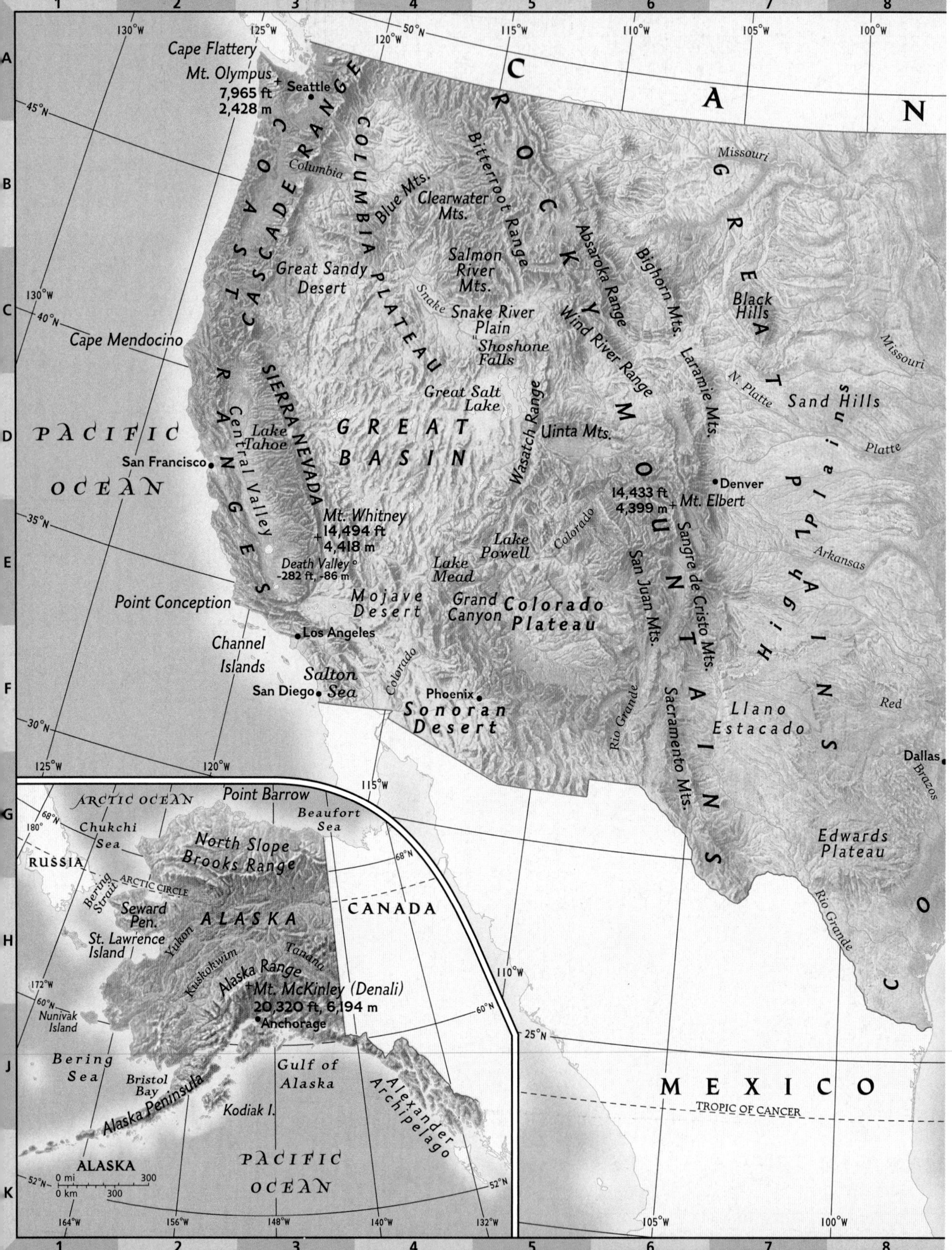

Cape Flattery
Mt. Olympus 7,965 ft 2,428 m
Seattle
COAST RANGES
CASCADE RANGE
COLUMBIA PLATEAU
ROCKY MOUNTAINS
CANADA
Bitterroot Range
Columbia
Blue Mts.
Clearwater Mts.
Salmon River Mts.
Absaroka Range
Bighorn Mts.
GREAT PLAINS
Missouri
Black Hills
Great Sandy Desert
Snake
Snake River Plain
Shoshone Falls
Wind River Range
Laramie Mts.
Cape Mendocino
SIERRA NEVADA
Great Salt Lake
Wasatch Range
N. Platte
Sand Hills
PACIFIC OCEAN
Central Valley
Lake Tahoe
GREAT BASIN
Uinta Mts.
Platte
San Francisco
Denver
14,433 ft 4,399 m Mt. Elbert
Mt. Whitney 14,494 ft 4,418 m
Colorado
Lake Powell
Sangre de Cristo Mts.
San Juan Mts.
High Plains
Arkansas
Death Valley -282 ft, -86 m
Lake Mead
Point Conception
Mojave Desert
Grand Canyon
Colorado Plateau
Los Angeles
Channel Islands
Salton Sea
San Diego
Phoenix
Sonoran Desert
Rio Grande
Sacramento Mts.
Llano Estacado
Red
Dallas
Brazos
Edwards Plateau
MEXICO
TROPIC OF CANCER
ARCTIC OCEAN
Point Barrow
Beaufort Sea
Chukchi Sea
North Slope
Brooks Range
RUSSIA
Bering Strait
ARCTIC CIRCLE
Seward Pen.
ALASKA
CANADA
St. Lawrence Island
Yukon
Tanana
Kuskokwim
Alaska Range
Mt. McKinley (Denali) 20,320 ft, 6,194 m
Anchorage
Nunivak Island
Bering Sea
Bristol Bay
Gulf of Alaska
Alaska Peninsula
Kodiak I.
Alexander Archipelago
PACIFIC OCEAN
ALASKA
0 mi 300
0 km 300

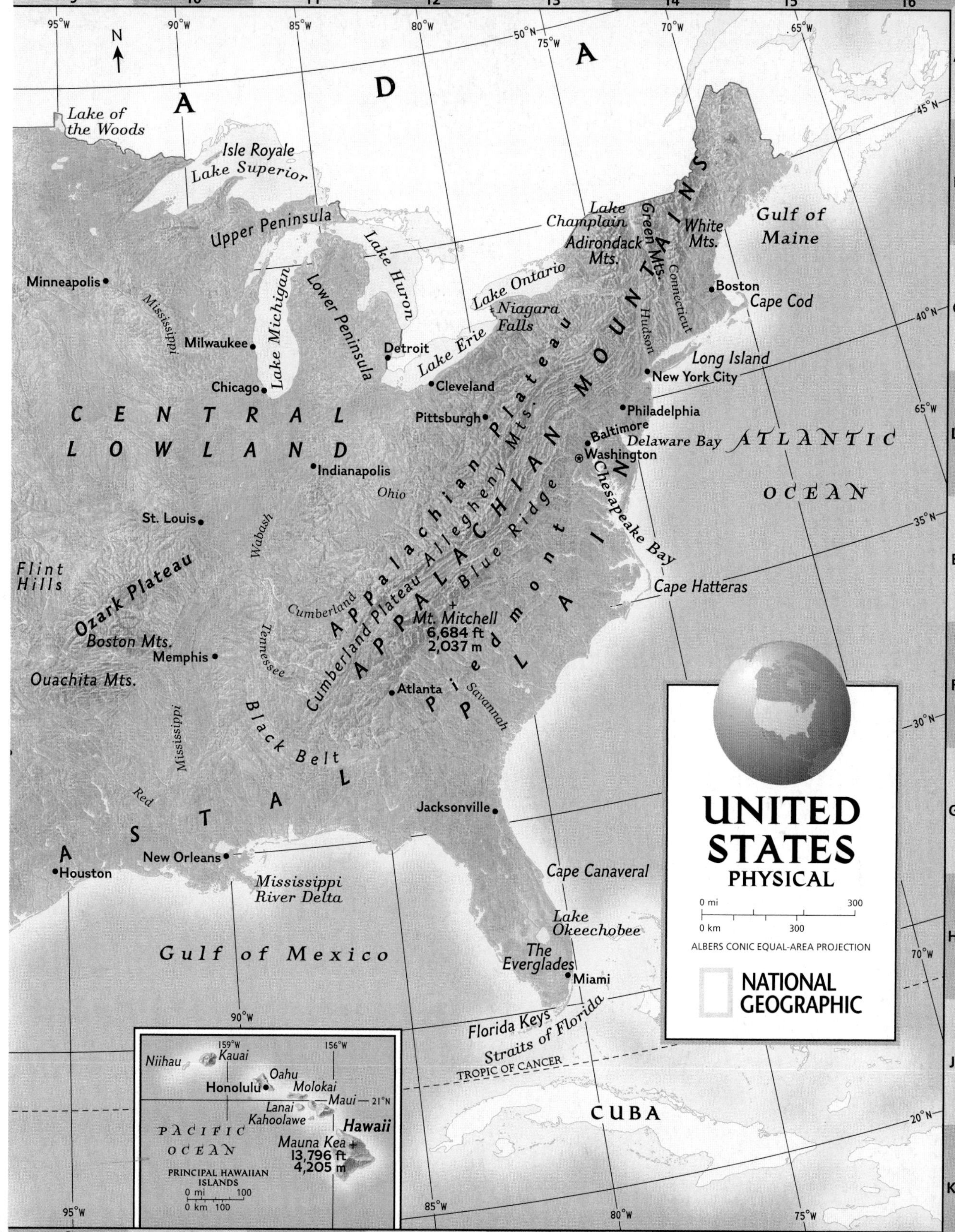
UNITED STATES
PHYSICAL
0 mi 300
0 km 300
ALBERS CONIC EQUAL-AREA PROJECTION
NATIONAL GEOGRAPHIC
C A N A D A
Lake of the Woods
Isle Royale
Lake Superior
Upper Peninsula
Lake Michigan
Lower Peninsula
Lake Huron
Lake Ontario
Niagara Falls
Lake Erie
Lake Champlain
Adirondack Mts.
Green Mts.
White Mts.
Gulf of Maine
Boston
Cape Cod
Connecticut
Hudson
Long Island
New York City
Philadelphia
Baltimore
Delaware Bay
Washington
Chesapeake Bay
Cape Hatteras
ATLANTIC OCEAN
APPALACHIAN MOUNTAINS
Appalachian Plateau
Allegheny Mts.
Blue Ridge
Cumberland Plateau
Piedmont
Mt. Mitchell 6,684 ft 2,037 m
Minneapolis
Mississippi
Milwaukee
Chicago
Detroit
Cleveland
Pittsburgh
CENTRAL LOWLAND
Indianapolis
Ohio
St. Louis
Wabash
Flint Hills
Ozark Plateau
Boston Mts.
Cumberland
Tennessee
Memphis
Ouachita Mts.
Atlanta
Savannah
Black Belt
Red
COASTAL PLAIN
Jacksonville
Houston
New Orleans
Mississippi River Delta
Gulf of Mexico
Cape Canaveral
Lake Okeechobee
The Everglades
Miami
Florida Keys
Straits of Florida
TROPIC OF CANCER
CUBA
Niihau
Kauai
Oahu
Honolulu
Molokai
Maui
Lanai
Kahoolawe
Hawaii
PACIFIC OCEAN
Mauna Kea 13,796 ft 4,205 m
PRINCIPAL HAWAIIAN ISLANDS
0 mi 100
0 km 100
159°W
156°W
21°N
95°W
90°W
85°W
80°W
75°W
70°W
65°W
50°N
45°N
40°N
35°N
30°N
20°N

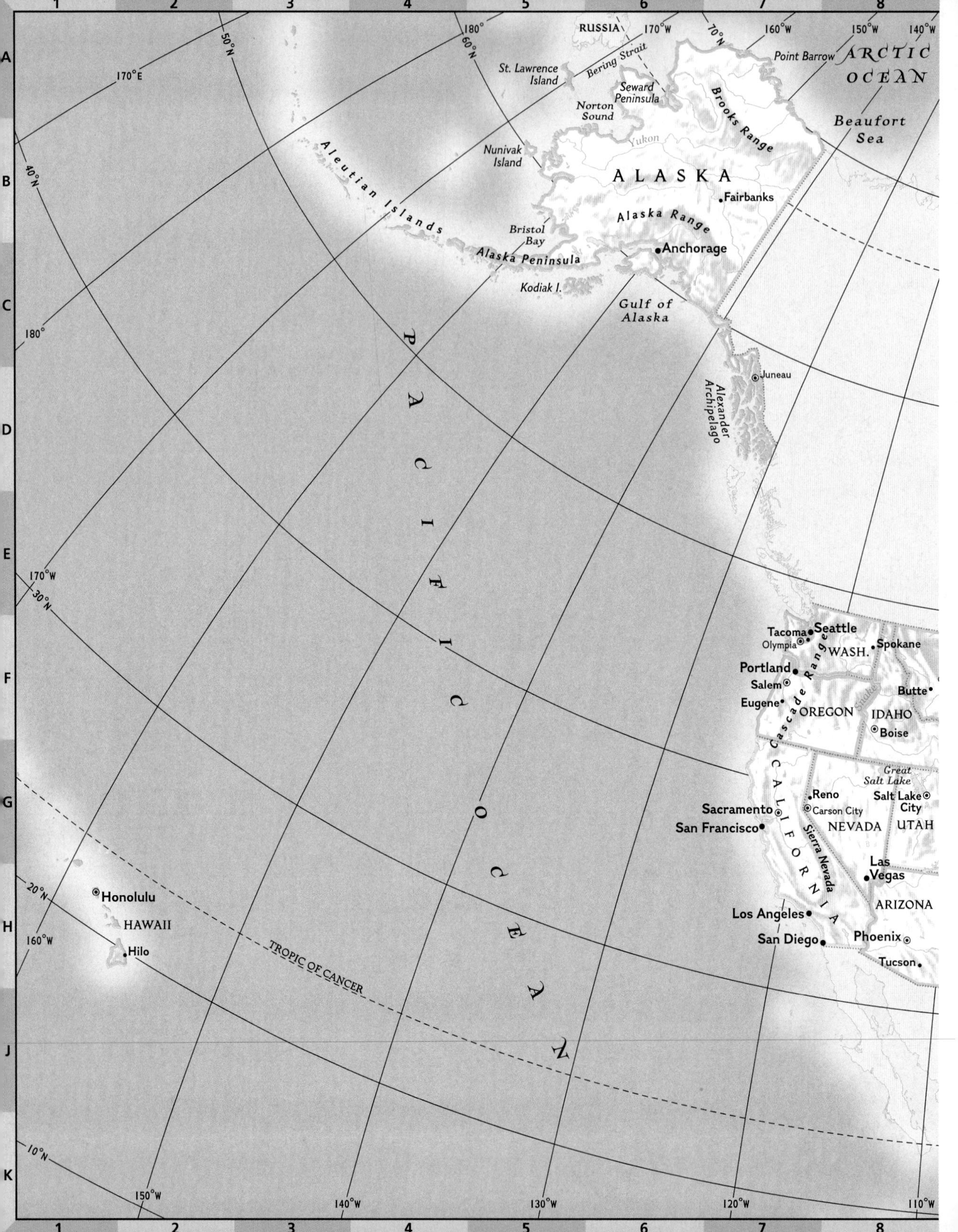

RUSSIA
Bering Strait
St. Lawrence Island
Seward Peninsula
Norton Sound
Point Barrow
ARCTIC OCEAN
Beaufort Sea
Brooks Range
Yukon
Nunivak Island
ALASKA
Fairbanks
Aleutian Islands
Alaska Range
Bristol Bay
Anchorage
Alaska Peninsula
Kodiak I.
Gulf of Alaska
Juneau
Alexander Archipelago
PACIFIC OCEAN
Tacoma
Seattle
Olympia
WASH.
Spokane
Portland
Salem
Cascade Range
Butte
Eugene
OREGON
Snake
IDAHO
Boise
Great Salt Lake
Salt Lake City
Reno
Carson City
Sacramento
San Francisco
NEVADA
UTAH
CALIFORNIA
Sierra Nevada
Las Vegas
ARIZONA
Los Angeles
San Diego
Phoenix
Tucson
Honolulu
HAWAII
Hilo
TROPIC OF CANCER
70°N
60°N
50°N
40°N
30°N
20°N
10°N
170°E
180°
170°W
160°W
150°W
140°W
130°W
120°W
110°W

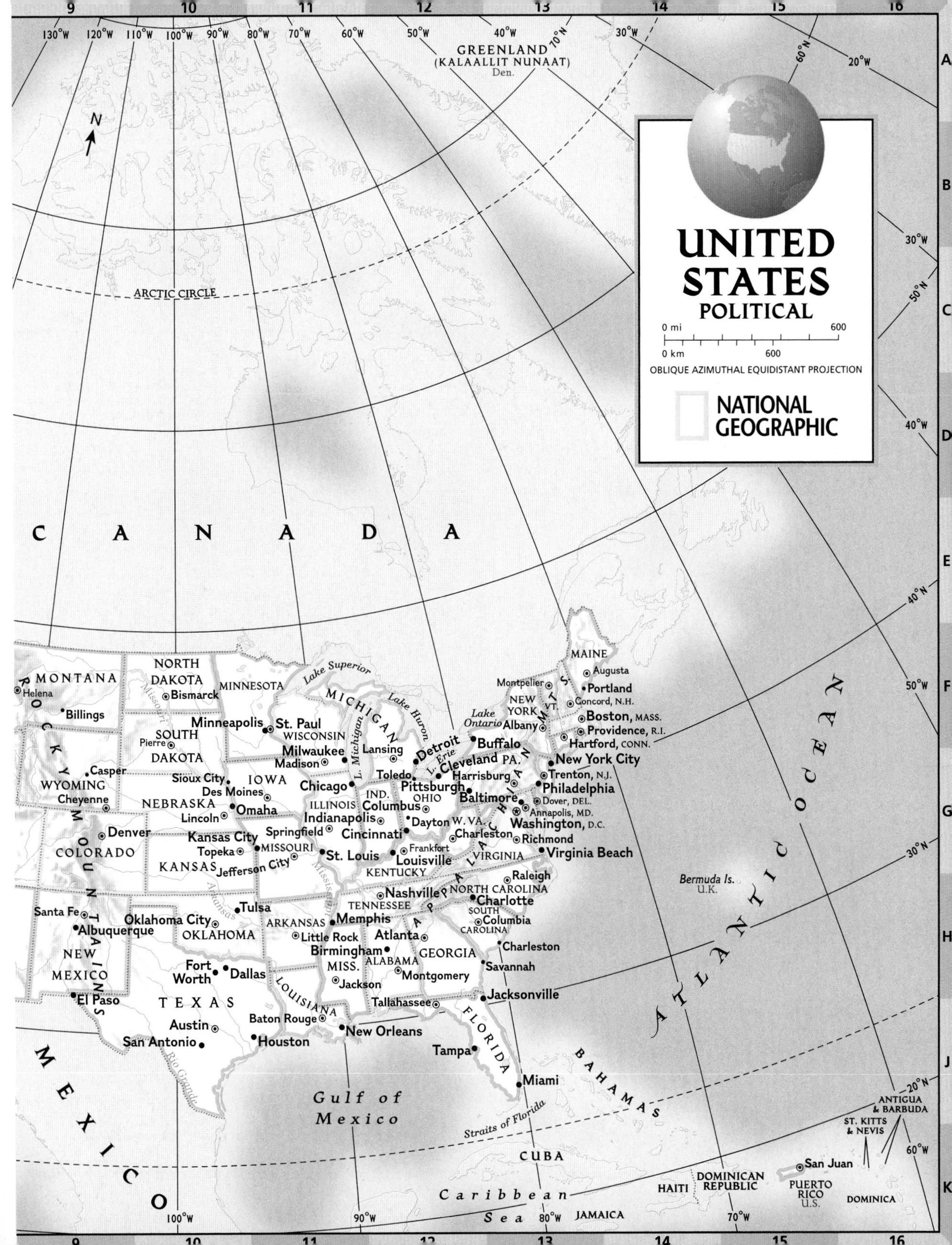

UNITED STATES
POLITICAL
0 mi 600
0 km 600
OBLIQUE AZIMUTHAL EQUIDISTANT PROJECTION
NATIONAL GEOGRAPHIC
GREENLAND (KALAALLIT NUNAAT) Den.
ARCTIC CIRCLE
CANADA
ATLANTIC OCEAN
Gulf of Mexico
Caribbean Sea
MEXICO
BAHAMAS
CUBA
HAITI
DOMINICAN REPUBLIC
PUERTO RICO U.S.
San Juan
JAMAICA
DOMINICA
ANTIGUA & BARBUDA
ST. KITTS & NEVIS
Bermuda Is. U.K.
Straits of Florida
ROCKY MOUNTAINS
APPALACHIAN MTS.
MONTANA
Helena
Billings
NORTH DAKOTA
Bismarck
SOUTH DAKOTA
Pierre
MINNESOTA
Minneapolis
St. Paul
WISCONSIN
Milwaukee
Madison
MICHIGAN
Lansing
Detroit
WYOMING
Casper
Cheyenne
NEBRASKA
Lincoln
Omaha
IOWA
Sioux City
Des Moines
ILLINOIS
Chicago
Springfield
IND.
Indianapolis
OHIO
Columbus
Toledo
Cleveland
Dayton
Cincinnati
COLORADO
Denver
KANSAS
Topeka
Kansas City
MISSOURI
Jefferson City
St. Louis
KENTUCKY
Frankfort
Louisville
W. VA.
Charleston
VIRGINIA
Richmond
Virginia Beach
PA.
Pittsburgh
Harrisburg
Philadelphia
NEW YORK
Albany
Buffalo
New York City
Trenton, N.J.
Baltimore
Dover, DEL.
Annapolis, MD.
Washington, D.C.
MAINE
Augusta
Portland
Montpelier
VT.
Concord, N.H.
Boston, MASS.
Providence, R.I.
Hartford, CONN.
NORTH CAROLINA
Raleigh
Charlotte
SOUTH CAROLINA
Columbia
Charleston
TENNESSEE
Nashville
Memphis
ARKANSAS
Little Rock
Tulsa
OKLAHOMA
Oklahoma City
NEW MEXICO
Santa Fe
Albuquerque
El Paso
TEXAS
Fort Worth
Dallas
Austin
San Antonio
Houston
LOUISIANA
Baton Rouge
New Orleans
MISS.
Jackson
ALABAMA
Birmingham
Montgomery
GEORGIA
Atlanta
Savannah
FLORIDA
Tallahassee
Jacksonville
Tampa
Miami
Lake Superior
Lake Huron
L. Michigan
L. Erie
Lake Ontario
Missouri
Mississippi
Arkansas
Rio Grande
130°W 120°W 110°W 100°W 90°W 80°W 70°W 60°W 50°W 40°W 30°W 20°W
70°N 60°N 50°N 40°N 30°N 20°N
9 10 11 12 13 14 15 16
A B C D E F G H J K

RUSSIA
ARCTIC OCEAN
Queen
Elizabeth
Islands
North Magnetic Pole
Prince Patrick I.
Melville Island
Bathurst Island
Banks Island
Somerset Island
Prince of Wales I.
Victoria Island
Boothia Peninsula
ARCTIC CIRCLE
Beaufort Sea
ALASKA
U.S.
Inuvik
YUKON TERRITORY
Mackenzie Mts.
Mackenzie
Mt. Logan
19,551 ft
5,959 m
Yukon Plateau
Great Bear Lake
NORTHWEST TERRITORIES
NUNAVUT
Whitehorse
Coast Mountains
Virginia Falls
Yellowknife
Great Slave Lake
ROCKY MOUNTAINS
Slave
Peace
CANADIAN
Lake Athabasca
Queen Charlotte Islands
BRITISH COLUMBIA
ALBERTA
Churchill
Fraser Plateau
Prince George
Athabasca
GREAT PLAINS
Nelson
Fraser
Columbia Mts.
PACIFIC OCEAN
Edmonton
SASKATCHEWAN
MANITOBA
Saskatchewan
Vancouver Island
Lake Winnipeg
Vancouver
Victoria
Calgary
Saskatoon
Lake Winnipegosis
Regina
Winnipeg
Lake of the Woods
UNITED STATES
170°W
160°W
150°W
140°W
130°W
120°W
110°W
100°W
80°N
70°N
60°N
50°N
40°N

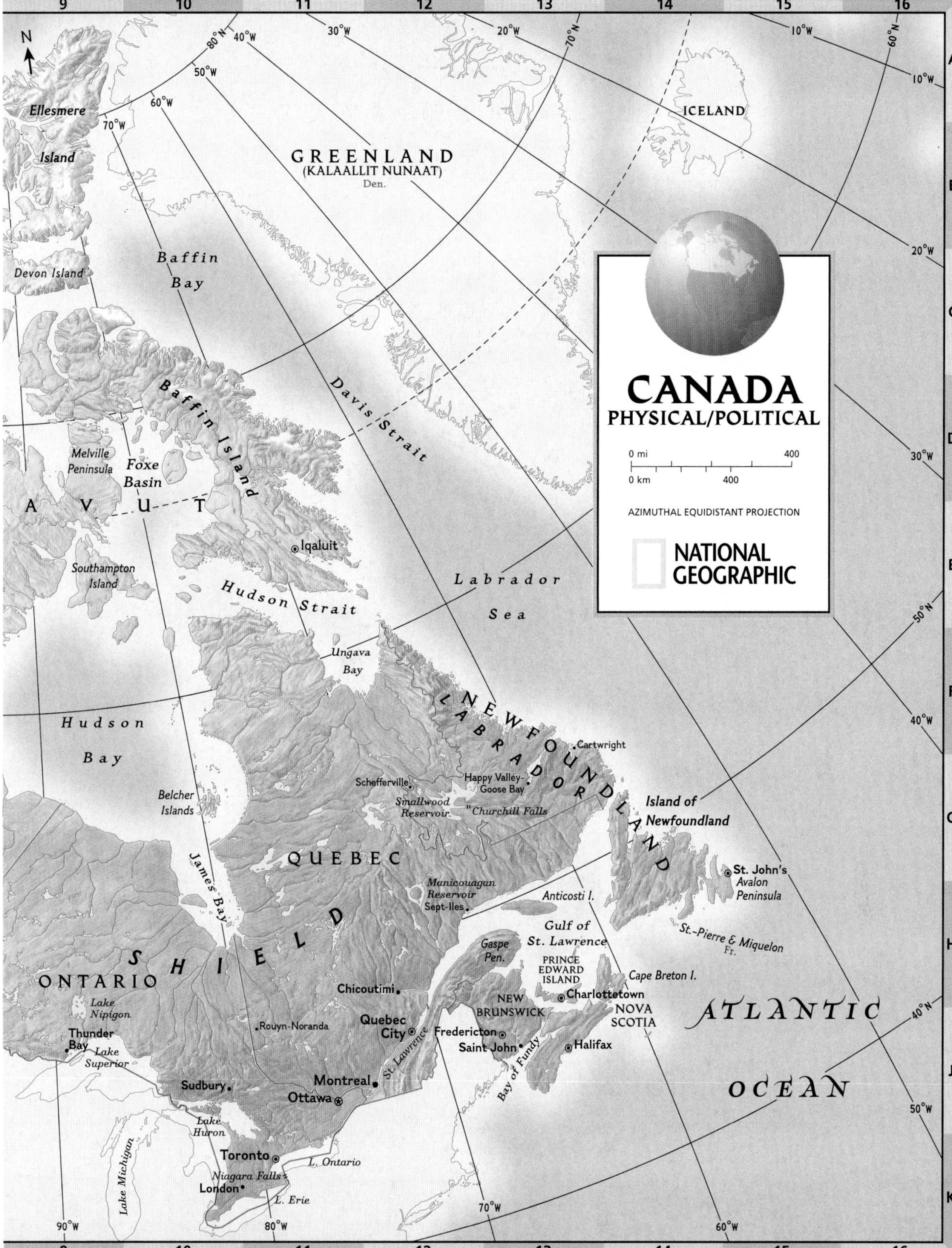
CANADA
PHYSICAL/POLITICAL
0 mi 400
0 km 400
AZIMUTHAL EQUIDISTANT PROJECTION
NATIONAL GEOGRAPHIC
GREENLAND
(KALAALLIT NUNAAT)
Den.
ICELAND
Ellesmere Island
Devon Island
Baffin Bay
Baffin Island
Davis Strait
Melville Peninsula
Foxe Basin
NUNAVUT
Iqaluit
Southampton Island
Hudson Strait
Labrador Sea
Ungava Bay
Hudson Bay
Belcher Islands
James Bay
NEWFOUNDLAND
LABRADOR
Cartwright
Schefferville
Happy Valley-Goose Bay
Smallwood Reservoir
Churchill Falls
Island of Newfoundland
St. John's
Avalon Peninsula
QUEBEC
Manicouagan Reservoir
Sept-Iles
Anticosti I.
Gulf of St. Lawrence
St.-Pierre & Miquelon
Fr.
Gaspe Pen.
PRINCE EDWARD ISLAND
Charlottetown
Cape Breton I.
NEW BRUNSWICK
NOVA SCOTIA
Fredericton
Saint John
Halifax
Bay of Fundy
ATLANTIC OCEAN
SHIELD
ONTARIO
Lake Nipigon
Thunder Bay
Lake Superior
Rouyn-Noranda
Chicoutimi
Quebec City
St. Lawrence
Montreal
Ottawa
Sudbury
Lake Huron
Lake Michigan
Toronto
L. Ontario
Niagara Falls
London
L. Erie
80°N
70°N
60°N
50°N
40°N
40°W
30°W
20°W
10°W
50°W
60°W
70°W
80°W
90°W
9 10 11 12 13 14 15 16
A B C D E F G H J K

UNITED STATES
Tijuana
Mexicali
Sonoran Desert
Baja California
Gulf of California
Ciudad Juarez
Chihuahua
Rio Grande
MEXICO
Sierra Madre Occidental
Sierra Madre Oriental
Nuevo Laredo
Monterrey
Matamoros
Gulf of Mexico
La Paz
False Cape
Mazatlan
Ciudad Madero
Tampico
San Luis Potosi
Leon
Guadalajara
Revillagigedo Islands
Mex.
Merida
Cozumel Island
Yucatan Peninsula
Bay of Campeche
Mexico City
Orizaba
18,855 ft
5,747 m
Popocatepetl
17,802 ft
5,426 m
Veracruz
Sierra Madre del Sur
Isthmus of Tehuantepec
Belize City
Belmopan
BELIZE
Gulf of Honduras
Acapulco
Sierra Madre
Gulf of Tehuantepec
HON
GUATEMALA
Guatemala City
Tegucigalpa
EL SALVADOR
San Salvador
Leon
CENTRAL AMERICA
PACIFIC OCEAN
Cocos Island
C.R.
MIDDLE AMERICA
PHYSICAL/POLITICAL
0 mi 400
0 km 400
AZIMUTHAL EQUIDISTANT PROJECTION
NATIONAL GEOGRAPHIC
N
30°N
20°N
10°N
0°
110°W
100°W
90°W

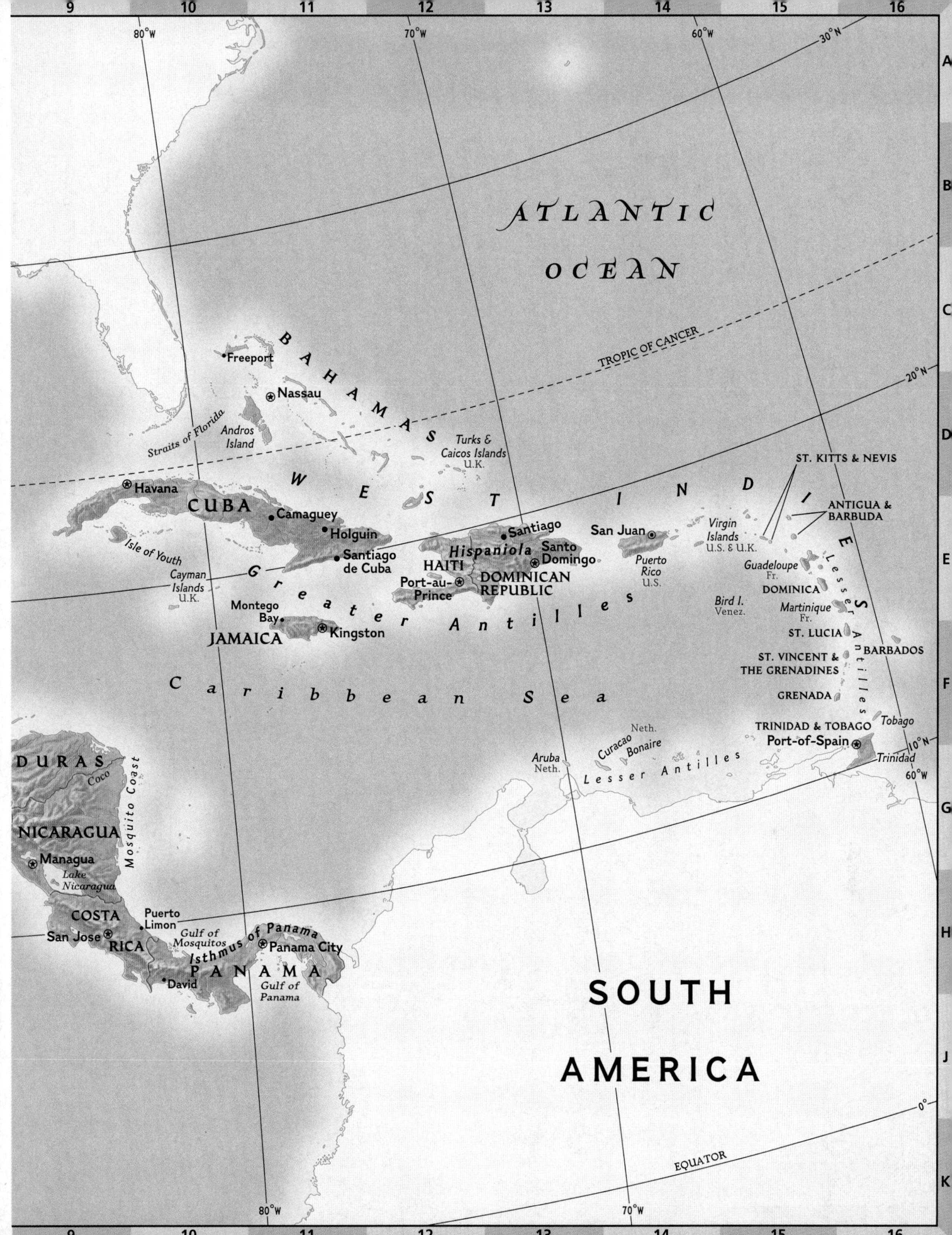

ATLANTIC OCEAN
BAHAMAS
Freeport
Nassau
Andros Island
Straits of Florida
Turks & Caicos Islands U.K.
TROPIC OF CANCER
WEST INDIES
Havana
CUBA
Camaguey
Holguin
Santiago de Cuba
Isle of Youth
Cayman Islands U.K.
Greater Antilles
Santiago
Hispaniola
HAITI
Port-au-Prince
Santo Domingo
DOMINICAN REPUBLIC
San Juan
Puerto Rico U.S.
Virgin Islands U.S. & U.K.
ST. KITTS & NEVIS
ANTIGUA & BARBUDA
Guadeloupe Fr.
DOMINICA
Martinique Fr.
ST. LUCIA
BARBADOS
ST. VINCENT & THE GRENADINES
GRENADA
Lesser Antilles
Tobago
TRINIDAD & TOBAGO
Port-of-Spain
Trinidad
Bird I. Venez.
Montego Bay
Kingston
JAMAICA
Caribbean Sea
Aruba Neth.
Curacao
Bonaire
Neth.
Lesser Antilles
DURAS
Coco
Mosquito Coast
NICARAGUA
Managua
Lake Nicaragua
COSTA RICA
San Jose
Puerto Limon
Gulf of Mosquitos
Isthmus of Panama
Panama City
PANAMA
David
Gulf of Panama
SOUTH AMERICA
EQUATOR
80°W
70°W
60°W
30°N
20°N
10°N
0°

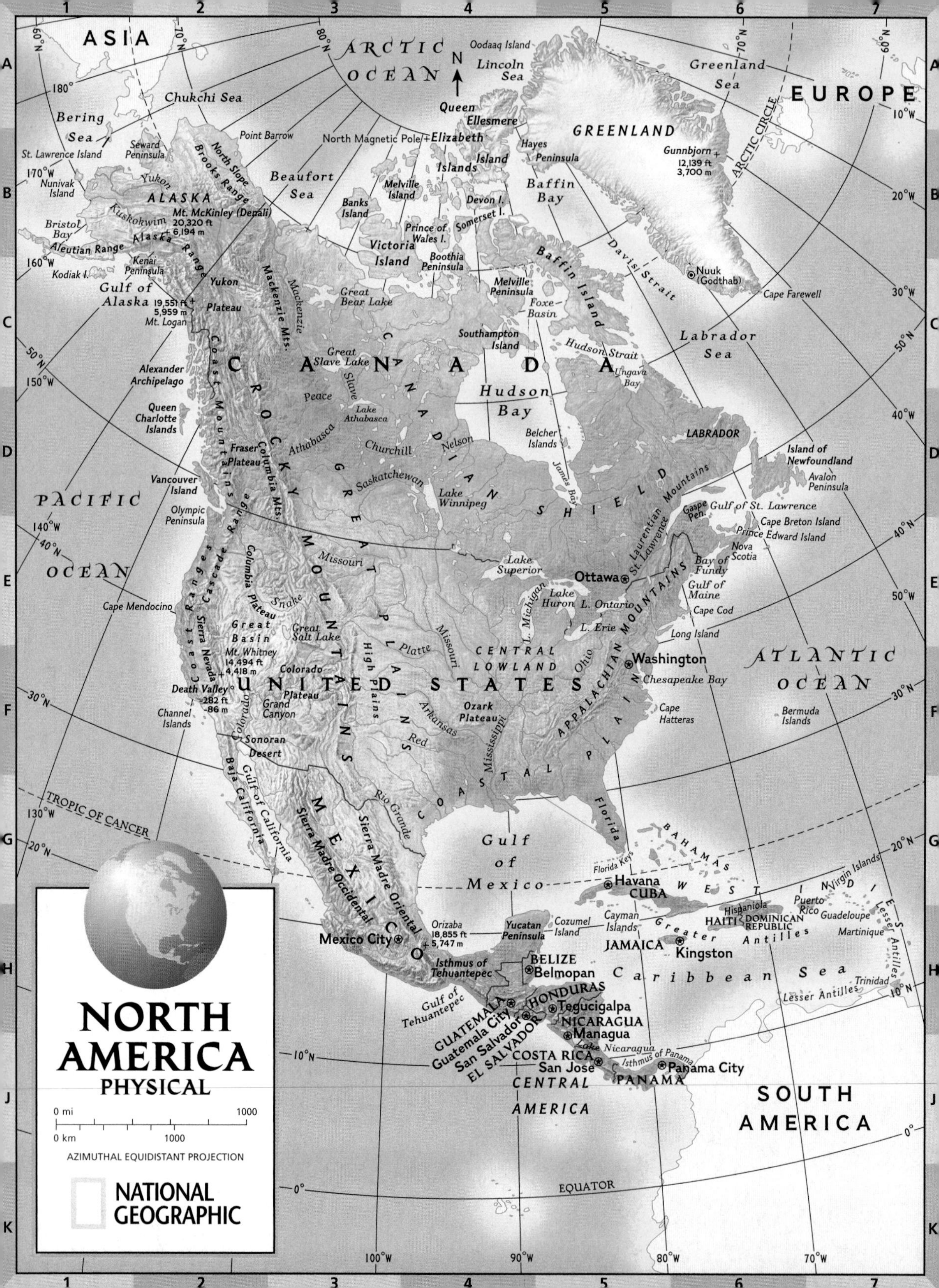
NORTH AMERICA
PHYSICAL
0 mi
1000
0 km
1000
AZIMUTHAL EQUIDISTANT PROJECTION
NATIONAL GEOGRAPHIC
ASIA
EUROPE
SOUTH AMERICA
CENTRAL AMERICA
ARCTIC OCEAN
PACIFIC OCEAN
ATLANTIC OCEAN
N
Chukchi Sea
Bering Sea
St. Lawrence Island
Seward Peninsula
Point Barrow
North Slope
Brooks Range
Yukon
ALASKA
Nunivak Island
Kuskokwim
Mt. McKinley (Denali)
20,320 ft
6,194 m
Bristol Bay
Alaska Range
Aleutian Range
Kenai Peninsula
Kodiak I.
Gulf of Alaska
Yukon Plateau
19,551 ft
5,959 m
Mt. Logan
Mackenzie Mts.
Mackenzie
Beaufort Sea
Banks Island
Victoria Island
Melville Island
Prince of Wales I.
Boothia Peninsula
Queen Elizabeth Islands
Ellesmere Island
North Magnetic Pole
Oodaaq Island
Lincoln Sea
Devon I.
Somerset I.
Hayes Peninsula
GREENLAND
Greenland Sea
Gunnbjorn
12,139 ft
3,700 m
ARCTIC CIRCLE
Baffin Bay
Baffin Island
Davis Strait
Nuuk (Godthab)
Cape Farewell
Melville Peninsula
Foxe Basin
Great Bear Lake
Great Slave Lake
Southampton Island
Hudson Strait
Labrador Sea
Ungava Bay
CANADA
Hudson Bay
Alexander Archipelago
Queen Charlotte Islands
Peace
Slave
Lake Athabasca
Athabasca
Churchill
Nelson
Belcher Islands
James Bay
LABRADOR
Island of Newfoundland
Avalon Peninsula
CANADIAN SHIELD
Laurentian Mountains
Coast Mountains
ROCKY MOUNTAINS
Fraser Plateau
Columbia Mts.
GREAT PLAINS
Saskatchewan
Lake Winnipeg
Vancouver Island
Olympic Peninsula
Cascade Range
Gaspe Pen.
Gulf of St. Lawrence
Cape Breton Island
Prince Edward Island
Nova Scotia
St. Lawrence
Missouri
Lake Superior
Ottawa
Bay of Fundy
Gulf of Maine
Columbia Plateau
Snake
Cape Mendocino
Coast Ranges
Sierra Nevada
Great Basin
Great Salt Lake
L. Michigan
Lake Huron
L. Ontario
L. Erie
Cape Cod
Long Island
APPALACHIAN MOUNTAINS
Mt. Whitney
14,494 ft
4,418 m
Colorado
High Plains
Platte
CENTRAL LOWLAND
Ohio
Washington
Chesapeake Bay
UNITED STATES
Death Valley
-282 ft
-86 m
Colorado Plateau
Grand Canyon
Ozark Plateau
Arkansas
Mississippi
Cape Hatteras
Bermuda Islands
Channel Islands
Sonoran Desert
Red
COASTAL PLAIN
TROPIC OF CANCER
Baja California
Gulf of California
Rio Grande
Florida
MEXICO
Sierra Madre Occidental
Sierra Madre Oriental
Gulf of Mexico
BAHAMAS
Florida Keys
Havana
CUBA
WEST INDIES
Virgin Islands
Puerto Rico
Guadeloupe
Martinique
Lesser Antilles
HAITI
Hispaniola
DOMINICAN REPUBLIC
Greater Antilles
Orizaba
18,855 ft
5,747 m
Mexico City
Yucatan Peninsula
Cozumel Island
Cayman Islands
JAMAICA
Kingston
Isthmus of Tehuantepec
BELIZE
Belmopan
Caribbean Sea
Trinidad
Gulf of Tehuantepec
GUATEMALA
Guatemala City
HONDURAS
Tegucigalpa
NICARAGUA
Managua
San Salvador
EL SALVADOR
Lake Nicaragua
COSTA RICA
San Jose
Isthmus of Panama
Panama City
PANAMA
EQUATOR
60°N
70°N
80°N
50°N
40°N
30°N
20°N
10°N
0°
180°
170°W
160°W
150°W
140°W
130°W
100°W
90°W
80°W
70°W
60°W
10°W
20°W
30°W
40°W
50°W
1 2 3 4 5 6 7
A B C D E F G H J K

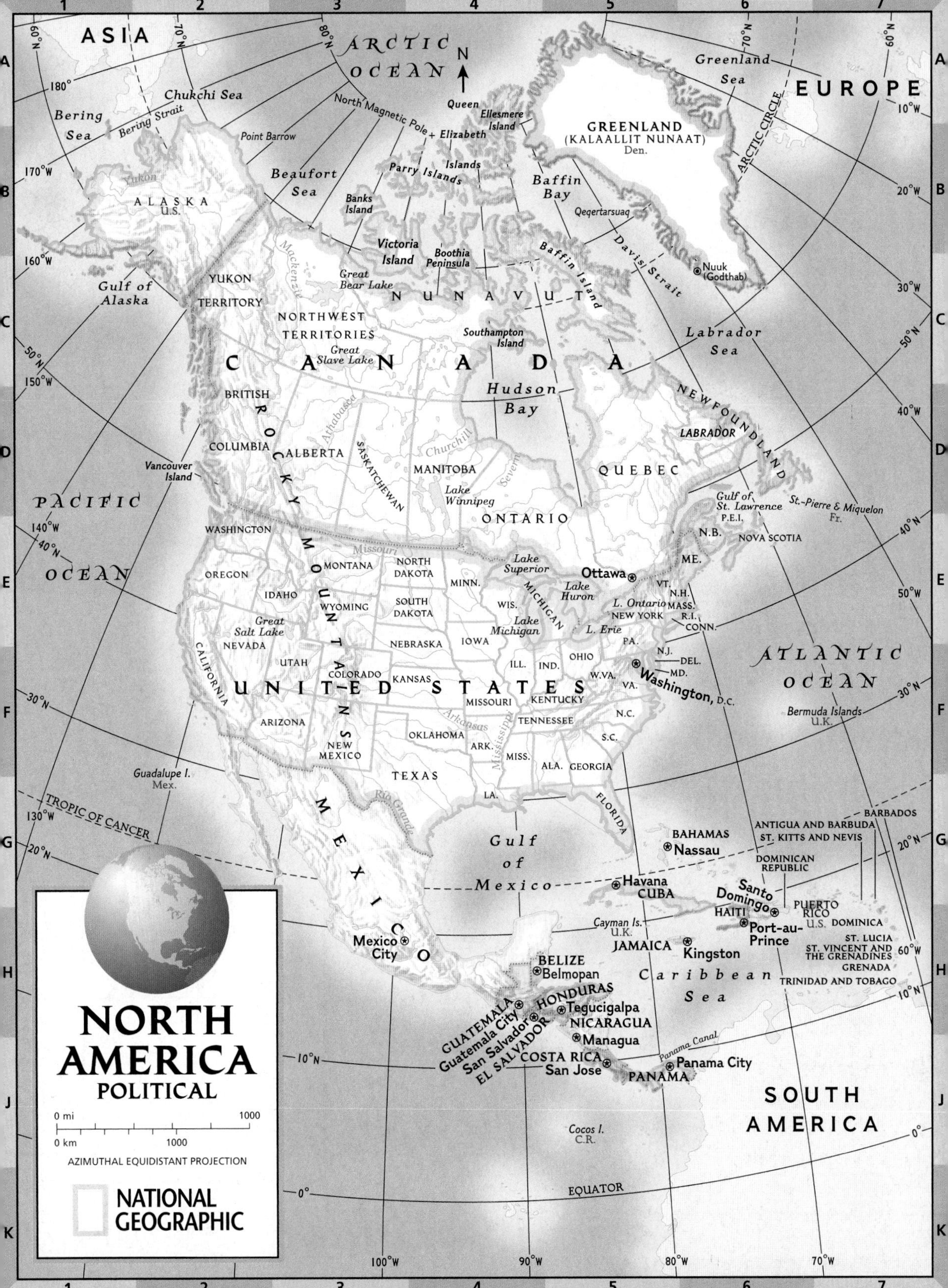
NORTH AMERICA
POLITICAL
AZIMUTHAL EQUIDISTANT PROJECTION
NATIONAL GEOGRAPHIC
ASIA
ARCTIC OCEAN
EUROPE
PACIFIC OCEAN
ATLANTIC OCEAN
SOUTH AMERICA
CANADA
UNITED STATES
MEXICO
GREENLAND (KALAALLIT NUNAAT) Den.
ALASKA U.S.
Chukchi Sea
Bering Sea
Bering Strait
Beaufort Sea
Point Barrow
North Magnetic Pole
Queen Elizabeth Islands
Ellesmere Island
Parry Islands
Banks Island
Victoria Island
Boothia Peninsula
Baffin Bay
Baffin Island
Davis Strait
Greenland Sea
ARCTIC CIRCLE
Qeqertarsuaq
Nuuk (Godthab)
Gulf of Alaska
YUKON TERRITORY
NORTHWEST TERRITORIES
NUNAVUT
Great Bear Lake
Great Slave Lake
Southampton Island
Labrador Sea
Hudson Bay
NEWFOUNDLAND
LABRADOR
BRITISH COLUMBIA
ALBERTA
SASKATCHEWAN
MANITOBA
ONTARIO
QUEBEC
Vancouver Island
Lake Winnipeg
Gulf of St. Lawrence
St.-Pierre & Miquelon Fr.
P.E.I.
N.B.
NOVA SCOTIA
ROCKY MOUNTAINS
Ottawa
WASHINGTON
OREGON
IDAHO
MONTANA
WYOMING
NORTH DAKOTA
SOUTH DAKOTA
MINN.
WIS.
MICHIGAN
Lake Superior
Lake Huron
Lake Michigan
L. Ontario
L. Erie
ME.
VT.
N.H.
MASS.
R.I.
CONN.
NEW YORK
PA.
N.J.
DEL.
MD.
Washington, D.C.
CALIFORNIA
NEVADA
Great Salt Lake
UTAH
COLORADO
NEBRASKA
IOWA
KANSAS
MISSOURI
ILL.
IND.
OHIO
W.VA.
VA.
KENTUCKY
TENNESSEE
N.C.
S.C.
ARIZONA
NEW MEXICO
OKLAHOMA
ARK.
MISS.
ALA.
GEORGIA
TEXAS
LA.
FLORIDA
Bermuda Islands U.K.
Guadalupe I. Mex.
TROPIC OF CANCER
Gulf of Mexico
BAHAMAS
Nassau
Havana
CUBA
Cayman Is. U.K.
JAMAICA
Kingston
HAITI
Port-au-Prince
Santo Domingo
DOMINICAN REPUBLIC
PUERTO RICO U.S.
ANTIGUA AND BARBUDA
ST. KITTS AND NEVIS
BARBADOS
DOMINICA
ST. LUCIA
ST. VINCENT AND THE GRENADINES
GRENADA
TRINIDAD AND TOBAGO
Caribbean Sea
Mexico City
BELIZE
Belmopan
GUATEMALA
Guatemala City
HONDURAS
Tegucigalpa
San Salvador
EL SALVADOR
NICARAGUA
Managua
COSTA RICA
San Jose
PANAMA
Panama City
Panama Canal
Cocos I. C.R.
EQUATOR
0 mi 1000
0 km 1000

SOUTH AMERICA
PHYSICAL
Caribbean Sea
ATLANTIC OCEAN
PACIFIC OCEAN
Caracas
Lake Maracaibo
Orinoco
VENEZUELA
LLANOS
Angel Falls
Total drop=
3,212 ft 979 m
GUYANA
Georgetown
SURINAME
Paramaribo
Cayenne
FRENCH GUIANA
GUIANA HIGHLANDS
Boundary claimed by Suriname
Bogota
COLOMBIA
Malpelo I.
Quito
ECUADOR
Negro
Amazon
Marajo Island
EQUATOR
AMAZON BASIN
Selvas
Madeira
Tapajos
Xingu
Purus
Ucayali
Teles Pires
Tocantins
BRAZIL
Sao Francisco
ANDES
PERU
Lima
Machu Picchu
Lake Titicaca
BOLIVIA
La Paz
Altiplano
Sucre
Salar de Uyuni
MATO GROSSO PLATEAU
BRAZILIAN HIGHLANDS
Brasilia
Paraguay
PARAGUAY
GRAN CHACO
Asuncion
Iguazu Falls
TROPIC OF CAPRICORN
San Felix I.
San Ambrosio I.
Parana
Uruguay
Aconcagua
22,834 ft
6,960 m
Santiago
Juan Fernandez Is.
Buenos Aires
URUGUAY
Montevideo
Rio de la Plata
PAMPAS
ARGENTINA
Negro
Chiloe Island
-131 ft
-40 m
Valdes Peninsula
PATAGONIA
Gulf of San Jorge
Taitao Peninsula
Wellington I.
Falkland Islands (Islas Malvinas)
Stanley
Strait of Magellan
Tierra del Fuego
Cape Horn
South Georgia I.
0 mi 800
0 km 800
AZIMUTHAL EQUIDISTANT PROJECTION
NATIONAL GEOGRAPHIC

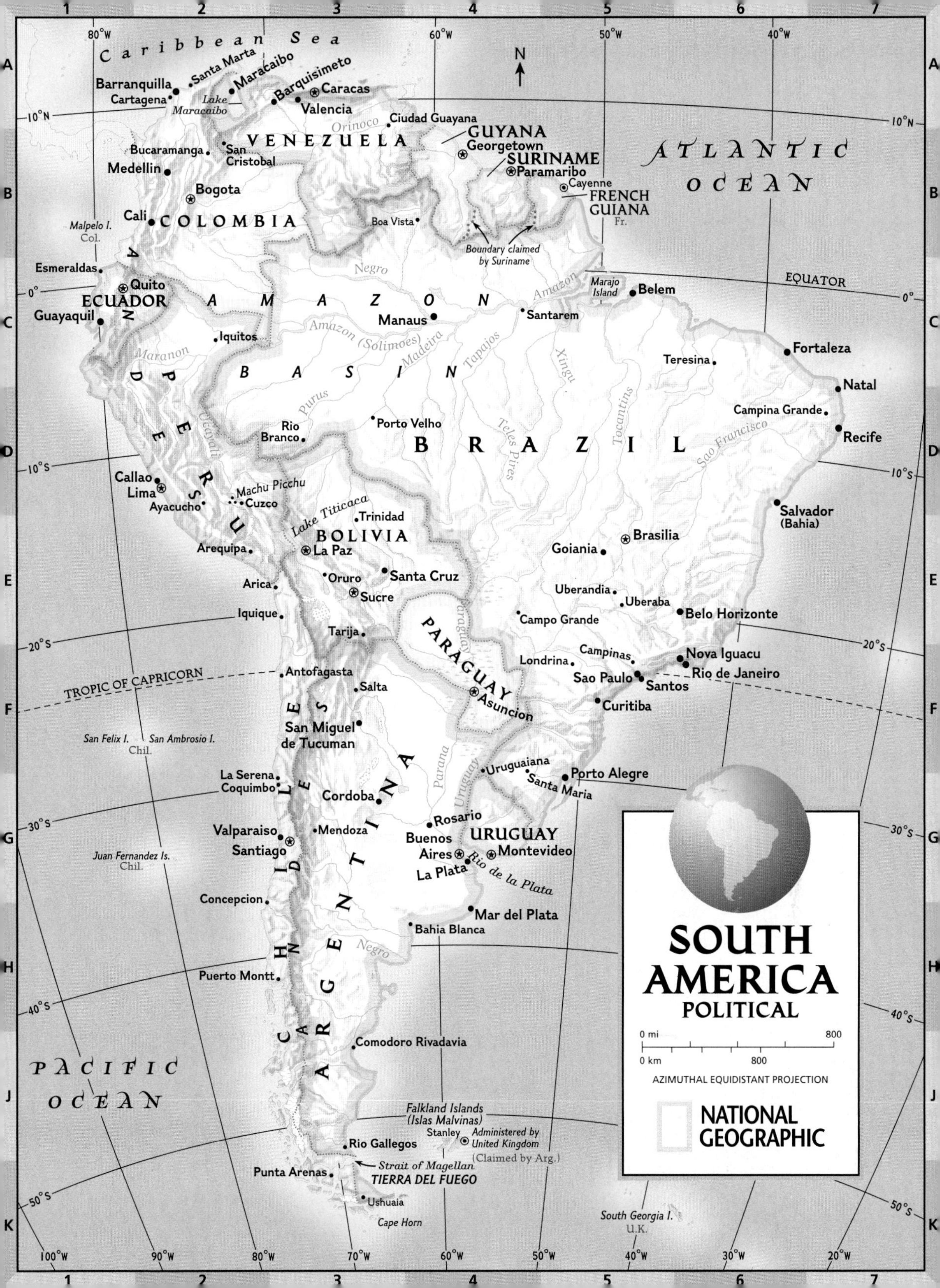

SOUTH AMERICA
POLITICAL
0 mi 800
0 km 800
AZIMUTHAL EQUIDISTANT PROJECTION
NATIONAL GEOGRAPHIC
Caribbean Sea
ATLANTIC OCEAN
PACIFIC OCEAN
EQUATOR
TROPIC OF CAPRICORN
N
Barranquilla
Cartagena
Santa Marta
Maracaibo
Lake Maracaibo
Barquisimeto
Caracas
Valencia
VENEZUELA
Orinoco
Ciudad Guayana
GUYANA
Georgetown
SURINAME
Paramaribo
Cayenne
FRENCH GUIANA
Fr.
Boundary claimed by Suriname
Bucaramanga
San Cristobal
Medellin
Bogota
Cali
COLOMBIA
Malpelo I.
Col.
Boa Vista
Esmeraldas
Quito
ECUADOR
Guayaquil
AMAZON BASIN
ANDES
Negro
Amazon
Marajo Island
Belem
Manaus
Santarem
Amazon (Solimoes)
Iquitos
Maranon
Madeira
Tapajos
Xingu
Fortaleza
Teresina
Natal
Campina Grande
Recife
Purus
Rio Branco
Porto Velho
Teles Pires
Tocantins
Sao Francisco
BRAZIL
PERU
Ucayali
Callao
Lima
Machu Picchu
Ayacucho
Cuzco
Lake Titicaca
Trinidad
BOLIVIA
La Paz
Salvador (Bahia)
Brasilia
Goiania
Arequipa
Arica
Oruro
Santa Cruz
Sucre
Uberandia
Uberaba
Belo Horizonte
Iquique
Campo Grande
Paraguay
PARAGUAY
Tarija
Campinas
Nova Iguacu
Rio de Janeiro
Londrina
Sao Paulo
Santos
Antofagasta
Salta
Asuncion
Curitiba
San Miguel de Tucuman
San Felix I.
San Ambrosio I.
Chil.
Parana
Uruguay
Uruguaiana
Porto Alegre
Santa Maria
La Serena
Coquimbo
Cordoba
ARGENTINA
Rosario
URUGUAY
Valparaiso
Mendoza
Santiago
Buenos Aires
Montevideo
La Plata
Rio de la Plata
Juan Fernandez Is.
Chil.
CHILE
Concepcion
Mar del Plata
Bahia Blanca
Negro
Puerto Montt
Comodoro Rivadavia
Falkland Islands (Islas Malvinas)
Stanley
Administered by United Kingdom
(Claimed by Arg.)
Rio Gallegos
Strait of Magellan
Punta Arenas
TIERRA DEL FUEGO
Ushuaia
Cape Horn
South Georgia I.
U.K.
80°W 60°W 50°W 40°W
100°W 90°W 80°W 70°W 60°W 50°W 40°W 30°W 20°W
10°N 0° 10°S 20°S 30°S 40°S 50°S
1 2 3 4 5 6 7
A B C D E F G H J K

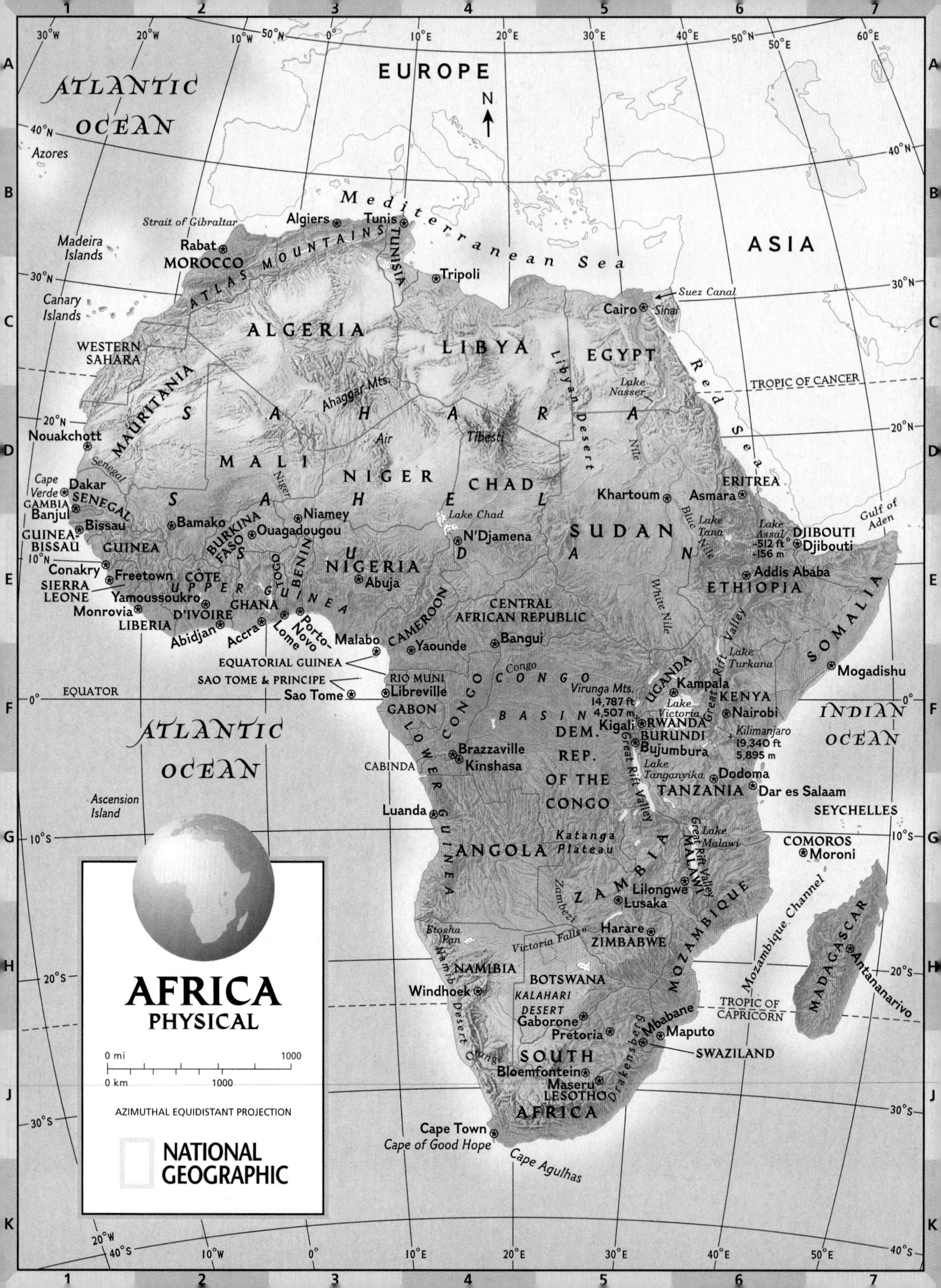

AFRICA
PHYSICAL
0 mi
1000
0 km
1000
AZIMUTHAL EQUIDISTANT PROJECTION
NATIONAL GEOGRAPHIC
EUROPE
ASIA
ATLANTIC OCEAN
ATLANTIC OCEAN
INDIAN OCEAN
Mediterranean Sea
Red Sea
Gulf of Aden
Mozambique Channel
N
Azores
Madeira Islands
Canary Islands
Strait of Gibraltar
Rabat
MOROCCO
ATLAS MOUNTAINS
Algiers
Tunis
TUNISIA
Tripoli
WESTERN SAHARA
ALGERIA
LIBYA
EGYPT
Cairo
Suez Canal
Sinai
Libyan Desert
Lake Nasser
TROPIC OF CANCER
MAURITANIA
S A H A R A
Ahaggar Mts.
Air
Tibesti
Nile
Nouakchott
MALI
NIGER
CHAD
Senegal
Niger
Cape Verde
Dakar
SENEGAL
GAMBIA
Banjul
Bissau
GUINEA-BISSAU
GUINEA
Conakry
Freetown
SIERRA LEONE
Monrovia
LIBERIA
Bamako
BURKINA FASO
Ouagadougou
Niamey
S A H E L
Lake Chad
N'Djamena
Khartoum
Asmara
ERITREA
SUDAN
Blue Nile
Lake Tana
Lake Assal -512 ft -156 m
DJIBOUTI
Djibouti
CÔTE D'IVOIRE
Yamoussoukro
Abidjan
GHANA
Accra
TOGO
Lome
BENIN
Porto-Novo
UPPER GUINEA
NIGERIA
Abuja
ETHIOPIA
Addis Ababa
White Nile
Rift Valley
SOMALIA
Mogadishu
Lake Turkana
CENTRAL AFRICAN REPUBLIC
Bangui
CAMEROON
Yaounde
Malabo
EQUATORIAL GUINEA
SAO TOME & PRINCIPE
Sao Tome
RIO MUNI
Libreville
GABON
EQUATOR
Congo
CONGO BASIN
CONGO
Virunga Mts. 14,787 ft 4,507 m
UGANDA
Kampala
Lake Victoria
Great Rift Valley
KENYA
Nairobi
Kilimanjaro 19,340 ft 5,895 m
Kigali
RWANDA
BURUNDI
Bujumbura
DEM. REP. OF THE CONGO
Brazzaville
Kinshasa
CABINDA
LOWER GUINEA
Lake Tanganyika
Dodoma
TANZANIA
Dar es Salaam
SEYCHELLES
Luanda
ANGOLA
Katanga Plateau
Lake Malawi
MALAWI
COMOROS
Moroni
ZAMBIA
Lilongwe
Lusaka
Zambezi
MOZAMBIQUE
MADAGASCAR
Antananarivo
Etosha Pan
Victoria Falls
Harare
ZIMBABWE
NAMIBIA
Namib Desert
Windhoek
BOTSWANA
KALAHARI DESERT
Gaborone
Pretoria
TROPIC OF CAPRICORN
Mbabane
Maputo
SWAZILAND
Drakensberg
Orange
SOUTH AFRICA
Bloemfontein
Maseru
LESOTHO
Cape Town
Cape of Good Hope
Cape Agulhas

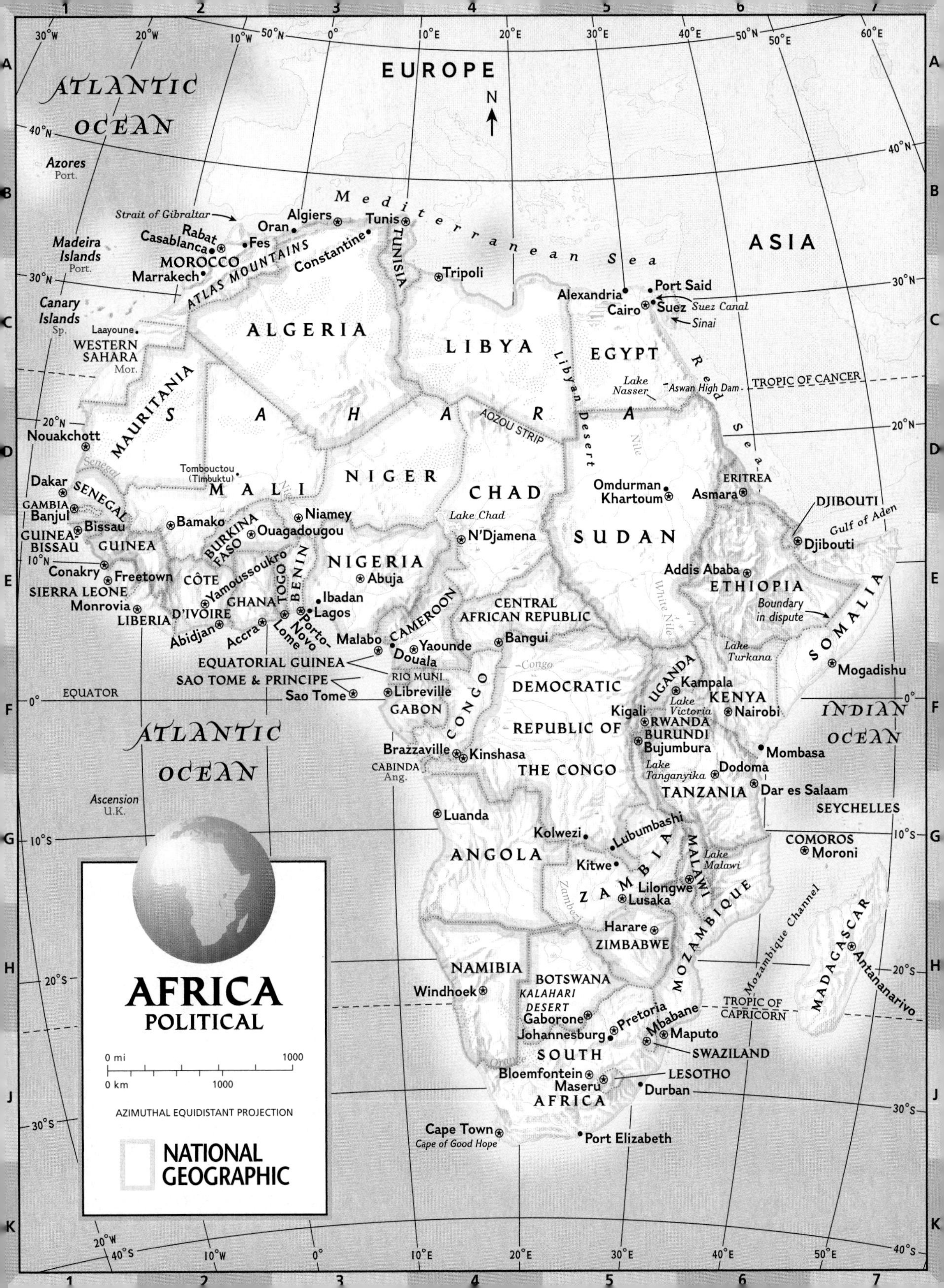

AFRICA
POLITICAL
0 mi 1000
0 km 1000
AZIMUTHAL EQUIDISTANT PROJECTION
NATIONAL GEOGRAPHIC
EUROPE
ASIA
ATLANTIC OCEAN
INDIAN OCEAN
Mediterranean Sea
Red Sea
Gulf of Aden
Mozambique Channel
Azores Port.
Madeira Islands Port.
Canary Islands Sp.
Ascension U.K.
Strait of Gibraltar
Suez Canal
Sinai
TROPIC OF CANCER
EQUATOR
TROPIC OF CAPRICORN
SAHARA
ATLAS MOUNTAINS
Libyan Desert
AOZOU STRIP
KALAHARI DESERT
Cape of Good Hope
Lake Nasser
Aswan High Dam
Lake Chad
Lake Turkana
Lake Victoria
Lake Tanganyika
Lake Malawi
Nile
White Nile
Congo
Zambezi
Orange
Senegal
Niger
Boundary in dispute
MOROCCO
Rabat
Casablanca
Fes
Marrakech
WESTERN SAHARA Mor.
Laayoune
ALGERIA
Algiers
Oran
Constantine
TUNISIA
Tunis
LIBYA
Tripoli
EGYPT
Alexandria
Cairo
Port Said
Suez
MAURITANIA
Nouakchott
MALI
Tombouctou (Timbuktu)
Bamako
NIGER
Niamey
CHAD
N'Djamena
SUDAN
Omdurman
Khartoum
ERITREA
Asmara
DJIBOUTI
Djibouti
ETHIOPIA
Addis Ababa
SOMALIA
Mogadishu
SENEGAL
Dakar
GAMBIA
Banjul
GUINEA-BISSAU
Bissau
GUINEA
Conakry
SIERRA LEONE
Freetown
LIBERIA
Monrovia
CÔTE D'IVOIRE
Yamoussoukro
Abidjan
BURKINA FASO
Ouagadougou
GHANA
Accra
TOGO
Lome
BENIN
Porto-Novo
NIGERIA
Abuja
Ibadan
Lagos
CAMEROON
Yaounde
Douala
EQUATORIAL GUINEA
Malabo
RIO MUNI
SAO TOME & PRINCIPE
Sao Tome
GABON
Libreville
CENTRAL AFRICAN REPUBLIC
Bangui
CONGO
Brazzaville
DEMOCRATIC REPUBLIC OF THE CONGO
Kinshasa
Kolwezi
Lubumbashi
CABINDA Ang.
UGANDA
Kampala
KENYA
Nairobi
Mombasa
RWANDA
Kigali
BURUNDI
Bujumbura
TANZANIA
Dodoma
Dar es Salaam
SEYCHELLES
COMOROS
Moroni
ANGOLA
Luanda
ZAMBIA
Kitwe
Lusaka
MALAWI
Lilongwe
MOZAMBIQUE
Maputo
ZIMBABWE
Harare
NAMIBIA
Windhoek
BOTSWANA
Gaborone
SOUTH AFRICA
Pretoria
Johannesburg
Bloemfontein
Cape Town
Durban
Port Elizabeth
SWAZILAND
Mbabane
LESOTHO
Maseru
MADAGASCAR
Antananarivo
N

EUROPE
PHYSICAL
0 mi 400
0 km 400
AZIMUTHAL EQUIDISTANT PROJECTION
NATIONAL GEOGRAPHIC
ATLANTIC OCEAN
ICELAND
Reykjavik
ARCTIC CIRCLE
MERIDIAN OF GREENWICH (LONDON)
Norwegian Sea
Faroe Islands
Shetland Islands
Orkney Islands
Outer Hebrides
British Isles
Highlands
Edinburgh
Belfast
UNITED KINGDOM
IRELAND
Dublin
Irish Sea
Great Britain
Cardiff
London
North Sea
English Channel
SCANDINAVIA
NORWAY
SWEDEN
Oslo
Stockholm
Gulf of
Jutland
DENMARK
Copenhagen
Zealand
Baltic
Amsterdam
NETH.
BELGIUM
Brussels
LUX.
Rhine
GERMANY
Berlin
NORTH
POLAND
Oder
Elbe
Prague
CZECH REP.
Danube
Bratislava
SLOVAKIA
Vienna
LIECH.
AUSTRIA
Budapest
HUNGARY
Brittany
Seine
Paris
Loire
FRANCE
Bay of Biscay
Mont Blanc
15,771 ft
4,807 m
Bern
SWITZ.
ALPS
Massif Central
Rhone
SLOVENIA
Ljubljana
Zagreb
Drava
CROATIA
Sava
Po
MONACO
Riviera
SAN MARINO
BOSNIA &
Sarajevo
HERZEGOVINA
Adriatic Sea
Cantabrian Mountains
Pyrenees
ANDORRA
Douro
PORTUGAL
IBERIAN PENINSULA
Madrid
SPAIN
Ebro
Tagus
Lisbon
Corsica
APENNINES
ITALY
VATICAN CITY
Rome
Tirana
ALBANIA
Sardinia
Tyrrhenian Sea
Balearic Islands
Baetic Mountains
GIBRALTAR
Strait of Gibraltar
Mediterranean
Ionian Sea
Sicily
Etna
10,902 ft
3,323 m
Valletta
MALTA
AFRICA
60°N
50°N
40°N
30°N
70°N
40°W
30°W
20°W
10°W
0°
10°E
N

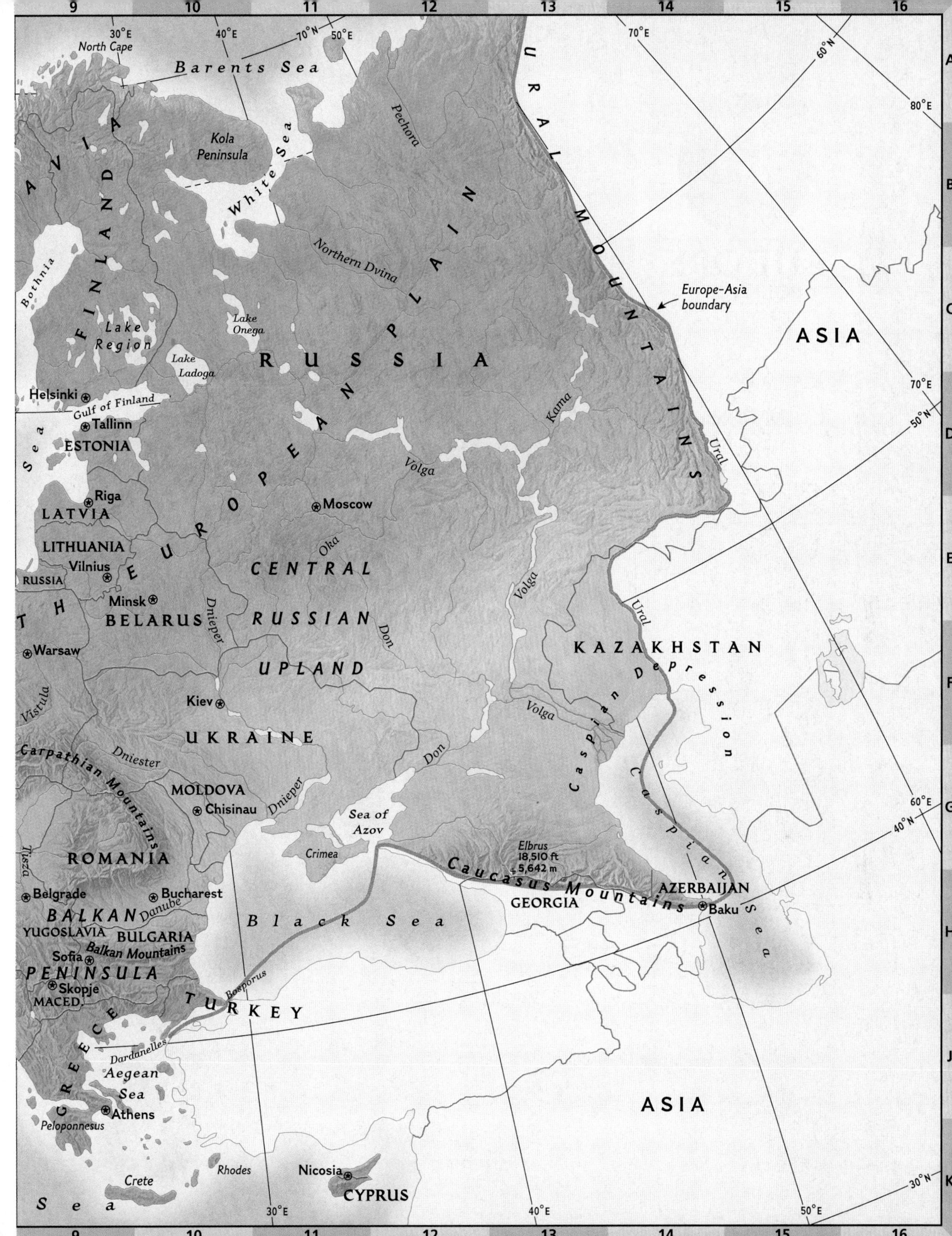

North Cape
Barents Sea
Kola Peninsula
White Sea
Pechora
URAL MOUNTAINS
Northern Dvina
Europe-Asia boundary
FINLAND
Lake Region
Lake Onega
Lake Ladoga
Bothnia
RUSSIA
ASIA
Helsinki
Gulf of Finland
Tallinn
ESTONIA
Kama
Ural
Volga
Riga
LATVIA
Moscow
LITHUANIA
Oka
Vilnius
RUSSIA
CENTRAL RUSSIAN UPLAND
THE EUROPEAN PLAIN
Minsk
BELARUS
Dnieper
Don
KAZAKHSTAN
Warsaw
Caspian Depression
Vistula
Kiev
UKRAINE
Carpathian Mountains
Dniester
MOLDOVA
Chisinau
Dnieper
Sea of Azov
Crimea
Elbrus 18,510 ft 5,642 m
Caucasus Mountains
Caspian Sea
ROMANIA
Tisza
Belgrade
Bucharest
Danube
AZERBAIJAN
GEORGIA
Baku
BALKAN PENINSULA
YUGOSLAVIA
BULGARIA
Balkan Mountains
Sofia
Black Sea
Skopje
MACED.
Bosporus
TURKEY
GREECE
Dardanelles
Aegean Sea
Athens
Peloponnesus
Rhodes
Nicosia
CYPRUS
Crete
Sea

EUROPE
POLITICAL
0 mi 400
0 km 400
AZIMUTHAL EQUIDISTANT PROJECTION
NATIONAL GEOGRAPHIC
ATLANTIC OCEAN
ICELAND
Reykjavik
Akureyri
ARCTIC CIRCLE
MERIDIAN OF GREENWICH (LONDON)
Norwegian Sea
Faroe Islands Den.
Torshavn
Shetland Islands
Lerwick
Orkney Islands
Rockall U.K.
Isle of Lewis
NORWAY
SWEDEN
Tromso
Trondheim
Are
Alesund
Sundsvall
Gulf of
Bergen
Oslo
Stavanger
Uppsala
Stockholm
Skagerrak
Goteborg
Gotland
Baltic
UNITED KINGDOM
SCOTLAND
Inverness
Aberdeen
Glasgow
Edinburgh
NORTHERN IRELAND
Belfast
IRELAND
Dublin
Cork
Irish Sea
North Sea
Liverpool
Manchester
WALES
ENGLAND
Birmingham
Cardiff
London
Celtic Sea
Land's End
Southampton
English Channel
DENMARK
Arhus
Copenhagen
Malmo
Kiel
Hamburg
Gdansk
NETH.
The Hague
Amsterdam
Berlin
Bydgoszcz
GERMANY
POLAND
Brussels
BELGIUM
Bonn
Rhine
Frankfurt
LUX.
Lodz
Wroclaw
Oder
Prague
CZECH REP.
Le Havre
Brest
Rennes
Paris
Nantes
Strasbourg
FRANCE
La Rochelle
Bay of Biscay
Limoges
Bordeaux
Munich
Bratislava
SLOVAKIA
Vienna
Zurich
LIECH.
Bern
SWITZERLAND
AUSTRIA
Budapest
HUNGARY
Geneva
Lyon
ALPS
SLOVENIA
Ljubljana
Zagreb
CROATIA
Milan
Turin
Venice
Genoa
BOSNIA & HERZEGOVINA
Sarajevo
MONACO
Nice
Marseille
SAN MARINO
Adriatic Sea
MONTENEGRO
ITALY
Corsica Fr.
VATICAN CITY
Rome
Tirana
ALBANIA
Naples
Sardinia It.
Cagliari
Tyrrhenian Sea
Ionian Sea
Palermo
Messina
Sicily
Catania
Valletta
MALTA
A Coruna
Vigo
Porto
Bilbao
Donostia-San Sebastian
Toulouse
Pyrenees
Valladolid
ANDORRA
Coimbra
PORTUGAL
Zaragoza
Madrid
Lisbon
Barcelona
SPAIN
Valencia
Cape St. Vincent
Cordoba
Seville
Palma
Balearic Islands Sp.
Murcia
Cadiz
Cartagena
GIBRALTAR U.K.
Malaga
Strait of Gibraltar
Mediterranean
AFRICA
N
60°N
50°N
40°N
30°N
40°W
30°W
20°W
10°W
0°
10°E
70°N
1 2 3 4 5 6 7 8
A B C D E F G H J K

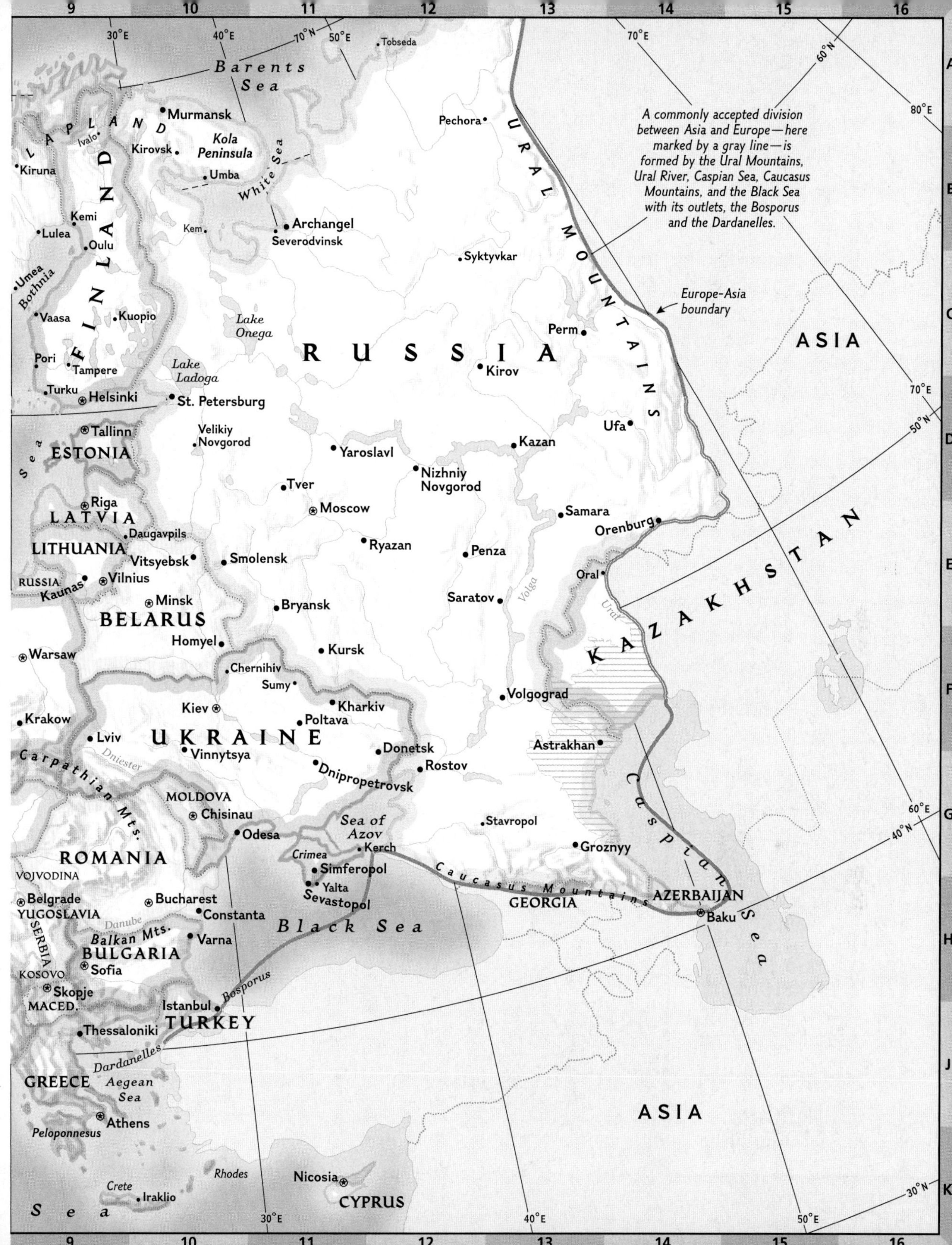

Barents Sea
Murmansk
Kola Peninsula
White Sea
LAPLAND
Ivalo
Kirovsk
Umba
Kiruna
FINLAND
Kemi
Lulea
Oulu
Kem
Archangel
Severodvinsk
Tobseda
Pechora
URAL MOUNTAINS
A commonly accepted division between Asia and Europe—here marked by a gray line—is formed by the Ural Mountains, Ural River, Caspian Sea, Caucasus Mountains, and the Black Sea with its outlets, the Bosporus and the Dardanelles.
Syktyvkar
Europe-Asia boundary
Umea
Bothnia
Vaasa
Kuopio
Lake Onega
Perm
ASIA
RUSSIA
Pori
Tampere
Lake Ladoga
Kirov
Turku
Helsinki
St. Petersburg
Ufa
Tallinn
Velikiy Novgorod
Kazan
ESTONIA
Yaroslavl
Sea
Nizhniy Novgorod
Tver
Riga
Moscow
Samara
LATVIA
Orenburg
Daugavpils
Ryazan
KAZAKHSTAN
LITHUANIA
Vitsyebsk
Smolensk
Penza
RUSSIA
Oral
Kaunas
Vilnius
Minsk
Saratov
Volga
Bryansk
Ural
BELARUS
Homyel
Kursk
Warsaw
Chernihiv
Sumy
Volgograd
Kiev
Kharkiv
Krakow
Poltava
Lviv
UKRAINE
Astrakhan
Dniester
Donetsk
Carpathian Mts.
Vinnytsya
Rostov
Dnipropetrovsk
Caspian Sea
MOLDOVA
Chisinau
Sea of Azov
Stavropol
Odesa
Kerch
Groznyy
ROMANIA
Crimea
Simferopol
Caucasus Mountains
VOJVODINA
Yalta
AZERBAIJAN
Belgrade
Sevastopol
GEORGIA
Bucharest
Baku
YUGOSLAVIA
Constanta
Danube
Black Sea
SERBIA
Balkan Mts.
Varna
BULGARIA
Sofia
KOSOVO
Bosporus
Skopje
MACED.
Istanbul
TURKEY
Thessaloniki
Dardanelles
GREECE
Aegean Sea
ASIA
Athens
Peloponnesus
Rhodes
Nicosia
Crete
Iraklio
CYPRUS
Sea
30°E
40°E
50°E
70°E
80°E
60°E
70°N
60°N
50°N
40°N
30°N
9
10
11
12
13
14
15
16
A
B
C
D
E
F
G
H
J
K

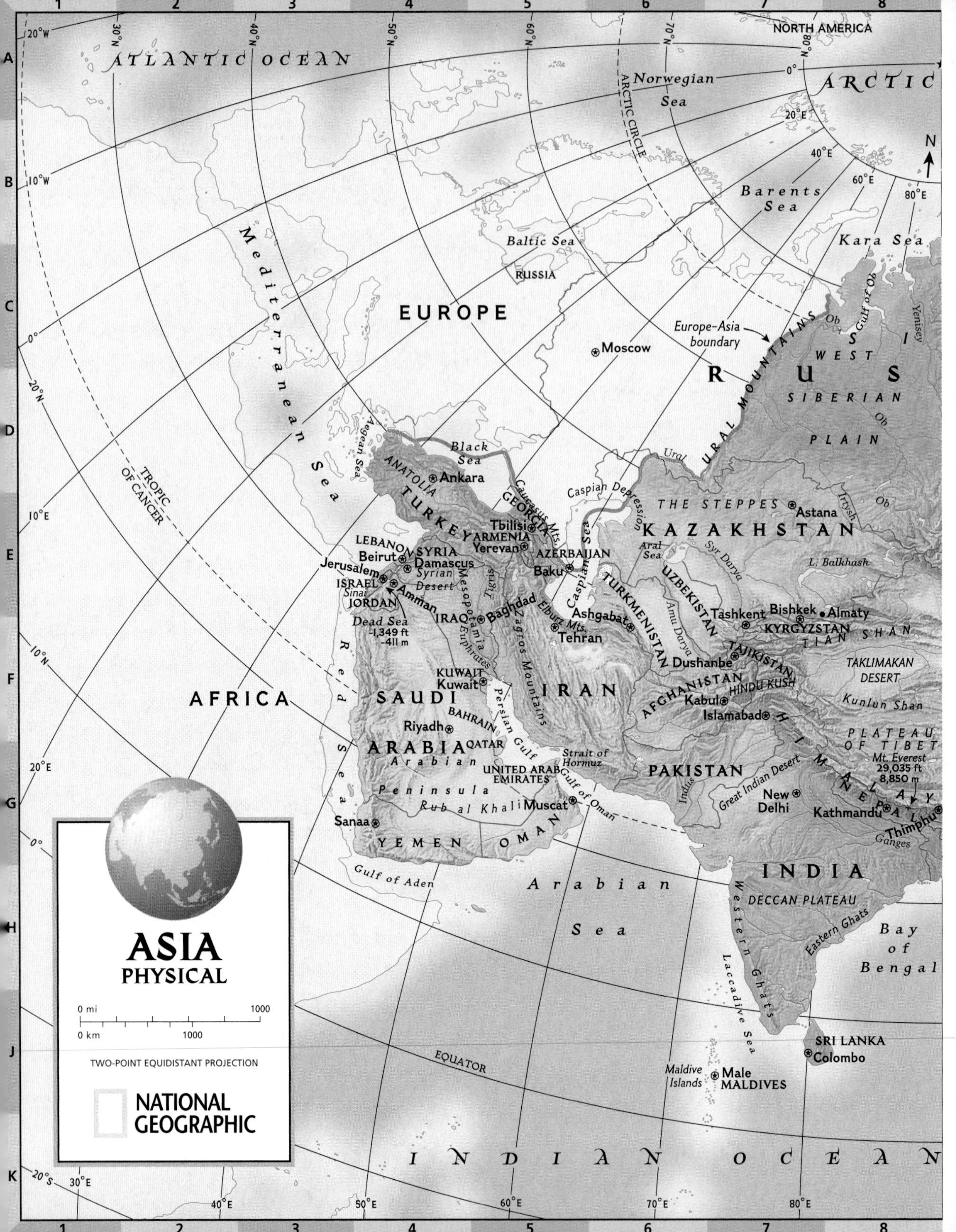

ASIA
PHYSICAL
0 mi 1000
0 km 1000
TWO-POINT EQUIDISTANT PROJECTION
NATIONAL GEOGRAPHIC
ATLANTIC OCEAN
NORTH AMERICA
Norwegian Sea
ARCTIC
ARCTIC CIRCLE
Barents Sea
Kara Sea
Baltic Sea
RUSSIA
EUROPE
Mediterranean Sea
Moscow
Europe-Asia boundary
URAL MOUNTAINS
R U S
WEST SIBERIAN PLAIN
Black Sea
Aegean Sea
ANATOLIA
Ankara
TURKEY
GEORGIA
Caucasus Mts.
Tbilisi
ARMENIA
Yerevan
AZERBAIJAN
Baku
Caspian Depression
Caspian Sea
THE STEPPES
Astana
KAZAKHSTAN
Aral Sea
Syr Darya
L. Balkhash
UZBEKISTAN
TURKMENISTAN
Amu Darya
Tashkent
Bishkek
Almaty
KYRGYZSTAN
TIAN SHAN
TAJIKISTAN
Dushanbe
Ashgabat
Elburz Mts.
Tehran
LEBANON
SYRIA
Beirut
Damascus
Syrian Desert
Jerusalem
ISRAEL
Sinai
Amman
JORDAN
Dead Sea -1,349 ft -411 m
IRAQ
Baghdad
Mesopotamia
Tigris
Euphrates
Zagros Mountains
IRAN
KUWAIT
Kuwait
Persian Gulf
BAHRAIN
QATAR
SAUDI ARABIA
Riyadh
Arabian Peninsula
Rub al Khali
UNITED ARAB EMIRATES
Strait of Hormuz
Gulf of Oman
Muscat
OMAN
YEMEN
Sanaa
Gulf of Aden
Red Sea
AFRICA
TROPIC OF CANCER
EQUATOR
AFGHANISTAN
Kabul
HINDU KUSH
Islamabad
PAKISTAN
Indus
Great Indian Desert
New Delhi
HIMALAYA
Kathmandu
NEPAL
Thimphu
Mt. Everest 29,035 ft 8,850 m
TAKLIMAKAN DESERT
Kunlun Shan
PLATEAU OF TIBET
Ganges
INDIA
DECCAN PLATEAU
Western Ghats
Eastern Ghats
Bay of Bengal
Arabian Sea
Laccadive Sea
SRI LANKA
Colombo
Maldive Islands
Male
MALDIVES
INDIAN OCEAN
Ob
Gulf of Ob
Yenisey
Irtysh
Ural

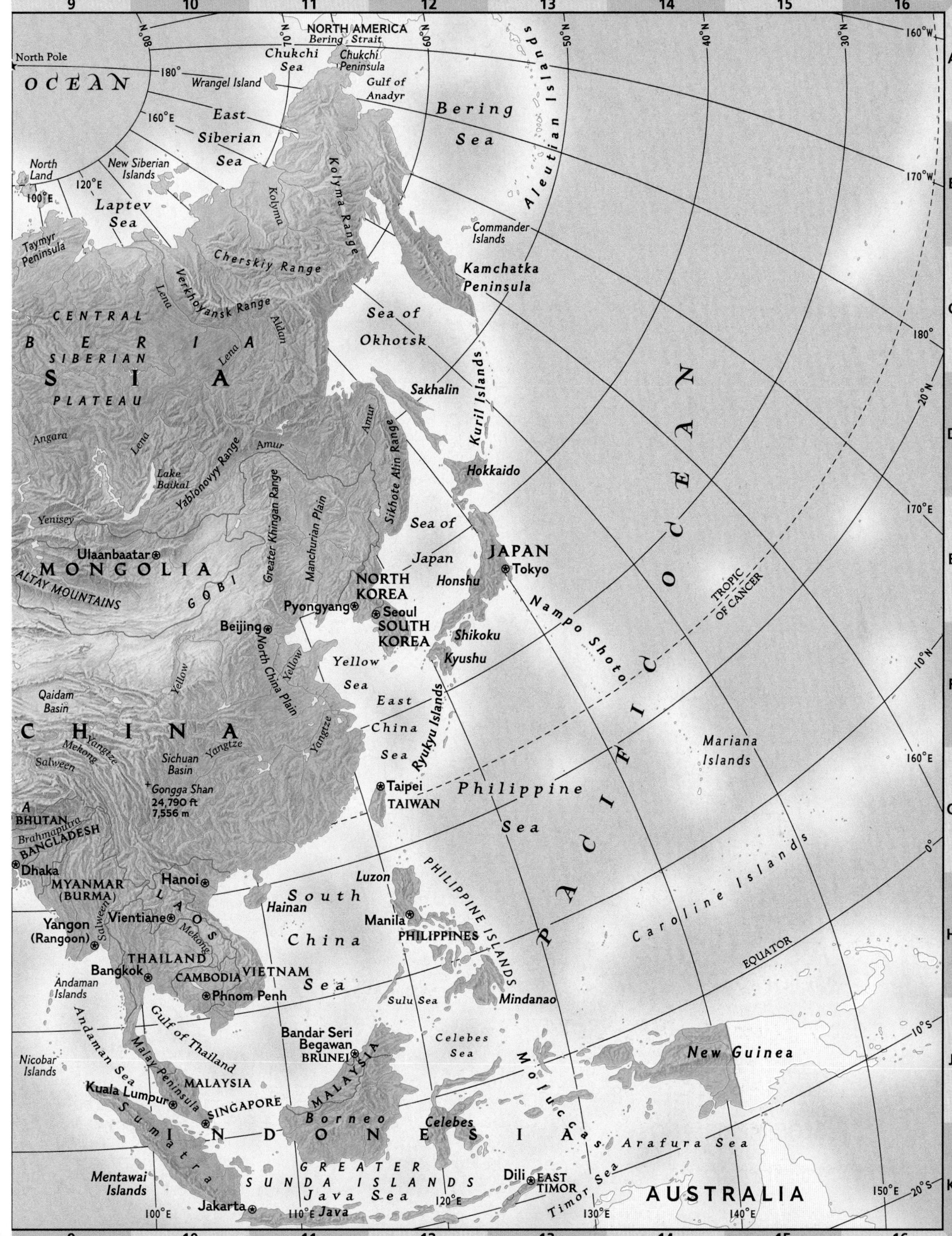

NORTH AMERICA
Bering Strait
North Pole
OCEAN
Chukchi Sea
Chukchi Peninsula
Wrangel Island
Gulf of Anadyr
East Siberian Sea
Bering Sea
Aleutian Islands
North Land
New Siberian Islands
Laptev Sea
Taymyr Peninsula
Kolyma
Kolyma Range
Commander Islands
Cherskiy Range
Kamchatka Peninsula
Verkhoyansk Range
Lena
Aldan
CENTRAL SIBERIAN PLATEAU
SIBERIA
ASIA
Sea of Okhotsk
Sakhalin
Kuril Islands
Amur
Sikhote Alin Range
Angara
Yablonovyy Range
Lake Baikal
Hokkaido
Yenisey
Greater Khingan Range
Manchurian Plain
Sea of Japan
JAPAN
Tokyo
Ulaanbaatar
MONGOLIA
ALTAY MOUNTAINS
GOBI
NORTH KOREA
Pyongyang
Seoul
SOUTH KOREA
Honshu
Shikoku
Kyushu
Beijing
North China Plain
Yellow
Yellow Sea
Nampo Shoto
PACIFIC OCEAN
TROPIC OF CANCER
Qaidam Basin
East China Sea
Ryukyu Islands
CHINA
Yangtze
Mekong
Salween
Sichuan Basin
Mariana Islands
Gongga Shan 24,790 ft 7,556 m
Taipei
TAIWAN
Philippine Sea
BHUTAN
Brahmaputra
BANGLADESH
Dhaka
MYANMAR (BURMA)
Hanoi
LAOS
Luzon
PHILIPPINE ISLANDS
Caroline Islands
Hainan
South China Sea
Manila
PHILIPPINES
Yangon (Rangoon)
Vientiane
THAILAND
Bangkok
CAMBODIA
VIETNAM
EQUATOR
Andaman Islands
Phnom Penh
Sulu Sea
Mindanao
Andaman Sea
Gulf of Thailand
Bandar Seri Begawan
BRUNEI
Celebes Sea
Moluccas
New Guinea
Nicobar Islands
Malay Peninsula
MALAYSIA
Kuala Lumpur
SINGAPORE
Borneo
Celebes
INDONESIA
Sumatra
Arafura Sea
Mentawai Islands
GREATER SUNDA ISLANDS
Java Sea
Dili
EAST TIMOR
Timor Sea
AUSTRALIA
Jakarta
Java
180°
160°E
120°E
100°E
80°N
70°N
60°N
50°N
40°N
30°N
20°N
10°N
0°
10°S
20°S
160°W
170°W
170°E
160°E
150°E
140°E
130°E
120°E
110°E
100°E
9
10
11
12
13
14
15
16
A
B
C
D
E
F
G
H
J
K

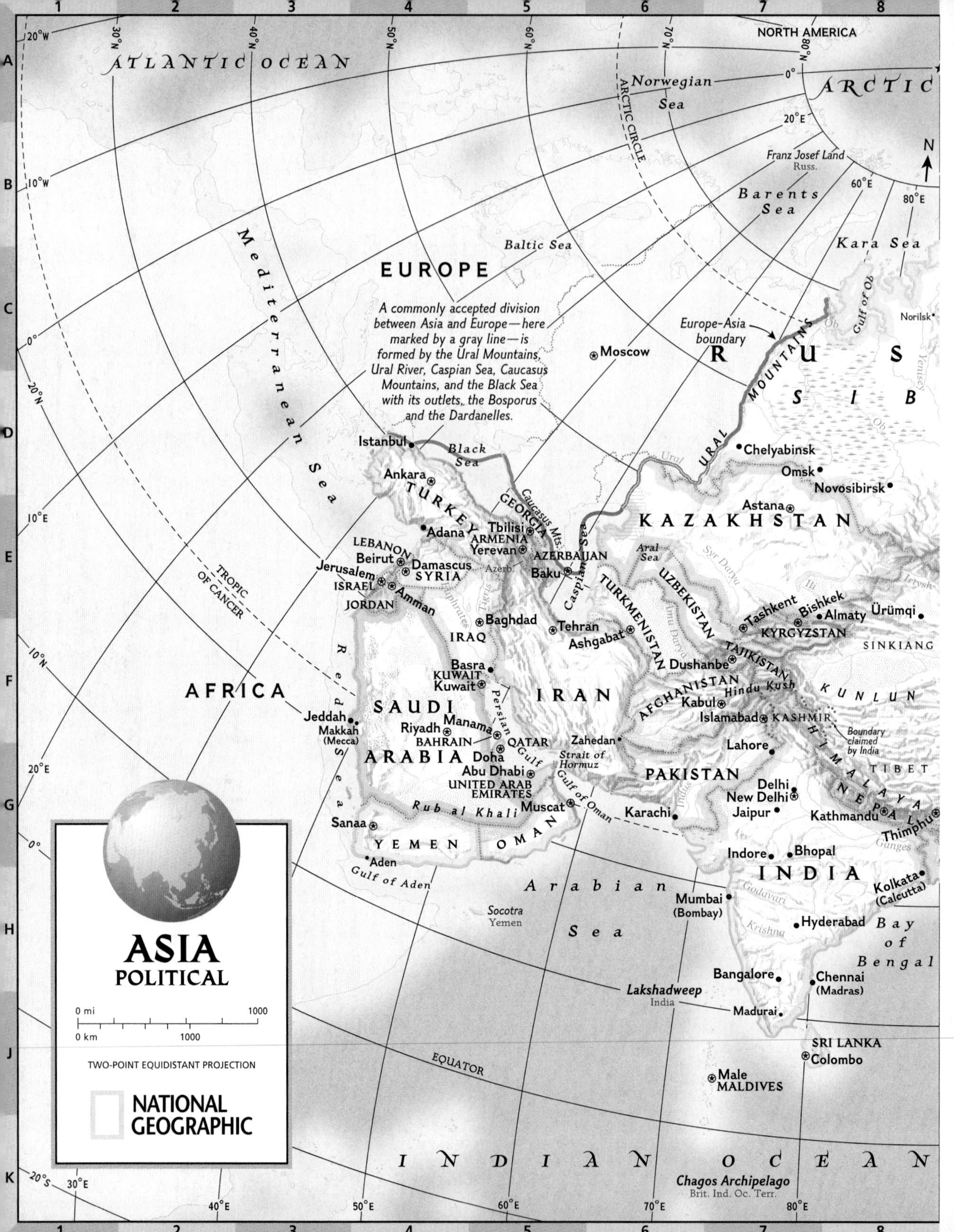

ATLANTIC OCEAN
NORTH AMERICA
Norwegian Sea
ARCTIC CIRCLE
ARCTIC
Franz Josef Land
Russ.
Barents Sea
Kara Sea
Baltic Sea
Mediterranean Sea
EUROPE
A commonly accepted division between Asia and Europe—here marked by a gray line—is formed by the Ural Mountains, Ural River, Caspian Sea, Caucasus Mountains, and the Black Sea with its outlets, the Bosporus and the Dardanelles.
Europe-Asia boundary
Moscow
URAL MOUNTAINS
RUSSIA
SIBERIA
Norilsk
Gulf of Ob
Ob
Yenisey
Istanbul
Black Sea
Ankara
TURKEY
GEORGIA
Caucasus Mts.
Tbilisi
ARMENIA
Yerevan
Adana
AZERBAIJAN
Azerb.
Baku
Caspian Sea
Ural
Chelyabinsk
Omsk
Novosibirsk
Astana
KAZAKHSTAN
Aral Sea
Syr Darya
LEBANON
Beirut
Damascus
SYRIA
Jerusalem
ISRAEL
Amman
JORDAN
TROPIC OF CANCER
Euphrates
Tigris
Baghdad
IRAQ
Tehran
Ashgabat
TURKMENISTAN
UZBEKISTAN
Amu Darya
Tashkent
Bishkek
Almaty
Ürümqi
KYRGYZSTAN
SINKIANG
Irtysh
Ili
TAJIKISTAN
Dushanbe
Basra
KUWAIT
Kuwait
IRAN
AFGHANISTAN
Hindu Kush
Kabul
Islamabad
KASHMIR
KUNLUN
AFRICA
Red Sea
Jeddah
Makkah (Mecca)
SAUDI ARABIA
Riyadh
Manama
BAHRAIN
Persian Gulf
QATAR
Doha
Zahedan
Strait of Hormuz
Lahore
Boundary claimed by India
HIMALAYA
TIBET
Abu Dhabi
UNITED ARAB EMIRATES
PAKISTAN
Indus
Delhi
New Delhi
Jaipur
Muscat
Gulf of Oman
Karachi
Rub al Khali
Kathmandu
NEPAL
Thimphu
Sanaa
YEMEN
OMAN
Aden
Gulf of Aden
Indore
Bhopal
Ganges
INDIA
Kolkata (Calcutta)
Arabian Sea
Mumbai (Bombay)
Godavari
Socotra
Yemen
Hyderabad
Krishna
Bay of Bengal
Bangalore
Chennai (Madras)
Lakshadweep
India
Madurai
SRI LANKA
Colombo
EQUATOR
Male
MALDIVES
INDIAN OCEAN
Chagos Archipelago
Brit. Ind. Oc. Terr.
ASIA
POLITICAL
0 mi 1000
0 km 1000
TWO-POINT EQUIDISTANT PROJECTION
NATIONAL GEOGRAPHIC

NORTH AMERICA
Bering Strait
Chukchi Sea
North Pole
OCEAN
Wrangel I.
Gulf of Anadyr
Anadyr
Bering Sea
East Siberian Sea
North Land
New Siberian Islands
Laptev Sea
Kolyma Range
Commander Is.
Kamchatka Peninsula
Cherskiy Range
Verkhoyansk Range
Magadan
Sea of Okhotsk
Yakutsk
SIA
ERIA
Sakhalin
Kuril Islands
Lake Baikal
Irkutsk
MANCHURIA
Hokkaido
Sapporo
Sea of Japan
Vladivostok
JAPAN
Tokyo
Ulaanbaatar
MONGOLIA
ALTAY MTS.
GOBI
Changchun
Shenyang
NORTH KOREA
Pyongyang
Seoul
SOUTH KOREA
Honshu
Kyoto
Osaka
Hiroshima
Kyushu
Beijing
Qingdao
Shijiazhuang
Yellow Sea
Marcus I. Jap.
TROPIC OF CANCER
Bonin Is. Jap.
Volcano Is. Jap.
PACIFIC OCEAN
SHAN
Lanzhou
Xi'an
Xuzhou
Nanjing
Shanghai
East China Sea
Ryukyu Islands
Okinawa
CHINA
Chengdu
Nanchang
Fuzhou
Taipei
TAIWAN
Parece Vela Jap.
Boundary claimed by China
Changsha
Guiyang
BHUTAN
BANGLADESH
Dhaka
Kunming
Guangzhou
Hong Kong
Macau
The People's Republic of China claims Taiwan as its 23rd province.
Philippine Sea
MYANMAR (BURMA)
Hanoi
Haiphong
South China Sea
Hainan
Luzon
Quezon City
Manila
Samar
PHILIPPINES
Leyte
Mindoro
Panay
Negros
Palawan
Mindanao
Yangon (Rangoon)
Vientiane
LAOS
Da Nang
THAILAND
Bangkok
CAMBODIA
VIETNAM
Phnom Penh
Ho Chi Minh City
Andaman Islands India
Gulf of Thailand
Andaman Sea
EQUATOR
Bandar Seri Begawan
BRUNEI
SABAH
MALAYSIA
SARAWAK
Halmahera
Morotai
Biak
Jayapura
New Guinea
Kepi
Merauke
Nicobar Islands India
Kuala Lumpur
Medan
SINGAPORE
Borneo
Celebes
Moluccas
Buru
Ceram
Aru Is.
Dolak
Tanimbar Is.
INDONESIA
Sumatra
Jambi
GREATER SUNDA ISLANDS
Mentawai Islands
Java Sea
Java
Jakarta
Dili
EAST TIMOR
Timor Sea
Kupang
AUSTRALIA
Lena
Aldan
Amur
Yellow
Yangtze
Mekong
Irrawaddy
Herlen
180°
160°E
120°E
100°E
80°N
70°N
60°N
50°N
40°N
30°N
20°N
10°N
0°
10°S
20°S
160°W
170°W
180°
170°E
160°E
100°E
110°E
120°E
130°E
140°E
150°E
9
10
11
12
13
14
15
16
A
B
C
D
E
F
G
H
J
K

NORTH PACIFIC OCEAN
TROPIC OF CANCER
A S I A
M I C R O N E S I A
NORTHERN MARIANA ISLANDS
U.S.
Saipan
GUAM
Hagatna
U.S.
Wake Island
U.S.
Bikini Atoll
MARSHALL ISLANDS
Ratak Chain
Ralik Chain
Majuro
PALAU
Koror
Yap Islands
Truk Islands
Caroline Islands
Palikir
Pohnpei (Ponape)
FEDERATED STATES OF MICRONESIA
Tarawa
Gilbert Islands
EQUATOR
Yaren
NAURU
M E L A N E S I A
New Guinea
Mt. Wilhelm
14,793 ft
4,509 m
PAPUA NEW GUINEA
New Britain
Port Moresby
Torres Strait
Solomon Is.
SOLOMON ISLANDS
Honiara
TUVALU
Funafuti
Santa Cruz Is.
Darwin
Gulf of Carpentaria
Coral Sea
CORAL SEA ISLANDS TERRITORY
Austral.
VANUATU
Port-Vila
Suva
FIJI ISLANDS
NEW CALEDONIA
Fr.
Noumea
Kimberley Plateau
GREAT DIVIDING RANGE
TROPIC OF CAPRICORN
Macdonnell Ranges
AUSTRALIA
Brisbane
GREAT VICTORIA DESERT
Lake Eyre
-52 ft
-16 m
Norfolk Island
Austral.
Perth
Darling
Lord Howe Island
Austral.
Great Australian Bight
Sydney
Adelaide
Murray
Canberra
Mt. Kosciuszko
7,310 ft
2,228 m
Melbourne
North Island
Auckland
NEW ZEALAND
Tasman Sea
Wellington
INDIAN OCEAN
Tasmania
Hobart
Mt. Cook
12,316 ft
3,754 m
Christchurch
South Island
Stewart Island
120°E
130°E
140°E
150°E
160°E
170°E
20°N
10°N
0°
10°S
20°S
30°S
40°S
50°S
1
2
3
4
5
6
7
8
A
B
C
D
E
F
G
H
J
K

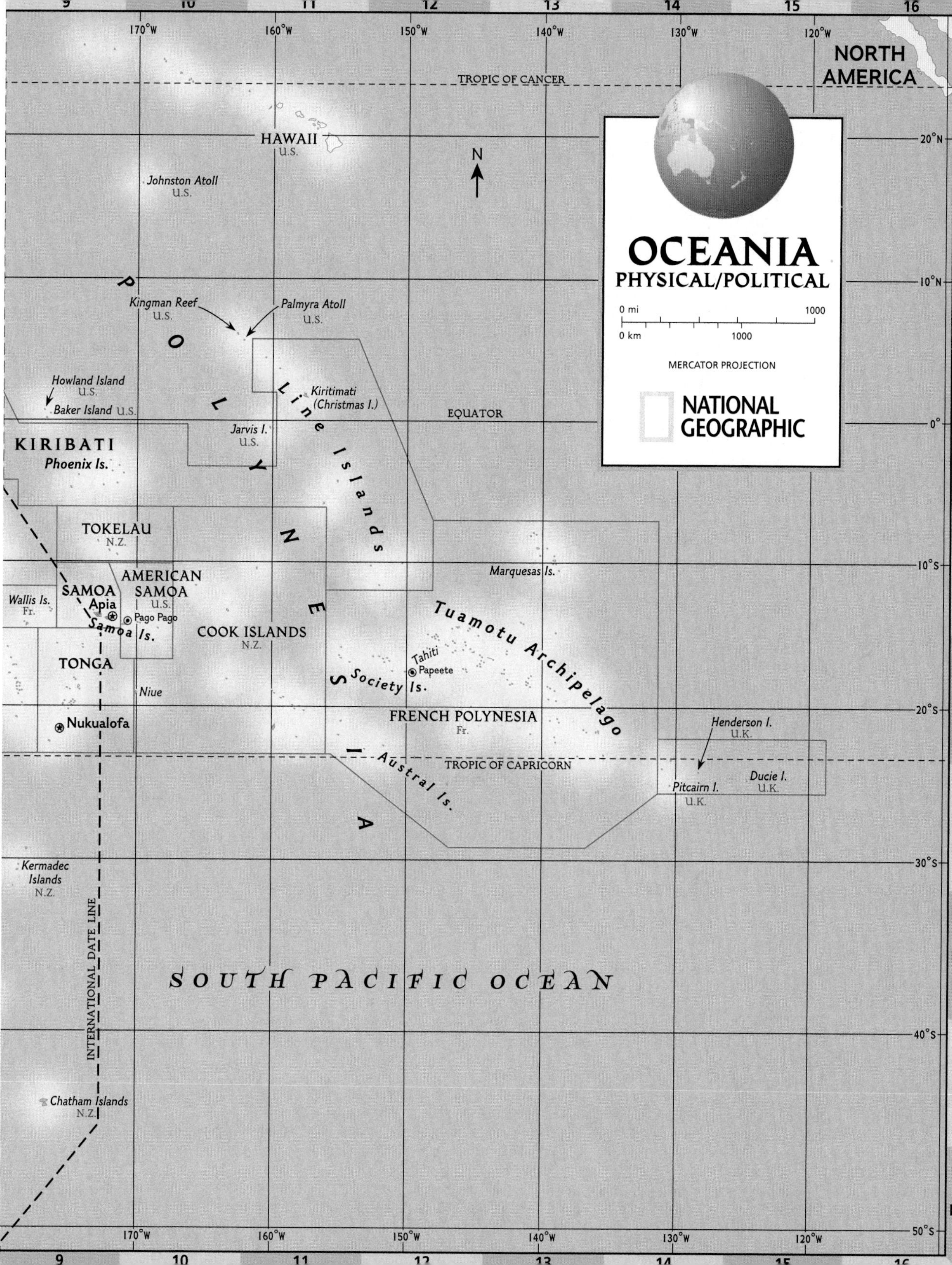

OCEANIA
PHYSICAL/POLITICAL
0 mi
1000
0 km
1000
MERCATOR PROJECTION
NATIONAL GEOGRAPHIC
NORTH AMERICA
TROPIC OF CANCER
HAWAII
U.S.
Johnston Atoll
U.S.
Kingman Reef
U.S.
Palmyra Atoll
U.S.
Howland Island
U.S.
Baker Island U.S.
Kiritimati (Christmas I.)
EQUATOR
Jarvis I.
U.S.
KIRIBATI
Phoenix Is.
POLYNESIA
Line Islands
TOKELAU
N.Z.
AMERICAN SAMOA
U.S.
SAMOA
Apia
Pago Pago
Wallis Is.
Fr.
Samoa Is.
COOK ISLANDS
N.Z.
TONGA
Niue
Nukualofa
Marquesas Is.
Tuamotu Archipelago
Tahiti
Papeete
Society Is.
FRENCH POLYNESIA
Fr.
Henderson I.
U.K.
TROPIC OF CAPRICORN
Austral Is.
Pitcairn I.
U.K.
Ducie I.
U.K.
Kermadec Islands
N.Z.
INTERNATIONAL DATE LINE
SOUTH PACIFIC OCEAN
Chatham Islands
N.Z.
170°W
160°W
150°W
140°W
130°W
120°W
20°N
10°N
0°
10°S
20°S
30°S
40°S
50°S

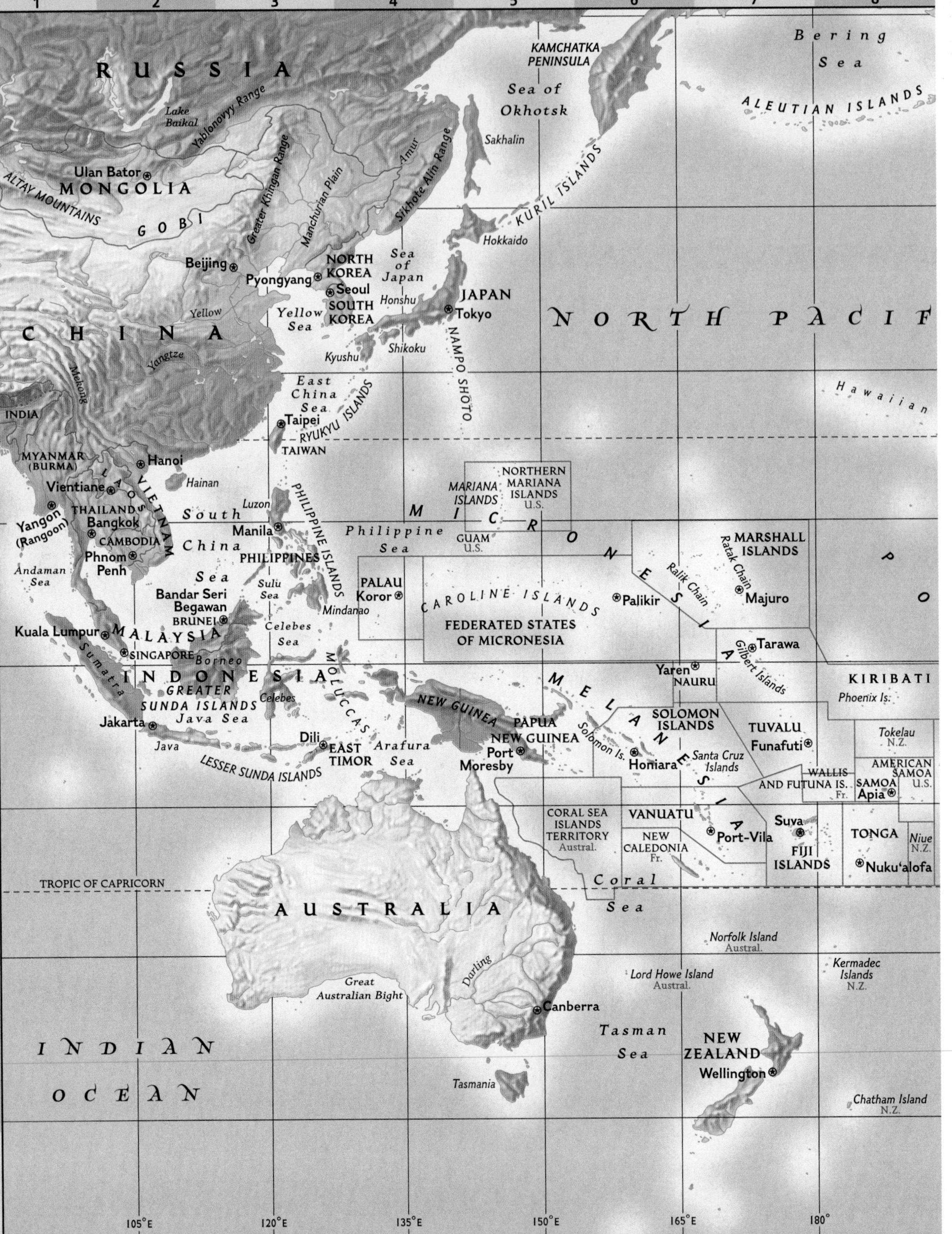

1
2
3
4
5
6
7
8
RUSSIA
KAMCHATKA PENINSULA
Bering Sea
Sea of Okhotsk
ALEUTIAN ISLANDS
Lake Baikal
Yablonovyy Range
Sakhalin
KURIL ISLANDS
Ulan Bator
MONGOLIA
ALTAY MOUNTAINS
GOBI
Greater Khingan Range
Manchurian Plain
Amur
Sikhote Alin Range
Hokkaido
Beijing
Pyongyang
NORTH KOREA
Seoul
SOUTH KOREA
Sea of Japan
Honshu
JAPAN
Tokyo
NORTH PACIF
Yellow
Yellow Sea
CHINA
Yangtze
Mekong
Kyushu
Shikoku
NAMPO SHOTO
East China Sea
RYUKYU ISLANDS
Taipei
TAIWAN
Hawaiian
INDIA
MYANMAR (BURMA)
Hanoi
LAOS
VIETNAM
Hainan
Vientiane
THAILAND
Bangkok
Yangon (Rangoon)
CAMBODIA
Phnom Penh
Andaman Sea
South China Sea
Luzon
Manila
PHILIPPINES
PHILIPPINE ISLANDS
Philippine Sea
NORTHERN MARIANA ISLANDS U.S.
MARIANA ISLANDS
GUAM U.S.
MICRONESIA
MARSHALL ISLANDS
Ratak Chain
Ralik Chain
Majuro
PO
Sulu Sea
Mindanao
PALAU
Koror
CAROLINE ISLANDS
Palikir
FEDERATED STATES OF MICRONESIA
Bandar Seri Begawan
BRUNEI
Kuala Lumpur
MALAYSIA
SINGAPORE
Borneo
Celebes Sea
Sumatra
INDONESIA
GREATER SUNDA ISLANDS
Celebes
MOLUCCAS
Tarawa
Gilbert Islands
Yaren
NAURU
KIRIBATI
Phoenix Is.
Jakarta
Java Sea
Java
NEW GUINEA
PAPUA NEW GUINEA
Port Moresby
MELANESIA
Solomon Is.
SOLOMON ISLANDS
Honiara
Santa Cruz Islands
TUVALU
Funafuti
Tokelau N.Z.
Dili
EAST TIMOR
Arafura Sea
LESSER SUNDA ISLANDS
WALLIS AND FUTUNA IS. Fr.
AMERICAN SAMOA U.S.
SAMOA
Apia
CORAL SEA ISLANDS TERRITORY Austral.
VANUATU
Port-Vila
NEW CALEDONIA Fr.
Suva
FIJI ISLANDS
TONGA
Nuku'alofa
Niue N.Z.
TROPIC OF CAPRICORN
Coral Sea
AUSTRALIA
Norfolk Island Austral.
Lord Howe Island Austral.
Kermadec Islands N.Z.
Darling
Great Australian Bight
Canberra
Tasman Sea
NEW ZEALAND
Wellington
INDIAN OCEAN
Tasmania
Chatham Island N.Z.
105°E
120°E
135°E
150°E
165°E
180°

PACIFIC
RIM
PHYSICAL/POLITICAL
0 mi
1500
0 km
1500
MILLER CYLINDRICAL PROJECTION
NATIONAL GEOGRAPHIC
Gulf of Alaska
Kodiak I.
Alexander Archipelago
Queen Charlotte Islands
Vancouver Island
Coast Mountains
ROCKY MOUNTAINS
Cascade Range
GREAT PLAINS
CANADIAN SHIELD
CANADA
Hudson Bay
Ottawa
Great Lakes
Missouri
CENTRAL LOWLAND
APPALACHIAN MTS.
Washington
UNITED STATES
Mississippi
COASTAL PLAIN
ATLANTIC OCEAN
IC OCEAN
Baja California
Gulf of California
Sierra Madre Occidental
Sierra Madre Oriental
MEXICO
Gulf of Mexico
Nassau
BAHAMAS
Havana
CUBA
JAMAICA
HAITI
DOMINICAN REPUBLIC
Santo Domingo
Mexico City
BELIZE
HONDURAS
Tegucigalpa
GUATEMALA
Guatemala City
San Salvador
EL SALVADOR
NICARAGUA
Managua
Caribbean Sea
Caracas
Panama City
San Jose
COSTA RICA
PANAMA
LLANOS
VENEZ.
Bogota
COLOMBIA
Islands
HAWAII
U.S.
TROPIC OF CANCER
Kiritimati
Line Islands
EQUATOR
Galapagos Islands
Ecua.
Quito
ECUADOR
AMAZON BASIN
BRAZIL
ANDES
PERU
Lima
BOLIVIA
La Paz
Marquesas Is.
Tuamotu Archipelago
COOK ISLANDS
N.Z.
Society Is.
FRENCH POLYNESIA
Fr.
Austral Is.
Pitcairn Island U.K.
POLYNESIA
SOUTH PACIFIC OCEAN
Santiago
CHILE
ARGENTINA
PATAGONIA
Chiloe Island
45°N
30°N
15°N
0°
15°S
30°S
45°S
165°W
150°W
135°W
120°W
105°W
90°W
75°W
9
10
11
12
13
14
15
16
D
E
F
G
H
J
K

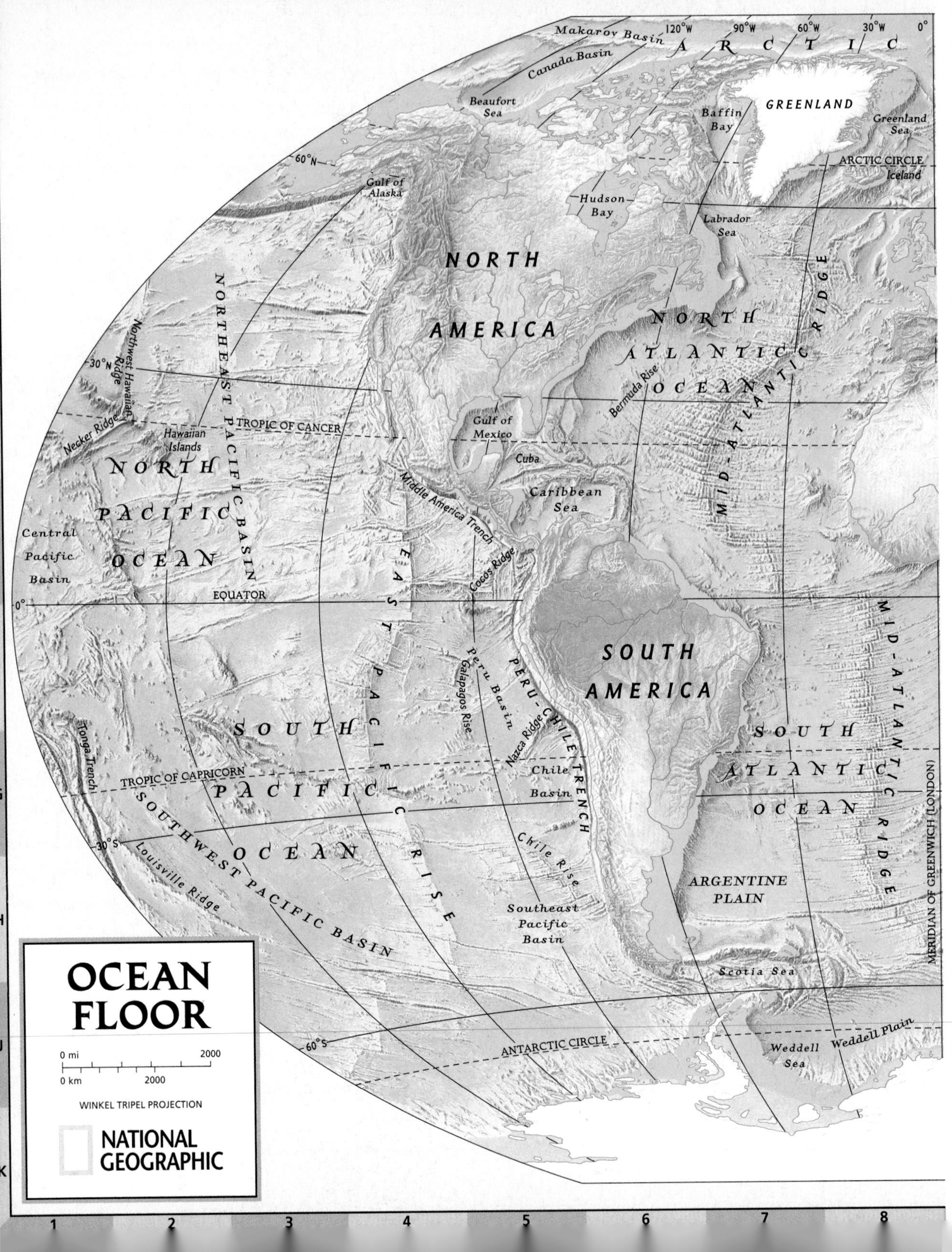

OCEAN FLOOR
NORTH AMERICA
SOUTH AMERICA
GREENLAND
ARCTIC
NORTH ATLANTIC OCEAN
SOUTH ATLANTIC OCEAN
NORTH PACIFIC OCEAN
SOUTH PACIFIC OCEAN
MID-ATLANTIC RIDGE
NORTHEAST PACIFIC BASIN
SOUTHWEST PACIFIC BASIN
EAST PACIFIC RISE
PERU-CHILE TRENCH
ARGENTINE PLAIN
Makarov Basin
Canada Basin
Beaufort Sea
Baffin Bay
Greenland Sea
Iceland
ARCTIC CIRCLE
Gulf of Alaska
Hudson Bay
Labrador Sea
Bermuda Rise
Northwest Hawaiian Ridge
Necker Ridge
Hawaiian Islands
TROPIC OF CANCER
Gulf of Mexico
Cuba
Caribbean Sea
Middle America Trench
Cocos Ridge
Central Pacific Basin
EQUATOR
Peru Basin
Galapagos Rise
Nazca Ridge
Chile Basin
Tonga Trench
TROPIC OF CAPRICORN
Louisville Ridge
Chile Rise
Southeast Pacific Basin
Scotia Sea
Weddell Sea
Weddell Plain
ANTARCTIC CIRCLE
MERIDIAN OF GREENWICH (LONDON)
120°W
90°W
60°W
30°W
0°
60°N
30°N
0°
30°S
60°S
0 mi 2000
0 km 2000
WINKEL TRIPEL PROJECTION
NATIONAL GEOGRAPHIC

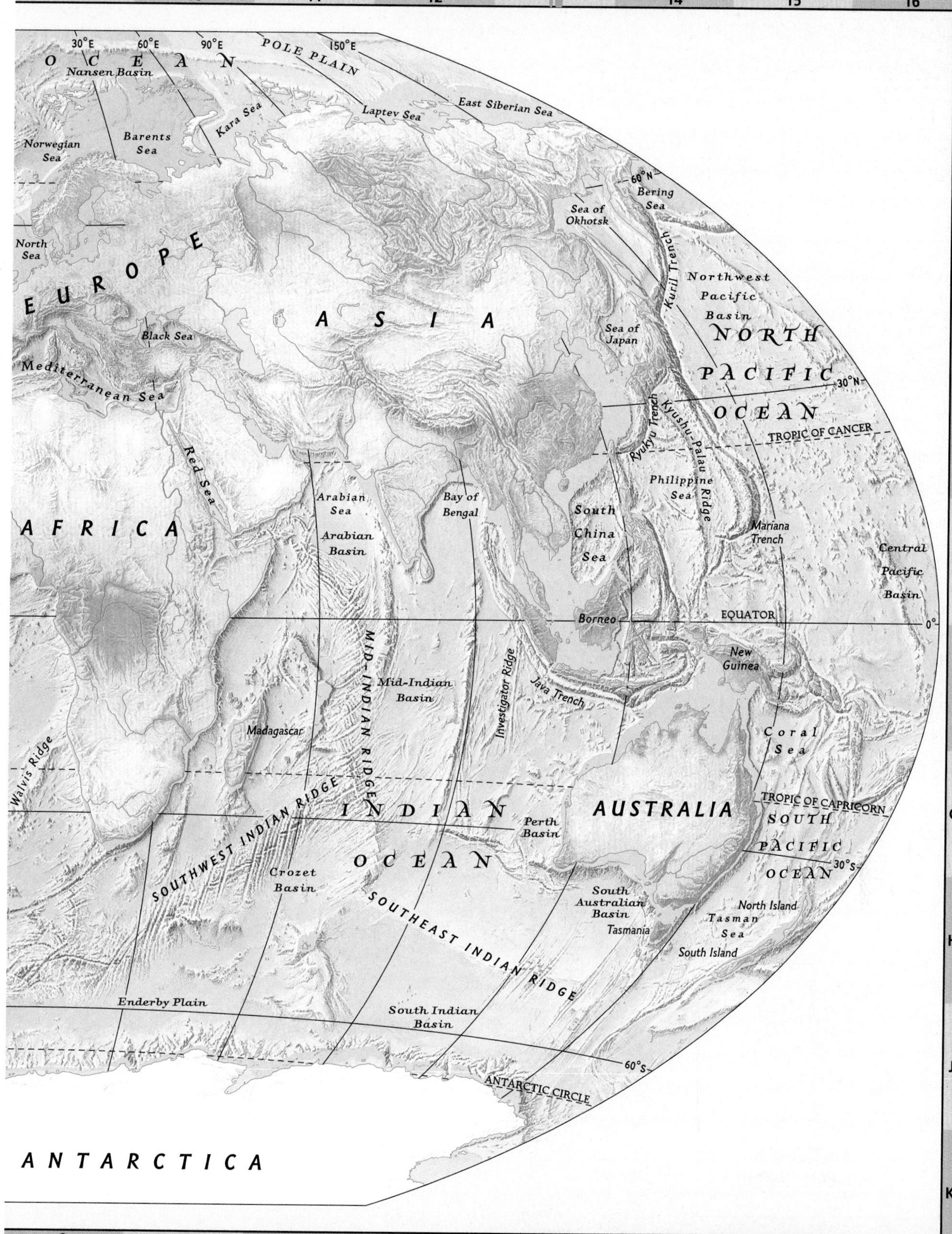

30°E
60°E
90°E
150°E
O C E A N
POLE PLAIN
Nansen Basin
Laptev Sea
East Siberian Sea
Kara Sea
Barents Sea
Norwegian Sea
60°N
Bering Sea
Sea of Okhotsk
North Sea
EUROPE
Kuril Trench
Northwest Pacific Basin
ASIA
Sea of Japan
NORTH PACIFIC OCEAN
Black Sea
Mediterranean Sea
30°N
Ryukyu Trench
Kyushu-Palau Ridge
TROPIC OF CANCER
Red Sea
Philippine Sea
Arabian Sea
Bay of Bengal
South China Sea
AFRICA
Mariana Trench
Arabian Basin
Central Pacific Basin
EQUATOR
Borneo
0°
New Guinea
MID-INDIAN RIDGE
Mid-Indian Basin
Investigator Ridge
Java Trench
Madagascar
Coral Sea
Walvis Ridge
TROPIC OF CAPRICORN
INDIAN OCEAN
AUSTRALIA
SOUTH PACIFIC OCEAN
SOUTHWEST INDIAN RIDGE
Perth Basin
30°S
Crozet Basin
South Australian Basin
North Island
Tasman Sea
Tasmania
South Island
SOUTHEAST INDIAN RIDGE
Enderby Plain
South Indian Basin
60°S
ANTARCTIC CIRCLE
ANTARCTICA
9
10
11
12
14
15
16

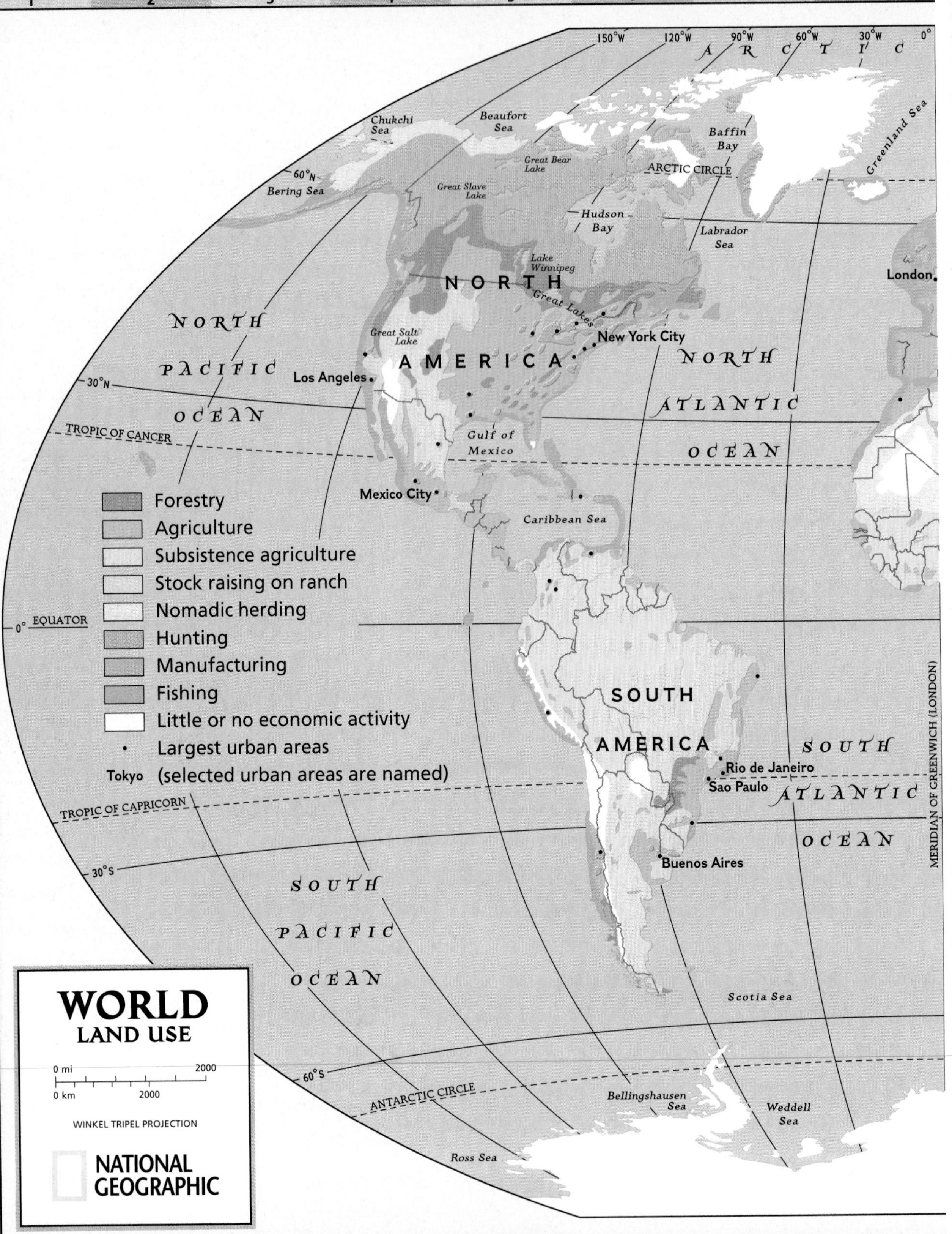

WORLD
LAND USE
Forestry
Agriculture
Subsistence agriculture
Stock raising on ranch
Nomadic herding
Hunting
Manufacturing
Fishing
Little or no economic activity
Largest urban areas
Tokyo (selected urban areas are named)
0 mi 2000
0 km 2000
WINKEL TRIPEL PROJECTION
NATIONAL GEOGRAPHIC
NORTH AMERICA
SOUTH AMERICA
NORTH PACIFIC OCEAN
SOUTH PACIFIC OCEAN
NORTH ATLANTIC OCEAN
SOUTH ATLANTIC OCEAN
ARCTIC
Chukchi Sea
Beaufort Sea
Bering Sea
Great Bear Lake
Great Slave Lake
Baffin Bay
Greenland Sea
ARCTIC CIRCLE
Hudson Bay
Labrador Sea
Lake Winnipeg
Great Lakes
Great Salt Lake
Los Angeles
New York City
London
Gulf of Mexico
Mexico City
Caribbean Sea
Rio de Janeiro
Sao Paulo
Buenos Aires
Scotia Sea
Bellingshausen Sea
Weddell Sea
Ross Sea
TROPIC OF CANCER
TROPIC OF CAPRICORN
EQUATOR
ANTARCTIC CIRCLE
MERIDIAN OF GREENWICH (LONDON)
150°W
120°W
90°W
60°W
30°W
0°
60°N
30°N
0°
30°S
60°S
1 2 3 4 5 6 7 8

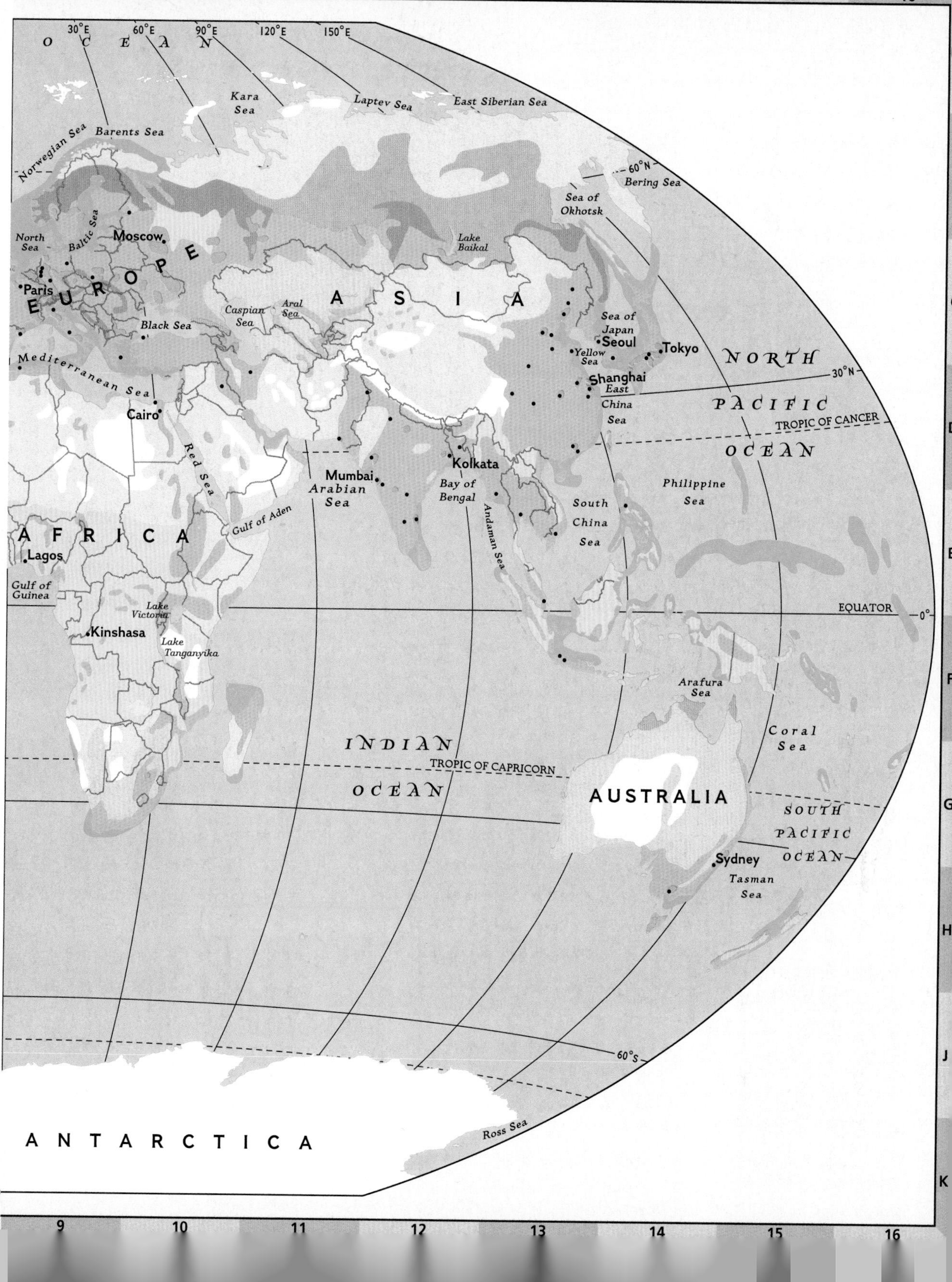

OCEAN
30°E
60°E
90°E
120°E
150°E
Kara Sea
Laptev Sea
East Siberian Sea
Norwegian Sea
Barents Sea
60°N
Bering Sea
Sea of Okhotsk
North Sea
Baltic Sea
Moscow
EUROPE
Lake Baikal
Paris
ASIA
Caspian Sea
Aral Sea
Black Sea
Sea of Japan
Seoul
Tokyo
Yellow Sea
NORTH
Mediterranean Sea
Shanghai
East China Sea
30°N
PACIFIC
Cairo
TROPIC OF CANCER
OCEAN
Red Sea
Kolkata
Mumbai
Arabian Sea
Bay of Bengal
Philippine Sea
South China Sea
Andaman Sea
Gulf of Aden
AFRICA
Lagos
Gulf of Guinea
Lake Victoria
EQUATOR
0°
Kinshasa
Lake Tanganyika
Arafura Sea
Coral Sea
INDIAN
TROPIC OF CAPRICORN
OCEAN
AUSTRALIA
SOUTH PACIFIC OCEAN
Sydney
Tasman Sea
60°S
Ross Sea
ANTARCTICA
9
10
11
12
13
14
15
16
D
E
F
G
H
J
K

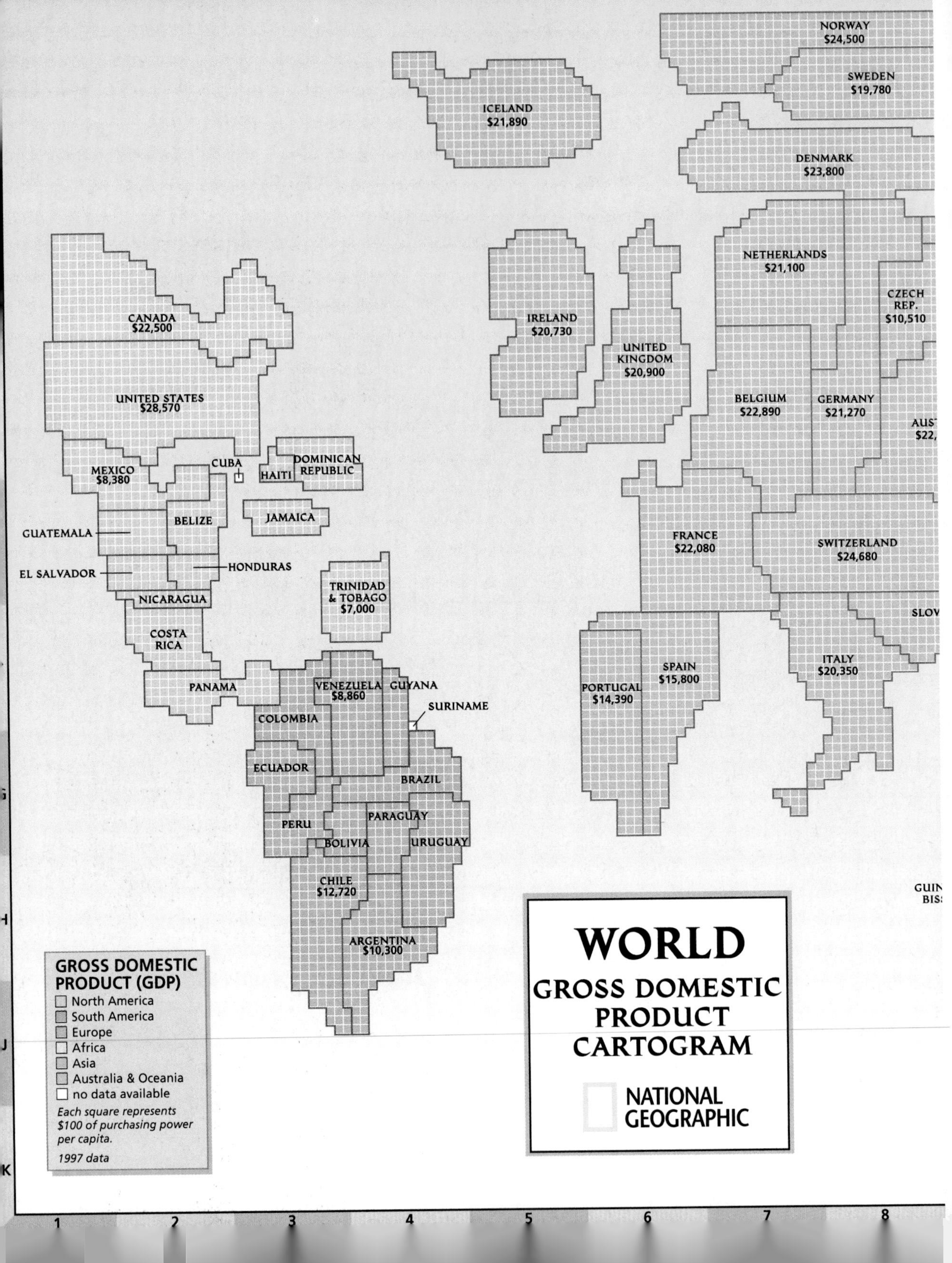

NORWAY
$24,500
SWEDEN
$19,780
ICELAND
$21,890
DENMARK
$23,800
NETHERLANDS
$21,100
CZECH REP.
$10,510
IRELAND
$20,730
UNITED KINGDOM
$20,900
BELGIUM
$22,890
GERMANY
$21,270
CANADA
$22,500
UNITED STATES
$28,570
MEXICO
$8,380
CUBA
HAITI
DOMINICAN REPUBLIC
JAMAICA
BELIZE
GUATEMALA
EL SALVADOR
HONDURAS
NICARAGUA
TRINIDAD & TOBAGO
$7,000
COSTA RICA
PANAMA
FRANCE
$22,080
SWITZERLAND
$24,680
SPAIN
$15,800
PORTUGAL
$14,390
ITALY
$20,350
VENEZUELA
$8,860
GUYANA
SURINAME
COLOMBIA
ECUADOR
BRAZIL
PERU
PARAGUAY
BOLIVIA
URUGUAY
CHILE
$12,720
ARGENTINA
$10,300
GROSS DOMESTIC PRODUCT (GDP)
North America
South America
Europe
Africa
Asia
Australia & Oceania
no data available
Each square represents $100 of purchasing power per capita.
1997 data
WORLD
GROSS DOMESTIC PRODUCT CARTOGRAM
NATIONAL GEOGRAPHIC

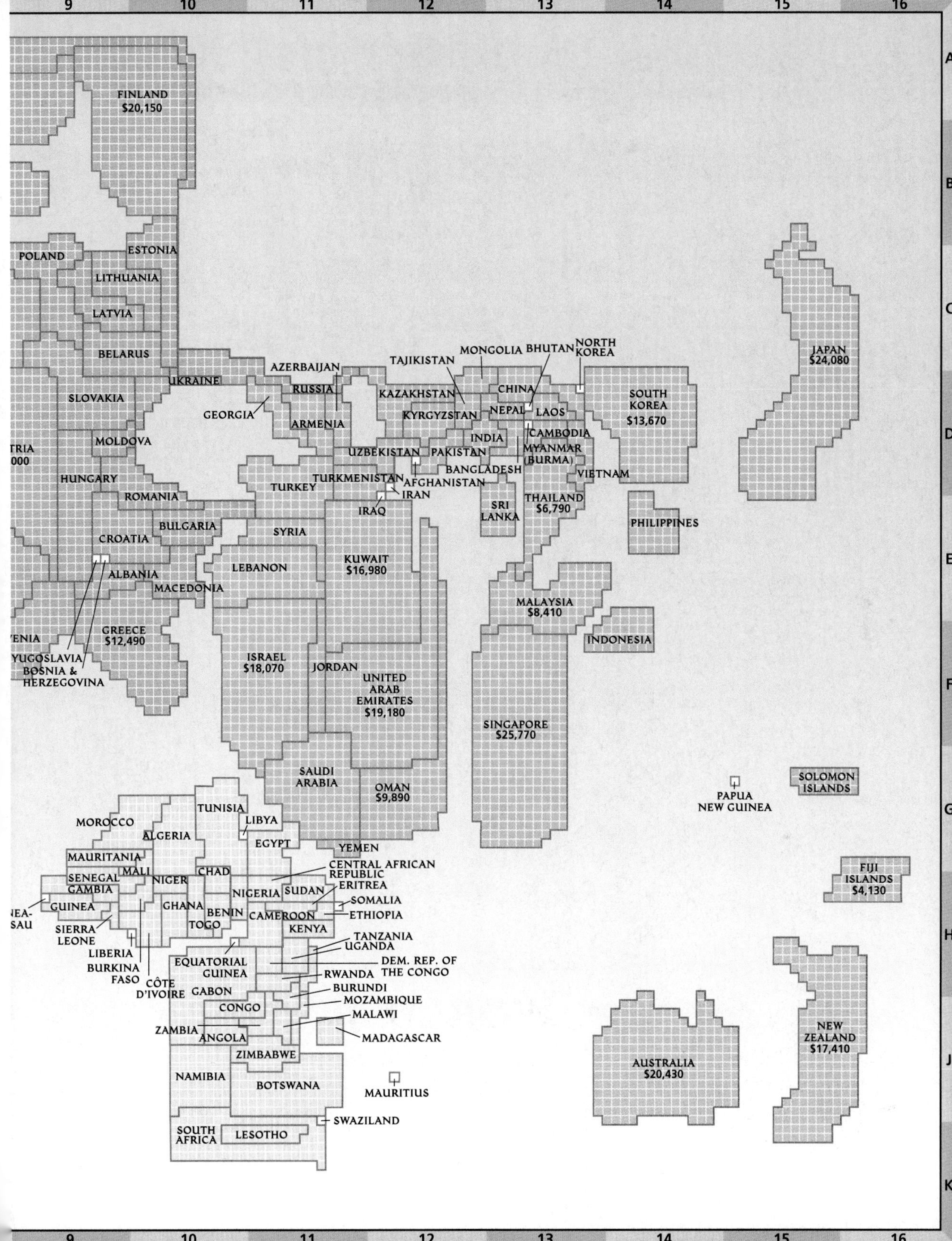
9
10
11
12
13
14
15
16
A
B
C
D
E
F
G
H
J
K
FINLAND
$20,150
POLAND
ESTONIA
LITHUANIA
LATVIA
BELARUS
UKRAINE
SLOVAKIA
MOLDOVA
TRIA
000
HUNGARY
ROMANIA
CROATIA
BULGARIA
ALBANIA
MACEDONIA
VENIA
GREECE
$12,490
YUGOSLAVIA
BOSNIA &
HERZEGOVINA
GEORGIA
AZERBAIJAN
RUSSIA
ARMENIA
TURKEY
SYRIA
LEBANON
ISRAEL
$18,070
JORDAN
IRAQ
IRAN
KUWAIT
$16,980
UNITED
ARAB
EMIRATES
$19,180
SAUDI
ARABIA
OMAN
$9,890
YEMEN
TAJIKISTAN
MONGOLIA
BHUTAN
NORTH
KOREA
KAZAKHSTAN
CHINA
KYRGYZSTAN
NEPAL
LAOS
INDIA
CAMBODIA
UZBEKISTAN
PAKISTAN
MYANMAR
(BURMA)
BANGLADESH
VIETNAM
TURKMENISTAN
AFGHANISTAN
SRI
LANKA
THAILAND
$6,790
SOUTH
KOREA
$13,670
JAPAN
$24,080
PHILIPPINES
MALAYSIA
$8,410
INDONESIA
SINGAPORE
$25,770
PAPUA
NEW GUINEA
SOLOMON
ISLANDS
FIJI
ISLANDS
$4,130
NEW
ZEALAND
$17,410
AUSTRALIA
$20,430
MOROCCO
TUNISIA
LIBYA
ALGERIA
EGYPT
MAURITANIA
MALI
SENEGAL
GAMBIA
NIGER
CHAD
NIGERIA
SUDAN
CENTRAL AFRICAN
REPUBLIC
ERITREA
SOMALIA
ETHIOPIA
NEA-
SAU
GUINEA
GHANA
BENIN
CAMEROON
TOGO
KENYA
SIERRA
LEONE
LIBERIA
BURKINA
FASO
CÔTE
D'IVOIRE
TANZANIA
UGANDA
DEM. REP. OF
THE CONGO
EQUATORIAL
GUINEA
RWANDA
BURUNDI
GABON
MOZAMBIQUE
CONGO
MALAWI
ZAMBIA
ANGOLA
MADAGASCAR
ZIMBABWE
NAMIBIA
BOTSWANA
MAURITIUS
SWAZILAND
SOUTH
AFRICA
LESOTHO

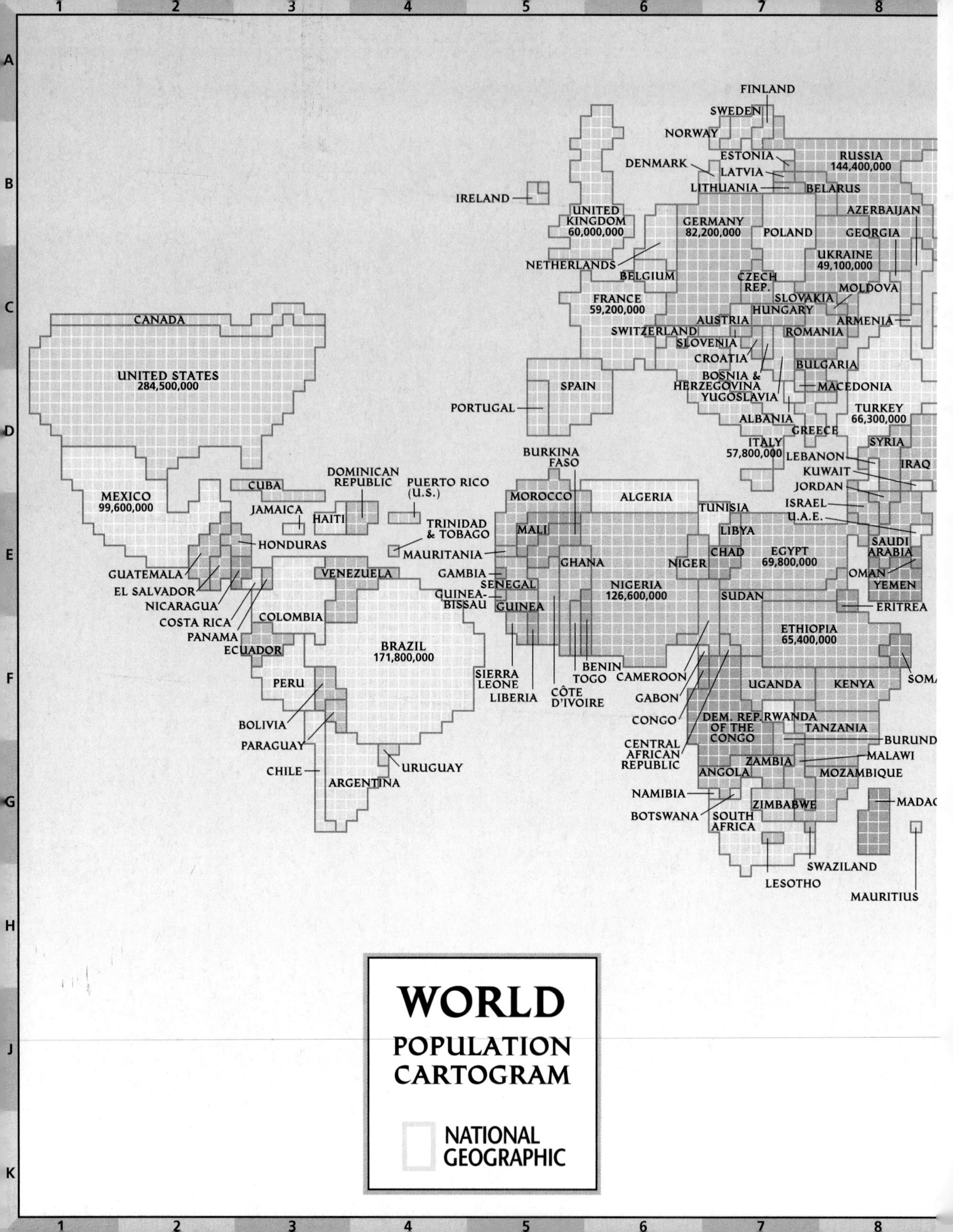

WORLD
POPULATION
CARTOGRAM
NATIONAL
GEOGRAPHIC
CANADA
UNITED STATES
284,500,000
MEXICO
99,600,000
CUBA
JAMAICA
HAITI
DOMINICAN
REPUBLIC
PUERTO RICO
(U.S.)
TRINIDAD
& TOBAGO
HONDURAS
GUATEMALA
EL SALVADOR
NICARAGUA
COSTA RICA
PANAMA
VENEZUELA
COLOMBIA
ECUADOR
PERU
BRAZIL
171,800,000
BOLIVIA
PARAGUAY
CHILE
ARGENTINA
URUGUAY
IRELAND
UNITED
KINGDOM
60,000,000
NETHERLANDS
BELGIUM
FRANCE
59,200,000
SWITZERLAND
PORTUGAL
SPAIN
FINLAND
SWEDEN
NORWAY
DENMARK
ESTONIA
LATVIA
LITHUANIA
RUSSIA
144,400,000
BELARUS
GERMANY
82,200,000
POLAND
AZERBAIJAN
GEORGIA
UKRAINE
49,100,000
CZECH
REP.
SLOVAKIA
MOLDOVA
HUNGARY
AUSTRIA
ARMENIA
SLOVENIA
ROMANIA
CROATIA
BOSNIA &
HERZEGOVINA
BULGARIA
MACEDONIA
YUGOSLAVIA
ALBANIA
TURKEY
66,300,000
GREECE
ITALY
57,800,000
SYRIA
LEBANON
IRAQ
KUWAIT
JORDAN
ISRAEL
U.A.E.
SAUDI
ARABIA
OMAN
YEMEN
ERITREA
BURKINA
FASO
MOROCCO
ALGERIA
TUNISIA
MALI
LIBYA
MAURITANIA
CHAD
EGYPT
69,800,000
NIGER
GHANA
GAMBIA
SENEGAL
GUINEA-
BISSAU
GUINEA
NIGERIA
126,600,000
SUDAN
ETHIOPIA
65,400,000
SIERRA
LEONE
LIBERIA
CÔTE
D'IVOIRE
BENIN
TOGO
CAMEROON
GABON
CONGO
CENTRAL
AFRICAN
REPUBLIC
UGANDA
KENYA
SOMA
DEM. REP.
OF THE
CONGO
RWANDA
TANZANIA
BURUND
ZAMBIA
MALAWI
ANGOLA
MOZAMBIQUE
NAMIBIA
BOTSWANA
ZIMBABWE
SOUTH
AFRICA
MADAG
SWAZILAND
LESOTHO
MAURITIUS

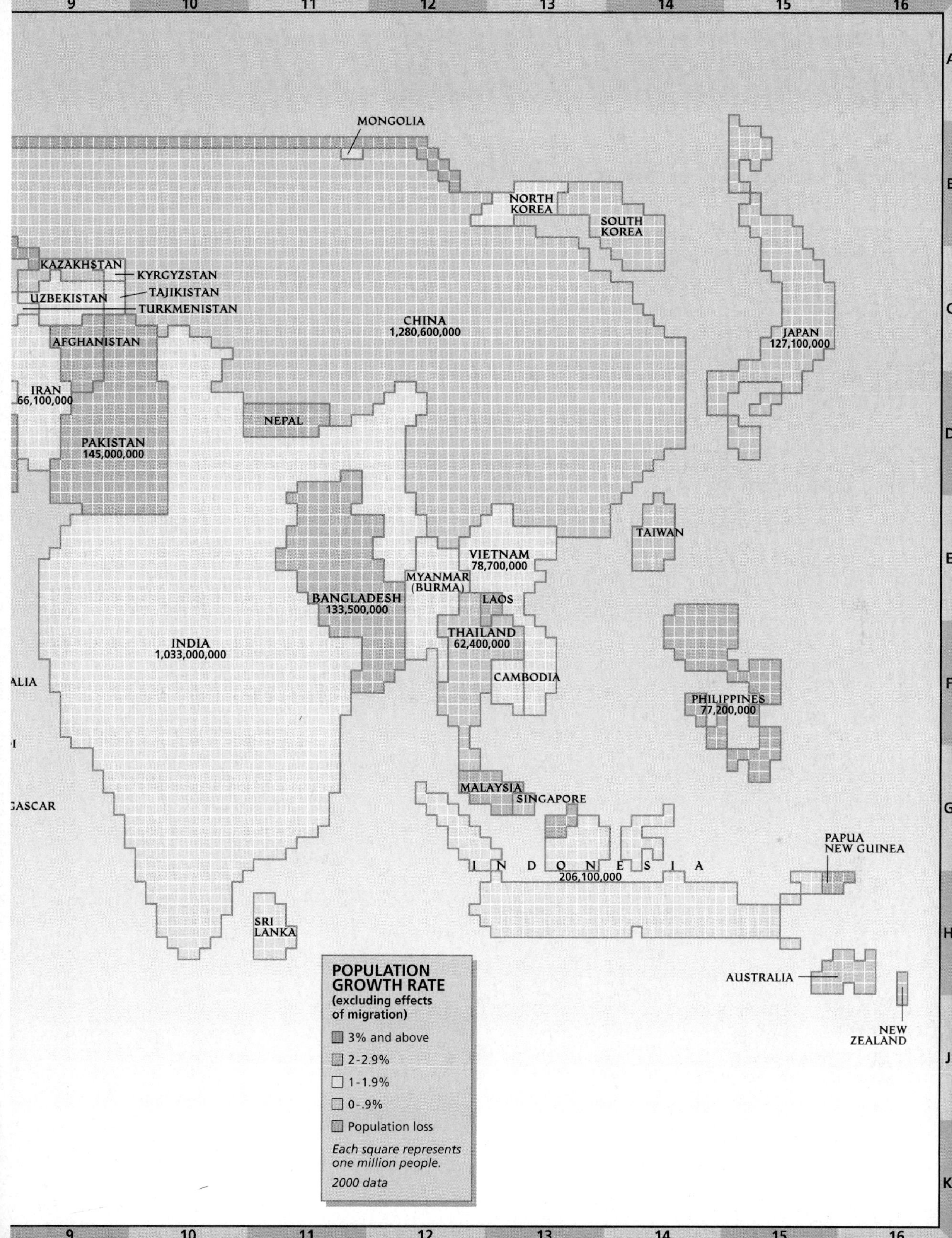

9
10
11
12
13
14
15
16
A
B
C
D
E
F
G
H
J
K
MONGOLIA
NORTH KOREA
SOUTH KOREA
KAZAKHSTAN
KYRGYZSTAN
TAJIKISTAN
UZBEKISTAN
TURKMENISTAN
CHINA
1,280,600,000
JAPAN
127,100,000
AFGHANISTAN
IRAN
66,100,000
NEPAL
PAKISTAN
145,000,000
TAIWAN
VIETNAM
78,700,000
MYANMAR
(BURMA)
BANGLADESH
133,500,000
LAOS
THAILAND
62,400,000
INDIA
1,033,000,000
ALIA
CAMBODIA
PHILIPPINES
77,200,000
I
MALAYSIA
SINGAPORE
GASCAR
PAPUA
NEW GUINEA
I N D O N E S I A
206,100,000
SRI
LANKA
AUSTRALIA
NEW
ZEALAND
POPULATION GROWTH RATE
(excluding effects of migration)
3% and above
2-2.9%
1-1.9%
0-.9%
Population loss
Each square represents one million people.
2000 data

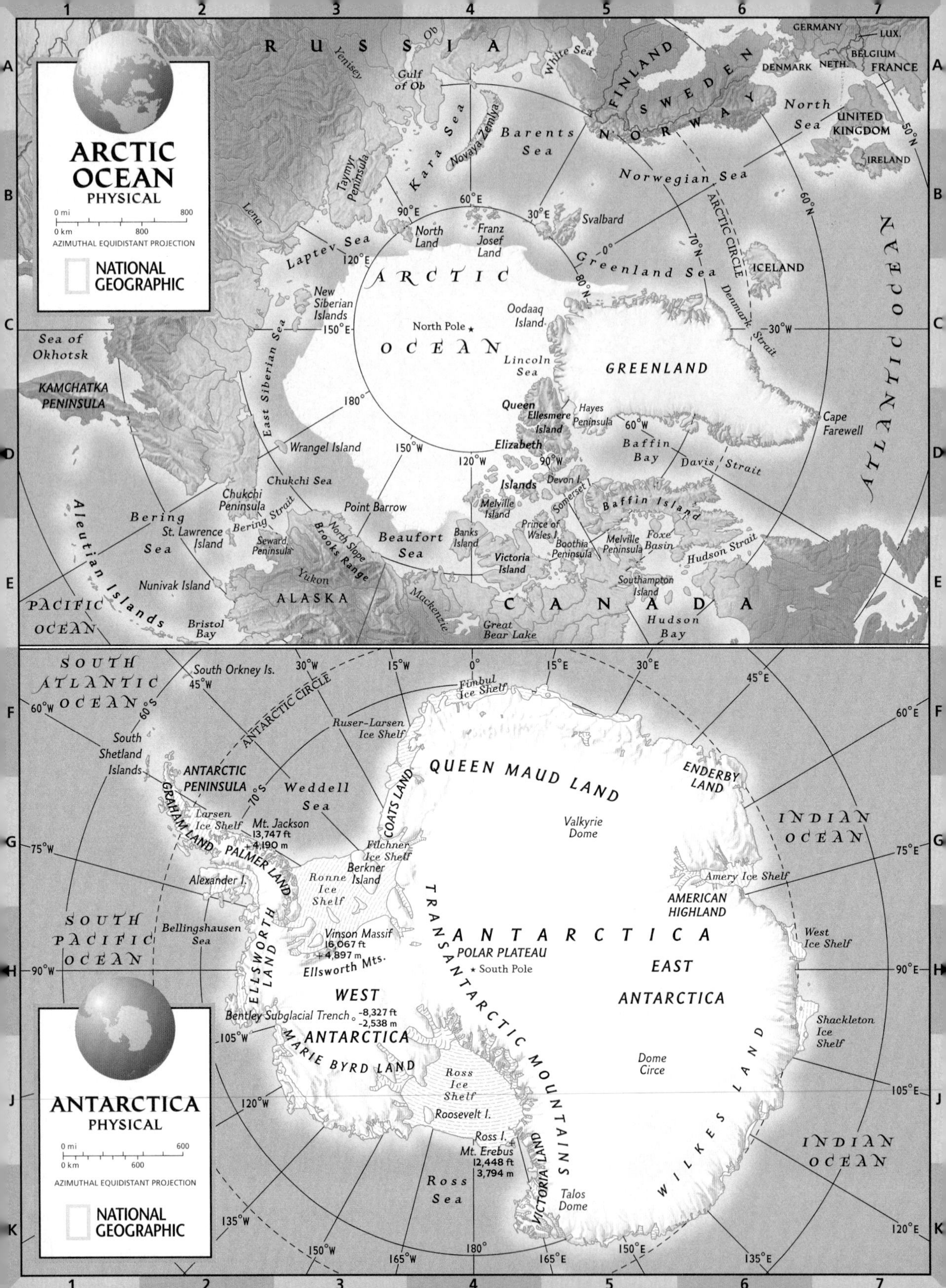

ARCTIC OCEAN
PHYSICAL
AZIMUTHAL EQUIDISTANT PROJECTION
NATIONAL GEOGRAPHIC
RUSSIA
FINLAND
SWEDEN
NORWAY
GERMANY
LUX.
BELGIUM
FRANCE
DENMARK
NETH.
UNITED KINGDOM
IRELAND
North Sea
Barents Sea
Kara Sea
Novaya Zemlya
Gulf of Ob
White Sea
Norwegian Sea
Svalbard
Franz Josef Land
North Land
Taymyr Peninsula
Laptev Sea
New Siberian Islands
ARCTIC OCEAN
North Pole
Greenland Sea
ICELAND
Denmark Strait
GREENLAND
Oodaaq Island
Lincoln Sea
Queen Elizabeth Islands
Ellesmere Island
Hayes Peninsula
Baffin Bay
Davis Strait
Cape Farewell
ATLANTIC OCEAN
Devon I.
Baffin Island
Somerset I.
Melville Island
Prince of Wales I.
Banks Island
Victoria Island
Boothia Peninsula
Melville Peninsula
Foxe Basin
Hudson Strait
Southampton Island
Hudson Bay
CANADA
Great Bear Lake
Mackenzie
Beaufort Sea
Point Barrow
North Slope
Brooks Range
Yukon
ALASKA
Chukchi Sea
Chukchi Peninsula
Bering Strait
Seward Peninsula
St. Lawrence Island
Bering Sea
Nunivak Island
Bristol Bay
Aleutian Islands
PACIFIC OCEAN
Wrangel Island
East Siberian Sea
Sea of Okhotsk
KAMCHATKA PENINSULA
Lena
Yenisey
Ob
ARCTIC CIRCLE
ANTARCTICA
PHYSICAL
AZIMUTHAL EQUIDISTANT PROJECTION
NATIONAL GEOGRAPHIC
SOUTH ATLANTIC OCEAN
South Orkney Is.
ANTARCTIC CIRCLE
South Shetland Islands
ANTARCTIC PENINSULA
GRAHAM LAND
PALMER LAND
Larsen Ice Shelf
Mt. Jackson 13,747 ft 4,190 m
Weddell Sea
Fimbul Ice Shelf
Riiser-Larsen Ice Shelf
COATS LAND
QUEEN MAUD LAND
ENDERBY LAND
INDIAN OCEAN
Valkyrie Dome
Filchner Ice Shelf
Berkner Island
Ronne Ice Shelf
Alexander I.
Amery Ice Shelf
AMERICAN HIGHLAND
SOUTH PACIFIC OCEAN
Bellingshausen Sea
ELLSWORTH LAND
Vinson Massif 16,067 ft 4,897 m
Ellsworth Mts.
ANTARCTICA
POLAR PLATEAU
South Pole
West Ice Shelf
EAST ANTARCTICA
WEST ANTARCTICA
Bentley Subglacial Trench -8,327 ft -2,538 m
TRANSANTARCTIC MOUNTAINS
Shackleton Ice Shelf
MARIE BYRD LAND
Ross Ice Shelf
Dome Circe
Roosevelt I.
Ross I.
Mt. Erebus 12,448 ft 3,794 m
WILKES LAND
Ross Sea
VICTORIA LAND
Talos Dome
INDIAN OCEAN

NATIONAL GEOGRAPHIC

Geography Skills Handbook

Geography skills provide the tools and methods for us to understand the relationships between people, places, and environments. We use geographic skills when we make daily personal decisions—where to buy a home; where to get a job; how to get to the shopping mall; where to go on vacation. Community decisions, such as where to locate a new school or how to solve problems of air and water pollution, also require the skillful use of geographic information.

This **Geography Skills Handbook** introduces you to the basic geographic tools—globes, maps, graphs—and explains how to use them. From this foundation, you will gain more reinforcement and practice in the **SkillBuilder** features located throughout the textbook. These resources will help you get the most out of your geography course—and provide you with skills you will use for the rest of your life.

Contents

Thinking Like a Geographer

Geographers use a wide array of tools and technologies—from basic globes to high-tech global positioning systems—to understand the earth. These help them collect and analyze a great deal of information. However, the study of geography is more than knowing a lot of facts about places. Rather, it has more do with asking questions about the earth, pursuing their answers, and solving problems. Thus, one of the most important geographic tools is inside your head: the ability to think geographically.

Skills for Learning Geography

Geography educators have identified a set of five skills that are key to geographic understanding. These skills, highlighted in the *Geography for Life* national geography standards, are listed in the chart below. Maps, globes, charts, graphs, satellite photos, global positioning systems, geographic information systems, library materials, the Internet, and this textbook are some of the resources available to help you in your study of geography.

Skill	Examples	Tool and Technologies
Asking Geographic Questions involves posing questions about your surroundings.	• Ask questions about why traffic has increased along a particular road. • Determine what factors should be considered in order to build a new community sports facility.	• Maps • Globes • Internet • Remote sensing • News media
Acquiring Geographic Information helps you answer geographic questions.	• Compare aerial photographs of a region taken over time. • Design a survey to determine who might use a community facility.	• Field observation • Interviews • GPS • Reference works • Satellite imagery • Historical records
Organizing Geographic Information helps you analyze and interpret the information you have collected.	• Compile a map that shows the spread of housing development over a period of time. • Summarize information obtained from interviews.	• Field maps • Databases • Statistical tables • Graphs • Diagrams • Summaries
Analyzing Geographic Information involves looking for patterns, relationships, and connections.	• Draw conclusions about the effects of road construction on traffic patterns. • Compare the information from different maps that show available land and zoning districts.	• Maps • Charts • Graphs • GIS • Spreadsheets
Answering Geographic Questions involves applying the information to real-life situations and problem-solving.	• Present a report that conveys the results of a case study. • Suggest locations for a new facility based on the geographic and community data gathered.	• Sketch maps • Reports • Research papers • Oral or multimedia presentations

Latitude, Longitude, and Location

Geography is often said to begin with the question: *Where?* The answer can be described in many ways, including direction, distance, country, or region. However, the basic tool for answering the question is **location.** Lines on globes and maps provide information that can help you locate places. These lines cross one another, forming a pattern called a **grid system.** This system helps you find exact places on the earth's surface.

Latitude

Lines of **latitude,** or **parallels,** circle the earth parallel to the Equator and measure the distance north or south of the Equator in degrees. The Equator is measured at 0° latitude, while the Poles lie at latitudes 90° N (north) and 90° S (south). Parallels north of the Equator are called **north latitude,** and parallels south of the Equator are called **south latitude.**

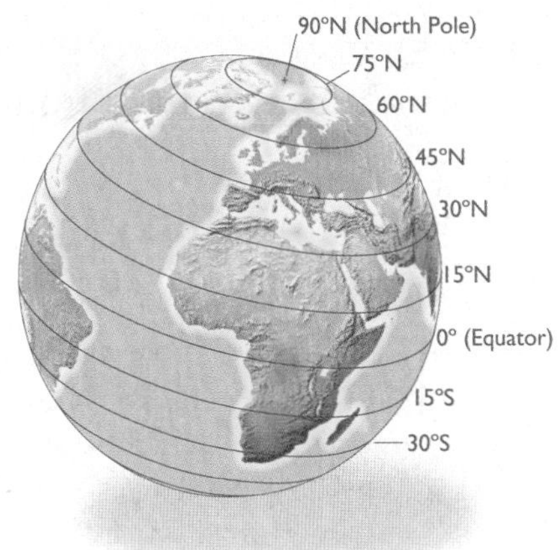

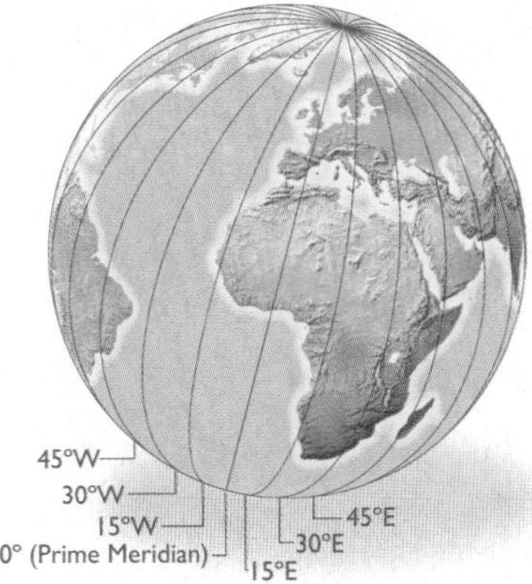

Longitude

Lines of **longitude,** or **meridians,** circle the earth from Pole to Pole. These lines measure distances east or west of the starting line, which lies at 0° longitude and is called the **Prime Meridian.** By international agreement, the Prime Meridian is the line of longitude that runs through the Royal Observatory in Greenwich, England. Places east of the Prime Meridian are known as **east longitude,** and places west of the Prime Meridian are known as **west longitude.**

The Global Grid

Every place has a global address, also called its absolute location (see page 9). You can identify the absolute location of a place by naming the longitude and latitude lines that cross exactly at that place. For example, the city of Tokyo, Japan, is located at 36°N latitude and 140°E longitude. For more precise readings, each degree of latitude and longitude is subdivided into 60 units called minutes.

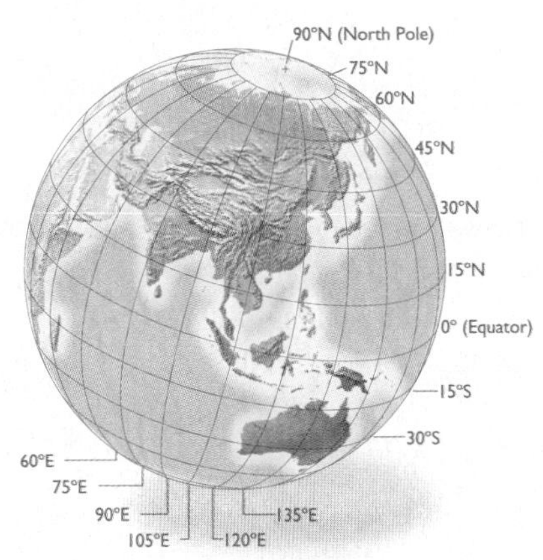

From Globes to Maps

A **globe** is a scale model of the earth. Because the earth is round, a globe presents the most accurate depiction of geographic information such as area, distance, and direction. However, globes show little close-up detail.

A printed **map** is a symbolic representation of all or part of the planet on a flat piece of paper. Unlike globes, maps can show small areas in great detail. Another advantage of printed maps is that they can be folded, stored, and easily carried from place to place.

From 3-D to 2-D

Think about the surface of the earth as the peel of an orange. To flatten the peel, you might have to cut it like the globe shown here. To create maps that are not interrupted, mapmakers, or **cartographers,** use mathematical formulas to transfer information from the three-dimensional globe to a two-dimensional map. However, when the curves of a globe become straight lines on a map, distortion of size, shape, distance, or area occurs. The purpose of the map usually dictates which projection is used.

How Map Projections Work

To create maps, cartographers project the round earth onto a flat surface—making a **map projection.** There are more than a hundred kinds of map projections, some with general names and some named for the cartographers who developed them. Three basic categories of map projections are shown here: **planar, cylindrical,** and **conic.**

Planar Projection

A planar projection shows the earth centered in such a way that a straight line coming from the center to any other point represents the shortest distance. Also known as an azimuthal projection, it is most accurate at its center. As a result, it is often used for maps of the Poles.

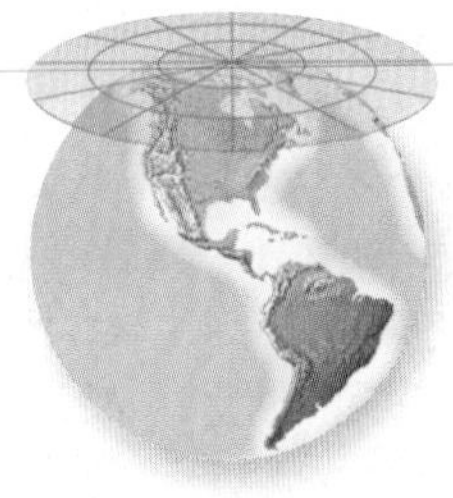

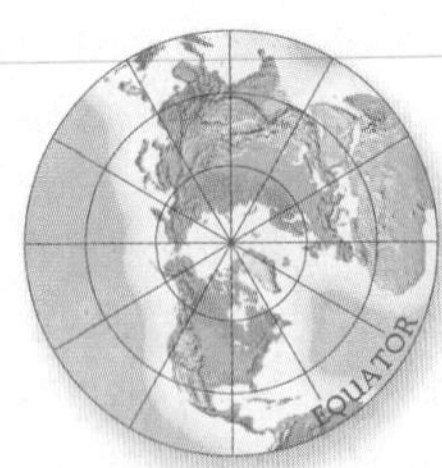

Great Circle Routes

A straight line of true direction—one that runs directly from west to east, for example—is not always the shortest distance between two points on Earth. This is due to the curvature of the earth. To find the shortest distance between any two places, stretch a piece of string around a globe from one point to the other. The string will form part of a *great circle,* or imaginary line that follows the curve of the earth. Traveling along a great circle is called following a **great circle route.** Ship captains and airline pilots use great circle routes to reduce travel time and save fuel.

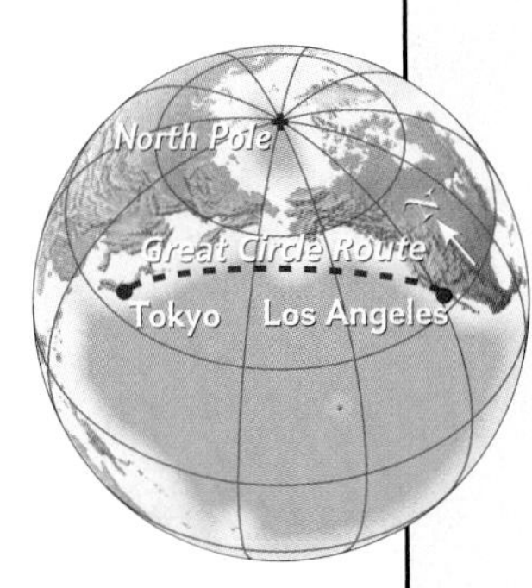

The idea of a great circle shows one important difference between globes and maps. Because a globe is round, it accurately shows great circle routes, as indicated on the partial globe shown (top). However, on a flat map, such as the Mercator projection (right), the great circle distance (dotted line) between Tokyo and Los Angeles appears to be far longer than the true direction distance (solid line). In fact, the great circle distance is 345 miles (555 km) shorter.

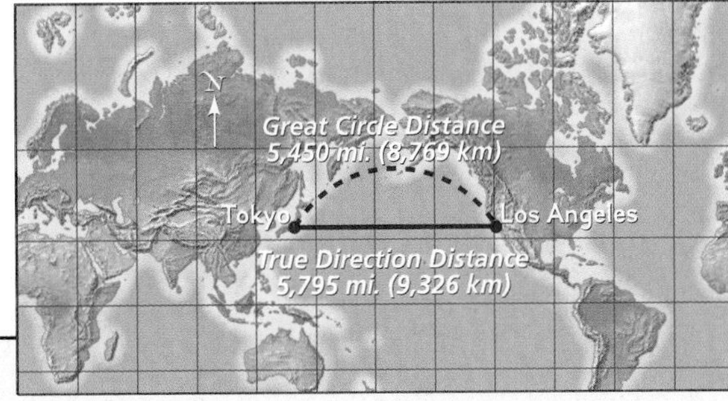

Cylindrical Projection

A cylindrical projection is based on the projection of the globe onto a cylinder. This projection is most accurate near the Equator, but shapes and distances are distorted near the Poles.

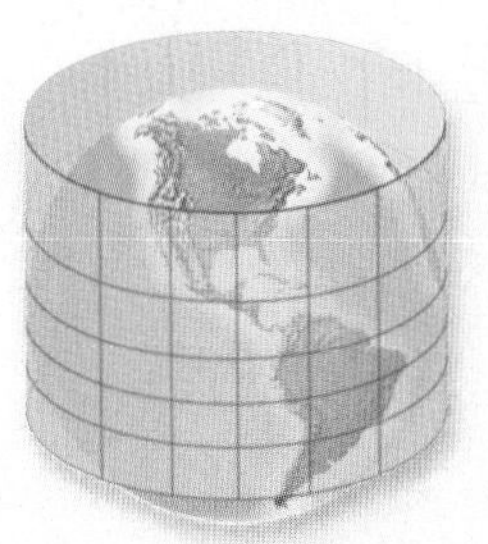

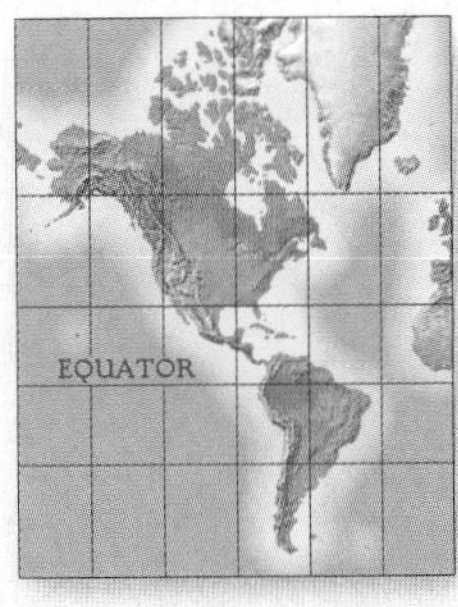

Conic Projection

A conic projection comes from placing a cone over part of a globe. Conic projections are best suited for showing limited east–west areas that are not too far from the Equator. For these uses, a conic projection can indicate distances and directions fairly accurately.

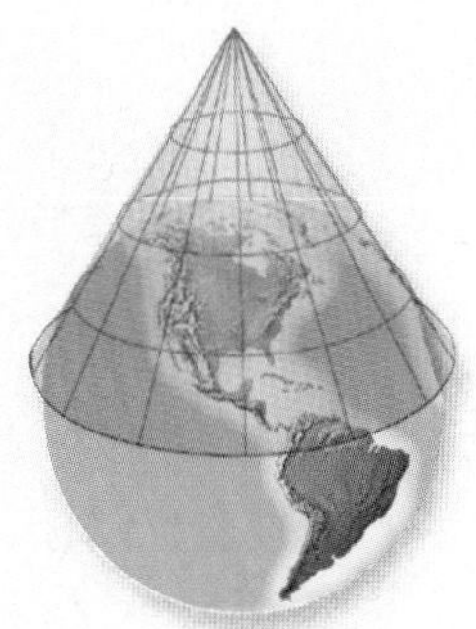

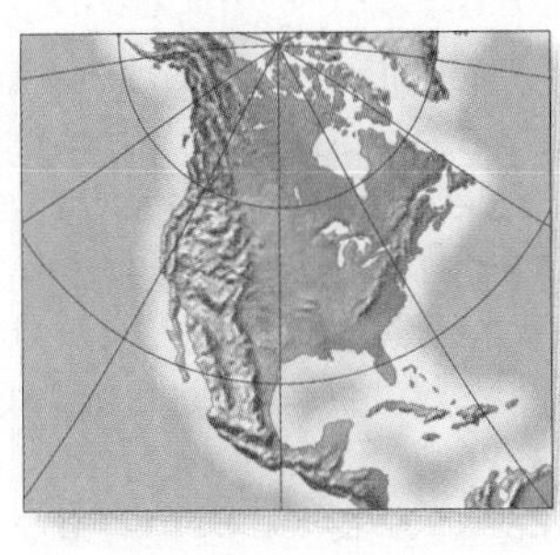

Common Map Projections

The curved surface of the earth cannot be shown accurately on a flat map. Every map projection stretches or breaks the curved surface of the planet in some way as it is flattened. Distance, direction, shape, or area may be distorted.

Cartographers have developed many map projections, each with some advantages and some degree of inaccuracy. Four of the most popular map projections, named for the cartographers who developed them, are shown on these pages.

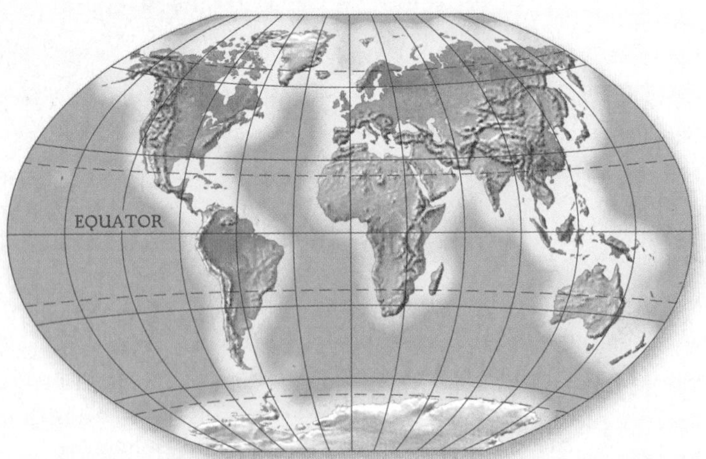

Winkel Tripel Projection

Most general reference world maps use the Winkel Tripel projection. Adopted by the National Geographic Society in 1998 for use in most maps, the Winkel Tripel projection provides a good balance between the size and shape of land areas as they are shown on the map. Even the polar areas are depicted with little distortion of size and shape.

Robinson Projection

The Robinson projection has minor distortions. The sizes and shapes near the eastern and western edges of the map are accurate, and the outlines of the continents appear much as they do on the globe. However, the shapes of the polar areas appear somewhat distorted.

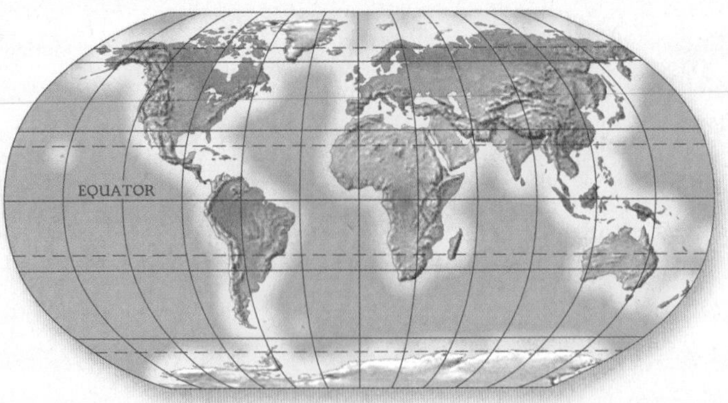

Goode's Interrupted Equal-Area Projection

An **interrupted projection** map looks something like a globe that has been cut apart and laid flat. Goode's Interrupted Equal-Area projection shows the true size and shape of the earth's landmasses, but distances are generally distorted.

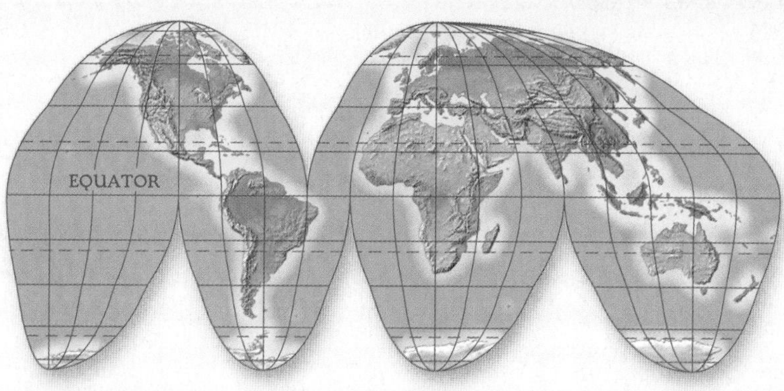

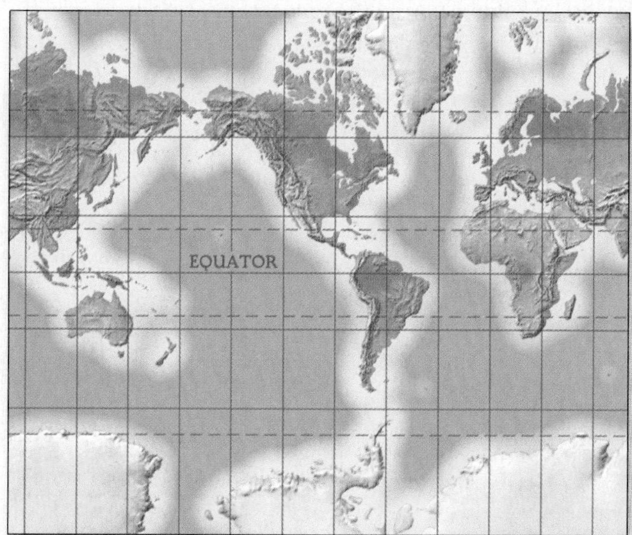

Mercator Projection

The Mercator projection, once the most commonly used projection, increasingly distorts size and distance as it moves away from the Equator. This makes areas such as Greenland and Antarctica look much larger than they would appear on a globe. However, Mercator projections do accurately show true directions and the shapes of landmasses, making these maps useful for sea travel.

Geographic Information Systems

Modern technology has changed the way maps are made. Most cartographers use computers with software programs called **geographic information systems (GIS).** A GIS is designed to accept data from many different sources, including maps, satellite images, and printed text and statistics. The GIS converts the data into a digital code, which arranges it in a database. Cartographers then program the GIS to process the data and produce the maps they need. With GIS, each kind of information on a map is saved as a separate electronic "layer" in the map's computer files. Because of this modern technology, cartographers are able to make maps—and change them—quickly and easily.

A cartographer uses GIS to make a map.

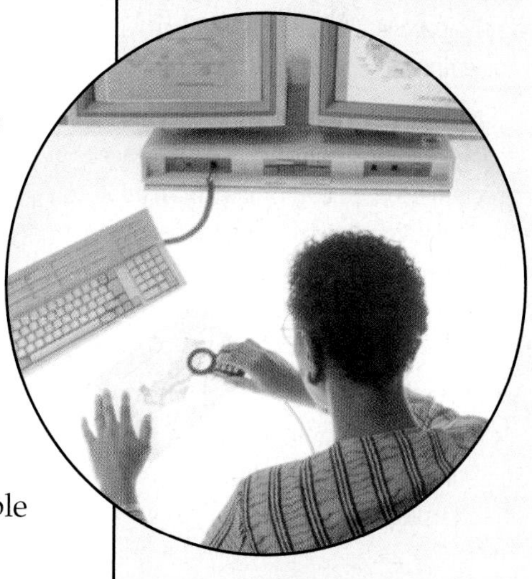

Reading a Map

In addition to scale and the lines of latitude and longitude, maps feature other important tools to help you interpret the information they contain. Learning to use these map tools will help you read the symbolic language of maps more easily.

Key

Cartographers use a variety of symbols to represent map information. Graphic symbols are easily understood by people around the world. To be sure that the symbols are clear, however, every map contains a **key**—a list that explains what the symbols stand for. This key shows symbols commonly used on a political map.

Boundary Lines

On political maps of large areas, boundary lines highlight the borders between different countries, states, or counties.

Compass Rose

Most maps feature a **compass rose,** a marker that indicates directions. The four **cardinal directions**—north, south, east, and west—are usually indicated with arrows or points of a star. The **intermediate directions**—northeast, northwest, southeast, southwest—may also be shown, usually with smaller arrows or star points.

Sometimes a compass rose may point in only one direction because the other directions can be determined in relation to the given direction. The compass rose on this map indicates north only.

Europe Before World War I

Cities

Cities are represented by a dot. Sometimes the relative sizes of cities are shown using dots of different sizes.

Scale Bar

The **scale bar** shows the relationship between map measurements and actual distances. By laying a ruler along the scale bar, you can calculate how many miles or kilometers are represented per inch or centimeter.

Capitals

National capitals are often represented by a star within a circle.

Using Scale

All maps are drawn to a certain scale. **Scale** is a consistent, proportional relationship between the measurement shown on the map and the measurement of the earth's surface. The scale of a map varies with the size of the area shown.

Use the scale bar to find actual distances on a map. The **scale bar** gives the relationship between map measurements and actual distances. Most scale bars are graphic representations, allowing you to use a ruler to calculate actual distances.

Small-Scale Maps

A small-scale map, like this political map of Mexico, can show a large area but little detail. Note that the scale bar for this map indicates that about ½ of an inch is equal to 300 miles and a little more than ½ of a centimeter is equal to 300 kilometers.

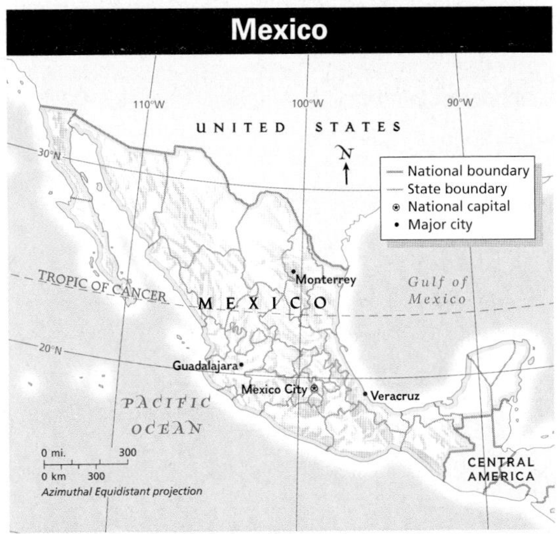

Large-Scale Maps

A large-scale map, like this map of Mexico City, can show a small area on the earth's surface with a great amount of detail. Study the scale bar. Note that the map measurements correspond to much smaller distances than on a small-scale map.

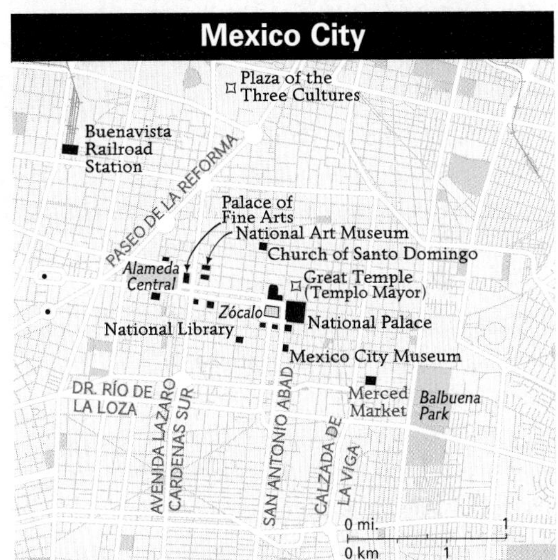

Absolute and Relative Location

As you learned on page 3, a place's **absolute location** is found at the precise point where one line of latitude crosses a line of longitude. Another way that people indicate location is by relative location. You may be told, for example, to look for a street that is "two blocks north" of another street. **Relative location** is the location of one place in relation to another place.

To find relative location, find a reference point—a location you already know—on a map. Then look in the appropriate direction for the new location. For example, locate Vienna (your reference point) on this map. The relative location of Budapest can be described as southeast of Vienna.

Types of Maps

Maps are prepared for many uses. The use for which a map is intended determines the kinds of information it contains. Learning to recognize a map's purpose will help you make the best use of its content.

General-Purpose Maps

Maps that show a wide range of information about an area are called **general-purpose maps.** General-purpose maps are typically used for reference, education, and travel. Two common forms of general-purpose maps are **physical maps** and **political maps.**

Physical Maps

A physical map shows the location and the **topography,** or shape, of the earth's physical features. Physical maps use colors or patterns to indicate **relief**—the differences in **elevation,** or height, of landforms. Some physical maps have **contour lines** that connect all points of land of equal elevation. Physical maps may show mountains as barriers to transportation. Rivers and streams may be shown as routes into the interior of a country. These physical features often help to explain the historical development of a country.

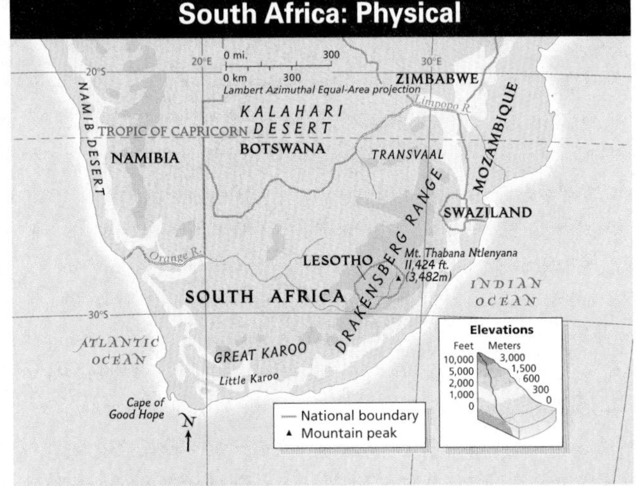

Political Maps

A political map shows the boundaries between countries. Smaller internal divisions, such as states or counties, may also be indicated by different symbols. Political maps often show human-made features such as capitals, cities, roads, highways, and railroads.

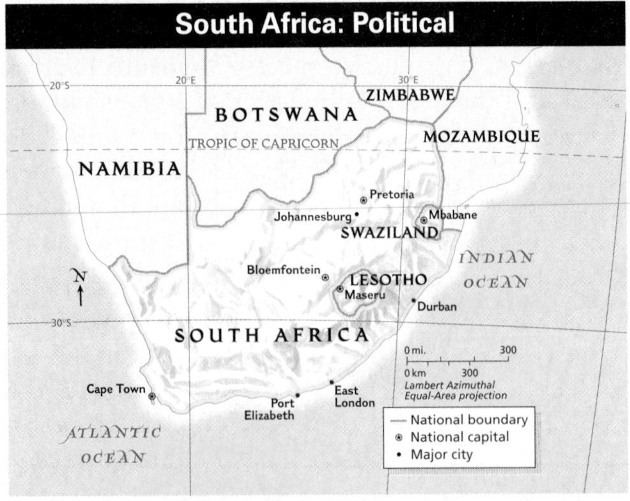

Special-Purpose Maps

Maps that emphasize a single idea or a particular kind of information about an area are called **special-purpose maps.** There are many kinds of special-purpose maps, each designed to serve a different need. You can learn more about several types of special-purpose maps in the SkillBuilder features in this textbook: relief maps (page 126), climate maps (page 172), population density maps (page 232), vegetation maps (page 432), elevation profiles (page 580), economic activity maps (page 680), and cartograms (page 754).

Some special-purpose maps—such as economic activity maps and natural resource maps—show the distribution of particular activities, resources, or products in a given area. Colors and symbols represent the location or distribution of activities and resources.

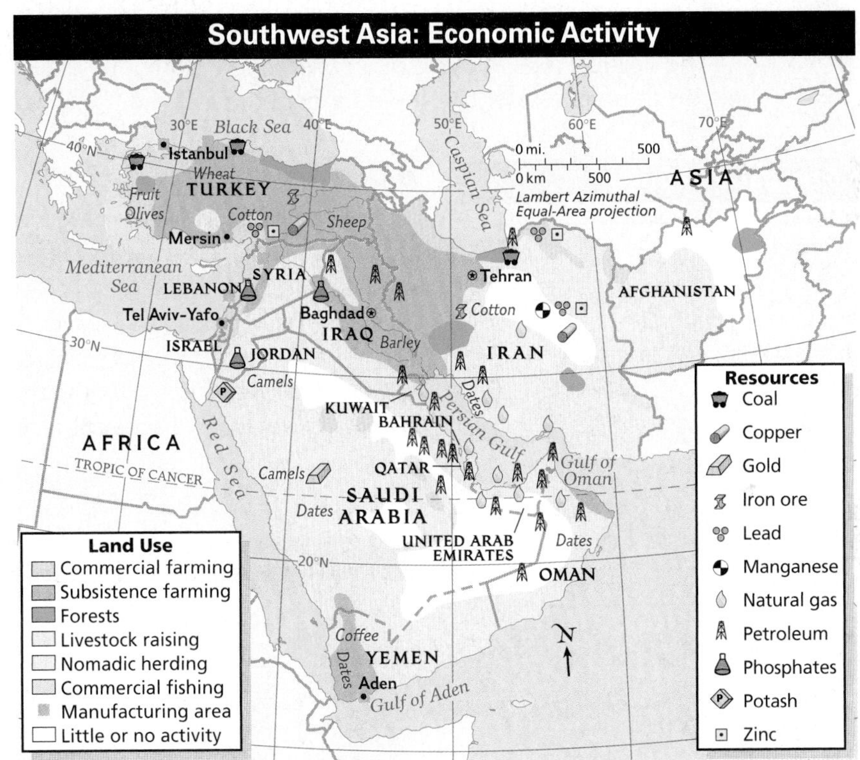

An Economic Activity Map

The special-purpose map above shows the distribution of land use and natural resources in Southwest Asia. Geographers use maps like this one to study the distribution of natural resources. Governments and industry leaders use land use maps and natural resource maps to monitor the economic activities of countries and regions.

Graphs, Charts, and Diagrams

In addition to globes and maps, geographers use other visual representations to display and interpret data. Graphs, charts, and diagrams provide valuable information in forms that are well organized and easy to read.

Graphs

A **graph** is a visual presentation of information. There are many kinds of graphs, each suitable for certain purposes. Most graphs show two sets of data, one displayed along the vertical axis and the other displayed along the horizontal axis. Labels on these axes identify the data being displayed.

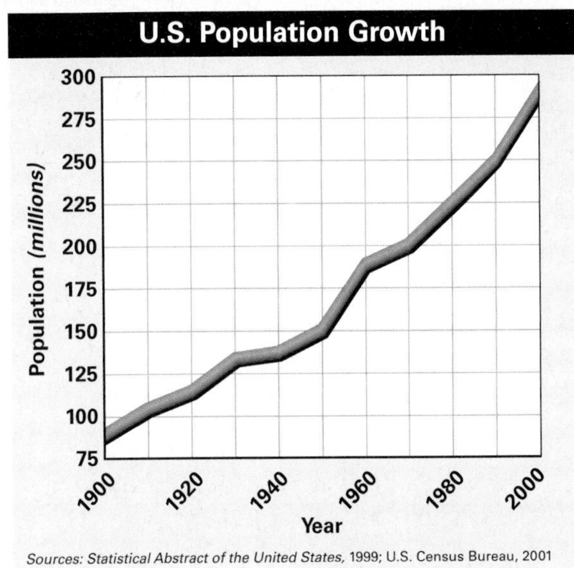

◀ Line Graphs

A **line graph** shows changes in two variables, or changing sets of circumstances over periods of time. To analyze data on a line graph, study the changes and trends as shown by the line. Then draw conclusions based on the information. This line graph shows U.S. population growth between 1900 and 2000. The vertical axis lists population, and the horizontal axis indicates the passage of time.

Bar Graphs ▶

A **bar graph** shows comparisons. To analyze a bar graph, note the differences in quantities. Then make generalizations or draw conclusions based on the data. This bar graph shows lumber production among the top five lumber-producing countries in the world. The vertical axis shows the amount of lumber produced.

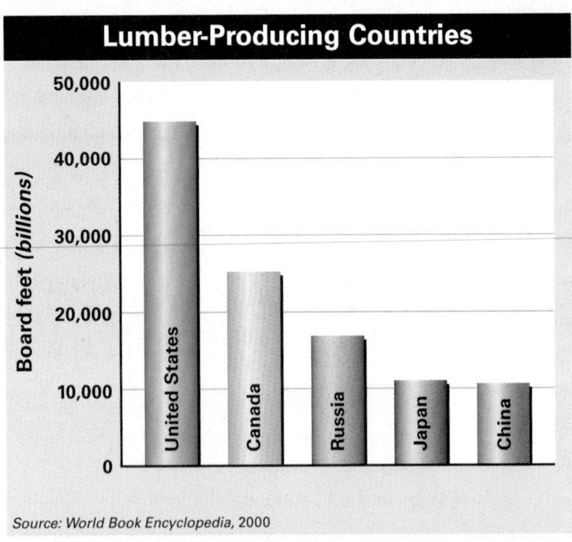

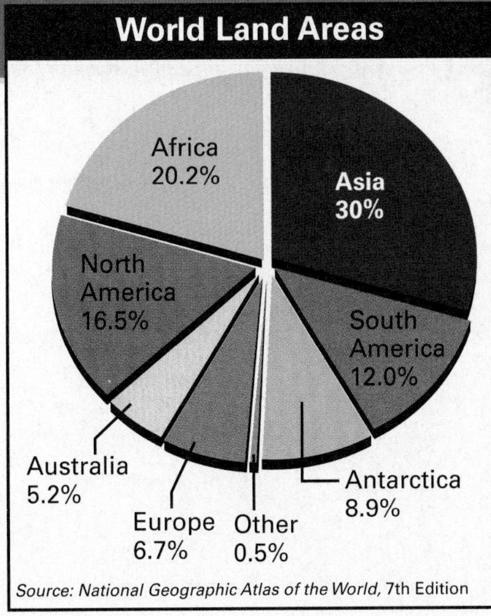

Circle Graphs

A **circle graph,** or **pie graph,** shows the relationship of parts to a whole. Percentages are indicated by relative size and sometimes by color. To analyze a circle graph, study the relationships of areas to one another and to the whole. This circle graph shows the land areas of the world's continents and other landmasses, such as islands, expressed as percentages of Earth's total landmass.

Charts and Tables

Data are arranged in columns and rows in a **chart** or **table.** Charts and tables display facts in an organized manner and make comparisons easy. To find key information in a chart or table, look for the intersections of columns and rows.

The table at right displays information about the population and land area of the world's continents.

Continents of the World

Continent	Population	Land Area
Africa	818,000,000	11,698,111 sq. mi. 30,298,107 sq. km
Antarctica	No permanent inhabitants	5,500,000 sq. mi. 13,209,000 sq. km
Asia	3,720,000,000	12,262,691 sq. mi. 31,760,369 sq. km
Australia	19,400,000	2,988,888 sq. mi. 7,741,220 sq. km
Europe	727,000,000	8,875,867 sq. mi. 22,988,495 sq. km
North America	491,000,000	8,747,613 sq. mi. 22,656,317 sq. km
South America	350,000,000	6,898,579 sq. mi. 17,867,319 sq. km

Source: World Population Data Sheet, 2001

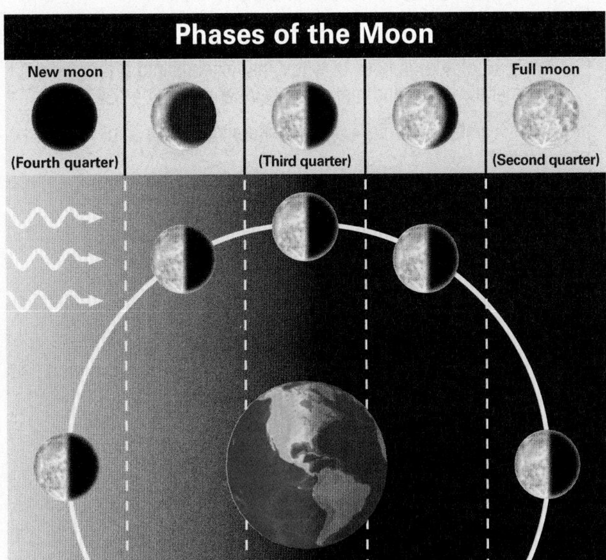

Diagrams

A **diagram** is a drawing that shows what something is or how something is done. Many diagrams feature several drawings or sections that show the steps in a process.

The diagram at left shows the way the moon seems to change shape as it goes through its phases each month. Note that as the moon revolves around the earth, it goes from the new moon phase, when it is almost invisible, to the full moon phase, when it appears as a giant globe.

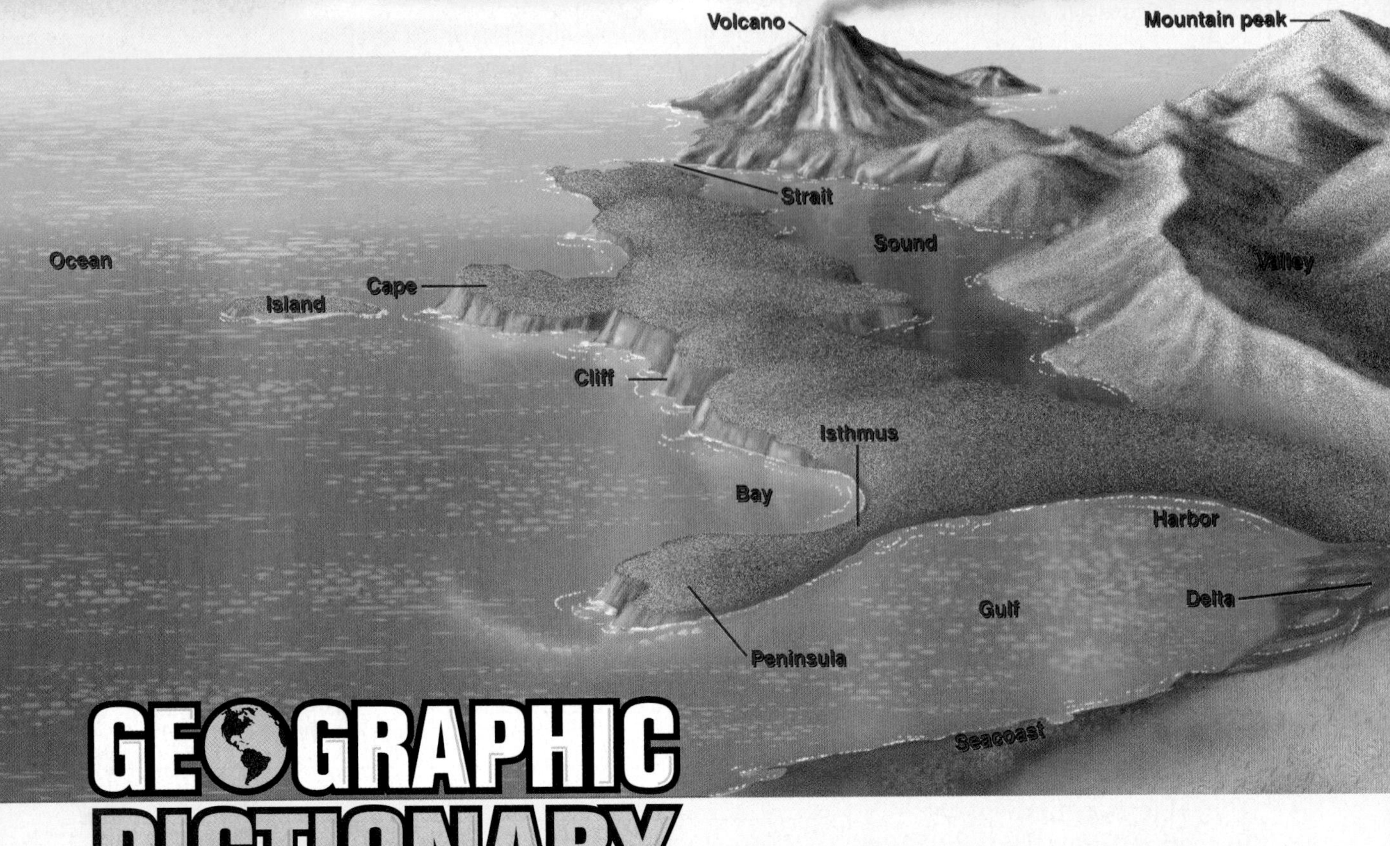

GEOGRAPHIC DICTIONARY

As you read about the world's geography, you will encounter the terms listed below. Many of the terms are pictured in the diagram.

absolute location exact location of a place on the earth described by global coordinates

basin area of land drained by a given river and its branches; area of land surrounded by lands of higher elevations

bay part of a large body of water that extends into a shoreline, generally smaller than a gulf

canyon deep and narrow valley with steep walls

cape point of land that extends into a river, lake, or ocean

channel wide strait or waterway between two landmasses that lie close to each other; deep part of a river or other waterway

cliff steep, high wall of rock, earth, or ice

continent one of the seven large landmasses on the earth

delta flat, low-lying land built up from soil carried downstream by a river and deposited at its mouth

divide stretch of high land that separates river systems

downstream direction in which a river or stream flows from its source to its mouth

elevation height of land above sea level

Equator imaginary line that runs around the earth halfway between the North and South Poles; used as the starting point to measure degrees of north and south latitude

glacier large, thick body of slowly moving ice

gulf part of a large body of water that extends into a shoreline, generally larger and more deeply indented than a bay

harbor a sheltered place along a shoreline where ships can anchor safely

highland elevated land area such as a hill, mountain, or plateau

hill elevated land with sloping sides and rounded summit; generally smaller than a mountain

island land area, smaller than a continent, completely surrounded by water

isthmus narrow stretch of land connecting two larger land areas

lake a sizable inland body of water

latitude distance north or south of the Equator, measured in degrees

longitude distance east or west of the Prime Meridian, measured in degrees

lowland land, usually level, at a low elevation

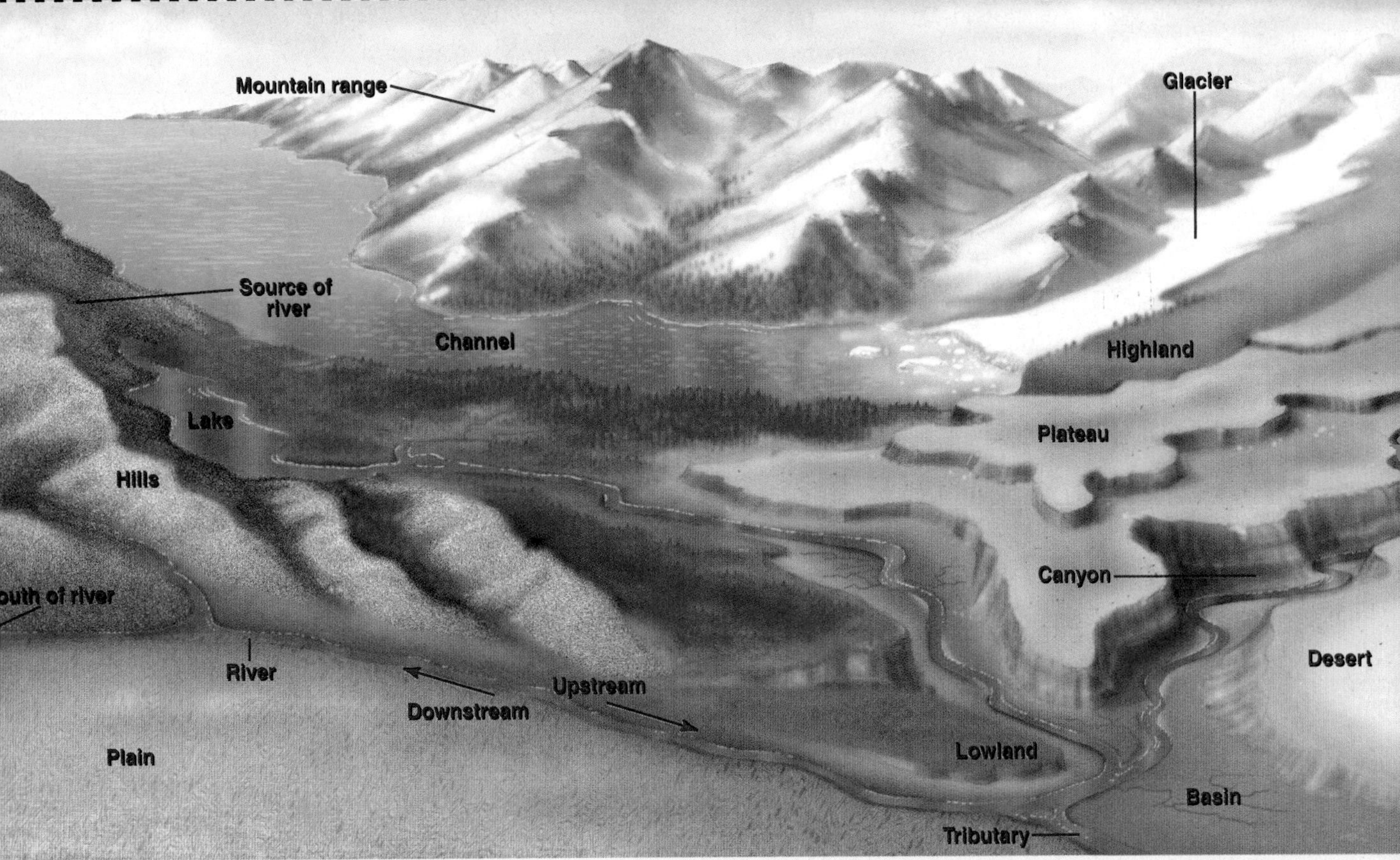

map drawing of the earth shown on a flat surface

meridian one of many lines on the global grid running from the North Pole to the South Pole; used to measure degrees of longitude

mesa broad, flat-topped landform with steep sides; smaller than a plateau

mountain land with steep sides that rises sharply (1,000 feet or more) from surrounding land; generally larger and more rugged than a hill

mountain peak pointed top of a mountain

mountain range a series of connected mountains

mouth (of a river) place where a stream or river flows into a larger body of water

ocean one of the four major bodies of salt water that surround the continents

ocean current stream of either cold or warm water that moves in a definite direction through an ocean

parallel one of many lines on the global grid that circles the earth north or south of the Equator; used to measure degrees of latitude

peninsula body of land jutting into a lake or ocean, surrounded on three sides by water

physical feature characteristic of a place occurring naturally, such as a landform, body of water, climate pattern, or resource

plain area of level land, usually at low elevation and often covered with grasses

plateau area of flat or rolling land at a high elevation, about 300 to 3,000 feet (90 to 900 m) high

Prime Meridian line of the global grid running from the North Pole to the South Pole at Greenwich, England; starting point for measuring degrees of east and west longitude

relief changes in elevation over a given area of land

river large natural stream of water that runs through the land

sea large body of water completely or partly surrounded by land

seacoast land lying next to a sea or an ocean

sound broad inland body of water, often between a coastline and one or more islands off the coast

source (of a river) place where a river or stream begins, often in highlands

strait narrow stretch of water joining two larger bodies of water

tributary small river or stream that flows into a large river or stream; a branch of the river

upstream direction opposite the flow of a river; toward the source of a river or stream

valley area of low land usually between hills or mountains

volcano mountain or hill created as liquid rock and ash erupt from inside the earth

The World

WHY IT'S IMPORTANT—

Entering the twenty-first century, the world is a much smaller place than it was at the time of your great-grandparents a hundred years ago. Advances in technology, communication, and transportation are responsible for much of this change. They have narrowed vast distances and made neighbors of the world's people. The Internet, for example, now puts you in immediate touch with people in other parts of the world. In the years to come, you and your generation—here and elsewhere—will be challenged to work together to use this and other technology to make the world a better place for everyone.

World Regions Video

To learn more about the physical and human geography of the world, view the *World Regions* video "Looking at the World."

NGS ONLINE

www.nationalgeographic.com/education

Skydivers in formation above patchwork fields, California

CHAPTER 1

How Geographers Look at the World

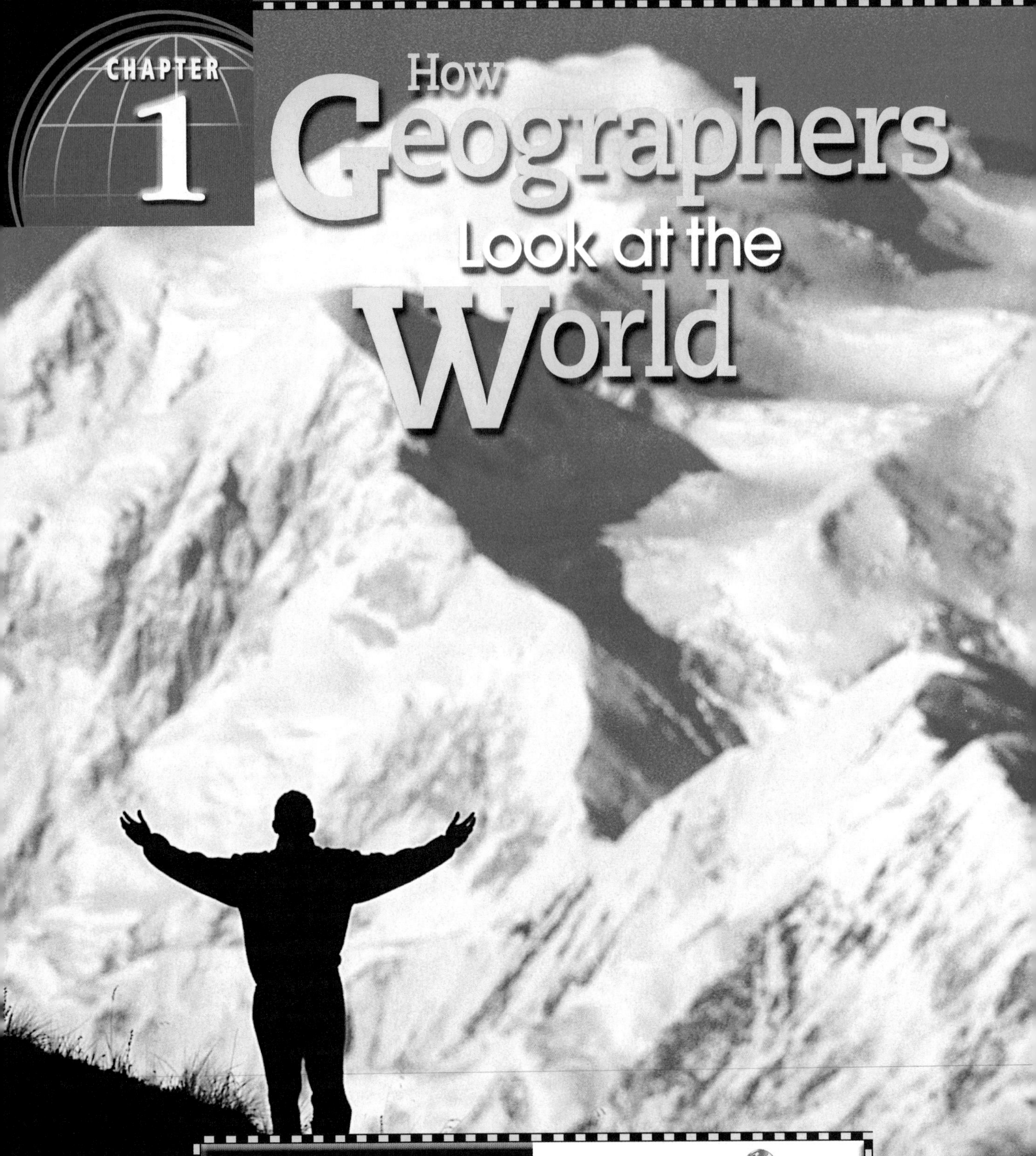

GeoJournal

Write a journal entry describing the part of the world in which you live—its physical features, plant and animal life, and people. Think about how your observations are similar to and different from the ways a geographer looks at the world.

GEOGRAPHY Online

Chapter Overview Visit the **Glencoe World Geography** Web site at geography.glencoe.com and click on Chapter Overviews—Chapter 1 to preview information about how geographers look at the world.

Exploring Geography

Guide to Reading

Consider What You Know

Think about where your school is located and the ways in which a place's location can be described. How many different ways can you think of to describe your school's location to another person?

Read to Find Out

- What are the physical and human features geographers study?
- How do geographers describe the earth's features and their patterns?
- How is geography used?

Terms to Know

- location
- absolute location
- hemisphere
- grid system
- relative location
- place
- region
- formal region
- functional region
- perceptual region
- ecosystem
- movement
- human-environment interaction

Places to Locate

- Equator
- North Pole
- South Pole
- Prime Meridian

◀ *Mt. McKinley, Alaska, United States*

NATIONAL GEOGRAPHIC

A Geographic View

Earth's Variety

A small planet in a modest solar system, a tumbling pebble in the cosmic stream, and yet . . . [t]his home is built of many mansions, carved by wind and the fall of water, lush with living things beyond number, perfumed by salt spray and blossoms. Here cool in a cloak of mist or there steaming under a brazen sun, Earth's variety excites the senses and exalts the soul.

—*Stuart Franklin, "Celebrations of Earth,"* National Geographic, *January 2000*

Labrador coast, Canada

How would you describe the world around you? Would it be in terms of people, places, things, or all of these? Geography is the study of the earth's physical features and the living things—humans, animals, and plants—that inhabit the planet. Geography looks at where all of these elements are located and how they relate to one another. In this section you will gain an understanding of what geography is and why it is important to study it.

The Elements of Geography

The root of the word *geography* is an ancient Greek word meaning "earth description." Geographers are specialists who describe the earth's physical and human features and the interactions of people, places, and environments. They not only describe but also search for patterns in these features and interactions, seeking to explain how and why they exist or occur. For example, geographers may study volcanoes and why they erupt, or they may analyze a city's location in relation to climate, landscape, and available transportation. In their work, geographers consider:

- *The world in spatial terms (location)*
- *Places and regions*
- *Physical systems*
- *Human systems*
- *Environment and society*
- *The uses of geography*

The World in Spatial Terms

Spatial relations refer to the links that places and people have to one another because of their locations. For geographers, **location**, or a specific place on the earth, is a reference point in the same way that dates are reference points for historians.

Absolute Location

One way of locating a place is by describing its **absolute location**—the exact spot at which the place is found on the globe. To determine absolute location, geographers use a network of imaginary lines around the earth. The **Equator**, the line circling the earth midway between the **North** and **South Poles**, divides the earth into **hemispheres**, or two halves (Northern and Southern). The **Prime Meridian**, the 0° north-south line that runs through Greenwich, England, and the 180° north-south line running through the mid-Pacific Ocean also divide the earth into hemispheres (Eastern and Western).

The Equator, the Prime Meridian (also called the Meridian of Greenwich), and other lines of latitude and longitude cross one another to form a pattern called a **grid system**. Using the grid, you can name the absolute, or precise, location of any place on Earth. This location is generally stated in terms of *latitude*, degrees north or south of the Equator, and *longitude*, degrees east or west of the Prime Meridian. For example, Dallas, Texas, is located at latitude 32°N (north) and longitude 96°W (west).

Relative Location

Although absolute location is useful, most people locate a place in relation to other places, or by its **relative location**. For example, New Orleans is located near the mouth of the Mississippi River. Knowing the relative location of a place helps you orient yourself in space and develop an awareness of the world around you.

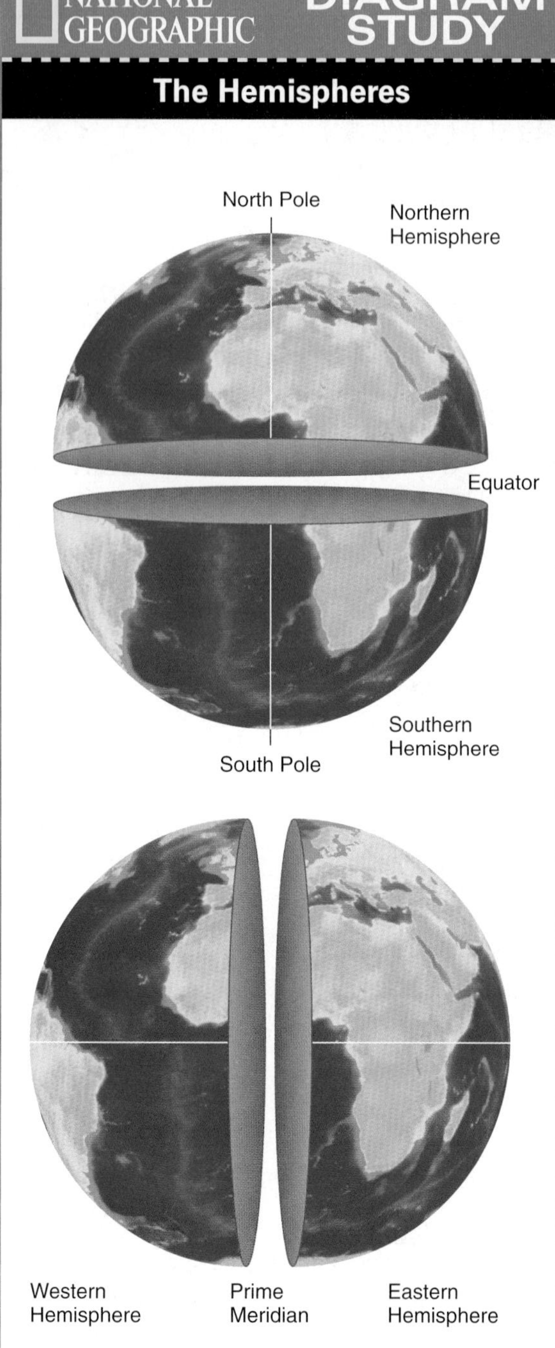

Geography Skills for Life

1. **Interpreting Diagrams** What lines of latitude and longitude divide the globe into hemispheres?
2. **Applying Geography Skills** In which hemispheres do you live?

NATIONAL GEOGRAPHIC **DIAGRAM STUDY**

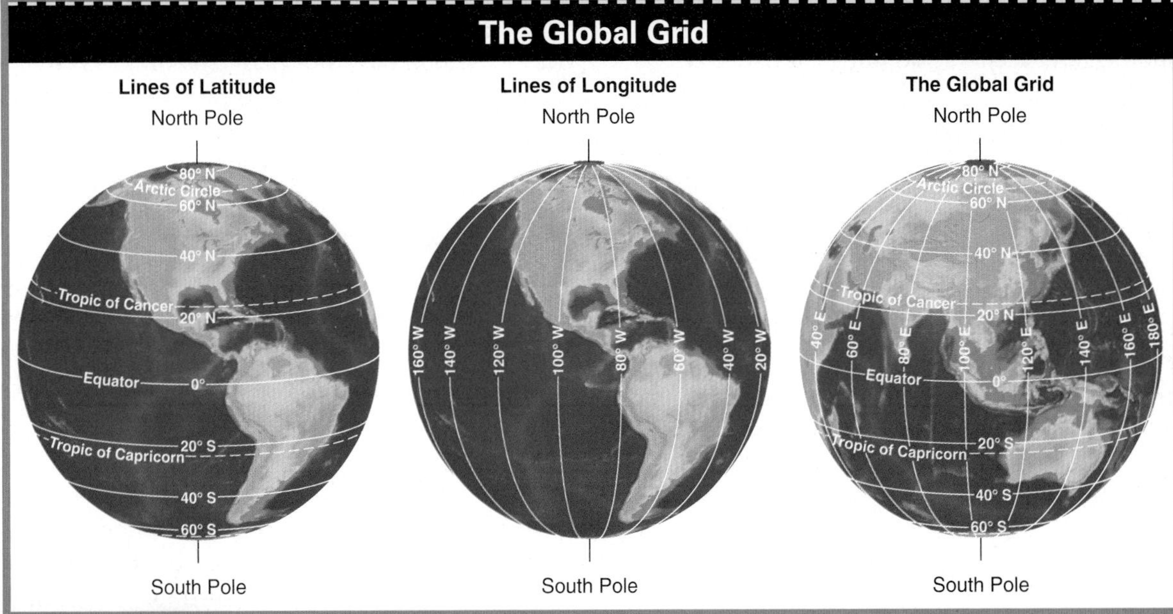

Geography Skills for Life

1. **Interpreting Diagrams** Between which lines of latitude is the Arctic Circle located?

2. **Applying Geography Skills** Approximate the absolute location of the state of Florida. Explain how you determined the answer.

Places and Regions

A **place** is a particular space with physical and human meaning. Every place on Earth has its own unique characteristics, determined by the surrounding environment and the people who live there. One task of geographers is to understand and explain how places are similar to and different from one another. To interpret the earth's complexity, geographers often group places into **regions**, or areas united by specific factors. The defining factors of a region may be physical, such as soil type, vegetation, river systems, and climate. A region may also have human factors that help define it. These may include language, religion, cultural traditions, forms of government, and trade networks.

Geographers identify three types of regions: formal, functional, and perceptual. A **formal**, or uniform, **region** is defined by a common characteristic, such as a product produced there. The Corn Belt—the Iowa-Illinois area in the United States—is a formal region because corn is its major crop. A **functional region** is a central place and the surrounding area linked to it, for example, by a highway system. Metropolitan areas such as Los Angeles and Tokyo are functional regions. A **perceptual region** is defined by popular feelings and images rather than by objective data. For, example, the term "heartland" refers to a central area in which traditional values are believed to predominate.

Physical Systems

In their work geographers analyze how certain natural phenomena, such as volcanoes, hurricanes, and floods, shape the earth's surface. The earth's systems are endlessly fascinating.

> *"Every astronaut loves to take pictures of the Earth. To me, that's the best part of flying in space."*
>
> Rick Searfoss, quoted in "Geographica," *National Geographic,* November 1996

Geographers look at how physical features interact with plant and animal life to create, support, or change ecosystems. An **ecosystem** is a community of plants and animals that depend upon one another, and their surroundings, for survival.

Human Systems

Geographers also examine how people shape the world—how they settle the earth, form societies, and create permanent features. A recurring theme in geography is the ongoing **movement** of people, goods, and ideas. For example, migrants entering a long-established society usually bring different ideas and practices that may transform that society's traditional culture. In studying human systems, geographers also look at how people compete or cooperate to change or control aspects of the earth to meet their needs.

Environment and Society

Human-environment interaction, or the study of the interrelationship between people and their physical environment, is another theme of geography. Geographers examine the ways people use their environment, how and why they have changed it, and what consequences result from these changes. In some cases the physical environment affects human activities. For example, mountains and deserts often pose barriers to human movement. In other instances human activities, such as building a dam, cause changes in the physical environment. By understanding how the earth's physical features and processes shape and are shaped by human activity, geographers help societies make informed decisions.

The Uses of Geography

Geography can provide insight into how physical features and living things developed in the past. It can also interpret present-day trends to plan for future needs. Governments, businesses, and individuals use geographic information in planning and decision making. Data on physical features and processes can determine whether a site is suitable for human habitation or has resources worth developing. Geographic information on human activities, such as population trends, can help planners decide whether to build new schools or highways in a particular place. Geographic information helps determine where to locate fire stations and shopping malls. As geographers learn more about the relationships among people, places, and the environment, their knowledge can help us plan and build a better future.

SECTION  ASSESSMENT

Checking for Understanding

1. **Define** location, absolute location, hemisphere, grid system, relative location, place, region, formal region, functional region, perceptual region, ecosystem, movement, human-environment interaction.
2. **Main Ideas** In a web diagram, list six elements in the study of geography (hint: use the headings in this section). Then explain how each is applied.

Critical Thinking

3. **Categorizing Information** Consider the physical and human factors that constitute a region. Identify the differences among formal, functional, and perceptual regions.
4. **Drawing Conclusions** How might geographers' knowledge of human systems benefit people?
5. **Making Generalizations** Explain how knowing about the geography of a particular city might influence your decision to move there.

Analyzing Diagrams

6. **Location** Study the diagram of the hemispheres on page 20. In which hemispheres is Africa located?

Applying Geography

7. **Relative Location** Write a paragraph that describes the relative location of your school in at least five ways. In what instances might relative location be more useful than absolute location? In what instances might absolute location be more useful?

SECTION 2

The Geographer's Craft

Guide to Reading

Consider What You Know

People use different types of maps when they need to move from place to place or learn where something is located. What kinds of maps have you used and for what purposes?

Read to Find Out

- What are the major branches of geography and the topics each branch studies?
- What research methods do geographers use at work?
- How is geography related to other subject areas?
- What kinds of geographic careers are available today?

Terms to Know

- physical geography
- human geography
- meteorology
- cartography
- geographic information systems (GIS)

NATIONAL GEOGRAPHIC

A Geographic View

The Power of Maps

Guyana [in 1966] . . . agreed to give Indians title to lands traditionally recognized as theirs. But in 1982 a tally of "village lands" using out-of-date maps reduced Indian holdings to a few fragments. . . . Local Earth Observation turned Indian villagers into digital mappers. Armed with handheld [GPS] . . . units that determine location using satellites, villagers named and located more than 4,000 . . . territorial landmarks. The data they collected were combined with drainage patterns to produce a large-scale map. . . . [T]he power of maps that merge ancient knowledge and modern technology has vastly strengthened their case.

—Allen Carroll, "CartoGraphic," National Geographic, *March 2000*

Mapping Guyana with GPS

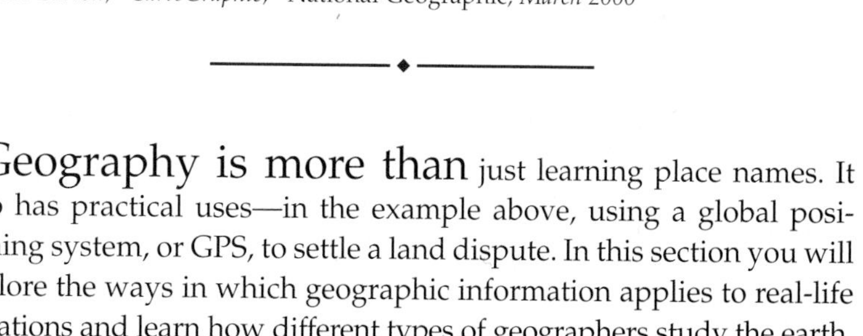

Geography is more than just learning place names. It also has practical uses—in the example above, using a global positioning system, or GPS, to settle a land dispute. In this section you will explore the ways in which geographic information applies to real-life situations and learn how different types of geographers study the earth.

Branches of Geography

Geography is a discipline that covers a broad range of topics. To make their work easier, geographers divide their subject area into different branches. Two major branches are physical geography and human

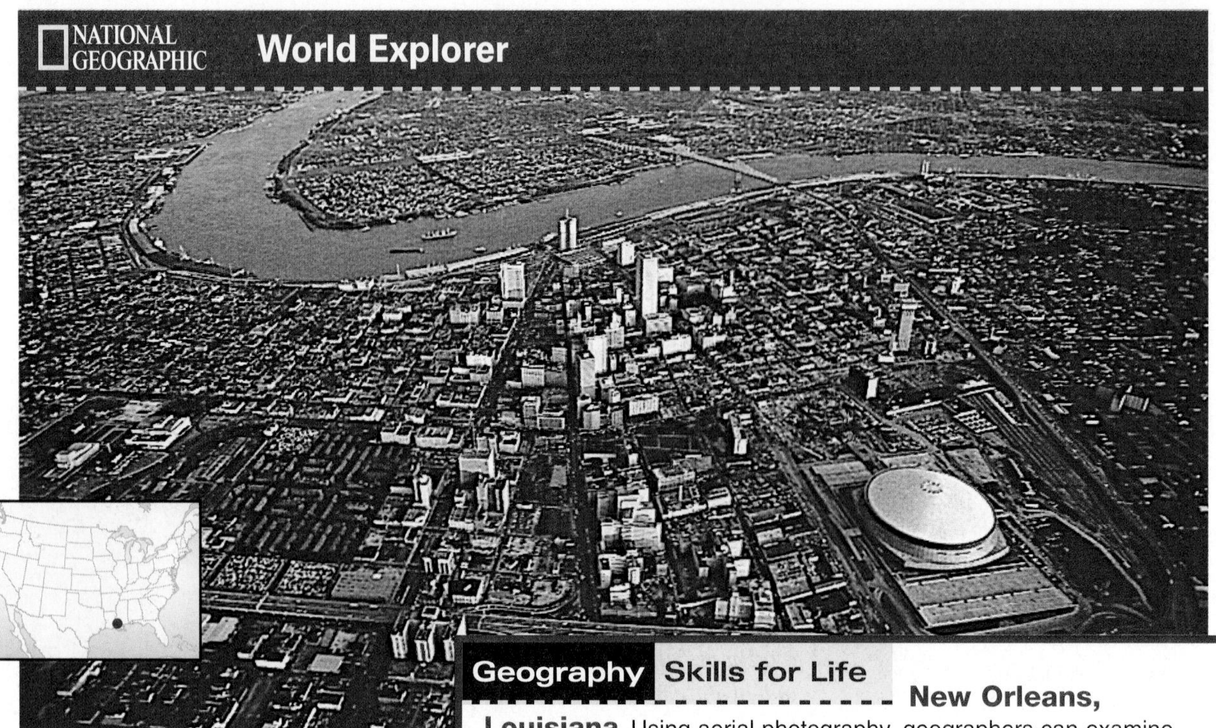

Geography Skills for Life **New Orleans, Louisiana** Using aerial photography, geographers can examine how large cities affect the physical environment.

Place How would you describe New Orleans based on the features shown in this photograph?

geography. **Physical geography** focuses on the study of the earth's physical features. It looks at climate, land, water, plants, and animal life in terms of their relationships to one another and to humans. **Human geography**, or cultural geography, is the study of human activities and their relationship to the cultural and physical environments. It concentrates on political, economic, and cultural factors, such as population density, urban development, economic production, and ethnicity.

Physical geography and human geography are further divided into smaller subject areas. Examples are **meteorology**, the study of weather and weather forecasting, and historical geography, the study of places and human activities over time and the various geographic factors that have shaped them.

Geographers at Work

Geographers use specialized research methods in their work. These methods include direct observation, mapping, interviewing, statistics, and the use of technology.

Direct Observation

Geographers use direct observation in studying the earth and the patterns of human activities that take place on its surface. They will often visit a region to gather specific information about the region and its geographic features. Geographers also employ remote sensing to study the earth, using aerial photographs and satellite images. For example, geographers may use aerial photographs or satellite images to locate mineral deposits or to determine the size of freshwater sources. They also might observe a forest that has been damaged by air pollution.

Mapping

Making and using maps are basic activities of geographers. Geographic specialists who make and design maps are known as cartographers; their area of work, known as **cartography**, involves studying and making maps.

Many geographic research findings can be shown on maps better than they can be explained in written text. Cartographers select complicated pieces of information about an area and present them in a more understandable form on a map. In this way they easily can show the location, features, patterns, and relationships of people, places, and things. In addition, maps allow a visual comparison between places and regions. For example, a geographer might compare population density maps of two counties in order to determine where to build new schools.

Interviewing

To answer a geographic question, geographers often must go beyond mere observation. In many cases geographers want to find out how people think or feel about certain places. They also may want to examine the ways in which people's beliefs and attitudes have led to changes in the physical environment. This kind of information is obtained by interviewing. Geographers choose a particular group of people for study. Instead of contacting everyone in that group, however, geographers talk to a carefully chosen sample whose answers represent the whole group.

Statistics

Some of the information that geographers use is numerical. Temperature and rainfall data point to a region's climate, for example. Geographers use computers to organize this information and present it in clear, understandable ways. They also analyze the data to find patterns and trends. For example, census data can be studied to learn about rates of population growth; the age, ethnic, and gender makeup of the population; and income levels. After identifying these patterns and trends, geographers use statistical tests to see whether their ideas are valid.

Technology

Geographers often use scientific instruments in their work. They especially depend on advanced technological tools, such as satellites and computers. Satellites orbiting the earth carry remote sensors, high-tech cameras, and radar that gather data and images related to the earth's environment, weather, human settlement patterns, and vegetation. **Geographic information systems (GIS)** are computer tools that process and organize data and satellite images with other pieces of information gathered by geographers and other scientists. GIS technology is valuable to urban planners, retailers, and local government officials who use this technology to help them determine where to build roads, stores, and parks.

The development of computer technology has also transformed the process of mapmaking. Allen Carroll, chief cartographer of the National Geographic Society, describes the changes in cartography as "revolutions in mapping." Technology has created

> *". . . computers that store vast archives of map data and render lines with superhuman precision, software programs that turn maps into analytical tools, satellite imagery that combines photographic beauty with cartographic precision, global electronic networks that enable maps to stream across our ever shrinking globe."*
>
> Allen Carroll, *National Geographic Atlas of the World,* 1999

NATIONAL GEOGRAPHIC **World Explorer**

Geography Skills for Life

Indonesian Surveyors These surveyors are helping plan a road in Borneo, Indonesia.

Human-Environment Interaction How do geographers play a part in our everyday lives?

Today, most cartographers rely on computers and computer software to make maps. Each type of data on a map is kept as a separate "layer" in the map's digital files. This method allows cartographers to make and change maps more quickly and easily.

NATIONAL GEOGRAPHIC **CHART STUDY**

Jobs in Geography

Geography Field	Description	Applications/Careers
Physical Geographer	Studies Earth's features and the geographic forces shaping them	Forecasting weather, tracing causes and effects of pollution, conserving wilderness areas
Human Geographer	Analyzes human aspects of culture—population, language, ethnicity, religion, government	Developing cultural policies for international organizations, such as the United Nations
Economic Geographer	Examines human economic activities and their relationship to the environment	Urban planning, focusing on the location of industries or transportation routes
Regional Geographer	Studies geographic features of a particular place or region	Assisting government and business in making decisions related to a region
Environmental Specialist	Focuses on the two-way interaction between humans and the physical environment	Advising government and business on ways of protecting the environment
Geographic Educator	Teaches about geography	Teaching geography at all educational levels; serving as consultant to business and government

Geography Skills for Life

1. **Interpreting Charts** What does a human geographer study?
2. **Applying Geography Skills** How might human geographers studying the effects of population growth gather information for their research?

Geography and Other Disciplines

Geographers study both the physical and human features of the earth and also analyze the patterns and relationships among geography and other disciplines. Studying subjects such as history, government, sociology, and economics helps geographers to understand how each of these subjects affects and is affected by geography.

History and Government

Geographers use history to help them understand what places looked like in the past. For example, geographers might want to know how Boston, Massachusetts, looked during the colonial years. They might also wish to look at the changes that have occurred in Boston over the past two centuries. Geographers may begin by gathering information about time periods in the city's history. This information could be used to answer such geographic questions as: How have human activities changed the natural vegetation in the area? Are the waterways different than they were in the past? Answers to these questions can help people make better decisions and plans for the future.

Geographers study political science to help them see how people in different places are governed. They look at how political boundaries have formed and how they have been changed. Geographers are interested in how the natural environment has influenced political decisions and how governments change natural environments. For example, the Egyptian government, helped by financing from abroad, built the massive

Aswan High Dam on the Nile River. The dam altered the surface of the earth in profound ways and so has had an impact on the region's people.

Culture

Human geographers use the tools of sociology and anthropology to understand the culture of societies throughout the world. They study the relationships between the physical environment and social structures. They examine people's ways of life in different parts of the world. Human geographers also seek to understand how the activities of different groups affect their physical environments and how the environment affects culture groups differently.

Economics

Geographers use economics to help them understand how the locations of resources affect the ways people make, transport, and use goods, and how and where services are provided.

Student Web Activity Visit the **Glencoe World Geography** Web site at geography.glencoe.com and click on Student Web Activities—Chapter 1 for an activity about careers in geography.

Geographers are interested in how locations are chosen for various economic activities, such as farming, mining, manufacturing, and selling. A desirable location usually includes plentiful resources and good transportation routes. Geographers are also interested in the interdependence of people's economic activities throughout the world. New developments in communications and transportation make the movement of information and goods faster and more efficient than ever before. A business can operate globally without depending on any one specific place to fill all of its needs.

Geography as a Career

Although people trained in geography are in great demand in the workforce, many of them do not have *geographer* as a job title. Geography skills are useful in so many different situations that geographers have more than a hundred different job titles. Geographers often combine the study of geography with other areas of study. For example, a salesperson must know the geographic characteristics of the region in which he or she is selling products. Also, a travel agent must have some knowledge of other places in order to plan trips for clients. Still, as the chart on page 26 shows, because geography itself has many specialized fields, there are many different kinds of geographers.

SECTION  ASSESSMENT

Checking for Understanding

1. **Define** physical geography, human geography, meteorology, cartography, geographic information systems (GIS).
2. **Main Ideas** Copy the table below on your paper, and fill in the ways geographers study the earth and use geography.

Geography Branches	
Geography Methods	
Other Disciplines	
Jobs in Geography	

Critical Thinking

3. **Predicting Consequences** What might happen if an economic geographer did not interview citizens when preparing a city transportation plan?
4. **Making Inferences** What kinds of geographers might be employed by a manufacturing company?
5. **Making Generalizations** How does the study of other disciplines help geographers in their work as countries become increasingly interdependent?

Analyzing Maps

6. **Place** Study the map of the United States in the Reference Atlas on pages RA6–RA7. What kinds of information can you learn from this map? How does the information on this map differ from the map on pages RA8–RA9?

Applying Geography

7. **Research Methods** As a geographer working on a plan for a new community center, what research methods would you use? Explain your choices in a paragraph.

Understanding Graphs

Graphs are visual representations of statistical data. Large amounts of information can be condensed when presented in graphs. Studying graphs allows readers to see relationships clearly.

Learning the Skill

The three main types of graphs present numerical information. **Line graphs** record changes in data over time. The vertical axis (*y*-axis) shows units of measurement, and the horizontal axis (*x*-axis) shows intervals of time. **Bar graphs** use bars of different lengths to compare different quantities. **Circle graphs** show the relationship of parts to a whole as percentages. To understand a graph:

- **Read the graph title to identify the subject.**
- **Study the labels to understand the numerical information presented.**
- **Study the information presented and the use of colors and patterns.**
- **Compare the lines, bars, or segments, and look for relationships in order to draw conclusions.**

Practicing the Skill

Study the graphs to answer these questions.

1. **Line graph** What is the difference in population between the low and high projections?
2. **Bar graph** In which decade did migration cause the least change in population?
3. **Circle graph** What percent of immigrants to the United States in the 1990s came from Europe?
4. What general population trends do the three graphs show?

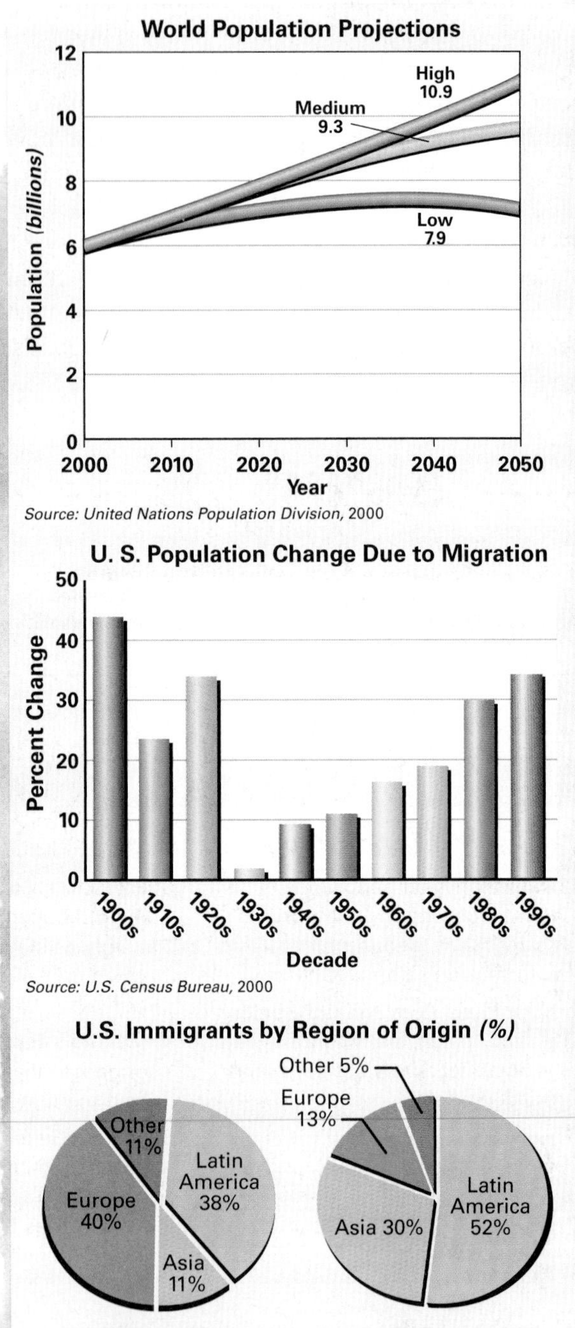

Source: United Nations Population Division, 2000

Source: U.S. Census Bureau, 2000

Source: U.S. Immigration and Naturalization Service, 1999

Applying the Skill

Take a poll of your classmates about a geographical topic. Design and draw a graph using the data. Consider geographic features, distributions, and relationships.

The **Glencoe Skillbuilder Interactive Workbook, Level 2** provides instruction and practice in key social studies skills.

CHAPTER 1 SUMMARY & STUDY GUIDE

SECTION 1 Exploring Geography (pp. 19–22)

Terms to Know
- location
- absolute location
- hemisphere
- grid system
- relative location
- place
- region
- formal region
- functional region
- perceptual region
- ecosystem
- movement
- human-environment interaction

Key Points
- Geographers study the earth's physical and human features and their interrelationships.
- Geographers use absolute and relative locations as reference points.
- Geographers identify three types of regions—formal, functional, and perceptual.
- Geography contributes knowledge about the relationships among human activities, the earth's physical systems, and the environment in order to develop a better future.

Organizing Your Notes
Create an outline using the format below to help you organize information about how geographers study the earth.

Exploring Geography

I. Elements of Geography
  A. World in spatial terms
    1. Absolute location

SECTION 2 The Geographer's Craft (pp. 23–27)

Terms to Know
- physical geography
- human geography
- meteorology
- cartography
- geographic information systems (GIS)

Key Points
- Geographers use special research skills, such as direct observation, mapping, interviewing, statistics, and technology.
- Studying other social sciences helps geographers analyze the patterns and relationships among these different fields.
- Geographers can specialize and may work in government, business, science, planning, or education.

Organizing Your Notes
Use a graphic organizer like the one below to help you organize your notes for this section.

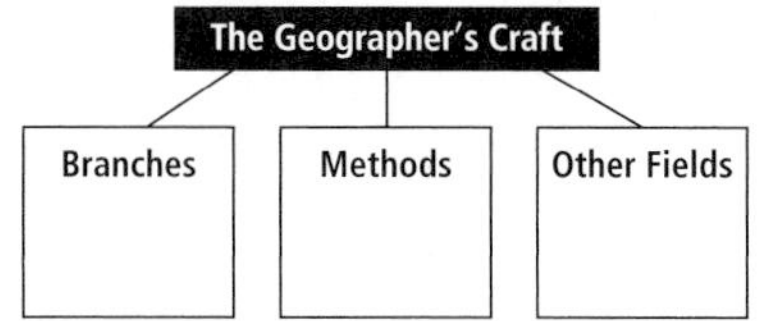

▲ *A scientist uses a global positioning system.*

CHAPTER 1

ASSESSMENT & ACTIVITIES

Reviewing Key Terms

Write the key term that best completes each of the following sentences. Refer to the Terms to Know in the Summary & Study Guide on page 29.

1. Plants and animals depend on one another in a(n) ________.
2. Geographers use a(n) ________ formed by lines of latitude and longitude to determine ________.
3. ________ is the study of the human aspects of geography.
4. A(n) ________ has boundaries determined by a common characteristic.
5. Another name for mapmaking is ________.
6. ______ is expressed in relation to other places.
7. Computer tools that process data and satellite images with other pieces of geographic information are called ________.
8. ________ focuses on the study of the earth's physical features.

Reviewing Facts

SECTION 1

1. How do geographers determine the locations of places?
2. What are the three types of regions identified by geographers?
3. Why do geographers study human systems and human-environment relationships?
4. What are two ways that every place on the earth can be located?

SECTION 2

5. How do physical and human geography differ?
6. What research methods do geographers use?
7. What other subjects do geographers study?

Critical Thinking

1. **Summarizing the Main Idea** How do geographers use the elements of geography to study the earth?
2. **Making Inferences** What subjects might you study in order to become an urban planner? Explain.
3. **Predicting Consequences** Consider the many ways that technology has affected the way people live and work. Then imagine that you have become a geographer of the future. How do you think technology will change the way you work?
4. **Categorizing Information** Use a web diagram like the one below to show five methods of geographic research.

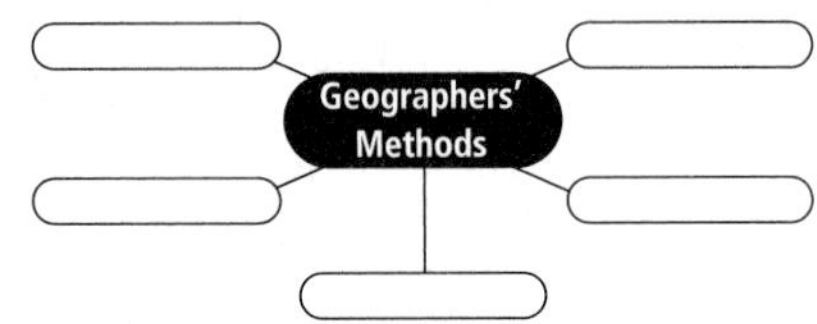

NATIONAL GEOGRAPHIC

Locating Places

The World: Physical Geography

Match the letters on the map with the places and physical features of the earth. Write your answers on a sheet of paper.

1. North America
2. South America
3. Africa
4. Asia
5. Europe
6. Antarctica
7. Australia
8. Atlantic Ocean
9. Indian Ocean
10. Pacific Ocean

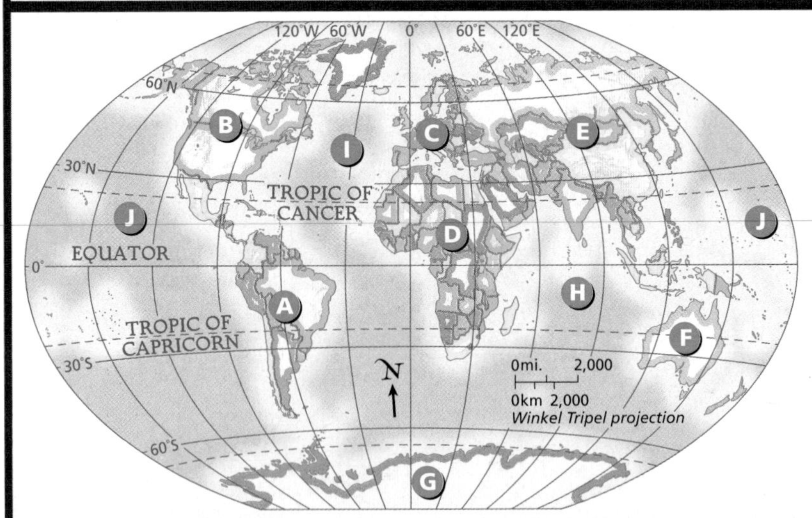

Self-Check Quiz Visit the **Glencoe World Geography** Web site at geography.glencoe.com and click on Self-Check Quizzes—Chapter 1 to prepare for the Chapter Test.

Thinking Like a Geographer

Imagine that you are an urban planner. What kinds of data might you want to assemble in order to plan a location for a new school? What methods would you use to collect the data? How would you use the data to determine the location for the school?

Problem-Solving Activity

Contemporary Issues Case Study Look at newspapers and magazines to identify one of the following issues:

- a local issue that involves land use.
- a local issue that involves economic development.
- a national issue that involves water resources.

Choose one issue, and research to learn more about its history, the various points of view surrounding the issue, and the final outcome. Use this information to prepare an outline. Then write an essay describing the influence of physical and human geography on the issue.

GeoJournal

Descriptive Writing Refer to the entry you wrote in your GeoJournal and the information in this chapter. Then imagine you are a physical geographer and write a paragraph describing another part of the earth's surface. For example, if you live in a plains area, describe how the geography of the mountains or the seashore would be different from your location. Include as many concrete details as you can to describe the physical and human geography of the place you chose.

Technology Activity

Using the Internet for Research Search the Internet for Web sites that provide information about geography to the public. Sponsors may include government agencies, scientific organizations, or special-interest magazines. Prepare a list of the five best sites, write a brief description of the kinds of information each one contains, and explain why you included it.

Standardized Test Practice

Use the circle graph below and your knowledge of geography to answer questions 1 and 2.

EARTH'S LAND AND WATER

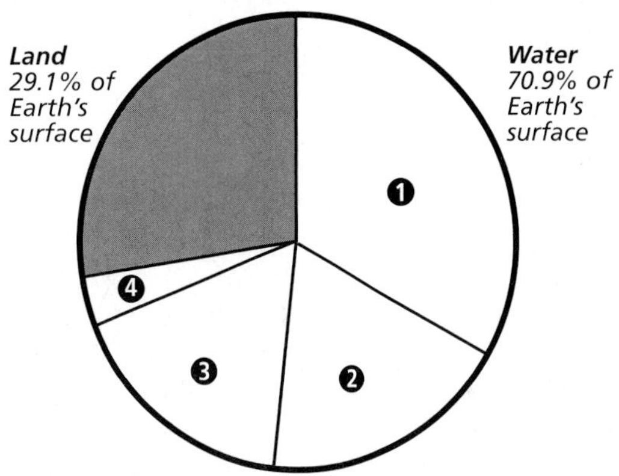

1 Pacific Ocean 64,169,000 sq. mi. (166,241,000 sq. km)
2 Atlantic Ocean 33,411,00 sq. mi. (86,557,000 sq. km)
3 Indian Ocean 28,342,800 sq. mi. (73,427,000 sq. km)
4 Arctic Ocean 3,661,200 sq. mi. (9,485,000 sq. km)

1. **Which ocean covers the smallest area of the earth's surface?**

   A Atlantic
   B Indian
   C Pacific
   D Arctic

2. **Which ocean covers about as much of the earth's surface as land does?**

   F Indian
   G Pacific
   H Arctic
   J Atlantic

Study the information shown on the circle graph for the areas of the earth covered by land and by oceans. Then compare the relative sizes of the different graph segments. By comparing the segments you will be able to determine the correct answers.

CHAPTER 2

The Earth

GeoJournal

As you read this chapter, use your journal to note facts about our planet. Include descriptive details about Earth's physical features, its structure, and the forces of change that shape its surface.

GEOGRAPHY *Online*

Chapter Overview Visit the **Glencoe World Geography** Web site at geography.glencoe.com and click on Chapter Overviews—Chapter 2 to preview information about Planet Earth.

Planet Earth

Guide to Reading

Consider What You Know

What do you know about Planet Earth? How does it compare to other planets and to other objects, such as stars and moons? How would you describe the surface of the earth?

Read to Find Out

- Where is Earth located in our solar system?
- How is Earth shaped?
- What is Earth's structure?
- What types of landforms are found on Earth?

Terms to Know

- hydrosphere
- lithosphere
- atmosphere
- biosphere
- continental shelf

Places to Locate

- Australia
- Antarctica
- Europe
- Asia
- North America
- South America
- Africa
- Mount Everest
- Dead Sea
- Mariana Trench

◀ *Earth viewed from space*

NATIONAL GEOGRAPHIC

A Geographic View

Patterns of Life

As we fly over the Sahara, we all strain to see Lake Chad, which is between the northern deserts and the green portion of equatorial Africa. Its level is a good indicator of how wet or dry this boundary region, called the Sahel, has been. It was mostly cloudy during my first flight: a good sign for those living below. On my second flight, 17 months later, we could see that the level had risen: The drought in the Sahel had abated [ended] for the moment.

—*Jay Apt, "Orbit: The Astronauts' View of Home,"* National Geographic, *November 1996*

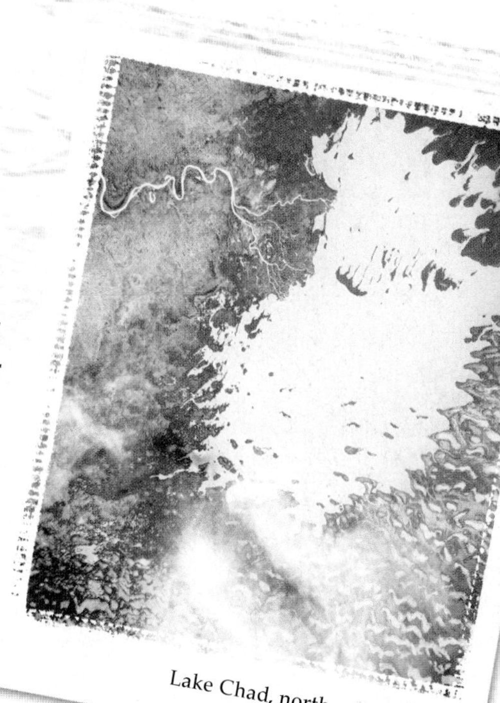

Lake Chad, northern Africa

An astronaut, seeing Earth from the blackness of space, described it as "piercingly beautiful." From the vantage point of space, the earth's great beauty resembles a blue and white marble, with contrasts of water and land beneath huge swirls of white clouds. Together these features form the physical environment of the earth. In this section you will discover what humans know about the physical nature of our planet, Earth.

Our Solar System

Earth is part of our solar system, which is made up of the sun and all of the countless objects that revolve around it. At our solar system's center is the sun—a star, or ball of burning gases. About 109 times wider than Earth, the sun's enormous mass—the amount of matter it contains—creates a strong pull of gravity. This basic physical force keeps the earth and the other objects revolving around the sun.

DIAGRAM STUDY

The Solar System

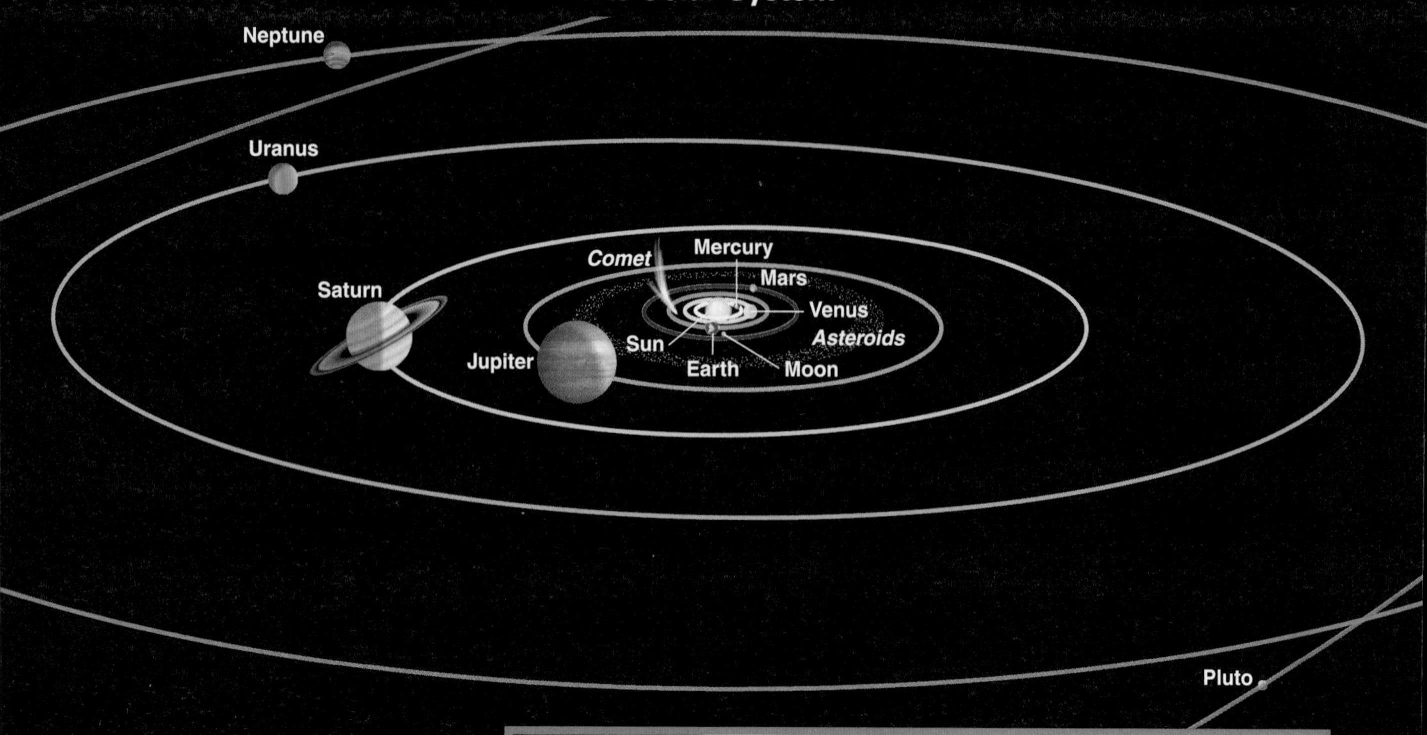

Geography Skills for Life

1. **Interpreting Diagrams** Which four planets are closest to the sun?
2. **Applying Geography Skills** What characteristics do the inner planets share? What characteristics do the outer planets share?

The Planets: Neighbors in Space

Except for the sun, spheres called planets are the largest objects in our solar system. At least nine planets exist, and each is in its own orbit around the sun. The diagram above shows that the planets vary in distance from the sun. Mercury, Venus, Earth, and Mars are the inner planets, or those nearest the sun. Earth, the third planet from the sun, is about 93 million miles (about 150 million km) away. Farthest from the sun are the outer planets—Jupiter, Saturn, Uranus, Neptune, and Pluto.

The planets vary in size. Jupiter is the largest. Earth ranks fifth in size among the planets, and distant Pluto is the smallest. All of the planets except Mercury and Venus have moons, smaller spheres or satellites that orbit them. The number of moons each planet has also differs. Earth has 1 moon, and Saturn has at least 18 moons.

All of the planets except Pluto are grouped into two types—terrestrial planets and gas giant planets. Mercury, Venus, Earth, and Mars are called *terrestrial planets* because they have solid, rocky crusts. Mercury and Venus are scalding hot, and Mars is a cold, barren desert. Only Earth has liquid water at the surface and can support varieties of life.

Farther from the sun are the *gas giant planets*—Jupiter, Saturn, Uranus, and Neptune. They are much more gaseous and less dense than the terrestrial planets, even though they are larger in diameter. Each gas giant planet is itself like a miniature solar system, with orbiting moons and thin, encircling rings. Only

Saturn's rings, however, are easily seen from Earth by telescope. Pluto, the exception among the planets, is a ball of ice and rock.

Asteroids, Comets, and Meteoroids

In addition to the planets, thousands of smaller objects, including asteroids, comets, and meteoroids, revolve around the sun. Asteroids are small, irregularly shaped, planetlike objects. They are found mainly between the orbits of Mars and Jupiter in a region called the *asteroid belt*. A few asteroids follow paths that cross the earth's orbit.

Comets, made of icy dust particles and frozen gases, look like bright balls with long, feathery tails. Their orbits are inclined at every possible angle to the earth's orbit. They may approach from any direction.

Meteoroids are pieces of space debris—chunks of rock and iron. When they occasionally enter Earth's gravitational field, friction usually burns them up before they reach the earth's surface. Those that collide with Earth are called meteorites. Meteorite strikes, though rare, can significantly affect the landscape, leaving craters and causing other devastation. In 1908 a huge area of forest in the remote Russian region of Siberia was flattened and burned by a "mysterious fireball." Scientists theorize it was a meteorite or comet. As a writer describes the effects:

> *"The heat incinerated herds of reindeer and charred tens of thousands of evergreens across hundreds of square miles. For days, and for thousands of miles around, the sky remained bright with an eerie orange glow—as far away as western Europe people were able to read newspapers at night without a lamp."*
>
> Richard Stone, "The Last Great Impact on Earth," *Discover*, September 1996

Getting to Know Earth

Although globes show our planet as a perfect sphere, the earth is really a rounded object wider around the center than from top to bottom. Earth has a larger diameter at the Equator—about 7,930 miles (12,760 km)—than from Pole to Pole, but the difference is less than 1 percent. With a circumference of about 24,900 miles (40,070 km), Earth is the largest of the inner planets.

Water, Land, and Air

The surface of the earth is made up of water and land. About 70 percent of our planet's surface is water, which gives the planet the deep blue appearance that astronauts see from space. Oceans, lakes, rivers, and other bodies of water make up a part of the earth called the **hydrosphere**.

About 30 percent of the earth's surface is land, including continents and islands. Land makes up a part of the earth called the **lithosphere**, the earth's crust. The lithosphere also includes the ocean basins, or the land beneath the oceans.

The air we breathe is part of Earth's **atmosphere**, a layer of gases extending about 6,000 miles (9,700 km) above the planet's surface. The atmosphere is composed of 78 percent nitrogen, 21 percent oxygen, and small amounts of argon and other gases.

All people, animals, and plants live on the earth's surface, close to the earth's surface, or in the atmosphere. The part of the earth that supports life is the **biosphere**. Life outside the biosphere, such as on a space station orbiting Earth, exists only with the assistance of mechanical life-support systems.

Landforms

Landforms are the natural features of the earth's surface. The diagram on pages 14–15 shows many of the earth's landforms, made up of physical features with a particular shape or elevation. Four major landforms are mountains, hills, plateaus, and plains. Others include valleys, canyons, and basins. Landforms often contain rivers, lakes, and streams.

Underwater landforms are as diverse as those found on dry land. In some places the ocean floor is a flat plain. Other parts of the seabed feature mountain ranges, cliffs, valleys, and deep trenches.

Seen from space, Earth's most visible landforms, however, are the seven large landmasses called continents. Two continents, **Australia** and **Antarctica**, stand alone, while the others are joined in some way. **Europe** and **Asia** are actually parts of one huge landmass called Eurasia. A narrow neck of land called the Isthmus of Panama links **North America** and **South America**. At the Sinai

NATIONAL GEOGRAPHIC **World Explorer**

Geography Skills for Life

Planet Earth Parts of the atmosphere (left), lithosphere (center), and hydrosphere (right) form the biosphere, the part of the Earth where life exists.

Human-Environment Interaction How does human activity impact the biosphere?

Peninsula, the human-made Suez Canal separates **Africa** and **Asia**.

The part of a continent that extends underwater is called a continental shelf. Continental shelves are narrow in some places and wide in others. They slope out from land for as much as 800 miles (1,287 km) and descend gradually to a depth of about 660 feet (200 m) where a sharp drop marks the beginning of the continental slope. This area drops more sharply to the ocean floor.

Earth's Heights and Depths

Great contrasts exist in the heights and depths of the earth's surface. The highest point on Earth is in South Asia at the top of **Mount Everest**, which is 29,035 feet (8,852 m) above sea level. The lowest dry land point, at 1,349 feet (411 m) below sea level, is the shore of the **Dead Sea** in Southwest Asia. Earth's deepest known depression lies under the Pacific Ocean southwest of Guam in the **Mariana Trench**, a long, narrow, underwater canyon about 35,827 feet (10,923 m) deep. These and other natural features have developed as the earth has changed over millions of years. In the next section you will learn about the forces of change that have shaped the earth.

SECTION 1 ASSESSMENT

Checking for Understanding

1. **Define** hydrosphere, lithosphere, atmosphere, biosphere, continental shelf.
2. **Main Ideas** Copy the diagram below on your paper, filling in information about the structures of the solar system and the earth.

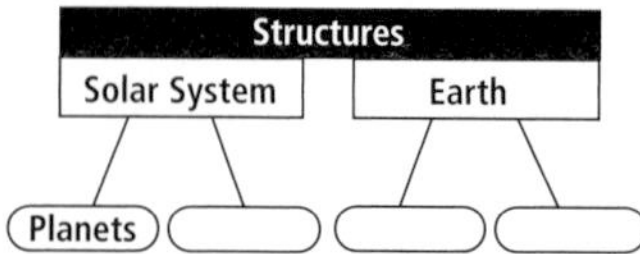

Critical Thinking

3. **Drawing Conclusions** Recently NASA has launched space probes to explore Mars. Why might Mars have been chosen for these explorations?
4. **Making Inferences** Think about Earth's surface and list conditions that must be present in a space station in order to support life.
5. **Comparing and Contrasting** How do the inner planets differ from the outer planets?

Analyzing Diagrams

6. **Location** Study the diagram of the solar system on page 34. How is the size of a planet's orbit influenced by its distance from the sun?

Applying Geography

7. **A Delicate Balance** Think about the ratio of water and land on Earth. Write a description of how Earth's physical features would be different if the proportions were reversed.

SECTION 2

Forces of Change

Guide to Reading

Consider What You Know

Through observation you are familiar with some features of the earth's surface. These features have been shaped by forces above and beneath the earth's surface. How many of these forces can you name?

Read to Find Out

- How do Earth's layers contribute to the planet's physical characteristics?
- What internal forces operate to affect Earth's surface, the setting for human life?
- What external forces affect Earth's surface?

Terms to Know

- mantle
- continental drift
- magma
- plate tectonics
- subduction
- accretion
- spreading
- fold
- fault
- weathering
- erosion
- loess
- glacier
- moraine

Places to Locate

- San Andreas Fault
- Ring of Fire

NATIONAL GEOGRAPHIC

A Geographic View

A Planet in Motion

Scientists now know that the earth is dynamic and enormously complex. At the [earth's] surface ride more than a dozen huge, stiff fragments, or plates. They move at a slower-than-snail's pace—only inches a year—but cover thousands of miles over millions of years. As they collide and separate, they change the face of the globe by deforming and rearranging its features. The engine that propels [these] plates lies below: hot inner layers that churn like thick soup simmering in very slow motion.

—Keay Davidson and A.R. Williams, "Under Our Skin: Hot Theories on the Center of the Earth," National Geographic, *January 1996*

San Andreas Fault, California

Although we cannot look into the center of the earth, scientists have concluded from available evidence that it is a dynamic interior of intense heat and pressure. Movements deep within the earth drive numerous changes that renew and enrich the earth's surface. In this section you will learn about the earth's structure and the natural forces that continually act upon our planet.

Earth's Structure

For hundreds of millions of years, the surface of the earth has been in slow but constant motion. Some forces that change the earth, such as wind and water, occur on the earth's surface. Others, such as volcanic eruptions and earthquakes, originate deep in the earth's interior.

NATIONAL GEOGRAPHIC **DIAGRAM STUDY**

Inside the Earth

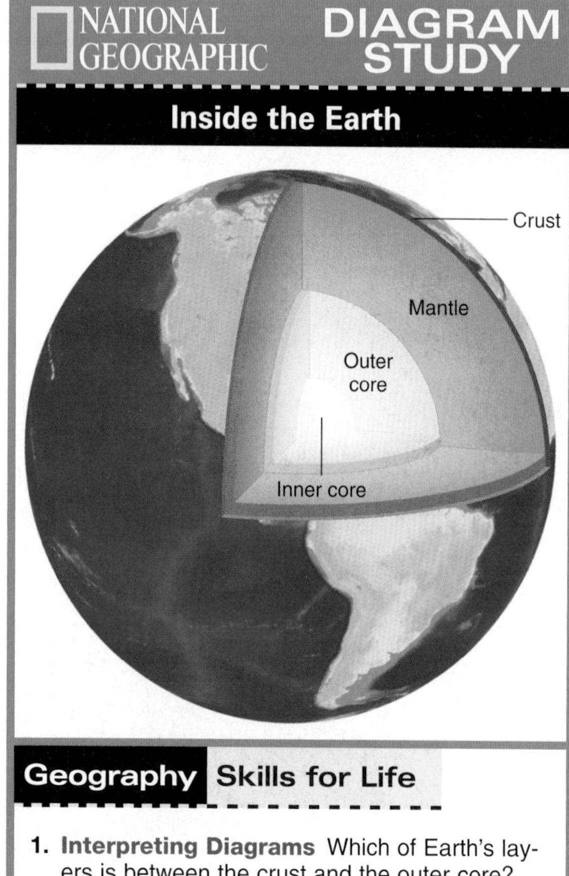

Geography Skills for Life

1. **Interpreting Diagrams** Which of Earth's layers is between the crust and the outer core?
2. **Applying Geography Skills** How do the plates of the earth's crust shape the physical landscape?

A Layered Planet

As you see in the diagram above, the earth is composed of three layers—the core, the mantle, and the crust. At the very center of the planet is a super-hot but solid inner core. It lies about 4,000 miles (6,430 km) below the surface of the earth. Scientists believe that the inner core is made up of iron and nickel under enormous pressure.

Surrounding the inner core is a liquid outer core, about 1,400 miles (2,250 km) thick. A band of melted iron and nickel, the outer core begins about 1,800 miles (2,900 km) below the surface of the earth. Temperatures there reach a scalding 8,500°F (about 4,700°C).

Next to the outer core is a thick layer of hot, dense rock called the **mantle**. The mantle consists of silicon, aluminum, iron, magnesium, oxygen, and other elements. This mixture continually rises, cools, sinks, warms up, and rises again, releasing 80 percent of the heat generated from the earth's interior.

The outer layer is the crust, a rocky shell forming the earth's surface. This relatively thin layer of rock ranges from about 2 miles (3.2 km) thick under oceans to about 75 miles (121 km) thick under mountains. The crust is broken into more than a dozen great slabs of rock called plates that rest—or more accurately, float—on a partially melted layer in the upper mantle. The plates carry the earth's oceans and continents. The map on page 39 shows the boundaries of the various plates.

Plate Movement

If you had seen the earth from space 500 million years ago, the planet probably would not have looked at all like it does today. Many scientists believe that most of the landmasses forming our present-day continents were once part of one gigantic supercontinent called *Pangaea* (pan•JEE•uh). Over millions of years, this supercontinent has broken apart into smaller continents. These continents in turn have drifted and, in some places, recombined. The theory that the continents were once joined and then slowly drifted apart is called **continental drift**.

Many scientists theorize that plates moving slowly around the globe have produced Earth's largest features—not only continents, but also oceans and mountain ranges. Most of the time, plate movement is so gradual—only about 4 inches (10 cm) a year—that it cannot be felt. As they move, the plates may crash into each other, pull apart, or grind and slide past each other. Whatever their actions, plates are constantly changing the face of the planet. They push up mountains, create volcanoes, and produce earthquakes. When the plates spread apart, **magma**, or molten rock, is pushed up from the mantle and ridges are formed. When the plates bump together, one may slide under another, forming a trench.

Scientists use the term **plate tectonics** to refer to all of these activities, which create many of the earth's physical features. Many scientists estimate that plate tectonics have been shaping the earth's surface for 2.5 to 4 billion years. According to some scientists, plate tectonics will have sculpted a whole new look for our planet millions of years from now.

Scientists, however, have not yet determined exactly what causes plate tectonics. They theorize that heat rising from the earth's core may create slow-moving currents within the mantle. Over millions of years, these currents of molten rock may shift the plates around, but the movements are extremely slow and difficult to detect.

Internal Forces of Change

The surface of the earth has changed greatly over time. Scientists believe that some of these changes come from internal forces associated with plate tectonics. One of these internal forces relates to the slow movement of magma within the earth. Other internal forces involve movements that can fold, lift, bend, or break the rock along the earth's crust.

Colliding and Spreading Plates

Mountains are formed in areas where giant continental plates collide. For example, the Himalaya ranges in South Asia were thrust upward when the Indian landmass rammed into Eurasia. Himalayan peaks are still getting higher as the Indian landmass continues to push against them.

Mountains also are created when a sea plate collides with a continental plate. In a process called **subduction** (suhb•DUHK•shuhn), the heavier sea plate dives beneath the lighter continental plate. Plunging into the earth's interior, the sea plate becomes molten material. Then, as magma, it bursts through the crust to form volcanic mountains. The Andes, for example, were formed over millions of years as a result of the process of subduction.

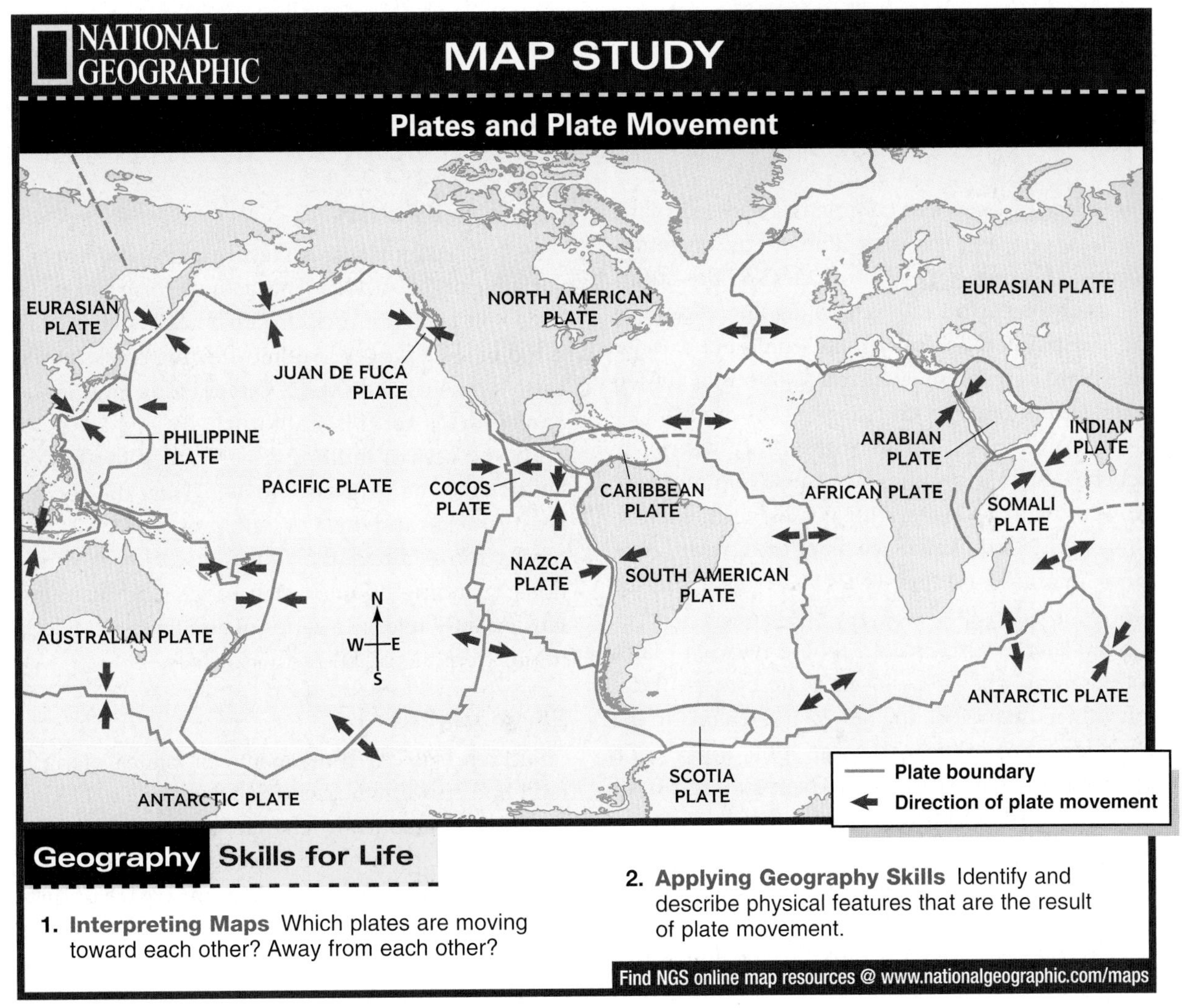

Geography Skills for Life

1. **Interpreting Maps** Which plates are moving toward each other? Away from each other?

2. **Applying Geography Skills** Identify and describe physical features that are the result of plate movement.

Find NGS online map resources @ www.nationalgeographic.com/maps

NATIONAL GEOGRAPHIC

DIAGRAM STUDY

Forces of Change

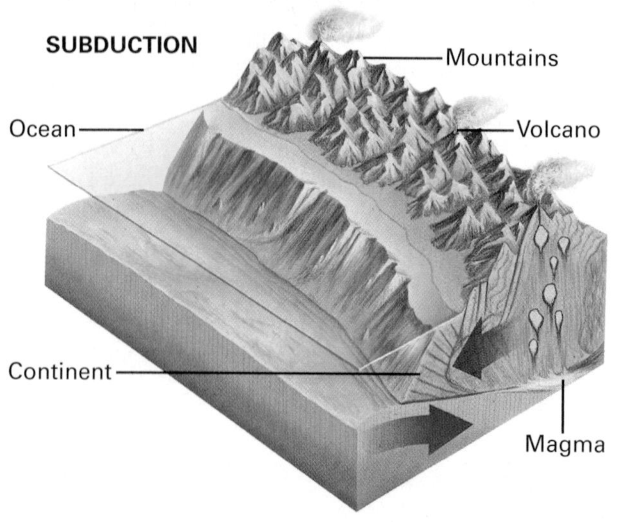

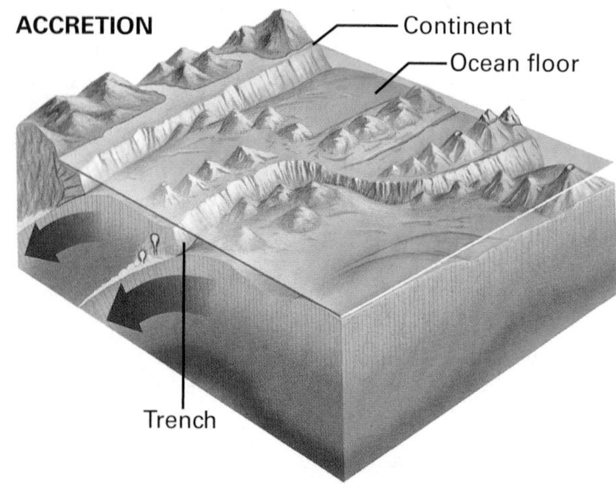

In other cases where continental and sea plates meet, a different process, known as accretion, occurs. During **accretion** (uh•KREE•shuhn), pieces of the earth's crust come together slowly as the sea plate slides under the continental plate. This plate movement levels off seamounts, underwater mountains with steep sides and sharp peaks, and piles up the resulting debris in trenches. Such a buildup can cause continents to grow outward. Most scientists believe that much of western North America expanded outward over more than 200 million years as a result of the process of accretion.

New land is also created where two sea plates converge. In this process one plate moves under the other, often forming an island chain at the boundary. Sea plates also can pull apart in a process known as **spreading**. The resulting rift, or deep crack, allows magma from within the earth to well up between the plates. The magma hardens to build undersea volcanic mountains or ridges. This spreading activity occurs down the middle of the Atlantic Ocean's floor, pushing Europe and North America away from each other.

Folds and Faults

Moving plates sometimes squeeze the earth's surface until it buckles. This activity forms **folds**, or bends, in layers of rock. In other cases plates may grind or slide past each other, creating cracks in the earth's crust called **faults**. One famous fault is the **San Andreas Fault** in California.

The process of faulting occurs when the folded land cannot be bent any further. Then the earth's crust cracks and breaks into huge blocks. The blocks move along the faults in different directions, grinding against each other. The resulting tension may release a series of small jumps, felt as minor tremors on the earth's surface.

Earthquakes

Sudden, violent movements of plates along a fault line are known as earthquakes. These shaking activities dramatically change the surface of the land and the floor of the ocean. For example, during a severe earthquake in Alaska in 1964, a portion of the ground lurched upward 38 feet (11.6 m).

Earthquakes often occur where different plates meet one another. Tension builds up along fault

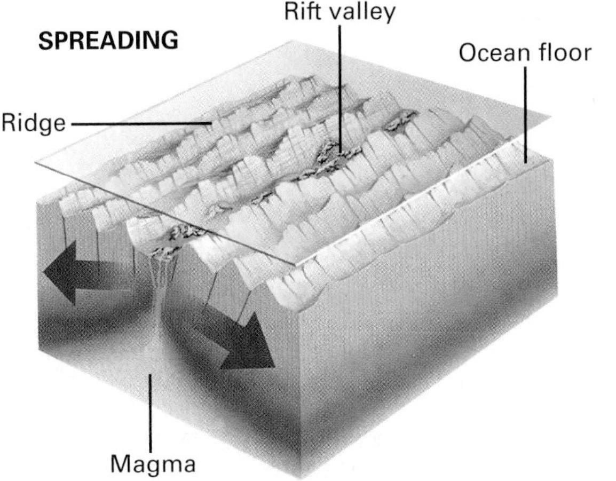

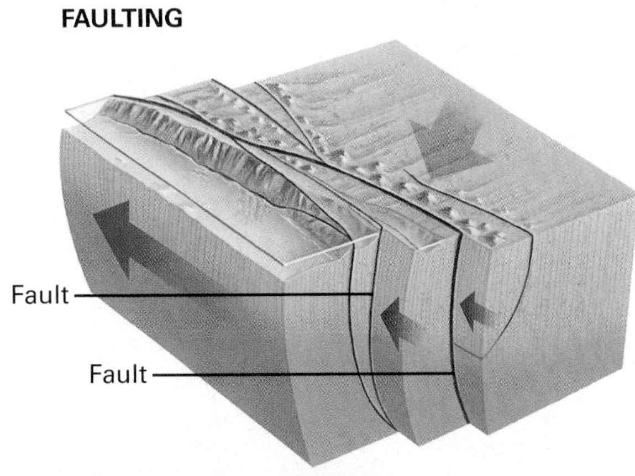

Geography Skills for Life

1. **Interpreting Diagrams** How does the process of accretion create deep trenches in the earth's surface?
2. **Applying Geography Skills** How do you think human settlement is affected by the process of faulting?

lines as the plates stick. The strain eventually becomes so intense that the rocks suddenly snap and shift. This movement releases stored-up energy along the fault. The ground then trembles as shock waves surge through it away from the area where the rocks snapped apart.

In recent years disastrous earthquakes have occurred in Kobe, Japan, and in Los Angeles and San Francisco, California. These cities are located along the **Ring of Fire**, one of the most earthquake-prone areas on the planet. The Ring of Fire is a zone of earthquake and volcanic activity surrounding the Pacific Ocean. It marks the boundary where the plates that cradle the Pacific meet the plates that hold the continents surrounding the Pacific.

Volcanic Eruptions

Volcanoes are mountains formed by lava or by magma that breaks through the earth's crust. Volcanoes often rise along plate boundaries where one plate plunges beneath another. This kind of volcanic formation occurs, for example, along the Ring of Fire. In such a process the rocky plate melts as it dives downward into the hot mantle. If the molten rock is too thick, its flow is blocked and pressure builds. A cloud of ash and gas may then spew forth, creating a funnel through which the red-hot magma rushes to the surface. There the lava flow may eventually form a large volcanic cone topped by a crater, a bowl-shaped depression at a volcano's mouth.

Volcanoes also arise in areas away from plate boundaries. Some areas deep in the earth are hotter than others, and magma often blasts through the surface as volcanoes. As a moving plate passes over these hot spots, molten rock flowing out of the earth's surface may create volcanic island chains. An example of this type of formation is the Hawaiian Islands in the Pacific Ocean. At various hot spots, molten rock may also heat underground water, causing hot springs or geysers. Yellowstone National Park in Wyoming has many spectacular geysers formed by this process, such as Old Faithful, which regularly sends water and steam into the air.

▲ Water erosion in the Grand Canyon

External Forces of Change

External forces, such as wind and water, also change the earth's surface. They have transformed the planet's appearance over millions of years and continue to do so today. Wind and water movements involve two processes: weathering and erosion. **Weathering** is the process that breaks down rocks on the earth's surface into smaller pieces. **Erosion** is the wearing away of the earth's surface by wind, glaciers, and moving water.

Weathering

The earth is changed by two basic kinds of weathering—physical weathering and chemical weathering. Physical weathering occurs when large masses of rock are physically broken down into smaller pieces. For example, water seeps into the cracks in a rock and freezes, then expands and causes the rock to split.

The process of chemical weathering changes the chemical makeup of rocks, transforming their minerals or combining them with new elements. For example, water mixed with carbon dioxide from the air easily dissolves certain rocks, such as limestone. Many of the world's caves have been and continue to be formed by this process.

Wind Erosion

Wind erosion involves the movement of dust, sand, and soil from one place to another. Plants help protect the land from wind erosion; however, in dry places where people have cut down trees and plants, winds pick up large amounts of soil and blow it away. Serious wind erosion devastated the Great Plains in the central United States during the 1930s. Fierce winds swept up dry, overworked soil from exposed farmland and carried it away in dust storms.

Wind erosion, however, also provides benefits. The dust carried by wind often forms large deposits of mineral-rich soil. These deposits provide fertile farmland in various parts of the world. China's Yellow River basin, for example, is thickly covered with **loess** (LEHS), a fertile, yellow-gray soil deposited by wind.

Glacial Erosion

Another cause of erosion is **glaciers**, or large bodies of ice that slowly move across the earth's surface. Glaciers form over a long period of time as layers of snow press together and turn to ice. Their great weight causes them to move slowly downhill or spread outward. As they move, glaciers pick up rocks and soil in their paths. Glacial movements change the landscape, destroying forests, carving out valleys, altering the courses of rivers, and wearing down mountaintops.

When glaciers melt and recede, in some places they leave behind large piles of rocks and debris called **moraines**. Some moraines form long ridges of land, while others form dams that hold water back and create glacial lakes.

There are two types of glaciers—sheet glaciers and mountain glaciers. Sheet glaciers are flat, broad sheets of ice. Today sheet glaciers cover most of Greenland and all of Antarctica. They advance a few feet each winter and recede during the summer. Large blocks of ice often break off from the coastal edges of sheet glaciers to become icebergs floating in the ocean.

Mountain glaciers are more common than sheet glaciers. They are located in high mountain valleys where the climate is cold, such as in the Rocky Mountains and Cascade Range of North America. As they move downhill, mountain glaciers gouge out round, U-shaped valleys. As these glaciers melt, rock and soil are deposited in new locations.

Water Erosion

Fast-moving water—rain, rivers, streams, and oceans—is the most significant cause of erosion. Water erosion begins when springwater and rainwater flow off the land downhill in streams. As the water flows, it cuts into the land, wearing away the soil and rock. The resulting sediment—small particles of soil, sand, and gravel—acts like sandpaper, grinding away the surface of rocks along the stream's path. Over time, the eroding action of water forms first a gully and then a V-shaped valley. Sometimes valleys are eroded even further to form valleys with high, steep walls, called canyons. The Grand Canyon of the Colorado River is a good example of the eroding power of water. Peter Miller describes the canyon's various layers as he hiked from the canyon rim to its floor:

> *"[A]s we hiked down the twisting Kaibab Trail to the Colorado River at the bottom of the canyon, we were hoping . . . to share a . . . great adventure. The canyon opened up before us with all the drama we had imagined. After we dropped below the ponderosa pine forests of the South Rim, we descended narrow ridgelines past crumbling cliffs of white sandstone, red limestone, chocolate sandstone, and maroon shale, descending 4,700 feet to the river."*
>
> Peter Miller, "John Wesley Powell: Vision for the West," *National Geographic*, April 1994

In addition to streams and rivers, oceans play an important role in water erosion. Pounding waves continually erode coastal cliffs, wear rocks into sandy beaches, and move sand away to other coastal areas. Violent storms speed up this process. In the next section, you will learn about the oceans and other water features of our planet and how they affect the earth's surface.

SECTION 2 ASSESSMENT

Checking for Understanding

1. **Define** mantle, continental drift, magma, plate tectonics, subduction, accretion, spreading, fold, fault, weathering, erosion, loess, glacier, moraine.
2. **Main Ideas** Copy the organizer below, and fill in information about forces that shape Earth's features. Then choose one force that shapes the earth's surface and explain this process.

Forces of Change	
Composition (layers)	
Internal forces	
External forces	

Critical Thinking

3. **Predicting Consequences** Based on your understanding of plate tectonics, what changes would you predict to the earth's appearance millions of years from now?
4. **Making Inferences** Think about water erosion associated with rivers and streams. Based on that information, what would you expect the areas where rivers empty into the oceans to be like? Explain.
5. **Drawing Conclusions** In what ways can erosion be both beneficial and harmful to agricultural communities?

Analyzing Maps

6. **Region** Study the map of plates and plate movement on page 39. Which plates are responsible for the earthquakes that have occurred in California in the United States?

Applying Geography

7. **Consequences of Earth's Forces** Review how internal forces shape the surface of the earth. Now imagine that the mantle ceased to circulate molten rock. Write a description of how land formation on the surface of the earth would be different.

Viewpoint

CASE STUDY on the Environment

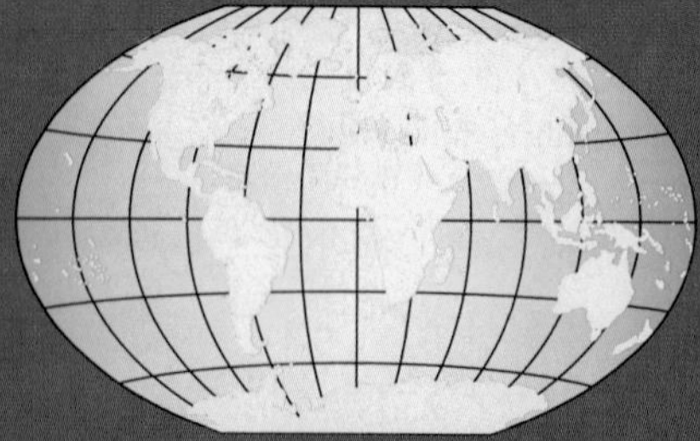

A Global Concern: Invasive Species

Until a few centuries ago, plants and animals living in one part of the world rarely mixed with those on other continents. This changed, however, as new ways of transport allowed more and more people to travel the planet. A turning point was the Columbian Exchange, the transfer of plants, animals, and diseases between Europe and the Americas that began with the voyages of Columbus. Since then, the biological exchange has become global. As a result, thousands of "alien" plants and animals have been introduced into areas where the species do not occur naturally. Some introduced species are beneficial, while others are invasive and pose environmental threats.

If you live near a wetland, you've probably seen the purple or pinkish flowers of the purple loosestrife (left). Native to Europe, purple loosestrife was brought to the United States as a garden plant. It soon spread to wetlands, where it now threatens native plants. Invasive species such as purple loosestrife are a global problem.

Sometimes invasive species arrive in new environments by accident—as seeds or as eggs or tiny insects hidden in packing materials, in food or soil, or in the ballast water of visiting ships. Without natural predators in their new habitats, alien species can multiply at alarming rates and crowd out native species. These invaders also can cause widespread environmental damage. The gypsy moth, for example, was accidentally introduced to the United States in the late 1800s. The insect, during its caterpillar stage, feeds on the leaves of several kinds of trees. Gypsy moth caterpillars have destroyed huge swaths of American forests.

Workers (below) cut invasive water hyacinths clogging waters in India. Yet some imported plants are beneficial. In Turkey (right), women harvest wheat to feed their families. ▼

Plants and animals are also introduced into new environments intentionally. For example, an insect might be imported to combat a crop pest. Some introduced species have turned out to be great successes. Wheat, for example, originated in the Middle East and is now grown on almost every continent. Spanish explorers brought horses to the Americas.

Yet for every helpful introduction, there have been many destructive ones. In 1859 British settlers released two dozen rabbits in Australia. Now millions of rabbits roam the continent. They devour crops and grasslands, creating deserts in the hot, dry climate. Water hyacinth, another invasive species, went from South America to Africa for use in ornamental ponds. The fast-growing plant soon spread to lakes and rivers, forming dense mats that interfere with boating and fishing.

◀ A gypsy moth caterpillar kills trees by eating their leaves.

Some scientists argue that alien species are serious threats to biodiversity. These people want laws to limit trade and to halt the spread of plants and animals from place to place. Many biologists think using foreign species to control pests is too risky. Scientists cite numerous examples of introduced species that have damaged their new environments.

Some farmers and economists point out that in a world where trade and transportation bring countries together, a global redistribution of species is inevitable. Such people, fearing financial losses, oppose laws that limit trade. Farmers generally support importing foreign species to control pests. These biological controls cause less environmental damage than the use of pesticides.

What's Your Point of View?

Should international trade be restricted to limit the risks of spreading invasive species around the world?

Earth's Water

Guide to Reading

Consider What You Know

Think of the ways you use water every day. Where does that water come from? Do you think it will always be available?

Read to Find Out

- How does the amount of water on Earth remain fairly constant?
- How is the water that makes up 70 percent of Earth's surface distributed?
- Why is freshwater important to humans?

Terms to Know

- water cycle
- evaporation
- condensation
- precipitation
- desalination
- groundwater
- aquifer

Places to Locate

- Pacific Ocean
- Atlantic Ocean
- Indian Ocean
- Arctic Ocean

NATIONAL GEOGRAPHIC

A Geographic View

Under the Arctic Ocean

In a world that's been almost completely mapped, it's easy to forget why cartographers used to put monsters in the blank spots. Today we got a reminder. The submarine captain had warned us that we were in uncharted waters. . . . Yet the first days of our cruise through this ice-covered ocean, Earth's least explored frontier, were . . . smooth. . . . Even when we passed over a mile-high mountain that no one on the planet knew existed, the reaction was one of quiet enthusiasm—"Neat."

—*Glenn Hodges, "The New Cold War: Stalking Arctic Climate Change by Submarine,"* National Geographic, *March 2000*

Navy submarine in the Arctic

A submarine crew investigating the Arctic Ocean can still experience the thrill of exploring uncharted territory—one of Earth's last frontiers. Although humans live mostly on land, water is important to our lives, and all living things need water to survive. Rivers, lakes, and oceans contain water in liquid form. The atmosphere holds water vapor, or water in the form of a gas. Glaciers and ice sheets are water in solid form. In this section you will learn about Earth's water and its importance to human life.

The Water Cycle

As you recall, oceans, lakes, rivers, and other bodies of water make up a part of the earth called the hydrosphere. Almost all of the hydrosphere is salt water found in the oceans, seas, and a few large saltwater lakes. The remainder is freshwater found in lakes, rivers, and springs.

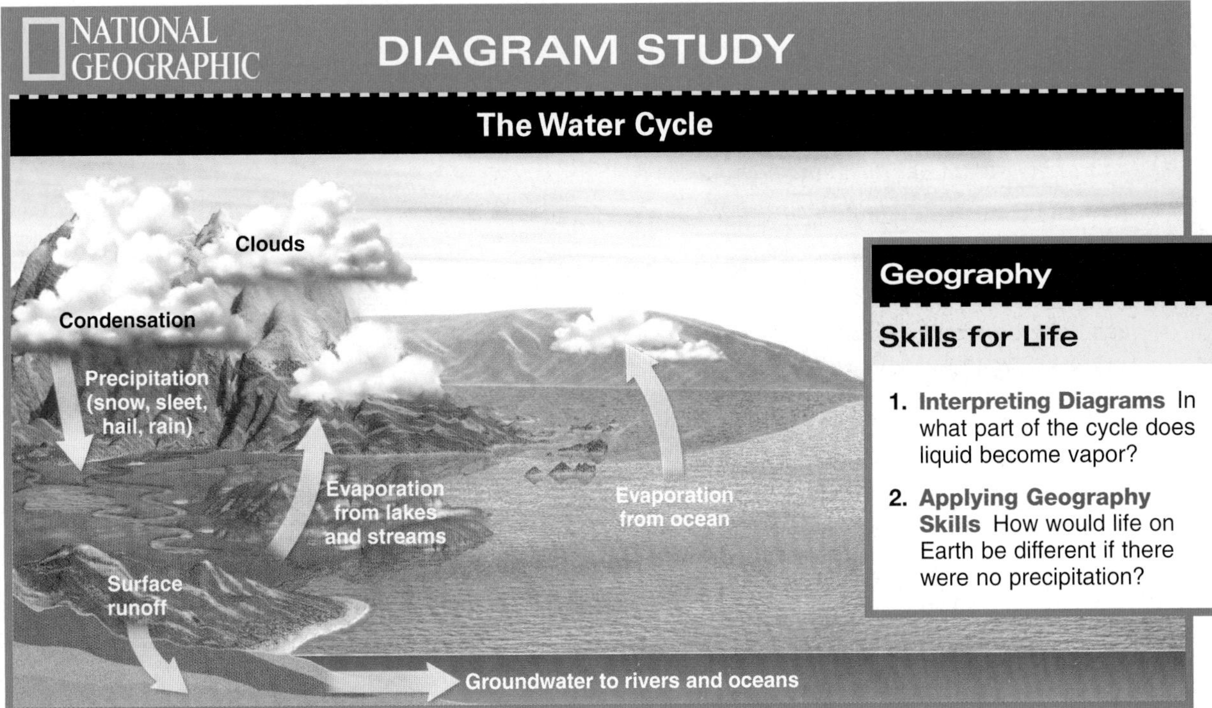

Geography Skills for Life

1. **Interpreting Diagrams** In what part of the cycle does liquid become vapor?
2. **Applying Geography Skills** How would life on Earth be different if there were no precipitation?

The total amount of water on the earth does not change, but the earth's water is constantly moving—from the oceans to the air to the ground and finally back to the oceans. The **water cycle** is the name given to this regular movement of water. The diagram above shows the major parts of the water cycle.

The sun drives the cycle by evaporating water from the surfaces of oceans, lakes, and streams. **Evaporation** is the changing of liquid water into vapor, or gas. The sun's heat causes evaporation. Water vapor rising from the oceans, other bodies of water, and plants is gathered in the air. The amount of water vapor the air holds depends on its temperature. Warm, less dense air holds more water vapor than does cool air.

When warm air cools, it cannot retain all of its water vapor, so the excess water vapor changes into liquid water—a process called **condensation**. Tiny droplets of water come together to form clouds. When clouds gather more water than they can hold, they release moisture, which falls to the earth as **precipitation**—rain, snow, or sleet, depending on the air temperature and wind conditions. This precipitation sinks into the ground and collects in streams and lakes to return to the oceans. Soon most of it evaporates, and the cycle begins again.

The amount of water that evaporates is approximately the same amount that falls back to the earth. This amount varies little from year to year. Thus, the total volume of water in the water cycle is more or less constant.

Bodies of Salt Water

Seen from space, the earth's oceans and seas are more prominent than the landmasses. As mentioned earlier, about 70 percent of the earth's surface is water, but almost all of it is salt water. Freshwater makes up only a small percentage of Earth's water.

Oceans

About 97 percent of the earth's water consists of a huge, continuous body of water that circles the planet. Geographers divide this enormous expanse into four oceans: the **Pacific**, the **Atlantic**, the **Indian**, and the **Arctic**. The Pacific, the largest of the oceans, covers more area than all the earth's land combined. The Pacific Ocean is also deep enough in some places to cover Mount Everest, the world's highest mountain, with more than 1 mile (1.6 km) to spare. The immense size of Earth's oceans continues to inspire awe and fascination in humans.

> *"It is useless to speculate at great length about why the sea has such a hold on us. Its mystery, its seeming infinity must be part of the reason. . . . And along with its vastness comes a visible, enduring wildness. We can raze the Amazonian rain forest if we like; we can settle the Alaskan tundra or the Arabian desert given the right economic incentives. We cannot, to the same extent, tame the ocean. . . ."*
>
> Robert Kunzig, *The Restless Sea: Exploring the World Beneath the Waves*, 1999

Seas, Gulfs, and Bays

Seas, gulfs, and bays are bodies of salt water smaller than oceans. These bodies of water are often partially enclosed by land. The Mediterranean Sea, one of the world's largest seas, is almost entirely encircled by southern Europe, northern Africa, and southwestern Asia. The Gulf of Mexico is nearly encircled by the coasts of the United States and Mexico. Scientists have identified 66 separate seas, gulfs, and bays, and many smaller divisions.

Economics

Ocean Water to Drinking Water

Although 97 percent of the world's water is found in oceans, the water is too salty for drinking, farming, or manufacturing. Today efforts focus on ways to meet the world's increasing need for freshwater, such as turning ocean water into freshwater by removing the salt. This process, known as **desalination**, is still in the early stages of development. Because desalination is expensive, only a small amount of freshwater is obtained this way. Even so, certain countries in Southwest Asia and North Africa use desalination because other freshwater sources are scarce.

Student Web Activity Visit the **Glencoe World Geography** Web site at geography.glencoe.com and click on Student Web Activities—Chapter 2 for an activity about the earth's oceans.

NATIONAL GEOGRAPHIC **World Explorer**

Geography Skills for Life

Kuwait Towers These towers—Kuwait's most famous landmark—have restaurants, an observation deck, and reservoirs storing 2 million gallons (7.6 million l) of desalinated water.

Human-Environment Interaction Why are scientists looking for ways to increase the amount of freshwater?

Bodies of Freshwater

Only about 3 percent of the earth's total water supply is freshwater, and most is not available for human consumption. More than 2 percent of Earth's total water supply is frozen in glaciers and ice caps. The Antarctic ice cap, for example, contains more freshwater than the rest of the world's regions combined. Another 0.5 percent is found beneath the earth's surface. Lakes, streams, and rivers contain far less than 1 percent of the earth's water.

Lakes, Streams, and Rivers

A lake is a body of water completely surrounded by land. Most lakes contain freshwater, although some, such as Southwest Asia's Dead Sea and Utah's Great Salt Lake, are saltwater remnants of ancient seas. Most lakes are found where glacial movement has cut deep valleys and built up dams of glacial soil and rock that held back melting ice-water. North America has thousands of glacial lakes.

Flowing water forms streams and rivers. Melt-water, an overflowing lake, or a spring may be the source, or beginning, of a stream. Streams may combine to form a river, a larger stream of higher volume that follows a channel along a particular course. When rivers join, the major river systems that result may flow for thousands of miles. Rain, runoff, and water from tributaries or branches swell rivers as they flow toward a lake, gulf, sea, or ocean. The place where the river empties into another body of water is its mouth.

Although lakes, streams, and rivers hold only a small part of the earth's water, they meet important needs. Most large urban areas began as settlements along the shores of lakes and rivers, where people would have a constant supply of water.

Geography Skills for Life

Freshwater The Snake River in southern Idaho provides vital freshwater for agriculture.

Human-Environment Interaction How have fresh-water sources affected the development of human settlements?

Groundwater

Groundwater, freshwater which lies beneath the earth's surface, comes from rain and melted snow that filter through the soil and from water that seeps into the ground from lakes and rivers. Wells and springs tap into groundwater and are important sources of freshwater for people in many rural areas and in some cities. An underground porous rock layer often saturated with water in the form of streams is called an **aquifer** (A•kwuh•fuhr). Aquifers and groundwater are important sources of freshwater.

SECTION 3 ASSESSMENT

Checking for Understanding

1. **Define** water cycle, evaporation, condensation, precipitation, desalination, groundwater, aquifer.
2. **Main Ideas** Use a web like the one below to organize information about Earth's water features.

Critical Thinking

3. **Drawing Conclusions** Use your knowledge of the water cycle to explain how droughts might occur.
4. **Making Inferences** Why might salt water someday provide water for drinking, farming, and manufacturing?
5. **Identifying Cause and Effect** How is drinking water contaminated by hazardous substances released on land or into rivers and lakes?

Analyzing Diagrams

6. **Physical Geography** Look at the diagram of the water cycle on page 47. What source of water supplies wells and springs?

Applying Geography

7. **Importance of Rivers** Many large urban areas developed in river basins. Write a description of how a river or rivers contributed to your community's development.

Comparing and Contrasting

Do you have a friend or relative who lives in a different state? You may have talked about the similarities and differences in the places where you each live. If so, you have compared and contrasted those states.

Learning the Skill

Comparing and contrasting help you identify similarities and differences between two or more things. Understanding similarities and differences can help you make better judgments about those things. When you compare, you look for similarities. Things that share at least one common quality can be compared. When you contrast, you look for differences. Things that differ from one another in at least one way can be contrasted.

To compare and contrast, apply the following steps:

- **Decide which items you will compare and contrast.** In the example above, you might compare and contrast the states' populations or their climates.
- **Determine which characteristics you will use to compare and contrast items.** In the example, you might decide to focus on the population sizes or the average temperatures of the two states.
- **Identify the similarities and differences in these characteristics.** In the example, you might compare populations and contrast climates.
- **If possible, identify causes for the similarities and differences.** For example, a state may have a larger population because it has a warmer climate.

Country	Languages	2001 Population (millions)	Government
Argentina	Spanish	37.5	Republic
Australia	English	19.4	Parliamentary democracy
Canada	English, French	31	Parliamentary democracy
France	French	59.2	Republic
Japan	Japanese	127.1	Constitutional monarchy
Kenya	English, Swahili	29.8	Republic

Source: Population Reference Bureau, 2001

Practicing the Skill

Study the chart above to answer the following questions.

1. What characteristics are used to compare and contrast the countries in the chart?
2. Which are the two smallest countries in population size? Which country is the largest?
3. How do Argentina and Japan differ in their population sizes? How do they differ in their governments?
4. How are Kenya, Canada, and Australia similar in the languages spoken?
5. Which two countries can you infer probably share similar cultural characteristics? Explain your answer.

Applying the Skill

With a partner, select four U.S. cities to research. Decide on at least three characteristics, such as population or land area, to compare and contrast. Collect the information for each of the three characteristics for each city. Design and draw a chart using your information. Develop three questions based on the chart. Exchange your work with another pair of students to answer the questions.

The **Glencoe Skillbuilder Interactive Workbook, Level 2** provides instruction and practice in key social studies skills.

CHAPTER 2

SUMMARY & STUDY GUIDE

SECTION 1 Planet Earth (pp. 33–36)

Terms to Know
- hydrosphere
- lithosphere
- atmosphere
- biosphere
- continental shelf

Key Points
- Planet Earth is located in our solar system.
- The hydrosphere (water), lithosphere (land), and atmosphere (air) make Earth's biosphere suitable for plant and animal life.
- Great contrasts exist in the heights and depths of the earth's surface.

Organizing Your Notes
Use a web like the one below to help you organize information about Planet Earth.

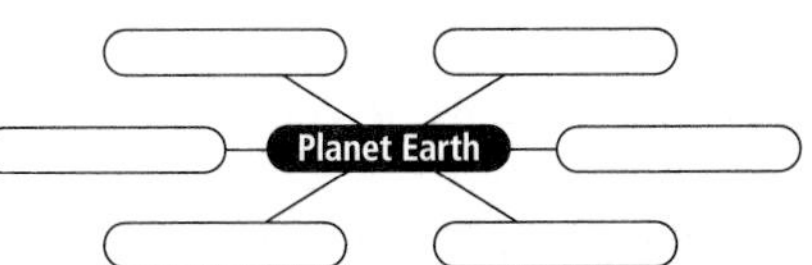

SECTION 2 Forces of Change (pp. 37–43)

Terms to Know
- mantle
- continental drift
- magma
- plate tectonics
- subduction
- accretion
- spreading
- fold
- fault
- weathering
- erosion
- loess
- glacier
- moraine

Key Points
- Planet Earth is composed of three layers—the core, the mantle, and the crust.
- Plates that move slowly around the globe produced Earth's largest features—continents, oceans, and mountain ranges.
- Mountains and islands are created by internal forces called subduction and accretion.
- Internal forces, such as earthquakes and volcanoes, also shape the surface of the earth.
- External forces, such as weathering and wind, glacial, and water erosion, also shape the surface of the earth.

Organizing Your Notes
Create a series of flowcharts like the ones below to show which forces or physical processes shape Earth's various and distinctive landforms.

Effects of Earth's Forces

Landform		Shaped by Force(s)
	→	
	→	
	→	
	→	

SECTION 3 Earth's Water (pp. 46–49)

Terms to Know
- water cycle
- evaporation
- condensation
- precipitation
- desalination
- groundwater
- aquifer

Key Points
- The amount of water on Earth remains fairly constant and moves in the water cycle.
- Water makes up 70 percent of the earth's surface.
- Earth's water features are classified as salt water or freshwater.
- Freshwater is necessary to sustain human life.

Organizing Your Notes
Create an outline using the format below to organize this section's notes.

Earth's Water

I. The Water Cycle
  A. Evaporation
    1.
    2.

CHAPTER 2 ASSESSMENT & ACTIVITIES

Reviewing Key Terms

Write the key term that best completes each of the following sentences. Refer to the Terms to Know in the Summary & Study Guide on page 51.

1. Four parts of the earth's surface are ________, ________, ________, and ________.
2. Underwater trenches are created through the process of ________.
3. The theory that continents are slowly moving is called ________.
4. An ________ is an underground porous rock layer often saturated with water.
5. A ________ is a bend in layers of rock.
6. Water vapor changes into liquid through ________.
7. ________ is molten rock within the earth.
8. Moisture that falls from the clouds is ________.
9. A ________ is a break in the earth's crust.
10. The activity of the earth's moving plates is called ________.
11. Liquid water changes into water vapor through the process of ________.
12. ________ wears away the earth's surface.

Reviewing Facts

SECTION 1

1. What are two types of planets?
2. What are the four major types of landforms found on Earth?

SECTION 2

3. Describe the earth's layers.
4. What produced some of Earth's largest landforms?

SECTION 3

5. What process keeps the amount of Earth's water constant?
6. Describe the forms that water takes throughout the water cycle.

Critical Thinking

1. **Comparing and Contrasting** How do internal and external forces of change affect Earth's surface differently?
2. **Making Inferences** How might the relationship among climate, vegetation, soil, and geology affect distribution of plants and animals in different regions?
3. **Finding and Summarizing the Main Idea** Copy the graphic organizer below, and write the main idea of each section in the outer ovals and the main idea of the chapter in the center oval. Write a summary using appropriate vocabulary.

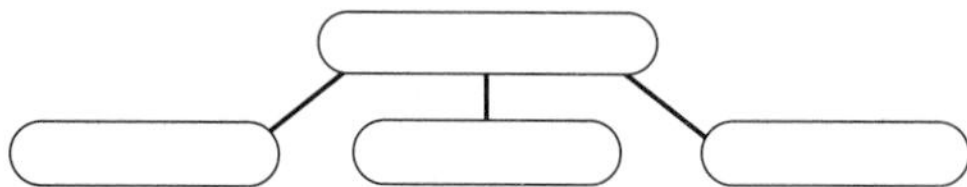

NATIONAL GEOGRAPHIC

Locating Places

The Earth: Physical Geography

Match the letters on the map with the physical features of Earth. Write your answers on a sheet of paper.

1. **Rocky Mountains**
2. **Isthmus of Panama**
3. **Gulf of Mexico**
4. **Andes**
5. **Himalaya**
6. **Ural Mountains**
7. **Arctic Ocean**
8. **Mediterranean Sea**
9. **Bay of Bengal**
10. **Europe**

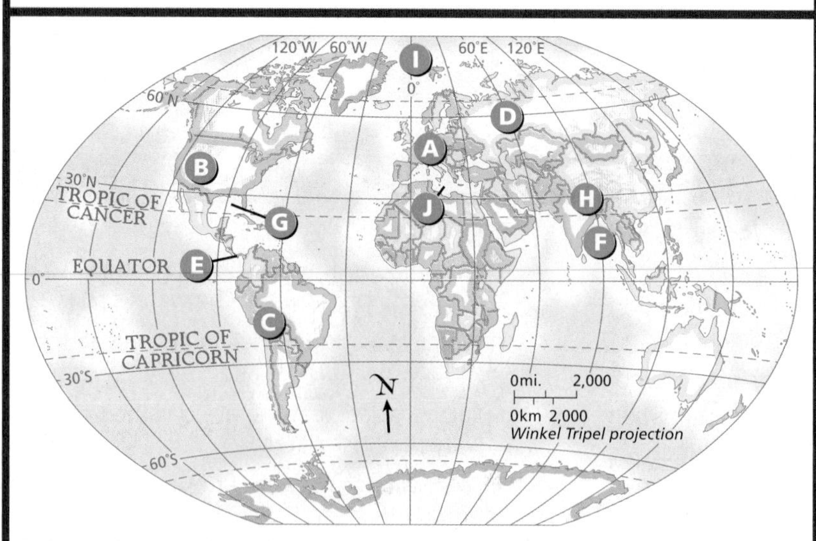

Self-Check Quiz Visit the **Glencoe World Geography** Web site at geography.glencoe.com and click on Self-Check Quizzes—Chapter 2 to prepare for the Chapter Test.

Thinking Like a Geographer

Think about the ways in which geographers classify the earth's physical features. As a geographer, in what other ways might you classify those features? What would your reasons be for classifying features differently? Identify different classification systems that may be helpful to geographers.

Problem-Solving Activity

Problem-Solution Proposal Using the Internet and library resources, learn more about a specific location where freshwater is scarce and the reasons for the scarcity. Then write a proposal identifying the problem and explaining possible solutions for the scarcity. Choose one solution, and give reasons why you think it is the best alternative. Include charts, graphs, or other data that will help readers understand the basis for your conclusions.

GeoJournal

Descriptive Writing Use the journal entries you wrote in your GeoJournal as you read this chapter. Pick one feature, structure, or force of change, and write an expanded description of it. Paint a word picture using concrete, specific details. Consider sights, sounds, smells, and textures associated with the feature. Be sure to organize your description in an order that guides your reader through your composition.

Technology Activity

Using the Internet for Research Choose one of the earth's features, such as mountains, lakes, or rivers. Use the Internet to find specific examples of that feature in different parts of the world. Note the measurement for each and the continent where it is located. Organize your findings in a chart. If, for example, you choose rivers as a feature to research, you might include headings such as longest river, widest river, swiftest river, highest volume river, and so on. Then write a summary, ranking the continents by the sizes of their features.

Standardized Test Practice

Use the information in the chart below to answer the questions. If you have trouble answering the questions, use the process of elimination to narrow your choices.

Notable Volcanic Eruptions

Year	Volcano	Location	Deaths (est.)
1631	Mt. Vesuvius	Italy	4,000
1783	Laki	Iceland	9,350
1883	Krakatau	Indonesia	36,000
1902	Mt. Pelée	Martinique	28,000
1980	Mt. St. Helens	United States	57
1991	Mt. Pinatubo	Philippines	800

1. **Based on the information shown in the chart, in which century did the deadliest eruption occur?**

   A seventeenth
   B eighteenth
   C nineteenth
   D twentieth

2. **The chart probably contains data from the past 300+ years because**

   F more volcanoes erupted then.
   G more information is available.
   H no volcanoes erupted before 1631.
   J eruptions are getting closer together.

Look at the chart title to see what kind of information is presented. For question 1, find out what categories are being compared. Identify which eruption caused the greatest number of deaths. Then, identify in which century it occurred. Question 2 asks you to infer causes. Note the chart's title, which gives a clue to the answer.

CHAPTER 3

Climates of the Earth

GeoJournal

As you read this chapter, note factors that affect climate. Then write a description of the climate in your community. List three factors that contribute to climate, and sketch a landscape showing one kind of common weather in this climate.

GEOGRAPHY Online

Chapter Overview Visit the **Glencoe World Geography** Web site at geography.glencoe.com and click on Chapter Overviews—Chapter 3 to preview information about Earth's climates.

SECTION 1

Earth-Sun Relationships

Guide to Reading

Consider What You Know

News reports sometimes feature unusual events such as solar eclipses or solar flares, intense bursts of energy from the sun's surface. In what ways does the sun affect human activities?

Read to Find Out

- How does Earth's position in relation to the sun affect temperatures on Earth?
- How does Earth's rotation cause day and night?
- What is Earth's position in relation to the sun during each season?
- How might global warming affect Earth's air, land, and water?

Terms to Know

- weather
- climate
- axis
- temperature
- revolution
- equinox
- solstice
- greenhouse effect
- global warming

Places to Locate

- Tropic of Cancer
- Tropic of Capricorn

◀ *Flamingoes enjoy an oasis at the foot of sand dunes in Namibia.*

NATIONAL GEOGRAPHIC

A Geographic View

Our Home Star

Through a small occulting telescope [equipped to block the sun's surface from view] . . . I stared at the darkened face of the sun, ringed by its glowing, gauzy corona. From that blazing disk high in the Hawaiian sky comes the endless power that drives and rules all life on earth: its plant growth and the food chains of all its creatures; the winds, rains, and churning weather of the planet; the ocean currents, forests, prairies, and deserts.

—*Samuel W. Matthews, "Under the Sun: Is Our World Warming?"* National Geographic, *October 1990*

Sunset at the beach

From atop the Mauna Loa Observatory on the island of Hawaii, scientists gather data about changes in the earth's atmosphere. Their research reveals information about the dynamic relationships between the earth and the sun, which influence all life on Earth. In this section you will explore how earth-sun relationships affect climate.

Climate and Weather

Climate is often confused with weather, which is a short-term aspect of climate. **Weather** is the condition of the atmosphere in one place during a limited period of time. When people look out the window or watch the news to see whether they need umbrellas or sunscreen, they are checking the weather. **Climate** is the term for the weather patterns that an area typically experiences over a long period of time. People who live in Seattle, Washington, for example, frequently use umbrellas because of the rainy, wet climate. In contrast, people who live in the dry, desert climate of Arizona must protect themselves from the sun.

Whether the climate in a particular region is cool and wet or hot and dry is determined by many factors, the most important of which is the earth's position in relation to the sun.

The sun's heat and light reach the earth as warmth and sunlight, but they do not reach all parts of the earth at the same time or with the same intensity.

Geography Skills for Life

Arizona Lightning

Lightning and thunder are related weather events that result from the powerful air currents of thunderstorms during the warm-weather months.

Place What factor distinguishes weather from climate?

Earth's Tilt and Rotation

Earth's tilt is one reason for variations in sunlight. The earth's axis—an imaginary line running from the North Pole to the South Pole through the planet's center—is currently tilted at an angle of about $23^1/2^{\circ}$. Because of the tilt of this axis, not all places on the planet receive the same amount of direct sunlight at the same time.

For this reason the angle of tilt affects the temperature—the measure of how hot or cold a place is. Areas that receive a large amount of direct sunlight have warmer temperatures than places that receive little direct sunlight. Temperature is usually measured in degrees on a set scale. The most common scales for measuring air temperature are Fahrenheit (°F) and Celsius (°C).

Whether or not a particular place on Earth receives light also depends on the side of the planet that is facing the sun. Earth rotates on its axis, making one complete rotation every 24 hours. Rotating from west to east, the earth turns first one hemisphere and then the other toward the sun, alternating between the light of day and the darkness of night.

Earth's Revolution

While planet Earth is rotating on its axis, it also is traveling in an orbit around the sun, our nearest star. It takes the earth a few hours more than 365 days—1 year—to complete one revolution, or trip around the sun.

The earth's revolution and its tilt cause changes in the angle and amount of sunlight that reach different locations on the planet. These changes follow a regular progression known as the seasons. During the course of a year, people on most parts of the earth experience distinct differences in the length of days and the daily temperature as the seasons change.

The seasons are reversed north and south of the Equator. When it is spring in the Northern Hemisphere, it is fall in the Southern Hemisphere. When it is winter in the Southern Hemisphere, it is summer in the Northern Hemisphere. Around March 21, the sun's rays fall directly on the Equator. This day is called an equinox (meaning "equal night") because daylight and nighttime hours are equal. In the Northern Hemisphere, the day on which this equinox falls marks the beginning of spring.

The Tropics of Cancer and Capricorn

As the earth continues its revolution around the sun, it moves so that eventually the sun's rays directly strike the latitude $23^1/2^{\circ}$N. This latitude is known as the **Tropic of Cancer**—the northernmost point on the earth to receive the direct rays of the sun. These direct rays reach the Tropic of Cancer about June 21, bringing the Northern Hemisphere its longest day of sunlight. This date, known as the summer solstice, marks the beginning of summer in the Northern Hemisphere.

By about September 23, the earth has moved so that the sun's rays directly strike the Equator again. This equinox marks the beginning of fall in the Northern Hemisphere. Gradually the sun's direct rays strike farther south, reaching their southernmost latitude—23½°S, or the **Tropic of Capricorn**—about December 22. The winter solstice is the day of shortest daylight in the Northern Hemisphere, beginning the season of winter. This cycle repeats itself each year as the earth revolves around the sun.

The Poles

The amount of sunlight at the Poles varies most dramatically as the earth's revolution and tilt cause the changing seasons. For six months of the year, one Pole is tilted toward the sun and receives continuous sunlight, while the other Pole is tilted away from the sun and receives little to no sunlight. An Arctic explorer describes the surreal quality of the winter day:

> *"The winter sun had completely disappeared below the horizon, and the 'days' were now only a few hours long. . . . I saw the Arctic in a whole new way, with its twilight days and auroral nights."*
>
> Keith Nyitray, "Alone Across the Arctic Crown," *National Geographic*, April 1993

At the North Pole, the sun never sets from about March 20 to September 23. At the South Pole, continuous daylight lasts from about September 23 to March 20. The tilt of the earth's axis as it revolves around the sun causes this natural phenomenon, known as the midnight sun. The occurrence of the

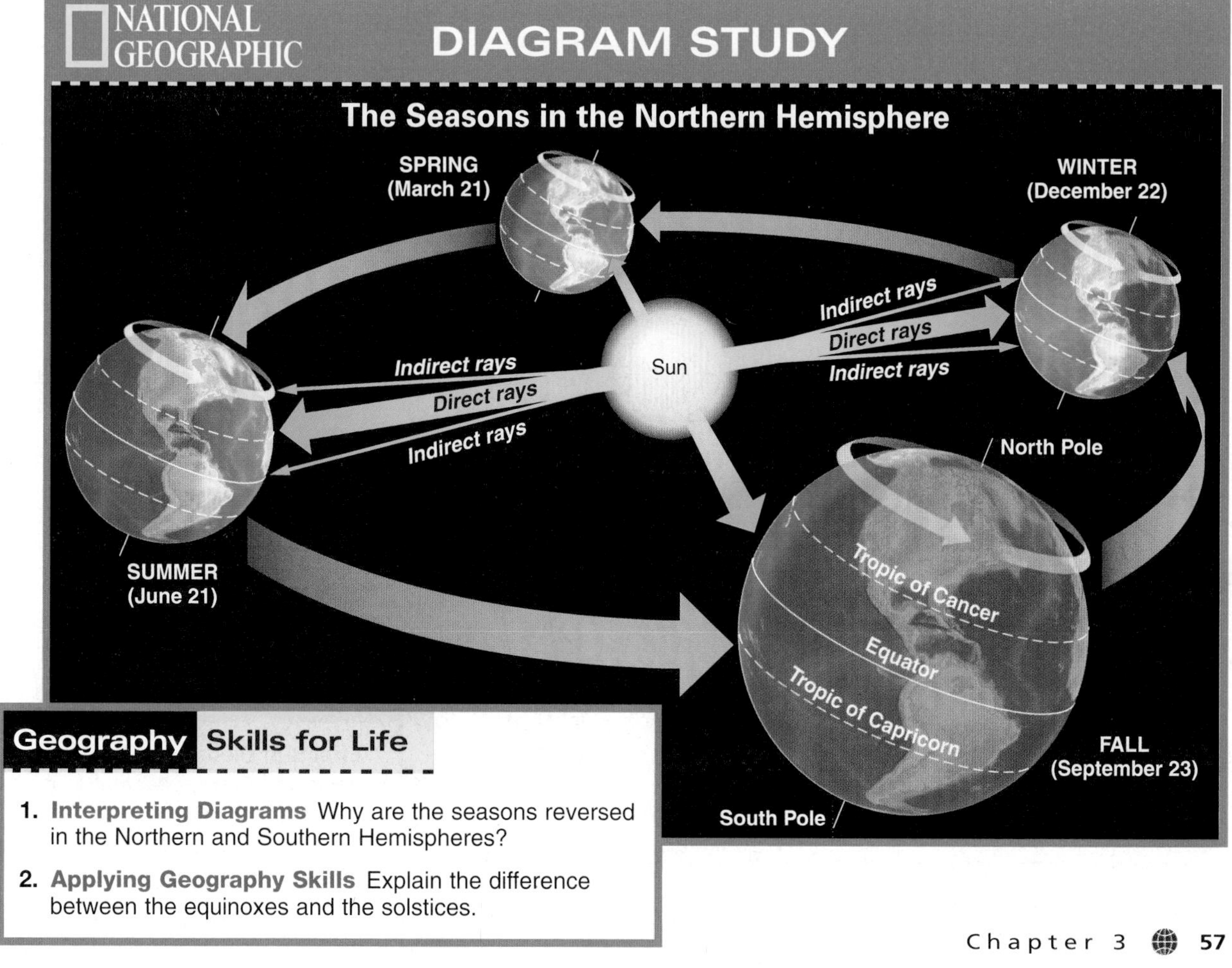

Geography Skills for Life

1. **Interpreting Diagrams** Why are the seasons reversed in the Northern and Southern Hemispheres?
2. **Applying Geography Skills** Explain the difference between the equinoxes and the solstices.

midnight sun goes almost unnoticed in sparsely populated Antarctica. Parts of northern North America (including Alaska) and northern Europe in the Arctic, however, have become popular tourist destinations as "lands of the midnight sun."

The Greenhouse Effect

Even on the sunniest days in the warmest climates, only part of the sun's radiation passes through the earth's atmosphere. The atmosphere reflects some radiation back into space. Enough radiation, however, reaches the earth to warm the air, land, and water.

Because the atmosphere traps some heat and keeps it from escaping back into space too quickly, Earth's atmosphere is like the glass in a greenhouse—it traps the sun's warmth for growing plants even in cold weather. Without this **greenhouse effect**, the earth would be too cold for most living things. The diagram on page 70 shows the greenhouse effect.

In order to support plant growth, conditions in a greenhouse must be regulated. If too much heat escapes, the plants will freeze. If too much heat is trapped, the plants will wilt or dry out.

The greenhouse effect of Earth's atmosphere follows some of the same general rules. Normally, the atmosphere provides just the right amount of insulation to promote life on the planet. The 50 percent of the sun's radiation that reaches the earth is converted into infrared radiation, or heat. Clouds and greenhouse gases—atmospheric components such as water vapor and carbon dioxide (CO_2)—absorb the heat reflected by the earth and radiate it back again so that a balance is created.

Many scientists, however, claim that in recent decades a rise in atmospheric CO_2 levels has coincided with a general rise in global temperatures. This trend—known as **global warming**—is believed to be caused in part by human activities, such as the burning of coal, oil, and natural gas. As more fossil fuels are burned, greenhouse gases enter the atmosphere and trap more heat.

Using computer models, some scientists predict that global warming will make weather patterns more extreme. Water, for example, will evaporate more rapidly from oceans, increasing humidity and rainfall generally. Rapid water evaporation from soil, however, will cause land to dry out more quickly between rains. Some areas may even become drier than before.

Scientists do not all agree on global warming and its effects. Some claim that a natural cycle, not human activity, is causing rising temperatures. Others claim that evidence for global warming is inconclusive and that it is too early to forecast future effects.

Student Web Activity Visit the **Glencoe World Geography** Web site at geography.glencoe.com and click on **Student Web Activities—Chapter 3** for an activity about global warming.

SECTION 1 ASSESSMENT

Checking for Understanding

1. **Define** weather, climate, axis, temperature, revolution, equinox, solstice, greenhouse effect, global warming.
2. **Main Ideas** On a chart, list characteristics of the earth-sun relationship and describe their effects on climate.

Earth-Sun Relationships	Effects on Climate

Critical Thinking

3. **Comparing and Contrasting** What differences in the weather would you expect in Alaska and in Florida? Explain.
4. **Drawing Conclusions** What effects does the earth's tilt on its axis have on your daily life?
5. **Analyzing Information** What would you pack if you were visiting Argentina in December?

Analyzing Diagrams

6. **Location** Study the diagram of the seasons on page 57. In what months do the sun's rays strike the Equator directly? The Tropics of Cancer and Capricorn?

Applying Geography

7. **Effects of Global Warming** Review the text in Section 1 about global warming. In what ways might agriculture be affected? Explain.

Factors Affecting Climate

Guide to Reading

Consider What You Know

You are familiar with the climate of the area where you live, such as the usual weather and temperature range for each season. What geographic factors do you think affect your climate?

Read to Find Out

- How do latitude and elevation affect climate?
- What role do wind patterns and ocean currents play in Earth's climates?
- How do landforms and climate patterns influence each other?

Terms to Know

- prevailing wind
- Coriolis effect
- doldrums
- current
- El Niño
- windward
- leeward
- rain shadow

Places to Locate

- low latitudes
- high latitudes
- Arctic Circle
- Antarctic Circle
- mid-latitudes

NATIONAL GEOGRAPHIC

A Geographic View

The Stormy Sea

As I survey the wreckage [from my yacht]—broken steering wheel, patched sails, ruined winches, life rails ripped away by bounding seas . . . I can see it's been a harrowing 4,600 miles since the start [of this leg of the Whitbread race] in Cape Town, South Africa. For eight of the twelve men aboard, including Paul Cayard, the skipper, the run east has been their first encounter with the ocean latitudes known as the roaring forties and furious fifties, where gales blow year-round and seas build to towering peaks, "the liquid Himalayas," as one Kiwi [New Zealand] broadcaster calls them.

Sailing in the southern Indian Ocean

—*Angus Phillips, "The Whitbread—Race Into Danger,"* National Geographic, *May 1998*

Sailing in the Whitbread race can be frightening. The part of this journey from South Africa to Australia is dangerous because of high winds and strong ocean currents at these latitudes. In this section you will learn how latitude, wind and water patterns, and landforms combine with the earth-sun relationship to influence Earth's climates.

Latitude and Climate

The influence of latitude on climate is part of the earth-sun relationship. During the earth's annual revolution around the sun, the sun's direct rays fall upon the planet in a regular pattern. This pattern can be

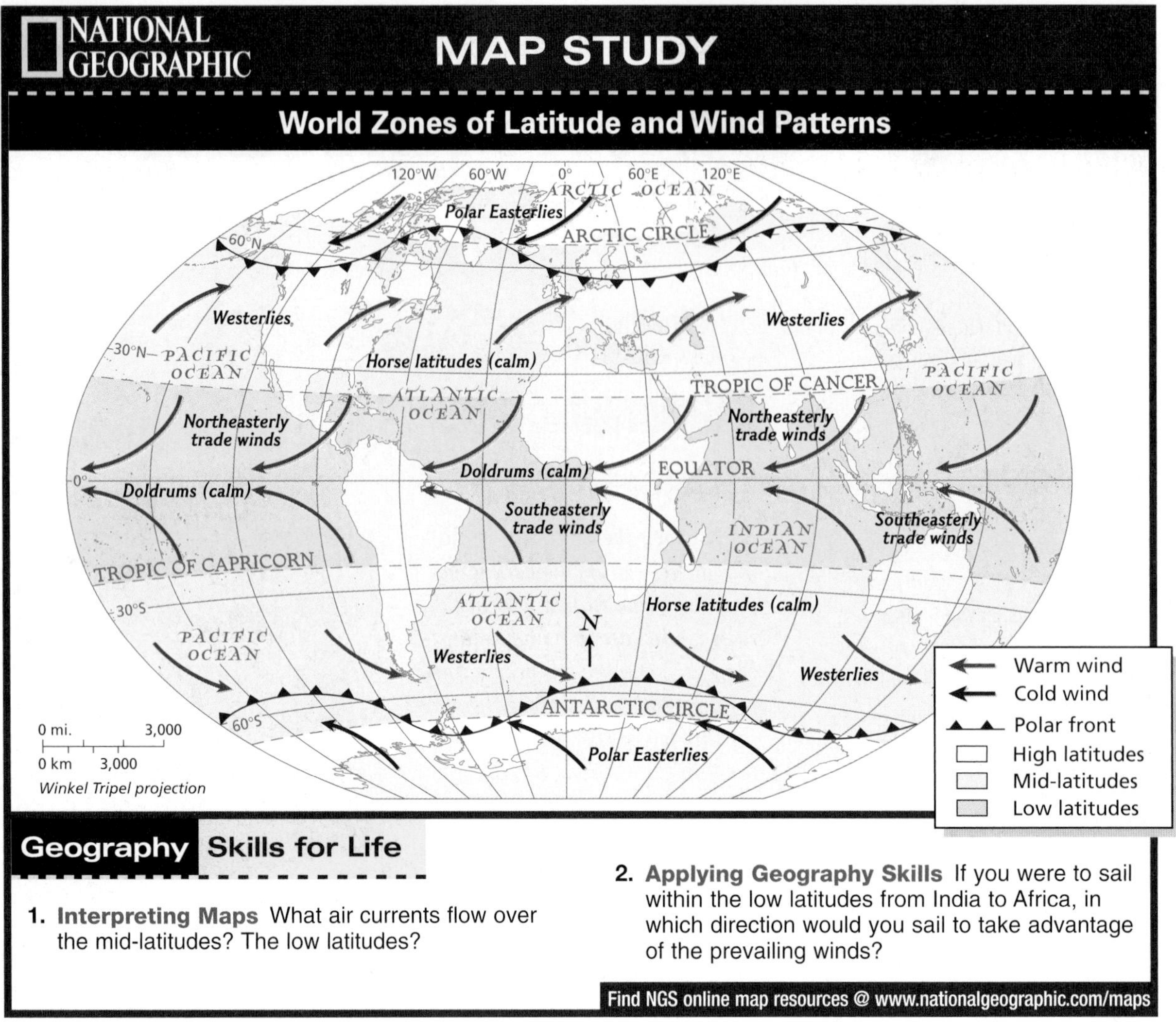

correlated with bands, or zones, of latitude to describe climate regions. Within each latitude zone, the climate follows general patterns.

Low Latitudes

Between the Tropic of Cancer and the Tropic of Capricorn is a zone known as the **low latitudes**. This zone includes the Equator. Portions of the low latitudes receive the direct rays of the sun year-round. Places located in the low latitudes have warm to hot climates. Because of the latitudes that form its boundaries, this zone is called the Tropics.

High Latitudes

The earth's polar areas are called the **high latitudes**. When either the Northern or the Southern Hemisphere is tilted toward the sun, its polar area receives continuous, but indirect, sunlight. From about March 20 to about September 23, the polar area north of the **Arctic Circle** (latitude 66°N) experiences continuous daylight or twilight. The polar area south of the **Antarctic Circle** (latitude 66°S) experiences continuous daylight or twilight for the other six months of the year.

Mid-Latitudes

The most variable weather on Earth is found in the **mid-latitudes** between the Tropic of Cancer and the Arctic Circle in the Northern Hemisphere and between the Tropic of Capricorn and the Antarctic Circle in the Southern Hemisphere. In summer the mid-latitudes receive warm masses of air from the Tropics. In winter, cold masses of air move into the mid-latitudes from the high latitudes. The

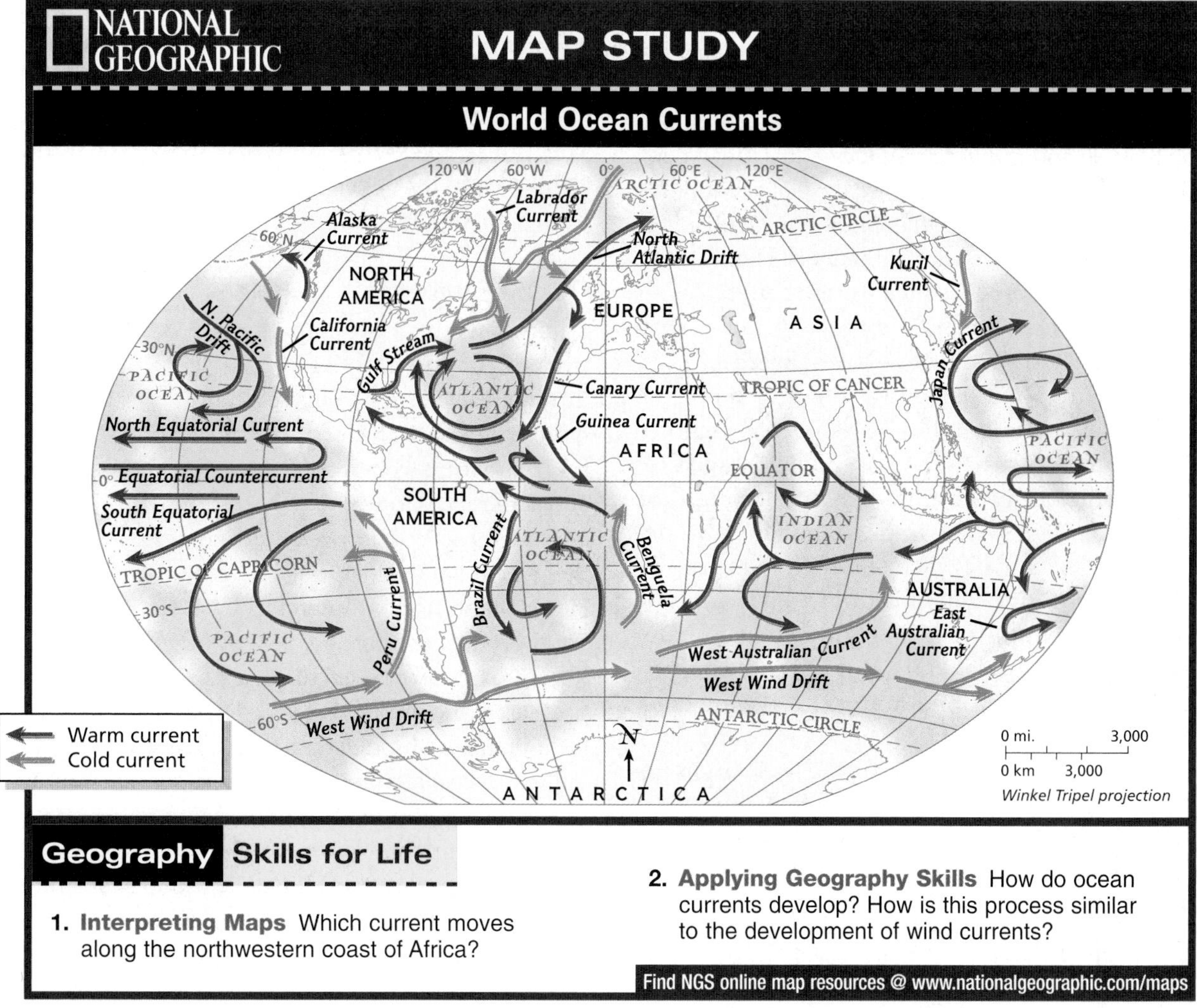

Geography Skills for Life

1. **Interpreting Maps** Which current moves along the northwestern coast of Africa?

2. **Applying Geography Skills** How do ocean currents develop? How is this process similar to the development of wind currents?

Find NGS online map resources @ www.nationalgeographic.com/maps

mid-latitudes generally have a temperate climate—one that ranges from fairly hot to fairly cold—with dramatic seasonal weather changes.

Elevation and Climate

At all latitudes elevation influences climate because of the relationship between the elevation of a place and its temperature. The earth's atmosphere thins as altitude increases. Thinner air retains less heat. As elevation increases, temperatures decrease by about 3.5°F (1.9°C) for each 1,000 feet (305 m). This effect occurs at all latitudes. For example, in Ecuador, the city of Quito (KEE•toh) is nearly on the Equator. However, Quito lies in the Andes at an elevation of more than 9,000 feet (2,743 m), so average temperatures are about 32°F (17°C) cooler than in the nearby lowlands.

Sunlight is bright in Quito and other places with high elevation because the thinner atmosphere filters fewer rays of the sun. Even in bright sunlight, the world's highest mountains are cold, snowy places year-round.

Wind and Ocean Currents

Wind and water combine with the effects of the sun to influence Earth's weather and climate. Air moving across the face of the earth is called wind. Winds occur because the sun heats up the earth's atmosphere and surface unevenly. Rising warm air creates areas of low pressure, and falling cool air causes areas of high pressure. The cool air then flows in to replace the warm rising air. These movements over the earth's surface cause winds, which distribute the sun's heat around the planet.

Wind Patterns

Global winds blow in fairly constant patterns called prevailing winds, shown on the map on page 60. The direction of prevailing winds is determined by latitude and is affected by the earth's movement. Because Earth rotates to the east, the global winds are displaced clockwise in the Northern Hemisphere and counterclockwise in the Southern Hemisphere. This phenomenon, called the Coriolis effect, causes prevailing winds to blow diagonally rather than along strict north-south or east-west lines.

Winds are often named for the direction from which they blow, but they sometimes were given names from the early days of sailing. Named for their ability to move trading ships through the region, the prevailing winds of the low latitudes are called *trade winds*. They blow from the northeast toward the Equator from about latitude 30°N and from the southeast toward the Equator from about latitude 30°S. *Westerlies* are the prevailing winds in the mid-latitudes, blowing diagonally west to east between about 30°N and 60°N and between about 30°S and 60°S. In the high latitudes, the *polar easterlies* blow diagonally east to west, pushing cold air toward the mid-latitudes.

History

The Horse Latitudes

At the Equator, global winds are diverted north and south, leaving a narrow, generally windless band called the doldrums. Two other narrow bands of calm air encircle the globe just north of the Tropic of Cancer and just south of the Tropic of Capricorn. In the days of wind-powered sailing ships, crews feared being stranded in these windless areas. With no moving air to lift the sails, ships were stranded for weeks in the hot, still weather. Meanwhile, food supplies dwindled, and perishable cargoes spoiled as the ships sat. The English poet Samuel Taylor Coleridge described this frightening experience:

> "*Day after day, day after day*
> *We struck, nor breath nor motion;*
> *As idle as a painted ship*
> *Upon a painted ocean.*"
>
> Samuel Taylor Coleridge, *The Rime of the Ancient Mariner*, 1798

To lighten the load so the ships could take advantage of the slightest breeze, sailors would toss excess cargo and supplies overboard, including livestock being carried to colonial settlements. This practice gave rise to the name by which the calm areas at the edges of the Tropics are known—the *horse latitudes*.

Ocean Currents

Just as winds move in patterns, cold and warm streams of water, known as currents, move through the oceans. Ocean currents are caused by many of the same factors that cause winds, including the earth's rotation, changes in air pressure, and differences in water temperature. The Coriolis effect is observed in ocean currents, too, causing them to move in clockwise circles in the Northern Hemisphere and counterclockwise circles in the Southern Hemisphere.

As ocean currents circulate, cold water from the polar areas moves slowly toward the Equator, warming as it moves through the Tropics. This water forms the warm ocean currents. The warm water, in turn, moves away from the Equator, cooling to become a cold ocean current.

Ocean currents affect climate in the coastal lands along which they flow. Cold ocean currents cool the lands they pass. Warm ocean currents bring warmer temperatures. For example, the North Atlantic Drift, a warm-water extension of the Gulf Stream current, flows near western Europe. This current gives western Europe a relatively mild climate in spite of its northern latitude.

Weather and the Water Cycle

Wind and water work together to affect weather in another important way. Driven by temperature, condensation creates precipitation, moisture falling to the earth in the form of rain, sleet, hail, or snow. The sudden cloudburst that cools a steamy summer day is an example of how precipitation both affects and is affected by temperature. Water vapor forms in the atmosphere from evaporated surface water. As colder temperatures cool the rising moist air, the vapor condenses into liquid droplets, forming clouds. Further cooling causes rain to fall, which can help lower the temperature on warm days.

El Niño

Climate is also affected by recurring phenomena, or events, that alter weather patterns. The most famous of these recurring climatic events is the **El Niño** (ehl NEE•nyoh) phenomenon. El Niño is a periodic change in the pattern of ocean currents and water temperatures in the mid-Pacific region.

El Niño does not occur every year, but its frequency appears to have increased in the latter half of the 1900s. In an El Niño year, the normally low atmospheric pressure over the western Pacific rises, and the normally high pressure over the eastern Pacific drops. This reversal causes the trade winds to diminish or even to reverse direction. The change in wind pattern reverses the equatorial ocean currents, drawing warm water from near Indonesia east to Ecuador, where it spreads along the South American coast.

The changed air pressures resulting from El Niño influence climates around the world in a kind of domino effect. Precipitation increases along the coasts of North and South America, warming winters and increasing the risk of floods. Hawaii experiences reduced winds like those in the doldrums, and also drier weather. In Southeast Asia and Australia, drought and occasional massive forest fires occur.

Scientists are not sure what causes El Niño or why it appears to be occurring more frequently. Preliminary studies have linked this climatic event to global warming. The costs in human and economic terms of the weather catastrophes associated with El Niño make learning more about this climatic event vitally important.

Landforms and Climate

The surface features of the earth, such as bodies of water and mountains, also can affect and be affected by climate. The climates of places located at the same latitude can be very different, depending on the presence or absence of certain landforms.

Large bodies of water, for example, are slower to heat and to cool than land. As a result, water temperatures are more uniform and constant than land temperatures. Coastal lands receive the benefit of this moderating influence and experience less changeable weather than do inland areas. Large lakes, such as the Great Lakes in the United States and Canada, also create this effect.

NATIONAL GEOGRAPHIC **DIAGRAM STUDY**

El Niño

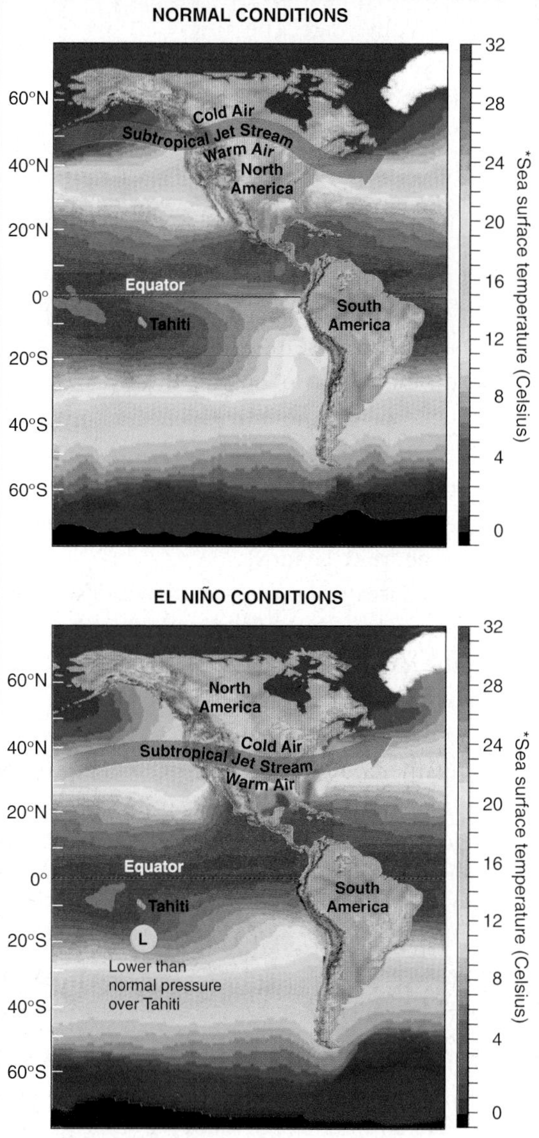

*Sea surface temperature data taken at one-degree intervals.

Geography Skills for Life

1. **Interpreting Diagrams** At what latitude does the subtropical jet stream flow under normal conditions?
2. **Applying Geography Skills** Describe the impact of El Niño on the lives of people in different parts of the world.

Mountain ranges also influence precipitation and affect climate. Winds that blow over an ocean are pushed upward when they meet a mountain range. The rising air cools and releases most of its moisture in the form of precipitation on the **windward** side—the side of the mountain range facing the wind. After the precipitation is released, winds become warmer and drier as they descend on the opposite, or **leeward**, side of the mountains. The hot, dry air produces little precipitation in an effect known as a **rain shadow**. The rain shadow effect often causes dry areas—and even deserts—to develop on the leeward sides of mountain ranges.

NATIONAL GEOGRAPHIC **DIAGRAM STUDY**

The Rain Shadow Effect

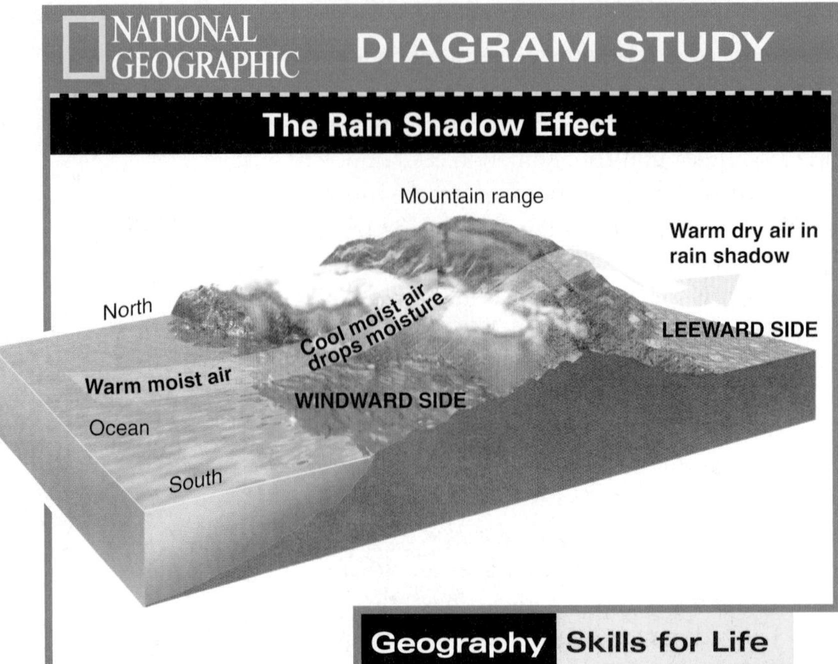

Geography Skills for Life

1. **Interpreting Diagrams** On which side of a mountain does precipitation fall?
2. **Applying Geography Skills** How do landforms cause the formation of a rain shadow?

From the interaction of landforms, wind and water currents, latitude, and elevation arise a remarkable variety of climates. In the next section, you will learn about the earth's climate regions and the kinds of natural vegetation that grows in each climate region. You will also learn about the ways that climates have changed over millions of years and explore the many natural causes of climate change. As you explore the ways that human activities may cause climate change, you will also learn about ways that humans can protect their environment.

SECTION 2 ASSESSMENT

Checking for Understanding

1. **Define** prevailing wind, Coriolis effect, doldrums, current, El Niño, windward, leeward, rain shadow.
2. **Main Ideas** On a web diagram, fill in information about each of the factors affecting climate. Indicate whether these factors affect the climate where you live.

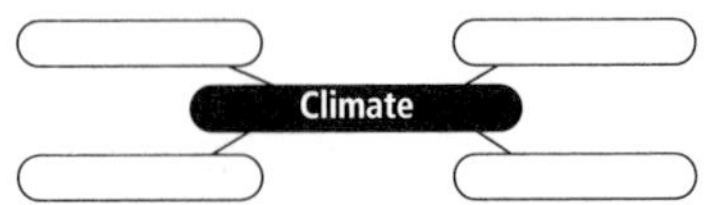

Critical Thinking

3. **Comparing and Contrasting** Describe the general differences in climate between the low latitudes and the mid-latitudes.
4. **Predicting Consequences** Without the Coriolis effect, how might the earth's climates be different?
5. **Identifying Cause-and-Effect Relationships** How does the presence of mountain ranges influence climate?

Analyzing Maps

6. **Location** Study the map of wind patterns on page 60. Where are the high latitudes located?

Applying Geography

7. **Movement of Ocean Currents** Study the map on page 61. If your ship were drifting from west to east in the Equatorial Countercurrent, what might happen as you drifted past longitude 120°W?

World Climate Patterns

Guide to Reading

Consider What You Know

Think about the types of vegetation that grow in the region where you live. What does the vegetation reveal about the regional climate there?

Read to Find Out

- How do geographers classify the climate regions of the world?
- Which kinds of vegetation are characteristic of each climate region?
- How do recurring phenomena influence climate patterns?

Terms to Know

- natural vegetation
- oasis
- coniferous
- deciduous
- mixed forest
- chaparral
- prairie
- permafrost
- hypothesis
- smog

Places to Locate

- Tropics
- Sahara
- Mediterranean Sea

NATIONAL GEOGRAPHIC

A Geographic View

Damage from El Niño

From South America, cradle of El Niño, came reports of appalling flood damage. I headed south, past drought-stricken Mexico, and alighted on the soggy soil of Guayaquil, Ecuador. Three months before Christmas a colossal slab of warm water 450 feet thick had arrived off Ecuador and Peru, smothering the cool ocean surface. Coastal fishing ceased, and the simmering sea brought rains that threatened to wash Ecuador into the sea. "Areas that normally measure precipitation in inches have had ten feet," said Jimmy Aycart, . . . director of emergency relief for the Guayaquil area.

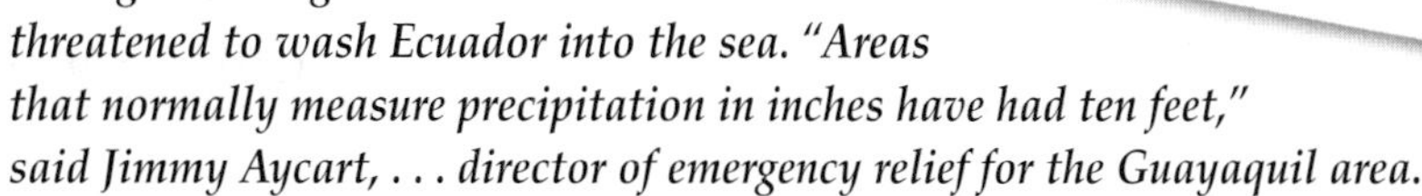

Flooding in Ecuador caused by El Niño

—*Thomas Y. Canby, "El Niño's Ill Wind,"* National Geographic, *February 1984*

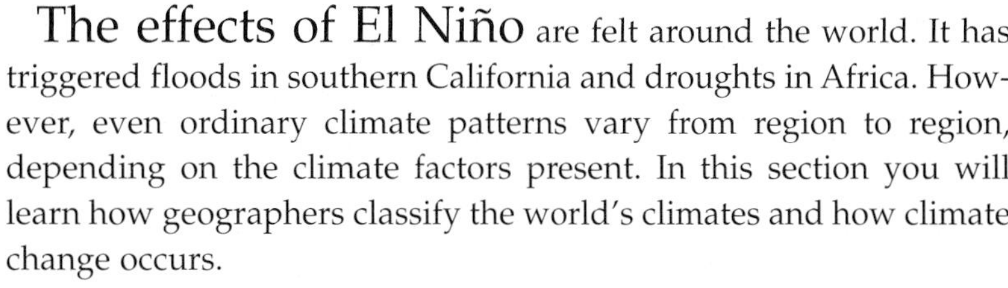

The effects of El Niño are felt around the world. It has triggered floods in southern California and droughts in Africa. However, even ordinary climate patterns vary from region to region, depending on the climate factors present. In this section you will learn how geographers classify the world's climates and how climate change occurs.

Climate Regions

Geographers often divide the earth into climate regions—tropical, dry, mid-latitude, high latitude, and highlands. Because climates vary within these broad regions, geographers further divide the major regions into smaller ones. Each of these divisions has its own

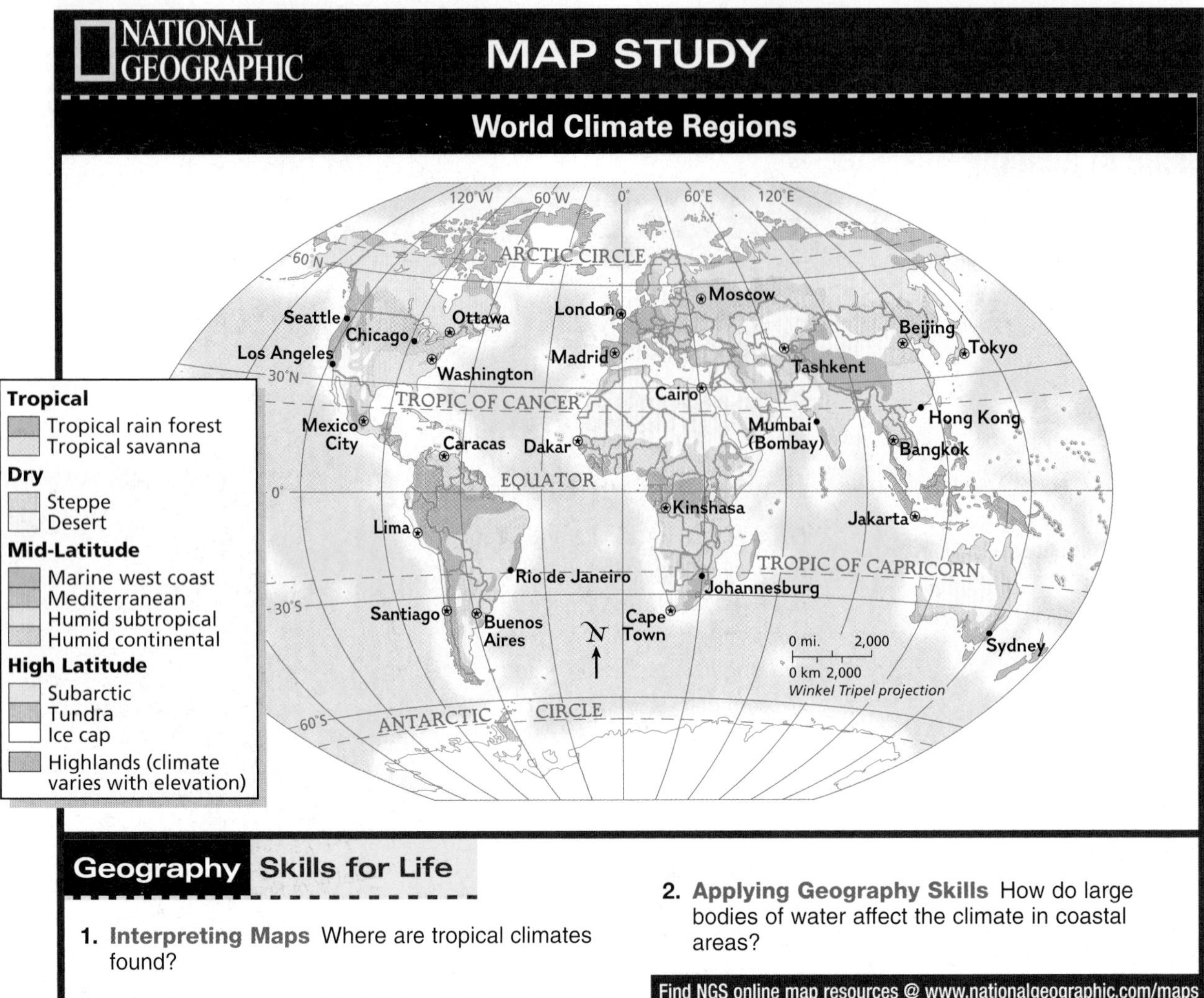

Geography Skills for Life

1. **Interpreting Maps** Where are tropical climates found?

2. **Applying Geography Skills** How do large bodies of water affect the climate in coastal areas?

Find NGS online map resources @ www.nationalgeographic.com/maps

characteristic soils and **natural vegetation**—the plant life that grows in an area where the natural environment is unchanged by human activity.

Tropical Climates

Tropical climates are found in or near the low latitudes—the **Tropics**. The two most widespread kinds of tropical climate regions are tropical rain forest and tropical savanna.

Hot and wet throughout the year, *tropical rain forest* climates have an average temperature of 80°F (27°C). The warm, humid air is saturated with moisture, producing rain almost daily. Yearly rainfall averages about 80 inches (203 cm). Lush vegetation is common in tropical climates, although the continual rain tends to leach, or draw out, nutrients from the soil in these climates. Wildlife is also abundant.

> *"Like an undiscovered continent encircling the globe, tropical rain forests shelter an astonishing abundance of organisms—probably more than half the Earth's plant and animal species."*
>
> Edward O. Wilson, "Rain Forest Canopy: The High Frontier," *National Geographic*, December 1991

Tropical rain forest vegetation grows thickly in layers. Tall teak or mahogany trees form a canopy over shorter trees and bushes. Vines and shade-loving plants grow on the forest floor. The world's largest tropical rain forest is in South America's Amazon River basin. Similar climate and vegetation exist in other parts of South America, in the Caribbean area, and Asia and Africa.

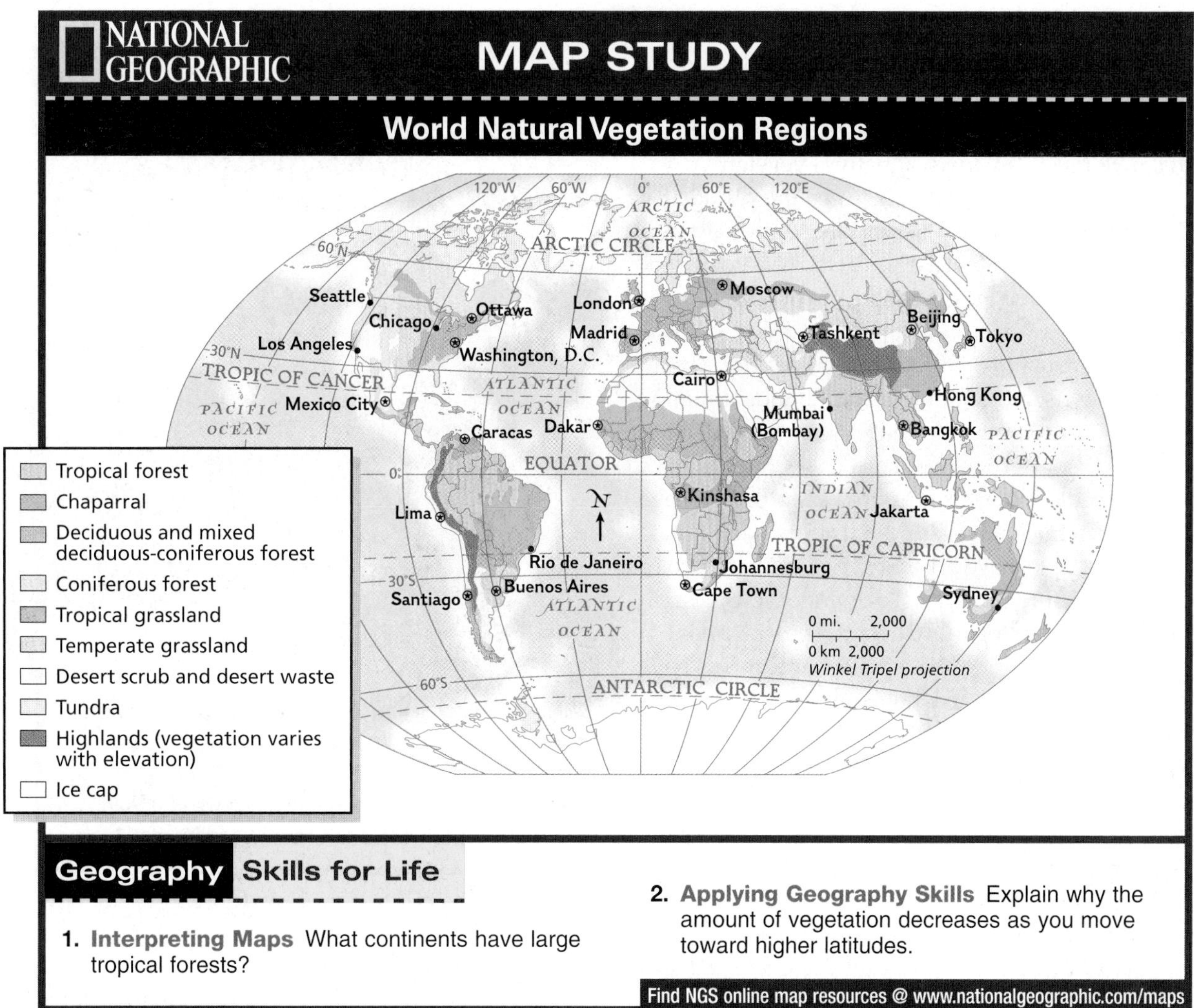

Geography Skills for Life

1. **Interpreting Maps** What continents have large tropical forests?

2. **Applying Geography Skills** Explain why the amount of vegetation decreases as you move toward higher latitudes.

Find NGS online map resources @ www.nationalgeographic.com/maps

Tropical savanna climates have dry winters and wet summers, accompanied by high year-round temperatures. In the dry season, the ground is covered with clumps of coarse grass. Fewer trees exist in savanna regions than in the rain forests. Tropical savannas are found in Africa, Central and South America, Asia, and Australia.

Dry Climates

Geographers have identified two types of dry climates, based on the vegetation in each. Both *desert* and *steppe* climates occur in many parts of the world.

Dry areas with sparse plant life are called deserts. Yearly rainfall in deserts seldom exceeds 10 inches (about 25 cm), and temperatures vary widely from the heat of day to the cool of night and from season to season. Desert climates occur in just under one-third of the earth's total land area. The **Sahara** alone extends over almost the entire northern one-third of the African continent.

The natural vegetation of deserts consists of scattered scrub and cactus, plants that tolerate low humidity and wide temperature ranges. In some desert areas, underground springs may support an oasis, an area of lush vegetation. Some deserts have dunes or rocky surfaces, and others have fertile soil that can yield crops through irrigation.

Often bordering deserts are dry, largely treeless grasslands called steppes. Yearly rainfall in steppe areas averages 10 to 20 inches (25 to 51 cm). The world's largest steppe stretches across eastern Europe and western and central Asia. Steppes are also found in North America, South America, Africa, and Australia.

Mid-Latitude Climates

The world's mid-latitudes include four temperate climate regions. Mid-latitude climates experience variable weather patterns and seasonal changes that give rise to a variety of natural vegetation.

Along western coastlines, between the latitudes of 30° and 60° north and south, are regions with a *marine west coast* climate. The Pacific coast of North America, much of Europe, and parts of South America, Africa, Australia, and New Zealand have marine west coast climates. In these areas ocean winds bring cool summers and damp winters. Abundant rainfall supports the growth of both coniferous and deciduous trees. **Coniferous** trees such as evergreens have cones, needle-shaped leaves, and keep their foliage throughout the winter. **Deciduous** trees, such as oak, elm, and maple, have broad leaves that change color and drop in autumn. Typical of marine west coast climates are **mixed forests** with both kinds of trees.

Lands surrounding the **Mediterranean Sea** have mild, rainy winters and hot, sunny summers. The natural vegetation includes **chaparral** (SHA•puh•RAL), thickets of woody bushes and short trees. Geographers classify as *Mediterranean* any coastal mid-latitude areas with similar climate and vegetation. Such areas also include Southern California and parts of southern Australia.

In the southeastern United States and in southeastern parts of South America and Asia, a *humid subtropical* climate brings short, mild winters and nearly year-round rain. The wind patterns and high pressure related to nearby oceans keep humidity levels high in these areas. Vegetation consists of **prairies**, or inland grasslands, and forests of evergreen and deciduous trees.

In some mid-latitude regions of the Northern Hemisphere, such as southern Canada, western Russia, and northeastern China, landforms influence climate more than winds, precipitation, or ocean temperatures do. These *humid continental* climate regions do not experience the moderating effect of ocean winds because of their northerly continental, or inland, locations. The farther north one travels in these humid continental areas, the longer and more severe are the snowy winters and the shorter and cooler are the summers. Vegetation in humid continental regions is similar to that found in marine west coast areas, with evergreens outnumbering deciduous trees in the northernmost areas of the region.

NATIONAL GEOGRAPHIC **World Explorer**

Geography Skills for Life

Mid-latitude Vegetation Chaparral, which includes shrubs and olive trees, grows in Mediterranean climate zones.

Region What kind of vegetation grows in humid subtropical climate zones? Humid continental?

High Latitude Climates

In high latitude climates, freezing temperatures are common throughout much of the year because of the lack of direct sunlight. As a result, the amount and variety of vegetation is limited.

Just south of the Arctic Circle lie the *subarctic* climate regions. Winters here are bitterly cold, and summers are short and cool. Subarctic regions have the world's widest temperature ranges, varying from winter to summer by as much as 120°F (49°C). In parts of the subarctic, only a thin layer of surface soil thaws each summer. Below it is permanently frozen subsoil, or **permafrost**. Brief summer growing seasons may support needled evergreens.

Closer to the polar regions, *tundra* climate regions are very cold. Here the winter darkness and bitter cold last for half the year, and the sun's indirect rays bring constant summer light but little heat. In tundra regions, most of which lie in the far north of the Northern Hemisphere, the layer of thawed soil is even thinner than in the subarctic.

Trees cannot establish roots on these frigid plains, so tundra vegetation is limited to low bushes, very short grasses, mosses, and lichens (LY•kuhns).

Snow and ice, often more than 2 miles (3 km) thick, constantly cover the surfaces of *ice cap* regions. Lichens are the only form of vegetation that can survive in these areas, where monthly temperatures average below freezing. Earth's largest polar ice cap covers almost all of Antarctica. Greenland's interior also has an ice cap climate.

Highlands Climates

Elevation can determine a climate region, regardless of latitude. High mountain areas, even along the Equator, share some of the same characteristics of high latitude climates because of the thinning of the atmosphere at high altitudes. The higher the elevation, the cooler the temperatures. The natural vegetation of highlands climates also varies with elevation. Mixed forests generally lie at the bases of mountain ranges. Higher up, meadows with small trees, shrubs, and wildflowers line the mountainsides.

Climatic Changes

Climates change gradually over time, although the causes of these changes are unclear. Scientists search for answers by studying the interrelationships among ocean temperatures, greenhouse gases, wind patterns, and cloud cover.

During the last 1 to 2 million years, for example, the earth passed through four ice ages, eras when glaciers covered large areas of the planet's surface. One **hypothesis**, or scientific explanation, for these ice ages is that the earth absorbed less solar energy because of variations in the sun's output of energy or because of variations in the earth's orbit. Another hypothesis suggests that dust clouds from volcanic activity reflected sunlight back into space, cooling the atmosphere and lowering surface temperatures.

Human interaction with the environment also affects climate. Burning fossil fuels releases gases that mix with water in the air, forming acids that fall in rain and snow. Acid rain can destroy forests. Fewer forests may result in climatic change. The exhaust released from burning fossil fuels in automobile engines and factories is heated in the atmosphere by the sun's ultraviolet rays, forming **smog**, a visible chemical haze in the atmosphere that endangers people's health. Other human-driven changes result from dams and river diversions. These projects, intended to supply water to dry areas, may cause new areas to flood or to dry out and may affect climate over time.

SECTION 3 ASSESSMENT

Checking for Understanding

1. **Define** natural vegetation, oasis, coniferous, deciduous, mixed forest, chaparral, prairie, permafrost, hypothesis, smog.
2. **Main Ideas** Create a table like the one below, adding information and a brief description about each of the world's climate regions.

Earth's Climates	
Climate Region	Features

Critical Thinking

3. **Analyzing Information** What patterns of vegetation are typical of tropical climates? Explain.
4. **Comparing and Contrasting** What factors account for the similarities and differences between the subdivisions in tropical climate zones?
5. **Categorizing Information** How are the five major climate regions related to the three zones of latitude?
6. **Drawing Conclusions** What are the two main categories of factors causing climate change?

Analyzing Maps

7. **Region** Study the map of world natural vegetation regions on page 67. What vegetation type dominates Europe? Canada and the United States?

Applying Geography

8. **Climate and Settlement Patterns** On the map of world climate regions on page 66, locate the climate regions for Tashkent, Cape Town, Lima, Chicago, London, and Jakarta. What can you conclude about the relationship between climate and settlement?

Reading a Diagram

Have you ever assembled a model airplane or car? Kits and how-to books give detailed instructions that often include diagrams. Because they present information visually, diagrams can help explain ideas and processes easily.

Learning the Skill

A diagram is a graphic design that shows a process or event. Diagrams are extremely useful in communicating information clearly and quickly. A diagram can show placement, relationships, cycles, and movement, using symbols and drawn objects. Diagrams can be very useful for showing changes over time or for comparing two or more actions or relationships. Presenting information visually can make complex events or ideas more understandable.

Newspapers, magazines, and the Internet use diagrams to supplement written information. Geographers use diagrams to explain complex processes such as climate and weather phenomena. A diagram can be a very effective way of communicating an idea.

Follow these steps to understand a diagram:

- **Note its title, caption, and labels.** These features provide information that is important for understanding the diagram.
- **Study carefully the objects and symbols used.** Sometimes a diagram includes a key to the symbols.
- **Look at the relationships or actions shown.** If the diagram compares two or more things, look for details that show how the relationship or action changes under different circumstances.

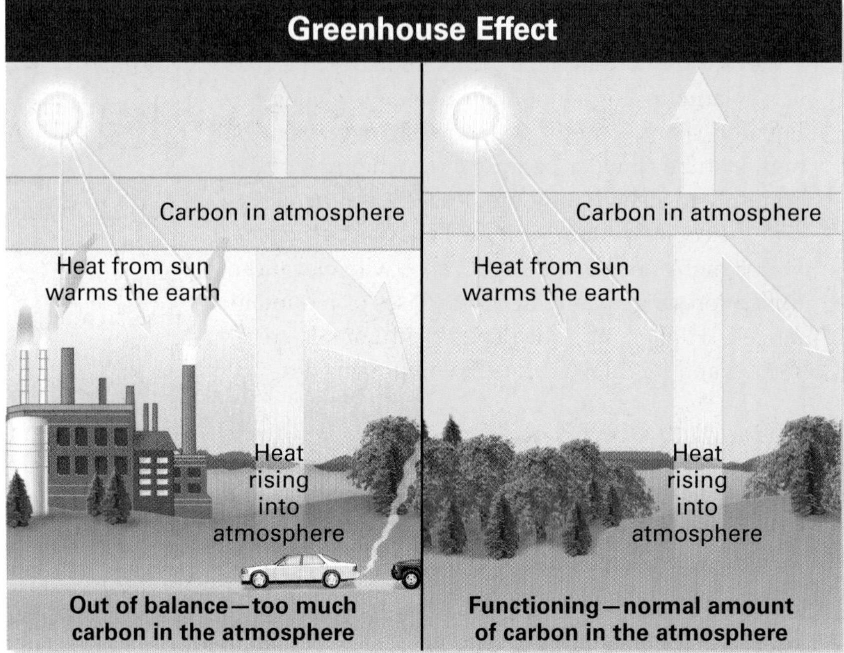

Practicing the Skill

The diagrams above explain and compare two aspects of the greenhouse effect. Use the diagrams to answer the following questions.

1. Describe the two aspects of the greenhouse effect shown above.
2. How is plant life different in the two diagrams?
3. What objects are shown contributing more carbon to the atmosphere?
4. How is the amount of heat retained in the Earth's atmosphere related to the amount of carbon in the atmosphere?

Design and draw a diagram that explains a natural process. Look through newspapers or magazines for ideas relating to geography. In the diagram, include drawn objects, symbols, a title, and labels.

The **Glencoe Skillbuilder Interactive Workbook, Level 2** provides instruction and practice in key social studies skills.

CHAPTER 3 SUMMARY & STUDY GUIDE

SECTION 1 Earth-Sun Relationships (pp. 55–58)

Terms to Know

- weather
- climate
- axis
- temperature
- revolution
- equinox
- solstice
- greenhouse effect
- global warming

Key Points

- The earth's position in relation to the sun affects temperatures on Earth.
- The rotation of the earth causes day and night.
- The earth's revolution and its tilt in relation to the sun produce the seasons.
- Global temperatures may be increasing as a result of human activity.

Organizing Your Notes

Use a graphic organizer like the one below to help you organize your notes for this section.

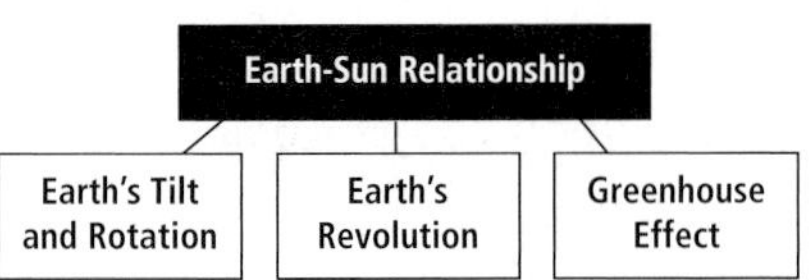

SECTION 2 Factors Affecting Climate (pp. 59–64)

Terms to Know

- prevailing wind
- Coriolis effect
- doldrums
- current
- El Niño
- windward
- leeward
- rain shadow

Key Points

- Latitude and elevation affect climate.
- Wind patterns and ocean currents play a key role in the earth's climates.
- Climate is affected by recurring phenomena such as El Niño.
- Landforms shape and are shaped by climate patterns.

Organizing Your Notes

Use a table like the one below to help you organize information about climate.

Climate Factors	
Location	
Currents	
Surface Features	

SECTION 3 World Climate Patterns (pp. 65–69)

Terms to Know

- natural vegetation
- oasis
- coniferous
- deciduous
- mixed forest
- chaparral
- prairie
- permafrost
- hypothesis
- smog

Key Points

- Geographers divide the world into major climate regions.
- Each climate region has its own characteristic natural vegetation.
- Climate patterns change over time as a result of both natural processes and human activity.

Organizing Your Notes

Use an outline like the one below to help you organize your notes for this section.

Climate Patterns

I. Tropical climates
  A. Tropical rain forest
    1. High temperatures
    2. Rain all year
  B.

CHAPTER 3

ASSESSMENT & ACTIVITIES

Reviewing Key Terms

On a sheet of paper, classify these key terms under the correct heading: earth-sun relationship, climate factors, or climate patterns.

greenhouse effect	temperature	El Niño
mixed forest	prevailing wind	oasis
equinox	chaparral	prairie
revolution	doldrums	current
deciduous	permafrost	solstice
Coriolis effect	rain shadow	axis

Reviewing Facts

SECTION 1

1. What are the effects of the earth's tilt, rotation, and revolution?
2. What are the differences in sunlight at the Tropic of Cancer, the Tropic of Capricorn, and the Poles?
3. How is CO_2 related to the greenhouse effect?

SECTION 2

4. List three key factors that affect climate.
5. Describe the changes in air pressure that occur during an El Niño year. How do these changes affect wind patterns?
6. How do large bodies of water and mountains affect climate?

SECTION 3

7. What are the major climate regions into which geographers divide the earth?
8. What main types of vegetation grow in the earth's major climate regions?
9. In what ways might the burning of fossil fuels affect a region's vegetation?

Critical Thinking

1. **Predicting Consequences** How might increased global warming affect the earth's climates? Give examples to support your answer.
2. **Making Generalizations** Why do the mid-latitudes have a temperate climate?
3. **Categorizing Information** Create a web diagram like the one below. Connect the items with a line to show that they are related. Describe the geographical events and phenomena associated with each category.

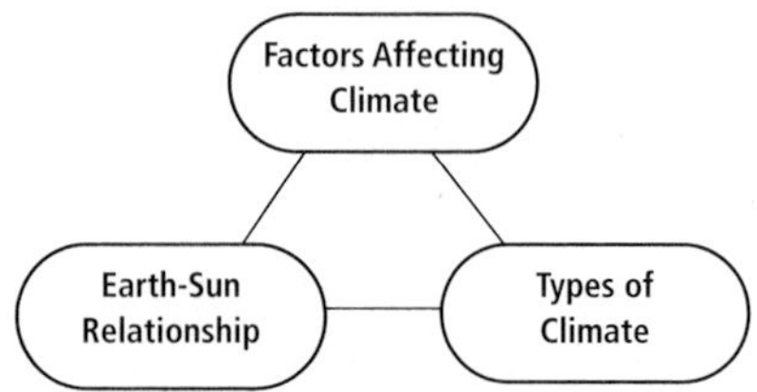

NATIONAL GEOGRAPHIC

Locating Places

The World: Physical Geography

Match the letters on the map with the appropriate places and physical features. Write your answers on a separate sheet of paper.

1. **Equator**
2. **Arctic Circle**
3. **Tropic of Capricorn**
4. **Warm current, Atlantic**
5. **Tropic of Cancer**
6. **Antarctic Circle**
7. **Cold current, Pacific**

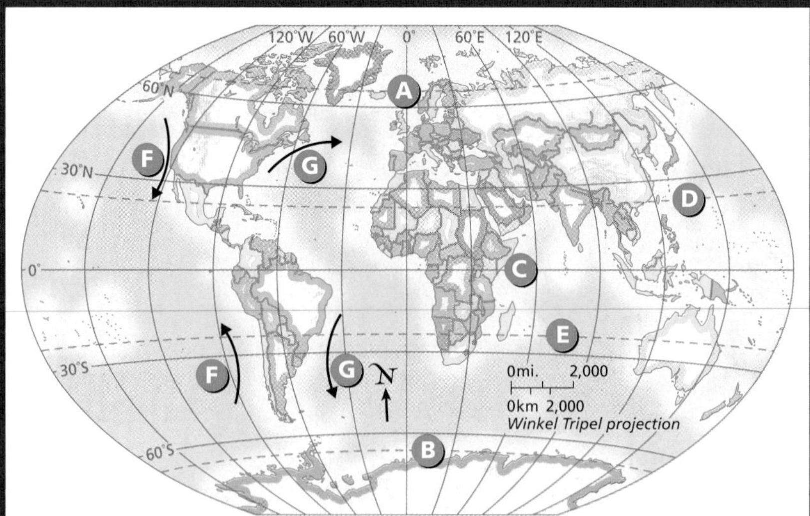

Self-Check Quiz Visit the **Glencoe World Geography** Web site at geography.glencoe.com and click on Self-Check Quizzes—Chapter 3 to prepare for the Chapter Test.

Thinking Like a Geographer

Imagine that you have been asked to speak about global warming to a group in your community. Your goal is to persuade the audience to take steps to reduce global warming. Write an outline of the remarks you will make to the group, and use visuals to support your arguments.

Problem-Solving Activity

Problem-Solution Proposal Research and analyze the effects of a physical geographic pattern, such as El Niño, on world economies. What actions did countries take in response? Determine what preparations might lessen the impact of this pattern in the future. Write a proposal in which you outline the problem, present several possible solutions, and recommend a course of action. Include diagrams, charts, maps, and other visual elements.

GeoJournal

Cause and Effect Review your journal entries about factors affecting climate. Write a story that takes place in the future. The climate in your community has changed. Include details about how this change occurred and how it has affected vegetation, economic activities, and human and animal populations. Use specific examples as well as illustrations and concrete language to make your story interesting and engaging.

Technology Activity

Using the Internet for Research Work with a team to search the Internet for information about the climate in your community. Include information such as climate region, local factors that influence climate, seasonal changes, and any unusual climatic events. Then choose one aspect of your area's climate and write a paragraph explaining its effect on people in your community. For example, if your community has experienced a weather-related disaster, such as a hurricane or tornado, what effects did the event have on your community?

Standardized Test Practice

Use the graph below and your knowledge of geography to answer the question. If you have trouble answering the question, use the process of elimination to narrow your choices.

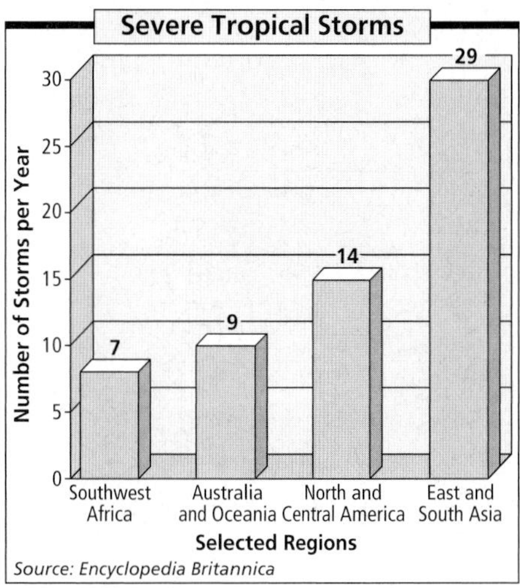

1. **Which of the following statements can be inferred from the graph?**

   F Severe tropical storms are rare in Oceania.

   G East and South Asia have about four times as many severe tropical storms as Southwest Africa.

   H North and Central America never go through a month without a storm.

   J South America does not have tropical storms.

Read the title and labels on the graph carefully to determine what is being presented. For example, the label on the bottom of the graph tells you that only selected regions are shown. The label on the left indicates the number of storms per year, but not the month in which they occur.

CHAPTER 4

The Human World

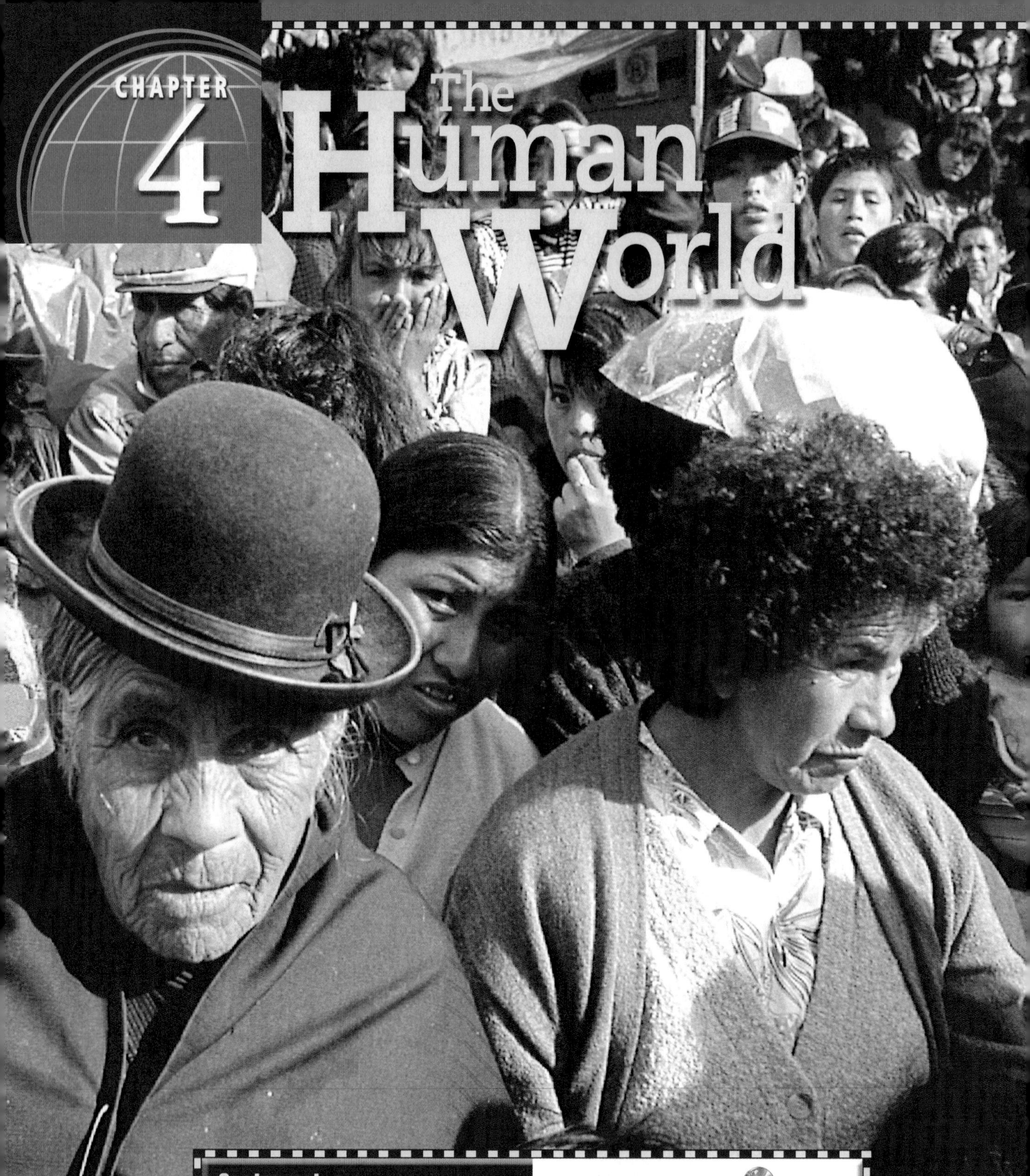

GeoJournal

As you read this chapter, use your journal to record information about aspects of the human world. Include details that show how geographers describe population, culture, government, resources, and the environment.

GEOGRAPHY Online

Chapter Overview Visit the **Glencoe World Geography** Web site at geography.glencoe.com and click on Chapter Overviews—Chapter 4 to preview information about the human world.

SECTION 1

World Population

Guide to Reading

Consider What You Know

The world's largest cities are often in the news. Think about where these cities are located and what their locations have in common. Why do you think people have settled in these places?

Read to Find Out

- What factors affect a country's population growth rate?
- What challenges does population growth pose for the planet?
- Why is the world's population unevenly distributed?

Terms to Know

- death rate
- birthrate
- natural increase
- doubling time
- population distribution
- population density
- migration

Places to Locate

- Nile Delta
- Hungary
- Germany
- Canada
- Bangladesh
- Mexico City

◀ *Young and old flock to a carnival in Oruro, Bolivia.*

NATIONAL GEOGRAPHIC

A Geographic View

Nile Delta in Peril

The black soil of the Nile Delta has made it the foundation stone of seven millennia of human history. . . . Today Egypt's battle is to preserve the soil and water that have always given life to the delta. One hundred fifty years ago this nation had five million acres of farmland and five million citizens; now it has seven million acres of farmland and 60 million citizens. And every nine months there are nearly a million more Egyptians to feed. . . . The Nile Delta . . . has survived many challenges from without. Now the challenges it must survive come from its own population. . . .

—*Peter Theroux, "The Imperiled Nile Delta,"* National Geographic, *January 1997*

Farmland in the Nile Delta

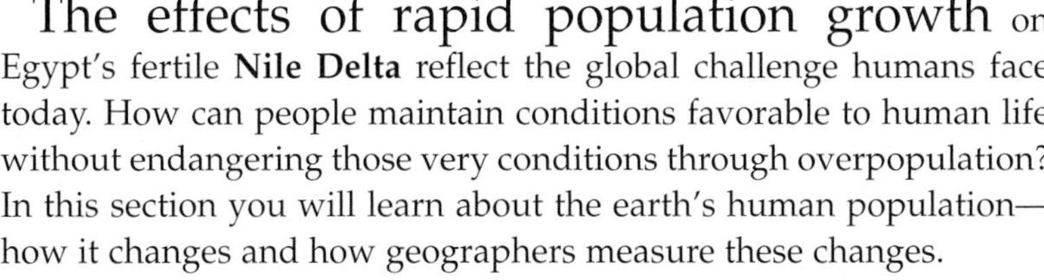

The effects of rapid population growth on Egypt's fertile **Nile Delta** reflect the global challenge humans face today. How can people maintain conditions favorable to human life without endangering those very conditions through overpopulation? In this section you will learn about the earth's human population—how it changes and how geographers measure these changes.

Population Growth

About 6.2 billion people now live on Earth, inhabiting about 30 percent of the planet's land. Global population is growing rapidly and is expected to reach about 7.8 billion by the year 2025. Such rapid growth was not always the case. The graph of population growth on page 76 shows that from the year 1000 until 1800, the world's population increased slowly. Then the number of people on Earth more than doubled between 1800 and 1950. By 2000 the world's population had soared to more than 6 billion. If the population continues to grow at its current rate, it will pass 9 billion by the year 2050.

Growth Rates

Global population is growing rapidly because birthrates have not declined as fast as death rates. The **death rate** is the number of deaths per year for every 1,000 people. The **birthrate** is the number of births per year for every 1,000 people. Scientists in the field of *demography,* the study of populations, calculate the **natural increase**, or growth rate, of a population as the difference between an area's birthrate and its death rate.

Population growth occurs at different rates in various parts of the world. Over the past 200 years, death rates have gone down in many places as a result of improved health care, more abundant food supplies, advances in technology, and better living conditions. In many wealthy industrialized countries, a declining death rate has been accompanied by a low birthrate. These countries have reached what is known as *zero population growth,* in which the birthrate and death rate are equal. When this balance occurs, a country's population does not grow.

In many countries in Asia, Africa, and Latin America, however, the birthrate is high. Families in these regions traditionally are large because of cultural beliefs about marriage, family, and the value of children. For example, a husband and wife in a rural agricultural area may choose to have several children who will help farm the land. A high number of births combine with low death rates to greatly increase population growth in these areas. As a result, the **doubling time**, or the number of years it takes a population to double in size, has been reduced to only 25 years in some parts of Asia, Africa, and Latin America. In contrast, the average doubling time of a wealthy, industrialized country can be more than 300 years.

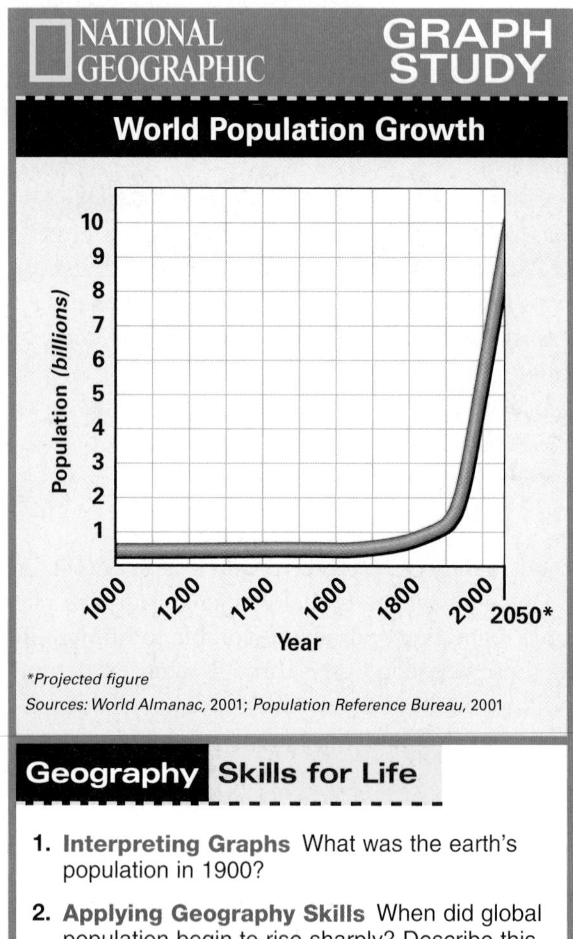

Geography Skills for Life

1. **Interpreting Graphs** What was the earth's population in 1900?
2. **Applying Geography Skills** When did global population begin to rise sharply? Describe this trend in population growth.

Challenges of Population Growth

Rapid population growth presents many challenges to the global community. As the number of people increases, so does the difficulty of producing enough food to feed them. Fortunately, since 1950 world food production has risen on all continents except Africa. Because so many people in various parts of Africa need food, warfare or severe weather conditions that ruin crops can bring widespread famine.

In addition, populations that grow rapidly use resources more quickly than populations that do not grow as rapidly. Some countries face shortages of water, housing, and clothing, for example. Rapid population growth strains these limited resources. Another concern is that the world's population is unevenly distributed by age, with the majority of some countries' populations being infants and young children who cannot contribute to food production.

While some experts are pessimistic about the long-term effects of rapid population growth, others are optimistic that, as the number of humans increases, the levels of technology and creativity also will rise. For example, scientists continue to study and develop ways to boost agricultural productivity. Fertilizers can improve crop yields. Irrigation systems can help increase the amount of land available for farming. New varieties of plants such as wheat and rice have been created to withstand severe conditions and yield more food.

Economics

Negative Population Growth

In the late 1900s, some countries in Europe began to experience negative population growth, in which the annual death rate exceeds the annual birthrate. **Hungary** and **Germany**, for example, show growth rates of –0.4 and –0.1, respectively. This situation has economic consequences different from, but just as serious as, those caused by high growth rates. In countries with negative population growth, it is difficult to find enough workers to keep the economy going. Labor must be recruited from other countries, often by encouraging immigration or granting temporary work permits. Although the use of foreign labor has helped countries with negative growth rates maintain their levels of economic activity, it also has created tensions between the "host" population and the communities of newcomers.

Geography Skills for Life

War Refugees War can cause population shifts by forcing refugees to flee to safety in other countries.

Movement How might population growth rates affect a country's economy?

Population Distribution

Not only do population growth rates vary among the earth's regions, but the planet's **population distribution**, or the pattern of human settlement, is uneven as well. Population distribution is related to the earth's geography. Only about 30 percent of the earth's surface is made up of land, and much of that land is inhospitable. High mountain peaks, barren deserts, and frozen tundra make human activity very difficult. As the population density map on page 78 shows, almost everyone on Earth lives on a relatively small portion of the planet's land—a little less than one-third. Most people live where fertile soil, available water, and a climate without harsh extremes make human life sustainable.

Of all the continents, Europe and Asia are the most densely populated. Asia alone contains more than 60 percent of the world's people. Throughout the world, where populations are highly concentrated many people live in *metropolitan areas*—cities and their surrounding urbanized areas. Today most people in Europe, North America, and Australia live in or around urban areas.

Population Density

Geographers determine how crowded a country or region is by measuring **population density**—the average number of people living on a square mile or square kilometer of land. To determine population density in a country, geographers divide the total population of the country by its total land area.

Student Web Activity Visit the **Glencoe World Geography** Web site at geography.glencoe.com and click on Student Web Activities—Chapter 4 for an activity on world population.

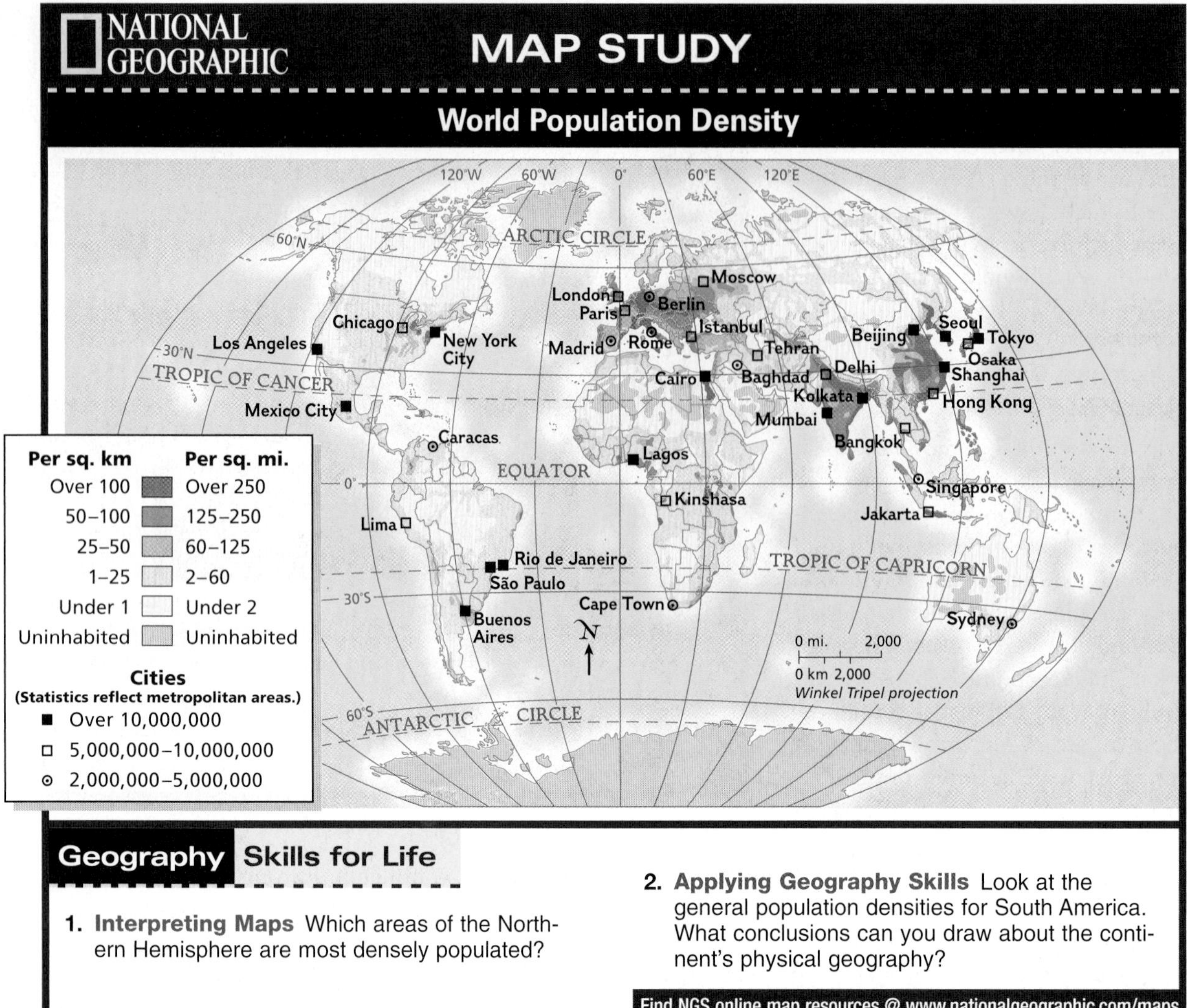

Geography Skills for Life

1. **Interpreting Maps** Which areas of the Northern Hemisphere are most densely populated?

2. **Applying Geography Skills** Look at the general population densities for South America. What conclusions can you draw about the continent's physical geography?

Find NGS online map resources @ www.nationalgeographic.com/maps

Population density varies widely from country to country. **Canada**, with a low population density of about 8 people per square mile (3 people per sq. km), offers wide-open spaces and the opportunity to choose between living in thriving cities or quiet rural areas. The country of **Bangladesh**, at the other extreme, has one of the highest population densities in the world—about 2,401 people per square mile (927 people per sq. km). In tiny Bangladesh even the rural areas are more crowded than many of the world's cities.

Countries that have populations of nearly the same size do not necessarily have similar population densities. For example, Niger and Belgium have about the same number of people, roughly 10.3 million. With a smaller land area, Belgium has 872 people per square mile (292 people per sq. km). However, Niger has an average of only 21 people per square mile (8 people per sq. km). Belgium, then, is more densely populated than Niger.

Because population density is an average, it does not account for uneven population distribution within a country—a common occurrence. In Egypt, for example, overall population density is 181 people per square mile (70 people per sq. km). In reality, about 99 percent of Egypt's people live within 20 miles of the Nile River. The rest of Egypt is desert. Thus, some geographers prefer to describe a country's population density in terms of land that can be used to support the population rather than total land area. When Egypt's population density is measured this way, it equals about 5,807 people per square mile (2,242 people per sq. km)!

Population Movement

Migration is the movement of people from place to place. The earth's human population is moving in great numbers. Some people are moving from city to city or from suburb to suburb. Large numbers of people are migrating from rural villages to cities.

> *"Migration is the dynamic undertow of population change. . . . It is, as it has always been, the great adventure of human life. Migration helped create humans, drove us to conquer the planet, shaped our societies, and promises to reshape them again."*
>
> Michael Parfit, "Human Migration," *National Geographic*, October 1998

The resulting growth of city populations brought about by migration and the changes that come with this increase in population are called *urbanization*. Urbanization has many causes. The primary cause is the desire of rural people to find jobs and a better life in more prosperous urban areas. Rural populations certainly have grown, but the amount of land that can be farmed has not increased to meet the growing number of people who need to work and to eat. As a result, many rural migrants find urban jobs in manufacturing or in service industries, such as tourism.

About half of the world's people live in cities—a far higher percentage than ever before. Between 1960 and 2000, the population of metropolitan **Mexico City** rose from about 5 million to about 18 million. Other cities in Latin America, as well as cities in Asia and Africa, have seen similar growth. Some of these cities contain a large part of their country's entire population. For example, about one-third of Argentina's people live in the city of Buenos Aires.

Population movement also occurs between countries. Some people emigrate from the country of their birth and move to another. They are known as emigrants in their homeland and are called immigrants in their new country. In the past 40 years, millions of people have left Africa, Asia, and Latin America to find jobs in the wealthier countries of Europe, North America, and Australia. Some people were forced to flee their country because of wars, food shortages, or other problems. They are *refugees*, or people who flee to another country to escape persecution or disaster. In the next section, you will learn how the movement of peoples has influenced the development of cultures.

SECTION 1 ASSESSMENT

Checking for Understanding

1. **Define** death rate, birthrate, natural increase, doubling time, population distribution, population density, migration.
2. **Main Ideas** On a table like the one below, fill in the main points about population growth and population distribution from this section.

Population Growth	Population Distribution
•	•
•	•
•	•

Critical Thinking

3. **Comparing and Contrasting** How do the effects of zero population growth and negative population growth differ? How are they similar?
4. **Drawing Conclusions** How might the population growth rates of developing countries be affected as they become increasingly industrialized?
5. **Predicting Consequences** What will happen to the standard of living in cities as urbanization increases? How might the standard of living differ between cities in the developing world and cities in the developed world?

Analyzing Maps

6. **Human-Environment Interaction** Look at the population density map on page 78. Identify three of the most densely populated areas on Earth. What physical features do they have in common?

Applying Geography

7. **Influences of Physical Geography** What geographic features might be present in countries that have large numbers of people concentrated in relatively small areas? Write a paragraph with supporting details to explain your answer.

SECTION 2

Global Cultures

Guide to Reading

Consider What You Know

You probably know that, as a country of immigrants, the United States includes people from a great variety of cultural backgrounds. How do you define your own cultural background?

Read to Find Out

- What factors define a culture?
- What are the major culture regions of the world?
- What developments have affected interaction between cultures in recent years?

Terms to Know

- culture
- language family
- ethnic group
- culture region
- cultural diffusion
- culture hearth

Places to Locate

- Egypt
- Iraq
- Pakistan
- China
- Mexico

NATIONAL GEOGRAPHIC

A Geographic View

Beijing Construction Boom

[T]oday's Beijing is awash with change, where the old Confucian ideals of personal cultivation and family values clash with a new emphasis on money and the market, . . . where a construction boom is reshaping Beijing's low-slung profile and cramped alleys with soaring skyscrapers of glass and steel, where car traffic clogs streets that once rang with bicycle bells. . . . Over the centuries the people of Beijing have become expert at adjusting. Like the willows planted around the capital, people have survived by being flexible, yielding to strong winds, then springing back as stillness returns.

—*Todd Carrel, "Beijing: New Face for the Ancient Capital,"* National Geographic, *March 2000*

Beijing, China

The Chinese people have shown a remarkable ability to adapt to changes over time. The evidence of change in Chinese and other societies is apparent in such areas as architecture, family customs, and economic activities. These factors and many others express the values that a group of people share and pass down from one generation to another. In this section you will read about how the world's people organize communities, develop their ways of life, and cope with their differences and similarities.

Elements of Culture

In addition to population trends, geographers study **culture**, the way of life of a group of people who share similar beliefs and customs. A particular culture can be understood by looking at various elements: what languages the people speak, what religions they follow, and what smaller groups form as parts of their society. The study of a culture also includes examining people's daily lives. Still other factors are the history the people have shared and the art forms they have created. Finally, culture includes how people govern their society and how they make a living.

Language

Language is a key element in a culture's development. Through language, people communicate information and experiences and pass on cultural values and traditions. Sharing a language is one of the strongest unifying forces for a culture. Even within a culture, however, there are language differences. Some people may speak a dialect, or a local form of a language that differs from the main language. These differences may include variations in pronunciation and the meaning of words.

Linguists, scientists who study languages, organize the world's languages into **language families**—large groups of languages having similar roots. Seemingly diverse languages may belong to the same language family. For example, English, Spanish, Russian, and Hindi (spoken in India) are all members of the Indo-European language family. The world map below shows where languages from the different language families are spoken.

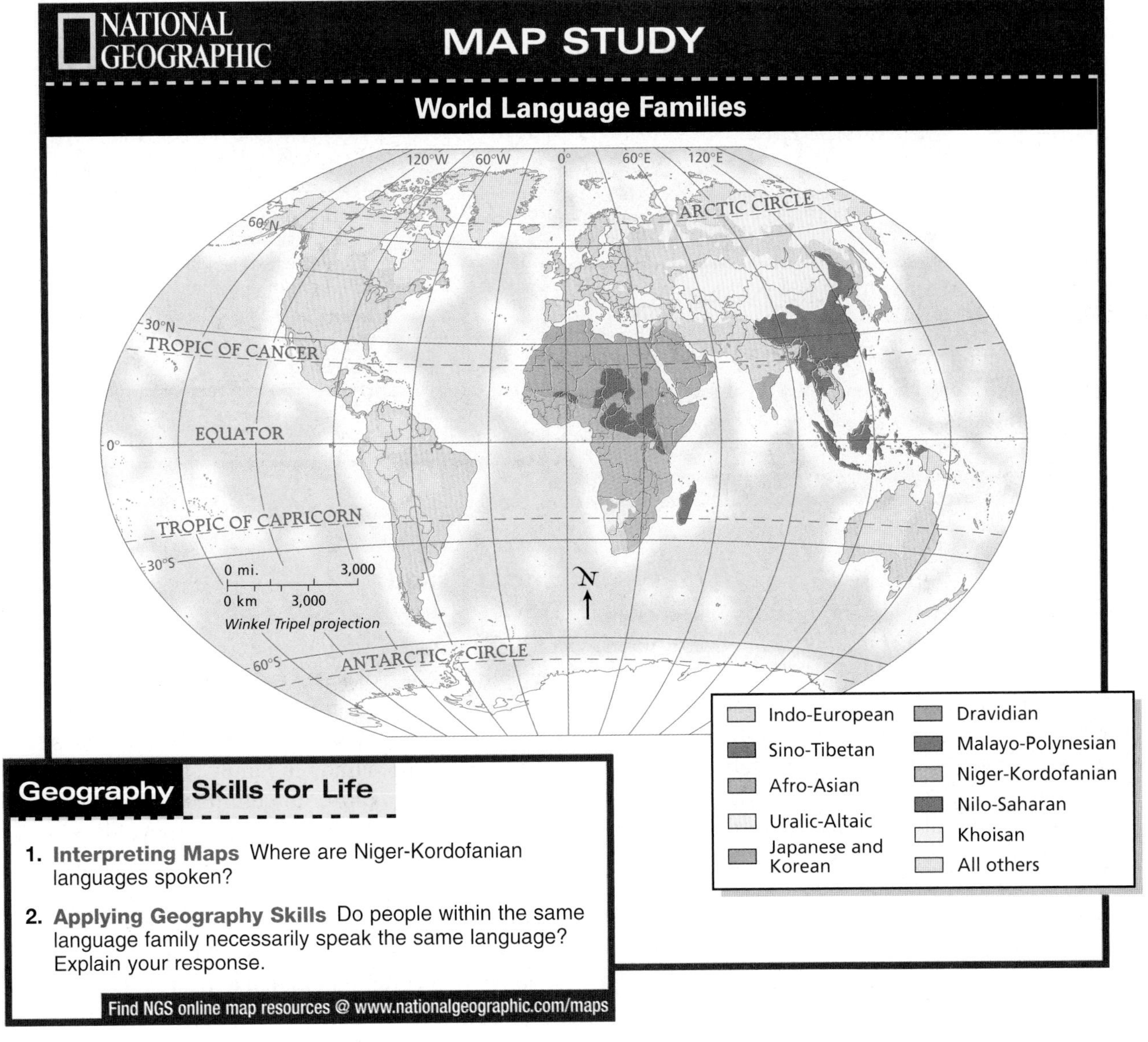

Geography Skills for Life

1. **Interpreting Maps** Where are Niger-Kordofanian languages spoken?
2. **Applying Geography Skills** Do people within the same language family necessarily speak the same language? Explain your response.

Find NGS online map resources @ www.nationalgeographic.com/maps

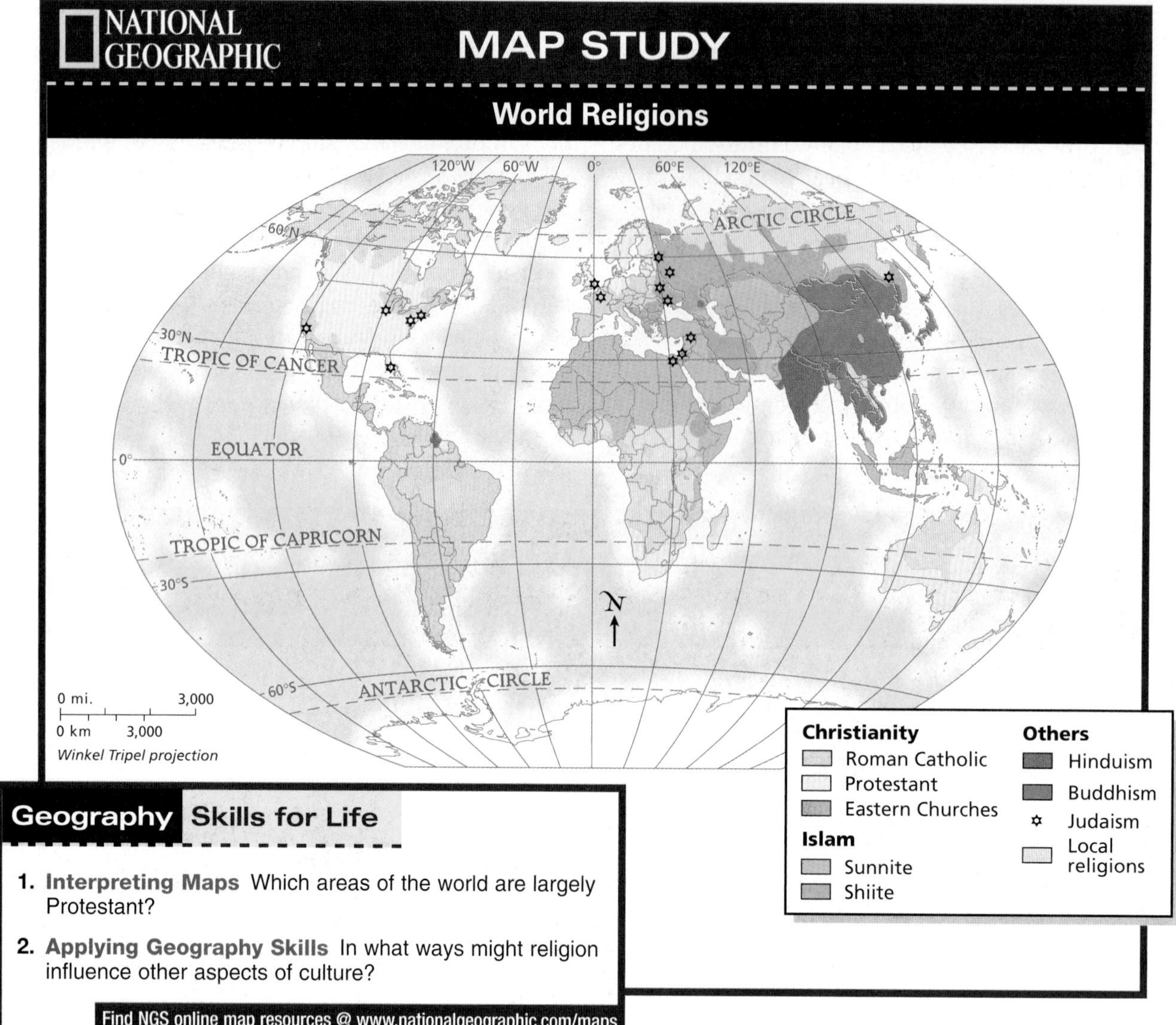

Geography Skills for Life

1. **Interpreting Maps** Which areas of the world are largely Protestant?
2. **Applying Geography Skills** In what ways might religion influence other aspects of culture?

Find NGS online map resources @ www.nationalgeographic.com/maps

Religion

Another important part of culture is religion. Religious beliefs vary significantly around the world, and struggles over religious differences are a source of conflict in many countries. In many cultures, however, religion enables people to find a sense of identity. It also influences aspects of daily life, from the practice of moral values to the celebration of holidays and festivals. Throughout history, religious symbols and stories have shaped cultural expressions such as painting, sculpture, architecture, music, and dance. Some of the major world religions are Hinduism, Buddhism, Judaism, Christianity, and Islam. The map above shows the areas of the world where these religions are practiced.

Social Groups

Every culture includes a social system in which the members of the society fall into various smaller groups. A social system develops to help the members of a culture work together to meet basic needs. In all cultures the family is the most important group, although family structures vary somewhat from culture to culture. Most cultures are also made up of social classes, groups of people ranked according to ancestry, wealth, education, or other criteria. Moreover, cultures may include people who belong to different ethnic groups. An **ethnic group** is made up of people who share a common language, history, place of origin, or a combination of these elements.

Government

A society's government reflects the uniqueness of its culture. Despite differences, governments of the world share certain features. Each government, for example, maintains order within the country, provides protection from outside dangers, and supplies other services to its people. Governments are organized according to levels of power—national, regional, and local—and by type of authority—a single ruler, a small group of leaders, or a body of citizens or their representatives.

Economic Activities

People in every kind of culture must make a living, whether in farming or in industry or by providing services such as preparing food or designing Web pages. In examining cultures, geographers look at economic activities. They study how a culture utilizes its natural resources to meet such human needs as food and shelter. They also analyze the ways in which people produce, obtain, use, and sell goods and services.

Culture Regions

To organize their understanding of cultural development, geographers divide the earth into specific areas called culture regions. Each **culture region** includes many different countries that have certain traits in common. They may share similar economic systems, forms of government, and social groups. Their histories, religions, and art forms may share similar influences. The food,

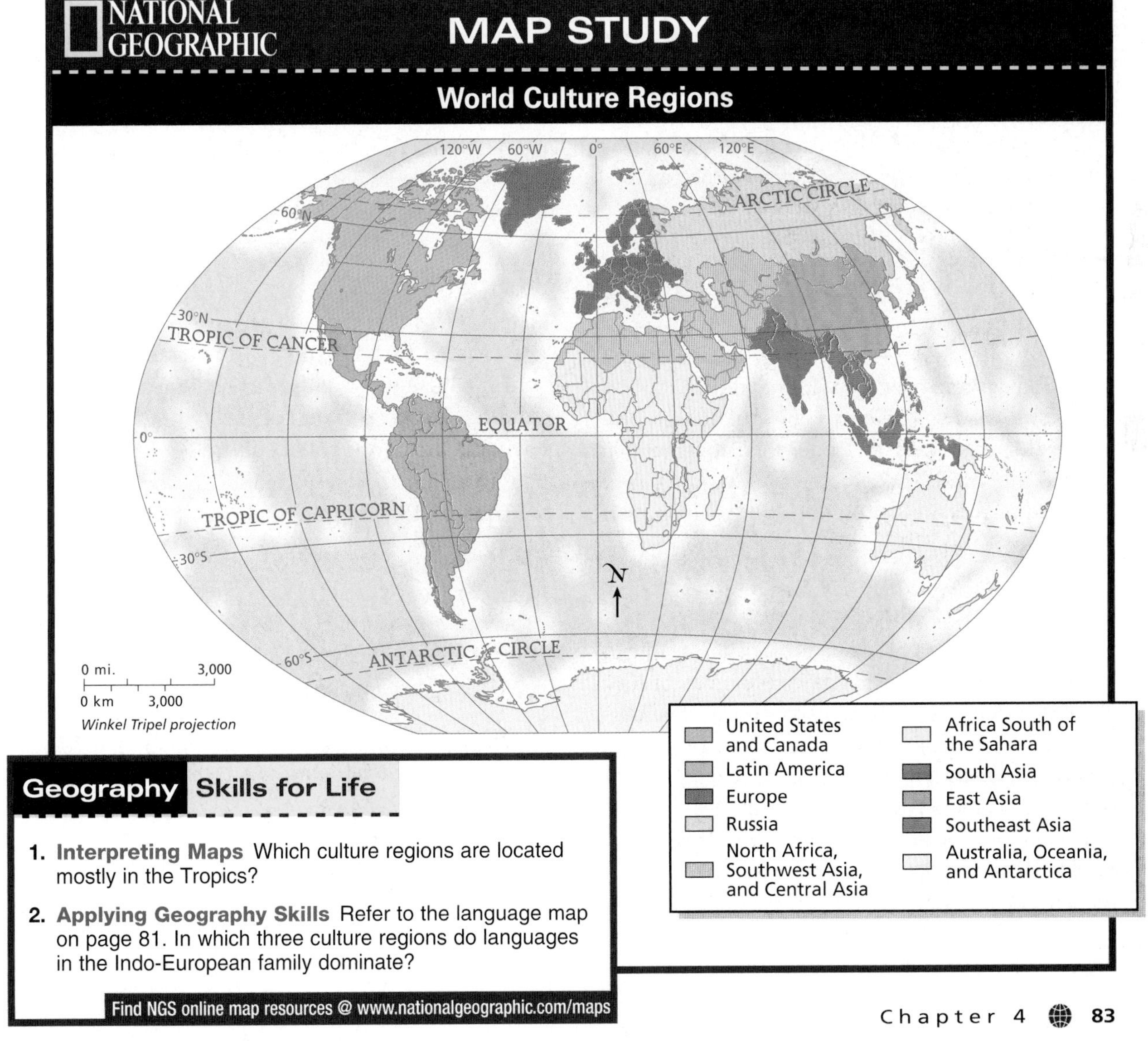

Geography Skills for Life

1. **Interpreting Maps** Which culture regions are located mostly in the Tropics?
2. **Applying Geography Skills** Refer to the language map on page 81. In which three culture regions do languages in the Indo-European family dominate?

Find NGS online map resources @ www.nationalgeographic.com/maps

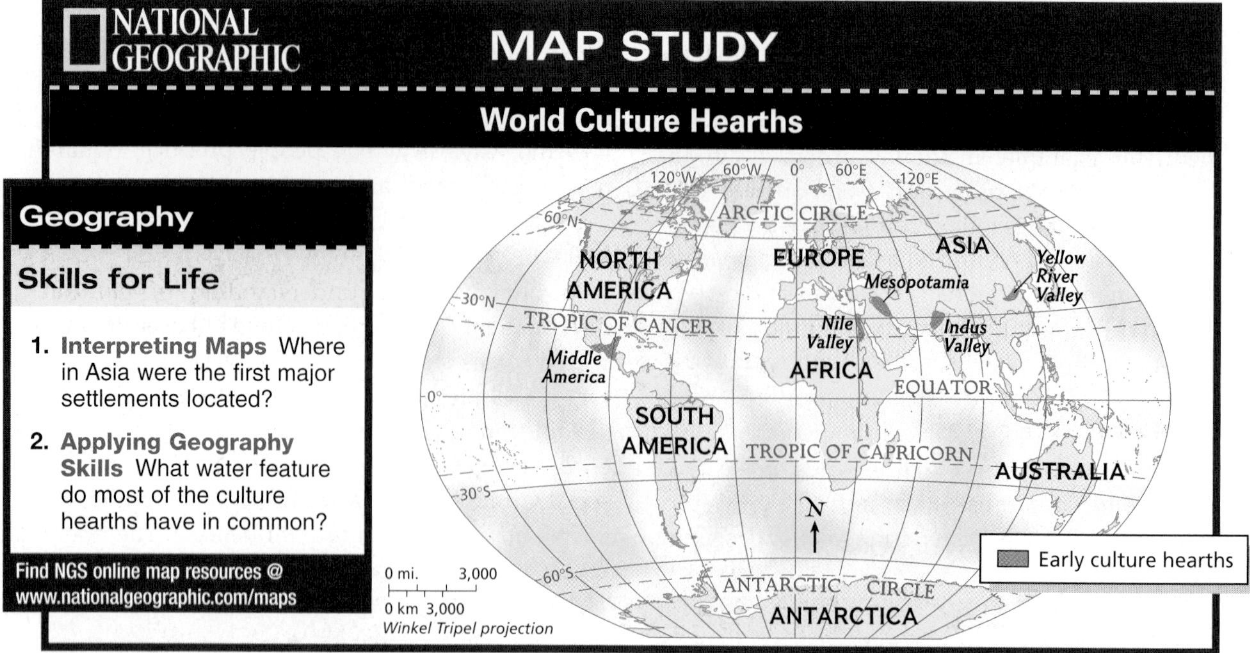

clothing, and housing of people in these countries may all have common characteristics as well. The map on page 83 shows the various culture regions that you will study in this textbook.

Cultural Change

No culture remains the same over the course of time. Internal factors—new ideas, lifestyles, and inventions—create change within cultures.

Change can also come to a culture through outside influences, such as trade, the movement of people, and war. The process of spreading new knowledge and skills from one culture to another is called **cultural diffusion**.

The Agricultural Revolution

Cultural diffusion has been a major factor in cultural development since the dawn of human history. The earliest humans were nomads, groups of herders who had no fixed home but who moved from place to place in search of food, water, and grazing land. As the earth's climate warmed about 10,000 years ago, many of these nomads settled in river valleys and on fertile plains. They became farmers who lived in permanent villages and grew crops on the same land every year. This shift from gathering food to producing food is known as the Agricultural Revolution.

By about 3500 B.C. some of these early farming villages had evolved into *civilizations*, highly organized, city-based societies with an advanced knowledge of farming, trade, government, art, and science.

Culture Hearths

The world's first civilizations arose in what are known as **culture hearths**, early centers of civilization whose ideas and practices spread to surrounding areas. As you can see from the map above, the most influential culture hearths developed in areas that now make up the modern countries of **Egypt**, **Iraq**, **Pakistan**, **China**, and **Mexico**. In Mexico the Olmec culture

> *". . . flourished along Mexico's Gulf Coast between 1200 and 400 B.C. . . . Because of early achievements in art, politics, religion, and economics, the Olmec stand for many as a kind of 'mother culture' to all the civilizations that came after, including the Maya and the Aztec."*
>
> George E. Stuart, "New Light on the Olmec," *National Geographic*, November 1993

These five culture hearths had certain geographic features in common. They all emerged from farming settlements in areas with a mild climate and fertile land and were located near a major river or source of water. The peoples of the culture hearths made use of these favorable environments. They dug canals and ditches in order to use the rivers to irrigate the land. All of these factors enabled people to grow surplus crops.

Economics

Specialization and Civilization

Surplus food set the stage for the rise of cities and civilizations. With more food available, there was less need for everyone in a settlement to farm the land. People were able to develop other ways of making a living. They created new technology and carried out specialized economic activities, such as metalworking and shipbuilding, that spurred the development of long-distance trade.

In turn, the increased wealth from trade led to the rise of cities and complex social systems. The ruler of a city needed a well-organized government to coordinate harvests, plan building projects, and manage an army for defense. Perhaps most importantly, officials and merchants created writing systems that made it possible to record and transmit information.

Cultural Contacts

Cultural contact among different civilizations promoted cultural change as ideas and practices spread through trade and travel. Permanent migration, in which people leave one land to seek a new life in another, also has fostered cultural diffusion. People migrate to avoid harsh governments, wars, persecution, and famines. In some cases, such as that of enslaved Africans brought to the Americas, mass migrations have been forced. Conversely, positive factors—a favorable climate, better economic opportunities, and religious or political freedoms—may draw people from one place to another. Migrants carry their cultures with them, and their ideas and practices often blend with those of the people already living in the migrants' adopted countries.

Industrial and Information Revolutions

Cultural diffusion has increased rapidly during the last 250 years. In the 1700s and 1800s, some countries began to industrialize, using power-driven machines and factories to mass-produce goods. New production methods dramatically changed these countries' economies, since goods could be produced quickly and cheaply. This development, known as the Industrial Revolution, also led to social changes. As people left farms for jobs in factories and mills, cities grew larger. Harsh working and living conditions at the outset of the Industrial Revolution eventually improved.

At the end of the 1900s, the world experienced a new turning point—the Information Revolution. Computers now make it possible to store huge amounts of information and to send information all over the world in an instant, thus linking the cultures of the world more closely than ever before.

SECTION 2 ASSESSMENT

Checking for Understanding

1. **Define** culture, language family, ethnic group, culture region, cultural diffusion, culture hearth.
2. **Main Ideas** Fill in the main features of global cultures on a web diagram like the one below.

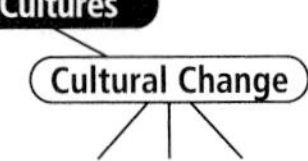

Critical Thinking

3. **Making Generalizations** Explain the factors that influence a country's power to control territory.
4. **Identifying Cause and Effect** What cultural changes have resulted from the Information Revolution?
5. **Analyzing Information** How do factors, such as trade, war, migration, and inventions, affect cultural change?

Analyzing Maps

6. **Place** Study the map of world religions on page 82. What factors are related to the diffusion of world religions?

Applying Geography

7. **Culture and Environment** Research the Internet to make a list of examples in which varying cultures view particular places or features differently.

SECTION 3

Political and Economic Systems

Guide to Reading

Consider What You Know

Political and economic systems help define a people's culture, or way of life. Think about the political and economic systems in your own region. How do they impact your culture?

Read to Find Out

- What are the various levels of government?
- What are the major types of governments in the world today?
- What are the major types of economic systems in the world?

Terms to Know

- unitary system
- federal system
- autocracy
- oligarchy
- democracy
- traditional economy
- market economy
- mixed economy
- command economy

Places to Locate

- United States
- Saudi Arabia
- United Kingdom
- China
- Vietnam

NATIONAL GEOGRAPHIC

A Geographic View

Global Connections

Geographic location of resources, labor, and capital means less as scattered countries use information technologies to work together. Many cars have parts made in a half dozen countries; stores sell look-alike clothes sewn on four continents. . . . Money moves most easily. Stocks, currency, and bonds traded on worldwide electronic markets amount to an estimated three trillion dollars each day, twice the annual U.S. budget.

—*Joel L. Swerdlow, "Information Revolution,"* National Geographic, *October 1995*

Mercantile Exchange, Chicago

As information technology continues to link the world's cultures, the governments and economies of countries around the globe become increasingly interconnected. Government is the institution through which a society maintains social order, provides public services, ensures national security, and supports its economic well-being. An economy is the way a society produces, distributes, and uses goods and services. In this section you will study the major political and economic systems found in the world today.

Features of Government

Today the world is made up of nearly 200 independent countries that vary in size, military might, natural resources, and world influence. Each country is defined by characteristics such as its territory, its population, and its sovereignty, or freedom from outside control. All of these elements are brought together under a government. In carrying out its tasks, a government must make and enforce policies and laws that are binding on all people living within its territory.

Levels of Government

The government of each country has unique characteristics that relate to that country's historical development. To carry out their functions, governments have been organized in a variety of ways. Most large countries have several different levels of government. These usually include a national or central government, as well as the governments of smaller internal divisions such as provinces, states, counties, cities, towns, and villages.

Unitary System

A unitary system of government gives all key powers to the national or central government. This structure does not mean that only one level of government exists. Rather, it means that the central government creates state, provincial, or other local governments and gives them limited sovereignty. The United Kingdom and France both developed unitary governments as they gradually emerged from smaller territories during the late Middle Ages and early modern times.

Federal System

A federal system of government divides the powers of government between the national government and state or provincial governments. Each level of government has sovereignty in some areas. The **United States** developed a federal system after the thirteen colonies became independent.

Another similar type of government structure is a confederation, a loose union of independent territories. The United States at first formed a confederation, but this type of political arrangement failed to provide an effective national government. As a result, the U.S. Constitution made the national government supreme, while preserving some state government powers. Today other countries with federal or confederal systems include Canada, Switzerland, Mexico, Brazil, Australia, and India.

Types of Governments

The governments of the world's countries also differ in the way they exercise authority. Governments can be classified by asking the question: "Who governs the state?" Under this classification system, all governments belong to one of the three major groups: (1) autocracy—rule by one person; (2) oligarchy—rule by a few people; or (3) democracy—rule by many people.

Autocracy

Any system of government in which the power and authority to rule belong to a single individual is an autocracy (aw•TAH•kruh•see). Autocracies are the oldest and one of the most common forms of government. Most autocrats achieve and maintain their position of authority through inheritance or by the ruthless use of military or police power.

Several forms of autocracy exist. One is an absolute or totalitarian dictatorship. In a totalitarian dictatorship, the decisions of a single leader determine government policies. The government under such a system can come to power through revolution or an election. The totalitarian dictator seeks to control all aspects of social and economic life. Examples of totalitarian dictatorships include Adolf Hitler's government in Nazi Germany (from 1933 to 1945), Benito Mussolini's rule in Italy (from 1922 to 1943),

NATIONAL GEOGRAPHIC **World Explorer**

Geography Skills for Life

Ultimate Authority Though an absolute monarch, King Fahd of Saudi Arabia is assisted by a cabinet, or group of advisers.

Place Describe the powers of an absolute monarch.

and Joseph Stalin's regime in the Soviet Union (from 1924 to 1953). In such dictatorships, the government is not responsible to the people, and the people have no power to limit their rulers' actions.

Monarchy (MAH•nuhr•kee) is another form of autocratic government. In a monarchy, a king or queen exercises the supreme powers of government. Monarchs usually inherit their positions. Absolute monarchs have complete and unlimited power to rule their people. The king of **Saudi Arabia**, for example, is an absolute monarch. Absolute monarchs are rare today, but from the 1400s to the 1700s kings or queens with absolute powers ruled most of Western Europe.

Today some countries, such as the United Kingdom, Sweden, Japan, Jordan, and Thailand, have constitutional monarchies. Their monarchs share governmental powers with elected legislatures or serve as ceremonial leaders.

Oligarchy

An **oligarchy** (AH•luh•GAHR•kee) is any system of government in which a small group holds power. The group derives its power from wealth, military power, social position, or a combination of these elements. Sometimes religion is the source of power. Today the governments of communist countries, such as China, are mostly oligarchies. In such countries, leaders in the Communist Party and the armed forces control the government.

Both dictatorships and oligarchies sometimes claim they rule for the people. Such governments may try to give the appearance of control by the people. For example, they might hold elections but offer only one candidate or control the election results in other ways. Such governments may also have some type of legislature or national assembly elected by or representing the people. These legislatures, however, only approve policies and decisions already made by the leaders. As in a dictatorship, oligarchies usually suppress all political opposition—sometimes ruthlessly.

Democracy

A **democracy** is any system of government in which leaders rule with the consent of the citizens. The term *democracy* comes from the Greek *demos* (meaning "the people") and *kratia* (meaning "rule"). The ancient Greeks used the word *democracy* to mean government by the many in contrast to government by the few. The key idea of democracy is that people hold sovereign power.

Direct democracy, in which citizens themselves decide on issues, exists in some places at local levels of government. No country today has a national government based on direct democracy. Instead, democratic countries have representative democracies, in which the people elect representatives with the responsibility and power to make laws and conduct government. An assembly of the people's representatives may be called a council, a legislature, a congress, or a parliament.

Many democratic countries, such as the United States and France, are republics. In a republic, voters elect all major officials, who are responsible to the people. The head of state—or head of government—is usually a president elected for a specific term. Not every democracy is a republic. The **United**

Geography Skills for Life

The Right to Vote In the United States, voters such as this man in Mississippi elect their officials.
Place How do direct democracies and representative democracies differ?

Kingdom, for example, is a democracy with a monarch as head of state. This monarch's role is ceremonial, and elected officials hold the power to rule.

Economic Systems

Governments around the world deal with many kinds of economic systems. All economic systems, however, must make three basic economic decisions: (1) what and how many goods and services should be produced, (2) how they should be produced, (3) who gets the goods and services that are produced. The three major types of these economic systems—traditional, market, and command—make decisions differently.

Geography Skills for Life **California Entrepreneur**

The owner of a bicycle shop uses a telephone and laptop computer to conduct business.

Region How does a market economy affect the economic activities in a region?

Traditional Economy

In a traditional economy, habit and custom determine the rules for all economic activity. Individuals are not free to make decisions based on what they would like to have. Instead their behavior is defined by the customs of their elders and ancestors. For example, it was a tradition in the Inuit society of northern Canada that a successful hunter would share the spoils of the hunt with the other families in the village. This custom allowed the Inuit to survive the Arctic climate for thousands of years. Today, traditional economic systems exist in very limited parts of the world.

Market Economy

In a market economy, individuals and private groups make decisions about what to produce. People, as shoppers, choose what products they will or will not buy, and businesses make more of what they believe consumers want. A market economy is based on *free enterprise*, the idea that private individuals or groups have the right to own property or businesses and make a profit with only limited government interference. In a free enterprise system, people are free to choose what jobs they will do and for whom they will work. Another term for an economic system organized in this way is *capitalism*.

No country in the world, however, has a pure market economy system. Today the U.S. economy and others like it are described as mixed economies. A mixed economy is one in which the government supports and regulates free enterprise through decisions that affect the marketplace. In this arrangement the government's main economic task is to preserve the free market by keeping competition free and fair and by supporting the public interest. Governments in modern mixed economies also influence their economies by spending tax revenues to support social services such as health care, education, and housing.

Command Economy

In a command economy, the government owns or directs the means of production—land, labor, capital (machinery, factories), and business managers—and controls the distribution of goods. Believing that such economic decision making benefits all of society and not just a few people, countries with command economies try to distribute goods and services equally among all citizens. Public taxes, for example, are used to support social services, such as housing and health care, for all citizens. However, citizens have no voice in how this tax money is spent.

Government

Socialism and Communism

A command economy is called either socialism or communism, depending on how much the government is involved. In theory, communism requires strict government control of almost the entire society, including its economy. The government decides how much to produce, what to produce, and how to distribute the goods and services produced. One political party—the Communist Party—makes decisions and may even use various forms of coercion to ensure that the decisions are carried out at lower political and economic levels.

Supporters of the market system claim, however, that without free decision making and incentives, businesses will not innovate or produce products that people want. Customers will be limited in their choices and economies will stagnate. As a result of these problems, command economies often decline. An example is the former Soviet Union, as described by a Russian observer.

> *"In 1961 the [Communist] party predicted . . . that the Soviet Union would have the world's highest living standard by 1980. . . . But when that year came and went, the Soviet Union still limped along, burdened by . . . a stagnant economy."*
>
> Dusko Doder, "The Bolshevik Revolution," *National Geographic*, October 1992

By 2000, Russia and the other countries that were once part of the Soviet Union were developing market economies. **China** and **Vietnam** have allowed some free enterprise to promote economic growth, although their governments tightly control political affairs.

An economic system called socialism allows an even wider range of free enterprise alongside government-run activities. Socialism has three main goals: (1) the equal distribution of wealth and economic opportunity; (2) society's control, through its government, of all major decisions about production; and (3) public ownership of most land, factories, and other means of production. Politically, some socialist countries, especially those in western Europe, are democracies. Under democratic socialism, people have basic human rights and elect their political leaders, even though the government controls certain industries.

SECTION 3 ASSESSMENT

Checking for Understanding

1. **Define** unitary system, federal system, autocracy, oligarchy, democracy, traditional economy, market economy, mixed economy, command economy.
2. **Main Ideas** Copy the outline below, and complete it with information from the section.

Political and Economic Systems
I. Levels of Government A. Unitary System 1.

Critical Thinking

3. **Comparing and Contrasting** What different roles might local citizens have in government decision making under a unitary system, a federal system, and a confederation?
4. **Making Generalizations** What functions do all types of governments carry out?
5. **Categorizing Information** Describe the characteristics of traditional, command, and market economies.

Analyzing Maps

6. **Region** Study the map of world religions on page 82. Then write two generalizations about the distribution of the world's religions.

Applying Geography

7. **Political Systems** Research political systems and geography. What geographic factors influence a country's foreign policy? Use Iraq, Israel, Japan, and the United Kingdom as examples.

Resources, Trade, and the Environment

Guide to Reading

Consider What You Know

People are dependent on the world's resources for survival. Yet you may have heard that certain economic activities threaten humans' future access to these resources. What are some actions people can take to preserve the world's natural treasures?

Read to Find Out

- What types of energy most likely will be used in societies in the future?
- What factors determine a country's economic development and trade relationships?
- How do human economic activities affect the environment?

Terms to Know

- natural resource
- developed country
- developing country
- industrialization
- free trade
- pollution

Places to Locate

- Malaysia
- European Union

NATIONAL GEOGRAPHIC

A Geographic View

Globalization in High Gear

Humans have been weaving commercial and cultural connections since before the first camel caravan ventured afield. In the 19th century the postal service, newspapers, transcontinental railroads, and great steam-powered ships [brought about] fundamental changes. . . . Now computers, the Internet, cellular phones, cable TV, and cheaper jet transportation have accelerated and complicated these connections. Still, the basic dynamic remains the same: Goods move. People move. Ideas move. And cultures change. The difference now is the speed and scope of these changes.

—*Erla Zwingle, "A World Together,"* National Geographic, *August 1999*

Buddhist monks in California restaurant

At the start of the twenty-first century, technological advances such as the Internet were connecting people around the globe. These connections continue to make the world's peoples increasingly interdependent, or reliant on each other. In this section you will learn about the growth of a global economy and the ways in which the world's peoples use—and misuse—natural resources.

Resources

Earth provides all the elements necessary to sustain life. The elements from the earth that are not made by people but can be used by them for food, fuel, or other necessities are called **natural resources**. People can use some natural resources as much as they want. These

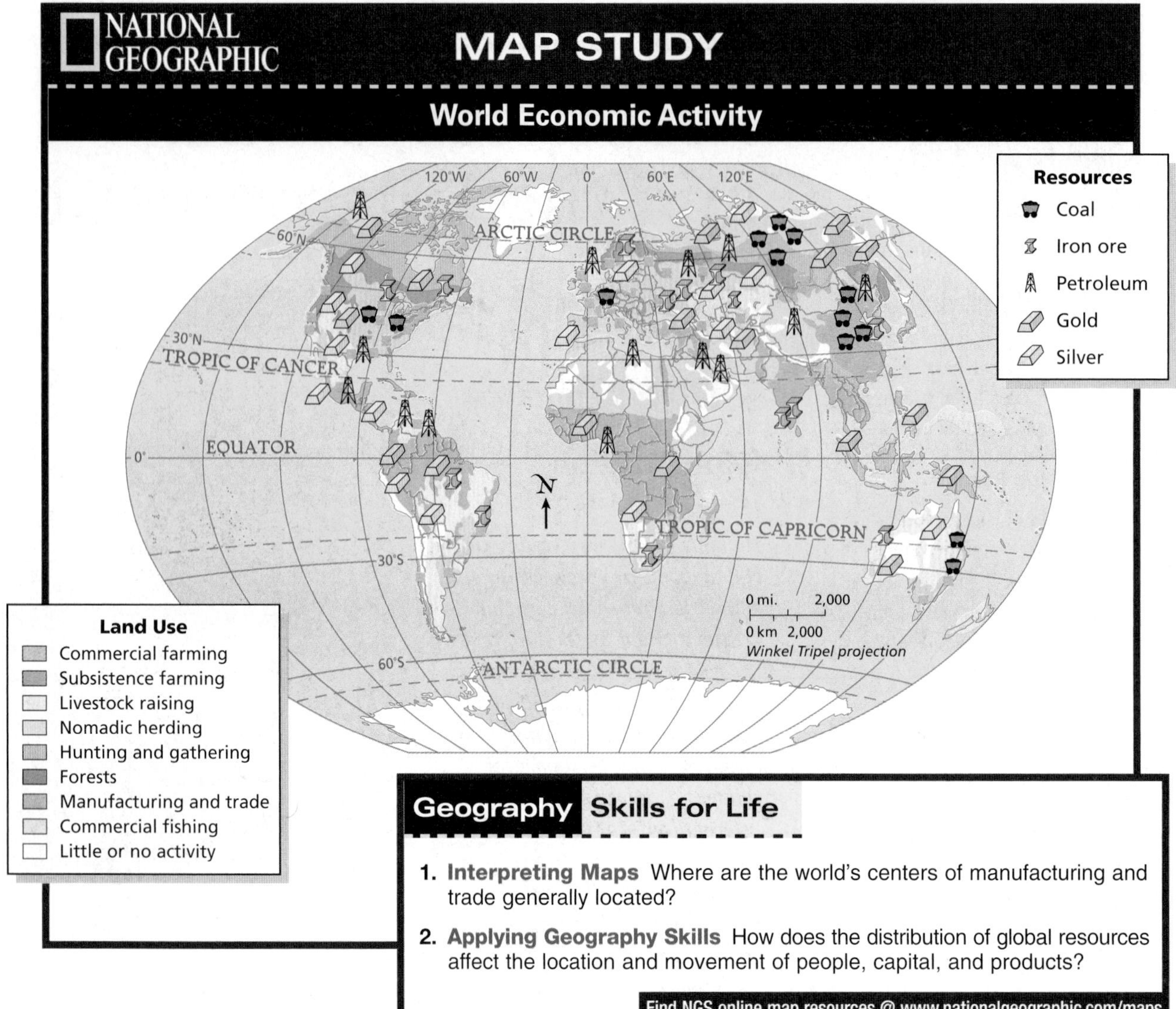

renewable resources cannot be used up or can be replaced naturally or grown again in a relatively short amount of time. Wind, sun, water, forests, and animal life are examples of renewable resources. The earth's crust, however, contains many *nonrenewable resources* that cannot be replaced, such as minerals and fossil fuels.

Resource Management

Because fossil fuels, like coal and oil, and other nonrenewable resources cannot be replaced, they must be conserved. The immediate goal of conservation is to manage vital resources carefully so that people's present needs are met. An equally important long-term goal is to ensure that the needs of future generations are met.

With these future needs in mind, environmental experts have encouraged people to replace their dependence on fossil fuels with the use of renewable energy sources. Many countries, for example, already produce hydroelectric power—a renewable energy source generated from falling water. Another renewable energy source is solar energy—power produced by the sun's heat. Unfortunately, harnessing solar energy requires large, expensive equipment, so it is not yet an economical alternative to other energy sources.

Still another renewable source is electricity created by nuclear energy, the power made by creating a controlled atomic reaction. Many concerns, however, surround the use of nuclear power because of the dangerous waste products it produces.

Economic Development

Most natural resources are not evenly distributed throughout the earth. This uneven distribution affects the global economy, as you see from the economic activities map on page 92. As a result, countries specialize in the economic activities best suited to their resources. Those having much technology and manufacturing, such as the United States, are called **developed countries**. There, most people work in manufacturing or service industries and enjoy a high standard of living. Farmers in developed countries engage in commercial farming, raising crops and livestock to sell in the market. Because of modern techniques, only a small percentage of these countries' workers is needed to grow food to feed entire populations. Those countries working toward greater manufacturing and technology use are called **developing countries**. In many developing countries, which are mainly in Africa, Asia, and Latin America, agriculture remains dominant. Despite much commercial farming, most farmers in these countries engage in subsistence farming, growing only enough food for family needs. As a result, most people in developing countries remain poor. **Industrialization**, or the spread of industry, however, has transformed once largely agricultural countries, such as China and **Malaysia**.

Despite advances, the global influence of developed countries has sparked resentment in some developing countries. Feeding on this discontent, militant groups have tried to strike back by engaging in terrorism, or the use of violence to create fear in a given population. Small in size and often limited in resources, these groups seek to use the fear unleashed by violence to heighten their influence to promote change.

Geography Skills for Life

Importing Goods

Many multinational companies import clothes for sale in the United States.

Place Why do some U.S. firms locate factories in developing countries?

World Trade

The unequal distribution of natural resources promotes a complex network of trade among countries. Countries export their specialized products, trading them to other countries that cannot produce those goods. When countries cannot produce as much as they need of a good, they import it, or buy it from another country. That country, in turn, may buy the first country's products, making the two countries trading partners.

A major stimulus to world trade has come from multinational companies. A multinational company is a firm that does business in many places throughout the world. Multinationals are usually headquartered in a developed country and often locate their manufacturing or assembly operations in developing countries with low labor costs. In recent decades many developing countries have allowed multinationals to buy property and build factories or form partnerships with local companies.

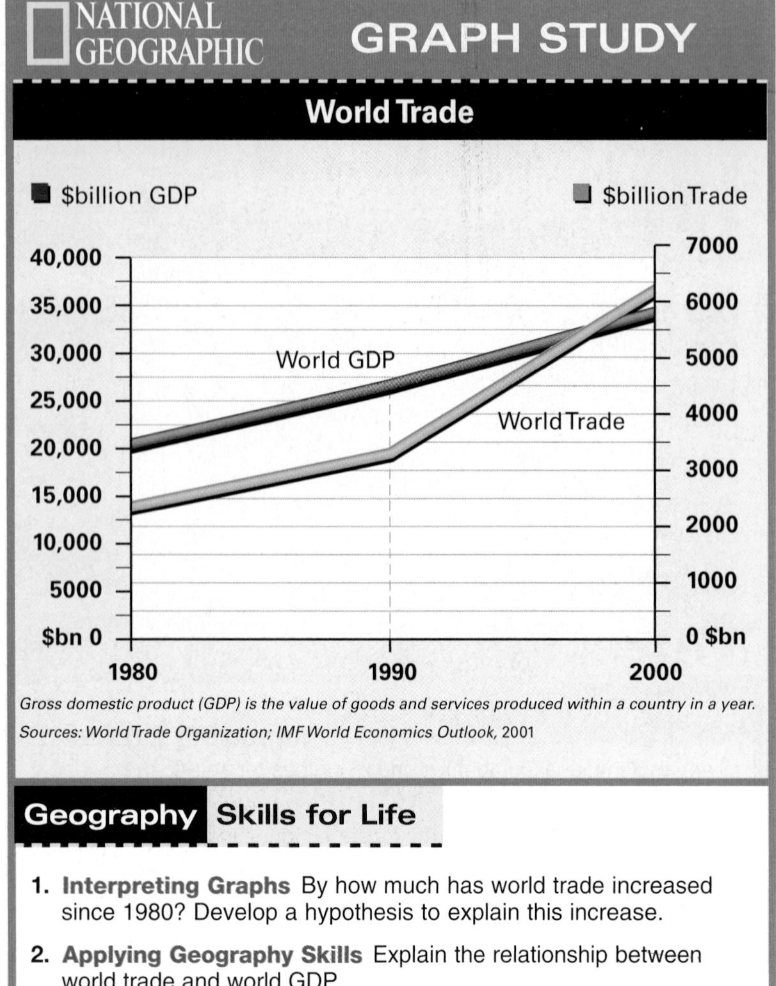

Gross domestic product (GDP) is the value of goods and services produced within a country in a year.

Sources: World Trade Organization; IMF World Economics Outlook, 2001

Geography Skills for Life

1. **Interpreting Graphs** By how much has world trade increased since 1980? Develop a hypothesis to explain this increase.
2. **Applying Geography Skills** Explain the relationship between world trade and world GDP.

Free Trade

In recent years governments around the world have moved toward **free trade**, the removal of trade barriers so that goods can flow freely among countries. The General Agreement on Tariffs and Trade (GATT) was the first international agreement to promote free trade. In 1995 GATT became the World Trade Organization (WTO), to which most countries now belong.

In various parts of the world, several countries have joined together to create regional free trade agreements. For example, the United States, Mexico, and Canada have set up the North American Free Trade Agreement (NAFTA) to eliminate all trade barriers to one another's goods. The **European Union** (EU), the largest trading bloc, includes many of the countries of Europe. Many members of the European Union have adopted a regional currency, the euro, to extend their cooperative efforts. Referring to decreasing trade restrictions, a U.S. trade official observed that

> *"... an opening world economy has allowed trade to expand fifteen-fold, sparking a six-fold increase in world economic production and a three-fold increase in global per capita incomes."*
>
> Charlene Barshefsky, *remarks on trade policy at the National Press Club,* Washington, D.C., October 19, 2000

Economics

Barriers to Trade

A government tries to manage its country's trade to benefit its own economy. Some governments add a tariff, or a tax, to the price of goods that are imported. Because tariffs make imported goods more costly, governments often use them to influence people to buy products made in their home country instead of imported goods.

Governments sometimes create other barriers to trade. They might put a strict quota, or number limit, on the quantity of a particular product that can be imported from a particular country. A government may even impose an embargo, banning trade with another country altogether as a way to punish that country for political or economic differences.

People and the Environment

In recent decades human economic activities have drastically affected the environment. A major environmental challenge today is **pollution**—the release of unclean or impure elements into the air, water, and land.

Water and Land Pollution

Earth's bodies of water are normally renewable, purifying themselves over time, but this natural cycle can be interrupted by human activity. Tankers and offshore rigs can cause oil spills and industries may dump chemical waste that enters and pollutes the water supply. Fertilizers and pesticides from farms can seep into groundwater and cause harm, as can animal waste and untreated sewage.

Land pollution occurs where chemical waste poisons fertile topsoil or solid waste is dumped in landfills. Radioactive waste from nuclear power plants and toxic runoff from chemical processing plants can also leak into the soil and cause contamination.

Air Pollution

The main source of air pollution is the burning of fossil fuels by industries and vehicles. Burning fuel gives off poisonous gases that can seriously damage people's health. Acidic chemicals in air pollution also combine with precipitation to form acid rain. Acid rain eats away at the surfaces of buildings, kills fish, and can even destroy entire forests.

Forests provide animal habitats, prevent soil erosion, and conduct photosynthesis (FOH•toh•SIHN•thuh•suhs)—the process by which plants take in carbon dioxide and, in the presence of sunlight, produce carbohydrates. The oxygen released during photosynthesis is vital for human and animal survival. Decreasing pollution-causing acid rain will preserve a region's environmental balance.

Some scientists believe that rising levels of pollutants in the atmosphere are contributing to a general increase in the earth's temperatures, a trend they call global warming. Although not all experts agree that global warming is occurring, scientists who study it warn that the increase in temperature may have disastrous effects, causing glaciers and ice caps to melt and raising the level of the world's oceans. Higher water levels in oceans, they claim, could flood coastal cities and submerge smaller islands.

The Fragile Ecosystem

As humans expand their communities, they threaten natural ecosystems, places where the plants and animals are dependent upon one another and their environment for survival. Ecosystems can be found in every climate and vegetation region of the world. Because the earth's land, air, and water are interrelated, what harms one part of the system harms all the other parts—including humans and other living things. As people become more aware of how their actions affect this delicate balance of life, they are starting to manage resources more wisely, by improving water treatment, preserving wilderness areas, and developing alternatives to fossil fuels.

SECTION  ASSESSMENT

Checking for Understanding

1. **Define** natural resource, developed country, developing country, industrialization, free trade, pollution.
2. **Main Ideas** Copy the chart below onto a sheet of paper. Fill in the challenges that each category presents to the world's countries.

Economic World			
Resources	Development	Trade	Environment

Critical Thinking

3. **Evaluating Information** Evaluate the impact of innovations, such as fire, steam power, diesel machinery, and electricity, on the environment.
4. **Making Inferences** What might be the advantages and disadvantages to a developing country of joining a free trade agreement?
5. **Making Comparisons** Research and compare two countries in the ways they depend on the environment for products that they export.

Analyzing Maps

6. **Region** Look at the world economic activity map on page 92. What regions of the world produce the most oil? The most coal?

Applying Geography

7. **Effects of Trade Policies** Think about the reasons countries use quotas and embargoes. How are quotas and embargoes different? What unintended consequence do you think quotas and embargoes might have on a country's economy?

Creating an Electronic Database

A computerized database program can help you organize and manage a large amount of information. Once you enter data in a database, you can quickly locate a record according to key information.

Learning the Skill

An electronic database is a collection of facts that are stored in a file on the computer. The information is organized into different fields. The table, for example, contains three fields: *Language, Speakers (in millions),* and *Main Areas Where Spoken.*

A database can be organized and reorganized in any way that is useful to you. By using special software developed for record keeping—a database management system (DBMS)—you can easily add, delete, change, or update information. You give commands to the computer that tell it what to do with the information, and it follows your commands. When you want to retrieve information, the computer searches through the files, finds the information, and displays it on the screen.

Follow these steps to create a database:

- **Determine what facts you want to include in your database.**
- **Follow the instructions in the DBMS you are using to set up fields.**
- **Determine how you want to organize the facts in the database—alphabetically, chronologically, or numerically.**

Language	Speakers (in millions)	Main Areas Where Spoken
Han Chinese (Mandarin)	874	China, Taiwan, Singapore
Hindi	366	Northern India
Spanish	358	Spain, Latin America, southwestern United States
English	341	British Isles, Anglo-America, Australia, New Zealand, South Africa, former British colonies in tropical Asia and Africa, Philippines
Bengali	207	Bangladesh, eastern India
Arabic	206	Southwest Asia, North Africa
Portuguese	176	Portugal, Brazil, southern Africa
Russian	167	Russia, Kazakhstan, parts of Ukraine and other former Soviet republics
Japanese	125	Japan
German	100	Germany, Austria, Switzerland, Luxembourg, eastern France, northern Italy

Source: National Geographic Desk Reference

- **Follow the instructions in the DBMS to sort the information in order of importance.**

Practicing the Skill

Enter the data in the table above into an electronic database. Then use the DBMS commands to answer the following questions.

1. In what order is the information in the table displayed?
2. Sort the data alphabetically by language. Which record appears first?
3. Request the database to display only those languages with more than 200 million speakers. Which records will *not* appear?
4. Sort your records using Africa as the main area where spoken. How many languages appear? What are they?

Study the world religion and world cultures maps on pages 82 and 83. Combine the information into an electronic database showing which religions are practiced in the world's culture regions. Write three questions that require sorting these records.

SECTION 1 World Population (pp. 75–79)

Terms to Know
- death rate
- birthrate
- natural increase
- doubling time
- population distribution
- population density
- migration

Key Points
- Population growth rates vary, posing different problems for different countries.
- The world's population is unevenly distributed.
- Large numbers of people are migrating from rural villages to cities.
- People emigrate because of wars, food shortages, persecution, lack of jobs, or other problems.

Organizing Your Notes
Use a graphic organizer like the one below to help you organize your notes for this section.

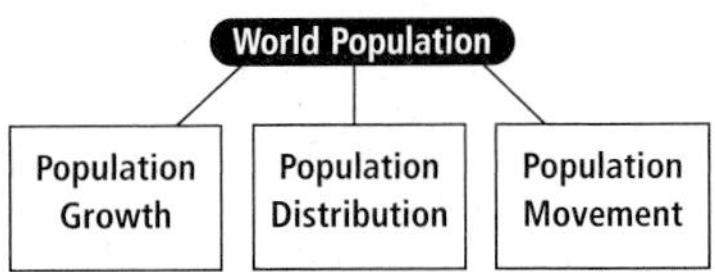

SECTION 2 Global Cultures (pp. 80–85)

Terms to Know
- culture
- language family
- ethnic group
- culture region
- cultural diffusion
- culture hearth

Key Points
- Language, religion, social groups, government, and economic activities define cultures.
- Geographers divide the earth into specific culture regions.
- Trade, migration, and war change cultures.
- The world's first civilizations arose in culture hearths.

Organizing Your Notes
Create an outline using the format below to organize your notes.

Global Cultures

I. Elements of Culture
   A. Language
      1.

SECTION 3 Political and Economic Systems (pp. 86–90)

Terms to Know
- unitary system
- federal system
- autocracy
- oligarchy
- democracy
- traditional economy
- market economy
- mixed economy
- command economy

Key Points
- A country's different levels of government may be organized as a unitary system, a federal system, or a confederation.
- An autocracy, an oligarchy, and a democracy differ in the way they exercise authority.
- The three major economic systems are traditional economy, market economy and command economy.

Organizing Your Notes
Create a chart like the one below to organize your notes for this section.

Political and Economic Systems	Characteristics
Features of Government	

SECTION 4 Resources, Trade, and the Environment (pp. 91–95)

Terms to Know
- natural resource
- developed country
- developing country
- industrialization
- free trade
- pollution

Key Points
- Peoples are increasingly interdependent.
- Because natural resources are not evenly distributed, countries must trade.
- Governments can create or eliminate trade barriers.
- Human economic activities have led to pollution.

Organizing Your Notes
Create a chart like the one below to organize your notes.

Trade	Resources	Development	Environment

ASSESSMENT & ACTIVITIES

Reviewing Key Terms

On a sheet of paper, classify each of the terms below into one of the following categories:

- **Population**
- **Political and Economic Systems**
- **Cultures**
- **Resources, Trade, and Environment**

a. death rate
b. culture
c. population distribution
d. free trade
e. language family
f. culture region
g. democracy
h. culture hearth
i. developed country
j. autocracy
k. cultural diffusion
l. market economy
m. mixed economy
n. federal system
o. developing country
p. birthrate

Reviewing Facts

SECTION 1

1. How does population growth affect the global community?

2. Why are large numbers of people moving to cities?

SECTION 2

3. What are the elements of a culture?

4. What influences may change a culture?

SECTION 3

5. Name two forms of autocratic government.

6. What kinds of benefits do people receive in a market economy system? In a command economy system?

SECTION 4

7. What is the difference between a renewable and a nonrenewable resource?

8. What factors make the world's countries increasingly interdependent?

Critical Thinking

1. Analyzing Information Explain the operation of a traditional economy, using Canada's Inuit as an example.

2. Making Inferences Why do you think geographers find it useful to divide the world into culture regions? Identify the human factors that constitute a region.

3. Predicting Consequences On a sheet of paper, create a graphic organizer like the one below to list the possible challenges faced by the citizens of a country whose government has changed from an autocracy to a democracy. Then suggest ways that people might address these challenges.

Autocracy to Democracy	
Challenges	Ways to Address Them
•	•
•	•
•	•
•	•

NATIONAL GEOGRAPHIC — Locating Places

The World: Cultural Geography

Match the letters on the map with the appropriate world culture regions. Write your answers on a sheet of paper.

1. **United States and Canada**
2. **Latin America**
3. **Europe**
4. **Russia**
5. **North Africa, Southwest Asia, and Central Asia**
6. **Africa South of the Sahara**
7. **South Asia**
8. **East Asia**
9. **Southeast Asia**
10. **Australia, Oceania, and Antarctica**

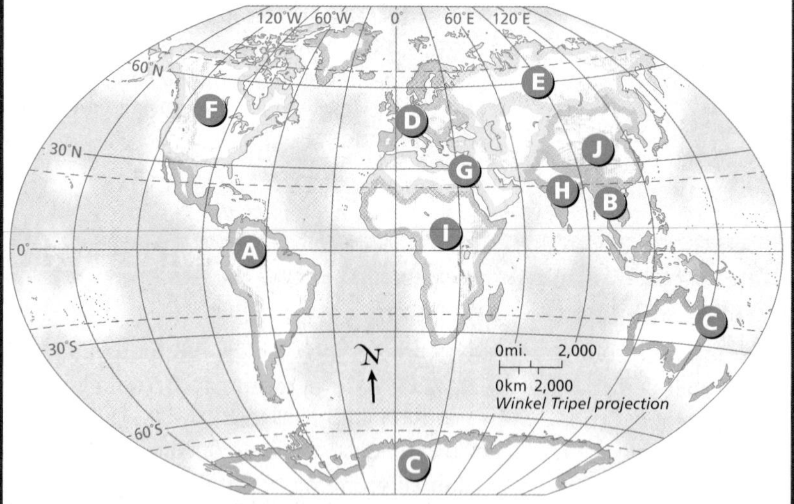

Self-Check Quiz Visit the **Glencoe World Geography** Web site at geography.glencoe.com and click on Self-Check Quizzes—Chapter 4 to prepare for the Chapter Test.

Thinking Like a Geographer

Based on what you know about cultural diffusion, research the role that diseases such as the bubonic plague have played in this process over the course of time. Create a map that traces the disease's spread from its point of origin to other areas. Write a paragraph that describes the disease's effects on regions of contact.

Problem-Solving Activity

Group Research Project Work with a group to find out more about the North American Free Trade Agreement (NAFTA). Research the important elements of the agreement, the supporting and opposing opinions, and the costs and benefits to participating countries. Decide whether NAFTA might be used as a model agreement for other regions, and present your decision and supporting reasons as a letter to a national news magazine. Include charts, graphs, or tables to support your ideas.

GeoJournal

Creative Writing Choose one of the world's culture regions. Use the notes in your GeoJournal and other sources to research and analyze the effects of human geographic patterns on the region's environment. Write a description of specific human activities that have positively or negatively changed physical features and natural resouces there.

Technology Activity

Creating an Electronic Database Choose several developed and developing countries and create a database of their trading activities. Include data about products they import and export and the amount of income each country earns from trade. Then write a paragraph explaining what the data show about developed and developing countries. Consider the differences among countries related to the kinds of products each category of country produces and the amount of income each kind produces. What accounts for the differences?

Standardized Test Practice

Choose the best answer for the following multiple-choice questions. If you have trouble answering the questions, use the process of elimination to narrow your choices.

1. Which of the following is NOT a challenge that rapid population growth presents to the global community?

A Shortages of food
B Shortages of metropolitan areas
C Shortages of water
D Shortages of housing

2. What is the most accurate description of an autocratic government?

F Power is divided among the national government and state or provincial governments.
G A small group of people have the power to govern, often because of wealth, military power, or social position.
H Leaders rule with the consent of the citizens.
J One person holds the power to rule and may use military or police power to maintain authority.

Read the questions carefully to determine what is being asked. Look for the key words and phrases that will help you eliminate incorrect answers, such as the word *not* in question 1 and the word *autocratic* in question 2. Then compare the answer choices you have not eliminated. Sometimes more than one answer choice seems correct. You must find the answer choice that is the *best.*

UNIT 2

The United States and Canada

WHY IT'S IMPORTANT—

The United States and Canada are peaceful neighbors, sharing the longest undefended border in the world. These two countries have many things in common, including similar ways of life and a democratic heritage. In recent years, free trade has brought their economies closer together. In each country, one finds an increasing number of products that were made in the other country.

World Regions Video

To learn more about the United States and Canada and their impact on your world, view the World Regions video "The United States and Canada."

Golden Gate Bridge across the entrance to San Francisco Bay

2

What Makes the United States and Canada a Region?

The United States and Canada span most of North America, stretching from the Pacific Ocean to the Atlantic. These two huge countries share many physical features. Mountains frame their eastern and western edges, cradling a central region of vast plains.

When people first arrived on these plains, they found an immense sea—not of water, but of grass. Beneath the gently rolling landscape lay dark, fertile soil. In time, the grasslands were transformed into some of the world's most productive farmland.

To the east of the plains stand the ancient, rounded Appalachian Mountains. To the west are the much younger Rocky Mountains, a majestic ribbon of jagged, snowcapped peaks. Still farther west are the Pacific Ranges, which run along the Pacific coast.

Almost every imaginable type of climate—from tundra to desert to tropical savanna—can be found within the borders of these two diverse countries.

3

1 **Six-foot-tall sunflowers** thrive on this farm in North Dakota, in the heart of the Great Plains. North Dakota leads the United States in the production of sunflowers. The protein-rich seeds are turned into margarine and cooking oil.

2 Boats line the harbor of a fishing village in Nova Scotia, along Canada's Atlantic coast. Both Canadians and Americans harvest fish and other types of seafood from the Atlantic's bountiful waters.

3 Snow-dusted peaks surround a climber in the Canadian Rockies. The backbone of North America, the Rockies extend from the farthest reaches of Alaska and the Yukon Territory down into the southwestern United States.

4 The only deserts in the region are found in the southwestern United States, in an area of low basins and high, windswept plateaus sandwiched between the Pacific Ranges and the Rocky Mountains. These rippled dunes lie in Utah.

2

Region of Immigrants

Even the ancestors of Native Americans came from a distant shore. These ancient people may have crossed from Asia to North America by way of a land bridge that spanned what is now the Bering Strait.

Immigrants began arriving from Europe in the 1500s. In the centuries that followed, others came from Africa, Asia, and Latin America. Many made this land their home by choice. Others were forced to come as exiles or slaves.

Today, most people in the United States and Canada live in urban areas. Major cities are ethnically diverse, reflecting an immigrant heritage. The economic strength of both countries was built on the bounty of agriculture. Manufacturing, technology, and service industries have joined agriculture as the region's primary economic activities.

1 **The white walls** of a Spanish mission, or religious settlement, stand out against a blue New Mexico sky. Hoping to convert the area's native inhabitants to Christianity, the Spanish built many missions in what is now the southwestern United States.

2 **Red lights and blues music** illuminate a musician's face. The blues, a distinctively American musical style, was developed by African Americans. It sprang from spiritual music and from the wails and calls used by Southern plantation workers.

3 Lights of Toronto, Ontario, stretch toward the horizon, brightening the night sky. With 4.7 million inhabitants in its metropolitan area, Toronto is Canada's largest city. It is a thriving center for service industries such as finance and communications.

4 Freshly caught fish chill in a snow bank outside an Inuit village in Canada's Northwest Territories. The Inuit have lived in the northern parts of Canada and Alaska for about a thousand years. Many still pursue traditional activities such as fishing and hunting.

2 The United States and Canada

PHYSICAL

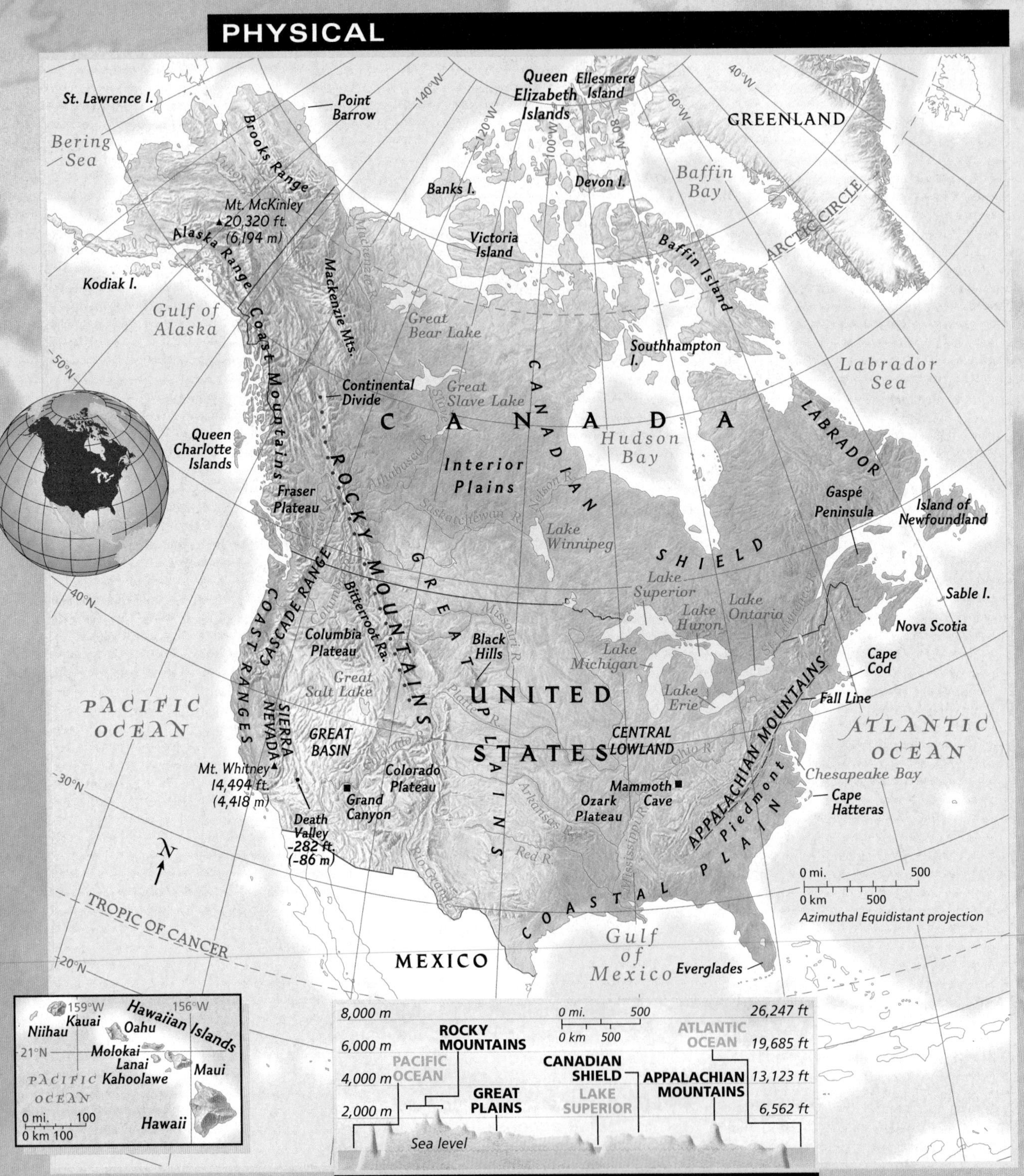

Elevation Profile

POLITICAL

RUSSIA
ARCTIC OCEAN
Bering Sea
Chukchi Sea
Beaufort Sea
ALASKA
Anchorage
Gulf of Alaska
YUKON TERRITORY
NORTHWEST TERRITORIES
NUNAVUT
BRITISH COLUMBIA
ALBERTA
SASK.
MANITOBA
ONTARIO
QUEBEC
NEWFOUNDLAND AND LABRADOR
P.E.I.
N.B.
NOVA SCOTIA
CANADA
Banks Island
Victoria Island
Ellesmere Island
Baffin Island
Southampton Island
Baffin Bay
Davis Strait
Hudson Strait
Hudson Bay
Labrador Sea
Greenland Sea
GREENLAND (KALAALLIT NUNAAT) Den.
ARCTIC CIRCLE
Edmonton
Calgary
Vancouver
Winnipeg
Montreal
Ottawa
Toronto
ROCKY MOUNTAINS
HAWAII
Honolulu
PACIFIC OCEAN
0 mi. 100
0 km 100
159°W
156°W
21°N
PACIFIC OCEAN
UNITED STATES
WASH.
OREGON
CALIFORNIA
IDAHO
NEVADA
UTAH
ARIZ.
MONT.
WYO.
COLO.
NEW MEXICO
N. DAK.
S. DAK.
NEBR.
KANSAS
OKLA.
TEXAS
MINN.
IOWA
MO.
ARK.
LA.
WIS.
MICHIGAN
ILL.
IND.
OHIO
KY.
TENN.
MISS.
ALA.
GA.
FLORIDA
S.C.
N.C.
VA.
W. VA.
PA.
N.Y.
VT.
N.H.
ME.
MASS.
R.I.
CONN.
N.J.
DEL.
MD.
Seattle
San Francisco
Los Angeles
Phoenix
Houston
St. Louis
Chicago
Atlanta
Miami
New York City
Philadelphia
Washington, D.C.
MEXICO
Gulf of Mexico
TROPIC OF CANCER
ATLANTIC OCEAN
Yukon R.
Mackenzie R.
Saskatchewan R.
Nelson R.
Missouri R.
Colorado R.
Arkansas R.
Red R.
Rio Grande
Mississippi R.
Ohio R.
St. Lawrence R.
180°
160°W
140°W
120°W
100°W
80°W
60°W
40°W
20°W
80°N
70°N
60°N
50°N
40°N
30°N

⊛ National capital
• Major city

0 mi. 500
0 km 500
Azimuthal Equidistant projection

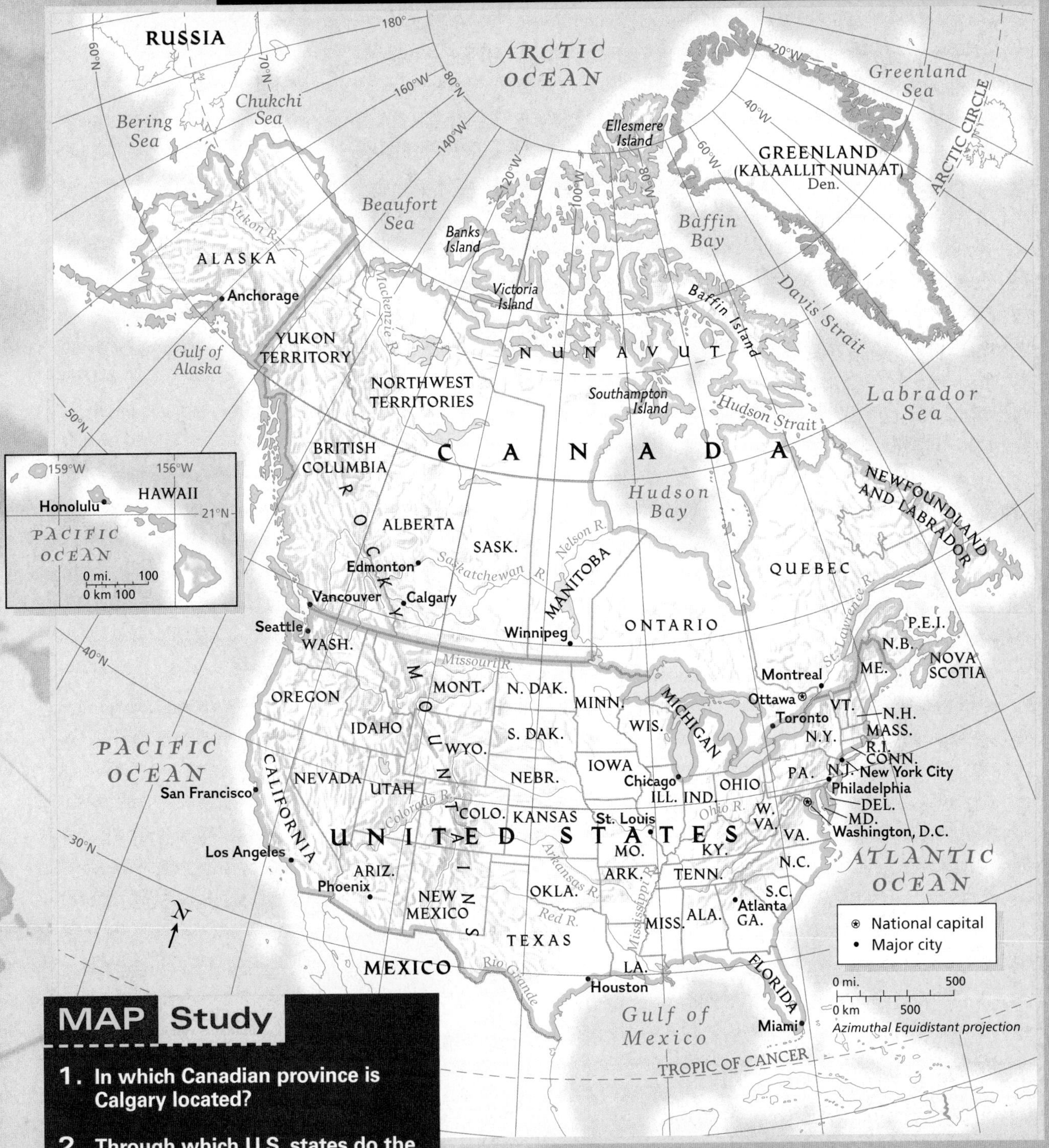

MAP Study

1. In which Canadian province is Calgary located?
2. Through which U.S. states do the Coast Ranges run?

2 The United States and Canada

POPULATION DENSITY

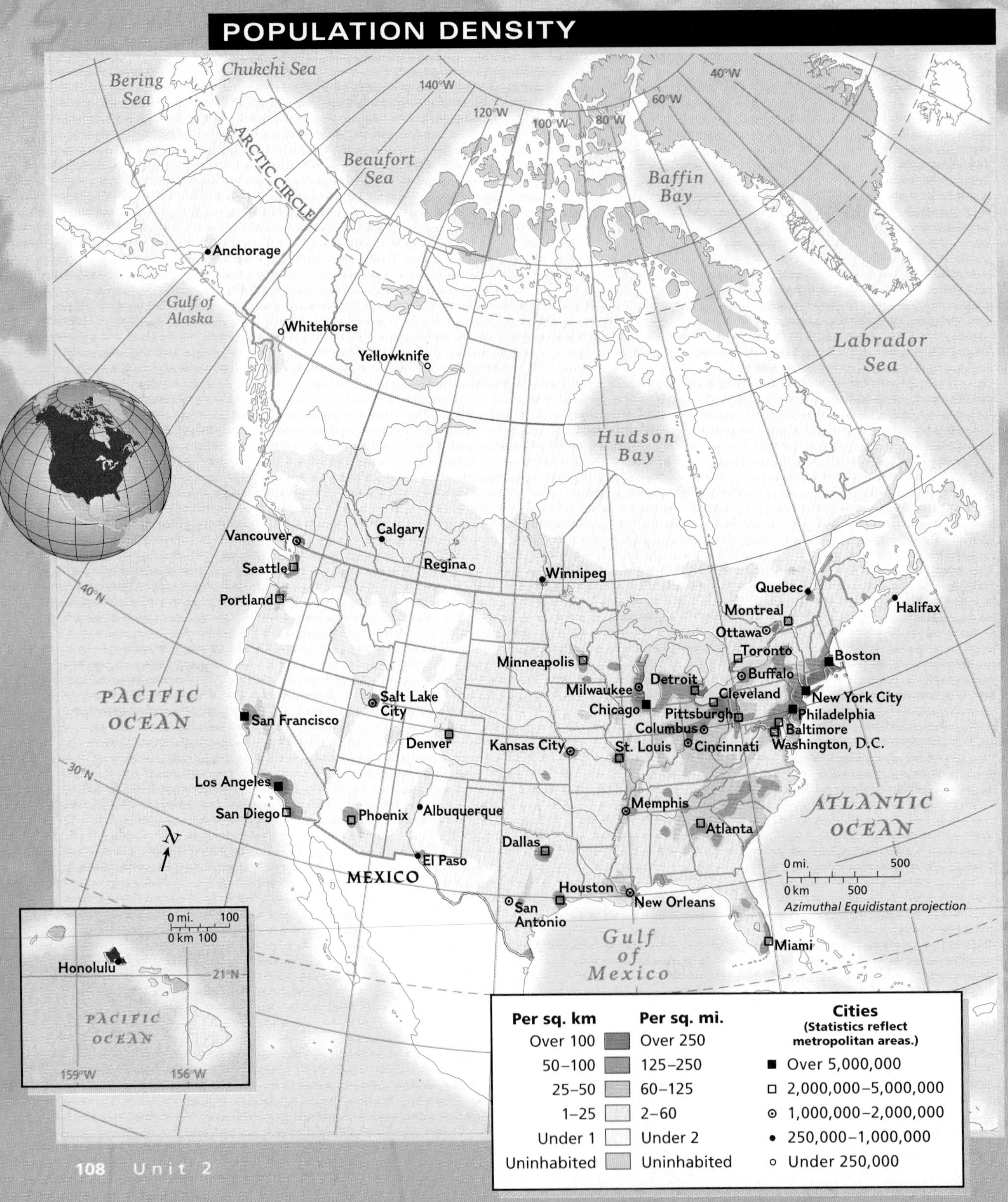

ECONOMIC ACTIVITY

GREENLAND
Beaufort Sea
Baffin Bay
Davis Strait
Labrador Sea
Hudson Strait
Hudson Bay
ARCTIC CIRCLE
Gulf of Alaska
Anchorage
CANADA
UNITED STATES
MEXICO
PACIFIC OCEAN
ATLANTIC OCEAN
Gulf of Mexico
TROPIC OF CANCER
60°N
50°N
30°N
120°W
100°W
80°W

0 mi. 500
0 km 500
Azimuthal Equidistant projection

Vancouver, Calgary, Seattle, Portland, Winnipeg, Ottawa, Montreal, Green Bay, Toronto, Boston, Minneapolis, Milwaukee, Detroit, Buffalo, New York City, Philadelphia, Pittsburgh, Baltimore, Washington, D.C., Norfolk, Raleigh, Chicago, Columbus, Indianapolis, Des Moines, Kansas City, St. Louis, Nashville, Memphis, Columbia, Atlanta, Birmingham, Dallas, Houston, New Orleans, Miami, San Francisco, Salt Lake City, Denver, Los Angeles, Phoenix

Wheat, Potatoes, Sheep, Cattle, Corn, Fruit, Tobacco, Cotton, Pecans

HAWAII
0 mi. 100
0 km 100
Sugarcane
Honolulu
21°N
PACIFIC OCEAN
Fruit
159°W
156°W

Resources

- Iron ore
- Petroleum
- Natural gas
- Coal
- Copper
- Zinc
- Gold
- Silver

Land Use

- Commercial farming
- Subsistence farming
- Livestock raising
- Nomadic herding
- Hunting and gathering
- Forests
- Manufacturing and trade
- Commercial fishing
- Little or no activity

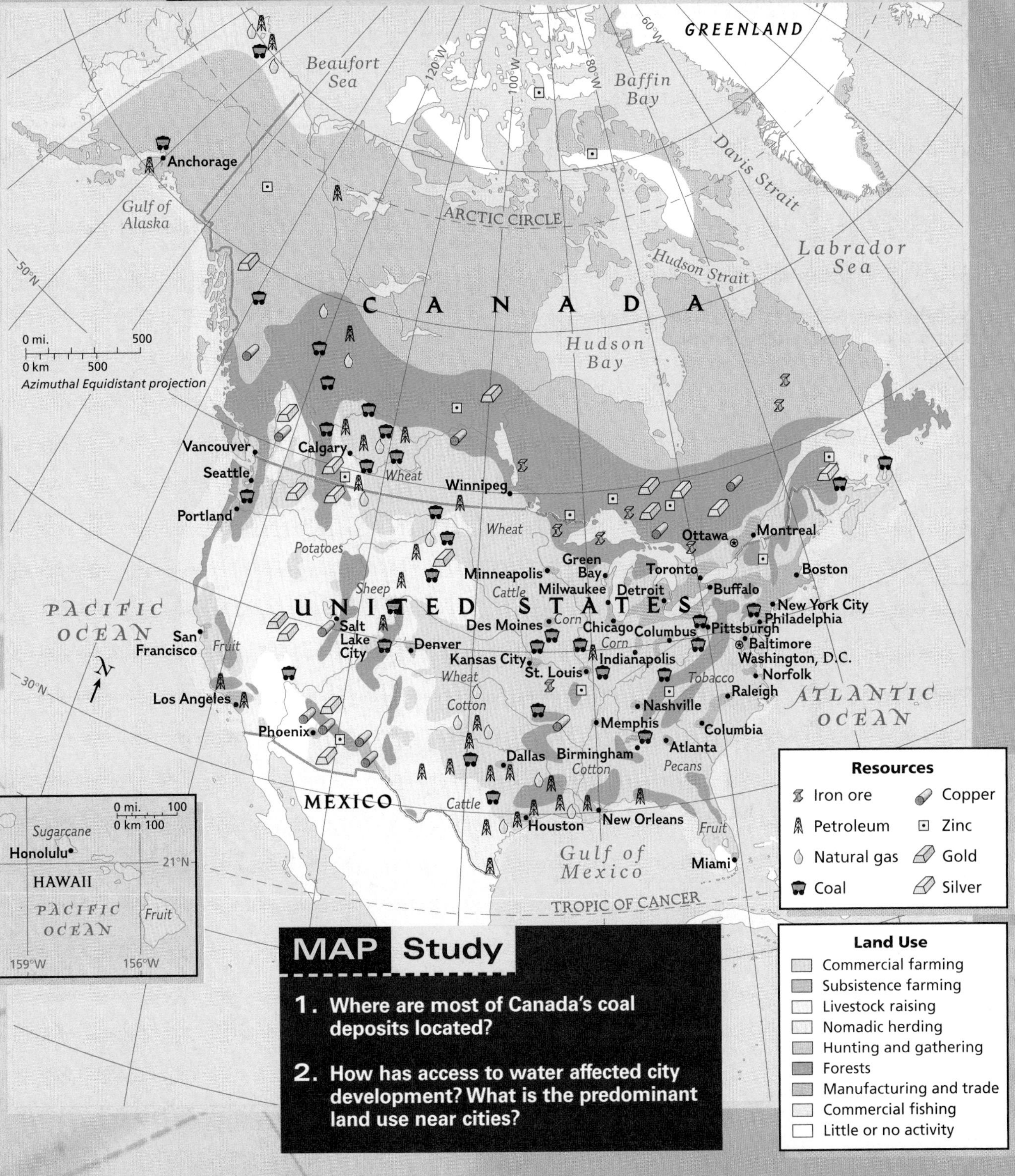

MAP Study

1. Where are most of Canada's coal deposits located?
2. How has access to water affected city development? What is the predominant land use near cities?

2

The United States and Canada

COUNTRY PROFILES

COUNTRY * AND CAPITAL	FLAG AND LANGUAGE	POPULATION AND DENSITY	LANDMASS	MAJOR EXPORT	MAJOR IMPORT	CURRENCY	GOVERNMENT
UNITED STATES Washington, D.C.	English	284,500,000 77 per sq. mi. 30 per sq. km	3,717,796 sq. mi. 9,629,091 sq. km	Machinery	Crude Oil	U.S. Dollar	Federal Republic
CANADA Ottawa	English, French	31,000,000 8 per sq. mi. 3 per sq. km	3,849,670 sq. mi. 9,970,610 sq. km	Newsprint	Crude Oil	Canadian Dollar	Parliamentary Democracy

FOR AN ONLINE UPDATE OF THIS INFORMATION, VISIT GEOGRAPHY.GLENCOE.COM AND CLICK ON "TEXTBOOK UPDATES."

U.S. State Names: Meaning and Origin

State	Capital	Meaning and Origin
ALABAMA	Montgomery	"thicket clearers" (Choctaw)
ALASKA	Juneau	"the great land" (Aleut)
ARIZONA	Phoenix	"little spring" (Papago), or "dry land" (Spanish)
ARKANSAS	Little Rock	"downstream people" (Quapaw)
CALIFORNIA	Sacramento	unknown meaning (Spanish)
COLORADO	Denver	"red" (Spanish)
CONNECTICUT	Hartford	"beside the long tidal river" (Native American)
DELAWARE	Dover	named for Virginia's colonial governor, Baron De La Warr
FLORIDA	Tallahassee	"feast of flowers" (Spanish)
GEORGIA	Atlanta	named for England's King George II
HAWAII	Honolulu	unknown meaning (Native Hawaiian)
IDAHO	Boise	unknown meaning (Native American)
ILLINOIS	Springfield	"tribe of superior men" (Native American)
INDIANA	Indianapolis	"land of Indians" (European American)
IOWA	Des Moines	unknown meaning (Native American)
KANSAS	Topeka	"people of the south wind" (Sioux)
KENTUCKY	Frankfort	"land of tomorrow" (Iroquoian)
LOUISIANA	Baton Rouge	named for France's King Louis XIV
MAINE	Augusta	named for an ancient French province
MARYLAND	Annapolis	named in honor of the wife of England's King Charles I
MASSACHUSETTS	Boston	"great mountain place" (Native American)
MICHIGAN	Lansing	"great lake" (Ojibway)
MINNESOTA	Saint Paul	"sky-tinted water" (Sioux)
MISSISSIPPI	Jackson	"father of the waters" (Native American)

* COUNTRIES, FLAGS, STATES, AND PROVINCES NOT DRAWN TO SCALE

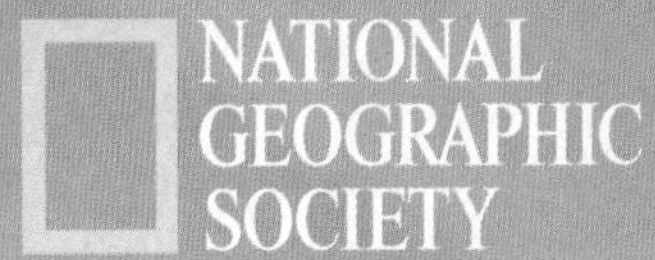

MISSOURI
Jefferson City
"town of the large canoes" (Native American)

MONTANA
Helena
"mountainous" (Spanish)

NEBRASKA
Lincoln
"flat water" (Native American)

NEVADA
Carson City
"snowcapped" (Spanish)

NEW HAMPSHIRE
Concord
named for Hampshire, a county in England

NEW JERSEY
Trenton
named for Isle of Jersey, a British territory

NEW MEXICO
Santa Fe
named for the state's former colonial ruler, Mexico

NEW YORK
Albany
named in honor of the English Duke of York

NORTH CAROLINA
Raleigh
named in honor of England's King Charles I

NORTH DAKOTA
Bismarck
named for the Dakota, a Native American group

OHIO
Columbus
"great river" (Native American)

OKLAHOMA
Oklahoma City
"red people" (Choctaw)

OREGON
Salem
unknown meaning and origin

PENNSYLVANIA
Harrisburg
"Penn's woodland," named for the father of Pennsylvania's founder, William Penn

RHODE ISLAND
Providence
unknown meaning and origin

SOUTH CAROLINA
Columbia
named for England's King Charles I

SOUTH DAKOTA
Pierre
named for the Dakota, a Native American group

TENNESSEE
Nashville
named for Tanasi, "Cherokee villages" (Cherokee)

TEXAS
Austin
"friends" (Tejas)

UTAH
Salt Lake City
"people of the mountains" (Ute)

VERMONT
Montpelier
"green mountain" (French)

VIRGINIA
Richmond
named for the unmarried Queen Elizabeth I of England, known as "the Virgin Queen"

WASHINGTON
Olympia
named in honor of George Washington

WEST VIRGINIA
Charleston
began as the western part of Virginia before becoming a state in 1863

WISCONSIN
Madison
"grassy place" (Chippewa)

WYOMING
Cheyenne
"upon the great plain" (Delaware)

Canadian Province and Territory Names: Meaning and Origin

ALBERTA
Edmonton
named for the daughter of England's Queen Victoria

BRITISH COLUMBIA
Victoria
named for Christopher Columbus and the province's British heritage

MANITOBA
Winnipeg
"strait of the great spirit" (Algonquian)

NEW BRUNSWICK
Fredericton
named for English royal family of Brunswick-Luneburg

NEWFOUNDLAND
St. John's
"new found land," named by European explorer John Cabot in 1497

NORTHWEST TERRITORIES
Yellowknife
named for lands north and west of Lake Superior

NOVA SCOTIA
Halifax
Latin term for "New Scotland," based on province's Scottish heritage

NUNAVUT
Iqaluit
"our land" (Inuktitut)

ONTARIO
Toronto
meaning unknown (Iroquoian)

PRINCE EDWARD ISLAND
Charlottetown
named for the son of England's King George III

QUEBEC
Quebec
"place where the river narrows" (Algonquian)

SASKATCHEWAN
Regina
"fast flowing river" (Cree)

YUKON TERRITORY
Whitehorse
"great river" (Native American)

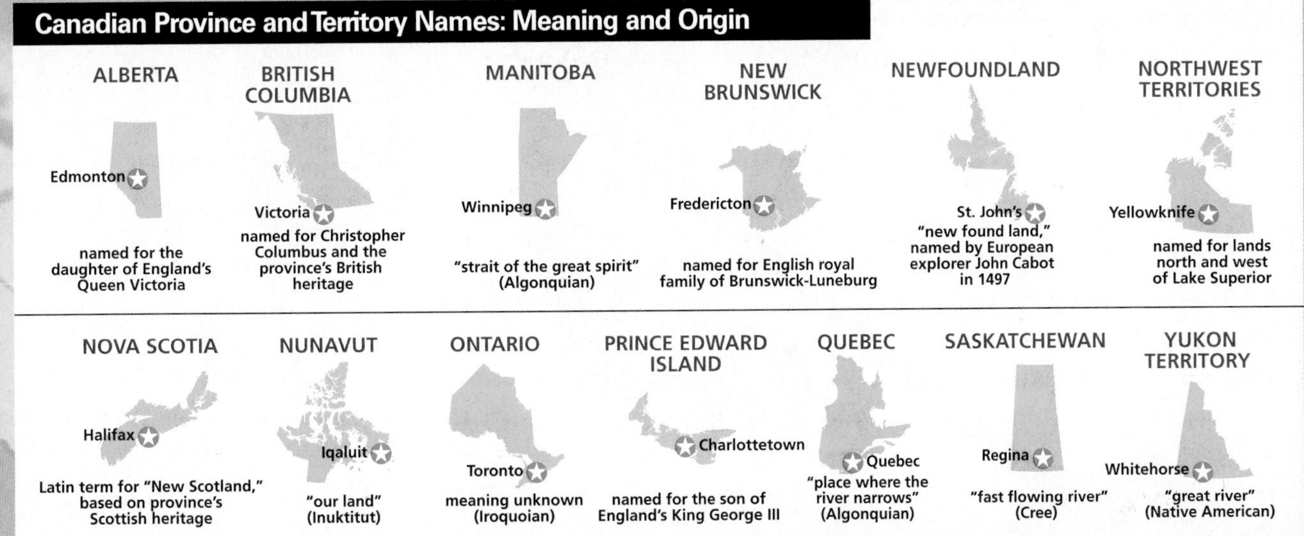

2

GLOBAL CONNECTION

CANADA AND THE UNITED STATES

ICE HOCKEY!

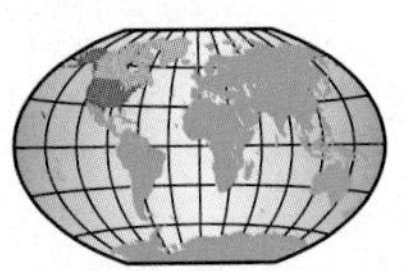

Historians still debate the details, but everyone agrees that ice hockey was invented in Canada. The game seems to have originated in the early 1800s in Nova Scotia, one of Canada's easternmost provinces. Not content to spend long winters indoors, some of Nova Scotia's inhabitants began tinkering with an Irish game, similar to field hockey, that was played with sticks and a ball. The eager sportsmen realized that the slick surface of a frozen pond was a worthy alternative to a grassy playing field. They traded their shoes for skates, and ice hockey was born.

As the new game gained popularity, it spread west and north across Canada. By the turn of the century, ice hockey had become Canada's national sport. And it was not just for men. The first all-female

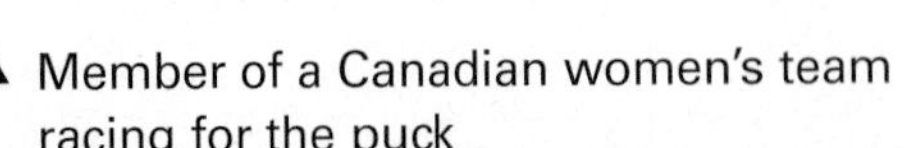

▲ Member of a Canadian women's team racing for the puck

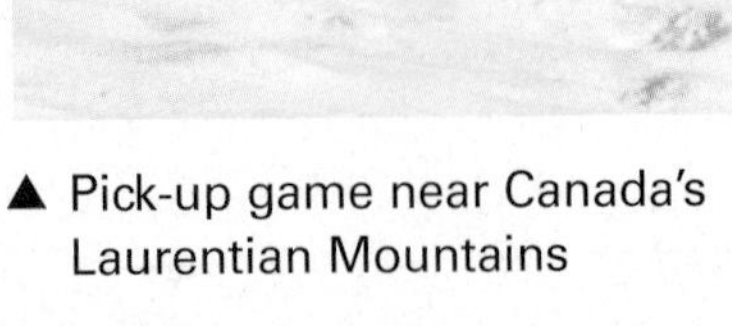

▲ Pick-up game near Canada's Laurentian Mountains

▼ Getting a feel for skating on wheels

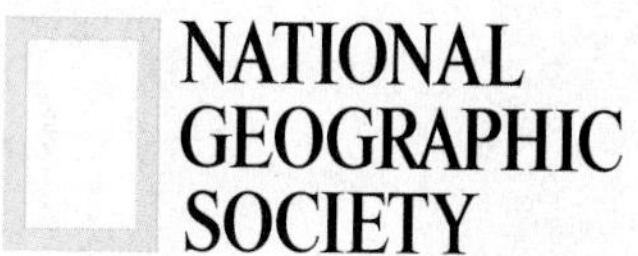

ice hockey game on record was played in Ontario in 1892.

Hockey fever spread southward, too, crossing the U.S.-Canadian border into northern states such as Minnesota, Michigan, Massachusetts, and New York. In 1924 the Boston Bruins became the first U.S. team to join Canada's National Hockey League. Other northern cities, including Chicago and Detroit, soon had teams on the League's roster as well.

At this point, geography checked ice hockey's southward spread. For nearly a quarter century, the game remained a northern pastime. It just didn't catch on in southern states where cold weather was rare and lakes never froze.

Eventually, however, indoor ice rinks, televised hockey games, and a steady influx of Canadian players into the United States overcame the geographic barriers, and the sport found a foothold in nearly every state. Hockey made headlines in 1980 when America's team beat the heavily favored Soviets in the Olympic Winter Games and then went on to win the gold medal. When Canadian superstar Wayne Gretzky came to play in the United States a few years later, hockey's popularity surged again.

▲ Hockey legend Wayne Gretzky entering a rink

Now in-line skates and roller hockey make it possible for would-be players to get a feel for the game no matter where they live or what the season. Many major U.S. cities have professional hockey teams, including Phoenix, Dallas, and Miami. In fact, America's Sun Belt alone is home to more hockey teams than there are in Canada! But ice hockey in the United States isn't just for professionals. It's played by kids—both boys and girls—and amateurs throughout the nation.

CHAPTER 5

The Physical Geography of the United States and Canada

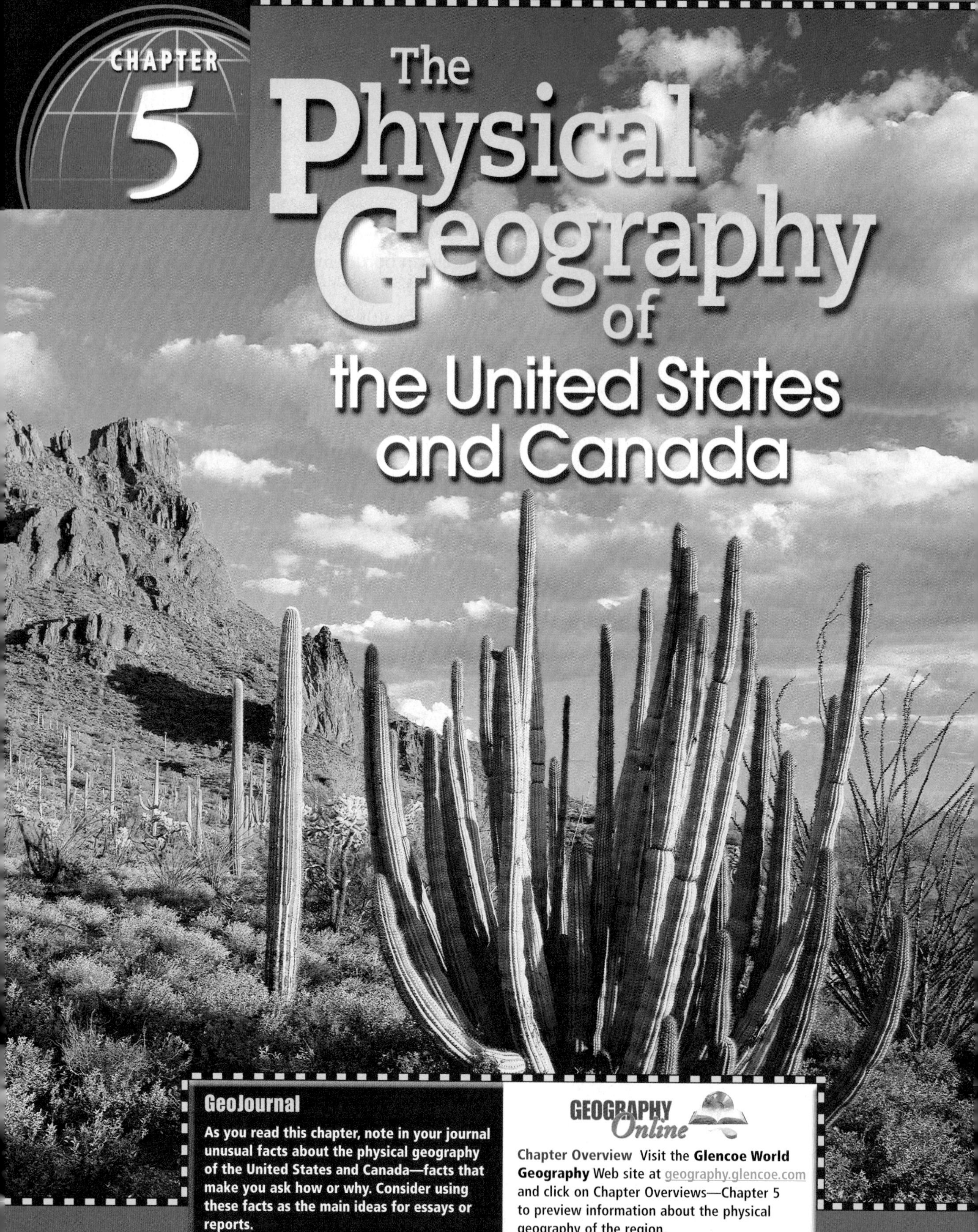

GeoJournal

As you read this chapter, note in your journal unusual facts about the physical geography of the United States and Canada—facts that make you ask how or why. Consider using these facts as the main ideas for essays or reports.

GEOGRAPHY Online

Chapter Overview Visit the **Glencoe World Geography** Web site at geography.glencoe.com and click on Chapter Overviews—Chapter 5 to preview information about the physical geography of the region.

SECTION 1

The Land

Guide to Reading

Consider What You Know

The United States and Canada share the world's longest undefended border. What famous natural feature do both the United States and Canada claim as a tourist attraction?

Read to Find Out

- What are some key similarities and differences in the physical geography of the United States and Canada?
- Why have rivers played such an important role in this region's development?
- What geographic factors have made the United States and Canada so rich in natural resources?

Terms to Know

- divide
- headwaters
- tributary
- fall line
- fishery

Places to Locate

- Mount McKinley
- Rocky Mountains
- Canadian Shield
- Appalachian Mountains
- Colorado River
- Rio Grande
- Mackenzie River
- Mississippi River
- St. Lawrence River
- Great Lakes

◀ *Organ Pipe Cactus National Monument, Arizona*

NATIONAL GEOGRAPHIC

A Geographic View

Carving Their Own Way

Grain fields spill their color across the badlands of the Missouri Breaks, a lonesome swatch of eastern Montana where the Great Plains roll to an abrupt and wild end. The Missouri River and its tributaries have cut deep paths through the underlying sandstone and shale, fracturing the open country. Rough and remote spaces rule the Breaks—perfect for folks who insist on carving their own way.

—*John Barsness, "The Missouri Breaks," National Geographic, May 1999*

Missouri Breaks, Montana

The rugged terrain of the Missouri Breaks bears witness to the geologic forces that have shaped the North American continent. The United States and Canada share the northern part of the continent. They form a geographic region of enormous physical variety and natural wealth. Together, Canada and the continental United States cover more than 7 million square miles (18 million sq. km), about 12 percent of Earth's land surface. In this section you will explore the physical geography of these two countries.

Landforms

Mountains rise at the eastern and western edges of both the United States and Canada. In the west young, sharp-edged mountain ranges tower above plateaus that descend to vast, rolling central plains. Mighty rivers and enormous lakes satisfy the thirst of cities, wildlife, and natural vegetation in the two countries' midsections. The fertile plains extend across the continent until they meet the lower, more eroded mountains in the east.

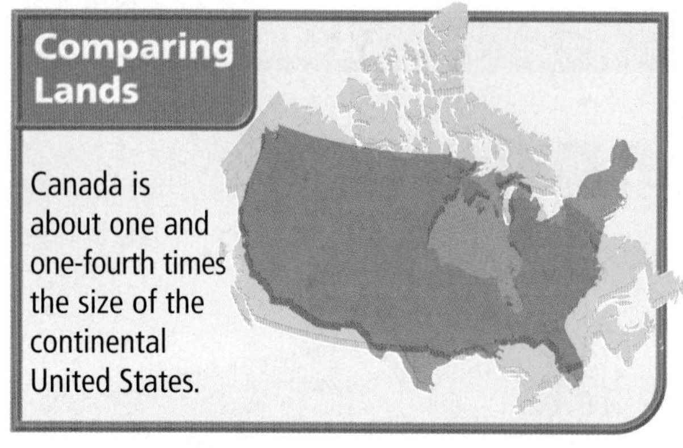

Canada is about one and one-fourth times the size of the continental United States.

Western Mountains and Plateaus

Collisions between the Pacific and the North American tectonic plates millions of years ago thrust up a series of impressive, sharp-peaked mountain ranges called the Pacific Ranges. Still young in geologic terms, the Pacific Ranges include the Sierra Nevada, the Cascade Range, the Coast Range, and the Alaska Range. The Alaska Range gives rise to the highest point on the continent, **Mount McKinley**, at 20,320 feet (6,194 m).

Like the Pacific Ranges, the **Rocky Mountains** grew as geologic forces heaved slabs of rock upward. The map on p. 117 shows that the snow-covered Rocky Mountains link the United States and Canada and stretch more than 3,000 miles (4,828 km) from New Mexico to Alaska. Some peaks of the Rockies soar to more than 14,000 feet (4,267 m).

Dry basins and plateaus fill the area between the Pacific Ranges and the Rockies. The Columbia Plateau in the north was formed by lava that seeped from cracks in the earth. The heavily eroded Colorado Plateau displays flat-topped mesas and the majestic Grand Canyon of the Colorado River. At its deepest the canyon's steep walls plunge 6,000 feet (1,829 m). The Great Basin cradles Death Valley, the hottest and lowest place in the United States. Canada's Nechako Plateau and Fraser Plateau are colder and narrower than the plateau areas in the United States.

Interior Landforms

East of the Rockies, the land falls and flattens into the Great Plains, which extend 300 to 700 miles (483 to 1,126 km) across the center of the region. The Great Plains are sometimes called the Interior Plains or the High Plains because of their location and elevation, which reaches up to 6,000 feet (1,829 m). Although the Great Plains appear flat, the land slopes gradually downward at about 10 feet per mile (about 2 m per km) to the heart of the Central Lowlands along the Mississippi River.

Eastern Mountains and Lowlands

East of the Mississippi, the land rises slowly into the foothills of the Appalachian Mountains. At the edge of the Canadian plains, the **Canadian Shield**, a giant core of rock centered on the Hudson and James Bays, anchors the continent. The stony land of the Shield makes up the eastern half of Canada and the northeastern United States. In northern Quebec the Canadian Shield descends to the Hudson Bay.

The heavily eroded **Appalachian Mountains** are North America's oldest mountains and the continent's second-longest mountain range. They extend about 1,500 miles (2,414 km) from Quebec to central Alabama. Coastal lowlands lie east and south of the Appalachians. Between the mountains and the coastal lowlands lies a wide area of rolling hills. Many rivers cut through the Piedmont and flow across to the Atlantic Coastal Plain in the Carolinas. In the southeast the Gulf Coastal Plain extends westward to Texas.

Islands

Islands are important in the region. New York City's Manhattan Island, at the mouth of the Hudson River, is a major United States and world economic center. Volcanic mountaintops emerging from the Pacific Ocean formed Hawaii, creating 8 major and 124 smaller islands with a land area of 6,471 square miles (16,760 sq. km). Newfoundland, Prince Edward Island, and Cape Breton Island in the east and Vancouver Island in the west are Canada's most important islands. Near the coast of Canada's Ellesmere Island lies the world's largest island, Greenland. An overseas territory of Denmark, Greenland spans 840,325 square miles (2.2 million sq. km), an area about the size of Alaska and Texas combined.

A Fortune in Water

Freshwater lakes and rivers have helped make the United States and Canada wealthy. Abundant water satisfies the needs of cities and rural areas, provides power for homes and industries, and moves resources across the continent.

NATIONAL GEOGRAPHIC

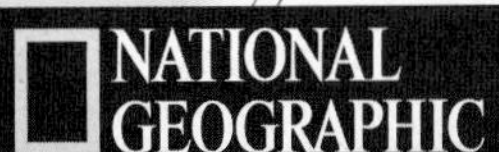

MAP STUDY

The United States and Canada: Physical-Political

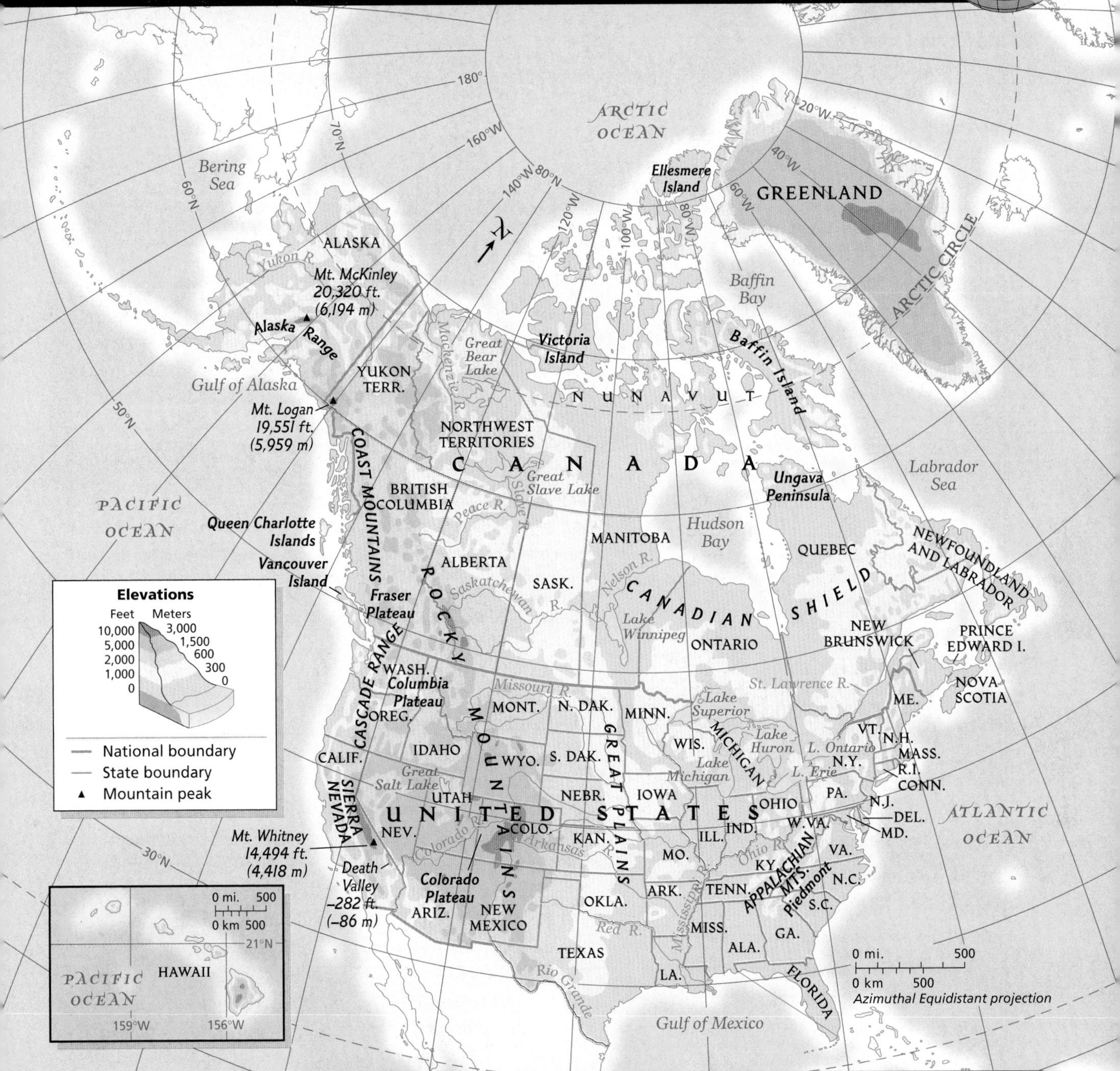

Geography Skills for Life

1. **Interpreting Maps** Which Canadian provinces border Hudson Bay?
2. **Applying Geography Skills** How do the Rocky Mountains affect the rivers in the United States and Canada?

Find NGS online map resources @ www.nationalgeographic.com/maps

Rivers from the Rockies

In North America the high ridge of the Rockies is called the Continental Divide, or the Great Divide. A **divide** is a high point or ridge that determines the direction that rivers flow. East of the Continental Divide, waters flow toward the Arctic Ocean, Hudson Bay, the Atlantic Ocean, and the Mississippi River system into the Gulf of Mexico; to the west, waters flow into the Pacific Ocean. Rivers—such as the **Colorado** and the **Rio Grande**—have their **headwaters**, or source, in the Rockies, and many **tributaries**, or brooks, rivers, and streams, connect with one of these rivers. Northeast of the Rockies, the **Mackenzie River**—which flows from the Great Slave Lake to the Arctic Ocean—drains much of Canada's northern interior.

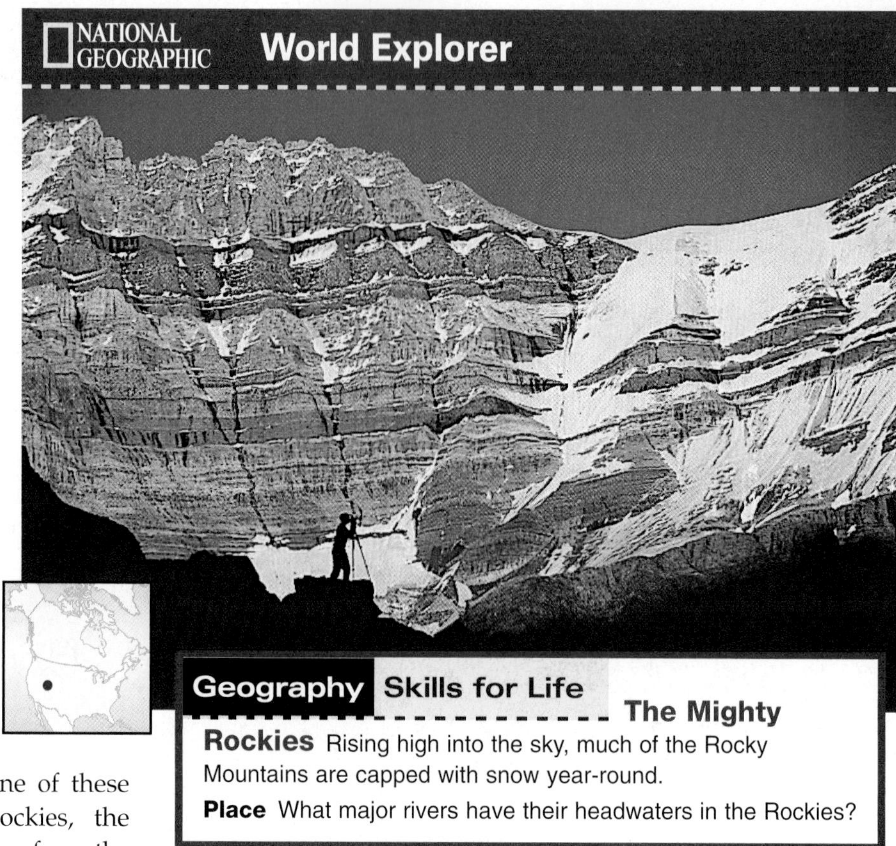

Geography Skills for Life

The Mighty Rockies Rising high into the sky, much of the Rocky Mountains are capped with snow year-round.

Place What major rivers have their headwaters in the Rockies?

The Mighty Mississippi

One of North America's longest rivers, the **Mississippi River**, flows 2,350 miles (3,782 km) from its source. It begins in Minnesota as a stream so narrow that a person can jump across it.

> *"When I was nine years old, I jumped across the Mississippi. . . . My parents let me know this modest stream I'd taken in stride was actually one of the Earth's great corridors, dominion of paddleboats and Huck Finn, prime mover of food, fertility, and commerce across our land."*
>
> Barbara Kingsolver, "San Pedro River: the Patience of a Saint," *National Geographic*, April 2000

The Mississippi, swelled to a width of a mile and a half (2.4 km), empties into the Gulf of Mexico. The Mississippi drains 1,200,000 square miles (3,108,000 sq. km) of land, including all or part of 31 U.S. states and 2 Canadian provinces. It is one of the world's busiest commercial waterways.

Eastern Rivers

The **St. Lawrence River**, one of Canada's most important rivers, flows for 750 miles (1,207 km) from Lake Ontario to the Gulf of St. Lawrence in the Atlantic, forming part of the border between Canada and the United States. The Canadian cities of Quebec, Montreal, and Ottawa grew up along the St. Lawrence River and its tributaries and depend on these waters as a transportation resource.

In the eastern United States, a boundary called the **fall line** marks the place where the higher land of the Piedmont drops to the lower Atlantic Coastal Plain. Along the fall line, eastern rivers break into rapids and waterfalls, blocking ships from traveling farther inland. Many key U.S. cities, such as Philadelphia, Pennsylvania; Baltimore, Maryland; and Washington, D.C., grew up along the fall line. They offer port facilities for oceangoing trading vessels. Smaller towns along the fall line, especially in New England and in the South, tap the waterpower of the falls for textile mills and factories.

Niagara Falls is a popular tourist attraction on the Niagara River, which forms part of the border between Ontario, Canada, and New York State in the United States. Niagara Falls is also a major source of hydroelectric power for both countries. Two separate drops form the falls, the Horseshoe Falls adjoining the Canadian bank of the river, and the American Falls adjoining the U.S. bank.

From Glaciers to Lakes

In northern Canada glacial dams created Great Bear Lake and Great Slave Lake. Glaciers also gouged the Canadian Shield and tore at the central section of the continent, leaving glacial basins that became the **Great Lakes**. Lake Superior, Lake Huron, Lake Erie, Lake Ontario, and Lake Michigan have had their current shapes for only about the last 14,000 years.

Providing a link between inland and coastal waterways has been crucial to the economic development of North America. The greatest of these connections is the Great Lakes–St. Lawrence Seaway, a series of canals, rivers, and other inland waterways linking the Great Lakes with the Atlantic Ocean. The seaway helped make cities along the Great Lakes, such as Chicago, powerful trade and industrial centers. Other important inland waterways include the Gulf Intracoastal Waterway, which connects cities from Florida to Texas with the Mississippi River, and the Atlantic Intracoastal Waterway, which provides sheltered inland channels for navigation between Norfolk, Virginia, and Key West, Florida.

NATIONAL GEOGRAPHIC **World Explorer**

Geography Skills for Life

Copper Mining During the 1990s, this copper mine in Utah produced over 300,000 tons of copper annually.

Place What minerals are mined in the United States and Canada?

Natural Resources

Ample freshwater is only one of the many natural resources of the United States and Canada. The same geologic processes that shaped the North American landscape left the region rich in a wide variety of resources. Access to this natural wealth has helped speed the industrialization of this region.

Fuels

The United States and Canada have important energy resources such as petroleum and natural gas. Texas and Alaska rank first and second in oil reserves in the United States. Texas also has the greatest reserves of natural gas. Most of Canada's oil and natural gas reserves lie in or near Alberta. Coal in the Appalachians, Wyoming, and British Columbia has been mined for more than 100 years.

Minerals

Mineral resources are also plentiful in the region. The Rocky Mountains yield gold, silver, and copper. Parts of the Canadian Shield are rich in iron and nickel. Deposits of low-grade iron ore exist in northern Minnesota and Michigan. Canada's minerals include 28 percent of the world's supply of potash (mineral salt used in fertilizers), 18 percent of its copper, 14 percent of its gold, and 12 percent of its silver.

Timber

Timber is vital for both countries. Forests and woodlands once covered much of the United

Geography Skills for Life

Catch of the Day This fisherman earns his livelihood fishing on Lake Michigan.

Place Why is the Grand Banks important to Canada?

States and Canada. Today, however, forests cover less than 50 percent of Canada and about one-third of the United States. Commercial lumbering operations face the challenge of harvesting the region's timber resources responsibly. Positive efforts to preserve the forests include planting new trees to replace those cut for lumber, cooperating to protect the 1,000 species of native animals in the forests, and preserving old-growth forests in areas set aside as national forests.

Economics

Fishing

The coastal waters of the Atlantic and Pacific Oceans and the Gulf of Mexico are important to the region's economy. Rich with fish and shellfish, these waters are important **fisheries**, or places for catching fish and other sea animals. The Grand Banks, once one of the world's richest fishing grounds, covers about 139,000 square miles (360,000 sq. km) off Canada's southeast coast. Fishers have harvested cod from the Grand Banks for at least 500 years. As these waters were overfished, however, stocks decreased, and the Canadian government banned cod fishing in 1992.

SECTION 1 ASSESSMENT

Checking for Understanding

1. **Define** divide, headwaters, tributary, fall line, fishery.
2. **Main Ideas** On a sheet of paper, fill in a chart like the one below. Then choose one example of landforms, water, of natural resources and describe its impact on the United States and Canada.

	Landforms	Water	Natural Resources
United States			
Canada			

Critical Thinking

3. **Drawing Conclusions** Why might fishing disputes arise in the region?
4. **Identifying Cause and Effect** How did the Great Lakes–St. Lawrence Seaway influence the development of cities in the region?
5. **Drawing Conclusions** In what ways did the actions of glaciers alter the physical geography of this region, and what effects did those alterations have on the region's development?

Analyzing Maps

6. **Location** Study the physical-political map on page 117. Describe the landscapes found in the following places: Montana, Texas, and Ontario.

Applying Geography

7. **Effects of Location** Write a paragraph describing the effects of a physical process, such as weather or gravity, on the flow of rivers in the United States and Canada.

Guide to Reading

Consider What You Know

Think about the climate differences between the United States and Canada. Why do you think Canada is so much colder than the United States?

Read to Find Out

- Which climate zones are found in the United States and Canada?
- In what ways do winds, ocean currents, latitude, and landforms affect the region's climates?
- What kinds of weather hazards affect the United States and Canada?
- How has human settlement affected the natural vegetation of the United States and Canada?

Terms to Know

- timberline
- chinook
- prairie
- supercell
- hurricane
- blizzard

Places to Locate

- Death Valley
- Great Plains
- Everglades
- Newfoundland
- Yukon Territory

Climate and Vegetation

NATIONAL GEOGRAPHIC

A Geographic View

Life Amid the Glaciers

The diversity of species on nunataks [mountains surrounded by glacial ice] takes patience to grasp. Only the showiest, such as moss campion and orange lichens, grab the eye. Wait and you might glimpse an alert wolf spider or resting butterfly. How did life reach these isolated peaks? Winds bore most pioneers over the glaciers. Plants were carried as seeds. Young spiders sailed in on strands of silk.

—*Kevin Krajick, "Nunataks,"* National Geographic, *December 1998*

Moss campion, Yukon Territory, Canada

The ice fields of Canada's northwestern Yukon Territory seem at first to be Arctic wastelands. Studding the glaciers, though, are craggy summits encased in glacial ice. Although temperatures there can fall below zero, mini-climates shelter an amazing variety of life forms. Similar diversity characterizes the whole of Canada and the United States. In this section you will learn about the climate regions and natural vegetation of the United States and Canada.

A Varied Region

Much of the United States and Canada experiences exactly the types of climate one might expect from the countries' latitudes. Two thirds of Canada and the U.S. state of Alaska lie in higher latitudes and experience long, cold winters and brief, mild summers. Most of the continental United States and the southern one third of Canada lie within more temperate latitudes, where climate regions vary with elevation. Hawaii, the only non-continental U.S. state, has a tropical climate.

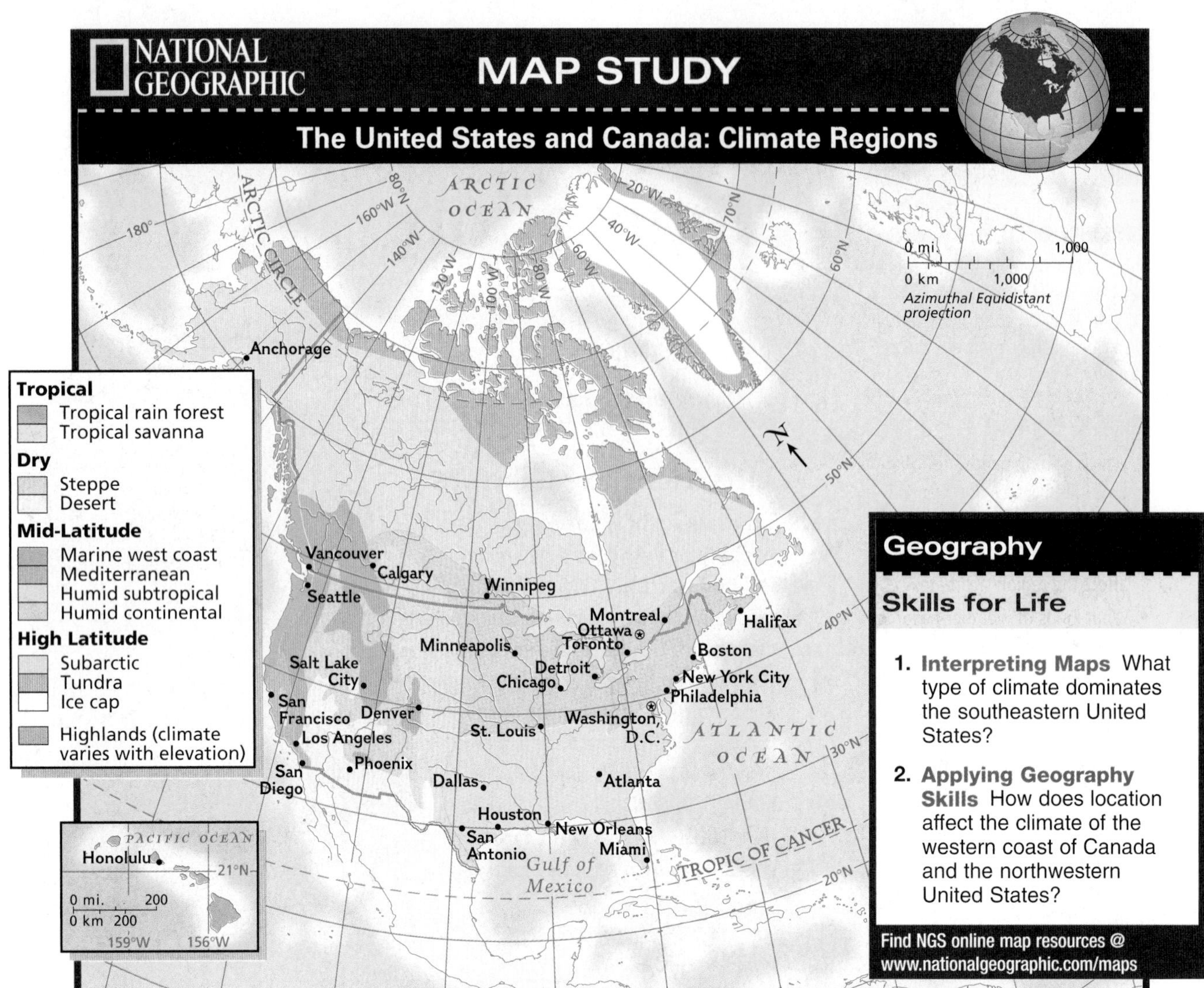

Geography Skills for Life

1. **Interpreting Maps** What type of climate dominates the southeastern United States?
2. **Applying Geography Skills** How does location affect the climate of the western coast of Canada and the northwestern United States?

Find NGS online map resources @ www.nationalgeographic.com/maps

Northern Climates

Large parts of Canada and Alaska lie in a subarctic climate zone with very cold winters and extensive coniferous forests. Two-thirds of Canada has January temperatures that average below 0°F (–18°C). In winter, temperatures can fall to –70°F (–57°C) in some places. A persistently high atmospheric pressure area over the Canadian subarctic spawns the cold winds that chill much of the central United States during the winter.

Lands along the Arctic coastline fall into the tundra climate zone. Bitter winters and cool summers in this vast expanse of wilderness make it inhospitable for most plants, and few people live there. Greenland's tundra vegetation consists of sedge, cotton grass, and lichens. The island's small ice-free areas have few trees, but some dwarfed birch, willow, and alder scrubs do survive. As in other northern climate areas, few people inhabit Greenland because of its harsh climate conditions.

Western Climates

From the cool, wet coast of British Columbia to the hot, dry deserts of California and the snow-capped peaks of the Rocky Mountains, the climate and vegetation patterns in the western areas of the United States and Canada vary widely. This variation in climate and vegetation is the result of the combined effects of latitude, elevation, ocean currents, and rainfall.

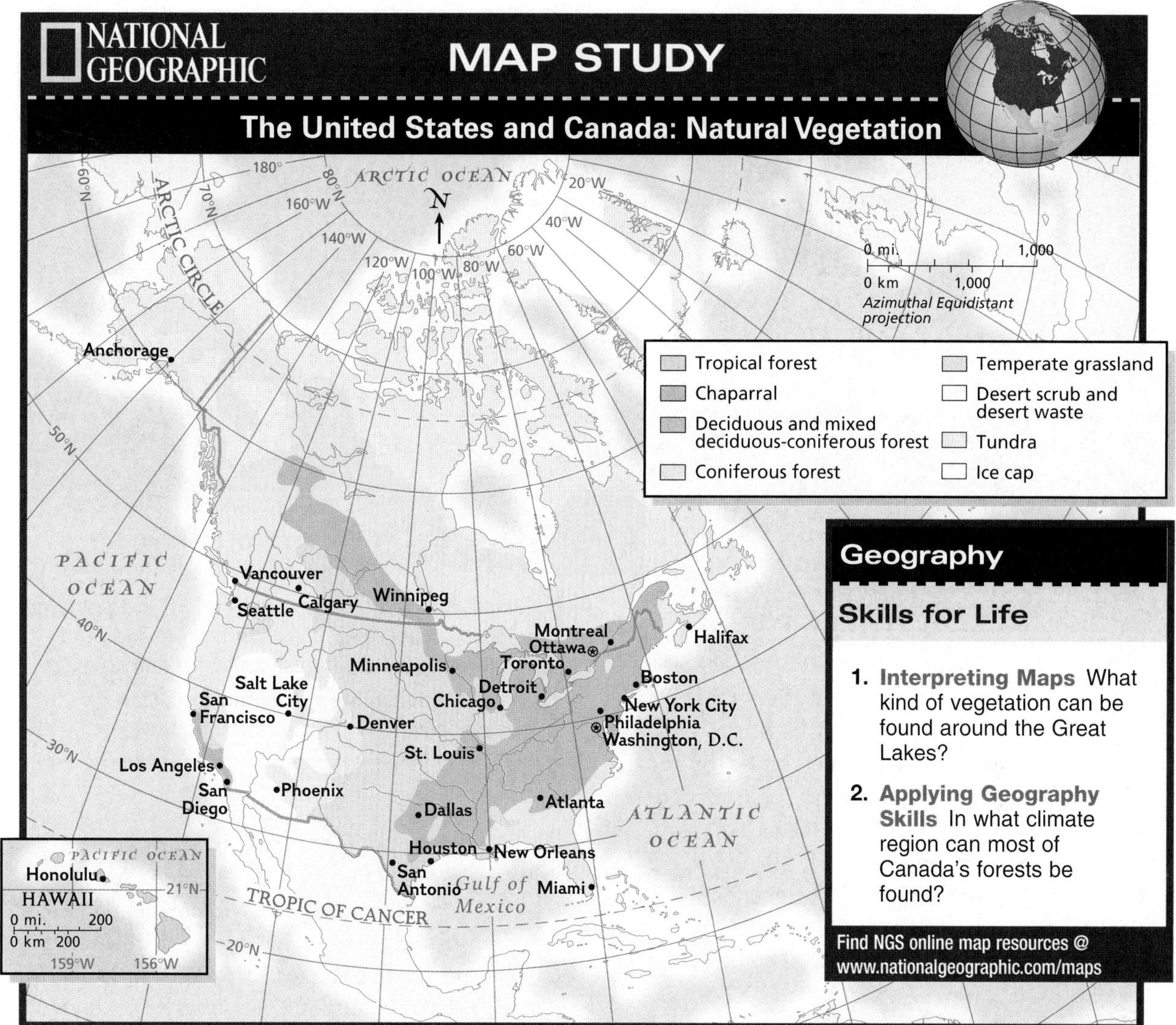

Marine West Coast

The interplay of ocean currents and winds with the Pacific Ranges gives the Pacific coast from California to southern Alaska a marine west coast climate. The mountains force the wet ocean air upward, where it cools and releases its moisture. As a result, more than 100 inches (254 cm) of rain soaks parts of this region each year. Coniferous forests, ferns, and mosses are common there. Southern California has a mild Mediterranean climate.

Plateaus, Basins, and Deserts

The rain shadow effect keeps the plateaus and basins that lie between the Pacific Ranges and the Rocky Mountains hot and dry. Much of the area has a steppe or desert climate. U.S. desert lands in this area, including the Great Salt Lake Desert, Death Valley, the Mojave (moh•HAH•vee) Desert, and the Chihuahuan (chee•WAH•wahn) Desert, bake in the relentless sun. **Death Valley** had the highest temperature ever recorded in the United States, 134°F (57°C). In the western deserts, cacti and hardy wildflowers bloom in the brief spring rains. The areas adjacent to these deserts usually experience a steppe climate with a mixture of desert scrub, grasslands, or coniferous forest, depending on latitude.

Elevation, not latitude, gives the higher reaches of the Rocky Mountains and Pacific Ranges their highlands climate. Coniferous forests cover the middle elevations of the western mountains, but

Geography Skills for Life

Dust Bowl to Recovery Farmland that turned to desert during the 1930s has been revived and today produces crops such as wheat and sorghum.

Human-Environment Interaction How did human activity bring about the Dust Bowl?

beyond the timberline, the elevation above which trees cannot grow, only lichens and mosses brave the ever-present cold. In late winter and early spring, a warm, dry wind called the chinook (shuh•NUK) may blow down the eastern slopes of the Rockies. Warming at a rate of about 1°F for every 180 feet (or about 1°C for every 99 meters) that it descends, the chinook rapidly melts and evaporates the snow at the base of the mountains.

Interior Climates

Far from large bodies of water that tend to moderate climate, the **Great Plains**, in the center of the continent, have a humid continental climate with bitterly cold winters and hot summers. Although western mountains do block moisture-bearing Pacific winds, the Great Plains benefit from moist winds that blow north along the Rockies from the Gulf of Mexico and south from the Arctic. The humid continental climate extends into southern Canada.

Student Web Activity Visit the **Glencoe World Geography** Web site at geography.glencoe.com and click on Student Web Activities—Chapter 5 for an activity about the physical features of North America.

Prairies

In the Great Plains of the United States and Canada, prairies, or naturally treeless expanses of grasses, spread across the continent's midsection. Each year, rainfall ranging from 10 to 30 inches (26 to 76 cm) waters tall prairie grasses, such as switchgrass and bluestem. Towering 6 to 12 feet (1.8 to 3.7 m) high, these grasses can grow as much as half an inch (1.3 cm) a day. In the Great Plains and the eastern United States, violent spring and summer thunderstorms called supercells spawn tornadoes, twisting funnels of air whose winds can reach 300 miles (483 km) per hour.

History

The Dust Bowl

The tangled roots of prairie grasses once formed dense, solidly packed layers of sod on the Great Plains. Then settlers broke up the sod to grow crops. When dry weather blanketed the plains in the 1930s, the wind eroded unprotected topsoil, reducing farmlands across several U.S. states to a barren wasteland called the Dust Bowl. The resulting economic hardships, made worse by the Great

Depression, caused mass migrations of people. Since the 1930s, improved farming and conservation methods have restored this region's soil.

Eastern Climates

The humid subtropical climate of the southeast has long, muggy summers and mild winters. Deciduous forests extend as far south as Louisiana, but land has been cleared for farming along the Mississippi River. Wetlands and swamps like Florida's **Everglades** shelter a great variety of vegetation and wildlife. In late summer and early autumn, hurricanes—ocean storms hundreds of miles wide with winds of 74 miles per hour (119 km per hour) or more—can pound the region's coastlines.

A humid continental climate extends from the northeastern United States into southeastern Canada. In Canada, a band of deciduous and mixed deciduous-coniferous forestland more than 1,375 miles (2,213 km) wide sweeps from **Newfoundland** into the subarctic **Yukon Territory**. In the United States, deciduous forests grow at lower elevations in the south. In winter, much of northern North America experiences blizzards with winds of more than 35 miles per hour (56 km per hour), heavy or blowing snow, and visibility of less than 1,320 feet (402 m) for three hours or more. On the East coast hazardous winter weather may disrupt travel.

Geography Skills for Life

New Hampshire Forest During autumn in the northeastern United States, deciduous forests show a dazzling display of colors.

Place Where are humid continental climate regions located in the United States and Canada?

Tropical Climates

Within the continental United States, only the extreme southern tip of Florida has a tropical savanna climate. Hawaii, 2,400 miles (3,862 km) west of the mainland, and the Caribbean island of Puerto Rico have tropical rain forests. The wide variety of climates and vegetation in the United States and Canada has helped shape the region's history.

SECTION 2 ASSESSMENT

Checking for Understanding

1. **Define** timberline, chinook, prairie, supercell, hurricane, blizzard.
2. **Main Ideas** Use a Venn diagram to compare the climate and vegetation of the United States and Canada.

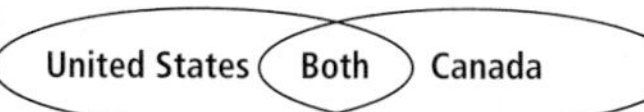

Critical Thinking

3. **Making Comparisons** How do the Pacific winds and the Arctic winds differ in their impact on climate?
4. **Problem Solving** How might the conditions that caused the 1930s Dust Bowl disaster have been avoided?
5. **Comparing and Contrasting** How do hurricanes and tornadoes differ?

Analyzing Maps

6. **Region** Study the maps on pages 122 and 123. Identify the three largest climate regions and the vegetation common in each.

Applying Geography

7. **Effects of Climate** Describe and explain the environmental factors that have affected human migration in the region.

Reading a Relief Map

When you plan a walk, do you prefer an easy stroll along flat ground, or do you look for a challenging hike up and down steep hills? By using a relief map, you can determine the elevation of the terrain you are going to cover.

Learning the Skill

A **relief map** is a special purpose map that shows variation in height, or elevation, of land areas. All elevation is measured from sea level, the average level of water in the world's oceans. Mapmakers label this elevation level zero feet (0 m). The actual elevation of some places is shown as a negative number because they lie below sea level.

It is not possible for a relief map to show the elevation of every single inch of land. As a result, areas are grouped together. A map may show all areas with an elevation between sea level and 1,000 feet (305 m) colored green. Within that area no hill will be higher than 1,000 feet (305 m) and no valley lower than sea level.

Follow these steps to read a relief map:

- **Note the title of the map.**
- **Study the map key.** Relief maps generally use colors or shaded areas to identify elevation.
- **Compare the relief map with other maps.** Observe how elevation affects climate, population distribution, and economic activity in an area.

Practicing the Skill

Refer to the relief map shown here to answer these questions.

1. What is the color of the map's highest elevation?
2. What elevation range does the color green indicate in feet? In meters?
3. What color is the elevation range of 2,000 to 5,000 feet (600 m to 1,500 m)?
4. At what elevation is the state of Mississippi?
5. What are the elevation levels as you travel west from New Jersey to Ohio?

Appplachian Region: Physical–Political

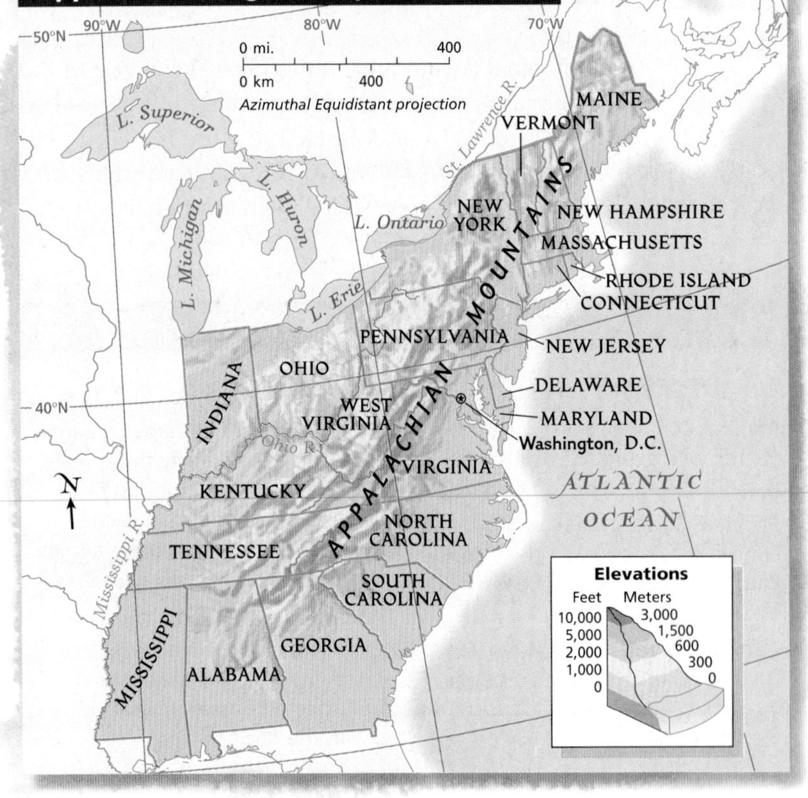

50°N 90°W 80°W 70°W 40°N
0 mi. 400
0 km 400
Azimuthal Equidistant projection
L. Superior, L. Michigan, L. Huron, L. Erie, L. Ontario, St. Lawrence R., Ohio R., Mississippi R.
MAINE, VERMONT, NEW HAMPSHIRE, MASSACHUSETTS, RHODE ISLAND, CONNECTICUT, NEW YORK, NEW JERSEY, PENNSYLVANIA, DELAWARE, MARYLAND, Washington, D.C., OHIO, INDIANA, WEST VIRGINIA, VIRGINIA, KENTUCKY, TENNESSEE, NORTH CAROLINA, SOUTH CAROLINA, GEORGIA, ALABAMA, MISSISSIPPI
APPALACHIAN MOUNTAINS
ATLANTIC OCEAN
N

Elevations
Feet: 10,000; 5,000; 2,000; 1,000; 0
Meters: 3,000; 1,500; 600; 300; 0

Compare the relief map of the United States and Canada on page 117 with the population density map on page 109. Then write a paragraph explaining how elevation affects population distribution.

The **Glencoe Skillbuilder Interactive Workbook, Level 2** provides instruction and practice in key social studies skills.

CHAPTER 5
SUMMARY & STUDY GUIDE

SECTION 1 The Land (pp. 115–120)

Terms to Know
- divide
- headwaters
- tributary
- fall line
- fishery

Key Points
- Canada and the continental United States have similar landforms, shaped by similar geologic processes. Both have high, sharp mountains and dry plateaus in the west; rolling, grassy plains in the center; and lower, older mountains and coastal lowlands in the east.
- The region's waterways, including rivers, lakes, coastal waters, and intracoastal channels, played a vital role in settling the land and continue to serve as commercial highways.
- The Continental Divide divides the region into two large drainage areas. To the east of the Divide, waters flow to the Arctic Ocean, to Hudson Bay, to the Atlantic Ocean, or to the Gulf of Mexico. To the west, they flow into the Pacific Ocean.
- Glacial movement shaped much of the North American landscape.
- The geologic factors that shaped the United States and Canada also provided the region with a wealth of natural resources.

Organizing Your Notes

Use a table like the one below to help you organize the notes you took as you read this section.

Physical Feature	Location
Cascade Range	
Great Plains	
Canadian Shield	
Appalachian Mountains	

SECTION 2 Climate and Vegetation (pp. 121–125)

Terms to Know
- timberline
- chinook
- prairie
- supercell
- hurricane
- blizzard

Key Points
- The region encompassing the United States and Canada experiences a great variety of climates.
- Some climate regions of the United States and Canada are influenced primarily by latitude.
- Wind, ocean currents, rainfall patterns, and elevation moderate the effects of latitude in other climate zones of the United States and Canada.
- Climatic factors cause hazardous seasonal weather patterns in the United States and Canada, including spring and summer tornadoes, and summer and fall hurricanes, and winter blizzards.
- The region's natural vegetation reflects its climatic variety, but human interaction with the environment has greatly altered natural vegetation.

Organizing Your Notes

Use diagrams like the one below to organize your notes under the following headings: Climate Regions, Seasonal Weather Patterns, and Vegetation.

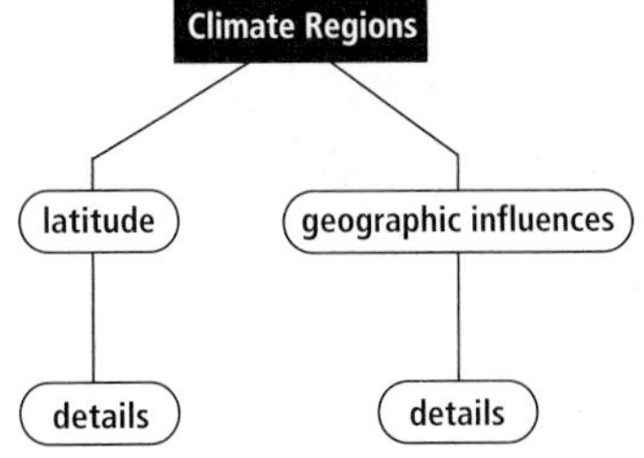

CHAPTER 5 ASSESSMENT & ACTIVITIES

Reviewing Key Terms

Write the key term that best completes each sentence. Refer to the Terms to Know in the Summary & Study Guide on page 127.

1. __________ supply great quantities of fish and other sea animals to North America.
2. The warm, dry wind, or __________, melts snow at the base of the Rockies.
3. Lichens and mosses grow above the __________.
4. Spring and summer tornadoes are spawned by a violent thunderstorm called a(n) __________.
5. Farmers on the wide grasslands, or __________, of the Great Plains broke up sod to grow crops.
6. Many North American rivers have their __________, or source, in the Rocky Mountains, where a(n) __________ determines the direction of the rivers' flow.
7. Important cities grew up along the __________, where the Piedmont drops to the Atlantic Coastal Plain.
8. A(n) __________ of the Mississippi River may be a stream or small river.

Reviewing Facts

SECTION 1

1. How were the Pacific Ranges formed?
2. What effect does the Continental Divide have on the direction rivers flow?

SECTION 2

3. What kind of climate is common in most of the United States and southern Canada?
4. Name two types of vegetation in this region.

Critical Thinking

1. **Analyzing Information** What geologic processes shaped much of this region?
2. **Drawing Conclusions** Why should the United States and Canada protect their natural vegetation?
3. **Classifying Information** On a web diagram, fill in information about the kinds of vegetation found in each of the region's climate zones.

Climate Zones and Vegetation

NATIONAL GEOGRAPHIC

Locating Places

The United States and Canada: Physical Geography

Match the letters on the map with the physical features of the United States and Canada. Write your answers on a sheet of paper.

1. Rocky Mountains
2. Great Plains
3. Appalachian Mountains
4. Canadian Shield
5. Great Lakes
6. Mississippi River
7. Hudson Bay
8. Great Bear Lake
9. Pacific Ranges
10. Mackenzie River
11. Rio Grande
12. Great Slave Lake

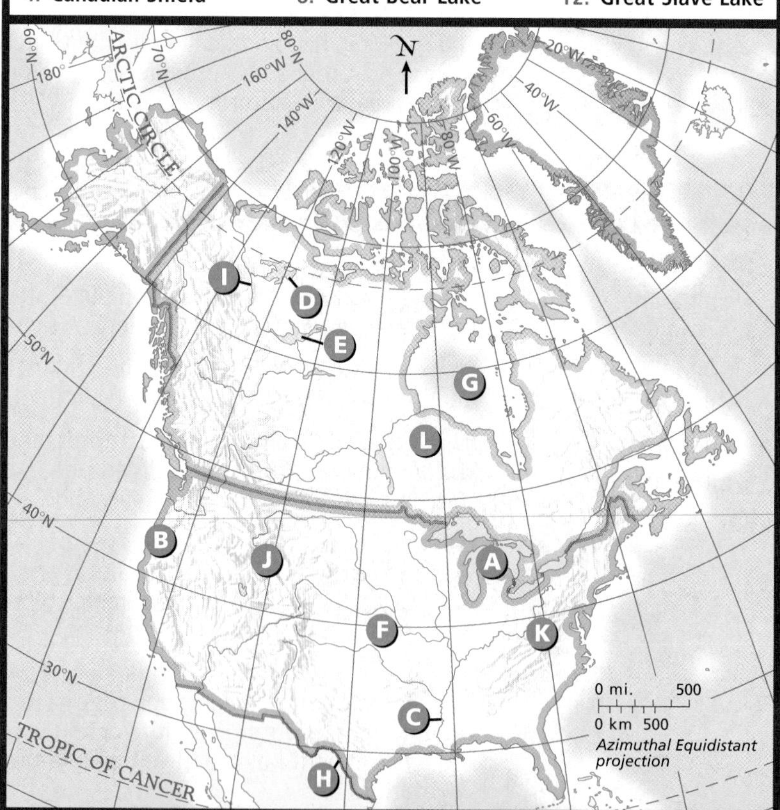

Self-Check Quiz Visit the **Glencoe World Geography** Web site at geography.glencoe.com and click on **Self-Check Quizzes—Chapter 5** to prepare for the Chapter Test.

Using the Regional Atlas

Refer to the Regional Atlas on pages 106–109.

1. **Region** How are the eastern and western halves of the United States and Canada different?
2. **Location** On the physical map, locate rivers that flow into the Mississippi. Then use the economic activity map to make a list of products that might be shipped using these rivers.

Thinking Like a Geographer

The region of the United States and Canada possesses natural resources that people depend on for survival. Choose one section of the region, and write a paragraph explaining how people depend on a natural resource in that area.

Problem-Solving Activity

Group Research Project The flooding of the Mississippi River floodplain in 1993 caused billions of dollars worth of damage and raised questions about the wisdom of controlling the flow of major rivers with dams and levees. Should rivers be allowed to take their natural course? In your group, choose who will argue for controlling rivers and who will argue against it. Be sure to give a fair presentation of the data, including supportive evidence on the pros and cons.

GeoJournal

Expository Writing Using the information you logged in your GeoJournal as you read, write a paragraph explaining how one of the region's physical features affects its inhabitants. Use your textbook and Internet resources to make your explanation clear and accurate.

Technology Activity

Using the Internet for Research Think about the effects of physical processes, landforms, and climate. Then use reliable Internet resources to find out more about one way in which life in your area is shaped by physical geography. Write a report, and share it with the class.

Standardized Test Practice

Choose the best answer for the following multiple-choice question. If you have trouble answering the question, use the process of elimination to narrow your choices.

Elevations: Selected U.S. Cities

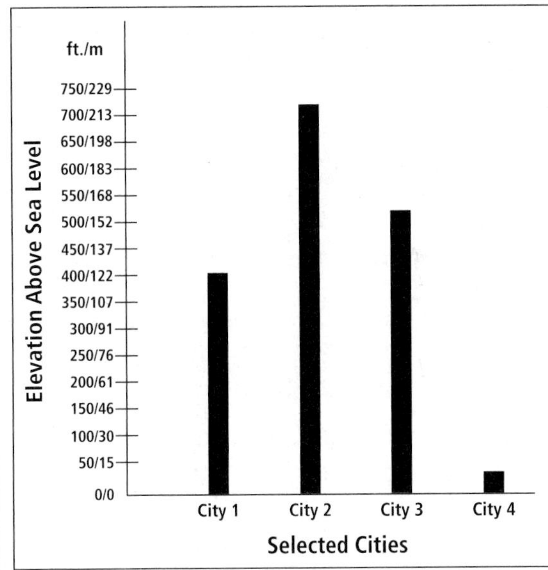

1. **Given the information shown in the bar graph, which city is most likely located east of the fall line in the eastern United States?**

   A City 1 **C** City 3
   B City 2 **D** City 4

Test-Taking Tip To determine which city is east of the fall line, remember that the fall line is where the higher land of the Piedmont drops to the lower Atlantic Coastal Plain to the east. Eliminate those choices that do not indicate a city on the coast, near sea level.

Geography Lab Activity

Comparing Soils

▲*Students collect soil samples for experiments.*

You may think that all soil is alike, but there are many different varieties. Several factors account for soil differences. The parent rock, or the type of rock from which soil is formed, is one factor. Weathering breaks down parent rock to produce different types of soil. For example, if limestone is the parent rock, it will produce a different soil than if sandstone were the parent rock. Climate, types of vegetation, and the slope of the land surface also affect soils.

The color of the soil indicates the presence of certain minerals or other substances. Sandy soil is usually light in color. Soil rich in humus is dark in color because of the presence of decaying plant and animal matter. Red soils are colored by large amounts of iron-bearing minerals. Different types of soil are found in the United States, including mountain soils, prairie soils, river soils, glacial soils, and desert soils.

1 Materials

- Computers with Internet access
- Large map of your state, with counties outlined and identified
- Small plastic envelopes or bags for soil samples
- Labels for the plastic envelopes or bags
- Pushpin or thumbtacks

2 Procedures

In this activity, you will use the Internet and other resources to compare soils in your state and explain why differences exist among them.

1. Using the Internet, locate e-mail addresses for as many other schools throughout your state as you can. Save the e-mail addresses in your program's address book.
2. Collect the e-mail addresses and postal addresses of your friends and relatives throughout your state.
3. Coordinate all of the addresses to ensure that there is full coverage and that counties or regions are not duplicated.
4. Send e-mail messages or handwritten notes to all your partners, asking each to send you a small soil sample of his or her area. Ask partners to identify exactly where the soil came from (for example, "from my yard," or "from the hillside behind my house"). Explain that you are trying to identify why soil samples are different within an area. Ask that the soil samples be sent as soon as possible.

5. Ask partners to sterilize the soil in a 350° oven for about 15 minutes before sealing the cooled soil sample in a small plastic bag.
6. As soil samples arrive, put each one in a separate plastic envelope or bag. Label each envelope or bag with the name and location of the person who sent it. Using the map below, try to identify the type of soil your partners sent.
7. Research to find the characteristics of that area's soils. Does the soil sample reflect information found in your research?
8. Using the pushpins, place each soil sample on the map in the area of its origin.
9. Describe how the location of where these soils were found affects how they are similar to or different from each other.

3 Lab Report

1. How many soil samples were collected?
2. What were some of the factors that accounted for the differences in the soils?
3. **Drawing Conclusions** Write a paragraph explaining how the differences in the soils collected do or do not reflect the economic activity of at least two areas of the state. Give reasons for your conclusions.

4 Find Out More

Identify a climate area in Canada or the United States, and research to find out, for example, how soils in that climate differ from each other. How has the climate affected the soil? The area's vegetation?

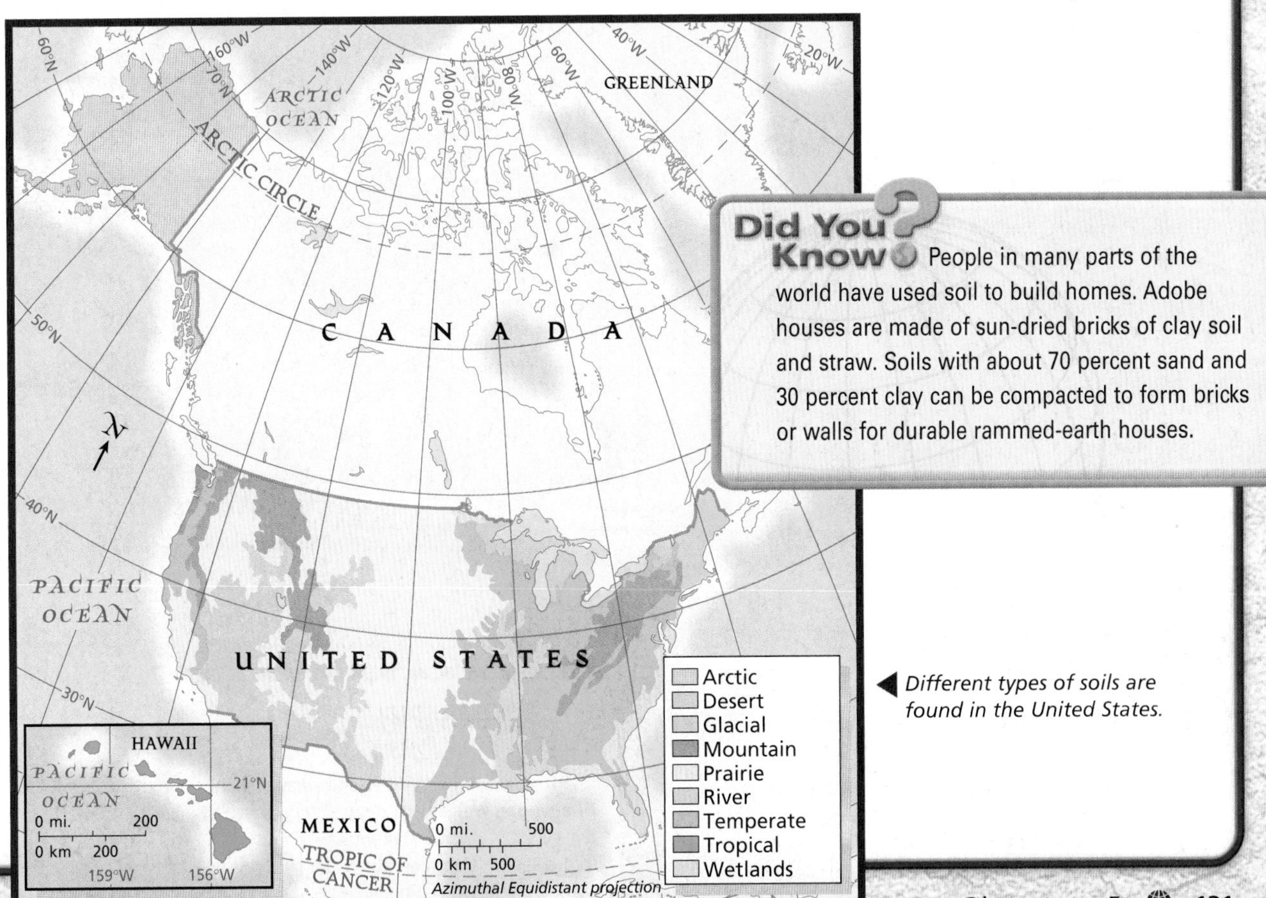

Did You Know? People in many parts of the world have used soil to build homes. Adobe houses are made of sun-dried bricks of clay soil and straw. Soils with about 70 percent sand and 30 percent clay can be compacted to form bricks or walls for durable rammed-earth houses.

Different types of soils are found in the United States.

CHAPTER 6

The Cultural Geography of the United States and Canada

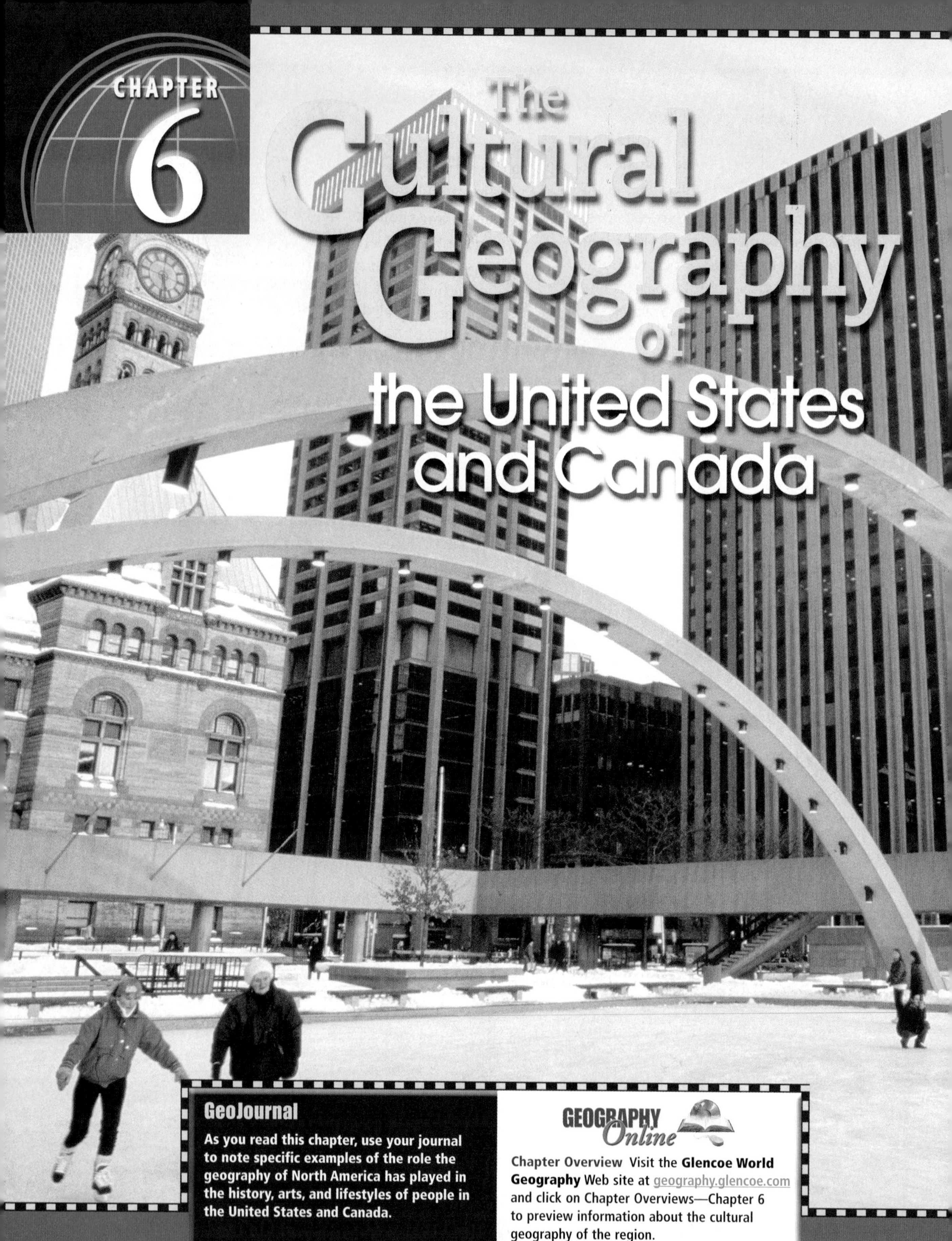

GeoJournal

As you read this chapter, use your journal to note specific examples of the role the geography of North America has played in the history, arts, and lifestyles of people in the United States and Canada.

GEOGRAPHY Online

Chapter Overview Visit the **Glencoe World Geography** Web site at geography.glencoe.com and click on Chapter Overviews—Chapter 6 to preview information about the cultural geography of the region.

Population Patterns

Guide to Reading

Consider What You Know

Think about news reports, movies, and songs you know that feature large, densely populated cities as topics or settings. What factors make urban living appealing to people?

Read to Find Out

- Who are the peoples of the United States and Canada?
- How are population patterns in the United States and Canada influenced by the region's physical geography?
- What geographic factors encouraged the industrialization and urbanization of the United States and Canada?

Terms to Know

- immigration
- Native American
- Sunbelt
- urbanization
- metropolitan area
- suburb
- megalopolis
- mobility

Places to Locate

- Washington, D.C.
- Miami
- New Orleans
- Houston
- Los Angeles
- Vancouver
- Ottawa
- Detroit

NATIONAL GEOGRAPHIC

A Geographic View

The Next Wave

For a century and a half this part of lower Manhattan has functioned as a catchment [holding place] for successive waves of poor immigrants, including Irish, Germans, Italians, and East European Jews. Each enclave [separate cultural community] dissolved as second and third generations seized the opportunities that education afforded them and then moved on to better neighborhoods, greener suburbs, or distant cities. But lower Manhattan, with its inexpensive housing, remained, ready to absorb the next wave.

—*Joel L. Swerdlow, "New York's Chinatown,"* National Geographic, *August 1998*

Chinatown, New York City

Lower Manhattan, a part of New York City, has a history that reflects population patterns throughout the United States and Canada. Both of these North American countries have been shaped by **immigration**, the movement of people into one country from another. In this section you will read about the peoples of the United States and Canada—who they are, why they came to the region, and where they live.

The People

About 5 percent of the world's population lives in the United States and Canada. The 285 million people of the United States and the 31 million people of Canada all are immigrants or descendants of immigrants. Some arrived only recently. Others belong to families whose ancestors came to North America centuries ago.

◀ *The city skyline rises behind an outdoor ice rink in Toronto, Ontario, Canada.*

Geography Skills for Life

Northwest California The Yurok Native American group is only one of many groups that help define the populations of the United States and Canada.

Place About how many Native Americans are there in the United States? Canada?

History

Waves of Immigrants

North America's first immigrants probably moved into the region from Asia thousands of years ago. Today their descendants, known as **Native Americans**, number 2.5 million in the United States and 700,000 in Canada. Other peoples—Europeans, Asians, Africans, and Latin Americans—came later. As a result of these waves of immigrants, the populations of the United States and Canada are among the world's most diverse.

Some immigrants came to the United States and Canada to seek political and religious freedom and to find better economic opportunities. Others fled wars or natural disasters. For example, the Irish potato famine of the 1840s caused about 1.5 million Irish people to immigrate to the United States.

Rich natural resources and the region's rapid industrial and economic development made the United States and Canada attractive destinations. Popular songs among European immigrants in the 1800s referred to the United States as the land "where the streets are paved with gold." Chinese immigrants nicknamed it "Gold Mountain." The rumors of gold were exaggerated, but the opportunities were real. Some immigrants faced discrimination at first, but they offered hard work, talent, enthusiasm, and diverse cultural practices. Throughout their histories, the United States and Canada have benefited from the contributions of immigrants.

Population Density and Distribution

Although the United States and Canada are "nations of immigrants," their populations differ in terms of density and distribution. Slightly larger than the United States in land area, Canada has an average population density of only 8 people per square mile (3 people per sq. km). Much of Canada's vast territory is inhospitable to human settlement because of rugged terrain and a bitterly cold climate. About 90 percent of Canadians live in a narrow strip of land along Canada's border with the United States. The poor soil of the Canadian Shield steered settlement toward the fertile land and industrial resources of the Great Lakes–St. Lawrence lowlands. Other population centers include the farming and ranching areas along the southern sweep of the Prairie Provinces of Manitoba, Saskatchewan, and Alberta and the Pacific coast of British Columbia.

Compared with Canada, the United States, with an average population density of 77 people per

Student Web Activity Visit the **Glencoe World Geography** Web site at geography.glencoe.com and click on Student Web Activities—Chapter 6 for an activity about the history of immigration to the United States.

square mile (30 people per sq. km), may seem relatively crowded. Outside large urban areas, however, the population is widely distributed. The Northeast and the Great Lakes regions are the most densely populated areas and the historic centers of American commerce and industry. Another population cluster lies on the Pacific coast, where pleasant climate, abundant natural resources, and economic opportunities attract residents. More people live in California than in any other state.

Since the 1970s the American South and Southwest, including California, have become the country's fastest growing areas. Nicknamed the **Sunbelt** for its mild climate, the southern United States draws employees to its growing manufacturing, service, and tourism industries. Retirees choose the Sunbelt for its mild winters. The area's geographic closeness to Mexico and the Caribbean also draws immigrants from those two regions.

The least densely populated areas of the United States include the subarctic region of Alaska, the parched Great Basin, and parts of the arid or semi-arid Great Plains. These areas owe their sparse populations to difficult climate conditions.

The Cities

Although both the United States and Canada began as agricultural societies, they have experienced **urbanization**, the concentration of population in cities. Cities grew as the use of machines in agriculture gave rise to large commercial farms. As a result, fewer agricultural laborers were needed, sending people to urban areas to search for work. Jobs, education, health care, and cultural opportunities also have drawn people to large cities.

Today most people in the United States and Canada live in metropolitan areas. A **metropolitan area** includes a city with a population of at least 50,000 and outlying communities called **suburbs**. More than 80 percent of the population of the United States lives in the country's 276 metropolitan areas. Canada's 25 metropolitan areas are home to about 60 percent of the Canadian population.

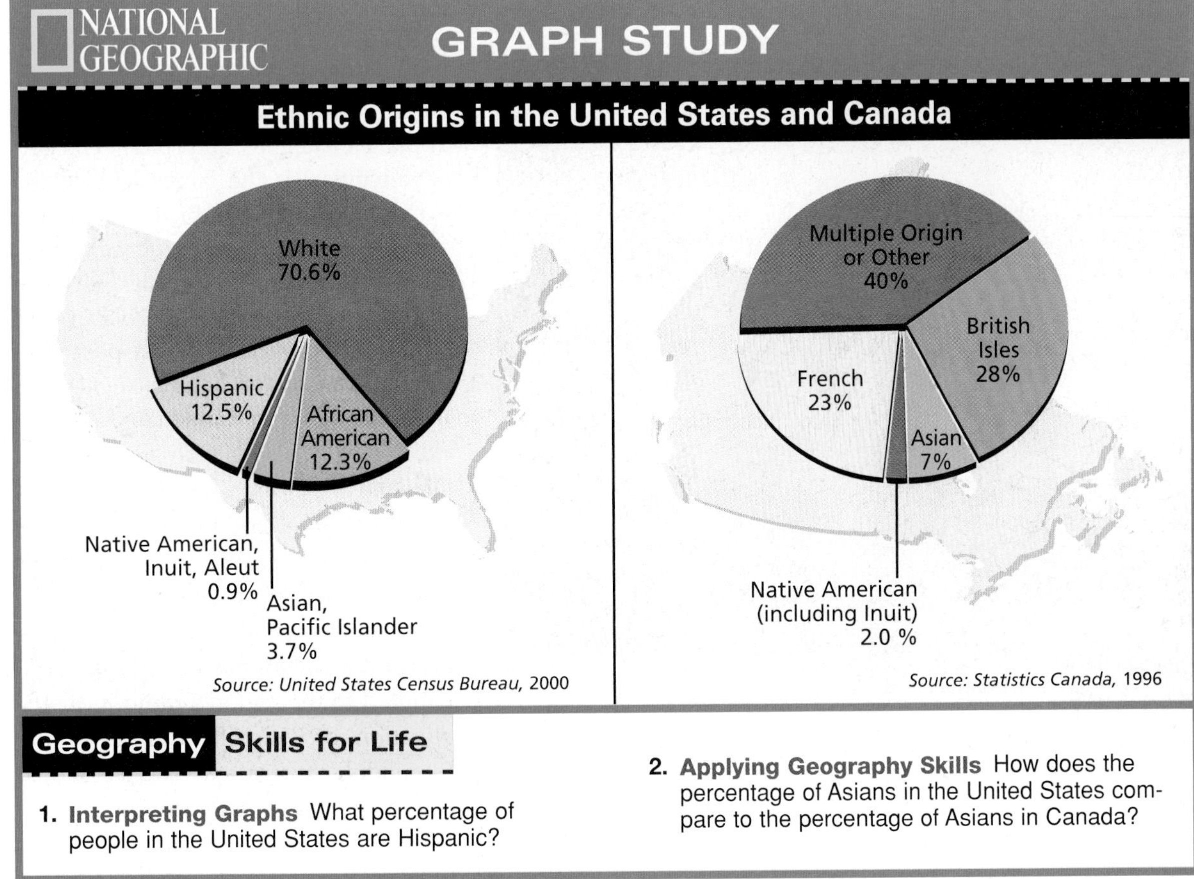

Geography Skills for Life

1. **Interpreting Graphs** What percentage of people in the United States are Hispanic?

2. **Applying Geography Skills** How does the percentage of Asians in the United States compare to the percentage of Asians in Canada?

Coastal Cities

Many population centers in the United States and Canada lie in coastal areas where healthy economies support large populations. Along the northern Atlantic coast of the United States, for example, a chain of closely linked metropolitan areas forms a **megalopolis**, or "great city." Home to about 42 million people, this megalopolis—nicknamed Boswash—includes the cities of Boston, New York, Philadelphia, Baltimore, and Washington, D.C. Four of the cities—Boston, New York, Philadelphia, and Baltimore—are important world trade centers because of their coastal or near coastal locations. The planned city of **Washington, D.C.**, established on the Potomac River near the Chesapeake Bay, is the country's capital. On the capital's 200th birthday, a native Washingtonian commended the city's chief designer:

> *"Most American cities grew haphazardly . . . with little overall planning. . . . But our nation's capital became one of the most attractive low-rise cities . . . in the world . . . [because of] the vision of Pierre L'Enfant who conceived the plan for the capital. . . ."*
>
> Gilbert M. Grosvenor, "Washington, D.C., Reaches Its 200th Birthday," *National Geographic*, August 1991

Other important U.S. coastal cities include the busy ports of **Miami**, on the Atlantic coast, and **New Orleans** and **Houston**, on the Gulf of Mexico. Houston, connected to the Gulf of Mexico by the Houston Ship Channel, is the southern end of a developing megalopolis that stretches north to the Dallas/Fort Worth metropolitan area.

Pacific coast cities also provide important commercial links to the rest of the world, especially to the growing Asian economies of the Pacific Rim. A developing megalopolis stretches from San Francisco south through **Los Angeles** to San Diego. All three cities have major ports. Another western port city, Seattle, as well as San Francisco and the neighboring area nicknamed the Silicon Valley, features innovative computer and Internet industries. These latter two areas also have developed aerospace industries, enterprises that design and manufacture airplanes, satellites, and space vehicles.

Vancouver is the largest city in the Canadian province of British Columbia and an important shipping center for western Canada. Despite its northern location, Vancouver's harbor never freezes, so ships use the busy port year-round. Vancouver handles nearly all the trade between Canada and Asia.

NATIONAL GEOGRAPHIC **World Explorer**

Geography Skills for Life **The United States at Night** City lights as seen at night from space reveal dense population clusters in the east and on the west coast.

Region What pattern do you observe in the distribution of cities in the United States and Canada?

Inland Cities

Rivers, lakes, and inland waterways promoted the growth of the region's inland cities. North America's waterways offered both natural resources and transportation routes that contributed to the region's rapid industrialization, or the shift from agriculture to manufacturing and service industries as the basis of an economy.

In Canada, ships reach the cities of Quebec, Montreal, Toronto, and **Ottawa** through the St. Lawrence River, the Ottawa River, and the Great Lakes. **Detroit**, the center of the United States automobile industry, uses the Great Lakes for shipping goods. A megalopolis links the U.S. Great Lakes cities of Chicago, Milwaukee, and Cleveland with Pittsburgh, a freshwater port on the Ohio River. Other U.S. river cities include Cincinnati, on the Ohio River, and Minneapolis and St. Louis, on the Mississippi River. Winnipeg, on the Red River, and Saskatoon and Edmonton, on the Saskatchewan River, are inland population centers in western Canada.

Other inland cities, such as Atlanta, Denver, Dallas, and San Antonio in the United States and Regina and Calgary in Canada, grew from agricultural and trading centers.

NATIONAL GEOGRAPHIC **World Explorer**

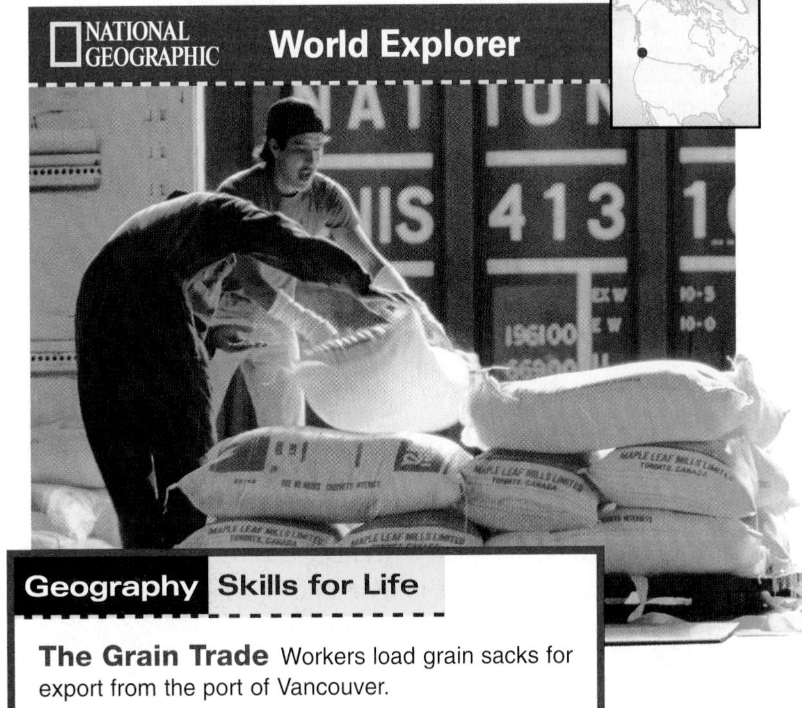

Geography Skills for Life

The Grain Trade Workers load grain sacks for export from the port of Vancouver.

Movement How do port cities sustain an economy?

Future Trends

Like most developed countries, the United States and Canada have low birthrates, which increase the population by only 0.5 percent annually. Immigration accounts for most of the region's population growth. In 1998 more than 9 percent of the population of the United States was born in another country. Like earlier immigrants, however, the people of the United States and Canada cherish their **mobility**, the freedom to move from place to place. In a typical year, one in six U.S. residents of the United States relocates, often to cities. As immigration adds to population diversity, living with cultural differences and managing urban congestion are ongoing challenges.

SECTION 1 ASSESSMENT

Checking for Understanding

1. **Define** immigration, Native American, Sunbelt, urbanization, metropolitan area, suburb, megalopolis, mobility.
2. **Main Ideas** Create a word web like the one below, listing the various peoples of North America.

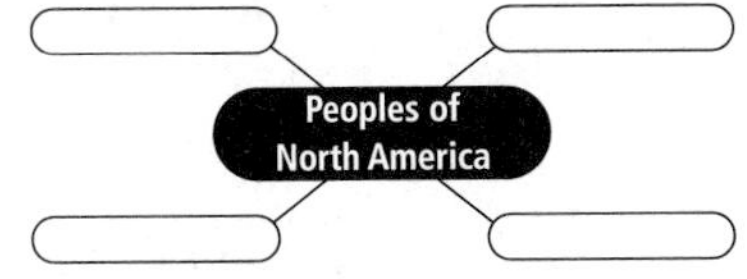

Critical Thinking

3. **Comparing and Contrasting** How are the population patterns of Canada and the United States similar? How do they differ?
4. **Making Inferences** What are some of the advantages and disadvantages of living in a megalopolis?
5. **Categorizing Information** Select three coastal and three inland cities and indicate the economic activities important to each city.

Analyzing Maps

6. **Human-Environment Interaction** Study the population density map on page 108. How many cities with populations over 1,000,000 lie along waterways? Explain.

Applying Geography

7. **Choosing a Destination** Imagine that you are an immigrant writing a letter to relatives about your new home in the United States. Explain your reasons for settling where you live.

GIVE-AND-TAKE ACROSS THE BORDER

CANADA AND THE UNITED STATES share the longest undefended border in the world. The 3,987-mile (6,416-km) border runs through the middle of rivers and lakes, crosses fields and forests, and slices through towns and farms. Each year more than a hundred million tourists, truck drivers, sports fans, and other visitors pass through 96 official border crossings and thousands of unofficial ones. Even more impressive, though, is the value of goods and services that flow between the United States and Canada—a total that exceeds trade between any other two countries in the world.

A History of Trade

Trade has played an important role in the growth of both the United States and Canada. The fur trade in Canada began in the 1500s. Native Americans gave Europeans furs in exchange for such items as tools, weapons, and kettles. During the 1700s, colonists traded numerous raw materials—timber and furs—for Europe's manufactured goods. In the early 1800s, the quest for furs by American and Canadian companies pushed frontiers westward as trading posts sprang up in the wilderness.

Cars line up to enter Canada at Fort Erie, Ontario. ▶

Each year, Canadians and Americans exchange products and people. Investors seek profits in real estate, mines, oil, and other ventures.

NATIONAL GEOGRAPHIC SOCIETY

As the neighboring economies grew and prospered, so did cooperation. Since the late 1800s, trade has flourished between Canada and the United States. Most goods pass freely across the border, without tariffs of any kind. Two major agreements have sought to eliminate remaining tariffs and other trade barriers: the United States-Canada Free Trade Agreement (FTA) in 1989 and the North American Free Trade Agreement (NAFTA), which includes Mexico, in 1994.

Cooperation and Conflict

Today each country has a major stake in the other's economy. Canadian companies operate plants in the United States and vice versa. Joint business ventures proliferate. The open border and long history of cooperation between the United States and Canada have led to good relations and a friendly give-and-take between neighbors.

Trade disputes do occur, however. Many Canadians dislike the effect free trade with the United States has on their culture and way of life. Canadians struggle to maintain a separate identity while they're bombarded by American music and movies. Moreover, differences arise over shared resources, such as fishing grounds, and over solutions to joint problems, such as pollution.

Looking Ahead

Will the spirit of cooperation between the United States and Canada prevail in this century? Or will trade disputes, disagreements over cultural issues, and other problems lead to conflict?

SECTION 2

History and Government

Guide to Reading

Consider What You Know

You may know that the early history of Canada and the United States featured the movement of people from east to west. In what ways did that movement affect the place where you live?

Read to Find Out

- What was life like for the earliest Americans and for European settlers?
- How did industrialization and technology enable westward expansion in North America?
- How do the governments of the United States and Canada differ?

Terms to Know

- republic
- Underground Railroad
- dry farming
- Constitution
- amendment
- Bill of Rights
- cabinet
- dominion
- Parliament

Places to Locate

- Hudson Bay
- Quebec
- Ontario
- Nova Scotia
- New Brunswick
- Yukon Territory
- Northwest Territories
- Nunavut
- Texas
- Alaska
- Hawaii
- Pennsylvania
- Ohio

NATIONAL GEOGRAPHIC

A Geographic View

Personality and History

History has bred the caricatures. The United States was born of rebellion and the cult of independence. It spread west two hops ahead of the law. Canada was formed by consensus among public servants. On its way west the law went first. Canadians never had a Wild West.

—*Priit J. Vesilind, "Common Ground, Different Dreams: The U.S.–Canada Border,"* National Geographic, *February 1990*

Monument Valley, Arizona

The United States and Canada share much in terms of geography, but they have taken different historical and cultural paths. In this section you will learn how the vast northern part of North America, originally inhabited by Native Americans, then colonized by Europeans, eventually developed into these two independent countries. You also will discover the key role physical geography played in the emergence of the United States and Canada and their development as industrialized countries.

History

Archaeologists generally believe that nomads crossing a land bridge from Asia to Alaska first settled North America thousands of years ago. Recent evidence suggests, however, that nomads from Central and South America may have populated North America at the same time as—or even before—those from Asia. Whatever theory proves correct, we know that as of 10,000 years ago, people lived in almost every part of what is now the United States and Canada.

Native Americans

Location and climate shaped the various cultures later known as Native American. For the peoples of the cold Arctic tundra, scarce resources and lack of farmland prompted them to hunt caribou and other animals for food and fur. By contrast, Pacific Coast peoples enjoyed a mild climate and abundant resources. They harvested salmon with fiber nets and used stone and copper tools to split cedar, fir, and redwood trees into planks for building houses and canoes.

In the high deserts of the Southwest, Native Americans used irrigation to farm the dry land. On the Great Plains, other groups hunted the buffalo, parts of which were used for food, clothing, shelter, and tools. Native Americans in the woodlands east of the Mississippi River built ceremonial mounds, hunted game, grew crops, and traded for shells and freshwater pearls. In the northeastern woodlands, Native American peoples hunted deer, turkeys, geese, and squirrels. These northeastern peoples lived in closely knit villages, developed systems of government, and traded throughout the region.

European Colonies

European migration had begun by the late 1500s. Europeans came to North America in search of land to farm, valuable minerals, and political and religious freedom. Most European migrants came from Spain, France, and England and settled in colonies.

The Spaniards controlled Florida and a large area west of the Mississippi River. Many Spanish settlements were founded as military posts or as missions—religious communities founded to convert Native Americans to Christianity. Spanish colonists also set up farms and huge cattle ranches.

The French came to North America primarily for the fur trade. French trappers canoed down rivers such as the St. Lawrence and the Mississippi. They set up trading posts, collecting beaver pelts and other furs from Native Americans to ship to Europe. Those who settled permanently lived along the St. Lawrence River and the Mississippi River near the Gulf of Mexico.

By the 1700s England had colonies or controlled land along much of the Atlantic coast and around **Hudson Bay**. The settlers in the northern English colonies found that the thin, rocky soil and short growing season made farming difficult. The area, however, had excellent harbors as well as good timber and fishing. Shipbuilding, trade, and fishing became important industries in this region. The middle English colonies had wide river valleys, level land, and fertile soil. They also had mild winters; long, warm summers; and an extended growing season. Many settlers there raised cash crops to be exported. The southern English colonies had mild climates, rich soils, and open land that encouraged plantation agriculture. Most

Geography Skills for Life

Fort Ticonderoga

French, British, and American forces fought over Fort Ticonderoga from 1758 through the American Revolution.

Human-Environment Interaction How did the physical geography of the United States influence settlement in the colonies?

plantation owners used enslaved Africans to provide the labor such large-scale farming required.

Two New Countries

In 1763 France was forced to give up much of its North American empire to Great Britain (formed by the union of England and Scotland in 1707). Conflicts soon arose between Native Americans and colonial settlers. Occupying the land, many settlers pushed out Native American communities and nearly destroyed their cultures.

During the 1760s the British government aroused the American colonists' anger by imposing new taxes and limiting their freedoms. Beginning in 1775, the thirteen British colonies, all of them along or near the Atlantic coast, fought a war for independence. The outcome was a new country—the United States of America. Rejecting monarchy, the Americans set up a **republic**, a government in which people elect their own officials, including their head of state. They elected George Washington as the first president of the United States.

Some American colonists, however, did not want to break ties with the British monarch. As many as 100,000 of these people, known as Loyalists, left the new country. Most settled in French-populated **Quebec,** which Great Britain controlled. During the early 1800s, English- and French-speaking communities in British North America constantly feuded about colonial government policies, but fears of a United States takeover forced them together. In 1867, under Prime Minister John A. Macdonald, four of the colonies—Quebec, **Ontario**, **Nova Scotia**, and **New Brunswick**—united as provinces of the Dominion of Canada, a new country within the British Empire. Neighboring areas—Manitoba, British Columbia, Alberta, and Saskatchewan in the west and Prince Edward Island and Newfoundland along the Atlantic coast—became provinces of Canada during the next 100 years. Today Canada encompasses these 10 provinces and 3 additional territories, the **Yukon Territory**, the **Northwest Territories**, and **Nunavut** (NOO•nuh•vut).

From Sea to Shining Sea

During the 1800s the United States and Canada expanded into western North America. In 1803, for example, the United States bought from France nearly all the land between the Mississippi River and the Rocky Mountains. This agreement, known as the Louisiana Purchase, nearly doubled the size of the country and gave the United States control of the Mississippi River and access to the Far West.

The western lands were rich in natural resources. **Texas,** a former Mexican territory that became an independent republic in 1836 and joined the United States in 1845, was valued for cotton production and cattle ranching. In the late 1840s, as a result of a war with Mexico, the United States gained all of the present-day states of California, Utah, and Nevada and parts of Colorado, Wyoming, Arizona, and New Mexico. The discovery of gold and silver boosted settlement in the region. A traveler on the California Trail during the Gold Rush of 1849 captured the attraction of the West in his journal:

> *"On, on, stay not for those who linger,*
> *on, on, look not for those behind. . . .*
> *America with one heave throws her life*
> *toward Sacramento!"*
>
> C. B. Darwin, quoted in "The Way West," *National Geographic*, September 2000

By trade or treaty, the United States eventually gained control of land from the Atlantic to the Pacific coasts, and from the Canadian border in the north to the Rio Grande in the south. In 1867 the United States purchased its last great frontier, **Alaska**, from Russia. Later it acquired **Hawaii** and some other islands located in the Pacific and the Caribbean. During this period Canada also acquired western lands, spreading from the Atlantic to the Pacific Ocean and from the Arctic region in the north to the United States border to the south. For Native Americans, however, westward expansion by both countries signaled the steady loss of their lands and restrictions on their traditional ways of life.

Economics

Growth, Division, and Unity

In the 1800s industrialization transformed the United States and Canada. The first factories in North America arose in the northeastern United

States, which had many waterfalls that could be harnessed to produce power to run machines. Because waterpower was limited to a few places, people in industry later used steam as a source of power. Large supplies of coal in **Pennsylvania** and **Ohio**, which were used to power steam engines, made steam power cheap and manufacturing very profitable. The Midwestern United States and the Canadian provinces of Ontario and Quebec became leading centers of industry and business. The many rivers and lakes in these areas, improved by the building of canals, were used to transport goods from factories to port cities.

Geography Skills for Life **Underground Railroad** Harriet Tubman (inset) helped enslaved people escape through the Underground Railroad. This was a system of safe houses like John Rankin's home in Ripley, Ohio.

Place How did geographic and economic factors help lead to the Civil War?

A growing demand for cotton by the textile industry in the northeastern United States made cotton production highly profitable. Cotton became the South's major cash crop. Swamps were drained and pine forests cleared for more cotton plantations. For plantation owners, the labor of enslaved Africans became more important than ever before.

Other people, however, worked to end slavery by enabling enslaved people to escape from bondage. The **Underground Railroad**, an informal network of safe houses, helped thousands of escaping enslaved people make their way north to freedom. Many escapees found shelter in Canada, which never practiced slavery and refused to honor U.S. laws that punished those who escaped.

Disputes over slavery, along with economic and political differences between Northern and Southern states, led to the American Civil War of 1861–1865. Under President Abraham Lincoln, the Northern states defeated the Southern states that had left the Union. After the war the United States abolished slavery and gave formerly enslaved African Americans citizenship, equal protection under the law, and the right to vote. Reunited, the United States set about rebuilding itself.

Technological and Social Change

During the late 1800s, the United States and Canada both encouraged settlement of the Great Plains. The United States and Canadian governments wanted to ease the crowding in eastern cities caused by immigration from Europe. They also wanted people to farm the region, thus providing more food for urban populations. Thousands of people from Atlantic coastal areas, as well as immigrants from eastern Europe and Scandinavia, started farms on the Great Plains.

Because of dry conditions, settlers on the Great Plains developed a special farming method, called **dry farming**, cultivating the land so that it caught and held rainwater. Strong steel plows, better able to break the hard prairie sod, soon replaced iron plows. Steam tractors made it possible to plant and harvest large areas of land faster and easier.

The late 1800s also saw the completion of transcontinental railroads in the United States and Canada. These made it possible to transport manufactured goods from east to west, as well as food products—especially beef cattle—from west to east. Immigrants from China, Ireland, Mexico, and other countries were recruited to help build the railroads.

During the early 1900s, the introduction of assembly lines for mass production cut the cost and the time needed to make many industrial products. Perhaps the most influential mass-produced item was the automobile. At this time people were becoming increasingly mobile, and more of them lived in urban areas than in rural areas.

Two world wars during the 1900s spurred economic growth in the United States and Canada. After 1940 both countries were linked in a close partnership that included increased trade between them. By the 1990s certain economic activities, mining and steel production, for example, were less important than rising high-tech industries. Social changes also took place. Immigration increased from Latin America and Asia. Women, African Americans, Hispanics, and other groups began to participate in business and the political process. In both Canada and the United States, Native Americans have negotiated with governments over land claims, mineral rights, and other issues. In 1999 the Inuit, one of Canada's native peoples, won the right to their own territory, called Nunavut, carved from the eastern half of the Northwest Territories.

NATIONAL GEOGRAPHIC **World Explorer**

Geography Skills for Life

Growth of Railroads Thousands of Chinese workers, such as these on the Northern Pacific Railway, helped build the railroads across western North America in the late 1800s.

Movement How did the completion of transcontinental railroads affect the United States?

Government

The United States and Canada both are democracies with federal systems, in which the national government shares power with state or provincial governments. To create a strong national government while preserving the rights of individual states and citizens, United States leaders in 1787 drafted a plan of government called the **Constitution**. Over the years, changes in the Constitution, called **amendments**, have been made to meet the country's changing needs. The first 10 amendments, called the **Bill of Rights**, guarantee the basic rights of citizens, including the freedoms of speech, religion, and the press.

The national government of the United States has three branches: executive, legislative, and judicial. The executive branch includes the president, the vice president, and the executive departments that administer various divisions of the national government. The heads of these departments form the president's **cabinet**—a group of special advisers. Congress, consisting of elected state representatives to both the Senate and the House of Representatives, is the legislative branch. The Supreme Court and lower federal courts make up the judicial branch.

Canada was created as a **dominion**, a partially self-governing country with close ties to Great Britain. It gained full independence in 1931, but the British government kept the right to approve changes to Canada's constitution. In 1982 this legislative link to Great Britain finally ended. Canada at its founding had a strong central government with only minor powers given to the individual provinces. Over the years, the power of the provinces has increased.

Today the executive part of Canada's government includes the governor-general, the prime minister, and the cabinet. The British monarch still serves as the head of state, appointing a governor-general to act in his or her place. The national legislature, called **Parliament**, is made up of the Senate and the House of Commons. Canada's prime minister, who is leader of the majority political party in Parliament, is the actual head of government. Nine judges sit on the Supreme Court of Canada, the country's highest court.

In the next section, you will learn about the culture and lifestyles in the United States and Canada as they enter the new millennium.

Geography Skills for Life

House of Parliament

Ottawa's location on the border between Quebec and Ontario helped determine its role as Canada's capital city.

Place How does Canada's prime minister derive his or her position?

SECTION 2 ASSESSMENT

Checking for Understanding

1. **Define** republic, Underground Railroad, dry farming, Constitution, amendment, Bill of Rights, cabinet, dominion, Parliament.
2. **Main Ideas** Create a time line like the one below, and label major events in the settlement and development of the United States and Canada.

Native Americans settled in region

circa 8000 B.C.

Critical Thinking

3. **Identifying Cause and Effect** How did physical geography influence the cultures of the region's first settlers?
4. **Drawing Conclusions** Why is the influence of French culture more pronounced in Canada than in the United States?
5. **Making Generalizations** Trace the spread of railroads in the United States and Canada. Describe the effects of the railroad on cultural sharing and national unity in both the countries.

Analyzing Maps

6. **Region** Study the political map on page 107. Then, without looking at this map, label the U.S. states and Canadian provinces/territories on an outline map.

Applying Geography

7. **Effects of Technology** How did people meet the challenges of settling the West through innovation and change? Write a brief essay describing these changes and how they affected the region's physical and human geography.

Cultures and Lifestyles

Guide to Reading

Consider What You Know

The arts and popular entertainment of the United States and Canada influence other culture regions of the world. What cultural trends can you think of that began in this region and spread to other parts of the world?

Read to Find Out

- How do the religious practices and languages of the region reflect the immigrant history of the United States and Canada?
- How do the arts of the United States and Canada reflect the region's colonial past?
- What kinds of educational and health care systems serve the people of the region?

Terms to Know

- bilingual
- jazz
- socioeconomic status
- literacy rate
- patriotism

Places to Locate

- New Mexico
- Hollywood

NATIONAL GEOGRAPHIC

A Geographic View

The Art of Everyday Life

Such attention to reality was at odds with artistic convention in [painter Winslow] Homer's time, as was his choice of subjects—barefoot boys, farm girls, working men, freed slaves, North Woods guides, ordinary soldiers, and women of leisure, all of whom represented everyday life in America. Early critics complained about it. . . . But like other American originals of his time—Walt Whitman and Mark Twain—Homer kept to his own path.

—*Robert M. Poole, "Winslow Homer: American Original,"* National Geographic, *December 1998*

Fresh Eggs by Winslow Homer

Winslow Homer, known for his naturalistic style, was one of the greatest American painters of the 1800s. His paintings express the independent thinking and the enthusiasm for new frontiers that mark the cultures of the United States and Canada. The immigrant roots of these countries also gives them a respect for diversity. In this section you will read about the cultures and lifestyles of the United States and Canada.

Cultural Characteristics

The United States and Canada are countries of many cultures. Like the threads of a brightly colored blanket, the cultures of these countries blend into a new pattern without losing their individual qualities.

music of NORTH AMERICA

The music of North America stems from Native American, European, and African influences. Music from this region is extremely varied and has greatly influenced other types of music around the world.

Instrument Spotlight

The **Native American flute** originated among the peoples of the Great Plains, and it was often played by men to express their feelings of love to women. Each flute, made individually by hand, has its own unique look and sound. Traditional flutes are made from a piece of cedar, cut the same length as the distance between the armpit and the longest finger of the musician. In addition to five or six playing holes, four "direction" holes are added to send the sound in all directions.

World Music: A Cultural Legacy Hear music of this region on Disc 1, Tracks 1–6.

History

Religious Freedom

Freedom of religion has always been valued in the United States and Canada. Many of the people who migrated to the region did so to worship freely. As early as 1774, the British Parliament passed a law recognizing the religious rights of Roman Catholic French Canadians. In 1791 the Bill of Rights, which became part of the United States Constitution, guaranteed Americans religious freedom in addition to a number of other rights.

Today most Americans and Canadians who are members of an organized religion are Christians. In the United States, the majority of Christians are Protestant, while in Canada most Christians are Roman Catholic. Judaism, Islam, and Buddhism are among other religions practiced in the United States and Canada.

Languages

English is the main language in the United States. In Canada, English and French are the official languages. Because of immigration from all over the world, however, people in the United States and Canada also speak or use various words and phrases of other languages. For example, street signs in ethnic neighborhoods of the region's port cities—New York, Los Angeles, San Francisco, and Vancouver—may be printed in Chinese, Korean, Russian, Arabic, or Hindi. A writer describes this mix of cultures and languages in New York:

> *"Store owners [on Third Avenue] are often Asian. Corner groceries are run by families from the Dominican Republic . . . Arabs operate the candy stores . . . Koreans run vegetable stands. . . . [Near 118th Street] Robert Kosches finished talking in Spanish to a young couple. . . . 'My grandfather, who came from Austria, started this [furniture] business,' he said, switching to English. . . ."*
>
> Jere Van Dyk, "Growing Up in East Harlem," *National Geographic*, May 1990

Geography Skills for Life

Unity Rally

Supporters of Canadian unity rally in Hull, Quebec, prior to a 1995 referendum on Quebec independence.

Place Why do some Quebecois desire independence?

Immigrants from Great Britain brought the English language to the United States and much of Canada. In the Canadian province of Quebec, however, French is the official language because most of the province's population are descended from French settlers who arrived from the 1500s to the 1700s.

French-speaking Canadians in Quebec and some other provinces want greater protection for their language and culture. To achieve this goal, many Quebecois (kay•beh•KWAH) want Quebec's independence and support a movement for *separatism*—the breaking away of one part of a country to create a separate, independent country.

The Southwestern United States since colonial times has had a large Spanish-speaking population. In **New Mexico**, any communications with the state government or with local governments may be in Spanish or English. Thus, New Mexico is **bilingual**, meaning "having two languages." In California the presence of Asian communities is evident in the signs written in Chinese, Japanese, Korean, and other Asian languages.

The Arts

The arts of the United States and Canada go back to the first Americans, who interwove art and music into daily life. Native Americans made detailed carvings from shell and stone, used clay to produce pottery, and wove baskets, sandals, and mats from local plants. After European settlement the arts of the region were dominated by European traditions. By the mid-1800s, however, Americans and Canadians had begun to create art forms that reflected their own lives as North Americans.

Music

In their music Native Americans used drums, flutes, whistles, and vocal chanting. Europeans later brought European folk and religious music to the region. At the beginning of the 1900s, a distinctive form of music known as **jazz** developed in African American communities throughout the United States. Jazz blended African rhythms with

European harmonies. By the end of the century, country music and rock 'n' roll had become popular musical forms, not only in North America but around the world. In classical music, dancers and choreographers created a modern form of ballet.

The Visual Arts

Painting and sculpture in the United States and Canada moved away from their European roots and explored new themes. In the early 1900s, a group of American artists known as the Ashcan School painted the grim realities of urban life. A group of Canadian painters called the Group of Seven showed the rugged landscape of Canada's far north in bright, dynamic colors. American artist Georgia O'Keeffe gave the world new visions of the American West. In the mid-1900s many artists in the region adopted from Europe the abstract style, which expresses the artist's emotions and attitudes without depicting recognizable images.

Architects in the United States and Canada also developed innovations. The skyscraper, a tall building with many floors, first appeared in the United States. The architects Frank Lloyd Wright in the United States and Arthur Erickson in Canada were noted for designing buildings that harmonized with the region's natural environments.

performing arts of **THE UNITED STATES**

WORLD CULTURE • WORLD CULTURE • WORLD CULTURE

Modern Dance Modern dance combines the techniques of social dance and ballet. One of modern dance's best known performers, Twyla Tharp, uses movement to interpret everyday activities.

Literature

Literature in the United States and Canada at first dealt mainly with European historical and religious themes. Later writers, such as James Fenimore Cooper, Nathaniel Hawthorne, and Edgar Allan Poe, wrote stories about life in North America. Since the late 1800s, many American and Canadian authors have written about different parts of the region. Mark Twain described life on the Mississippi River, Margaret Laurence focused on the prairies of central Canada, and Willa Cather described life on the Great Plains.

More recently, writers have concentrated on highlighting aspects of the region's cultures. For example, writers such as Richard Wright and Toni Morrison depict the African American experience, Maxine Hong Kingston and Amy Tan write about the experience of Asian immigrants, Isaac Bashevis Singer's stories reflect the world of Jewish Americans, and Rudolfo Anaya and Sandra Cisneros focus on Hispanic American lives and issues.

Popular Entertainment

The cultural influence of the United States and Canada on the rest of the world is strongest in the area of popular entertainment. During the 1900s the United States became the world's dominant source for entertainment and popular fashion, from jeans and T-shirts to rock stars, movies, and television programs. The motion picture industry began in New York City and later moved to southern California. Today the name of a Los Angeles district, **Hollywood**, has become synonymous with the movie business. Canada's film industry, supported by the government, is known for its innovative documentaries. In the performing arts, Canada is noted for its Shakespeare Festival, held annually in Stratford, Ontario. Broadway, a New York City street name, is internationally identified with popular theater. The musical, combining elements of drama with music, became a popular form of theater in the United States.

Lifestyles

As citizens of two of the world's wealthiest countries, most people in the United States and Canada enjoy a high standard of living. Their **socioeconomic status**, or level of income and education, means having the advantage of many personal choices and opportunities. Because the region has an agricultural surplus, foods are relatively inexpensive. Housing varies to suit the needs of individuals and families, whether it be high-rise apartments, multifamily row houses, or suburban houses in a variety of styles.

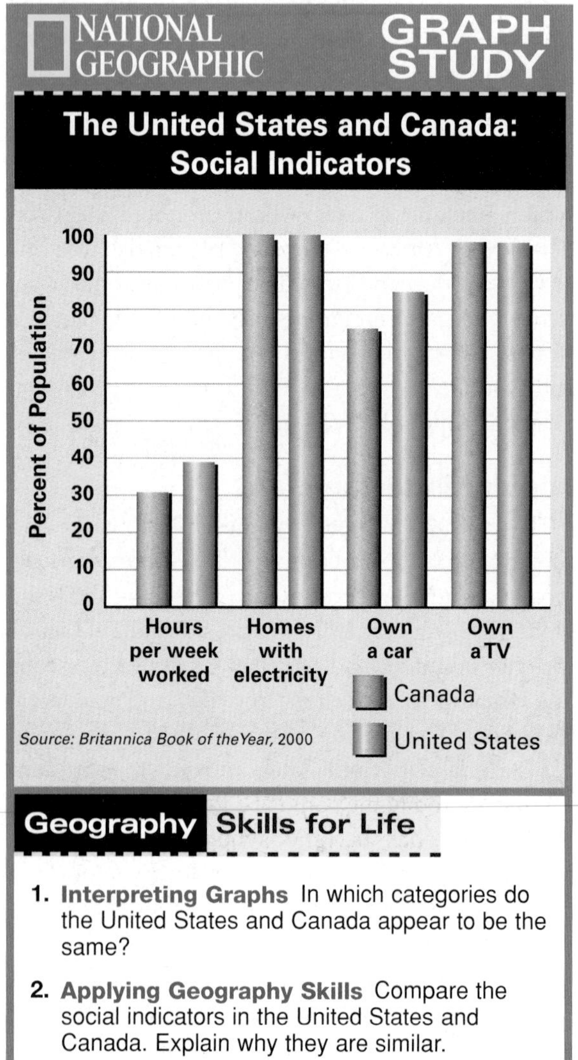

Geography Skills for Life

1. **Interpreting Graphs** In which categories do the United States and Canada appear to be the same?
2. **Applying Geography Skills** Compare the social indicators in the United States and Canada. Explain why they are similar.

Economics

Health Care

People in both the United States and Canada can expect to live longer, generally healthier lives than people in many other parts of the world. The region's high level of economic development enables governments to devote substantial resources to health care. Health care is administered differently in the two countries. In Canada, the government pays for health care. In the United States, most people are expected to pay for their own health care through health insurance provided by employers or other organizations. Federal and state governments, however, pay for some health insurance for people who are older, people with disabilities, or low-income families. Still, many people in the United States are unable to purchase insurance, and others cannot afford health care even with insurance. In the United States, the role of the government in providing health care for all citizens, regardless of their socioeconomic status, is currently under debate.

Education

The United States and Canada have similar educational systems, including networks of public and private schools. Both countries maintain compulsory education requirements. In the United States and most Canadian provinces, school systems have 12 grades. Colleges and universities exist in every state and province. In the United States, the **literacy rate**, the percentage of people who can read and write, is 97 percent; Canada's literacy rate is also 97 percent.

Sports and Recreation

Although a strong work ethic is woven into the culture of both the Americans and Canadians, they also enjoy plenty of leisure time. Some of the most popular activities involve watching and participating in sports. Most people associate baseball and football with the United States and ice hockey with Canada. Fans and players of these sports, however, come from both sides of the border. Basketball, soccer, golf, tennis, and competitive ice-skating also have their supporters in both countries.

Television transformed sports during the late 1900s. Many people in the United States and Canada share in exciting sports telecasts throughout the year: baseball during the spring and summer and its World Series in the early fall; football during the fall, capped by the Super Bowl in January; and the National Basketball Association (NBA) championships in the spring. As a result, sports heroes, such as baseball's Derek Jeter, football's Brett Favre, and basketball's Michael Jordan, have become household names.

The vast North American landscape is ideal for camping, canoeing, and hiking. The first U.S. national park was created in 1872. Located in parts of Wyoming, Montana, and Idaho, Yellowstone National Park covers more than two million acres, and dazzles visitors from around the world with its spectacular physical features. Since then, the United States and Canada have set aside millions of acres as national parks for conservation and recreation.

Celebrations

Holiday celebrations in the United States and Canada are essentially similar. Both countries celebrate many of the same religious holidays, and many civic observances are similar although held on different dates. Celebrations such as American Independence Day (July 4) and Canada Day (July 1) are occasions for public displays of **patriotism**, or loyalty to one's country.

NATIONAL GEOGRAPHIC World Explorer

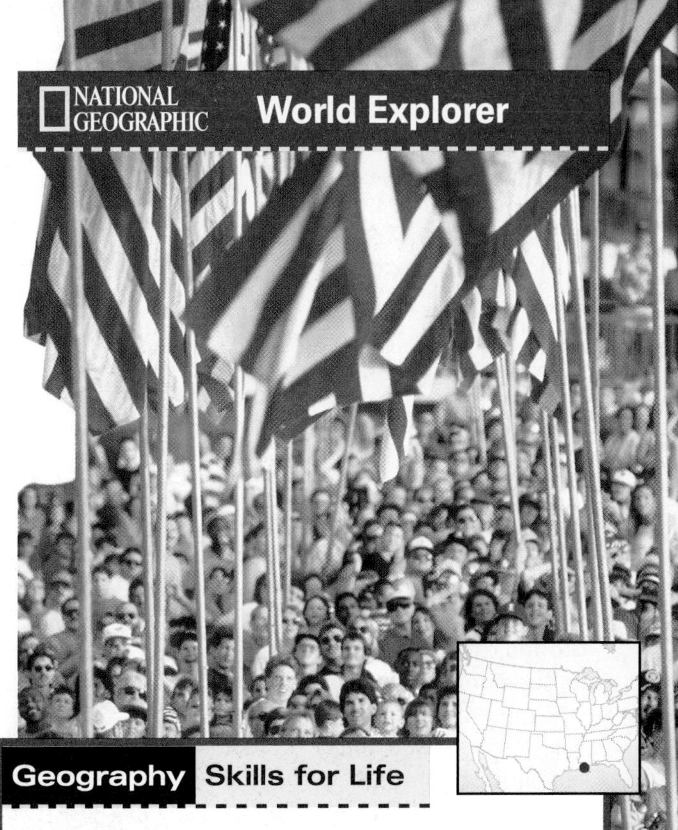

Geography Skills for Life

United States Patriotism A crowd enjoys a Fourth of July celebration in Baton Rouge, Louisiana.

Place What Canadian holiday is associated with patriotism?

SECTION 3 ASSESSMENT

Checking for Understanding

1. **Define** bilingual, jazz, socioeconomic, literacy rate, patriotism.
2. **Main Ideas** Create a diagram like the one below, listing aspects of the cultures, arts, and lifestyles of the region.

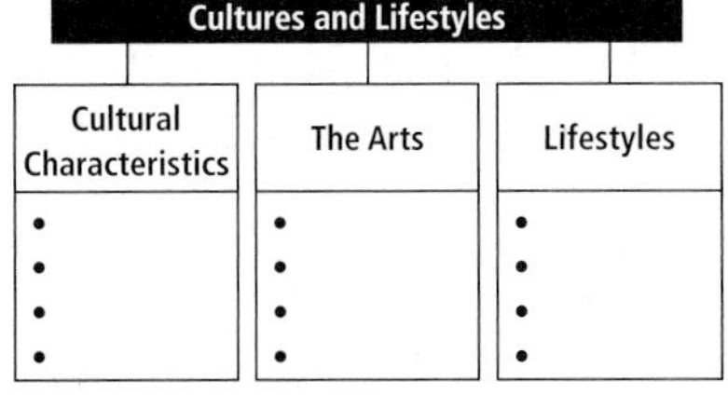

Critical Thinking

3. **Making Generalizations** What challenges are created for government, education, and business when a country has two official languages?
4. **Making Inferences** Why do you think immigrants to the United States and Canada did not develop new styles of art, music, and literature at first?
5. **Identifying Cause and Effect** How has the region's history of religious freedom contributed to the development of culturally diverse societies in the United States and Canada?

Analyzing Maps

6. **Place** Study the physical map in the Regional Atlas on page 106. Why does the physical geography of California make it ideal for both surfers and mountain climbers?

Applying Geography

7. **Sports as Culture** Think about the popularity of various sports in the region. Write a paragraph explaining how sports can increase cultural understanding among the peoples of Canada and the United States.

Understanding GIS and GPS

A Global Positioning System (GPS) can accurately determine a position on the earth to within .08 inches (2 mm). Geographic information systems (GIS) are computer tools for handling, processing, and analyzing geographic data. Both systems help us understand information about location.

Learning the Skill

GPS satellites in space continually broadcast signals to Earth. By tracking the signals from several satellites, a GPS receiver on the ground can determine its current latitude, longitude, and altitude. GPS measurements showed that Mt. Everest is actually 7 feet (2.1 m) higher than its official recorded height! The receiver can also report current time and the direction and speed of travel. The unit even has a feature that tells exact sunrise, sunset, and moon phase based on location and time.

Originally developed by the United States military, GPS is now available for many different uses:

- **Hikers** use GPS receivers with physical maps that show an area's surface features and elevations.
- **Drivers** use GPS receivers in cars to obtain digital street maps and plot travel routes.
- **Sailors** use GPS technology to plot a ship's course.

GPS receivers often feed data into geographic information systems (GIS). GIS are computer tools that gather, combine, and display information relevant to a specific geographic location. After information about an area is entered into the GIS database, the computer can create maps showing any combination of the data.

▲ *Soldier using GPS receiver*

Businesses use GIS to find prime locations to open franchises. Creating a database with information such as traffic patterns, competitors' locations, average income, and vacant lots for sale helps pinpoint the best location for a new store. Using data from various sources, GIS technology might display a map that shows the factories, air pollution count, and cancer rates in a particular neighborhood. GIS aid in information analysis by visually presenting the interaction among various factors in a given location.

GPS allows people to locate themselves inside the map, while GIS creates maps that highlight the elements affecting a location. GPS and GIS technology represents the state of the art in geography and mapmaking.

Practicing the Skill

Using the data you have read, answer the following questions.

1. What do GPS receivers use to plot exact locations?
2. Why was the military the original developer of GPS?
3. How does seeing various factors on a map help people make decisions?
4. How is GPS information used with a GIS database?
5. How could GPS and GIS improve traffic safety?

Applying the Skill

Your community is planning to build a recreation center. If you have access to a GIS program, use the program to help determine the best location. Use three types of data—such as roads, housing estates, and high school gyms—to create maps. Then analyze the maps and write your proposal.

CHAPTER 6 SUMMARY & STUDY GUIDE

SECTION 1 Population Patterns (pp. 133–137)

Terms to Know
- immigration
- Native American
- Sunbelt
- urbanization
- metropolitan area
- suburb
- megalopolis
- mobility

Key Points
- Both Canada and the United States are home to various groups of native peoples and descendants of immigrants.
- Physical geography impacts the distribution and density of population in the U.S. and Canada.
- North America's settlements and its largest cities developed along waterways.
- Natural resources and waterways for transportation helped North America industrialize.

Organizing Your Notes
Use a table like the one below to help you organize your notes about the region's population patterns.

Peoples	Population Patterns

SECTION 2 History and Government (pp. 140–145)

Terms to Know
- republic
- Underground Railroad
- dry farming
- Constitution
- amendment
- Bill of Rights
- cabinet
- dominion
- Parliament

Key Points
- Native Americans are North America's earliest people.
- Europeans set up colonies in North America for trading, conquest, and religious freedom.
- The thirteen British colonies won their independence from Britain in 1776 and formed their own republic, the United States of America.
- In 1867 the eastern provinces combined to form the Dominion of Canada. Canada today encompasses 10 provinces and 3 territories; it became an independent country in 1931.
- Industrialization and technology enabled westward expansion and spurred social change.

Organizing Your Notes
Create an outline using the format below to help you organize your notes for this section.

History and Government

I. History
   A. Native Americans
      1.
      2.
   B. European Colonies
      1.
      2.

SECTION 3 Cultures and Lifestyles (pp. 146–151)

Terms to Know
- bilingual
- jazz
- socioeconomic status
- literacy rate
- patriotism

Key Points
- The immigrant roots of the United States and Canada make these two countries diverse.
- Both countries have a heritage of religious freedom.
- Musical and artistic expression began with immigrants and gradually became uniquely North American.
- Health care is supported by the governments of both countries but in different ways.
- Both countries in the region have high standards of living.

Organizing Your Notes
Use a cluster map like the one below to help you organize your notes for this section.

ASSESSMENT & ACTIVITIES

Reviewing Key Terms

On a sheet of paper, write the key term that matches the definition. Refer to the Terms to Know in the Summary & Study Guide on page 153.

1. level of income and education
2. loyalty to one's country
3. percentage of people who can read and write
4. the movement of people into one country from another
5. a central city and outlying communities
6. a government in which people elect their own officials
7. a chain of closely linked urban areas and suburbs
8. partially self-governing country with close British ties
9. ability to use two languages
10. cultivating land so that it catches and holds rainwater

Reviewing Facts

SECTION 1

1. Most settlers in the United States and Canada came from what region of the world?
2. Where is most of Canada's population concentrated?

SECTION 2

3. Explain how the Underground Railroad helped enslaved African Americans escape to freedom.
4. What three technological innovations led to the expansion and development of the United States?

SECTION 3

5. What is the most widely practiced religion in Canada? In the United States?
6. How did the literature of the United States begin to change in the late 1800s?

Critical Thinking

1. **Making Generalizations** How did physical features influence population patterns and urbanization in the United States and Canada?
2. **Drawing Conclusions** Explain how diverse ethnic groups influenced the development of the region's arts.
3. **Comparing and Contrasting** Use a Venn diagram to compare and contrast the governments of the United States and Canada.

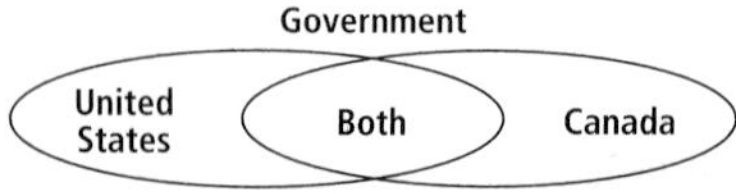

NATIONAL GEOGRAPHIC

Locating Places

The United States and Canada: Physical-Political Geography

Match the letters on the map with the places and physical features of the United States and Canada. Write your answers on a sheet of paper.

1. **Texas**	5. **Alberta**	9. **Detroit**
2. **Great Salt Lake**	6. **New Mexico**	10. **Nunavut**
3. **Nova Scotia**	7. **Miami**	11. **British Columbia**
4. **Quebec**	8. **Pennsylvania**	12. **Hudson Bay**

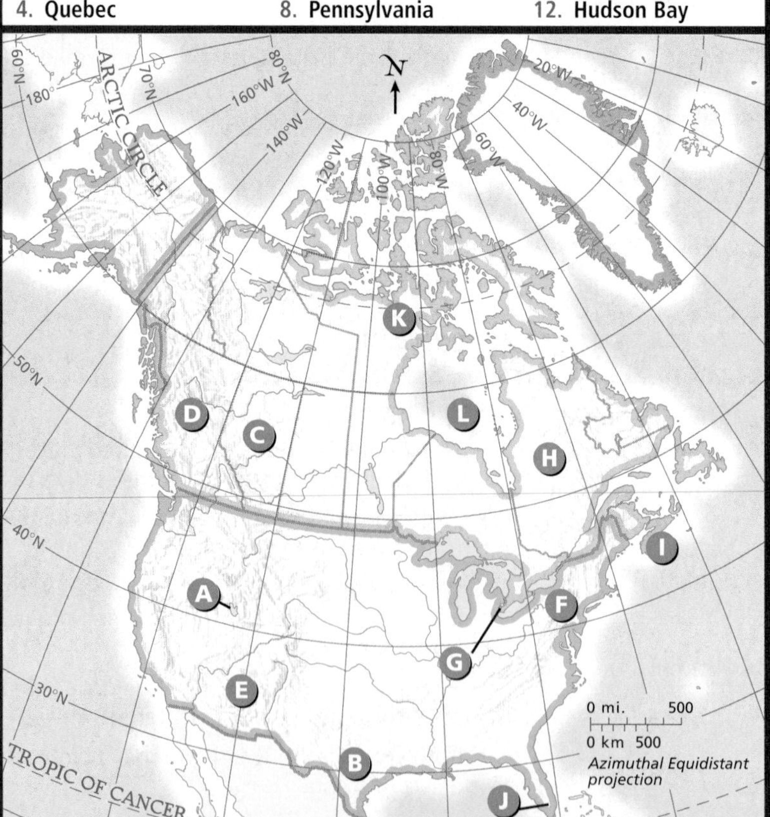

Self-Check Quiz Visit the **Glencoe World Geography** Web site at geography.glencoe.com and click on Self-Check Quizzes—Chapter 6 to prepare for the Chapter Test.

Using the Regional Atlas

Refer to the Regional Atlas on pages 106–109.

1. **Human-Environment Interaction** Compare the political map to the economic activity map. What commercial crops are grown in the southern United States?
2. **Place** Compare the population density map to the political map. What states and provinces have the highest population density?

Thinking Like a Geographer

Describe the effects of cultural diffusion between the United States and Canada and other parts of the world in recent decades. Trace this process, using specific examples, such as films, music, foods, and American slang.

Problem-Solving Activity

Group Research Project Research U.S. voting patterns and the distribution of political power. Study a map showing the outcome of the latest congressional election, district by district. Then compare it to a map of a previous race. Write a report explaining the political changes from one election to the next as well as the factors shaping the formation of congressional voting districts.

GeoJournal

Expository Writing Using your GeoJournal data, write an essay analyzing the effects of processes, such as migration, on the territorial growth of the United States and Canada.

Technology Activity

Developing Multimedia Presentations Choose one Native American or immigrant group and describe its influences on the region's cultures and lifestyles. Include contributions such as religion, language, the arts, food, clothing, and celebrations. Create a multimedia presentation that displays examples of these contributions and explains their origins. To enhance your presentation, play music appropriate to the group you choose for the class.

Standardized Test Practice

Choose the best answer for each of the following multiple-choice questions. If you have trouble answering the questions, use the process of elimination to narrow your choices.

1. **Researchers using GIS and GPS technology to correlate water quality indexes, zoning maps, census figures, and maps of area rivers and aquifers are most likely trying to determine which of the following?**

   A The relationship between the location of industrial plants and water quality
   B The water pressure for new fire hydrants for a developing community
   C The location of scenic hiking trails
   D The lung disease rates for various areas in the region

Several of the data elements correlated by the researchers relate to water. Eliminate those answers that do not relate to water.

2. **Ships at sea use GPS technology**

   F as a communication device.
   G as a navigational aid.
   H for inventory control.
   J to maintain personnel files.

If you know that GPS technology deals with precise positioning data, you can eliminate choices H and J, and choose the best answer from the answer choices that remain.

CHAPTER 7

The United States and Canada Today

GeoJournal

As you read this chapter, take notes in your journal on the economic activities, transportation, communications, and environmental concerns of the United States and Canada.

GEOGRAPHY Online

Chapter Overview Visit the **Glencoe World Geography** Web site at geography.glencoe.com and click on Chapter Overviews—Chapter 7 to preview information about the region today.

SECTION 

Living in the United States and Canada

Guide to Reading

Consider What You Know

People in the United States and Canada depend heavily on the automobile for personal travel. How is travel by automobile different from travel by bus or train? How do you think dependence on automobiles has influenced the development of new neighborhoods?

Read to Find Out

- What are the effects of physical geography on the region's agriculture?
- What kinds of transportation and communications systems does the region have?
- How are the economies of the United States and Canada dependent on each other and interdependent with those in other parts of the world?

Terms to Know

- market economy
- post-industrial
- commodity
- retooling
- pipeline
- monopoly
- trade deficit
- tariff
- trade surplus

Places to Locate

- Corn Belt
- New York State
- Minnesota
- Seattle
- Research Triangle
- Pittsburgh
- Trans-Canada Highway

◀ *The skyline of Houston, Texas*

NATIONAL GEOGRAPHIC

A Geographic View

An Occasion to Celebrate

County fairs endure as an occasion to celebrate our agrarian traditions, to honor family, inventiveness, and hard work. More important, perhaps, they allow us, as communities, to come together and get to know one another. County fairs also give us a chance to glimpse the American past. Yet they have lasted not by being annual historical reenactments but by evolving as American society evolves and becomes more urban.

—*John McCarry, "County Fairs,"* National Geographic, *October 1997*

County fair, Vermont

Urban lifestyles predominate in the United States and Canada, yet people in both countries continue to respect traditional rural values, such as inventiveness and hard work. Adhering to these values, Americans and Canadians have utilized their rich natural resources and technological skills, placing their countries among the world's top economic powers. In this section you will learn how people in the United States and Canada make their livings and how their economies are interrelated with each other and with the rest of the world.

Economic Activities

The United States and Canada both have free **market economies**, which allow people the freedom to own, operate, and profit from their own businesses. Businesses can hire employees and pay them for their work. Laws protect private property rights, employment opportunities, and the health and safety of workers. Although their economies are similar, the United States and Canada take different approaches to the ownership of some corporations and the administration of some services. In Canada the government owns and administers many services, such as broadcasting and health care, that tend to be handled by private, nongovernmental corporations in the United States.

Like other developed countries, the United States and Canada have moved from primarily agricultural to primarily industrialized economies. As technology transforms the workplace, both countries are developing **post-industrial** economies, which place less emphasis on heavy industry and traditional manufacturing and more emphasis on service and high-tech businesses. Agriculture and manufacturing continue to play significant roles in the region's economic life, however.

Agriculture

As in most developed countries, farming in the United States and Canada is overwhelmingly commercial, with agricultural **commodities**, or goods, produced for sale. Large commercial corporations, however, account for only 5 percent of farm ownership. Most farms in the United States and Canada, no matter what their size, are still owned by farming families, many of whom have formed cooperative operations.

The United States devotes about 1 billion acres (405,000,000 ha) of land to agriculture. A little less than half of that total is cropland—the largest cropland area of any country in the world—and the rest is used for the grazing of livestock. Canada, with much less arable land than the United States, still devotes 167 million acres (67,583,000 ha) to agriculture, evenly divided between crops and livestock.

Since the 1950s, although the average size of farms in the United States and Canada has risen, the number of people employed as farmers in the region has decreased. Today only 2 percent of Americans and 4 percent of Canadians work in agriculture. Among the factors contributing to this decline is the high cost of farming. Successful agriculture requires investing in expensive machinery, fertilizers, and chemical pesticides—all of which make farming easier but drive up costs and impact profits. Another factor is unpredictable consumer demand. If a farm product produced in large quantities—such as hogs or cranberries—does not sell well on the market, farmers have to lower their prices for the product and may lose money as a result. The risk of natural disasters and the time and hard work needed to run a farm also contribute to fewer farmers entering this segment of the economy.

Geography Skills for Life

Wheat Harvest The Peace River area of British Columbia produces most of the grain harvested in the province.

Place Where is wheat grown in the United States?

Key Agricultural Products

The United States and Canada rank among the world's leading producers of beef, milk, and eggs, and of corn, wheat, and other grains. These agricultural products are shipped to markets across the country and around the world.

Cattle ranches operate mostly in the western, southern, and midwestern United States and in Canada's western Prairie Provinces. Other important livestock-producing areas include the north-central parts of the United States and the Canadian provinces of Quebec and Ontario. Hogs, chickens, and dairy products also lead the list of farm products from these areas.

Wheat is grown in the Prairie Provinces of Canada and on the Great Plains of the United States, a region often called the Wheat Belt. The type of wheat grown depends on the climate. Farmers in the northern plains, with their short growing season, concentrate on spring wheat, which is planted in the spring and harvested in the fall. Farther south, farmers plant winter wheat, which germinates before the ground freezes, grows with the spring rains, and is harvested in the early summer.

The **Corn Belt** of the United States consists of a band of farmland stretching from Ohio to Nebraska. Corn is also grown in the Canadian provinces of Quebec, Ontario, and Manitoba. About 50 percent of the corn crop is used to feed livestock; the rest is processed to make sweeteners and corn oil, used in industrial manufacturing, or used for food by people.

Geography Skills for Life

Advanced Technologies Automation and robotics have revolutionized heavy industry in the United States and Canada.

Place What kinds of industries might use robots for welding?

Fruits and vegetables are grown in many parts of the United States and Canada. Apples, peaches, and cherries flourish in the Great Lakes region and in the St. Lawrence River Valley. Potatoes are an important crop in the U.S. states of Maine, North Dakota, and Idaho as well as in the Canadian provinces of Prince Edward Island and New Brunswick. California ranks first among U.S. states in the production of tomatoes, lettuce, peas, asparagus, okra, avocados, grapes, and strawberries. Citrus fruits are grown in central and southern Florida, in the lower Rio Grande Valley of Texas, and in southern California. Sugarcane, pineapples, and bananas are grown in Hawaii.

Economics

Breaking Geographic Boundaries

Throughout much of the region's history, geographic factors often limited the type of agriculture that could be carried out in a particular area. Cattle ranching, for example, needed the wide open spaces and natural grasses of the western prairies and plains. Most American dairy farms were concentrated in a belt of land stretching from upper **New York State** to **Minnesota**. This region, known as America's Dairyland, has cooler summers and native grasses ideal for dairy cattle.

Today, however, advances in agricultural technology have changed or widened the traditional growing areas. The development of breeds of cattle that need less room to roam has opened the southern United States to cattle ranching. Because of improved feed sources and automation, large productive dairy farms can now be found in every American state and many Canadian provinces.

Manufacturing and Service Industries

Manufacturing makes up about 20 percent of both the United States and Canadian economies and employs about 20 percent of the region's workforce. Advanced technologies, such as robotics and computerized automation, have transformed manufacturing in the region. As with farming, the region's factories produce greater quantities of goods with fewer workers than in the past.

Transportation equipment and machinery are large export categories of both countries. In the United States, aircraft and aerospace equipment are produced in California and Washington, and factories in the Midwest produce most of the country's automobiles. United States auto manufacturers also operate plants in Quebec and Ontario. Food processing is another important economic activity in California and in the northeastern United States. Canada, especially Quebec, manufactures and exports a variety of wood-based products drawn from its timber resources.

Post-Industrial Economies

The largest area of economic growth in the United States and Canada is in service industries. About 75 percent of the region's workers are employed in service jobs, such as government, education, health care, tourism, entertainment, banking, and real estate.

The rising post-industrial economy is best reflected in the region's high-tech and biotechnology industries. Both countries produce high-tech equipment for use in computer sciences and telecommunications. California's Silicon Valley, for example, is home to around 20 of the world's 100 largest high-tech companies. Led by the development of high-tech industries in **Seattle**, the state of Washington has the sixth-highest concentration of high-tech businesses in the United States. Texas boasts more than 1,000 software companies in its capital city of Austin, and some of the fastest-growing high-tech companies are based in Dallas. Raleigh, Durham, and Chapel Hill, known as North Carolina's **Research Triangle**, have attracted prestigious biotechnology companies. Boston is a leading area in software, telecommunications, and media technology. In Canada, Ontario is home to many thriving telecommunications and Internet businesses.

Retooling the Rust Belt

During the last third of the 1900s, the switch from heavy industry and traditional manufacturing to service industries left cities in the east and near the Great Lakes, such as Buffalo, Pittsburgh, Cleveland, and Detroit, without their major economic bases. As corporations began to move

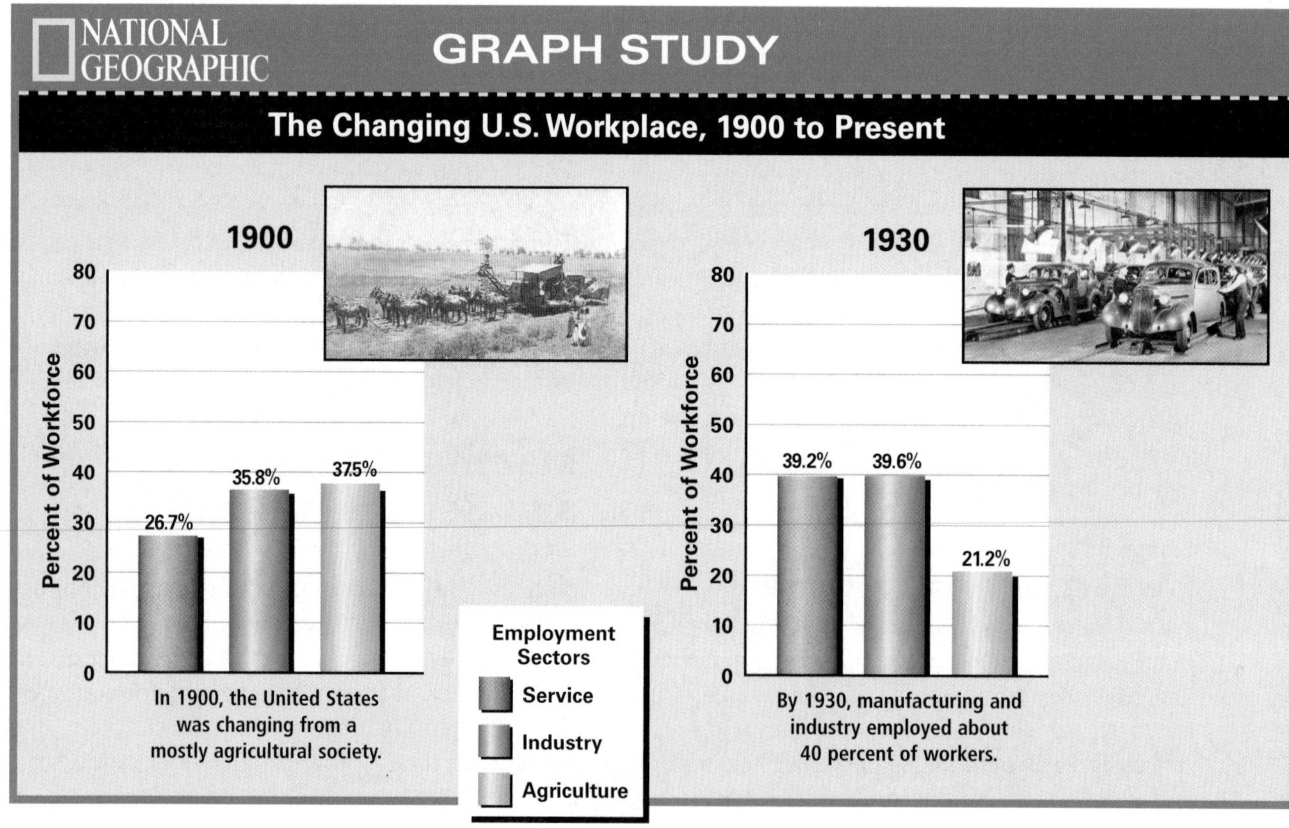

south to the Sunbelt, some older industrial areas were left with abandoned factories and rusting steel mills. Together they acquired the derogatory nickname "the Rust Belt." Today, however, many of these cities are converting old factories for use in new industries, a process called **retooling**, and transforming run-down areas into tourist attractions and public spaces. For cities such as **Pittsburgh**, this change has brought renewed energy:

> *"Shaken by the collapse of the steel industry, which had provided them with an unshakable sense of identity for more than a century, Pittsburghers hunkered down and built a new economy based on services, medicine, education, and technology. In the process, they transformed their community from one driven by quantity of production into one devoted to quality of life."*
>
> Peter Miller, "Pittsburgh: Stronger than Steel," *National Geographic*, December 1991

Transportation and Communications

Good transportation and reliable communications systems are the backbone of economic success in the United States and Canada. Both also contribute to the quality of life in the region today.

The Automobile

Since World War II, the most popular means of personal transportation in the United States and Canada has been the automobile. Extensive automobile use in the region has required heavy investment in the building and maintenance of highways, roads, and bridges. In the United States, more than 3,900,000 miles (6,276,442 km) of streets, roads, and highways carry about 208 million motor vehicles each year. Canada's smaller, more concentrated population relies on about 15 million motor vehicles and 550,000 miles (885,139 km) of roads. The **Trans-Canada Highway**, a well-maintained modern roadway, runs 4,860 miles (7,821 km) from Victoria, British Columbia, to St. John's, Newfoundland.

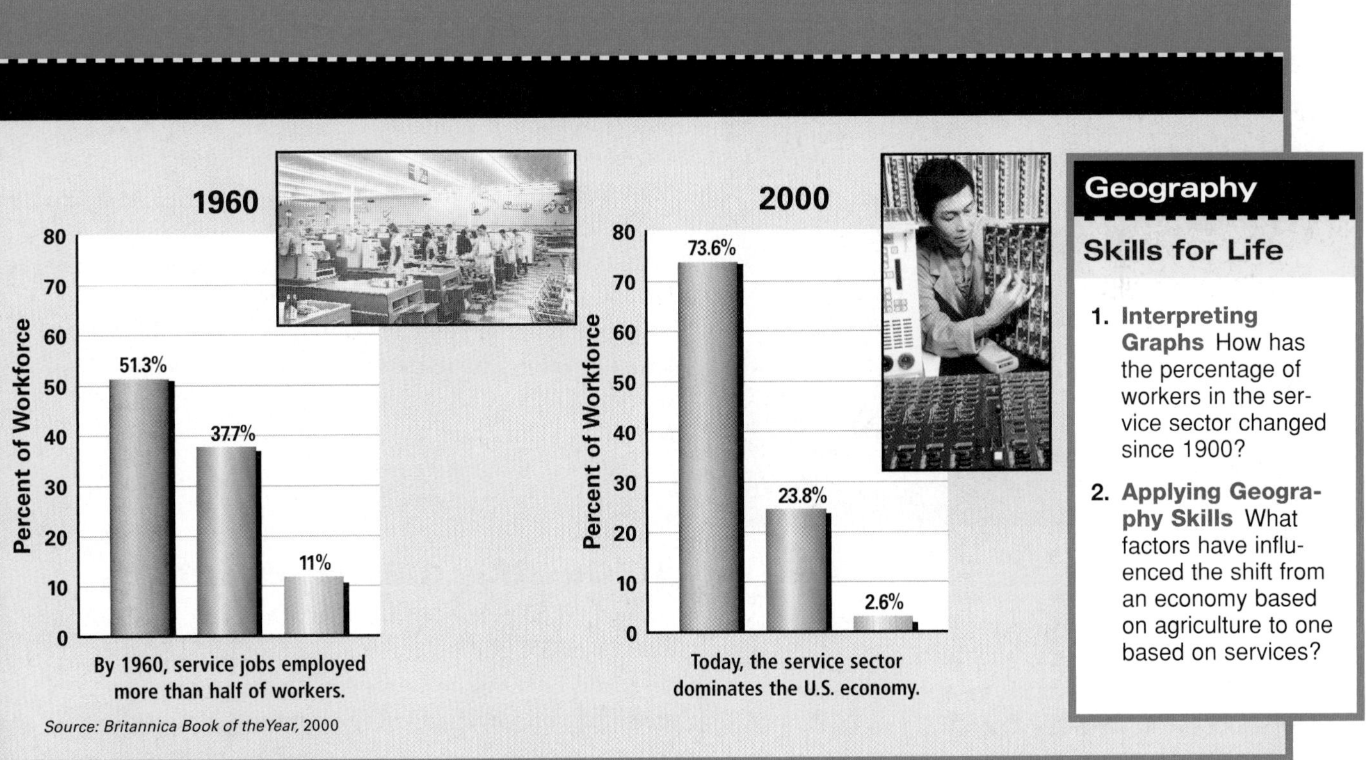

Geography Skills for Life

1. **Interpreting Graphs** How has the percentage of workers in the service sector changed since 1900?
2. **Applying Geography Skills** What factors have influenced the shift from an economy based on agriculture to one based on services?

More than a simple means of getting from one place to another, the automobile has become a status symbol for many North Americans. Cars are marketed for the image they represent, and obtaining a driver's license has become an unofficial rite of passage for most teenagers in the United States and Canada. Reliance on the automobile, however, creates many challenges because automobile-related pollution affects most urban areas. Automakers and government agencies are working together to reduce the use of automobiles in certain urban districts and to find clean, efficient ways to use fuel.

Another challenge posed by automobile use is traffic congestion in the region's cities, where traffic jams can last for hours. Mass public transportation can help ease such congestion. Cities such as Montreal, New York, San Francisco, and Boston now have well-established subway systems. Los Angeles, which has some of the world's largest traffic jams, is completing a transport system that will combine subways with elevated trains, and Seattle and Dallas both use monorail systems. Urban areas also use buses and commuter trains to ease some of the congestion.

Geography Skills for Life

Transporting Goods Long-haul trucks are used heavily in the United States and Canada.

Region What other means of transporting goods are important to the region?

Other Means of Transportation

For long-distance travel, many people in the United States and Canada use the region's busy network of airports. In the United States, Atlanta's Hartsfield and Chicago's O'Hare International Airports vie for the title of the busiest airport in the country and in the world. Toronto's Pearson International Airport is Canada's busiest. Passenger railroads and long-distance buses account for only a small portion of the region's passenger travel.

The transport systems of the region move goods as well as people. Railroads move about 35 percent of the region's freight, and about 15 percent continues to be carried along inland waterways. More modern means of transport include long-haul trucks, which carry about 20 percent of the region's freight. Airplanes carry only a small portion of the region's heavy freight but do a growing amount of overnight delivery business. Finally, **pipelines**, long networks of underground or aboveground pipes, carry almost one-fourth of the region's freight in the form of gas and oil.

Communications

In the United States and Canada, telephone and mail services are the primary means of communication. While Canada's broadcasting and telephone services are publicly owned, private companies operate the same services in the United States. Federal government regulations, however, make sure that there is no **monopoly**, the total control of a type of industry by one person or one company.

Wireless microwave and satellite relays are increasingly used for long-distance contacts. Cellular and digital services have made telephone communication more mobile. Computer use is high, although efforts are underway to make this technology available to all people. In the midst of these advances, Americans and Canadians still rely on newspapers and magazines.

Trade and Interdependence

The United States and Canada are among the world's major trading countries. The United States is second only to the European Union in exports, providing more than 10 percent of all world exports. The U.S. economy supplies chemicals,

Empire State Building Completed in New York City in 1931, the Empire State Building was the tallest building in the world at 1,250 feet (381 m). It was created in the streamlined art deco style of geometric patterns that was popular in the 1920s and 1930s.

agricultural and manufactured goods, and raw materials, such as metals, iron ore, and cotton fiber. Canada exports many of the same goods, as well as large quantities of seafood and timber products. In 2000, the United States granted China—one of the world's largest potential markets—permanent normal trade relation status (PNTR, formerly called most-favored nation status), which gives China the same trading opportunities granted to other trading partners.

Exports and Imports

Despite its many resources, the United States spends more on imports than it earns from exports. The resulting **trade deficit**, or difference in value between a country's imports and its exports, is hundreds of billions of dollars. The U.S. trade deficit results from the country's large population and its growing industries that require costly energy purchases. Also, some countries charge high **tariffs**, or taxes, on imports, thus raising the price of U.S. products and reducing their sales abroad. As a result, growth rates for U.S. exports are very slow.

Canada, by contrast, enjoys a **trade surplus**, earning more from exports than it spends for imports. Canada's smaller population makes its energy needs less costly. Although both countries are spending more on imports, Canada's export revenues have grown yearly at a higher rate than those of the United States.

NAFTA

The United States and Canada are each other's largest trade partners. In 1989, the two countries signed an agreement that removed trade restrictions between them. A 1994 pact—the North American Free Trade Agreement (NAFTA)—included these two countries and Mexico. Unlike the European Union, however, NAFTA prohibits the free flow of labor among member countries.

In recent years, businesses in the United States and other developed countries have sought lower production and labor costs by *outsourcing*, or setting up plants abroad to produce parts and products for domestic use or sale. Outsourcing provides cheaper goods for home markets, while offering jobs to foreign workers. Because of NAFTA, more American companies have set up assembly plants in Mexico, where labor costs are less expensive than in the United States.

United Against Terrorism

On September 11, 2001, terrorists hijacked four passenger planes, crashing two of them into New York City's World Trade Center and the third into the Pentagon, the defense department headquarters near Washington, D.C. A fourth plane plummeted into a Pennsylvania field. The devastation and loss of so many lives made the United States firmly resolved to rid the world of terrorism.

Student Web Activity Visit the **Glencoe World Geography** Web site at geography.glencoe.com and click on Student Web Activities—Chapter 7 for an activity on the economic interdependence of Canada and the United States.

Geography Skills for Life

American Heroes New York City firefighters raise the U.S. flag on the ruins of the World Trade Center.

Region What was the U.S. response to the terrorist attacks of September 11, 2001?

Although the attacks stunned Americans, they responded quickly to aid victims and rescue workers. To show their unity, Americans across the country put up flags and held candlelight vigils and prayer services. The United States government also acted swiftly in dealing with the crisis. Military forces were put on high alert, security was increased at airports and other public places, and the FBI began a massive investigation. President George W. Bush announced the creation of the Office of Homeland Security to organize efforts to protect Americans from further terrorist attacks.

In October, another crisis arose when traces of anthrax, a type of bacteria used in deadly biological weapons, were found in mail sent to major news offices and federal government buildings. Several people contracted the anthrax disease, and some died. The FBI began investigating who had made the anthrax and distributed it.

Meanwhile, the September 11th attacks had shocked other nations, some of whom had lost citizens in the World Trade Center. As a result, the United States won much international support for a massive, wide-ranging response to terrorism. The first military operation of the war on terrorism began in the Southwest Asian country of Afghanistan, which harbored Saudi exile Osama bin Laden, the leader of the terrorist network believed to have carried out the attacks. President Bush warned that Afghanistan was only the beginning of a struggle that would not end "until every terrorist group of global reach has been found, stopped, and defeated."

SECTION 1 ASSESSMENT

Checking for Understanding

1. **Define** market economy, postindustrial, commodity, retooling, pipeline, monopoly, trade deficit, tariff, trade surplus.
2. **Main Ideas** Use a table to organize details about agriculture, manufacturing, and trade in the United States and Canada.

Economic Activity	United States	Canada

Critical Thinking

3. **Analyzing Information** Describe how recent technological changes have affected the location and pattern of economic activities in the United States and Canada.
4. **Identifying Cause and Effect** What factors caused technological growth in the region? How did technology affect agriculture?
5. **Drawing Conclusions** Why do the United States and Canada have strong economies?

Analyzing Maps

6. **Region** Study the economic activity map on page 109. How are the locations of manufacturing centers related to the region's lakes and rivers?

Applying Geography

7. **Public Policies** Research an issue related to global trade and the United States and Canada. Identify different points of view in each country that affect public policy and decision making on the issue.

SECTION 2

People and Their Environment

Guide to Reading

Consider What You Know

Pollution of the air, water, and land in the United States and Canada is a well-known problem. Why does reducing pollution require regional cooperation?

Read to Find Out

- How are the United States and Canada learning to manage their natural resources more responsibly?
- What are the causes and effects of pollution in the region? How can pollution be prevented?
- What environmental challenges face the United States and Canada in the twenty-first century, both as individual countries and as a region?

Terms to Know

- clear-cutting
- acid rain
- smog
- groundwater
- eutrophication

Places to Locate

- Sudbury
- Banks Island

NATIONAL GEOGRAPHIC

A Geographic View

From Waste to Wetland

Gold mine in Nevada

This is gold mining today, the ads proclaim—beautiful hills, waving fields of grass, prancing mule deer, a glimmering lake. . . . I saw waste rock piles shaped into eye-pleasing mounds, the milling operation that recycles and contains all processed water, and the huge [residue-collecting] pond that, over time, will become a 600-acre wetland. I saw the sophisticated monitoring system for the early detection of contamination in the groundwater. I even saw the gate placed over the mouth of a tunnel to protect the maternity roost for a local population of Townsend's big-eared bats.

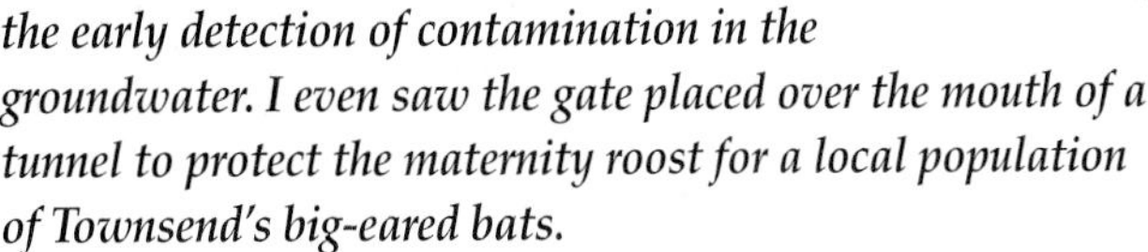

—*T. H. Watkins, "Hard Rock Legacy,"* National Geographic, *March 2000*

Strip mining made the American West rich, but it also left deep scars on the landscape. Today mining and other resource-based western industries are working to control ecological damage. In this section you will discover how people in the United States and Canada are managing scarce resources and seeking ways to overcome the effects of pollution.

Human Impact

The rich natural resources of the United States and Canada have not always been managed responsibly. The practice of **clear-cutting**, or taking out whole forests when harvesting timber, has destroyed many of the region's old-growth forests, endangered wildlife, and left the land subject to erosion and flooding. Overfishing has depleted many of the region's freshwater and ocean fisheries. Although efforts to reverse the damage have begun, the region has a long way to go toward the sustainable use of its natural resources.

Natural resource management also includes evaluating the impact of human activity on the environment. In some cases, policies that appear to make good environmental sense must be rethought. For example, in the dry western regions of the United States and Canada, it used to be standard practice to extinguish wildfires as quickly as possible. The vegetation in these areas, however, needs periodic wildfires to clear overgrowth and to germinate seeds. Without burning, grasses and scrub grow thick and underbrush dries out. Too much burning, however, can be devastating. For example, when lightning from summer storms ignited brushfires in 2000, the result was explosive infernos that raged across several states, endangering human and animal life and destroying agricultural and grazing lands. One solution may be to follow the practice of the early Native American inhabitants of these dry areas, who set deliberate fires, known today as controlled burns, to clear dry brush before it became too dense.

Pollution

One of the unfortunate consequences of industrial development in the United States and Canada has been the increase in human-made pollution. Pollution, the introduction of harmful materials into the environment, damages the quality of water, air, and land. The kinds of pollution that trouble the United States and Canada are directly related to the region's physical geography and economic activities.

Acid Rain

Acid rain, precipitation that carries abnormally high amounts of acidic material, affects plants and fish in a large area of the eastern United States and Canada. Acid rain forms when chemical emissions from cars, power plants, factories, and refineries react with water vapor in the air. The reaction turns the chemicals, chiefly sulfur dioxide and nitrogen oxide, into their acidic forms. As the acid rain falls to the ground, it corrodes stone and metal buildings, damages crops, and pollutes the soil. Acid rain is especially damaging to the region's waters, however. Plant life and fish cannot survive in highly acidic waters. Over time, lakes may become biologically dead, unable to support most organisms.

The winds that carry acid rain do not respect local or national boundaries. The source of the pollution may be quite distant from the place where acid rain falls. Carried by eastward winds, acid rain from the U.S. Midwest's coal-burning plants falls on the Adirondack Mountains, where it mixes with the runoff from melting acid snow. The result is that 26 percent of all lakes in the region are

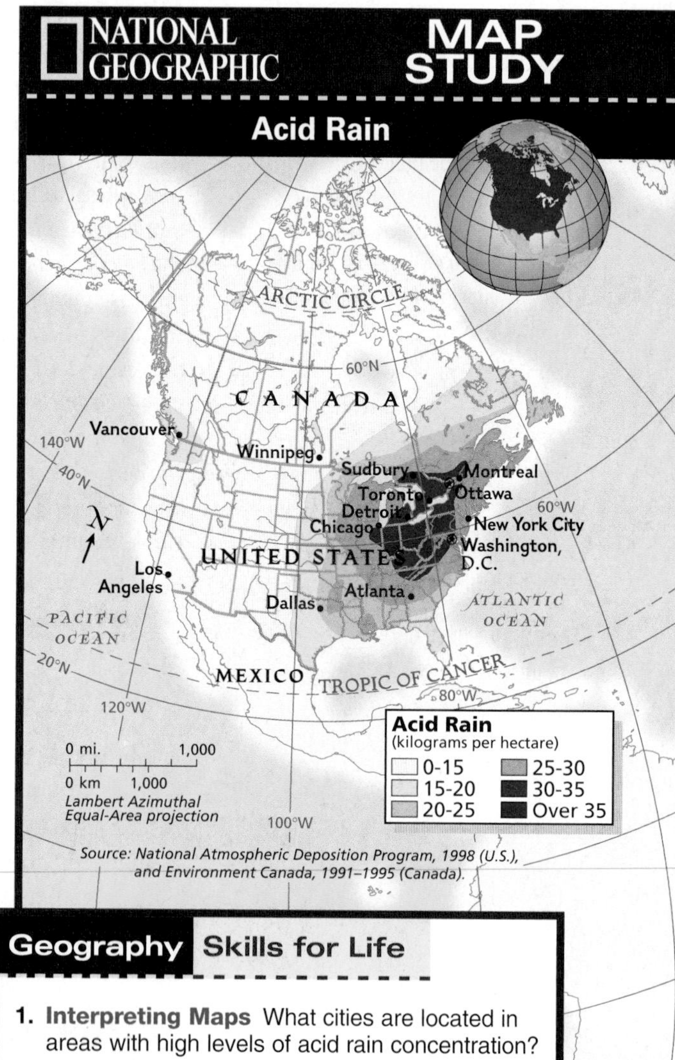

Geography Skills for Life

1. **Interpreting Maps** What cities are located in areas with high levels of acid rain concentration?
2. **Applying Geography Skills** In which state of the southeastern United States are trees least likely to show the effects of acid rain? Why?

Find NGS online map resources @ www.nationalgeographic.com/maps

acidic, and hundreds are unsuitable for the survival of sensitive fish species. Emissions from the United States also result in acid rain in Canada, threatening important timber and water resources. Canada's eastern provinces—Ontario, Quebec, New Brunswick, and Nova Scotia—are the most vulnerable. Thousands of lakes throughout Canada, including 100 in Ontario alone, are so acidic that they are biologically dead.

About half of the acid rain in Canada comes from the United States. As a result, the two countries have begun cooperating to improve air quality. Improvement has already been noted. In Canada, 33 percent of the acidified lakes studied since the 1980s show reduced acid levels. In the **Sudbury** region of Ontario, for example, fish populations are rising, as are the number of fish-eating birds, such as loons.

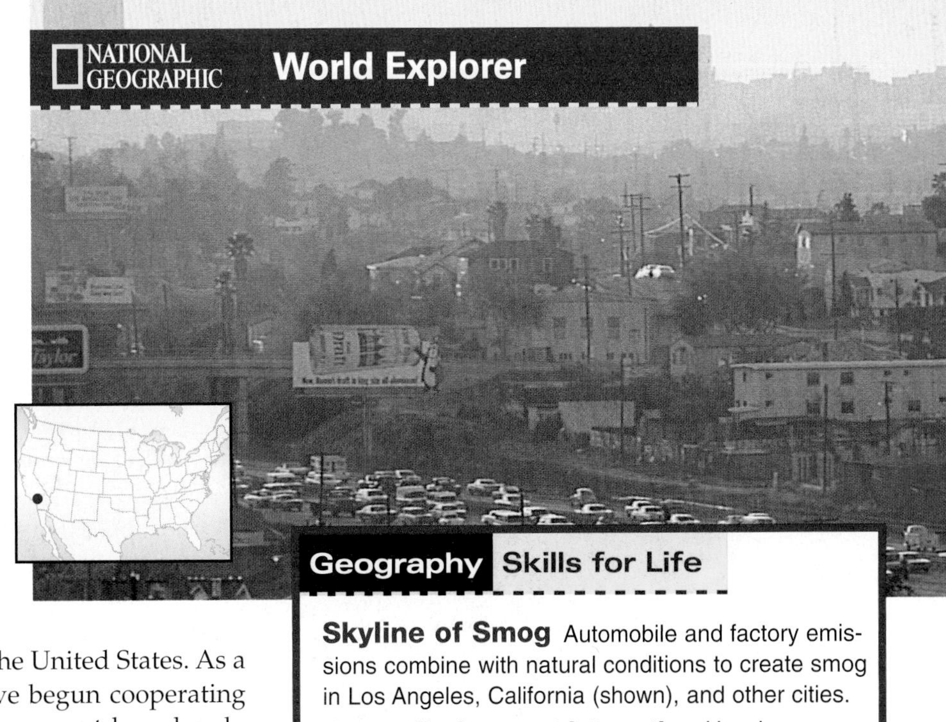

Geography Skills for Life

Skyline of Smog Automobile and factory emissions combine with natural conditions to create smog in Los Angeles, California (shown), and other cities.

Human-Environment Interaction How has industrialization affected the environment?

Smog

The sulfur and nitrogen oxides that create acid rain also contribute to the type of air pollution known as smog. As the sun's rays interact with automobile exhaust gases and industrial emissions, a visible haze forms, damaging or killing plants and irritating people's eyes, throats, and lungs.

Health officials in many of the region's metropolitan areas now measure air quality on a daily basis. When emissions interact with climate conditions and create dangerous levels of smog, officials issue alerts, urging children, the elderly, and people with respiratory problems to stay indoors. Under these conditions authorities may prohibit nonessential driving and the use of lawnmowers, chainsaws, and other devices with gas-powered engines. Industrial activity may be restricted, and industries with excessive emissions may be fined. Some local and state governments in the United States require emissions testing as part of the automobile licensing process. In many parts of the United States and Canada, fuel pumps in service stations must have special nozzles that reduce the leakage of petroleum vapors into the air.

Clean-air practices have substantially reduced air pollution in Los Angeles and other major cities, and still more is being done. Some car manufacturers are producing vehicles that run on electricity instead of fossil fuels. Engineers also continue to research air-, water-, and solar-powered cars. In the United States, proposed legislation would require the reduction of the sulfur content in diesel fuel by 97 percent. By 2007, officials hope to make all new diesel vehicles, such as trucks and buses, smoke free. Smog can also be reduced by encouraging the use of alternatives to automobiles, such as walking, bicycling, or using public transportation.

Water Pollution

Water systems in the United States and Canada become polluted not only by acid rain but also by the introduction of sewage and industrial and agricultural wastes into the water supplies. Industrial wastes, including toxic substances, may be illegally dumped into rivers and streams or may find their way through small, unnoticed leaks into the groundwater, freshwater that lies beneath the earth's surface. Industries also cause thermal pollution by releasing heated industrial waste water into cooler lakes and rivers. Runoff from agricultural

▲ The polluted Nashua River, Massachusetts, in the 1960s

▼ The present-day Nashua River, after cleanup

chemicals, such as fertilizers and pesticides, also pollutes the water resources of the region.

Water pollution has disastrous effects on marine life and on the birds and other animals that feed on fish or breed in wetlands. The toxic chemicals and wastes that pollute the water supply also endanger humans. In 2000 seven people in the Canadian farming town of Walkerton died and thousands became ill after being infected by *E. coli* bacteria in their drinking water. Groundwater contaminated with animal waste and other toxins had entered Walkerton's water supply through corroded pipes.

Water pollution also speeds **eutrophication** (yu•TROH•fuh•KAY•shuhn), the process by which a lake or other body of water becomes rich in dissolved nutrients, encouraging overgrowth of small plants, especially algae. In extreme cases the algae growth depletes the water's oxygen, leaving none for fish. Algae overgrowth can also turn a lake first into a swamp and later into dry land. Normally, eutrophication takes thousands of years, but pollution greatly speeds the process.

History

Back from the Brink

In the 1960s the region's waterways were under assault from pollution. The Cuyahoga River near Cleveland, Ohio, was so fouled by industrial chemicals that it burst into flames. Oil from a spill off the coast of Santa Barbara, California, coated beaches and wildlife. Eutrophication threatened Lake Erie. A biologist warned of serious consequences:

> *"The most alarming of all man's assaults upon the environment is the contamination of air, earth, rivers, and sea with dangerous and even lethal materials. . . . The poisons . . . kill vegetation, sicken cattle and . . . travel from link to link of the food chain. . . ."*
>
> Rachel Carson, *Silent Spring*, 1962

In 1972 the United States and Canada signed the Great Lakes Water Quality Agreement to combat pollution in the lakes. The United States also passed the Clean Water Act, mandating measures to restore the quality of the country's waters.

In New England, the Act forced an end to asbestos dumping in the Nashua River and spurred the construction of waste-water treatment plants. The facilities protected the river from paper pulp, chemical dyes, and other industrial wastes. Like many of the country's waterways, the Nashua slowly regained its health. Today it is once again safe for wildlife and people.

The passage of the North American Free Trade Agreement (NAFTA), however, has shifted some environmental concerns south to the U.S.–Mexico border. Along the Rio Grande, rapid industrial growth threatens the environment. The Commission for Environmental Cooperation, a nongovernmental agency with representation from Canada, the United States, and Mexico, is monitoring the environmental effects of NAFTA and suggesting ways to reduce pollution.

Challenge for the Future

Like people worldwide, those who live in the United States and Canada are concerned about the possible effects of global warming. The slight but steady rise in the earth's temperatures over the past century is not easily explained, nor are its consequences completely understood. Some effects of global warming, however, are easy to see, especially in the Arctic regions of Alaska and Canada. In these areas, the melting of polar ice is accelerating, a phenomenon with potentially disastrous consequences. In one Inuit community on the western tip of Canada's **Banks Island**, thinning sea ice has forced caribou, polar bears, and seals, on which the hunting lifestyle of the Inuit depends, to move farther north. More disturbing, the permafrost—the frozen soil of the tundra—is beginning to thaw, buckling the land and weakening the foundations of houses.

Global warming has a chain reaction of effects that threaten to alter life throughout the United States and Canada. When polar ice melts, ocean levels rise, increasing the danger of coastal flooding. For example, the city of New Orleans, much of which lies below sea level, is in danger of being completely submerged because of the combined effects of rising ocean waters and more frequent Mississippi River floods. Warmer, higher seas also alter climate patterns, leading to increased frequency and severity of weather events such as El Niño, which has been responsible for both flooding and drought. Monitoring and responding appropriately to the effects of global warming remains a critical challenge for the future of the region and for the world.

SECTION 2 ASSESSMENT

Checking for Understanding

1. **Define** clear-cutting, acid rain, smog, groundwater, eutrophication.
2. **Main Ideas** Copy the flowchart below on a sheet of paper. Complete the chart by listing causes, effects, and possible solutions for a regional environmental problem.

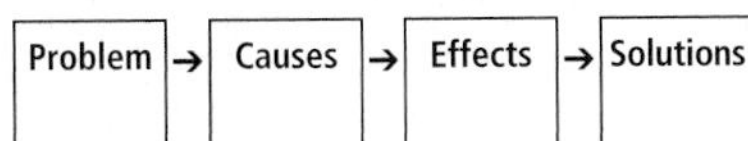

Critical Thinking

3. **Analyzing Information** Why is it important for Canada and the United States to work together to reduce pollution?
4. **Drawing Conclusions** Why are more metropolitan areas of the United States and Canada beginning to experience smog?
5. **Identifying Cause and Effect** What are the short-term and long-term effects of water pollution on people and the environment?

Analyzing Maps

6. **Place** Study the map of acid rain on page 166. Which parts of the region have the greatest concentration of acid rain? Why?

Applying Geography

7. **Regional Cooperation** Think about the cooperation among the United States, Canada, and Mexico in NAFTA to promote free trade. Identify the human factors involved in the trade network created by this agreement.

Viewpoint

CASE STUDY on the Environment

Source: USGS Water Supply Paper 2425

United States's Wetlands: Under Siege

Marsh. Bog. Swamp. Different words, but they all describe wetlands. A wetland is an area where water covers the soil, or lies just beneath its surface, for at least part of the year. For centuries, wetlands were regarded as smelly, insect-choked wastelands. They were places to eliminate. Across America, wetlands were filled in or drained. But in the 1970s, research confirmed what many had suspected—wetlands are valuable ecosystems that link water, life, and land. Laws were passed to protect wetlands. However, balancing wetlands preservation with development is controversial.

NATIONAL GEOGRAPHIC SOCIETY

A tricolored heron stands in the still, shallow water (left). It is early morning in Florida's Everglades National Park, one of the largest wetlands in the United States.

Wetlands teem with life. Spend a few hours in the Everglades and what might you see? Egrets, herons, and dozens of other birds. Scores of insects, fish, frogs, and snakes. Perhaps an alligator or two.

Wetlands rank among Earth's most productive ecosystems. They are nurseries, where fish and shellfish come to spawn. They are rich feeding grounds, where tiny plants and aquatic insects form the base of complex food webs. They are also sanctuaries, home to living things found nowhere else. About a third of all endangered or threatened species in the United States live in wetlands.

Wetlands are valuable in other ways, too. They slow erosion, thus stabilizing shorelines and riverbanks. They filter out pollutants that would otherwise end up in lakes and rivers. Coastal wetlands buffer the impact of storm tides. Inland wetlands slow fast-moving floodwaters.

◀ In its wetland habitat, a bullfrog eyes the world.

However, wetlands often occur where people want to grow crops. Before wetlands were recognized as valuable, the United States government encouraged farmers to drain these ecosystems to create cropland. In addition, wetlands frequently lie in the path of new housing developments, shopping malls, airports, roads, and reservoirs.

When Europeans arrived in North America, some 220 million acres (89 million ha) of wetlands existed in what are now the lower 48 states. Today slightly more than 100 million acres (40 million ha) remain. Wetlands are protected by law, but some wetlands development is allowed.

Supporters of wetlands development think the land where wetlands are found has more profitable uses. As cities and communities grow, people need more housing, schools, businesses, and roads. Developers insist they often have no choice but to build in wetlands. In many instances, it is now legal for developers to destroy natural wetlands as long as they create "new" wetlands as replacements.

Opponents of wetlands development argue that wetlands are too valuable to lose. Without these unique habitats, some kinds of animals may become extinct. Opponents are concerned that human-made wetlands are not true replacements for existing ecosystems.

What's Your Point of View?

Do you think it's acceptable to build on natural wetlands, as long as "new" wetlands are created? Or do you think wetlands should be completely protected?

A worker in an all-terrain tractor (below) digs ditches in wetlands in Florida. A waterfront housing development (right) abuts wetlands on Long Island in New York State. ▼

Interpreting a Climate Map

Climate helps determine how people live, work, dress, and play in a particular region. People on different continents may share similar climates. By reading a climate map, you can discover these similarities and differences among regions.

Learning the Skill

A climate map shows the climate zones of a region. Latitude, temperature, precipitation, altitude, wind patterns, and nearness to oceans help determine the climate of a region. Variation in precipitation also creates different types of climates, such as *rain forest* (very wet), *desert* (very dry), and *savanna* (wet and dry seasons).

On a climate map, colors represent different climate regions. The map key explains the color code. To interpret a climate map:

- **Identify the area covered by the map.**
- **Study the key to identify the climate regions on the map.**
- **Locate the regions in each climate zone.**
- **Draw conclusions about the climate similarities and differences among regions.**

Practicing the Skill

Study the climate map of eastern Canada. Use the information to answer the following questions.

1. What climate dominates the far northeast part of Canada?
2. Which area shown has a humid continental climate?
3. What climate does the coast of Newfoundland and Labrador have?
4. Why are so few major cities located in Nunavut and northern Quebec?
5. Why are there only three climate regions represented in eastern Canada? What factors of physical geography may account for this?
6. Compare the climate map on this page to the natural vegetation map on page 123. What is the relationship between climate patterns and vegetation patterns in eastern Canada?

Applying the Skill

Research to learn more about the climate of a place in the United States or Canada. Then write a paragraph describing how you think the location of the area affects its climate. Include examples of agricultural products or vegetation found in the area.

The **Glencoe Skillbuilder Interactive Workbook, Level 2** provides instruction and practice in key social studies skills.

SECTION 1 Living in the United States and Canada (pp. 157–164)

Terms to Know

- market economy
- post-industrial
- commodity
- retooling
- pipeline
- monopoly
- trade deficit
- tariff
- trade surplus

Key Points

- The region's economy has shifted from reliance on agriculture and traditional manufacturing to emphasis on service and high-tech industries.
- Agriculture is a key economic activity of the region, although it employs only a small percentage of the workforce.
- Technology and improved agricultural methods have helped farmers overcome the limitations of physical geography and climate.
- Dependable transportation and advanced communications systems help make the region an economic leader.
- The United States and Canada are among the world's leading exporters.
- The region's two countries are each other's largest trade partners. The region also trades with countries and trade blocs around the world.

Organizing Your Notes

Use an outline format similar to the one below to help you organize your notes for this section.

Living in the U.S. and Canada

I. Economic Activities
  A. Agriculture
    1. Livestock
      a.
      b.
      c.
      d.
    2. Crops
  B. Manufacturing and Services

SECTION 2 People and Their Environment (pp. 165–169)

Terms to Know

- clear-cutting
- acid rain
- smog
- groundwater
- eutrophication

Key Points

- The United States and Canada are working to manage their rich natural resources responsibly.
- Acid rain, smog, and water pollution cause damage to the region's environment and affect human health.
- Cooperative efforts to address environmental concerns are making a difference in the region.

Organizing Your Notes

Use a web diagram like the one below to help you organize your notes for this section.

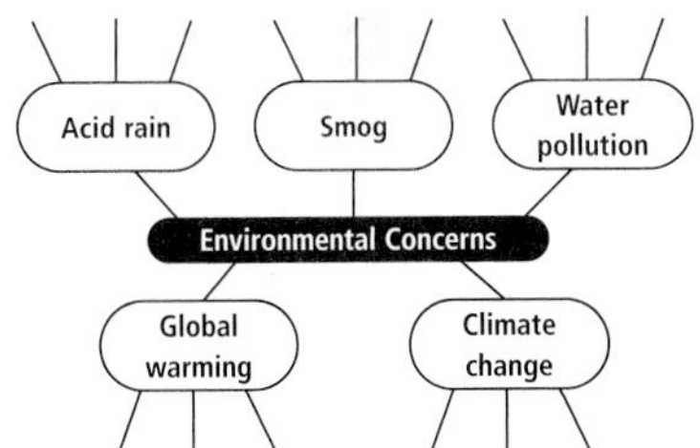

Workers guide a barge down the Erie Canal.

ASSESSMENT & ACTIVITIES

Reviewing Key Terms

Write the letter of the key term that best matches each definition below.

a. trade surplus
b. retooling
c. clear-cutting
d. acid rain
e. trade deficit
f. market economy
g. post-industrial
h. tariff

1. an economic system in which people can own and profit from their own businesses
2. reduced emphasis on heavy industry
3. converting old factories to new uses
4. loss of income through trade
5. a tax on imported trade goods
6. taking out whole forests when harvesting timber
7. precipitation that carries high amounts of acids
8. earning money through export sales

Reviewing Facts

SECTION 1

1. What type of economic system do the United States and Canada have?
2. What economic activity employs the most people in both the United States and Canada?

SECTION 2

3. What solutions have the United States and Canada implemented to deal with air pollution?
4. What factors contribute to water pollution in the region?
5. What part of the region is experiencing increased environmental problems as a result of NAFTA?

Critical Thinking

1. Making Generalizations What challenges will industrial cities face as the economy becomes more dependent on high-technology industries?
2. Analyzing Information Explain the connection between transportation patterns and air pollution.
3. Identifying Cause and Effect Use a chart like the one below to analyze the causes of acid rain and its effects on the environment.

Acid Precipitation	
Causes	Effects

Locating Places

The United States and Canada: Physical-Political Geography

Match the letters on the map with the places and physical features of the United States and Canada. Write your answers on a sheet of paper.

1. Midwest
2. Prairie Provinces
3. California
4. New York
5. Toronto
6. St. Lawrence River
7. Ohio River
8. Alaska
9. Texas
10. Pacific Northwest

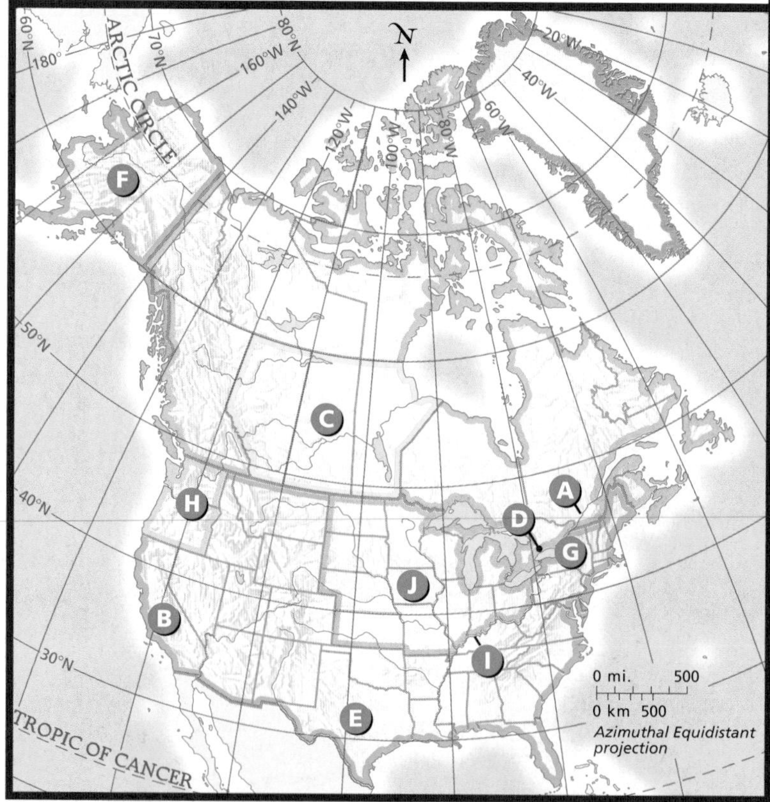

Self-Check Quiz Visit the **Glencoe World Geography** Web site at geography.glencoe.com and click on Self-Check Quizzes—Chapter 7 to prepare for the Chapter Test.

Using the Regional Atlas

Refer to the Regional Atlas on pages 106–109.

1. **Region** Describe the relationship between areas where livestock is raised and the population density in these areas.
2. **Human-Environment Interaction** What types of natural resources are clustered around large cities and manufacturing areas in the United States?

Thinking Like a Geographer

Study the economic activity map on page 109. Identify an activity that is represented in your area. Then, in geographic terms, explain why your area is suited to this activity and what other related activities might be developed there.

Problem-Solving Activity

Problem–Solution Proposal Identify a transportation problem in your community or state. Find examples of different points of view that affect decision making and the development of public policies on the problem. Then devise your own solution and present it to the class.

GeoJournal

Descriptive Writing Use your GeoJournal to write an essay describing how varying physical and cultural patterns in the region influenced the development and spread of new ideas and technologies. Use your textbook and the Internet as resources to make your essay accurate and interesting.

Technology Activity

Creating an Electronic Database Choose a region of the United States or Canada. Research that region and create an electronic database. Include types of industries, jobs, trading partners, major transportation routes, communications, land use, and environmental problems. Share your findings with the class, using charts, maps, and other graphics.

Standardized Test Practice

Study the bar graph below. Then choose the best answer for the following multiple-choice question. If you have trouble answering the question, use the process of elimination to narrow your choices.

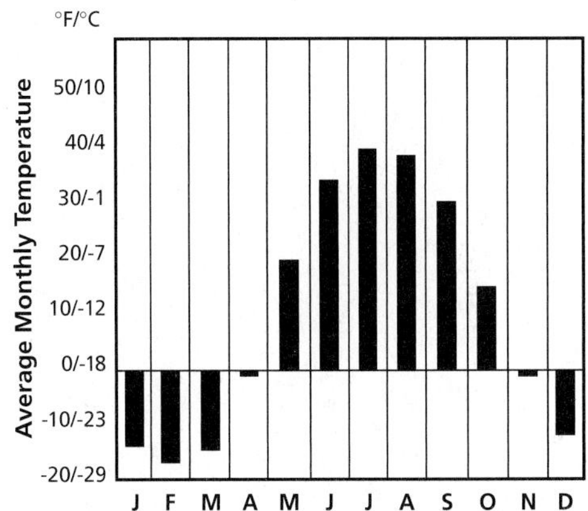

1. **As a regional geographer for an oil company, you need to determine the best time for a survey team to work near Barrow, Alaska. Given the information on the bar graph, during which three-month period should the survey take place?**

   A January, February, March
   B September, October, November
   C March, April, May
   D June, July, August

Study the information shown on the bar graph for average monthly temperature. Look for three consecutive months in which temperatures would be the most favorable for people and equipment to function outside.

UNIT 3

Latin America

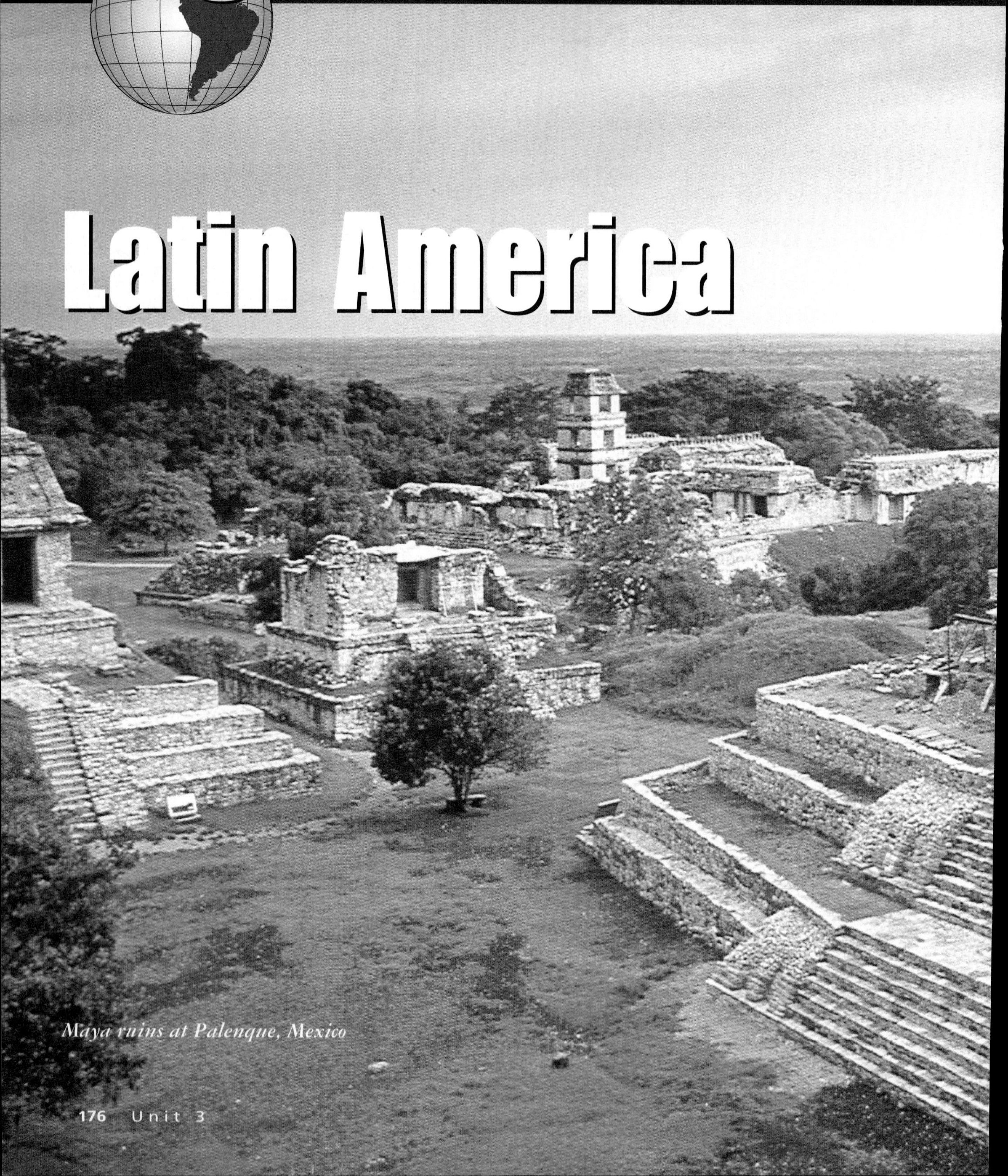

Maya ruins at Palenque, Mexico

NGS ONLINE

www.nationalgeographic.com/education

WHY IT'S IMPORTANT—

Latin America reflects a unique blend of world cultures, including Native American, European, and African. In turn, Latin America's diverse cultures have spread to other parts of the world. For example, the languages, music, foods, and arts of Latin America have profoundly influenced life in the United States. Today, many Americans are of Latin American descent and maintain close ties to their heritage. In addition, Americans and Latin Americans are close trading partners. They share democratic values based on human rights and revolt from European rule.

World Regions Video

To learn more about Latin America and its impact on your world, view the World Regions video "Latin America."

3

What Makes Latin America a Region?

Spanning more than 85 degrees of latitude, Latin America encompasses Mexico, Central America, the Caribbean Islands, and South America. It is a region of startling contrasts.

High mountains run from northern Mexico through the heart of Central America. The higher peaks of the Andes course down South America's western side.

Elsewhere, broad plateaus span huge areas. At still lower elevations, plains dominate the landscape. These great grasslands, such as the pampas in Argentina and the llanos in Venezuela and Colombia, are ideal for grazing cattle and sheep.

But when people think of Latin America, it's often rain forests that come to mind. Eternally wet, intensely green, and bursting with life, rain forests cover parts of many Caribbean islands and Central American countries. Yet none of these compare to the Amazon rain forest of Brazil. Drained by the Amazon River, this lowland forest covers one-third of South America and is home to nearly half of the world's plant and animal species.

3

1

2

1 Like a cartoon come to life, a brightly colored toucan calls out from a leafy perch in a Costa Rican rain forest. A toucan's enormous beak has serrated edges that help the bird get a good grip on slippery-skinned fruits.

2 Rust-red terraces curve around the Carajás iron mine, in Brazil. This mine boasts the world's largest deposit of high-quality iron ore. Tin, copper, silver, oil, and natural gas are among Latin America's other important natural resources.

3 Like jagged teeth in some enormous jaw, the snow-covered peaks of the Andes guard South America's western flank. The world's longest mountain chain, the Andes stretch the entire length of the continent.

4 On the plains of Paraguay, cowboys known as vaqueros round up cattle before driving them to fresh pastures. The western part of Paraguay, the Gran Chaco, is a grassland area where cattle roam on large ranches.

3

Mix of Old and New

3

Latin America is a region where cultures have collided and combined. Maya, Aztec, and Inca civilizations flourished here long ago. Then Europeans arrived in the late 1400s. For more than 300 years, Spain and Portugal controlled most of Latin America. They forced new laws, new languages, and a new religion onto the region's inhabitants. Yet native cultures survived by blending with those of the conquerors. Today, the faces, costumes, and customs of many Latin Americans reveal their mixed heritage.

This is a region of developing nations—countries in the process of becoming industrialized. Latin America's urban population is increasing rapidly as people flock from the countryside to modern, bustling cities.

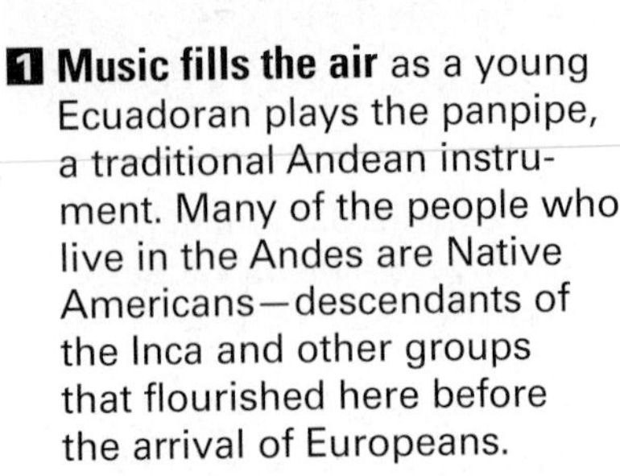

1

1 Music fills the air as a young Ecuadoran plays the panpipe, a traditional Andean instrument. Many of the people who live in the Andes are Native Americans—descendants of the Inca and other groups that flourished here before the arrival of Europeans.

2

2 **Villages and farms** dot the highlands near Cuzco, Peru. The economies of many Latin American countries are still based largely on agriculture, though manufacturing and other industries play an increasingly important role.

3 **Arms outstretched** as if in blessing, a huge statue of Jesus keeps watch over the sprawling city of Rio de Janeiro, Brazil. Most Latin Americans are Catholic, a legacy from the days of Spanish and Portuguese rule.

4 **A glass-covered bridge** links a hotel to a convention center, part of a new business complex in Monterrey, Mexico. Among Latin American countries, Mexico has been one of the quickest to modernize and industrialize.

3

Latin America

PHYSICAL

POLITICAL

120°W 100°W 80°W 60°W 40°W
30°N 20°N 10°N 0° 10°S 20°S 30°S

NORTH AMERICA
Rio Grande
Gulf of Mexico
Monterrey
MEXICO
Guadalajara
Mexico City
BERMUDA U.K.
BAHAMAS
Nassau
Havana
CUBA
Cayman Is. U.K.
HAITI
Port-au-Prince
DOMINICAN REPUBLIC
Santo Domingo
Puerto Rico U.S.
BELIZE
Belmopan
JAMAICA
Kingston
GUATEMALA
Guatemala City
HONDURAS
Tegucigalpa
San Salvador
EL SALVADOR
NICARAGUA
Managua
San José
COSTA RICA
Panama City
PANAMA
ST. KITTS AND NEVIS
Caribbean Sea
ATLANTIC OCEAN
TROPIC OF CANCER
Virgin Islands U.S. & U.K.
ANTIGUA AND BARBUDA
Guadeloupe Fr.
DOMINICA
Martinique Fr.
ST. LUCIA
ST. VINCENT AND THE GRENADINES
BARBADOS
GRENADA
TRINIDAD AND TOBAGO
Caracas
Port-of-Spain
VENEZUELA
Georgetown
GUYANA
Paramaribo
Cayenne
SURINAME
FRENCH GUIANA Fr.
Medellín
Bogotá
Cali
COLOMBIA
EQUATOR
Galápagos Islands Ecua.
Quito
ECUADOR
Negro R.
Amazon R.
AMAZON BASIN
Selvas
Madeira R.
PACIFIC OCEAN
PERU
Lima
ANDES
Lake Titicaca
La Paz
BOLIVIA
Sucre
MATO GROSSO PLATEAU
BRAZIL
BRAZILIAN HIGHLANDS
Brasília
Recife
Salvador
Belo Horizonte
PARAGUAY
Paraguay R.
Paraná R.
TROPIC OF CAPRICORN
Asunción
Rio de Janeiro
São Paulo
Curitiba
CHILE
Porto Alegre
ARGENTINA
Valparaíso
Santiago
Rosario
URUGUAY
Buenos Aires
Montevideo
PAMPAS
Río de la Plata
ATLANTIC OCEAN
ANDES
PATAGONIA
Falkland Islands (Islas Malvinas) U.K.
South Georgia Island U.K.

- ⊛ National capital
- ⊙ Territorial capital
- • Major city

0 mi. 1,000
0 km 1,000
Lambert Azimuthal Equal-Area projection

N

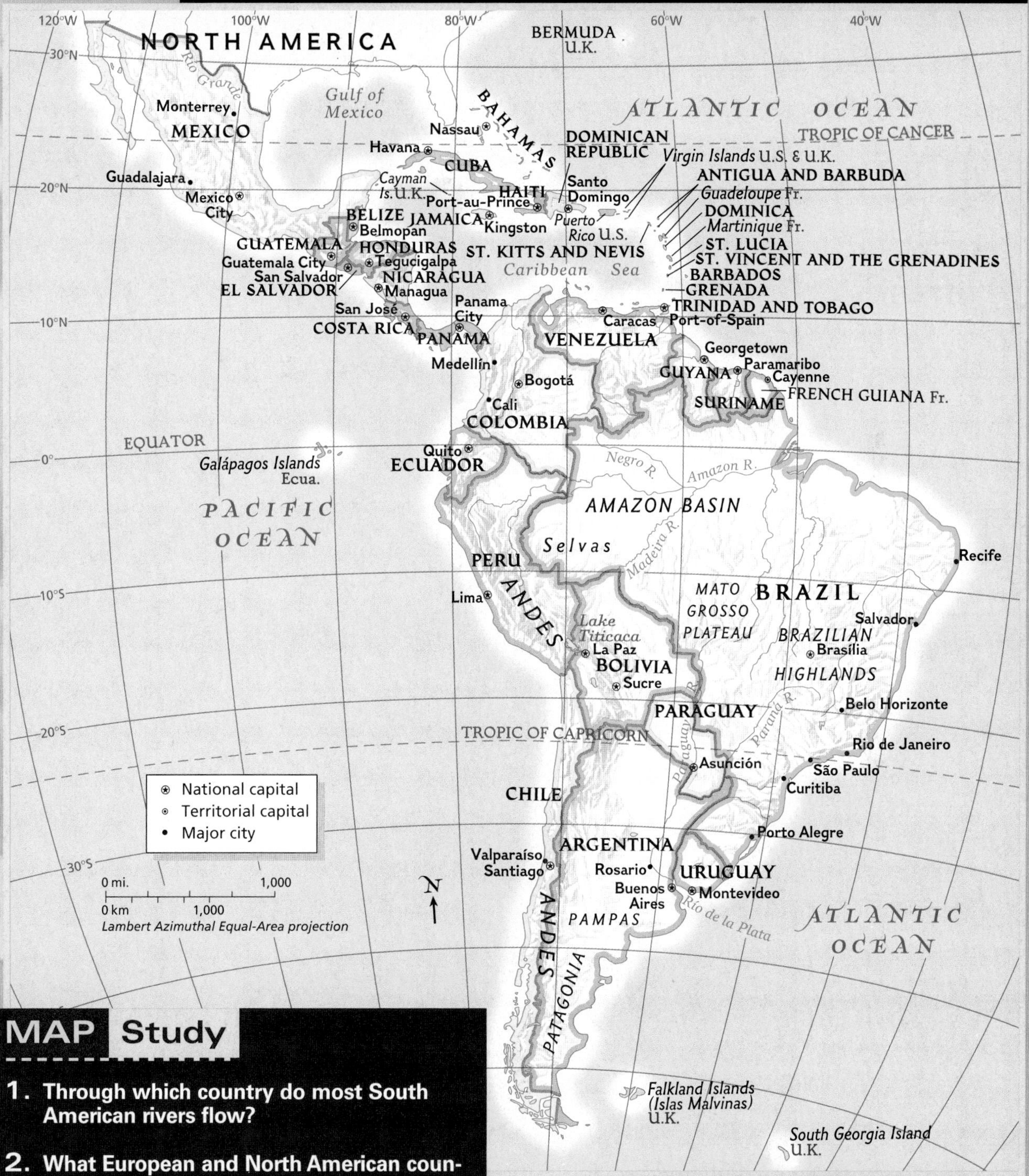

MAP Study

1. Through which country do most South American rivers flow?
2. What European and North American countries have territories in Latin America?

3

Latin America

ECONOMIC ACTIVITY

Land Use

- Commercial farming
- Subsistence farming
- Livestock raising
- Forests
- Manufacturing and trade
- Commercial fishing
- Little or no activity

Resources

Coal	Bauxite
Petroleum	Nickel
Natural gas	Copper
Uranium	Lead
Iron ore	Manganese
Tin	Gold
Zinc	Silver

0 mi. 1,000
0 km 1,000
Lambert Azimuthal Equal-Area projection

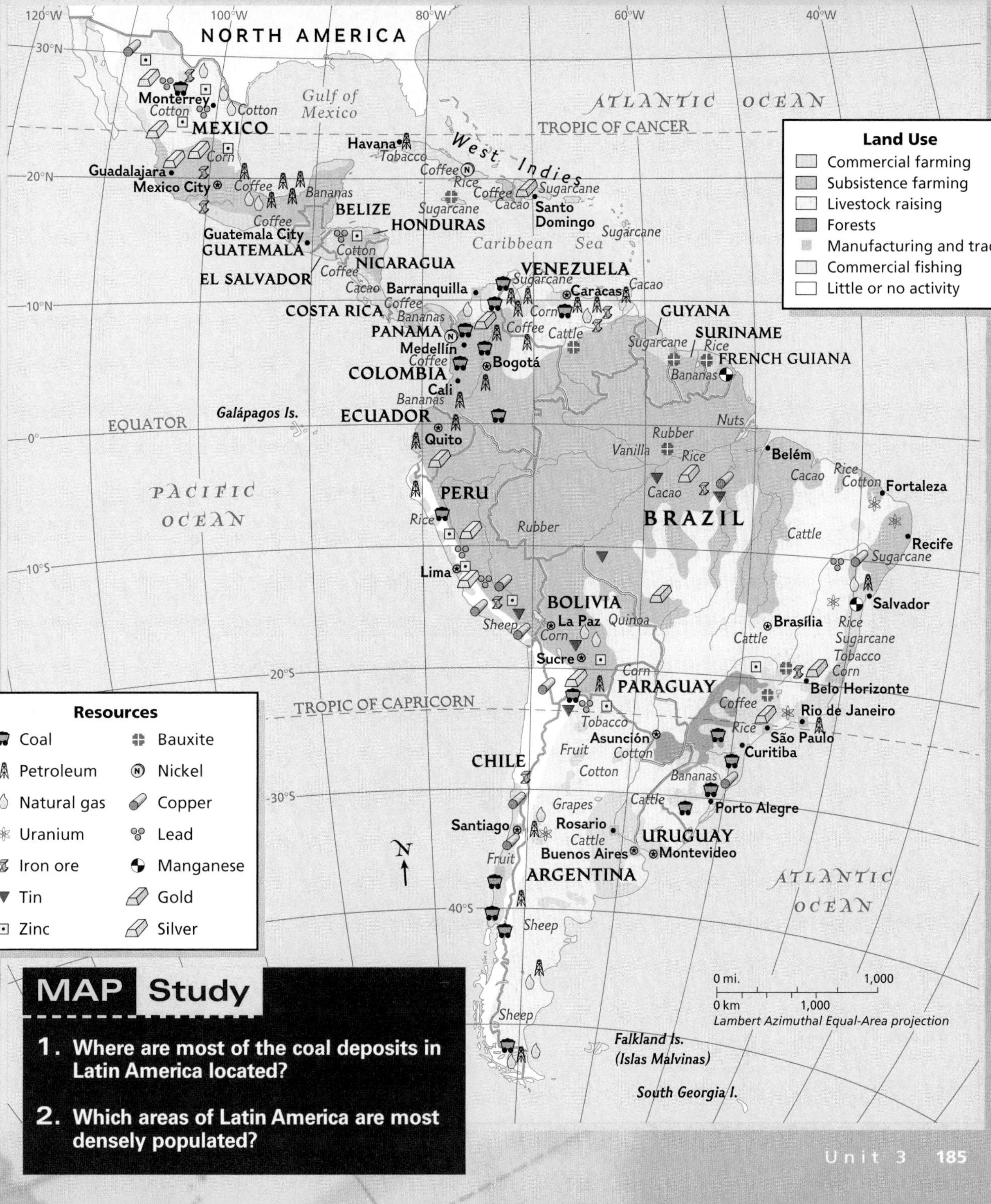

MAP Study

1. Where are most of the coal deposits in Latin America located?
2. Which areas of Latin America are most densely populated?

3 Latin America

COUNTRY PROFILES

COUNTRY * AND CAPITAL	FLAG AND LANGUAGE	POPULATION AND DENSITY	LANDMASS	MAJOR EXPORT	MAJOR IMPORT	CURRENCY	GOVERNMENT
ANTIGUA AND BARBUDA St. John's	English	100,000 394 per sq.mi. 152 per sq.km	170 sq.mi. 440 sq.km	Petroleum Products	Foods and Livestock	East Caribbean Dollar	Parliamentary Democracy
ARGENTINA Buenos Aires	Spanish	37,500,000 35 per sq.mi. 14 per sq.km	1,073,514 sq.mi. 2,780,401 sq.km	Meat	Machinery	Peso	Republic
BAHAMAS Nassau	English, Creole	300,000 58 per sq.mi. 23 per sq.km	5,359 sq.mi. 13,880 sq.km	Pharma-ceuticals	Foods	Bahamian Dollar	Parliamentary Democracy
BARBADOS Bridgetown	English	300,000 1,620 per sq.mi. 626 per sq.km	166 sq.mi. 430 sq.km	Sugar	Manufactured Goods	Barbados Dollar	Parliamentary Democracy
BELIZE Belmopan	English	300,000 29 per sq.mi. 11 per sq.km	8,865 sq.mi. 22,960 sq.km	Sugar	Machinery	Belize Dollar	Parliamentary Democracy
BOLIVIA La Paz Sucre	Spanish, Quechua, Aymara	8,500,000 20 per sq.mi. 8 per sq.km	424,162 sq.mi. 1,098,580 sq.km	Metals	Machinery	Boliviano	Republic
BRAZIL Brasília	Portuguese	171,800,000 52 per sq.mi. 20 per sq.km	3,300,154 sq.mi. 8,547,399 sq.km	Iron Ore	Crude Oil	Real	Federal Republic
CHILE Santiago	Spanish	15,400,000 53 per sq.mi. 20 per sq.km	292,135 sq.mi. 756,626 sq.km	Copper	Machinery	Peso	Republic
COLOMBIA Bogotá	Spanish	43,100,000 98 per sq.mi. 38 per sq.km	439,734 sq.mi. 1,138,511 sq.km	Petroleum	Machinery	Peso	Republic

* COUNTRIES AND FLAGS NOT DRAWN TO SCALE

COUNTRY * AND CAPITAL	FLAG AND LANGUAGE	POPULATION AND DENSITY	LANDMASS	MAJOR EXPORT	MAJOR IMPORT	CURRENCY	GOVERNMENT
COSTA RICA San José	Spanish	3,700,000 188 per sq. mi. 73 per sq. km	19,730 sq. mi. 51,100 sq. km	Coffee	Raw Materials	Colón	Republic
CUBA Havana	Spanish	11,300,000 264 per sq. mi. 102 per sq. km	42,803 sq. mi. 110,860 sq. km	Sugar	Petroleum	Peso	Communist State
DOMINICA Roseau	English, French	100,000 262 per sq. mi. 101 per sq. km	290 sq. mi. 751 sq. km	Bananas	Manufactured Goods	East Caribbean Dollar	Republic
DOMINICAN REPUBLIC Santo Domingo	Spanish	8,600,000 456 per sq. mi. 176 per sq. km	18,815 sq. mi. 48,731 sq. km	Ferronickel	Foods	Peso	Republic
ECUADOR Quito	Spanish, Quechua	12,900,000 118 per sq. mi. 46 per sq. km	109,483 sq. mi. 283,561 sq. km	Petroleum	Transport Equipment	Sucre	Republic
EL SALVADOR San Salvador	Spanish	6,400,000 788 per sq. mi. 304 per sq. km	8,124 sq. mi. 21,041 sq. km	Coffee	Raw Materials	Colón	Republic
FRENCH GUIANA (FRANCE) Cayenne	French	200,000 6 per sq. mi. 2 per sq. km	34,749 sq. mi. 89,999 sq. km	Shrimp	Foods	French Franc	Overseas Department of France
GRENADA St. George's	English, French	100,000 678 per sq. mi. 262 per sq. km	131 sq. mi. 339 sq. km	Bananas	Foods	East Caribbean Dollar	Parliamentary Democracy
GUATEMALA Guatemala City	Spanish, Mayan Languages	13,000,000 309 per sq. mi. 119 per sq. km	42,042 sq. mi. 108,889 sq. km	Coffee	Petroleum	Quetzal	Republic

* COUNTRIES AND FLAGS NOT DRAWN TO SCALE

FOR AN ONLINE UPDATE OF THIS INFORMATION, VISIT GEOGRAPHY.GLENCOE.COM AND CLICK ON "TEXTBOOK UPDATES."

3 Latin America

COUNTRY PROFILES

COUNTRY * AND CAPITAL	FLAG AND LANGUAGE	POPULATION AND DENSITY	LANDMASS	MAJOR EXPORT	MAJOR IMPORT	CURRENCY	GOVERNMENT
GUYANA Georgetown	English	700,000 8 per sq. mi. 3 per sq. km	83,000 sq. mi. 214,969 sq. km	Sugar	Manufactured Goods	Guyana Dollar	Republic
HAITI Port-au-Prince	French, Creole	7,000,000 650 per sq. mi. 251 per sq. km	10,714 sq. mi. 27,750 sq. km	Manufactured Goods	Machinery	Gourde	Republic
HONDURAS Tegucigalpa	Spanish	6,700,000 155 per sq. mi. 60 per sq. km	43,278 sq. mi. 112,090 sq. km	Bananas	Machinery	Lempira	Republic
JAMAICA Kingston	English, Creole	2,600,000 624 per sq. mi. 241 per sq. km	4,243 sq. mi. 10,989 sq. km	Alumina	Machinery	Jamaican Dollar	Parliamentary Democracy
MEXICO Mexico City	Spanish, Native American Languages	99,600,000 132 per sq. mi. 51 per sq. km	756,062 sq. mi. 1,958,201 sq. km	Crude Oil	Machinery	Peso	Federal Republic
NICARAGUA Managua	Spanish	5,200,000 104 per sq. mi. 40 per sq. km	50,193 sq. mi. 129,999 sq. km	Coffee	Manufactured Goods	Cordoba	Republic
PANAMA Panama City	Spanish	2,900,000 100 per sq. mi. 39 per sq. km	29,158 sq. mi. 75,519 sq. km	Bananas	Machinery	Balboa	Republic
PARAGUAY Asunción	Spanish, Guaraní	5,700,000 36 per sq. mi. 14 per sq. km	157,046 sq. mi. 406,749 sq. km	Cotton	Machinery	Guaraní	Republic
PERU Lima	Spanish, Quechua, Aymara	26,100,000 53 per sq. mi. 20 per sq. km	496,224 sq. mi. 1,285,220 sq. km	Copper	Machinery	Nuevo Sol	Republic

* COUNTRIES AND FLAGS NOT DRAWN TO SCALE

COUNTRY * AND CAPITAL	FLAG AND LANGUAGE	POPULATION AND DENSITY	LANDMASS	MAJOR EXPORT	MAJOR IMPORT	CURRENCY	GOVERNMENT
PUERTO RICO (U.S.) San Juan	Spanish, English	3,900,000 1,139 per sq. mi. 440 per sq. km	3,456 sq. mi. 8,951 sq. km	Pharma-ceuticals	Chemical Products	U.S. Dollar	U. S. Commonwealth
ST. KITTS AND NEVIS Basseterre	English	40,000 281 per sq. mi. 108 per sq. km	139 sq. mi. 360 sq. km	Machinery	Electronic Goods	East Caribbean Dollar	Parliamentary Democracy
ST. LUCIA Castries	English, French	200,000 656 per sq. mi. 253 per sq. km	239 sq. mi. 619 sq. km	Bananas	Foods	East Caribbean Dollar	Parliamentary Democracy
ST. VINCENT AND THE GRENADINES Kingstown	English, French	100,000 757 per sq. mi. 297 per sq. km	151 sq. mi. 391 sq. km	Bananas	Foods	East Caribbean Dollar	Parliamentary Democracy
SURINAME Paramaribo	Dutch	400,000 7 per sq. mi. 3 per sq. km	63,039 sq. mi. 163,271 sq. km	Bauxite	Machinery	Suriname Guilder	Republic
TRINIDAD AND TOBAGO Port-of-Spain	English	1,300,000 656 per sq. mi. 253 per sq. km	1,981 sq. mi. 5,131 sq. km	Petroleum	Machinery	Trinidad and Tobago Dollar	Republic
URUGUAY Montevideo	Spanish	3,400,000 49 per sq. mi. 19 per sq. km	68,498 sq. mi. 177,410 sq. km	Wool	Machinery	Peso	Republic
VENEZUELA Caracas	Spanish	24,600,000 70 per sq. mi. 27 per sq. km	352,143 sq. mi. 912,050 sq. km	Petroleum	Raw Materials	Bolivar	Federal Republic
VIRGIN ISLANDS (U.S.) Charlotte Amalie	English	120,917 713 per sq. mi. 276 per sq. km	134 sq. mi. 347 sq. km	Chemical Products	Crude Oil	U.S. Dollar	U.S. Territory

* COUNTRIES AND FLAGS NOT DRAWN TO SCALE

FOR AN ONLINE UPDATE OF THIS INFORMATION, VISIT GEOGRAPHY.GLENCOE.COM AND CLICK ON "TEXTBOOK UPDATES."

3

GLOBAL CONNECTION

LATIN AMERICA AND THE UNITED STATES

FOOD CROPS

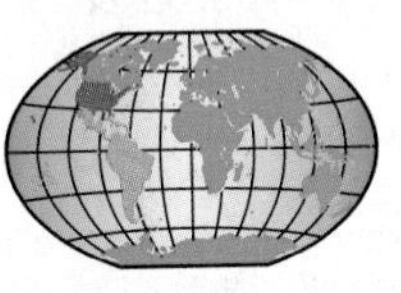

If you've had cornflakes, French fries, or a chocolate bar recently, you can thank Latin America. That's because these foods all are made from crops that originated there.

About 10,000 years ago, Native Americans in what is now Mexico gathered ears of wild corn for food. Between 5000 B.C. and 3500 B.C., they domesticated the plants and began raising them. Corn became a staple in the diets of the Maya and Aztec peoples. Gradually, corn cultivation spread from Mexico, eventually reaching the northeastern part of North America.

When European colonists arrived on America's eastern shores, Native Americans taught them how to grow this crop. It's been an important part of American agriculture ever since. In fact, the United States now leads the world in corn production.

Roughly 2,000 years ago, people in South America's Andes began cultivating potatoes, which are native to that area. When Spanish and English explorers arrived in the 1500s, they sampled potatoes and then carried some back to their homelands. It took a while for Europeans to develop a taste for these strange-looking tubers. But by the 1700s, potatoes were widely grown, especially in Ireland. Immigrants from Europe brought potatoes to the American colonies.

Chocolate is made from the seeds of cacao, a tree native to the Amazon River basin. How and when cacao seeds arrived in Central

▲ Guatemalan farmer carrying cornstalks

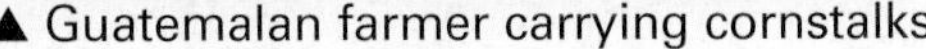

NATIONAL GEOGRAPHIC SOCIETY

America remains a mystery, but we do know that cacao came to play a major role in Maya and Aztec cultures. The Aztec believed that cacao seeds were a gift from heaven. The seeds were ground up to make a rich beverage called *xocoatl* (shoh•KOH•ahtl). However, the drink wasn't sweet. It was rather bitter and spiced with chili peppers!

In 1519 the Spanish explorer Hernán Cortés was served a cup of *xocoatl* by an Aztec ruler. When Cortés returned home, he took cacao seeds with him and introduced the drink to Spain. The Spanish made a few alterations. Their "chocolate" was sweetened with sugar and flavored with cinnamon and vanilla. For about 80 years, the Spanish kept their new beverage a secret. Once word got out, though, a chocolate craze spread across Europe.

When chocolate first arrived in the American colonies, it was an expensive European delicacy that only the wealthy could afford. Then in 1765, cacao seeds began to be imported directly, and relatively cheaply, from the West Indies. Finally, the average American was able to afford what the Aztec believed was the "food of the gods."

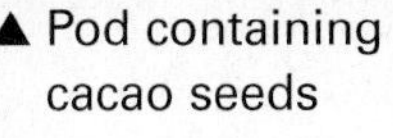

▲ Pod containing cacao seeds

CHAPTER 8

The Physical Geography of Latin America

GeoJournal

As you read this chapter, use your journal to describe the geographic features of Latin America. Choose strong, vivid terms to capture the beauty, grandeur, and economic importance of the physical features of the region.

GEOGRAPHY Online

Chapter Overview Visit the **Glencoe World Geography** Web site at geography.glencoe.com and click on Chapter Overviews—Chapter 8 to preview information about the physical geography of the region.

SECTION 1

The Land

Guide to Reading

Consider What You Know

News accounts of natural disasters in Latin America describe the destruction caused by hurricanes, earthquakes, and volcanic eruptions. What geographic factors might make the region vulnerable to such natural disasters?

Read to Find Out

- How do geographers divide the large region known as Latin America?
- What factors have shaped the formation of Latin America's landforms?
- How has the Latin American landscape influenced patterns of human settlement?
- What natural resources make Latin America an economically important region?

Terms to Know

- cordillera
- altiplano
- escarpment
- llano
- pampa
- gaucho
- hydroelectric power
- estuary

Places to Locate

- Amazon River
- Middle America
- Central America
- West Indies
- South America
- Sierra Madre
- Andes
- Mexican Plateau
- Patagonia
- Mato Grosso Plateau
- Rio Grande
- Río de la Plata

◀ *View from the top of the Iguaçu Falls, Brazil*

NATIONAL GEOGRAPHIC

A Geographic View

On the Amazon

I watched the river. Each boat carried a tiny cross-section of Amazon society. . . . Canoes drifted past. Small wooden passenger boats or traders mumbled their smoky way downstream. No matter how blue the sky, the river never caught the color in its reflection; it was loaded with sediment carved from the Andes. Logs and brush and whirlpools moved past in endless flow, and river dolphins rolled ahead of us.

—*Jere van Dyk, "Amazon: South America's River Road,"* National Geographic, *February 1995*

Brazilian riverboat

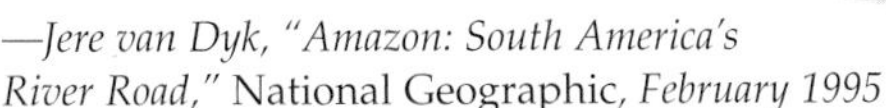

From the headwaters of the Peruvian Andes to the Atlantic coast of Brazil, the **Amazon River** winds about 4,000 miles (6,400 km) through the heart of South America. This mighty river, the world's second longest, is only one prominent feature of Latin America's large and varied landscape. In this section you will explore the region's physical geography: mountains, islands, coastal lowlands, plains, and waterways.

A Vast Region

Located in the Western Hemisphere south of the United States, Latin America has a land area of about 8 million square miles (20,720,000 sq. km)—nearly 16 percent of Earth's land surface. The countries of the region share a heritage of settlement by Europeans, especially those from Spain and Portugal. Most of these settlers spoke Spanish or Portuguese—languages based on Latin, the language of the Roman Empire, which gives the region its name.

Geographers usually divide Latin America into three areas—Middle America, the Caribbean, and South America. **Middle America** consists of Mexico and the seven countries of **Central America**, the stretch of land that links the landmasses of North and South America. The Caribbean islands, also known as the **West Indies**, fall into three groups—the Bahamas, the Greater Antilles, and the Lesser Antilles. The continent of **South America** is by far the largest land area of Latin America. Among South America's 13 countries, Brazil is the largest in both land area and population.

Mountains and Plateaus

One of Latin America's most distinctive landforms is its towering mountains. Thrusting upward in countless folds and ridges, this mountainous profile begins in North America as the Rocky Mountains and extends all the way to South America's southern tip. The mountains' names change as you move south. In Mexico they are the **Sierra Madre**; in Central America, the Central Highlands; and in South America, the **Andes**.

Latin America has such a rugged landscape because much of the region sits along the Pacific Ring of Fire, where plates of the earth's crust have collided for billions of years. These collisions have formed mountains and volcanoes and have caused tremendous earthquakes. They continue to change the landscape today. In 1999, for example, a strong earthquake reduced to rubble many towns and villages in northwestern South America.

Despite obstacles, the mountains and plateaus of Latin America have been places of human settlement for thousands of years. People wanting to escape the heat of the lowland areas have been drawn to cooler mountain climates. They also have been attracted by the mountains' rich natural resources—water, volcanic soil, timber, and minerals. Historically, Latin America's rugged terrain has tended to block movement and trade and to isolate regions and peoples. In recent decades radio, television, air transport, and the Internet have begun to break down old physical barriers.

Comparing Lands

Latin America is about three times the size of the continental United States.

Mountains of Mexico, Central America, and the Caribbean

Look at the physical-political map on page 195. Notice that Mexico's Sierra Madre consists of two mountain ranges—the Sierra Madre Oriental ("Eastern") and the Sierra Madre Occidental ("Western")—that meet near Mexico City to form the sharp-peaked Sierra Madre del Sur ("of the South"). These ranges surround the densely populated **Mexican Plateau**, which covers much of central Mexico. In the plateau's southern area, the mild climate, fertile volcanic soil, and adequate rainfall have attracted human settlement for thousands of years.

Farther south, the Central Highlands, a chain of volcanic mountains, rise like a backbone across Central America. Many Caribbean islands are also part of this mountain range, which extends across the bed of the Caribbean Sea. The islands are actually volcanic peaks that rise above sea level. Some of these volcanoes are still active, which can make living on these islands hazardous.

Andes of South America

None of Latin America's other mountains compare with the 4,500-mile (7,242-km) stretch of the Andes along the western edge of South America. Their extent makes the Andes the world's longest mountain range, as well as one of the highest, with some peaks rising to more than 20,000 feet (6,096 m) above sea level. The Andes consist of several ranges that run parallel to one another like deep folds in a carpet. Such parallel ranges are called **cordilleras** (KAWR•duhl•YEHR•uhs).

In Peru and Bolivia, the spectacular Andes peaks encircle a region called the **altiplano**, which means "high plain." In southern Argentina, hills and lower flatlands form the plateau of **Patagonia**.

MAP STUDY

Latin America: Physical-Political

ATLANTIC OCEAN
PACIFIC OCEAN
ATLANTIC OCEAN
Gulf of Mexico
Caribbean Sea
TROPIC OF CANCER
EQUATOR
TROPIC OF CAPRICORN
30°N
20°N
10°N
0°
10°S
20°S
30°S
40°S
50°S
120°W
100°W
80°W
60°W
40°W
Bermuda Is.
Rio Grande
Baja California
Sierra Madre Occidental
Mexican Plateau
Sierra Madre Oriental
MEXICO
Sierra Madre del Sur
Yucatán Pen.
BAHAMAS
WEST INDIES
CUBA
Greater Antilles
HAITI
DOMINICAN REPUBLIC
Virgin Is.
Puerto Rico
Lesser Antilles
JAMAICA
ANTIGUA AND BARBUDA
Guadeloupe
DOMINICA
Martinique
ST. LUCIA
ST. VINCENT AND THE GRENADINES
BARBADOS
GRENADA
TRINIDAD AND TOBAGO
BELIZE
HONDURAS
GUATEMALA
EL SALVADOR
NICARAGUA
COSTA RICA
PANAMA
Isthmus of Panama
VENEZUELA
Orinoco R.
LLANOS
GUYANA
SURINAME
FRENCH GUIANA
GUIANA HIGHLANDS
COLOMBIA
ECUADOR
Galápagos Islands
AMAZON BASIN
Amazon R.
PERU
ANDES
BRAZIL
MATO GROSSO PLATEAU
BRAZILIAN HIGHLANDS
São Francisco R.
Lake Titicaca
BOLIVIA
Paraguay R.
Paraná R.
GRAN CHACO
PARAGUAY
CHILE
ARGENTINA
Aconcagua 22,834 ft. (6,960 m)
URUGUAY
PAMPAS
Río de la Plata
PATAGONIA
Tierra del Fuego
Cape Horn
Falkland Islands (Islas Malvinas)
South Georgia I.
N
0 mi. 1,000
0 km 1,000
Lambert Azimuthal Equal-Area projection

Elevations

Feet	Meters
10,000	3,000
5,000	1,500
2,000	600
1,000	300
0	0

National boundary
▲ Mountain peak

Geography Skills for Life

1. **Interpreting Maps** What physical features surround the Mexican Plateau?
2. **Applying Geography Skills** Which South American rivers flow through highlands areas? Lowlands areas?

Find NGS online map resources @ www.nationalgeographic.com/maps

Wildlife expert William Franklin describes the windswept Patagonia region at the southern end of South America:

> *"The sky is full of mountains in this country. I often get a sore neck from admiring these Andean peaks as we trek on foot and horseback over the plains and hills. The wind is our constant companion; locals advise that if you want to see Patagonia, just stand still and it will all blow past you."*
>
> William Franklin, "Patagonia Puma: The Lord of Land's End," *National Geographic,* January 1991

Highlands of Brazil

Eastern South America is marked by broad plateaus and valleys. The **Mato Grosso Plateau**, a sparsely populated plateau of forests and grasslands, spreads over much of Brazil and across the west to Bolivia and Peru. East of the Mato Grosso Plateau lie the Brazilian Highlands, a plateau so vast that it spans several climate and vegetation zones. On the eastern edge of the Brazilian Highlands, the plateau plunges sharply to the Atlantic Ocean, forming a steep cliff or slope called an **escarpment**.

Lowlands and Plains

Narrow coastal lowlands wind their way along the Gulf of Mexico and the Caribbean and also hem the Atlantic and Pacific coasts of South America. One of the longest strips of coastal plain in Latin America lies along Brazil's Atlantic coast. In northeastern Brazil, this plain is about 40 miles (60 km) wide but narrows considerably as it winds southward. Between Rio de Janeiro and the southeastern seaport of Santos, the plain disappears entirely, only to reappear and widen again near Brazil's borders with Uruguay and Argentina. Hemmed in by highland escarpment, Brazil's coastal plain has been a major area of settlement and economic activity since the 1500s.

Inland areas of South America hold vast grasslands: the **llanos** (LAH•nohs) of Colombia and Venezuela, and the **pampas** of Argentina and

Geography Skills for Life

The Coast of Brazil Ipanema Beach in Rio de Janeiro is a popular tourist area.

Place Describe the physical environment in which Rio de Janeiro is located.

NATIONAL GEOGRAPHIC **World Explorer**

Geography Skills for Life

The Gaucho Argentine cowhands known as gauchos ride the pampas herding livestock, the major agricultural product of Argentina.

Human-Environment Interaction How does terrain in Argentina support cattle ranching?

Uruguay. Both plains areas provide wide grazing lands for beef cattle. Ranchers on large estates employ cowhands, called *llaneros* in the llanos and **gauchos** in the pampas, to drive great herds of cattle across the rolling terrain. Known for its fertile soil, the pampas region is one of the world's major "breadbaskets," producing an abundance of wheat and corn. Many people in the pampas region grow crops on small- and medium-sized farms.

Water Systems

Like a massive circulatory system, Latin America's many waterways serve as arteries that transport people and goods to different parts of the region and the world. Most of the region's major rivers are in South America. One important exception is the **Rio Grande**, or *Río Bravo del Norte* ("Wild River of the North"), which forms part of the long border between Mexico and the United States.

Economics

Rivers of South America

Middle America's rivers are generally small, but the rivers that cross South America are gigantic. The Amazon is the Western Hemisphere's longest river and carries ten times the water volume of the Mississippi River. Hundreds of smaller rivers join the Amazon as it journeys from the Andes to the Atlantic Ocean. These rivers together form the Amazon Basin, which drains parts of Bolivia, Peru, Ecuador, Colombia, and Venezuela, as well as Brazil. Despite the tremendous force of water at its mouth, the Amazon is navigable. Oceangoing ships can travel upstream as far as 2,300 miles (3,701 km) from the Atlantic coast.

The Paraná, Paraguay, and Uruguay Rivers together form the second-largest river system in Latin America. This system drains the rainy eastern half of South America. Important commercial highways, these three rivers provide inland water routes and **hydroelectric power**—electricity generated from the energy of water—for Argentina, Bolivia, Brazil, Paraguay, and Uruguay. After coursing through inland areas, the three rivers flow into a broad **estuary**, an area where the tide meets a river

Student Web Activity Visit the **Glencoe World Geography** Web site at geography.glencoe.com and click on Student Web Activities—Chapter 8 for an activity about the physical geography of Costa Rica.

current. This estuary, the **Río de la Plata** ("River of Silver"), meets the Atlantic Ocean. Buenos Aires, the capital of Argentina, and Montevideo, the capital of Uruguay, lie along the Río de la Plata.

Lakes

Latin America has few large lakes. The region does include the world's highest navigable lake, Lake Titicaca (TEE•tee•KAH•kah), in the Andes of Bolivia and Peru. Lake Titicaca lies about 12,500 feet (3,810 km) above sea level. The area surrounding Lake Titicaca was one of the centers of early Native American civilization. It holds many architectural remains from the distant past. Lake Maracaibo (MAH•rah•KY•boh) in Venezuela is regarded as South America's largest lake, even though it is actually an inlet of the Caribbean Sea. Lake Maracaibo and the surrounding area contain the most important oil fields in Venezuela. The largest lake in Central America is Lake Nicaragua, which lies between Nicaragua and Costa Rica.

Gold mask from Ecuador, about 500 B.C.–A.D. 500

Natural Resources

Latin America has significant natural resources, including minerals, forests, farmland, and water. Major deposits of oil and natural gas lie in rock beds located in mountain valleys and in offshore areas, especially along the Gulf of Mexico and in the southern Caribbean Sea. These deposits help make Mexico and Venezuela leading oil producers.

Latin America's mineral wealth was first mined by Native American peoples and later by European colonists. The foothills along Venezuela's Orinoco River contain large amounts of gold. Brazil also is rich in gold, while Peru and Mexico are known for silver. Mines in Colombia have been producing the world's finest emeralds—precious green stones composed of beryllium—for more than 1,000 years. Even Latin America's nonprecious minerals have great economic value. Chile is the world's largest exporter of copper, and Jamaica is a leading source of bauxite, the main ore of aluminum. Bolivia and Brazil have large reserves of tin.

Not all of Latin America's countries share equally in this bounty. Geographic inaccessibility, lack of capital for development, and deep social and political divisions keep many of the region's natural resources from being developed fully or distributed evenly. The challenge for Latin Americans in the future is how to overcome these obstacles and make the best use of the region's natural resources.

SECTION 1 ASSESSMENT

Checking for Understanding

1. **Define** cordillera, altiplano, escarpment, llano, pampa, gaucho, hydroelectric power, estuary.
2. **Main Ideas** Use a table like the one below to describe Latin America's three main geographic areas. Then choose one area, and explain how it differs from the other two.

Geographic Area	Physical Features

Critical Thinking

3. **Identifying Cause and Effect** How do the physical features of Latin America affect everyday life? Give examples.
4. **Drawing Conclusions** Why does much of South America have the potential to produce hydroelectric power?
5. **Making Inferences** What factors make Latin America important to the global economy?

Analyzing Maps

6. **Region** Study the physical-political map on page 195. What part of South America is dominated by mountains?

Applying Geography

7. **Effects of Landforms** Think about the physical features of South America. Write a descriptive paragraph explaining how landforms affect the course of South America's water systems.

SECTION 2

Climate and Vegetation

Guide to Reading

Consider What You Know

Most of Latin America's people live in an area between the Tropic of Cancer and the Tropic of Capricorn, an area that includes the Equator. What types of climate and vegetation would you expect to find in this broad area?

Read to Find Out

- Which climate regions are represented in Latin America?
- How do Latin America's location and landforms affect climates even within particular regions?
- How are the natural vegetation and agriculture of Latin America influenced by climatic factors?

Terms to Know

- canopy
- *tierra caliente*
- *tierra templada*
- *tierra fría*

Places to Locate

- Amazon Basin
- Colombia
- Venezuela
- Argentina
- Uruguay
- Atacama Desert

NATIONAL GEOGRAPHIC

A Geographic View

Exploring Chile's Mountains

Green was the color we least expected when we landed on [Chile's] Sarmiento [ranges]. . . . Mosses and lichens carpeted the rocks above an iceberg-littered bay. . . . After several days of exploring, our progress thwarted by glacial canyons and snarly ice-falls, we discovered a route to the peaks. . . . To reach the ridge, we had to hack through rain forest, our skis catching on limbs, our boots slipping off logs.

—*Jack Miller, "Chile's Uncharted Cordillera Sarmiento,"* National Geographic, *April 1994*

Sarmiento peak and bay

Diverse climates make Latin America a region of sharp contrasts. To reach the glacial peaks of Chile's Cordillera Sarmiento, for example, climbers must trek through thick, nearly impenetrable vegetation. Steamy rain forests, arid deserts, grassy plains, and sandy beaches are all part of the region. In this section you will learn about Latin America's various climate regions and how the region's climates and landforms together influence natural vegetation and the growing of crops.

Climate and Vegetation Regions

Much of Latin America lies between the Tropic of Cancer and the Tropic of Capricorn. As a result, vast areas of the region have some form of tropical climate with lush green vegetation. Yet, even within the Tropics, mountain ranges and wind patterns create a variety of climates and natural vegetation in Latin America. The

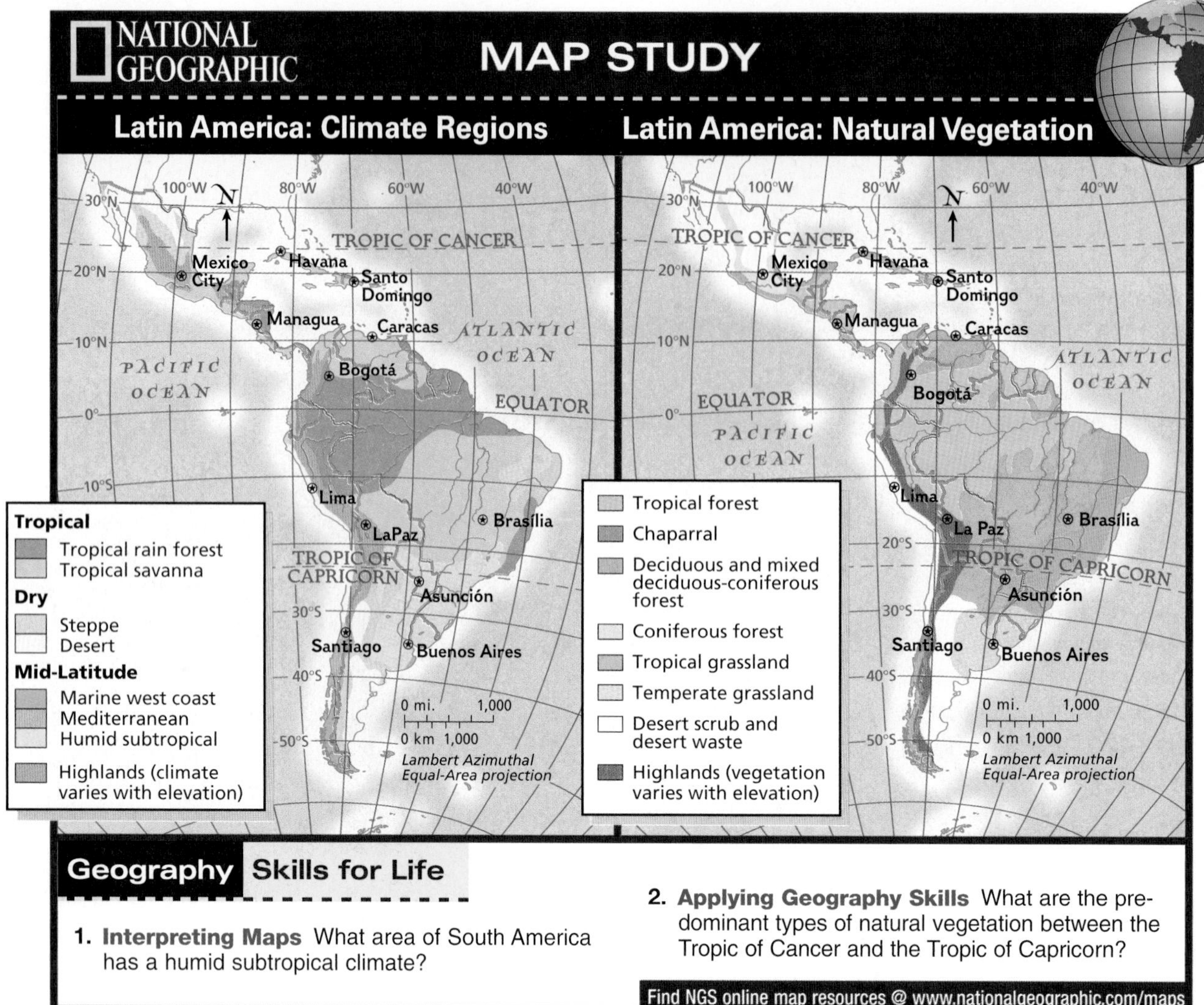

maps above show Latin America's climate regions and natural vegetation zones.

Tropical Regions

A tropical rain forest climate and vegetation dominate southern Mexico, eastern Central America, some Caribbean islands, and parts of South America. Hot temperatures and abundant rainfall occur year-round. In the **Amazon Basin**, this combination results from the area's location on the Equator and the patterns of the prevailing winds.

The Rain Forest

Wet tropical areas of Latin America have a dense cover of rain forest, or *selva* as it is called in Brazil. Latin American rain forests contain a variety of trees, including tropical hardwoods, palms, tree ferns, and bamboos. In Latin America's tropical rain forest areas, broad-leafed and needle-leafed evergreen trees are so close together that their crowns form a dense **canopy**, or a continuous layer of leaves. The canopy may soar to 130 feet (40 m) and is so dense that sunlight seldom reaches the forest floor. Plants beneath the canopy must be shade tolerant.

The Amazon Basin, with Earth's largest rain forest, covers about one-third of South America. It is also the world's wettest tropical plain. Heavy rains drench much of the densely forested lowlands throughout the year, but especially between January and June. During the months of heavy rainfall, large areas crossed by the Amazon River are often severely flooded. In Brazil, the width of the river ranges between 1 and 6 miles (1.6 and

10 km) but enlarges to 30 miles (48 km) or more during annual flooding.

The Amazon rain forest shelters more species of plants and animals per square mile than anywhere else on Earth. One journalist described a recent survey by scientists from the Smithsonian Institution in Washington, D.C.:

> *"Here at this one site on the Equator, in about 1,500 acres, scientists have counted 3,000 species of plants, 530 species of birds, nearly 80 species of bats, and 11 species of primates. There are jaguars and other wild cats, tapir, deer, otters, capybaras, and agoutis. . . ."*
>
> Virginia Morell, "The Variety of Life," *National Geographic,* February 1999

The Amazon rain forest is also a habitat for many reptiles. The snakes there include boas and anacondas. Iguanas and crocodiles also are found in many rain forests. Rivers and streams teem with varied and abundant freshwater fish.

Tropical Savanna

A tropical savanna climate is typical of the coast of southwestern Mexico, most Caribbean islands, and north-central South America. These areas have hot temperatures and abundant rainfall but also experience an extended dry season. In many tropical savanna areas, vast grasslands flourish. Some of these grasslands, such as the llanos of **Colombia** and **Venezuela**, are covered with scattered trees and are considered transition zones between grasslands and forests.

History

The Humid Subtropics

A humid subtropical climate prevails over much of southeastern South America, from Rio de Janeiro, Brazil, to the pampas of **Argentina** and **Uruguay**. In this area, winters are short and mild, and summers are long, hot, and humid. Summers occasionally bring short dry periods.

The vast pampas today consist primarily of short grasses but once had scattered trees. Spanish settlers brought cattle and horses to the pampas and cut down trees to set up ranches. Overgrazing

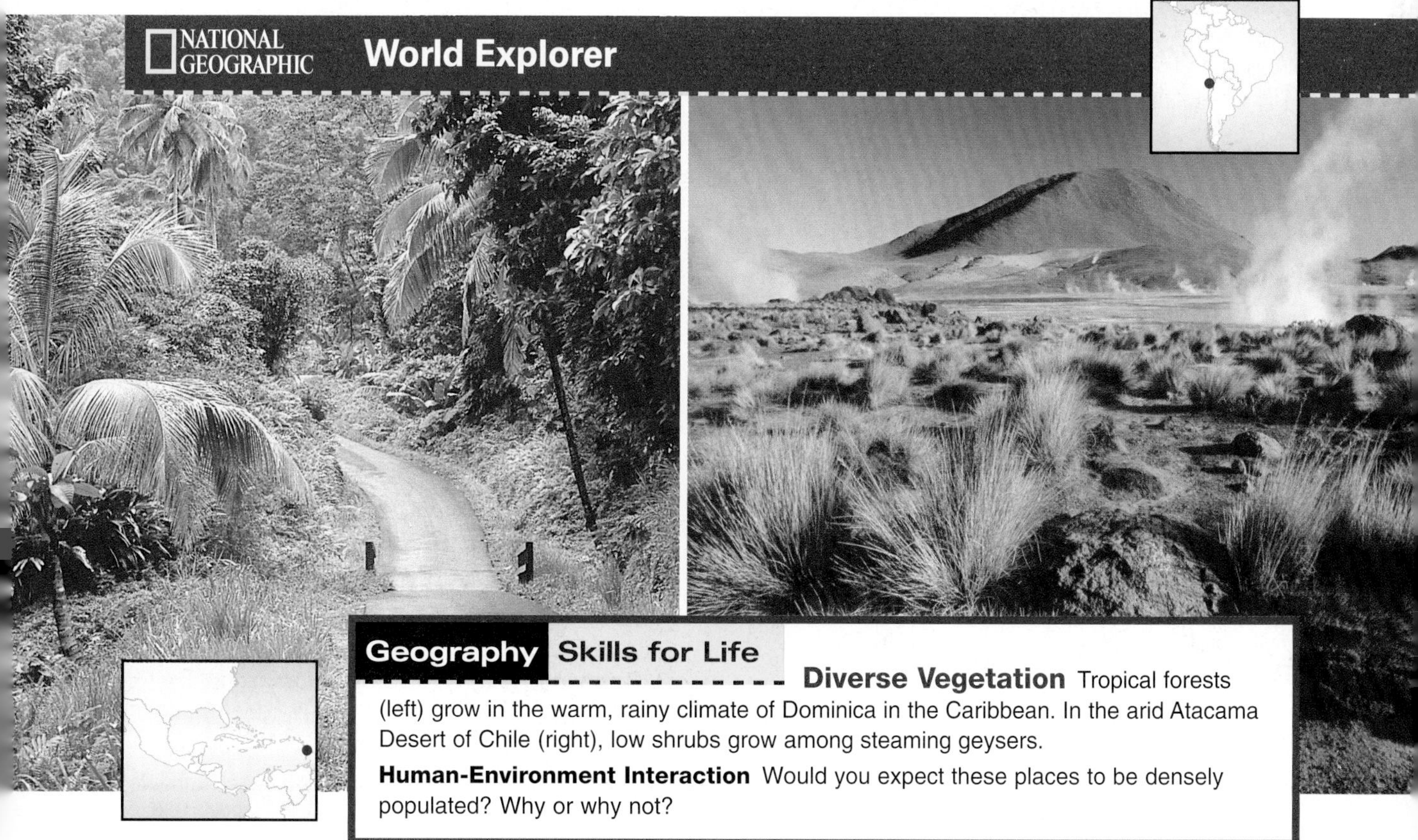

Geography Skills for Life

Diverse Vegetation Tropical forests (left) grow in the warm, rainy climate of Dominica in the Caribbean. In the arid Atacama Desert of Chile (right), low shrubs grow among steaming geysers.

Human-Environment Interaction Would you expect these places to be densely populated? Why or why not?

eventually left only short clumps of grass to anchor the pampas soil, and dust storms periodically swept over the region. Argentine farmers now plant alfalfa, corn, and cotton to hold the topsoil in place.

Desert and Steppe Areas

Parts of northern Mexico, coastal Peru and Chile, and the southeastern coast of Argentina have desert climates and vegetation. In Chile the rain shadow effect of the Andes has produced the **Atacama Desert**, a region so arid that in some places no rainfall has ever been recorded. In the desert areas of Latin America, vegetation is sparse. Prickly cacti and drought-resistant shrubs, however, have adapted to the harsh environment.

Parts of Latin America—northern Mexico, northeastern Brazil, and south central South America—receive little rainfall but do not have desert climates and vegetation. Instead, they have steppe climates—hot summers, cool winters, and light rainfall—and grassy or lightly forested vegetation.

Elevation and Climate

Although Latin America lies in the Tropics, its varied climates are more affected by elevation than by distance from the Equator. Throughout the region, Spanish terms are used to describe three different vertical climate zones that occur as elevation increases. Each of these three zones has its own characteristic natural vegetation and crops.

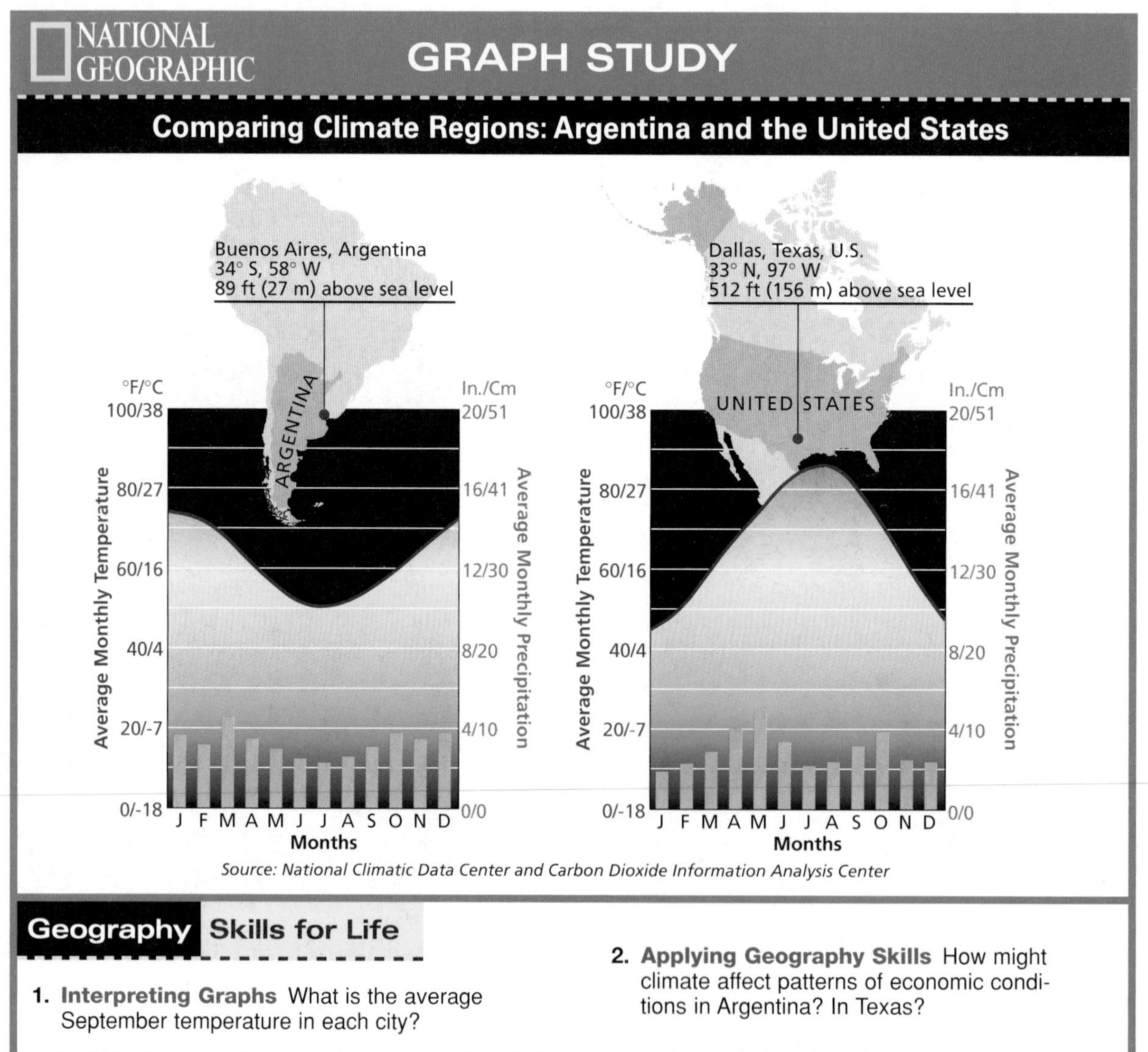

Geography Skills for Life

1. **Interpreting Graphs** What is the average September temperature in each city?
2. **Applying Geography Skills** How might climate affect patterns of economic conditions in Argentina? In Texas?

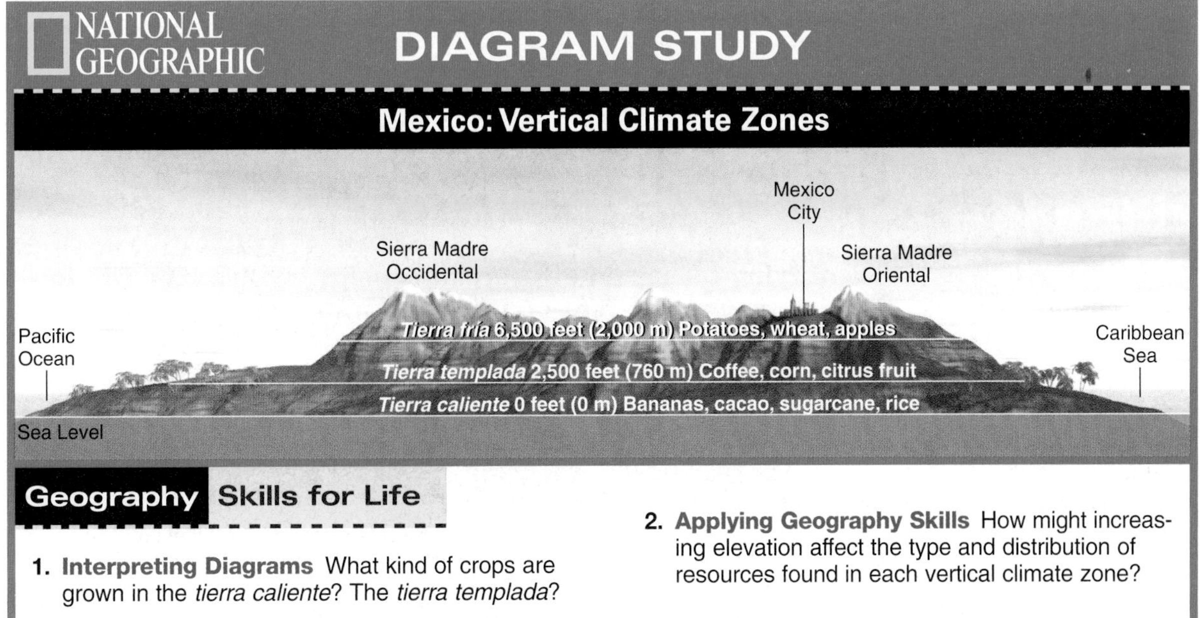

Geography Skills for Life

1. **Interpreting Diagrams** What kind of crops are grown in the *tierra caliente*? The *tierra templada*?
2. **Applying Geography Skills** How might increasing elevation affect the type and distribution of resources found in each vertical climate zone?

The *tierra caliente*, or "hot land," lies at elevations between sea level and 2,500 feet (760 m). Average annual temperatures in these coastal areas and foothills range from 68° to 91°F (20° to 33°C). In the rain forests of the *tierra caliente*, the main crops include bananas, sugar, rice, and cacao.

The *tierra templada*, or "temperate land," lies between 2,500 and 6,500 feet (760 and 2,000 m). In this zone temperatures range between 60° and 72°F (16° and 22°C). Broad-leafed evergreen trees at lower levels give way to needle-leafed, cone-bearing evergreens at upper levels. In the *tierra templada*, the most densely populated of the vertical climate zones, coffee and corn are the main crops.

Land at 6,500 to 10,000 feet (2,000 to 3,048 m) is known as the *tierra fría*, or "cold land." At this elevation, frosts are common during winter months. However, crops such as potatoes and barley grow well here. Above the *tierra fría*, conditions are more difficult for agriculture or human habitation.

SECTION 2 ASSESSMENT

Checking for Understanding

1. **Define** canopy, *tierra caliente*, *tierra templada*, *tierra fría*.
2. **Main Ideas** Create a table to identify, locate, and describe Latin America's climate regions. Then write a sentence that describes each zone's features and vegetation.

Climate Region	Location	Characteristics
Humid Subtropical		

Critical Thinking

3. **Making Inferences** Why might some Latin Americans live in areas in which climate and agriculture are unfavorable?
4. **Determining Cause and Effect** How does elevation affect climate and vegetation in Latin America?
5. **Comparing and Contrasting** Compare the pampas to your region. How do climate and vegetation help define the economic activities in each place?

Analyzing Maps

6. **Place** Study the vegetation map on page 200. Which two types of vegetation cover most of South America?

Applying Geography

7. **Effects of Climate** Write a paragraph describing the effects of climate on economic activities in a particular Latin American country. Then map the locations of these activities.

Identifying Cause-and-Effect Relationships

Identifying cause-and-effect relationships involves considering how and why an event occurred. A *cause* is the action or situation that leads to the event. An *effect* is the result or consequence of an action or situation.

Learning the Skill

Cause-and-effect relationships may be simple or complex. Several causes can produce a single effect. For example, a forest fire may be caused by a series of events or conditions. Hot weather and lack of rain make grass and wood dry and flammable. The day the fire started might have been windy, and the wind might have blown sparks from a camper's fire to some dry grass.

Similarly, one event can produce several effects. A large forest fire can destroy animal habitats. It can also suddenly reduce plant cover, making the land more susceptible to erosion from wind and rain. A large fire can also be expensive to fight and damaging to homes and businesses, harming the economy of an entire region.

Sometimes one event causes several other events in a chain reaction. A traffic accident on a highway may cause another accident, which causes another accident, and so on. Strings of causal relationships are called *cause-and-effect chains.*

Follow these steps to identify cause-and-effect relationships:

- **Ask questions about why events occur.**
- **Identify the outcomes of events.**
- **Look for clues that indicate a cause-and-effect relationship.** Words and phrases such as *because, as a result of, brought about, as a consequence, therefore,* and *thus* can help you identify cause-and-effect relationships.

Causes → Forest Fire → Effects

Causes
- Hot weather and lack of rain make grass and wood dry and flammable.
- Natural factors such as lightning strikes or wind can ignite or fuel a fire.
- Careless campers and hikers may leave a fire unattended.

Effects
- Animal habitats destroyed.
- Soil erosion increases due to loss of plant cover.
- Expense of fighting the fire and the loss of homes and businesses harms the region's economy.

Practicing the Skill

Identify one cause and one effect associated with each of the events or conditions listed below.

1. The 1999 earthquake in Colombia
2. The formation of several Caribbean islands
3. Limited access to the rich mineral resources of the Amazon Basin
4. Cold temperatures in the *tierra fría*

Applying the Skill

Use the library or the Internet to research volcanic activity in Latin America. Then explain the causes and effects of a volcanic eruption by creating a graphic like the one above.

The **Glencoe Skillbuilder Interactive Workbook, Level 2** provides instruction and practice in key social studies skills.

8

SECTION 1 The Land (pp. 193–198)

Terms to Know
- cordillera
- altiplano
- escarpment
- llano
- pampa
- gaucho
- hydroelectric power
- estuary

Key Points
- Latin America includes Middle America, the Caribbean, and South America.
- Latin America's physical features include high mountain ranges, less rugged highlands, vast central plains, and volcanic islands.
- The water systems of Latin America, especially the mighty rivers of South America, are key to human activity in the region.
- Although the region is rich in natural resources, geographic, political, and economic obstacles have kept resources from being developed fully or shared equally.

Organizing Your Notes
Create a table like the one below to help you organize information about the physical features of Latin America.

Physical Feature	Location
Mexican Plateau	
Andes	
Rio Grande	
Amazon River	
Río de la Plata	

SECTION 2 Climate and Vegetation (pp. 199–203)

Terms to Know
- canopy
- *tierra caliente*
- *tierra templada*
- *tierra fría*

Key Points
- Much of Latin America lies in the Tropics; however, landforms and wind patterns give the region great climatic diversity.
- Tropical climates such as tropical forest and tropical savanna are the most common climates in Latin America.
- The natural vegetation of Latin America consists mainly of rain forests and grasslands.
- The tropical highlands in Latin America include three vertical climate zones that are based on latitude and elevation.

Organizing Your Notes
Create an outline using the format below to help you organize your notes for this section.

Climate and Vegetation

I. Climate and Vegetation Regions
  A. Tropical Regions
    1. The Rain Forest
II.

Andean peaks in northern Chile

CHAPTER 8

ASSESSMENT & ACTIVITIES

Reviewing Key Terms

Write the key term that best completes each of the following sentences. Refer to the Terms to Know in the Summary & Study Guide on page 205.

1. The Andes consist of parallel mountain ranges, or _________.
2. The high plain encircled by the Andes of Bolivia and Peru is known as the _________.
3. The plateau of the Brazilian Highlands plunges sharply to the Atlantic Ocean, forming a steep cliff called an _________.
4. Cattle are raised on the broad grasslands called _________ in Colombia and Venezuela and _________ in Argentina and Uruguay.
5. The Río de la Plata is typical of an _________, an area where the tide meets a river current.
6. Highlands climates are divided into vertical zones, including the hot _________, the temperate _________, and the cold _________.

Reviewing Facts

SECTION 1

1. What are the three major geographic areas within Latin America?
2. What three island groups make up the West Indies?
3. Which three rivers flow into the Río de la Plata?

SECTION 2

4. What are the eight climate regions of Latin America?
5. What factors determine why Latin America's highlands climate is divided into three zones?
6. Where is the world's largest rain forest located?
7. What are South America's two main grassland areas called?

Critical Thinking

1. **Making Generalizations** Write a generalization that describes the kinds of economic activities you would expect to find in grasslands areas, using Latin America as an example.
2. **Analyzing Information** Identify and explain the factors affecting the location of different types of economic activities in Latin American countries.
3. **Comparing and Contrasting** Use a Venn diagram to compare the climate and vegetation found in Latin America's tropical areas.

Tropical Savanna | Both | Tropical Rain Forest

NATIONAL GEOGRAPHIC

Locating Places

Latin America: Physical Geography

Match the letters on the map with the physical features of Latin America. Write your answers on a sheet of paper.

1. Amazon River
2. Lake Titicaca
3. Rio Grande
4. Hispaniola
5. Lake Maracaibo
6. Río de la Plata
7. Gulf of Mexico
8. Pampas
9. Caribbean Sea
10. Orinoco River
11. Mexican Plateau

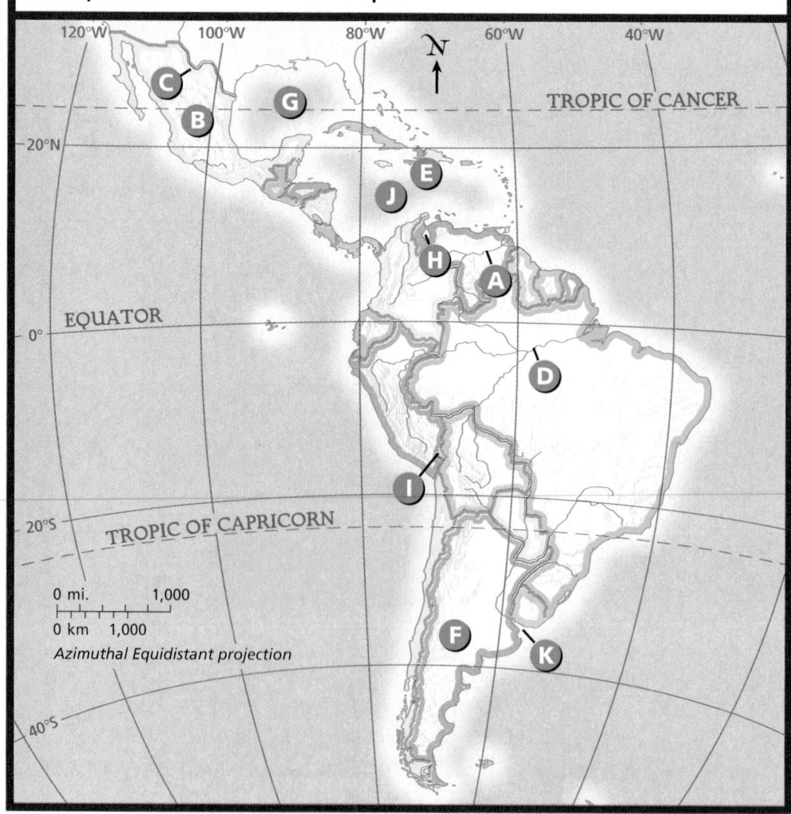

Self-Check Quiz Visit the **Glencoe World Geography** Web site at geography.glencoe.com and click on Self-Check Quizzes—Chapter 8 to prepare for the Chapter Test.

Using the Regional Atlas

Refer to the Regional Atlas on pages 182–185.

1. **Location** What river makes up a major part of the boundary between Mexico and the United States?
2. **Place** In terms of land use, why is there little to no activity along much of the Pacific coast of South America?

Thinking Like a Geographer

Review the economic activity map on page 187. Analyze the effects of physical and human geographic processes on the development of Latin America's resources. Make three practical suggestions for improving resource development in the region.

Problem-Solving Activity

Problem-Solution Proposal Working with a group, contact media services to find out more about a recent natural disaster in Latin America. Investigate accounts of the disaster to determine whether human activity made the disaster worse. In a report, describe the disaster's impact and propose ways to reduce the potential for damage in the future.

GeoJournal

Descriptive Writing Using the information you logged in your GeoJournal as you read, write a descriptive paragraph about one of the physical features of the region. Use your textbook and the Internet as resources to make your descriptions vivid, accurate, and interesting.

Technology Activity

Building an Electronic Database Collect facts about the countries of Latin America, such as natural resources, climate, average annual temperature, average annual rainfall, and natural vegetation. Create a database to organize and analyze the data. From the database, develop a table that presents your analysis.

Standardized Test Practice

Use the climograph below and your knowledge of geography to answer the question.

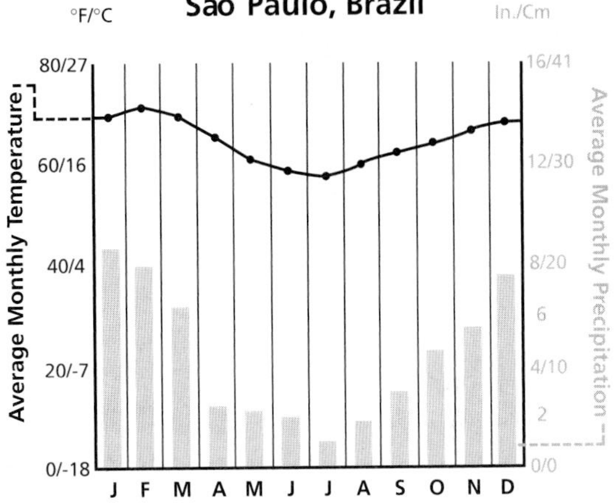

1. **Based on the information shown in the climograph, which statement about the months of April and November is accurate?**

   A The average temperature and amount of rainfall are about the same.

   B It is hotter and drier in November.

   C The average temperature is about the same, but it is wetter in November.

   D The amount of rainfall is about the same, but it is hotter in November.

Study the information shown on the climograph for average temperature and monthly precipitation. Then look carefully at the data for the months of April and November. Compare the amount of precipitation for the two months. As a result, you will be able to eliminate some of the statements.

Geography Lab Activity

Simulating Geographic Information Systems

▲City planning is made easier with GIS technology.

Geographic information systems (GIS) use computer software to create specialized maps that display a range of geographic information about an area. The user first creates a database with fields for images, such as digitized maps and satellite photos, and statistical information, such as census figures or property taxes. The GIS software can then generate maps to display the data, either separately or in combination.

GIS technology makes maps that can help people manage resources, select sites for buildings, and plan transportation routes. If Mexico City's government wants to add a new bus route, for example, an urban planner might input such data as population distribution, traffic patterns, congested areas, and existing bus routes. After examining these data, displayed in "layers" on a computerized map, the planner can analyze relationships and make an informed decision about the new bus route.

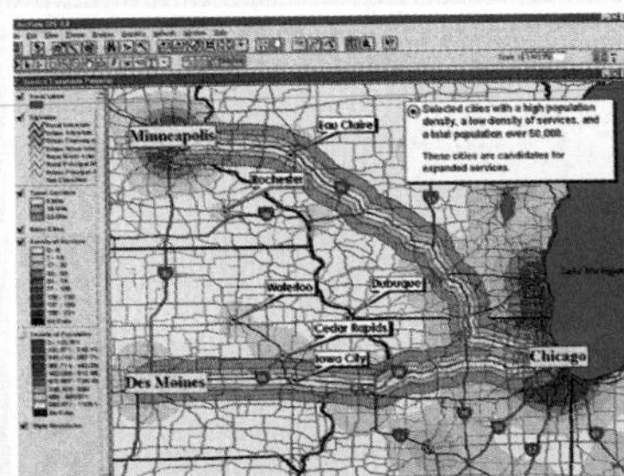

GIS technology allows layers of data drawn on a map to be turned on or off. ▶

1 Materials

- Overhead projector
- Transparency markers
- Seven blank $8\frac{1}{2}$″ × 11″ transparencies
- Street map of your community
- Computer

2 Procedures

In this activity, you will simulate a GIS map, using data about your community. As a class, you will analyze the data to determine the best emergency evacuation routes for your community.

1. Form three groups (A, B, and C). Each group will collect a different type of data.
2. Each group should copy a street map of the town or community onto a blank transparency. These will be the base maps. (Be sure that each group uses the same map.)
3. **Group A:** Determine the locations of major roads. Draw these roads on the blank overlay transparency placed on top of the base map. Color-code primary roads in black and secondary roads in purple. Be sure to note specific features such as bridges, railroads, and high-water crossings.
4. **Group B:** Gather information on the population distribution in the community. Locate residential areas, business districts, and shopping centers. Identify single-family homes and clusters of apartment buildings or college dormitories, and mark them on the overlay transparency. Color-code the areas or use symbols.

List peak business hours and traffic rush hours, when roads may be crowded.

5. **Group C:** Gather information on the location of emergency shelters in the community. Use a symbol to indicate these shelters on the overlay transparency. Also, identify physical features of the town, such as rivers, creeks, or mountains.
6. As a class, layer the overlay transparencies of each group over one base map on the overhead projector. Analyze the map and the data to determine the best emergency evacuation routes. Using a blank transparency taped over the other layers, draw the suggested emergency evacuation route(s) in red.

3 Lab Report

1. Which of the steps was the most time-consuming? Why?
2. Which of the overlays provided you with the most useful information? Explain.
3. What layers of information might you need to determine the best location for a new shopping center in your town?
4. **Drawing Conclusions** How do you think GIS technology might help scientists monitor earthquakes and perhaps prevent heavy earthquake damage in the future?

4 Find Out More

If you have access to an actual GIS program, use the program to determine the locations of new bridges, roads, or schools in your community. Is the information you gathered in questions 3 and 4 similar to or different from that of the actual GIS program?

Did You Know? In Portland, Maine, researchers are working with the Federal Emergency Management Agency's Project Impact to help minimize flood damage. They are using GIS and handheld GPS (global positioning systems) to create a database of the city's drainage systems. The project is part of a nationwide effort to minimize damage from natural disasters.

Growing populations and limited roads may indicate a need for more emergency routes. City planners look at traffic patterns in the community and use GIS technology to help identify solutions.

CHAPTER 9

The Cultural Geography of Latin America

GeoJournal

As you read this chapter, use your journal to list and describe the cultural influences that have shaped life in Latin America. Note how both native and imported cultures have formed a uniquely Latin American way of life.

GEOGRAPHY Online 

Chapter Overview Visit the **Glencoe World Geography** Web site at geography.glencoe.com and click on Chapter Overviews—Chapter 9 to preview information about the cultural geography of the region.

Population Patterns

Guide to Reading

Consider What You Know

Think about what you have read about the physical geography of Latin America. What geographic factors influence where people have settled in this region?

Read to Find Out

- What ethnic groups make up the population of Latin America?
- How have geography and economics influenced the distribution of Latin America's population?
- How has migration affected the Latin American culture region?
- In what ways does Latin America's cultural diversity present both benefits and challenges for its people?

Terms to Know

- indigenous
- dialect
- patois
- urbanization
- megacity
- primate city

Places to Locate

- Ecuador
- Peru
- Bolivia
- Guyana
- Buenos Aires
- Caracas
- Santiago
- Patagonia
- Rio de Janeiro
- Barbados

◀ *Woman at a Guatemalan market*

NATIONAL GEOGRAPHIC

A Geographic View

A Flavorful Mix

More than any other Caribbean island, Trinidad is a multiethnic stew. Africans and East Indians, each with about 40 percent of the population, make up the base, while smaller groups add their own flavor. Spanish and French families trace their roots to the 18th century, when their ancestors came to clear the land for plantations or to trade. . . . Portuguese, Chinese, and Syrian immigrants became merchants and shopkeepers. Today Trinidadians compare the resulting mix to **callaloo,** *a soup with many ingredients.*

—*A. R. Williams, "The Wild Mix of Trinidad and Tobago,"* National Geographic, *March 1994*

Trinidad's Laventille neighborhood

The island country of Trinidad and Tobago reflects in miniature Latin America's diverse population. In this section you will learn how Latin America's multiethnic population came about, how the land shaped patterns of human migration, and what benefits and challenges population growth and diversity bring to the region.

Human Characteristics

Latin America's 525 million people—about 9 percent of the world's population—live in 33 countries that span more than half of the Western Hemisphere. The region's population includes Native Americans, Europeans, Africans, Asians, and mixtures of these groups. The bar graph on page 212 shows you the ethnic diversity that characterizes Latin America today.

History

A Blending of Peoples

The ancestors of Native Americans were the first people to settle Latin America. As a result, Native Americans today are known as an **indigenous** (ihn•DIH•juh•nuhs) group, people descended from an area's first inhabitants. Centuries ago three Native American groups—the Maya of the Yucatán Peninsula and parts of Central America, the Aztec of Mexico, and the Inca of Peru's highlands—developed great civilizations with important cities and ceremonial centers.

Today many Native American cultural features still remain in parts of Latin America. Most of Latin America's present-day Native Americans live in Mexico, Central America, and the Andes region of **Ecuador**, **Peru**, and **Bolivia**. In areas where they are a large part of the population, Native American peoples have worked to preserve their traditional cultures while adopting features of other cultures.

Europeans first arrived in what is now Latin America in the late 1400s. Since that time millions of European immigrants have come to the region. Most of these settlers were Spanish and Portuguese. Over the years other European groups—Italians, British, French, and Germans—came as well. In modern times so many Europeans settled in Argentina and Uruguay that these countries became known as *immigrant nations.* In Latin America today, descendants of European immigrants continue to follow many of the ways of life their ancestors brought with them.

Africans first came to Latin America in the 1500s. They arrived as enslaved people, brought forcibly by Europeans to work sugar and other cash crop plantations in Brazil and the Caribbean islands. The labor of enslaved Africans helped build Latin American economies. By the late 1800s, slavery had finally ended in the region. Many Africans whose families had been in Latin America for generations remained in parts of the region. They added their rich cultural

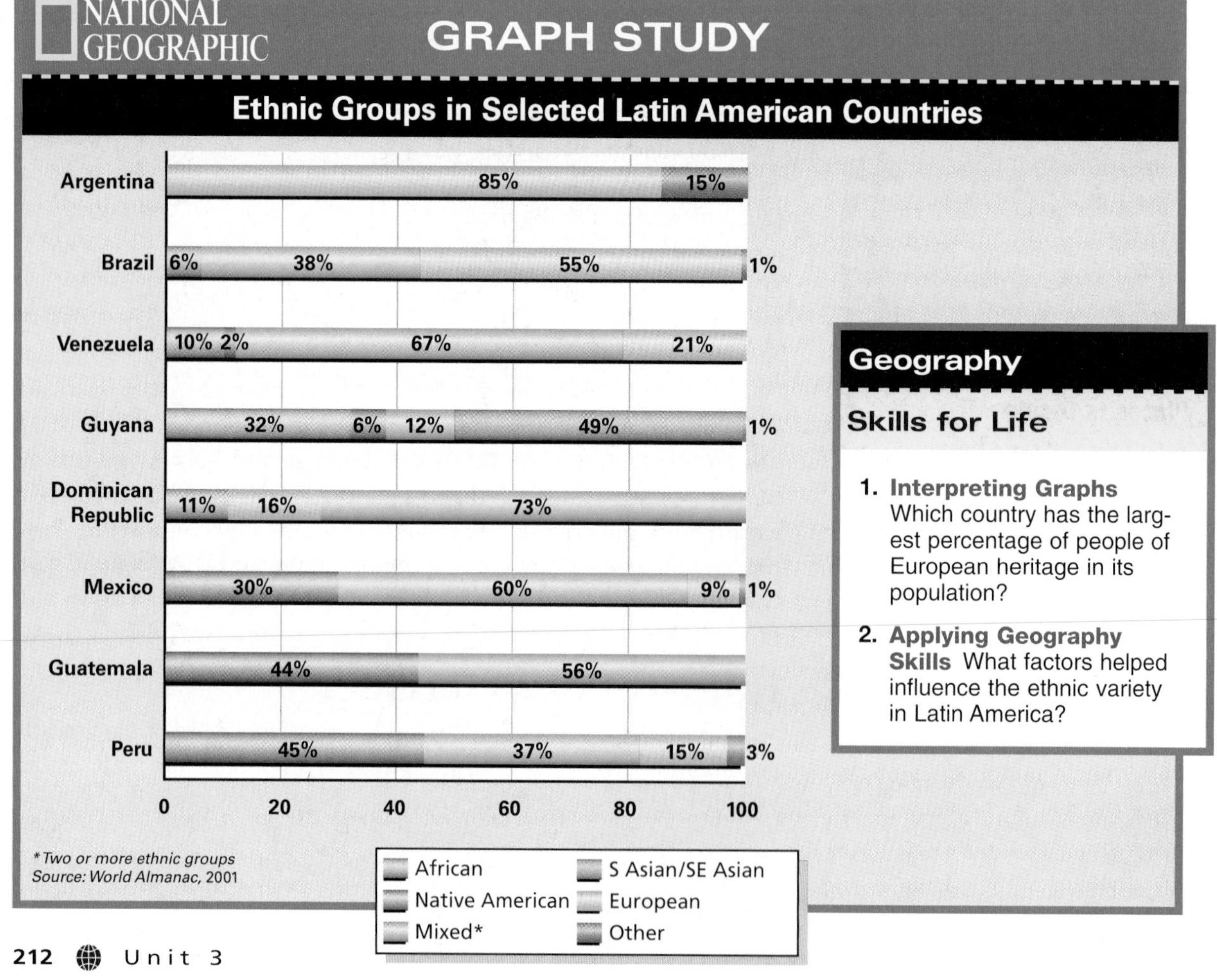

*Two or more ethnic groups
Source: World Almanac, 2001

Geography Skills for Life

1. **Interpreting Graphs** Which country has the largest percentage of people of European heritage in its population?
2. **Applying Geography Skills** What factors helped influence the ethnic variety in Latin America?

influences to the food, music, arts, and religions of Latin America.

Asians first settled in Latin America during the 1800s. They labored as temporary workers, and many remained to form ethnic communities. Today the Caribbean islands and some countries of South America have large Asian populations. In **Guyana** about one-half of the population is of South Asian or Southeast Asian descent. Many people of Chinese descent make their homes in Peru, Mexico, and Cuba, and many people of Japanese descent live in Brazil and Peru.

Over the centuries there has been a blending of these different ethnic groups throughout Latin America. For example, in countries such as Mexico, Honduras, and El Salvador, people of mixed Native American and European descent make up the largest part of the population. In other countries, such as Cuba and the Dominican Republic, people of mixed African and European descent form a large percentage of the population.

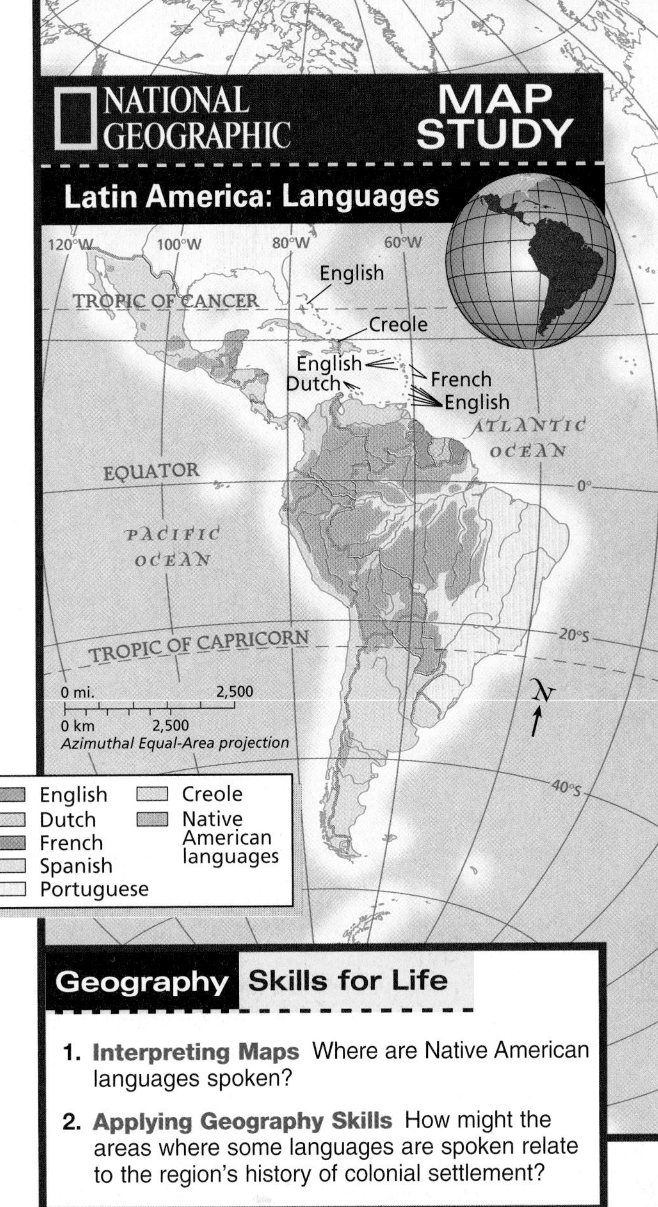

Geography Skills for Life

1. **Interpreting Maps** Where are Native American languages spoken?
2. **Applying Geography Skills** How might the areas where some languages are spoken relate to the region's history of colonial settlement?

Find NGS online map resources @ www.nationalgeographic.com/maps

Language

Language is a major factor in bringing together the diverse ethnic groups of Latin America. Most people in the region have adopted the languages of the European countries that once colonized the region. Today Spanish is the primary language of most countries of Latin America. However, other languages also are spoken. For example, the official language of Brazil is Portuguese; of Haiti and Martinique, French; and of Jamaica, Belize, and Guyana, English.

Not all Latin Americans, however, speak these European languages the same way as, or even in a way similar to, the original European colonists. Each country has its own **dialects**, forms of a language unique to a particular place or group. Meanings of words and the words themselves often differ from one place to another.

In addition, millions of Latin Americans speak Native American languages. In Central America, Mayan dialects such as K'iche' (kee•CHAY) are common. Tupi-Guarani predominates in Paraguay and Brazil. Aymara is spoken in Bolivia, and Quechua (KEH•chuh•wuh) in Ecuador, Peru, and Bolivia.

Many Latin Americans are bilingual, speaking two languages—a European language and another language, either indigenous, African, or Asian. Other Latin Americans speak one of many Latin American forms of **patois** (PA•TWAH), dialects that blend elements of indigenous, European, African, and Asian languages.

Where Latin Americans Live

In addition to having a diverse population, Latin America today has a high rate of population growth. By most estimates the region's population will soar to about 800 million by the year 2050—an increase of 55 percent. This high growth rate magnifies the challenges to human patterns of

Geography Skills for Life **Latin American Immigration** These floral arrangers, in the state of São Paulo, Brazil, are part of a community of Japanese Brazilians.

Movement Why have Asian and European immigrants settled in Latin America?

settlement already presented by Latin America's physical geography.

Latin America's varied climates and landscapes have an impact on where Latin Americans live. Temperature extremes, dense rain forests, towering mountains, and arid deserts limit human habitation in many parts of Latin America. In fact, most of Latin America's population lives on only one-third of the region's land.

About 350 million people live in South America, generally along the coasts. Another 138 million people live in Central America and Mexico, either along Central America's Pacific coast or on the inland Mexican Plateau and Central Highlands. The Caribbean island countries are home to 37 million people.

South America's Populated Rim

Rain forests, deserts, and mountains dominate South America's interior. In these areas harsh living conditions and poor soil discourage human settlement. As a result, most South Americans live on the continent's edges, an area sometimes known as the "populated rim." The coastal regions provide favorable climates, fertile land, and easy access to transportation systems.

South America's eastern coast, from the mouth of the Amazon River in Brazil to the pampas around **Buenos Aires**, Argentina, is Latin America's largest populated area. A narrower strip of densely populated land stretches along the continent's northern and western coast from **Caracas**, Venezuela, to **Santiago**, Chile.

South America's populated rim does not encircle the entire continent, however. For example, the eastern coast between the Amazon's mouth and Caracas has a hot, rainy climate and is sparsely populated. Another area of low population density lies to the far south in the Andes and **Patagonia**, where the climate and land are harsh.

With the exception of Native Americans, few South Americans live in the continent's inland areas. To draw people away from the densely populated coast, the Brazilian government in 1960 moved the capital from coastal **Rio de Janeiro** to Brasília, a planned city built in the country's interior.

Population Density

As the population density map on page 184 shows, population density varies greatly throughout Latin America. One important factor in a country's population density is its area. South American countries, with their relatively large land areas, tend to have low population densities. In Ecuador, the most densely populated country in South America, an average of only 118 people share a square mile (46 people per sq. km). Brazil has a large population, but its enormous land area, over 3.3 million square miles (8.5 million sq. km), results in a population density averaging only 52 people per square mile (20 people per sq. km).

Caribbean countries, in contrast, combine small land areas with large populations that tend to grow at rapid rates. These factors make the Caribbean countries some of the most densely populated in Latin America. The tiny island nation of **Barbados** has the highest population density in the Caribbean, with an average of 1,620 people per square mile (698 people per sq. km).

Population density also varies within countries. With 99.6 million people, Mexico is the world's most populous Spanish-speaking country, and it is the second most populous country in Latin America, after Brazil. Mexico's population and its land area of 756,000 square miles (1.9 million sq. km) give it a population density of 132 people per square mile (51 people per sq. km), making Mexico seem relatively uncrowded. This overall density rate is only an average, however. In metropolitan Mexico City, more than 18 million people live within an area of 597 square miles (1,547 sq. km). That makes the population density of Mexico City a staggering 30,150 people per square mile (11,641 people per sq. km)!

Geography Skills for Life

Market Street in La Paz A crowded street in La Paz, Bolivia, shows the effects of rapid urbanization.

Movement How does internal migration contribute to urbanization in the region?

Migration

Migration has been a major force shaping population patterns in Latin America. As a geography writer recently observed,

> "*Migration is . . . everyone's solution, everyone's conflict. . . . Unlike the flight of refugees, which is usually chaotic, economic movement is a chain that links the world. Migration . . . continues to push us toward change.*"
>
> Michael Parfit, "Human Migration," *National Geographic,* October 1998

In past centuries Europeans, Africans, and Asians migrated to Latin America in large numbers, either voluntarily or involuntarily. Today people from places such as Korea, Armenia, Lebanon, and Syria come to Latin America seeking economic and political opportunities.

Migrating North

In addition to receiving an inflow of migrants from foreign countries, Latin America also experiences an outflow of people to different parts of the world. For many Latin Americans, the desire for improved living conditions, political freedom, or an escape from political unrest leads them to move north to the United States. Latin Americans come to the United States primarily from Mexico, Central America, and the Caribbean islands. Immigrants from Latin America live in every state of the Union, with large numbers in California, Texas, New York, Illinois, and Florida. Many Latin American immigrants go through the process of legally entering the United States; others enter illegally. All of these immigrants bring elements of their culture with them. Most retain close ties with family and friends in their home countries, and many intend to return when economic conditions there improve.

Internal Migration

Internal migration, or movement within a region or country, also has shaped Latin America

NATIONAL GEOGRAPHIC

MAP STUDY

Latin America: Migration and Urbanization

People in Latin America continue to look to urban areas, both inside and outside the region, for work and a better life. Most often people migrate to areas of higher economic prosperity. For example, people migrate from Mexico to the United States. However, migration also occurs within Latin America. Migration has caused an increase in the levels of urbanization throughout the region.

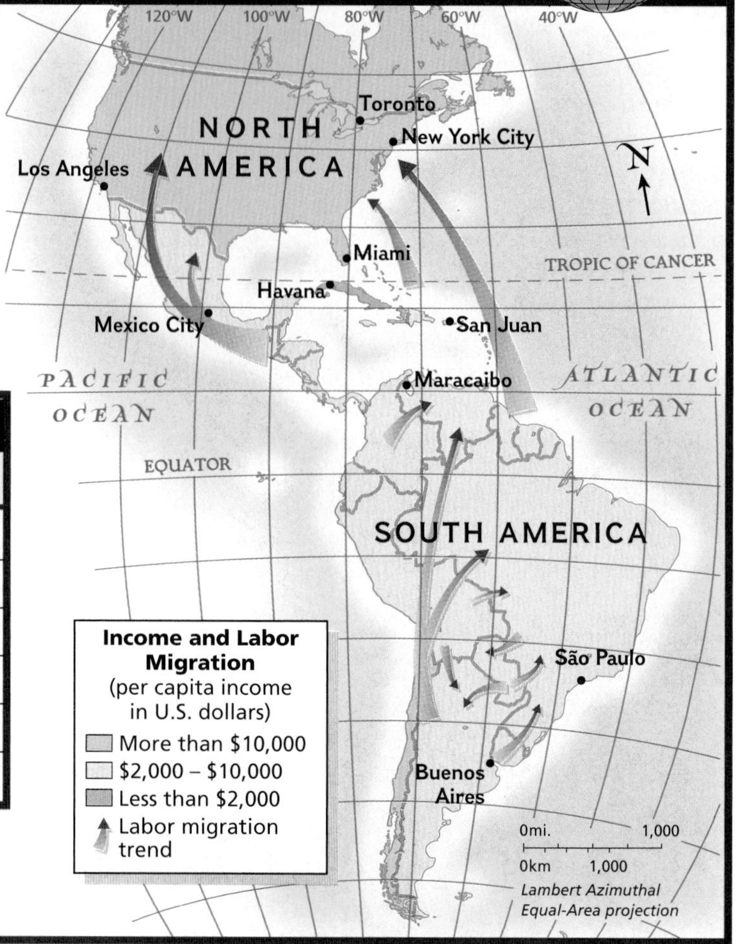

Level of Urbanization

	1970	2001	2025*
Argentina	78%	90%	93%
Bolivia	36%	63%	76%
Brazil	56%	81%	86%
Guatemala	37%	39%	42%
Mexico	59%	74%	82%
Latin America	**57%**	**74%**	**81%**

*Projected figures.
Sources: World Population Data Sheet, 2001; United Nations Population Division; Economic Commission for Latin America and the Caribbean

Geography Skills for Life

1. **Interpreting Maps** Which countries receive a significant inflow of labor migration?

2. **Applying Geography Skills** Study the levels of urbanization in Argentina and Guatemala. How do they compare with the region as a whole?

Find NGS online map resources @ www.nationalgeographic.com/maps

in recent decades. As in many parts of the world, migrants within Latin America usually move from rural to urban areas because of better job opportunities in the cities. This one-way migration also occurs because in many rural areas fertile land is in short supply or a small portion of the population controls access to the land. As the rural population rises, there is less fertile land to go around. Smaller farms can no longer support families. The result is continuing, rapid **urbanization**—the migration of people from the countryside to cities as well as the change from a rural to an urban society that accompanies this movement.

Growth of Cities

In the past most Latin Americans lived in the countryside and worked the land. Today most live in urban areas. Four cities of Latin America—Mexico City, Mexico; São Paulo and Rio de Janeiro, Brazil;

and Buenos Aires, Argentina—now rank among the world's 20 largest urban areas in population.

The Urban Setting

In some Latin American countries, as cities have grown they have absorbed surrounding cities and suburbs to create **megacities**, cities with more than 10 million people. The region's largest megacity is Mexico City, with a current population of more than 18 million. By 2015, the city is expected to have 19.2 million people. Mexico City's rapidly growing population already stresses the city's ability to provide safe drinking water, underground sewers, and utilities for new arrivals. Although the city has many areas with comfortable homes, its challenge for the future is to provide adequate housing for many who now live in cardboard shacks or makeshift houses made from sheets of metal.

Because of its size and influence, Mexico City is a **primate city**, an urban area that dominates its country's economy, culture, and political affairs. Other primate cities in Latin America include Caracas, Venezuela; Montevideo, Uruguay; Santiago, Chile; Buenos Aires, Argentina; and Havana, Cuba. Many primate cities began near waterways during the colonial era. Today these cities serve as central locations for gathering, collecting, and shipping resources overseas. They are especially powerful magnets for rural migrants seeking a higher standard of living.

Urban Challenges

Most rural Latin Americans migrate to cities to find a better life—higher incomes, more educational opportunities, better housing, and increased access to health care. In many cases people do not find what they seek. As a city's resources are strained by rapid population growth, jobs and housing become scarce. At the same time, many rural people lack the education and skills to obtain urban employment. Schools and health care centers are overwhelmed.

Despite disappointments, most rural migrants do not have the resources to return to their villages. They remain in the cities, forced by poverty to live in neighborhoods with substandard housing, poor sanitation, and little opportunity for improvement. Families sometimes split apart under the stress, leaving large numbers of homeless children to fend for themselves on the streets.

Many of Latin America's urban challenges arise from modern developments, such as the growth of cities. Others, however, stem from social and economic issues deeply rooted in the past. In the next section you will read about the historical factors that still shape current ways of life in Latin America.

SECTION 1 ASSESSMENT

Checking for Understanding

1. **Define** indigenous, dialect, patois, urbanization, megacity, primate city.
2. **Main Ideas** Create a web diagram like the one below, and fill in important information about Latin America's people.

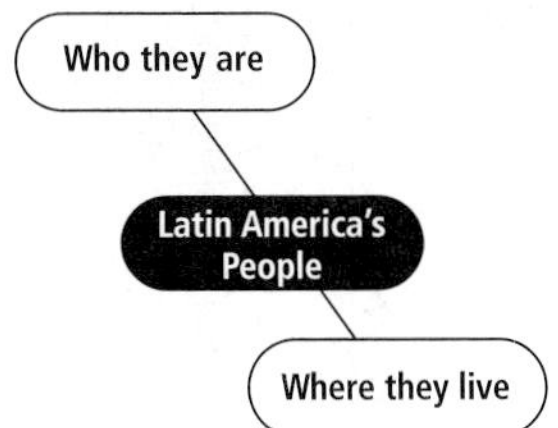

Critical Thinking

3. **Analyzing Cause and Effect** What factors account for the differences in the way Spanish is spoken in various Latin American countries?
4. **Drawing Conclusions** Develop a hypothesis describing probable population patterns in Latin America in the year 2050. Defend your hypothesis, using present trends as evidence.
5. **Making Inferences** In what ways might physical geography influence the development of megacities in Latin America?

Analyzing Maps

6. **Region** Study the language map on page 213. What is the most widely spoken language in Central America?

Applying Geography

7. **Population Density** Consider the physical geography of Latin America. Write a paragraph suggesting suitable locations for constructing new cities to relieve population pressures in Latin America's existing cities. Consider the kinds of resources required to sustain large populations.

PASSAGE THROUGH PANAMA

THE PANAMA CANAL, a vital waterway connecting the Atlantic and Pacific Oceans, has been an important trade route since the day it opened. About 14,000 ships pass through the canal's system of locks and lakes each year. Using the canal, ships can avoid the treacherous waters around Cape Horn at the southern tip of South America and can shave 7,000 miles (11,265 km) off their trip. For most of the twentieth century, the United States controlled the canal. This changed on the last day of 1999, when control passed to the nation of Panama. Today the United States and other countries anxiously watch how Panama operates this international shortcut between the world's largest oceans.

A freighter winds its way through the Panama Canal. ▶

Big Dreams and Political Shenanigans

Spanish explorer Vasco Núñez de Balboa was the first to grasp the unique geographic features of the land in Central America known today as Panama. In 1513, while exploring the isthmus, he climbed a peak and discovered a body of water as vast as the Atlantic, the ocean he had left behind. It wasn't long before thoughts turned to building a waterway to breach the slender neck of land that connects Central and South America. The limitations of manual labor, however, kept the idea in

◀ **Canal enthusiast Theodore Roosevelt operates a steam shovel at a canal work site.**

the realm of dreams for more than 300 years.

By the late 1800s, technology had caught up with the imagination. The first to try to build a waterway was Frenchman Ferdinand de Lesseps, who masterminded the Suez Canal. Cutting through the mountainous terrain proved extremely difficult, and de Lesseps failed. But one of his engineers, Philippe Jean Bunau-Varilla, refused to quit. In 1901 he pitched the idea to U.S. President Theodore Roosevelt, who was willing to pay the engineer's price if Colombia, of which Panama was a part, relinquished control of the proposed canal route. When Colombia refused, Bunau-Varilla supported Panamanian revolutionaries and persuaded the United States to intervene. The presence of American gunboats was enough to make Colombia give in. Panama became an independent country. Bunau-Varilla, Panama's new minister to the United States, negotiated a treaty giving the United States control of the land along the proposed route. In 1904 construction began.

Engineering Wonder of the World

Nearly 75,000 laborers from around the world built what is still regarded as one of the engineering wonders of the world. Instead of trying to make a cut through the rugged hills to carry ships across at sea level, the American solution was to build a system of locks to lift ships up to a newly created lake, and in the same way, lower them down the other side. The volcanic soil, heat and rain, dense vegetation, and disease-spreading insects conspired to make progress painfully slow. Thousands of workers lost their lives, and costs grew to more than $380 million. The canal was completed in 1914.

Looking Ahead

In acquiring the Panama Canal, the Panamanians gained a sizable investment. Do you think Panama will find the resources to maintain and operate the canal? How might the United States be affected by the change in command?

NATIONAL GEOGRAPHIC SOCIETY

1500s First road built across isthmus

1855 American business interests build railroad across isthmus

1881 French company begins building a sea level canal; project abandoned in 1887

1903 Backed by U.S. President Theodore Roosevelt (cartoon above), Panama becomes independent country and signs treaty to create Panama Canal Zone

1904 Work begins on the Panama Canal (background photo)

1914 First ship passes through canal

1977 U.S. and Panama sign Panama Canal Treaty, gradually transferring ownership to Panama

1999 Canal ownership transfers to Panama

History and Government

Guide to Reading

Consider What You Know

Latin American politics and social conflicts often make news in the United States. How are Latin American governments similar to or different from the United States government?

Read to Find Out

- What contributions have Latin America's Native American empires made to the region's cultural development?
- How has colonial rule influenced Latin America's political and social structures?
- How did most Latin American countries make the transition from colonialism to democracy?
- What political and social factors continue to challenge the Latin American culture region?

Terms to Know

- glyph
- *chinampas*
- quipu
- conquistador
- viceroy
- caudillo

Places to Locate

- Mexico
- Tikal
- Tenochtitlán
- Cuzco
- Haiti
- Cuba

NATIONAL GEOGRAPHIC

A Geographic View

Native Rights Protest

Drawn machetes slapped against trouser legs. Dark eyes stared in anger. About 40 Tojolabal Indian men and women surrounded two men. . . . They talked angrily, and the phrase that came through was, "This is our land." The sharp edges of the machetes gleamed. . . . These Indians might be Zapatistas, rebel Indian farmers named for Mexico's revolutionary war hero Emiliano Zapata. . . .

—Michael Parfit, "Chiapas: Rough Road to Reality," National Geographic, *August 1996*

Zapatista protesters, Chiapas, Mexico

In 1994 the Zapatistas attacked government troops and captured several towns in southern Mexico. One of their aims was to recover lands that Spanish conquerors had seized from their ancestors four centuries earlier. They finally succeeded in pressuring the Mexican government to introduce reforms giving Native Americans more power in Mexico's political system. Throughout Latin America today people struggle with unresolved issues rooted in the past. In this section you will learn about Latin America's long and often violent history, which includes ancient Native American civilizations, European colonial rule, and struggles for independence.

Native American Empires

Years before Christopher Columbus arrived in the Americas in 1492, three Native American empires—the Maya, the Aztec, and the Inca—flourished in the area that is present-day Latin America.

The civilization of each empire left enduring marks on Latin American cultures.

The Maya

The Maya dominated southern **Mexico** and northern Central America from about A.D. 250 to 900. They established many cities, the greatest of which was **Tikal**, located in what is today Guatemala. Terraces, courts, and pyramid-shaped temples stood in these cities. Priests and nobles ruled the cities and surrounding areas. The Maya based their economy on agriculture and trade.

Skilled in mathematics, the Maya developed accurate calendars and used astronomical observations to predict solar eclipses. They used **glyphs**, picture writings carved in stone, on temples to honor their deities and record their history.

For reasons that are still a mystery, the Maya eventually abandoned their cities, which over time became lost beneath the vegetation of the rain forest. Archaeologists continue to search for more information about the ancient Maya. Researchers have uncovered the ruins of over 40 Mayan cities, but most of the glyphs remain untranslated. Today many temple ruins are popular tourist attractions. Descendants of the Maya still live in villages in southern Mexico and northern Central America, where they practice subsistence farming.

The Aztec

The Aztec civilization arose in central Mexico, in the A.D. 1300s. The Aztec founded their capital, **Tenochtitlán** (tay•NAWCH•teet•LAHN), today the site of Mexico City, on an island in a large lake. To feed the growing population, Aztec farmers grew beans and maize on *chinampas*—floating "islands" made from large rafts covered with mud from the lake bottom.

The Aztec developed a highly structured class system headed by an emperor and military officials. High-ranking priests performed rituals to win the deities' favor and to guarantee good harvests. At the bottom of Aztec society were the majority—farmers, laborers, and soldiers.

Culture

Gifts to the World's Tables

Several foods grown by the Aztec have become worldwide favorites. Corn, a staple food of Latin America, came from the maize cultivated by the Aztec. The tomato, later used in Mediterranean cuisine, was unknown in Europe until the European conquest of Latin America. From bitter cacao beans, the Aztec made a concoction called *xocoatl* (chocolate), or "food of the gods."

The Inca

During the time of the Aztec, the Inca established a civilization in the Andes mountain ranges of South America. At its height the Incan Empire stretched from what is today Ecuador to central

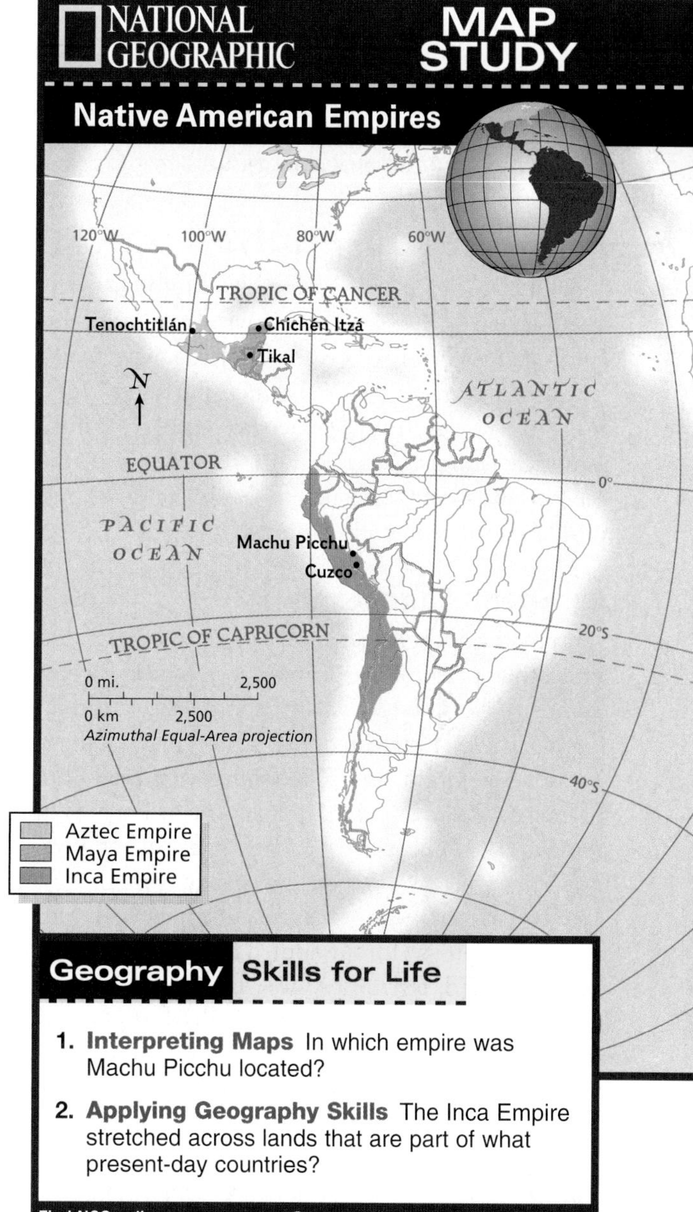

Geography Skills for Life

1. **Interpreting Maps** In which empire was Machu Picchu located?
2. **Applying Geography Skills** The Inca Empire stretched across lands that are part of what present-day countries?

Find NGS online map resources @ www.nationalgeographic.com/maps

Chile. The Inca built their capital, **Cuzco**, in what is now Peru and ruled their lands through a central government headed by an emperor.

Using precisely cut stones, Incan builders constructed massive temples and fortresses. They laid out a network of roads that crossed high mountain passes and penetrated dense forests. To keep soil from washing away, Incan farmers cut terraces into the steep slopes of the Andes and built irrigation systems to bring water to Pacific coast deserts. The Inca also domesticated the alpaca and the llama, which they used for wool. With no written language, the Inca used oral storytelling to pass on knowledge to each generation. To keep track of financial records, Incan traders used a **quipu** (KEE•poo), a series of knotted cords of various colors and lengths. Each knot represented a different item or number.

Empires to Nations

Beginning with Christopher Columbus's voyages from 1492 to 1504, Europeans explored and colonized vast areas of the Americas. The major European powers of Spain and Portugal ruled huge territories from Mexico to southern South America. Later Great Britain, France, and the Netherlands colonized in the Caribbean area and parts of northern South America.

European Conquests

From the West Indies, the Spaniards expanded into other parts of the Americas. Desiring riches, Spanish **conquistador**, or conqueror, Hernán Cortés in 1521 defeated the Aztec and claimed Mexico for Spain. In 1535 another conquistador, Francisco Pizarro, destroyed the Incan Empire in Peru and began Spain's South American empire. The Portuguese settled on the coast of Brazil.

As a result of these conquests, European colonies gradually arose throughout Latin America. In Spanish-ruled territories, for example, the conquerors set up highly structured political systems under royally appointed officials known as **viceroys**. The Roman Catholic Church became the major unifying institution in both Spanish and Portuguese colonies. Missionaries from Europe converted the Native Americans to Christianity and set up schools and hospitals.

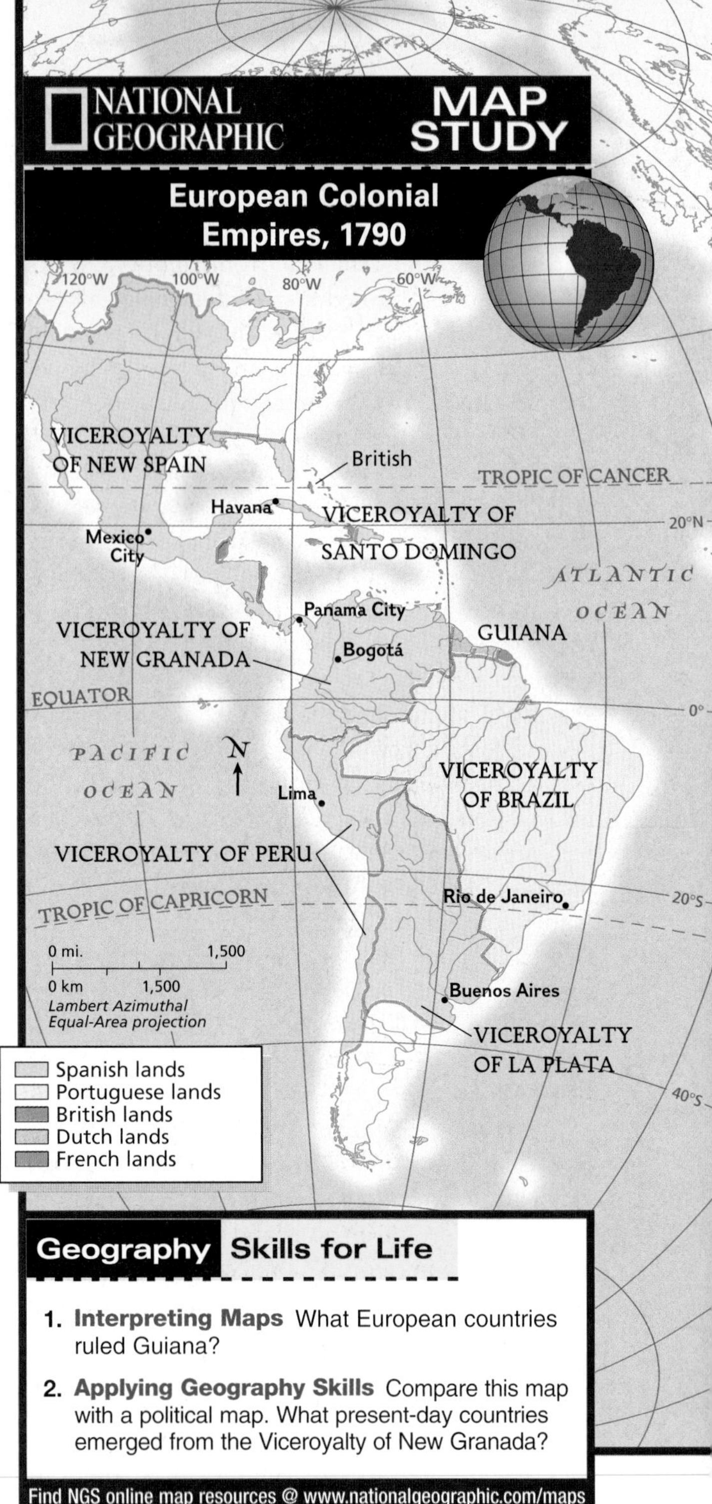

European Colonial Empires, 1790

Geography Skills for Life

1. **Interpreting Maps** What European countries ruled Guiana?
2. **Applying Geography Skills** Compare this map with a political map. What present-day countries emerged from the Viceroyalty of New Granada?

Find NGS online map resources @ www.nationalgeographic.com/maps

Colonial Economies

The European colonies in the Americas became sources of wealth for the home countries. Some Spanish settlers prospered from the mining of gold and silver. The Portuguese discovered precious metals in Brazil and made use of brazilwood,

WORLD CULTURE • WORLD CULTURE • WORLD CULTURE • WORLD CULTURE

architecture of LATIN AMERICA

Colonial Architecture Built in the 1500s, the chapel Nuestra Señora de los Remedios (below) in Cholula, Mexico, is typical of Spanish colonial architecture. Colonial buildings often reflect a mixture of European styles with Native American and African influences.

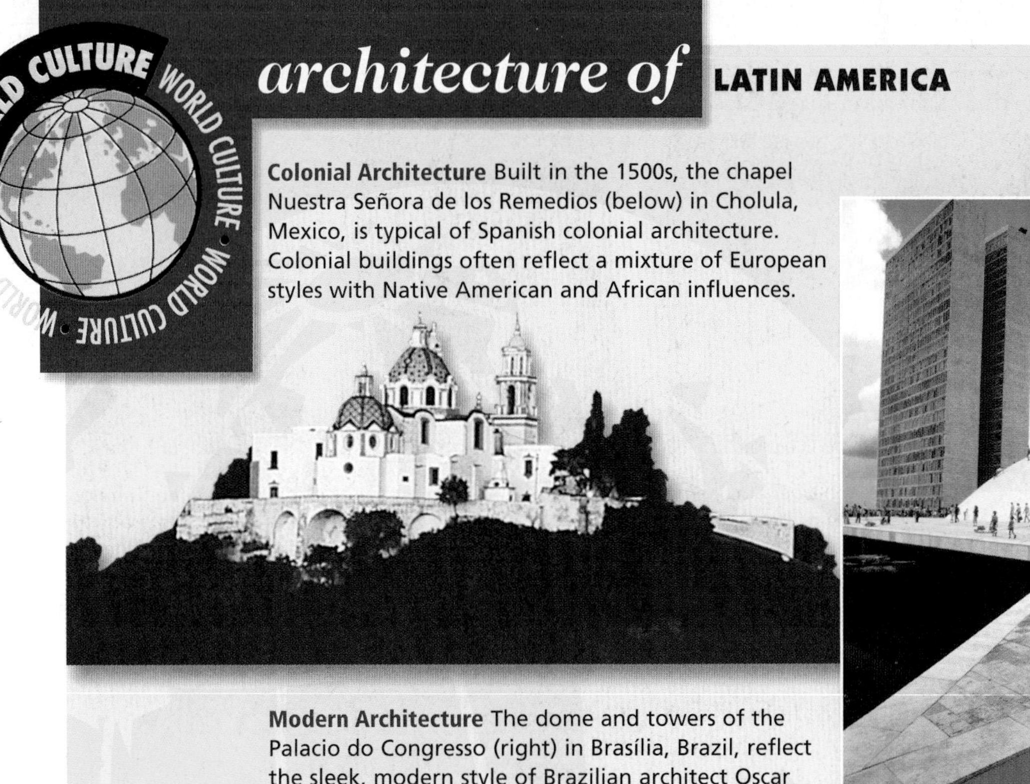

Modern Architecture The dome and towers of the Palacio do Congresso (right) in Brasília, Brazil, reflect the sleek, modern style of Brazilian architect Oscar Niemeyer. The major government buildings in Brasília were planned and built in the 1950s and 1960s.

a tree used to make red dye. Spanish and Portuguese colonists also built cities and towns that served as trade centers and seats of government. In the tropics their plantations grew coffee, bananas, and sugarcane for export to Europe. In cool highlands areas, they established farms and cattle ranches.

The Spaniards and Portuguese used Native Americans to work on the plantations and ranches. As epidemic diseases and hardships drastically reduced the numbers of Native Americans, the European colonists imported enslaved Africans to meet the labor shortage. Despite European dominance many aspects of the Native American and African ways of life survived, creating a blend of the cultures of three continents in Latin America.

Gaining Independence

In the late 1700s, resentment against European rule spread throughout Latin America. Wealthy colonists of European origin wanted self-rule. Those Europeans lower on the social scale demanded more rights. Native Americans and Africans simply yearned for freedom from servitude. Encouraged by the revolutions in North America and France, many Latin Americans joined together to end European colonial rule.

The first Latin American country to gain its independence was **Haiti**, located on the Caribbean island of Hispaniola. In the 1790s François Toussaint-Louverture (frahn•SWAH TOO•SAN•LOO•vuhr•TYUR), a soldier born of enslaved parents, led a revolt by enslaved Africans. By 1804 Haiti had won its independence from France. The first Spanish-ruled country in Latin America to win independence was Mexico. The independence movement there began in 1810 and was led by a parish priest, Father Miguel Hidalgo. After a long struggle, Mexico became independent in 1821.

Other territories in Latin America also sought independence. By the mid-1800s most of them had achieved their goal under such leaders as Simón Bolívar of Venezuela and José de San Martín of Argentina. However, only one country—Brazil—became independent without a violent upheaval.

Except for Haiti, Caribbean island countries were the last territories in Latin America to achieve

Latin America's Independence Leaders

François Toussaint-Louverture led enslaved Haitians in a violent revolt against French rule. He died in a French prison in 1803.

Called "the Liberator," Simón Bolívar of Venezuela won freedom for the present-day countries of Venezuela, Colombia, Ecuador, Peru, and Bolivia.

José de San Martín of Argentina led his Latin American forces across the Andes to win independence for Chile and Peru.

Father Miguel Hidalgo called on Mexicans to fight for "Independence and Liberty" from Spain. He was executed in 1811.

independence. **Cuba**, for example, did not win its freedom from Spain until 1898. British-ruled islands, such as Jamaica and Barbados, did not gain independence until well into the 1900s. Even today some islands remain under foreign control; for example, Martinique is a possession of France, the Cayman Islands of Great Britain, and Curaçao of the Netherlands. In addition, Puerto Rico and some of the Virgin Islands have political links to the United States.

Era of Dictatorships

Latin America's wars for independence ushered in a period of political and economic instability. During the 1800s some leaders in the region wanted to build democratic institutions and prosperous economies. However, they had to contend with the legacy of indigenous and European class structures, which stressed rank and privilege. As a result, political and economic power often remained in the hands of a small group of wealthy landowners, army officers, and clergy. Written constitutions were ignored, public dissatisfaction led to revolts, and governments relied on the military to keep order.

In this chaotic situation, a new kind of leader emerged—the **caudillo** (kow•DEE•yoh), or dictator. With the backing of military forces and wealthy landowners, caudillos became absolute rulers with sole authority to make decisions.

Movements for Change

During the 1900s Latin America experienced dramatic political, social, and economic changes. As European rule declined, the influence of the United States increased in the region. For example, after Panama became an independent country in 1903, the United States and Panama signed a treaty creating the Panama Canal Zone. The formation of industries, the building of railroads, and the expansion of trade all brought new wealth to the upper classes. These developments also created new middle and working classes in the cities. However, for the vast majority of Latin Americans, especially rural dwellers, progress was limited.

As the gap between the rich and the poor widened, unrest spread among farmers and workers. Conservative dictators and military governments resisted demands for reform and crushed uprisings. In Cuba, however, a revolution in 1959 set up a communist state under Fidel Castro.

During the 1990s communism remained entrenched in Cuba, but military dictatorships gave way to democratically elected governments in a

Student Web Activity Visit the **Glencoe World Geography** Web site at geography.glencoe.com and click on Student Web Activities—Chapter 9 for an activity about the Panama Canal.

number of countries. Today Latin American countries are struggling to end corrupt politics and bring economic benefits to all their citizens. In Mexico, for example, nearly 70 years of one-party rule ended in the year 2000 when the candidate of the ruling party PRI (*Partido Revolucionario Institucional*) lost the presidency to Vicente Fox of the opposition party PAN (*Partido Acción Nacional*) in a genuinely democratic election.

As Latin America entered the 2000s, Native Americans, farmers, and workers demanded more political power and greater economic benefits. A spokeswoman for Guatemala's modern-day Maya people, Rigoberta Menchú, discusses the need for greater inclusion in political processes:

> *"National unity must be defined in the context of the right of the whole society to diversity, protected by and reflected in a democratic state. Eventually governments will have to tackle the issue of the self-determination of diverse peoples within national boundaries. . . .We must accept that humanity is a beautiful multicolored garden."*
>
> Rigoberta Menchú (Ann Wright, trans.), *Crossing Borders*, 1998

NATIONAL GEOGRAPHIC **World Explorer**

Geography Skills for Life

New President In his campaign, Vicente Fox (shown here with Rigoberta Menchú) promised better public education and more attention to the poor.

Region What political issues are important in Latin America today?

SECTION 2 ASSESSMENT

Checking for Understanding

1. **Define** glyph, *chinampas*, quipu, conquistador, viceroy, caudillo.
2. **Main Ideas** Create a web diagram like the one below for each Native American culture, and show its major achievements. Then choose one achievement and explain why it was important.

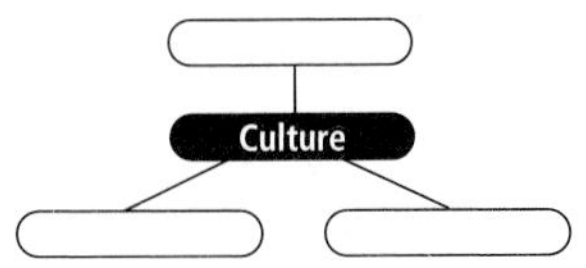

Critical Thinking

3. **Making Comparisons** How was the social structure of the Aztec Empire similar to the social structures of Latin America under European colonialism?
4. **Drawing Conclusions** Was the plantation system beneficial or harmful? Explain.
5. **Analyzing Information** According to Rigoberta Menchú, how can diversity bring unity? Do you agree or disagree with her assessment, and what steps would you take to bring about unity?

Analyzing Maps

6. **Region** Compare the maps of Latin America and the colonial empires on pages 195 and 222. Which Spanish viceroyalty was named for a geographic feature of Latin America?

Applying Geography

7. **Development and History** On a time line trace the development of indigenous and European empires in Latin America. Include at least one achievement that occurred during each empire.

SECTION 3

Cultures and Lifestyles

Guide to Reading

Consider What You Know

Latin American foods and music are popular in the United States and around the world. How do people today discover and learn about the cultural traditions of Latin America?

Read to Find Out

- What role does religion play in Latin American culture?
- How have Latin Americans used the arts to express their history, their social struggles, and their cultural diversity?
- How is Latin America's cultural diversity reflected in family life, leisure activities, and public celebrations?

Terms to Know

- syncretism
- mural
- mosaic
- extended family
- malnutrition
- *fútbol*
- jai alai

Places to Locate

- West Indies
- Dominican Republic
- Guatemala
- Brasília
- Chile

NATIONAL GEOGRAPHIC

A Geographic View

Shadows of the Ancients

Doffing his mask, a member of Los Panchitos dance troupe takes a breather from the vigorous street dancing. . . . With its origins deep in the past, the dance pokes fun at figures of the present. . . . The finger of ridicule points to a landowner who abuses peasant workers, a judge who decides a case in favor of the rich. . . .

—*Michael E. Long, "Enduring Echoes of Peru's Past,"* National Geographic, *June 1990*

Masked dancer at a Peruvian festival

The past and present intermingle in the lives of Latin Americans. Here, along Peru's northern coast, a masked dance blends Native American and European influences, music and visual arts, religion and social criticism. This interweaving of diverse elements is a hallmark of Latin American culture. In this section you will learn how Latin Americans express their culture through religion, the arts, and everyday life.

Religion

Religion has long played an important role in Latin American society. During the colonial era, most Latin Americans became Christians, and Christianity still has the most followers. In addition, other faiths are found in the region. For example, scores of traditional Native American and African religions thrive, often mixed with Christianity and other faiths. Islam, Hinduism, and Buddhism, brought by Asian immigrants, are practiced in the **West Indies** and coastal areas of South America. Judaism has followers in the largest Latin American cities.

Roman Catholicism

Most Christians in Latin America are Roman Catholics, and Roman Catholic traditions influence daily life in the region. During colonial times Roman Catholicism was the official religion of the Spanish colonies and Brazil. Roman Catholic clergy had accompanied European conquerors and colonists to the Americas. They established Roman Catholicism throughout Latin America, converting many Native Americans to their faith. When European settlers arrived, the priests saw to their spiritual needs as well.

Before long, church leaders were playing an important role in political affairs in the region, and the Roman Catholic Church had become wealthy. When the fight for independence came, church officials backed the wealthy and powerful classes. During the late 1900s, however, Roman Catholics in Latin America began to support the concerns of the poor and the oppressed. In recent years many Roman Catholic clergy and laypeople have opposed dictatorships and worked to improve the lives of disadvantaged groups. For example, the Church has been active in movements for land reform and for improvements in education and health care.

Protestantism

Various forms of Protestant Christianity came to Latin America with British and Dutch settlers in the 1800s. In time American Protestant missionaries came and built hospitals, schools, and colleges. Protestants in the region were few in number until the late 1900s, when Protestantism grew rapidly. According to religious observers, many Latin Americans were drawn to Protestantism because it gave laypeople a major role in religious life and emphasized personal religious experience.

A Mixing of Religions

Throughout Latin America a mixing of religions has occurred since the colonial era. Many Latin

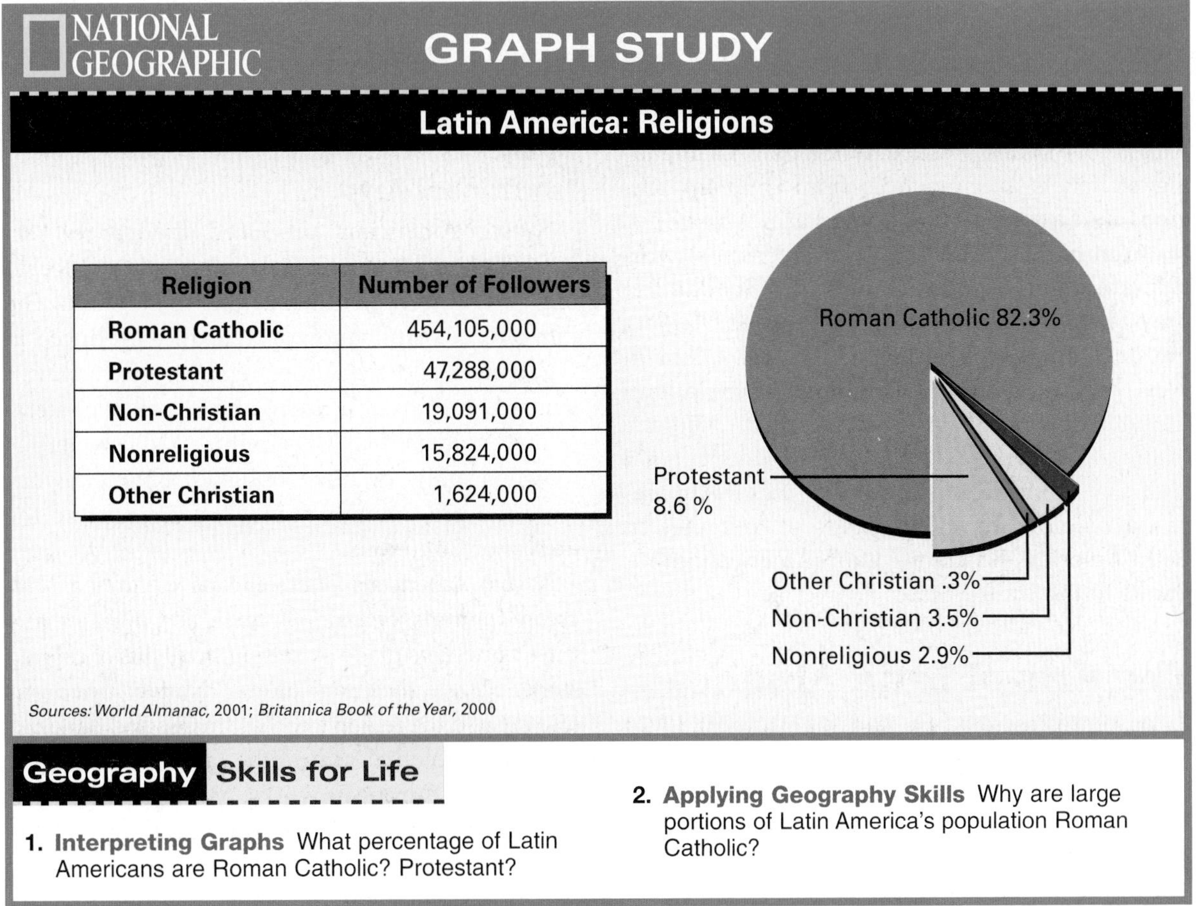

Religion	Number of Followers
Roman Catholic	454,105,000
Protestant	47,288,000
Non-Christian	19,091,000
Nonreligious	15,824,000
Other Christian	1,624,000

Sources: World Almanac, 2001; Britannica Book of the Year, 2000

Geography Skills for Life

1. **Interpreting Graphs** What percentage of Latin Americans are Roman Catholic? Protestant?
2. **Applying Geography Skills** Why are large portions of Latin America's population Roman Catholic?

music of LATIN AMERICA

A wide mix of music traditions in Latin America comes from the native inhabitants (wind and percussion instruments), the Europeans (strings, vocal harmonies), and the Africans (drums, varied rhythms).

Instrument Spotlight

Panpipes are one of the most common musical instruments from the Andean region of South America, dating from before the arrival of the Europeans. Often called *zampona* or *siku*, panpipes are made of bamboo in varying sizes and pitches. Individual bamboo stalks are cut precisely and lashed together in rows with strips of bamboo and string. The notes of a given scale often alternate from one set of pipes to another. For a complete melody to be played, the two rows of pipes are stacked one on top of the other.

Go To **World Music: A Cultural Legacy** Hear music of this region on Disc 1, Tracks 7–12.

Americans today practice syncretism—a blending of beliefs and practices from different religions into a single faith. Some Latin Americans, for example, especially Native Americans, worship at Roman Catholic churches on Sunday but pray to nature deities during the week. Among the descendants of enslaved Africans, belief in West African deities is combined with Roman Catholic devotion to the saints. Called *condomblé* in Brazil, Santería in Cuba, and voodoo in Haiti and the **Dominican Republic**, these African-based religions have thousands of followers in Latin America and among Latin American immigrants to the United States.

The Arts of Latin America

For centuries, the arts and literature of Latin America were shaped by European styles. Today's Latin American artists and writers have developed styles that often reflect their diverse ethnic heritages, blending European styles with those of Native American cultures.

History

Traditional Arts

Native Americans produced the earliest art forms in Latin America. They left a legacy of weaving, woodcarving, pottery, and metalwork. The intricate, colorful handwoven textiles produced in **Guatemala** and the Andes regions reflect Mayan symbols and Incan weaving. The work of contemporary goldsmiths, silversmiths, and jewelers is matched only by the sophisticated metalwork from the pre-Columbian era, the time before the arrival of Columbus.

Native Americans built temples decorated with colored murals, or wall paintings, and mosaics, pictures or designs made by setting small bits of colored stone, tile, or shell into mortar. Native Americans also created the region's earliest music and dance.

During colonial times the arts were largely inspired by European works. Most paintings had Christian themes. Murals, however, mixed the brightly colored abstract designs of the Native Americans with the more realistic European styles.

Churches built in Spanish and Portuguese designs often were enlivened by the ethnic details added by Native American and African artists. Meanwhile, Africans brought to the region the rhythms, songs, and dances that evolved into today's Latin American musical styles and dances, such as calypso, reggae, and samba.

Modern Arts

During the 1900s Latin American artists mixed European, Native American, and African artistic traditions. Many of them also focused on social and political subjects. Diego Rivera, a well-known Mexican artist, created huge murals that illustrated key events in Mexico's history, especially the struggles of impoverished farmers to win social justice. Other noted Latin American painters included Mexico's Frida Kahlo, known for her self-portraits, and Colombia's Fernando Botero, who satirized the lifestyles of Latin America's upper classes.

Latin American music combines Native American, European, and African influences to create unique styles. These musical styles include Brazilian samba, Cuban salsa, and Mexican mariachi.

During the past 50 years, Latin American architects, dancers, and writers also have won international recognition. The Brazilian architect Oscar Niemeyer is known for the buildings he designed in the Brazilian capital of **Brasília**. Dance companies such as the Ballet Folklórico of Mexico fascinate audiences worldwide with their performances of traditional Native American and Spanish dances. Latin America also has produced outstanding novelists, such as Colombia's Gabriel García Márquez and Chile's Isabel Allende, who skillfully blend everyday reality with the mythical and fantastic in their writings. A continuing theme of Latin American literature is cultural identity. The Argentine poet Jorge Luis Borges wrote of this theme in his life:

> "*From a lineage of Protestant ministers and South American soldiers who fought, with their incalculable dust, against the Spaniards and the desert lances, I am and am not . . .*"
>
> "Yesterdays," *Jorge Luis Borges: Selected Poems,* Alexander Coleman, ed., Stephen Kessler, trans., 1999

▲ *Diego Rivera's mural* Teatro Insurgentes *depicts leaders of the Mexican Revolution.*

Everyday Life

Latin Americans place great emphasis on social status and family life. They also cherish values such as personal honor and individual freedom.

Families

Most Latin Americans have a strong sense of loyalty to family. Each person is part of an **extended family** that includes grandparents, aunts, uncles, and cousins as well as parents and children. Latin American parents and children often share their home with grandparents and sometimes other members of the extended family. *Compadres*, or godparents, play an important role in family life. Godparents are people chosen by the mother and father to sponsor their new baby. Godparents are concerned with the child's religious and moral upbringing and help take care of the child if something happens to the parents.

Latin American society still displays traces of machismo, a Spanish and Portuguese tradition of male supremacy, although women have made rapid advances in public life in recent decades. Latin American women are in charge of home life, making important financial and family decisions. Each year more women attend universities and hold jobs in a variety of professions. Many have been elected as national legislators, as mayors of large cities, and as country leaders. For example, in 1999, Panama elected Mireya Elisa Moscoso as president.

Education and Health Care

The quality of education varies throughout Latin America. Children generally are required to complete elementary school, but they often do not because of long distances to school and lack of money for clothing and supplies. Also, many children drop out to help with family farming or to find jobs.

Despite such realities many Latin American countries have made gains in education. Adult literacy rates have risen steadily, governments now devote more funds to schools, and some countries have seen impressive gains in school attendance. University enrollment also is rising, as some public universities provide higher education at little or no cost to students. Although Latin America has lagged behind some other regions in computer literacy, Internet usage is beginning to transform education in countries such as **Chile** and Mexico.

In Latin America, as in other regions, health care is linked to standards of living. As people become employed and better educated, health concerns linked to poverty, lack of sanitation, and **malnutrition**, a serious condition caused by a lack of proper food, become much less severe. Today, despite a wide gap between the rich and the poor, Latin America overall is improving the health of its people. Infant mortality rates for the region have fallen dramatically in recent years, and most people now have access to clean, treated water for drinking.

Still, health conditions vary from country to country. In lands with prosperous economies and high standards of living, such as Chile, people have access to better health care systems and are able to live healthier, longer lives. By contrast, countries with less developed economies, such as Haiti, have little money to spend on health care. Consequently, disease is more prevalent and life expectancy is low. In most Latin American countries, the quality of health care falls between these two extremes.

Sports and Leisure

Throughout Latin America fans are as passionate about ***fútbol***, or soccer, as fans in the United States are about American football. In many Latin American countries it is the national sport. Thousands of dedicated spectators crowd into huge stadiums to watch their teams play. Baseball, basketball, and volleyball also have large followings, especially in the West Indies. Many Latin

Geography Skills for Life

Extended Families Family celebrations such as birthdays and weddings are important traditions within extended families.

Region How do Latin Americans view families?

American baseball stars, including home-run hitter Sammy Sosa from the Dominican Republic, have gone on to play in the North American major leagues. A favorite sport among many Mexicans and Cubans is jai alai (HY•LY), a fast-paced game much like handball, played with a ball and a long, curved basket strapped to each player's wrist.

Watching television, listening to the radio, and attending movies, concerts, and plays are leisure activities as popular in Latin America as they are around the world. The most popular Latin American leisure activity of all, however, may be celebrating. From impromptu gatherings of friends to special family dinners to religious feast days, and patriotic events, almost any social occasion is a party—a *fiesta*, or festival.

Perhaps the best-known festival is Carnival, celebrated in the week before the Roman Catholic observance of Lent, a 40-day period of fasting and prayer before Easter. In Rio de Janeiro, home of one of the largest Carnival celebrations, teams from different parts of the city compete to win the prize for the best hand-decorated float. People make their own brightly colored masks and elaborate costumes and then parade to samba music through the streets. Today Carnival draws people from around the world to Latin America.

NATIONAL GEOGRAPHIC **World Explorer**

Geography Skills for Life

Fútbol The Brazilian star Ronaldo breaks through the Italian defense during a tournament game.

Region What other sports have large followings in Latin America?

SECTION 3 ASSESSMENT

Checking for Understanding

1. **Define** syncretism, mural, mosaic, extended family, malnutrition, *fútbol*, jai alai.
2. **Main Ideas** Create a chart like the one below, and fill in the influences that contributed to each aspect of Latin American culture.

Cultures and Lifestyles		
Influences		Aspects of Culture
	→	
	→	
	→	

Critical Thinking

3. **Making Inferences** Why do you think Roman Catholicism has remained the predominant religion in Latin America?
4. **Drawing Conclusions** Why do you think Latin American arts imitated the arts of Europe?
5. **Making Generalizations** On an outline map, label the countries of South America. What factors do you think determine their political boundaries?
6. **Making Inferences** Why might parties—fiestas and festivals—be so popular in Latin America?

Analyzing Charts

7. **Place** Study the graph showing religions on page 227. Which religion in Latin America is second to Roman Catholicism in its number of followers?

Applying Geography

8. **Cultural Influences** Make a sketch map to show where the region's arts originated. Include representative examples of various art forms and examples from Africa, Europe, and Latin America. Provide notes about each example's ethnic origins.

Reading a Population Density Map

Population density measures how many people live within a certain unit area, such as a square mile or square kilometer. Population density may vary from place to place within a country or region. A population density map shows you these variations.

Learning the Skill

To determine a country's overall population density, divide the number of people within a country's boundaries by its land area in square miles or square kilometers. The map at right shows how population density differs within Brazil.

- **Study the map keys to determine what the colors and symbols represent.** Notice that the map uses colors to show population densities and symbols to show the populations of cities.
- **Look for patterns that might explain population density patterns.** Ask yourself what geographical features are shared by areas with high or low population densities.
- **Compare the map with other regional information,** such as natural resources and physical geography, to draw conclusions about the possible causes and effects of population density patterns.

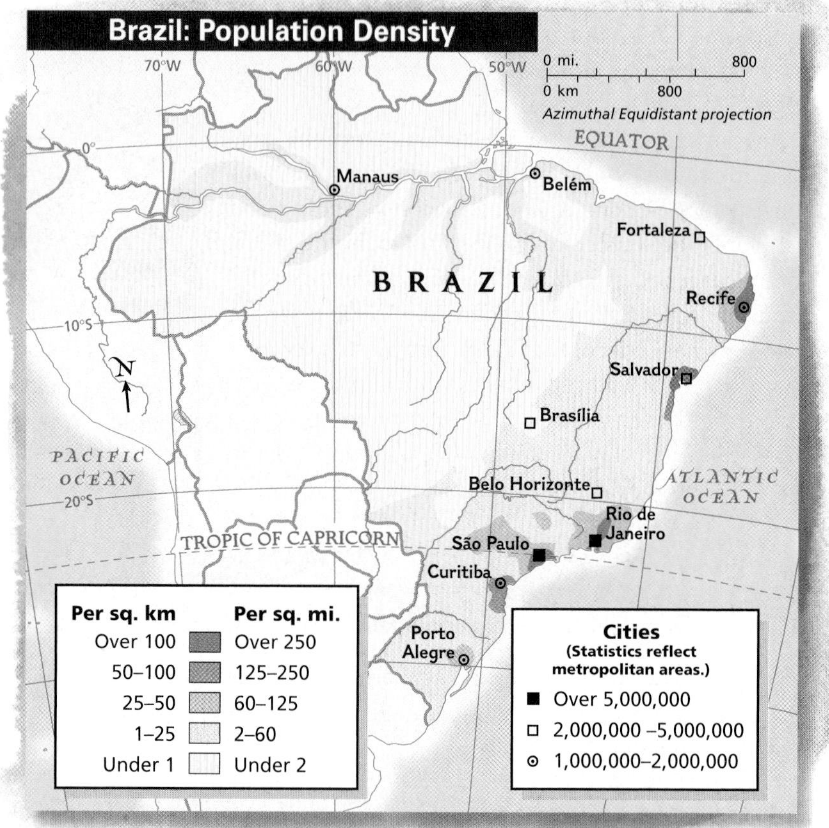

Practicing the Skill

Use the population density map to answer the questions.

1. What does the dark orange color represent?
2. What symbol represents cities of more than 5,000,000 people?
3. Which areas of Brazil have low population densities?
4. Which areas have the highest population densities?
5. Which two cities have the most people?
6. Which cities have fewer than 2 million people?
7. Why do you think the east coast of Brazil is more densely populated?

Compare the physical map of Latin America with the population density map. Write a paragraph explaining how physical geography affects population density.

The **Glencoe Skillbuilder Interactive Workbook, Level 2** provides instruction and practice in key social studies skills.

9

SECTION 1 Population Patterns (pp. 211–217)

Terms to Know

- indigenous
- dialect
- patois
- urbanization
- megacity
- primate city

Key Points

- Latin America's people descended from indigenous peoples, Europeans, Africans, and Asians.
- Latin Americans speak Spanish, Portuguese, other European languages, indigenous languages, and mixed dialects or patois.
- Latin America's population is mostly concentrated in coastal areas.
- Urbanization has created an imbalance in Latin America's population density.
- The region has some of the world's largest cities.

Organizing Your Notes

Use a graphic organizer like the one below to help you organize important details from this section.

Peoples	Population Patterns	Migration

SECTION 2 History and Government (pp. 220–225)

Terms to Know

- glyph
- *chinampas*
- quipu
- conquistador
- viceroy
- caudillo

Key Points

- The Maya, the Aztec, and the Inca developed complex civilizations before Europeans arrived.
- Spanish and Portuguese colonization had lasting effects on Latin America's culture.
- Most Latin American countries achieved independence during the 1800s.
- Most Latin American countries developed democratic self-rule in the twentieth century.
- The political, economic, and cultural legacy of colonialism still challenges Latin America.

Organizing Your Notes

Use a time line like the one below to help you organize your notes on key historical events discussed in this section.

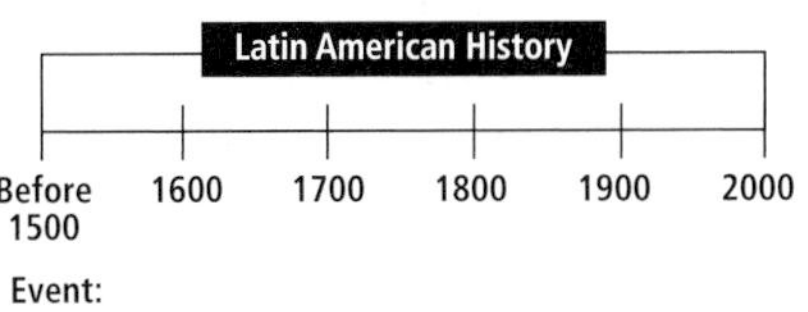

SECTION 3 Cultures and Lifestyles (pp. 226–231)

Terms to Know

- syncretism
- mural
- mosaic
- extended family
- malnutrition
- *fútbol*
- jai alai

Key Points

- Religion plays an important role in Latin American life.
- Educational quality varies throughout the region.
- As each country improves its economy, nutrition, and sanitation, people's health improves.
- Latin American traditional arts, music, and literature reflect the region's cultural diversity.
- Deep divisions between economic and social classes still characterize Latin American life.
- Latin Americans value family activities, sports such as *fútbol* and jai alai, and holidays and festivals.

Organizing Your Notes

Create an outline using the format below to help you organize your notes for this section.

Cultures and Lifestyles

I. Religion
  A. Roman Catholicism
  B. Protestantism

CHAPTER 9

ASSESSMENT & ACTIVITIES

Reviewing Key Terms

Write the key term that best matches each description. Refer to the Terms to Know in the Summary & Study Guide on page 233.

1. native; original inhabitant
2. two popular sports in Latin America
3. designs made by setting small pieces of colored stone, tile, or shell into mortar
4. grandparents, aunts, uncles, cousins
5. two language variations
6. a city that dominates its country's economy and government
7. knotted cords used for keeping accounts
8. Spanish or Portuguese conqueror
9. government officials appointed by European monarchs
10. a city with more than 10 million inhabitants
11. mixing of diverse religious traditions
12. wall painting
13. Mayan picture writing
14. migration from rural areas to cities
15. condition caused by lack of food

Reviewing Facts

SECTION 1

1. Where is most of South America's population located?
2. Why is the region's population density unbalanced?

SECTION 2

3. Name three indigenous Latin American empires.
4. What fueled the movement for Latin American independence?

SECTION 3

5. What ancient art form inspired the region's painters?
6. What sports are most popular in Latin America?

Critical Thinking

1. **Categorizing Information** Define the types of migration that occur in the region.
2. **Making Comparisons** Compare social and family life in Latin America and the United States.
3. **Identifying Cause and Effect** Use a diagram like the one below to fill in three lasting effects of colonialism.

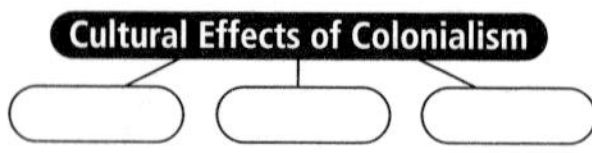

NATIONAL GEOGRAPHIC

Locating Places

Latin America: Political Geography

Match the letters on the map with the places of Latin America. Write your answers on a sheet of paper.

1. Caracas	5. Montevideo	9. Mexico City
2. Brasília	6. Bogotá	10. La Paz, Sucre
3. Port-au-Prince	7. Quito	11. Buenos Aires
4. Santiago	8. Havana	12. Lima

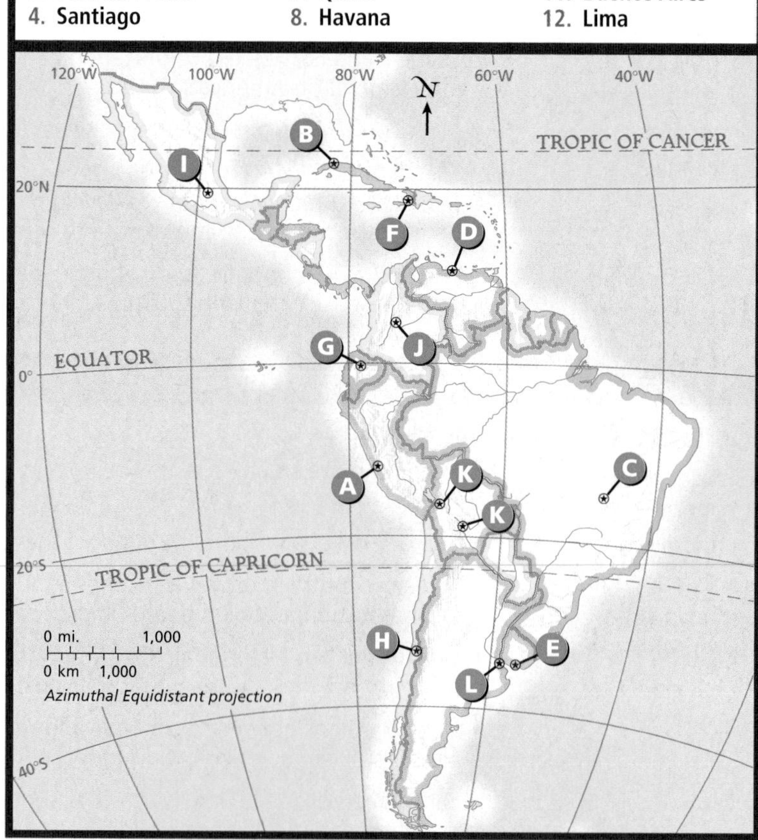

Self-Check Quiz Visit the **Glencoe World Geography** Web site at geography.glencoe.com and click on Self-Check Quizzes—Chapter 9 to prepare for the Chapter Test.

Using the Regional Atlas

Refer to the Regional Atlas on pages 182–185.

1. **Place** What features draw a large population to the Buenos Aires area?
2. **Human-Environment Interaction** Study the physical and population density maps. Why are parts of Argentina and Bolivia uninhabited?

Thinking Like a Geographer

Trace the diffusion and exchange of foods between the Americas and other parts of the world. Describe the foods involved, their place of origin, and their effects on the places to which they spread.

Problem-Solving Activity

Contemporary Issues Case Study Using the Internet, research a democratic country in Latin America. Then write a report that discusses the spread and adaptation of democracy to that country. Also, explain how other countries in the region might learn from its experience. Use photos, charts, and other graphics in your report.

GeoJournal

Descriptive Writing Using the information you logged in your GeoJournal, write a paragraph describing European or African influences on the art or religion of a particular Latin American country. Use additional resources to make your descriptions as vivid and accurate as possible.

Technology Activity

Creating an Electronic Database

Use reliable sources to gather population data for the past 10 years for three Latin American countries. Choose one category of information, such as literacy rates, population under age 18, or male/female ratio. Create an electronic computer database, and then use computer software to design and draw a graph or chart. Present your conclusions orally to the class, using the graph or chart to illustrate your findings.

Standardized Test Practice

Choose the best answer for each of the following multiple-choice questions. If you have trouble answering the questions, use the process of elimination to narrow your choices.

1. Latin American peoples speak a variety of languages. Which of the following statements is NOT true?

A Millions of Latin Americans speak Native American languages.
B Portuguese is the official language of most Latin American countries.
C French is the official language in some Latin American countries.
D Many Latin Americans are bilingual.

Do not quickly choose the first answer that makes sense—your answer will most likely be incorrect. This question asks you to identify which statement is *not* true. Eliminate any answer choices you know to be true before selecting the correct answer.

2. Diego Rivera was a Mexican artist who was well known for his creation of

F folk dramas.
G woven tapestry.
H political and social satires in poetry.
J large murals of historic events.

This question is factual. Try to recall what you know about Rivera, considering that he was a popular modern artist and important political activist.

CHAPTER 10

Latin America Today

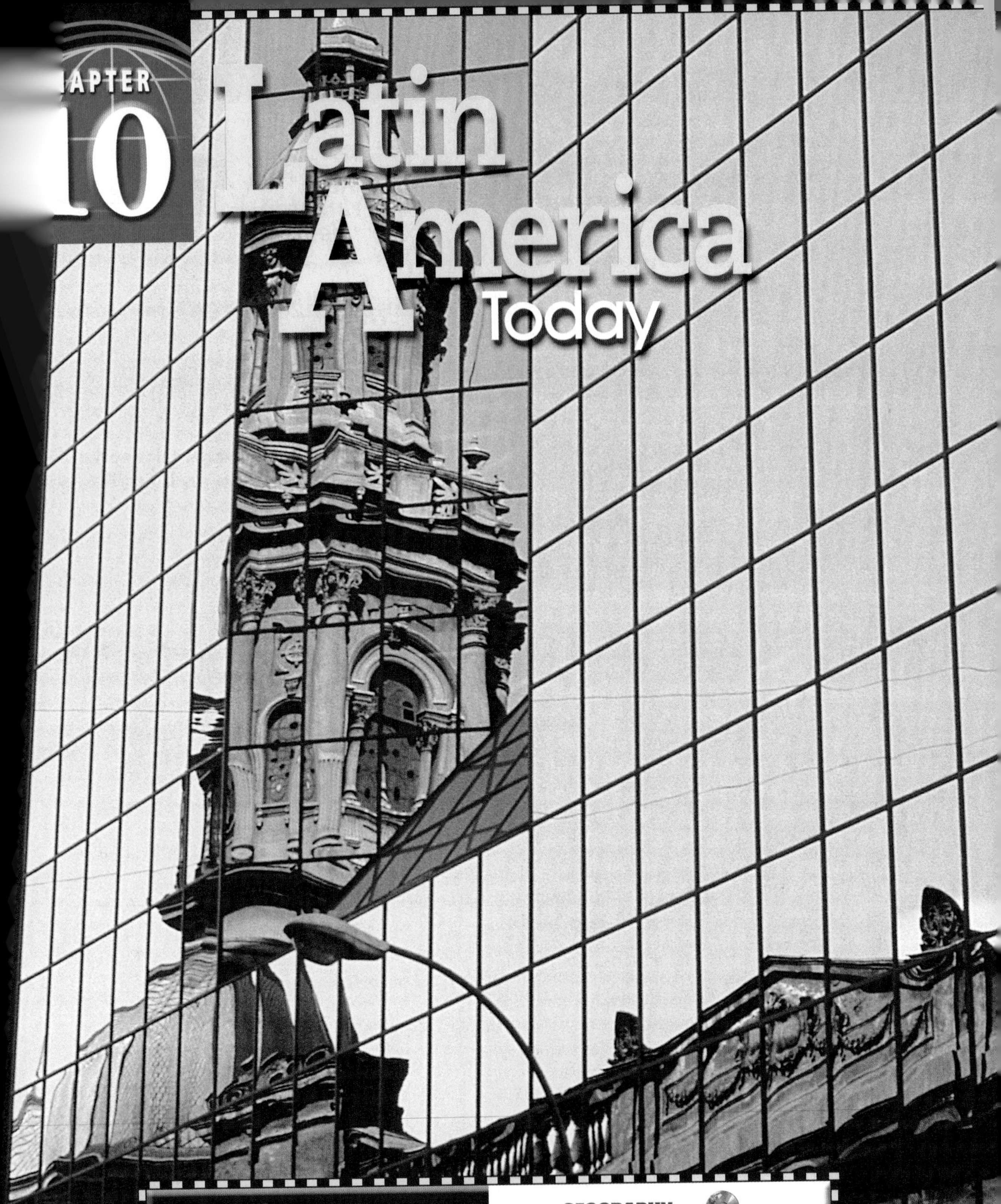

GeoJournal

As you read this chapter, use your journal to note examples of how geography affects life in Latin America and how the people of this region interact with their environment.

GEOGRAPHY Online

Chapter Overview Visit the **Glencoe World Geography** Web site at geography.glencoe.com and click on Chapter Overviews—Chapter 10 to preview information about the region today.

Living in Latin America

Guide to Reading

Consider What You Know

The North American Free Trade Agreement (NAFTA) has been the focus of economic and political news in recent years. What do you know about this agreement? How does NAFTA affect your community?

Read to Find Out

- What is the basis of the economies of many Latin American countries?
- What are the advantages and disadvantages of the North American Free Trade Agreement (NAFTA) for Mexico?
- What are the causes and consequences of Latin America's economically dependent status?
- How has the region's physical geography affected transportation and communications?

Terms to Know

- export
- campesino
- *latifundia*
- *minifundia*
- cash crop
- developing country
- service industry
- maquiladora
- North American Free Trade Agreement (NAFTA)

Places to Locate

- Honduras
- Tijuana

Modern Santiago, Chile, reflects the past.

NATIONAL GEOGRAPHIC

A Geographic View

Unlikely Neighbors

Tijuana's character changes from street to street. In one **colonia**, *or neighborhood, people wash laundry in tubs. . . . Up another road you pass dozens of modest homes built of concrete block and metal. Across town in wealthy Colonia Chapultepec, magnificent homes are built like fortresses right to the edge of the sidewalk. . . . Upper-middle-class sections abut neighborhoods of shacks with chickens, stray dogs, winding dirt roads, and crumbling embankments.*

—Michael Parfit, "Tijuana and the Border," National Geographic, *August 1996*

Tijuana, Mexico

Like Mexico's Tijuana, many Latin American cities reveal the sharp divisions between the wealthy and the poor. These class differences stem from social, political, and economic factors, but they are also shaped by physical geography. In this section you will learn about the ways in which Latin America's physical environment relates to the region's economic development and quality of life.

Agriculture

Although about three-fourths of Latin America's people live in cities, most of the region's countries still depend on agriculture to supply a major portion of their incomes. Latin American countries **export**, or sell to other countries, much of what their farms produce, such as bananas, sugarcane, and coffee.

History

Latifundia and Minifundia

For centuries farmland in Latin America has been unevenly distributed between a small group of wealthy landowners and a much larger force of **campesinos** (KAM•puh•SEE•nohs), or rural farmers and workers. Large agricultural estates owned by wealthy families or corporations are called ***latifundia***. Today's *latifundia* are highly mechanized commercial operations that yield high returns for low investment in labor. All other farms are called ***minifundia***, small plots of land intensively farmed by campesinos to feed their families. Campesinos, though, rarely own these plots, which are held by either wealthy landowners or the government.

The centuries-old system of *latifundia* and *minifundia* is gradually breaking down, however. As *latifundia* become more mechanized, farmworkers are leaving the land for the cities. In addition, reform-minded governments are passing laws to distribute farmland more fairly. Many campesinos have formed agricultural cooperatives, combining *minifundia* into large, jointly run farms. The legacy of economic inequality, however, is difficult to overcome completely, and Latin America's campesinos remain very poor.

NATIONAL GEOGRAPHIC **World Explorer**

Geography Skills for Life

Mexican Countryside Farmers harvest limes on a *latifundium* near the Mexico-U.S. border.

Human-Environment Interaction What is the difference between subsistence farming and cash-crop farming?

Cash Crops and Livestock

Latin America's physical geography makes it a suitable region for growing **cash crops**, crops produced in large quantities to be sold or traded. Fertile highlands help make Brazil, Mexico, Guatemala, and Colombia among the world's leading coffee producers. Lush, tropical coastal areas enable Central America, as well as Jamaica, Honduras, Ecuador, and Brazil, to be major producers of bananas. Tropical climates and fertile soil also help make Brazil and Cuba the world's leading producers of sugarcane. These export crops, all grown most efficiently on *latifundia*, usually benefit large-scale commercial producers more than individual farmers. In addition to growing cash crops, some Latin American countries—Argentina, Mexico, and Brazil—raise cattle for export on large ranches located in grassland areas.

Countries run great risks, however, when they depend on just one or two export products. If droughts, floods, or volcanic eruptions destroy a country's cash crop, the damage to that country's economy causes great hardship. In 1998 Hurricane Mitch devastated parts of Central America and destroyed about 90 percent of the banana crop, the main export, in **Honduras**. Tragically, the storm hit Honduras just as it was beginning to make some economic progress.

Industry

Most of Latin America's countries are **developing countries**, or countries that are working toward greater manufacturing and technology use. Countries with skilled workforces, good energy supplies, efficient transportation networks, and many natural resources are industrializing more rapidly than countries without these advantages. In many Latin American countries, **service industries**, such as banking, which provide services rather than goods, have grown rapidly in recent decades.

Industrial Growth

Several factors have limited industrial growth in Latin America. Physical geography may present obstacles. The high Andes and the dense Amazon rain forest, for example, limit access to natural resources. Ties to more developed regions also have limited growth. Foreigners have brought new technology to the region, but many have drained local resources and profits. Finally, political instability in many Latin American countries has made investors wary of investing in Latin American enterprises.

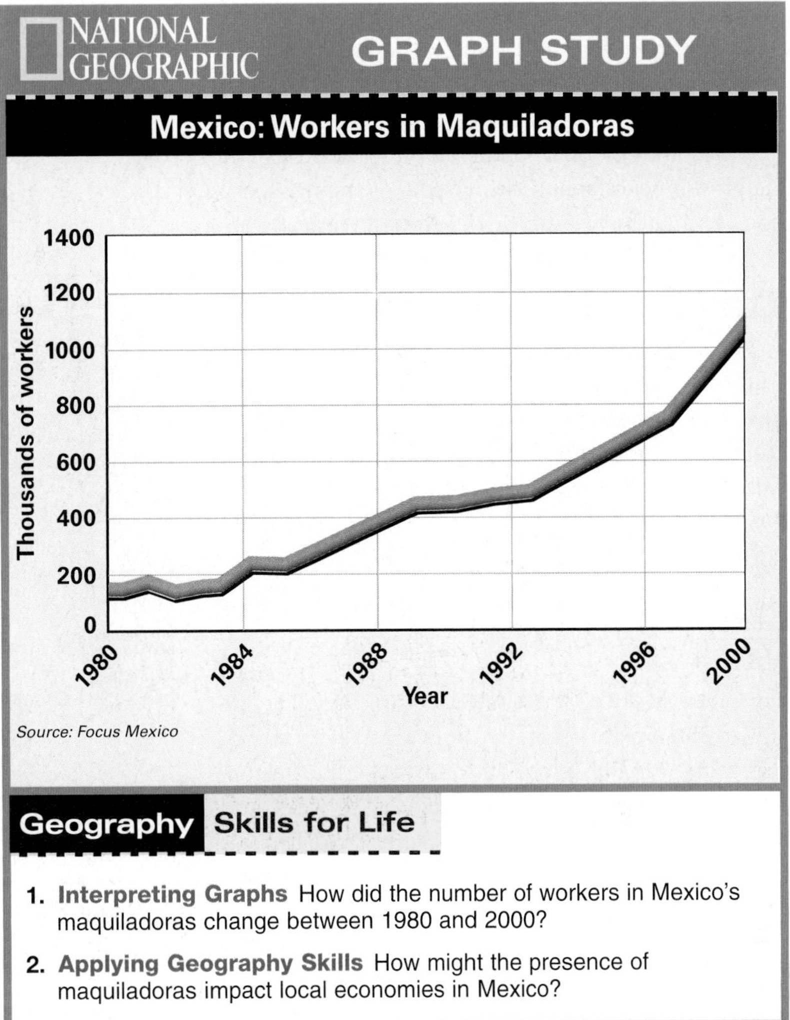

Geography Skills for Life

1. **Interpreting Graphs** How did the number of workers in Mexico's maquiladoras change between 1980 and 2000?
2. **Applying Geography Skills** How might the presence of maquiladoras impact local economies in Mexico?

Some Latin American countries, however, are overcoming these barriers. They combine the necessary human and natural resources with relatively stable governments and active business communities. Mexico, for example, is a major producer of motor vehicles, textiles, and processed foods. Brazil is a leading producer of iron and steel, cars, airplanes, textiles, and electrical goods. After weathering serious financial crises in the 1990s, both countries emerged with stronger economies because of their expanding global trade.

Other countries in the region also are developing industries. Argentina produces cars and processed meats. Venezuela refines oil, and Chile, Costa Rica, and Nicaragua all produce foods and textiles. Bolivia mines and refines tin, and Barbados, Cuba, and Saint Kitts and Nevis refine sugar.

Economics

Maquiladoras

During the past 50 years, American and Japanese firms have built manufacturing plants in Latin American countries. Most of these factories, known as **maquiladoras** (muh•KEE•luh•DOHR•uhs), lie along the Mexico–United States border—especially near the Mexican cities of Ciudad Juárez and **Tijuana**—where they employ many Mexicans. Maquiladoras benefit foreign corporations by allowing them to hire low-cost labor and to produce duty-free exports. They also offer the host country and its people employment opportunities and investment income. As one observer noted:

> *"... Tijuana lures foreign investors with cheap labor and proximity to U.S. markets, while beckoning workers from across Mexico with the chance for a new beginning. Here their dreams converge and sometimes collide, pulled hard by the magnet of the north."*
>
> Michael Parfit, "Tijuana and the Border," *National Geographic*, August 1996

Critics of maquiladoras charge that the system often ignores labor and environmental protection laws, damaging the environment and encouraging low-paying or dangerous jobs. As the world's economy becomes globalized, developing countries will weigh the benefits and drawbacks of their associations with industries of the developed world.

Trade and Interdependence

Like other countries of the world, Latin American countries depend on trade to obtain the goods and food that they cannot produce. Some Latin American countries, for example, import raw materials and expensive technology. In recent years Latin America has begun to promote trade within the region and with the rest of the world.

Economics

NAFTA

In 1992 Mexico, the United States, and Canada signed the **North American Free Trade Agreement (NAFTA)**. NAFTA gradually reduced trade restrictions and increased the flow of goods, services, and people among these countries. After NAFTA was implemented, trade among the three countries grew by 10 to 15 percent annually.

NAFTA, however, has been controversial in the United States. American labor groups fear the loss of jobs to generally lower-paid Mexican workers. Still, U.S. companies have not yet relocated south of the border in large numbers because certain production costs, such as electricity, are higher in Mexico. From Mexico's viewpoint, NAFTA has helped boost exports and create thousands of new jobs. Other Latin American countries are watching Mexico's economy to see if an agreement like NAFTA could work for them.

Foreign Debts

Many Latin American countries borrow funds from foreign sources to finance industrial development. During the 1980s a worldwide economic slowdown caused a sharp decline in demand for Latin America's products. When their incomes fell, many Latin American governments threatened to default, or not pay back their loans on time. Lenders then rescheduled the loans, which lengthened the time allowed to repay them and decreased monthly payments. However, this remedy also raised the total amount of interest on the debt. Repaying large foreign debts has halted needed domestic programs in some countries. Now international agencies are looking for other ways to offer debt relief.

NATIONAL GEOGRAPHIC **World Explorer**

Geography Skills for Life

Traditional Handicrafts Local potters in Peru produce clay jars and jugs to sell.

Region What other goods are produced in Latin America?

Transportation

In Latin America building roads and railroads is difficult. Many governments cannot afford building projects that must cross rugged mountains, dense rain forests, and arid deserts. Even so, some Latin American countries do have good roads. The region's major road system, the Pan-American Highway, stretches from northern Mexico to southern

▲ *An Andean highway echoes the past when Inca roads crossed the area.*

Chile and links more than a dozen Latin American capitals. A trans-Andean highway runs through the Andes and links cities in Chile and Argentina. To develop the Amazon Basin's mineral resources, Brazil is building the Trans-Amazonian Highway.

Although physical barriers limit railroad use, Mexico, Panama, Argentina, and Brazil have well-developed rail systems. In some places, however, railways have fallen into disrepair. As a result, inland waterways such as the Amazon River, the Paraná-Paraguay Rivers, and the Panama Canal remain important. As air travel becomes more affordable it will help overcome geographic barriers. All Latin American capitals and most major cities receive domestic and international flights. Mexico City, Buenos Aires, Rio de Janeiro, and São Paulo have the region's busiest airports. Many private and military landing strips serve remote locations.

Communications

Latin America's developing communications networks include newspapers, radio, and television, but all may be censored by governments during political unrest. Millions of Latin Americans use telephones, but few have them in their homes. Some countries cannot afford the equipment needed to provide residential phone service. In larger cities though, many people, especially young people, use cellular phones.

Although computer technology is slowly changing communications in Latin America, most people cannot afford personal computers. In 1998, on average, only 34 of every 1,000 Latin Americans owned computers. Innovative ways to offer computer access, such as Peru's public Internet centers, are helping Latin Americans go online, however.

SECTION 1 ASSESSMENT

Checking for Understanding

1. **Define** export, campesino, *latifundia*, *minifundia*, cash crop, developing country, service industry, maquiladora, North American Free Trade Agreement (NAFTA).
2. **Main Ideas** On a table, fill in examples of how five Latin American countries produce income.

Latin American Country	Chief Source of Income

Critical Thinking

3. **Drawing Conclusions** Why might political instability in a country discourage investors?
4. **Identifying Cause and Effect** What effects might defaulting on debt repayments have?
5. **Making Comparisons** How are *latifundia* and *minifundia* systems of farming alike? Different?

Analyzing Maps

6. **Place** Study the economic activity map on page 185. Which countries produce the most petroleum?

Applying Geography

7. **Industrialization** In a paragraph, discuss why industrialization requires good transportation and communications systems. Describe the impact of new technologies.

People and Their Environment

Guide to Reading

Consider What You Know

People around the world are concerned about the destruction of Latin America's rain forests. Why is the preservation of the rain forests important, and what makes this issue so complex?

Read to Find Out

- How has development affected Latin America's forest resources?
- How are Latin American governments working to balance forest conservation with human and economic development?
- What challenges are posed by the growth of Latin America's urban population?
- What regional and international issues continue to pose challenges for Latin American countries?

Terms to Know

- sustainable development
- deforestation
- slash-and-burn
- reforestation
- shantytown

Places to Locate

- São Paulo
- El Salvador

NATIONAL GEOGRAPHIC

A Geographic View

A Crucial Decision

Taking life slow, a three-toed sloth hangs out on an **ambaibo** *tree along the Río Tuichi in Madidi National Park. A planned hydroelectric dam may permanently [flood] this area—claiming one of South America's most biologically diverse rain forests even before it has been fully explored.*

—*Steve Kemper, "Madidi,"*
National Geographic, *March 2000*

Three-toed sloth

Bolivia and other Latin American countries face a difficult choice: whether to preserve large tracts of wilderness, such as Madidi National Park in northwestern Bolivia, or develop these areas for the purpose of raising peoples' standard of living. One way of resolving this dilemma is to work toward **sustainable development**—technological and economic growth that does not deplete the human and natural resources of a given area. In this section you will learn about the interrelationship of Latin Americans and their environment, and how the region is working to protect the environment while meeting human needs.

Managing Rain Forests

As you recall, extensive rain forests cover South America's Amazon River basin and the coastal areas of the Caribbean region. Like rain forests in other regions of the world, those in Latin America are disappearing as a result of **deforestation**, the clearing or destruction of forests. Although the threats to the world's rain forests are well known, the proposed strategies for preserving them are hotly debated.

Brazil's experience serves as an example of the complexity of the deforestation issue. During the past several decades, Brazil has worked to boost its economy by tapping the rain forest's vast mineral resources, such as petroleum, iron, copper, and tin. Roads have been carved out of the rain forest to open up Brazil's interior to settlement and development. For example, a 3,400-mile (5,472-km) east-west segment of the Trans-Amazonian Highway now crosses the region. This development of Brazil's interior, however, has proved disastrous for the Amazon rain forest and its indigenous human and animal inhabitants. The indigenous people of Brazil's interior have seen their homes and traditional ways of life disappear along with trees, other forms of vegetation, and animal life.

More than 13 percent of the Amazon rain forest has already been destroyed. (See the map below.) As the Amazon rain forest is depleted, the diversity of Earth's plant and animal species is threatened. Many of the world's key medicines are derived from rain forest plants and organisms, and deforestation risks the loss of compounds that have the potential to treat cancer and other illnesses. Because plants use carbon dioxide and produce oxygen, destroying rain forest plants could result in less carbon dioxide being used and more of it remaining in the atmosphere. Since carbon dioxide is a greenhouse gas that helps trap heat, catastrophic global warming, climate change, and rising ocean levels could result. Traditional wisdom has another way of expressing the value of rain forests, as one journalist notes:

> *"'Who cuts the trees as he pleases cuts short his own life.' The Maya adage . . . is spoken in a language that uses the same word for both 'blood' and 'tree sap.'"*
>
> George E. Stuart, "Maya Heartland Under Siege," *National Geographic*, November 1992

Brazil and other rain forest countries are listening to the advice of scientists and environmentalists, but they still face pressing social and economic

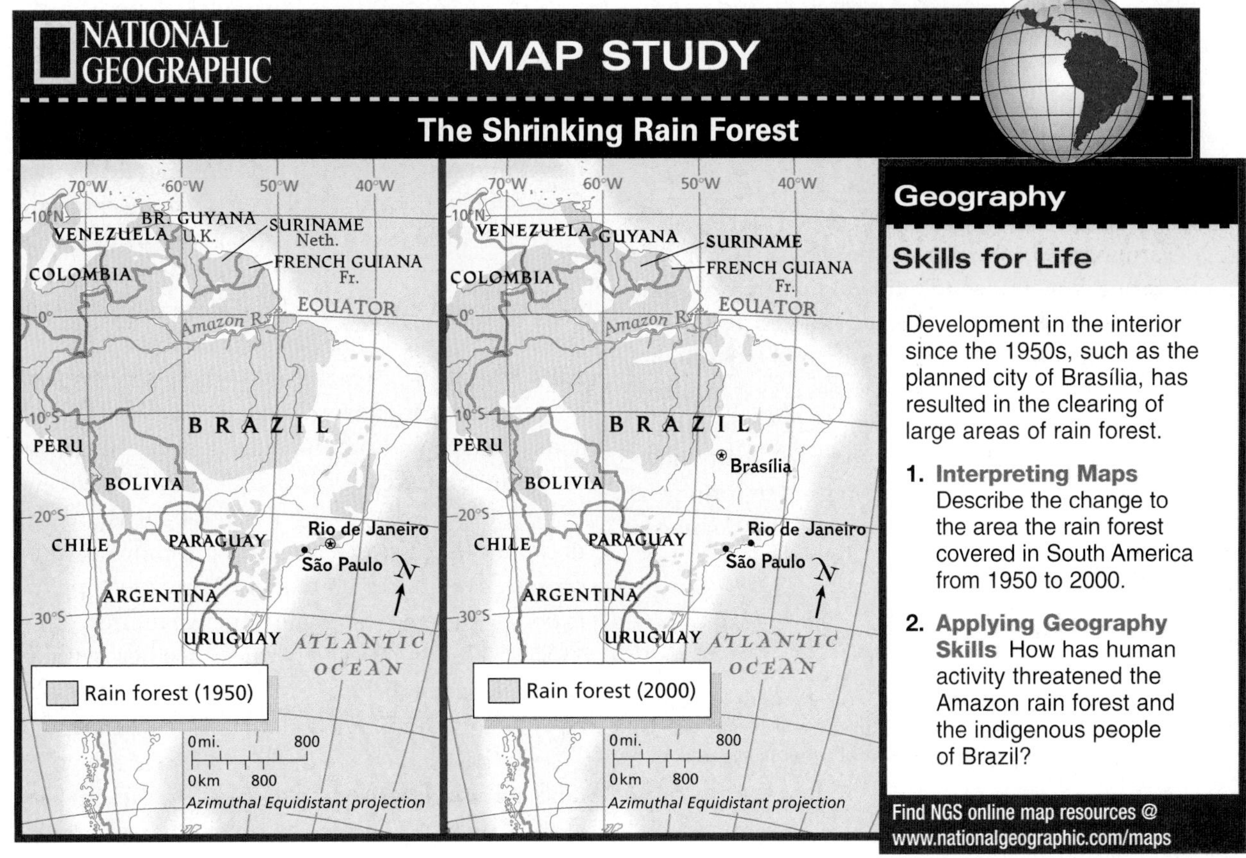

Geography Skills for Life

Development in the interior since the 1950s, such as the planned city of Brasília, has resulted in the clearing of large areas of rain forest.

1. **Interpreting Maps** Describe the change to the area the rain forest covered in South America from 1950 to 2000.
2. **Applying Geography Skills** How has human activity threatened the Amazon rain forest and the indigenous people of Brazil?

Find NGS online map resources @ www.nationalgeographic.com/maps

realities. If Brazil were to ban the use of rain forest lands, for example, how would it provide for all the people who would no longer have a way to support themselves? How would the country handle population growth in coastal areas if vast stretches of the interior were off-limits to its people?

Economics

Farms and Ranches Versus Forests

One of the most widespread activities in the Amazon Basin is the clearing of rain forest to provide more land for farming. To prepare the land, settlers use an ancient technique called **slash-and-burn** farming—but practiced on a larger scale. Farmers cut down all plants and strip any trees of bark. After the plants and trees have dried out, they are set on fire. The ash from the fire puts nutrients into the soil. Unfortunately, frequent rains leach away the benefits, and within one or two years, the soil loses its fertility or is washed away. Crop yields decline, and farmers move on to clear new parts of the forest.

Slash-and-burn methods are also used to carve huge cattle ranches out of the forest. Ranchers plant grasses in the charred ground for cattle grazing. After about four years, the grasses dry up and the ranchers, like the farmers, move on. The spent land supports little growth, and centuries-old rain forests have disappeared in just a few years.

Planting for the Future

Farming and ranching are not the only activities that contribute to deforestation in the Amazon area. Commercial logging operations harvest trees for timber and other products. Some estimates indicate that for every tree cut, two-thirds of the wood is wasted. Since colonial times, few attempts have been made to regulate the profitable logging industry. Today, however, the importance of conserving and restoring forest resources in Brazil and other Latin American countries has become increasingly clear. Brazil has set aside about 10 percent of its Amazon rain forest for national forests or parks in which logging is banned. In Costa Rica concerned citizens are buying abandoned, burned-over tropical forests that were once home to old-growth mahogany and other trees. The citizens then donate the land to a conservation district for restoration.

Given time, rain forests will regenerate on their own but with a considerable loss of biodiversity. Laws requiring **reforestation**—the planting of young trees or the seeds of trees on the land that has been stripped—can help, especially if the laws are rigorously enforced. Developing new methods of farming, mining, and logging and combining conservation with responsible tourism can protect the forests while boosting local economies.

Urban Environments

Latin America also has environmental challenges in its urban areas. More of the region's people live in megacities or towns than in rural areas. In 2000 Mexico City and **São Paulo** (sown POW•loo) ranked as the world's second- and fourth-largest metropolitan areas.

Overcrowded Cities

As Latin America's rural workers migrate to cities, they often cannot find jobs or adequate housing. Some are forced to live in slums or **shantytowns**, makeshift communities on the edges of cities. Known in different cities by different names—the *favelas* of São Paulo, the *barriadas* of Bogotá, and the *villas miserias* of Buenos Aires—these shantytowns often rest on dangerous slopes and wetlands. Mudslides, floods, and other natural disasters can wipe out entire communities. Lacking running water and underground sewage systems, these areas are also unsanitary, so diseases can spread rapidly. Because people have little or no money to buy food, malnutrition is common, especially among children.

Air pollution affects people in cities without adequate clean air laws. Millions of vehicles clog city streets and release massive amounts of exhaust gases into the air. Added to that are pollutants from industrial smokestacks. In Mexico City, air pollution can become so severe that authorities

GEOGRAPHY Online

Student Web Activity Visit the **Glencoe World Geography** Web site at geography.glencoe.com and click on Student Web Activities—Chapter 10 for an activity about economic development in Brazil.

periodically order cars off the roads and children are not allowed to play outside.

Building a Better Life

Rapid urbanization creates environmental challenges for Latin American cities. Cities experience rapid urbanization when their rates of population growth far exceed the available resources for housing, sanitation, employment, education, and government services.

The trend toward urbanization is global and is not likely to be reversed. Governments and international agencies, however, are beginning to address the needs of Latin America's urban areas. For example, Mexico City's recent improvements include a new water supply system and expanded public transportation. The World Bank has targeted cities in Venezuela, Peru, Brazil, and Guatemala for intensive neighborhood improvements. Grassroots efforts are even more promising. Groups of homeless people in cities such as Buenos Aires, Rio de Janeiro, and Santiago have successfully turned abandoned city buildings into affordable housing and commercial space.

Regional and International Issues

The quality of life in Latin America today continues to be shaped by geographic, economic, and historical realities that reach beyond national borders. Regional cooperation in addressing international issues will help move the region forward.

History

Disputed Borders

During the past 150 years, Latin America has faced a number of territorial conflicts. These

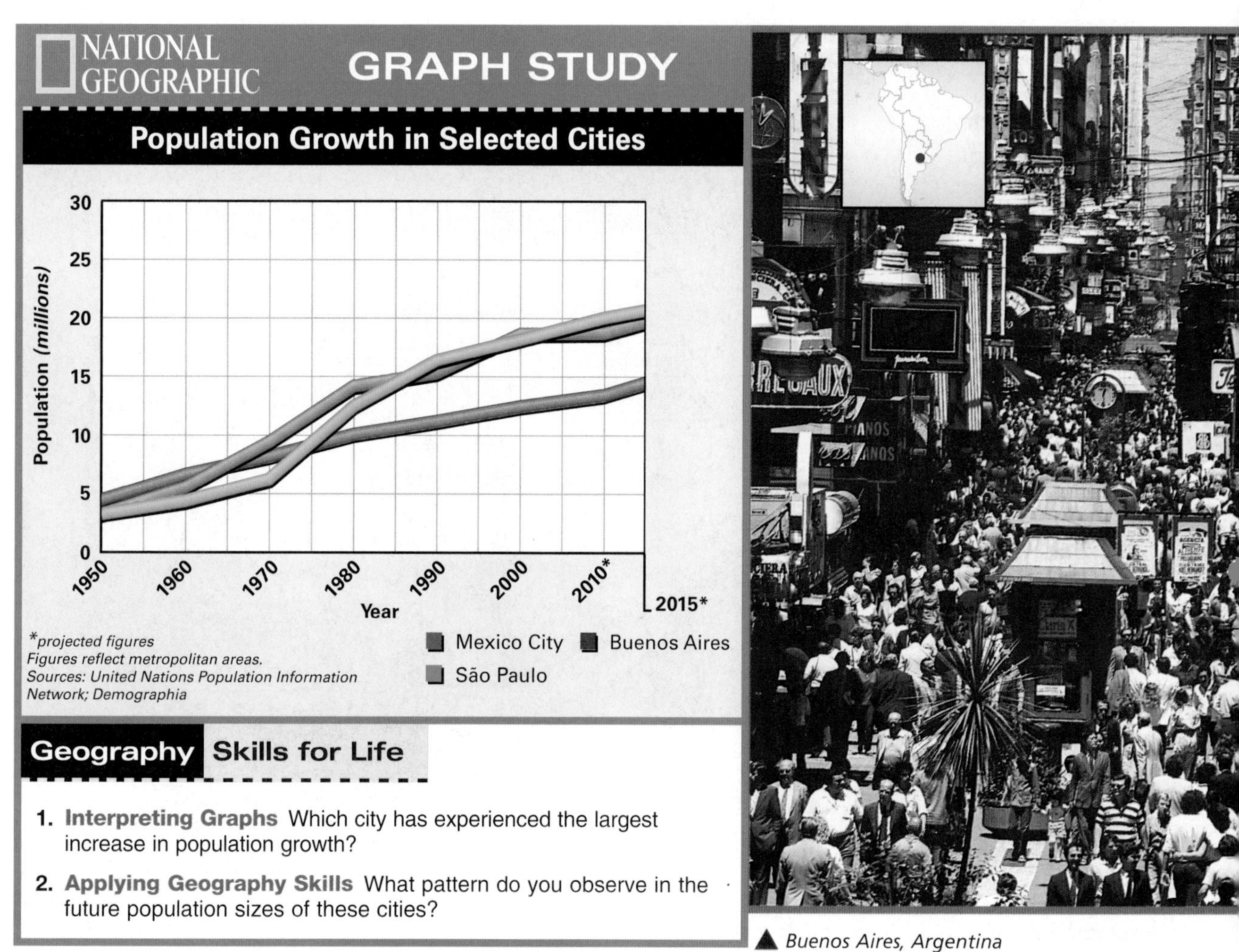

▲ *Buenos Aires, Argentina*

conflicts occur over disputed regions involving strategic locations or rights to valuable natural resources. Nicaragua, Honduras, and **El Salvador**, for example, have quarreled over fishing rights in the Gulf of Fonseca. Venezuela and Guyana battle over petroleum holdings along their shared border. Border wars divert resources that might better be used for development, but economic incentives can encourage countries to resolve their differences. After going to war three times, Peru and Ecuador finally settled a 60-year-old border dispute in 1998. During the negotiations international investors offered more than $3 billion in aid to develop economies and human services on both sides of the border.

Population Growth and Migration

Through education and economic improvement, most Latin American countries have begun to lower the high birthrates that have led to overpopulation. In the 2000s Latin America's population challenges will likely involve balancing the distribution of goods and services. Migration within the region—for economic or political reasons, or to escape the devastation of natural disasters—will continue to strain the resources of overcrowded cities.

In addition, growing numbers of Latin Americans have migrated abroad, especially to the United States. Many of these migrants have come to the United States to find a better way of life, some entering without visas. Others are well-educated people or skilled workers who could make important contributions to their home countries. For example, if scientists or researchers leave Latin America, its countries may ultimately lack the human resources necessary to solve environmental problems. To stem this out-migration, Latin American leaders are seeking to create jobs for their people by attracting more foreign investment.

Latin American migrants, meanwhile, have brought many changes to the United States. Immigration from Latin America has made the United States the country with the fifth largest Spanish-

NATIONAL GEOGRAPHIC **World Explorer**

Geography Skills for Life

Mexico's Smoking Mountain

The volcano Popocatépetl smokes while children play in central Mexico.

Human-Environment Interaction How are Latin American governments using technology to better plan for natural disasters?

speaking population in the world. Latin Americans have contributed much to the many communities in which they have settled—Mexicans in Texas, California, and Illinois; Cubans in Florida; and Puerto Ricans and Dominicans in New York.

Disaster Preparedness

Latin America's physical geography makes the region especially vulnerable to natural disasters, such as earthquakes, volcanic eruptions, and hurricanes, which take a huge toll in human life and economic resources. In order to increase the region's emergency preparedness, Latin American governments are cooperating in the use of sophisticated technology, such as satellite imaging and computer modeling, to forecast the direction and severity of hurricanes, for example. Such cooperative efforts can help Latin Americans anticipate emergencies rather than reacting after the fact. The savings in lives and economic resources are sure to be significant.

Industrial Pollution

Air and water pollution do not respect national boundaries. Multinational firms and free-trade agreements have increased industrial growth in some countries. Environmental laws, however, have not reduced the risks of increased pollution from new factories. Similarly, runoff from chemical fertilizers and pesticides used on commercial farms may cross borders to damage health or endanger lives. Governments and international agencies are cooperating to help Latin America address these challenges.

NATIONAL GEOGRAPHIC **World Explorer**

Geography Skills for Life

Water Pollution Bauxite mining runoff covers a cove in Jamaica.

Human-Environment Interaction How can Latin America overcome the challenges of industrial pollution?

SECTION 2 ASSESSMENT

Checking for Understanding

1. **Define** sustainable development, deforestation, slash-and-burn, reforestation, shantytown.
2. **Main Ideas** Use a diagram similar to the one below to identify activities that have contributed to deforestation in Latin America.

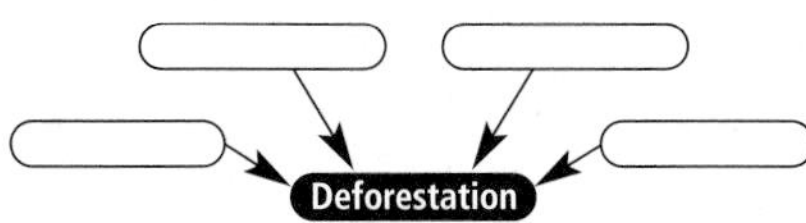

Critical Thinking

3. **Predicting Consequences** How might the destruction of the Amazon rain forest affect your life?
4. **Making Comparisons** Compare the ways urban populations in Latin America and those in your state have modified their physical environments.
5. **Drawing Conclusions** What circumstances might make environmental protection a low priority for some Latin American governments?

Analyzing Maps

6. **Human-Environment Interaction** Study the maps of the Amazon rain forest on page 243. What kinds of activities are responsible for these changes?

Applying Geography

7. **Mental Mapping** Without consulting a map, identify the Latin American countries most at risk from hurricanes. Write a description of ways that a hurricane is a threat to these countries.

Viewpoint

CASE STUDY on the Environment

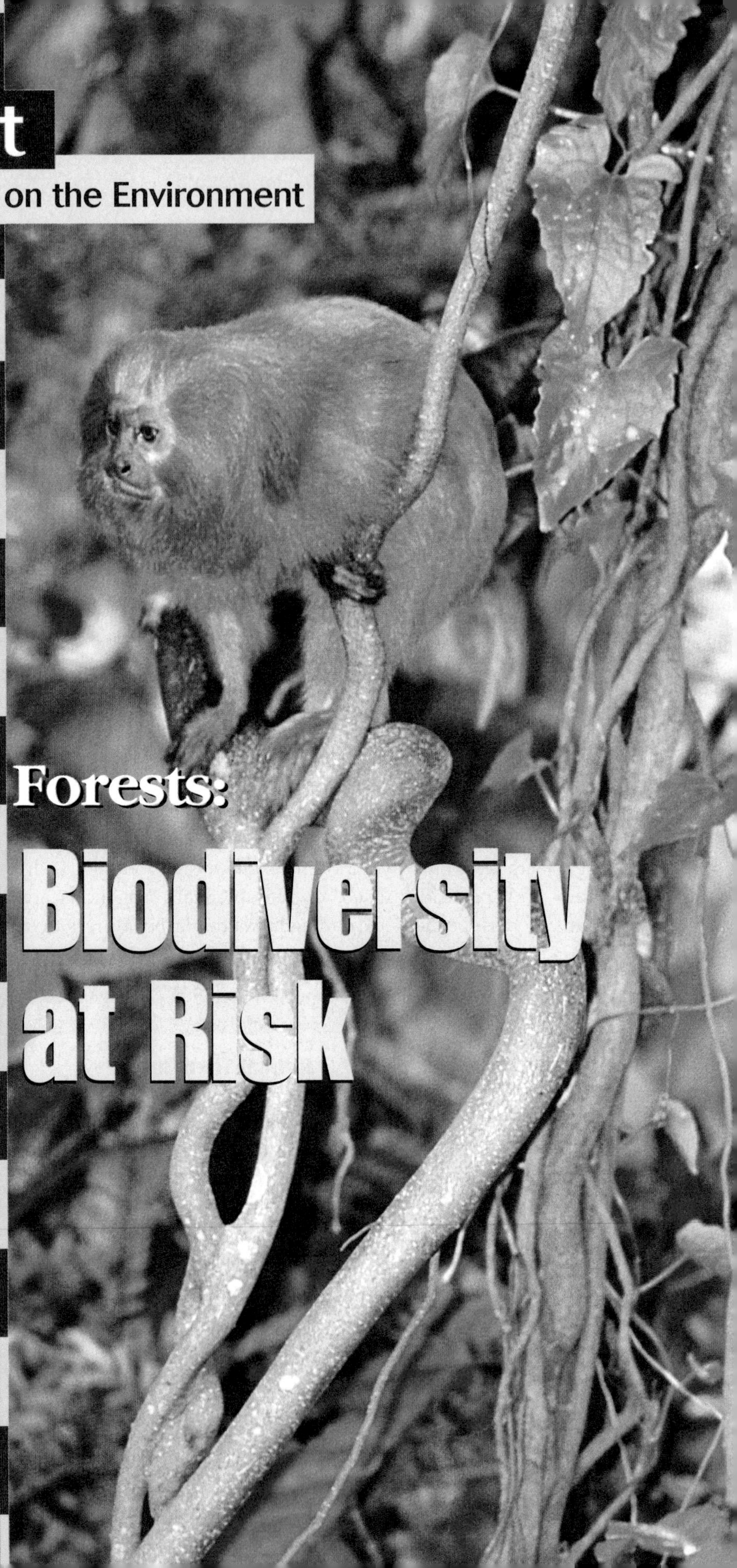

Brazil's Rain Forests: Biodiversity at Risk

Nowhere is "biodiversity"—the term biologists coined to describe our planet's bountiful variety of living things—more apparent than in a tropical rain forest. Rain forests harbor at least half of all species on Earth. However, deforestation severely threatens these biologically rich ecosystems. Worldwide, roughly 150 acres (61 ha) of rain forest are destroyed every minute. As the forests disappear, habitats are lost and Earth's biodiversity dwindles. The world's largest remaining expanses of tropical rain forest are in Brazil. But the fate of these forests, and their astounding array of plants and animals, is uncertain.

Small enough to fit in the palm of your hand, a golden lion tamarin (left) is a blaze of orange against vivid rain-forest green. A tiny primate, the golden lion is one of four tamarin species that live in Brazil's Amazon and less-well-known Atlantic rain forests. All tamarins are endangered, but a successful captive-breeding program recently brought the golden lion tamarin back from the brink of extinction. The future of thousands of other plant and animal species also is in jeopardy as pressures on Brazil's rain forests intensify.

These forests boast the richest variety of plant life on the planet. A few acres might contain 450 different species of trees. Thousands of other types of plants—orchids, bromeliads, ferns, and vines—grow on, among, or beneath the trees.

Brazil's rain forests reverberate with the hum of countless insect species. A single tree might harbor 650 kinds of beetles. Sharing the forest with these insect multitudes are snakes, such as anacondas and jararacas, and other animals, such as poison dart frogs, fruit bats, jaguars, and spider monkeys. But this biodiversity is rapidly disappearing.

Large areas of Brazil's rain forests are burned or cut down by farmers and ranchers to make way for cropland and cattle pastures. Loggers cut the fine hardwoods and export them for a profit. Growing cities, new roads, and industries encroach on the forests. Each year hundreds, perhaps thousands, of species are lost as their habitats are destroyed.

Rain forest preservationists want future generations to enjoy Earth's biodiversity. They point out that rain forest plants provide us with many things: foods, such as bananas and Brazil nuts; medicines, such as quinine and muscle relaxants; and substances, such as dyes and waxes. Scientists have identified only a fraction of rain forest species. Could a cure for cancer or AIDS lurk in a plant that has not yet been studied?

◀ The flowering four-o'clock contains substances used to treat rheumatism.

Rain forest developers argue that people living in or near Brazil's rain forests need to feed their families. For many, farming or raising cattle in clear-cut areas is the only way they can survive. Harvesting rain forest timber allows workers to lift their standard of living above the poverty line. With a growing population of more than 160 million, Brazil needs room for urban growth and industrial development. Some argue that Brazil's rain forests belong to Brazilians. Shouldn't they manage their forests as they see fit?

What's Your Point of View?
Only about 8 percent of Brazil's Atlantic rain forest remains. Do you think this remnant is worth saving? Why or why not?

Brazilian farmers (below) clear a field in the Amazon rain forest. Fire (right) destroys large areas of forest. ▼

Creating an Outline

Outlining may be used as a starting point for a reader who wants to understand and organize information. The reader begins with the rough shape of the material and gradually fills in the details in a logical manner.

Learning the Skill

Outlining can be used as a method of note taking and organizing information. There are two types of outlines—informal and formal. An informal outline is similar to taking notes—you write words and phrases needed to remember main ideas. A formal outline has a standard format.

To make a formal outline, begin by thinking about big ideas and dividing them into units of information. Give each of these major ideas a *heading*—a word or phrase that will identify the concept. Each major idea will be followed by two or more *subtopics*, or parts of main ideas. Include *supporting details* within each subtopic.

To create a formal outline, follow these steps:

- **Identify the general topic of the outline, and write the topic as a question.** Refer to the topic question as you work to be sure you are recording the most important ideas.
- **Write the main ideas that answer this question.** Label these with Roman numerals.
- **Write subtopics under each main idea.** Label these with capital letters.
- **Write supporting details for each subtopic.** Label these with Arabic numerals and lowercase letters.

Topic as a question:

______________________________?

I. Managing Rain Forests
  A. Rain Forests
    1. Location
    2. Brazil's experience
      a. Trans-Amazonian Highway
      b. Indigenous peoples
    3. World concerns
    4. Local challenges
  B. Farms and Ranches Versus Forests
    1. Slash-and-burn farming
    2. Future land value
  C. Planting for the Future
    1. ____________
    2. ____________
II. ____________
  A. Overcrowded Cities
    1. Shantytowns
    2. ____________
  B. Building a Better Life
    1. Rapid urbanization
    2. ____________
III. Regional and International Issues
  A. ____________
    1. ____________
    2. ____________
  B. Population Growth and Migration
    1. ____________
    2. ____________
  C. ____________
    1. ____________
    2. ____________
  D. ____________
    1. ____________
    2. ____________

Practicing the Skill

Study the incomplete outline of Chapter 10, Section 2, above. The main ideas generally correspond to the section headings in the chapter. Copy this outline on a sheet of paper, and fill in the missing information for Section 2 of Chapter 10.

When you have completed your outline, answer the following questions:

1. What are the most important topics in Chapter 10, Section 2?
2. What are the four main subtopics under the heading "Regional and International Issues"?
3. What are two situations in which an outline such as this might be useful?
4. In addition to being useful to readers, how would an outline help writers?

Create an outline for Chapter 10, Section 1. Use the section headings for your main ideas. Remember to include at least two subtopics for each main heading. When you have finished, use your outline to identify the main ideas of the section.

Go To The **Glencoe Skillbuilder Interactive Workbook, Level 2** provides instruction and practice in key social studies skills.

CHAPTER 10

SUMMARY & STUDY GUIDE

SECTION 1 Living in Latin America (pp. 237–241)

Terms to Know

- export
- campesino
- *latifundia*
- *minifundia*
- cash crop
- developing country
- service industry
- maquiladora
- North American Free Trade Agreement (NAFTA)

Key Points

- Latin America's economy is based on the export of agricultural products.
- A small group of wealthy families or businesses owns a large percentage of the agricultural land in Latin America.
- The economy of many Latin American countries is linked to one or two cash crops.
- The maquiladora system, trade agreements, and international borrowing are attempts to speed the industrialization of many Latin American countries.
- Geographic and economic realities have presented obstacles to developing transportation and communications in the region.

Organizing Your Notes

Use a graphic organizer like the one below to summarize your notes for this section.

Living in Latin America	
Agriculture	
Industry	
Trade and Interdependence	
Transportation and Communications	

SECTION 2 People and Their Environment (pp. 242–247)

Terms to Know

- sustainable development
- deforestation
- slash-and-burn
- reforestation
- shantytown

Key Points

- A key challenge for the Latin American region is sustainable development.
- Damage to the Amazon rain forest has both local and global consequences.
- Slash-and-burn cultivation contributes to Latin America's environmental challenges.
- Latin America's urban environmental problems are a result of rapid urbanization.
- Solutions to the region's environmental concerns will come through cooperation among local, national, regional, and international governments and organizations.

Organizing Your Notes

Create graphic organizers like the one below for each of the following topics: deforestation, population growth, and international issues.

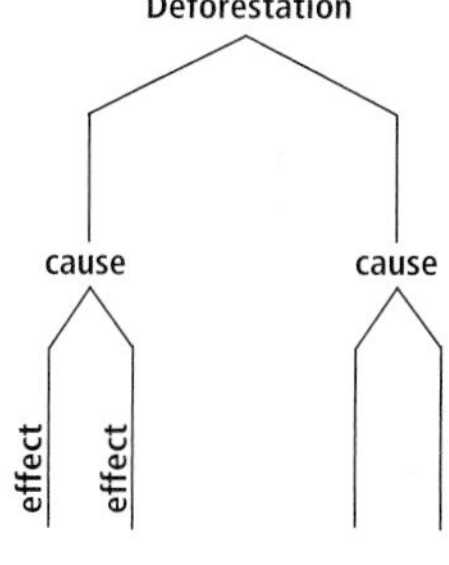

Signing of NAFTA, 1992

CHAPTER 10

ASSESSMENT & ACTIVITIES

Reviewing Key Terms

Examine the sets of terms below. Then write a sentence explaining how each set is related.

a. export — cash crop
b. *latifundia* — *minifundia*
c. developing country — sustainable development
d. maquiladora — service industries — North American Free Trade Agreement (NAFTA)
e. deforestation — slash-and-burn — reforestation
f. shantytown — rapid urbanization

Reviewing Facts

SECTION 1

1. What three cash crops supply much of Latin America's income?
2. How can dependence on a single crop affect a country's economy?
3. How do unstable governments prevent industrial development?
4. What obstacles have slowed the development of Latin America's transportation and communications systems?

SECTION 2

5. What are the environmental effects of slash-and-burn cultivation?
6. Why do some people believe that preservation of the Amazon rain forest is a global concern?
7. How are Latin American countries addressing the problems resulting from rapid urbanization?
8. Why do border disputes slow the economic development of the Latin American region?

Critical Thinking

1. **Making Generalizations** Has the maquiladora system had a positive or a negative effect on Mexico's people? Explain.
2. **Predicting Consequences** How might the North American Free Trade Agreement (NAFTA) change migration patterns in Latin America? What are the implications?
3. **Identifying Cause and Effect** Complete a diagram by giving examples of how rapid urbanization in Latin America affects housing, employment, education, sanitation, and government services.

Rapid urbanization affects . . .

NATIONAL GEOGRAPHIC

Locating Places

Latin America: Political Geography

Match the letters on the map with the places in Latin America. Write your answers on a sheet of paper.

1. **Panama**
2. **Belém**
3. **Brazil**
4. **São Paulo**
5. **Santiago**
6. **Bogotá**
7. **Costa Rica**
8. **Lima**
9. **Mexico**
10. **Rio de Janeiro**
11. **Mexico City**

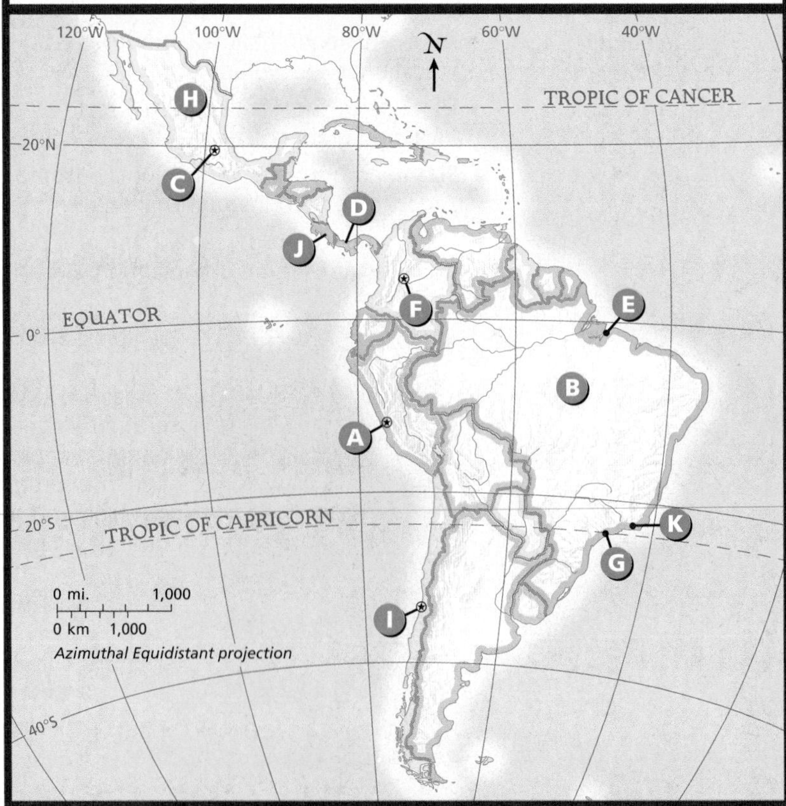

Self-Check Quiz Visit the **Glencoe World Geography** Web site at geography.glencoe.com and click on Self-Check Quizzes—Chapter 10 to prepare for the Chapter Test.

Using the Regional Atlas

Refer to the Regional Atlas on pages 182–185.

1. **Region** What do the economic activity and population density maps suggest about where commercial farming occurs in Latin America?
2. **Location** Compare the economic activity and population density maps. What industry probably attracts settlers around Córdoba, Argentina?

Thinking Like a Geographer

Why do you think Latin American countries often depend on a single cash crop? As a geographer, what suggestions would you make to help governments make their agricultural output more varied?

GeoJournal

Descriptive Writing Review the data in your GeoJournal. Then write a paragraph comparing the ways Latin Americans in rural, urban, coastal, and highlands areas depend on or adapt to their environment. Focus on specific human activities.

Problem-Solving Activity

Problem-Solution Proposal In Latin America thousands of children spend most of their lives roaming the streets. Pick one Latin American city with street children, and learn more about them: how they survive, where they come from, how they spend their time, and where their families are. Then write a proposal in which you present your solution to the problem.

Technology Activity

Using the Internet for Research Search the Internet for photographs showing the destruction of a rain forest in Latin America. Narrow your search by using words such as *Amazon rain forest* or *Brazil*. Print the photographs, write captions to explain what is happening, and display your work on the classroom bulletin board. If possible, include charts, graphs, and maps.

Standardized Test Practice

Choose the best answer for the following multiple-choice question. If you have trouble answering the question, use the process of elimination to narrow your choices.

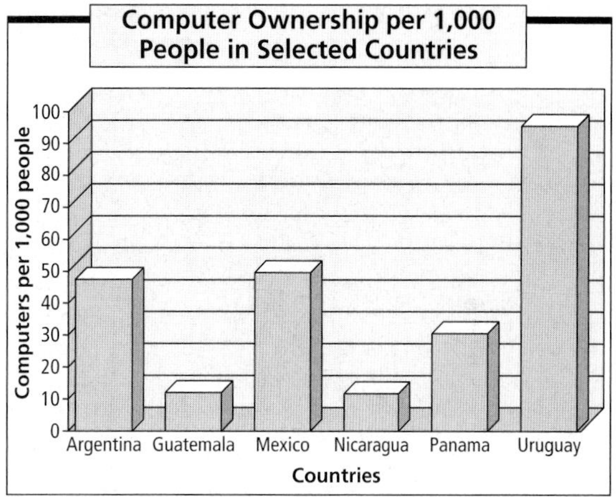

1. **On average, 34 of every 1,000 Latin Americans owned computers in 1998. Which countries had higher rates of computer ownership than the regional average?**

   A Nicaragua and Panama
   B Argentina, Mexico, Panama, and Uruguay
   C Argentina, Mexico, and Uruguay
   D Guatemala, Nicaragua, and Panama

Reread the title and x- and y-axis labels of the graph to determine the information the graph shows. Notice that Guatemala, Nicaragua, and Panama all fall short of the regional average. Tackle each answer choice one by one, eliminating those that contain even one country that has a lower rate than the regional average.

UNIT 4

Europe

WHY IT'S IMPORTANT—

In the 1990s, several nations of Europe formed the European Union, an alliance that works for the region's economic and political unity. Many European countries are replacing their national currencies with a common currency—the euro. As one of the world's leading economic powers, Europe has long had close political, cultural, and trading ties with the United States. Because of this important relationship, European ideas and practices have shaped your life and will continue to do so in the years ahead.

World Regions Video

To learn more about Europe and its impact on your world, view the World Regions video "Europe."

NGS ONLINE

www.nationalgeographic.com/education

Gondolier in Venice, Italy

4

What Makes Europe a Region?

Europe is a small continent with a long, jagged coastline. With watery fingers, the sea reaches deep into the land, embracing peninsulas and carving out bays, gulfs, and channels. Warm Atlantic winds and currents bathe European shores, helping to give this northerly landmass an unexpectedly mild climate. They also bring abundant rain that nurtures lush, green landscapes.

Fertile plains extend across much of northern Europe. Farther south, the plains become rugged hills, then mountains. The Alps are the continent's highest mountain range. They stretch across south-central Europe, forming a barrier that shelters the sunny Mediterranean area from moist northern winds.

Great rivers wind their way through Europe's landscapes, linking inland areas with the sea. The Danube flows through or along more countries than any other river in the world. The Rhine, with its source high in the Swiss Alps, is the continent's most important waterway.

1 **Guernsey cows** tread a familiar track on one of the Channel Islands, between England and France. Warmed by ocean currents that originate in the tropical Atlantic, these islands have a mild, moist climate—perfect for cattle and crops.

NATIONAL GEOGRAPHIC SOCIETY

2 Rows of bright umbrellas shelter beachgoers at Positano, Italy. Europe's unusually long coastline borders many seas. Countries along the Mediterranean Sea enjoy what is called a Mediterranean climate, with mild winters and hot, dry summers.

3 Ships and barges follow the snaking curves of the Rhine River in Germany. For centuries, the Rhine has provided an important transportation route through western Europe. Some of Germany's largest cities—and many medieval castles—lie along the Rhine and its tributaries.

4 Icy peaks in the Swiss Alps reflect the colors of the setting sun. The Alpine mountain system forms a broad arc that reaches from Spain to the Balkan Peninsula. The range's highest peak, Mont Blanc, lies in France, near the border with Switzerland and Italy.

Cultural Colossus

A mosaic of more than 40 countries, Europe enjoys a rich cultural heritage. Western traditions of art, architecture, science, and mathematics had their start in ancient Greece and Rome. In the centuries that followed, European culture spread far beyond the continent, aided by easy access to the sea. Modern European cities remain thriving centers for education and the arts.

Europe is home to people of many ethnic groups. Differences among those groups have led to frequent conflicts throughout European history. Toward the end of the twentieth century, political reforms greatly changed the face of Europe and brought new unity—as well as new challenges—to the region's inhabitants.

1 Crates of cargo stand on docks lining the harbor of Aberdeen, Scotland. The cargo awaits loading onto oceangoing ships. Europe's long coastline is dotted with busy ports. Access to the sea has helped spread European goods and culture worldwide.

2 West Berliners batter the wall that once separated the city into eastern and western sectors—and represented Europe's division into Communist and non-Communist camps. In 1989 several Communist governments were toppled, and the Berlin Wall began to come down.

3 The extravagant Opéra Garnier, one of the largest theaters in the world, stands near the center of Paris. Sometimes compared to a gilded wedding cake, this ornate structure was built in the mid-1800s. Originally an opera house, it now features mostly ballet.

4 A young Basque boy dons his father's cap and will carry on the elder's ethnic traditions. Three million Basques inhabit a wedge-shaped homeland that straddles the border between France and Spain. Basques speak an ancient tongue that is unrelated to any other known language.

4 Europe

PHYSICAL

Elevation Profile

POLITICAL

⊛ National capital
• Major city

0 mi. 500
0 km 500
Lambert Azimuthal Equal-Area projection

MAP Study

1. Which European countries border the Black Sea?

2. Which European countries are land-locked, or have no coastline?

4 Europe

POPULATION DENSITY

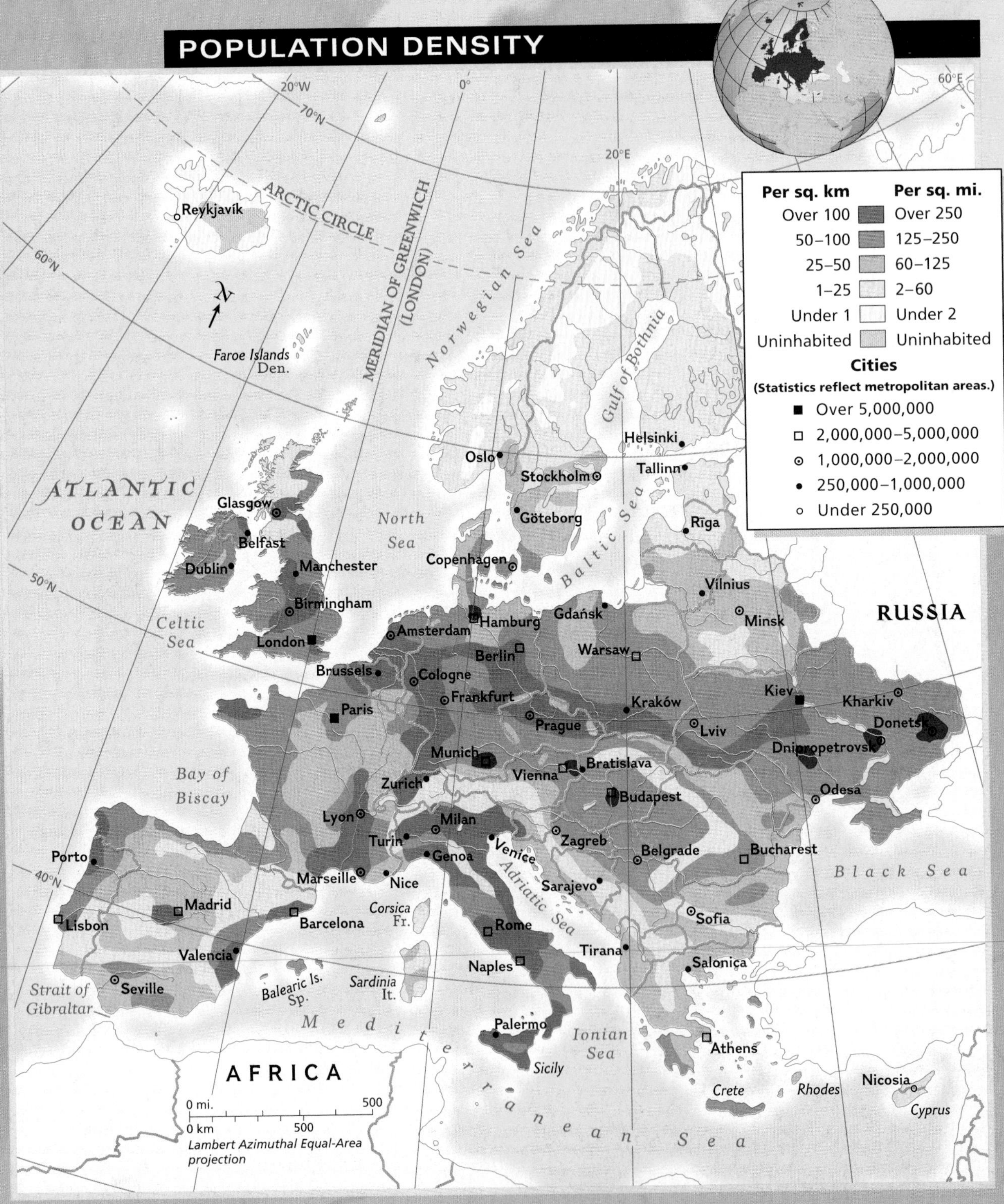

ECONOMIC ACTIVITY

Resources
- Coal
- Iron ore
- Petroleum
- Bauxite
- Copper
- Lead
- Zinc
- Silver
- Uranium
- Phosphate
- Nickel
- Natural gas
- Manganese

Land Use
- Commercial farming
- Subsistence farming
- Nomadic herding
- Forests
- Manufacturing and trade
- Commercial fishing
- Little or no activity

0 mi. 500
0 km 500
Lambert Azimuthal Equal-Area projection

MAP Study

1. Which European countries appear to have the highest population densities?
2. Describe the relationship between coal deposits and population density in Europe.

UNIT 4

REGIONAL ATLAS

Europe

COUNTRY PROFILES

COUNTRY * AND CAPITAL	FLAG AND LANGUAGE	POPULATION AND DENSITY	LANDMASS	MAJOR EXPORT	MAJOR IMPORT	CURRENCY	GOVERNMENT
ALBANIA Tirana	Albanian	3,400,000 310 per sq. mi. 120 per sq. km	11,100 sq. mi. 28,748 sq. km	Asphalt	Machinery	Lek	Republic
ANDORRA Andorra la Vella	Catalan, French, Spanish	100,000 380 per sq. mi. 147 per sq. km	174 sq. mi. 451 sq. km	Electricity	Manufactured Goods	French Franc, Spanish Peseta	Parliamentary Democracy
AUSTRIA Vienna	German	8,100,000 251 per sq. mi. 97 per sq. km	32,378 sq. mi. 83,859 sq. km	Machinery	Petroleum	Schilling, Euro	Federal Republic
BELARUS Minsk	Belarussian, Russian	10,000,000 125 per sq. mi. 48 per sq. km	80,154 sq. mi. 207,598 sq. km	Machinery	Fuels	Belarussian Ruble	Republic
BELGIUM Brussels	Flemish, French	10,300,000 872 per sq. mi. 337 per sq. km	11,787 sq. mi. 30,528 sq. km	Iron and Steel	Fuels	Belgian Franc, Euro	Constitutional Monarchy
BOSNIA AND HERZEGOVINA Sarajevo	Serbo-Croatian	3,400,000 173 per sq. mi. 69 per sq. km	19,741 sq. mi. 51,129 sq. km	N/A	N/A	Convertible Mark	Republic
BULGARIA Sofia	Bulgarian	8,100,000 190 per sq. mi. 73 per sq. km	42,822 sq. mi. 110,909 sq. km	Machinery	Fuels	Lev	Republic
CROATIA Zagreb	Serbo-Croatian	4,700,000 197 per sq. mi. 76 per sq. km	21,830 sq. mi. 56,540 sq. km	Transport Equipment	Machinery	Kuna	Republic
CYPRUS Nicosia	Greek, Turkish	900,000 247 per sq. mi. 95 per sq. km	3,571 sq. mi. 9,249 sq. km	Citrus Fruits	Manufactured Goods	Cyprus Pound	Republic
CZECH REPUBLIC Prague	Czech, Slovak	10,300,000 337 per sq. mi. 130 per sq. km	30,448 sq. mi. 78,860 sq. km	Machinery	Crude Oil	Koruna	Republic
DENMARK Copenhagen	Danish	5,400,000 322 per sq. mi. 124 per sq. km	16,637 sq. mi. 43,090 sq. km	Machinery	Machinery	Krone	Constitutional Monarchy

* COUNTRIES AND FLAGS NOT DRAWN TO SCALE

COUNTRY * AND CAPITAL	FLAG AND LANGUAGE	POPULATION AND DENSITY	LANDMASS	MAJOR EXPORT	MAJOR IMPORT	CURRENCY	GOVERNMENT
ESTONIA Tallinn	Estonian	1,400,000 78 per sq. mi. 30 per sq. km	17,413 sq. mi. 45,099 sq. km	Textiles	Machinery	Kroon	Republic
FINLAND Helsinki	Finnish, Swedish	5,200,000 40 per sq. mi. 15 per sq. km	130,560 sq. mi. 338,150 sq. km	Paper	Foods	Markka, Euro	Republic
FRANCE Paris	French	59,200,000 278 per sq. mi. 107 per sq. km	212,934 sq. mi. 551,499 sq. km	Machinery	Crude Oil	French Franc, Euro	Republic
GERMANY Berlin	German	82,200,000 597 per sq. mi. 231 per sq. km	137,830 sq. mi. 356,978 sq. km	Machinery	Machinery	Deutsche Mark, Euro	Federal Republic
GREECE Athens	Greek	10,900,000 214 per sq. mi. 83 per sq. km	50,950 sq. mi. 131,960 sq. km	Foods	Machinery	Drachma	Republic
HUNGARY Budapest	Hungarian	10,000,000 278 per sq. mi. 107 per sq. km	35,919 sq. mi. 93,030 sq. km	Machinery	Crude Oil	Forint	Republic
ICELAND Reykjavík	Icelandic	300,000 7 per sq. mi. 3 per sq. km	39,768 sq. mi. 102,999 sq. km	Fish	Machinery	Icelandic Króna	Republic
IRELAND Dublin	English, Irish Gaelic	3,800,000 142 per sq. mi. 55 per sq. km	27,135 sq. mi. 70,280 sq. km	Chemicals	Foods	Irish Pound, Euro	Republic
ITALY Rome	Italian	57,800,000 497 per sq. mi. 192 per sq. km	116,320 sq. mi. 301,269 sq. km	Metals	Machinery	Lira, Euro	Republic
LATVIA Ríga	Latvian, Russian	2,400,000 95 per sq. mi. 37 per sq. km	24,942 sq. mi. 64,599 sq. km	Wood	Fuels	Lat	Republic
LIECHTENSTEIN Vaduz	German	30,000 534 per sq. mi. 206 per sq. km	62 sq. mi. 160 sq. km	Machinery	Machinery	Swiss Franc	Constitutional Monarchy

* COUNTRIES AND FLAGS NOT DRAWN TO SCALE

FOR AN ONLINE UPDATE OF THIS INFORMATION, VISIT GEOGRAPHY.GLENCOE.COM AND CLICK ON "TEXTBOOK UPDATES."

4 Europe

COUNTRY PROFILES

COUNTRY * AND CAPITAL	FLAG AND LANGUAGE	POPULATION AND DENSITY	LANDMASS	MAJOR EXPORT	MAJOR IMPORT	CURRENCY	GOVERNMENT
LITHUANIA Vilnius	Lithuanian, Polish, Russian	3,700,000 147 per sq.mi. 57 per sq.km	25,174 sq.mi. 65,200 sq.km	Foods and Livestock	Minerals	Litas	Republic
LUXEMBOURG Luxembourg	Luxembourgian, German, French	400,000 446 per sq.mi. 172 per sq.km	999 sq.mi. 2,587 sq.km	Steel Products	Minerals	Luxembourg Franc, Euro	Constitutional Monarchy
MACEDONIA Skopje	Macedonian, Albanian	2,000,000 205 per sq.mi. 79 per sq.km	9,927 sq.mi. 25,711 sq.km	Manufactured Goods	Fuels	Denar	Republic
MALTA Valletta	Maltese, English	400,000 3,157 per sq.mi. 1,219 per sq.km	124 sq.mi. 321 sq.km	Machinery	Foods	Maltese Lira	Republic
MOLDOVA Chişinău	Moldovan, Russian	4,300,000 328 per sq.mi. 127 per sq.km	13,012 sq.mi. 33,701 sq.km	Foods	Petroleum	Moldovan Leu	Republic
MONACO Monaco	French	30,000 45,333 per sq.mi. 17,503 per sq.km	1 sq.mi. 2.6 sq.km	N/A	N/A	French Franc	Constitutional Monarchy
NETHERLANDS Amsterdam	Dutch	16,000,000 1,018 per sq.mi. 393 per sq.km	15,768 sq.mi. 40,839 sq.km	Manufactured Goods	Raw Materials	Guilder, Euro	Constitutional Monarchy
NORWAY Oslo	Norwegian	4,500,000 36 per sq.mi. 14 per sq.km	125,050 sq.mi. 323,880 sq.km	Petroleum	Machinery	Krone	Constitutional Monarchy
POLAND Warsaw	Polish	38,600,000 310 per sq.mi. 120 per sq.km	124,807 sq.mi. 323,250 sq.km	Manufactured Goods	Machinery	Zloty	Republic
PORTUGAL Lisbon	Portuguese	10,000,000 282 per sq.mi. 109 per sq.km	35,514 sq.mi. 91,981 sq.km	Clothing	Machinery	Escudo, Euro	Republic

* COUNTRIES AND FLAGS NOT DRAWN TO SCALE

COUNTRY * AND CAPITAL	FLAG AND LANGUAGE	POPULATION AND DENSITY	LANDMASS	MAJOR EXPORT	MAJOR IMPORT	CURRENCY	GOVERNMENT
ROMANIA Bucharest	Romanian, Hungarian	22,400,000 243 per sq. mi. 94 per sq. km	92,042 sq. mi. 238,389 sq. km	Textiles	Fuels	Leu	Republic
SAN MARINO San Marino	Italian	30,000 1,166 per sq. mi. 450 per sq. km	23 sq. mi. 60 sq. km	Building Stone	Manufactured Goods	Italian Lira	Republic
SLOVAKIA Bratislava	Slovak, Hungarian	5,400,000 286 per sq. mi. 110 per sq. km	18,923 sq. mi. 49,011 sq. km	Transport Equipment	Machinery	Koruna	Republic
SLOVENIA Ljubljana	Slovene, Serbo-Croatian	2,000,000 256 per sq. mi. 99 per sq. km	7,819 sq. mi. 20,251 sq. km	Transport Equipment	Machinery	Slovenian Tolar	Republic
SPAIN Madrid	Spanish, Catalan, Galician, Basque	39,800,000 204 per sq. mi. 79 per sq. km	195,363 sq. mi. 505,990 sq. km	Cars and Trucks	Machinery	Peseta, Euro	Constitutional Monarchy
SWEDEN Stockholm	Swedish	8,900,000 51 per sq. mi. 20 per sq. km	173,730 sq. mi. 449,961 sq. km	Paper Products	Crude Oil	Krona	Constitutional Monarchy
SWITZERLAND Bern	German, French, Italian	7,200,000 453 per sq. mi. 175 per sq. km	15,942 sq. mi. 41,290 sq. km	Precision Instruments	Machinery	Swiss Franc	Federal Republic
UKRAINE Kiev	Ukrainian, Russian	49,100,000 211 per sq. mi. 81 per sq. km	233,089 sq. mi. 603,701 sq. km	Metals	Machinery	Hryvnya	Republic
UNITED KINGDOM London	English, Welsh, Scottish Gaelic	60,000,000 635 per sq. mi. 245 per sq. km	94,548 sq. mi. 244,879 sq. km	Manufactured Goods	Foods	Pound Sterling	Constitutional Monarchy
VATICAN CITY	Italian, Latin	1,000	0.2 sq. mi. 0.4 sq. km	N/A	N/A	Lira	Sovereign State Under the Pope
YUGOSLAVIA Belgrade	Serbo-Croatian, Albanian	10,700,000 270 per sq. mi. 104 per sq. km	39,448 sq. mi. 102,170 sq. km	Manufactured Goods	Machinery	Yugoslav New Dinar	Federal Republic

* COUNTRIES AND FLAGS NOT DRAWN TO SCALE

FOR AN ONLINE UPDATE OF THIS INFORMATION, VISIT GEOGRAPHY.GLENCOE.COM AND CLICK ON "TEXTBOOK UPDATES."

4

GLOBAL CONNECTION

EUROPE AND THE UNITED STATES

ARCHITECTURE

Wander through any city in the United States, and you'll see European influences—not just in foods and fashions, but in brick, wood, and stone. From churches to country homes, many American buildings reflect our connection to European cultures.

The Capitol is a national landmark in the heart of Washington, D.C. Its great dome, or large arched roof, dominates the structure. Roman architects favored arching shapes, and domes are their legacy.

▲ Queen Anne house in Washington State

In 1792 President George Washington asked architects to submit designs for a "federal Capitol" to house the U.S. Congress. William Thornton, an amateur draftsman, won the competition with a neo-Roman design. Thornton modeled the Capitol dome after the one that crowns the Pantheon, an ancient Roman temple built in the A.D. 100s.

The Gothic style of architecture originated in France in the 1100s and became the style of choice for cathedrals. Gothic cathedrals are huge and soaring, with

▲ Interior of Pantheon, in Rome

pointed arches, large stained glass windows, and towers and spires that seem to point toward heaven. Such cathedrals were built across western Europe during the Middle Ages.

Hundreds of years later, American architect James Renwick designed St. Patrick's Cathedral, which was built in New York City starting in 1858. It is considered one of the best examples of Gothic architecture in the United States. True to the Gothic style, the cathedral has pointed arches, stained glass windows, and a pair of enormous towers.

The Queen Anne style developed in England in the 1860s and 1870s. Queen Anne buildings tend to be asymmetrical and quirky, with prominent chimneys, steep roofs, dormer windows, and corner turrets jutting out. Ornamental details such as fancy brickwork and contrasting trim help give Queen Anne buildings their characteristic look.

Queen Anne houses became popular throughout the United States in the late 1800s. You probably wouldn't have to travel far to see a Queen Anne house. There might even be one in your neighborhood.

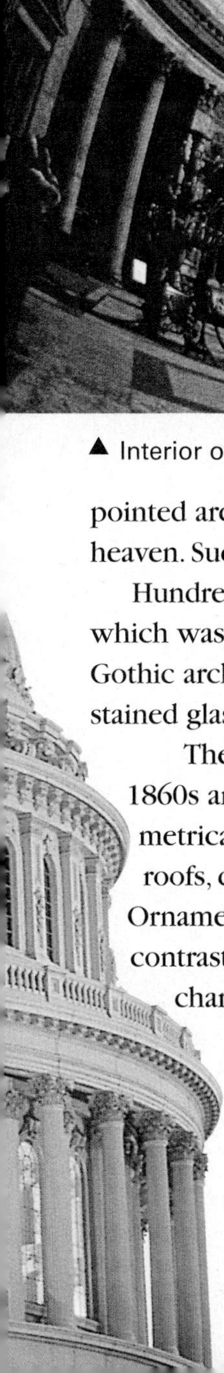

▲ St. Patrick's Cathedral

◀ Dome of U.S. Capitol

CHAPTER 11

The Physical Geography of Europe

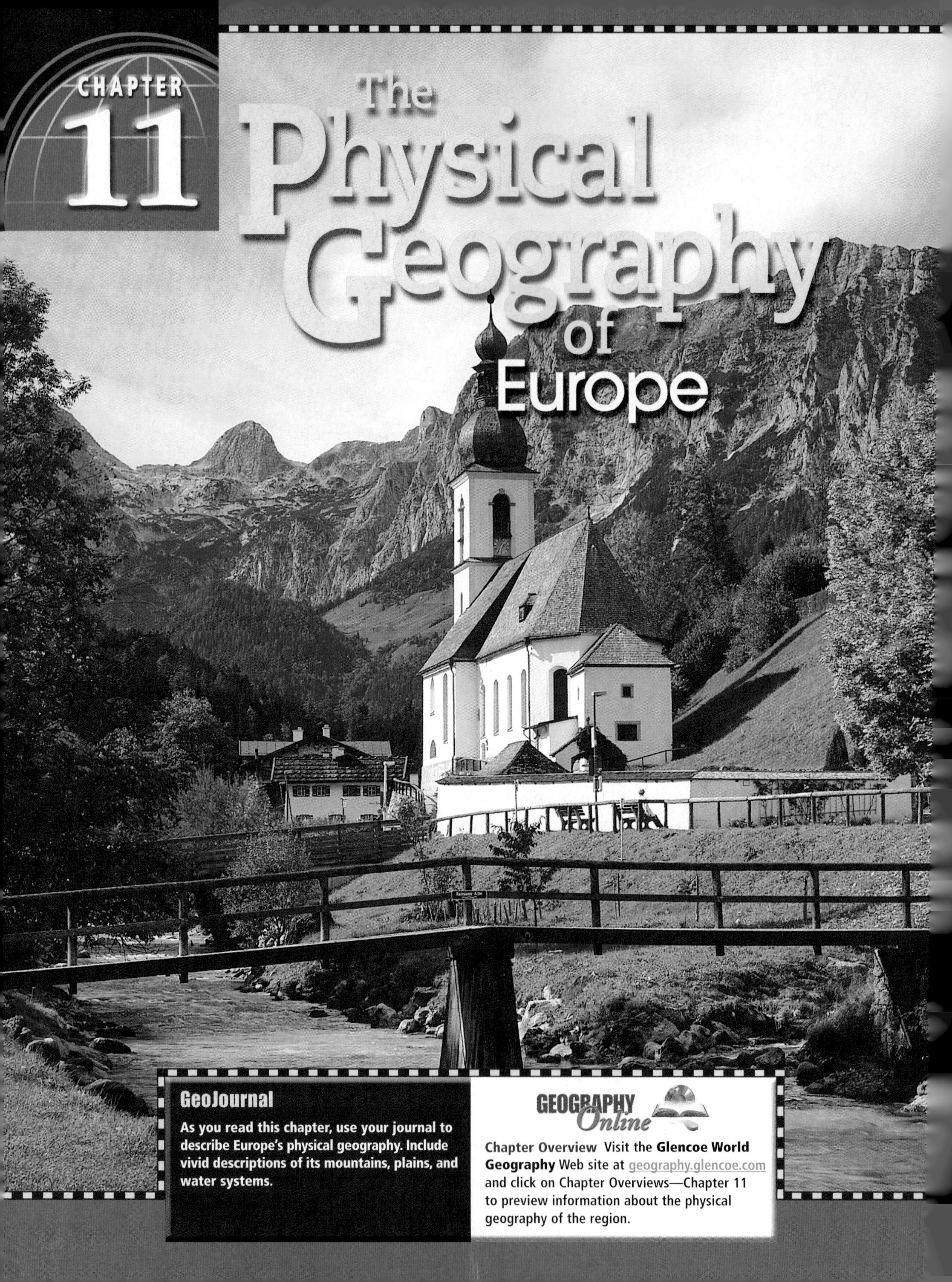

GeoJournal

As you read this chapter, use your journal to describe Europe's physical geography. Include vivid descriptions of its mountains, plains, and water systems.

GEOGRAPHY Online

Chapter Overview Visit the **Glencoe World Geography** Web site at geography.glencoe.com and click on Chapter Overviews—Chapter 11 to preview information about the physical geography of the region.

SECTION 1

The Land

Guide to Reading

Consider What You Know

References to Europe are often in the news. What physical features come to mind when you think of Europe? What do you know about them?

Read to Find Out

- Why is Europe sometimes called a "peninsula of peninsulas"?
- What are some of the numerous islands surrounding the continent of Europe?
- Why are rivers vital to Europe's economy?
- What are some of Europe's most important natural resources?

Terms to Know

- dike
- polder
- glaciation
- fjord
- loess

Places to Locate

- North Sea
- Iberian Peninsula
- Balkan Peninsula
- Alps
- Rhine River
- Po River
- North European Plain

◀ *Church in the Alps, Bavaria, Germany*

NATIONAL GEOGRAPHIC

A Geographic View

Fire in Iceland

. . . [O]ne of the largest volcanic eruptions to hit Iceland this century rumbled to life beneath the country's biggest ice cap. For two weeks ash and steam billowed skyward as elemental forces clashed in thermal battle. . . . [A]sh-laden runoff rushed from the eruption site, carving an ice canyon 500 feet deep and more than two miles long.

—*Glenn Oeland, "Iceland's Trial by Fire,"* National Geographic, *May 1997*

Ice canyon, Iceland

Though few natural occurrences are as dramatic as Iceland's volcanic eruptions, physical forces continue to shape the landscape of Europe. In this section you will learn about the variety of Europe's landforms, water systems, and natural resources.

Seas, Peninsulas, and Islands

Unlike the world's other continents, Europe and Asia share a common landmass called Eurasia. Yet Europe, the second smallest of the continents after Australia, is a distinct region. Jutting westward from Asia, Europe has an unusually long, irregular coastline that touches a number of bodies of water, including the Atlantic Ocean and the Baltic, North, Mediterranean, and Black Seas.

History

Struggle With the Sea

Most of Europe lies within 300 miles (483 km) of a seacoast. This closeness to the sea has shaped the lifestyles of its peoples. In the

Netherlands, water can be friend or foe. About 25 percent of the Netherlands lies below sea level. Coastal dunes have not always been helpful in keeping out North Sea waters, so the Dutch since the Middle Ages have built **dikes**, large banks of earth and stone, to hold back water. With the dikes for protection, they have reclaimed new land from the sea. These reclaimed lands, called **polders**, once were drained and kept dry by the use of windmills. Today, other power sources run pumps to remove seawater. Polders provide hundreds of thousands of acres for farming and settlement. Still, from time to time, stormy seas breach the dikes, creating devastating floods.

The Northern Peninsulas

Europe is a large peninsula made up of smaller peninsulas. In the far north of Europe lies the scenic Scandinavian Peninsula. During the last Ice Age, in a process known as **glaciation**, glaciers formed and spread over the peninsula. They carved out long, narrow, steep-sided inlets called **fjords** (fee•AWRDZ) on the Atlantic coastline. The map on page 273 shows Norway's jagged coastal strip, where many fjords provide fine harbors.

Much of Norway and northern Sweden is mountainous, but in southern Sweden, lowlands slope gently to the Baltic Sea. In both countries, and in Finland, Ice Age glaciers left behind thousands of sparkling lakes.

The peninsula of Jutland forms the mainland part of Denmark and extends into the **North Sea** toward Norway and Sweden. Glaciers deposited sand and gravel on Jutland's flat western side and carved fjords into the slightly higher coastline on the east. Flat plains or low hills make up most of Jutland's interior.

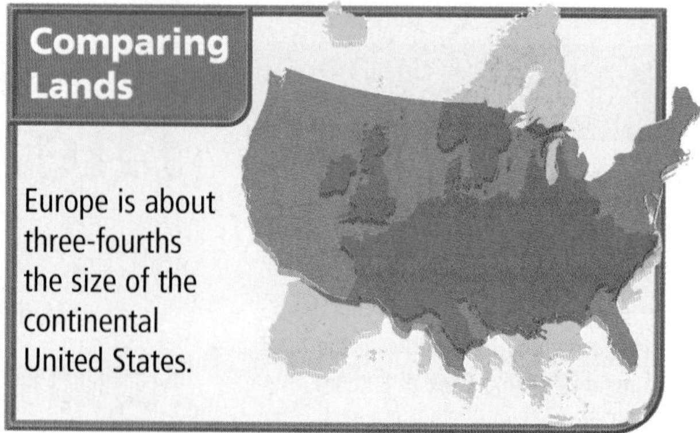

Comparing Lands

Europe is about three-fourths the size of the continental United States.

Geography Skills for Life

Dutch Windmills By the 1800s the Dutch had built about 9,000 windmills to pump seawater from low-lying areas.

Place Why do you think the Dutch shifted to other power sources to drain flooded areas?

The Southern Peninsulas

The **Iberian Peninsula** extends off the southwestern edge of Europe. Home to Spain and Portugal, the peninsula separates the Atlantic Ocean from the Mediterranean Sea. Only 20 miles (32 km) of water at the Strait of Gibraltar, however, separates the peninsula's southern tip from Africa.

Most of the Iberian Peninsula is a semiarid plateau, rising above slender coastal plains. To the north, the Pyrenees (PIHR•uh•NEEZ) Mountains cut off the peninsula from the rest of Europe. Because of this rugged barrier, the people of the Iberian Peninsula until modern times were relatively isolated from the rest of Europe and were oriented toward the sea.

The Apennine (A•puh•NYN) Peninsula, where Italy is located, extends like a giant boot into the Mediterranean Sea. Its long coastline varies from high, rocky cliffs to long, sandy beaches. Forming the peninsula's spine are the Apennines, a geologically young mountain chain that includes an active

Europe: Physical-Political

Elevations

Feet: 10,000 / 5,000 / 2,000 / 1,000 / 0

Meters: 3,000 / 1,500 / 600 / 300 / 0

National boundary

▲ Mountain peak

ICELAND
Faroe Is.
Shetland Is.
Orkney Is.
Ben Nevis 4,406 ft. (1,343 m)
ATLANTIC OCEAN
UNITED KINGDOM
IRELAND
Great Britain
Thames R.
North Sea
Norwegian Sea
ARCTIC CIRCLE
MERIDIAN OF GREENWICH (LONDON)
Lapland
SCANDINAVIA
NORWAY
SWEDEN
FINLAND
ESTONIA
LATVIA
LITHUANIA
Baltic Sea
DENMARK
Jutland
NETH.
BELGIUM
LUX.
GERMANY
NORTH EUROPEAN PLAIN
Elbe R.
POLAND
Vistula R.
BELARUS
UKRAINE
Dnieper R.
Dniester R.
MOLDOVA
CZECH REP.
SLOVAKIA
Carpathian Mountains
HUNGARY
Great Hungarian Plain
ROMANIA
Danube R.
Black Sea
AUSTRIA
LIECH.
SWITZ.
L. Geneva
ALPS
Mt. Blanc 15,771 ft. (4,807 m)
FRANCE
Seine R.
Loire R.
Rhine R.
Rhone R.
Massif Central
Bay of Biscay
PORTUGAL
IBERIAN PENINSULA
Ebro R.
Meseta
SPAIN
ANDORRA
Pyrenees
MONACO
GIBRALTAR
Strait of Gibraltar
Balearic Islands
Corsica
Sardinia
SAN MARINO
Po R.
ITALY
Apennines
Vesuvius 4,190 ft. (1,277 m)
Adriatic Sea
SLOV.
CROATIA
BOSN. & HERZG.
YUG.
Balkan Mts.
Balkan Peninsula
BULGARIA
MACED.
ALBANIA
GREECE
TURKEY
Bosporus
Dardanelles
Aegean Sea
Ionian Sea
Sicily
MALTA
Crete
Rhodes
CYPRUS
Mediterranean Sea
20°W
0°
20°E
40°E
70°N
60°N
50°N
40°N
N

0 mi. 500
0 km 500
Lambert Azimuthal Equal-Area projection

Geography Skills for Life

1. **Interpreting Maps** What body of water separates the United Kingdom from Denmark?
2. **Applying Geography Skills** Which country has areas of land below sea level? How might people live in these areas?

Find NGS online map resources @ www.nationalgeographic.com/maps

volcano—Mount Vesuvius, near the city of Naples. Plains cover only about one-third of the Apennine Peninsula, the largest being the fertile plain of Lombardy along the Po River in the north.

In southeastern Europe lies the **Balkan Peninsula**. Bounded by the Adriatic and Ionian Seas on the west and the Aegean and Black Seas on the east, the Balkan Peninsula holds a tangle of mountain ranges and valleys that stretch southward from the Danube River. Because of the region's craggy landscape, overland travel is difficult. Historically people moved along rivers and seas in this mountainous region.

Europe's Islands

In addition to peninsulas, Europe includes many islands. Iceland is located south of the Arctic Circle in the North Atlantic Ocean. Lying astride the Mid-Atlantic Ridge, Iceland has volcanoes, hot springs, and geysers. Because of Iceland's far northern location, glaciers are found next to the volcanoes and hot springs. Most of the homes and industries in the area of the capital, Reykjavík (RAY•kyah•VEEK), pipe in water from hot springs for heat. Grassy lowlands stretch along Iceland's coast, but the land rises sharply to form a large inland plateau.

The British Isles lie northwest of the European mainland. They consist of two large islands, Great Britain and Ireland, and thousands of smaller islands. Mountain ranges, plateaus, and deep valleys make up most of northern and western Great Britain, and low hills and gently rolling plains dominate in the south. Ireland, often called the Emerald Isle, is a lush green land of cool temperatures and abundant rainfall. In many places the rugged coastline of the British Isles features rocky cliffs that drop to deep bays. One visitor to the British coast writes:

> *"We hiked past . . . plenty of farms, and mile after mile of rocky cliffs, their long faces carved raw and craggy by the ocean's dull knife. All day we stayed close to Cornwall's serrated edge, weaving in and out like a conga line."*
>
> Alan Mairson, "Saving Britain's Shore," *National Geographic,* October 1995

Islands also lie south of the European mainland, in the Mediterranean Sea. Rugged mountains form the larger islands of Sicily, Sardinia, Corsica, Crete, and Cyprus. Volcanic and earthquake activity are characteristic of the region. Mount Etna, Europe's highest active volcano, rises over Sicily. Smaller island groups in the Mediterranean area are Spain's Balearic Islands, Malta's 5 islands, and Greece's nearly 2,000 islands in the Aegean Sea. The scenic, rugged landscape and the sunny climate of Europe's Mediterranean islands draw tourists from around the world.

Mountains and Plains

Europe's mainland, in essence, consists of plains interrupted by mountains running through its interior and along its northern and southern edges. The map on page 273 shows the names and locations of some of these landforms.

Mountain Regions

Europe's northwestern mountains have some of the earth's most ancient rock formations. Rounded by eons of erosion and glaciation, these ranges feature relatively low peaks, such as Ben Nevis, the highest mountain in the British Isles at 4,406 feet (1,343 m). Extending from the Iberian Peninsula to eastern Europe, the central uplands consist of low, rounded mountains and high plateaus with scattered forests. This region includes the Meseta, Spain's central plateau, and the Massif Central, France's central highlands.

By contrast, southern Europe's geologically younger mountains are high and jagged. As the earth's crust lifted and folded, the Pyrenees Mountains were thrust upward to more than 11,000 feet (3,354 m). Created by glaciation and folding, the mountain system known as the **Alps** forms a crescent from southern France to the Balkan Peninsula. The highest peak in the Alps, Mont Blanc, stands at

Student Web Activity Visit the **Glencoe World Geography** Web site at geography.glencoe.com and click on Student Web Activities—Chapter 11 for an activity about the physical geography of the Netherlands.

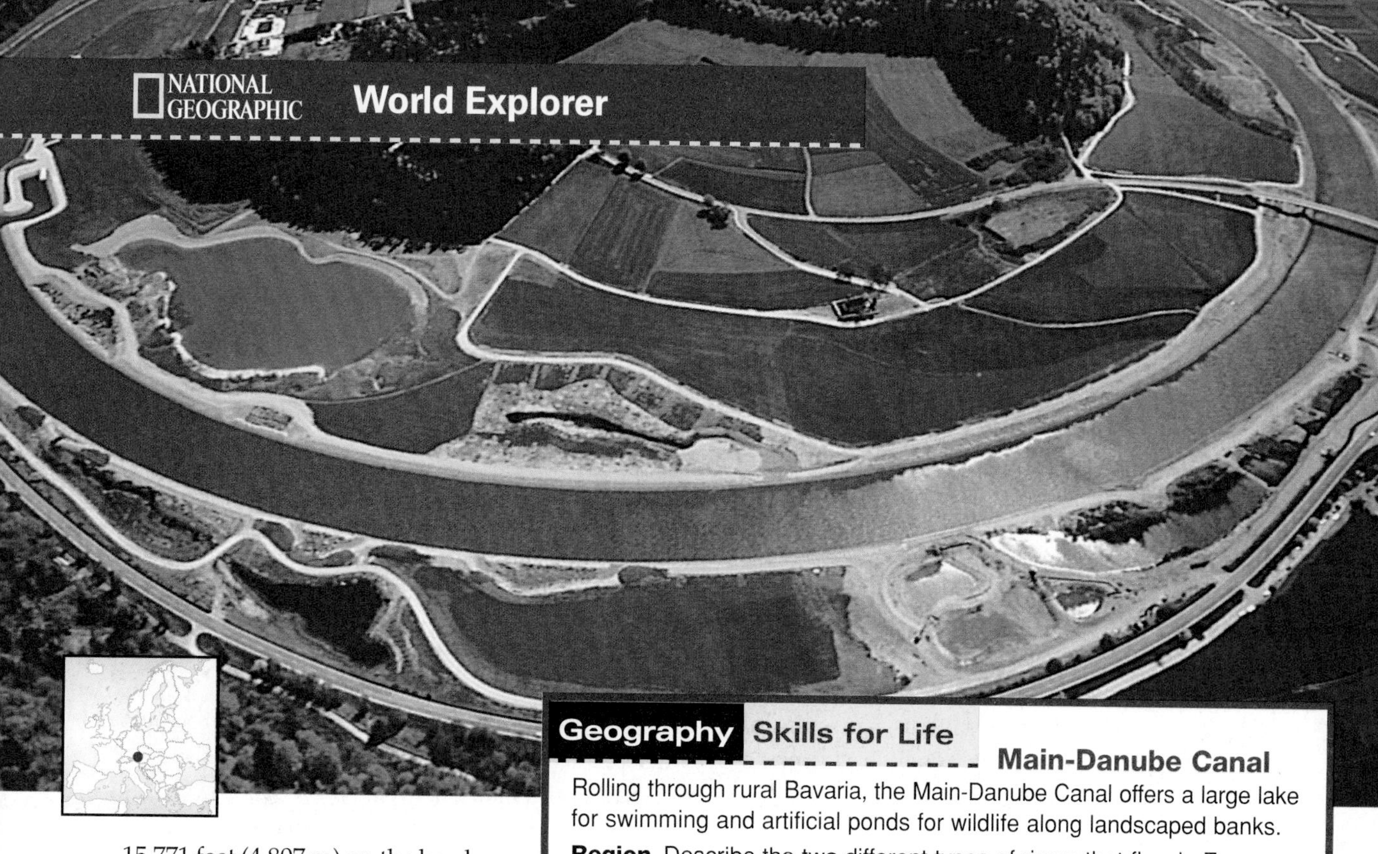

Geography Skills for Life

Main-Danube Canal

Rolling through rural Bavaria, the Main-Danube Canal offers a large lake for swimming and artificial ponds for wildlife along landscaped banks.

Region Describe the two different types of rivers that flow in Europe.

15,771 feet (4,807 m) on the border of France and Italy. Some of Europe's major rivers, such as the **Rhine** and the **Po**, originate in the Alps. The Alps also form a barrier that separates the warm, dry climate of the Mediterranean region from the cooler climates of the north. Another towering mountain chain, the Carpathians, runs through eastern Europe from Slovakia to Romania.

Plains Regions

Europe's broad plains curve around the highlands. Scoured by Ice Age glaciers, the **North European Plain** stretches from southeastern England and western France eastward to Poland, Ukraine, and Russia. The plain's fertile soil and wealth of rivers originally drew farmers to the area, and the plain is still a major agricultural region. The southern edge is especially fertile because deposits of loess, a fine, rich, wind-borne soil, cover it.

Deposits of coal, iron ore, and other minerals found on the North European Plain led to western Europe's industrial development during the 1800s. Today many of Europe's largest cities, such as Paris and Berlin, are located on the plain.

Another fertile plains area, the Great Hungarian Plain, extends from Hungary to Croatia, Serbia, and Romania. Farmers cultivate grains, fruit, and vegetables and raise livestock in the lowlands along the Danube River.

Water Systems

Many of Europe's water systems flow from inland mountain and highlands areas to the coasts. By connecting navigable rivers with canals, Europeans have greatly enhanced their natural waterways as transportation links. Europe's rivers and canals also provide water to irrigate farmland and to produce electricity.

Europe's rivers have differing characteristics. The rivers in Scandinavia are short and do not provide easy connections between cities. In the Iberian Peninsula, main rivers generally are too narrow and shallow for large ships. England's Thames (TEHMZ) River, on the other hand, allows ocean-going ships to reach the port of London.

In the heartland of Europe, however, relatively long rivers provide links between inland areas as well as to the sea. The Rhine is the most important river in western Europe. It flows from the Swiss

Alps through France and Germany and into the Netherlands, connecting many industrial cities to the busy port of Rotterdam on the North Sea.

The Danube, which flows from Germany's Black Forest to the Black Sea, is eastern Europe's major waterway. Each year ships and barges carry millions of tons of cargo on the Danube. In 1992 the Main (MYN) River, a tributary of the Rhine, became connected to the Danube when the Main-Danube Canal was completed, thereby linking the North Sea with the Black Sea.

Other major European rivers include the Seine, Rhône, and Loire in France; the Elbe and Weser in Germany; the Vistula in Poland; the Po in Italy; and the Dnieper in Ukraine.

▲ *An Irish farmer digging peat*

Natural Resources

Europe has a long history of utilizing its natural resources, including energy sources, agricultural areas, water, and especially minerals. Europe's abundant supply of coal and iron ore fueled the development of modern industry.

Major reserves of coal lie in the United Kingdom, Germany, Ukraine, and Poland as well as other European countries. Although coal is still an important fuel source, many coalfields in western Europe are depleted or are too expensive to mine. Large deposits of iron ore lie in northern Sweden, northeastern France, and southeastern Ukraine. Europe's other mineral resources include bauxite, zinc, and manganese.

In places where other fuels are scarce, Europeans burn peat, a kind of vegetable matter found in swamps and usually composed of mosses. Peat is dug up, chopped into blocks, and dried so it can be burned. Europeans, however, largely rely on coal, oil, gas, and nuclear and hydroelectric power. Vast oil and natural gas deposits under the North Sea contribute greatly to Europe's energy needs. France, which lacks large oil or gas reserves, has invested heavily in nuclear power.

SECTION 1 ASSESSMENT

Checking for Understanding

1. **Define** dike, polder, glaciation, fjord, loess.
2. **Main Ideas** Re-create the table below on a sheet of paper, and fill in examples of the physical features and natural resources of Germany, Norway, Ukraine, Italy, and France.

Country	Physical Features	Natural Resources

Critical Thinking

3. **Comparing and Contrasting** How does the landscape of the Jutland peninsula differ from that of the Balkan Peninsula?
4. **Making Generalizations** Europe's Mediterranean islands are popular vacation destinations. What physical features make these islands attractive to tourists?
5. **Drawing Conclusions** How does Europe's network of rivers and canals contribute to industrial development in the region?

Analyzing Maps

6. **Location** Study the physical-political map of Europe on page 273. What part of Europe has the lowest elevation? The highest?

Applying Geography

7. **Conflict Over Resources** Use the economic activity map on page 263 to identify three areas in which natural resources cross international boundaries. Describe the areas in which conflict could arise because of the management of these resources.

Climate and Vegetation

Guide to Reading

Consider What You Know

Much of Europe borders oceans and seas. What kinds of climates would you expect in Europe?

Read to Find Out

- What are the climate regions in Europe?
- What physical features influence Europe's climates?
- Why are most of Europe's original forests gone?

Terms to Know

- timberline
- foehn
- avalanche
- mistral
- sirocco
- chaparral
- permafrost

Places to Locate

- Gulf Stream
- North Atlantic Drift

NATIONAL GEOGRAPHIC

A Geographic View

Power of the Wind

Sailboats on Als Sund

I stood on the shore of Als Sund, a saltwater sound . . . with [Flemming] Rieck, a maritime archaeologist at the National Museum of Denmark. . . . After a week of rain the spring sun radiated intense light but little heat. Chins tucked in our windbreakers, Rieck and I stared at sailboats bobbing on the whitecaps.

"At some point . . . ," he said, "Scandinavians began using sails." He spread his arms into the stiff breeze. "No one can say why it took so long for them to use the power of all this wind."

—*Michael Klesius, "Mystery Ships From a Danish Bog,"* National Geographic, *May 2000*

Wind is only one of the factors affecting Europe's climates. Latitude, mountain barriers, ocean currents, and the distance from large bodies of water all help determine Europe's varied climates. In this section you will read about Europe's climate regions—from the sunny, dry Mediterranean climate to the frozen subarctic zone. You will also study the patterns of vegetation growth found in each region of Europe.

Water and Land

The climates and vegetation of Europe vary from the cold, barren tundra and subarctic stretches of Iceland, Norway, Sweden, and Finland to the warm, shrub-covered Mediterranean coasts of Italy, Spain, and Greece. What factors account for such variety in a relatively small area?

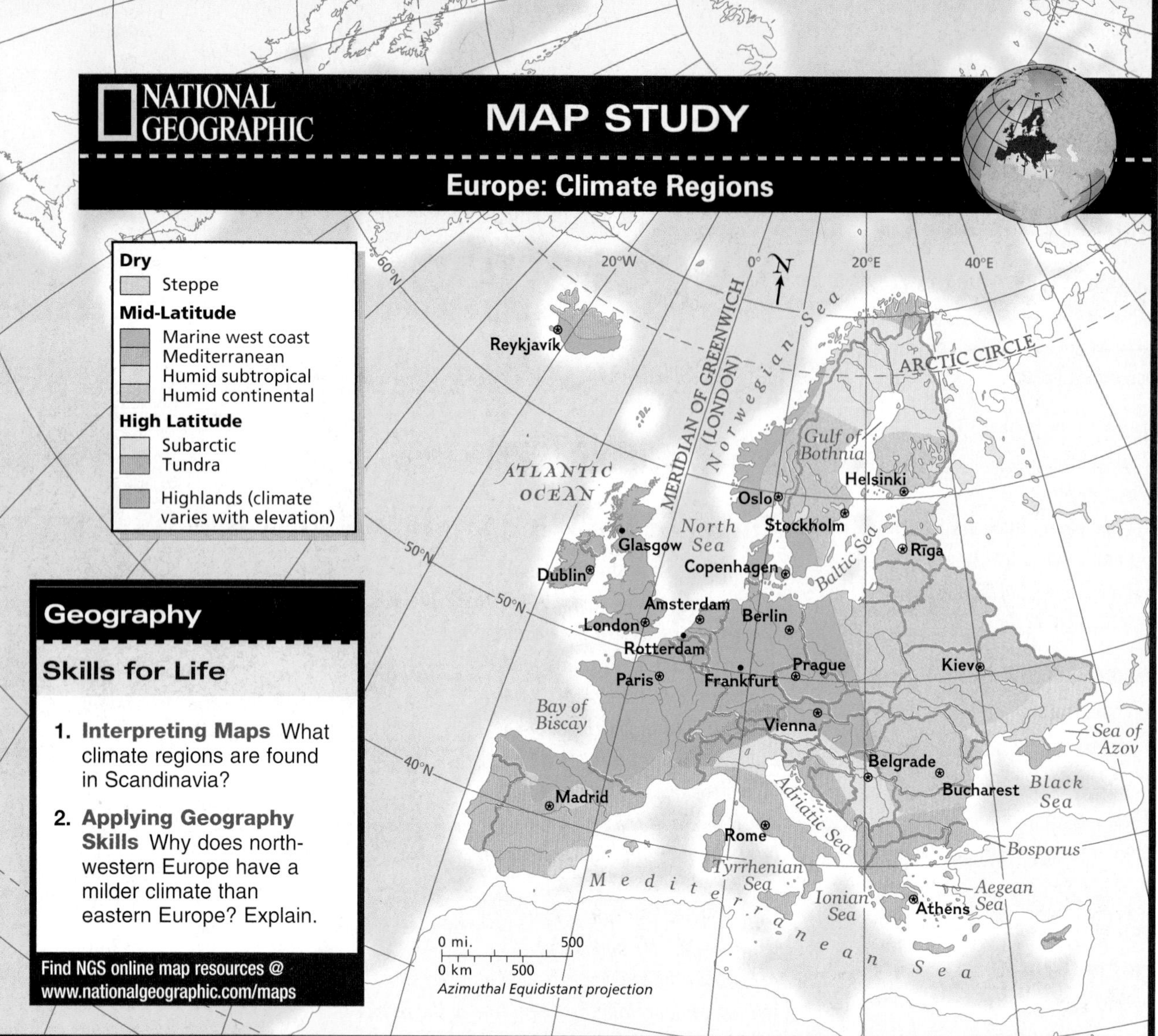

Europe's northern latitude and its relationship to the sea influence its climates and vegetation. Western and southern parts of Europe, which lie near or along large bodies of water, benefit from warm maritime winds. These areas have a generally mild climate compared with other places in the world at the same latitude. Frankfurt, Germany, as well as Paris, France, and Boston, Massachusetts, are about the same distance from the Arctic Circle, yet January temperatures in Paris are milder than those in Boston. By contrast, parts of eastern and northern Europe have a colder climate than most of western and southern Europe because of their distance from the warming effects of the Atlantic Ocean.

As in other areas of the world, location influences vegetation patterns in Europe. Natural vegetation in the region varies from forests and grasslands to tundra plants and small shrubs. Compare the natural vegetation map on page 279 with the climate map above. Notice that the types of vegetation found in Europe are closely linked to the climate regions.

Western Europe

As the climate map on this page shows, much of western Europe has a marine west coast climate—mild winters, cool summers, and abundant rainfall. The Atlantic Ocean's **Gulf Stream** and its northern extension, the **North Atlantic Drift**, bring warm waters to this part of Europe from the Gulf of Mexico and regions near the Equator (see map on page 61). Prevailing westerly winds blowing over these currents carry warm, moist air across the surface of the European landmass.

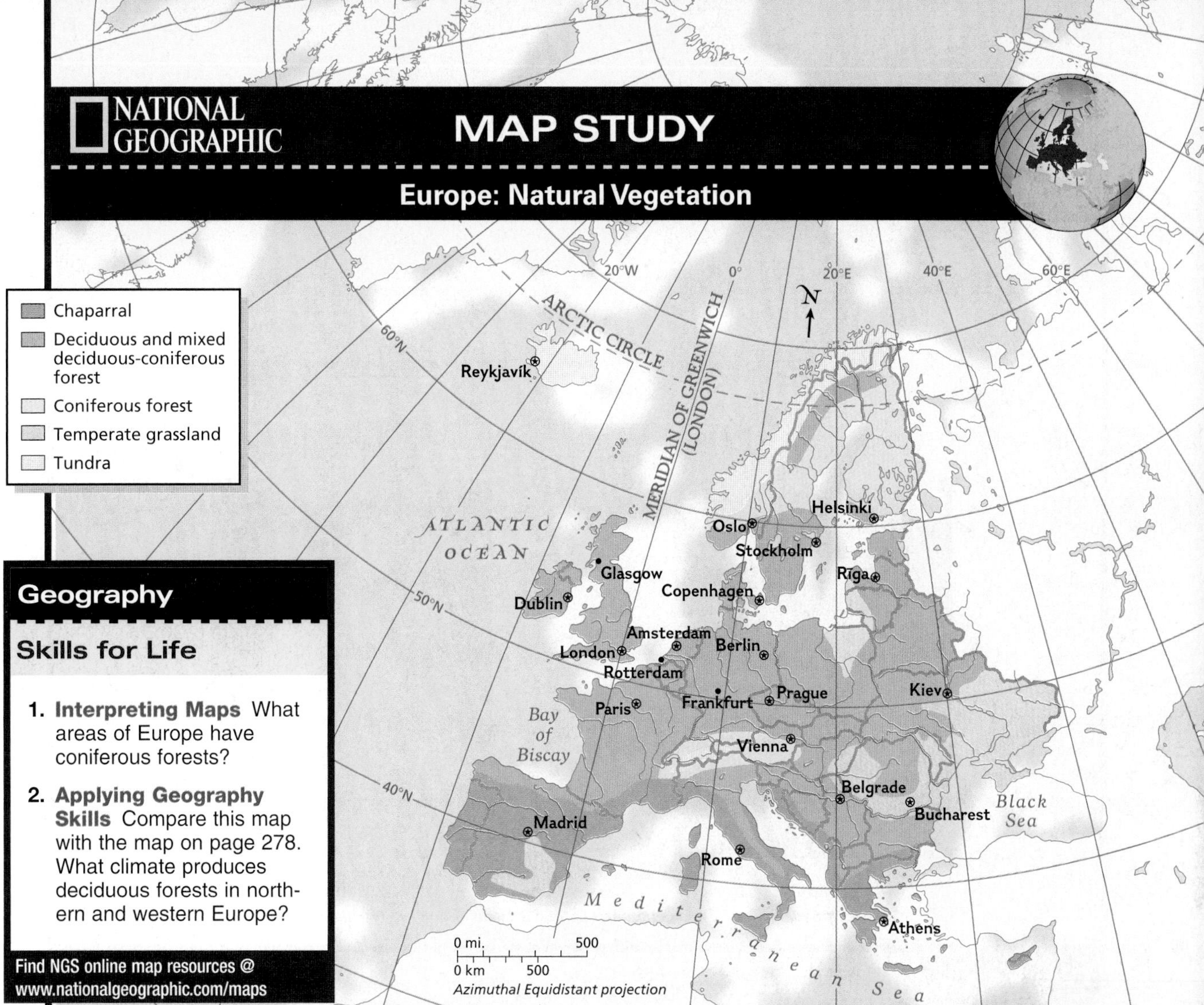

Geography Skills for Life

1. **Interpreting Maps** What areas of Europe have coniferous forests?
2. **Applying Geography Skills** Compare this map with the map on page 278. What climate produces deciduous forests in northern and western Europe?

Find NGS online map resources @ www.nationalgeographic.com/maps

Trees and Highlands

Western Europe's natural vegetation includes varieties of deciduous (dih•SIH•juh•wuhs) and coniferous (koh•NIH•fuh•ruhs) trees. Deciduous trees, those that lose their leaves, such as ash, maple, and oak, thrive in the area's marine west coast climate. Coniferous trees, cone-bearing fir, pine, and spruce, are found in cooler Alpine mountain areas up to the **timberline**, the elevation above which trees cannot grow.

The Alps have a highlands climate with generally colder temperatures and more precipitation than nearby lowland areas. Sudden changes can occur, however, when dry winds called **foehns** (FUHNZ) blow down from the mountains into valleys and plains. Foehns can trigger **avalanches**, destructive masses of ice, snow, and rock sliding down mountainsides. Avalanches threaten skiers and hikers, and often carry away everything in their paths. They represent a serious natural hazard in the Alps.

History

Ireland's Forests

Much of Europe was orginally covered by forest, but over the centuries human settlement and clearing of the land have transformed the vegetation. For example, prior to the 1600s, much of the midlands region of Ireland was covered with forests of broadleaved trees. However, pressure from agriculture and the large-scale harvest of native lumber for firewood depleted the country's forests. By 1922, when Ireland gained independence, only 1 percent of the

NATIONAL GEOGRAPHIC

GRAPH STUDY

Comparing Climate Regions: France and the United States

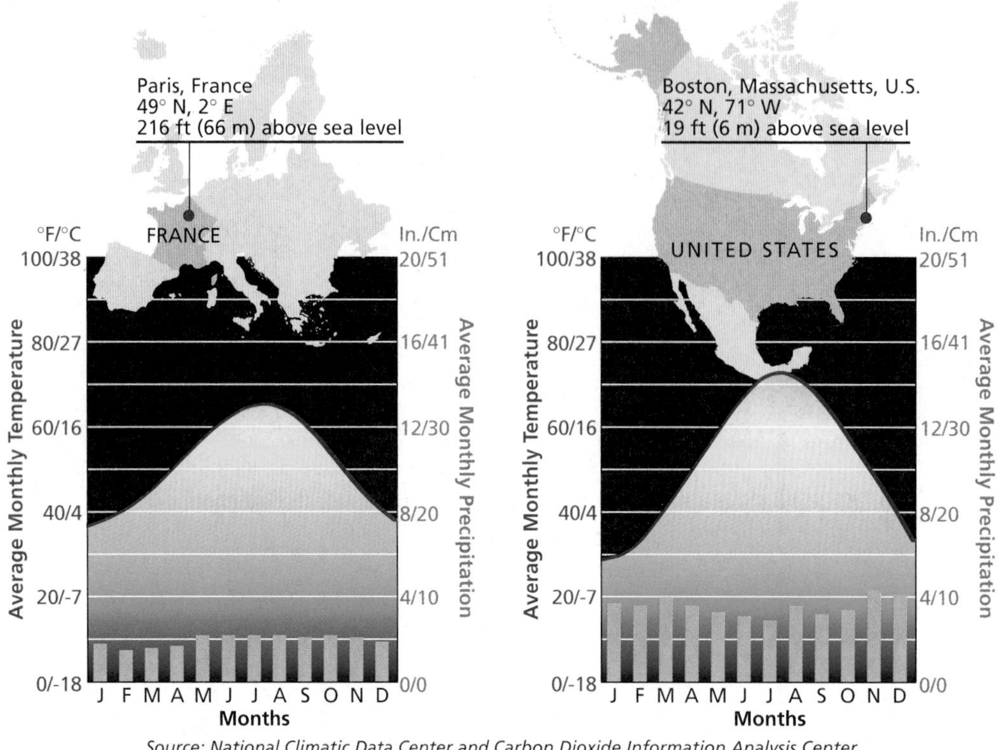

Source: National Climatic Data Center and Carbon Dioxide Information Analysis Center

Geography Skills for Life

1. **Interpreting Graphs** What is the average March temperature and precipitation in each city?
2. **Applying Geography Skills** Why does Paris have a cooler summer than Boston? (Refer to the map of world climate regions on page 66.)

country was woodland. Searching for old-growth forests can be challenging, as one traveler notes:

> *"'Of course Tomies Wood is all second growth.' . . . The real thing, Padraig told me, was far more remote, far from the trails, in the heights of MacGillycuddy's Reeks, where even now few people ventured."*
>
> Rebecca Solnit, "The Lost Woods of Killarney," *Sierra*, March/April 1997

State-sponsored reforestation efforts since World War II have increased Ireland's woodland areas.

Southern Europe

Most of southern Europe has a Mediterranean climate—warm, dry summers and mild, rainy winters. Several other climates, however, are found in small areas of the region. For example, a humid subtropical climate stretches from northern Italy to the central part of the Balkan Peninsula. In addition, parts of Spain's Meseta have a drier steppe climate.

The Alps block moist Atlantic winds, so less precipitation falls in southern Europe than in northwestern Europe. Local winds in the region sometimes cause changes in the normal weather pattern. The **mistral**, a strong north wind from the Alps, sometimes sends gusts of bitterly cold air into southern France. By contrast, **siroccos** (suh•RAH•kohs), high, dry winds from North Africa, may bring high temperatures to the region. The hot, dry summers in much of southern Europe support the growth of **chaparral**, or shrubs and small trees, such as the cork oak tree and the olive tree.

Eastern and Northern Europe

Eastern and certain northern areas of Europe have a generally humid continental climate—cold, snowy winters and hot summers. Warm Atlantic currents have less influence on climate in these areas farther from the Atlantic Ocean. As a result, summer and winter temperatures vary more widely in eastern and northern Europe than in the rest of Europe.

In eastern Europe the vegetation is generally a mix of deciduous and coniferous forests. Coniferous trees, which are able to survive long, cold winters, are found in parts of Scandinavia and the region around the Baltic Sea. Grasslands cover parts of eastern Europe, especially in Hungary, Yugoslavia, and Romania.

Europe's far north—for example, Iceland, northern Scandinavia, and Finland—has subarctic and tundra climates of bitterly cold winters and short, cool summers. Tundra and subarctic regions have **permafrost**, soil that is permanently frozen below the surface. Tundra areas support little vegetation, with the exception of mosses, small shrubs, and wildflowers that bloom during the brief summer. The subarctic supports a vast coniferous forest that broadens in the eastern part where Europe and Russia share a border.

Geography Skills for Life

Land of Lakes Inari, in the far north of Finland, is one of some 60,000 lakes that dot the Finnish countryside.

Region What climates dominate Europe's far north?

SECTION 2 ASSESSMENT

Checking for Understanding

1. **Define** timberline, foehn, avalanche, mistral, sirocco, chaparral, permafrost.
2. **Main Ideas** Create an outline like the one below, showing the climates and vegetation found in three European countries.

Climate and Vegetation

I. Iceland
   A. Climates: subarctic, tundra, and permafrost
   B. Vegetation: conifers, lichens, moss

Critical Thinking

3. **Predicting Consequences** Prevailing westerly winds bring warm air from the North Atlantic Drift to the European continent. What do you think happens when the winds temporarily change course?
4. **Analyzing Information** What geographic factors contribute to vegetation differences between highlands and tundra climate regions?
5. **Identifying Cause and Effect** How has human interaction with the environment caused changes in Europe's vegetation patterns?

Analyzing Maps

6. **Location** Study the map of Europe's climate regions on page 278. Where are highlands climate regions found? What are their physical features?

Applying Geography

7. **Physical Processes** Describe the physical processes that affect Europe's climate and vegetation. Provide specific examples related to the variety of climates and vegetation found in the region.

Finding and Summarizing the Main Idea

Finding and summarizing the main idea in an article or book will help you organize information. It will also help you identify the most important concepts to remember.

. . . Patterdale lies within the 885 square miles of the Lake District National Park. . . . It is the largest of ten national parks in England and Wales (Scotland has none), but, as with the others, the designation is really a misnomer [incorrect name] since the land is neither owned by the nation nor is it in any conventional sense a park. It is, rather, a lived-in landscape, full of towns and farms, with a resident population of 40,000. All but a small fraction of the land is in private hands.

Unlike U.S. national parks, which often aim to preserve wilderness, British parks inevitably include residents. These parks were created so there could be a way to exert some control over the speed and nature of change, not to prevent it altogether. Unfortunately, the various authorities have little power, relying primarily on persuasion to resolve myriad [numerous] demands.

—Bill Bryson, "England's Lake District,"
***National Geographic**, August 1994*

Learning the Skill

To identify the main idea, you may need to "read between the lines" and interpret the facts and evidence that are presented. Review the important details, and decide which ones are central to the message. By looking closely at important details, you can infer an author's main meaning.

When looking for a main idea, follow these steps:

- **Skim the material to identify its general subject.** Look at any headings and subheadings.
- **Read the information to pinpoint the ideas that the details support.** Why is the author presenting these facts and this evidence?
- **Identify the main idea.** Ask yourself: How can I state the main idea in my own words?

Practicing the Skill

Read the passage above. Then answer the following questions.

1. What is the general subject of the passage?
2. What important facts and details does the passage include?
3. What is the main idea of the passage? State the main idea in your own words.

Applying the Skill

Bring to class a news article about an issue facing Europe. Summarize the main idea of the article, and explain why it is important.

The **Glencoe Skillbuilder Interactive Workbook, Level 2** provides instruction and practice in key social studies skills.

11 SUMMARY & STUDY GUIDE

SECTION 1 The Land (pp. 271–276)

Terms to Know

- dike
- polder
- glaciation
- fjord
- loess

Key Points

- Europe is a huge peninsula extending westward from the Eurasian landmass.
- Europe has a long coastline with many peninsulas and islands.
- Europe has a large plains region in its northern areas; mountains are found along the continent's eastern and southern boundaries.
- Rivers provide important transportation in Europe, linking the interior of the continent with coastal ports.
- Europe has important deposits of minerals, oil, and natural gas.

Organizing Your Notes

Use a table like the one below to help you organize the notes you took as you read the chapter.

Country	Mountains	Rivers and Lakes	Other Features

SECTION 2 Climate and Vegetation (pp. 277–281)

Terms to Know

- timberline
- foehn
- avalanche
- mistral
- sirocco
- chaparral
- permafrost

Key Points

- Warm ocean currents give much of Europe a milder climate than other areas at similar latitudes.
- Areas of western Europe with a marine west coast climate have generally moderate temperatures.
- Much of southern Europe has a Mediterranean climate, with mild, rainy winters and warm, dry summers.
- Europe's interior has more extreme seasonal temperatures than do areas nearer the sea.
- Both climate and human activity affect the natural vegetation of Europe.

Organizing Your Notes

Create graphic organizers like the one below to help organize your notes about each of Europe's climate regions.

Marine West Coast
-
-
-

Reindeer herding, northern Sweden

ASSESSMENT & ACTIVITIES

Reviewing Key Terms

Write the letter of the key term that best matches each definition below.

a. sirocco
b. fjord
c. foehn
d. polder
e. timberline
f. mistral

1. elevation above which trees cannot grow
2. dry wind that blows in the Alps
3. hot wind that blows from North Africa to Europe's Mediterranean coast
4. drained area reclaimed from the sea
5. deep, water-filled valley carved by glaciers
6. strong north wind from the Alps that brings cold air to southern France

Reviewing Facts

SECTION 1

1. Why is Europe a "peninsula of peninsulas"?
2. What geographic area in Europe has rich, fertile farmland and is a center of industry?
3. How have human actions over the centuries changed Europe's waterways?

SECTION 2

4. How do the Gulf Stream and the North Atlantic Drift affect Europe's climate?
5. What kinds of climate regions are found in Iceland and the Scandinavian Peninsula?

Critical Thinking

1. **Drawing Conclusions** How did geographic features help shape European cultures? Provide examples to support your answers.
2. **Identifying Cause and Effect** Why did the North European Plain develop into a densely populated industrial center?
3. **Drawing Conclusions** Copy the diagram of European rivers, seas, and waterways below onto a sheet of paper. In each oval, write the name of a city that is located on or beside the body of water. Then draw lines to show how cities are linked by waterways.

North Sea –	Danube River –
Baltic Sea –	Thames River –
Rhine River –	Main-Danube Canal –

NATIONAL GEOGRAPHIC

Locating Places

Europe: Physical Geography

Match the letters on the map with the physical features of Europe. Write your answers on a sheet of paper.

1. British Isles
2. Rhine River
3. Sicily
4. Apennines
5. Baltic Sea
6. Mediterranean Sea
7. Scandinavia
8. Crete
9. Iberian Peninsula
10. Balkan Peninsula

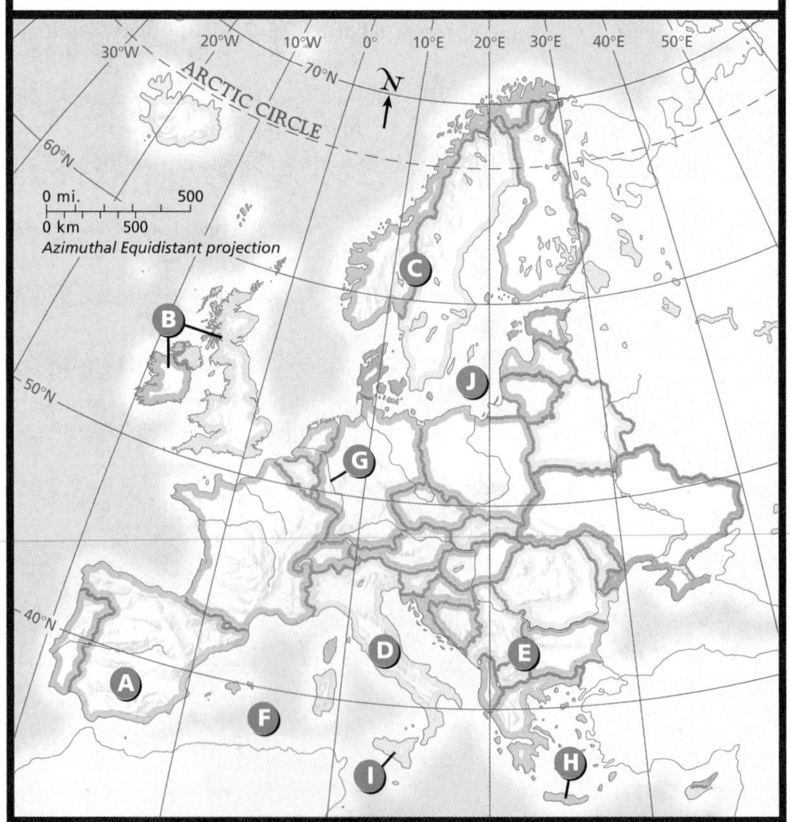

Self-Check Quiz Visit the **Glencoe World Geography** Web site at geography.glencoe.com and click on Self-Check Quizzes—Chapter 11 to prepare for the Chapter Test.

Using the Regional Atlas

Refer to the Regional Atlas on pages 260–263.

1. **Location** Through what country do the Seine, Loire, and Rhône Rivers flow?
2. **Place** What are three major agricultural products of the North European Plain?

Thinking Like a Geographer

Think about the physical geography of Europe. Identify Europe's energy resources, and where they are located. Which of these are nonrenewable resources? What future energy sources would you advise European countries to pursue?

Problem-Solving Activity

Group Research Activity People in Europe face many weather-related challenges, from avalanches in the mountains to flooding in the lowlands. Using the Internet and other resources, research an area in Europe that has successfully coped with weather-related events. Then report to the class on the solutions to these challenges. Include photos, charts, graphs, or any other visual elements to enhance your report.

GeoJournal

Creative Writing Using the information in your GeoJournal, describe an imaginary trip through a European country of your choice. Describe the country's physical features and the climate and natural vegetation you find. Use what you have learned in your reading to make your account detailed and colorful.

Technology Activity

Using an Electronic Spreadsheet Choose a city in each of Europe's climate regions, and find the average rainfall for each city. Use a spreadsheet program to organize your information, listing the cities in the first column and the rainfall amounts in the next column. Use the program's graphics feature to make a bar graph. Write a paragraph summarizing the variations in rainfall among the cities.

Standardized Test Practice

Choose the best answer for the following multiple-choice question. If you have trouble answering the question, use the process of elimination to narrow your choices.

"And so I have finally come to understand that while I am hopelessly American, accustomed to (and dependent on) the relentless pressures and fierce energies of the New World, . . . there are moments when I want to escape to a different place with a beauty and a beat of its own. And when that happens, when I want to disappear from who I am, and where I live, the place I think of is Paris."

—David Halberstam, "Paris," *National Geographic Traveler,* October 1999

1. **What kind of place does the author want to escape to sometimes?**

   A He wants a place where there is a lot of pressure and energy.

   B He wants a beautiful place halfway around the world.

   C He wants a unique, beautiful place that is different from where he lives.

   D He wants a place where he can disappear into the crowds.

When choosing an answer for a multiple-choice question, sometimes more than one option may seem correct. Read the question carefully, and then look in the reading for information about the kind of place. Compare each answer with that information.

CHAPTER 12

The Cultural Geography of Europe

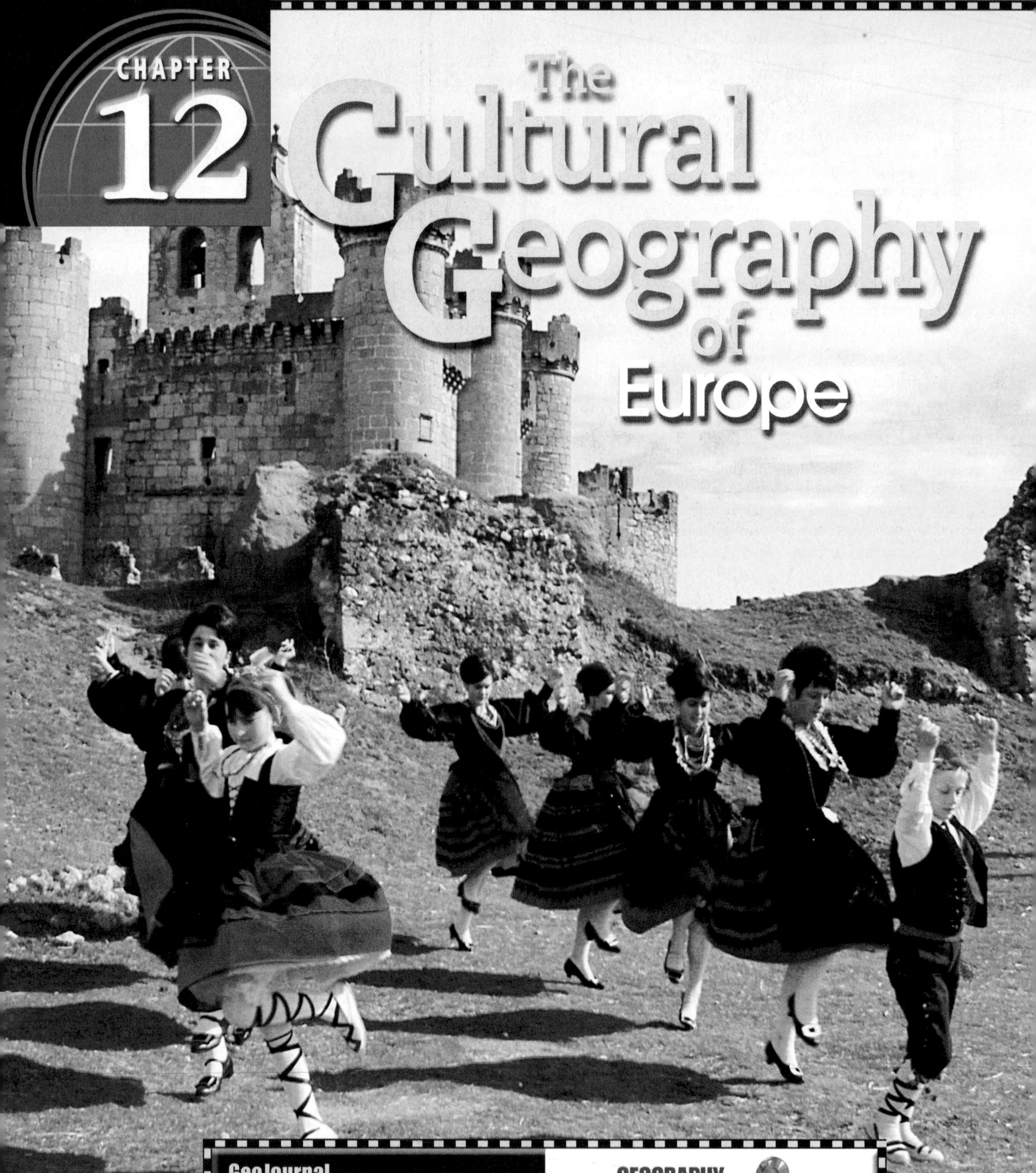

GeoJournal

As you read this chapter, list the differences among European peoples and the similarities that bind them into one cultural region. Which of these differences and similarities might affect Europe's future?

GEOGRAPHY Online

Chapter Overview Visit the **Glencoe World Geography** Web site at geography.glencoe.com and click on Chapter Overviews—Chapter 12 to preview information about the cultural geography of the region.

Population Patterns

Guide to Reading

Consider What You Know

Recent conflicts in Europe's Balkan Peninsula frequently make newspaper headlines. How do these conflicts affect everyday life for people in the region today?

Read to Find Out

- How does the physical geography of Europe influence its population density and distribution?
- What effects have industrialization and urbanization had on Europe's people?
- How have recent patterns of migration influenced the region's cultures?

Terms to Know

- ethnic group
- ethnic cleansing
- refugee
- urbanization

Places to Locate

- Sweden
- Belgium
- Bosnia-Herzegovina
- Kosovo
- Germany
- Vatican City
- United Kingdom
- France
- Czech Republic
- Poland
- Paris
- London
- Naples

◀ *Turegano folk dancers, Spain*

NATIONAL GEOGRAPHIC

A Geographic View

Search for a New Life

In a world of shifting demographics—where the poor, the dispossessed, and the war-ravaged are on the move—Western Europe has become the migrant's preferred destination. . . . There are nearly 20 million legal immigrants there—plus an estimated two million illegal aliens. . . . In 1992 more than 750,000 political asylum seekers crowded into Europe, more than half of them into Germany. Almost all become economic wards [dependents] of their adopted nations.

—Peter Ross Range, "Europe Faces an Immigrant Tide," National Geographic, *May 1993*

Immigrants shop at a London market.

Europe is home to more than 40 countries, whose peoples belong to many different cultural groups and speak a variety of languages. This diversity stems from centuries of migration, cultural diffusion, conflict, and changing borders. In this section you will learn about Europe's peoples, their ethnic characteristics, and where they live.

Ethnic Diversity

Europe's diverse population reflects a long history of migrations throughout the continent. Most Europeans are descended from various Indo-European and Mediterranean peoples who settled the continent centuries ago. Europe's population today also includes more recent immigrants from Asia, Africa, and the Caribbean area who arrived during the past 100 years. Many of these immigrants have come from areas of the world once ruled by European countries.

Ethnic Groups

Today Europe is home to more than 160 separate ethnic groups—groups of people with a shared ancestry, language, customs, and, often, religion. Some European countries have one major ethnic group. In **Sweden**, for example, 89 percent of the population are Swedes, descendants of Germanic and other groups that settled the peninsula of Scandinavia centuries ago. They share a common culture, the Swedish language, and a Lutheran religious heritage.

In other countries the population consists of two or more major ethnic groups. For example, **Belgium** has two leading ethnic groups—the Flemings and the Walloons. The Flemings make up about 56 percent of Belgium's population and the Walloons about 32 percent. The Flemings, closely related to the Dutch, are descended from Germanic groups who invaded the area of present-day Belgium during the A.D. 400s. The Walloons trace their ancestry to the Celts who lived in the area during the Germanic invasions. Flemings and Walloons are both Roman Catholic, but language differences have often led to bitter relations between them. Both groups, however, have been able to keep their disputes from endangering Belgium's national unity.

Ethnic Tensions

Tensions among some European ethnic groups have led to armed conflict. The Balkan Peninsula has long been a shatterbelt, a region caught between external and internal rivalries. In the 1990s, the Balkans was a battleground among Serbs, Croats, Bosnian Muslims, and Kosovar Albanians. Following World War II, these and other Balkan peoples had belonged to a communist-ruled land called Yugoslavia. For a time, hatreds were muted. But after the communist system's fall in the early 1990s, ethnic tensions erupted, resulting in Yugoslavia's breakup into separate independent republics.

Within some of the new republics, ethnic hatreds were serious enough to spark the worst fighting in Europe since World War II. The republic of **Bosnia-Herzegovina** (BAHZ•nee•uh HERT•seh•GAW•vee•nah) and the Serb-ruled territory of **Kosovo** (KAW•saw•VAW) were centers of the most brutal warfare. Following a policy called ethnic cleansing, Serb leaders expelled or killed rival ethnic groups in these areas. As a result, many people became refugees—people who flee to a foreign country for safety. International peacemaking efforts, however, enabled many of these refugees to later return to their homes.

Sources of Unity

Although division and conflict have characterized much of Europe's history, Europeans in recent years have been working toward greater unity. Their efforts at cooperation rest on common attitudes and values. For example, most Europeans value the importance of the past and the cultural achievements of their ancestors. They also take pride in their families, which they place at the center of their social lives.

Despite having varying forms of government, the peoples of Europe generally share a commitment to

NATIONAL GEOGRAPHIC **World Explorer**

Geography Skills for Life

Paris Shopping A world center of fashion, Paris, France, is known for its elegant shops and department stores.

Region What cultural factors unite Europeans? What cultural factors divide them?

democracy and free markets. Their sense of individualism, however, is combined with the belief that government should regulate economies and provide for social welfare. These similarities make it easier for residents to think of themselves as Europeans as well as members of ethnic or national groups.

Population Characteristics

Europe is smaller in land area than any other continent except Australia. Yet it is the third most populous continent, after Asia and Africa. In the year 2001, Europe's population (excluding Russia) was about 583 million. **Germany**, with 82.2 million people, is Europe's largest country in population, and **Vatican City** is the smallest, with only 1,000 people.

Population Density

Europe's large numbers of people are crowded into a relatively small space. In fact, Europe's population density is greater than that of any other continent except Asia. If Europe's population were distributed evenly throughout the continent, the average population density would be 255 people per square mile (98 people per sq. km). In Europe, however, as in other continents, the population is not distributed evenly. Most of Europe has far less than the average population density. The region's highly industrialized urban centers, however, are among the world's most densely populated areas.

Population Distribution

As in other parts of the world, Europe's population distribution is closely related to its physical geography. Compare the population density map on page 262 with the physical map on page 260. Notice that mountainous areas and cold northern areas in Europe are less populated than plains areas. In fact, the parts of Europe with average or higher than average population densities share one or more of the following features: favorable climates, plains, fertile soil, mineral resources, and inland waterways. One of the most densely populated parts of Europe extends from the **United Kingdom** into **France** and across the fertile North European Plain into the **Czech Republic** and **Poland**. Another densely populated area extends from southeastern France into northern Italy. In addition to having rich farmland, these regions contain densely populated, industrial cities.

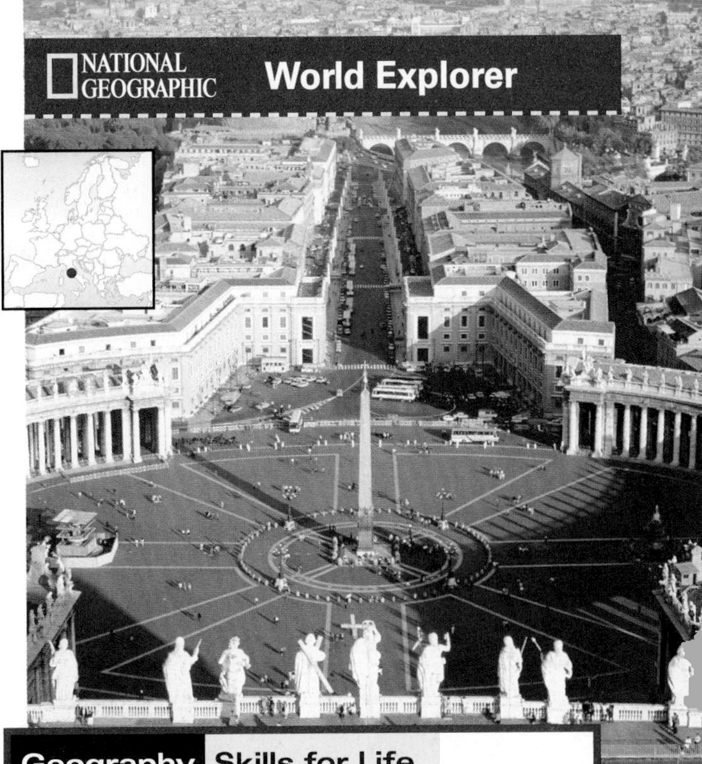

Geography Skills for Life

Vatican City This view of St. Peter's Square is from the top of St. Peter's Basilica, one of the world's largest Christian churches and a gathering place for many Roman Catholics.

Place What other European countries have small populations?

Urbanization

Beginning in the late 1700s, the Industrial Revolution transformed Europe from a rural, agricultural society to an urban, industrial society. Rural villagers moved in large numbers to urban areas

Student Web Activity Visit the **Glencoe World Geography** Web site at geography.glencoe.com and click on Student Web Activities—Chapter 12 for an activity on researching the cultural similarities and differences among Scandinavian countries.

and became factory workers. This concentration of populations in towns and cities is known as **urbanization**. The growth of industries and cities began first in western Europe during the late 1700s. Later, after World War II, this process spread to eastern Europe.

Today about 75 percent of Europeans live in cities. **Paris** and **London** rank among the world's largest urban areas. Other European cities with large populations include Rome, Italy; Madrid, Spain; Berlin, Germany; Stockholm, Sweden; Budapest, Hungary; Athens, Greece; and Kiev, Ukraine.

Urban Features

Europe's largest cities, like cities everywhere, face the challenges of overcrowding and pollution. In spite of these problems, European cities provide a unique combination of old and new ways of life. Landmarks that date back hundreds of years stand near fast-food restaurants and shopping malls. **Naples**, Italy, is one city that reflects this coming together of past and present in modern Europe:

> *"One morning I went on a . . . walk through . . . the historic center of Naples. Here the grid plan remains from the original Greek settlement, with laundry-festooned streets barely the width of an average driveway. Lack of space has never presented any serious problem to the Neapolitan. . . . At any given moment there will be at least one car on the street, along with two motorbikes (coming from opposite directions), three girls walking arm in arm, and a family with a baby carriage, all of whom unaccountably manage to avoid collision."*
>
> Erla Zwingle, "Naples Unabashed," *National Geographic*, March 1998

NATIONAL GEOGRAPHIC **World Explorer**

Geography Skills for Life

Historic Prague Prague, in the Czech Republic, is a city of churches and palaces.

Place How does Prague compare with major cities in your state?

Economics

Population Movements

Population movements have been a constant aspect of European life. During the 1800s and early 1900s, many Europeans migrated to the Americas and parts of Africa and the South Pacific region. Since the mid-1900s, far fewer Europeans have permanently left the region, but large numbers of foreigners have migrated to Europe.

When western Europe's economy boomed during the 1950s and 1960s, labor shortages developed. Many European countries invited guest workers from other countries to fill available jobs. Soon guest workers and immigrants began arriving, seeking the social and economic opportunities that western Europe had to offer. In France, for example, immigrants came from North African countries newly independent from France. In Germany guest workers from Turkey, Greece, and the Balkan countries of southeastern Europe filled industrial jobs. The United Kingdom also saw increased immigration from countries in South Asia and the Caribbean areas that had once been British but were now independent.

By the time Europe's economy had slowed in the 1970s, many guest workers had moved their families and established homes in host countries. Tensions rose as the immigrants and local residents competed for jobs, housing, and social services. As a result, many immigrants felt unwelcome in their new countries. Since the 1970s, European governments have tried to limit further immigration while protecting the rights of their immigrant communities.

Despite its growing immigrant populations and abundant resources, Europe's overall population is shrinking. Italy and Germany, for example, have the world's lowest birthrates. Experts predict that Italy's population will fall from 57.8 million today to about 41 million by 2050. In addition, older people are making up a larger percentage of Europe's population.

Europe's population continues to change even as it maintains and honors its historic traditions. In the next section you will learn how Europe's physical geography affected the settlement of its peoples and their cultural and economic development.

Geography Skills for Life

A Village in Ruins During the 1990s ethnic violence uprooted many people in the Balkan Peninsula.

Movement What major European country has hosted Balkan migrants since the 1950s?

SECTION 1 ASSESSMENT

Checking for Understanding

1. **Define** ethnic group, ethnic cleansing, refugee, urbanization.
2. **Main Ideas** On a sheet of paper, create a web diagram like the one below to show important details about Europe's population.

Population of Europe
- Ethnic Diversity
  - •
  - •
  - •
  - •
- Population Characteristics
  - •
  - •
  - •
  - •

Critical Thinking

3. **Making Generalizations** Why have people migrated to Europe from various parts of the world?
4. **Drawing Conclusions** What factors contribute to the patterns of population density and distribution in Europe?
5. **Identifying Cause and Effect** What effect has migration had on modern Europe?
6. **Predicting Consequences** What might be the consequences of falling birthrates in some European countries?

Analyzing Maps

7. **Region** Look at the population density map page 262. Find three regions in which population density varies within a country. Explain how climate and physical features contribute to these differences.

Applying Geography

8. **Geography and Population** Consider how physical geography has influenced population patterns in Europe. In an essay, describe one population group, and explain the impact of physical geography on the settlement of this group.

GEOGRAPHY AND HISTORY

YUGOSLAVIA: THEN AND NOW

CAN YOU IMAGINE waking in the night to the sounds of soldiers and gunfire? Quickly you grab a few belongings and flee with your family from your home. This scenario may sound far-fetched, but if you lived on the Balkan Peninsula, it might be more believable. There, many people have been forced from their homes during the last decade. The peninsula has long been a region of instability and conflict. But since the breakup of the former Yugoslavia in 1991–1992, long-held resentments among various ethnic groups have erupted into full-scale wars.

A New Nation Emerges

The Balkan Peninsula lies in southeast Europe, between the Black and Adriatic Seas. Towering mountain ranges—the Carpathian and Dinaric—dominate the area. Long ago, Slavic peoples moved south into the region from what are now southern Poland and Russia. Slavic groups established independent states—Croatia, founded by the Croats; Serbia, founded by the Serbs; and Slovenia, founded by the Slovenes. Foreign nations ruled these lands for centuries. The Ottomans, who were based in what is now

Families flee Rača, Yugoslavia, when fighting erupts in their village. Minority ethnic groups throughout the region have been forced to leave their homes. ▶

AUSTRIA
ITALY
SLOVENIA
Ljubljana
HUNGARY
Zagreb
CROATIA
ROMANIA
VOJVODINA
SERBIA
Belgrade
BOSNIA AND HERZEGOVINA
Sarajevo
YUGOSLAVIA
MONTENEGRO
KOSOVO
BULGARIA
ALBANIA
Skopje
MACEDONIA
GREECE

Turkey, controlled much of the region and gave the peninsula its name—*Balkan*, or "mountains."

The Ottoman Turks were defeated in the Balkan Wars of 1912–1913, ending their 500-year reign. Following their departure, a movement to unite the Slavs into one country gained strength. In 1918 the Slavs formed the Kingdom of Serbs, Croats, and Slovenes. It was later renamed Yugoslavia, "Land of the South Slavs."

During World War II, Nazi Germany's occupation of Yugoslavia divided the country. Throughout the war, an underground group headed by Croatian Josip Broz, who was code-named Tito, worked against Germany. At war's end, Yugoslavia emerged as a Communist country with Tito as its leader. It consisted of six republics—Croatia, Bosnia and Herzegovina, Slovenia, Montenegro, Serbia, and Macedonia.

The Nation Splinters

Tito ruled Yugoslavia with an iron hand and succeeded in holding the ethnically mixed republics together. One of his challenges was to prevent Serbia, the largest of the republics, from dominating the central government. To dilute Serb power, Tito gave greater autonomy to two provinces within Serbia—Kosovo and Vojvodina.

Yugoslavia's economy began to crumble in the 1970s. When Tito died in 1980, the country began to fracture along ethnic lines. In 1991 Serbia's president, Slobodan Milosevic, tried to assert Serb leadership over the republics. Slovenia, Macedonia, and Croatia declared independence. Fighting erupted in Croatia between Serbs and non-Serbs. When Bosnia and Herzegovina tried to secede, civil war broke out. By 1992 only two of the six republics—Serbia and Montenegro—remained united. They make up present-day Yugoslavia.

Looking Ahead

The hostilities are far from over. Minority ethnic groups throughout the region may be tomorrow's targets. How do the historic and geographic roots of the area shed light on the conflict?

NATIONAL GEOGRAPHIC SOCIETY

1400s Ottoman Empire gains control of nearly all lands in the Yugoslav region

1912–1913 Balkan Wars end Ottoman rule

1918 Slavs unite to form Kingdom of Serbs, Croats, and Slovenes

1929 Kingdom is renamed Yugoslavia

1945 World War II ends; Tito (photo above) and the Communists control Yugoslavia

1980 Tito dies

1991–1992 Several Yugoslav republics declare independence; civil wars ravage cities (background photo)

1992 Serbia and Montenegro remain united as Yugoslavia

History and Government

Guide to Reading

Consider What You Know

Europe, a relatively small continent, is home to a great variety of ethnic groups. How do you think being exposed to many different cultures has affected the way Europeans live?

Read to Find Out

- What contributions did early Europeans make to world culture?
- In what ways has Europe's geography shaped its history?
- What were the effects of world wars and economic and political revolutions in Europe?

Terms to Know

- city-state
- Middle Ages
- feudalism
- Crusades
- Renaissance
- Reformation
- Enlightenment
- industrial capitalism
- communism
- reparations
- Holocaust
- Cold War
- European Union

Places to Locate

- Greece
- Rome
- Athens
- Italy
- Constantinople
- Spain
- Portugal

NATIONAL GEOGRAPHIC

A Geographic View

Layers of Culture

Bosnia and Herzegovina spreads across the gnarled reaches of the Dinaric Alps, a region possessed of enough bracing mountain beauty, enterprise, and gusto to have landed the 1984 Winter Olympics at its capital, Sarajevo. Even in this rugged corner of the Balkan Peninsula, the wash of empires—Roman, Byzantine, Ottoman, Austro-Hungarian—deposited layer upon layer of culture.

—*Priit J. Vesilind, "In Focus: Bosnia,"* National Geographic, *June 1996*

Sarajevo, Bosnia and Herzegovina

The layering of cultures in the Balkans area is typical of Europe as a whole. Throughout the region, buildings, monuments, and local customs reflect the different periods of Europe's long history and the peoples that dominated its stage at these times. Through empire-building, immigration, and trade, Europe's cultures also have influenced other parts of the world. In this section you will learn about the contributions Europeans have made in learning, the arts, and technology.

The Rise of Europe

Physical geography in part has shaped Europe's history. The physical map on page 260 shows that several large bodies of water touch Europe. This closeness to the sea enabled Europeans to move beyond their own borders to other parts of the world. In addition, European mountain ranges contained passes and so did not severely hinder contacts within the region as did mountain ranges in other parts of the world. Also, Europe's river-crossed fertile plains encouraged peaceful settlement as well as invasions and conflicts.

Early Peoples

Fossils found by archaeologists suggest that early humans lived in Europe more than a million years ago. Prehistoric Europeans moved from place to place in search of food. By about 6000 B.C., farming spread from Southwest Asia to southeastern Europe and then to all but the densely forested areas in the northern part of the continent. With the introduction of farming, early Europeans settled in agricultural villages, some of which later developed into Europe's first cities.

Ancient Greece and Rome

Two civilizations in the Mediterranean world laid the foundations of European—and Western—civilization. The first was ancient **Greece**, which reached its peak during the 400s and 300s B.C. The second civilization, whose capital was **Rome**, ruled a vast empire that reached its height of power between 27 B.C. and A.D. 180.

Greece's mountainous landscape and its closeness to the sea influenced the ancient Greeks to form separate communities called **city-states**. Each city-state was independent, but was linked to other city-states by Greek language and culture. Fleeing overpopulated areas and desiring new wealth, Greek merchants and sailors eventually colonized many parts of the Mediterranean coast.

The ancient Greeks laid the foundations of European government and culture. The city-state of **Athens** introduced the Western idea of democracy. Although women and enslaved persons could not vote, more people had a voice in Athens' government than in any earlier civilization. Greek art, literature, drama, and philosophy as well as mathematics and medicine also left a lasting impression on the Western world.

In **Italy** around 500 B.C., another Mediterranean people, the Romans, founded a republic. From the city of Rome, Roman armies went forth to conquer an empire that spanned much of Europe, some of Southwest Asia, and North Africa. The Romans imitated Greek art and literature, and borrowed from Greek science and architecture.

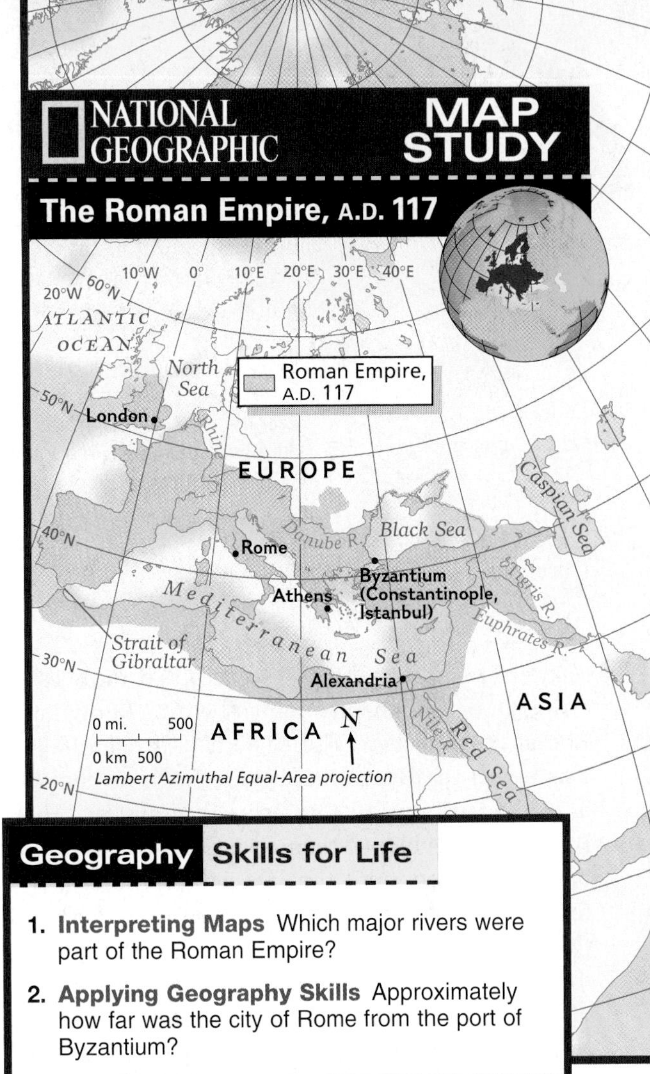

Geography Skills for Life

1. **Interpreting Maps** Which major rivers were part of the Roman Empire?
2. **Applying Geography Skills** Approximately how far was the city of Rome from the port of Byzantium?

Find NGS online map resources @ www.national geographic.com/maps

Roman developments in government, law, and engineering, however, influenced other cultures. Throughout the Roman Empire, for example, engineers built a vast network of roads, bridges, and aqueducts—artificial channels for carrying water.

A Christian Europe

In the late A.D. 300s, Christianity became the official religion of the Roman Empire and, later, one of the world's major religions. Although united in name, the empire came to be ruled by two emperors, one in the eastern half and the other in the western half. Eventually the two parts developed into eastern and western Europe, each with its own political, cultural, and religious traditions. During the 400s, Germanic groups from the north overthrew Roman rule in the western half and founded

architecture of EUROPE

Notre Dame de Paris Built between 1163 and 1250, Notre Dame Cathedral in Paris, France, is one of the best known Gothic cathedrals in the world. Gothic architecture is characterized by pointed arches and flying buttresses—external stone beams that support the outer walls. High ceilings, beautiful stained glass windows, and open interiors are also typical elements of Gothic architecture. The western front of Notre Dame is adorned with fine carvings of biblical figures and events.

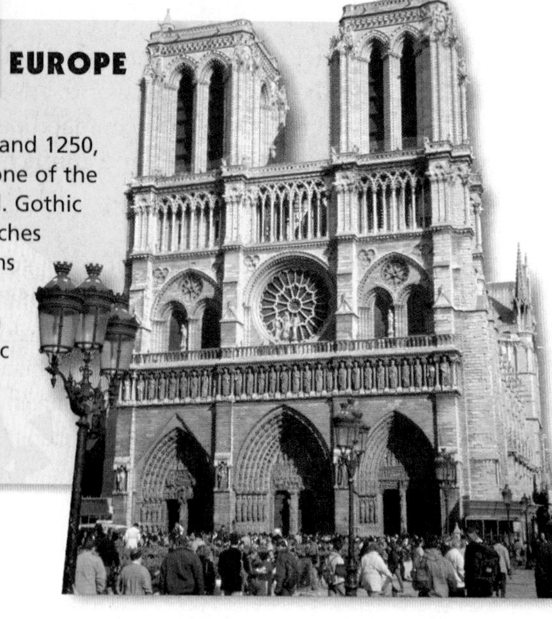

separate kingdoms. They also accepted the western form of Christianity, which became known as Roman Catholicism. The eastern half eventually was called the Byzantine Empire, with its capital at **Constantinople**, formerly Byzantium. The Byzantines developed their own Christian civilization that lasted until the late 1400s. The eastern form of Christianity became known as Eastern Orthodoxy. During the 500s Slavic peoples migrated from Ukraine into eastern and central Europe. Some of them later accepted western European ways and Roman Catholicism, and others adopted Byzantine traditions and Eastern Orthodoxy.

The Middle Ages

After the fall of Rome, western Europe entered the **Middle Ages**, the period between ancient and modern times. During this era, from about A.D. 500 to 1500, **feudalism**—a system in which monarchs or lords gave land to nobles in return for pledges of loyalty—replaced centralized government. The Roman Catholic Church, western Europe's major religious body, brought Roman culture and principles of government and law to the region's Germanic peoples. During the Middle Ages, religious centers, such as cathedrals and monasteries, were major centers of learning. Meanwhile, in eastern Europe, the Byzantine Empire preserved ancient Greek and Roman cultures, and Byzantine missionaries spread Eastern Orthodoxy among many of the Slavic peoples.

Although generally Christian, Europe was influenced by other religious groups during the Middle Ages. Cities and towns in western Europe were home to Jewish communities that made contributions to European society. Many Christians, however, saw the Jews as outsiders and persecuted or discriminated against them. Expelled from much of western Europe, many Jews settled in eastern Europe, where they developed new communities based on their religious traditions.

Another influence on Christian Europe was Islam, a religion based on belief in one God and the preachings of Muhammad, a prophet who lived in Southwest Asia during the 600s. Within a century of Muhammad's death, Islam had spread from Southwest Asia through North Africa and into **Spain**. Muslims, the followers of Islam, developed a culture in Spain that passed on to Europeans many achievements in science, mathematics, and medicine.

Expansion of Europe

Beginning in the 1000s, western European armies fought the **Crusades**—a series of brutal religious wars—to win Palestine, the birthplace of Christianity, from Muslim rule. Europeans failed to win permanent control of the area but did extend trade routes to the eastern Mediterranean world. Spices and other products that came with increased trade sparked the interest of the small number of educated Europeans in other parts of the world. Beginning in the 1300s, the **Renaissance** (REH•nuh•SAHNTS)—a 300-year period of discovery and learning—brought about great advances in European civilization.

The Renaissance

During the Renaissance, educated Europeans developed a new interest in the cultures of ancient Greece and Rome. They stressed the importance of people and their place in this world. Writers described human feelings, and artists created lifelike paintings and sculptures. In addition to religious structures, architects designed buildings, such as palaces and villas, for private use. The Renaissance also led to scientific advances. For example, the invention of movable type in printing spread new ideas more quickly and easily.

The increased production of books and pamphlets aided a religious movement called the **Reformation**, which lessened the power of the Roman Catholic Church and led to the beginnings of Protestantism. By the mid-1500s, Protestant churches were dominant in northern Europe, but Roman Catholicism retained its hold on the southern, central, and northeastern parts of the region. Religious wars soon engulfed Europe, and European monarchs were able to strengthen their power over nobles and church leaders.

European Explorations

During the Middle Ages, Europe lagged behind the Chinese and Muslim empires in economic development. In the 1400s, however, western Europe began to emerge as a significant force in world affairs. At that time seafarers from **Portugal** developed new trade routes around Africa to Asia. Spanish rulers financed the Italian-born explorer Christopher Columbus, who reached the Americas in the late 1400s. England, France, and the Netherlands also sent out expeditions of explorers. These voyages resulted in conquests of foreign lands, often destroying the cultures already thriving there. Trade with colonies in the Americas, Asia, and Africa brought great wealth and power to western Europe.

A Changing Europe

During the late 1600s and early 1700s, many educated Europeans emphasized the importance of reason and began to question long-standing traditions and values. This movement, known as the **Enlightenment**, was followed by political and economic revolutions that swept the entire region.

Revolutions

At this time Europeans wanting a voice in government began political revolutions. In the late 1600s, the English Parliament, or lawmaking body, passed a Bill of Rights that limited the power of the monarch. The French Revolution in the late 1700s overthrew France's monarchy and spread the ideals of democracy. The 1800s saw many uprisings throughout the rest of Europe that challenged the power of monarchs and nobles. By 1900 most European countries had constitutions that limited rulers' powers and guaranteed at least some political rights to citizens.

During this time of political change, the Industrial Revolution began in England and rapidly spread to other countries. Power-driven machinery and new methods of production transformed life in Europe. Industrial cities and improved transportation and communication developed. These sweeping changes led to the rise of **industrial capitalism**, an economic system in which business leaders used profits to

NATIONAL GEOGRAPHIC **World Explorer**

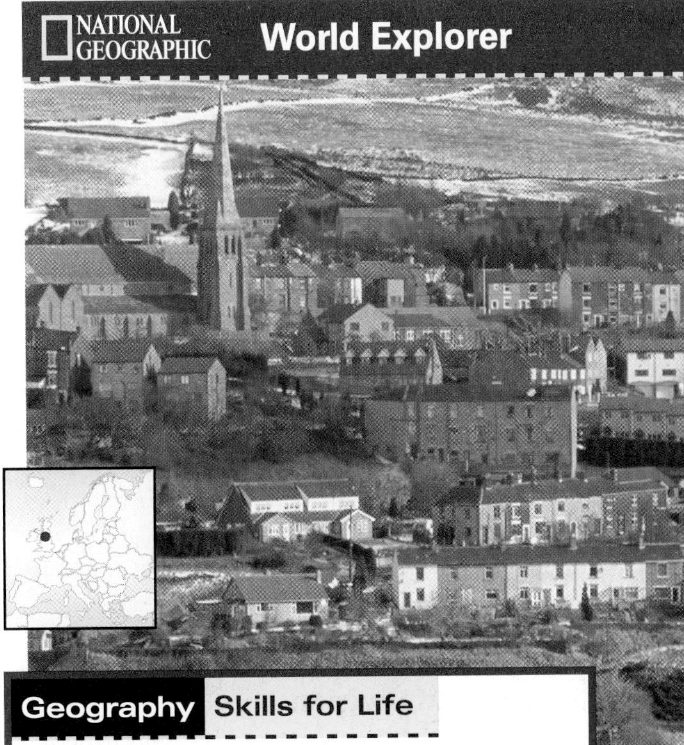

Geography Skills for Life

Industrial Revolution Industrial cities, such as Mossley, England, developed as the Industrial Revolution spread across Europe.

Region How did the Industrial Revolution affect life in Europe?

Geography **Skills for Life**

Auschwitz Memorial In Auschwitz, Poland—site of infamous Nazi death camps—a Holocaust memorial exhibit shows piles of footwear worn by Jewish and other prisoners who lost their lives.

Region How did World War II impact the political landscape of Europe?

expand their companies. Under this system, new social groups emerged: a middle class of merchants and factory owners, and a working class of factory laborers. Although the middle class prospered, factory workers at first were poorly paid and lived in crowded, unhealthy conditions.

These social problems led in the mid-1800s to the birth of **communism**—a philosophy that called for a society based on economic equality in which the workers would control the factories and industrial production. By the end of the 1800s, various European governments began passing laws to improve conditions for workers in the workplace and to expand education, housing, and health care.

Conflict and Division

In the first half of the 1900s, two world wars resulted in major changes in Europe. Rivalries among European powers for colonies and economic power led to World War I, which lasted from 1914 to 1918. An American journalist, Richard Harding Davis, described the German invasion of Belgium in these words:

> "... For three days and three nights the column of gray, with fifty thousand bayonets and fifty thousand lances, with gray transport wagons, gray ammunition carts, gray ambulances, gray cannons, like a river of steel cut Brussels in two."
>
> quoted in John N. Chettle, "When War Called, Davis Answered," *Smithsonian*, April 2000

As a result of World War I, monarchies collapsed in Germany, Austria-Hungary, and Russia, and several central and eastern European countries won independence. The Versailles peace treaty in 1919 found Germany guilty of starting the war and demanded that Germany make **reparations**, or payment for damages.

The large number of unresolved political problems from World War I and a worldwide economic depression enabled dictators Benito Mussolini and Adolf Hitler to gain control of Italy and Germany, respectively. Following aggressive territorial expansion by these two countries, World War II broke out in 1939. By the time this conflict ended in 1945, most of Europe and much of the rest of the world were involved. A major horror of World War II in Europe was the **Holocaust**, the mass killing of more than 6 million European Jews and others by Germany's Nazi leaders.

World War II left Europe ruined and divided. Most of eastern Europe came under communist control of the Soviet Union, and most of western Europe backed democracy and received economic and military support from the United States. This division of Europe brought about the **Cold War**, a power struggle between the communist world, led by the Soviet Union, and the noncommunist world, led by the United States. A divided Germany—communist East Germany and democratic West Germany—became the "hot point" of the Cold War in Europe.

History

The Cold War in Europe

At the end of World War II, the victorious Allies, including the United States, the Soviet Union, the

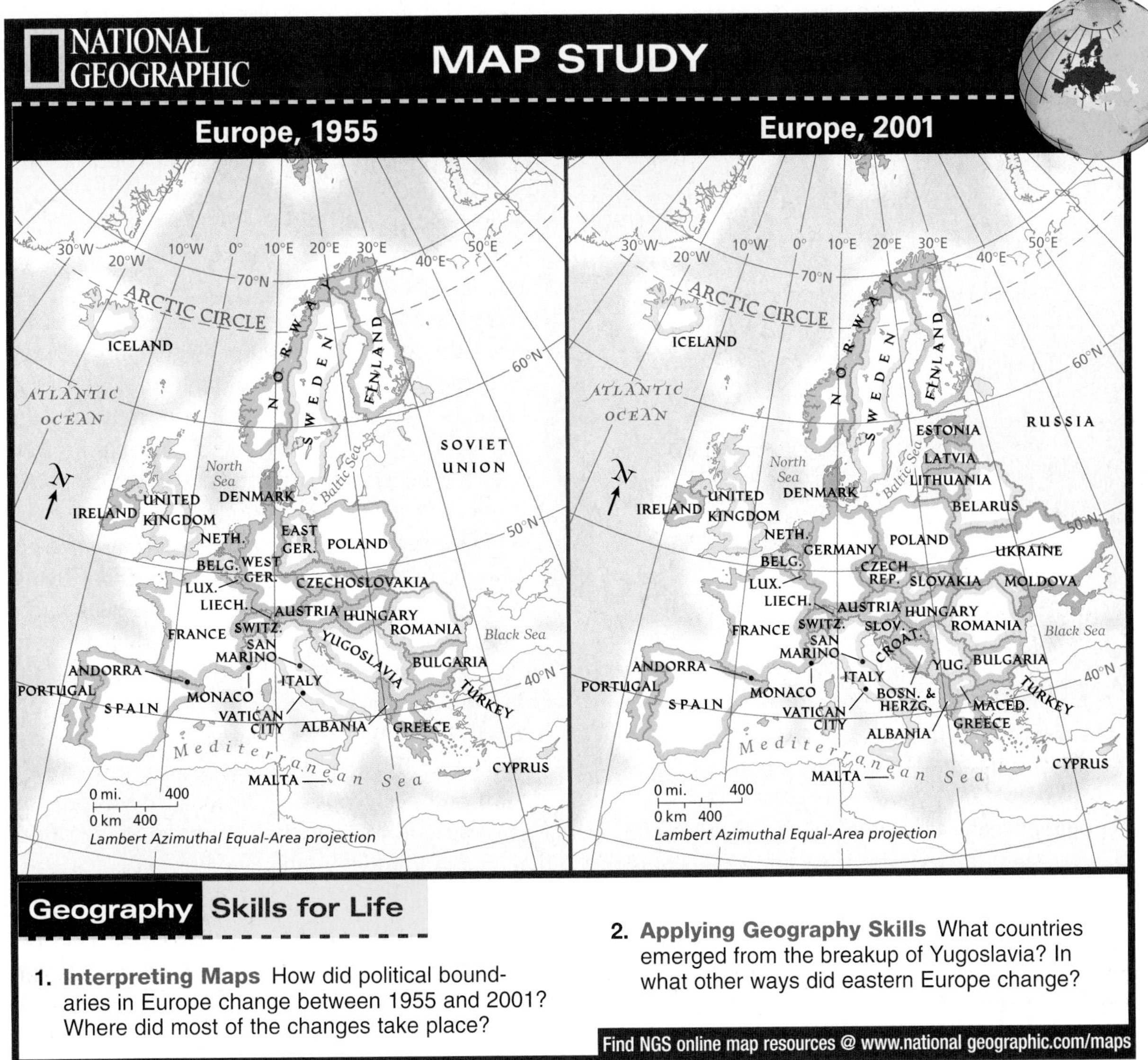

Geography Skills for Life

1. **Interpreting Maps** How did political boundaries in Europe change between 1955 and 2001? Where did most of the changes take place?

2. **Applying Geography Skills** What countries emerged from the breakup of Yugoslavia? In what other ways did eastern Europe change?

Find NGS online map resources @ www.national geographic.com/maps

United Kingdom, and France, divided Germany into four zones. By 1949 the three western zones of Germany were combined into West Germany, with Bonn as its capital. The eastern zone, occupied by the Soviets, became East Germany with East Berlin as its capital. Throughout the decade following the division, many East Germans fled to the West to escape communism. In the 1960s, East Germany built the Berlin Wall and other barriers to stop this movement of people.

During the Cold War era, most western European democracies became more productive and economically secure than they had been before World War II. In contrast, communist governments in eastern Europe allowed people little voice in government or the economy. Although eastern European communist countries pushed for industrial growth, their economies and standards of living lagged behind those of western Europe.

A New Era for Europe

From the 1950s to the 1980s, revolts against communist rule periodically swept eastern Europe. In Poland, Hungary, Czechoslovakia, East Germany, Romania, and Bulgaria, citizens demanded freedom and a better way of life. In the early 1980s, Polish workers founded Solidarity, the first free labor union in the communist world. In 1989, public demonstrations—and the refusal of reform-minded Soviet leaders to intervene—

Geography Skills for Life

European Union A car in Germany is painted to display the logo of the European Union during a celebration of the new organization.

Region What is the goal of the European Union?

swiftly led to the fall of eastern Europe's communist governments. Dramatic changes followed. The infamous Berlin Wall came down, and in 1990, the two parts of Germany reunited. Three years later, Czechoslovakia split into two separate countries: the Czech Republic and Slovakia. Throughout much of eastern Europe during the 1990s, free elections installed democratic leaders, who encouraged the rise of market economies.

Changes also occurred in western Europe. During the 1950s, Belgium, France, Italy, Luxembourg, the Netherlands, and West Germany banded closer together economically and politically. By the 1990s this growing movement toward unity had led to the **European Union (EU)**, an organization whose goal was a united Europe in which goods, services, and workers could move freely among member countries. The Maastricht Treaty, signed in 1992, set goals for a central bank and a common currency. Launched in 1999, that currency, the euro, replaced national currencies, such as the Italian lira and the German mark, in 2002. Currently comprising 15 member countries, the European Union plans to expand its membership to include eastern European countries. In the next section, you will learn about the variety of cultures and lifestyles that are found today in the new Europe.

SECTION 2 ASSESSMENT

Checking for Understanding

1. **Define** city-state, Middle Ages, feudalism, Crusades, Renaissance, Reformation, Enlightenment, industrial capitalism, communism, reparations, Holocaust, Cold War, European Union.
2. **Main Ideas** Create a section outline like the one below. Fill in headings from this section, and add supporting details.

European History and Government

I. The Rise of Europe
  A. Early peoples
    1. Prehistoric Europeans
    2.
  B. Ancient Greece and Rome

Critical Thinking

3. **Identifying Cause and Effect** How did the rise of communism and working-class movements affect the lives of workers in the 1800s?
4. **Drawing Conclusions** How were the countries of eastern Europe affected by World War I? By World War II?
5. **Predicting Consequences** How might an organization such as the European Union encourage unity among the various European countries? Provide examples to support your answer.

Analyzing Maps

6. **Region** Study the map of the Roman Empire on page 295. Compare it with the map of Europe 2001 on page 299. Which areas of Europe were not conquered by the Romans?

Applying Geography

7. **Causes of Political Change** Study the maps of Europe in 1955 and in 2001 on page 299. Choose one area whose boundaries changed between 1955 and 2001, and write a paragraph explaining the causes and effects of these changes.

Cultures and Lifestyles

Guide to Reading

Consider What You Know

The cultures of Europe have had a profound influence on the rest of the world. How many European artworks—paintings, sculptures, literary works, musical compositions, or works of architecture—can you list? Can you name the artist who created each work, and the country where the artist worked or lived?

Read to Find Out

- How has religion influenced the cultural development of Europe?
- Why has European art and culture been so influential throughout the world?
- How do European governments meet the educational and health-care needs of their peoples?

Terms to Know

- dialect
- language family
- Good Friday Peace Agreement
- romanticism
- realism
- impressionist
- welfare state

Places to Locate

- Switzerland
- Northern Ireland
- the Netherlands
- Ukraine

NATIONAL GEOGRAPHIC

A Geographic View

Denmark's Two Seasons

Life in Denmark is divided into two parts, the Golden Summer and the Great Murk, which extends from late fall to mid-spring. The months of youth and beauty [are] when the sky is light until almost 11 p.m. and Danes take to the beaches, eat in their gardens, soak up the sun, feel sleek and smart. . . . [T]he other months [are] when they go to and from work in the dark and the rain and just try to keep putting one foot in front of the other and not get too glum.

—*Garrison Keillor, "Civilized Denmark,"* National Geographic, *July 1998*

Sidewalk café in Denmark

In Denmark, as in other countries of Europe, people have developed distinct ways of life in response to their physical environment. At the same time that Europe becomes more united politically and economically, its peoples struggle to maintain their separate cultural identities. In this section you will learn about the cultural characteristics that both unite and divide Europeans as well as impact their everyday lives.

Expressions of Culture

Like people in other regions, Europeans express their values through language, religion, and the arts. A study of European languages, religions, and art forms reveals the rich diversity of culture in Europe today.

Languages

In Europe there are about 50 different languages and more than 100 **dialects**, or local forms of languages. At times dialects are so different that even people speaking the same language have difficulty understanding one another. For example, a dialect of English called Orkney is spoken in the Orkney Islands off Scotland's northern coast:

> *"[The Orkney dialect] combines Old Norse words with many unique or archaic English expressions in a way that leaves outsiders—known as 'eens fae aff,' literally 'ones from off'—hopelessly befuddled. [An Orkney resident observes:] "When someone says to you, 'We hid a quey caff yistreen' (We had a female calf last night), you know you are on the fringes of the English-speaking world."*
>
> Bill Bryson, "Orkney: Ancient North Sea Haven," *National Geographic,* June 1998

Almost all of Europe's languages and dialects, however, belong to the Indo-European language

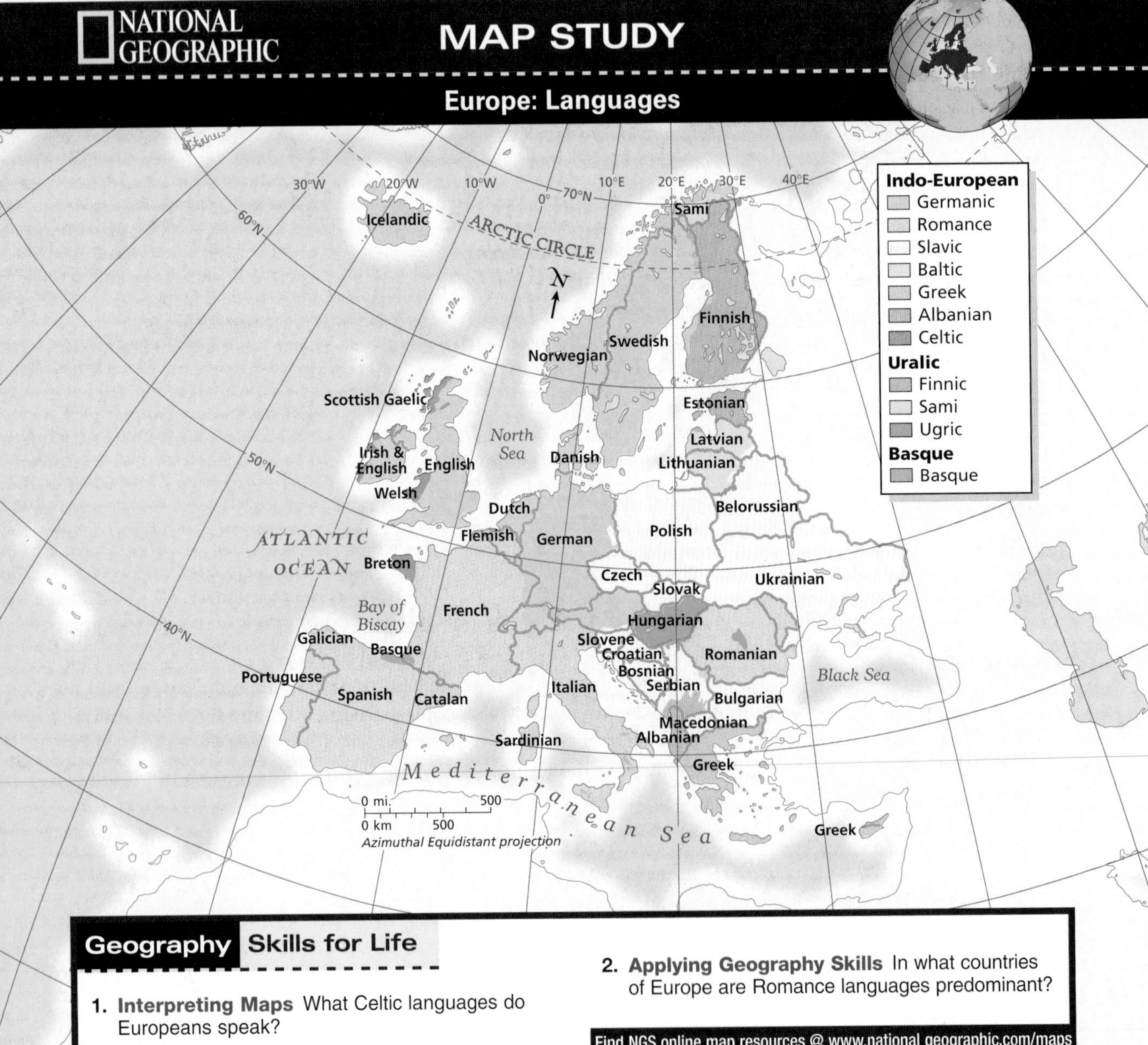

Geography Skills for Life

1. **Interpreting Maps** What Celtic languages do Europeans speak?

2. **Applying Geography Skills** In what countries of Europe are Romance languages predominant?

Find NGS online map resources @ www.national geographic.com/maps

NATIONAL GEOGRAPHIC

GRAPH STUDY

Europe: Religions

Religion	Number of Followers
Roman Catholic	286,124,000
Eastern Orthodox	134,000,000
Protestant	86,000,000
Other Christian*	26,000,000
Muslim	17,000,000
Jewish	1,900,000
Other religions	5,000,000
Nonreligious	26,000,000

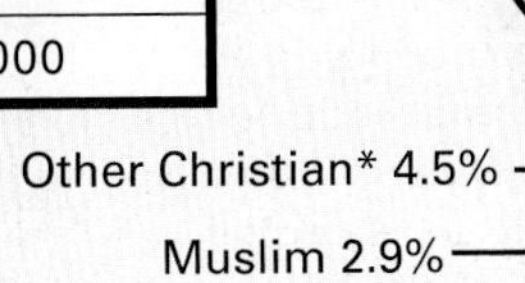

*mainly Anglican (Church of England)
Sources: World Almanac, 2001; Britannica Book of the Year, 2000

Geography Skills for Life

1. **Interpreting Graphs** How does the percentage of Europeans who are Roman Catholic compare with the percentage of those who are Protestant? Eastern Orthodox?

2. **Applying Geography Skills** Why do you think religion has both united and divided Europeans throughout their history? Provide examples to support your answer.

family. A **language family** is a group of related languages that developed from an earlier language. The Indo-European family has several major branches in Europe. Most people in eastern Europe speak Slavic languages—including Bulgarian, Czech, Polish, Slovak, Ukrainian, Belorussian, and Serbo-Croatian—or Baltic languages, such as Latvian and Lithuanian. In northern Europe, most people speak Germanic languages—German, Dutch, English, Danish, Swedish, and Norwegian. The Romance languages, which come from Latin, the language of the Roman Empire, are widely spoken in southern Europe. They include Italian, Spanish, Portuguese, French, and Romanian. Other Indo-European branches are Greek, Albanian, and the Celtic languages. Two European language groups are not Indo-European—the Uralic languages (Finnish, Estonian, and Hungarian) and Basque, one of the world's few languages that is not related to any other.

Many European countries have one or more official languages, those that are recognized by the government, and a smaller number of the other languages. For example, Romanian is the official language of Romania, but Hungarian and German are also spoken there. **Switzerland** has three official languages—German, French, and Italian. A fourth language—Romansch, closely related to Latin—is spoken by a small number of Swiss.

Religion

Religion—primarily Christianity—has deeply shaped European values, societies, and cultures. Today many Europeans are not practicing members of a religious body, but they still maintain cultural links to the faiths of their ancestors, especially in celebrating religious holidays. Although many European countries have a largely Christian heritage, others are Muslim or have a diversity of faiths.

music of EUROPE

European music has a history descending from the ancient Greeks through the Christian era and into Western classical music. Traditional music from Europe is based strongly on melody, with less emphasis on percussion and rhythm.

Instrument Spotlight

It is believed that **bagpipes** originated with early European shepherds. The bag traditionally is made from an animal's hide or stomach. Hollow sticks or bones are connected to it so the air can escape in a controlled way to produce different notes. With all bagpipes, sound is produced either by blowing air through an intake tube or by pumping with a bellows held under one arm. Most bagpipes create one or more steady drone notes, produced by pipes called "drones." The other pipe, which plays the melody, is called a "chanter." The most well-known types of bagpipes are from Scotland and Ireland.

Go To **World Music: A Cultural Legacy** Hear music of this region on Disc 1, Tracks 13–19.

Most of Europe's Christians are Roman Catholics, who live in southern Europe, parts of western Europe, and the northern part of eastern Europe. Protestants, who generally belong to the Anglican, Lutheran, and Reformed churches, are dominant in northern and northwestern Europe. Eastern Orthodox churches are strongest in the southern part of eastern Europe. Many Muslims live in Albania, Bosnia-Herzegovina, and Bulgaria. Jewish communities are found in all major European cities.

Religious leaders, such as Pope John Paul II, head of the Roman Catholic Church, inspired religious believers in eastern Europe in the struggle against communist controls. With the fall of communism and its antireligious policies, religious freedom came to eastern Europeans during the 1990s.

Although religion unites some Europeans, it divides others. For years, hostility between Catholics and Protestants led to conflict in **Northern Ireland**, a part of the United Kingdom. Roman Catholics there wanted to become part of the largely Catholic Republic of Ireland, and Protestants favored keeping ties with the mostly Protestant United Kingdom. In 1998 the **Good Friday Peace Agreement** paved the way for Protestant and Roman Catholic communities to share political power. Though hopes for peace run high, the political situation in Northern Ireland remains unstable.

Religious and ethnic differences were at the heart of conflict in the Balkan Peninsula. During the early 1990s, Roman Catholic Croats, Eastern Orthodox Serbs, and Muslim Bosnians fought over land and political power in Bosnia-Herzegovina. Later in the decade, Eastern Orthodox Serbs fought the Albanian Muslim majority in the Serb province of Kosovo.

The Arts

As a result of Europe's global influence in the 1800s and 1900s, European art forms have spread around the world and influenced other cultures. The arts of Europe reflect its history as well as the ideas and values of its people.

Europe's temples and churches show the close relationship of religion and architecture. The

Parthenon in Athens and the Pantheon in Rome are examples of temples built by the ancient Greeks and Romans. The Roman Catholic cathedral in Córdoba, Spain, once was a mosque built by North African Muslims who brought Islam to Spain. The Church of the Holy Apostles in Salonica, Greece, is an example of Byzantine art that reflects Eastern Orthodox spirituality. Notre Dame Cathedral in Paris is an example of the Gothic architecture that flourished in Roman Catholic western Europe from the mid-1100s to the 1400s.

During the 1500s and 1600s, European artists and writers began to work with everyday subjects as well as religious themes. The paintings of Leonardo da Vinci and Michelangelo Buonarotti influenced generations of artists. England's William Shakespeare wrote numerous plays, and Spain's Miguel de Cervantes penned *Don Quixote*, a classic novel about a landowner who imagines himself a knight called to perform heroic deeds.

In the 1600s and 1700s, new music forms, such as opera and the symphony, emerged in Europe. In the 1800s artists such as French painter Eugène Delacroix, British writer Sir Walter Scott, and German composer Ludwig van Beethoven reflected the style of **romanticism**, which focused on the emotions, stirring historical events, and the exotic. During the mid-1800s, **realism**—an artistic style that focused on accurately depicting the details of everyday life—became prominent. Later in the century, a group of French painters called **impressionists** moved outdoors from their studios to capture immediate experiences, or "impressions," of the natural world.

During the 1900s, European artists and writers explored a variety of new forms and styles. Abstract painting and sculpture, which emphasized form and color over realistic content, became dominant. An important European artist who influenced modern art was the Spanish painter Pablo Picasso. In architecture, Germany's Bauhaus school of design emphasized clean geometric forms and the use of glass and concrete.

Quality of Life

Today western Europe, with its heritage of industrial and urban growth, generally enjoys a higher standard of living than southern and eastern Europe. Many eastern European countries especially struggle with problems inherited from the communist past or are rebuilding economies damaged by recent warfare or internal unrest. The

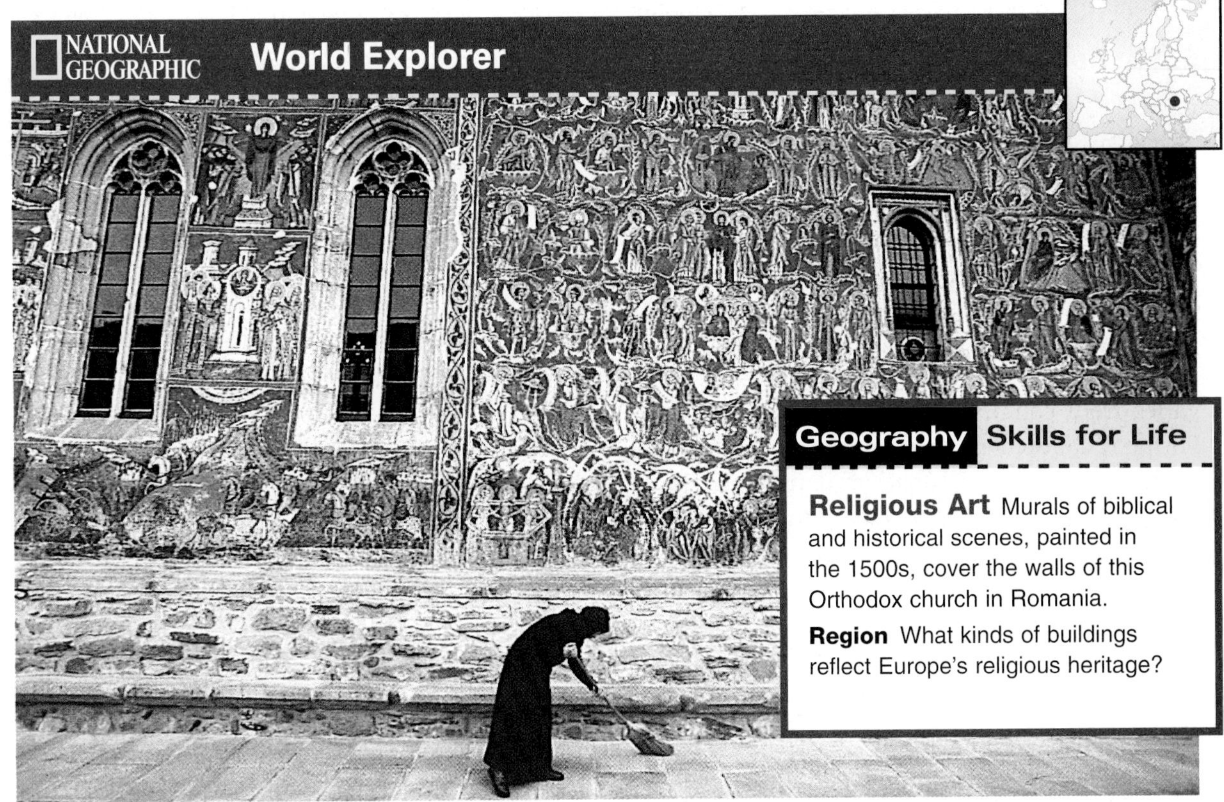

Geography Skills for Life

Religious Art Murals of biblical and historical scenes, painted in the 1500s, cover the walls of this Orthodox church in Romania.

Region What kinds of buildings reflect Europe's religious heritage?

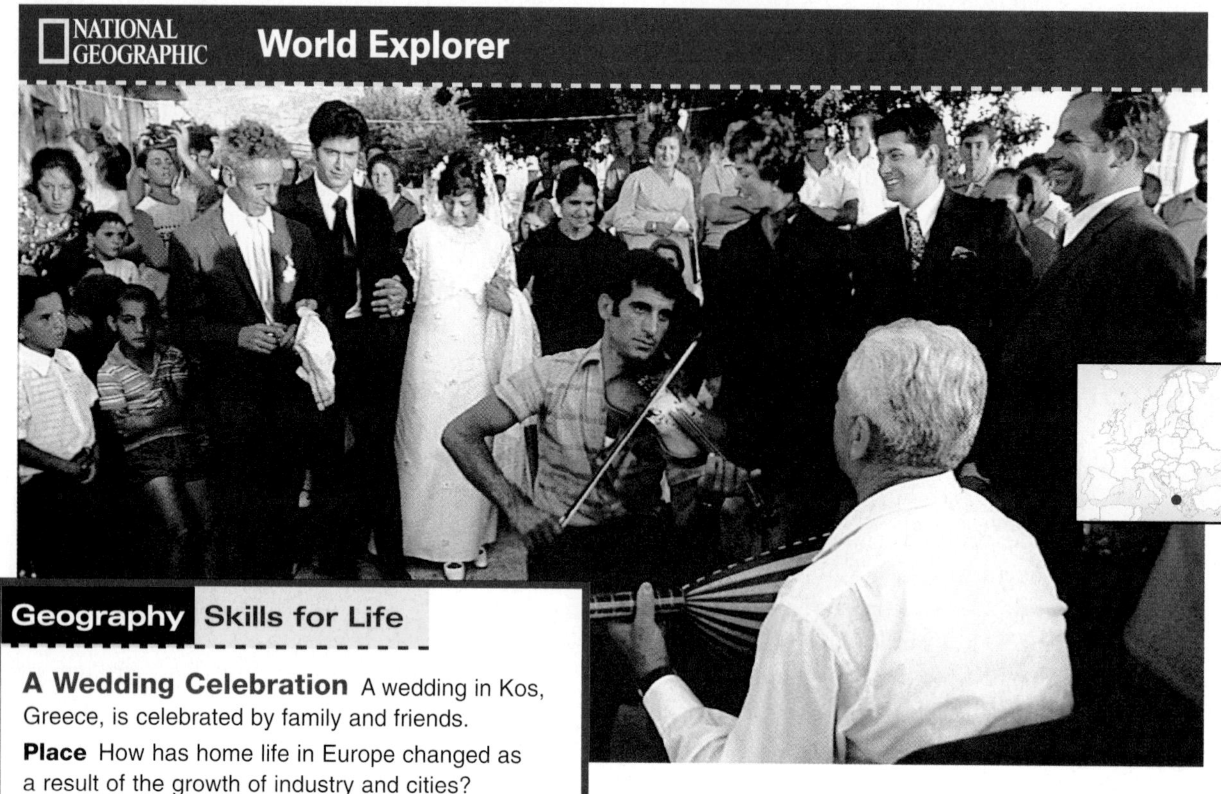

Geography Skills for Life

A Wedding Celebration A wedding in Kos, Greece, is celebrated by family and friends.

Place How has home life in Europe changed as a result of the growth of industry and cities?

gap in the quality of life among various parts of Europe poses an obstacle to full European unity.

Education

Respect for education is a traditional European value. The quality of education in Europe, whose people are among the world's best educated, is linked to economic performance. Countries with a high standard of living can afford to improve schools and provide specialized training for students. With the exception of war-torn Balkan countries, European countries have literacy rates above 90 percent.

The number of years of required schooling varies from country to country. For example, in Portugal children must attend school for only 6 years, but the United Kingdom requires 12 years. Many European school systems provide preparation for either college or vocational training.

Economics

State-Sponsored Human Services

Some European countries, such as Sweden and the United Kingdom, offer complete social welfare programs to their citizens. These countries, known as **welfare states**, have tax-supported programs for higher education, health care, and social security. The government of Sweden is Europe's most wide-ranging provider of human services. For example, each Swedish family receives an allowance for every child under 16 years of age and for secondary or university students. Also, single parents with low incomes can obtain allowances for family vacations.

Funding social programs is expensive for European governments. Many countries, such as Sweden and France, have spent large portions of their national budgets to provide social services. In recent years European governments have had to tighten their budgets and limit human services to citizens most in need. This cutback has met with opposition from trade unions and voters.

Lifestyles

Cultural and economic differences have produced a variety of lifestyles in Europe. These differences, however, have lessened as industrialization, urbanization, and technological advances have brought a common culture to many places.

Home Life

Extended families often shared homes and economic resources in Europe before the Industrial Revolution. As the number of Europeans moving to the cities increased, this traditional pattern changed. Today women in most European countries have entered the workforce, families are more mobile, and government agencies tend to many social concerns once handled by families. Still, family life remains important in Europe. In many European cultures, life revolves around the extended family. Even when young people move away from home, they often maintain close family ties.

Sports and Recreation

Soccer is a major sport in Europe, and many countries have professional soccer teams. Rugby football is a popular team sport, especially in the United Kingdom, France, and Ireland. Many Europeans play tennis for recreation, and the British tennis tournament at Wimbledon is a major international championship.

Some European sports evolved in response to a country's climate, landscape, or culture. In Spain, soccer's popularity only recently surpassed that of bullfighting. In **the Netherlands**, the *Elfstedentocht*, or Eleven Cities Tour, is a Dutch ice-skating marathon along frozen rivers and canals. Winter sports, such as downhill skiing in the Alpine regions, cross-country skiing in Scandinavia, and ice-skating in **Ukraine** (yoo•KRAYN), have made European athletes famous in the Winter Olympics.

Celebrations

Europeans celebrate many of the same religious holidays observed in other parts of the world, although their celebrations are marked by distinctive traditions. Greeks celebrate Easter with a feast of roast lamb, and Ukrainians share intricately decorated eggs called *pysanky*. European Jews make potato pancakes called latkes to eat during the eight-day festival of religious freedom known as Hanukkah. Muslim families gather for family feasting at the end of Ramadan, a month-long period of fasting during daylight hours.

European Roman Catholics celebrate local festivals in honor of patron saints. Many festivals blend Christian symbolism with customs that date back to pre-Christian times.

Other European holidays mark the change of seasons or patriotic events. In the British Isles, for example, Yule logs and mistletoe decorate homes at the winter solstice. On July 14 the French celebrate Bastille Day to commemorate the storming of the Bastille prison in 1789 and the start of the French Revolution. Countries such as Denmark and the Netherlands celebrate the birthdays of their reigning monarchs as national holidays. Celebrations help Europe's peoples maintain their cultural heritages even as they move toward greater unity.

SECTION 3 ASSESSMENT

Checking for Understanding

1. **Define** dialect, language family, Good Friday Peace Agreement, romanticism, realism, impressionist, welfare state.
2. **Main Ideas** Use a web to organize major cultural influences on the people of Europe.

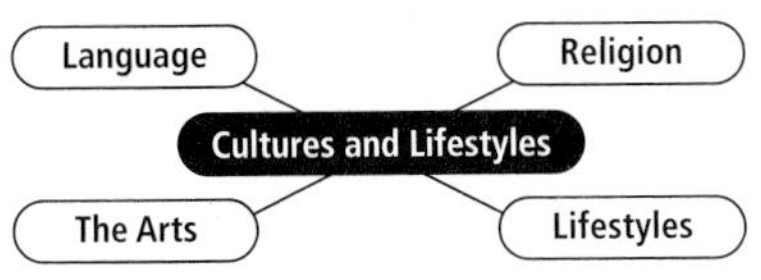

Critical Thinking

3. **Making Generalizations** How has Europe influenced the arts of other cultures?
4. **Comparing and Contrasting** How do European welfare states differ from the social welfare systems in the United States?
5. **Predicting Consequences** How will recent political and economic developments in Europe produce greater similarities or differences in lifestyles and quality of life?

Analyzing Maps

6. **Region** Study the map of Europe's languages on page 302. Where are Germanic languages predominant?

Applying Geography

7. **Historical Geography** Research daily life in Rome today and as it was in ancient times. Describe the human characteristics of the city then and now. What are the similarities and the differences?

Reading a Political Map

Lines on a map that indicate counties, states, and countries are called political boundaries because they divide areas controlled by different governments. A political map illustrates these divisions.

Learning the Skill

Unlike physical maps, which remain fairly constant over time, political maps change as political relationships shift. By comparing political maps from different historical periods, you can observe changes in political relationships over time.

On political maps of large areas, lines indicate boundaries between countries, or *national boundaries*. Political maps may also include cities, counties, or provinces. A map key can show the symbols for national boundaries, national capitals, and state or provincial capitals.

To interpret a political map:

- **Read the map's title to identify the geographic area.**
- **Note the time period the map reflects.**
- **Identify the countries or other political units named on the map.**
- **Use the information from the map to make generalizations about the history, government, and political geography of the region.**

Practicing the Skill

Use the political map of present-day Scandinavia to answer the following questions.

1. Which Scandinavian capital lies farthest south?
2. Which Scandinavian countries have port cities on the Baltic Sea?
3. Which Scandinavian country owns the Faroe Islands? Which country owns Jan Mayen Island?
4. Which Scandinavian country probably has the strongest historical and cultural ties to Germany? Explain your answer.

Use an encyclopedia or the Internet to find a political map showing Scandinavia in the mid-1600s. Write a paragraph comparing this map to the present-day political map.

The **Glencoe Skillbuilder Interactive Workbook, Level 2** provides instruction and practice in key social studies skills.

CHAPTER 12 SUMMARY & STUDY GUIDE

SECTION 1 Population Patterns (pp. 287–291)

Terms to Know
- ethnic group
- ethnic cleansing
- refugee
- urbanization

Key Points
- Europe's cultures and ethnic groups are diverse.
- Physical features, climate, and resources have affected the region's population density and distribution.
- Industrialization, urbanization, and patterns of migration have helped define Europe as a region.

Organizing Your Notes
Create a table like the one below. Fill in details about each aspect of Europe's population patterns.

Ethnic Diversity	Population Characteristics	Urbanization

SECTION 2 History and Government (pp. 294–300)

Terms to Know
- city-state
- Middle Ages
- feudalism
- Crusades
- Renaissance
- Reformation
- Enlightenment
- industrial capitalism
- communism
- reparations
- Holocaust
- Cold War
- European Union

Key Points
- The contributions of Greek and Roman civilizations have influenced much of European history.
- During the Middle Ages, Christianity played a major role in shaping European societies.
- Trade, colonization, and immigration spread European cultures to other continents.
- After World War II, the Cold War divided communist-controlled eastern Europe from noncommunist western Europe.
- The European Union was formed to promote economic unity and stability among European countries.

Organizing Your Notes
Create a time line that shows key dates in European history. The time line below has been started for you.

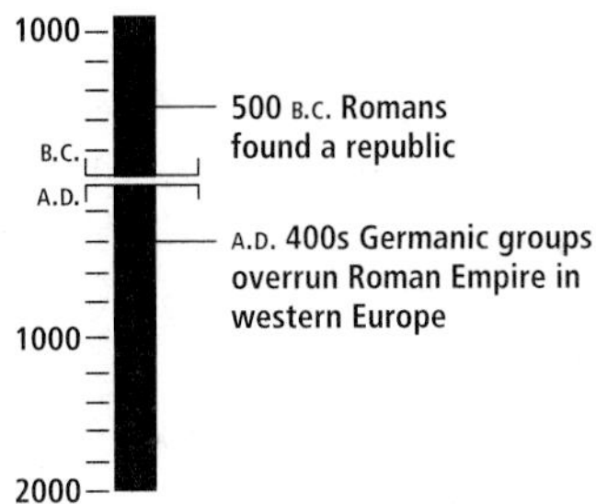

SECTION 3 Cultures and Lifestyles (pp. 301–307)

Terms to Know
- dialect
- language family
- Good Friday Peace Agreement
- romanticism
- realism
- impressionist
- welfare state

Key Points
- Most of Europe's various languages belong to one language family.
- Religion has influenced European values and has sometimes contributed to conflicts.
- Because of colonialism, European art and culture have profoundly influenced the Western world.
- Eastern and western European countries have differences in standards of living.
- Some European governments provide comprehensive social services to their citizens.

Organizing Your Notes
Create a table like the one below to help you organize your notes for this section. Give examples of each aspect of European culture.

European Culture	Examples
Language	
Religion	
The Arts	

ASSESSMENT & ACTIVITIES

Reviewing Key Terms

On a sheet of paper, classify each of the lettered terms below into the following categories. (Some key terms may apply to more than one category.)

- **European Peoples**
- **Art and Ways of Life**
- **Historical Development**

a. romanticism
b. ethnic cleansing
c. city-states
d. Middle Ages
e. feudalism
f. Renaissance
g. urbanization
h. reparations
i. welfare state
j. European Union
k. Reformation
l. refugees
m. realism
n. Cold War

Reviewing Facts

SECTION 1

1. From which world regions have many of Europe's more recent immigrants arrived?
2. Which parts of Europe are the most densely populated?
3. What has contributed to the rapid rise of urbanization in Europe?

SECTION 2

4. What civilizations shaped early Europe?
5. What important revolutions changed Europe in the 1700s and 1800s?

SECTION 3

6. What political changes swept through Europe during the second half of the twentieth century?
7. What language family includes most of the European languages? Give examples.
8. Name the religion that most Europeans practice. What are its three major branches?

Critical Thinking

1. **Drawing Conclusions** How did Europe's Christians relate to Jews and Muslims during the Middle Ages?
2. **Predicting Consequences** How might the European Union affect the social services provided by European governments?
3. **Identifying Cause and Effect** Create a diagram that shows factors that led to conflicts in Northern Ireland. Then describe what has been done to resolve them.

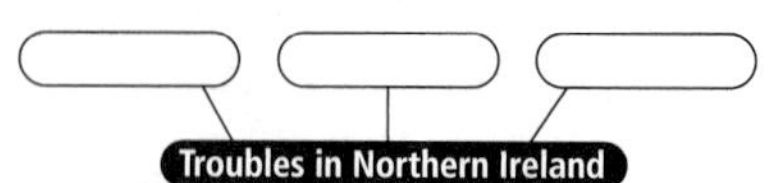

NATIONAL GEOGRAPHIC

Locating Places

Europe: Political Geography

Match the letters on the map with these capitals of European countries. Write your answers on a sheet of paper.

1. **Berlin**	5. **Lisbon**	9. **London**
2. **Oslo**	6. **Belgrade**	10. **Rome**
3. **Brussels**	7. **Kiev**	11. **Paris**
4. **Warsaw**	8. **Sarajevo**	12. **Vienna**

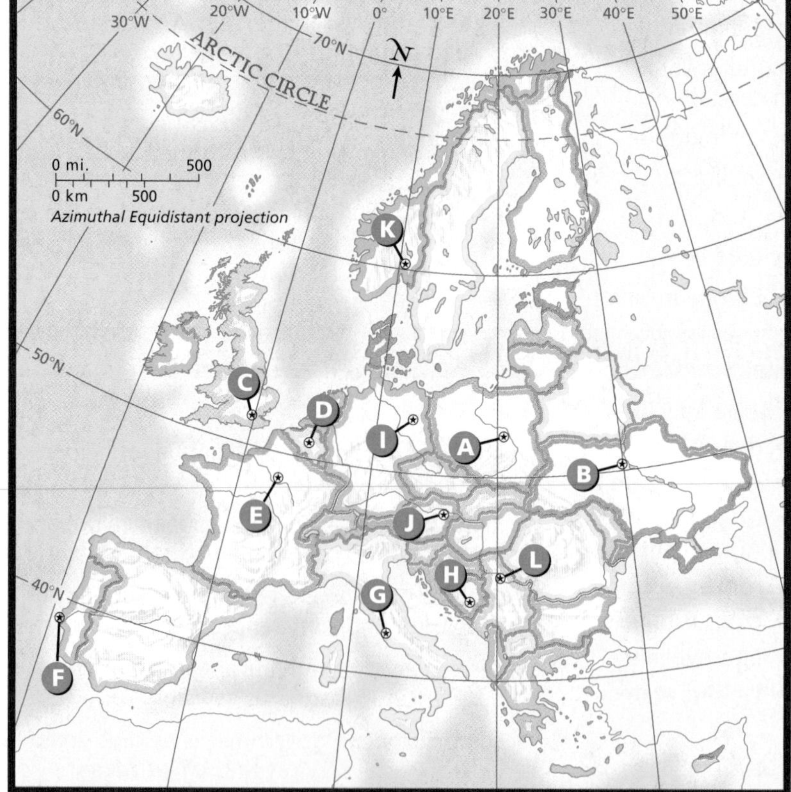

Self-Check Quiz Visit the **Glencoe World Geography** Web site at geography.glencoe.com and click on Self-Check Quizzes—Chapter 12 to prepare for the Chapter Test.

Using the Regional Atlas

Refer to the Regional Atlas on pages 260–263.

1. **Human-Environment Interaction** Compare the population density map and the economic activity map. Describe the population patterns in manufacturing areas.
2. **Location** What capitals are located on the North European Plain? What are the advantages of this location?

Thinking Like a Geographer

Use your textbook, library sources, and the Internet to answer the following: How did Europe's culture spread overseas? How did physical geography affect Europe's expansion?

Problem-Solving Activity

Contemporary Issues Case Study Membership in the European Union offers many benefits but also many challenges. Choose one EU member country, and research the issues that were raised before it joined the EU. Then analyze the situation in the country today, highlighting the benefits and difficulties of EU membership. Present your findings as a case study.

GeoJournal

Expository Writing Using the information you logged in your GeoJournal, explain in a paragraph how two groups of Europeans are attempting to overcome the cultural differences that exist between them.

Technology Activity

Using the Internet for Research Search the Internet for the Web sites of European museums, such as the Louvre in Paris, France, and the Uffizi Gallery in Florence, Italy. Look for information about European architecture, sculpture, paintings, or historical artifacts. Choose one work that expresses the spirit of a particular European country or historical period, and explain your choice to the class. Provide photographs or illustrations of the work you chose.

Standardized Test Practice

Choose the best answer for the following multiple-choice question. If you have trouble answering the question, use the process of elimination to narrow your choices.

Selected Countries in Eastern Europe

1. **Study the locations of the Czech Republic, Slovakia, and Hungary. How might the locations of these three countries affect their role in world trade?**

   A They are smaller.
   B They are landlocked.
   C They are communist.
   D They have many resources.

Notice that the question asks you to base your answer on location. Three of the choices deal with resources, form of government, and size. Choice B, however, focuses on the countries' locations away from seas, which would influence their role in world trade.

CHAPTER 13

Europe Today

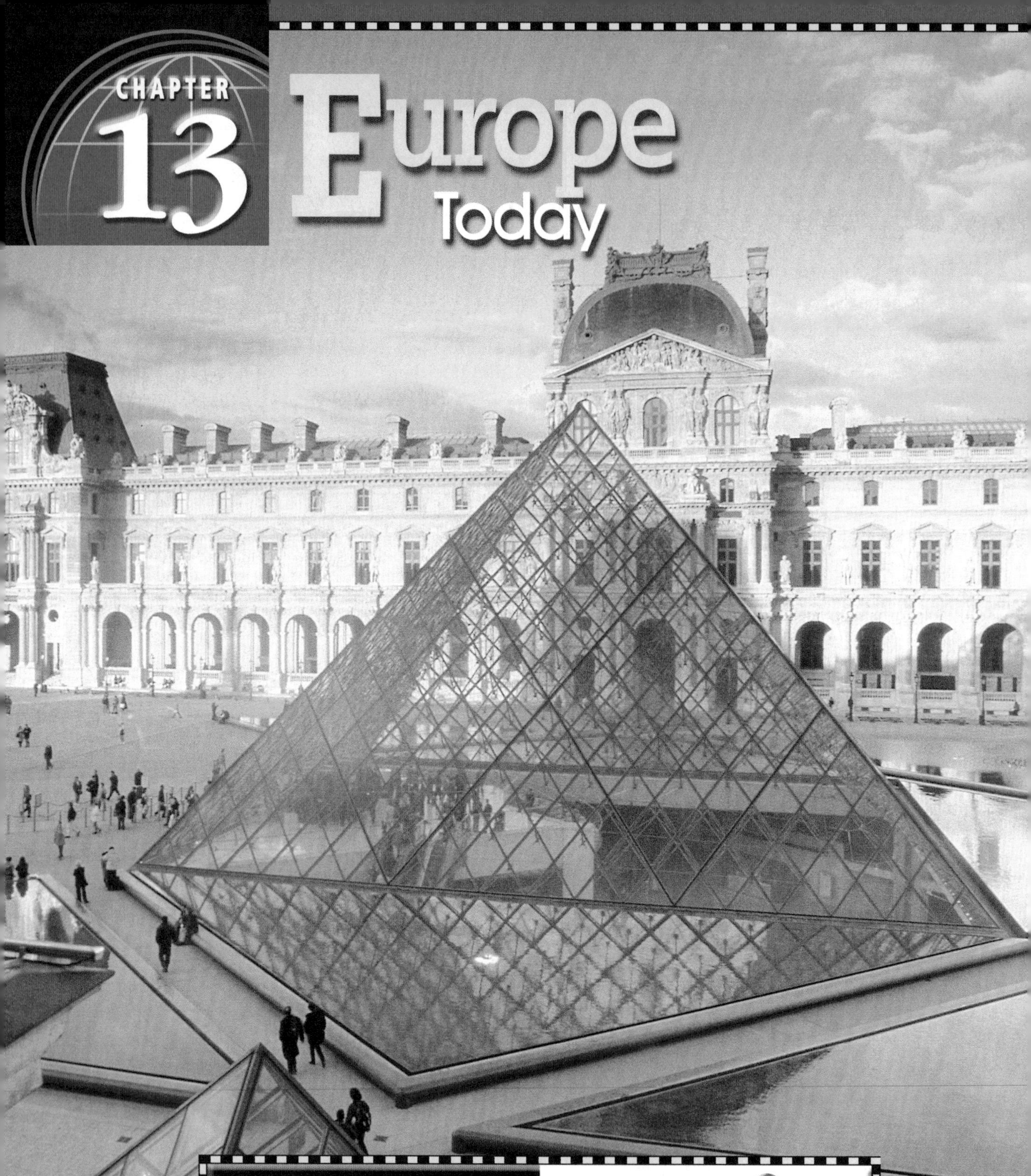

GeoJournal

As you read this chapter, note the ways Europeans are striving to care for their environment. Choose one environmental challenge, and write a short essay comparing Europeans' solutions to measures in your community.

GEOGRAPHY Online

Chapter Overview Visit the **Glencoe World Geography** Web site at geography.glencoe.com and click on Chapter Overviews—Chapter 13 to preview information about the region today.

SECTION 1

Living in Europe

Guide to Reading

Consider What You Know

Based on what you learned in Chapter 12, how do you think Europeans are adjusting to the changes caused by the fall of communism?

Read to Find Out

- What economic systems are found in Europe?
- Why are economic changes taking place in Europe?
- How do transportation and communications systems link European countries to each other and to the rest of the world?

Terms to Know

- European Union (EU)
- Maastricht Treaty
- heavy industry
- light industry
- mixed farming
- farm cooperative
- collective farm
- state farm
- genetically modified food
- organic farming

Places to Locate

- Ruhr
- Denmark

◀ *Pyramid entrance of the Louvre Museum, Paris, France*

NATIONAL GEOGRAPHIC

A Geographic View

A New Europe

By 5:30 A.M., 3,000 or 4,000 workers of the first shift are pouring through the gates of the iron and steel works on the Danube River island of Csepel. Expanded by ardent communists in the 1950s, it became Hungary's largest industrial site. . . . Today's worker wants to become part of the middle class, to own a car and a weekend cottage in the country. "That's what I want," says Gábor Szabó, a young welder, "to become a European."

—*Tad Szulc, "Dispatches from Eastern Europe,"* National Geographic, *March 1991*

Eastern European miners

Like many eastern Europeans during the early 1990s, this factory worker in Hungary was eager to leave behind the dark legacy of communism and share in the prosperity that democratic western Europe enjoyed. Today, despite many difficulties, the countries of eastern Europe are building democracies and market economies. As standards of living rise, people in these countries also are developing closer ties to western Europe. Throughout Europe, people still remain proud of their individual national identities, but they are also beginning to identify with the region as a whole. In this section you will learn about Europe's social, political, and economic systems and the recent changes that are transforming them.

Changing Economies

Europe's economies, like its peoples, are diverse and changing. Today Europe is one of the world's major manufacturing and trading regions. The **European Union (EU)**, which unites much of western Europe into one trading community, enjoys a greater volume of trade than any single country in the world. Meanwhile, the former communist countries of eastern Europe are trying to build free market economies. Some also seek to eventually become part of the European Union.

Economics

The European Union

The movement for European unity arose from the ashes of World War II, as western European countries struggled to rebuild their ruined economies. In 1950 France proposed closer links among Europe's coal and steel industries, a move seen as the first step toward a united Europe.

Over the years more steps were taken toward that goal, but not until the 1990s did most Europeans agree that such a goal could ever be reached. In 1992 representatives from various European governments met in Maastricht, the Netherlands, and signed the **Maastricht Treaty**, which set up the European Union (EU). This new body aimed to make Europe's economies competitive with those of the rest of the world by getting rid of restrictions on the movement of goods, services, and people across its members' borders. It also paved the way for a single European currency, a central bank, and a common foreign policy.

Since the EU was formed, member countries have worked to boost trade and to make their economies more efficient and more productive. They have also tried to control government spending for many costly social welfare programs. Many Europeans, however, oppose scaling down the welfare state, believing that such a step would increase hardships on people during times of rising unemployment. The EU continues to work toward the goal of a stronger single economy in spite of the difficulties brought on by change. In the years ahead, the European Union plans to extend its membership to include a number of additional countries, mainly in eastern Europe.

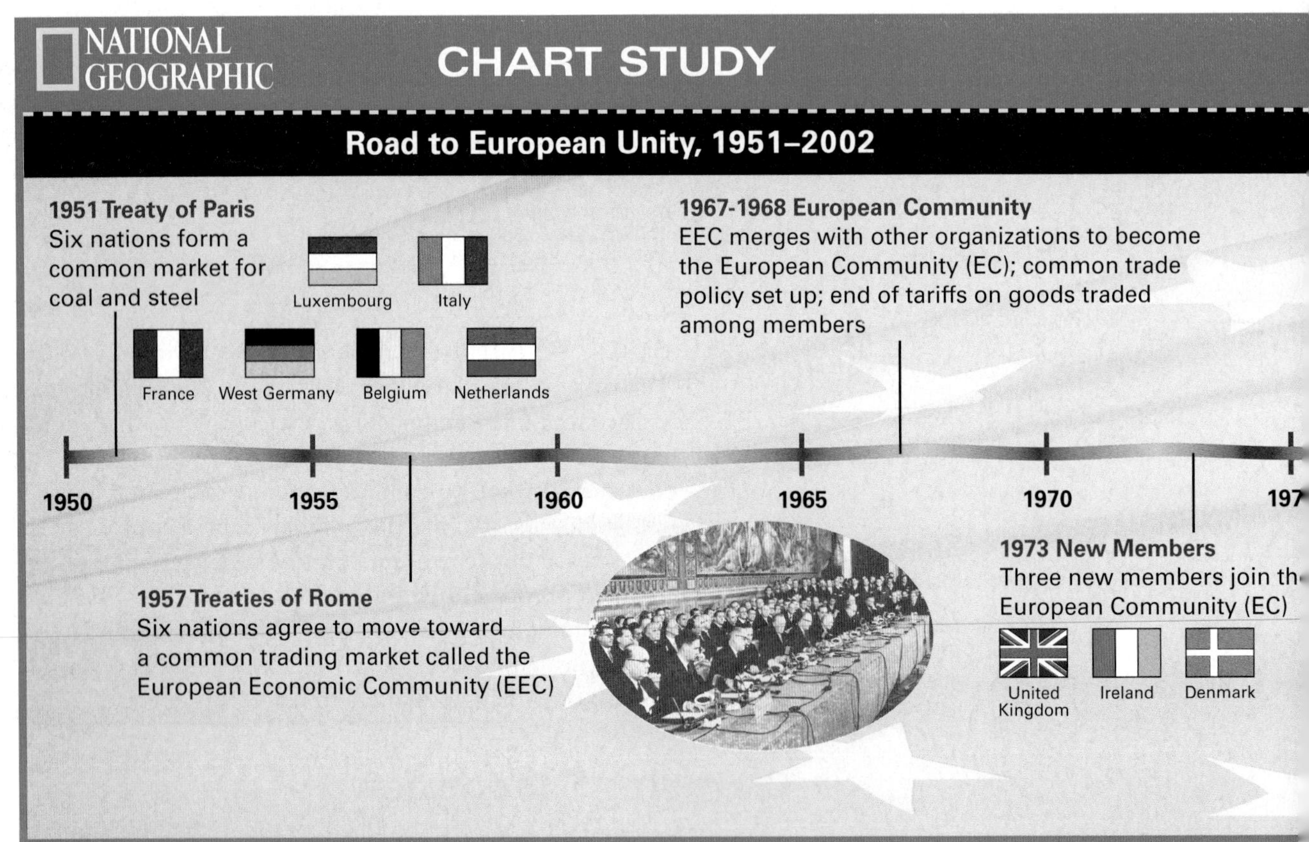

Eastern Europe

For more than 40 years after World War II, communist governments loyal to the Soviet Union ran eastern Europe's command economies. Under these systems, government planners made decisions about what goods to produce and how to produce them. Industries employed many more workers and managers than they needed, and many factories lacked modern technology.

Since the fall of communism in 1989, eastern European countries have been moving from command economies to market economies. To compete in global markets, eastern European industries are working to overcome the obstacles of outdated equipment and inefficient production methods. Many laid-off workers are being retrained, as industries try to acquire new technology and to adopt energy conservation measures to reduce pollution. Eastern European governments are seeking to attract investments and financial aid from western Europe and other parts of the world.

Eastern Europeans have realized, however, that change is often costly and difficult. Workers have lost part of their social "safety net"—the free health care, child care, lifetime jobs, and other social benefits—provided by the communist system. With reduced benefits, death rates among newborns have risen in some parts of eastern Europe, and life expectancy levels have declined. Despite these difficulties, however, people in eastern Europe are slowly adjusting to a new way of life. For example, Germany—reunited in 1990—has faced challenges in improving industrial and living conditions in its eastern part, once under

Student Web Activity Visit the **Glencoe World Geography** Web site at geography.glencoe.com and click on Student Web Activities—Chapter 13 for an activity about the European Union.

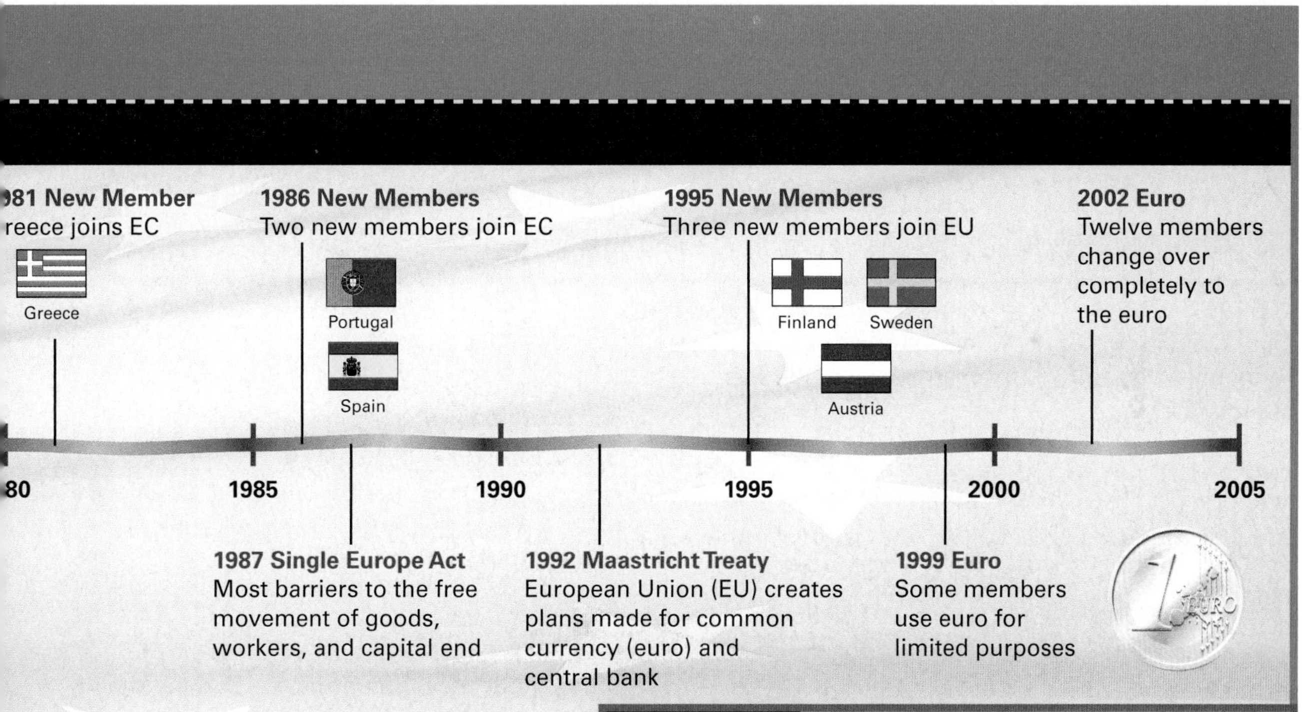

981 New Member
reece joins EC
Greece

1986 New Members
Two new members join EC
Portugal
Spain

1995 New Members
Three new members join EU
Finland
Sweden
Austria

2002 Euro
Twelve members change over completely to the euro

80 | 1985 | 1990 | 1995 | 2000 | 2005

1987 Single Europe Act
Most barriers to the free movement of goods, workers, and capital end

1992 Maastricht Treaty
European Union (EU) creates plans made for common currency (euro) and central bank

1999 Euro
Some members use euro for limited purposes

Geography Skills for Life

1. **Interpreting Charts** When did the United Kingdom join the European Community?
2. **Applying Geography Skills** What steps have European countries taken toward unity since 1985?

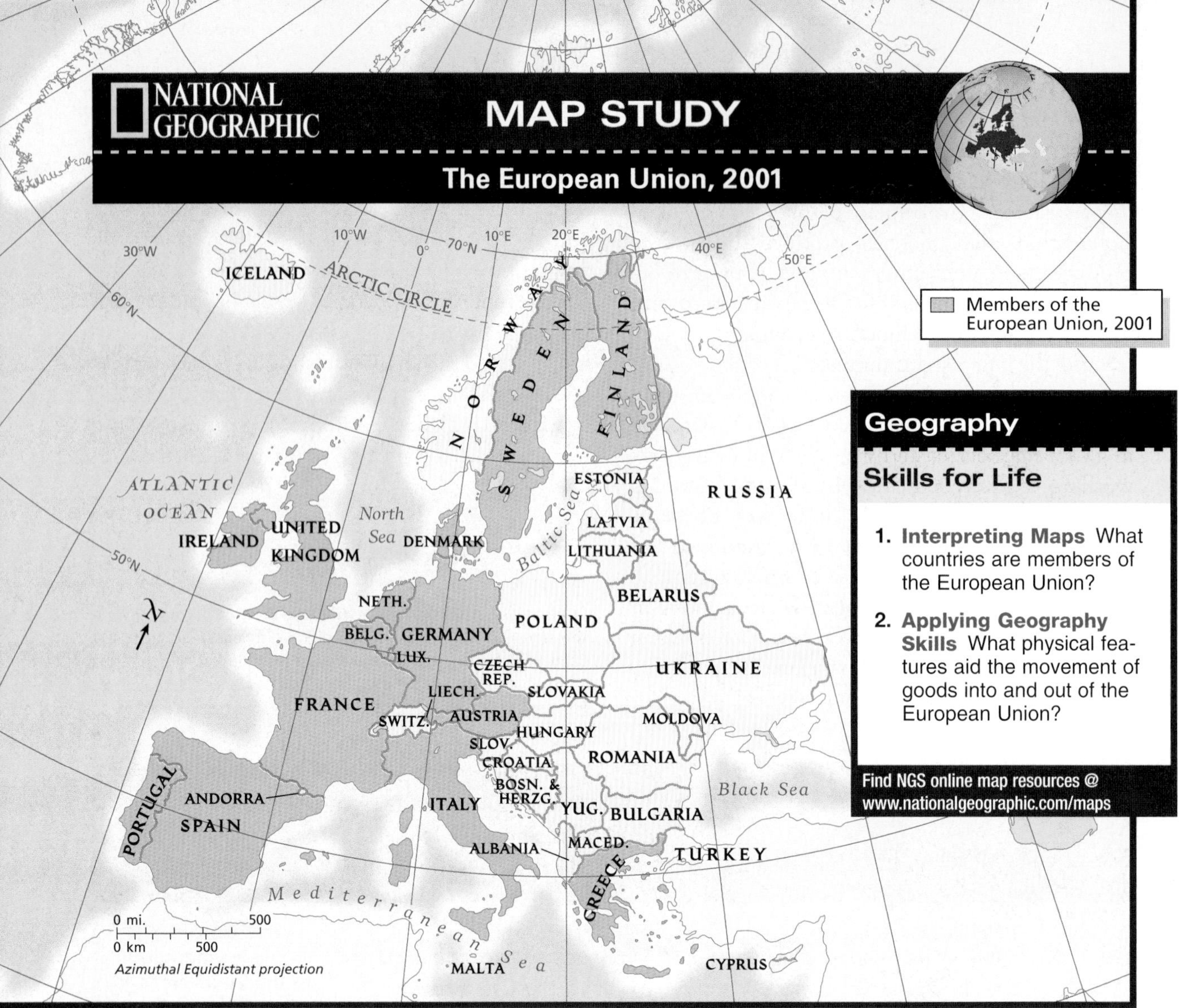

communist control. Yet for many people, life has improved:

> *"Despite the . . . joblessness in eastern Berlin, there are thousands of quiet success stories. . . . 'It was the fulfillment of a dream,' says Stefan Geissler, a 37-year-old former publishing clerk in the old East Berlin, who started the business with a partner, using their combined savings of $2,250. Despite a shaky beginning, Goethe & Co. is now making a small profit. 'Everybody I know is making it—somehow,' Geissler says with conviction. 'People who say "Bring back the Wall" are talking nonsense. Most people are better off today.'"*
>
> Peter Ross Range, "Reinventing Berlin," *National Geographic*, December 1996

Industry

The Industrial Revolution made Europe the birthplace of modern industry. Today large-scale manufacturing centers are found across Europe from the United Kingdom to Poland. In both heavy and light industries, Europe produces everything from computers and cellular phones to transportation equipment and packaged goods.

Manufacturing

The development of industry is often linked to the availability of raw materials. In the 1800s Europe's large deposits of coal and iron ore sparked the growth of heavy industry—the manufacture of machinery and industrial equipment. Today Europe's leading industrial centers include the **Ruhr** and the Middle Rhine districts in Germany, the Lorraine-Saar district in France, the Po basin in Italy, and the Upper Silesia-Moravia district in Poland and the Czech Republic. Vast mineral deposits help to make the United Kingdom, France, and Germany leaders in manufacturing. Countries lacking industrial raw materials, such as the Netherlands and Denmark, specialize in light industry, such as making textiles or processing food.

Service and Technology Industries

Service industries employ a large percentage of the workforce in most European countries—in fact, about 60 percent of workers in western Europe. International banking and insurance rank among Europe's top service industries. Switzerland and the United Kingdom are leaders in these fields. Belgium serves as the headquarters for hundreds of international companies. Tourism is another large service industry in Europe, especially in the United Kingdom, France, Germany, and Switzerland. As in the United States and Canada, high-technology industries are a growing sector of western Europe's economy. Ireland, for example, has become a leading manufacturer of computer products and software.

▲ *Harvesting wheat in Ukraine*

Agriculture

Although largely industrialized, Europe also has fertile farmland. More Europeans earn a living from farming than from any other single economic activity. Yet the percentage of farmers in each country varies widely. For example, about 50 percent of Albania's workers are farmers, but in the industrialized United Kingdom, fewer than 2 percent engage in agriculture.

Europe's crops vary from area to area. Olives, citrus fruits, dates, and grapes grow in warm Mediterranean areas. Farther north, in the cooler plains region, farmers raise wheat, rye, and other grains as well as livestock. Northern countries, such as Denmark and the Netherlands, are major producers of dairy products. The Scandinavian countries are among the world's leading suppliers of fish.

Farming Techniques

In western Europe, farmers use advanced technology to make the best use of limited agricultural space. Mixed farming—raising several kinds of crops and livestock on the same farm—is common. Most western European farmers own their own land, and the average farm covers about 30 acres (about 12 ha). In **Denmark** and some other countries, farm cooperatives, organizations in which farmers share in growing and selling products, reduce costs and increase profits.

In eastern Europe, the fall of communism has brought many changes to farming. Under communism, farmers worked either on government-owned collective farms, receiving wages plus a share of products and profits, or on state farms, not sharing in the profits but getting wages like factory workers. On both types of farms, outdated equipment and lack of incentive resulted in low crop yields. Since the shift to democracy, private ownership of land and food production has risen, and eastern European farmers are expected to increase yields and profits by using modern equipment and fertilizers.

Agricultural Issues

Throughout Europe, new farming methods have not escaped criticism. Many Europeans, for example, oppose genetically modified foods, foods with genes altered to make them grow bigger or faster or be more resistant to pests. Opponents claim that little is yet known about the safety of these foods. In addition, many consumers also avoid foods grown in fields treated with toxic chemicals to control insects or weeds. Because of the concern about chemical use, some farmers rely on **organic farming**, using natural substances instead of fertilizers and chemicals to increase crop yields.

Despite much agricultural success, western Europe today faces a livestock crisis. In 2001 an outbreak of foot-and-mouth disease in the United Kingdom required the killing of thousands of animals, severely crippling the country's livestock industry. The disease—highly contagious among animals but harmless to humans—then crossed to the European continent. As the disease threatened to spread across Europe, consumer panic led to plummeting beef sales. Fearing a global threat from the foot-and-mouth outbreak, the United States and other countries banned imports of animals, meat, and milk from Europe.

architecture of EUROPE

Eiffel Tower Built in Paris, France, the 984-foot (300-m) Eiffel Tower was an early example of modern architecture using wrought iron construction. French engineer Gustave Eiffel designed the tower for the Paris World's Fair in 1889. The lower section of the tower consists of four arched legs that curve inward until they rise together in a single tapered tower. Stairs and elevators allow people to visit the tower's three platform levels, each with an observation deck. Modern additions to the tower include a meteorological station and a radio station.

Transportation and Communications

Europe's network of highways, railroads, waterways, and airline routes is among the best in the world. Modern communications systems also link most parts of Europe to one another and to the rest of the world. Many of the continent's transportation and communications systems are government-owned, with standards and performance varying from one country to another. Eastern Europe, for example, is trying to improve its less advanced transportation and communications systems to match the quality of those in western Europe.

Railways and Highways

Throughout Europe, railroads move freight and passengers. Rail lines connect the region's major cities and airports as well as link natural resources to major industrial centers. Railroads provide easy access to downtown and suburban areas. Bridges and tunnels carry traffic over or through barriers posed by water, mountains, or valleys. For example, in 2000 Denmark and Sweden opened a rail and road bridge that links Sweden to western Europe for the first time since the last Ice Age.

France pioneered the use of high-speed trains with its introduction in 1981 of *trains à grand vitesse* (TGVs), which means "very fast trains." The fastest trains in the world, TGVs cause less damage to the environment than most other forms of transportation. High-speed rail lines, more economical than airline travel, now also operate in Germany, Italy, and Spain. A high-speed rail triangle links Paris, Brussels, and London, passing beneath the English Channel through the Chunnel, or Channel Tunnel.

A well-developed highway system also links Europe's major cities. Germany's four-lane superhighways, called *autobahnen*, are among Europe's best roads. Europe has the highest number of automobile owners in the world except for the United States. Bicycles and motorcycles also provide popular forms of transportation.

Seaports and Waterways

With its long coastline, Europe has a seafaring tradition. Europe handles more than half the world's international shipping at its bustling ports. Major ports include London, England; Antwerp, Belgium;

Genoa, Italy; Le Havre and Marseille in France; Odesa, Ukraine; and Gdańsk, Poland. Rotterdam, the Netherlands, is the world's largest port in surface area, amount of freight handled, and numbers of ships that it can dock at one time.

Europe has many navigable rivers and human-built canals. The Rhine River and its tributaries carry more freight than any other river system in Europe, providing access to the North Sea for five European countries. The Kiel Canal cuts across southern Denmark and shortens the route between the North Sea and the Baltic Sea. The Main-Danube Canal in Germany links hundreds of inland ports between the North Sea and the Black Sea.

Geography Skills for Life

Rail Travel At London's Victoria Station, travelers wait to board a transcontinental train.

Movement Why do some Europeans use high-speed rail lines rather than airline travel?

Communications Links

Communications systems bring information and programming to Europe. The International Telecommunications Satellite Organization uses a series of communications satellites (INTELSATs) to broadcast and receive television programs. The Eurovision network links most of western Europe, and the Intervision network operates in eastern Europe. The two networks sometimes exchange programming.

Telephone service and print media vary throughout Europe. High-quality telephone service is not as available in eastern Europe as it is in western Europe. Western European telephone systems include extensive cable and microwave radio relay, fiber optics, and satellite systems. A large percentage of western Europeans use cellular phones, electronic mail, and the Internet. Books, magazines, and newspapers continue to shape public opinion in Europe. As democracy has grown in eastern Europe, government censorship of printed materials has ended. In the next section, you will read how people in Europe interact with the region's physical environment.

SECTION 1 ASSESSMENT

Checking for Understanding

1. **Define** European Union (EU), Maastricht Treaty, heavy industry, light industry, mixed farming, farm cooperative, collective farm, state farm, genetically modified food, organic farming.
2. **Main Ideas** Use a chart to organize data about factors affecting Europe's economy.

Economic Impact		
Industry	Agriculture	Transportation and Communications

Critical Thinking

3. **Making Comparisons** What different challenges do eastern and western Europeans face as they move toward a more unified Europe?
4. **Identifying Cause and Effect** Explain how physical geography influenced Europe's economic development.
5. **Drawing Conclusions** What are the advantages and disadvantages of Europe's communications systems?

Analyzing Maps

6. **Region** Study the map of the European Union on page 316. What common cultural and geographic features aided the formation of the EU?

Applying Geography

7. **Agriculture in Europe** Imagine that you are a farmworker in eastern Europe. Write a description about how your work activities have changed since the fall of communism.

SECTION 2

People and Their Environment

Guide to Reading

Consider What You Know

Environmental issues are frequently in the news in the United States. What threats to the environment in European countries have made headlines in U.S. newspapers?

Read to Find Out

- How have industry and farming practices affected Europe's environment?
- What steps are being taken to protect Europe's environment?
- What successes have Europeans had in recent decades in reversing the effects of pollution?

Terms to Know

- dry farming
- acid rain
- meltwater
- acid deposition
- environmentalist
- greenhouse effect
- global warming
- biologist

Places to Locate

- Romania
- Mediterranean Sea
- Strait of Gibraltar
- Carpathian Mountains

NATIONAL GEOGRAPHIC

A Geographic View

A High Price

A Czechoslovak friend described what it was like in the 1950s. . . . "The bright future lay in industrializing as fast as possible. This way we would exploit all natural resources and gain mastery over nature. The technology was often out of date, but we were after short-term benefits—there was no thought of the future environmental consequences."

—*Jon Thompson, "East Europe's Dark Dawn,"* National Geographic, *June 1991*

Factory worker in Czechoslovakia

When communism ended in eastern Europe, the results of nearly 40 years of rapid industrialization were shockingly clear: polluted air and rivers, acres of destroyed forests, and soot-covered, decaying buildings. In this section you will learn about the interaction of Europeans with their environment. You will also discover how Europeans are working together to reverse the effects of pollution, a problem that crosses national borders.

Humans and the Environment

As in other parts of the world, people in Europe face challenges posed by the physical environment. In southern Europe, about 40 million years ago, two tectonic plates collided, thrusting up great mountain ranges, including the Alps and the Apennines. The frequent occurrence of earthquakes in countries such as Italy, Greece, and Macedonia indicates that tectonic changes are still taking place today, and earthquakes may strike with devastating effects. Like peoples in other areas, Europeans affect and are affected by their environment.

People in parts of southern Europe also have to cope with low rainfall. For example, Spain's Meseta is so arid that streams dry up, the ground becomes scorched, and drought is common. The arid climate makes dry farming necessary in this area. **Dry farming** is a way of farming in dry areas that produces crops without any irrigation and relies on farming methods that conserve soil moisture.

The Delta Project

In northwestern Europe, violent Atlantic and North Sea storms strike countries that border the sea, such as the Netherlands and Denmark. During these storms, sea travel is often hazardous along these countries' coasts. In 1953 a severe Atlantic storm, combined with the North Sea's heavy spring tide, flooded the southwest corner of the Netherlands, killing about 1,800 people. For nearly the next 30 years, Dutch engineers carried out the Delta Plan, a project that aimed to prevent such severe flooding. Under the plan, a system of dams and dikes was built to seal off and protect the Netherlands' southwestern coast.

Floods

In recent years heavy rains have lashed much of Europe, causing widespread floods and mudslides. This extreme weather has led to loss of life, property damage, and disruption of transportation networks. Some scientists claim that the natural climate cycle accounts for the rains. Others believe that global warming is responsible.

Pollution

In some ways, Europeans have not dealt wisely with their environment. Over the years, Europe's high concentration of industry and population has had a devastating impact on the land, air, and water. For example, in central Europe's "black triangle," a heavily industrialized area in Poland, eastern Germany, and the Czech Republic, soot covers the ground, and the air bears the smell of sulfur from smokestacks.

Before 1989 eastern European countries had practically no laws to control pollution. With the communist emphasis on rapid industrial growth—not environmental safety—the pollution of the air, water, and soil increased until it affected public health. Although efforts are now under way to clean up the environment, the "black triangle" still bears the scars of poorly considered development from the communist era. Western European countries also have experienced serious environmental damage from the dumping of industrial wastes into the air and water. The European Union (EU) now requires environmental protection and cleanup from its members.

Acid Rain

In the 1960s industries in several European countries built high smokestacks to carry pollution away from industrial sites. This method worked locally, but the pollution directed away from the factories drifted across national borders. The pollution, containing acid-producing chemicals, combined with moisture in the air and fell as **acid rain**. Polluted clouds drifting from the industrial belt of Europe, for example, wither forests in other areas, and increase the trees' vulnerability to insects and disease.

The effects of acid rain are especially severe in eastern Europe, where lignite coal continues to serve as a main fuel source. Also called brown coal, lignite is found close to the earth's surface, making the cost of production low. Lignite, however, burns inefficiently and pollutes heavily. As a result, acid rain has ravaged 35 percent of Hungary's forests, 82 percent of Poland's, and 73 percent of the forests in the Czech Republic and Slovakia.

Acid rain damage is not limited to forests. Acid rain also falls on lakes and rivers. In winter, snow carries the industrial pollution to the ground. In spring, **meltwater**—the result of melting snow and ice—carries the acid into lakes and rivers. As acid concentrations build, fish and other aquatic life die. Nearly 20 percent of Sweden's lakes have no fish. A third of the rivers in the Czech Republic and half of those in Slovakia cannot support aquatic life.

Automobile exhaust also adds acid-forming compounds to the atmosphere. **Acid deposition**, wet or dry acid pollution that falls to the ground, harms not only Europe's natural environment but also its historic buildings. The Acropolis in Athens, the Tower of London, and Cologne Cathedral in Germany all show damage from acid deposition. Statues, bridges, and stained glass windows also show the harmful effects of this type of pollution.

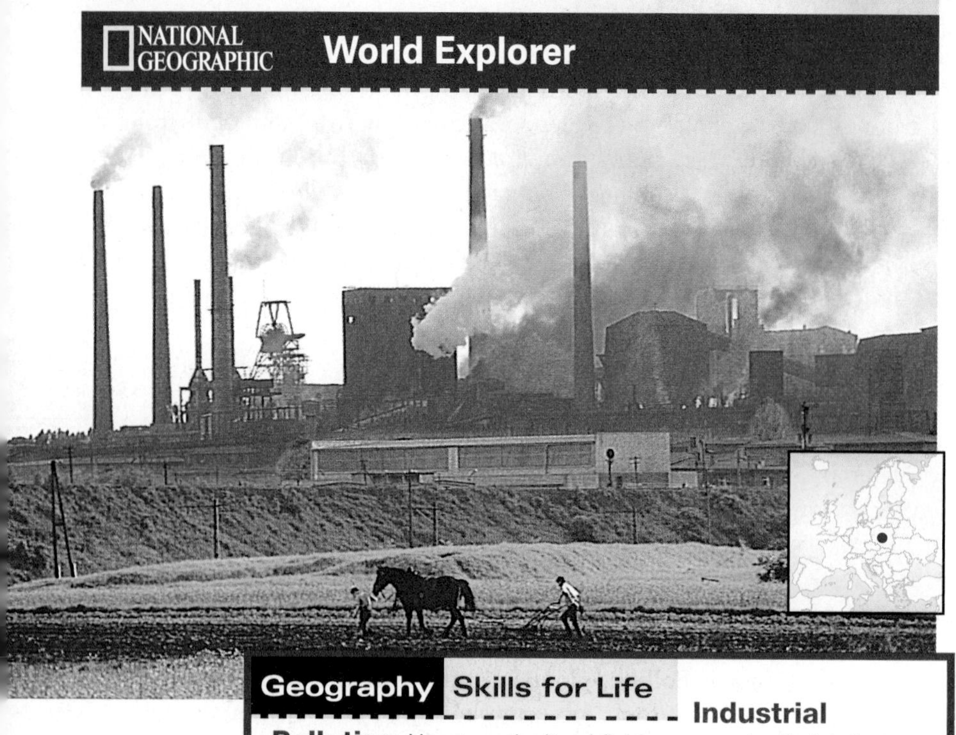

Geography Skills for Life

Industrial Pollution Hearty agricultural fields surround a Polish factory whose smokestacks may carry pollution north into Scandinavia.

Human-Environment Interaction How does industrial pollution contribute to global warming?

Air Pollution

Air pollution is a problem throughout Europe. Traffic exhausts and industrial fumes cause eye irritations and asthma, and make respiratory infections worse in people who live in the industrial areas of western Europe. In the Netherlands—where people drive the greatest number of cars per square mile in Europe—high levels of air pollution also affect public health. In 2000 Swiss researchers estimated that pollution from automobiles and trucks was responsible for about 6 percent of all deaths across Austria, France, and Switzerland.

In eastern Europe, factories built in the communist era belch soot, sulfur, and carbon dioxide into the air by the hundreds of tons. As a result, Poland, **Romania**, and the Czech Republic are among the world's most polluted countries. Life expectancy is lower in eastern Europe than in environmentally cleaner regions, and cancer rates and birth defects are higher. Air pollution also has poisoned crops.

Although steps are being taken to reduce pollution, Europe still faces many challenges. For example, some former communist countries are closing polluting factories. Yet they are also putting more cars on the road, increasing air pollution from traffic.

Global Warming

The problems of air quality in Europe, like those in other industrialized regions, may have global consequences. Many **environmentalists**—people concerned with the quality of the environment—are studying the effects of increased carbon dioxide in the earth's atmosphere. Carbon dioxide and other gases trap the sun's heat near the earth's surface, creating the **greenhouse effect**. Without this greenhouse effect, the earth would be so cold that even the oceans would freeze. Plants would not grow, and life would not exist.

The burning of fossil fuels such as coal, oil, and gasoline, however, has significantly raised the amounts of carbon dioxide in the atmosphere, increasing the greenhouse effect. Some scientists estimate that the earth's average temperature may rise 2.5° to 10.4°F (1.4° to 5.8°C) by the year 2100, a trend called **global warming**. A warmer global climate, they claim, will melt polar ice caps and mountain glaciers and cause oceans to submerge coastal areas. Weather patterns might change, producing new extremes of rainfall and drought.

Although the potential destruction from global warming is more widespread than are other environmental issues, governments give it less attention than they do other environmental issues that are more regional or local. Facing the threat of global warming requires international cooperation. However, because not all scientists agree that global warming is occurring, the international community so far has done little to reduce its possible causes.

Economics

Water Pollution

Water pollution is another issue facing Europe, particularly in the Mediterranean region. Countries bordering the **Mediterranean Sea** use the sea for transportation and recreation. They also use it for waste disposal, dumping sewage, garbage, and industrial waste there. In the past, bacteria in the Mediterranean Sea broke down most of the waste the sea received. In recent times, however, growing populations and tourism along the coast have increased the environmental problems of the Mediterranean. Small tides and weak currents tend to keep pollution where people discharge it. The Mediterranean Sea, open to the Atlantic only through the narrow **Strait of Gibraltar**, takes almost a century to renew itself completely.

Pollution contaminates marine and animal life and creates health hazards for people. The Mediterranean is overfished and cannot provide its former bounty. Only small schools of tuna enter from the Atlantic, and disease has claimed the last Mediterranean monk seals. Native species of seaweed and shellfish compete with foreign species carried into the Mediterranean by ships.

Water pollution affects Europe's rivers and lakes as well as its coastal waters. The Danube River, for example, is seriously affected by agricultural runoff. When fertilizers enter the river, they encourage algae growth. Algae, in turn, rob the river of so much oxygen that fish cannot survive. Another source of pollution is raw sewage, which is dumped into rivers in various places. In Warsaw, for example, only half of the sewage is treated. The other half is released untreated into the Vistula River. Industries in western Europe deposit wastes into the Meuse and Rhine Rivers; from there the pollutants flow into the North Sea. Consequently, pollution levels from the Netherlands to Denmark have doubled over the past few years.

Reducing Pollution

Europeans are working to solve environmental problems, such as pollution and waste disposal. They understand the economic impact of pollution, such as the loss of tourists and the high cost of cleanup. They also recognize the cultural effects, primarily the destruction of natural and historical sites.

Concern for the Environment

Today's Europeans feel responsible for protecting and preserving the environment and their national heritages for future generations. For example, many Europeans share a respect for nature. Those who live in densely populated areas value the opportunity to get away from urban areas and enjoy the natural landscape. Those who inhabit sparsely populated areas often depend on the natural environment to support their way of life.

The European concept of a natural environment is different from that in other parts of the world. Few

NATIONAL GEOGRAPHIC **World Explorer**

Geography Skills for Life

Water Pollution Children play in the Gulf of Gdańsk in Sopot, Poland. Nearby rivers dump agricultural and industrial waste into the gulf.

Human-Environment Interaction What happens when fertilizers enter lakes and rivers?

areas in Europe remain unchanged by the clearing of forests, the drainage of seas, or the building of canals. Although much of Europe has been greatly altered by human activity, Europeans want to preserve what little wilderness area is left. One of the largest areas of Europe still in its natural state is the Bialowieza (bee•ahl•lah•WEH•zhah) Forest in Belarus and Poland. Today this area is home to animal species such as the wolf, lynx, and European bison, all of which are now rarely seen elsewhere in Europe.

An effort to reintroduce wolves—which help reduce large herds of musk oxen, elk, reindeer, and other deer—is under way in some parts of Europe. Spain recolonized the animals in the northwest areas of the country, and their number has tripled to more than 2,000. Wolves now live within 25 miles of Rome, where their reintroduction succeeded in part because Italian farmers are paid for livestock lost to wolves. Wolves also are thriving in Romania's **Carpathian Mountains**. About 2,500 wolves—weighing up to 150 pounds (68 kg) each—live in the heavily forested mountains, preying on chamois, roe deer, and red deer.

Cleanup Efforts

In recent decades Europeans have made more concerted efforts to clean up the environment. Member countries of the EU can face legal action if they do not respect environmental protection laws. For example, France was cited for violating the European Union's guidelines on nitrate pollution, and Greece is being taken to court for failing to protect a rare Mediterranean sea turtle. European countries are also addressing the consequences of pollution. Cities in western Europe now protect buildings and statues with acid-resistant coatings. Lime added to lakes in Scandinavia reduces acid levels. **Biologists**—scientists who study plant and animal life—are researching the effects of acid levels on fish.

England's Thames River cleanup is a notable success story. Until the 1960s, the river was lifeless, its fish destroyed by sewage and industrial pollution. Factory closings and strict environmental controls have allowed the return of many fish and birds. After having disappeared for 150 years, even the giant conger eel, a traditional delicacy, returned in large numbers to the Thames.

Pollution that crosses national borders, however, presents a more complicated situation. For example, pollution in the Danube River, flowing through central and eastern Europe, threatens wildlife in its outlet—the Black Sea:

> *"In the past 50 years the number of dolphins in the Black Sea has declined from an estimated million to about 200,000. We must improve the water quality, but how will it be possible financially and administratively when the Danube flows through eight countries, and 70 million people live within its drainage area?"*
>
> Jon Thompson, "East Europe's Dark Dawn," *National Geographic*, June 1991

NATIONAL GEOGRAPHIC **World Explorer**

Geography Skills for Life **Preserving Wildlife**

A litter of wolves boosts hopes that fears of extinction are over as European countries seek to increase wolf populations.

Human-Environment Interaction Why have the wolf populations in Spain and Italy increased?

Solving wide-ranging pollution problems requires international cooperation. The United Nations' Mediterranean Action Plan, which involves 20 countries and the European Union, is a model of international joint effort. In 1999 the EU approved guidelines to protect endangered species, increase protection from industrial waste, and prevent the dumping of pollutants by ships and aircraft into the Mediterranean. The EU also required large companies to recycle a portion of their packaging waste. As a result, over 9 billion pounds (over 4 billion kg) of waste plastics were recycled in 1998.

Plans for the Future

The EU and European governments continue to develop ways to protect the environment. Many power plants now burn clean natural gas instead of lignite coal. By 2010 the EU wants all member countries to lower emissions to 15 percent below 1990 levels to reduce greenhouse gases.

In order to be admitted to the EU, countries in eastern Europe are expected to meet EU environmental standards. Because they will need to spend about $120 billion on cleanup, eastern Europeans are now seeking financial aid from EU countries in western Europe. Pollution from eastern Europe also threatens western Europeans, so they and U.S. companies are also providing technology, expertise, and investment to help modernize eastern Europe's industries. Such efforts highlight the global range of Europe's environmental concerns.

NATIONAL GEOGRAPHIC **World Explorer**

Geography Skills for Life

Inspecting the Baltic Divers in the Baltic Sea off Finland inspect the water quality and measure pollution from nearby factories.

Human-Environment Interaction How have European countries improved the conditions of their physical environments?

SECTION 2 ASSESSMENT

Checking for Understanding

1. **Define** dry farming, acid rain, meltwater, acid deposition, environmentalist, greenhouse effect, global warming, biologist.
2. **Main Ideas** On a table, list physical and human-made features affected by pollution. Then describe steps to counteract pollution's effects.

People and Their Environment		
Pollution Concern	Feature Affected	Steps Taken

Critical Thinking

3. **Identifying Cause and Effect** Why does eastern Europe have higher levels of industrial pollution than western Europe?
4. **Finding and Summarizing the Main Idea** Why do cleanup and preservation of the environment require cooperation in Europe?
5. **Making Generalizations** What are some of Europe's major challenges as its countries work to improve the environment?

Analyzing Maps

6. **Human-Environment Interaction** Refer to the population density map on page 262 and the map of Germany on page 326. In what ways are areas of dense population and areas affected by acid rain related?

Applying Geography

7. **Environmental Protection** Imagine that you live in a polluted area of Europe. Write a letter to the editor of a newspaper there, suggesting steps to halt environmental damage.

Viewpoint

CASE STUDY on the Environment

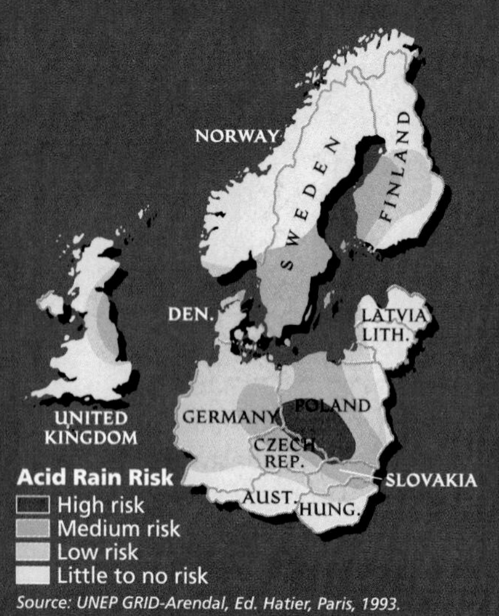

Source: UNEP GRID-Arendal, Ed. Hatier, Paris, 1993.

Germany's Forests: In the Path of Acid Rain

It looks harmless as it falls, pattering softly on the ground. Yet acid rain is a quiet killer. It can turn a forest into a patch of leafless trunks and a pond into a lifeless pool. Human activities are to blame for most acid rain. Chemical gases emitted from power plants, factories, and cars are the chief causes. Acid rain has damaged many European forests and lakes. Germany's once-picturesque forests have been especially hard hit. Acid rain can be reduced. But to do so requires balancing environmental protection with the needs of modern industrialized societies.

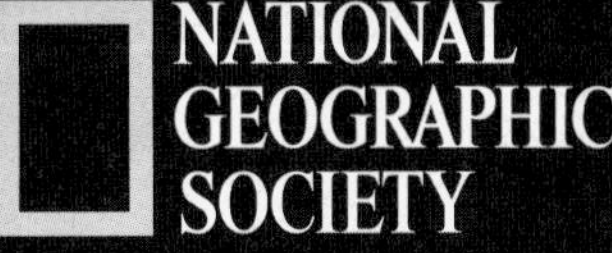

Like pale skeletons, dead evergreen trees (left) haunt the Ore Mountains near the border between Germany and the Czech Republic. What caused this destruction? Acid rain.

When fossil fuels, such as coal and gasoline, are burned, sulfur and nitrogen compounds are produced. Sulfur dioxide swirls from the smokestacks of coal-burning power plants. Nitrogen oxides escape in the exhaust of gasoline-powered cars and trucks.

Wind carries these compounds high into the sky. As the gases travel through air, they combine with moisture to form sulfuric acid and nitric acid. These acids make rainwater much more acidic than normal. The result is "acid rain"—a term applied to rain, snow, fog, or any form of precipitation that contains abnormal amounts of acid. This precipitation can be as acidic as battery acid!

Acid rain damages trees, especially evergreens. High levels of acidity in streams and lakes harm aquatic life, killing fish and plants. Acid rain also eats away at stone monuments and buildings.

Power plants, steel mills, and factories are found in or near most large European cities. Millions of cars and trucks travel European highways. The emissions from these industries and vehicles cause acid rain, which often falls hundreds of miles from its source. More than 25 percent of Germany's forests have been damaged by acid rain. Forests in Poland, the Czech Republic, Sweden, and Norway are also dying. Solutions to the problem do exist, but they often conflict with the needs of people.

◀ Industrial emissions contribute to acid rain.

Environmentalists want stricter emissions regulations for industries and vehicles. This often involves equipping smokestacks and vehicle exhaust systems with devices that remove sulfur and nitrogen compounds. Many people believe we should replace fossil fuels with alternative energy sources, such as solar and wind power.

Industrialists point out that modern societies cannot function without the electricity and material goods produced by power plants and factories. People need vehicles to get from place to place. Devices that reduce acid-causing emissions are expensive. Furthermore, solar power and other alternative energy sources are not yet realistic replacements for fossil fuels.

Despite these challenges, Germany is developing new technologies that will help reduce acid rain. German companies manufacture some of the world's most efficient gas turbines. Germans also built the first steel mill that does not burn coal to make steel.

What's Your Point of View?
Acid rain falling in Germany can be caused by another country's power plants. How does this complicate finding solutions to the acid rain problem?

A German power plant (below) emits chemicals that cause acid rain. But industries such as car companies (right) that use electricity make essential products. ▼

Using the Internet for Research

Using the Internet for research is both easier and harder than using the library. It is easier because you can look through many different sources at one time. Internet research can sometimes be difficult because of the large amounts of information and the lack of organization to it.

Learning the Skill

Fortunately, you can search for information on the Internet in several ways. You can start your search with a search engine, such as www.yahoo.com or a reference center, such as Internet Public Library, at www.ipl.org.

Once you find information, however, you need to consider its reliability.

- **Evaluate the source of the information.** Avoid sources that do not provide facts or that are heavily slanted toward a particular view.
- **Keep records.** Always record the Web site title and address, the date you viewed the Web site, and the author's name (if available) so you can cite it.

▲ *Information and photographs about European wildlife, such as these Alpine ibexes, can be found on the Internet.*

The top level domain (TLD) at the end of a Web site address tells you what kind of site you have accessed. These are the most common TLDs:

1. .gov—government agencies, such as the Library of Congress or the U.S. State Department
2. .edu—educational sites, such as universities or the Smithsonian Institution
3. .org—nonprofit organizations, such as the United Nations and the World Wildlife Fund
4. .com—business sites, such as the National Geographic Society or the Discovery Channel; search engines, such as Yahoo, are often .com sites

In 2000, seven new TLDs were introduced, including ".mus" for museums and ".biz", an additional TLD for businesses.

Practicing the Skill

Use an Internet search engine to search for information about the environmental policies of the European Union. Analyze the Web sites you find. Then choose three sites, and record the reference information about each site.

Search the Internet to find three sites that provide data and statistics on Europe's wildlife. Write a report analyzing and evaluating the sites' validity and usefulness. Then use reliable site data to answer the following questions: What kinds of wildlife does Europe have? How is wildlife being protected? What countries are the most committed to wildlife protection?

CHAPTER 13 SUMMARY & STUDY GUIDE

SECTION 1 Living in Europe (pp. 313–319)

Terms to Know

- European Union (EU)
- Maastricht Treaty
- heavy industry
- light industry
- mixed farming
- farm cooperative
- collective farm
- state farm
- genetically modified food
- organic farming

Key Points

- The countries of the European Union work toward bringing the continent economic and political unity.
- After years of communist rule, countries in eastern Europe face a difficult transition to market economies.
- Europe's economic activities include manufacturing, service and technology industries, and agriculture.
- Much of Europe has well-developed communications and transportation systems.

Organizing Your Notes

Create an outline using the format below to help you organize your notes for this section.

Living in Europe

I. Changing Economies
  A. The European Union
  B. Eastern Europe
II. Industry
  A.
  B.
  C.
III. Agriculture
IV.

SECTION 2 People and Their Environment (pp. 320–325)

Terms to Know

- dry farming
- acid rain
- meltwater
- acid deposition
- environmentalist
- greenhouse effect
- global warming
- biologist

Key Points

- Acid rain is damaging Europe's forests, waterways, wildlife, and buildings.
- Air pollution from Europe's factories endangers public health and the environment.
- Greenhouse gases contribute to global warming.
- Pollution threatens the water quality and wildlife in the Mediterranean Sea and eastern Europe.
- European countries are taking steps to reduce pollution and clean up the environment.

Organizing Your Notes

Use charts like the one below to help you organize the notes you took as you read this section.

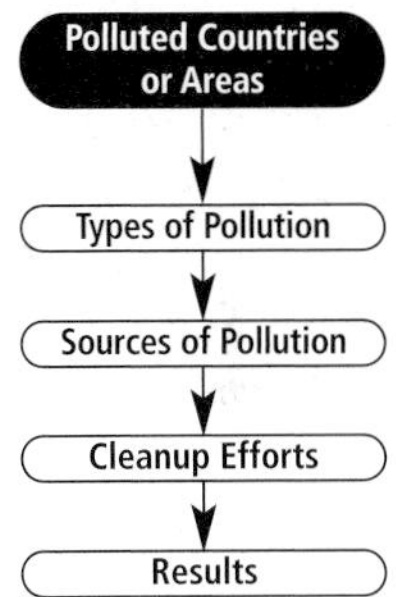

Bird rescued from oil-covered beach, Wales

CHAPTER 13

ASSESSMENT & ACTIVITIES

Reviewing Key Terms

On a sheet of paper, write the key term that best completes each sentence. Refer to the Terms to Know in the Summary & Study Guide on page 329.

1. Wet or dry pollution that falls directly to the ground is also known as _________.
2. _________ is damaging Europe's forests.
3. Raising several types of crops and livestock is called _________.
4. _________ may cause the ice caps to melt.
5. _________ produces machinery.
6. Soviet officials managed a(n) _________, but did not share profits with the farmers.
7. _________ carries the acid precipitation into rivers and lakes in the spring.
8. The _________ set up the European Union (EU).
9. The _________ causes the sun's heat to be trapped near the earth's surface.
10. _________ is the production of textiles or processed food.
11. _________ uses natural substances to increase crop yield.

Reviewing Facts

SECTION 1

1. How is the European Union working toward economic and political unity for its members?
2. How has eastern European agriculture changed since the communist era ended?

SECTION 2

3. Why is air pollution in Europe so widespread?
4. How has the EU encouraged environmental protection and cleanup?

Critical Thinking

1. **Drawing Conclusions** Why are pollution problems most severe in eastern Europe?
2. **Making Predictions** How might global warming affect Europe? How do you think the countries of Europe will address the issue of global warming in the future?
3. **Comparing and Contrasting** On a Venn diagram, compare and contrast pollution in western and eastern Europe. Explain the interrelationships among physical and human processes regarding environmental change.

Western Europe | Both | Eastern Europe

NATIONAL GEOGRAPHIC

Locating Places

Europe: Physical-Political Geography

Match the letters on the map with the places and physical features of Europe. Write your answers on a sheet of paper.

1. Mediterranean Sea
2. Po River
3. Italy
4. Athens
5. Madrid
6. Corsica
7. France
8. Adriatic Sea
9. Cyprus
10. Sardinia
11. Black Sea
12. Strait of Gibraltar

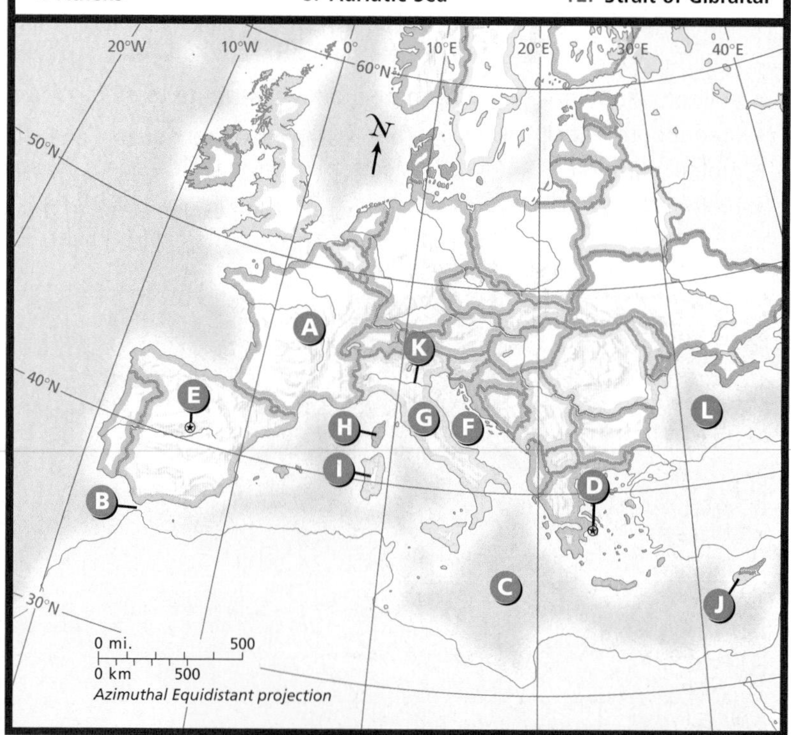

Self-Check Quiz Visit the **Glencoe World Geography** Web site at geography.glencoe.com and click on Self-Check Quizzes—Chapter 13 to prepare for the Chapter Test.

Using the Regional Atlas

Refer to the Regional Atlas on pages 260–263.

1. **Location** In what area of Europe is subsistence farming predominant?
2. **Place** Name three European capitals that have populations greater than 5,000,000.

Thinking Like a Geographer

Think about the physical and human geography of Europe. What factors helped establish the European Union? Research and identify different points of view that will shape the future structure and role of the European Union.

Problem-Solving Activity

Problem-Solution Proposal Choose a city located on the Mediterranean Sea. Imagine that you head a planning committee that wants to encourage tourism but also wants to reduce the pollution in the local bay. Research the industrial and tourism activities of the city, and then write a proposal suggesting ways to develop tourism while reducing pollution. Address the proposal to the city's industrial leaders, hotel managers, tourism directors, and water quality experts.

GeoJournal

Creative Writing Use the information in your GeoJournal to create an outline and storyboard for a television special on preserving the environment in Europe.

Technology Activity

Creating an Electronic Database Choose five countries from western Europe and five countries from eastern Europe. Using the Internet and other references, research the per capita income, or the average individual earnings in a year, for each country. Also find the percentage of the workforce involved in agriculture, manufacturing, and service industries for each country. Then use the information to create an electronic database. Write an essay explaining how these economic factors affect the standard of living in western and eastern Europe.

Standardized Test Practice

Choose the best answer for each of the following multiple-choice questions. If you have trouble answering the questions, use the process of elimination to narrow your choices.

1. Lignite, or brown coal, is easily and inexpensively mined. Why should European cities be discouraged from using lignite as a main fuel source?

A Mining of lignite creates unsightly open pits that are dangerous to children.
B Acid rain in European cities would be reduced by burning lignite.
C European cities, especially in the east, use natural gas more than lignite.
D Sulfur dioxide emissions from lignite cause high levels of air pollution.

2. How do prevailing winds affect the acid rain that falls in Europe?

F Prevailing winds disperse acid rain across national borders.
G Prevailing winds help clear away the acid rain, which results in less pollution.
H Acid rain is heavier than air, so prevailing winds do not affect acid rain at all.
J Europe's industrial belt lies in an area with no prevailing winds.

For multiple-choice questions, remember to read each answer choice carefully. Some answer choices may not answer what the question asks. Sometimes, more than one answer may seem correct. Therefore, closely study the question so that you are sure of what it is asking, and then choose the answer choice that best answers the question.

UNIT 5

Russia

WHY IT'S IMPORTANT—

For most of the last century, Russia was part of the vast Soviet Union. Ruled by a Communist government, the Soviet Union challenged the United States and other democracies for global influence. Then the Soviet Union collapsed, and Russia emerged as an independent republic. Now Russia is struggling to build a stable democracy and free-enterprise economy. Because Russia is a key player in world affairs, its success—or failure—will affect your world in the years to come.

World Regions Video

To learn more about Russia and its impact on your world, view the World Regions video "Russia."

Snow-covered trees in St. Petersburg

What Makes Russia a Region?

The world's largest country, Russia stretches almost halfway around the northern part of the globe, covering nearly half of two continents. It is a land of vast distances, bitter winters, and remote frontiers.

The ancient Ural Mountains separate European Russia in the west from Asian Russia, or Siberia, in the east. Most Russians live in the west, where several rivers course through rolling plains. East of the Urals, Siberia begins as a vast plain. It rises gradually to an immense plateau, then reaches higher still to rugged mountain ranges that border Russia's eastern shores. Siberia has abundant natural resources but few inhabitants.

North of the Arctic Circle, the land is treeless tundra where most of the ground is permanently frozen and winters are some of the coldest on Earth. South of the tundra lies the taiga, an enormous belt of dense coniferous forest. South of the trees is the Russian steppe, a rolling grassland with rich soil and a more hospitable climate.

1 Sparks flare against a dark January sky as lengths of pipe are welded in western Siberia. The pipeline will carry natural gas. Siberia is rich in natural resources, but its rugged terrain, harsh climate, and isolation make extracting and transporting those resources a tremendous challenge.

2 The splendid skyline of St. Petersburg, Russia's second largest city, forms a glowing backdrop for the Neva River. The Neva flows into the Gulf of Finland, an extension of the Baltic Sea. Like many of Russia's rivers, the Neva freezes over during the country's harsh winters.

3 Worn down over time, the rounded peaks of the Ural Mountains shelter a Russian village. The Urals extend 1,500 miles (2,400 km) from the Arctic Ocean south to Kazakhstan. They have long been a rich source of minerals, including gemstones such as emeralds, amethyst, and topaz.

4 At home in the snow, a trio of Siberian tigers roams through the frigid forests of eastern Siberia. Long, thick fur protects these big cats from the cold—but not from poaching and habitat loss, which have left Siberian tigers extremely endangered. Fewer than 500 survive in the wild.

From Empire to Free Enterprise

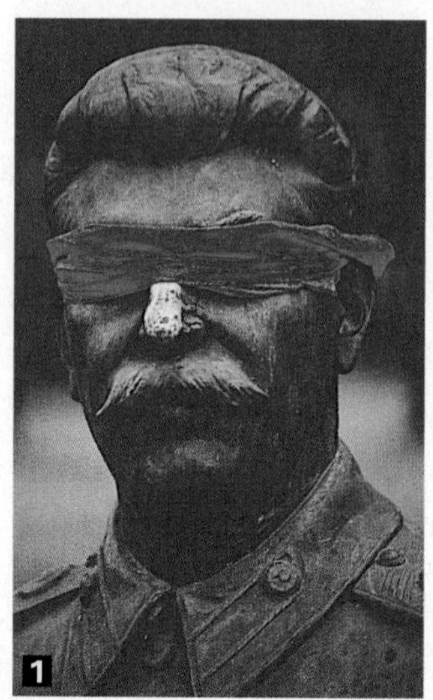

Most Russians are descended from Slavs, ancient European peoples who settled in western Russia. The settlements they established were eventually united under the rule of Ivan the Great, who ruled in the 1400s.

For centuries, autocratic czars governed what became a vast Russian empire. Revolution ended the czars' rule in 1917. Communism took its place, and Russia, along with neighboring republics, became part of the Soviet Union. Communist authorities controlled the Soviet economy. Everything from farms to steel mills was owned and operated by the government.

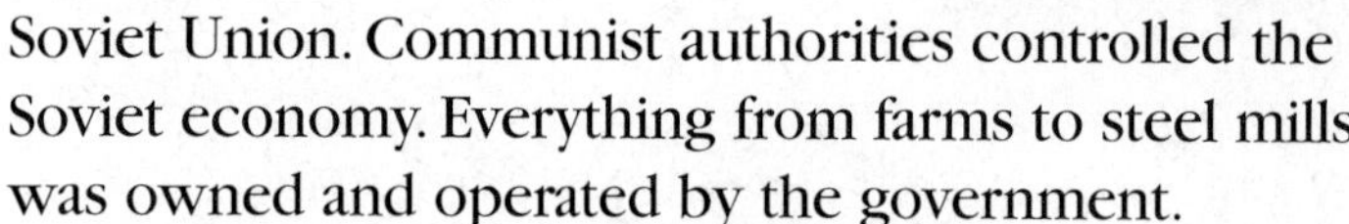

Since the Soviet Union disintegrated in 1991, Russians have struggled to establish a free-enterprise economy in their now independent nation.

1 A blindfolded statue of Stalin symbolizes the end of Communist rule in Russia. Stalin was a brutal dictator who ruled the Soviet Union for some 30 years. Millions of Russians starved or were put to death under his totalitarian regime. Stalin died in 1953; the Communist system would last almost four decades more.

2 Snow-dusted sculptures inspire an artist in Peter the Great's Summer Garden, in St. Petersburg. Czar Peter the Great founded St. Petersburg in 1703. His goal: to create a Russian capital that rivaled Western cities such as London and Amsterdam.

3 Onion-shaped domes cap Russian Orthodox churches beside a lake in northwestern Russia. The Russian Orthodox faith has its roots in the ancient Byzantine Empire, which was centered in what is now Turkey. The domes are characteristic of Byzantine architecture.

4 Eager for customers, vendors on a Moscow street wait anxiously by their produce. After the disintegration of the Soviet Union, the Russian economy entered a period of great instability. Wages fell, while prices for food and other necessities soared.

5

Russia

PHYSICAL

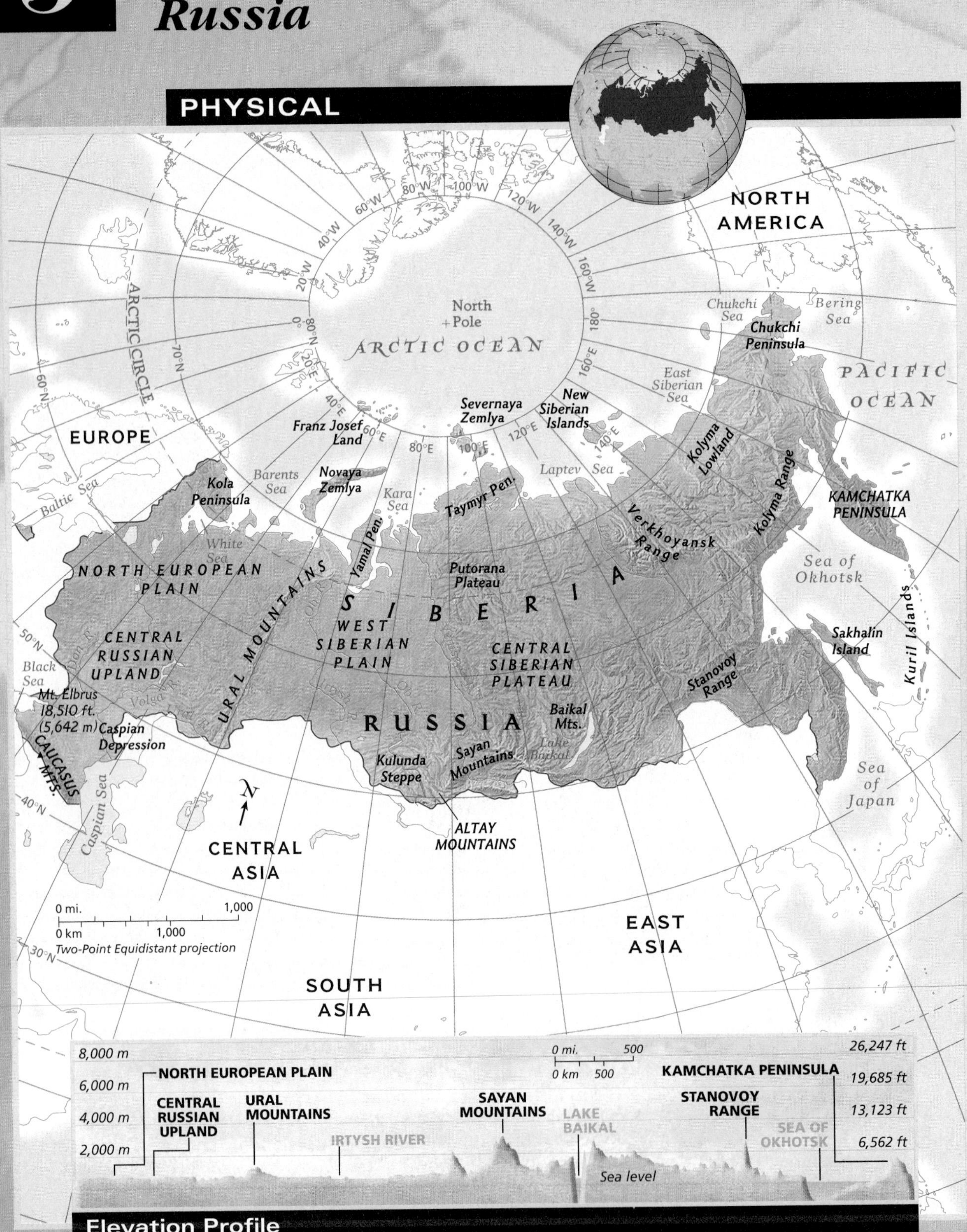

NATIONAL GEOGRAPHIC SOCIETY

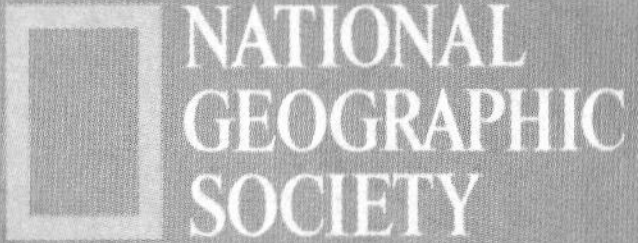

POLITICAL

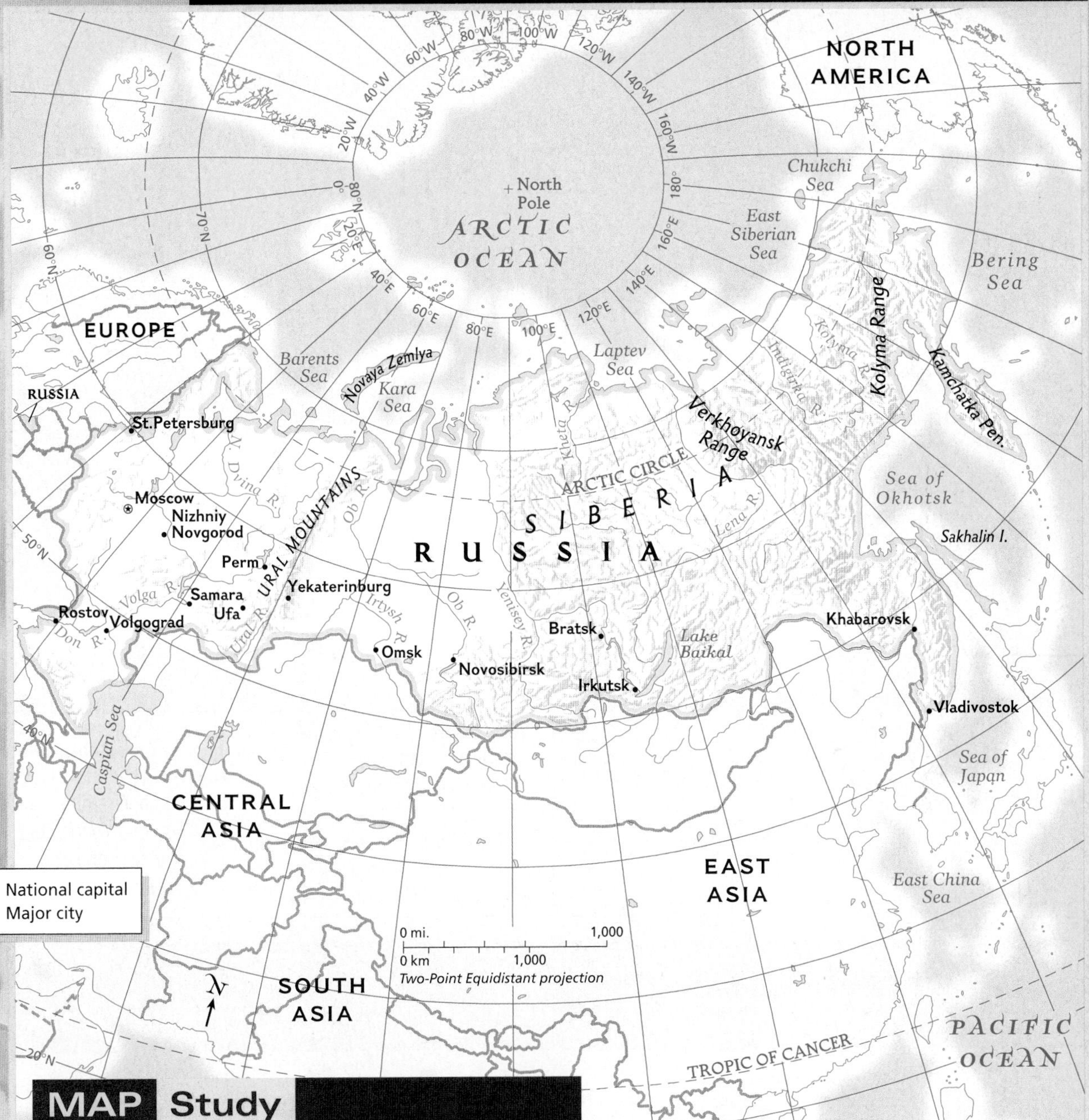

NORTH AMERICA
North Pole
ARCTIC OCEAN
Chukchi Sea
East Siberian Sea
Bering Sea
Kolyma Range
Kamchatka Pen.
Laptev Sea
Barents Sea
Novaya Zemlya
Kara Sea
EUROPE
RUSSIA
St. Petersburg
Moscow
Nizhniy Novgorod
Perm
URAL MOUNTAINS
Yekaterinburg
Samara
Ufa
Rostov
Volgograd
Omsk
Novosibirsk
Bratsk
Irkutsk
Lake Baikal
Khabarovsk
Vladivostok
Verkhoyansk Range
ARCTIC CIRCLE
SIBERIA
RUSSIA
Sea of Okhotsk
Sakhalin I.
Sea of Japan
Caspian Sea
CENTRAL ASIA
EAST ASIA
SOUTH ASIA
East China Sea
PACIFIC OCEAN
TROPIC OF CANCER
National capital
Major city
0 mi. 1,000
0 km 1,000
Two-Point Equidistant projection

MAP Study

1. What physical feature separates the North European Plain from the West Siberian Plain?
2. Which Russian cities are located on the North European Plain?

5

Russia

POPULATION DENSITY

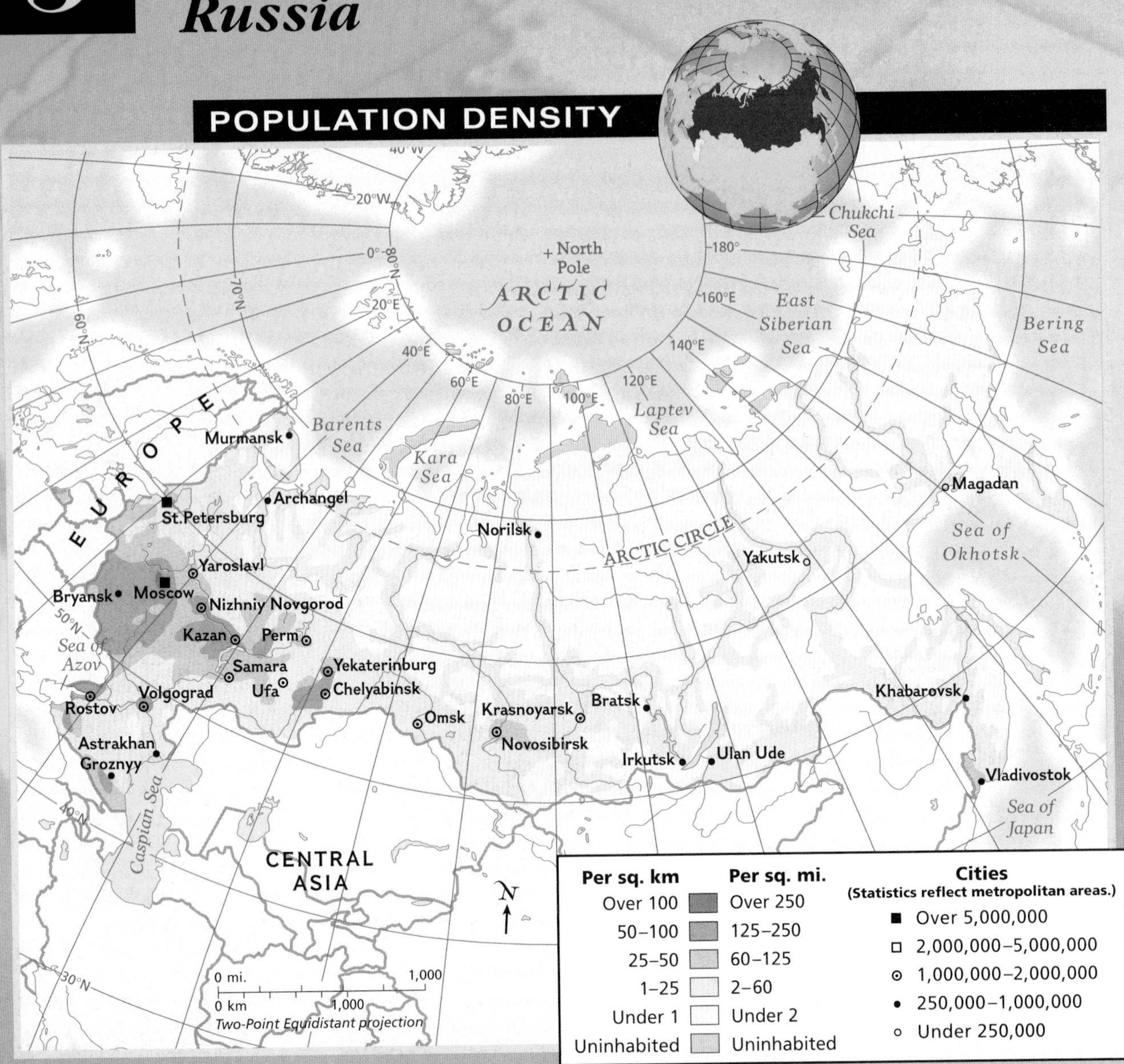

COUNTRY PROFILES

COUNTRY * AND CAPITAL	FLAG AND LANGUAGE	POPULATION AND DENSITY	LANDMASS	MAJOR EXPORT	MAJOR IMPORT	CURRENCY	GOVERNMENT
RUSSIA Moscow	Russian, Local Languages	144,400,000 22 per sq. mi. 9 per sq. km	6,592,819 sq. mi. 17,075,401 sq. km	Petroleum	Machinery	Ruble	Federated republic

*COUNTRY AND FLAG NOT DRAWN TO SCALE

FOR AN ONLINE UPDATE OF THIS INFORMATION, VISIT GEOGRAPHY.GLENCOE.COM AND CLICK ON "TEXTBOOK UPDATES."

ECONOMIC ACTIVITY

NORTH AMERICA
North Pole
ARCTIC OCEAN
Chukchi Sea
East Siberian Sea
Bering Sea
Laptev Sea
Barents Sea
Kara Sea
EUROPE
Baltic Sea
RUSSIA
Reindeer
St. Petersburg
Flax
Corn
Moscow
Nizhniy Novgorod
Oats
Barley
Kazan
Perm
Samara
Rostov
Wheat
Tyumen
Chelyabinsk
ARCTIC CIRCLE
Sea of Okhotsk
Novosibirsk
Irkutsk
Lake Baikal
Sheep
Vladivostok
Sea of Japan
Caspian Sea
CENTRAL ASIA
EAST ASIA
East China Sea
PACIFIC OCEAN
N

0 mi. 1,000
0 km 1,000
Two-Point Equidistant projection

Land Use

- Commercial farming
- Subsistence farming
- Livestock raising
- Nomadic herding
- Hunting and gathering
- Forests
- Manufacturing and trade
- Commercial fishing
- Little or no activity

Resources

Coal	Tungsten
Petroleum	Platinum
Natural gas	Gold
Iron ore	Copper
Nickel	Lead
Bauxite	Zinc
Manganese	Tin

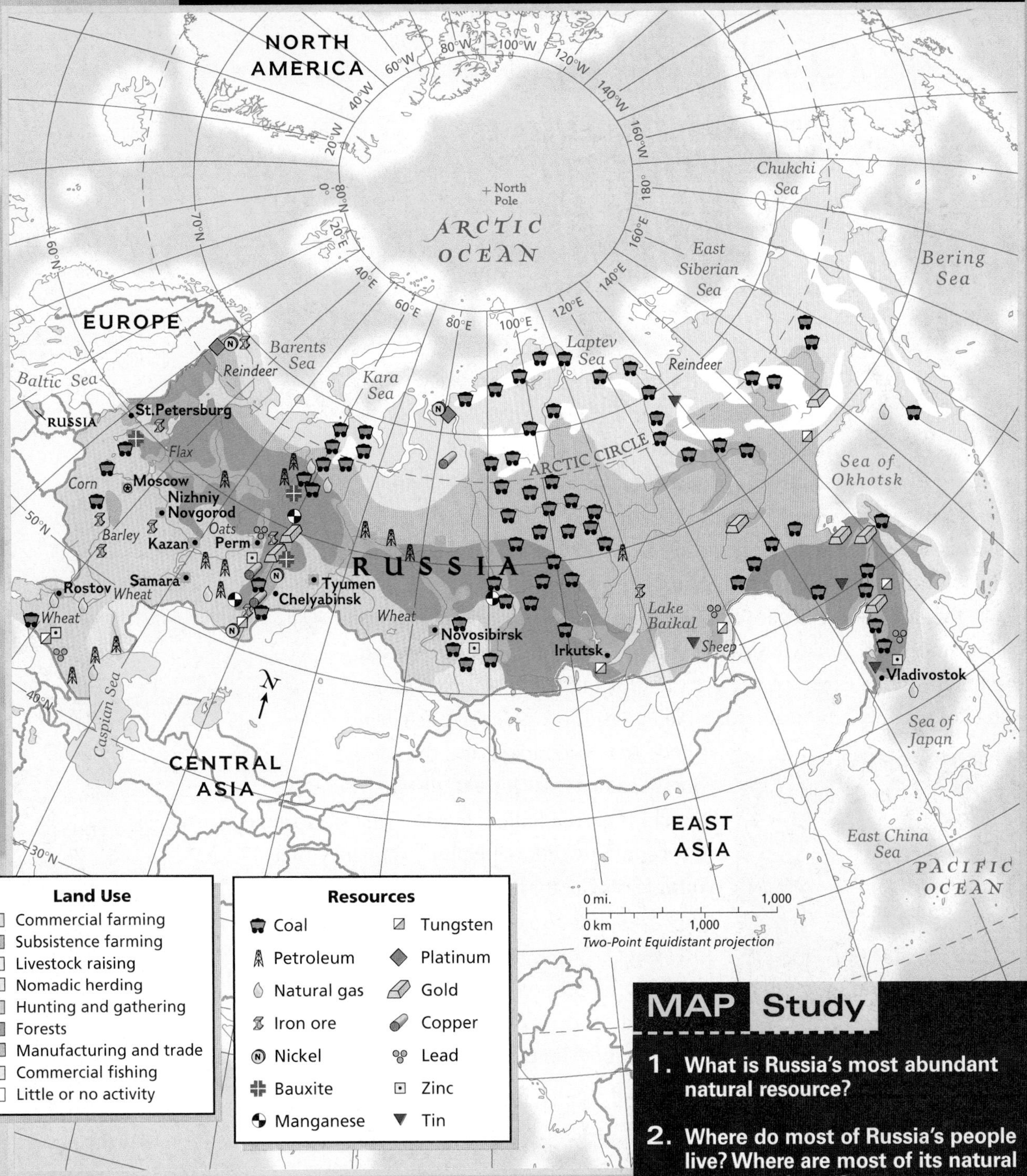

MAP Study

1. What is Russia's most abundant natural resource?
2. Where do most of Russia's people live? Where are most of its natural resources found?

GLOBAL CONNECTION

RUSSIA AND THE UNITED STATES

NUTCRACKER

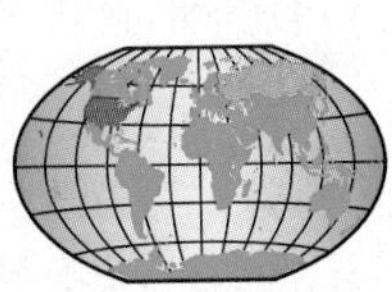

Clara. The Mouse King. The Sugarplum Fairy. Do these names sound familiar? All across America, children and adults alike would instantly recognize them as the names of characters from *The Nutcracker* ballet. Although *The Nutcracker* originated in Russia more than a century ago, it has become a beloved holiday tradition in the United States.

In the late 1800s, elegant St. Petersburg was the world center for ballet. Peter Ilich Tchaikovsky, the famous Russian composer, was at the height of his career. A theater director in St. Petersburg asked Tchaikovsky to write music for a ballet based on a German fairy tale called "The Nutcracker and the Mouse King." Tchaikovsky agreed, and in 1892 the Russian Imperial Ballet gave the first performance of this unconventional ballet—the story of a young girl's dream of Christmas presents that come to life in a magical kingdom of snowflakes and sweets.

The premiere of *The Nutcracker,* however, was far from a success. Critics sneered and audiences were unimpressed. The ballet wasn't seen outside Russia until 1934. Even then, it was only modestly popular.

◀ Peter Ilich Tchaikovsky

▲ Dancers performing *The Nutcracker* in Austin, Texas

In 1940, musical excerpts from *The Nutcracker* were included in the score of Walt Disney's animated film *Fantasia*. The movie was a hit, and as a result, Tchaikovsky's *Nutcracker* melodies were suddenly all the rage in the United States. The stage was set, so to speak, for a successful American production of the ballet.

It was George Balanchine, a Russian choreographer, who brought *The Nutcracker* to life for American audiences. Balanchine, whose given name was Georgi Melitonovitch Balanchivadze, was born in St. Petersburg and studied ballet there. As a boy, he danced the roles of the Mouse King and the Nutcracker Prince. In 1933 Balanchine came to the United States, where he helped found a ballet company that would become the famous New York City Ballet. In 1954 Balanchine directed the New York City Ballet in a lush and imaginative production of *The Nutcracker.* It was an immediate—and enormous—success.

In the years that followed, *The Nutcracker* became, and still remains, one of the most widely produced and widely attended ballets in American history—as much a part of December holiday celebrations as colored lights, carols, and presents under the tree.

CHAPTER 14

The Physical Geography of Russia

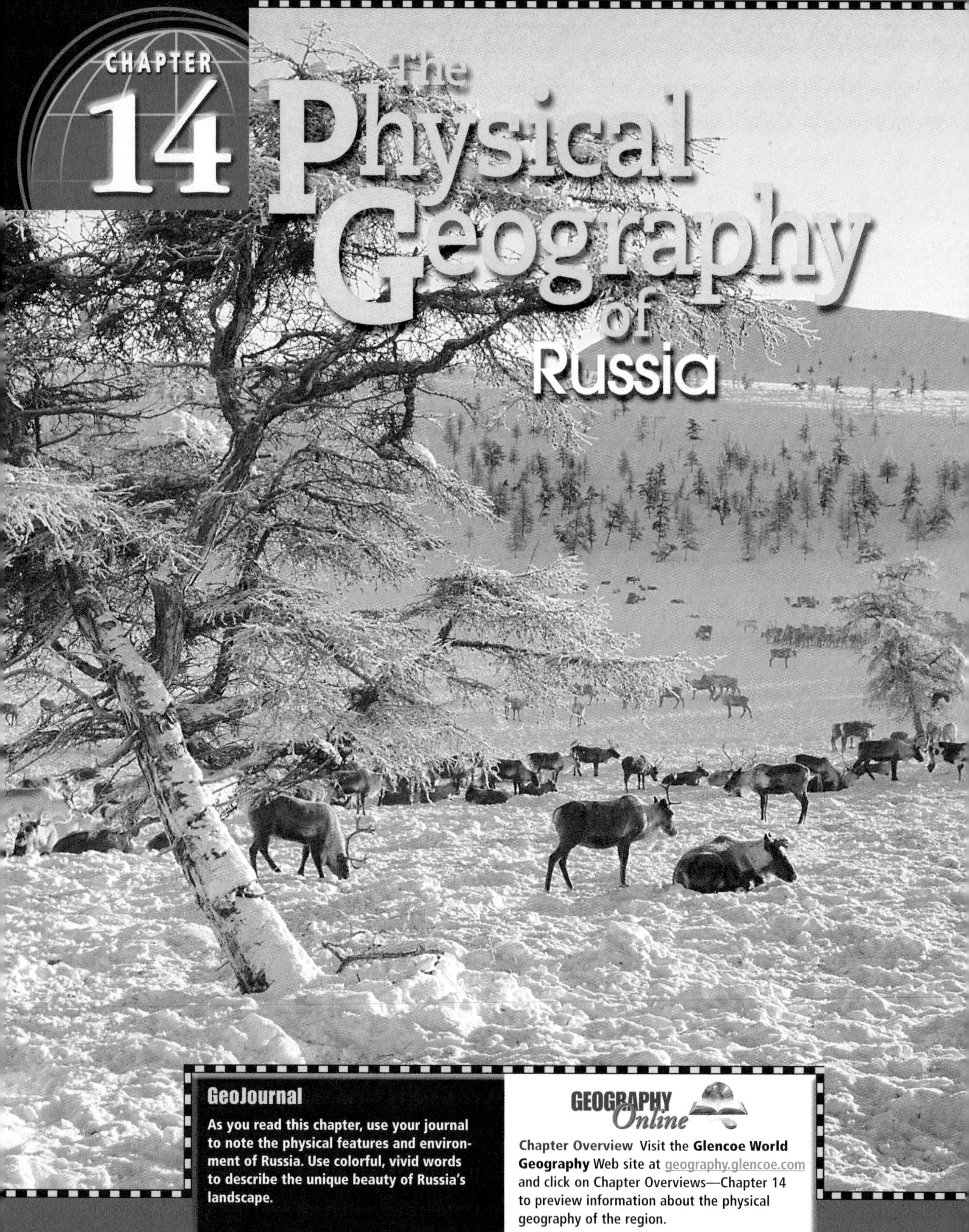

GeoJournal

As you read this chapter, use your journal to note the physical features and environment of Russia. Use colorful, vivid words to describe the unique beauty of Russia's landscape.

GEOGRAPHY Online

Chapter Overview Visit the **Glencoe World Geography** Web site at geography.glencoe.com and click on Chapter Overviews—Chapter 14 to preview information about the physical geography of the region.

SECTION 1

The Land

Guide to Reading

Consider What You Know

Like Canada, Russia is a far northern country. How might Russia's location affect its connections to other parts of the world?

Read to Find Out

- How large is the land area of Russia?
- How do Russia's interconnected plains and mountain ranges shape settlement in the country?
- What are Russia's natural resources?

Terms to Know

- chernozem
- hydroelectric power
- permafrost

Places to Locate

- Caucasus Mountains
- Central Siberian Plateau
- North European Plain
- West Siberian Plain
- Volga River

◀ *Reindeer in winter pastures, Siberia, Russia*

NATIONAL GEOGRAPHIC

A Geographic View

Swimming in the Volga

Crossing the Volga on a summer day, the ferry **Moskva-44** *took pale passengers to a sandy beach and brought away bathers red as lobsters. Upon this island Sahara were . . . boom boxes, sand castles, paper hats. . . . Yes, people still swim in the Volga, and I did, too. . . . You just forget about heavy metals and the 24 smokestacks vying on the horizon with the awesome Stalingrad memorial. The love of Mother Volga is real and has priority.*

—Mike Edwards, "Mother Russia on a New Course," National Geographic, *February 1991*

Sunbathers, Volga River

In 1991 the powerful Soviet Union broke up into 15 independent republics. Of these, Russia is by far the largest. In this section you will explore Russia, a gigantic and varied land of grassy and swampy plains divided and bordered by mountain ranges; tundra; subarctic forests; and wide, often frozen rivers and seas.

A Vast and Varied Land

In both total land area and geographic extent, Russia is the world's largest country. Covering about 6.6 million square miles (17.1 million sq. km), Russia stretches across parts of two continents—Europe and Asia. The country's greatest east-west extent is about 6,200 miles (about 9,980 km). This vast distance spans 11 time zones, contains 9 mountain ranges, and borders 13 seas, 3 oceans, and 14 other countries. The Russian landscape consists of an interrupted belt of rugged mountains and plateaus in the south and east and vast plains in the north and west.

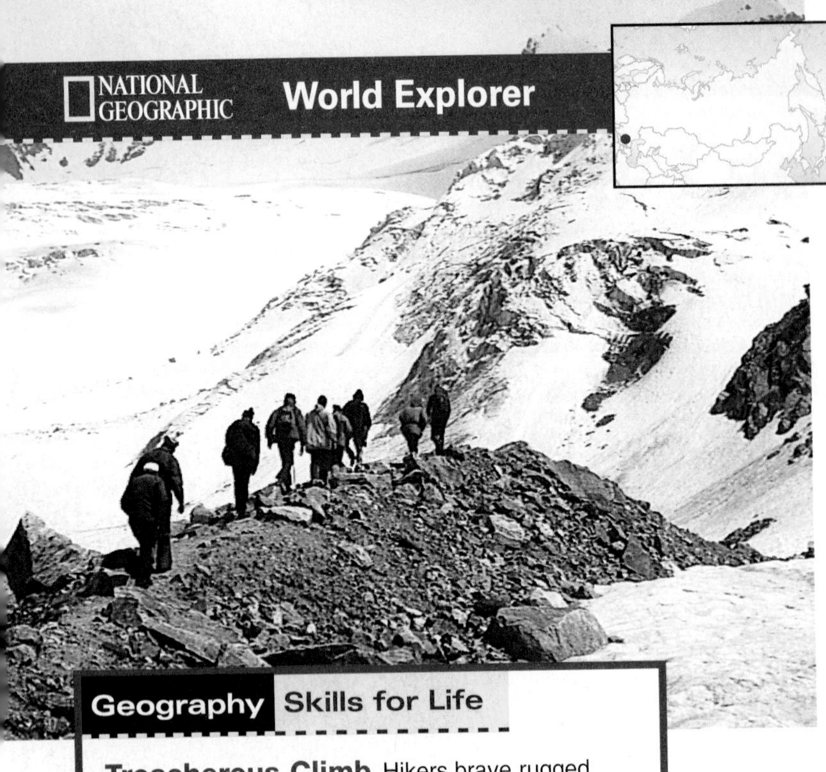

Geography Skills for Life

Treacherous Climb Hikers brave rugged terrain and glacial temperatures in the Caucasus Mountains of southern Russia.

Place How do mountain ranges help define Russia's territory?

Mountains and Plateaus

Mountains and plateaus punctuate the generally flat landscape of Russia. The Ural Mountains mark the traditional boundary between European Russia and Asian Russia. The Urals are an old, worn-down series of mountain ranges with an average height of about 2,000 feet (about 610 m). Though modest in height, the Urals are rich in iron ore and mineral fuels, such as oil and natural gas.

In southwestern Russia the rugged **Caucasus** (KAW•kuh•suhs) **Mountains** lie between the Black and Caspian Seas. The Caucasus reach their highest elevation at Mount Elbrus, an extinct volcano that reaches 18,510 feet (5,642 m), Russia's highest point.

Mountain ranges also form a rugged natural boundary between Russia and China. These mountains mark the southeastern edge of the **Central Siberian Plateau**. This rolling plateau has elevations ranging from 1,600 to 2,300 feet (480 to 700 m). Throughout the plateau's expanse, swiftly flowing rivers have carved out canyons.

Still farther east, mountains and basins extend to the Pacific Ocean. Temperatures in this remote area have plunged to a record –90°F (–68°C). In the easternmost part of Russia, on the Kamchatka Peninsula, there are more than 100 volcanoes, including 23 that are active.

Plains Areas

Vast plains span nearly half of Russia. Most of European Russia is part of the rolling **North European Plain** that sweeps across western and central Europe into Russia. In Russia, the northern part of this plain is very flat and poorly drained, resulting in many swamps and lakes. By contrast, the southern part has navigable waterways and a rich black soil, known as chernozem (cher•nuh•ZYAWM), that supports the production of wheat, barley, rye, oats, and other crops. About 75 percent of the Russian population lives on the North European Plain. This region holds Russia's most populous cities, including Moscow, the capital, and the port city of St. Petersburg.

Farther to the east, the Ural Mountains divide the North European Plain from another vast plains area—the **West Siberian Plain**. With almost 1 million square miles (2.6 million sq. km), the West Siberian Plain is one of the world's largest areas of flatland. At its widest this plain stretches from the Arctic Ocean in the north to the grasslands of central Asia in the south. Its lowland areas are poorly drained, with many swamps and marshes.

Coasts, Seas, and Lakes

Russia has the longest continuous coastline of any country in the world. Stretching 23,400 miles (37,650 km), Russia's coastline touches both the Arctic and Pacific Oceans. Other coasts lie along an arm of the Baltic Sea in the northwest and along the Black and Caspian Seas in the south. Because of Russia's far northern location, most of its coast lies along waters that are frozen for many months of the year. As a result, Russia has few ocean ports that are free of ice year-round.

In Russia's southwest corner, the Black Sea provides Russia with a warm-water outlet to the Aegean and Mediterranean Seas through three Turkish-controlled waterways—the Bosporus strait, the Sea of Marmara, and the Dardanelles (DAHRD•uhn•EHLZ). Despite sudden storms that sometimes strike the Black Sea, Russia's fishing industry has thrived in its waters.

The Caspian Sea, on Russia's southwestern border, is the largest inland body of water in the world. Although called a sea, it is actually a saltwater lake that occupies a deep depression. Rivers flow into the Caspian, but there is no outlet to the ocean. Water in the Caspian Sea evaporates over time, slowly shrinking the sea and leaving behind salts that accumulate and make the water saltier.

Another large body of water, Lake Baikal (by•KAHL), lies in southern Siberia. At nearly 400 miles (644 km) long, 40 miles (64 km) wide, and

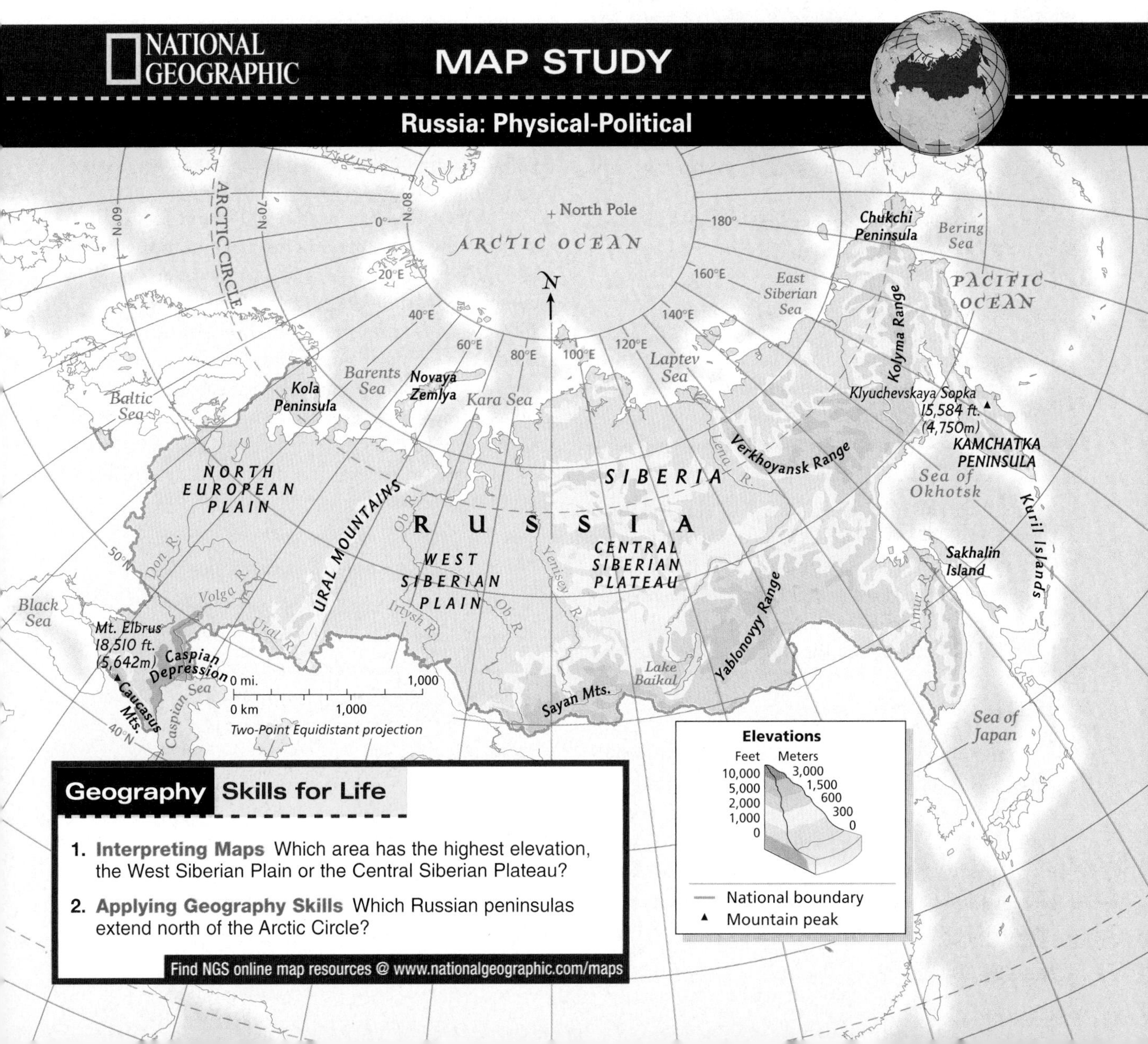

Geography Skills for Life

1. **Interpreting Maps** Which area has the highest elevation, the West Siberian Plain or the Central Siberian Plateau?
2. **Applying Geography Skills** Which Russian peninsulas extend north of the Arctic Circle?

Find NGS online map resources @ www.nationalgeographic.com/maps

Geography Skills for Life

Lake Baikal At an estimated 20 to 25 million years old, Russia's Lake Baikal is the world's oldest existing freshwater lake.

Human-Environment Interaction How has industrial development affected Lake Baikal?

over 1 mile (1.6 km) deep, Lake Baikal is the third largest lake in Asia and the deepest freshwater lake in the world. It is estimated to contain about 20 percent of the earth's total supply of freshwater. In recent years runoff from nearby pulp and paper factories has threatened the purity of the lake.

Rivers

Some of the world's longest rivers flow through Russia, draining a large portion of the land and providing water for irrigation. They also serve as transportation routes or sources of electric power for densely populated urban areas. Most of Russia's longest rivers—which supply 84 percent of the country's water—are located in Siberia, where only 25 percent of the population lives. Thus, people in Siberia enjoy a surplus of freshwater, but European Russians often face water shortages or problems with water quality.

The Volga River

The **Volga River** in European Russia is the fourth-longest river in Russia. Affectionately called *Matushka Volga*, or "Mother Volga," the river is vital to Russia. The Volga and its tributaries drain much of the eastern part of Russia's North European Plain. They connect Moscow to the Caspian Sea and, by way of the Volga-Don Canal, to the Sea of Azov and the Black Sea. Canals link the Volga to the Baltic Sea in the north, giving Russia a water route to northern Europe. Although frozen almost half of each year, the river provides **hydroelectric power**, or power generated by falling water, and water for drinking and irrigation.

Two-thirds of Russia's water traffic travels along the Volga. Heavy use of the river, however, has created challenges for Russia's people. Fed by melting snow, the wide, swift Volga supplies 33 percent of Russia's usable water, but half of it returns to the river carrying human and industrial waste. In addition, dams interrupt the river's flow, threatening wildlife and drinking water supplies.

Other rivers important to European Russia include the Western Dvina, the Dnieper, and the Don. Many fishing villages line the banks of the Don as it flows through rich farmland toward Rostov, where it empties into the Sea of Azov. A visitor to Veshenskaya, several hundred miles upstream, describes the river's peaceful flow:

> *"It is indeed a quiet Don there, flowing dreamily at sunup under a diaphanous mist. Rowboats move upon the surface like water spiders, as geese waddle down to the edge, honking joyously."*
>
> Mike Edwards, "A Comeback for the Cossacks," *National Geographic*, November 1998

Siberian Rivers

In Siberia, rivers such as the Ob, the Irtysh, the Yenisey, and the Lena rank among the world's largest river systems. They flow north through

Siberia to the Arctic Ocean. Temperatures are warmer at the rivers' sources in the south than at their mouths in the north. Frozen rivers melt and land along the rivers thaws earlier in the south than in the north. Blocked by ice as they head northward, the meltwaters often flood the land, creating vast inland swamps and marshes.

The Amur River, which drains eastward, forms the border between Russia and China for about 1,000 miles (1,610 km). Influenced by summer monsoon winds from the southeast, the Amur River valley is warmer than the rest of Siberia and is Siberia's main food-producing area.

Natural Resources

Russia's physical geography is both a blessing and a challenge. The country holds an abundance of natural resources. Much of this wealth, however, lies in remote and climatically unfavorable areas and is difficult to tap or utilize.

Minerals and Energy

Russia has huge reserves of mineral resources. It is especially rich in mineral fuels. The country holds large petroleum deposits and 16 percent of the world's coal reserves. Russia produces more dry natural gas than any other country in the world. It also leads the world in nickel production and ranks among the top three producers of aluminum, gemstones, platinum group metals, sulfur, and tungsten. Russia's rivers make it a leading producer of hydroelectric power.

Soil and Forest Land

Because of Russia's generally cold climate, only about 10 percent of Russia's land can support agriculture. In the far north there is very little farming because of permafrost, a permanently frozen layer of soil beneath the surface of the ground, which underlies much of Russia. However, a wide, fertile band called the Black Earth Belt covers about 250 million acres (100 million ha) and stretches from Ukraine to southwestern Siberia. The chernozem soils of this farmland produce crops such as wheat, rye, oats, barley, and sugar beets that feed much of Russia.

About one-fifth of the world's remaining forest lands lie in Russia—75 percent of them in eastern Siberia. Second only to the Amazon rain forest in terms of the amount of oxygen returned to the

NATIONAL GEOGRAPHIC **World Explorer**

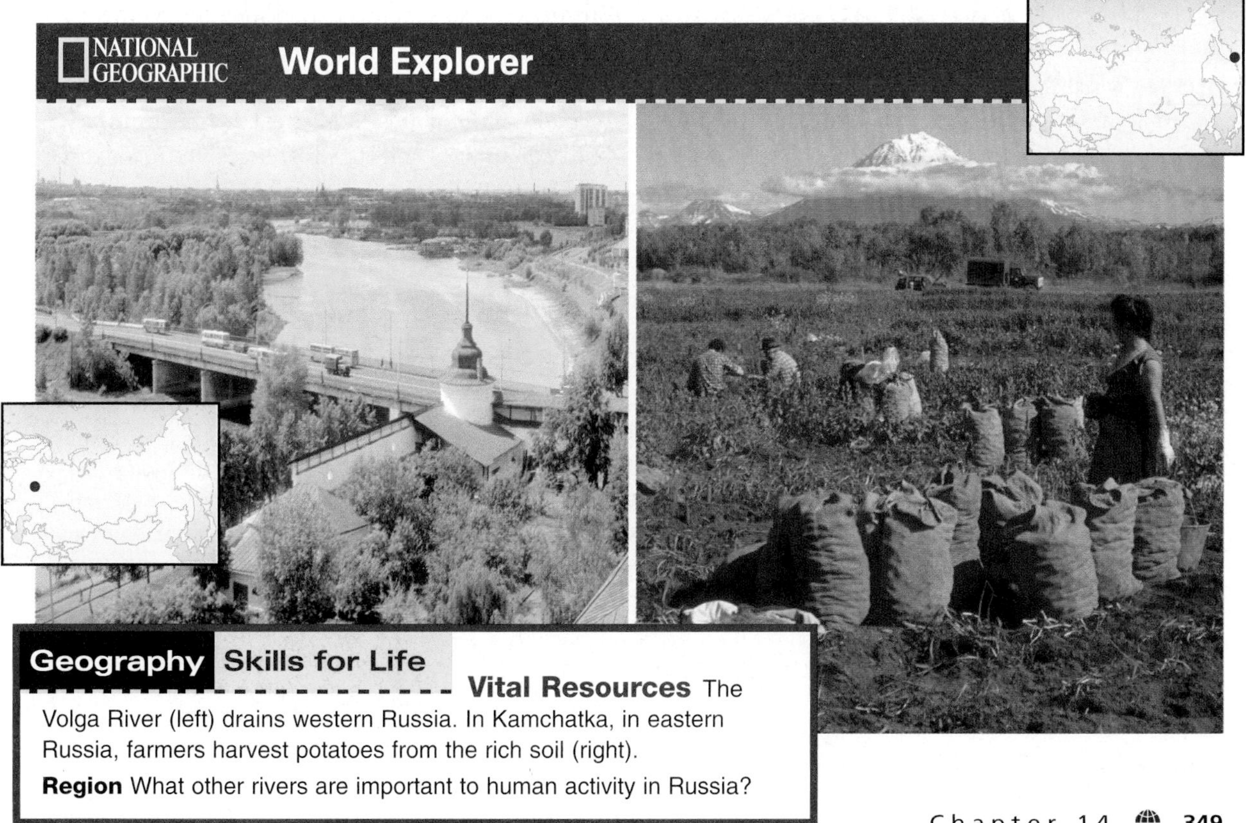

Geography Skills for Life

Vital Resources The Volga River (left) drains western Russia. In Kamchatka, in eastern Russia, farmers harvest potatoes from the rich soil (right).

Region What other rivers are important to human activity in Russia?

NATIONAL GEOGRAPHIC World Explorer

Geography Skills for Life

Russia's Forests Students gather plant samples in a Siberian forest.

Human-Environment At what rate are Russian forests being depleted each year?

atmosphere, Russian forests also supply much of the world's timber, mainly pine, fir, spruce, cedar, and larch. As a result of commercial logging, however, Russian forests shrink by almost 40 million acres (16 million ha) each year—a rate of loss higher than that of the Amazon Basin.

Economics

Russia's Fishing Industry

Fish remain important to the Russian diet and economy, even though many of Russia's rivers and seas are overfished or polluted. Salmon from the Pacific Ocean and herring, cod, and halibut from the Arctic Ocean support a flourishing fishing industry. However, the supply of world-famous Russian caviar, or salted fish eggs, has declined. Dams built on the Volga River have interrupted the migration of sturgeon, the fish that provide the eggs for caviar. Sturgeon is now being fished illegally to meet the global demand for this delicacy.

As you have learned, Russia's vast and varied land of plains, mountains, and great frozen tundra has both advantages and disadvantages. In the next section, you will learn in more detail how climate restricts access to the country's natural resources.

SECTION 1 ASSESSMENT

Checking for Understanding

1. **Define** chernozem, hydroelectric power, permafrost.
2. **Main Ideas** Use a Venn diagram like the one below to show how European Russia and Asian Russia are alike and different.

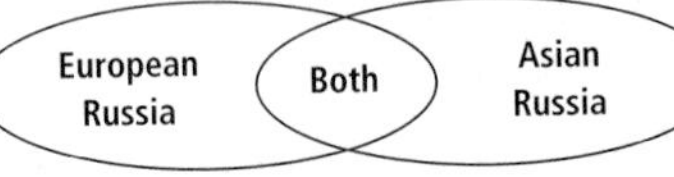

Critical Thinking

3. **Drawing Conclusions** Why do 75 percent of Russians live west of the Ural Mountains?
4. **Identifying Cause and Effect** What problems arise as a result of the large number of Russians living on the North European Plain?
5. **Making Generalizations** Explain how Russia's geography affects access to natural resources.

Analyzing Maps

6. **Region** Study the map on page 347. What types of physical features form Russia's boundaries?

Applying Geography

7. **Effect of Location** Think about the locations of Russia's seas. Then write a descriptive paragraph explaining how the locations of these seas affect Russia's economy.

SECTION 2

Climate and Vegetation

Guide to Reading

Consider What You Know

You can see on a map that Russia's vast landmass lies in the far northern latitudes. What effect do you think this location has on Russia's climate?

Read to Find Out

- What are Russia's major climates?
- What are the seasons like in Russia?
- How does climate affect the way Russians live?
- What types of natural vegetation are found in each of Russia's climate regions?

Terms to Know

- tundra
- taiga
- steppe

Places to Locate

- Siberia
- Arctic Circle

NATIONAL GEOGRAPHIC

A Geographic View

Ice or Mud

We had started on a fine summer day, but as we approached the coast, blue skies gave way to a bitter wind and soul-drenching fog. When Yuri stopped at a small outpost called Nizhne Kamchatsk . . . I was ready to accept his friend's offer of hot tea made from tree fungus ("Good for the kidneys!") and an all-purpose forecast: "Fickle, the weather," he growled. "In Kamchatka the earth is a piece of ice or a piece of mud."

—*Bryan Hodgson, "Kamchatka: Russia's Land of Fire and Ice,"* National Geographic, *April 1994*

Mountain range, Kamchatka Peninsula

Shifting extremes of weather in the Kamchatka Peninsula challenge the people who live there. Much of Russia experiences extreme cold and long winters. In a land where it is frigid and dark for long periods of time and where the rivers do not move for months, people learn patience.

Russia's Climates and Vegetation

Most of Russia has a harsh climate with long, cold winters and short, relatively cool summers. The country's climate is characterized by temperature extremes. The coldest winter temperatures occur in eastern **Siberia**. Warmer air from the Atlantic Ocean moderates temperature to some extent in certain areas of European Russia. Most of the country, however, lies well within the Eurasian landmass, far away from any moderating ocean influences. The Siberian city of Verkhoyansk, located at about 68°N latitude, has been called the "cold pole of the world" because of its bitter winters. January temperatures there have fallen to a low of –90°F (–68°C).

High Latitude Climates

Russia's high-latitude climates feature extremely cold winters and short summers. Seasonal temperatures across this broad landmass can vary greatly. In Yakutsk, in eastern Russia, for example, January temperatures often fall below –33°F (–36°C), and July temperatures average 64°F (18°C). Isolated from oceans and moisture-bearing air masses, Siberia's interior has very little precipitation.

Tundra

Far to the north, the **tundra**, a vast, treeless plain, dominates the Russian landscape. Hugging the edges of the Arctic seas, almost all of the tundra climate region lies north of the **Arctic Circle** ($66^1/_2$°N). An isolated patch of tundra in northeastern Siberia lies near the Sea of Okhotsk. *Tundra* in Finnish means "barren land," an apt term for a place where the average annual temperature is below freezing. In this region the sky stays dark for many weeks before and after December 22. Then, for several weeks during summer, there is continuous sunlight.

The tundra covers about 10 percent of Russia. Its short growing season and the thin, acidic soil lying just above the permafrost limit the kinds of plants

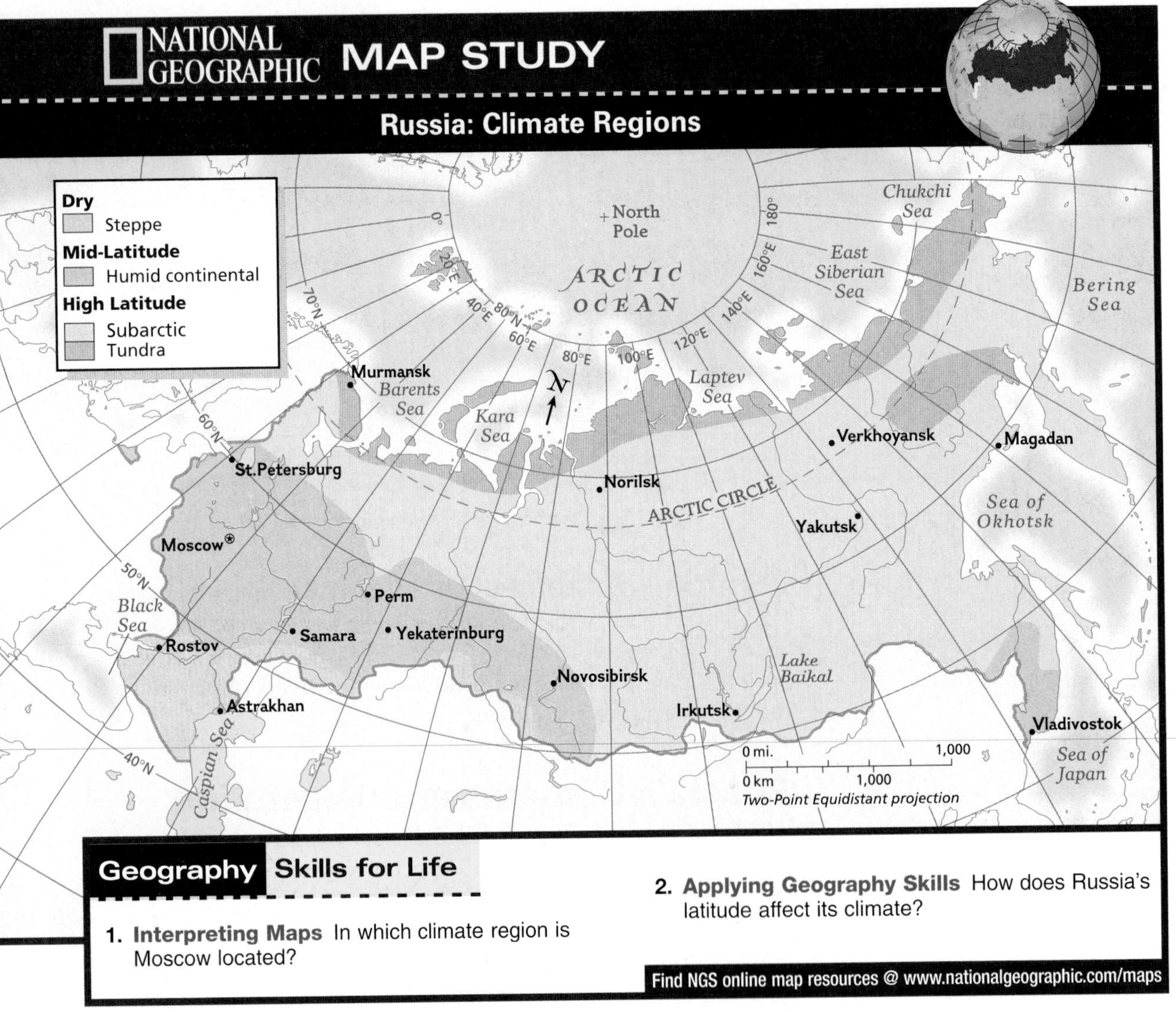

Geography Skills for Life

1. **Interpreting Maps** In which climate region is Moscow located?

2. **Applying Geography Skills** How does Russia's latitude affect its climate?

Find NGS online map resources @ www.nationalgeographic.com/maps

that can grow there. Only mosses, lichen, algae, and dwarf shrubs thrive in the tundra.

Subarctic

Russia's dominant climate region is the subarctic. Although the subarctic lies south of the tundra, some of the world's coldest temperatures occur there. For 120 to 250 days each year, snow covers the ground. The subarctic climate supports the **taiga** (TY•guh), a forest belt that covers two-fifths of European Russia and extends into much of Siberia. Roughly the size of the United States, the Russian taiga is the world's largest coniferous forest, containing about one-half of the world's softwood timber.

Culture

Living in a Cold Climate

Living in an extremely cold climate challenges Russians' creativity. Russians must make adjustments to the climate in all aspects of their lives—jobs, transportation, food and water supplies, heating, clothing, and plumbing. Keeping warm requires a great deal of energy from oil, gas, wood, or coal. Layers of clothing made of wool or fur protect those who brave the frigid outdoor temperatures.

Businesses and industries also must make adjustments to the extreme cold. Builders plan for the cold when they construct buildings. To make machinery and cars, manufacturers must use a special type of steel that will not crack in the cold.

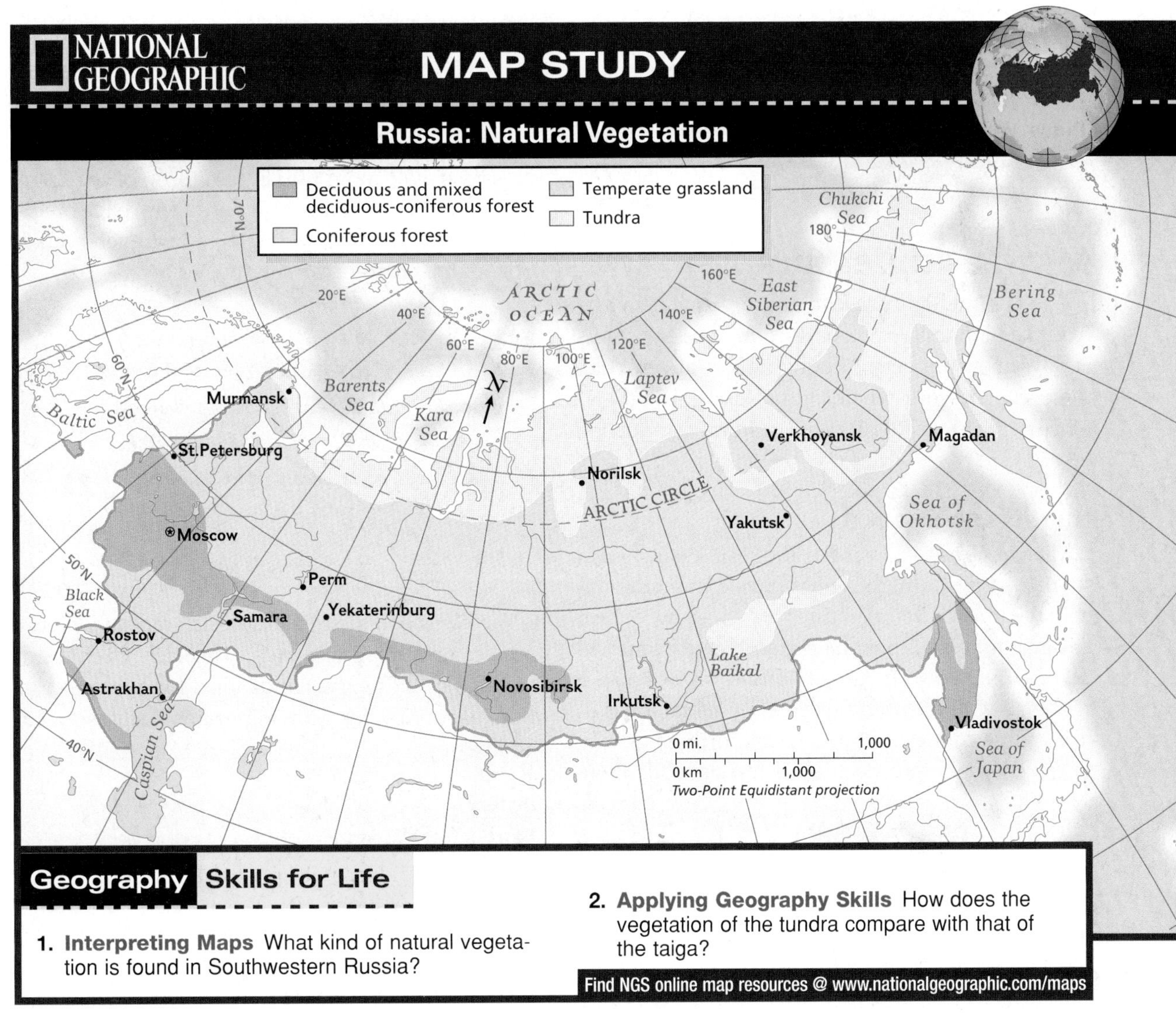

Geography Skills for Life

1. **Interpreting Maps** What kind of natural vegetation is found in Southwestern Russia?
2. **Applying Geography Skills** How does the vegetation of the tundra compare with that of the taiga?

Find NGS online map resources @ www.nationalgeographic.com/maps

NATIONAL GEOGRAPHIC **World Explorer**

Geography Skills for Life

Siberian Temperatures Arctic conditions are part of daily life in Siberia. The woman shown at right is carrying blocks of frozen milk to her home.

Place Why is the climate so cold in Siberia?

Mid-Latitude Climates

Russia's mid-latitude climates are much milder than the high-latitude climates, with milder winters and warmer summers. Although still relatively cold, these climates are where most Russians live and where much of Russia's agricultural production takes place.

Humid Continental

Most of Russia's North European Plain and a small part of southern Siberia have a humid continental climate. Temperatures in Moscow, which lies in a humid continental region, range from 9° to 21°F (–13° to –6°C) in January and from 56° to 75°F (13° to 24°C) in July. In humid continental areas of Russia, the coniferous taiga of the north gives way to mixed coniferous-deciduous forests. Soils there are somewhat more fertile than those of the taiga, and farming methods and fertilizers have made them very productive. Here, one traveler takes note of the crops he sees in the region:

> *"The train bored through a corridor of birch and pine interspersed every 15 minutes with a village. Each of these openings revealed, in the evening light, patches of cabbages, beets, onions, and tomatoes. . . ."*
>
> Mike Edwards, "Playing by New Rules," *National Geographic*, March 1993

Farther south the mixed forests gradually merge into temperate grasslands. The rich chernozem soil makes these grasslands ideal for crop production, especially for growing grains such as wheat and barley.

Student Web Activity Visit the **Glencoe World Geography** Web site at geography.glencoe.com and click on Student Web Activities—Chapter 14 for an activity about Siberia.

History

War and Winter

Russia's cold climate played an important role during World War II. When German troops invaded the former Soviet Union on June 22, 1941, German commanders thought they would win a quick victory. By November, German forces had surrounded Leningrad (now St. Petersburg) and were approaching Moscow. As the Germans advanced into the suburbs of Moscow in December, they unexpectedly encountered Russia's most effective weapon, its brutal winter.

In 1941 winter arrived early, blanketing the front lines with temperatures as low as –40°F (–40°C), the coldest temperatures in decades. Many of the German troops, who were still wearing summer uniforms, suffered frostbite. The frigid cold paralyzed the German tanks, mechanized vehicles, artillery, and aircraft. Russian soldiers, used to the harsh cold, fought well. Siberian troops in particular proved to be extremely effective cold-weather fighters. Under pressure, the Germans retreated, forced back by a combination of Russia's harsh winter and its military forces.

Geography Skills for Life

Russian Steppe Over the centuries, many nomadic invaders, including Attila the Hun, have crossed over Russia's steppe to invade territories to the west.

Region What allows plants to flourish in Russia's steppe region?

Steppe

A small area between the Black and Caspian Seas and a thin band along Russia's border with Kazakhstan make up Russia's **steppe** climate region. This temperate grassland area has dry summers and long, cold, dry winters with swirling, sparse snow. The steppe's chernozem soil is rich in organic matter that enables many plants to flourish. Rippling in the winds, seas of grass stretch to the horizon in every direction. Sunflowers, mint, and beans also flourish in the steppe. In recent years, however, the introduction of foreign plants and overgrazing by animals have damaged the steppe ecosystem. As the newly introduced plants crowd out native grasses, soil fertility declines.

As you have learned, frigid climates dominate large areas of Russia. Nonetheless, in the pockets of more moderate climates, vegetation and human life coexist quite comfortably.

SECTION 2 ASSESSMENT

Checking for Understanding

1. **Define** tundra, taiga, steppe.
2. **Main Ideas** On a sheet of paper, create a table like the one below. Complete the table by filling in information on the climate regions and vegetation of Russia.

Climate Region	Description	Vegetation

Critical Thinking

3. **Making Generalizations** What generalization can you make about Russia's climate regions?
4. **Making Inferences** How does Russia's climates and short growing season affect food production?
5. **Comparing and Contrasting** What are the differences between the tundra and subarctic climate regions? Between the humid continental and steppe climate regions?

Analyzing Maps

6. **Region** Study the maps on pages 347 and 352. Which type of climate characterizes the North European Plain?

Applying Geography

7. **Impact of Climate** Write a paragraph describing physical processes, such as freezing and thawing, and the effect they have on the land and the people of Siberia and other northern parts of Russia.

Understanding Climographs

Climate is the result of the complex interaction of latitude, wind patterns, temperature, and precipitation. Climographs allow us to compare and contrast different climates in different regions based on temperature and precipitation.

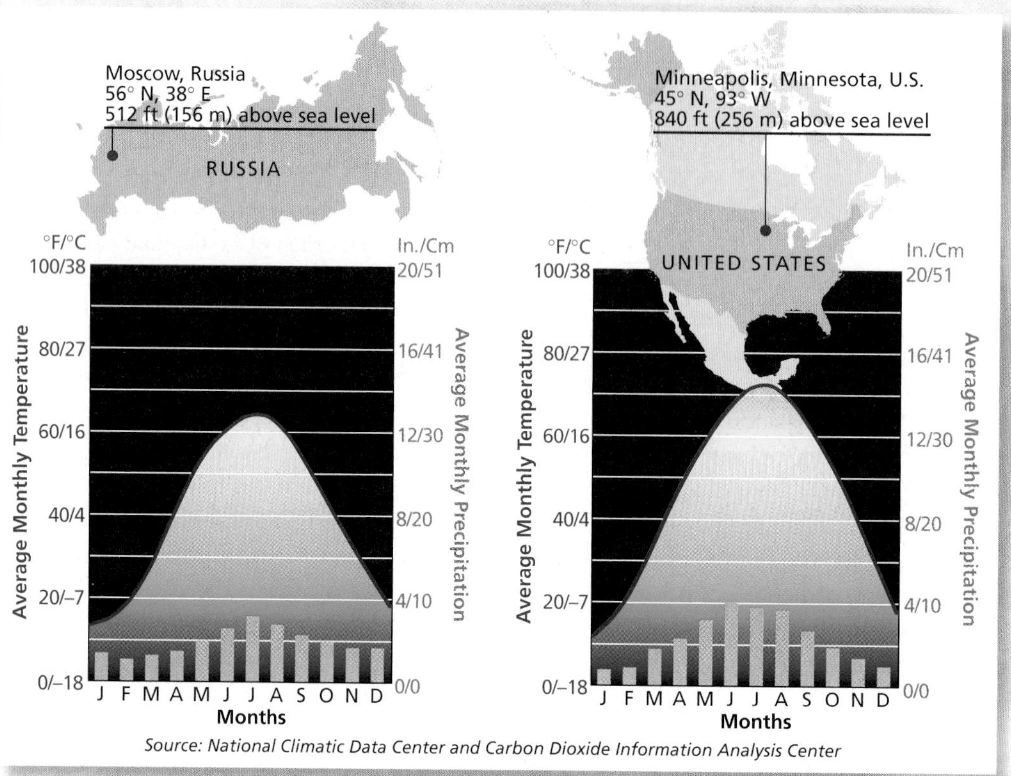

Source: National Climatic Data Center and Carbon Dioxide Information Analysis Center

Learning the Skill

A climograph combines a line graph and bar graph to show average variation in temperature and precipitation. In the graphs above, the months of the year are shown on the horizontal axis. Temperature appears on the left vertical axis as a line graph; precipitation appears on the right vertical axis as a bar graph.

To analyze the information in a climograph:

- **Identify highest and lowest temperatures.**
- **Determine the variation in annual precipitation.**
- **Use this information to describe and compare the two climates.**

Practicing the Skill

Answer the questions using the climographs above.

1. Which city is warmer year-round? Wetter?
2. Which city has the greater annual variation in temperature?
3. What kind of climate does Moscow have? Minneapolis?

Research the average monthly precipitation and temperature in your area using the library, a local newspaper, or the Internet. Use the data to make a climograph. How are the climates of your area and Moscow similar? Different?

The **Glencoe Skillbuilder Interactive Workbook, Level 2** provides instruction and practice in key social studies skills.

CHAPTER 14 SUMMARY & STUDY GUIDE

SECTION 1 The Land (pp. 345–350)

Terms to Know
- chernozem
- hydroelectric power
- permafrost

Key Points
- Russia is the largest country in the world, spanning Europe and Asia.
- Russia's land consists of interconnected plains and plateaus and is bordered on the south and east by mountain ranges.
- Most rivers in Russia flow northward and are frozen for much of the year.
- Russia is rich in resources, such as petroleum, coal, minerals and gems, and timber.

Organizing Your Notes
Use a diagram like the one below to help you organize important details about Russia's physical features.

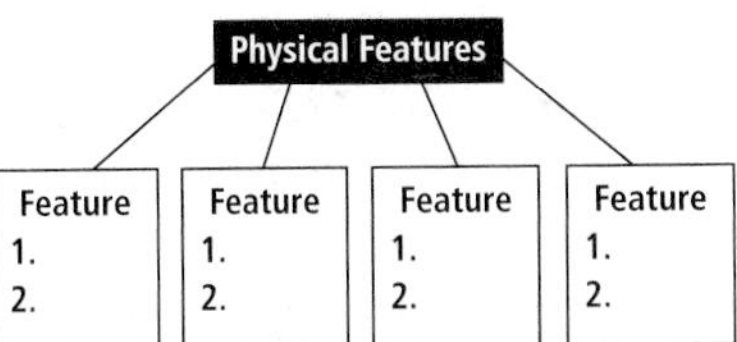

SECTION 2 Climate and Vegetation (pp. 351–355)

Terms to Know
- tundra
- taiga
- steppe

Key Points
- Most of Russia has a harsh climate with wide extremes of temperatures, which creates challenges in all aspects of Russian life.
- Russian winters are long and cold, and its summers are short and relatively cool.
- Permanently frozen subsoil, or permafrost, lies beneath much of Siberia.
- The vegetation in Russia is varied, with treeless tundra in the far north, densely wooded taiga in the north and central areas, and temperate steppe grasslands in the southwest.

Organizing Your Notes
Create an outline using the format below to help you organize your notes for this section.

Climate's and Vegetation

I. Russia's Climate and Vegetation
  A. High Latitude
    1. Tundra
      a. north of Arctic Circle
      b.
      c.

Birch tree forest in Siberia.

ASSESSMENT & ACTIVITIES

Reviewing Key Terms

Write the key term that best completes each of the following sentences. Refer to the Terms to Know in the Summary & Study Guide on page 357.

1. The permanently frozen _________ lies beneath much of northern Russia.
2. The frigid _________ stretches along Russia's northern boundary.
3. Many varieties of grasses grow in the _________ climate region.
4. The rich _________ soil of the North European Plain supports the production of grains.
5. Coniferous trees grow in the _________, a forest belt that covers most of Russia.
6. The Volga River provides western Russia with _________.

Reviewing Facts

SECTION 1

1. Which mountains form a natural dividing line between European Russia and Asian Russia?
2. What are Russia's two main plains?
3. Explain why the Volga River is so important to the people of Russia.
4. What are Russia's major natural resources?

SECTION 2

5. What are the main characteristics of Russian seasons?
6. What are the four climate regions in Russia?
7. Which climate region dominates Russia?
8. What kinds of vegetation are found in each of Russia's climate regions?

Critical Thinking

1. **Drawing Conclusions** Why do most Russians live on the North European Plain?
2. **Analyzing Information** Why is Russia's Volga River often called "Mother Volga"?
3. **Identifying Cause and Effect** Create a web diagram like the one below, and fill in effects of the cold climate on Russian life. Then write a paragraph describing one effect in detail.

NATIONAL GEOGRAPHIC

Locating Places

Russia: Physical Geography

Match the letters on the map with the places and physical features of Russia. Write your answers on a sheet of paper.

1. Ural Mountains
2. Caucasus Mountains
3. Verkhoyansk Range
4. Central Siberian Plateau
5. Arctic Ocean
6. Bering Sea
7. Caspian Sea
8. Volga River
9. Ob River
10. North European Plain

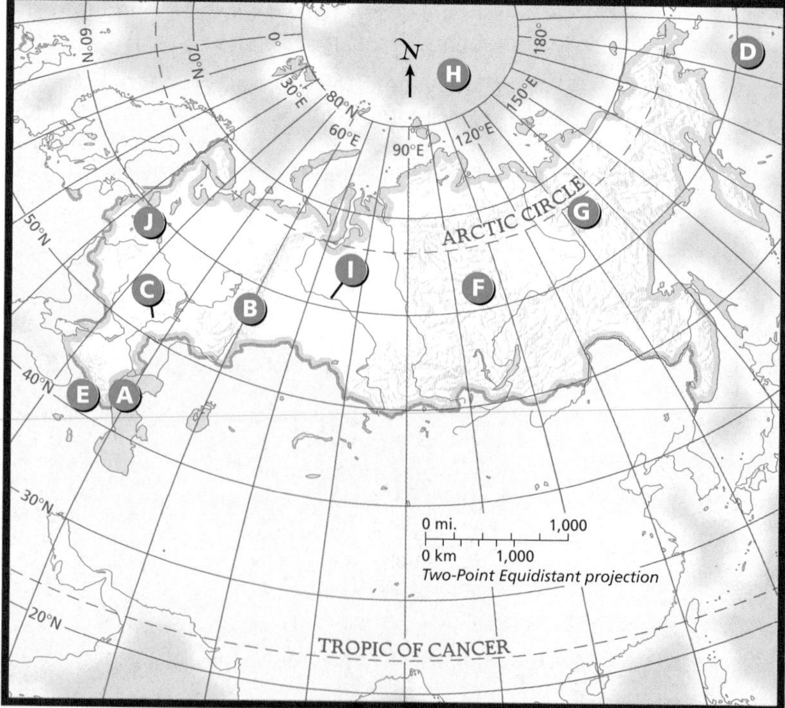

Self-Check Quiz Visit the **Glencoe World Geography** Web site at geography.glencoe.com and click on Self-Check Quizzes—Chapter 14 to prepare for the Chapter Test.

Using the Regional Atlas

Refer to the Regional Atlas on pages 338–341.

1. **Location** What challenges do Russia's physical features create for mining and transporting coal?
2. **Region** What is the major natural resource found on the Central Siberian Plateau?

Thinking Like a Geographer

Think about the physical geography of Russia. What factors prevent Russia from being a major shipping country with many ports? As a geographer, which form of transportation within Russia do you think would be the best choice for further development?

Problem-Solving Activity

Decision Making Imagine you are a Russian engineer. A foreign automobile manufacturing company has asked you to recommend a location within Russia to build a new plant. List the resources and services needed for the factory. Then use your text, the Internet, or other sources to select one or two ideal areas for the plant. Write your recommendations in a letter to the president of the company.

GeoJournal

Descriptive Writing Using the information you logged in your GeoJournal as you read this chapter, write a descriptive paragraph about one of Russia's unique or beautiful physical features. Describe the feature in detail, using your textbook and the Internet as resources to make your descriptions as specific and vivid as possible.

Technology Activity

Using E-mail Locate an e-mail address for a youth organization in Russia. Write a letter requesting information about the land and climate in the youths' area and the effects on their lifestyle. Share the response you receive with your class.

Standardized Test Practice

Using the table below and your knowledge of geography, choose the best answer for the following multiple-choice question. If you have trouble answering the question, use the process of elimination to narrow your choices.

Arable Land in Selected Countries

	Russia	Canada	United States	France
arable land	8%	5%	19%	33%
permanent crops	0%	0%	0%	2%
permanent pastures	4%	3%	25%	20%
forests and woodland	46%	54%	30%	27%
other	42%	38%	26%	18%

Source: CIA World Factbook 2000

1. **What factor may help explain why Russia and Canada have a lower percentage of arable land than do the United States and France?**

   A Russia and Canada have been settled longer.

   B Russia and Canada extend farther into cold northern regions.

   C Russia and Canada have larger land masses.

   D Russia and Canada are less industrialized.

Study the information shown in the table about land use. Then think about climate regions in the selected countries. Notice similarities or differences between figures for the four countries. Choice C is not relevant, so it can be eliminated.

Geography Lab Activity

Cleaning Up Oil Spills

▲*Approximately 30 tons (27,216 kg) of oil spilled into the Neva River in October 1999.*

Petroleum and water are both natural resources. However, when they mix, they pose a grave danger to the environment. In 1999 a massive oil spill on the Neva River in northwest Russia threatened to poison the water supply of St. Petersburg and damage the Gulf of Finland downstream. Russian officials immediately dispatched vessels equipped with floating barriers to contain the oil after the ship carrying it ran aground. Cleaning crews used a bio-absorbent, a substance that helps break down oil, to contain the spill and remove much of the oil from the water.

When oil mixes with water, a number of chemical changes may take place. Depending on the movement and temperature of the water and the presence of wind and sunlight, the oil may change its shape, density, and chemical composition. Cleanup crews have a limited number of tools to use against an oil spill. Pumps and sponges may suck up oil if it is not too heavy after mixing with water. Skimmers that look like conveyor belts draw oil off the water's surface. Oil-eating microbes and strong detergents are also used to dissolve oil.

1 Materials

- Large, shallow, rectangular pan about 11 inches (28 cm) long
- Water
- 1 tablespoon vegetable or cooking oil
- String or dental floss, at least 36 inches (91 cm) long
- Ruler
- Timer
- Drinking straw
- 2 paper towels or several cotton balls
- 1 or 2 drops of dishwashing detergent
- Measuring cup
- Writing materials

2 Procedures

In this activity, you will simulate an oil spill in order to observe how oil reacts in water. You also will experiment with two cleanup strategies.

1. Fill the pan about halfway with water, and set the pan on a level surface.
2. Gently pour the vegetable oil into the center of the water.
3. Use the string or dental floss to measure the size (the circumference) of the "oil slick." Record your measurement.
4. Wait 2 to $2^1/2$ minutes, and then measure the oil spill again. Record your measurements.
5. Repeat step 4 three more times, until you have five measurements.

6. Use the straw to blow gently across the surface of the water. Blow steadily for about 20 to 30 seconds. Record your observations.
7. Gently shake or vibrate the pan to create a wavelike motion. Do this for about 30 seconds to 1 minute. Record your observations.
8. Set a paper towel or cotton ball on the surface of the water in the pan until it is completely soaked, but do not let it sink.
9. Remove the paper towel or cotton ball, and repeat this step with a fresh paper towel or cotton ball. Record your observations.
10. Now mix the drops of dishwashing detergent into 4 ounces of water. Pour the mixture gently into the middle of the pan. Record your observations.

3 Lab Report

1. Which of the steps was most time-consuming? Why?
2. What did measuring the oil slick demonstrate about the way oil behaves in water?
3. What did blowing on and shaking the water demonstrate about the way oil behaves in rivers, lakes, and oceans?
4. **Drawing Conclusions** Based on your observations, what oil cleanup strategy would you recommend for future oil spills? Why?

4 Find Out More

Look in reference books or check Internet sources for more information about strategies for cleaning up oil spills. Take notes on the different strategies, including the way they work and how successful they are. Make a poster or multimedia presentation of the different strategies and their success rates, and then share your findings with the class.

Did You Know? Magnets may prove to be an effective tool against the effect of oil spills on wild birds. When oil coats birds' feathers, the birds lose their natural protection against the elements, and they cannot fly. Scientists have discovered that coating bird feathers with an iron powder and then applying magnets removed nearly all the oil in test cases.

Water movement and air flow can affect the shape and density of an oil spill.

CHAPTER 15

The Cultural Geography of Russia

GeoJournal

Create two columns in your journal. Label the first column "Questions" and the second "New Knowledge." First, list questions you have about Russian history and culture. Then, as you read this chapter, record the answers.

GEOGRAPHY Online

Chapter Overview Visit the **Glencoe World Geography** Web site at geography.glencoe.com and click on Chapter Overviews—Chapter 15 to preview information about the cultural geography of the region.

SECTION 1

Population Patterns

Guide to Reading

Consider What You Know

Throughout much of its history, Russia has been a land of many different peoples. The United States also has a diverse population. What do you think are the benefits and challenges of living in a country that has people of many different backgrounds?

Read to Find Out

- What ethnic groups make up Russia's population?
- Why is Russia's population unevenly distributed?
- How does the climate east of the Ural Mountains affect the population distribution in this region?

Terms to Know

- ethnic group
- nationality
- sovereignty

Places to Locate

- Ural Mountains
- Moscow

◀ *Children playing on ice, Lake Baikal*

NATIONAL GEOGRAPHIC

A Geographic View

Russian Heartland

I have come back to Mother Russia, the old heartland from which had sprung the Russian Empire and its successor, the Union of Soviet Socialist Republics. . . . A modern land in many ways, yet profoundly tied to the past. . . . There is [today] a new quest for the much trampled Russian culture, for the "soul" that writers lauded for its breadth and warmth. The old love of the gentle landscape that 'spreads out evenly across half the world,' as [the writer] Nikolay Gogol saw it in [his novel] **Dead Souls**, *blooms anew—in the form of anger over polluted rivers and smoky vistas.*

—*Mike Edwards, "Mother Russia on a New Course,"* National Geographic, *February 1991*

Historic church in Moscow

Over the centuries Russia's borders moved beyond Moscow to include vast territories inhabited by people of different ethnic backgrounds. Today, the citizens of Russia are not one people, but many. Each of the diverse groups within the country has its own cultural traditions, history, and language. In this section you will learn about the various culture groups of Russia—from the Arctic peoples in the north to the peoples of the Caucasus region in the south.

Russia's Ethnic Diversity

Russia has one of the widest varieties of ethnic groups in the world—in fact, more than a hundred! An **ethnic group** shares a common ancestry, language, religion, or set of customs, or a combination of these things. Despite Russia's ethnic diversity, more than 80 percent of the population are ethnic Russians, people who follow Russian customs and speak Russian as their first language. The percentage of

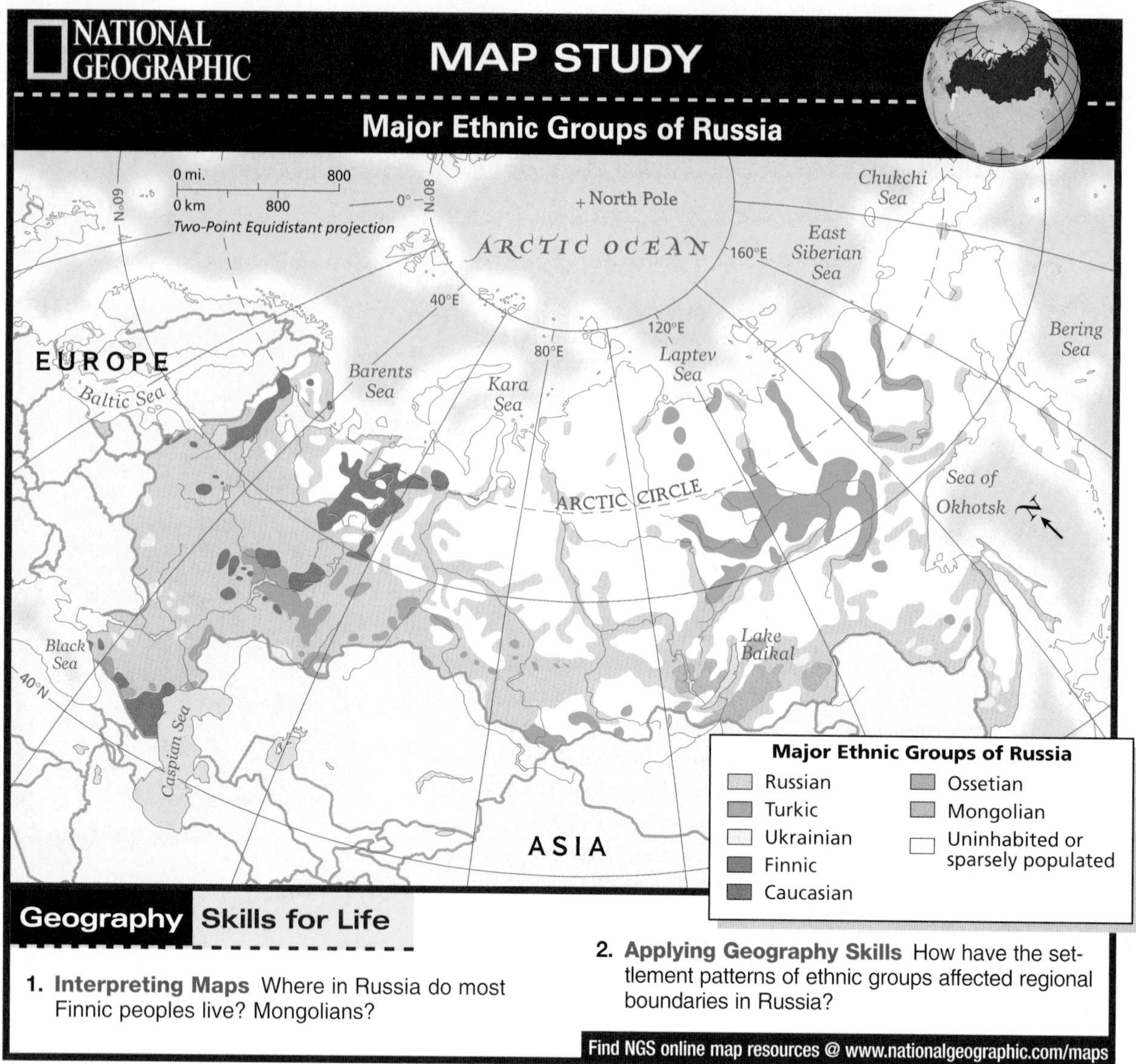

Geography Skills for Life

1. **Interpreting Maps** Where in Russia do most Finnic peoples live? Mongolians?

2. **Applying Geography Skills** How have the settlement patterns of ethnic groups affected regional boundaries in Russia?

Find NGS online map resources @ www.nationalgeographic.com/maps

the population that is not ethnic Russian usually became part of Russia's population as a result of conquest. This fact has made it difficult for some groups to consider themselves truly Russian.

Culture

Ethnic Regions

Over the centuries Russia grew from a small territory to a gigantic empire that stretched from the plains of Europe to the waters of the Pacific Ocean. In the process, many non-Russian ethnic groups came under its control. In some cases ethnic groups were concentrated in a single area. During the Soviet era, regional political boundaries often reflected the locations of major ethnic groups, or **nationalities**. In 1991, after the Soviet breakup, several of these larger republics, including Russia, became independent countries. Today 32 ethnic groups have their own republics or administrative territories within Russia.

The Slavs

Ethnic Russians are a part of a larger ethnic group known as Slavs, a family that also includes Poles, Serbs, Ukrainians, and other eastern Europeans.

Throughout Russia's history the Russian Slavs have dominated the country's politics and culture. Most Slavs practice Eastern Orthodoxy, a form of Christianity brought to Russia from the eastern Mediterranean area. Russian national identity has long been tied to the Slav, or ethnic Russian, culture.

Although more than 100 languages are spoken in Russia today, Russian is the country's official language. Ethnic Russians generally speak only this language, while people belonging to other ethnic groups speak both their own languages and Russian.

Turkic Peoples

Russia's second-largest family of ethnic groups, the Turkic peoples, live in the Caucasus area, in Siberia, and in the middle Volga area. Although Turkic peoples are mainly Muslims, their ethnicity is based primarily on language.

The Turkic peoples of Russia include the Tatars, Chuvash, Bashkirs, and Sakha. The most numerous of these groups are the Tatars, about one-third of whom live in Tatarstan (TA•tuhr•STAN) in east-central Russia. The Tatar population there, however, is growing rapidly, as this observer reveals:

> *"Tatars make up 48 percent of Tatarstan's 3.7 million population. Russians are 43 percent. The ratio is close, but the Russians are worried. . . . [T]he Tatar birthrate is 40 percent higher than the Russian, and efforts to revive Tatar ways . . . will surely erode Russian influence."*
>
> Mike Edwards, "Russia: Playing by New Rules," *National Geographic*, March 1993

Russia has ruled Tatarstan since the mid-1500s. In 1994, however, the Russian government granted Tatarstan a limited amount of **sovereignty** (SAH•vuh•ruhn•tee), or self-rule. The government hopes that this arrangement will dampen any desire the people of Tatarstan may have to separate from Russia.

Caucasian Peoples

Another large group of diverse peoples is classified as Caucasian (kaw•KAY•zhuhn) because they live in the Caucasus region of southeastern Russia. Mainly Muslims, the Caucasian peoples have similar languages and cultures, but local dialects often make communication among them difficult. Caucasian groups such as the Chechens, Dagestanis, and Ingushetians today are demanding independence or at least local self-rule.

Population Density and Distribution

Russia is the sixth most populous country in the world, after China, India, the United States, Indonesia, and Brazil. Russia does not, however, have a large population relative to its land area.

Population and the Environment

With a population of about 144.4 million people and an area of about 6.6 million square miles (about 17.1 million sq. km), Russia's average population density is about 22 people per square mile (9 per sq. km). Compare this figure with that of the United States, where an average of 77 people live within a square mile (30 per sq. km), and you can begin to appreciate how sparsely populated parts of Russia are.

Averages alone, however, can be misleading. About 75 percent of all Russians live in the area between the Belarus and Ukraine borders and the **Ural Mountains**, making the population density of European Russia about 120 people per square mile (46 per sq. km). Meanwhile, Russia's largest eastern republic, Sakha, averages less than 1 person per square mile.

NATIONAL GEOGRAPHIC **World Explorer**

Geography Skills for Life

Kazan Marketplace Consumers shop at the poultry counter in the marketplace of Kazan, the major economic center of Tatarstan.

Place What is Tatarstan's political relationship with Russia?

Geography Skills for Life

Russian Language A poster in St. Petersburg announces the opening of an opera.

Region What two cultural characteristics give ethnic Russians a sense of identity?

The uneven distribution of Russia's population relates to its physical environment. East of the Ural Mountains, the Siberian climate is harsh. Mountains, frozen tundra, and forests there are unsuitable for farming. Although Siberia makes up about 75 percent of Russia's land area, only 25 percent of Russia's people live there.

By contrast, the more densely settled European Russia includes the region's industrialized cities, many of which are connected by waterways. The major industrial city is **Moscow**, Russia's capital. Other industrial centers include St. Petersburg, Nizhniy Novgorod, Kazan, Perm, Volgograd, and Yekaterinburg. Since 1990, urban population growth in most industrialized centers has leveled off or decreased, particularly in cities with more than 500,000 inhabitants.

Population Trends

During the Soviet era, many ethnic Russians migrated to non-Russian republics of the Soviet Union. In the 1970s this trend began to reverse. Since the breakup of the Soviet Union in 1991, more ethnic Russians have returned to their homeland. Most have settled in Moscow, St. Petersburg, and southwestern Russia. Because of this trend, the number of immigrants to Russia has been greater than the number of Russians leaving the country.

Still, Russia is currently experiencing a population crisis because of a rise in illnesses as the quality and availability of health care have declined. Since 1992 the number of deaths has exceeded the number of births. During the 1990s, male life expectancy dropped from 64 years to 59 years, but is expected to rise slowly in the 2000s. During the same time period, female life expectancy decreased from 74 years to 72 years. Infant mortality during this period rose from 17.4 deaths per 1,000 births to 19 per 1,000 births. One of the tasks facing Russia in the years ahead is to improve health care.

SECTION 1 ASSESSMENT

Checking for Understanding

1. **Define** ethnic group, nationality, sovereignty.
2. **Main Ideas** Copy the web below, and use it to fill in current information about Russia's population.

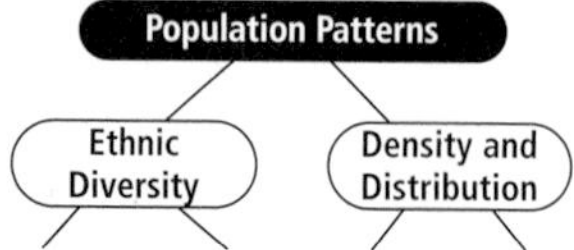

Critical Thinking

3. **Categorizing Information** What would be the advantages and disadvantages of an ethnic group forming an independent country?
4. **Making Generalizations** How might improved health care help solve Russia's current population crisis?
5. **Predicting Consequences** What are some likely effects of changes in Russia's population in the future?

Analyzing Maps

6. **Human-Environment Interaction** Study the map of Russia's ethnic groups on page 364. Explain the pattern of settlement east of the Ural Mountains.

Applying Geography

7. **Migration** Consider past population trends in Russia. What might have drawn immigrants to settle in an area like Moscow?

History and Government

Guide to Reading

Consider What You Know

Events in Russia are often in the news. What recent news events have helped you understand more about Russia's government and its challenges?

Read to Find Out

- Who were the ancestors of the ethnic Russians?
- Why did the rule of the czars end in revolution?
- What were the causes of the Soviet Union's collapse?
- Why does Russia face an uncertain future?

Terms to Know

- czar
- serf
- Russification
- socialism
- Bolshevik
- communism
- satellite
- Cold War
- perestroika
- glasnost

Places to Locate

- Baltic Sea
- Black Sea
- St. Petersburg

NATIONAL GEOGRAPHIC

A Geographic View

End of an Era

The Bolshevik dream finally ended with Mikhail Gorbachev's program of **glasnost**, *or openness, which allowed citizens to speak freely for the first time in decades. All the carefully constructed "truths" began to unravel, and there was no turning back. Gorbachev's era passed. Russia's President, Boris Yeltsin, outlawed the Communist Party by signing a few pieces of paper. The Bolsheviks surrendered without a shot.*

—*Dusko Doder, "The Bolshevik Revolution,"* National Geographic, *October 1992*

Mikhail Gorbachev

Mikhail Gorbachev saw firsthand both the costs and benefits of political change, even changes that come with democratic reforms. As the last Soviet leader, Gorbachev tried to reform the Soviet system, but his efforts failed to prevent its collapse. The history of Russia, once the dominant republic of the Soviet Union, is a story of the rise and fall of great empires. Monarchs, Communist Party officials, and democratic politicians—as well as foreign invaders—have all shaped Russia's national character.

Early Peoples and States

Russia's historical roots go back to the A.D. 600s, when Slav farmers, hunters, and fishers settled near the waterways of the North European Plain. Over time, the Slavs separated into distinct cultural groups. The West Slavs eventually became the Poles, Czechs,

MAP STUDY

Russia's Changing Borders, 1462–Present

ALASKA
Russian until 1867
Chukchi Sea
Bering Sea
North Pole
ARCTIC OCEAN
N
80°N
0°
60°N
40°E
80°E
120°E
160°E
East Siberian Sea
FINLAND
EUROPE
LATVIA
ESTONIA
LITHUANIA
Barents Sea
Kara Sea
Laptev Sea
St.Petersburg (Leningrad)
BELARUS
UKRAINE
Kiev
Moscow
ARCTIC CIRCLE
Sea of Okhotsk
RUSSIA
MOLDOVA
Black Sea
Omsk
GEORGIA
40°N
Caspian Sea
KAZAKHSTAN
Vladivostok
Sea of Japan
ARMENIA
AZERBAIJAN
TURKMENISTAN
UZBEKISTAN
KYRGYZSTAN
TAJIKISTAN
ASIA

0 mi. 1,000
0 km 1,000
Two-Point Equidistant projection

Russian and Soviet Expansion

1462–1505
1505–1689
1689–1917
1917–1945
Kievan Rus
Muscovy of Ivan III, 1462
Boundary of Soviet Union in 1945
Present boundaries

Geography Skills for Life

1. **Interpreting Maps** In what time period did Russia gain control of St. Petersburg? Omsk?
2. **Applying Geography Skills** What geographical factors influenced Russian leaders to expand their empire's boundaries?

Find NGS online map resources @ www.nationalgeographic.com/maps

and Slovaks. The South Slavs became the Bulgarians, Croats, Serbs, and Slovenes. The East Slavs became the Russians, Ukrainians, and Belarusians. These East Slav peoples remained settled along the Dnieper (NEE•puhr) River in the west and the Volga River in the east.

Kievan Rus

During the 800s Scandinavian warriors called the Varangians settled among the Slavs living near the Dnieper and Volga Rivers. Within a century the Varangians had adopted the Slav language and many Slav customs and had organized the Slav communities into a loose union of city-states known as Kievan Rus. Ruled by princes, the leading city-state, Kiev, controlled a prosperous trading route, using Russia's western rivers as a link between the **Baltic Sea** and the **Black Sea**.

Eventually, fighting among the city-states weakened Kievan Rus. Then, in the early 1200s, Mongol invaders from Central Asia conquered Kiev and many of the Slav territories. Although the Mongols allowed the Slavs self-rule, they continued to control the area for more than 200 years. During this period the Slav territories still remained in contact with western and central Europe. However, they followed their own distinctive cultural path based on the traditions of Eastern Orthodoxy.

The Rise of Russia

When the Mongols first overran Kiev, many Slavs fled into nearby forests, and some of them later settled along the Moskva River to the northeast. In time one of their settlements grew into the city of Moscow, which became the center of a territory called Muscovy (muh•SKOH•vee). Muscovy was linked by rivers to major trade routes and surrounded by lands good for farming and trapping fur-bearing animals.

For about two centuries, Muscovy's princes kept peace with the Mongols. Their territory grew in power as the princes helped the Mongols collect taxes from other Slav territories. By the late 1400s, however, the Muscovites became strong enough to refuse payments to the Mongols and to drive them out. Following this triumph, Muscovy's Prince Ivan III brought many Slav territories under his control, thus earning the title "the Great." Ivan's expanded realm eventually became known as Russia. In the heart of Moscow, Ivan built a huge fortress, called the Kremlin, and filled it with churches and palaces.

In 1533 Ivan the Great's grandson, Ivan IV, became Russia's first crowned **czar** (ZAHR), or supreme ruler. Called Ivan the Terrible, Ivan IV crushed all opposition to his power and expanded his realm's borders.

After Ivan's reign, however, the country faced foreign invasion, economic decline, and social upheaval. When the Romanov dynasty came to power in 1613, the government gradually tightened its grip on the people. By 1650 many peasants had become **serfs**, a virtually enslaved workforce bound to the land and under the control of nobility.

Romanov Czars

While Russia struggled through chaos and harsh rule, western Europe moved forward and left Russia behind, especially in the areas of science and technology. Then in the late 1600s, Czar Peter I—known as Peter the Great—came to power determined to modernize Russia. Under him, Russia enlarged its territory, built a strong military, and developed trade with Europe. To acquire seaports, Peter gained land along the Baltic Sea from Sweden. He also strengthened Russia's control of Siberia.

A new capital—**St. Petersburg** —was carved out of the wilderness. Built along the Gulf of Finland, St. Petersburg provided access to the Baltic Sea and gave Russia a "window to the West." Since most of Russia's other ports were icebound for almost half the year, St. Petersburg became a major port.

During the late 1700s, Empress Catherine the Great continued to expand Russia's empire and gained a long-sought-after warm-water port on the Black Sea. By that time the Russian nobility had adopted western European ways—for example, using French instead of Russian as their primary language. As a result, a cultural gap developed between the nobility and Russia's serfs, who followed traditional Russian ways. Meanwhile, poverty and heavy work fell even more harshly on the serfs. Russia's great expansion also brought

NATIONAL GEOGRAPHIC World Explorer

Geography Skills for Life

The Catherine Palace Along with other palaces, the summer residence of Catherine the Great is located outside of St. Petersburg.

Region How did Catherine the Great expand Russia's empire?

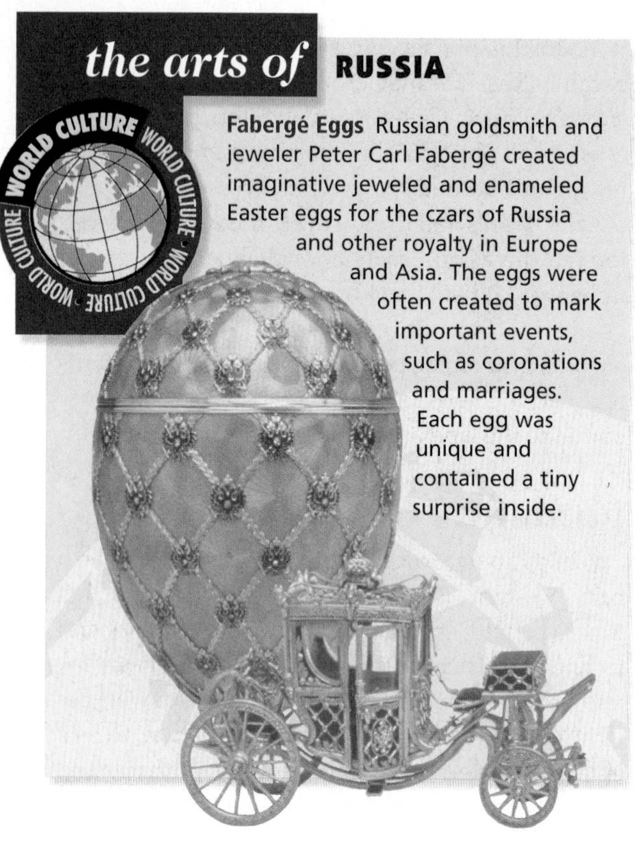

the arts of RUSSIA

Fabergé Eggs Russian goldsmith and jeweler Peter Carl Fabergé created imaginative jeweled and enameled Easter eggs for the czars of Russia and other royalty in Europe and Asia. The eggs were often created to mark important events, such as coronations and marriages. Each egg was unique and contained a tiny surprise inside.

under its rule many non-Russians, including Poles, Ukrainians, Estonians, Baltic Germans, Jews, and Tatars in Crimea near the Black Sea.

The Russian Revolution

The 1800s saw a long cycle of popular discontent, half-hearted political reforms, and governmental repression. Inspired by the American and French Revolutions, educated Russians wanted to make Russian society more open. The government, however, held on tightly to power, and reforms were limited. Czar Alexander II freed the serfs in 1861, but they had no education and few ways to earn a living. Industrialization drew some peasants from the country to the cities, where they worked long hours in poor conditions for meager wages.

At the same time, non-Russian peoples were facing prejudice and hostility. Spurred by increasing nationalism, the government introduced the policy of **Russification**, which required everyone to speak Russian and follow Eastern Orthodox Christianity. People who refused were often persecuted. Harsh treatment was directed especially toward Jews, who were often blamed for Russia's problems.

Frustrated and discontented, many Russian thinkers and workers were attracted to **socialism**, a belief that calls for greater economic equality in society. Some Russians especially liked the socialist ideas of German philosopher Karl Marx. Marx advocated public ownership of all land and a classless society with an equal sharing of wealth. He claimed that continual struggle between the wealthy and working classes would lead to a worldwide revolution. This revolution, he thought, would be led by workers and end the power of the wealthy.

In the early 1900s, discontent with the iron rule of the czars spilled into the streets. Strikes and demonstrations in 1905 nearly ended the reign of Czar Nicholas II. One event, called Bloody Sunday, began with a peaceful crowd of workers desiring better working conditions and personal freedoms marching toward the czar's palace in St. Petersburg. The march ended abruptly when soldiers fired into the marchers, killing nearly 1,000 people.

Twelve years later, in 1917, the hardships of World War I brought even larger numbers of workers into the streets of the capital. With soldiers joining them, the workers demanded "bread and freedom." Finally, Nicholas II was forced to give up his throne, ending the rule of the czars in Russia.

The Soviet Era

The Russian Revolution of March 1917 established a representative government, but it was too weak to control the passion for change that had swept Russia. In November of that year, the **Bolsheviks**, a revolutionary group led by Vladimir Ilyich Lenin, seized control. The victorious Bolsheviks believed in **communism**, a philosophy based on Karl Marx's ideas that called for the violent overthrow of government and the creation of a new society led by workers.

Promising the Russian people "Peace, Land, and Bread!" the Bolsheviks withdrew Russia from World War I, surrendering much territory to Germany. They used their complete hold on political power to take over industry, direct food distribution, establish an eight-hour workday, and reform the army.

Not all Russians supported the Bolsheviks. To maintain power, the Bolsheviks dealt harshly with their opponents. A civil war soon divided the country, pitting the Bolshevik Red Army against the anti-Bolshevik White Army.

History

The Soviet Union

In 1921 the Bolsheviks, now known as Communists, won the civil war. The following year they established a new country, the Union of Soviet Socialist Republics (USSR), or the Soviet Union, with Moscow as the capital. Under the Communists the Soviet Union gradually gained back Ukraine, Belorussia (now Belarus), much of the Caucasus region, and a large part of Central Asia.

After Lenin's death in 1924, Joseph Stalin, a leading Communist Party official, began a five-year climb to power. Defeating his rivals, Stalin set about making the Soviet Union into a powerful industrial giant by ruthlessly taking control of farms and factories. Millions either were killed or died as a result of hunger, physical hardships, or the brutal conditions in labor camps. Stalin also eliminated from the party and the military those people who might threaten his power.

A Superpower

During World War II, the growth of industry—and the fierce Russian winter—helped the Russians push out the invading Germans, but at great cost. More than 27 million Russian soldiers and civilians died as a result of the war. At the war's end in 1945, the Soviet Union controlled much of eastern Europe. By 1949 most of the countries in the region had become Soviet **satellites**, countries controlled by the Soviet Union. These satellite states, notably East Germany, Hungary, Poland, and Czechoslovakia, strengthened the Soviet Union's military and supplied critically needed raw materials, such as coal and iron ore, as well as manufactured goods.

For the next four decades, the Soviet Union and the United States were engaged in the **Cold War**, the struggle between the two competing systems—communist and capitalist—for world influence and power. Since each country had destructive nuclear weapons, outright conflict was avoided. Instead, the two countries used as "weapons" propaganda, the threat of force, and economic aid to developing countries.

The Soviet Breakup

During the Cold War, the Soviet economy weakened while many other economies grew. Soviet workers struggled with economic hardships, yet their leaders enjoyed great privileges. By the 1980s it was clear that communism was failing.

In 1985 Mikhail Gorbachev, a reform-minded official, assumed power in the Soviet Union. Gorbachev was keenly aware of the abuses of the past—Joseph Stalin had imprisoned both his grandfathers. Although Gorbachev remained a dedicated communist, he began a policy of economic restructuring called **perestroika** (PEHR•uh•STROY•kuh) and a policy of greater political openness called **glasnost** (GLAZ•nohst). Gorbachev's reforms, however, failed to save the Soviet Union.

Poland, Hungary, Czechoslovakia, and other communist countries overthrew their communist rulers in 1989. Meanwhile, nationalist fervor was rising in

NATIONAL GEOGRAPHIC **World Explorer**

Geography Skills for Life

Lenin's Plan Lenin and the Bolsheviks promised to build an economy in which each citizen shared equally in the wealth.

Place What conditions led many people to identify with the promises of Lenin?

Geography Skills for Life

Red Square, Moscow A retired Russian colonel confronts an anti-communist demonstrator in the early 1990s.

Place How did Gorbachev seek to reform the Soviet system?

the non-Russian Soviet republics. The Baltic states of Latvia, Lithuania, and Estonia were the first Soviet republics to declare independence.

Then in 1991 loyal Communists tried to overthrow Gorbachev and preserve the Soviet Union. Boris Yeltsin, who had recently become the first democratically elected president of the Russian republic, rallied the people of Moscow to defy the plotters. A reporter described how Yeltsin's cause was helped by financier Konstantin Borovoy:

> *"On the first day. . . , few Russians knew that Yeltsin was resisting. . . . [Yeltsin] sent faxes to Borovoy's office. . . . Brokers copied them and spread leaflets. . . . Citizens. . . threw up barricades as tanks took positions. . . . 'The [secret police] came to arrest the Xerox machines,' Borovoy said, 'but we had already taken them to a safer place.'. . ."*
>
> Mike Edwards, "Russia: Playing by New Rules," *National Geographic*, March 1993

The coup collapsed, and Gorbachev remained the leader of the Soviet Union. By year's end, however, all the remaining Soviet republics had declared independence. Boris Yeltsin remained president of Russia, the largest of the former Soviet republics. Russia, Belarus, and Ukraine formed the Commonwealth of Independent States (CIS) in 1991, and they eventually were joined by other former republics. On December 25, 1991, Gorbachev's presidency ended, and the Soviet Union ceased to exist.

A New Russia

Boris Yeltsin assumed the leadership of a devastated Russia. The economy was in shambles, and ethnic conflicts threatened the Caucasus region.

A Market Economy

On the economic front, Russia began moving from a command economy to a market economy. This transition caused massive unemployment as outdated and inefficient factories were closed and agriculture was restructured. By 2000, however, the Russian economy began to improve. The rate of inflation, at an unbelievable high of 1,500 percent in 1992, fell below 20 percent by 1997. In addition, Russia's currency, the ruble, which had been sharply losing value on international markets, began to stabilize.

Separatist Movements

After the fall of the Soviet Union, separatist movements and ethnic conflict threatened Russia's stability. Beginning in the 1990s, Tatarstan, Dagestan, Chechnya, and other Russian ethnic territories demanded greater self-rule or sought a complete break from Moscow. Although some conflicts have been settled by compromise, often violence or full-scale war has erupted. The bloody war between the Russian government and separatist forces in Chechnya is a tragic example. In 1991 the Chechens declared their independence. Fearing Russia's breakup if other groups did the same, Boris Yeltsin sent Russian troops into Chechnya in 1994. Under Yeltsin's successor, Vladimir Putin, Russia claimed to control much of the territory. Chechen resistance, however, continued in rural areas. By 2001 about 335,000 people had been displaced by the conflict, and Chechens faced severe food shortages.

▲ *Apartment building damaged during the Chechen conflict*

The years of warfare have ravaged Chechnya's civilian population and the area's oil-based economy. The conflict has also drained economic-development funds from the rest of Russia.

SECTION 2 ASSESSMENT

Checking for Understanding

1. **Define** czar, serf, Russification, socialism, Bolshevik, communism, satellite, Cold War, perestroika, glasnost.
2. **Main Ideas** List the key events in Russia or in the Soviet Union during each of the following time periods: Kievan Rus, Russian Empire, Soviet Union, and Russia.

Era	Dates and Key Events
Kievan Rus	

Critical Thinking

3. **Making Inferences** Why do you think Russians have almost always had a centralized government? What problems do you think the government had as Russia grew?
4. **Comparing and Contrasting** How was the government during czarist rule and the Soviet era similar? Different?
5. **Predicting Consequences** How might Russia be affected if separatist groups gain independence?

Analyzing Maps

6. **Human-Environment Interaction** Look at the map of Russia's changing borders on page 368. What geographic factors encouraged Russian expansion?

Applying Geography

7. **Geography and History** Think about ways that physical geography influenced the Russian people's history and culture. Write an essay explaining the impact of geography on one of Russia's ethnic groups.

GEOGRAPHY AND HISTORY

RUSSIA'S IRON ROAD

EASTWARD HO? Opening the Russian frontier meant traveling east—far east. But Conestoga wagons could not have crossed the frozen lands of Siberia. Encompassing more than half of Russia's total area, Siberia dwarfs the American West and ranks as one of Earth's coldest climates. Only the Trans-Siberian Railroad could accomplish Russia's eastward expansion.

Czar Alexander III approved plans for the railroad that would link the European and Asian parts of the Russian Empire and bring eastern lands under Russia's control. Construction of the world's longest railroad began in 1891. The builders hoped to connect Moscow to the port city of Vladivostok, on the Sea of Japan, by 1900. The distance between the cities is nearly 6,000 miles (9,650 km).

Get Me to Vladivostok on Time

Huge construction problems loomed from the start. Siberia's severe climate and rugged topography slowed progress. By the expected end date of 1900, two unfinished segments remained. The first was the section around Lake Baikal, the world's deepest freshwater lake. To lay track around the lake's southern tip, bridges spanning hundreds of gorges and 33 tunnels through

A train bound for Vladivostok rumbles over Trans-Siberian tracks. More passengers and freight move by train than by any other form of transportation in Russia. ▶

NATIONAL GEOGRAPHIC SOCIETY

Workers built more than 200 bridges to span rivers and gorges along the railway's route.

rock would have to be built. Farther east, the Shilka and Amur Rivers presented similar challenges.

As a temporary solution, an icebreaking steam ferry about the size of a football field carried rail cars and up to 800 passengers at a time across Lake Baikal. Farther down the line, passengers and freight were loaded onto riverboats—ice sledges in winter—for the 1,400-mile (2,250-km) trip along the rivers whose banks had yet to be conquered by rail.

Eager to complete an east-west railway, Russia negotiated an alternate route through Chinese-controlled Manchuria that bypassed the Shilka and Amur Rivers. Completion of this shortcut, along with the Lake Baikal segment in 1904, made travel by rail between Moscow and Vladivostok possible for the first time. Twelve years later, the original route within Russia was completed.

A Driving Force

The railroad opened Russia's interior to homesteaders and developers who exploited Siberia's vast store of raw materials—including coal, timber, and gold. During World War II, rail cars carried supplies to the front and moved hundreds of factories from western sites in the Soviet Union to safer sites east of the Ural Mountains.

Since the 1950s much of the line has been electrified. From start to finish, passengers can make the trip in slightly less than a week across seven time zones.

Looking Ahead

Today the Trans-Siberian Railroad shows signs of wear. With Russia's political turmoil and shaky economy, passenger and freight traffic have steadily declined. Worker morale is low. Will the railroad withstand Russia's upheavals? If the railroad falls apart, how might that affect Russia's future?

1891 Construction begins (photo above) on Trans-Siberian Railroad

1900 Line opens using boats to cross Lake Baikal and the Shilka and Amur Rivers

1903 Shortcut opens through Manchuria

1904 Workers finish section around Lake Baikal

1916 Train travel begins (background photo) along original route in Russia

1950s Electrification of rail line begins

SECTION 3

Cultures and Lifestyles

Guide to Reading

Consider What You Know

Think of books or movies about Russia that you may have read or seen. What images of life in Russia stand out in your mind?

Read to Find Out

- How has the role of religion changed in post-Soviet Russia?
- How are education and health care in Russia adjusting to the fall of communism?
- What role do art, music, and literature have in Russia's cultural heritage?

Terms to Know

- atheism
- patriarch
- icon
- pogrom
- intelligentsia
- socialist realism

Places to Locate

- Caspian Sea
- Lake Baikal

NATIONAL GEOGRAPHIC

A Geographic View

A Cultural Center

. . . [T]oday's Russian aristocracy of entrepreneurs and artists . . . [feel] nostalgia for a Russia long gone—an age of glittering accomplishment when St. Petersburg reigned as a world center of music, ballet, and literature. . . . Reflecting that legacy, the city counts some 30 theaters devoted to the performing arts. . . . Under communist rule, the arts . . . were lavishly subsidized. The Bolsheviks may have made Moscow the political capital of the Soviet Union, but St. Petersburg remained its cultural rival—a position Petersburgers are resolved to maintain.

—*Steve Raymer, "St. Petersburg: Capital of the Tsars,"* National Geographic, *December 1993*

A Russian family reciting poems

Russia's adjustment to a new government and economic system has had a profound effect on all Russians. As they move into a new era, Russia's people are also looking for a cultural renewal. Now that the Soviet state no longer dictates their personal lives, millions of Russians are rediscovering their faiths and traditions, reeducating themselves, and expressing themselves creatively.

Religion in Russia

The Eastern Orthodox Church had been central to Russian culture for a thousand years before the communist revolution in 1917. After acquiring power, the Soviet government strictly discouraged religious practices. It actively promoted **atheism** (AY•thee•IH•zuhm), or the belief that there is no God or other supreme being, in schools and other public institutions. In the late 1980s, however, the government began to relax its restrictions on religion.

NATIONAL GEOGRAPHIC

GRAPH STUDY

Russia: Religions

Religion	Number of Followers
Christian	25,100,000
Muslim	14,600,000
Jewish	600,000
Other religions	1,900,000
Nonreligious	103,800,000

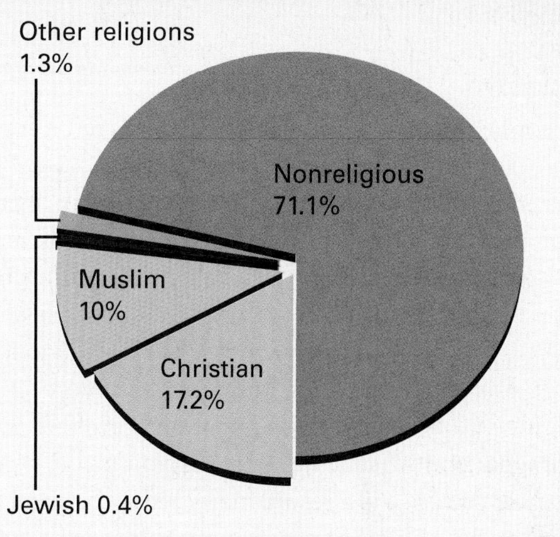

Sources: World Almanac, 2001; Britannica Book of the Year, 2000

Geography Skills for Life

1. **Interpreting Graphs** How many more Christians than Muslims live in Russia?

2. **Applying Geography Skills** Why do you think there is such a large percentage of nonreligious people in Russia?

After the Soviet breakup, many Russians returned to religious practices. However, the influx of many foreign missionaries from Western Christian denominations prompted lawmakers in 1997 to place restrictions on the activities of newly established religious groups. Only Christianity, Islam, Judaism, and Buddhism were allowed full liberty as traditional religions of Russia.

History

Christianity in Russia

In 988 Prince Vladimir, leader of Kievan Rus, adopted Eastern Orthodox Christianity as Russia's official religion. By 1453 the Byzantine Empire, the center of the Eastern Orthodox Church, had fallen, and Russia asserted its claim as leader of the Orthodox Christian world.

During the 1900s the Soviet government weakened Orthodoxy's influence, but today the Russian Orthodox Church is enjoying a resurgence. Most Russians who claim a religious affiliation belong to the Russian branch of the Orthodox Church. The faithful have repaired or rebuilt many of the churches that were looted or destroyed during Soviet times. Like some other Eastern Orthodox churches, the Russian Church has a spiritual leader called a **patriarch** (PAY•tree•AHRK) and uses **icons**, or religious images or symbols, in its religious practices.

Despite recent government efforts to restore the dominant position of Eastern Orthodoxy and restrict other denominations, Russia is also home to many other Christian groups, including Roman Catholics and Protestants. Persecuted during the Soviet era along with members of all other religions, these groups have reemerged since the 1980s.

Islam

Islam, the second-largest religion in Russia, is also enjoying a rebirth. Islam is practiced mostly by people living in the southern regions of Russia, particularly in the Caucasus region and in areas north of Kazakhstan. Most Russian Muslims belong to the Sunni branch of Islam. Sunni Islam is also practiced by people in most Arab countries of Southwest Asia as well as Turkey and Afghanistan. Some citizens of Russia also practice other forms of Islam, including Sufism, which is a deeply spiritual branch of Islam.

Judaism

People practicing Judaism in Russia have long been persecuted. In czarist times Jews could settle only in certain areas and could not own land. They were often the targets of organized persecution and massacres known as **pogroms**. Yet Jewish communities managed to thrive in many of Russia's cities.

Events in the twentieth century took a tragic toll on Russia's Jews. During most of the communist era, Jews experienced discrimination and were discouraged from practicing their religion or celebrating their culture. As a result, many Jews migrated to Israel or the United States, though the process was difficult in some cases. By 1995, however, 700,000 Jews still lived in Russia. Despite lingering prejudice, Jewish communities in Russia are restoring their religious practices and organizing schools.

Buddhism

Russia has two ethnic republics that are mainly Buddhist. Kalmykia (kal•MIH•kee•uh), near the **Caspian Sea** in the southwest, and Buryatia, near **Lake Baikal** in south-central Russia, together have nearly half a million Buddhists. For this reason Buddhism is accepted in today's Russia as a traditional religion.

Education

During the 1900s education in Russia showed significant advances. Today the country's literacy rate is nearly 100 percent in most urban areas populated by ethnic Russians. This high rate is largely the result of the Soviet emphasis on free but mandatory education. During the Soviet era, the education system favored military, science, and engineering studies rather than language, history, and literature. This educational focus produced generations of technology-focused government officials. They, along with prominent educators, writers, and artists, made up the Soviet **intelligentsia** (in•TEH• luh•JEHN•see•uh), or intellectual elite. In contrast, doctors and teachers were among the lowest-paid professionals in Soviet society.

When the Soviet Union collapsed, the curriculum in Russia's schools changed dramatically. Communist teachings disappeared, and schools emphasized a more objective and less authoritarian approach to learning. Today students in Russia have a choice of several different kinds of high schools, including traditional schools, schools specializing in elective studies such as languages, university preparatory schools, and alternative schools with experimental programs.

Unfortunately, Russia's unstable economy has severely limited budgets for schools. Many schools are overcrowded and in disrepair. Frustrated teachers

NATIONAL GEOGRAPHIC **World Explorer**

Geography Skills for Life

Russian Education

Students in Moscow School 1173 are linked by the Internet to students in more than 1,000 U.S. schools and 27 from other countries.

Place What is the main reason for Russia's high literacy rate?

often abandon teaching because of low pay, lack of respect, and low morale. In an unstable economy, many young people focus on earning money rather than getting an education. Still, Russian students and teachers are reexamining Russia's traditions in education and the arts.

Health Care

Disease, lifestyle choices such as smoking tobacco and drinking alcoholic beverages, and inefficient health care systems all threaten the well-being of Russia's people. Russian birthrates fell after World War II because of the massive loss of life in the war. This drop, coupled with an aging population in the 1990s, is shrinking Russia's population, but the trend may be slowing. Male life expectancy in Russia is expected to rise slowly in the 2000s, moving from a low of 59 years in the late 1990s, compared with 74 years in the United States during the same period. However, infertility in Russia is increasing by more than 3 percent a year, and 75 percent of all pregnant women develop serious health problems. Concern about increasing rates of infectious disease, such as tuberculosis, typhoid, and diphtheria, has led some countries to carefully screen Russian immigrants.

Today the Russian health care system is struggling to meet people's needs. Privatization has helped, but the government still owns and manages many clinics and hospitals, and these are often inefficient. Doctors and nurses are giving up their professions because they can earn more money as cab drivers or store clerks. Better insurance funding and wiser health care management are among the many reforms needed to improve health care in Russia.

The Arts

Russians revere their artists, musicians, and writers not only for their creativity but also for their courage in expressing themselves in the face of censorship. Modern Russians are still devoted to their long and rich cultural heritage.

Russia's Artistic Golden Age

Before the late 1600s, Russian architects and artists often found inspiration in religion. They built beautiful churches, crowned with onion-shaped domes and filled with icons of Jesus, Mary, and the saints as well as wall paintings of biblical stories. When Peter the Great introduced western European culture to Russia in the early 1700s, Russian arts began to focus on nonreligious themes. By the early 1800s, Russia had entered an artistic golden age that lasted into the 1900s.

NATIONAL GEOGRAPHIC **CHART STUDY**

State of Health

Country	Infant Mortality Rate (per 1000)	Life Expectancy Male	Life Expectancy Female	Rate of Natural Increase
Russia	16.0	59	72	–0.7
Poland	9.2	68	77	0.0
Czech Republic	4.1	71	78	–0.2
Albania	12.0	69	75	1.2
Ukraine	15.0	63	74	–0.7

Source: 2001 World Population Data Sheet

Geography Skills for Life

1. **Interpreting Charts** Which country has the highest infant mortality rate?
2. **Applying Geography Skills** How are infant mortality rates and life expectancy related to the general state of health in a country?

music of RUSSIA

Traditional Russian music includes many styles, from the rich folk traditions of the steppes to religious choral music performed in richly decorated urban churches. In the former Soviet Union, music and art were government-controlled, but individual artistic expression is very much alive today.

Instrument Spotlight

The **balalaika** appeared in Russia during the 1600s and was based on a two-string Tatar instrument from the 1200s. The instrument is crafted of pine in a rounded or triangular shape, and it has three strings, which are strummed or plucked. Balalaikas are often played in folk groups along with accordions, guitars, zithers, and percussion instruments. Today there is a wide range of balalaikas in different sizes, ranging from soprano to bass.

Go To **World Music: A Cultural Legacy** Hear music of this region on Disc 1, Tracks 20–23.

Russian painters such as Ilya Repin, Wassily Kandinksy, and Marc Chagall contributed to the wealth of Russian art. Composers Pyotr (Peter) Tchaikovsky, Nikolay Rimsky-Korsakov, and Modest Mussorgsky revolutionized Russian classical music and created memorable ballets. Many of their compositions used themes from Russian folk music. Today the Bolshoi and Kirov ballet companies are world famous for their stunning performances of traditional Russian ballet.

Russian literature owes a great debt to poets such as Alexander Pushkin, Boris Pasternak, and Anna Akhmatova, who wrote eloquently about their private lives and about historical events. Novelists of the 1800s, such as Leo Tolstoy and Fyodor Dostoyevsky, became known for epic works filled with vivid characters caught up in the struggle between good and evil or between love and hate. These two literary giants also focused on social and political injustices of life under the czars. Tolstoy's *War and Peace* and Dostoyevsky's *Crime and Punishment* still captivate readers today.

GEOGRAPHY Online

Student Web Activity Visit the **Glencoe World Geography** Web site at tx.geography.glencoe.com and click on Student Web Activities—Chapter 15 for an activity on nineteenth-century Russian painters.

Government

Culture and the Soviets

After 1917 the Soviet government severely limited individual artistic expression. It believed that all artists had the duty to glorify the achievements of Soviet communism in their works. This approach to art was called **socialist realism**. Writers, painters, and other artists who did not follow government guidelines were severely punished. The writer Alexander Solzhenitsyn, for example, was banished to a succession of labor camps and finally expelled from the country. He described the horrors of the labor camps in his famous work, *The Gulag Archipelago*.

Post-Soviet Arts

Beginning in the mid-1980s, activity in the arts renewed, as loosening government controls allowed the printing of previously unpublished works and new materials. During the height of Soviet repression, some of these works had been smuggled from Russia and printed in other countries. In 1989, a journalist from the United States noted the frenzy of cultural activity that came with the dawn of freedom:

> "*On [Moscow's] Arbat pedestrian mall, would-be Pushkins and Pasternaks peddle their autographed poetry for a ruble or more a page. . . . More than 200 experimental studio theaters have sprouted in Moscow alone. The cultural explosion has been felt as far away as the Pacific port of Nakhodka, where local artists set up a puppet theater workshop, and in Yaroslavl in the Soviet heartland, scene of a rollicking street festival celebrating the arts.*"
>
> John Kohan, "Freedom Waiting for Vision," *Time*, April 10, 1989

Life and Leisure

Daily life has always been difficult for ordinary people in Russia. During Soviet times apartment dwellers often found residential buildings crowded. Because of shortages of consumer goods, people spent many hours trying to purchase daily staples. Today, although some Russians are prospering and are building new homes in suburbs, others still live in crowded apartments and find it hard to pay the high prices charged for certain goods. Despite the frustrations, urban life offers many opportunities for people to enjoy the arts and culture. Reading, playing chess, and attending concerts, the ballet, and the theater all provide popular entertainment.

Both in cities and rural areas, Russians enjoy relaxing at mealtime with family and close friends. Sports, both amateur and professional, are quite popular with all age groups. Russia's tennis, track and field, and ice hockey athletes have had remarkable success in international events, as have figure skaters and gymnasts.

In the Soviet era, holidays were celebrated to honor Soviet workers or Soviet history. On May 1, the traditional workers' holiday known as May Day, great parades passed through Red Square, a large open area next to the Kremlin.

Today Russians observe May Day more as a spring festival than as a workers' holiday. Traditional religious holidays also have reemerged. In 1991, Christmas, celebrated by Eastern Orthodox Christians, became an official holiday in Russia for the first time since 1918.

SECTION 3 ASSESSMENT

Checking for Understanding

1. **Define** atheism, patriarch, icon, pogrom, intelligentsia, socialist realism.
2. **Main Ideas** Create a graphic organizer like the one below, and use it to fill in the key details for each aspect of Russian culture today.

Critical Thinking

3. **Making Inferences** Why do you think Russian lawmakers have restricted activity by religious groups other than Russia's four traditional religions?
4. **Comparing and Contrasting** What was the education system like during the Soviet era, and what is it like today?
5. **Making Generalizations** How have Russian artists, musicians, and writers inspired the Russian people during difficult times?

Analyzing Graphs

6. **Region** Study the graph of religions in Russia on page 377. What percentage of people living in Russia today is Muslim? What percentage is nonreligious?

Applying Geography

7. **Influence of Location** In which part of Russia do most Russian followers of Islam live? Why do you think this is so? Write a paragraph to explain your reasoning.

Analyzing Primary and Secondary Sources

When you read an account written by someone who witnessed an event, you are reading a primary source. If you read about the event based on a historian's research, you are reading a secondary source.

Learning the Skill

Knowing whether you are reading a primary or secondary source is important for evaluating the information. A primary source has the advantage of firsthand knowledge of an event. A secondary source often benefits from a broader perspective on the event.

Primary sources may include letters, interviews with eyewitnesses, photographs, and historical documents. Secondary sources rely on primary sources to create a broader picture. History books, encyclopedias, and documentary films are examples of secondary sources.

To analyze primary and secondary sources, ask yourself the following questions:

- **Did the source witness the event, or just gather information about it?**
- **When was the account written? At the time of, or after the event?**
- **Is the account valid? Do emotions, opinions, and biases influence the account?**
- **How useful is the source? What kind of information does the source provide, and what questions are left unanswered?**

▲ *This military hero who played a role in the 1917 Russian Revolution can be a primary source for historical research.*

Practicing the Skill

Read the following excerpt about the Bolshevik seizure of power in 1917, and then answer the questions.

"A tall iron gate surrounded the palace. One of the gates had not been locked. We saw this and opened the gate wide.... Like a wave of black lava, we moved into the palace, followed by workers and soldiers. There was no resistance, none at all. They surrendered their weapons. We arrested the members of the ... government."

—Karl G. Rianni, colleague of Lenin, quoted by Dusko Doder, "The Bolshevik Revolution," *National Geographic*, October 1992

1. What information does the source provide?
2. What is the writer's relationship to the information?
3. Is the source a primary or secondary source? How do you know?

Research a topic about Russia. Analyze your sources, and evaluate their validity and usefulness as primary or secondary sources.

The **Glencoe Skillbuilder Interactive Workbook, Level 2** provides instruction and practice in key social studies skills.

15

SECTION 1 Population Patterns (pp. 363–366)

Terms to Know
- ethnic group
- nationality
- sovereignty

Key Points
- More than 80 percent of Russia's population is ethnic Russian, and the remainder comprises about 100 different ethnic groups.
- Although more than 100 different languages are spoken in Russia, Russian is the official language.
- Russia is experiencing a population crisis, largely the result of health care problems.
- Russia's population is unevenly distributed, with 75 percent of Russians living west of the Urals.

Organizing Your Notes
Create an outline similar to the one started below to help you organize important details from this section.

Population Patterns

I. Russia's Ethnic Diversity
   A. Ethnic Regions
   B. Slavs
   C.

SECTION 2 History and Government (pp. 367–373)

Terms to Know
- czar
- serf
- Russification
- socialism
- Bolshevik
- communism
- satellite
- Cold War
- perestroika
- glasnost

Key Points
- Kievan Rus, an early Slavic state, grew out of settlements of Slavs and Varangians.
- Under the czars Russia expanded its territory and became an enormous empire.
- In 1917 a revolution overthrew Czar Nicholas II. Later that year, the Bolsheviks, under Lenin, seized power.
- In 1922 the Communists formed the Union of Soviet Socialist Republics, or Soviet Union.
- In December 1991 the Soviet Union collapsed and was replaced by Russia and other independent republics.

Organizing Your Notes
Organize your notes for this section by listing the important events under each century of Russian history.

Russian History

Before 1600	1601–1700	1701–1800	1801–1900	1901–Present

SECTION 3 Cultures and Lifestyles (pp. 376–381)

Terms to Know
- atheism
- patriarch
- icon
- pogrom
- intelligentsia
- socialist realism

Key Points
- Since the Soviet Union's collapse, many Russians have resumed their religious practices.
- Post-Soviet Russian schools are more open to new ideas and methods, but they face low budgets, overcrowding, and disrepair.
- Russia's artistic golden age began in the 1800s. After 1917 the Soviet government severely restricted certain kinds of artistic expression.
- Today Russians' respect for culture, traditions, and the arts has increased as a result of the new freedoms.

Organizing Your Notes
Create web diagrams like the one below to help you organize your notes for this section. Make separate diagrams for Religion, Education, Health Care, and the Arts.

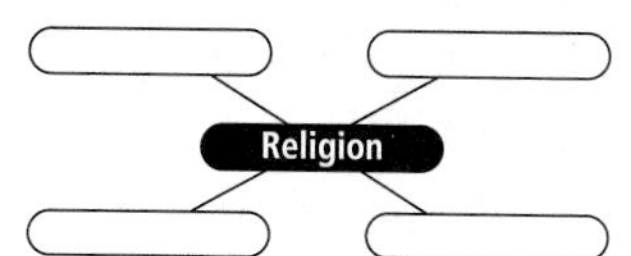

CHAPTER 15

ASSESSMENT & ACTIVITIES

Reviewing Key Terms

Write the key term that best completes each of the following sentences. Refer to the Terms to Know in the Summary & Study Guide on page 383.

1. A person who was part of the revolutionary group led by Lenin was called a(n) ________.
2. A(n) ________ ruled Russia at the time of the Russian Revolution.
3. The ________ is the head of the Russian Orthodox Church.
4. The Soviet Union's intellectual elite was called the ________.
5. A religious symbol is called a(n) ________.
6. The Russian term for restructuring is ________.
7. The Russian term for political openness is ________.
8. A peasant worker who farmed a plot of land that was owned by someone else was called a(n) ________.
9. ________ is the belief that there is no God or supreme being.

Reviewing Facts

SECTION 1

1. Which ethnic group forms the majority in Russia?
2. Where do most of Russia's people live?

SECTION 2

3. How did princes and czars change Russia's territory?
4. What were the major goals and events of the Soviet era?

SECTION 3

5. What major religions are found in Russia?
6. How have education and health care changed since the Soviet breakup?

Critical Thinking

1. **Drawing Conclusions** Explain why you agree or disagree with the following statement: "The Soviet Union was a 74-year-long experiment that failed."
2. **Making Inferences** Why do you think many people in Russia have returned to earlier traditions?
3. **Finding and Summarizing the Main Idea** Fill in four key events in Russian history in the order they occurred, on a flowchart. Then explain why each event was a turning point in Russia's history.

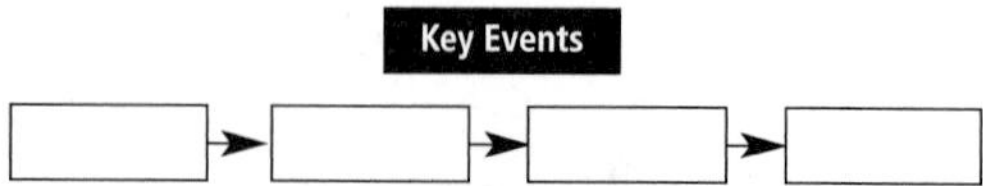

NATIONAL GEOGRAPHIC

Locating Places

Russia: Physical-Political Geography

Match the letters on the map with the places and physical features of Russia. Write your answers on a sheet of paper.

1. St. Petersburg
2. Baltic Sea
3. Barents Sea
4. Volga River
5. Moscow
6. Yenisey River
7. Yekaterinburg
8. Black Sea

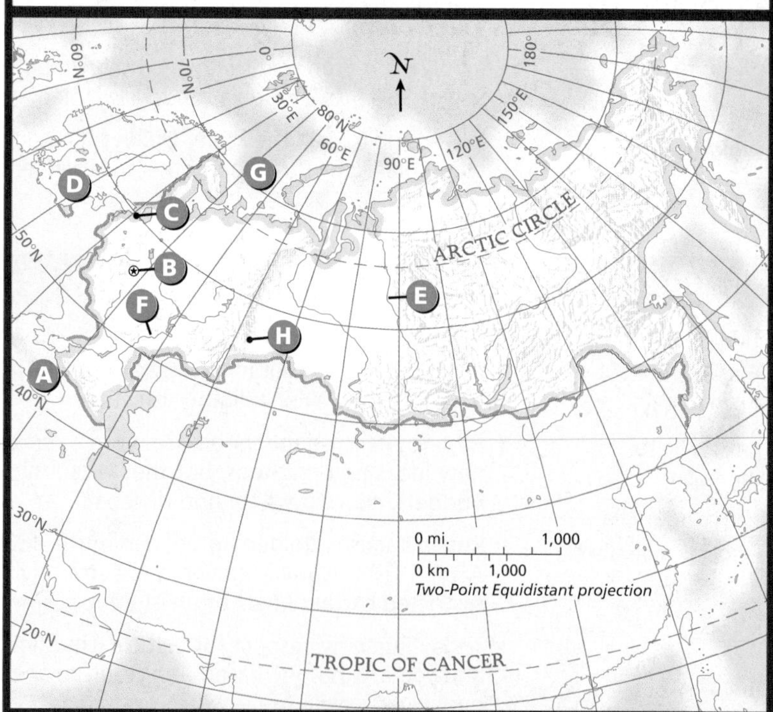

Self-Check Quiz Visit the **Glencoe World Geography** Web site at geography.glencoe.com and click on Self-Check Quizzes—Chapter 15 to prepare for the Chapter Test.

Using the Regional Atlas

Refer to the Regional Atlas on pages 338–341.

1. **Human-Environment Interaction** Which important rivers have helped in Russia's development?
2. **Place** What physical processes have affected migration and patterns of settlement in Russia?

Thinking Like a Geographer

Russia's population is spread unevenly across an enormous country. What physical features influence population density? How might human action affect population density? Design and draw a chart of elements that encourage population and those that discourage it.

Problem-Solving Activity

Contemporary Issues Case Study Choose one aspect of Russian culture today in which Russian and foreign cultural traits have converged, or come together. Research your topic in news magazines and newspapers or on the Internet to find a specific example, such as growth of U.S.-based fast-food restaurants or the spread of Western religions. Then write a one-page essay describing your example.

GeoJournal

Summarizing Return to the chart you made in your GeoJournal before you started reading this chapter. Write a brief summary of what you learned about Russia from reading the textbook. Use your chart, the textbook, and the Internet to prepare your summary.

Technology Activity

Using the Internet for Research The Soviet government required all artists to portray communism in a positive way. Use the Internet to locate examples of socialist realism in Russian art. Develop a brochure to educate people about this style of art. Download examples, and use them as illustrations in your brochure.

Standardized Test Practice

Read the quote by Zina Popova below, and then choose the best answer for each of the following multiple-choice questions. If you have trouble answering the questions, use the process of elimination to narrow your choices.

> *"That hero stuff was a millstone around [my mother's] neck. Mama told me she knew nothing about Lenin and Marxism when she joined the revolutionaries. They were spurred by hunger. My mother believed in the myth of the October Revolution but only for a few years. Then there was no exit. She put in her time, like most of the others."*
>
> —Zina Popova, in "The Bolshevik Revolution," *National Geographic,* October 1992

1. **The quote above could be used by a geographer to learn more about a country's**

   A cultural geography.
   B foreign policy.
   C ethnic minorities.
   D physical geography.

2. **Zina Popova's perspective on the Bolshevik Revolution comes from**

   F the Communist Party.
   G her own experience.
   H her mother.
   J a reference book.

In answering questions about quotations, make sure that you have a clear understanding of the quote. Also make sure that you understand the perspective of the person being quoted. Often, as in this case, you can find this information after the person's name.

CHAPTER 16

Russia Today

GeoJournal

As you read this chapter, make notes in your journal about life in Russia today. Use clear, specific language to explain how recent economic and political changes have affected the people of Russia.

GEOGRAPHY Online

Chapter Overview Visit the **Glencoe World Geography** Web site at geography.glencoe.com and click on Chapter Overviews—Chapter 16 to preview information about the region today.

SECTION 1

Living in Russia

Guide to Reading

Consider What You Know

Russia continues to adjust to dramatic and sometimes difficult political and economic changes. How do you think these changes probably affect Russians' attitudes toward the old Soviet government?

Read to Find Out

- How has Russia made the transition to a market economy?
- How have agriculture, industry, transportation, and communications in Russia changed since the breakup of the Soviet Union?
- What is Russia's relationship to the global community?

Terms to Know

- command economy
- consumer goods
- black market
- market economy
- privatization
- kolkhoz
- sovkhoz

Places to Locate

- Siberia
- Vladivostok

◀ *Teens walking through Red Square, Moscow*

NATIONAL GEOGRAPHIC

A Geographic View

The Price of Freedom

Not long ago . . . I met a woman named Larissa Pavlova. She was a teacher who now sold old clothes evenings and weekends to supplement her family's income. Countless thousands of Muscovites work second and third jobs to get by. . . . "Moscow is filled with what our good Comrade Lenin called contradictions," she said. "The rich get richer and the rest of us tread water or drown. I work much harder than I did in the old days, and sometimes that makes it hard to remember what we've gained. Freedom is sweet, but it's also a heavy, heavy load."

—*David Remnick, "Moscow: The New Revolution,"* National Geographic, *April 1997*

Muscovites selling food on the street

Russians hoped that the end of the Soviet-controlled economy and the birth of Russian independence would bring quick and painless economic change. As the teacher from Moscow discovered, however, shifting toward a freer economy could also bring hard times. Russia continues its efforts to create a working economy that will provide for its people and maintain its place in the global marketplace.

Changing Economies

Since the fall of communism, Russia has faced many economic challenges, such as providing more jobs for its citizens, increasing food production at home, and expanding trade internationally. As Russia works to strengthen its economy, its citizens also face ethnic unrest, rising crime, and declining health and social services.

The Soviet Command Economy

Under Communist leaders, the Soviet Union operated as a command economy. In a **command economy**, a central authority makes key economic decisions. The government owned banks, factories, farms, mines, and transportation systems. Members of the State Planning Committee, known as *Gosplan*, decided what and how much to produce, how to produce it, and who would benefit from the profits. *Gosplan* also controlled the pricing of most goods and decided where they would be sold.

The Soviet government emphasized heavy industry—the manufacture of goods such as tanks and other military hardware, machinery, and electric generators. As a result, the Soviet Union became an industrial giant and a world power, but its people could not buy many **consumer goods**, or goods needed for everyday life.

Unemployment in the Soviet Union was low, but so were wages, because most men and women worked at state-run factories and farms. People often could not afford the few consumer goods that factories produced. Even when people had enough money, such goods were hard to find. Some items could be bought on the **black market**, an illegal trade in which scarce or illegal goods are sold at prices even higher than those set by the government. Most workers, however, could not afford to pay such high prices with their limited incomes.

By the 1970s and 1980s, Western countries and some Asian countries had turned away from heavy industry to focus on computer technology and global communications. The Soviet system during this time, however, focused on increased industrial production and did not invest in developing new high-technology industries. As a result, the Soviet Union's economy stagnated, and its standard of living declined while the global economy entered a dynamic new era of change.

The Market Economy

When Mikhail Gorbachev came to power in 1985, the Soviet Union's economy was in serious trouble. To remedy the crisis, Gorbachev began to move away from a command economy toward a **market economy**, in which businesses are privately owned. Production and prices in a market economy depend on supply and demand. Things offered for sale are *supply;* people's desire to buy those things is *demand.* As part of his program of

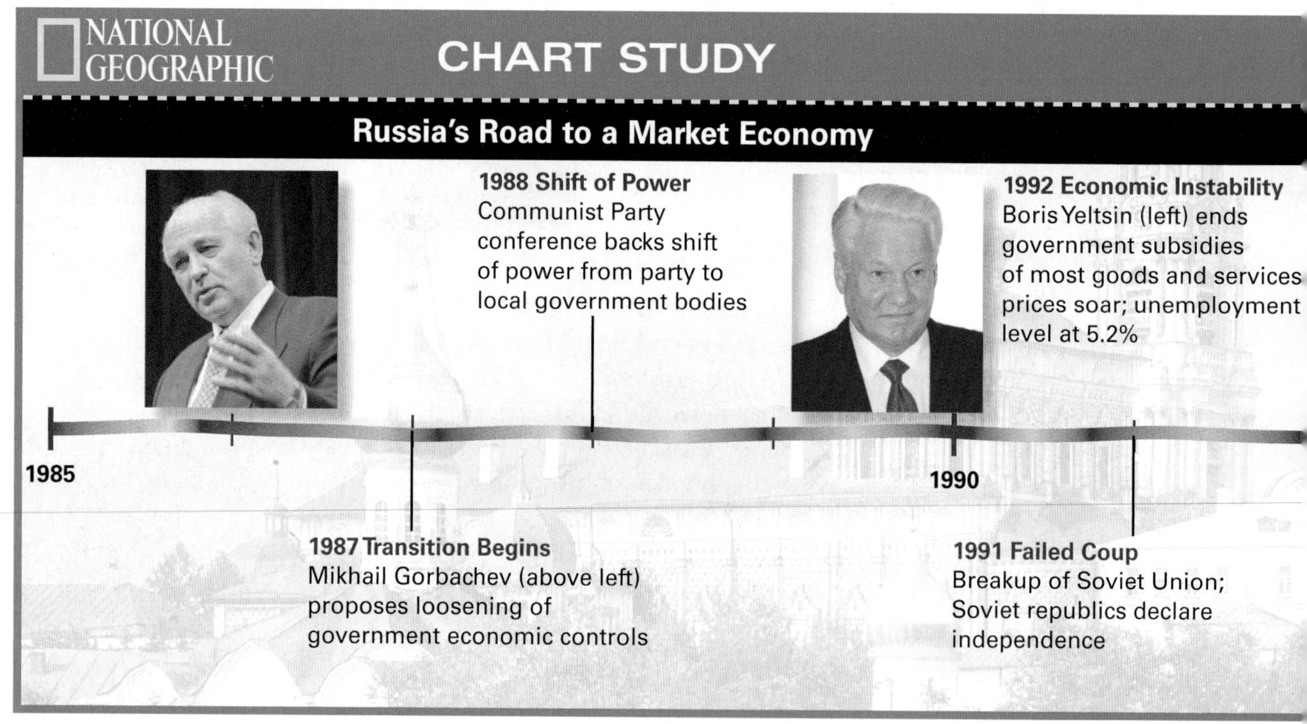

perestroika, or restructuring, Gorbachev reduced some government controls, allowed people to start small businesses, and encouraged foreign investment. Boris Yeltsin, Gorbachev's successor, expanded this process.

Economics
Privatization

Russia's economy continued to change after the Soviet Union officially ceased to exist in 1991. When Russia and the other Soviet republics became independent, they eliminated most remaining economic controls. Russian President Boris Yeltsin removed 90 percent of price controls and encouraged the mass **privatization**—a change to private ownership—of state-owned companies and industries, such as mining and oil extraction and processing. This process of privatization favored important businesspeople, political insiders, and foreign investors, all of whom could afford to purchase large companies. Rather than reinvest in Russia and its economy, many of these people invested their profits outside the country. Most average Russian workers did not benefit from this changing economic system: they neither earned nor were spending the new wealth.

> *"By 1995 privatization had gained a negative reputation with ordinary Russians, who coined a slang word* prikhvatizatsiya*, a combination of the Russian word for 'grab' and the Russianized English word 'privatize,' producing the equivalent of 'grabification.'"*
>
> Glenn E. Curtis (ed.), *Russia: A Country Study*, 1996

Widespread corruption complicated the privatization process in the new Russia. Organized crime groups and corrupt public officials operated throughout the country, especially in Moscow. Some people grew rich through special government favors that allowed them to buy property at far below its true value. This illegal behavior damaged the economy and absorbed investment funds that could have been used to rebuild the country.

The Transition Continues

The Russian economy experienced ups and downs throughout the 1990s. Russians could find

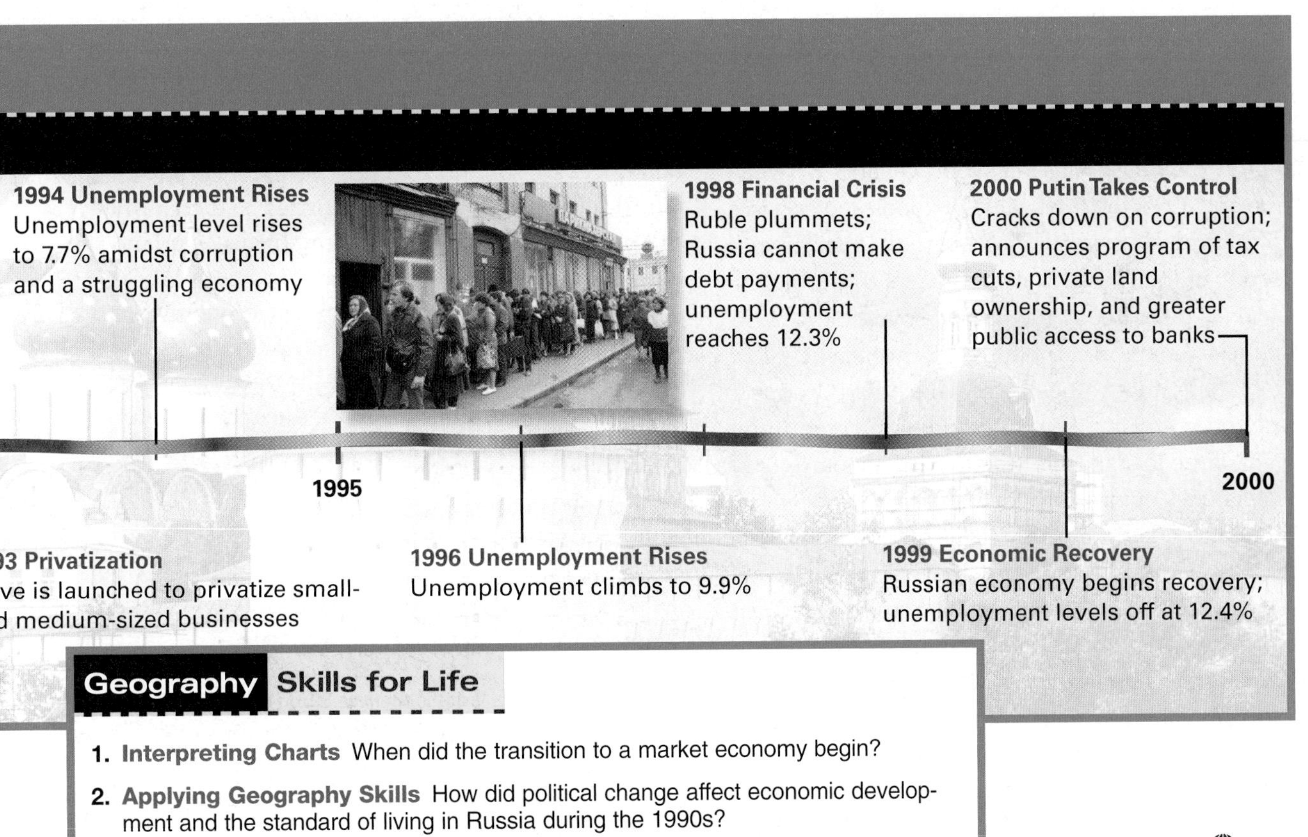

Geography Skills for Life

1. **Interpreting Charts** When did the transition to a market economy begin?
2. **Applying Geography Skills** How did political change affect economic development and the standard of living in Russia during the 1990s?

NATIONAL GEOGRAPHIC **World Explorer**

Geography **Skills for Life**

Kalininggrad, Russia To supplement their income, many Russians sell hard-to-find goods, such as car parts, in open-air markets.

Place Describe the characteristics of Russia's economy in 1991 and today.

more consumer goods in shops. However, without controls, prices soared, and many people could not afford to buy the goods that were available. Between 1990 and 1995, the total value of goods and services produced in Russia fell by 50 percent, a far greater drop than the United States experienced during the Great Depression of the 1930s. Following a 1998 financial crisis, the ruble, Russia's currency, lost 71 percent of its value. Prices, which had dropped, rose once again. The international community made large loans to aid the Russian economy.

Yeltsin resigned as president of Russia in 1999. His successor, Vladimir Putin, inherited an unstable economy. Russia's economy needed money and a stronger banking system, which would help keep more Russian money at home. Putin also needed to improve the Russian military. However, he needed to do so without overspending on the armed forces at the expense of overall economic growth, a problem experienced previously by Soviet leaders.

An inefficiently managed government, budget shortfalls, unclear property rights, an unstable currency, corruption, and organized crime all threaten Russia's economic stability. However, there is also potential for success. Russia can rebuild its economy by relying on its vast natural resources, developed industries, and well-educated citizens.

Agriculture and Industry

Under the Soviet system, farms were organized into state-controlled kolkhozes (kahl•KAW•zehz) and sovkhozes (sahf•KAW•zehz). The **kolkhozes** were small farms worked by farmers who shared, to a degree, in the farm's production and profits. **Sovkhozes** were large farms run more like factories, with the farm workers receiving wages. However, prices and production in both the agricultural and industrial sectors were controlled by the government. Both the agricultural and industrial sectors suffered because the system did not motivate workers. As a result, long before the 1980s Soviet agriculture did not produce enough food to feed its people, and the government had to import additional grain and other foods.

In 1991 President Yeltsin started to restructure state-run farms so they could function better in a market economy. However, Russian farmers—accustomed to the stability of Soviet controls—continued to operate many of Russia's farms as kolkhozes or sovkhozes. Most farmers could not afford to buy land, and they worried that wealthy Russians or foreign investors might use the land for nonagricultural development. Because of these concerns, progress toward a market economy for agriculture has been slow, and crop and livestock production has fallen. Recently, however, gains in farm productivity have helped reduce the need for agricultural imports.

Like agriculture, Russian industry has also been transformed since the early 1990s. For many years

Student Web Activity Visit the **Glencoe World Geography** Web site at geography.glencoe.com and click on Student Web Activities—Chapter 16 for an activity about living in Russia today.

Russia's state-owned aerospace industry and its military-industrial system were its economic and technical focus. Many of these components have become privately owned and provide export income. Russia has also encouraged foreign investment by selling shares of ownership in some Russian companies and by opening Russia's markets to Western companies. A popular American fast-food chain, for example, now has 52 restaurants in 17 Russian cities.

Russia's most important industry is petroleum extraction and processing, and the country is one of the world's largest producers of crude oil. Russia's domestic oil provides its other industries with vital energy at a reasonable cost. The country is also a major producer of iron ore, manganese, and nickel. Huge forests in Russia produce one-fifth of the world's softwood, and Russian fish-factory ships process catches from both the Atlantic and Pacific Oceans. Other major manufacturing industries include steel milling; auto and truck production; aircraft construction; and the manufacturing of chemicals, heavy machinery, and agricultural equipment. Most of these industries are in the Volga Valley, near Moscow and St. Petersburg, and in the Ural Mountains. Although Russia's industries still face difficulties, industrial production is now rising.

Transportation and Communications

Russian transportation and communications systems lag behind those of most of the world's developed countries. In an age of speedy transportation, the Internet, and a global economy, Russia struggles to find funds for new highways and high-tech communications.

Transporting Goods

Russia's transportation systems must move food and other resources great distances to reach consumers. A major highway system links Moscow with other major Russian cities, but many roads are in poor repair. Harsh winters in places like **Siberia** often make roads impassable.

Because of its great size and climate extremes, Russia depends on railroads and waterways for most of its transportation needs. Not surprisingly, Russia boasts the world's longest continuous railroad line. The Trans-Siberian Railroad is the greater part of the rail route from Moscow through the Siberian steppes to the Pacific port city of **Vladivostok**. Major cities are found where the Trans-Siberian Railroad crosses large rivers, such as the Ural, Irtysh, Ob, and Yenisey. Millions of tons of materials travel along thousands of miles of navigable inland waterways, which connect seaports and inland cities.

NATIONAL GEOGRAPHIC **World Explorer**

Geography Skills for Life

Russian Agriculture Outdated farm equipment makes farming labor-intensive for humans and animals.

Place Why have farmers in Russia been reluctant to accept a market economy system?

Transporting People

Most Russians live in cities and many do not own cars, so public transportation, such as trains, buses, and, in several large cities, subways, is common. Private car ownership doubled in the 1990s, but public transportation remains a practical option for Russians, in part because the government helps pay for it. The systems and equipment, however, need repair and improvements.

The Soviet Union used jet airplanes for passenger traffic, and the government

Geography Skills for Life

Russian Waterways Cargo cranes along the Pregolya River in Kaliningrad assist in shipping goods for export.

Place What goods does Russia export?

financially supported air travel for many years. The passenger airline Aeroflot was once the only one in the Soviet Union and at its peak carried 135 million people a year. After the fall of the Soviet Union, Aeroflot split into numerous smaller airlines. The high cost of fuel and reduced passenger traffic because of skyrocketing ticket prices have forced about 60 percent of Russia's airports to close.

Transporting Energy

Russia's large size also affects how it transports natural gas, crude oil, and other petroleum products. Pipelines are effective, although constructing and maintaining them can be difficult in areas of harsh climate. A complex maze of pipelines crisscrosses Russia, providing major Russian cities and parts of western Europe with fuel.

The oil pipelines run through Chechnya and Dagestan (DAH•guh•STAHN), ethnic republics in southwestern Russia. Because people in these republics are fighting for their independence from Russia, control of the area's oil reserves and working pipelines is a major concern.

History

Mass Communications

Under the Soviet Union, the state owned and controlled all the mass communications systems, including newspapers, magazines, television, the postal service, and the telegraph and telephone systems. State agencies reviewed all print and broadcast materials to make sure that they contained no criticism of the government. Since the breakup of the Soviet Union, Russians have heard and read new voices and fresh views. Most families own radios and television sets, and by 1995 Russians could choose from among 10,000 newspapers and journals.

Telephone service in Russia has also grown. As a result of Russia's vast size, only 22 percent of rural households have telephones, compared with 56 percent of urban households. However, communications companies are increasingly offering services such as the Internet, e-mail, and cellular phones. These advances in communications systems will make vital contributions to the successful transition to a market economy.

Global Interdependence

After independence Russia and the other former Soviet republics began to increase their interdependence with other countries. By expanding international trade and building political and financial relations, Russia has increasingly focused on becoming a full partner in the global community.

Trade

Russia has already established trade relations in world markets and is a major source of energy and fuels, which make up 48 percent of its exports. Lumber, metals, and chemicals are also important Russian exports. The United States, the European Union, the other former Soviet republics, China, and Japan are among Russia's major trading partners. These countries provide Russia with the consumer goods, medicines, meat, and sugar it needs.

Energy is expected to remain Russia's main item of international trade until its manufactured goods, such as machinery and light industrial products, improve in quality and become more competitively priced. Working to strengthen its industries, Russia became a member of the Asia-Pacific Economic Cooperation (APEC) forum in 1998. Negotiations are continuing for Russia's membership in the World Trade Organization (WTO). As Russian manufacturing makes further gains, these trading networks will become even more important for the Russian economy.

International Relations

Despite its political and economic challenges at home, Russia maintains its important role in world affairs. Russia benefits from occupying the former Soviet Union's seat in the United Nations Security Council. The country has also joined European organizations that support security and cooperation. Russia has helped settle conflicts and has supported peace efforts in several countries, especially in former Soviet republics. Even as Russia asserts itself internationally, however, economic problems have drained money from its military. As a result, military forces have old equipment, and soldiers' morale is low.

Adequate financial resources are vitally important to Russia's stability and progress in the global community. Other countries and world organizations have provided loans, and foreign investors have made funds available to Russian industry. With foreign help, Russia is trying to create secure and workable systems for banking, farming, manufacturing, transportation, and communications. Although Russia has a long way to go, the economic gains made in recent years are positive signs.

SECTION 1 ASSESSMENT

Checking for Understanding

1. **Define** command economy, consumer goods, black market, market economy, privatization, kolkhoz, sovkhoz.
2. **Main Ideas** On a chart like the one below, fill in details about agriculture, industry, transportation, and communications in the Soviet command economy and the Russian market economy.

	Soviet Command Economy	Russian Market Economy
Agriculture		

Critical Thinking

3. **Predicting Consequences** How might Russia's agricultural and industrial sectors be affected by Russia's growing global interdependence?
4. **Comparing and Contrasting** How did the Soviet command economy and the Russian market economy affect the Russian people?
5. **Making Inferences** What can you infer about Russia's goals, based on changes in Russia's trade and international relations since the Soviet breakup?

Analyzing Maps

6. **Human-Environment Interaction** Study the economic activity map on page 341. In what area is the raising of livestock concentrated? How is this related to the physical geography of the region?

Applying Geography

7. **Effects of Size and Distance** Think about the physical geography of Russia. Write a paragraph analyzing how Russia's vast size affects the availability of natural resources and the country's ability to develop them.

Viewpoint

CASE STUDY on the Environment

Russia's Supertrawlers: Factories at Sea

More than a million fishing vessels scour the oceans for fish. As a result, fish populations are shrinking worldwide. Enormous ships called supertrawlers are largely to blame. Towing huge trawl nets—some large enough to scoop up a whale—supertrawlers are floating fish factories. These ships can catch and process more than 400 tons (360 t) of fish a day. No one knows how many fish swim the oceans. However, if too many are caught, some species may not recover. Is sustainable fishing possible with super-trawlers harvesting the seas?

NATIONAL GEOGRAPHIC SOCIETY

State-of-the-art electronic gear allows supertrawlers to track schools of fish with pinpoint accuracy. Trawl nets may stretch half a mile behind the ships, engulfing everything in their paths.

The first factory trawler was built in Scotland in 1954. By 1970 the Soviet Union had 400 trawlers, the world's largest fleet at that time. Other countries, including the United States, China, and Japan, soon launched their own trawler fleets. Marine harvests soared as these enormous vessels went to work in the world's richest fishing grounds. But after a few years of bounty, there were signs of trouble at sea.

In the western Bering Sea, for example, Soviet (later Russian) supertrawlers initially harvested large numbers of sole, perch, herring, and especially pollock—the fish used in frozen fish sticks and fast-food fish sandwiches. As catches started to outpace reproduction rates, fish populations plummeted. Data gathered by Russian marine biologists show that pollock catches are declining by 10 percent every year. American scientists raise similar concerns about the impact of American supertrawlers working the eastern Bering Sea.

Supertrawlers are usually after certain kinds of fish. Everything else hauled up in the nets gets discarded. Millions of fish and other marine animals die unnecessarily every year. Many trawlers also drag nets along the seafloor, destroying countless organisms and their habitats. Animals higher on the food chain are affected, too. Seals, sea lions, and kittiwakes can starve if there are fewer fish to eat. Since the 1970s these mammal and bird populations in the Bering Sea have declined.

Opponents of supertrawlers argue that the ships are doing irreparable damage to fish stocks and marine habitats. Opponents feel the unnecessary slaughter of healthy marine organisms is wasteful. Even though there are some restrictions on supertrawlers, opponents maintain that the laws are hard to enforce. Furthermore, since smaller boats can't compete with supertrawlers, the big ships threaten traditional fishing cultures on every continent.

◀ Using a huge net, a Russian fisherman empties a load of fish into his boat.

Supporters of supertrawlers cite the growing global demand for fish and fish products. They point out that their catches supply high-protein food to millions of people. Some trawler operators dispute data that show a decline in fish populations. Others say that if trawlers reduce their catches, other ships will simply harvest what's left behind. Russian officials must balance the risk of destroying fish stocks with Russia's need for a profitable fishing industry.

What's Your Point of View?
Should further restrictions be placed on supertrawlers? Should they be banned worldwide?

A Russian supertrawler (below) searches for fish in the Sea of Okhotsk. Aboard these vessels, workers (right) prepare fish for a global market. ▼

People and Their Environment

Guide to Reading

Consider What You Know

You have read about the Soviet government's development of heavy industry and about Russia's development of a market economy after the breakup of the Soviet Union. How do you think these activities have affected the quality of Russia's environment?

Read to Find Out

- How does Russia manage its natural resources?
- How has pollution affected the lives of Russia's people?
- What are the environmental challenges in Russia's future?

Terms to Know

- radioactive material
- pesticide
- nuclear waste

Places to Locate

- Kamchatka
- Lake Baikal

NATIONAL GEOGRAPHIC

A Geographic View

The Aftermath

I found little likelihood that things [would] improve soon; the economies of Russia and most of the other . . . former Soviet republics are in shambles. "They used to show us films of the corrupted West with its polluted waters, like your Great Lakes," a Siberian environmental worker said. "Now the situation you had in the 1960s is here. But if the chaos continues, we will need two or three times as many years as you needed just to decide *it's necessary to clean up."*

—*Mike Edwards, "Lethal Legacy,"* National Geographic, *August 1994*

Steel plant in Siberia

The world's expanding industries and rapidly growing population often strain the natural environment. Careless management of natural resources for short-term gain destroys economic opportunities for future generations, damages the environment, threatens people's health, and jeopardizes people's quality of life. In this section you will learn how Russia is managing its resources and balancing economic growth with environmental conservation.

Managing Resources

Russia is trying to make the best use of its vast and abundant natural resources in order to strengthen its economy and improve its standard of living. Unfortunately, the country has inherited a legacy of environmental damage. Russia's main challenge is to manage its resources without repeating its past disregard for the environment.

A second challenge is to improve the environment and repair damage that has already been done. One target for improvement is the timber industry. Russia contains the world's largest forest reserve, and the World Bank's Sustainable Forestry Pilot Project is helping Russia manage its forests more effectively. Using land more wisely, protecting forests, planting new trees, and increasing private forestry investment all help Russia's environment and economy. Higher taxes paid by Russian citizens provide income for the government to help protect the environment. Increased employment opportunities in the forest industry and more stable local economies will be possible only if steps to conserve the forest are taken. It is in the best interest of the people to protect the forests because the timber industry provides jobs and economic resources for communities.

Individual Russians are becoming more aware of the value of good environmental management. People have banded together to oppose a mining operation located in remote **Kamchatka** (kuhm•CHAWT•kuh), a region of Siberia in eastern Russia. The Kamchatka Committee for the Protection of the Environment and Natural Resources has demanded that the mining company meet strict environmental standards. The possible threat to the area's salmon spawning grounds prompted the local fishing industry to support the effort. The mine also caused concern among local residents and environmentalists because it was close to a nature park that was recently named a United Nations World Heritage site.

Pollution

The Soviets' disregard for the environmental effects of industrialization damaged Russia's water, air, and soil. By the 1990s, 40 percent of Russia's vast territory was under "ecological stress," with the health of millions of Russians affected by unchecked pollution and radiation.

Water Quality

Although Russia has one of the world's largest supplies of freshwater, industrialization has polluted most of its lakes and rivers. Fertilizer runoff, sewage, metals such as aluminum, and radioactive material—material contaminated by residue from the generation of nuclear energy—all contribute to poor water quality. The waters of the Moskva and Volga Rivers, for instance, pose severe health risks. The many dams along the Volga have trapped contaminated water. Moscow has also reported cholera-causing bacteria in its water. Pollution even threatens the Caspian Sea.

Lying on the southeastern edge of the Central Siberian Plateau, **Lake Baikal** (by•KAWL) is the world's oldest and deepest lake. It contains one-fifth of the world's freshwater, and 1,500 native species of aquatic plants and animals make their home there. Calling it "the Pearl of Siberia," Russians consider the lake a natural wonder. A recent traveler learned from a local resident what Baikal means to Russians:

> *"Lake Baikal is a symbol, Sasha told me once, of all the things that give Siberian life its distinct sweetness—the natural beauty, the purity of open air, the hardy generosity of people and the poetry in their collective soul. 'This is what Russians mean when they talk about the Motherland,' he said. 'And nothing, nothing is more precious to us than that.'"*
>
> Don Belt, "Russia's Lake Baikal: The World's Great Lake," *National Geographic*, June 1992

NATIONAL GEOGRAPHIC **World Explorer**

Geography Skills for Life

Timber in Siberia Timber processing is a major economic activity along Siberia's Yenisey River.

Human-Environment Interaction How does the proper management of forests affect a country?

Geography Skills for Life

Lake Baikal Efforts to protect Lake Baikal include closing paper mills and installing wastewater treatment plants.

Human-Environment Interaction How has industrial development affected Lake Baikal?

In 1957 the Soviet Union announced a plan to build a paper pulp factory in Baikalsk along Lake Baikal's southwestern shores. Although this plan was opposed by citizens in the area, their protests were ignored and the factory was built. This factory and others that followed continue to dump industrial waste into the lake. However, in response to the ongoing protests of local residents, the most serious polluters either have been closed or are in the process of reducing pollution. Pollution levels in the lake are now relatively low compared with many lakes in Europe and the United States.

Soil and Air Quality

For decades toxic waste dumps and airborne pollution poisoned Russia's soil. Aging storage containers cracked and leaked toxic wastes into the soil. Petroleum pipelines also often broke and tainted the land. Overuse of fertilizers and pesticides—chemicals used to kill crop-damaging insects, rodents, and other pests—has damaged farmland.

Russian experts believe that during the 1990s only 15 percent of Russia's urban population lived with acceptable air quality. Industries, emissions from vehicles, and the soft coal burned for fuel are all sources of air pollution. In addition to releasing soot, sulfur, and carbon dioxide into the air, burning coal leads to another harmful agent—acid rain. Experts estimate that the combination of acid rain and chemical pollution has reduced Russian forests by about 1.5 million acres (607,500 ha) since the early 1970s.

Nuclear Wastes

Between 1949 and 1987, the Soviet Union set off more than 600 nuclear explosions. Soviets developed and then stockpiled nuclear weapons throughout the Cold War. Today, the condition and fate of those weapons concern Russia and the rest of the world.

Nuclear wastes are the by-products of producing nuclear power. Some of these wastes can remain radioactive for thousands of years, posing great dangers to people and the environment. The Soviets placed most nuclear wastes in storage facilities, but they also dumped some radioactive nuclear materials directly into Russia's northern waters, such as the Baltic and Bering Seas.

History

Chernobyl

During the Cold War, nuclear power generated much-needed electricity in the Soviet Union. It also provided power for building military weapons and vehicles. The urgency of keeping pace with the West during the Cold War often resulted in substandard nuclear plants and reactors that employed poorly trained workers who ignored proper safeguards. In 1986 a fire in a nuclear reactor in the town of Chernobyl (chuhr•NOH•buhl), 60 miles (97 km) north of Kiev, Ukraine, released tons of radioactive particles into the local environment. This radiation was then carried great distances by the wind, and contaminated other countries.

Thousands of people were exposed to deadly levels of radiation because Soviet officials were slow to alert the public to the crisis and did not evacuate people soon enough. By the mid-1990s over 8,000 people had died as a direct result of radiation poisoning. Millions more continue to suffer from cancer, stomach diseases, and immune system disorders. Radiation covered thousands of acres of farmland and forests in Belarus, Ukraine, and Russia. In Russia alone, radiation covered over 19,300 square miles (50,000 sq. km), where more than 30 million people lived. Because of prevailing winds, other countries suffered as well, most notably Finland, Sweden, Poland, and the former Czechoslovakia.

After the Chernobyl accident, international pressure prompted Soviet leaders to improve nuclear safety standards and to shut down dangerous plants. In response to these demands, Soviet officials never opened some newly built reactors and abandoned plans for building others. Despite concerns from other countries, 28 nuclear reactors continue to operate at nine sites throughout Russia. Much of Russia's electricity continues to come from these plants. In late 2000, however, the remaining reactor at Chernobyl was shut down. Experts in Western countries as well as in Russia and Ukraine think that many remaining Soviet-era reactors are poorly designed, unsafe, and should be made secure.

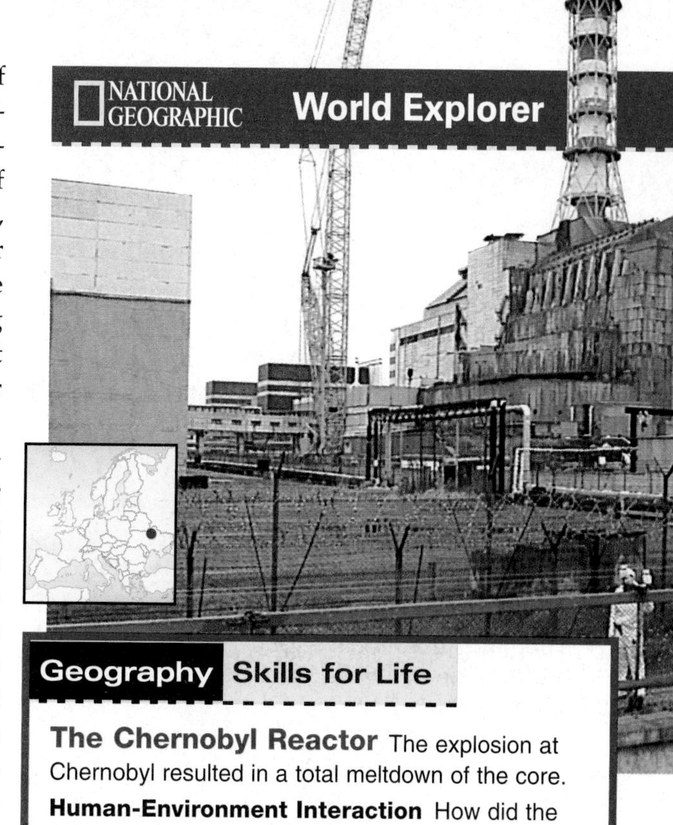

Geography Skills for Life

The Chernobyl Reactor The explosion at Chernobyl resulted in a total meltdown of the core.

Human-Environment Interaction How did the Soviet government improve nuclear safety standards after the Chernobyl accident?

SECTION 2 ASSESSMENT

Checking for Understanding

1. **Define** radioactive material, pesticide, nuclear waste.
2. **Main Ideas** Create a graphic organizer like the one below, and fill in information about each of the topics. Then choose one of the topics, and summarize efforts currently under way in Russia to address the situation.

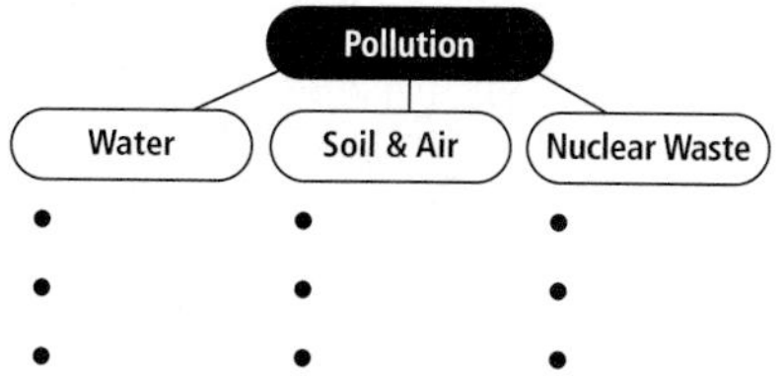

Critical Thinking

3. **Making Generalizations** What generalizations can you make about the relationship between economic development and the environment in Russia?
4. **Problem Solving** Assume the role of the Russian president, and identify an environmental problem in your country. What steps would you take to solve this problem?
5. **Predicting Consequences** Think about what you know about the Russian economy. What is the likelihood of a dramatic improvement in Russia's environmental problems in the near future?

Analyzing Maps

6. **Human-Environment Interaction** Study the economic activity map on page 341. Think about the regions of Russia in which pollution is a problem. Describe the relationship between the location of manufacturing centers and pollution.

Applying Geography

7. **Influence of Location** Think about the challenges Russia faces concerning water quality. Write a paragraph explaining why Russians do not use more water from Lake Baikal to supply their freshwater needs.

Categorizing Information

When you read a map, you make sense of the data you see—the symbols, words, and different-colored lines and shapes—by categorizing the information. Categorizing means grouping information and details together in a way that helps you understand and compare two or more ideas or concepts.

Learning the Skill

Categorizing information helps you make connections and retain information. This skill helps you answer questions such as *What is it? What parts does it have?* and *How is this like or unlike something else?*

When you categorize, you sort details into groups. You may be looking at a map, reading an informative article, or watching a basketball game. Once you understand how the details are grouped, you can make comparisons and draw conclusions. One way to keep track of the different details is to create a chart.

Follow these steps to categorize written information, using a chart.

- **As you read a section of a chapter, identify its main categories.** Make a two-column chart with one row for titles and one row for each category.
- **Spend a few minutes reading the section.** Record the title for each category in the first column of the chart. Then note some details and characteristics that you found for each category. List these in the second column of the chart.
- **Review the details in the second column of the chart.** Use them to write a summary statement about each category and to compare the categories with each other.

The Changing Economy of Russia

Type of Economy	Characteristics
Command (Soviet)	• Central authority owns banks, factories, farms, mines, and transportation systems. • Production and prices depend on decisions of the central authority. • Meets the basic needs of consumers but not designed to meet their wants.
Market	• Businesses are privately owned. • Production and prices depend on supply and demand. • A high degree of individual freedom allows producers to make whatever they think they will sell.

Practicing the Skill

Use the information about Russia on pages 390–392 and the chart on this page to answer the questions below.

1. What are the main categories of information on pages 390–392?
2. What are two other characteristics you could list in the chart?
3. How are these systems alike? How are they different?
4. What are two ways that a chart similar to the one above could help you?

Use library or Internet research and the information in Chapters 7 and 16 to categorize information about pollution in Russia, the United States, and Canada. Use a chart like the one on this page to list details about the sources of air, water, and soil pollution and proposed solutions for these challenges.

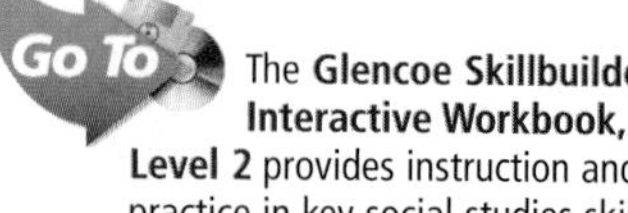

The **Glencoe Skillbuilder Interactive Workbook, Level 2** provides instruction and practice in key social studies skills.

16 SUMMARY & STUDY GUIDE

SECTION 1 Living in Russia (pp. 387–393)

Terms to Know

- command economy
- consumer goods
- black market
- market economy
- privatization
- kolkhoz
- sovkhoz

Key Points

- The Soviet economy was a command economy controlled by government agencies.
- Since the 1980s the Russians have been making the difficult transition from the Soviet command economy to a market economy.
- After the breakup of the Soviet Union, Boris Yeltsin encouraged privatization of state-owned farms and businesses.
- Transportation and communications systems must improve in order to support a strong market economy.
- To take its place as a full partner in the global community, Russia needs good international trade and strong political and economic relations.

Organizing Your Notes

Create an outline like the one below, using the section headings to help you organize your notes for this section.

Living in Russia

I. Changing Economies
  A. The Soviet Command Economy
    1.
    2.

SECTION 2 People and Their Environment (pp. 396–399)

Terms to Know

- radioactive material
- pesticide
- nuclear waste

Key Points

- Soviet leaders' drive for an industrial-based economy caused major and lasting damage to Russia's water, soil, and air.
- Russia needs to manage its use of natural resources properly in order to avoid more environmental damage.
- Radioactivity from nuclear waste, nuclear accidents, and aging nuclear weapons poses a grave danger to Russia's environment and its people's health.

Organizing Your Notes

Use a graphic organizer like the one below to help you organize information about the challenges facing Russia today.

Environmental Challenges

Water	Soil and Air	Nuclear Waste

Russian passengers wait to board a train.

CHAPTER 16

ASSESSMENT & ACTIVITIES

Reviewing Key Terms

On a sheet of paper, classify each of the lettered terms below into the following categories. (Some terms may apply to both categories.)

- **Soviet Era**
- **After Independence**

a. command economy
b. consumer goods
c. black market
d. market economy
e. privatization
f. kolkhoz
g. sovkhoz
h. radioactive material
i. pesticide
j. nuclear waste

Reviewing Facts

SECTION 1

1. What was the role of the government in the Soviet command economy?
2. How has the transition to a market economy challenged Russian society?
3. How did privatization impact daily life in Russia?
4. What steps has Russia taken to become part of the global community?

SECTION 2

5. What problems have been created by pollution in Russia?
6. What challenges with the environment and natural resources does Russia face today?
7. How did the Cold War contribute to Russia's environmental problems?

Critical Thinking

1. **Making Inferences** Study the chart on pages 388–389. How might political and economic reforms in Russia eventually affect the distribution of the country's resources?
2. **Problem Solving** Identify one kind of pollution affecting Russia. Describe the cause or origin of the pollution. What steps do you think would be necessary to reduce its effects? Explain the reasons for your answer.
3. **Categorizing Information** Complete a diagram like the one below to show the changes in Russian life after independence. Then write a paragraph explaining the change you think had the greatest impact on Russian life.

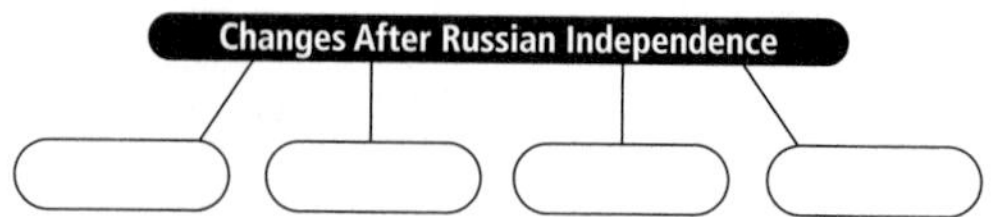

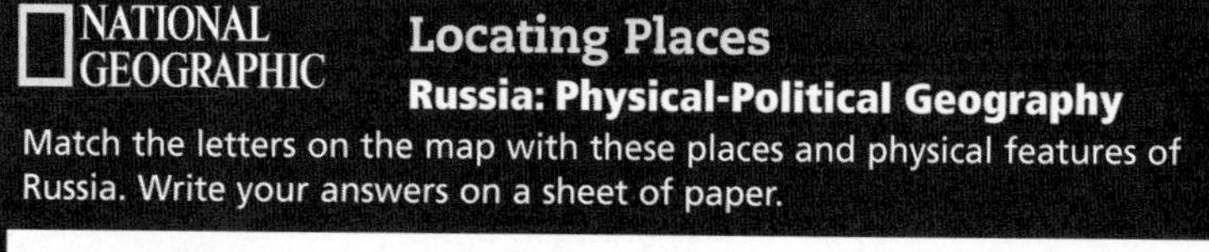

Locating Places

Russia: Physical-Political Geography

Match the letters on the map with these places and physical features of Russia. Write your answers on a sheet of paper.

1. **Don River**
2. **Caspian Sea**
3. **East Siberian Sea**
4. **West Siberian Plain**
5. **Lena River**
6. **Amur River**
7. **Vladivostok**
8. **Lake Baikal**

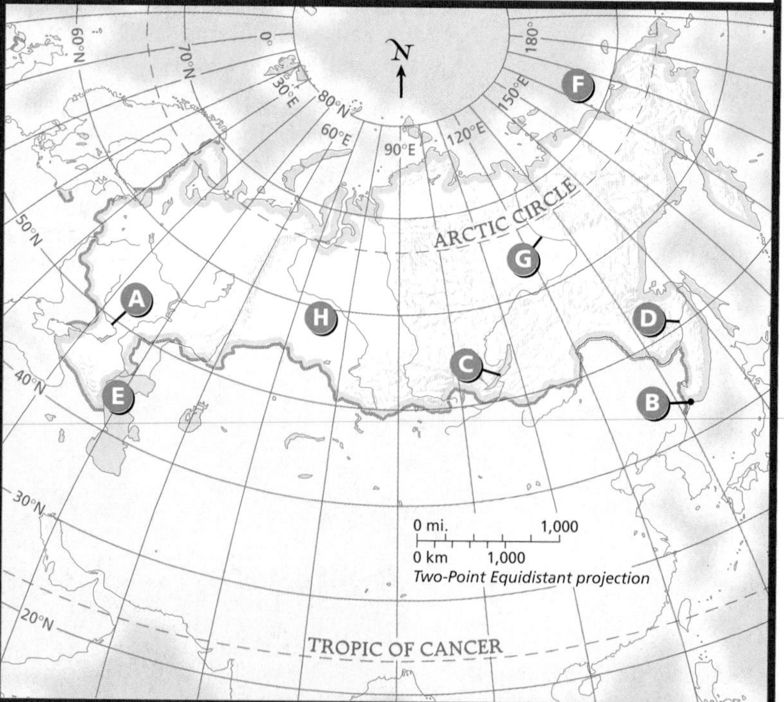

Self-Check Quiz Visit the **Glencoe World Geography** Web site at geography.glencoe.com and click on Self-Check Quizzes—Chapter 16 to prepare for the Chapter Test.

Using the Regional Atlas

Refer to the Regional Atlas on pages 338–341.

1. **Location** Which manufacturing areas are located along the Volga River? Along the Ob and Irtysh Rivers?
2. **Human-Environment Interaction** Compare the population density map and the economic activity map. Describe the correlation between commercial farmland and population density.

Thinking Like a Geographer

Think about the population distribution in Russia. How do you think the shift from a command economy to a market economy might affect migration patterns and population density? Write a paragraph about the most significant effect.

Problem-Solving Activity

Contemporary Issues Case Study Russia remains an influential international power despite its recent political and economic changes. Learn more about Russia's policies toward the expansion of the European Union. Then focus on ways Russian and European interests are similar and different. In an essay, present your conclusions about the future of Russian-European relations.

GeoJournal

Expository Writing Use your GeoJournal, your textbook, and the Internet to research and write an essay that analyzes how Moscow's character as a city is related to its political, social, economic, and cultural features.

Technology Activity

Developing Multimedia Presentations Compile information about people's work, school, community, home life, and leisure activities in Russia. Use various media to introduce Russian life to other students or people in your community. For example, you might play recordings of popular Russian music or show a film that presents an aspect of Russian culture, in addition to your oral report.

Standardized Test Practice

Use your knowledge of Russia to choose the best answer for each of the following multiple-choice questions. If you have trouble answering the questions, use the process of elimination to narrow your choices.

1. In Russia which of these challenges affects the transportation of both petroleum products and other goods?

A Poorly repaired roads
B Harsh weather and vast distances
C Frozen waterways
D Separatist movements

First determine what choices you can eliminate. Since petroleum products are transported through pipelines, choices A and C do not apply and can be eliminated. Choose the best answer from the remaining options.

2. In Russia, nuclear power plants built during the Soviet era

F have been shut down.
G provide much of Russia's electricity.
H are now safer than ever before.
J have been replaced by coal-fired generators.

Only one answer is completely true. Some reactors have been shut down, and some safety standards were improved. Choose the answer that is completely true.

North Africa, Southwest Asia, and Central Asia

WHY IT'S IMPORTANT—

Your modern lifestyle depends on oil. Without vehicles powered by gasoline, how would you get from one place to another, and how would goods be sent from warehouses to stores? Today, much of the world's oil comes from the region of North Africa, Southwest Asia, and Central Asia. Many American companies do business in the region. As a result, political, social, and economic changes there have a major impact on your daily life.

World Regions Video

To learn more about North Africa, Southwest Asia, and Central Asia and their impact on your world, view the World Regions video "North Africa, Southwest Asia, and Central Asia."

NGS ONLINE

www.nationalgeographic.com/education

Bedouin resting on roof of ancient stone building at Petra, Jordan

6

What Makes North Africa, Southwest Asia, and Central Asia a Region?

Arid and often forbidding, the region of North Africa, Southwest Asia, and Central Asia stretches from Morocco to Kazakhstan. Rugged mountain ranges surround vast, dry plateaus and some of Earth's greatest deserts. Through these parched landscapes flow a handful of life-sustaining rivers. The Nile, the world's longest river, slices northward through Egypt to the Mediterranean Sea. The Tigris and Euphrates flow southeast across Turkey, Syria, and Iraq. These two rivers cradle the "Fertile Crescent," an area of rich soil where some of the world's earliest agricultural societies took root.

Where slightly more rain falls, deserts give way to grass-covered steppes where nomadic herders roam with their flocks. Only coastal areas and highlands enjoy a moister, milder Mediterranean climate. On the whole, water, perhaps the most precious resource, is very scarce in this region. Oil, in contrast, is one of the region's most abundant resources.

1 A bedouin girl holds a baby goat on an arid plain in Jordan. Bedouins traditionally are nomadic herders of goats, sheep, and camels. Nomadic herding is common—and practical—in the vast parts of this region that are too dry for growing crops.

2 Gleaming pipes surround an oil refinery in Kuwait. This tiny country and its neighbors on the Arabian Peninsula produce much of the world's oil. An elaborate system of pipelines transports the oil from refineries to seaports where huge oil tankers dock.

3 Wind-carved sand dunes surround a small oasis in the Algerian Sahara. The world's largest hot desert, the Sahara covers most of North Africa. Surprisingly, sand dunes are relatively rare in the Sahara. Far more common are wind-swept expanses of rock and gravel.

4 Muslims pause to pray high in the mountains of Afghanistan's Hindu Kush, one of the region's many mountain ranges. Other ranges include the Atlas Mountains, which span Morocco and Algeria, and the glacier-crowned Tian Shan of Kyrgyzstan.

6

Cradle of Civilization

Along the banks of the Nile River, colossal stone statues are mute reminders of the ancient Egyptian civilization. Many other great civilizations, including those of the Sumerians, Persians, and Phoenicians, also arose in this region. So did three of the world's great religions—Judaism, Christianity, and Islam. Today, Islam claims the greatest number of followers here. Yet Georgia and Armenia remain Christian strongholds, and Israel is the Jewish state.

Ethnic diversity is a hallmark of this region, which has long been a cultural crossroads linking Europe, Africa, and Asia. Just as cultures mix in this part of the world, so tradition intermingles with the newest technology. Ancient customs persist even in the most modern cities.

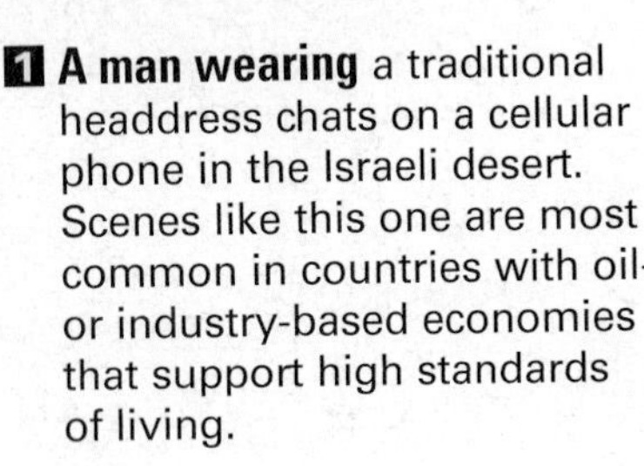

1 A man wearing a traditional headdress chats on a cellular phone in the Israeli desert. Scenes like this one are most common in countries with oil- or industry-based economies that support high standards of living.

4

2 Draped in flowing *chadris,* or body veils, women shop for shoes in a market in Kabul, Afghanistan. The women practice a conservative form of Islam, which encourages women to conceal their bodies under these traditional full-length garments.

3 Straddling the Bosporus Strait, Istanbul, Turkey, is the only major city to stand on two continents—Asia and Europe. The magnificent, domed Hagia Sophia was initially built as a Christian cathedral. Nearly a thousand years later, it was converted into a mosque. It now serves as a museum.

4 Standing guard for centuries, giant stone figures flank the entrance to an ancient temple in Egypt. Pharaoh Ramses II built this and a neighboring temple beside the Nile River during the 1200s B.C. When the Aswan High Dam was built in the 1960s, the temples were moved to higher ground.

6

North Africa, Southwest Asia, and Central Asia

PHYSICAL

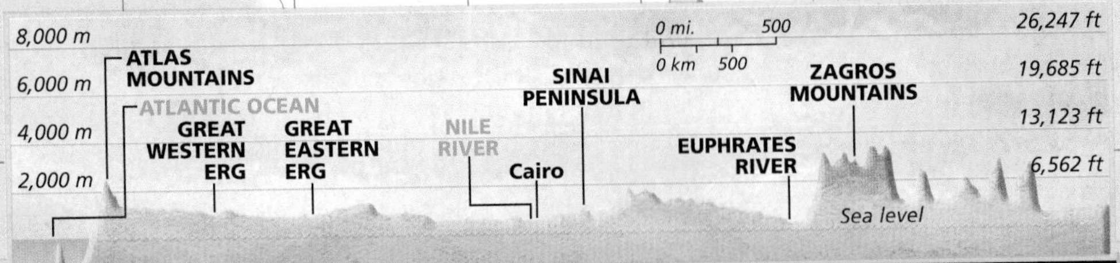

ARCTIC CIRCLE
RUSSIA
EUROPE
ATLANTIC OCEAN
THE STEPPES
Kazakh Uplands
KAZAKHSTAN
Caspian Depression
Aral Sea
Lake Balkhash
KYRGYZSTAN
Ustyurt Plateau
Turan Lowland
Qizilqum
TIAN SHAN
UZBEKISTAN
TAJIKISTAN
GEORGIA
CAUCASUS MTS.
Caspian Sea
GARAGUM
Ismail Samani Peak 24,590 ft. (7,495 m)
Black Sea
Mt. Ararat 16,945 ft. (5,165 m)
TURKMENISTAN
Hindu Kush
ANATOLIA
TURKEY
AZERBAIJAN
ARMENIA
Taurus Mts.
ZAGROS MOUNTAINS
Strait of Gibraltar
Mediterranean Sea
SYRIA
MESOPOTAMIA
Dasht e Kavir
AFGHANISTAN
Madeira Is.
MOROCCO
ATLAS MOUNTAINS
TUNISIA
LEBANON
Dead Sea
IRAN
Great Western Erg
Great Eastern Erg
ISRAEL
IRAQ
Qattara Depression
Sinai Pen.
JORDAN
KUWAIT
Persian Gulf
Canary Is.
ALGERIA
LIBYA
BAHRAIN
Makran Ra.
SAHARA
EGYPT
Hejaz
SAUDI ARABIA
QATAR
Gulf of Oman
WESTERN SAHARA
TROPIC OF CANCER
AHAGGAR MOUNTAINS
Marzuq Desert
Gilf Kebir Plateau
Red Sea
ARABIAN PENINSULA
Akhdar Mts.
UNITED ARAB EMIRATES
OMAN
Rub' al Khali (Empty Quarter)
Asir
AFRICA
YEMEN
Hadramaut
Arabian Sea
Gulf of Aden
Socotra I.
N
0 mi. 1,000
0 km 1,000
Lambert Azimuthal Equal-Area projection
INDIAN OCEAN
EQUATOR
40°W 20°W 0° 20°E 40°E 60° 100°E
70°N 60°N 50°N 40°N 30°N 20°N 10°N 0° 10°S

8,000 m
6,000 m
4,000 m
2,000 m
26,247 ft
19,685 ft
13,123 ft
6,562 ft
0 mi. 500
0 km 500
ATLAS MOUNTAINS
ATLANTIC OCEAN
GREAT WESTERN ERG
GREAT EASTERN ERG
NILE RIVER
Cairo
SINAI PENINSULA
EUPHRATES RIVER
ZAGROS MOUNTAINS
Sea level

Elevation Profile

POLITICAL

40°W 20°W 0° 20°E 40°E 60°E 80°E 100°E

ARCTIC CIRCLE

60°N 50°N 40°N 30°N 20°N 10°N 0° 10°S

0 mi. 1,000
0 km 1,000
Lambert Azimuthal Equal-Area projection

ATLANTIC OCEAN

RUSSIA

EUROPE

Astana
Lake Balkhash
KAZAKHSTAN
Aral Sea
Bishkek
KYRGYZSTAN
UZBEKISTAN
Tashkent
Caspian Sea
Caucasus Mts.
Black Sea
GEORGIA
T'bilisi
Baku
TAJIKISTAN
ARMENIA
TURKMENISTAN
Dushanbe
Istanbul
Ankara
Yerevan
Ashgabat
Izmir
TURKEY
AZERBAIJAN
Kabul
Algiers
Tunis
Aleppo
Tehran
AFGHANISTAN
Madeira Is. Port.
Rabat
Atlas Mountains
TUNISIA
Mediterranean Sea
LEBANON
SYRIA
IRAN
Beirut
Damascus
Baghdad
Tripoli
ISRAEL
IRAQ
Isfahan
Canary Is. Sp.
MOROCCO
Jerusalem
Amman
KUWAIT
Alexandria
JORDAN
Kuwait
Persian Gulf
ALGERIA
Cairo
Nile R.
Laayoune
BAHRAIN
Manama
QATAR
WESTERN SAHARA Mor.
LIBYA
Riyadh
Doha
Abu Dhabi
EGYPT
Red Sea
SAUDI ARABIA
Muscat
Gulf of Oman
TROPIC OF CANCER
UNITED ARAB EMIRATES
Jeddah
Makkah (Mecca)
S A H A R A
OMAN
YEMEN
Sanaa
Arabian Sea
Gulf of Aden
N

National capital
Major city

AFRICA

EQUATOR

INDIAN OCEAN

ATLANTIC OCEAN

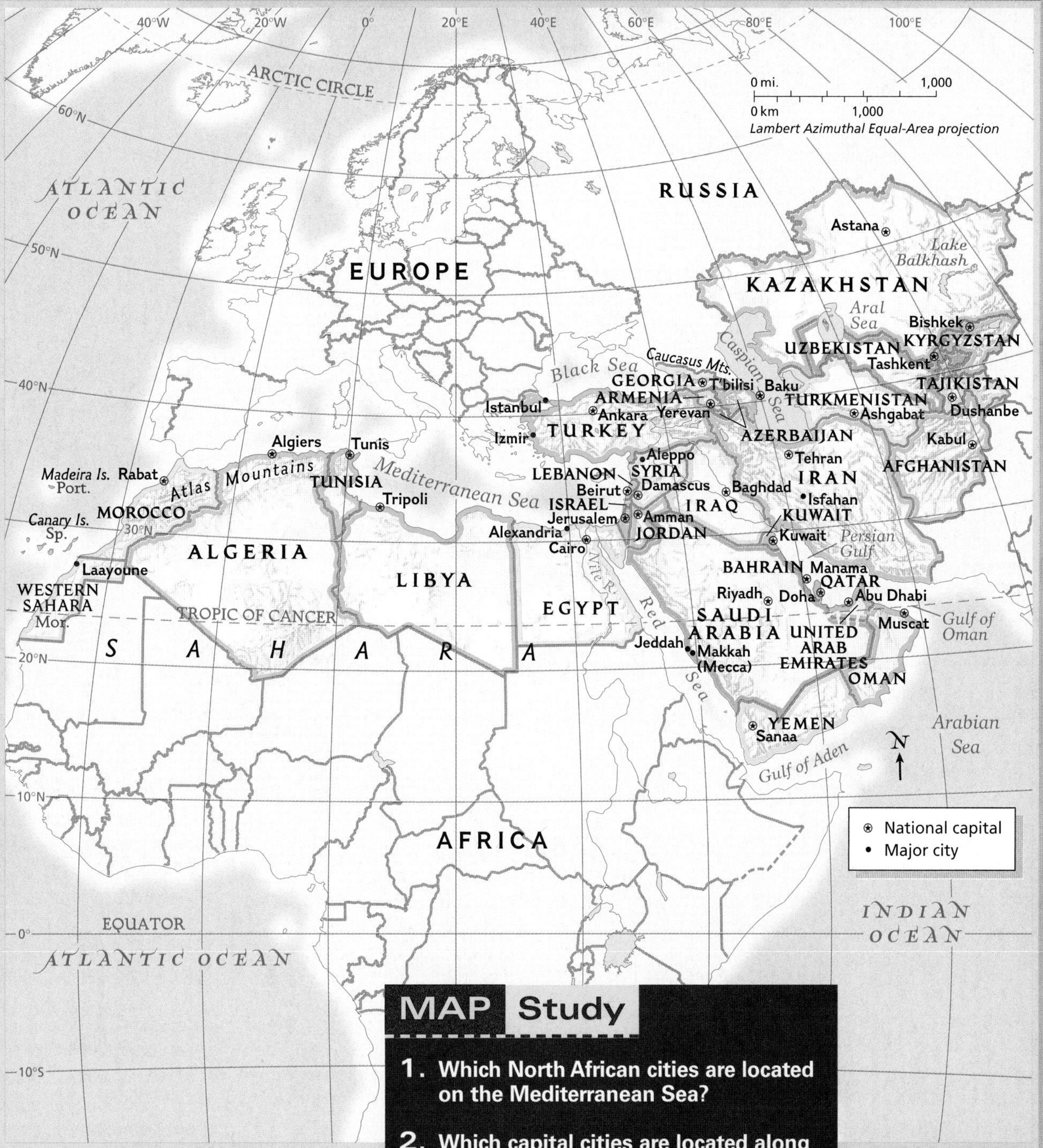

MAP Study

1. Which North African cities are located on the Mediterranean Sea?
2. Which capital cities are located along the Persian Gulf?

6

North Africa, Southwest Asia, and Central Asia

POPULATION DENSITY

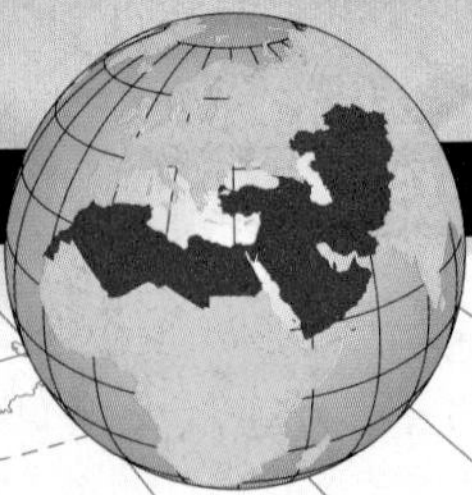

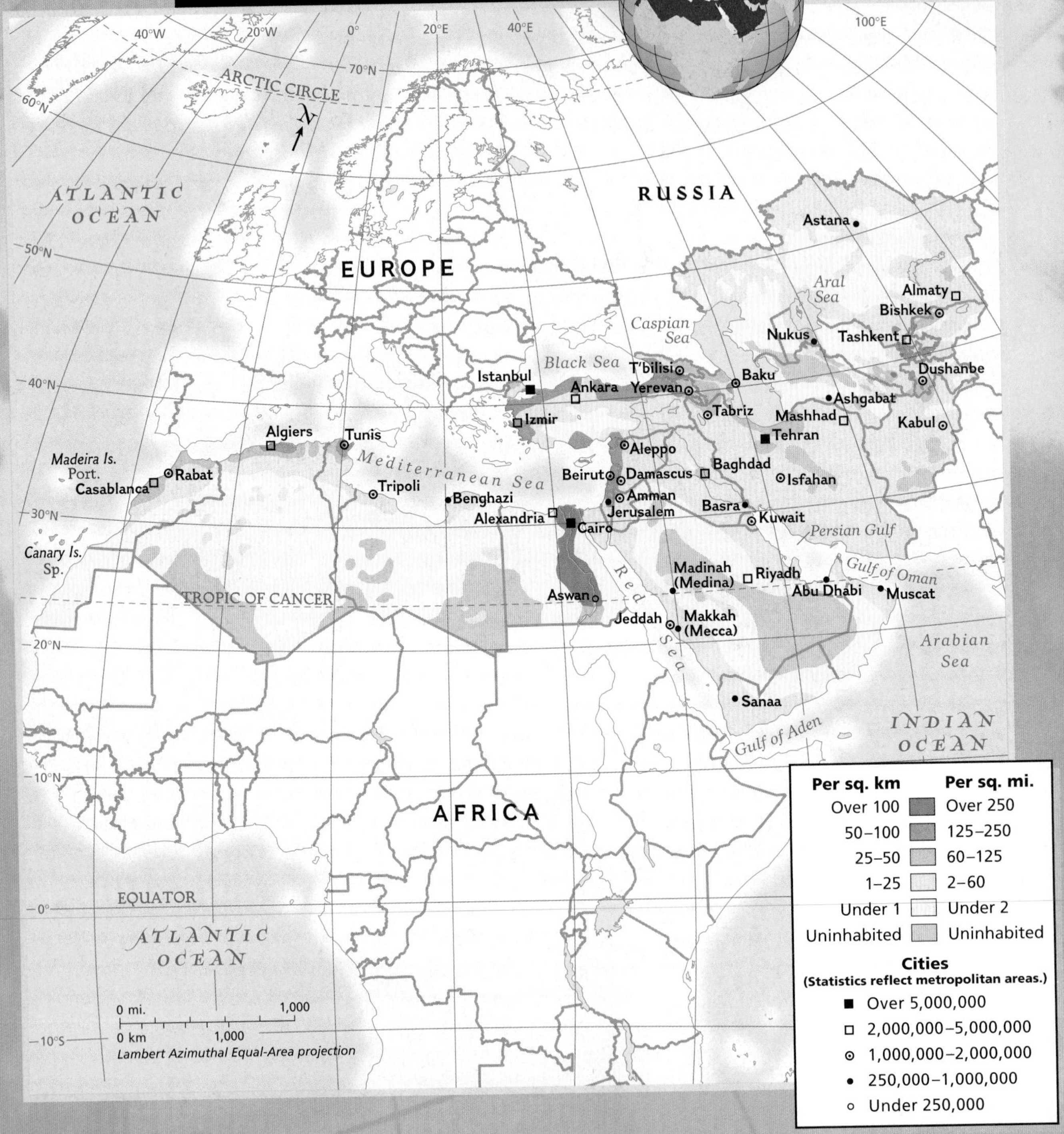

Per sq. km	Per sq. mi.
Over 100	Over 250
50–100	125–250
25–50	60–125
1–25	2–60
Under 1	Under 2
Uninhabited	Uninhabited

Cities
(Statistics reflect metropolitan areas.)

- ■ Over 5,000,000
- □ 2,000,000–5,000,000
- ⊙ 1,000,000–2,000,000
- • 250,000–1,000,000
- ○ Under 250,000

0 mi. 1,000
0 km 1,000

Lambert Azimuthal Equal-Area projection

ECONOMIC ACTIVITY

Land Use

- Commercial farming
- Subsistence farming
- Livestock raising
- Nomadic herding
- Hunting and gathering
- Manufacturing and trade
- Commercial fishing
- Little or no activity

Resources

- Coal
- Petroleum
- Natural gas
- Iron ore
- Copper
- Lead
- Zinc
- Bauxite
- Manganese
- Phosphate
- Tungsten
- Gold
- Silver
- Chromite

0 mi. 1,000
0 km 1,000
Lambert Azimuthal Equal-Area projection

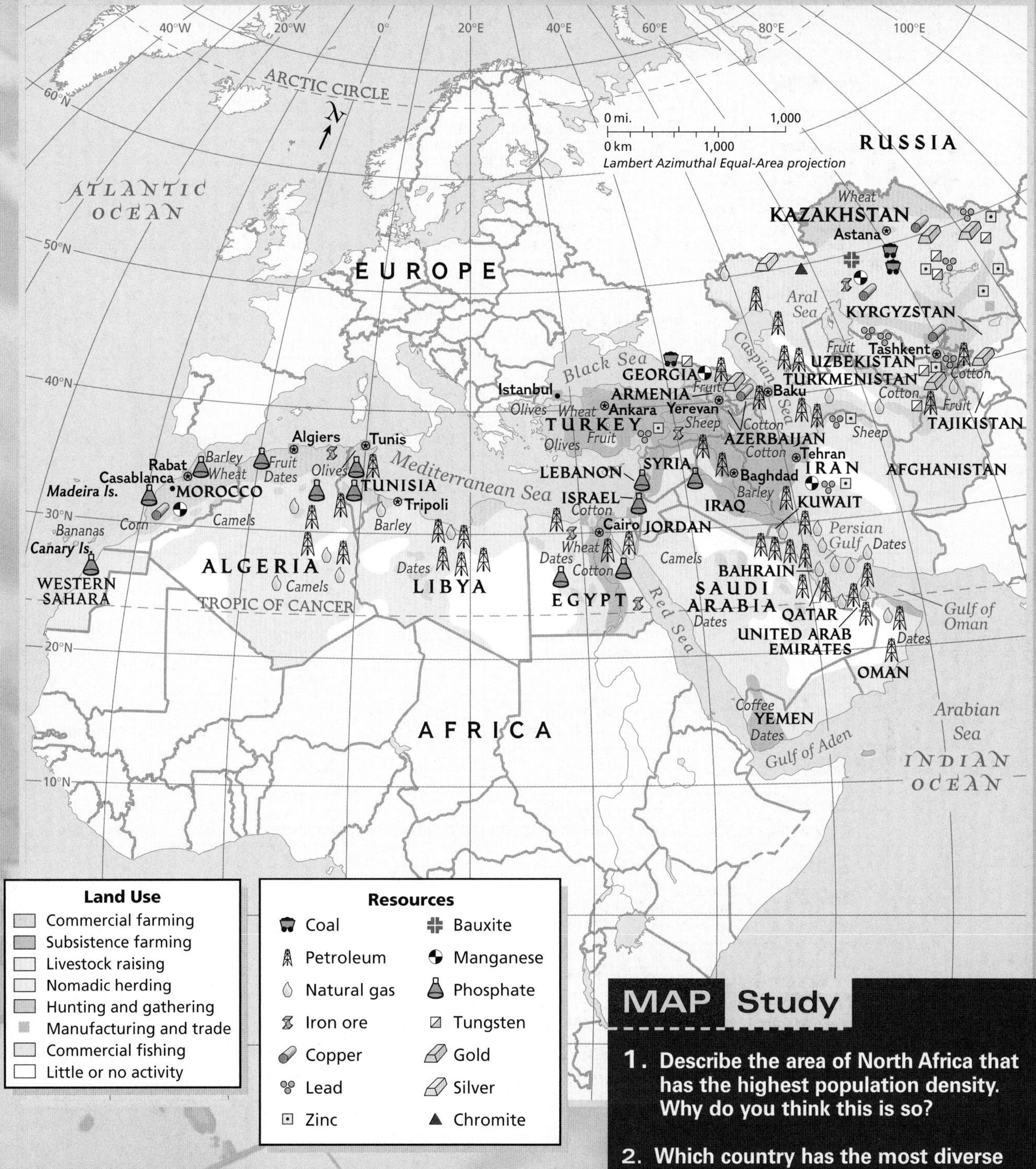

MAP Study

1. Describe the area of North Africa that has the highest population density. Why do you think this is so?
2. Which country has the most diverse natural resources?

6 North Africa, Southwest Asia, and Central Asia

COUNTRY PROFILES

COUNTRY * AND CAPITAL	FLAG AND LANGUAGE	POPULATION AND DENSITY	LANDMASS	MAJOR EXPORT	MAJOR IMPORT	CURRENCY	GOVERNMENT
AFGHANISTAN Kabul	Pashto, Dari	26,800,000 106 per sq. mi. 41 per sq. km	251,772 sq. mi. 652,089 sq. km	Fruits and Nuts	Foods	Afghani	Islamic Republic
ALGERIA Algiers	Arabic, French, Berber	31,000,000 34 per sq. mi. 13 per sq. km	919,591 sq. mi. 2,381,521 sq. km	Petroleum	Machinery	Algerian Dinar	Republic
ARMENIA Yerevan	Armenian, Russian	3,800,000 330 per sq. mi. 127 per sq. km	11,506 sq. mi. 29,801 sq. km	Gold	Grain	Dram	Republic
AZERBAIJAN Baku	Azeri, Russian, Armenian	8,100,000 243 per sq. mi. 94 per sq. km	33,436 sq. mi. 86,599 sq. km	Petroleum	Machinery	Manat	Republic
BAHRAIN Manama	Arabic	700,000 2,688 per sq. mi. 1,038 per sq. km	266 sq. mi. 689 sq. km	Petroleum	Machinery	Bahrain Dinar	Traditional Monarchy
EGYPT Cairo	Arabic	69,800,000 181 per sq. mi. 70 per sq. km	386,660 sq. mi. 1,001,449 sq. km	Crude Oil	Machinery	Egyptian Pound	Republic
GEORGIA T'bilisi	Georgian, Russian	5,500,000 139 per sq. mi. 54 per sq. km	26,911 sq. mi. 69,699 sq. km	Citrus Fruits	Fuels	Lari	Republic
IRAN Tehran	Persian, Kurdish	66,100,000 108 per sq. mi. 42 per sq. km	630,575 sq. mi. 1,633,189 sq. km	Petroleum	Machinery	Rial	Islamic Republic
IRAQ Baghdad	Arabic, Kurdish	23,600,000 139 per sq. mi. 54 per sq. km	169,236 sq. mi. 438,321 sq. km	Crude Oil	Machinery	Iraqi Dinar	Republic
ISRAEL Jerusalem**	Hebrew, Arabic	6,400,000 791 per sq. mi. 305 per sq. km	8,131 sq. mi. 21,059 sq. km	Polished Diamonds	Chemicals	Shekel	Republic

* COUNTRIES AND FLAGS NOT DRAWN TO SCALE

**Israel has proclaimed Jerusalem as its capital, but many countries' embassies are located in Tel Aviv. The Palestinian Authority has assumed all governmental duties in non-Israeli-occupied areas of the West Bank and Gaza Strip.

COUNTRY * AND CAPITAL	FLAG AND LANGUAGE	POPULATION AND DENSITY	LANDMASS	MAJOR EXPORT	MAJOR IMPORT	CURRENCY	GOVERNMENT
JORDAN Amman	Arabic	5,200,000 150 per sq. mi. 58 per sq. km	34,444 sq. mi. 89,210 sq. km	Phosphates	Crude Oil	Jordanian Dinar	Constitutional Monarchy
KAZAKHSTAN Astana	Kazakh, Russian	15,417,000 15 per sq. mi. 6 per sq. km	1,049,039 sq. mi. 2,716,998 sq. km	Petroleum	Machinery	Tenge	Republic
KUWAIT Kuwait	Arabic	2,300,000 297 per sq. mi. 115 per sq. km	6,880 sq. mi. 17,818 sq. km	Petroleum	Foods	Kuwaiti Dinar	Constitutional Monarchy
KYRGYZSTAN Bishkek	Kirghiz, Russian	5,000,000 65 per sq. mi. 25 per sq. km	76,641 sq. mi. 198,500 sq. km	Cotton	Grain	Som	Republic
LEBANON Beirut	Arabic, French	4,300,000 1,061 per sq. mi. 410 per sq. km	4,015 sq. mi. 10,399 sq. km	Paper	Machinery	Lebanese Pound	Republic
LIBYA Tripoli	Arabic	5,200,000 8 per sq. mi. 3 per sq. km	679,359 sq. mi. 1,759,540 sq. km	Crude Oil	Machinery	Libyan Dinar	Military Dictatorship
MOROCCO** Rabat	Arabic, French, Berber	29,500,000 105 per sq. mi. 41 per sq. km	279,757 sq. mi. 724,571 sq. km	Foods	Manufactured Goods	Dirham	Constitutional Monarchy
OMAN Muscat	Arabic	2,400,000 29 per sq. mi. 11 per sq. km	82,031 sq. mi. 212,460 sq. km	Petroleum	Machinery	Omani Rial	Traditional Monarchy
QATAR Doha	Arabic	600,000 139 per sq. mi. 54 per sq. km	4,247 sq. mi. 11,000 sq. km	Petroleum	Machinery	Qatari Riyal	Traditional Monarchy

* COUNTRIES AND FLAGS NOT DRAWN TO SCALE

**Morocco claims the Western Sahara area, but other countries do not accept this claim.

FOR AN ONLINE UPDATE OF THIS INFORMATION, VISIT GEOGRAPHY.GLENCOE.COM AND CLICK ON "TEXTBOOK UPDATES."

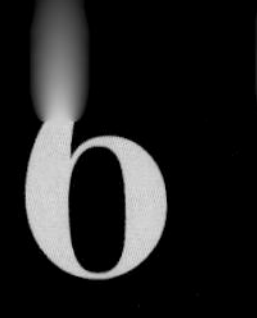

North Africa, Southwest Asia, and Central Asia

COUNTRY PROFILES

COUNTRY * AND CAPITAL	FLAG AND LANGUAGE	POPULATION AND DENSITY	LANDMASS	MAJOR EXPORT	MAJOR IMPORT	CURRENCY	GOVERNMENT
SAUDI ARABIA Riyadh	Arabic	21,100,000 25 per sq. mi. 10 per sq. km	829,996 sq. mi. 2,149,690 sq. km	Petroleum	Machinery	Riyal	Traditional Monarchy
SYRIA Damascus	Arabic, Kurdish, Armenian	17,100,000 231 per sq. mi. 89 per sq. km	71,498 sq. mi. 185,180 sq. km	Petroleum	Machinery	Syrian Pound	Republic
TAJIKISTAN Dushanbe	Tajik, Russian	6,200,000 112 per sq. mi. 43 per sq. km	55,251 sq. mi. 143,100 sq. km	Aluminum	Fuels	Tajik Ruble	Republic
TUNISIA Tunis	Arabic, French	9,700,000 154 per sq. mi. 59 per sq. km	63,170 sq. mi. 163,610 sq. km	Petroleum Products	Machinery	Tunisian Dinar	Republic
TURKEY Ankara	Turkish, Kurdish	66,300,000 221 per sq. mi. 85 per sq. km	299,158 sq. mi. 774,819 sq. km	Foods and Livestock	Machinery	Turkish Lira	Republic
TURKMENISTAN Ashgabat	Turkmen, Russian, Uzbek	5,500,000 29 per sq. mi. 11 per sq. km	188,456 sq. mi. 488,101 sq. km	Natural Gas	Machinery	Manat	Republic
UNITED ARAB EMIRATES Abu Dhabi	Arabic, Persian	3,300,000 103 per sq. mi. 40 per sq. km	32,278 sq. mi. 83,600 sq. km	Petroleum	Manufactured Goods	Emirian Dirham	Federal Monarchy
UZBEKISTAN Tashkent	Uzbek, Russian, Tajik	25,100,000 145 per sq. mi. 56 per sq. km	172,471 sq. mi. 446,700 sq. km	Cotton	Machinery	Som	Republic
YEMEN Sanaa	Arabic	18,000,000 88 per sq. mi. 34 per sq. km	203,849 sq. mi. 527,969 sq. km	Cotton	Textiles	Rial	Republic

* COUNTRIES AND FLAGS NOT DRAWN TO SCALE

FOR AN ONLINE UPDATE OF THIS INFORMATION, VISIT GEOGRAPHY.GLENCOE.COM AND CLICK ON "TEXTBOOK UPDATES."

▶ Traffic speeds by eastern harbor, Alexandria, Egypt

NATIONAL
GEOGRAPHIC
SOCIETY
سيراميكا كليوباترا
سيراميك وأطقم حمامات
إيجيبل
أبواب وشبابيك
EGYBEL
MI-BANK

6

GLOBAL CONNECTION

SOUTHWEST ASIA AND THE UNITED STATES

RELIGIONS

Chances are there's a church, a synagogue, or a mosque in your community. These places of worship represent three of the most widespread religions in the United States: Christianity, Judaism, and Islam. All three have their roots in Southwest Asia and profess belief in one God.

Jews trace their ancestry to a herder named Abraham, who lived at least 3,500 years ago in what is now Iraq. According to Jewish scripture, God instructed Abraham to settle in the area that became known as Israel and promised to bless Abraham's descendants if they worshiped one God.

Around 1000 B.C., Israel was united under a powerful king, David, who made Jerusalem his capital. Political strife later divided Israel into two parts, Israel and Judah, which were conquered by other nations. Many of the people of Judah—the Jews—left their homeland, and their descendants scattered around the world. The first Jews in North America arrived in the American colonies in the 1650s. Today, the United States is home to the world's largest Jewish population.

The Jews believed that God would send a Messiah to unite and lead them. Jesus was a Jew who was born in Judah when it was under Roman rule. Jesus interpreted Jewish teachings in a new way. His message made him unpopular with the authorities, and the Romans executed him around A.D. 30.

▲ Christmas celebration in Bethlehem, birthplace of Jesus

▲ Jewish worshiper at wall of ancient temple in Jerusalem

▲ Muslim women leaving mosque in Jerusalem

Jesus' followers, known as Christians, believed that Jesus was both the Messiah and the Son of God. The Christian faith eventually became the official religion of the Roman Empire, and then the dominant faith throughout Europe. European explorers and colonists carried it to the Western Hemisphere, and it became the most widely practiced religion in the United States.

More than 500 years after Jesus died, the prophet Muhammad was born on the Arabian Peninsula. According to Muslim tradition, Muhammad received revelations from God and began to teach lessons to his followers. The heart of his teachings form the basis of Islam, which revolves around belief in a single God who periodically communicates through prophets. For believers of Islam, Muhammad was the last in a series of prophets that included Abraham and Jesus.

After Muhammad's death in A.D. 632, Islam spread quickly. Unlike Judaism and Christianity, however, Islam remained the dominant faith in the region where it originated. Islam has more than a billion followers worldwide, and its numbers are growing in the United States.

CHAPTER 17

The Physical Geography of North Africa, Southwest Asia, and Central Asia

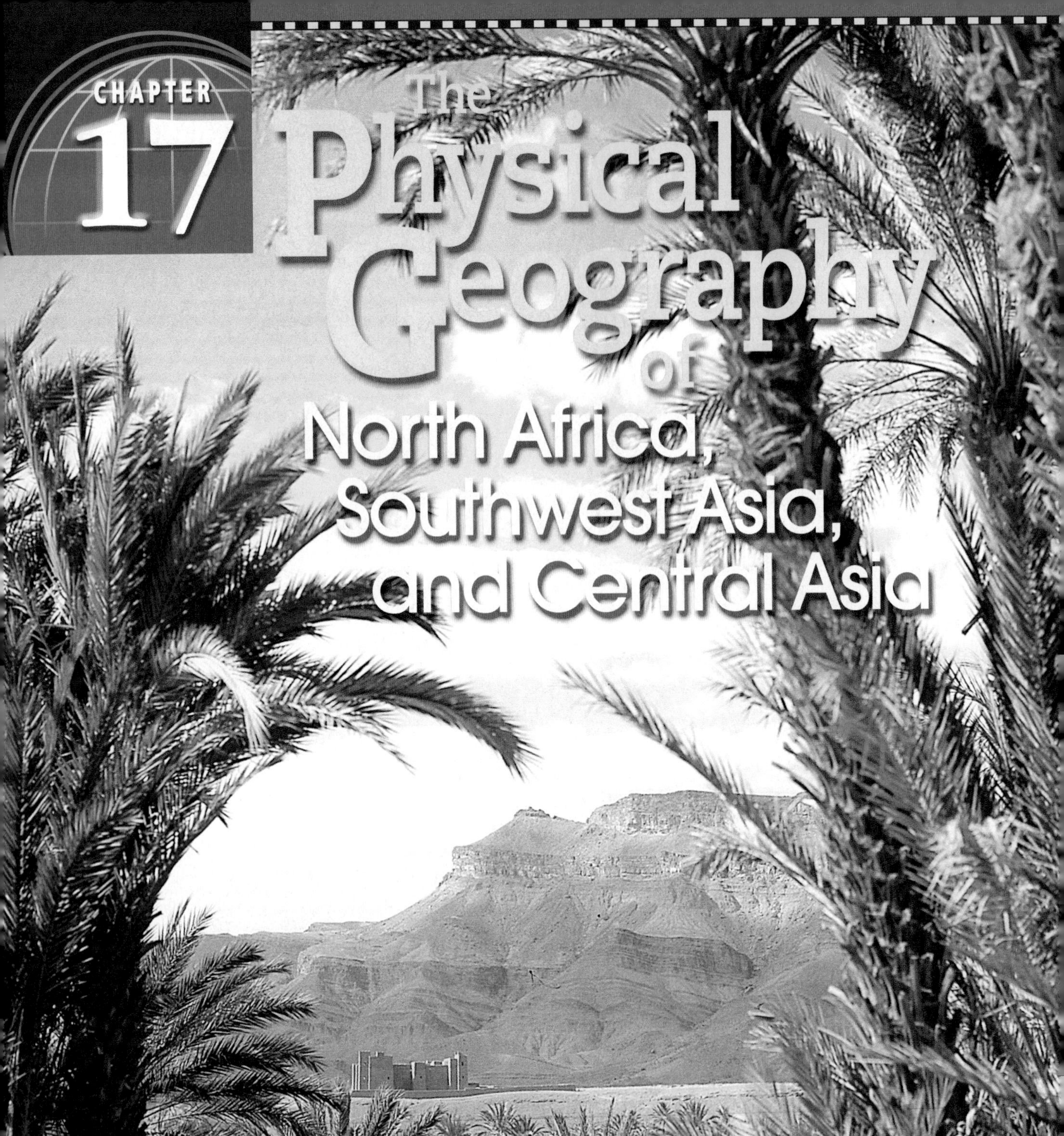

GeoJournal

As you read this chapter, list ways the physical geography of North Africa, Southwest Asia, and Central Asia shapes the lives of people in the region. Include examples you discover in media sources.

GEOGRAPHY Online

Chapter Overview Visit the **Glencoe World Geography** Web site at geography.glencoe.com and click on Chapter Overviews—Chapter 17 to preview information about the physical geography of the region.

The Land

Guide to Reading

Consider What You Know

The vast region of North Africa, Southwest Asia, and Central Asia spans portions of Africa and Asia. Considering this great expanse, what landforms would you expect to discover in the region?

Read to Find Out

- What land and water features dominate the region?
- Why are the region's major rivers important to its people?
- Why is much of the world economically dependent on the region?

Terms to Know

- alluvial soil
- wadi
- *kum*
- phosphate

Places to Locate

- Red Sea
- Arabian Peninsula
- Persian Gulf
- Sinai Peninsula
- Anatolia
- Dead Sea
- Caspian Sea
- Aral Sea
- Nile River
- Tigris River
- Euphrates River
- Atlas Mountains
- Caucasus Mountains

◀ *Desert fort and oasis, Morocco*

NATIONAL GEOGRAPHIC

A Geographic View

Timeless Travel

Men and boys of the caravan form a ragged rank, facing distant Mecca. . . . In unison the caravanners kneel, then bow, pressing their foreheads into the sand. In the cool shadows of morning they rejoin the line of beasts tethered head to tail and wait for a signal. . . . The **madougou**, *or caravan boss, raises his staff, jerks the rope halter on his lead camel, and, to shouts and the clanging of pans and bowls, the half-mile-long train grudgingly lurches forward.*

—*Thomas J. Abercrombie, "Ibn Battuta, Prince of Travelers,"* National Geographic, *December 1991*

Camel caravan, Sahara

Joining a camel caravan in the Sahara, writer Thomas J. Abercrombie followed in the footsteps of the Muslim traveler Ibn Battuta, who crisscrossed the lands of North Africa, Southwest Asia, and Central Asia more than five centuries ago.

People, goods, and ideas have come together in this part of the world for thousands of years because of its location on or near the Mediterranean Sea. This section examines the varied landscape and the wealth of natural resources of the region where the continents of Europe, Africa, and Asia meet.

Seas and Peninsulas

North Africa, Southwest Asia, and Central Asia form an intricate jigsaw puzzle of seas and peninsulas. Edging the coast of North Africa as far as the Strait of Gibraltar, the Mediterranean Sea separates Africa and Europe.

Comparing Lands

The region of North Africa, Southwest Asia, and Central Asia is about twice the size of the continental United States.

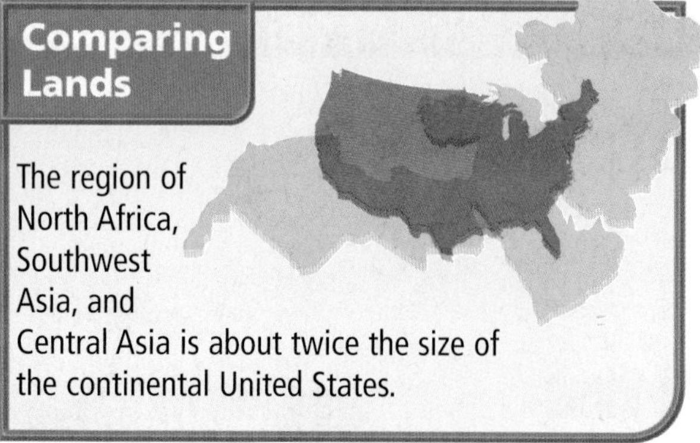

To the east, the **Red Sea** and the Gulf of Aden separate Southwest Asia's **Arabian Peninsula** from Africa. The **Persian Gulf** frames this peninsula on the east, and the Arabian Sea borders it on the south. Northwest of the Arabian Peninsula, the Gulf of Suez and the Gulf of Aqaba flank the smallest piece in the puzzle, the **Sinai Peninsula**.

To the north the peninsula of **Anatolia** points west to the Aegean Sea. Two more seas—the Black Sea and the Mediterranean Sea—lie at the peninsula's north and south. The Dardanelles, the Sea of Marmara, and the Bosporus strait, which together separate Europe and Asia, connect the Aegean and Black Seas.

Three landlocked bodies of salt water lie east of the Mediterranean Sea. The smallest of these, the **Dead Sea**, sits at the mouth of the Jordan River, forming part of the Israeli-Jordanian border. It is a source of chemical products such as potash. In Central Asia, the **Caspian Sea** is the largest inland body of water on Earth. Stretching for almost 750 miles (1,207 km), this landlocked sea laps the shores of both Asia and Europe. As you read in Unit 4, evaporation and decreased flow from feeder rivers have resulted in the Caspian Sea's lower water levels. Irrigation and industry also cut the flow of other rivers flowing into the Caspian Sea, further reducing water levels.

East of the Caspian Sea, in the heart of Central Asia, is the **Aral Sea**. Until the 1960s the Aral Sea was the world's fourth-largest inland sea, and it supported a healthy fishing community. Now it is just a fraction of its former size and looks more like a desert than a sea. The Aral Sea began to dry up when the Soviet Union diverted huge amounts of water for irrigation from the major rivers flowing into the sea. Today the Aral Sea seems to be coming back. By building small dams in parts of the former sea, local people plan to create smaller freshwater basins with water from the rivers.

Rivers

Rivers are the lifeblood of North Africa, Southwest Asia, and Central Asia. Their lush and productive valleys have always welcomed travelers

NATIONAL GEOGRAPHIC **World Explorer**

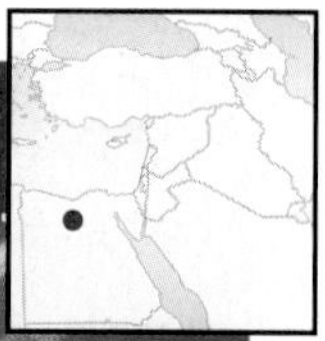

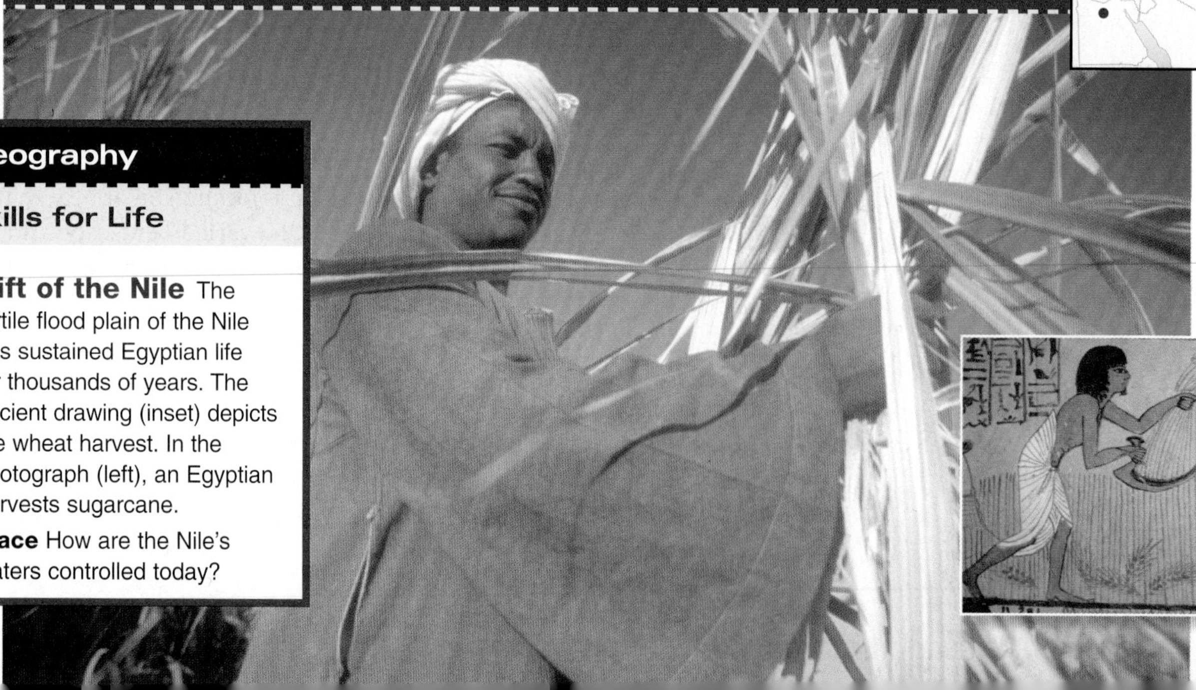

Geography

Skills for Life

Gift of the Nile The fertile flood plain of the Nile has sustained Egyptian life for thousands of years. The ancient drawing (inset) depicts the wheat harvest. In the photograph (left), an Egyptian harvests sugarcane.

Place How are the Nile's waters controlled today?

NATIONAL GEOGRAPHIC

MAP STUDY

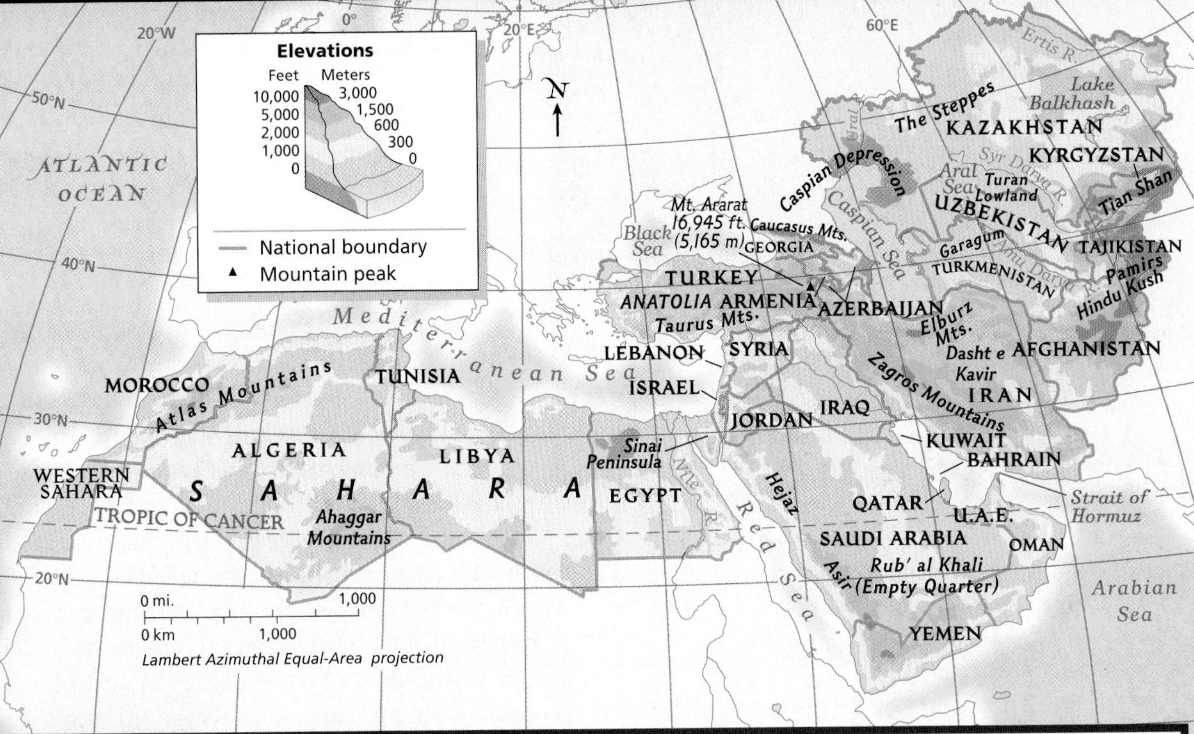

Geography Skills for Life

1. **Interpreting Maps** Where are the Zagros Mountains located? The Atlas Mountains?

2. **Applying Geography Skills** Which country in the region is dominated by areas of elevations of more than 5,000 feet (1,500 meters)?

Find NGS online map resources @ www.nationalgeographic.com/maps

and provided food for local peoples. Egypt's **Nile River** is the world's longest river at 4,160 miles (6,695 km). The Tigris (TY•gruhs) and Euphrates (yu•FRAY•teez) Rivers, which flow mainly through Iraq, are also important to the region.

Culture

Major Rivers: Cradles of Civilization

The Nile Delta and the fertile land along the river's banks gave birth to one of the world's earliest civilizations. Today more than 90 percent of Egypt's people live in the Nile Delta or along the course of the river on only 3 percent of Egypt's land. The Aswan High Dam and other modern dams farther up the Nile now control the river's flow, reducing both flooding and deposits of alluvial soil, rich soil made up of sand and mud deposited by moving water.

Early civilizations also thrived in the Tigris-Euphrates river valley, a fertile farming valley in Central Asia. Known by ancients as Mesopotamia, which is Greek for "land between two rivers," this valley owes its fertile character to the **Tigris** and **Euphrates Rivers**. A complex irrigation network has watered the valley and supported farming there for 7,000 years. Today the Tigris and Euphrates help irrigate farms throughout Syria, Turkey, and Iraq.

Originating only 50 miles (80 km) from each other in eastern Turkey, the Tigris and Euphrates Rivers join in Iraq to form the Shatt al Arab, which empties into the Persian Gulf. The Euphrates is the longer river, flowing 1,700 miles (2,736 km) toward the sea.

Geography Skills for Life

Snowy Desert Morocco's Atlas Mountains lie near the Sahara, but mountain travelers must be prepared for cold, snowy weather.

Place What are Morocco's most important economic activities?

The Tigris extends about 1,180 miles (1,899 km). Dams control the flow of both rivers.

Streambeds

Many streams in arid North Africa and Southwest Asia flow only intermittently, appearing suddenly and disappearing just as quickly. In the region's deserts, runoff from infrequent rainstorms creates wadis (WAH•dees)—streambeds that remain dry until a heavy rain. Irregular rainstorms often produce flash flooding. During a flash flood, wadis fill with so much sediment that they can rapidly become mud flows, or moving masses of wet soil, which are a danger to humans and animals.

Plains, Plateaus, and Mountains

A traveler in North Africa, Southwest Asia, and Central Asia could expect to see varied and dramatic landforms. Low plains extend to the horizon and sometimes rise to a plateau or mountains. Study the map on page 423 to see elevation patterns within the region.

Coastal Plains

In a region dominated by deserts and mountains, lush coastal plains stand out. The region's agricultural base is rooted in fertile plains along the Mediterranean Sea, such as those stretching east to west along the Moroccan and Algerian coasts and those along the Caspian Sea and Persian Gulf.

Highlands

Africa's longest mountain range, the **Atlas Mountains**, reaches across Morocco and Algeria, in the westernmost part of the region. Enough precipitation falls on the northern side of these mountains to water the coastal regions and make them hospitable to settlement and farming. Despite Morocco's generally rugged terrain, for example, the fertile farmlands of the Atlas's northern slopes produce an abundance of crops typical of the Mediterranean climate. About 50 percent of Morocco's people engage in agriculture, producing barley, oats, and wheat. In years of drought, as in 1999, the economy suffers. With more rain predicted, the economy is expected to grow by about 6 percent per year. Fishing and raising livestock also play a large role in Morocco's economy.

In Southwest Asia, two mountain ranges, the Hejaz and the Asir, stretch along the western coast of the Arabian Peninsula. The taller Asir Mountains receive more rainfall than the Hejaz, accumulating up to 19 inches (48 cm) annually. This precipitation makes the Asir region the most agriculturally productive on the Arabian Peninsula. In contrast, the Central Plateau to the east of the Asir Mountains averages between 0 and 4 inches (0 and 10 cm) of rain per year, mainly because of the rain shadow effect.

Student Web Activity Visit the **Glencoe World Geography** Web site at geography.glencoe.com and click on Student Web Activities—Chapter 17 for an activity about physical processes in North Africa, Southwest Asia, and Central Asia.

The Pontic Mountains and the Taurus Mountains rise from the Turkish landscape. Between these ranges, the Anatolian Plateau stands 2,000 to 5,000 feet (610 to 1,524 m) above sea level. East of the Pontic range, camel-backed Mount Ararat, at almost 17,000 feet (5,182 m), overlooks the Turkish-Iranian border.

As the map on page 423 shows, the **Caucasus Mountains** rise north of Mount Ararat between the Black Sea and Caspian Sea. The grandeur and beauty of this mountain range and surrounding country are captured in a journalist's words:

> "*To glimpse the landscape of the . . . Caucasus . . . is to imagine Eden. Beneath the icy summits of its mountain range, grapevines and pomegranate trees hang [heavy] with fruit.*"
>
> Mike Edwards, "The Fractured Caucasus," *National Geographic*, February 1996

West of the Tian Shan range, the Turan Lowland provides some irrigated farmland. To the south, dune-covered *kums* (KOOMZ), or deserts, offer a stark contrast to the cultivated fields of the lowland. The Garagum (Kara Kum), or black sand desert, covers most of Turkmenistan. The Qizilqum (Kyzyl Kum), or red sand desert, blankets half of Uzbekistan. Farther west, the Ustyurt Plateau has salt marshes, sinkholes, and caverns.

Earthquakes

The African, Arabian, and Eurasian plates come together in the lands of North Africa, Southwest Asia, and Central Asia. As the plates move, they build mountains, shift landmasses, and cause earthquakes. Tectonic movement built the Zagros Mountains of southern Iran and the Taurus Mountains of Turkey. The movement continues to shape the region. For example, the shifting of the African and Arabian plates causes the widening of the Red Sea.

Earthquakes rumble throughout the region regularly. Turkey, lying at the boundary of the Arabian and Eurasian plates, experienced a 1999 earthquake measuring 7.4 on the Richter scale. It toppled more than 76,000 buildings and killed nearly 20,000 people.

NATIONAL GEOGRAPHIC **World Explorer**

Geography Skills for Life

Nature's Wrath Survivors survey the destruction caused by an earthquake in Turkey.

Human-Environment Interaction What factor accounts for frequent earthquakes in this region?

▲ *Working in the oil fields of Azerbaijan*

Natural Resources

The lands of North Africa, Southwest Asia, and Central Asia contain many natural resources. Petroleum and natural gas, the region's most abundant resources, are important to the economies of countries around the world.

Economics

Oil and Natural Gas

Sixty-six percent of the world's known oil reserves and 33 percent of the world's known natural gas reserves lie beneath the region. Unmeasured reserves include newly discovered gas fields in the Gaza Strip and Egypt and under the Caspian Sea.

Although North Africa, Southwest Asia, and Central Asia produced little oil before World War II, production increased dramatically after 1945. Petroleum exports have enriched the region, but heavy reliance on petroleum exports is risky. When oil prices fluctuate on world markets, as they did between 1997 and 1999, the region's economies suffer. By the time oil prices rose from a low of $7 per barrel to about $30 per barrel in early 2000, oil-exporting countries' economies had been damaged.

Minerals

Minerals also provide revenue for the region. Turkmenistan has the world's largest deposits of sulfate used in paperboard, glass, and detergents, and the largest deposits of sulfur. Morocco ranks second in the production of **phosphate**—a chemical used in fertilizers. Deposits of chromium, gold, lead, manganese, and zinc are sprinkled across the region. Discoveries of iron ore and copper deposits indicate that the region may contain up to 10 percent of the world's iron ore reserves.

Building Diverse Economies

Some countries in the region are diversifying their economies to decrease their reliance on oil and minerals exports. The United Arab Emirates, for example, is investing oil earnings in banking, information technology, and tourism. Libya, which relies on oil for 98 percent of its export income, is investing in infrastructure, agriculture, and fisheries.

SECTION 1 ASSESSMENT

Checking for Understanding

1. **Define** alluvial soil, wadi, *kum*, phosphate.
2. **Main Ideas** Complete the table by listing physical features found in this region. Then describe how the physical features of one part of the region influence people's lives.

Region	Physical Features
North Africa	

Critical Thinking

3. **Comparing and Contrasting** How are the Caspian Sea and the Aral Sea alike? How are they different?
4. **Predicting Consequences** How might development of oil fields in the Caspian Sea affect the region of North Africa, Southwest Asia, and Central Asia?
5. **Analyzing Information** How has diversification affected the economies of countries in the region?

Analyzing Maps

6. **Place** Study the physical-political map on page 423. What physical feature dominates western Iran?

Applying Geography

7. **Benefits of Rivers** Write a descriptive paragraph explaining how the major rivers of North Africa, Southwest Asia, and Central Asia benefit people in the region.

SECTION 2

Climate and Vegetation

Guide to Reading

Consider What You Know

In much of North Africa, Southwest Asia, and Central Asia, rainfall averages 10 inches (25 cm) or less annually. How does lack of precipitation affect the growth of vegetation in this region?

Read to Find Out

- How do the climates of North Africa, Southwest Asia, and Central Asia differ?
- How have the needs of a growing population affected the natural vegetation of the region?

Terms to Know

- oasis
- pastoralism
- cereal

Places to Locate

- Sahara
- Rub' al Khali
- Garagum (Kara Kum)

NATIONAL GEOGRAPHIC

A Geographic View

Algeria's Desert Art

From the mouth of this cave Algeria stretches dry and desolate before me, but the paintings inside . . . tell of a time, perhaps 7,000 years ago, when this land was wet and green enough to support cattle and a community of herders. Today our only evidence of this rich life is an ancient artist's rendering of it. . . . Amazingly, even after thousands of years the colors are still vibrant.

—*David Coulson, "Ancient Art of the Sahara,"* National Geographic, *June 1999*

Desert scene, Algeria

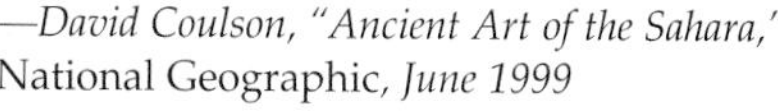

The North African landscape is commonly associated with images of vast stretches of sand, huge dunes, and the occasional watering hole. However, as David Coulson suggests, ancient cave paintings tell us that this part of the African continent was once wet and green. This section explores how differences and changes in climate across the region affect vegetation and human activities in North Africa, Southwest Asia, and Central Asia today.

Water: A Precious Resource

Water scarcity defines the region's climates. Rainfall in some areas is plentiful. The southern edge of the Caspian Sea receives more than 78 inches (198 cm) of rainfall per year. Elsewhere, however, water evaporation rates far exceed rainfall, making water very precious. Desert predominates, although steppe, Mediterranean, and highlands climates are also present in North Africa, Southwest Asia, and Central Asia.

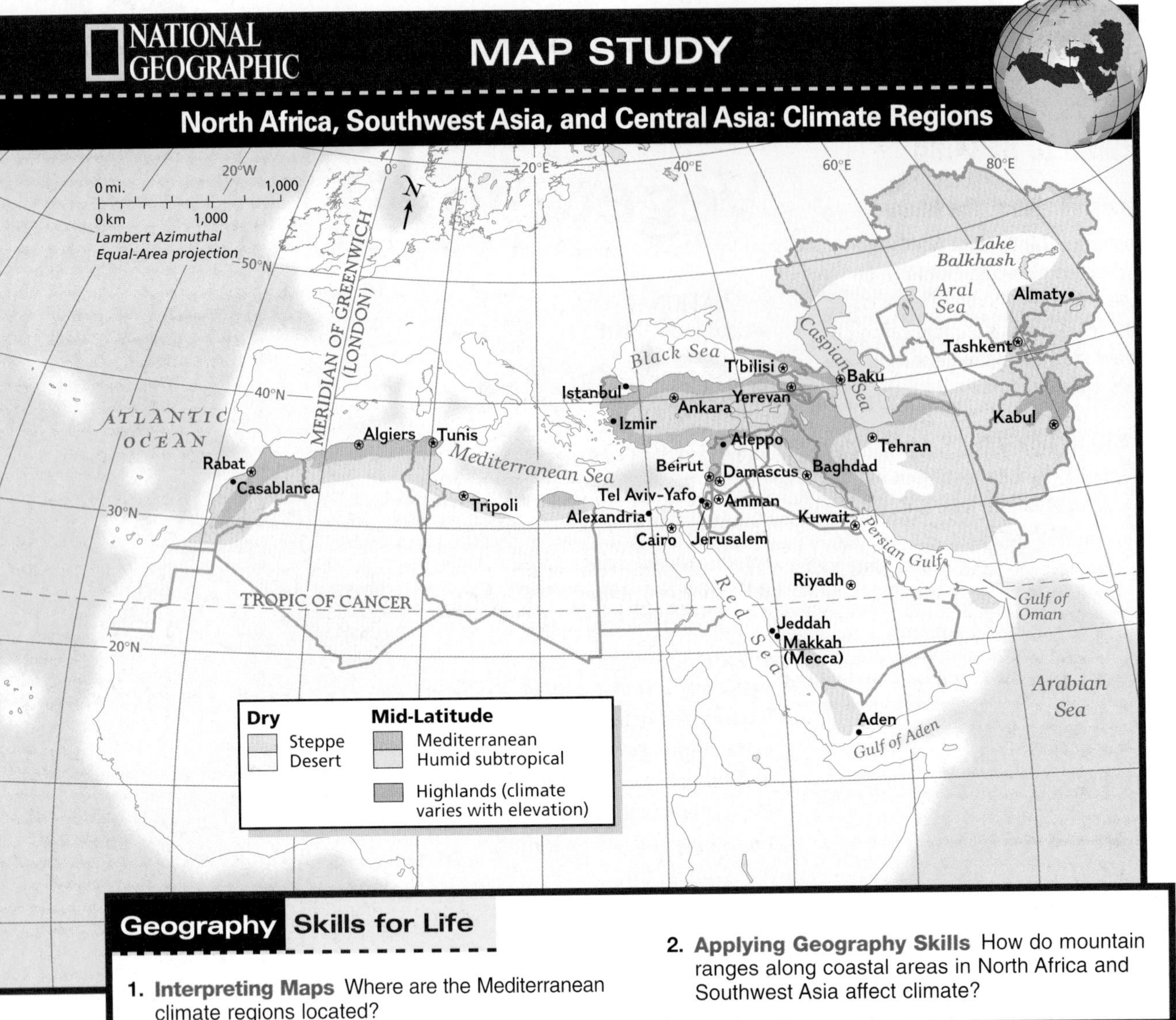

Geography Skills for Life

1. **Interpreting Maps** Where are the Mediterranean climate regions located?

2. **Applying Geography Skills** How do mountain ranges along coastal areas in North Africa and Southwest Asia affect climate?

Find NGS online map resources @ www.nationalgeographic.com/maps

Desert Climate

In prehistoric times a grassy plain extended across North Africa, and the climate was moderate. Today the climate in the area is hot and dry. The **Sahara**, the largest desert in the world at about 3.5 million square miles (about 9.1 million sq. km), covers most of North Africa. How much of the entire region is desert? Scientists define a desert climate as one in which precipitation averages 10 inches (25 cm) or less per year. By that definition deserts encompass almost 50 percent of the lands in North Africa, Southwest Asia, and Central Asia. In recent decades, droughts have expanded the Sahara.

Weather patterns in the desert tend to be extreme. The deserts of Central Asia and northern parts of the Sahara and the Arabian Desert have relatively cold winters with freezing temperatures. Winters in the southern Sahara and the Arabian Desert are generally milder. Summers in all these desert regions are long and hot. In July, daytime temperatures in the Central Asian deserts sometimes exceed 120°F (49°C) in the shade. At night, however, temperatures drop significantly because of the air's lack of moisture.

A traveler crossing any of the region's deserts would probably see only a few *ergs*, or sandy, dune-covered areas. *Regs*, stony plains covered with

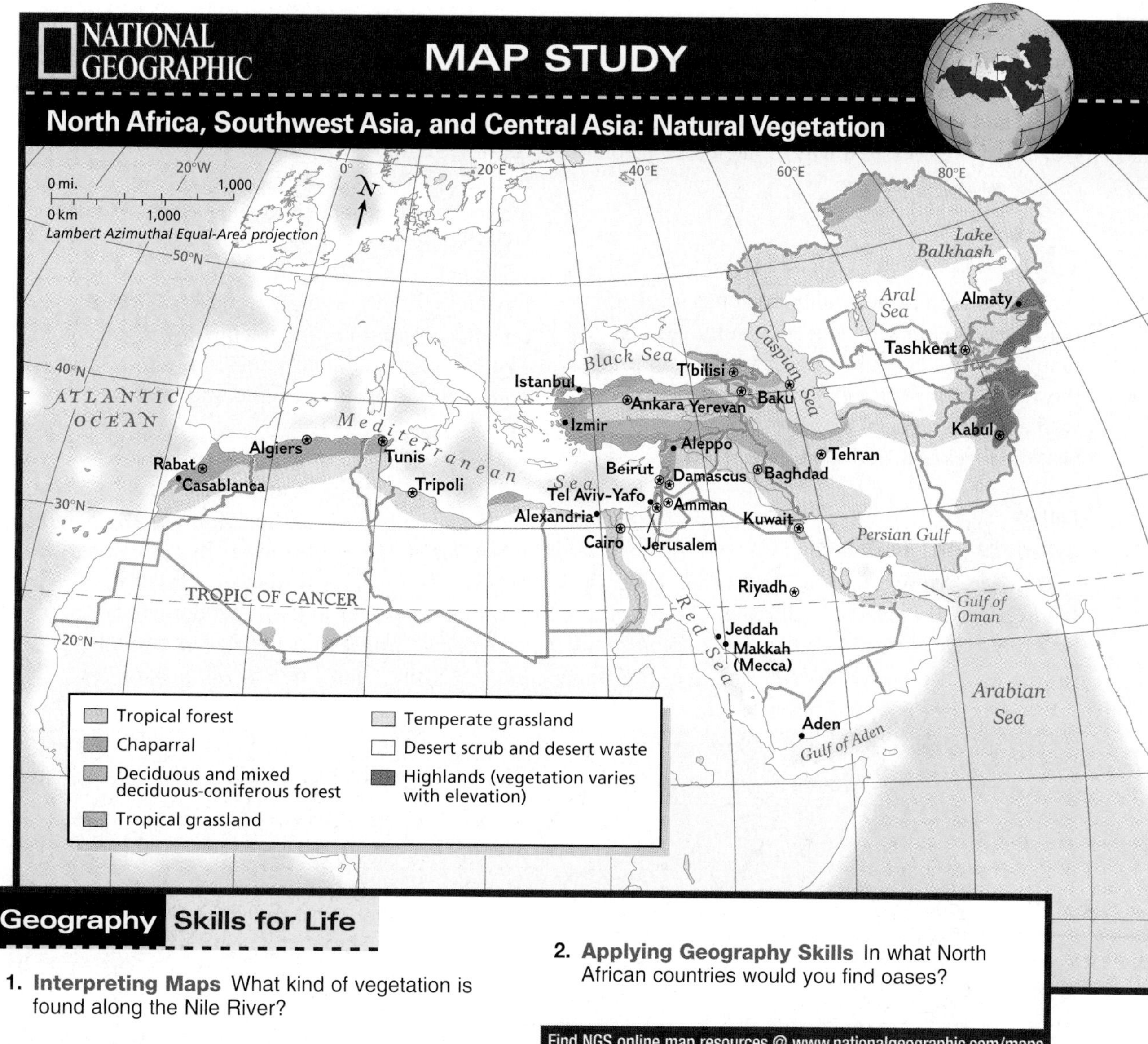

Geography Skills for Life

1. **Interpreting Maps** What kind of vegetation is found along the Nile River?

2. **Applying Geography Skills** In what North African countries would you find oases?

rocky gravel called "desert pavement," and an occasional *hamada*, or flat, sandstone plateau, would be more common. Sand covers less than 10 percent of the Sahara; desert pavement, mountains, and barren rock cover the rest.

The 250,000-square-mile (647,500-sq.-km) **Rub' al Khali**, or Empty Quarter, has the largest area of sand in the region. One of several deserts on the Arabian Peninsula, the Rub' al Khali covers almost the entire southern quarter of the peninsula.

Despite their arid conditions, the Sahara and other deserts in the region support vegetation such as cacti and drought-resistant shrubs. Nomadic herds of sheep, goats, and camels graze on brush in Central Asia's **Garagum (Kara Kum)**. Small-scale farming is possible in an **oasis**, a place in the desert where underground water surfaces. Villages, towns, and cities have risen around many Saharan oases.

Steppe Climate

Steppe is the second-largest climate region in the lands of North Africa, Southwest Asia, and Central Asia. The steppe borders the Sahara to the north and snakes between other climate regions from Turkey to eastern Kazakhstan. Precipitation in this semi-arid climate region usually averages less than 14 inches (36 cm) annually. This amount is enough to support short grasses in the steppe climate, providing pasture for sheep, goats, and camels, as well as shrubs and some trees. **Pastoralism**, the raising and grazing of livestock, is a way of life for the steppe's people, such as bedouins.

Climatic Variations

In the Mediterranean climate zones, cool, rainy winters alternate with hot, dry summers. As the map on page 428 shows, this climate is common in the Tigris-Euphrates valley and in uplands areas as well as on the coastal plains of the Mediterranean Sea, the Black Sea, and the Caspian Sea.

Culture

Exports and Tourists

Morocco, Tunisia, Syria, and other countries having Mediterranean climates boost their economies by exporting citrus fruits, olives, and grapes to Europe and North America. Some of these Mediterranean countries also benefit from tourism, as people from colder climates seek the sun and warmth. Morocco's city of Agadir, with 360 days of sunshine per year, attracts many of the country's 2 million tourists, who come mainly from Europe. Travelers in Morocco also visit the cultural attractions of ancient cities such as Fès, Marrakech, and Casablanca.

Higher areas, like the Caucasus Mountains, have a highlands climate, which is generally wetter and colder than other climates in the region. The highlands climate varies, however, with elevation and exposure to wind and sun.

Rainfall

Coastal and highlands areas near mountain ranges usually receive the most rainfall, as moist, warm air is driven off the sea by prevailing westerly winds. The North African coast near the Atlas Mountains, for example, averages more than 30 inches (76 cm) of rain each year, enough rain to support flourishing forests. More than twice that

NATIONAL GEOGRAPHIC **World Explorer**

Geography Skills for Life

Mediterranean Climate The bushes and short trees of the landscape around this Moroccan village typify the Mediterranean climate region.

Region What type of natural vegetation is found in Mediterranean climate regions?

amount falls each year at the foot of the Elburz Mountains. Batumi, in the Republic of Georgia, one of the region's wettest places, receives more than 100 inches (254 cm) of rain a year. In areas where more than 14 inches (36 cm) of rain falls yearly, farmers can raise **cereals**—food grains such as barley, oats, and wheat—without irrigation.

A Sign of Things to Come?

Landscapes can change with variations in climate and with people's activities. Under the pressure of climate changes, grassy plains in the region turned into desert, as explorer Thor Heyerdahl observed:

> *"The desert, encroaching upon the spring-green marshes from all sides, has swallowed up the former Sumerian homeland [in Mesopotamia] and all that it contained. . . . The landscape which once throbbed with life is today as silent and lifeless as the North Pole."*
>
> Thor Heyerdahl, *The Tigris Expedition*, 1981

Will other fertile lands give way to the desert as the grasslands of North Africa and Mesopotamia did? Will pollution threaten other bodies of water as it has the Aral and Caspian Seas? The answers depend on future world climate changes and the interactions of people with their environments.

Geography Skills for Life

Grape Harvest Grape vineyards, such as this one in Georgia, have been cultivated for food and wine for 8,000 years.

Human-Environment Interaction In what areas can farmers raise cereals without irrigation?

SECTION 2 ASSESSMENT

Checking for Understanding

1. **Define** oasis, pastoralism, cereal.
2. **Main Ideas** On a web diagram, fill in the climate regions found in North Africa, Southwest Asia, and Central Asia. Then describe the characteristics of one region.

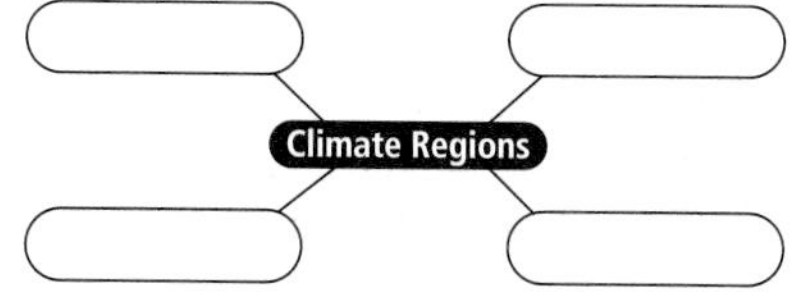

Critical Thinking

3. **Comparing and Contrasting** Compare and contrast agriculture in steppe climate regions with that of Mediterranean climate regions.
4. **Analyzing Cause and Effect** Why has natural vegetation declined in areas of North Africa, Southwest Asia, and Central Asia?
5. **Drawing Conclusions** How did climate changes in the Sahara centuries ago affect its people?

Analyzing Maps

6. **Region** Study the map of climate regions on page 428 and the map of natural vegetation on page 429. What kind of natural vegetation thrives in Mediterranean climates?

Applying Geography

7. **Climate and Population** Write a paragraph explaining the possible effects of climate on settlement patterns in North Africa, Southwest Asia, and Central Asia.

Reading a Vegetation Map

Geographers call the plant life that grows naturally in an area natural vegetation. Variations in vegetation can make areas of the same country look very different.

Learning the Skill

Climate greatly affects natural vegetation. For example, thick layers of plants that make up tropical forest vegetation grow only in tropical rain forest climates. Likewise, areas with less than 10 inches (25 cm) of rain support only desert scrub vegetation.

Elevation also affects vegetation. Forests grow at the bases of mountains. At higher elevations, grasses, small trees, and shrubs grow. Where elevation makes it too cold for trees and shrubs, only mosses thrive.

On a vegetation map, colors indicate different vegetation types. The map key explains the color code. To read a vegetation map:

- **Identify the area covered on the map.**
- **Study the key to identify the vegetation types that the map depicts.**
- **Locate the regions covered by each vegetation type.**
- **Draw conclusions about the similarities and differences between the types of vegetation found in different areas of the map.**

Practicing the Skill

Use the map showing the vegetation of Central Asia to answer the following questions.

1. What geographic area does this map show?
2. In which vegetation region is the capital of Kyrgyzstan located?
3. What kinds of vegetation are found along the coast of the Caspian Sea?
4. What factors would explain the distribution of vegetation throughout the region?
5. Of the areas shown on the vegetation map, where do you think irrigation is used for cultivating crops?

Applying the Skill

Look at the vegetation map on page 429. Compare the vegetation types of North Africa, Southwest Asia, and Central Asia with those of the United States and Canada, found on page 123. How are they similar? Different?

The **Glencoe Skillbuilder Interactive Workbook, Level 2** provides instruction and practice in key social studies skills.

CHAPTER 17

SUMMARY & STUDY GUIDE

SECTION 1 The Land (pp. 421–426)

Terms to Know

- alluvial soil
- wadi
- *kum*
- phosphate

Key Points

- North Africa, Southwest Asia, and Central Asia are located at the crossroads of Asia, Africa, and Europe.
- The region is a jigsaw puzzle of peninsulas and seas.
- Rivers feed the inland seas and supply irrigation to parched lands. Their alluvial soil deposits enrich the land, especially in the Nile River Valley and delta.
- The movement of tectonic plates forms mountains, moves landforms, and causes earthquakes in the region.
- The region contains much of the world's oil and natural gas reserves.

Organizing Your Notes

Use a table like the one below to help you organize the notes for this section. Complete the table by listing and describing the location of the region's important physical features.

Feature	Location
Sahara	
Atlas Mountains	
Nile River	

SECTION 2 Climate and Vegetation (pp. 427–431)

Terms to Know

- oasis
- pastoralism
- cereal

Key Points

- Rainfall in North Africa, Southwest Asia, and Central Asia varies widely. Most of the region contains arid areas.
- The four climate regions in North Africa, Southwest Asia, and Central Asia are desert, steppe, Mediterranean, and highlands.
- Natural vegetation in the region varies widely and is closely related to rainfall and irrigation patterns.

Organizing Your Notes

Create an outline using the format below to help you organize your notes for this section.

Climate and Vegetation

I. Water: A Precious Resource
  A. Desert Climate
    1. Sahara
    2.
  B. Steppe Climate

Mosque in Afghanistan

ASSESSMENT & ACTIVITIES

Reviewing Key Terms

Write the key term that best completes each of the following sentences. Refer to the Terms to Know in the Summary & Study Guide on page 433.

1. In the Sahara, a place where underground water surfaces is a(n) ________.
2. Runoff from infrequent rainstorms creates ________, or dry streambeds.
3. ________, or the raising and grazing of livestock, is a way of life on the steppe.
4. Morocco produces ________, which is used in fertilizers.
5. Much of the region is covered by sandy deserts, or ________.
6. Barley is an example of a ________ grain.
7. ________ is rich soil deposited by running water.

Reviewing Facts

SECTION 1

1. What physical features separate the Arabian Peninsula from the African continent?
2. What physical features separate Europe and Asia and connect the Aegean and Black Seas?
3. What desert covers most of Turkmenistan? What desert covers about half of Uzbekistan?

SECTION 2

4. About how much of North Africa, Southwest Asia, and Central Asia experience desert climate?
5. Describe the natural vegetation of steppe areas.
6. In what part of the region does tropical vegetation flourish? What climate factors allow this kind of vegetation to grow in that area?

Critical Thinking

1. **Drawing Conclusions** How do you think the region's resources affect the global economy?
2. **Analyzing Information** Compare the climate map on page 428 with the population density map on page 412. How does climate influence where people live in the region?
3. **Identifying Cause and Effect** On a sheet of paper, complete a chart like the one below to show how increased irrigation affected the region's inland seas.

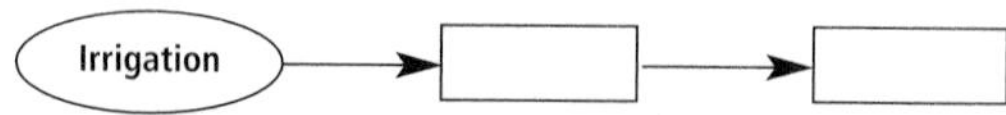

Locating Places

NATIONAL GEOGRAPHIC

North Africa, Southwest Asia, and Central Asia: Physical Geography

Match the letters on the map with the physical features of North Africa, Southwest Asia, and Central Asia. Write your answers on a sheet of paper.

1. **Arabian Peninsula**	5. **Aral Sea**	9. **Caspian Sea**
2. **Sahara**	6. **Red Sea**	10. **Black Sea**
3. **Atlas Mountains**	7. **Persian Gulf**	11. **Gulf of Aden**
4. **Nile River**	8. **Mediterranean Sea**	12. **Tian Shan**

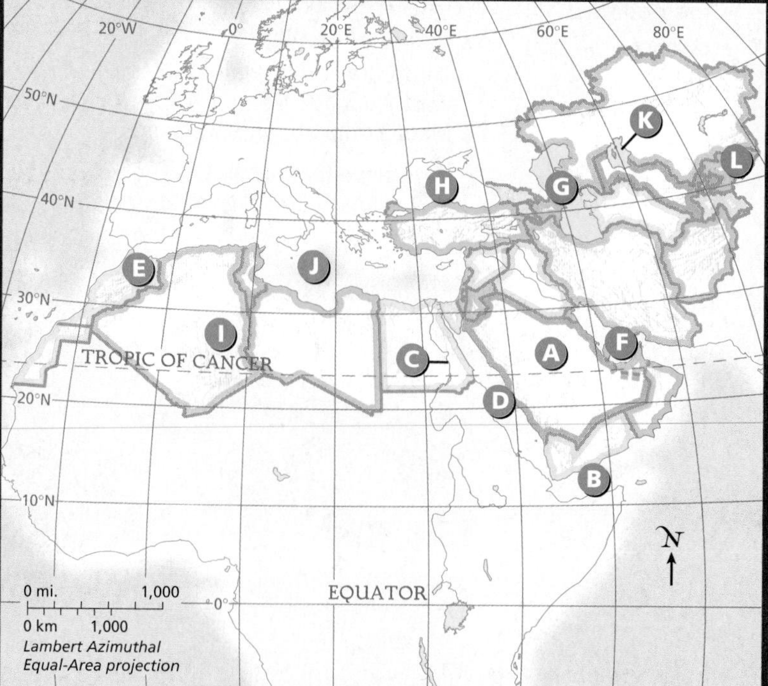

Self-Check Quiz Visit the **Glencoe World Geography** Web site at geography.glencoe.com and click on Self-Check Quizzes—Chapter 17 to prepare for the Chapter Test.

Using the Regional Atlas

Refer to the Regional Atlas on pages 410–413.

1. **Region** In which area of the region is livestock raising practiced? Subsistence farming?
2. **Place** Compare the physical map on page 410 with the population density map on page 412. What do the gray areas on the population map represent? How does the physical map help explain the distribution of the population in these areas?

Thinking Like a Geographer

Think about the areas in North Africa, Southwest Asia, and Central Asia that do not have enough freshwater. As a geographer, where would you recommend desalination plants to be built? Consider population centers, energy needs, and water sources.

Problem-Solving Activity

Group Research Project As a group, choose an oil-producing country from this region and investigate possible ways the country could diversify its economy. Present your research in a written report that gives reasons for your recommendations. Be sure to include photos, maps, charts, or graphs to help illustrate your findings.

GeoJournal

Descriptive Writing Select three physical features in North Africa, Southwest Asia, or Central Asia. Then, using your GeoJournal data, describe and analyze in writing how these physical features shape the distribution of culture groups in the region.

Technology Activity

Using the Internet for Research Use the Internet to research the natural resources of one of the countries in this region. Identify factors affecting the location of the economic activities there. Create a bulletin board display about the country, including a list of its primary imports and exports.

Standardized Test Practice

Choose the best answer for the following multiple-choice questions. If you have trouble answering the questions, use the process of elimination to narrow your choices.

1. **Part of Uzbekistan has a desert climate. What kind of vegetation can grow in a desert climate?**

   A No vegetation at all
   B Drought-resistant shrubs and cacti
   C Drought-resistant shrubs, cacti, and occasional small-scale farm crops in areas with underground water
   D Short grasses for grazing

Note that the directions ask you to choose the best answer to the question. The best answer will contain the most precise information for answering the question.

2. **In part of the region of North Africa, Southwest Asia, and Central Asia, people earn their living by growing citrus fruits, olives, and grapes, as well as from the tourist trade. This region probably has a**

   F highlands climate.
   G steppe climate.
   H Mediterranean climate.
   J desert climate.

Think about the conditions needed to grow the specific crops. Desert climates are too dry, as are steppe climates. Highlands climates are wet but may be too cold. Eliminating wrong choices helps you choose the correct answer.

Geography Lab Activity

Desalination

▲ *Students distill salt water.*

In spite of the location of large river systems in North Africa, Southwest Asia, and Central Asia, most of the usable water comes from regional river basins such as the Jordan and the Nile and from aquifers. Aquifers are underground layers of porous rock, gravel, or sand that contain water. Although abundant, seawater is not usable because of its salt content. Countries in the region are searching for new sources of water as well as increasing their use of desalination—the removal of salt from seawater. These countries produce about 75 percent of the world's desalinated water. Worldwide, more than 2 billion gallons (7.5 billion liters) of freshwater were produced daily at desalination plants at the end of the twentieth century.

Distillation is the most widely used desalination method. The process of distillation purifies water by imitating the way ocean water evaporates into clouds, condenses, and falls back to Earth as precipitation. The distillation process varies little whether producing one cup or millions of gallons of freshwater. Salt water is heated until the water evaporates. The vapor condenses into freshwater in a second container, while the salt remains in the first container.

1 Materials

- Table salt
- Water
- 1 flask
- Rubber stopper
- Plastic tubing
- Rubber tubing
- Scissors
- Cardboard
- Metal washers (for weight)
- Beaker
- Ice
- Shallow pan
- Hot plate
- Measuring cup
- Thermal mitt

CAUTION: Be careful when using the hot plate. It should be cool before the flask is moved.

2 Procedures

In this activity, you will distill salt water to make drinking water.

1. In the flask, dissolve 2 teaspoons (10 ml) of salt in 1 cup (237 ml) of water. Swish the salt and water mixture around until no salt crystals remain.
2. Insert the plastic tubing into the rubber stopper, and then insert the stopper into the flask. Be sure the plastic tubing is above the surface of the saltwater solution.

3. Attach one end of the rubber tubing to the plastic tubing. Insert the other end of the rubber tubing through a small hole cut in the cardboard. The hole in the cardboard should be small enough that the tubing fits snugly.
4. Place the cardboard over the beaker. Add several washers to the cardboard to hold it in place.
5. Place the beaker in the shallow pan filled with ice water to speed up the condensation process.
6. Set the flask on the hot plate. Bring the saltwater solution to a boil, and continue boiling until the solution is almost boiled away. You will notice a salt residue forming as the boiling water evaporates.
7. Turn off the hot plate. After letting it cool, remove the flask.
8. Pour the water you collected in the beaker into a measuring cup.
9. Taste the water in the measuring cup. Does the water still taste salty?

3 Lab Report

1. What happened to the water in the flask as you boiled the solution?
2. What happened inside the beaker?
3. Why did the water, and not the salt, move from the flask to the beaker?
4. **Drawing Conclusions** How could this process be used to extract minerals from seawater?
5. **Predicting Consequences** Based on your observations, what do you think might be the biggest drawback to using this process?

4 Find Out More

Research where desalination is used in the United States. What other places in the country would benefit from desalination plants? Create a map showing existing plants and areas where you would propose building new plants.

Did You Know? Today's desalination plants produce 15 times as much freshwater as they did 20 years ago. Saudi Arabia, a world leader in desalination projects, relies on about 30 desalination plants to change seawater to freshwater. One plant turns out 250 million gallons (950 million liters) of freshwater daily for human use!

As stagnant water evaporated, it left behind a crust of salt in this field in southern Iraq.

CHAPTER 18

The Cultural Geography of North Africa, Southwest Asia, and Central Asia

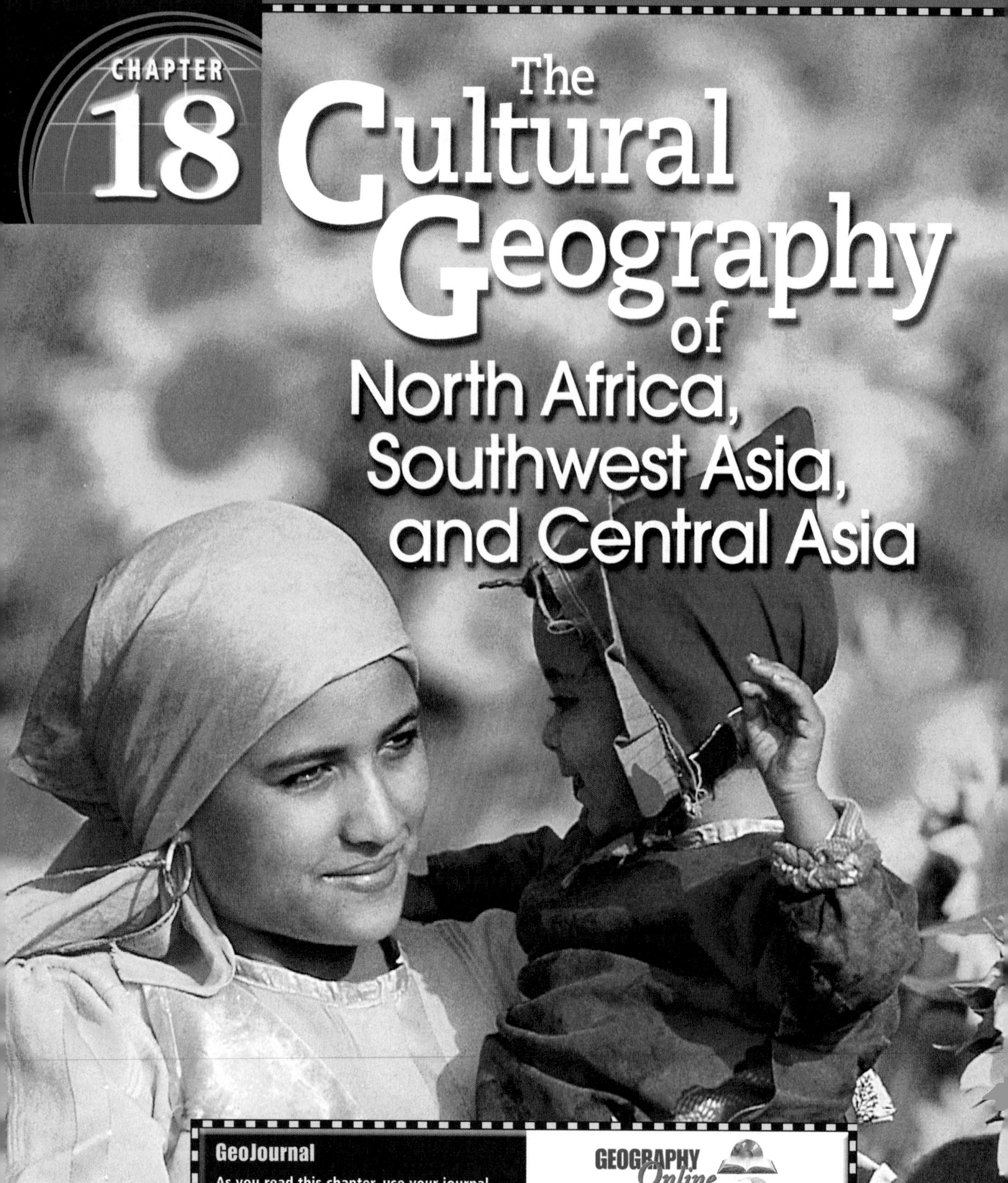

GeoJournal

As you read this chapter, use your journal to note the ethnic diversity of North Africa, Southwest Asia, and Central Asia. Record both similarities and differences among the peoples who inhabit this region.

GEOGRAPHY Online

Chapter Overview Visit the **Glencoe World Geography** Web site at geography.glencoe.com and click on Chapter Overviews—Chapter 18 to preview information about the cultural geography of the region.

Population Patterns

Guide to Reading

Consider What You Know

As you know, the region of North Africa, Southwest Asia, and Central Asia is made up of a variety of physical features and climates. Many different peoples live in the region. How does the diversity of ethnic groups affect life in North Africa, Southwest Asia, and Central Asia?

Read to Find Out

- How have movement and interaction of people in the region led to ethnic diversity?
- How do the region's seas, rivers, and oases influence where people live?
- What effect does the growing migration into the cities have on the region?

Terms to Know

- ethnic diversity
- infrastructure

Places to Locate

- Turkey
- Afghanistan
- Armenia
- Georgia
- Kazakhstan
- Tajikistan
- Uzbekistan
- Tehran

◀ *El Faiyum oasis, Egypt*

NATIONAL GEOGRAPHIC

A Geographic View

Refuge of Peoples

A refuge since the last period of Eurasian glaciation, the Caucasus region has been a gateway for travel, trade, and conquest. [Despite the numerous power struggles] the Caucasus has remained a [stronghold] of peoples whose identities are tied to the 50-some languages they speak. . . . The persistence of the enduring identities of ethnic groups has been aided by the rugged terrain and by societies whose loyalties are to clan and family as much as to nation or region.

—*Mike Edwards, "The Fractured Caucasus,"* National Geographic, *February 1996*

Family in the Caucasus

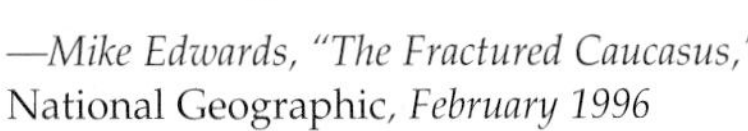

Like North Africa, Southwest Asia, and Central Asia as a whole, the Caucasus area has long been home to many peoples. Some of these peoples vanished long ago—defeated in wars, wiped out by famines, or absorbed by more powerful groups. Others have survived for hundreds of years and flourish today because of contact with travelers, merchants, and conquerors from distant places. The result is a tapestry as rich and varied as the region's much-sought-after carpets.

Many Peoples

The region of North Africa, Southwest Asia, and Central Asia has served as the crossroads for Asia, Africa, and Europe. As a result, the region has remarkable **ethnic diversity**, or differences among groups based on their languages, customs, and beliefs.

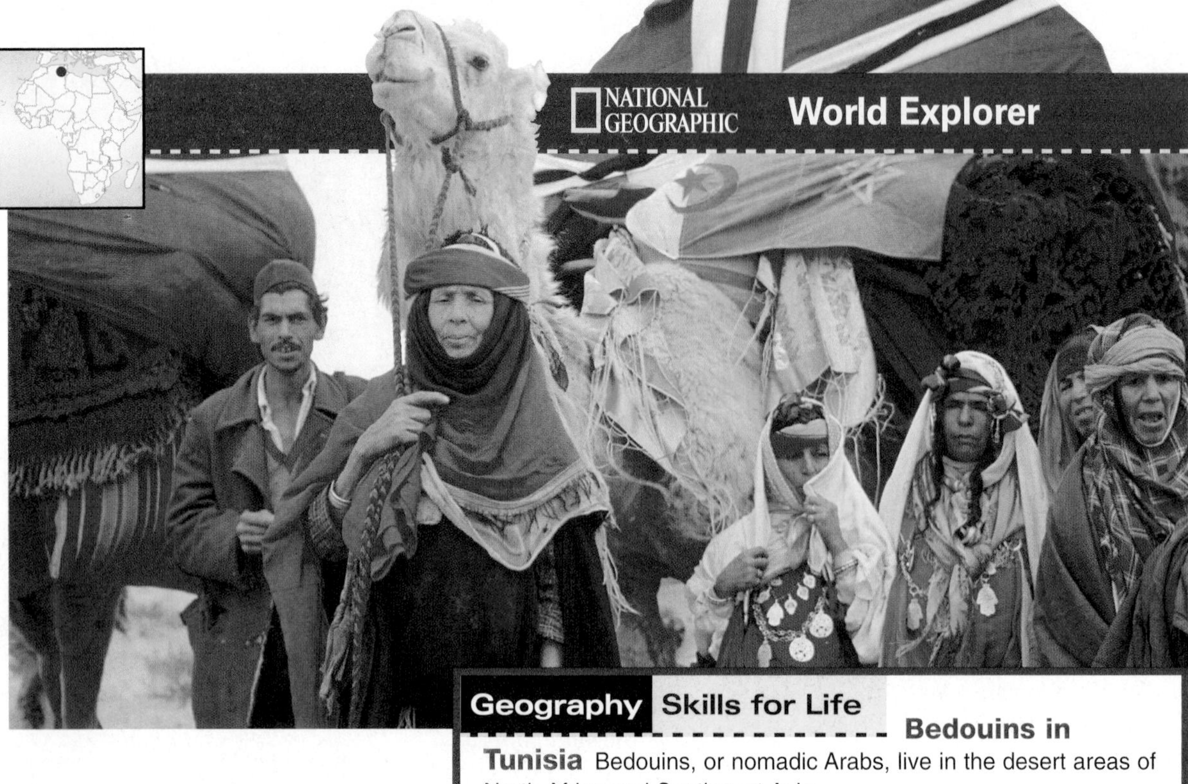

Geography Skills for Life

Bedouins in Tunisia Bedouins, or nomadic Arabs, live in the desert areas of North Africa and Southwest Asia.

Place Which countries are part of the Maghreb?

Arabs

Most people of the region—about 275 million—are Arabs. Most Arabs are Muslims, followers of the religion of Islam, but a small percentage follow Christianity or other religions. Both Islamic culture and Arabic, the language of the Arabs, have had a significant impact in this region.

Before the spread of Islam in the A.D. 600s, Arabic-speaking peoples inhabited the Arabian Peninsula and a few areas to its north. Many Arabic-speaking people today, however, descend from ancient groups such as the Egyptians, Phoenicians, Saharan Berbers, and peoples speaking Semitic languages. Currently, Arabs live in 16 countries, including Libya, Tunisia, Algeria, and Morocco—the countries known as the Maghreb—"the West" in Arabic.

Israelis

About 6.4 million people of the region are Israelis living in Israel. Of these, 82 percent are Jewish. The remaining 18 percent are mostly Arabs who are Muslim or Christian.

Jews living in Israel and elsewhere trace their religious heritage to the Israelites, who in ancient times settled Canaan, the land shared today by Israel and Lebanon. The Israelites believed that God had given them this area as a permanent homeland. Over the centuries, wars, persecution, and trade led many Jews—as the descendants of the Israelites are called—to settle in other countries. Their religious identity, however, kept alive their link to the ancestral homeland. Finally, in 1948, Israel was founded as a Jewish state. Today half of Israel's Jews were born in Israel, and half have emigrated from elsewhere.

The Arabs of the region, however, did not want a Jewish state in territory that had been their homeland for centuries. Tensions between Arabs and Jews resulted in four wars that brought severe hardship to all the people of the area, including the Palestinians—Arabs living in the territory in which Israel was established. During this period of conflict, many Palestinians were displaced from their homes and lived in refugee settlements in neighboring Arab countries.

Today agreements between Israeli and Palestinian leaders have led to greater Palestinian self-rule. Nevertheless, peace is still elusive. Issues such as the ownership of the Old City of Jerusalem, the return of Palestinian refugees, and ownership of water and other natural resources remain unresolved.

Turks

Over the past 8,000 years, many peoples have occupied Anatolia, the Asian part of what is today the country of **Turkey**. Each group added its own customs and beliefs to the cultural blend. Turkic peoples migrated to the peninsula in the A.D. 1000s from Central Asia. One Turkic group, known as the Ottoman Turks, later built the Ottoman Empire, which ruled much of the eastern Mediterranean world for more than 600 years. When a group of Turkish citizens was asked to define who a Turk is today, one of them responded this way:

> *"'I don't believe anybody is Turkish, whatever that means,' he said. Then, swinging his arms to take in the lunch crowd, he exclaimed, 'Look at us! A mix of Turks, Arabs, Jews, Greeks, Iranians, Armenians, Kurds.'"*
>
> Thomas B. Allen, "Turkey Struggles for Balance," *National Geographic*, May 1994

Most Turks practice Islam and speak the Turkish language. They have a culture that blends Turkish, Islamic, and Western elements.

Iranians and Afghanis

About 66 million people live in Iran, once called Persia. The word *Iran* means "land of the Aryans." Many Iranians believe they are descendants of the Aryans (AR•ee•uhnz), Indo-Europeans who migrated into the region from southern Russia about 1000 B.C. Iranians speak Farsi, and almost 90 percent of them are Shiite (SHEE•EYET) Muslims.

On the eastern border of Iran is **Afghanistan**. This mountainous country is home to many ethnic groups that reflect centuries of migrations and invasions by different peoples. People in Afghanistan speak many languages, and most practice Islam.

Caucasian Peoples

More than 50 ethnic groups and nationalities live in the Caucasus area. Armenians and Georgians are among the largest ethnic groups.

Armenians make up more than 90 percent of the population of the republic of **Armenia**, which became independent after the Soviet Union dissolved in 1991. The Armenians have had their own language and literature for more than 15 centuries, and in the A.D. 300s most accepted Christianity.

In ancient times the Armenians ruled a large, powerful kingdom. For much of their later history, however, the Armenians were ruled by others—Arabs, Persians, Turks, and Russians. In 1915 about 1 million Armenians in Turkey were massacred, were deported, or died of illness at the hands of the Ottoman Turks. Many survivors fled to Southwest Asia, Europe, and the United States.

The republic of **Georgia** also became independent after the fall of the Soviet Union in 1991. Like the Armenians, most Georgians became Christian in the A.D. 300s. Today they have their own Orthodox Christian Church. The Georgian language, with its unique alphabet, is related to other Caucasian languages, which suggests that the Georgians probably originated in the Caucasus region.

Turkic Peoples

Most Turkic peoples outside of Turkey, including Uzbeks and Kazakhs, live in the republics of Central Asia. All of these peoples speak Turkic languages, and almost all are Muslims.

The Uzbeks form the largest Turkic group in the Central Asian republics. Of the Central Asian Turkic peoples, only the Kazakhs are a minority in their own country, **Kazakhstan**. Under Russian and, later, Soviet rule, Kazakhstan was settled by large numbers of Russians, Ukrainians, and Germans. Since the end of the Soviet era, the proportion of Kazakhs has increased for two reasons: a high birthrate and the movement of many non-Kazakhs out of Kazakhstan.

The Tajiks (tah•JIHKS), a predominantly Muslim non-Turkic group in the Central Asian republics, make up most of the population of **Tajikistan**. Tajiks also live in **Uzbekistan** and Afghanistan and speak a language similar to Farsi.

Student Web Activity Visit the **Glencoe World Geography** Web site at geography.glencoe.com and click on Student Web Activities—Chapter 18 for an activity about visiting Egypt's cultural and historic sites.

Kurds

The Kurds also speak a language related to Farsi, and most Kurds are Muslims. They live in the border areas of Turkey, Iraq, Iran, Syria, and the Caucasian republics, in an area that is sometimes called Kurdistan. However, the Kurds have no country of their own. Their efforts to win self-rule have been repeatedly crushed by their Turkish and Arab rulers.

NATIONAL GEOGRAPHIC **World Explorer**

Geography Skills for Life

Water and Population A canal supplies water to farms near Luxor, Egypt.

Place How does the availability of water affect human settlement in Egypt?

Population and Resources

Geographic factors, especially the availability of water, help determine where the region's people have settled. Because water is scarce, people have for centuries settled along seacoasts and rivers, near oases, or in rain-fed highlands where drinking water is readily available. For example, many people live along the Nile River in Egypt or in the Tigris-Euphrates Valley in Iraq. Desert areas remain largely unpopulated except where oil is abundant. Nomadic herders live in or near the desert oases or where there is enough vegetation to support their herds.

Government

Control of a Vital Resource

Water has been a major issue in border disputes between Israel and Syria. As much as 30 percent of Israel's water comes from the Sea of Galilee, which is partly fed by streams beginning in the Golan Heights, a Syrian area that Israel conquered in the 1967 Arab-Israeli War. The Jordan River carries the water south, where Israeli farmers use it to irrigate their crops. Some 15,000 Israelis live in the Golan Heights. The area also has about 17,000 Arabs. Syria wants Israel to return the Golan Heights, but Israel is reluctant to give up needed water resources.

Population Growth

The region's most populous countries are Turkey, Egypt, and Iran, each with more than 65 million people. Morocco, Uzbekistan, Algeria, Iraq, Saudi Arabia, and Afghanistan each have between 20 million and 31 million people. Other countries each have about 18 million or fewer people.

Overall, the region's population is growing rapidly. The result is that many citizens in some countries, especially those in North Africa, are unemployed and must migrate to other countries to find work. This migration serves as a safety valve for some countries, helping to diffuse political discontent.

Urbanization

Large urban areas, such as Istanbul, Turkey; Cairo, Egypt; Tehran, Iran; and Baghdad, Iraq, dominate social and cultural life in their respective countries. Cities like these have been growing rapidly as villagers move there in search of a better life. Problems have arisen, however, because cities have grown too fast to supply enough jobs and housing or improve the **infrastructure**—basic urban necessities like streets and utilities. Poverty, snarled traffic, and pollution have resulted. Families moving to a city sometimes crowd into single rooms or live in makeshift shelters far from the city's center, and they overload public resources. For example, illegal developments without water or waste services have cropped up on the outskirts of Cairo, adding to the city's sanitation problems.

Some cities have tried to cope by installing traffic control systems and improving public transportation. Iran has tried another solution—decentralizing its government. It has set up many government offices in various towns and villages away from the capital, **Tehran**. By doing so, Iran hopes to improve services in outlying areas and slow Tehran's rapid growth.

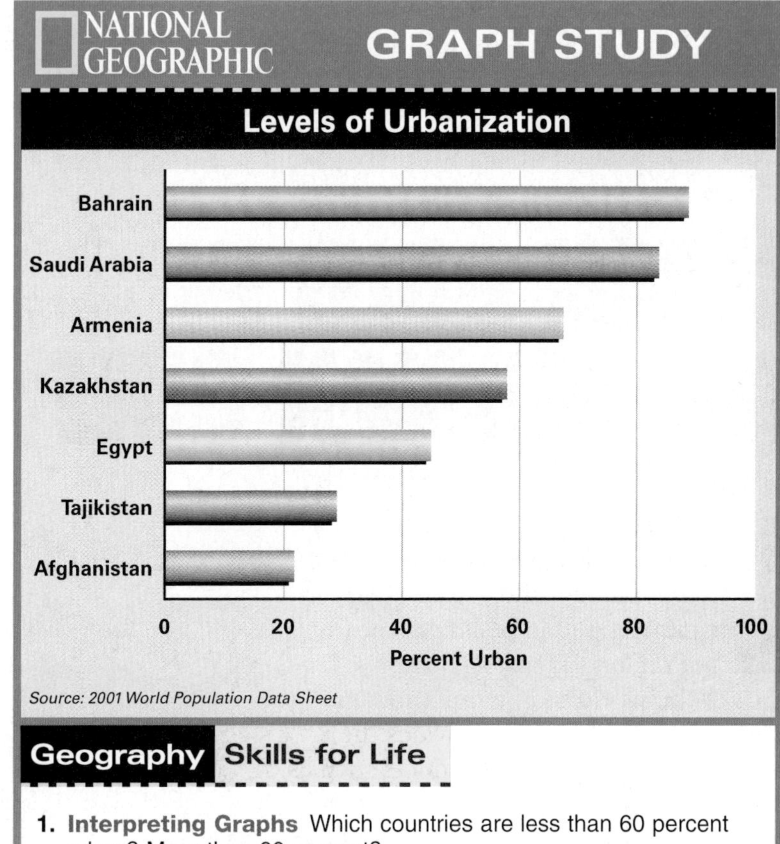

Geography Skills for Life

1. **Interpreting Graphs** Which countries are less than 60 percent urban? More than 60 percent?
2. **Applying Geography Skills** Why do you think some countries are more urbanized than others?

SECTION 1 ASSESSMENT

Checking for Understanding

1. **Define** ethnic diversity, infrastructure.
2. **Main Ideas** Create a table like the one below, and fill it in to show information about the diverse peoples, religions, and languages of this region.

	North Africa	Southwest Asia	Central Asia
Peoples			
Religions			
Languages			

Critical Thinking

3. **Comparing and Contrasting** In what ways is the population of Turkey similar to and different from the population of Iran? What may account for these differences and similarities?
4. **Identifying Cause and Effect** What historical event accounts for the large number of Armenians living outside their homeland?
5. **Predicting Consequences** What might happen if Israel returns the Golan Heights to Syria? How would this affect life in Israel? In Syria?

Analyzing Maps

6. **Location** Study the population density map on page 412. Where are the largest concentrations of people in the region? Why are they concentrated there?

Applying Geography

7. **Ethnic Diversity** Think about the diverse groups of people you have read about. Write a paragraph describing positive aspects of ethnic diversity in the region. Also mention any drawbacks to ethnic diversity.

GEOGRAPHY AND HISTORY

BLACK GOLD IN THE PERSIAN GULF

LIKE THE GENIE IN ALADDIN'S LAMP, oil has brought unimagined riches to the nations of the Persian Gulf. Trapped in pockets beneath the region's sandy soils are two-thirds of the world's known petroleum reserves. This "black gold" provides the raw material for everyday products such as compact discs, crayons, and house paint. In addition, oil supplies more than half of the energy used worldwide. Almost overnight, oil profits transformed villages in Saudi Arabia, Kuwait, Bahrain, and other Gulf countries from watering holes for camel caravans into gleaming, modern cities.

The discovery of oil in the early 1900s, however, did not immediately bring riches to the region. Nor did drilling wells to extract oil from the ground. The American and European companies who owned the wells paid host countries only about 20 cents a barrel, and the quantity of oil tapped was small.

Boom Times

Low oil prices in the late 1950s caused Western companies to cut payments to the oil-producing countries. In 1960 Venezuela joined with four Gulf states—Iran, Iraq, Kuwait, and Saudi Arabia—to form the

Massive pipelines carry tons of crude oil from wells in Saudi Arabia. ▶

◀ **Money from oil profits builds new schools for children in the Persian Gulf region.**

NATIONAL GEOGRAPHIC SOCIETY

Organization of Petroleum Exporting Countries (OPEC). The OPEC nations agreed to reduce oil production in an effort to cut supplies and increase prices. As demand grew, the group gradually assumed more power. They set their own prices for oil and mandated production quotas for each country. In 1973 the Arab oil embargo, sparked by the Arab-Israeli War, reduced supplies and further boosted prices. In less than a year, prices increased fourfold.

With money pouring in, Gulf countries took over ownership of their oil operations. Big budgets meant big spending. Billions were used to build highways, airports, and telecommunications systems. Hospitals and schools sprang up, and governments showered their citizens with free medical care, low-cost housing, and lifetime jobs.

Planning for Post-Oil Days

Beginning in the early 1980s, however, oil prices started to decline. Why? Reduced consumption and increased oil production outside the Middle East led to a surplus of oil. As oil profits shrank, collaboration among OPEC members began to break down. Quota disputes and other disagreements led Iraq to invade Kuwait in 1990, igniting the Persian Gulf War. Many Gulf countries have cut spending—an unpopular move among citizens accustomed to subsidies.

While OPEC members manipulate current oil prices, they also know they must prepare for the day their oil reserves will run out. Today Gulf countries are investing in foreign real estate and creating new businesses at home, from cement factories to theme parks.

> **Looking Ahead**
>
> Persian Gulf leaders expect their oil to run out within this century. Many are reconsidering their dependence on oil. Will oil prove to be a genie of good or bad fortune? What will be oil's legacy in the Persian Gulf?

1908	Workers discover oil in Persia (Iran)
1960	Four Gulf countries and Venezuela form OPEC
1960s	OPEC members press for oil price increases
1970s	Gulf countries acquire their oil production facilities (background photo)
1973	Arab oil embargo leads to gas rationing in United States (photo above)
1991	Persian Gulf War
2090s	Experts predict Persian Gulf oil supplies will be depleted

History and Government

Guide to Reading

Consider What You Know

The Egyptian civilization was one of several civilizations that arose in this region. Ancient Egypt is a popular subject in films and books. What can you recall about its history and government?

Read to Find Out

- What great civilizations arose in North Africa, Southwest Asia, and Central Asia?
- What three major world religions originated in the region?
- How did countries of the region gain independence in the modern era?

Terms to Know

- domesticate
- culture hearth
- cuneiform
- hieroglyphics
- *qanat*
- monotheism
- prophet
- mosque
- nationalism
- nationalize
- embargo

Places to Locate

- Mesopotamia
- Fertile Crescent
- Persian Empire
- Silk Road
- Samarqand
- Jerusalem
- Makkah (Mecca)
- Iraq
- Iran

NATIONAL GEOGRAPHIC

A Geographic View

A Long History

Tucked away at the bottom of the Arabian Peninsula, . . . Yemen [is] . . . [d]ivided by nature into three distinct geographical regions—coastal plains, highlands, and desert. . . . Yemen has for much of its long history been no less divided politically by the shifting fortunes of its fiercely independent inhabitants. Kingdoms and empires have risen and fallen here for more than 3,000 years.

—*Andrew Cockburn, "Yemen United,"* National Geographic, *April 2000*

Yemeni woman in traditional clothing

Yemen is only one of many young countries with a long history in the region of North Africa, Southwest Asia, and Central Asia. This region saw the rise of some of the world's greatest civilizations and the birth of three of the world's major religions. Sadly, the region also has a long history of intense conflicts.

Prehistoric Peoples

Hunters and gatherers settled throughout North Africa, Southwest Asia, and Central Asia by the end of the last Ice Age, about 10,000 years ago. By 6000 B.C. farming communities had arisen in areas along the Nile River, the Mediterranean Sea, and the Taurus and Zagros Mountains.

The region's farmers were among the first in the world to **domesticate** plants and animals, or take them from the wild and make them useful to people. These farmers captured and herded cattle, sheep, goats, pigs, and camels. Some of the animals were used for food. Farmers used the hides to make clothes and shelters.

Early Civilizations

Although much of North Africa, Southwest Asia, and Central Asia has dry land, important civilizations developed there. These civilizations began to grow in the region's most fertile areas about 6,000 years ago.

The civilizations that arose in **Mesopotamia**, the area between the Tigris and Euphrates Rivers, comprised one of the world's first **culture hearths**, or centers where cultures developed and from which ideas and traditions spread outward. Part of a larger, rich agricultural region known as the **Fertile Crescent**, the area was home to the Sumerian civilization. The Sumerians mastered farming by growing crops year-round and using canals to irrigate them. The Sumerians made great strides in soil science, mathematics, and engineering. They also established at least 12 cities and created a code of law to keep order. They kept records by using a writing system called **cuneiform** (kyu•NEE•uh•FAWRM), wedge-shaped symbols written on wet clay tablets that were then baked to harden them.

Egyptian civilization flourished along the Nile River. Annual floods from the Nile deposited rich soils on the flood plain. During dry seasons Egyptians used sophisticated irrigation systems to water crops, enabling farmers to grow two crops each year. The Egyptians also developed a calendar with a 365-day year, built impressive pyramids as tombs for their rulers, and invented a form of picture writing called **hieroglyphics** (HY•ruh•GLIH•fihks).

Empires and Trade

The Phoenician civilization, which arose along the eastern Mediterranean coast, developed an alphabet in which letters stood for sounds. It formed the basis for many alphabets used in much of the Western world today.

During the 500s B.C., the **Persian Empire** extended from the Nile River and the Aegean Sea in the west to Central Asia's Amu Darya in the east. Realizing that irrigation water would evaporate in surface canals, the Persians constructed a system of *qanats*, or underground canals, to carry water from the mountains across the desert to farmlands.

Beginning about 100 B.C., parts of Central Asia and Southwest Asia prospered from the **Silk Road**, a trade route connecting China with the Mediterranean Sea. Many cities in the region, such as **Samarqand** in present-day Uzbekistan, thrived as trading stations along the Silk Road. At these stations travelers and merchants traded Chinese silks and Indian cotton as well as ideas and inventions. Because of the Silk Road, with its cultural and commercial exchange, the region became known as the "crossroads of civilization."

Today, as they did hundreds of years ago, nomads travel across the steppes of Central Asia seeking grasslands for their herds. Sometimes nomadic peoples, including the Mongols, invaded these lands. During the late 1100s, a leader known as Genghis Khan united the nomadic Mongol tribes living north of China. In the 1200s they invaded Central Asia, establishing a vast empire. The Mongols killed tens of thousands of people to gain control, but later they brought many improvements to the region, such as paper money and safer trade routes.

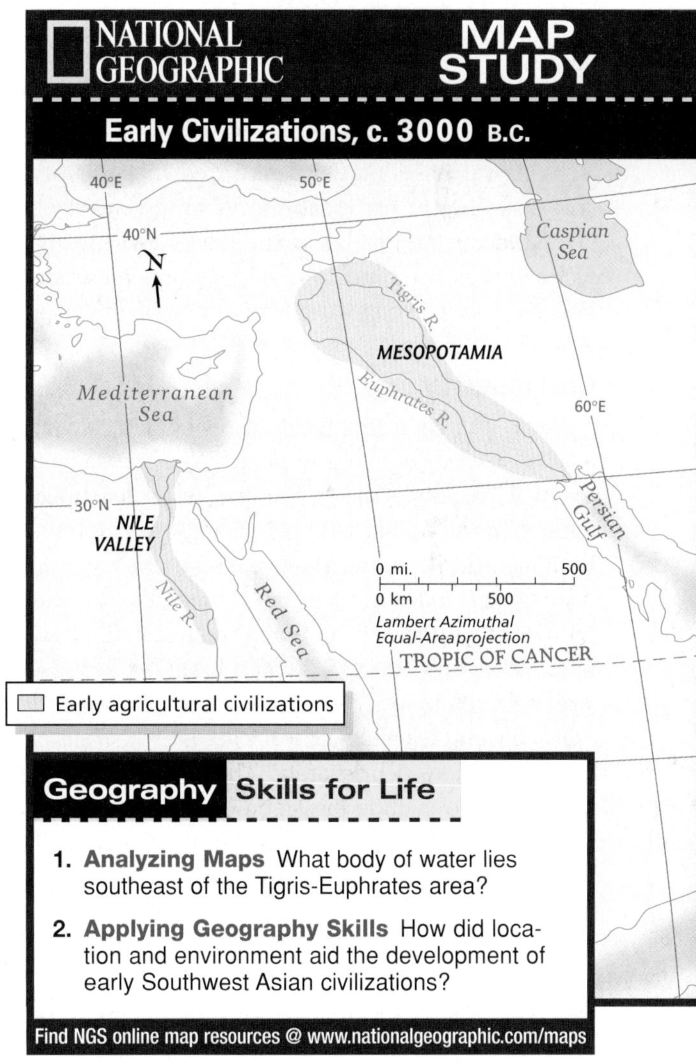

Three Major Religions

Three major religions began in the region: Judaism, Christianity, and Islam. All three share many beliefs, especially **monotheism**, or belief in one God.

Judaism

Judaism is the oldest of the monotheistic faiths. Followers of Judaism, known as Jews, trace their origin to the ancient Israelites, who set up the kingdom of Israel along the eastern Mediterranean coast. There they made **Jerusalem** their capital and religious center.

Despite political division, conquest, and exile to Mesopotamia, Jews and Judaism continued to survive and flourish. Many Jews eventually left Mesopotamia and returned to their homeland, now known as Judah. Others settled elsewhere in the Mediterranean. As they scattered, the Jews took their beliefs with them.

Judaism teaches obedience to God's laws and the creation of a just society. Believing that events have a divine purpose, the Jews recorded their history and examined it for meaning. Writings based on laws and on the history of the Jews make up the Hebrew Bible, or Torah. Worship services are traditionally held in synagogues, where a rabbi officiates.

Christianity

About A.D. 30, in the territory of Judah, a Jewish teacher named Jesus began preaching a message of renewal and God's mercy. Some of Jesus' teachings made him unpopular with people in power, and the Roman officials ruling the area had Jesus put to death. Jesus' followers soon proclaimed that he was the world's savior, alive in heaven, and that a new life in the world to come would be given to those who believed in Jesus and followed his teachings.

The life and teachings of Jesus became the basis of a new religion—Christianity. The Christian scriptures came to include the Hebrew Bible as the Old Testament, and writings on the life and teachings of Jesus as well as on the experiences of the earliest Christian communities as the New Testament. As the centuries passed, Christians spread the message of Jesus throughout the Mediterranean world and into Asia, Africa, and Europe, and eventually to the Americas.

Islam

Islam today is the major religion of Southwest Asia, North Africa, and Central Asia. Islamic tradition states that in A.D. 610, revelations from God came to Muhammad, a merchant in the city of **Makkah (Mecca)** in the Arabian Peninsula. Muhammad began preaching that people should turn away from sin and worship the one true God. Various groups in the peninsula accepted Muhammad's message, acknowledging him as the last in a line of **prophets**, or messengers, that included Abraham and Jesus.

By the 800s, Islam had spread to North Africa, Central Asia, South Asia, Southwest Asia, and parts of Europe. Islam had profound religious, political, and cultural influences in these areas. One of the new features seen in the region's cities was the **mosque**, a house of worship where Muslims pray. Muslim scholars also made important contributions:

> *"During Europe's [Middle Ages], the light of Islam shone, unifying, stimulating the cultures of many lands with the currents of trade and the bond of a common language, Arabic. Ibn Sina of Bukhara, known to the West as Avicenna, wrote his Canon, which remained Europe's medical textbook for more than 500 years. Mathematician al-Khwarizmi of Baghdad introduced 'Arabic' numerals and the decimal system from India and wrote the standard treatise on* al-jabr*—algebra."*
>
> Thomas J. Abercrombie, *Great Religions of the World*, 1971

The geographer Ibn Battuta traveled extensively throughout the Muslim world in the 1300s. He described the peoples and places of the region in his famous book, the *Rihlah*. Other Muslim scholars wrote about Islamic achievements and translated Greek writings into Arabic, works that later added to European knowledge about the ancient world.

Today around one-fifth of the world's population follows Islam and is called Muslim, a term meaning "those who submit to God's will." Muslims follow their faith's principles set down in the Quran, Islam's holy book. They also fulfill five duties known as the Five Pillars of Islam: professing faith in

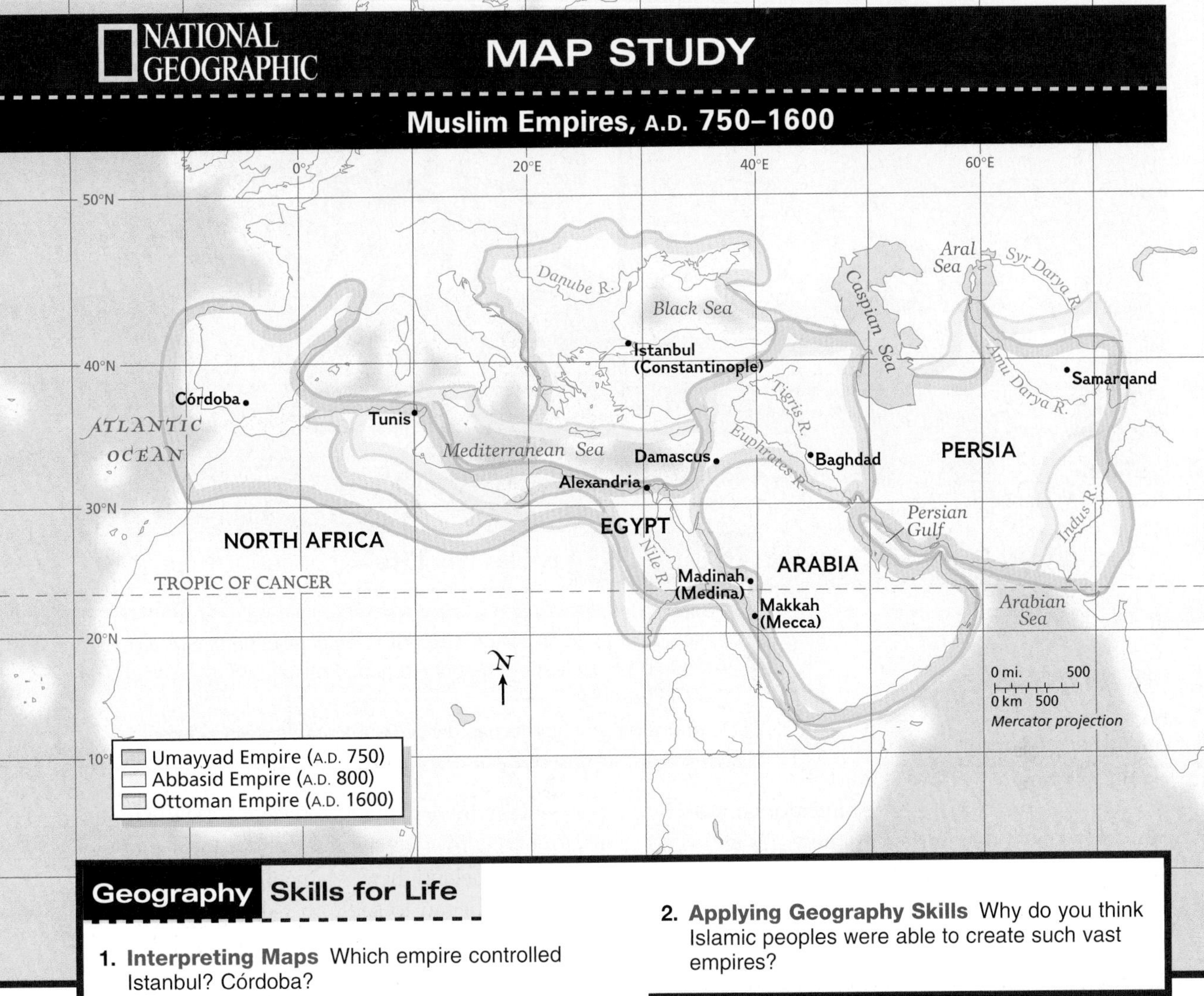

Geography Skills for Life

1. **Interpreting Maps** Which empire controlled Istanbul? Córdoba?

2. **Applying Geography Skills** Why do you think Islamic peoples were able to create such vast empires?

Find NGS online map resources @ www.nationalgeographic.com/maps

God and the prophet Muhammad, praying five times a day, helping the poor and needy, fasting during the ninth month of the Islamic calendar, and making a pilgrimage to Makkah, Islam's holiest city.

The Modern Era

As the centuries passed, Muslim empires in North Africa, Southwest Asia, and Central Asia rose and fell. Major conflicts, including the Crusades and the Mongol invasions, brought challenges to the region. Physical geography sometimes placed limits on economic development. For example, empires in North Africa and Southwest Asia lacked resources such as minerals, wood, and coal to fuel an industrial revolution like that of western Europe. By the late 1800s, western European powers controlled large areas of North Africa and Southwest Asia, and the Russian Empire took much of Central Asia.

Although the Caucasus area prospered under the Russians, peoples in other parts of the region were discontented under foreign control. During the 1800s a well-educated urban middle class developed in North Africa and Southwest Asia. Trained in European ways, this new middle class adopted European ideas about **nationalism**, or a belief in the right of an ethnic group to have its own independent country. This development stirred demands for self-rule that provided the basis for the modern countries that have emerged in the region.

Geography Skills for Life

War and Peace

From the late 1940s to the early 1970s, Arabs and Israelis fought a series of wars, such as the 1967 Six-Day conflict (left). Since then, Arab and Israeli leaders have held peace talks to try to resolve their differences.

Region What major issues divide Arabs and Israelis today?

Independence

In North Africa and Southwest Asia, the continuing rise of nationalism after World Wars I and II gradually ended direct European colonial rule. By the 1960s most territories in these regions had achieved political freedom. Independence has been a more recent development in Muslim Central Asia, where countries did not win their freedom until the breakup of the Soviet Union in 1991. Even after gaining independence, the regional economies of countries often remained under European control. Regional governments sometimes retaliated by seizing European property. Leaders in Iran, Iraq, and Libya **nationalized**, or placed under government control, the foreign-owned oil companies within their borders.

Arab-Israeli Conflict

Not all the independent countries in the region are Arab or Muslim. An exception is Israel, founded in 1948 as a Jewish state. About 1,900 years earlier, the Romans had expelled most Jews from their ancestral homeland, known as Palestine. These Jewish migrants eventually settled in communities scattered around the world. In their adopted countries, the Jews often faced persecution by the majority population around them. In the late 1800s, a fierce wave of persecution drove many European Jews to call for the return of the Jews to Palestine and for the creation of a Jewish homeland there. Many of these Jews, known as Zionists, began to settle in Palestine, which was then largely Arab and under Ottoman Turkish rule.

After World War I, the British gained control of Palestine. They supported a Jewish homeland there while claiming to give equal attention to the interests of the majority Arab population. These conflicting goals, as well as increasing Jewish immigration into Palestine, sparked conflict between Palestine's Arab and Jewish communities. Later, the murder of 6 million European Jews by the Nazis in the Holocaust increased Western sympathy for the Zionist cause.

After World War II, hostilities broke out in Palestine among Jews, Arabs, and British forces. Finally, the United Nations decided in 1947 to divide Palestine into separate Jewish and Arab states. When the British withdrew from Palestine, the Jews proclaimed the independent state of Israel in 1948. During the next 25 years, Arab opposition to Israel and Israel's concern for its security led to four major wars in the region. In the 1948 and 1967 Arab-Israeli conflicts, victorious Israeli forces took over Arab lands that had been part of Palestine. Since its formation, Israel has drawn many Jewish immigrants from around the world.

Israelis and Palestinians

The wars that followed the birth of Israel forced many Palestinian Arabs from their homes to live as refugees or settlers in other lands. The status of the Palestinian refugees is an ongoing issue in the Arab-Israeli dispute. In addition, the Palestinians—both refugees and those living in Israeli-occupied areas—want an independent state of their own in the West Bank and Gaza Strip areas. The West Bank lies west of the Jordan River, between Israel and Jordan. The Gaza Strip is a territory bordered on the south by Egypt, on the west by the Mediterranean Sea, and on the north and east by Israel.

The goal of Palestinian independence is complicated by the many Jewish settlements that have been built on the West Bank since the 1967 war. The challenge, says one West Bank resident, is straightforward: "Israelis and Palestinians claim the right of return to the same land."

Israel and the Palestinians finally agreed to the first stages of a peace settlement in 1993. Under its terms the Palestinians would gain limited self-rule in return for Arab recognition of Israel's right to exist as a nation. Another stage began with the Wye River Agreement, signed in 1998. It called for Israeli troop withdrawals from Israeli-held areas in the West Bank and Gaza Strip in order to increase Palestinian self-rule.

In 2000, peace talks stalled over the status of Jerusalem and over other issues. During the next year, renewed violence between Israeli forces and Palestinians had put the peace process in jeopardy.

War in Afghanistan

In past centuries, Hindu Kush mountain passes brought waves of invaders and traders to Afghanistan. Having an ethnically diverse population, Afghanistan in recent years has seen conflict involving foreign forces and rival Afghan groups. In the 1990s, radical Muslims known as the Taliban won control of most of the country. Taliban leaders were criticized internationally for human rights abuses, especially in limiting education and jobs for women, and for sheltering terrorists, such as wealthy Saudi exile Osama bin Laden.

In October 2001, American and British warplanes began bombing Afghan targets in the first military

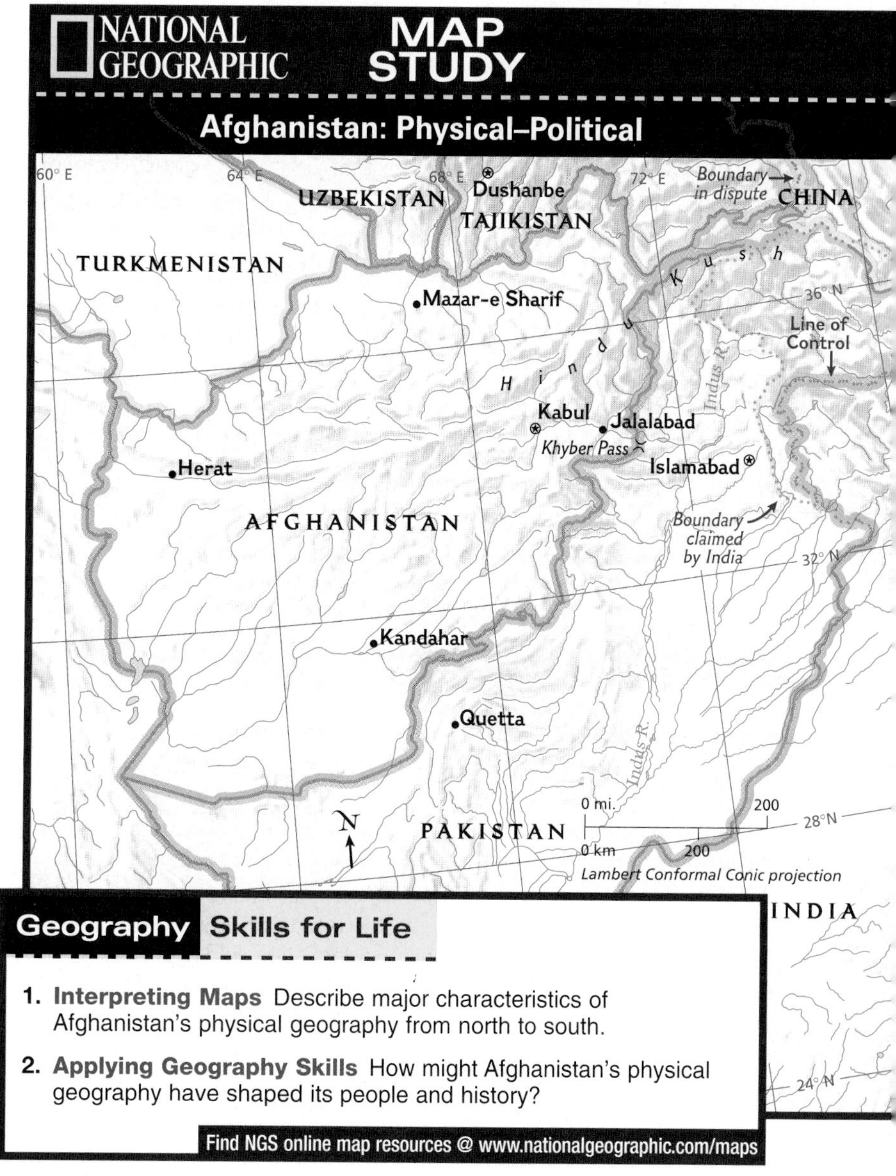

Geography Skills for Life

1. **Interpreting Maps** Describe major characteristics of Afghanistan's physical geography from north to south.
2. **Applying Geography Skills** How might Afghanistan's physical geography have shaped its people and history?

Find NGS online map resources @ www.nationalgeographic.com/maps

operation of the war on terrorism. The United States also gave ground and air support to the Northern Alliance, a group of Afghan rebels fighting the Taliban. With this help, the Northern Alliance in November captured major Afghan cities and routed most Taliban forces. Talks then began to form a new Afghan government. Meanwhile, bin Laden and his aides remained at large in the mountain caves that honeycomb Afghanistan. The United States and other nations expressed resolve to defeat them and bring them to justice.

Border Conflicts

Since World War II, various nations in Southwest Asia, North Africa, and Central Asia have fought each other over land and water resources. In 1980 a border dispute led to years of war between **Iraq** and **Iran**. Ten years later, Iraq's invasion of its oil-rich neighbor Kuwait forced the world community to impose an **embargo**, or a ban on trade, against Iraq. During the Persian Gulf War in early 1991, the United States and other countries forced Saddam Hussein, Iraq's leader, to withdraw his army from Kuwait. For years, the region's 20 million Kurds, most of whom live in border areas of Armenia, Iraq, Iran, Syria, and Turkey, have sought a country of their own. Political differences among the Kurds themselves and opposition by the governments ruling them have kept the Kurds from realizing this goal.

Government

Today's Governments

The countries of North Africa, Southwest Asia, and Central Asia have various forms of government. Traditionally the region was under the rule of dynasties. Today monarchs with varying degrees of power still rule in eight countries, including Saudi Arabia, the United Arab Emirates, Morocco, and Jordan.

The rest of the region's countries call themselves republics, although their republican governments differ greatly. Israel is a parliamentary democracy with a president as head of state and a prime minister as head of government. In the West Bank and Gaza Strip, a body known as the Palestinian National Authority is laying the foundation of statehood for Arab Palestinians.

Elsewhere, powerful presidents rule in Egypt, Syria, Kazakhstan, Turkmenistan, Uzbekistan, and Georgia. Military-based dictators govern Libya and Iraq. Iraq's leader Saddam Hussein, for example, remains in power despite a UN trade embargo that has crippled Iraq's economy.

In some countries, such as Algeria and Egypt, Islamist, or politically Islamic, groups have opposed secular, or non-religious, governments. Some of these movements have been successful. Under Shiite Muslim religious leaders, Iran's Islamic government was set up in 1979 after a revolution toppled the country's shah, or monarch.

SECTION 2 ASSESSMENT

Checking for Understanding

1. **Define** domesticate, culture hearth, cuneiform, hieroglyphics, *qanat*, monotheism, prophet, mosque, nationalism, nationalize, embargo.
2. **Main Ideas** Re-create a web diagram like the one below, and write in the features of one of the major religions that began in Southwest Asia.

Critical Thinking

3. **Drawing Conclusions** Why was the domestication of plants and animals so important for the early peoples in the region?
4. **Comparing and Contrasting** How are Judaism, Christianity, and Islam alike, and how do they differ? Describe the similarities and differences.
5. **Identifying Cause and Effect** What are the main causes of conflict in the region today?

Analyzing Maps

6. **Place** Study the map of Afghanistan on page 451. What challenges might military forces face in fighting a war there?

Applying Geography

7. **Expansion and Geography** Look at the map of Muslim empires on page 449. Consider the physical geography of the region. Then write a paragraph explaining why the locations of the three empires are similar.

SECTION 3

Cultures and Lifestyles

Guide to Reading

Consider What You Know

As you have learned, North Africa, Southwest Asia, and Central Asia have diverse geographic features, climate zones, and ethnic groups. How might these aspects of the region affect its culture?

Read to Find Out

- How have religion and language both unified and divided the peoples of North Africa, Southwest Asia, and Central Asia?
- What arts are popular in the region?
- What are some characteristics of everyday life in the region?

Terms to Know

- ziggurat
- bedouin
- bazaar

Places to Locate

- Qatar
- United Arab Emirates

NATIONAL GEOGRAPHIC

A Geographic View

City of Tradition Meets the Modern World

Alexandria, Egypt

Smoke and the fragrance of roasting quail float up from long charcoal grills lining the perimeter of Suq el-Attarine, the Market of Scents in Alexandria, Egypt. . . . Along sidewalks men sit on benches. . . . Some play dominoes. Above us hang the purple flowers of jacaranda trees.

The tranquil scene recalls earlier times in the city that Alexander the Great founded more than 2,300 years ago. But as I stroll from the marketplace toward the harbor, I am clearly in a modern city. Apartment buildings . . . surround me. Traffic jams the streets. Supermarkets, cell phones, motorcycles, and teenagers in baseball caps are everywhere.

—*Joel L. Swerdlow, "Tale of Three Cities,"* National Geographic, *August 1999*

Everyday scenes in Alexandria, Egypt, and elsewhere in the region reflect both tradition and change. In this section you will look at aspects of culture that have long shaped the lives and experiences of peoples in the region. You will also consider how the peoples of the region balance tradition and change in their daily lives.

Religion

Religion both unifies and divides the peoples of the region. The great majority of the people are Muslims. Most belong to the Sunni branch of Islam, which believes that leadership should be in the hands of the

NATIONAL GEOGRAPHIC

GRAPH STUDY

North Africa, Southwest Asia, and Central Asia: Religions

Religion	Number of Followers
Sunni Muslim	342,000,000
Shiite Muslim	87,000,000
Christian	16,900,000
Jewish	6,000,000
Other religions	17,000,000

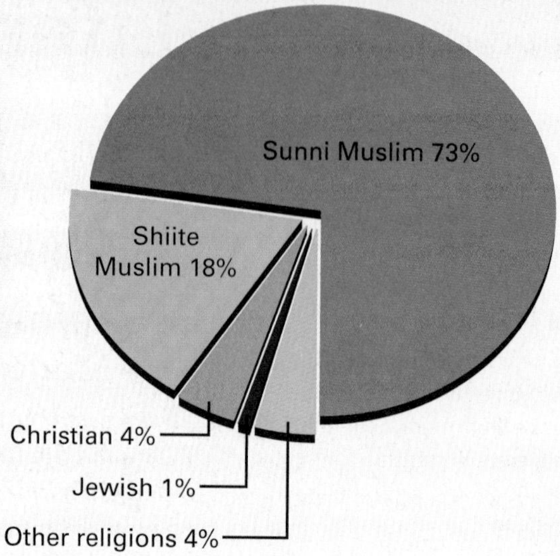

Sunni Muslim 73%
Shiite Muslim 18%
Christian 4%
Jewish 1%
Other religions 4%

Sources: World Almanac, 2001; Britannica Book of the Year, 2000

Geography Skills for Life

1. **Interpreting Graphs** How does the percentage of people who are Muslim compare with that of the followers of other religions?

2. **Applying Geography Skills** How are the beliefs of Sunni Muslims and Shiite Muslims similar? How are they different?

Islamic community at large. In Iran, Azerbaijan, Iraq, and parts of Syria and Lebanon, however, most Muslims follow the Shia branch of Islam. The Shia, or Shiites, believe that only Muhammad's descendants should lead the Islamic community.

Although Judaism and Christianity originated in the region, their followers make up only a small percentage of the population. Most Jews in the area live in Israel. Christians predominate in Armenia and Georgia, and large groups of Christians also live in Lebanon, Egypt, and Syria.

Languages

As Islam spread across the region, so did the Arabic language. Non-Arab Muslims learned Arabic in order to read Islam's holy book, the Quran. As more people became Muslims, Arabic became the region's main language. Other major languages in the region include Hebrew in Israel, Berber in southern Morocco and Algeria, and Turkish in Turkey. The languages of the Iranians, the Afghanis, and the Kurds include Farsi, Pashto, and Kurdish, respectively. Turkic languages are spoken in most of Central Asia.

The Arts

From earliest times, the peoples of the region have expressed themselves through the arts and architecture. Architects, artists, and writers later found inspiration in Judaism, Christianity, and Islam. Today the region's cultural expressions reflect the influence of both East and West.

Art and Architecture

The region's early civilizations created sculptures, fine metalwork, and buildings. In Mesopotamia the Sumerians built large, mud-brick temples called **ziggurats**, which were shaped like pyramids and rose above the flat landscape. The Egyptians built towering pyramids from massive stone blocks to serve as royal tombs. The Persians erected great stone palaces decorated with beautiful textiles.

Mosques and palaces are the best-known examples of Islamic architecture. Because Islam discourages depicting living figures in religious art, Muslim artists work in geometric patterns and floral designs. They also use calligraphy, or elaborate writing, for decoration. Passages from the Quran adorn the walls of many mosques.

Literature

Based on a strong oral tradition, epics and poetry are the region's dominant literary forms. The epic *Shahnameh (King of Kings)* describes heroic events in early Persian history. The *Rubaiyat* by the Persian poet Omar Khayyam is one of the few world masterpieces that has been translated into most languages. *The Knight in the Panther's Skin*, a Georgian epic by the writer Shota Rustaveli, paints a picture of brave warriors and their battles during the reign of Georgia's Queen Tamara. *The Thousand and One Nights*, a well-known collection of Arab, Indian, and Persian stories, reflects life in the early period of the Muslim empires.

Today rhythmic patterns in the region's poetry show an increased Western influence. Much modern literature has nationalistic themes. Many writers also focus on the challenges of change in traditional society. Kyrgyz writer Chingiz Aitmatov, for example, defends his homeland's traditional values against modernization. The Egyptian writer Naguib Mahfouz's novels about Cairo's recent past portray the conflicts between traditional village life and the new urban environment. In 1988 Mahfouz became the first winner of the Nobel Prize in literature whose native language is Arabic.

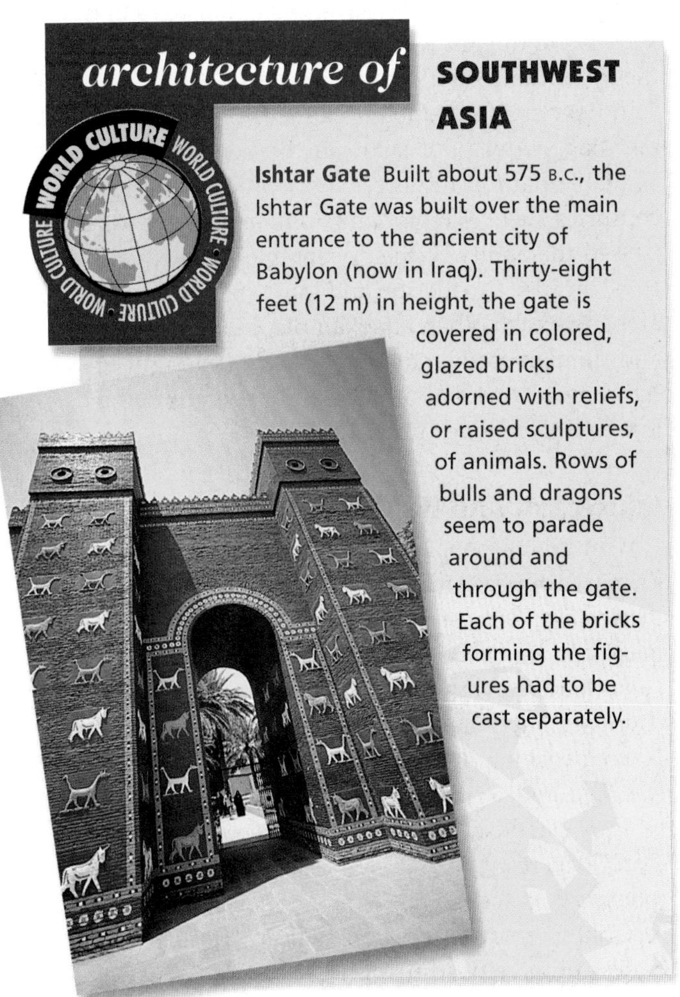

architecture of **SOUTHWEST ASIA**

Ishtar Gate Built about 575 B.C., the Ishtar Gate was built over the main entrance to the ancient city of Babylon (now in Iraq). Thirty-eight feet (12 m) in height, the gate is covered in colored, glazed bricks adorned with reliefs, or raised sculptures, of animals. Rows of bulls and dragons seem to parade around and through the gate. Each of the bricks forming the figures had to be cast separately.

Everyday Life

The lives of people in Southwest Asia, North Africa, and Central Asia have changed dramatically in the last century. The population has grown rapidly with improved health care and a high birthrate. In most countries more than one-third of the population is under 15 years of age. Many people also have moved to urban areas. For example, less than 50 percent of North Africans and Southwest Asians still cultivate the land, and only a small percentage are **bedouins** (BEH•duh•wuhnz), or desert nomads. Contact with other regions of the world through travel, trade, and the Internet is also changing lifestyles. Even so, cherished customs and traditions survive. Daily life still revolves around family, home, education, religion, and recreation.

Home and Community

In the region's largest cities, many people live in high-rise apartments. In the older parts of cities, however, people may live in stone or mud-brick buildings hundreds of years old. Similarly, many rural people in North Africa and Southwest Asia reside in stone or wooden structures. Some of these dwellings still lack running water or electricity.

Many families are very close-knit, often gathering at midday for their main meal. The menu might feature grains such as wheat and barley as well as

NATIONAL GEOGRAPHIC **World Explorer**

Geography Skills for Life

Pilgrims in Makkah This Asian couple eating a meal in front of the holy al-Haram Mosque is among the two million pilgrims that visit the city of Makkah each year.

Location Why is the direction they face when praying important to Muslims?

fruits, vegetables, and dairy products. Meat, especially lamb or mutton, is also a part of the diet of most of the region's peoples.

Rural dwellers often depend on their own farms or the village market for food. City dwellers can shop at supermarkets, but the **bazaar** is still popular. This traditional marketplace is a bustling area ranging from a single street of stalls to an entire district in a large city. The bazaar of Istanbul, for example, extends along miles of passageways:

> *"Sizzling hot kebabs give off an aroma. . . . The pounding of the hammers on copper pots assails our ears. . . . Merchants . . . wave and call out hoping to lure us into their shops. Lights make the glass, beads, and brass trays sparkle. The dimly lit twisting streets and alleys that curve off into the unknown add a . . . sense of adventure."*
>
> Stewart G. McHenry, "Markets, Bazaars, and Suqs," *Focus*, Summer 1993

Economics

Standards of Living

Standards of living vary widely across the region and even within countries. Urbanized countries with economies based on oil production or manufacturing and trade have relatively high standards of living. In Israel and **Qatar**, for example, the majority of people have access to the material goods they need. They can also afford additional goods that they want. Some oil-rich countries, such as the **United Arab Emirates**, are so prosperous that they have labor shortages and depend on foreign workers from India, Sri Lanka, the Philippines, and other countries.

In developing countries, however, much of the population does not share in the benefits of the available natural resources. Population growth in countries such as Egypt and Afghanistan has surpassed the ability of their economies to meet citizens' needs. Prosperity and poverty often exist alongside each other. For example, cellular phones and foreign cars may be a common sight in Azerbaijan's capital of Baku, but many other Azeris live in poverty.

Education and Health Care

Most young people in the region attend school. Primary education is free, and enrollment is increasing. Many students now complete both primary and secondary school, and a small percentage attend university. Eighteen of the region's 28 countries have literacy rates above 75 percent; in 10 countries, more than 90 percent of the people can read and write. Before 1979, when revolution in Iran established an Islamic government, less than 50 percent of Iranians could read or write; today, 79 percent can. Women have advanced especially in education, now making up fully half of new university admissions.

In recent decades health care also has improved and expanded in the region. People needing medical treatment usually go to government-owned hospitals. In wealthier countries, the hospital stay is often free, but doctor shortages in the rural areas of many countries mean that treatment is available mainly in large towns and cities. So despite improvements, average life expectancies have remained low in much of the region.

Celebrations and Leisure Time

Calls to worship occur five times each day in countries with large Muslim populations. A muezzin, or crier, calls the faithful to prayer from the minaret, or tower, of each local mosque. Men gather in rows on the mosque's mats or carpets after leaving their shoes at the entrance. Following the movements of the imam, or prayer leader, they bow and kneel, touching their foreheads to the ground in the direction of the holy city of Makkah in Saudi Arabia.

Religious holidays and observances often bring family and community together. Many Muslims mark Id al Adha, the Feast of Sacrifice, by making a pilgrimage to Makkah. They also observe Ramadan, a holy month of fasting from dawn to dusk ordained by the Quran. Yom Kippur, the Jews' most solemn holy day, is also a time of fasting and prayer. Passover and Hanukkah are other important holy days for Jews. Christians observe the holy days of Christmas and Easter, with special services at the places associated with Jesus' life.

People also visit with family members during their leisure time, often daily. Simple activities such as watching television or going to the movies bring young and old together. Soccer matches draw many spectators, and hunting and fishing are also popular. Board games such as backgammon and chess amount to unofficial national pastimes in countries like Armenia.

Interpretations of Islamic law have prevented Muslim women in some countries from fully participating in certain public activities such as sports. Some Muslim women, however, have begun to protest these restrictions. For example, women gather daily in Tehran's Mellat Park for a morning aerobic session, but in public places they must cover themselves completely. In sports where such dress is not practical, women perform in separate areas where the only spectators are female. Today Iranian women are active in many sports, including skiing, bodybuilding, shooting, and soccer. Their enthusiasm helped launch the first Islamic Women's Games in Tehran in 1993. Women competing in the games represented many predominantly Muslim countries, including Afghanistan, Azerbaijan, Kazakhstan, Kyrgyzstan, Oman, Syria, Turkmenistan, and Yemen.

SECTION 3 ASSESSMENT

Checking for Understanding

1. **Define** ziggurat, bedouin, bazaar.
2. **Main Ideas** Use a diagram like the one below to organize information about religion, language, the arts, and everyday life in the region.

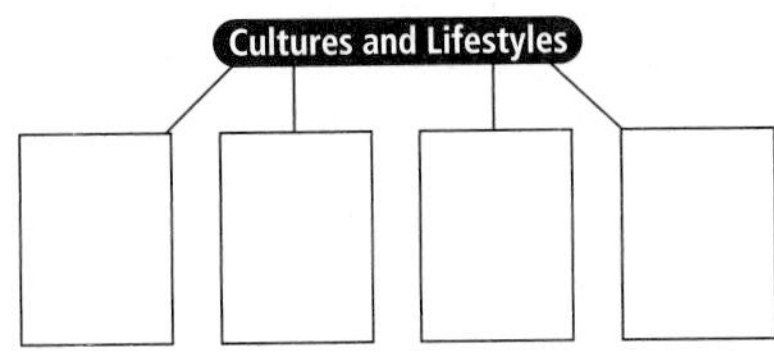

Critical Thinking

3. **Making Generalizations** How has religion been expressed in the arts from earliest times in North Africa, Southwest Asia, and Central Asia?
4. **Predicting Consequences** What are two possible effects of recent increases in literacy in Iran?
5. **Identifying Cause and Effect** Why does a large segment of the region's population live in poverty, even in oil-rich countries?

Analyzing Graphs

6. **Region** Study the graph of religions on page 454. How does the percentage of Sunni Muslims compare to the percentage of Shiite Muslims in the region?

Applying Geography

7. **Ways of Life** Think about the language, religion, systems of education, and customs in this region. Then write a paragraph comparing the ways of life there with your own.

Using an Electronic Spreadsheet

Electronic spreadsheets are used to manage numbers quickly and easily. Formulas may be used to add, subtract, multiply, and divide the numbers in the spreadsheet. If you make a change to one number, the totals are recalculated automatically.

Learning the Skill

An electronic spreadsheet is a worksheet for numerical information. All spreadsheet programs follow the same basic design of rows and columns. Columns, arranged vertically, are assigned letters. Rows, arranged horizontally, are assigned numbers. The point where a column and a row intersect is called a *cell*. The cell's position on the spreadsheet is labeled according to its column and row. For example, the cell at the intersection of Column A and Row 1 is labeled A1.

Spreadsheets use *standard formulas* to perform calculations using numbers in the cells. To create an equation using the standard formulas, you should first select the cell in which you want to display the results of your calculation. Here are some examples of equations you can build:

- **The equation = *B4* + *B5* applies a standard formula to add the values in cells *B4* and *B5*.**
- **The equation = *B5/B6* divides the value in cell *B5* by the value in cell *B6*.**
- **An asterisk (*) signifies multiplication.** The equation = (*B7* * *C4*) + *D4* means you want to multiply the value in cell *B7* by the value in cell *C4*, and then add the value in cell *D4* to the total.

Geneva 10 · C5 = 31

GWG 18 skill spreadsheet

	A	B	C	D	E
1	Country	GDP	Estimated	Life	Life
2		Per Capita	Population	Expectancy	Expectancy
3		(U.S. $)	(millions)	(Male)	(Female)
4	Morocco	3,500	29.5	67	71
5	Algeria	4,000	31.0	68	70
6	Tunisia	6,150	9.7	70	74
7	Libya	6,700	5.2	73	77
8	Egypt	4,400	69.8	65	68
9	Total Population (estimated)		145.2		
10					
11					

Sheet1 / Sheet2 / Sheet3

Because adding is the most common function of spreadsheets, most spreadsheet programs have an *AutoSum* key (Σ) that you can click on to place a sum in a highlighted cell.

Practicing the Skill

To practice using an electronic spreadsheet, follow these steps.

1. Open a new spreadsheet file.
2. Enter the information in Columns A through E as shown above.
3. In cell *C9*, use the *AutoSum* function (Σ) to calculate total population in millions for North Africa.
4. Print your results and share them with the class.

Applying the Skill

Use the information on pages 414–416 to develop a spreadsheet on the land area and population for all countries in the region. Use the *AutoSum* function to create calculations showing the total land area and the total number of people in the region. Then create an equation to calculate the population density of the entire region.

18

SECTION 1 Population Patterns (pp. 439–443)

Terms to Know
- ethnic diversity
- infrastructure

Key Points
- Movement and interaction of people have created the region's ethnic diversity.
- The largest concentrations of population are in coastal and river valley areas where water is readily available.
- Urbanization has caused increased pollution and overcrowding, challenges that cities and regional governments are addressing in many ways.

Organizing Your Notes
Use a cause-and-effect chart like the one below to help reinforce your understanding of how change affects population patterns in the region.

Cause	Effect
Movement of people →	
→	
→	

SECTION 2 History and Government (pp. 446–452)

Terms to Know
- domesticate
- culture hearth
- cuneiform
- hieroglyphics
- *qanat*
- monotheism
- prophet
- mosque
- nationalism
- nationalize
- embargo

Key Points
- Early peoples in the region were among the first to domesticate plants and animals.
- Two of the world's earliest civilizations arose in Mesopotamia and the Nile River valley.
- Three of the world's major religions—Judaism, Christianity, and Islam—trace their origins to Southwest Asia.
- After centuries of foreign rule, independent states arose in North Africa, Southwest Asia, and Central Asia during the 1900s.

Organizing Your Notes
Create an outline using the format below to help you organize important details from this section.

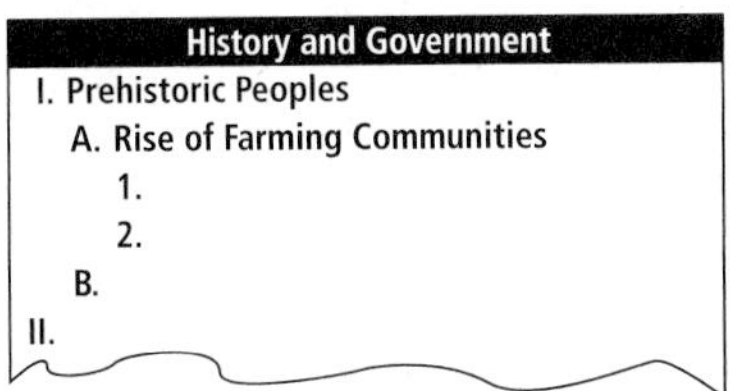

SECTION 3 Cultures and Lifestyles (pp. 453–457)

Terms to Know
- ziggurat
- bedouin
- bazaar

Key Points
- Islam and the Arabic language have been unifying forces in much of North Africa, Southwest Asia, and Central Asia.
- Many people in the region speak Arabic. Other major languages in the region include Hebrew, Berber, Greek, Farsi, Pushtu, Kurdish, and various Turkic languages.
- The peoples of North Africa, Southwest Asia, and Central Asia have expressed themselves from the earliest times through the arts and architecture.
- Tradition, especially religious observance, plays an important role in everyday life in the region.

Organizing Your Notes
Use a graphic organizer like the one below to fill in examples of the role tradition plays in different aspects of everyday life in the region.

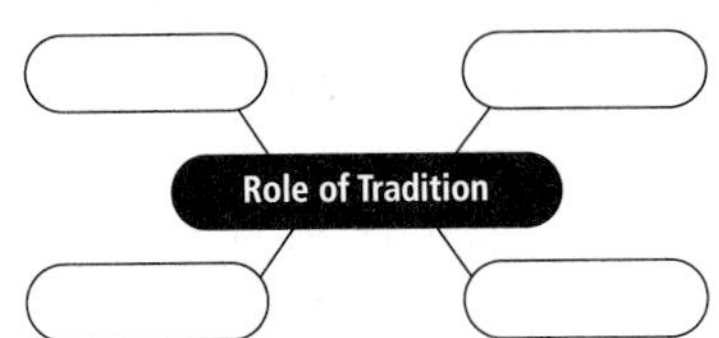

ASSESSMENT & ACTIVITIES

Reviewing Key Terms

Match the following terms with their definitions.

a. cuneiform
b. culture hearth
c. hieroglyphics
d. *qanat*
e. monotheism
f. ziggurat
g. bedouin
h. bazaar

1. center where cultures developed and from which ideas and traditions spread outward
2. form of picture writing
3. writing system developed by the Sumerians
4. belief in one God
5. large, mud-brick temple shaped like a pyramid
6. traditional public marketplace
7. desert nomad
8. underground canal

Reviewing Facts

SECTION 1

1. What groups of people live in the region?
2. How has urbanization affected cities in North Africa, Southwest Asia, and Central Asia?

SECTION 2

3. What physical features allowed areas in Mesopotamia and the Nile Valley to become culture hearths?
4. What basic idea is shared by Judaism, Christianity, and Islam?
5. In which areas of Israel do the Palestinians want an independent state of their own?

SECTION 3

6. How do religion and language influence the region's cultures?
7. How does tradition blend with modern ways in everyday life?

Critical Thinking

1. **Comparing and Contrasting** How are Armenians and Georgians similar? Different?
2. **Predicting Consequences** How might the impact of new technologies affect the region's ways of life?
3. **Categorizing Information** Create a web diagram like the one below to explain reasons for varying standards of living in the region.

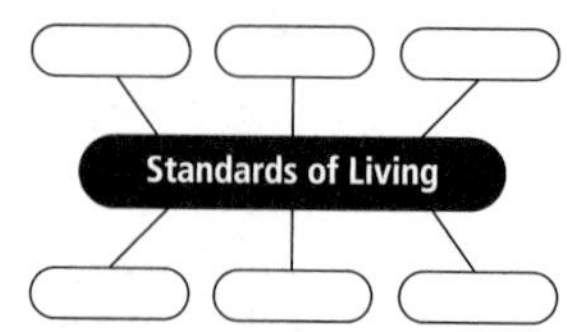

NATIONAL GEOGRAPHIC

Locating Places

North Africa, Southwest Asia, and Central Asia: Physical-Political Geography

Match the letters on the map with the places and physical features of North Africa, Southwest Asia, and Central Asia. Write your answers on a sheet of paper.

1. Tripoli
2. Iran
3. Istanbul
4. Casablanca
5. Riyadh
6. Israel
7. Tehran
8. Suez Canal
9. Astana
10. Cairo
11. Kabul
12. Uzbekistan

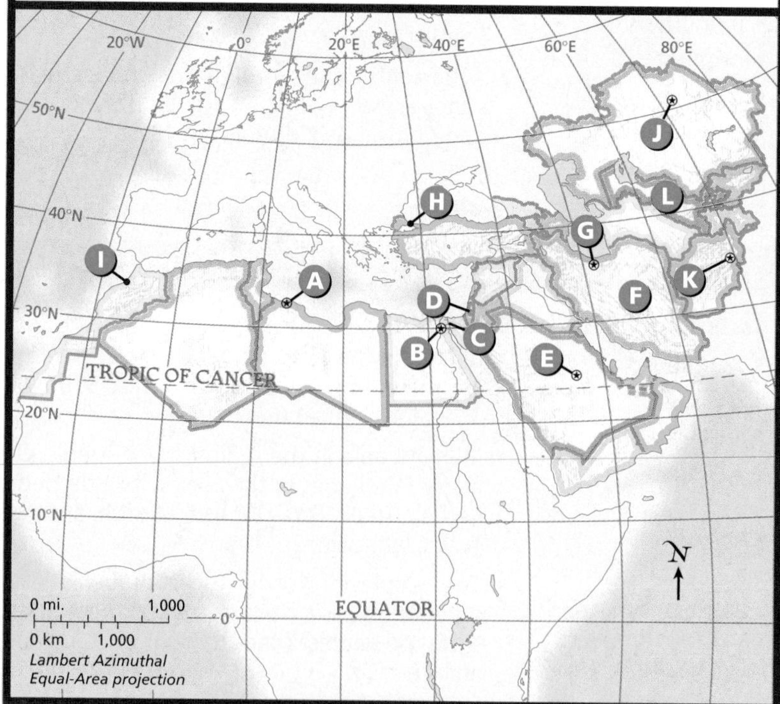

Self-Check Quiz Visit the **Glencoe World Geography** Web site at geography.glencoe.com and click on Self-Check Quizzes—Chapter 18 to prepare for the Chapter Test.

Using the Regional Atlas

Refer to the Regional Atlas on pages 410–413.

1. **Location** Where do most people in Turkey live? What accounts for this pattern?
2. **Place** Which countries' capitals have populations of more than 5 million?

Thinking Like a Geographer

Think about how language both unites and divides the region's peoples. Prepare a map of North Africa, Southwest Asia, and Central Asia, using different colors to show specific language areas. Use one color for areas where Arabic is the main language, another for Turkish and Turkic languages, and a third color for Farsi-related languages. As a geographer, what might you suggest to improve communication within the region?

Problem-Solving Activity

Group Research Project Working in a small group, simulate a meeting of delegates from four or five oil-producing countries. Each group member should research and report to the group on his or her country's oil production, oil revenues, and ways the revenues should be spent. Group members should then work together to create a chart or a graph to present the information to the class.

GeoJournal

Descriptive Writing Use details from your journal to write a descriptive paragraph about one of the culture groups of North Africa, Southwest Asia, or Central Asia. Share your paragraph with the class.

Technology Activity

Using E-Mail Search the Internet for the e-mail address of a museum or university in one of the region's countries. Compose and send an e-mail message requesting information about some aspect of the country's culture, such as architecture, religion, art, or language. Write a short report from the response you receive.

Standardized Test Practice

The following question refers to the accompanying quotation. Read the quotation carefully and then answer the question.

"Censorship in Saudi Arabia is even more overt. Under a system that took two years to develop, all Internet connections in the country have been routed through a hub outside Riyadh, where high-speed government computers block access to thousands of sites catalogued on a rapidly expanding blacklist."

—Douglas Jehl, "The Internet's 'Open Sesame' Is Answered Warily," *New York Times on the Web* (online), March 18, 1999

1. **Which of the following statements CANNOT be inferred about Saudi Arabia from the excerpt above?**

   F There is a great amount of censorship in Saudi Arabia.

   G People in Saudi Arabia are not interested in technology.

   H The Saudi Arabian government feels threatened by the impending technological revolution.

   J Many Internet sites are off-limits in Saudi Arabia.

Many questions ask you to identify information that CAN and CANNOT be inferred from a passage. In this example, you are asked to identify information NOT implied by the passage. Eliminating answers that are directly referred to in the passage helps narrow the possible choices. Ask yourself: Which of the statements are true about the passage, and which of the statements are false? Eliminate statements that you are certain can refer directly to the passage above.

CHAPTER 19

North Africa, Southwest Asia, and Central Asia Today

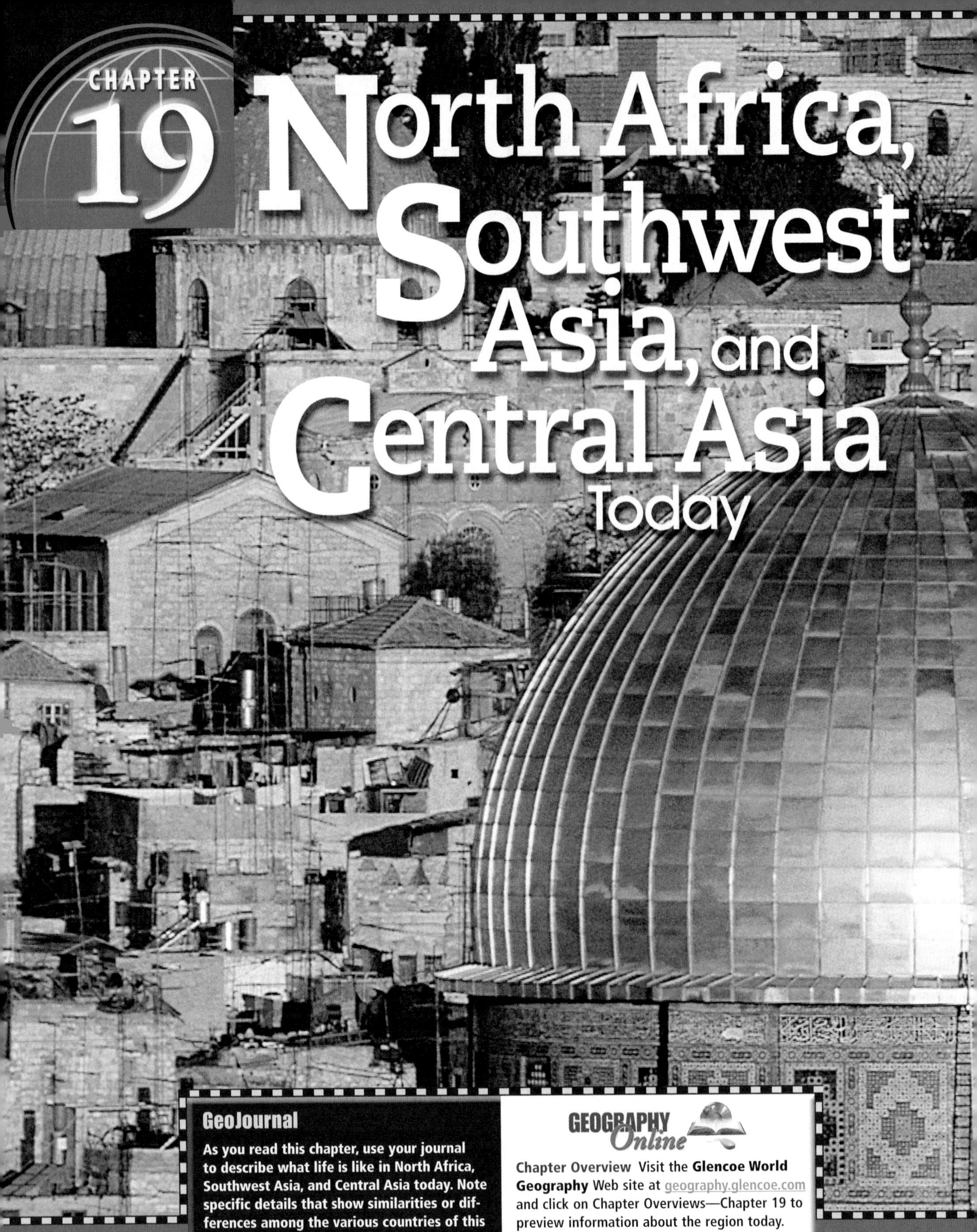

GeoJournal

As you read this chapter, use your journal to describe what life is like in North Africa, Southwest Asia, and Central Asia today. Note specific details that show similarities or differences among the various countries of this diverse region.

GEOGRAPHY Online

Chapter Overview Visit the **Glencoe World Geography** Web site at geography.glencoe.com and click on Chapter Overviews—Chapter 19 to preview information about the region today.

SECTION 1

Living in North Africa, Southwest Asia, and Central Asia

Guide to Reading

Consider What You Know

Reflect on what you have learned about the physical geography of North Africa, Southwest Asia, and Central Asia. Which countries in the region do you think have experienced the greatest economic development? Why?

Read to Find Out

- How does physical geography affect farming and fishing in North Africa, Southwest Asia, and Central Asia?
- What kinds of industries are important in the region?
- How are improvements in transportation and communications changing life in the region?

Terms to Know

- arable
- commodity
- petrochemical
- gross domestic product (GDP)
- hajj
- embargo

Places to Locate

- Saudi Arabia
- Israel
- Kuwait
- Morocco
- Istanbul
- Gulf of Aqaba
- Strait of Hormuz
- Baku

◀ *The Old City of Jerusalem*

NATIONAL GEOGRAPHIC

A Geographic View

Oil Boom

Baku oil derricks, Azerbaijan

On a clear, warm Sunday . . . Jamshid Khalilov, a 22-year-old student at the Azerbaijan State Oil Academy, rose early to study. Jamshid lives on the third floor of a dormitory a mile from the Caspian Sea in the Azerbaijani capital of Baku. In Baku Bay oil derricks spike the horizon like dead trees, and water seems to carry a gray, viscous film. . . . "As a boy I wanted to be a doctor," he said. "But then I decided there were better opportunities in oil."

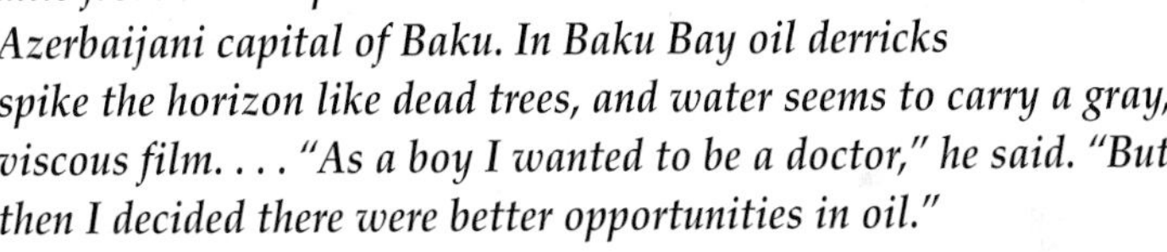

—*Robert Cullen, "The Rise and Fall of the Caspian Sea,"* National Geographic, *May 1999*

Jamshid wants a job in the oil industry. His future, however, depends on the Caspian Sea's oil potential. Like other areas in North Africa, Southwest Asia, and Central Asia, the Caspian Sea has great oil reserves that encourage economic activities such as oil production. Improved transportation and communications also link the region and its global neighbors.

Meeting Food Needs

Producing food for a rapidly growing population is a challenge in many parts of the region. More developed countries, such as

Saudi Arabia, buy food with oil profits. Less developed countries, such as Afghanistan, often grow their own food, but yields are usually small because of unreliable rainfall or poor soil. Farmers in some countries, such as **Israel**, however, take advantage of better climate and soils or use effective irrigation to grow food for export.

Agriculture

Only a small part of the region's land is arable, or suitable for farming, yet a large percentage of the population works in agriculture. In Afghanistan, for example, where only 12 percent of the land is arable, 67 percent of the people farm for a living. Agriculture plays a smaller role in many countries, such as **Kuwait**, that have economies based on oil.

Areas of North Africa and Southwest Asia that have a Mediterranean climate are best suited for cereal crops, citrus fruits, grapes, olives, and dates. When rainfall is below normal, however, harvests of major crops such as wheat, barley, and corn seldom meet people's needs. Countries such as Tunisia, Morocco, and Egypt that grow these crops often import grains to feed their people. Other crops of North Africa and Southwest Asia, like citrus fruits, are important exports. For example, Georgia, located at the eastern end of the Black Sea, has a subtropical climate that is good for producing citrus fruits, grapes, tobacco, and cotton.

Farmers in Central Asia raise both crops and livestock. Uzbekistan is one of the world's largest cotton producers. Both Uzbekistan and Turkmenistan are

NATIONAL GEOGRAPHIC

CHART STUDY

Land Use in Selected Countries

Country	Total Land Area sq. mi.	(sq. km)	Arable Land†	Forests and Woodlands†	Herding†
Afghanistan	251,772	(652,090)	12	3	46
Algeria	919,591	(2,381,741)	3	2	13
Egypt	386,660	(1,001,450)	2	*	*
Iran	630,575	(1,633,190)	10	7	27
Israel	8,131	(21,060)	17	6	7
Jordan	34,444	(89,210)	4	1	9
Lebanon	4,015	(10,399)	18	8	1
Morocco	279,757	(724,571)	21	20	47
Saudi Arabia	829,996	(2,149,690)	2	1	56
Tunisia	63,170	(163,610)	19	4	20
Turkey	299,158	(774,820)	32	26	16

* Less than 1 percent
† Data represent a percentage of the total land area of each country. Columns will not total 100 percent, as some land uses are omitted.
Sources: 2001 World Population Data Sheet; CIA World Fact Book, 2000

Geography Skills for Life

1. **Interpreting Charts** What percentage of Saudi Arabia's land is forested? What percentage of Egypt's land is suitable for farming?
2. **Applying Geography Skills** How might a country with relatively little arable land, forests and woodlands, or grasslands for herding make up for these deficiencies?

important centers for raising silkworms. Wheat, cotton, potatoes, and tea earn Azerbaijan substantial export income, even though less than one-eighth of its land is cultivated. Kazakhstan, which has fertile soil, is a major grain producer.

Fishing

Fish serve as an important food source in the region. Fishing boats ply the region's waters. Moroccan fishing boats bring in sardines and mackerel from the Atlantic Ocean. The majority of Israel's annual fish catch consists of freshwater fish raised in human-made ponds. Fishers from other countries harvest fish from the Persian Gulf, which is home to about 150 edible species. The size of fish catches has declined in the Caspian Sea because of overfishing and pollution. Still, Iran and several other countries have flourishing fishing industries.

NATIONAL GEOGRAPHIC **GRAPH STUDY**

World Oil Reserves (Billions of Barrels)

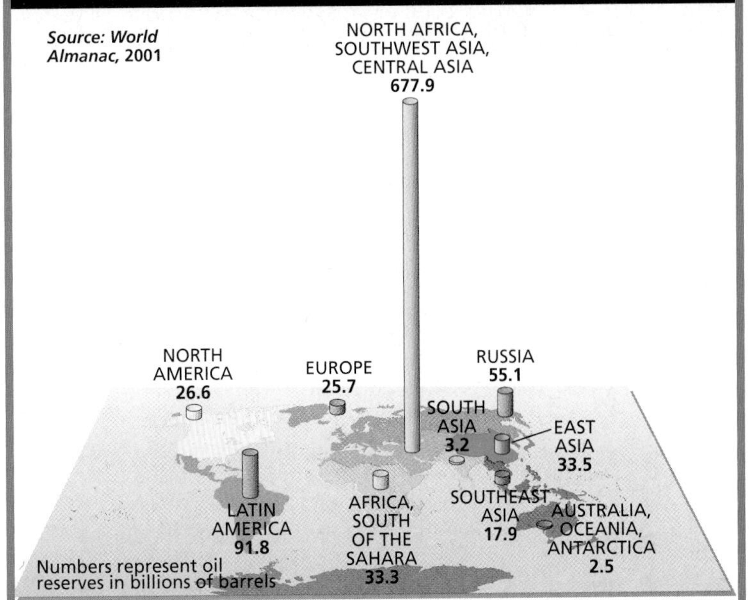

Source: World Almanac, 2001

NORTH AFRICA, SOUTHWEST ASIA, CENTRAL ASIA 677.9
NORTH AMERICA 26.6
EUROPE 25.7
RUSSIA 55.1
SOUTH ASIA 3.2
EAST ASIA 33.5
LATIN AMERICA 91.8
AFRICA, SOUTH OF THE SAHARA 33.3
SOUTHEAST ASIA 17.9
AUSTRALIA, OCEANIA, ANTARCTICA 2.5
Numbers represent oil reserves in billions of barrels

Geography Skills for Life

1. **Interpreting Graphs** About how much greater are oil reserves in North Africa, Southwest Asia, and Central Asia than those in the rest of the world?
2. **Applying Geography Skills** How might having large oil reserves affect a region's relations with other world regions?

Industrial Growth

Petroleum and oil products are the main export commodities, or economic goods, of North Africa, Southwest Asia, and Central Asia. The region holds about 67 percent of the world's oil and is likely to continue to supply much of the world's fossil fuels. In addition to significant oil reserves (the amount that can be recovered for use), the region also holds about 33 percent of the world's natural gas reserves.

Oil, Natural Gas, and Mining

Wealth from oil has helped build industry in the region. **Iran** and Saudi Arabia operate large oil-refining and oil-shipping facilities, and most other oil-producing countries export crude oil to industrialized countries. Natural gas has also advanced the region, powering steel, textile, and electricity production in various countries. Some countries have developed industries using petrochemicals—products derived from petroleum or natural gas—to make fertilizers, medicines, plastics, and paints. The economic growth brought by industries provides thousands of jobs and helps improve the region's standard of living.

Mining also contributes to the region's economic growth. Coal and copper mining and cement production are important in both Southwest Asia and Central Asia. In North Africa, Morocco is the world's largest exporter of phosphate, an essential ingredient in agricultural fertilizers.

Service Industries

Service industries—banking, real estate, insurance, financial services, and tourism—play significant roles in the region's economies. For example, the banking, real estate, and insurance industries amount to more than 60 percent of Bahrain's gross domestic product (GDP). GDP is the value of goods and services produced in a country in a year.

music of NORTH AFRICA, SOUTHWEST ASIA, AND CENTRAL ASIA

The region of North Africa, Southwest Asia, and Central Asia is home to a wide variety of music that is divided into three general cultural categories: Arabic, Turkish, and Persian. Islam is an important unifying influence of the music of this region.

Instrument Spotlight

The **oud** is the most popular stringed instrument of North Africa, Southwest Asia, and Central Asia. The body of the oud is pear-shaped, with a thin neck that bends sharply backward toward the player. The oud is made of various kinds of wood and is usually decorated with ebony, ivory, and other materials. Often used in solos, the oud is also an important ensemble instrument and is used to accompany classical pieces. It is said that the instrument owes its special tone to the birdsongs absorbed by the wood from which the oud is crafted.

Go To **World Music: A Cultural Legacy** Hear music of this region on Disc 1, Tracks 24–29.

Tourism also benefits some of the region's economies. North Africa and Southwest Asia are popular travel destinations because of their historical importance. Ancient monuments and religious sites have attracted millions of visitors, especially followers of the three major religions that began in the region. Christians and Jews tour Israel, Jordan, and other countries with deep roots in the heritage of the Bible. Muslims make a **hajj**, or pilgrimage, to Makkah in Saudi Arabia. Other visitors come to enjoy sunny Mediterranean beaches or the vibrant music and other cultural traditions of the region. Tourism is especially vital to **Morocco**:

> *"Tourism is Morocco's third largest industry. . . . Europeans come for hiking and skiing in the Atlas Mountains, or to the beaches around Agadir. . . . Americans come for the culture, . . . the medieval* medina *[quarter] of Fez, where they comb the market. . . . "*
>
> Erla Zwingle, "Morocco," *National Geographic*, October 1996

Some countries, however, discourage visitors in order to limit unwanted foreign influences. After the Islamic revolution in 1979, the Iranian government placed restrictions on tourists from non-Muslim countries. Regional conflicts and political instability in places such as Algeria, Syria, and Lebanon have also affected tourism.

Transportation and Communications

Advances in transportation and communications systems in the region are bringing the peoples closer together. Countries in the eastern Mediterranean area have experienced the region's greatest expansion in transportation and communications.

Roads, Railroads, and Airlines

Extensive road systems cross Iran, Turkey, and Egypt, connecting their major cities with oil fields and seaports. More than 200,000 miles (321,869 km) of roads span Turkey alone. In some countries of the region, mountains and deserts make road building difficult and costly. However, the growing number of vehicles and the need to link cities fosters highway development. In parts of the Caucasus area, roads provide the only access to the outside world. To ease traffic congestion in crowded urban areas and to improve urban-rural connections, some governments have built rapid transit systems and railroads. A new subway in **Istanbul**, Turkey, a city of some 9.5 million people, carries commuters to and from the city's center. National rail lines also connect urban areas and seaports. In 1998 Tajikistan unveiled part of a major railway system, which is designed to make trade and travel easier throughout Central Asia.

Since World War II, the growth in the air travel industry has benefited North Africa and Southwest Asia. In recent years Central Asia also has benefited from increased air traffic. Before the breakup of the Soviet Union, Central Asian countries relied on the Soviet airline Aeroflot, but now some Central Asian countries have their own airlines.

Waterways and Pipelines

Water transportation is vital to the region. Ships load and unload cargo at ports on the Mediterranean and Black Seas. The Strait of Tiran—between the **Gulf of Aqaba** and the Red Sea—and the **Strait of Hormuz**—linking the Persian Gulf with the Arabian Sea—are of strategic and economic importance. Oil tankers entering and leaving the Persian Gulf must pass through the Strait of Hormuz. The **Suez Canal**, a major human-made waterway lying between the Sinai Peninsula and the rest of Egypt, enables ships to pass from the Mediterranean Sea to the Red Sea.

An elaborate system of pipelines transports oil overland to ports on the Mediterranean and Red Seas and the Persian Gulf. In Central Asia, pipelines carry oil from **Baku**, Azerbaijan's capital, to Batumi, Georgia, on the Black Sea coast. The recent discovery of large oil and natural gas reserves in the Caspian Sea has prompted governments to plan for the building of more pipelines.

Communications

Throughout the region, television and radio broadcasting is expanding, although government control of the media in many places limits programming. Communication is difficult in some areas because of vast stretches of desert. Satellite technology, however, is helping countries improve communications services. Technologies such as wireless service and solar-powered radiophones are bringing telephone service to more people. Cellular phones are a common sight on the streets of major cities. Although service is limited, more and more people in the region have computer and Internet access. In Dubai, a territory of the United Arab Emirates, plans are in place to build a computer-based "cybercity" that will include a free trade zone, a research center, a science and technology park, and a university.

Economics

Two New Silk Roads

The year 1998 marked the opening of the world's longest telecommunications highway. The "highway" is actually a 16,767-mile (26,984-km) cable

NATIONAL GEOGRAPHIC **World Explorer**

Geography Skills for Life

Road Construction Workers build a bridge across a riverbed in Tajikistan.

Human-Environment Interaction What new development in Central Asia has made trade and travel easier in the area?

that follows the route of the Silk Road, the ancient trade route that linked Europe, Central Asia, and China. The cable provides the 20 countries along its path with digital circuits for voice, data, fax, and video transmissions.

Plans are also under way to build a network of road, rail, and air transportation systems tracing the Silk Road's path. The Transport Corridor Europe-Caucasus-Asia (TRACECA) will extend from Moldova in Europe eastward to Mongolia in East Asia. The more than 30 countries involved hope the project will promote peace in this vast area and provide access to newly discovered oil and gas deposits in the Caspian region.

Interdependence

Good transportation and communications networks go a long way toward increasing interaction between North Africa, Southwest Asia, and Central Asia and the rest of the world. Interdependence is also growing within the region, as developed countries provide foreign aid, trade deals, and development loans to less developed countries. Following the breakup of the Soviet Union, for example, Turkey, Iran, and Saudi Arabia helped smooth the new Central Asian republics' transition to independence.

Eight of the region's oil-producing countries—Algeria, Libya, Iran, Iraq, Kuwait, Qatar, Saudi Arabia, and United Arab Emirates—have become a majority in the 11-member Organization of Petroleum Exporting Countries (OPEC). Founded in 1960, OPEC has given member countries more control over oil production and prices. Because other countries depend heavily on the region's oil, OPEC has considerable influence in global affairs. It exercised political muscle by restricting oil shipments to the United States because of its aid to Israel during the 1973 Arab-Israeli war. OPEC raised oil prices during the 1970s. It also placed and later canceled an **embargo**, or restriction, on oil shipments to the United States and other industrialized countries. In 1999 and again in 2000, OPEC cut back oil production, forcing up oil prices around the world.

The countries of North Africa, Southwest Asia, and Central Asia and the rest of the world depend on one another. Industrialized countries, such as the United States, need oil from the region, and the region needs industrial products for its markets. For these reasons, both sides recognize that, despite political and economic disagreements, they must work together to ensure the well-being of all.

Student Web Activity Visit the **Glencoe World Geography** Web site at geography.glencoe.com and click on Student Web Activities—Chapter 19 for an activity about OPEC.

SECTION 1 ASSESSMENT

Checking for Understanding

1. **Define** arable, commodity, petrochemical, gross domestic product (GDP), hajj, embargo.
2. **Main Ideas** Use a graphic organizer like the one below to list ways that the activities listed help meet food needs in the region.

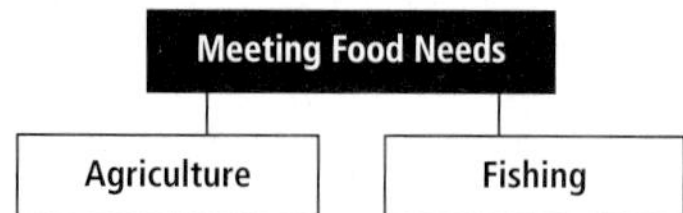

Critical Thinking

3. **Drawing Conclusions** Why does this region refine only a small amount of the oil it produces?
4. **Predicting Consequences** How might recent advances in communications technology help unify the region and change its cultures?
5. **Identifying Cause and Effect** Why do oil prices rise and fall? How do these changes affect global consumers?

Analyzing Maps

6. **Region** Study the economic activity map on page 413 of the Regional Atlas. In what areas of the region are oil deposits most abundant?

Applying Geography

7. **Effects of Transportation** List recent changes in global transportation and communications. Then create a graphic organizer showing how these changes have affected everyday life in the region.

People and Their Environment

Guide to Reading

Consider What You Know

North Africa, Southwest Asia, and Central Asia produce much of the world's oil. Because of this commodity, what particular environmental problems do you think people in this region face?

Read to Find Out

- How have peoples in the region dealt with scarce water resources?
- What are the causes and effects of environmental problems in the region?

Terms to Know

- aquifer
- desalination

Places to Locate

- Tripoli
- Aswan High Dam
- Elburz Mountains
- Dead Sea
- Aral Sea

NATIONAL GEOGRAPHIC

A Geographic View

Resources in Danger

An ocean of yellow sand covers Egypt, divided by the dark green vein of the Nile River. The river injects life into the bright green fan at its mouth, while the gray, man-made mass of Cairo eats away at the fan's delicate stem. . . . Cairo's commercial and residential sprawl has locked priceless soil beneath miles of concrete; the discharge of chemicals into delta lakes threatens the fishing industry and the supply of clean drinking water.

—Peter Theroux, "The Imperiled Nile Delta," National Geographic, *January 1997*

Satellite view of the Nile Delta

Human actions in North Africa, Southwest Asia, and Central Asia, like human actions in many places, often threaten the environment. These actions take many forms—oil spills, urban sprawl, and overuse of water supplies. The dilemma faced by people in the region is how to meet human needs while protecting the environment.

The Need for Water

Because more than 70 percent of the earth is covered by water, we often think of it as an abundant natural resource. However, about 2 percent of the earth's water is frozen, and 97 percent is salt water. According to the United Nations, about 1.2 billion people worldwide cannot obtain clean drinking water. About two-thirds of the world's households do not have a nearby source of freshwater. Some experts predict that by the year 2050 about 10 billion people will be living on the earth, producing an even greater strain on water resources.

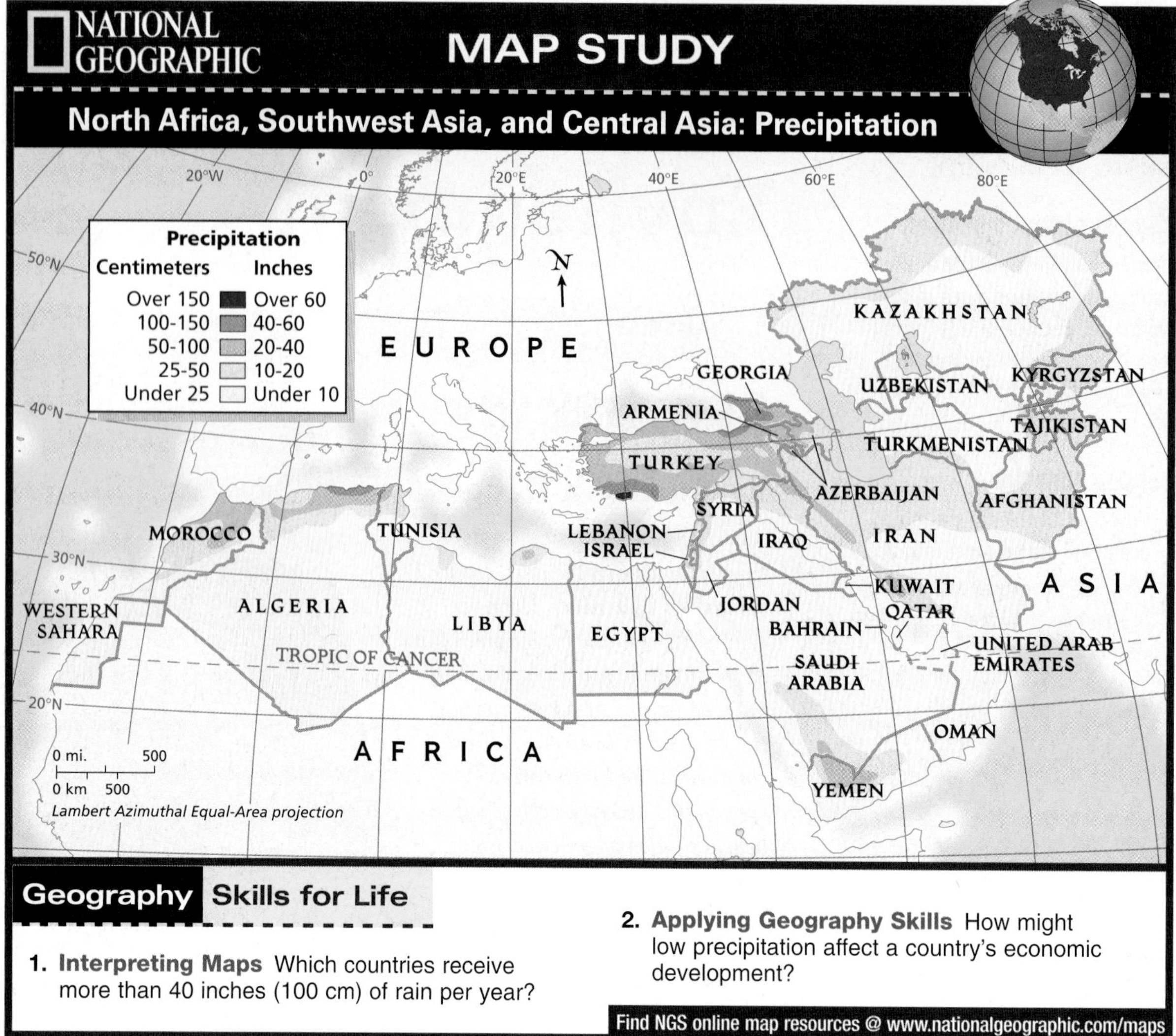

Geography Skills for Life

1. **Interpreting Maps** Which countries receive more than 40 inches (100 cm) of rain per year?

2. **Applying Geography Skills** How might low precipitation affect a country's economic development?

Find NGS online map resources @ www.nationalgeographic.com/maps

Water Resources

Much of the freshwater in North Africa, Southwest Asia, and Central Asia comes from rivers, oases, and **aquifers**—underground layers of porous rock, gravel, or sand that contain water. As the population grows, demand for water taxes these aquifers and other water resources.

The Nile, Tigris, Euphrates, Jordan, Amu Darya, and Syr Darya are the area's only major rivers, so only a few of the region's countries have enough freshwater for irrigation. Israel, for instance, uses an elaborate system of human-made canals to funnel the freshwater of the Jordan River from north to south. In the rest of the region, people turn to smaller rivers and other sources for water.

Desalination

Limited water resources have prompted scientists to develop ways to remove salt from seawater, a process called **desalination**. As the world's population increases and becomes more highly concentrated in urban areas, desalination helps meet the need for more freshwater. Within the region, Israel was the first country to attempt desalination. Other countries soon followed. The region now has about 60 percent of the world's freshwater-producing capacity, producing more than 2.4 billion gallons (9.1 billion l) a day. Many countries, particularly those near the Persian Gulf, depend on desalination plants. At the new Middle East Desalination Research Center, freshwater needs have brought Israeli and Arab scientists together.

History

An Ancient Solution

Creative solutions to the scarce water supply in North Africa, Southwest Asia, and Central Asia are an ongoing need. Scientists are now looking to the region's past for insights into possible solutions for the future. The ancient Nabataeans built the city of Petra, located in present-day Jordan, in a desert canyon that receives only about six inches of rain each year. To supply the 30,000 residents of Petra with the water they needed, the Nabataeans harvested rainwater, collecting and storing it in an amazingly intricate system of pipes, dams, terraces, and cisterns, or other artificial reservoirs.

> *"Hundreds of cisterns kept Petra from dying of thirst in times of drought, while masonry dams in the surrounding hills protected the city from flash floods after bursts of rain. . . . That kind of planning is called for again today."*
>
> Don Belt, "Petra: Ancient City of Stone," *National Geographic*, December 1998

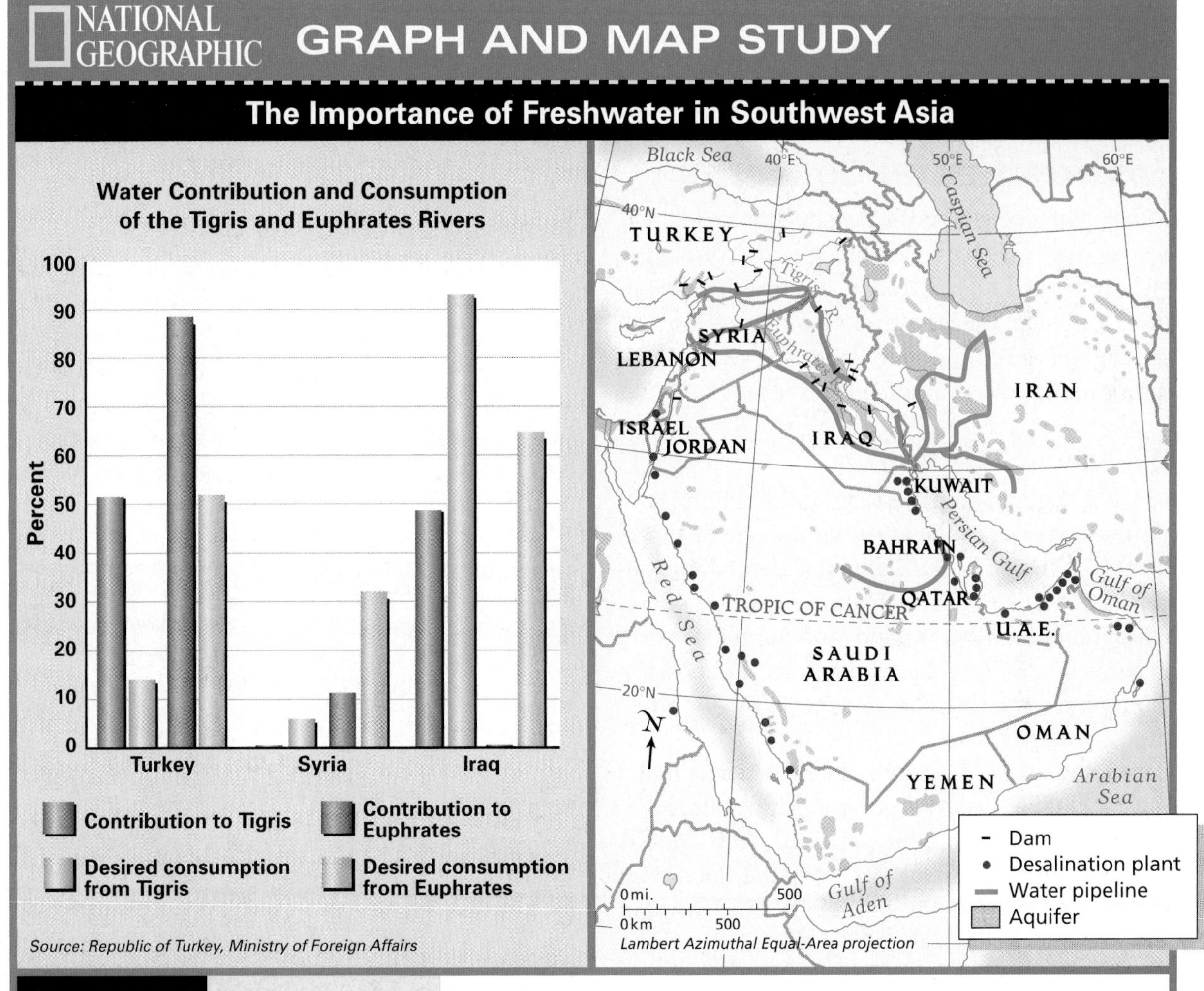

Geography Skills for Life

The graph above shows the countries through which the Tigris and Euphrates Rivers flow, and their contribution and consumption of water from the two rivers, which often is not equivalent. As shown on the map above, countries throughout Southwest Asia rely on various methods to acquire freshwater.

1. **Interpreting Graphics** Where are most of the larger aquifers in Southwest Asia located?
2. **Applying Geography Skills** Why do you think controlling freshwater is such a difficult issue in Southwest Asia?

Find NGS online map resources @ www.nationalgeographic.com/maps

The "Great Man-Made River"

Libya's "great man-made river" is an ambitious effort to supply freshwater. This multibillion-dollar project uses two pipelines to carry water from large aquifers beneath the Sahara to farms near the Mediterranean. The first pipeline, completed in 1991, brings freshwater across eastern Libya to the coast, and plans are under way to extend the pipeline to other areas. A second pipeline, completed in 1996, carries water to areas near **Tripoli** (TRIH•puh•lee), the country's capital, from an aquifer in the west. Yet pipelines may create environmental challenges. Scientists fear that the pipelines could drain aquifers in Libya and neighboring countries and that pumping aquifers near the Mediterranean could draw in salt water from the sea, contaminating the freshwater.

Environmental Concerns

In recent decades both new technologies and destructive wars have heightened environmental concerns in the region. Today countries must balance accessing their natural resources with preserving the environment. Egypt's **Aswan High Dam** provides an example of this struggle.

The Aswan High Dam

In 1970 Egypt completed the Aswan High Dam, located about 600 miles (966 km) south of Cairo. Started in the 1950s, the 364-foot (111-m) dam controls the Nile's floods, irrigates around 3 million acres (1.2 million ha) of land, and supplies nearly 50 percent of Egypt's electrical power. To boost the fishing industry, the dam also created the world's largest artificial lake.

In spite of these successes, the project also had a negative impact on the environment. Before the dam's construction, the annual Nile floods deposited fertile alluvial soil along the floodplain and washed away salt from the soil. Now the dam traps the soil, and Egyptian farmers must use expensive fertilizers. The land also retains salt because floodwaters no longer cleanse the soil.

The health of people and their livestock also suffers. After the dam was completed, parasite-related diseases and deaths around the dam and downriver increased. With aid from other countries and international organizations, however, Egypt is overcoming many of the difficulties created by the dam.

History

The Persian Gulf War

War in the region has also had a negative effect on the environment. During the Persian Gulf War, Iraqi troops retreating from Kuwait set fire to more than 700 oil wells. Huge black clouds of smoke polluted the area. Iraqi troops also dumped about 250 million gallons (947 million l) of oil into the Persian Gulf.

Scientists do not yet know what long-term effects these events will have. Thousands of fish and other marine life died when the oil spill spread 350 miles (563 km) along the Persian Gulf coastline. Smoke from oil-well fires threatened millions of birds. Oil pollution from routine shipping also adversely affects the Persian Gulf environment.

Nuclear and Chemical Dangers

Central Asia also inherited the Soviet era's environmental problems. Kazakhstan was once home to Soviet nuclear bases. During the Cold War, the Soviets tested nuclear, chemical, and biological weapons there. In 1989 it was found that this weapons testing had caused radiation leaks. Scientists think many years will pass before all the resulting contamination disappears.

Geography Skills for Life

Aswan High Dam The dam creates 10 billion kilowatt-hours of electricity annually.

Human-Environment Interaction How has the Aswan High Dam affected Egypt's water sources?

Soviet planners also chose Kazakhstan as a site for heavy industry, which polluted the air with toxic chemicals. Scientists have linked increased infant mortality in Kazakhstan directly to industrial pollution. The people of Kyrgyzstan, another site of Soviet heavy industry, have suffered similar effects.

Three Troubled Seas

The Caspian Sea, the Dead Sea, and the Aral Sea also face severe environmental challenges. Pollution at the Caspian Sea's southern end, near the **Elburz Mountains** of Iran, is especially severe. Pollution and overfishing threaten fish, like sturgeon, whose products are important exports.

The water level of the **Dead Sea** has dropped more than 262 feet (80 m) over the past 40 years. Ninety percent of the water from the sea's feeder rivers is diverted for irrigation and to hydroelectric plants. Scientists have suggested pumping water into the Dead Sea from the Gulf of Aqaba, but the $5 billion price is too high. To reduce the amount of water diverted from the Dead Sea, planners recommend building a desalination plant on Israel's Mediterranean coast.

Like the Dead Sea, the **Aral Sea** has had water diverted from feeder rivers to irrigate cropland. Once the world's fourth-largest body of inland water, by the year 2000 it had become separate, smaller lakes. These changes destroyed the sea's fishing industries, and dust storms have spread disease. People living by the Aral Sea are now working to revive their sea as a chain of lakes that can support fish.

Geography Skills for Life

1. **Interpreting Maps** How has the Aral Sea changed since 1960?
2. **Applying Geography Skills** What factors have affected the size of the Aral Sea?

Find NGS online map resources @ www.nationalgeographic.com/maps

SECTION 2 ASSESSMENT

Checking for Understanding

1. **Define** aquifer, desalination.
2. **Main Ideas** Re-create the web diagram below on a sheet of paper, and fill in ways the region meets its freshwater needs.

Critical Thinking

3. **Drawing Conclusions** Have the effects of the Aswan High Dam been mostly positive or mostly negative? Explain.
4. **Predicting Outcomes** What problems might occur if new sources of water are not found for the region?
5. **Comparing and Contrasting** How are the problems facing this region's seas similar? Different?

Analyzing Maps

6. **Region** Study the precipitation map on page 470. What kind of vegetation would you expect to find in most of Kazakhstan? Why?

Applying Geography

7. **Planning for the Future** Think about the needs of this region. Write a plan to address its water needs in the future.

Viewpoint

CASE STUDY on the Environment

Turkey's Atatürk Dam: Diverting a River's Flow

Rivers are the lifeblood of arid regions. As rivers wind through parched landscapes, they supply precious water to millions of people. Anything that disrupts a river's flow can be a threat to life itself. So it is with Turkey's Atatürk Dam on the Euphrates River. The Atatürk restrains the Euphrates and diverts some of its water for irrigation. For the Turks, the dam is turning barren plains into lush croplands. Downstream, however, Syria and Iraq worry that the Atatürk project will rob their countries of a vital lifeline.

Like a silver ribbon, the Euphrates River (left) winds down through Turkey's Anti-Taurus Mountains. It skirts the western edge of the Harran Plain, then crosses the border into Syria. From Syria, the Euphrates flows into Iraq, where it eventually joins the Tigris River and empties into the Persian Gulf.

In the early 1980s, Turkey embarked on the Southeastern Anatolia Project, a plan to bring water from both the Euphrates and Tigris Rivers to Turkey's arid southeast region. The Atatürk Dam is the centerpiece of this massive irrigation project. Completed in 1990, the Atatürk is one of the world's largest dams. Water held in the dam's reservoir is channeled into two huge irrigation tunnels that lead to the Harran Plain 40 miles (64 km) away.

Eventually 21 other dams will be constructed along the Euphrates and Tigris Rivers. Nineteen hydroelectric power plants associated with the dams will generate about 27 billion kilowatt-hours of electricity each year—about twice Turkey's current output.

Yet the project is creating tension downstream. People in Syria and Iraq worry that Turkey's dams will reduce the precious flow of water through their lands. They were outraged when Turkey stopped the flow of the Euphrates for a month in 1990 while filling the Atatürk reservoir. Despite Turkey's assurances that its southern neighbors will receive a fair share of river water, Syria and Iraq remain unconvinced.

◀ Woman in Turkey hauls water to her family.

Supporters of the Atatürk Dam claim that the Southeastern Anatolia Project will help Turkey expand its agricultural base and raise its standard of living. They note that every country has the right to control river water within its borders. Before the dam's reservoir was filled, Turkey increased water flow from the Euphrates for two months to prevent adverse effects to Syria and Iraq. Supporters say that while Syria and Iraq rely on petroleum deposits for energy, oil-poor Turkey needs hydroelectric power.

Opponents of the dam claim that once the Anatolia Project is complete, Syria's share of Euphrates waters could be reduced by 40 percent and Iraq's by 60 percent. Reduced flow will make it harder for Syria's hydroelectric plants to maintain current levels of production. Irrigation-based agriculture in both Syria and Iraq could also be jeopardized. Opponents believe that because the Turks stopped the flow of the Euphrates once, they may do it again—whenever it suits their needs. Archaeologists in the region also oppose the project because the dams are destroying unexplored ancient cities.

What's Your Point of View?
Does Turkey have the right to restrict the flow of river water to its downstream neighbors?

A Syrian farmer (below) relies on river water to grow crops. Turkey's Atatürk Dam (right) may reduce his share of irrigation waters. ▼

Creating Sketch Maps

While traveling, it is often much easier to follow visual images on a map than directions given in words. A simple sketch map can display a wide range of useful information.

Learning the Skill

Think about how you get from place to place each day. In your mind you have mentally mapped your route. You could probably draw sketch maps of many familiar places. Making mental maps and sketching them are also useful skills in the study of geography. They can help you remember and organize information about the regions you study.

To create a sketch map, follow these steps:

- **When a country or city name is mentioned, find it on a map to get an idea of where it is and what it is near.** Look for important features, such as highways, mountains, buildings, or bodies of water.
- **Draw a sketch map of the area.** Include a compass rose to show direction. Include these important features and political boundaries on your map.
- **As you read or hear information about the place, picture where on your sketch map you would fill in this information.** Add the information to your sketch map. Use colors and symbols to show different kinds of information, and add a legend, or key.
- **Compare your sketch to an actual map of the place.** Change your sketch if you need to by adjusting the locations of places or including additional information.

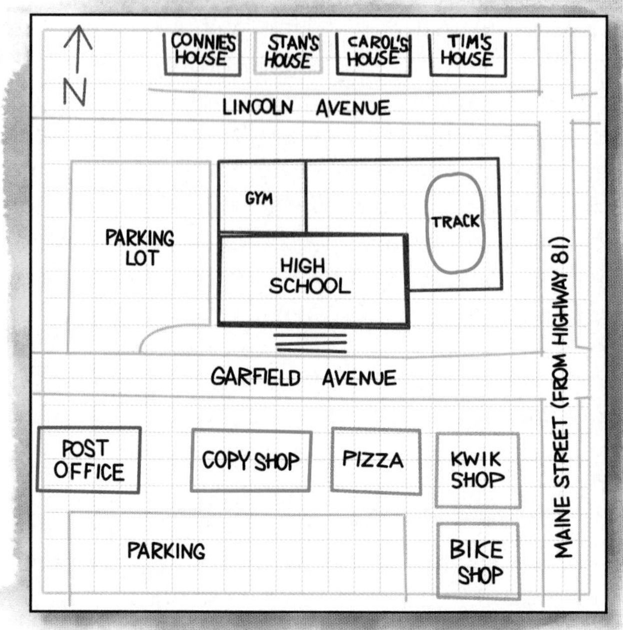

Practicing the Skill

Study the map above. Use it to answer the following questions.

1. If you are at the post office, what is the quickest route to Tim's house?
2. How many blocks apart are the bike shop and Carol's house?
3. Suppose your classroom is in the southeast corner of this high school. Could you have a view of the school's parking lot from the classroom?
4. Suppose you deliver pizzas for the pizza shop on Garfield Avenue. You have deliveries to the post office, Connie's house, the high school gym, and the bike shop. What is the best route?

Applying the Skill

Without looking back at this chapter, make a list of important physical features and cities in North Africa, Southwest Asia, and Central Asia. Design and draw a sketch map that shows the locations of each of these places. Compare your sketch map with an actual map of the region. Revise your sketch if you need to. Then create three questions based on your sketch map, and have another student use your map to answer them.

Go To The **Glencoe Skillbuilder Interactive Workbook, Level 2** provides instruction and practice in key social studies skills.

CHAPTER 19 SUMMARY & STUDY GUIDE

SECTION 1 Living in North Africa, Southwest Asia, and Central Asia (pp. 463–468)

Terms to Know
- arable
- commodity
- petrochemical
- gross domestic product (GDP)
- hajj
- embargo

Key Points
- Although North Africa, Southwest Asia, and Central Asia have limited arable land, a relatively large percentage of the region's people work in agriculture.
- The oil-producing countries in North Africa, Southwest Asia, and Central Asia have experienced greater economic growth than other countries in the region.
- Expanded and more advanced transportation and communications systems are helping connect the region's urban and economic centers with one another and with the world.
- Interdependence is increasing among the countries of the region, especially in controlling oil production and prices.

Organizing Your Notes
Use a chart like the one below to help you organize the notes you took as you read this section. Under each head, fill in the important supporting details.

Food Production	Industrial Growth	Transportation and Communications	Interdependence

SECTION 2 People and Their Environment (pp. 469–473)

Terms to Know
- aquifer
- desalination

Key Points
- Countries in the region have modified their environments to meet people's needs for water for drinking and irrigation.
- New technologies and destructive wars have subjected the region's environment to stress.
- People are working to revive areas damaged by past events.

Organizing Your Notes
Create an outline using the format below to list ways in which the people of North Africa, Southwest Asia, and Central Asia use their environment.

People and Their Environment

I. Need for Water
  A. Water Resources
    1. Population

Business district, Ankara, Turkey

ASSESSMENT & ACTIVITIES

Reviewing Key Terms

Write the key term that best completes each of the following sentences. Refer to the Terms to Know in the Summary & Study Guide on page 477.

1. Underground layers of porous rock, gravel, or sand that contain water are called _________.
2. _________ is the value of goods and services produced in a country in a year.
3. _________ is the process that removes salt from seawater.
4. Government restrictions on buying or selling certain goods are called _________.
5. _________ are products derived from petroleum or natural gas.
6. The _________ is the pilgrimage to Makkah made by many Muslims.
7. Land that is suitable for farming is _________.
8. Petroleum is one of the main economic goods, or _________, exported by the region.

Reviewing Facts

SECTION 1

1. How has natural gas helped advance the region's industrial growth?
2. Why do some countries in North Africa, Southwest Asia, and Central Asia discourage tourism?
3. Why is most freight carried by road in Armenia?

SECTION 2

4. What, and where, is the "great man-made river" project?
5. How did the Persian Gulf War affect the environment of the region?
6. What has caused the water levels of the Dead and Aral Seas to drop?

Critical Thinking

1. **Making Inferences** Why is industrial growth limited in some parts of the region?
2. **Analyzing Information** How does the region of North Africa, Southwest Asia, and Central Asia compare with other world regions in terms of economic opportunities for women?
3. **Identifying Cause and Effect** List examples of economic growth in the region and the effects of each on the environment. Then write a paragraph that explains the impact of one of them.

Examples	Effects on Environment

NATIONAL GEOGRAPHIC

Locating Places

North Africa, Southwest Asia, and Central Asia: Physical-Political Geography

Match the letters on the map with the places and physical features of North Africa, Southwest Asia, and Central Asia. Write your answers on a sheet of paper.

1. Libya
2. Turkey
3. Kuwait
4. Shatt al Arab
5. Jordan
6. Morocco
7. Persian Gulf
8. Ural River
9. Hindu Kush
10. Sinai Peninsula
11. Turkmenistan
12. Makkah

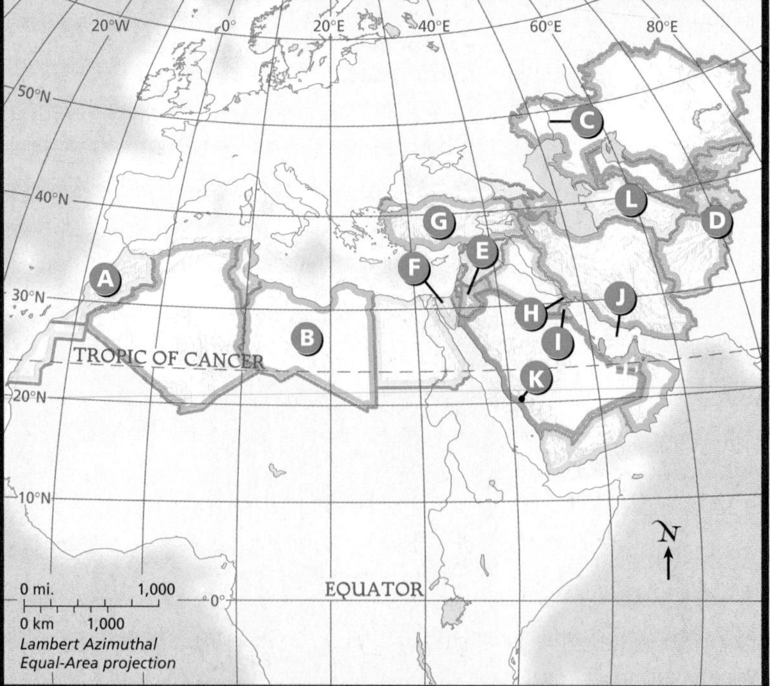

Self-Check Quiz Visit the **Glencoe World Geography** Web site at geography.glencoe.com and click on **Self-Check Quizzes—Chapter 19** to prepare for the Chapter Test.

Using the Regional Atlas

Refer to the Regional Atlas on pages 410–413.

1. **Location** For which densely populated areas in North Africa, Southwest Asia, and Central Asia are desalination plants practical?
2. **Human-Environment Interaction** What is the relationship between the locations of commercial farms and water resources in North Africa and Southwest Asia?

Thinking Like a Geographer

Consider human-environment interaction in the region. How have people's views about the environment there changed since ancient times? How have these changing viewpoints led to changes in lifestyles? Give examples.

Problem-Solving Activity

Contemporary Issues Case Study Research an OPEC member country. Find statistics that illustrate how important petroleum and natural gas are to the country's economy. Share a written summary of your findings with the class.

GeoJournal

Cause and Effect Using your GeoJournal data and other resources, write an essay that compares an environmental trouble spot in North Africa, Southwest Asia, and Central Asia with that in another region. Focus on the relationship between technology and environmental change.

Technology Activity

Using the Internet for Research On the Internet, search for an online news source with articles about North Africa, Southwest Asia, and Central Asia. Find a recent article about an environmental issue, and summarize the article. Be sure to cite source information on which the summary is based. Identify any biases, if present, in the article, and make sure the source is reliable. Then evaluate the impact of the issue on the region.

Standardized Test Practice

Choose the best answer for each of the following multiple-choice questions. If you have trouble answering the questions, use the process of elimination to narrow your choices.

1. **Imagine that you are hired to create a sketch map of Egypt to show the importance of the Nile River to Egypt's people. What combination of information would be the most useful to show?**

   A Coastal areas, mountains, and oil and phosphate resources
   B Population density, commercial farming, and deserts
   C Population density, subsistence farming, and city locations
   D Deserts, plateaus, and mountains

Use the Process of Elimination (POE) to answer this question. First, consider the physical features, land use, resources, and population patterns near the Nile River. Then, eliminate answer choices that contain even one feature or resource that is not likely to be found along the river. Choose your answer from those that remain.

2. **What set of latitude lines would be best to use on your sketch map of Egypt and the Nile River?**

   F 40°N, 50°N, 60°N
   G 20°S, 25°S, 30°S
   H 30°S, 0°N, 30°N
   J 20°N, 35°N, 70°N

Use POE to answer this question. First, visualize where Egypt lies in relation to the Equator on a map. Then, eliminate coordinates that are likely to be too far from Egypt.

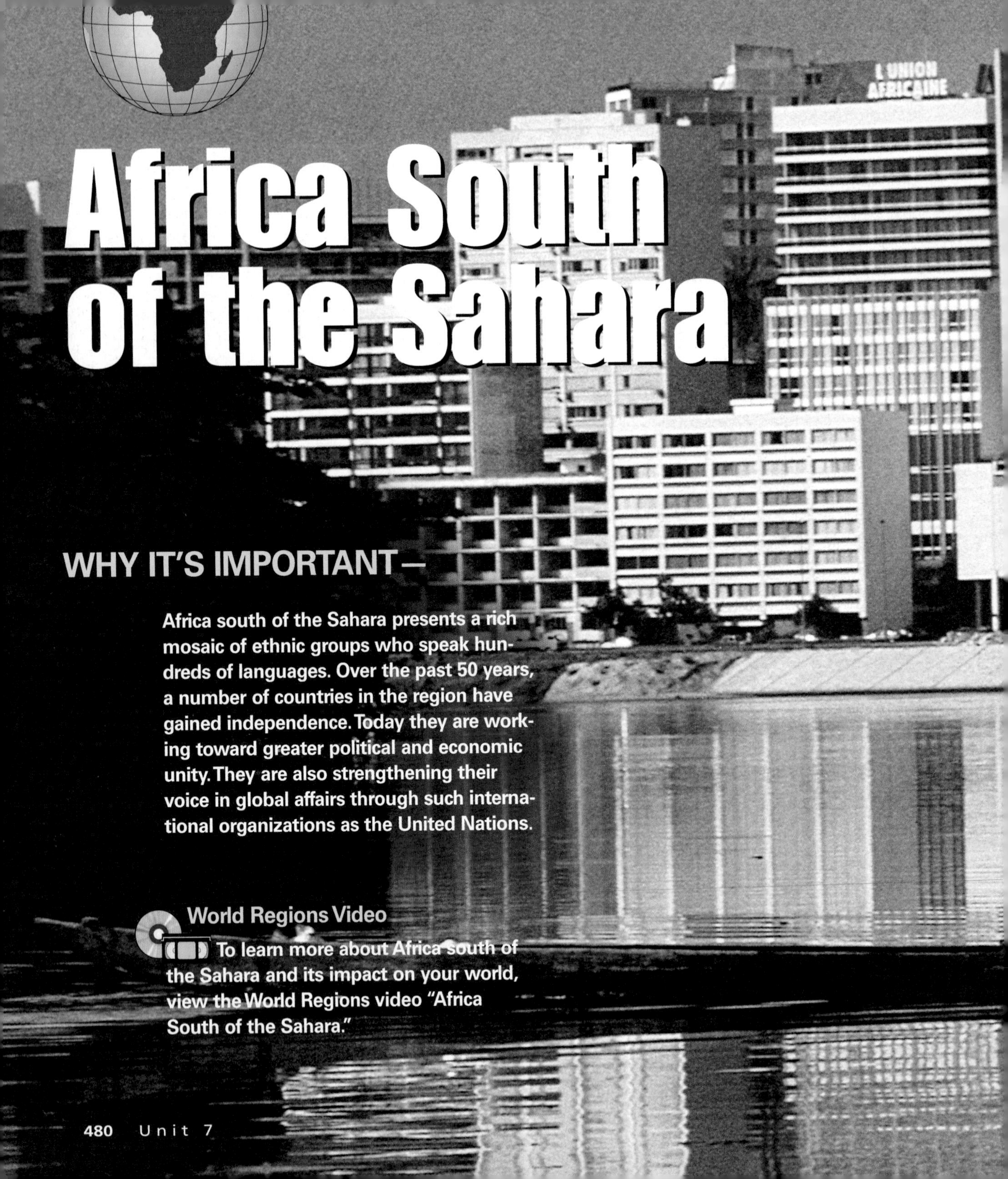

Africa South of the Sahara

WHY IT'S IMPORTANT—

Africa south of the Sahara presents a rich mosaic of ethnic groups who speak hundreds of languages. Over the past 50 years, a number of countries in the region have gained independence. Today they are working toward greater political and economic unity. They are also strengthening their voice in global affairs through such international organizations as the United Nations.

World Regions Video

To learn more about Africa south of the Sahara and its impact on your world, view the World Regions video "Africa South of the Sahara."

Port city of Abidjan, Côte d'Ivoire

What Makes Africa South of the Sahara a Region?

Straddling the Equator, Africa south of the Sahara encompasses about 9.5 million square miles (24.6 million sq. km) and nearly 50 countries. It is a region of immense plateaus that rise, like steps, from west to east across the continent. Several great rivers flow across this landscape. As the rivers journey to the sea, they cascade from one plateau to the next, creating spectacular waterfalls.

The Great Rift Valley, formed by the movement of the Earth's crust, slices through the plateaus of eastern Africa. Along the valley's rim stand some of the region's isolated mountain peaks, including the highest: snowcapped Kilimanjaro.

Most of this region lies within the Tropics. Closest to the Equator are steamy rain forests, second in size only to those of the Amazon River basin. At higher latitudes lie grasslands, home to many of Africa's famous wild animals. Beyond the grasslands, deserts stretch out under the fierce African sun.

1 Sculpted sand dunes rise in the Namib, one of Africa's deserts. The Namib lies in western Namibia, bordering the Atlantic Ocean. Eastern Namibia is home to another desert, the Kalahari, which stretches far into neighboring Botswana.

2 Covered in mineral-rich mud, a South African miner drills for gold. In the late 1800s, huge deposits of gold and diamonds were discovered in South Africa. Mining has made this country the wealthiest and most developed in the region.

3 A rainbow dances in the spray of Victoria Falls, on the Zambezi River. The river plummets 355 feet (108 m) as it spills over the edge of a steep cliff. The spray and the roar prompted local people to call the falls *Mosi oa Tunya*—"smoke that thunders."

4 Like regal lords, two male lions stride across an African savanna, or tropical grassland. Some savannas support huge herds of antelope, buffalo, wildebeests, and zebras, which are hunted by lions, cheetahs, and other predators.

Rich in Resources and Challenges

Many scientists believe that the human race originated in Africa millions of years ago. Ever since, the lands south of the Sahara have been home to diverse peoples, cultures, and empires. Europeans arrived in the 1400s and quickly began to exploit the region's abundant natural resources. By the 1800s, Africa was a patchwork of European colonies. Colonial rule ended in the twentieth century, leaving independent, but struggling, nations in its wake.

In Africa south of the Sahara, most of the people depend on small-scale agriculture or herding for their livelihood. Drought, disease, illiteracy, political instability, and poor transportation systems make economic development difficult in this region—the poorest of all world regions.

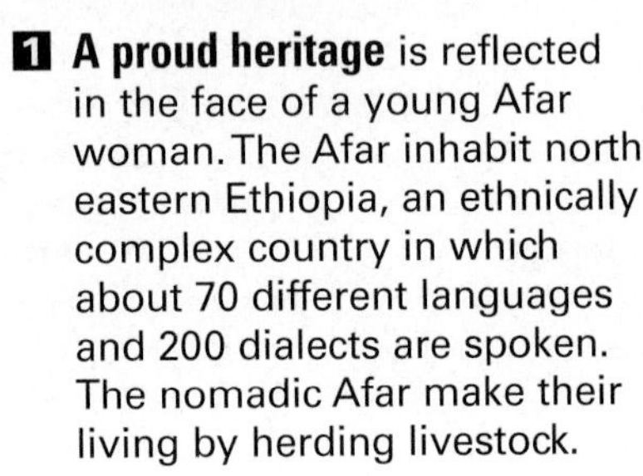

1 **A proud heritage** is reflected in the face of a young Afar woman. The Afar inhabit northeastern Ethiopia, an ethnically complex country in which about 70 different languages and 200 dialects are spoken. The nomadic Afar make their living by herding livestock.

2 Mud-brick walls of an old mosque rise behind a busy market in Djenné, Mali. Five centuries ago, Djenné was a center of commerce and Muslim scholarship in the Songhai Empire, a wealthy and powerful trading kingdom.

3 Table Mountain rises steeply behind Cape Town, in South Africa. The city was established in 1652 as a port of call for Dutch ships sailing from Europe to India. Today, Cape Town is an important shipping center as well as the legislative capital of South Africa.

4 Waist-deep in tea plants, a Kenyan man picks leaves that will go into making one of the world's most popular drinks. Most farms in Africa are small, but Kenya has several large plantations that grow cash crops of tea and coffee for export.

Africa South of the Sahara

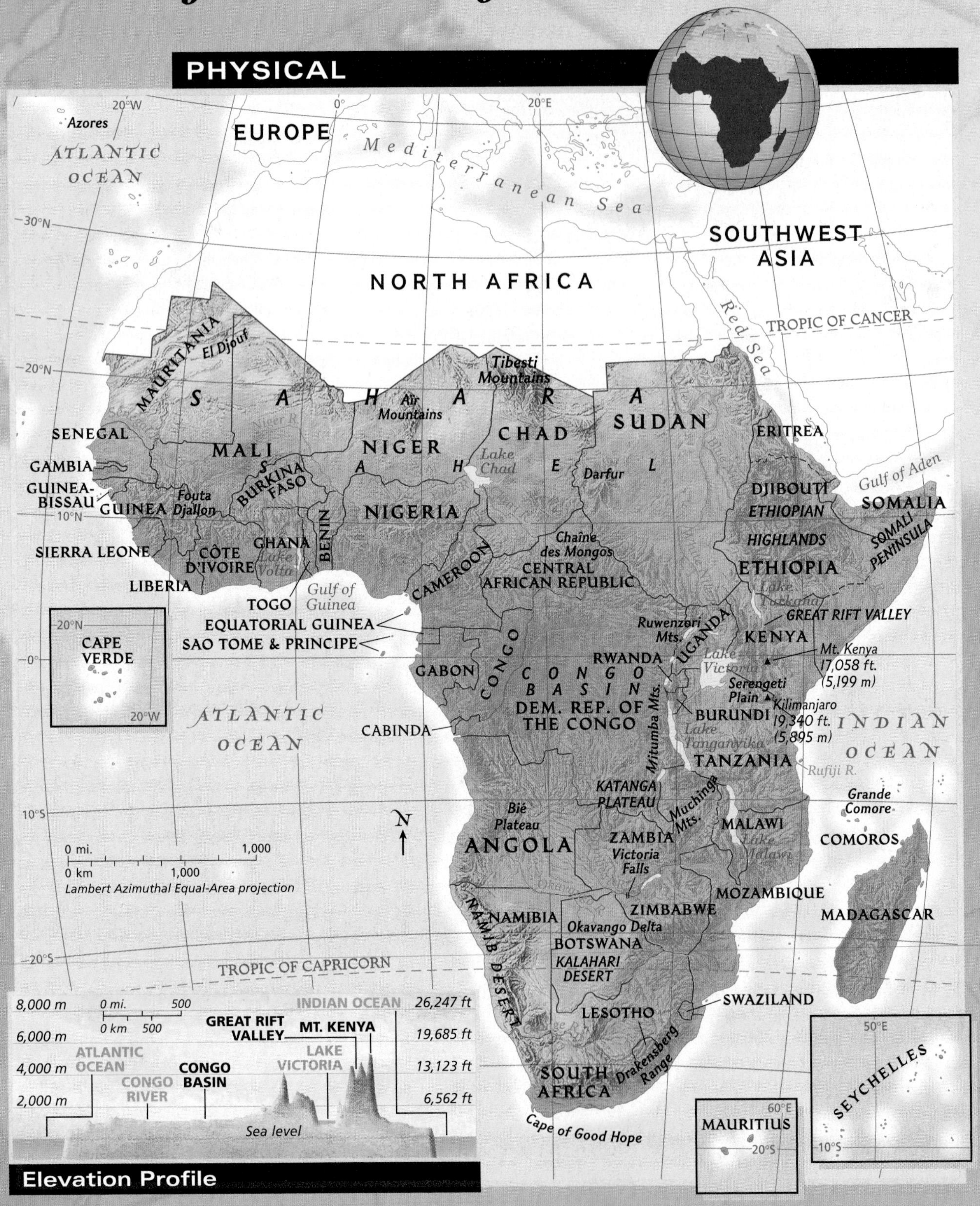

POLITICAL

⊛ National capital
• Major city

0 mi. 1,000
0 km 1,000
Lambert Azimuthal Equal-Area projection

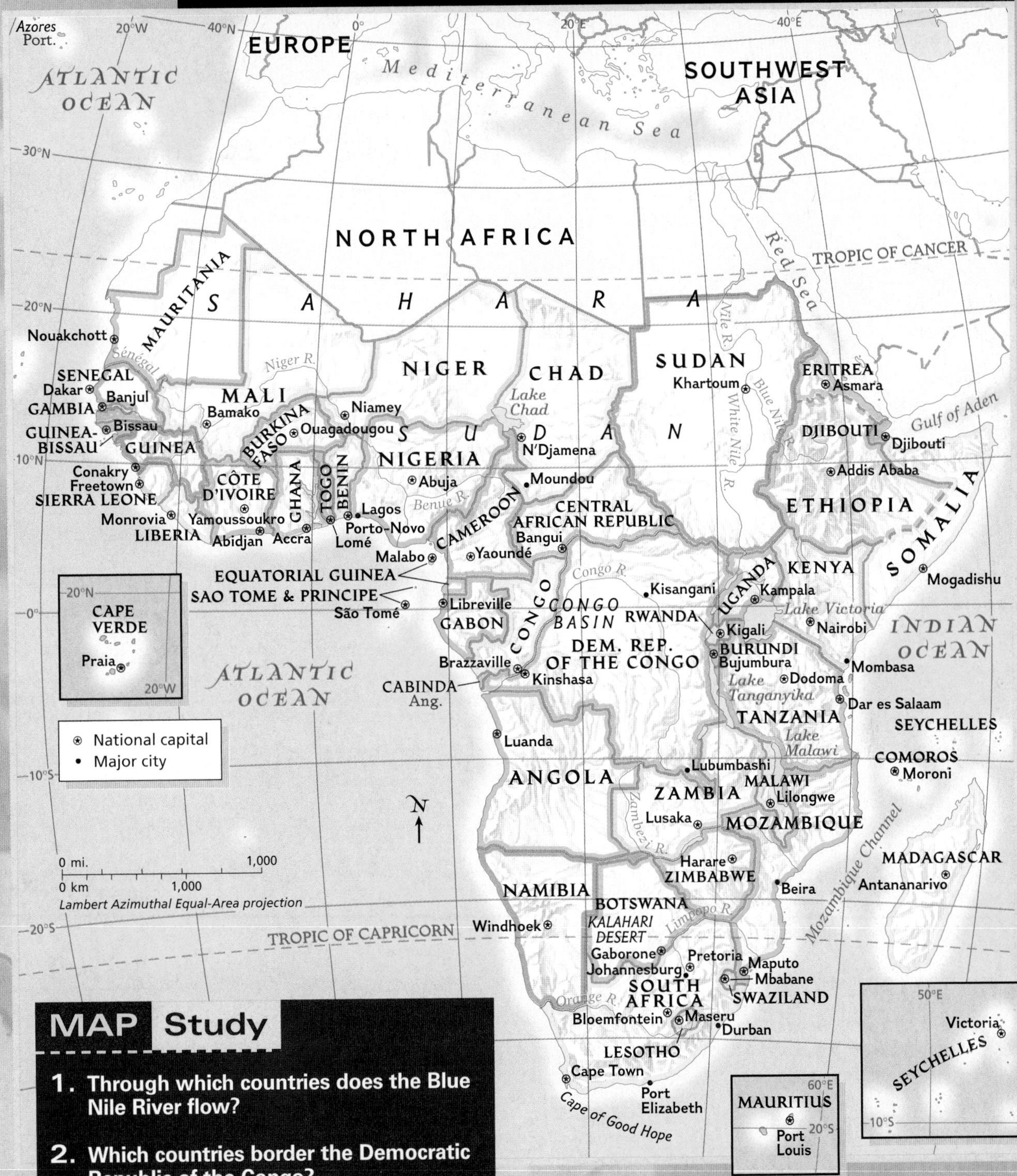

MAP Study

1. Through which countries does the Blue Nile River flow?
2. Which countries border the Democratic Republic of the Congo?

Africa South of the Sahara

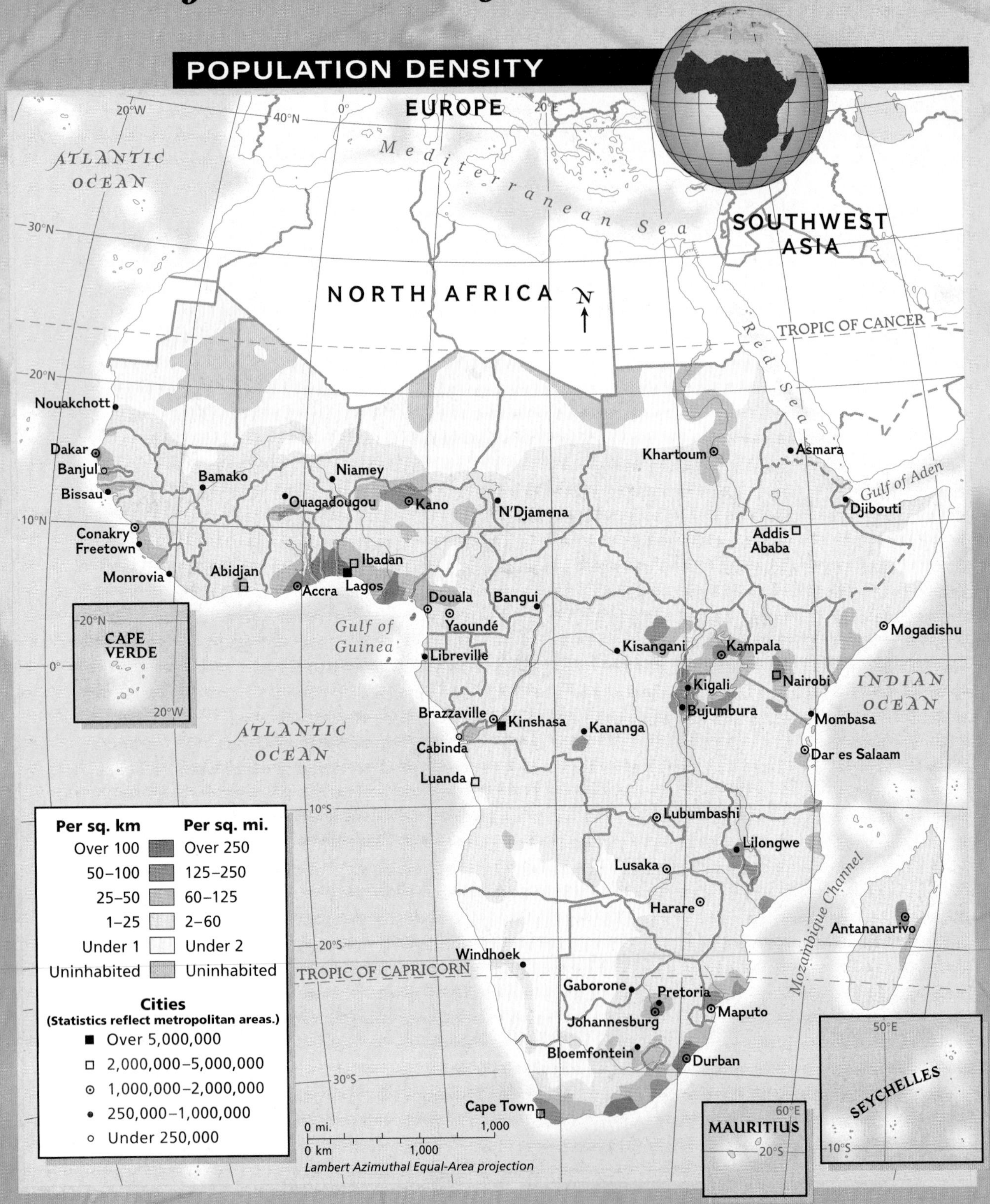

ECONOMIC ACTIVITY

Resources

- Diamonds
- Gold
- Copper
- Uranium
- Manganese
- Cobalt
- Zinc
- Petroleum
- Silver

Land Use

- Commercial farming
- Subsistence farming
- Livestock raising
- Nomadic herding
- Manufacturing and trade
- Commercial fishing
- Little or no activity

0 mi. 1,000
0 km 1,000
Lambert Azimuthal Equal-Area projection

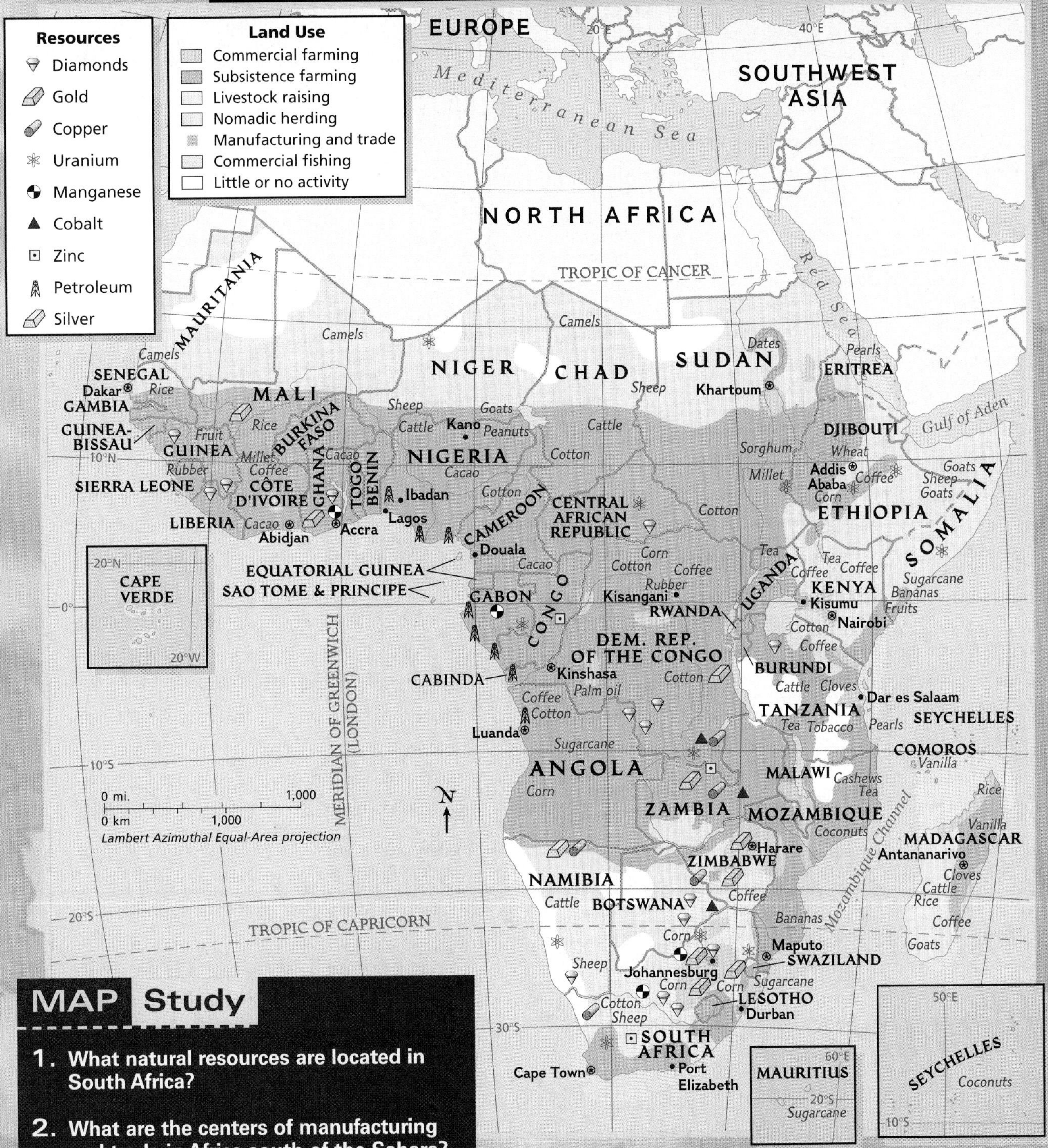

MAP Study

1. What natural resources are located in South Africa?
2. What are the centers of manufacturing and trade in Africa south of the Sahara?

Africa South of the Sahara

COUNTRY PROFILES

COUNTRY * AND CAPITAL	FLAG AND LANGUAGE	POPULATION AND DENSITY	LANDMASS	MAJOR EXPORT	MAJOR IMPORT	CURRENCY	GOVERNMENT
ANGOLA Luanda	Portuguese, Local Languages	12,300,000 26 per sq.mi. 10 per sq. km	481,351 sq. mi. 1,246,699 sq. km	Crude Oil	Machinery	Kwanza	Republic
BENIN Porto-Novo	French, Fon, Yoruba	6,600,000 152 per sq. mi. 59 per sq. km	43,483 sq. mi. 112,621 sq. km	Cotton	Foods	CFA Franc	Republic
BOTSWANA Gaborone	English, Setswana	1,600,000 7 per sq. mi. 3 per sq. km	224,606 sq. mi. 581,730 sq. km	Diamonds	Foods	Pula	Republic
BURKINA FASO Ouagadougou	French, Local Languages	12,300,000 116 per sq. mi. 45 per sq. km	105,792 sq. mi. 274,000 sq. km	Cotton	Machinery	CFA Franc	Republic
BURUNDI Bujumbura	Kirundi, French	6,200,000 519 per sq. mi. 200 per sq. km	10,745 sq. mi. 27,830 sq. km	Coffee	Machinery	Burundi Franc	Republic
CAMEROON Yaoundé	French, English, Local Languages	15,800,000 86 per sq. mi. 33 per sq. km	183,568 sq. mi. 475,442 sq. km	Crude Oil	Machinery	CFA Franc	Republic
CAPE VERDE Praia	Portuguese, Crioulo	400,000 287 per sq. mi. 111 per sq. km	1,556 sq. mi. 4,030 sq. km	Shoes	Foods	Cape Verdean Escudo	Republic
CENTRAL AFRICAN REPUBLIC Bangui	French, Sango, Arabic, Hunsa	3,600,000 15 per sq. mi. 7 per sq. km	240,533 sq. mi. 622,980 sq. km	Diamonds	Foods	CFA Franc	Republic
CHAD N'Djamena	French, Arabic, Sara, Sango	8,700,000 18 per sq. mi. 7 per sq. km	495,753 sq. mi. 1,283,948 sq. km	Cotton	Machinery	CFA Franc	Republic
COMOROS Moroni	Arabic, French, Comoran	600,000 692 per sq. mi. 267 per sq. km	861 sq. mi. 2,230 sq. km	Vanilla	Rice	CFA Franc	Republic

* COUNTRIES AND FLAGS NOT DRAWN TO SCALE

COUNTRY * AND CAPITAL	FLAG AND LANGUAGE	POPULATION AND DENSITY	LANDMASS	MAJOR EXPORT	MAJOR IMPORT	CURRENCY	GOVERNMENT
CONGO Brazzaville	French, Lingala, Monokutuba	3,100,000 24 per sq. mi. 9 per sq. km	132,046 sq. mi. 341,999 sq. km	Crude Oil	Machinery	CFA Franc	Republic
CONGO, DEMOCRATIC REPUBLIC OF THE Kinshasa	French, Lingala, Kingwana	53,600,000 59 per sq. mi. 23 per sq. km	905,351 sq. mi. 2,344,859 sq. km	Diamonds	Manufactured Goods	Congolese Franc	Republic
CÔTE D'IVOIRE Yamoussoukro Abidjan	French, Dioula	16,400,000 132 per sq. mi. 60 per sq. km	124,502 sq. mi. 322,460 sq. km	Cocoa	Foods	CFA Franc	Republic
DJIBOUTI Djibouti	French, Arabic	600,000 71 per sq. mi. 27 per sq. km	8,958 sq. mi. 23,200 sq. km	Hides and Skins	Foods	Djibouti Franc	Republic
EQUATORIAL GUINEA Malabo	Spanish, French, Fang, Bubi, Ibo	500,000 43 per sq. mi. 17 per sq. km	10,830 sq. mi. 28,050 sq. km	Petroleum	Machinery	CFA Franc	Republic
ERITREA Asmara	Afar, Amharic, Arabic, Tigre	4,300,000 95 per sq. mi. 37 per sq. km	45,405 sq. mi. 117,599 sq. km	Livestock	Processed Foods	Nakfa	Republic
ETHIOPIA Addis Ababa	Amharic, Tigrinya, Orominga	65,400,000 153 per sq. mi. 59 per sq. km	426,371 sq. mi. 1,104,301 sq. km	Coffee	Foods and Livestock	Birr	Federal Republic
GABON Libreville	French, Local Languages	1,200,000 12 per sq. mi. 4 per sq. km	103,347 sq. mi. 267,667 sq. km	Crude Oil	Machinery	CFA Franc	Republic
GAMBIA Banjul	English, Mandinka, Fula	1,400,000 323 per sq. mi. 125 per sq. km	4,363 sq. mi. 11,300 sq. km	Peanuts	Foods	Dalasi	Republic
GHANA Accra	English, Local Languages	19,900,000 216 per sq. mi. 83 per sq. km	92,100 sq. mi. 238,537 sq. km	Gold	Machinery	Cedi	Republic

* COUNTRIES AND FLAGS NOT DRAWN TO SCALE

FOR AN ONLINE UPDATE OF THIS INFORMATION, VISIT GEOGRAPHY.GLENCOE.COM AND CLICK ON "TEXTBOOK UPDATES."

Africa South of the Sahara

COUNTRY PROFILES

COUNTRY * AND CAPITAL	FLAG AND LANGUAGE	POPULATION AND DENSITY	LANDMASS	MAJOR EXPORT	MAJOR IMPORT	CURRENCY	GOVERNMENT
GUINEA Conakry	French, Local Languages	7,600,000 80 per sq. mi. 31 per sq. km	94,927 sq. mi. 245,861 sq. km	Bauxite	Petroleum Products	Guinean Franc	Republic
GUINEA-BISSAU Bissau	Portuguese, Crioulo, Fula	1,200,000 88 per sq. mi. 34 per sq. km	13,946 sq. mi. 36,120 sq. km	Cashews	Foods	CFA Franc	Republic
KENYA Nairobi	English, Swahili	29,800,000 133 per sq. mi. 51 per sq. km	224,081 sq. mi. 580,370 sq. km	Tea	Machinery	Kenyan Shilling	Republic
LESOTHO Maseru	English, Sesotho, Zulu, Xhosa	2,200,000 186 per sq. mi. 72 per sq. km	11,718 sq. mi. 30,350 sq. km	Clothing	Corn	Loti	Constitutional Monarchy
LIBERIA Monrovia	English, Local Languages	3,200,000 75 per sq. mi. 29 per sq. km	43,000 sq. mi. 111,369 sq. km	Diamonds	Natural Gas	Liberian Dollar	Republic
MADAGASCAR Antananarivo	French, Malagasy	16,400,000 71 per sq. mi. 27 per sq. km	226,656 sq. mi. 587,039 sq. km	Coffee	Machinery	Malagasy Franc	Republic
MALAWI Lilongwe	Chewa, English	10,500,000 231 per sq. mi. 89 per sq. km	45,745 sq. mi. 118,480 sq. km	Tobacco	Foods	Kwacha	Republic
MALI Bamako	French, Bambara	11,000,000 23 per sq. mi. 9 per sq. km	478,838 sq. mi. 1,240,307 sq. km	Cotton	Machinery	CFA Franc	Republic
MAURITANIA Nouakchott	Hasaniya Arabic, Wolof	2,700,000 7 per sq. mi. 3 per sq. km	395,954 sq. mi. 1,025,521 sq. km	Fish	Foods	Ouguiya	Islamic Republic
MAURITIUS Port Louis	English, Creole, Bhojpuri, French	1,200,000 1,520 per sq. mi. 587 per sq. km	788 sq. mi. 2,040 sq. km	Sugar	Foods	Mauritian Rupee	Republic

* COUNTRIES AND FLAGS NOT DRAWN TO SCALE

COUNTRY * AND CAPITAL	FLAG AND LANGUAGE	POPULATION AND DENSITY	LANDMASS	MAJOR EXPORT	MAJOR IMPORT	CURRENCY	GOVERNMENT
MOZAMBIQUE Maputo	Portuguese, Local Languages	19,400,000 63 per sq.mi. 24 per sq.km	309,494 sq.mi. 801,598 sq.km	Cashews	Foods	Metical	Republic
NAMIBIA Windhoek	English, Afrikaans, Local Languages	1,800,000 6 per sq.mi. 3 per sq.km	318,259 sq.mi. 824,291 sq.km	Diamonds	Construction Materials	Namibian Dollar	Republic
NIGER Niamey	French, Hausa, Djerma	10,400,000 21 per sq.mi. 8 per sq.km	489,189 sq.mi. 1,267,000 sq.km	Uranium	Manufactured Goods	CFA Franc	Republic
NIGERIA Abuja	English, Hausa, Yoruba, Igbo	126,600,000 355 per sq.mi. 137 per sq.km	356,668 sq.mi. 923,770 sq.km	Petroleum	Machinery	Naira	Federal Republic
RWANDA Kigali	Kinyarwanda, French, English	7,300,000 719 per sq.mi. 278 per sq.km	10,170 sq.mi. 26,340 sq.km	Coffee	Foods	Rwanda Franc	Republic
SAO TOME AND PRINCIPE São Tomé	Portuguese, Crioulo	260,000 445 per sq.mi. 172 per sq.km	371 sq.mi. 961 sq.km	Cocoa	Textiles	Dobra	Republic
SENEGAL Dakar	French, Wolof, Pulaar, Diola	9,700,000 127 per sq.mi. 49 per sq.km	75,954 sq.mi. 196,721 sq.km	Fish	Foods	CFA Franc	Republic
SEYCHELLES Victoria	English, French, Creole	100,000 449 per sq.mi. 173 per sq.km	174 sq.mi. 451 sq.km	Fish	Foods	Seychelles Rupee	Republic
SIERRA LEONE Freetown	English, Mende, Temne, Krio	5,400,000 196 per sq.mi. 76 per sq.km	27,699 sq.mi. 71,740 sq.km	Diamonds	Foods	Leone	Republic
SOMALIA Mogadishu	Somali, Arabic	7,500,000 30 per sq.mi. 12 per sq.km	246,201 sq.mi. 637,657 sq.km	Livestock	Textiles	Somali Shilling	Republic

* COUNTRIES AND FLAGS NOT DRAWN TO SCALE

FOR AN ONLINE UPDATE OF THIS INFORMATION, VISIT GEOGRAPHY.GLENCOE.COM AND CLICK ON "TEXTBOOK UPDATES."

7

Africa South of the Sahara

COUNTRY PROFILES

COUNTRY * AND CAPITAL	FLAG AND LANGUAGE	POPULATION AND DENSITY	LANDMASS	MAJOR EXPORT	MAJOR IMPORT	CURRENCY	GOVERNMENT
SOUTH AFRICA Pretoria Bloemfontein Cape Town	Afrikaans, English, Zulu	43,600,000 92 per sq. mi. 36 per sq. km	471,444 sq. mi. 1,221,038 sq. km	Gold	Transport Equipment	Rand	Republic
SUDAN Khartoum	Arabic, Nubian, Ta Bedawie	31,800,000 33 per sq. mi. 13 per sq. km	967,494 sq. mi. 2,505,809 sq. km	Cotton	Petroleum Products	Sudanese Pound	Republic
SWAZILAND Mbabane	English, Swazi	1,100,000 165 per sq. mi. 64 per sq. km	6,703 sq. mi. 17,361 sq. km	Soft Drink Concentrates	Machinery	Lilangeni	Monarchy
TANZANIA Dodoma Dar es Salaam	Swahili, English	36,200,000 99 per sq. mi. 38 per sq. km	364,900 sq. mi. 945,087 sq. km	Coffee	Machinery	Tanzanian Shilling	Republic
TOGO Lomé	French, Ewe, Mina, Kabye	5,200,000 235 per sq. mi. 91 per sq. km	21,927 sq. mi. 56,791 sq. km	Phosphates	Manufactured Goods	CFA Franc	Republic
UGANDA Kampala	English, Ganda	24,000,000 258 per sq. mi. 100 per sq. km	93,066 sq. mi. 241,041 sq. km	Coffee	Machinery	Ugandan Shilling	Republic
ZAMBIA Lusaka	English, Local Languages	9,800,000 34 per sq. mi. 13 per sq. km	290,583 sq. mi. 752,610 sq. km	Copper	Manufactured Goods	Kwacha	Republic
ZIMBABWE Harare	English, Shona, Sindebele	11,400,000 75 per sq. mi. 30 per sq. km	150,873 sq. mi. 390,761 sq. km	Gold	Machinery	Zimbabwean Dollar	Republic

* COUNTRIES AND FLAGS NOT DRAWN TO SCALE

FOR AN ONLINE UPDATE OF THIS INFORMATION, VISIT GEOGRAPHY.GLENCOE.COM AND CLICK ON "TEXTBOOK UPDATES."

▶ Crowd in Burundi market wearing colorful traditional clothing

NATIONAL GEOGRAPHIC SOCIETY

GLOBAL CONNECTION

AFRICA SOUTH OF THE SAHARA AND THE UNITED STATES

ROOTS OF JAZZ

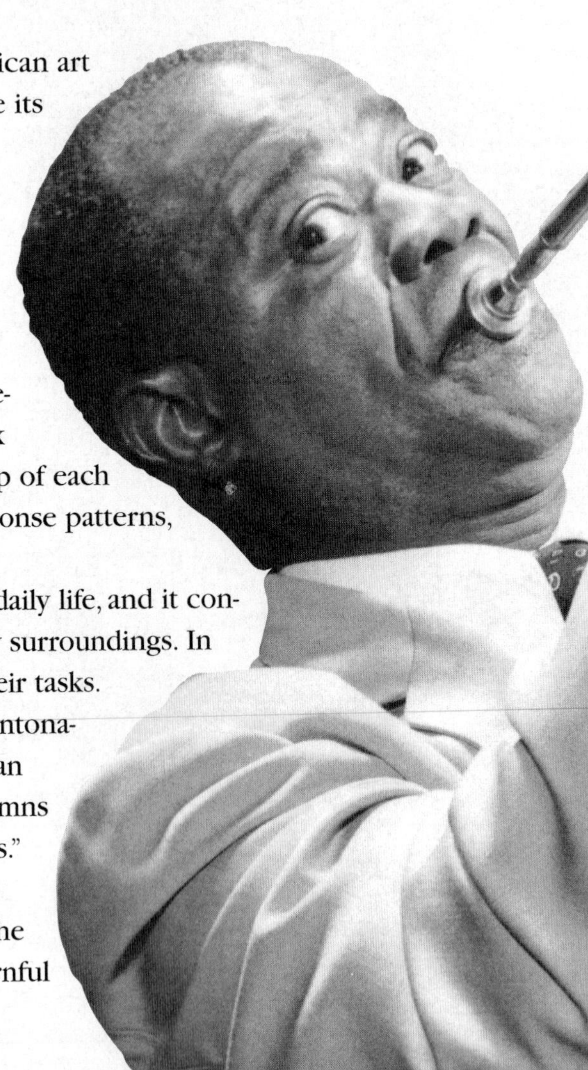

It's been called the only truly American art form. However, jazz music can trace its roots straight back to Africa.

Beginning in the 1600s, thousands of black Africans, mostly from West Africa, were forcibly brought to the American colonies as slaves. Torn from their homelands, these enslaved Africans held tightly to the only possessions they retained—their cultural traditions, especially their music. That music had strong, complex rhythms, with layers of different beats built on top of each other. Other characteristics included call-and-response patterns, sliding-pitch tones, and improvisation.

In Africa, music had been an essential part of daily life, and it continued to be for the enslaved people in their new surroundings. In the fields, they sang to relieve the drudgery of their tasks. Their work songs echoed with the rhythms and intonations of their homelands. Forced to adopt Christian beliefs, the enslaved people altered traditional hymns to suit their own tastes, creating soulful "spirituals."

After the Civil War, work songs, spirituals, and other influences came together to give birth to the blues—simple, repeated harmonies set to a mournful scale. Close on the heels of the blues came ragtime—a jaunty style of piano music with a distinctly African rhythmic undertone.

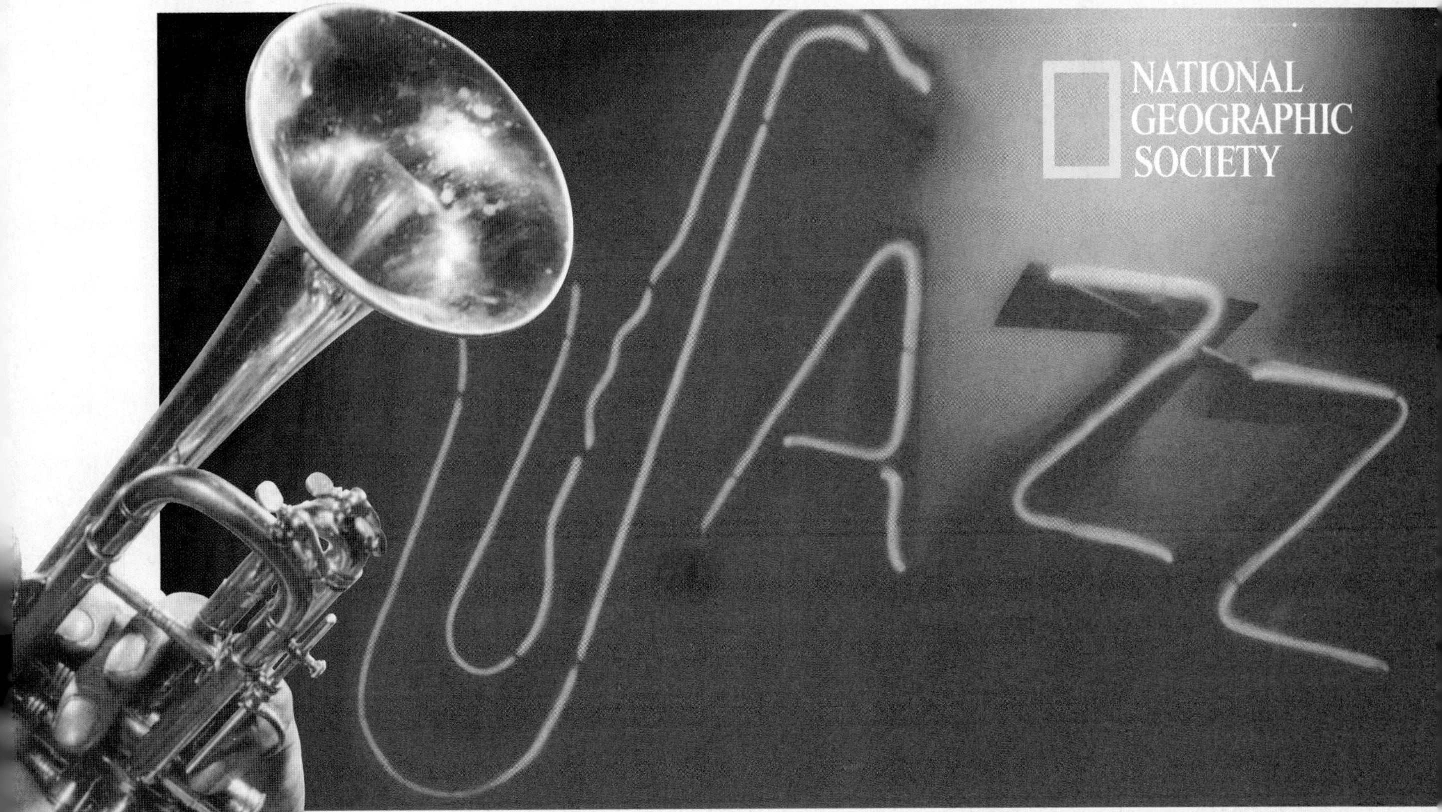

It was in New Orleans that blues and ragtime blended with Creole, European, and other influences to become jazz. By 1917, the new music echoed throughout New Orleans.

In the 1920s, the center of jazz began shifting northward, first to Chicago and then to New York. In the decades that followed, jazz underwent many changes. New forms of jazz emerged, such as big band swing, bebop, and cool jazz. Jazz continues to develop today. No matter what variations appear, however, the roots of jazz remain firmly planted in Africa.

▲ West African musicians

◀ American jazz innovator Louis Armstrong

CHAPTER 20

The Physical Geography of Africa South of the Sahara

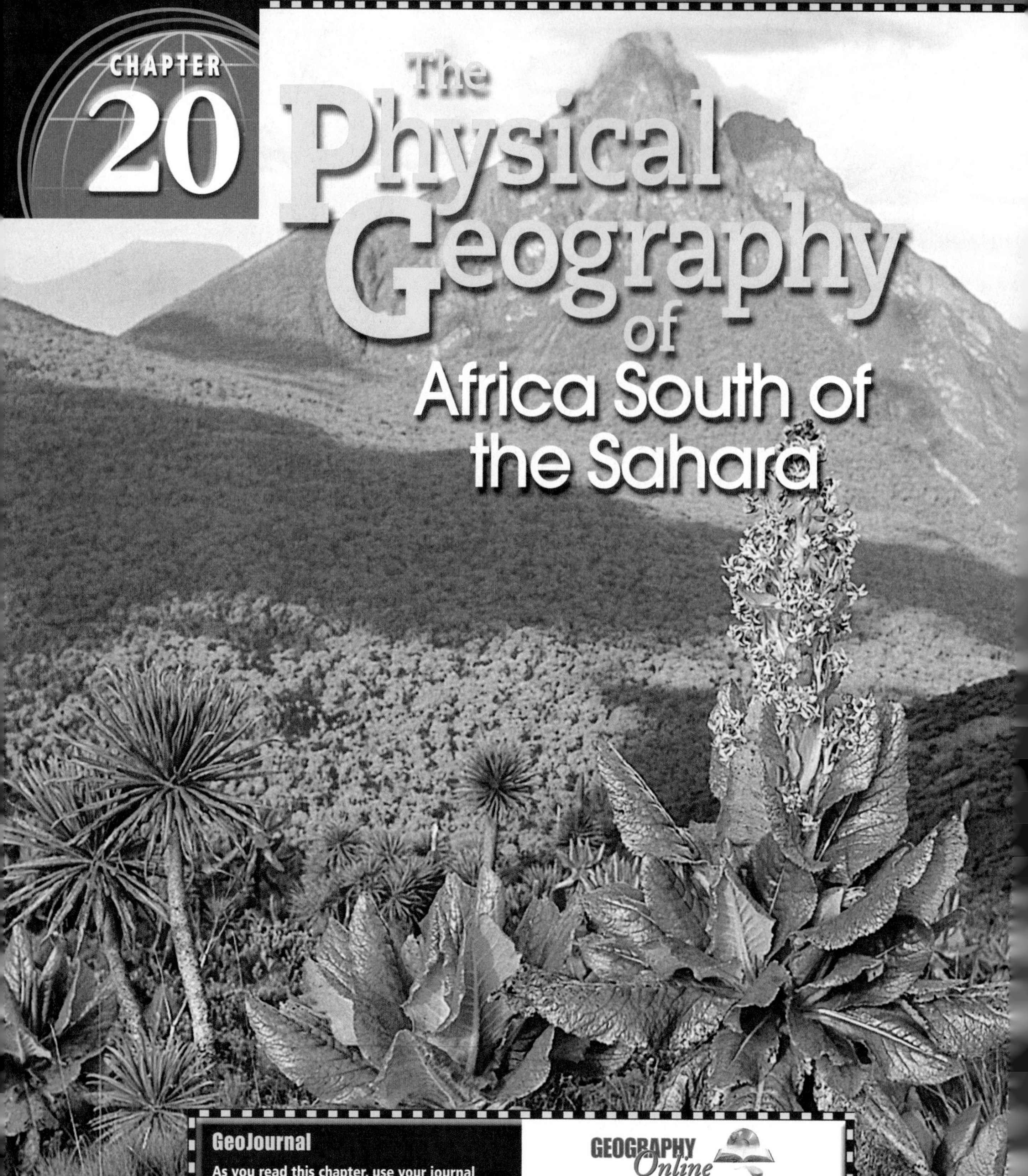

GeoJournal

As you read this chapter, use your journal to compare and contrast the physical geography of Africa south of the Sahara with that of Latin America. List similarities and differences in your journal.

GEOGRAPHY Online

Chapter Overview Visit the **Glencoe World Geography** Web site at geography.glencoe.com and click on Chapter Overviews—Chapter 20 to preview information about the physical geography of the region.

SECTION 1

The Land

Guide to Reading

Consider What You Know

Africa south of the Sahara is known throughout the world for its extraordinary physical geography as well as its wildlife. What animals do you associate with the region? What landforms do you think are present in the places where those animals live?

Read to Find Out

- What are the major landforms in Africa south of the Sahara?
- How does the land affect the water systems of Africa south of the Sahara?
- What are the region's most important natural resources?

Terms to Know

- escarpment
- cataract
- rift valley
- fault
- delta
- estuary

Places to Locate

- Ruwenzori Mountains
- Drakensberg Range
- Great Rift Valley
- Lake Victoria
- Niger River
- Zambezi River
- Congo River

◀ *Mt. Mikeno, Democratic Republic of the Congo*

NATIONAL GEOGRAPHIC

A Geographic View

Across the Great Rift

The road has risen swiftly as we leave Nakuru, a town that sits just south of the Equator in Kenya. . . . We drive beneath cool gray rain clouds. . . . Abruptly the clouds break, and before our windshield the red earth drops away. A gash 700 meters (2,300 feet) deep and 16 kilometers (ten miles) across—an offshoot of Africa's Great Rift—lies before us. Far to the left, stretching beyond the horizon, I see a glimmering expanse of water—Lake Victoria.

—*Curt Stager, "Africa's Great Rift,"* National Geographic, *May 1990*

Lake Victoria, East Africa

Nothing in his research prepared biologist Curt Stager for the breathtaking views on the road out of Nakuru, Kenya, in East Africa. Along the Great Rift, long troughs and deep lakes slash the land in the shadow of majestic volcanic mountains. In this section you will read about the dramatic physical features and the rich natural resources of Africa south of Sahara.

Landforms

Africa south of the Sahara is an immense region covering about 9.5 million square miles (24.6 million sq. km). Bounded on the north by the Sahara, the region extends to the sea in all other directions. To its northeast is the Red Sea, to its west the Atlantic Ocean, and to its east the Indian Ocean. On the southern edge of Africa south of the Sahara, the waters of the Atlantic and Indian Oceans meet at the Cape of Good Hope. This vast region embraces a broad variety of physical features.

Plateaus, Highlands, and Mountains

Seen from space, Africa south of the Sahara might be described as a series of steps. These steps are actually plateaus that rise in elevation from the coast inland and from west to east. Ranging in elevation from 500 feet (152 m) in the west to 8,000 feet (2,438 m) or more in the east, the plateaus are outcroppings of the solid rock that makes up most of Africa.

The edges of the continent's plateaus are marked by escarpments—steep, often jagged slopes or cliffs. Most of the escarpments are located less than 20 miles (32 km) from the coast. Rivers crossing the plateaus plunge suddenly down the sides of the escarpments in cataracts, or towering waterfalls.

Although Africa's overall surface is higher in average elevation than that of every other continent, it has relatively few mountains. Most African mountains dot the Eastern Highlands, an area that stretches from Ethiopia almost to the Cape of Good Hope. These highland areas include the Ethiopian Highlands as well as volcanic summits, such as Kilimanjaro and Mount Kenya.

West of the Eastern Highlands, the **Ruwenzori** (ROO•wuhn•ZOHR•ee) **Mountains** divide Uganda and the Democratic Republic of the Congo. Covered with snow and cloaked in clouds, these mountains, also called the "Mountains of the Moon," have fascinated observers since ancient times. The writer Christopher Ondaatje, who traveled to East Africa in the 1990s, recorded these impressions of the legendary range:

> *"One can feel the water in the air. It is damp, dank, grey, and cold. . . . [T]he clouds seldom clear so the upper slopes remain veiled. Still, we were constantly aware of them, a powerful presence brooding over the surrounding countryside."*
>
> Christopher Ondaatje, *Journey to the Source of the Nile*, 1999

Moist air from the Indian Ocean creates the clouds that wrap around the Ruwenzoris and give these mountains their wondrous appearance.

Farther south are the Cape Mountains, which include the **Drakensberg Range** in South Africa and Lesotho. These mountains rise to more than 11,000 feet (3,353 m) and form part of the sharp escarpment along the southern edge of the continent.

The Great Rift Valley

An amazing natural wonder known as the **Great Rift Valley** stretches from Syria in Southwest Asia to Mozambique (MOH•zahm•BEEK) in the southeastern part of Africa. A rift valley is a large crack in the earth's surface formed by shifting tectonic plates. Millions of years ago, plate movements created the system of faults or fractures in the earth's crust within which the Great Rift Valley lies. Volcanic eruptions as well as earthquakes helped create the valley's striking landscape, and they continue to shape it today.

In East Africa, the Great Rift Valley forms two branches, with volcanic mountains rising at its edges and deep lakes that run parallel to its length. The main volcanic cones, among them Kilimanjaro, are found along the eastern branch. Lake Tanganyika, one of the deepest and longest freshwater lakes in the world, lies on the western branch. To the south is Lake Malawi, a mountain-rimmed lake that looks much like a fjord. Like the glacier-cut valleys of seawater in northern Europe, Lake Malawi lies well below the land surrounding it. It is also very deep, its floor dropping to more than 2,300 feet (700 m) at its deepest point.

Water Systems

The land has influenced the water systems of Africa south of the Sahara in important ways. The lakes and rivers that drain the region are located in huge basins formed millions of years ago by the uplifting of the land. The great rivers of Africa

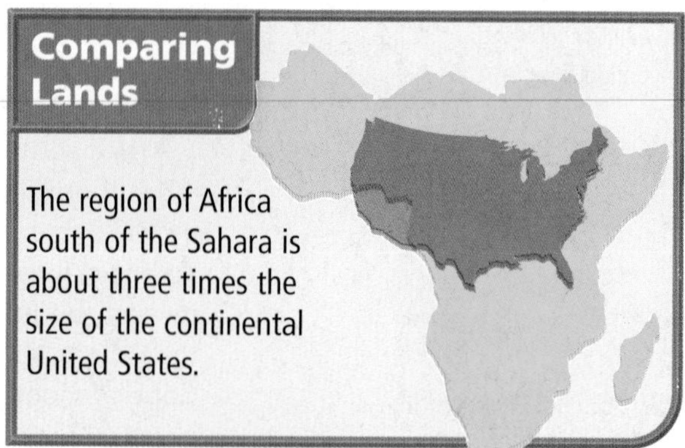

Comparing Lands

The region of Africa south of the Sahara is about three times the size of the continental United States.

MAP STUDY

Africa South of the Sahara: Physical/Political

Geography Skills for Life

1. **Interpreting Maps** What country is completely surrounded by South Africa?
2. **Applying Geography Skills** Where are areas of highest elevation in Africa south of the Sahara?

originate high in the plateaus and eventually make their way to the sea. Escarpments and ridges, also created long ago by movements of the earth's crust, frequently break the rivers' paths to the ocean with rapids, waterfalls, and cataracts. This broken landscape makes it impossible to navigate most of the region's rivers from mouth to source.

Land of Lakes

Most of the region's lakes, including Lakes Tanganyika and Malawi, are near the Great Rift Valley. **Lake Victoria**, the largest lake in Africa, lies between the eastern and western branches of the Great Rift. It is the world's second largest freshwater lake, after Lake Superior in North America. Lake Victoria is the source of the White Nile River. Despite its large size, Lake Victoria is comparatively shallow at only 270 feet (82 m) deep.

Lake Chad, outside the Great Rift Valley in west-central Africa, is threatened with extinction. Although fed by three large streams, landlocked Lake Chad is shrinking. Droughts in the 1970s completely dried up the northern portion of the lake, and the water level continues to be shallow even during years when rainfall is normal. Because of the arid climate, much of the lake's water evaporates or seeps into the ground.

Economics

A Lake Meets Many Needs

Lake Volta in West Africa ranks among the largest human-made lakes in the world. This artificial lake was created in the 1960s by damming the Volta River south of Ajena, Ghana. The new lake flooded more than 700 villages, forcing more than 70,000 people to find new places to live.

Although the dam was originally built as part of a hydroelectric project to provide power to an aluminum plant, the people of Ghana today benefit from the lake in many ways. Capable of storing 124 million acre-feet (153 billion cubic m) of water, Lake Volta supplies irrigation for farming in the

Geography Skills for Life

Niger River The Niger River spreads into a vast inland delta. The river is important for farming, travel, trade, and fishing (inset).

Place What are the longest rivers in Africa south of the Sahara?

plains below the dam and is well stocked with fish. In addition to supplying power to the aluminum industry in the port of Tema, the hydroelectric plant now generates electricity used throughout Ghana.

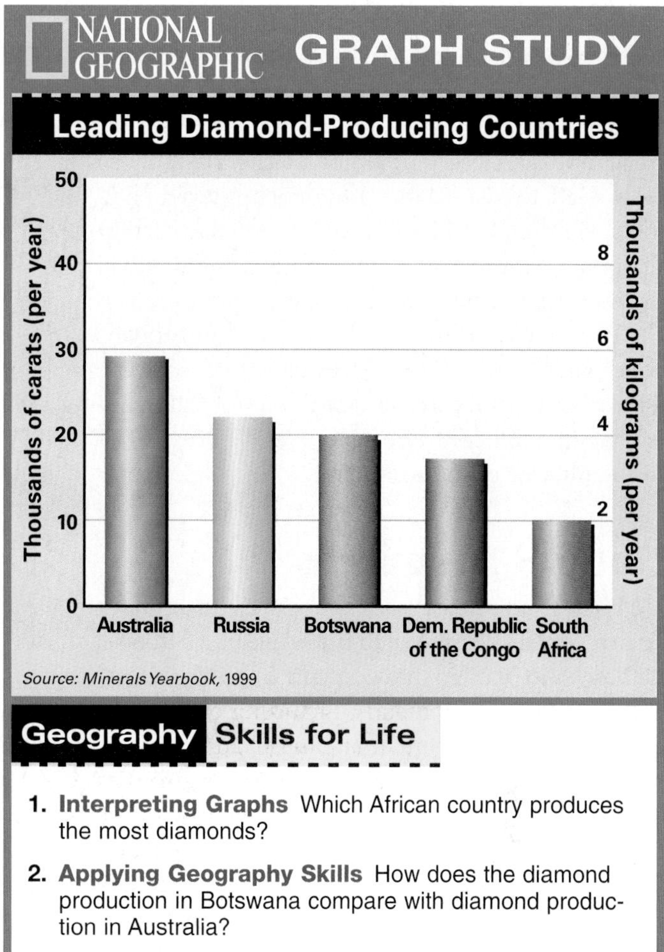

River Basins

The **Niger** (NY•juhr) **River** is known by many names along its course, but all its names have roughly the same meaning—"great river." The Niger is the main artery in western Africa. Originating in the highlands of Guinea only 150 miles (241 km) inland from the Atlantic Ocean, the river forms a great arc. It flows about 2,600 miles (4,184 km) northeast and then curves southeast to meet the Atlantic Ocean at the coast of Nigeria. In addition to being vitally important to agriculture, the Niger River is a major means of transportation for people in the region. It also provides a leisurely means of travel for tourists.

This great river does not flow as one well-defined stream into the sea. At Aboh in southern Nigeria, the Niger splits into a vast inland **delta**, a triangular section of land formed by sand and silt carried downriver. The Niger Delta stretches 150 miles (241 km) north to south and extends to a width of about 200 miles (322 km) along the shore of the Gulf of Guinea.

The **Zambezi River** of south-central Africa also meets the ocean in a delta. The Zambezi flows 2,200 miles (3,540 km) from its source near the Zambia-Angola border in the west to the Indian Ocean in the east, where it fans out in a delta that is 37 miles (60 km) wide. The Zambezi's course to the sea is interrupted in many places by waterfalls. At Victoria Falls, on the border of Zambia and Zimbabwe, the Zambezi plummets a sheer 355 feet (108 m), about twice the drop of Horseshoe Falls on the Niagara River between Canada and the United States. The water at Victoria Falls flows at 35,400 cubic feet (1,002 cubic m) per second.

Unlike the Niger, the Zambezi, and most other African rivers, the Congo River reaches the sea through a deep **estuary** (EHS•chuh•WEHR•ee), or passage where freshwater from a river meets seawater. The Congo's estuary is 6 miles (10 km) wide and is easily navigated by ocean vessels, making it an important waterway. The 2,900 miles (4,667 km) of the Congo form the largest network of navigable waterways on the continent. Some parts of the river, however, such as rapids and waterfalls, present serious obstacles to traffic. The river plunges up to almost 900 feet (274 m) in numerous cataracts not far from where it meets the Atlantic Ocean. The cataracts are a major barrier to travel from the estuary upriver.

History

Daunting Physical Barriers

Although North Africans enjoyed relatively easy access to Europe and Southwest Asia across the Mediterranean and Arabian Seas, the Sahara prevented most land travel to and from central and southern Africa. The daunting physical geography along the West African coast made travel inland by river very difficult for European traders, who began arriving in the late 1400s. Sand and silt deposits made navigation through the deltas treacherous. At certain times of the year, those who tried to sail inland often encountered shallows, sandbars, and even dry riverbeds. Farther upstream, rapids and waterfalls made travel upriver almost impossible. As a result, between the late 1400s and the late 1700s, most Europeans conducted trade with Africans from offshore islands or coastal forts, and regional African leaders maintained control of goods and trade routes in the interior of the continent.

Natural Resources

Mineral resources are abundant throughout Africa south of the Sahara. Angola, Nigeria, Gabon, and Congo have plentiful oil reserves. Deposits of various metals, including chromium, cobalt, copper, iron ore, manganese, and zinc, are scattered across the region. South Africa supplies about half the world's gold. Zimbabwe, the Democratic Republic of the Congo, Tanzania, and Ghana are additional sources of this precious metal. Uranium, usually found with gold, is abundant in South Africa as well as in Niger, Gabon, the Democratic Republic of the Congo, and Namibia. South Africa, Botswana, and the Congo River basin hold major diamond deposits. Diamonds also are mined in Angola, the Democratic Republic of the Congo, and Sierra Leone.

Water is an abundant resource in parts of Africa south of the Sahara, and it has tremendous potential for agricultural and industrial uses. Areas in the west of the region and near the Equator receive abundant rainfall. Controlling water for practical uses, such as irrigation and hydroelectric power, is difficult because rainfall often is irregular and unpredictable. Because of these physical challenges, combined with a lack of financial support, Africa has a great deal of unused hydroelectric power potential. The Congo River, for instance, has more potential hydroelectric power than all the lakes and rivers in the United States combined, but this resource has remained underdeveloped. Despite these challenges, some development has occurred, however. For example, most of the electricity generated in Kenya, Tanzania, Zambia, Ghana, and many other countries comes from hydroelectric power.

Solar power is another renewable energy source that has been harnessed in the region. In Kenya, rural electrification programs resulted in the installation of more than 20,000 small-scale solar power systems from 1986 to 1996. In the next section, you will read about the climate and vegetation of this vast region and their role in the development of Africa south of the Sahara.

SECTION  ASSESSMENT

Checking for Understanding

1. **Define** escarpment, cataract, rift valley, fault, delta, estuary.
2. **Main Ideas** Create a web like the one below to record and organize information about the region's physical geography.

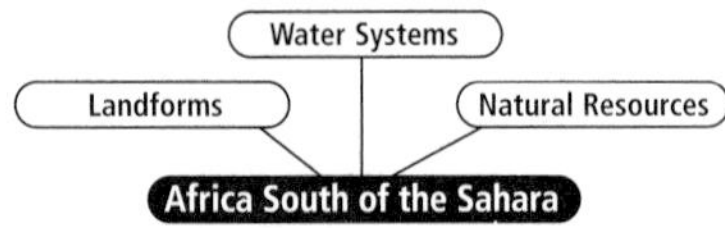

Critical Thinking

3. **Identifying Cause and Effect** Physical features such as the Sahara to the north and oceans to the east and west isolated Africa south of the Sahara from other regions. What effects did this isolation have?
4. **Making Inferences** Considering the region's physical geography, what advancement in transportation do you think has contributed the most to improved travel in Africa?

Analyzing Maps

5. **Place** Study the map on page 501. What do most of the countries with elevations of less than 1,000 feet (300 m) have in common?

Applying Geography

6. **Identifying Cause and Effect** As a geographer from Kenya, write a paper on the geological process that caused the formation of the Great Rift Valley to deliver to the National Council for Geographic Education.

Climate and Vegetation

Guide to Reading

Consider What You Know

The continent of Africa straddles the Equator. How do you think this location affects climate and vegetation in Africa south of the Sahara?

Read to Find Out

- What geographic factors affect climate in Africa?
- What kinds of climate and vegetation are found in Africa south of the Sahara?

Terms to Know

- leach
- harmattan
- savanna

Places to Locate

- Serengeti Plain
- Sahel
- Namib Desert
- Kalahari Desert

NATIONAL GEOGRAPHIC

A Geographic View

Desert Delta

The Kalahari spread out below us. . . . Thunderheads spread cobras' hoods on the horizon, and the air was heavy with the musk of rain-wet earth somewhere up the breeze. There would be lightning that night, flashing on the burnished hills, then wind and finally, perhaps, the water that the whole land craved like a kind of forgiveness, like a blessing long withheld.

—Douglas B. Lee, "Okavango Delta: Old Africa's Last Refuge," National Geographic, *December 1990*

Kalahari sand dune, South Africa

"Pula"—in Botswana's Okavango Delta, this word, meaning "rain," is also used as a greeting. Rain is so important to the area, in fact, that *pula* is also the word for the country's currency and the word for blood, or life. In many places in Africa south of the Sahara, water is such a precious resource that rain and life are considered one and the same. In this section you will discover how rain helps determine climate, and thus vegetation, in every part of the region—its deserts, steppes, savannas, and tropical forests.

Tropical Climate

In addition to rainfall, other factors—ocean currents, prevailing wind patterns, elevation, and latitude—cause great variations in climate and vegetation throughout Africa south of the Sahara. However, as the map on page 506 shows, much of the region lies in the Tropics and has tropical climate and vegetation areas.

Tropical Rain Forest

Tropical rain forest climate, located near the Equator, is the wettest climate region in Africa. Warm temperatures prevail in this zone. More than 60 inches (150 cm) of rainfall per year soak the dense forests. Rainfall amounts vary seasonally, but the tropical rain forests do not experience a truly dry season. Daily, rain falls on an amazing number and variety of life forms.

Shrubs, ferns, and mosses grow together at the lowest level of the rain forest, which rises 6 to 10 feet (2 to 3 m). A layer of trees and palms reaching as high as 60 feet (18 m) tops this undergrowth. Arching over all is a canopy of leafy trees with a maximum height of 150 feet (46 m). Orchids, ferns, and mosses grow among the branches of the canopy, and woody vines link the trees in a tangle.

Economics

Crops and Cutting at a Cost

Although heavy rains in the tropical rain forest leach, or dissolve and carry away, nutrients from the soil, various crops are still grown in this zone. Bananas, pineapples, cocoa, tea, coffee, palms for oil, rubber, and cotton are grown as cash crops on large plantations. As farmers clear more land,

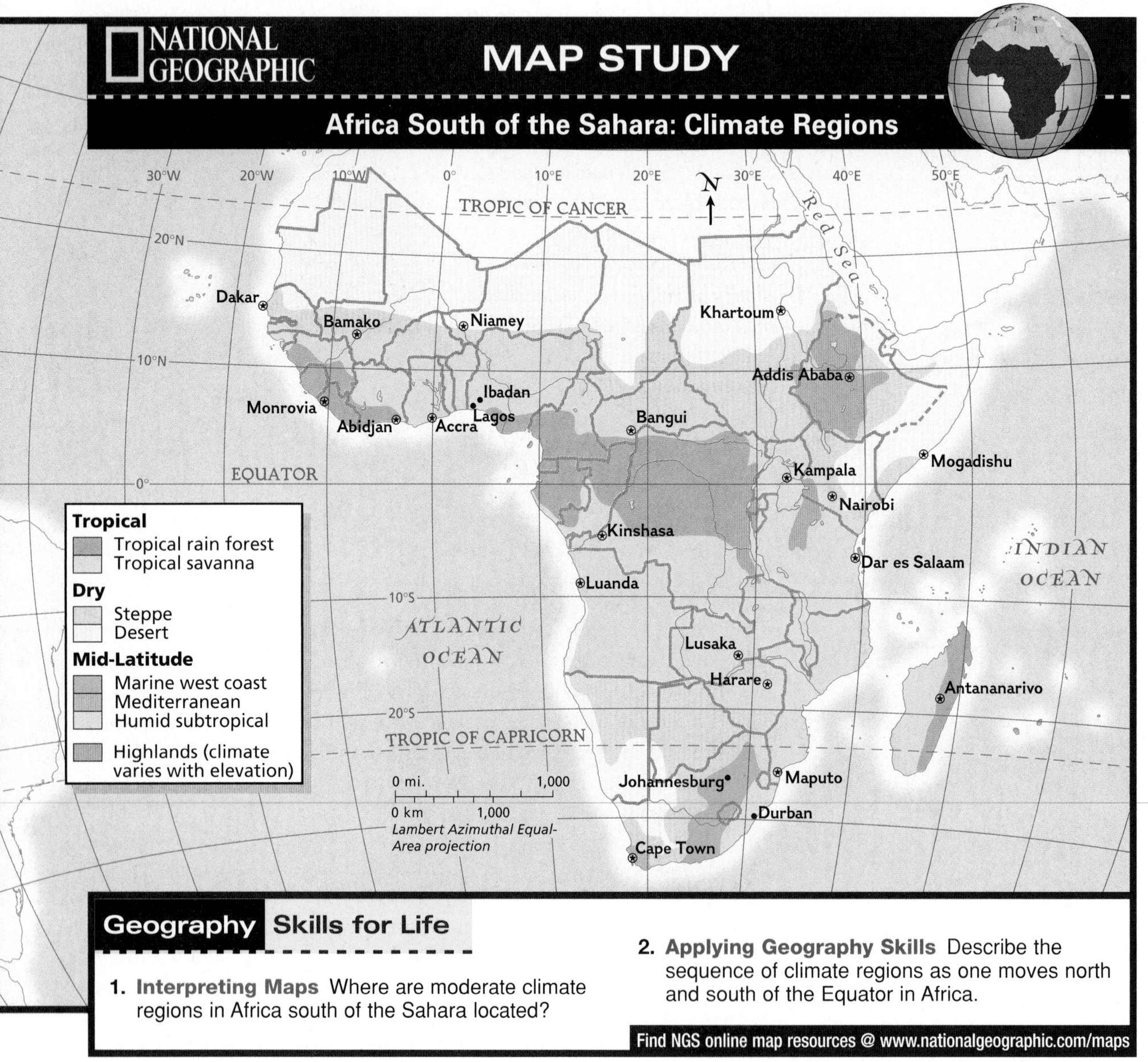

Geography Skills for Life

1. **Interpreting Maps** Where are moderate climate regions in Africa south of the Sahara located?
2. **Applying Geography Skills** Describe the sequence of climate regions as one moves north and south of the Equator in Africa.

Find NGS online map resources @ www.nationalgeographic.com/maps

agriculture seriously threatens the rain forests. In addition, commercial loggers diminish the rain forest by clear-cutting tropical timber. The deforestation of Africa's tropical rain forests concerns people worldwide, who fear that if the clear-cutting continues the rain forests may disappear. In Chapter 22 you will read about steps that governments, groups, and individuals are taking to protect Africa's rain forest environments.

Savanna

Tropical grassland with scattered trees—known as savanna—covers almost half of the continent of Africa. Rainfall is seasonal in this climate zone, with alternating wet and dry seasons. In the wettest areas, which are closest to the Equator, six months of almost daily rain is followed by a six-month dry season. Average annual rainfall in the savanna is about 35 to 45 inches (90 to 115 cm).

Dueling winds affect the savanna climate of western Africa. Hot, dry air streams in from the Sahara on a northeast trade wind known as a harmattan. Although dusty, a harmattan is welcome in the summer because it dries up moisture left by heavy summer rains. Around the same time of year, cool, humid air blows in from the southwest.

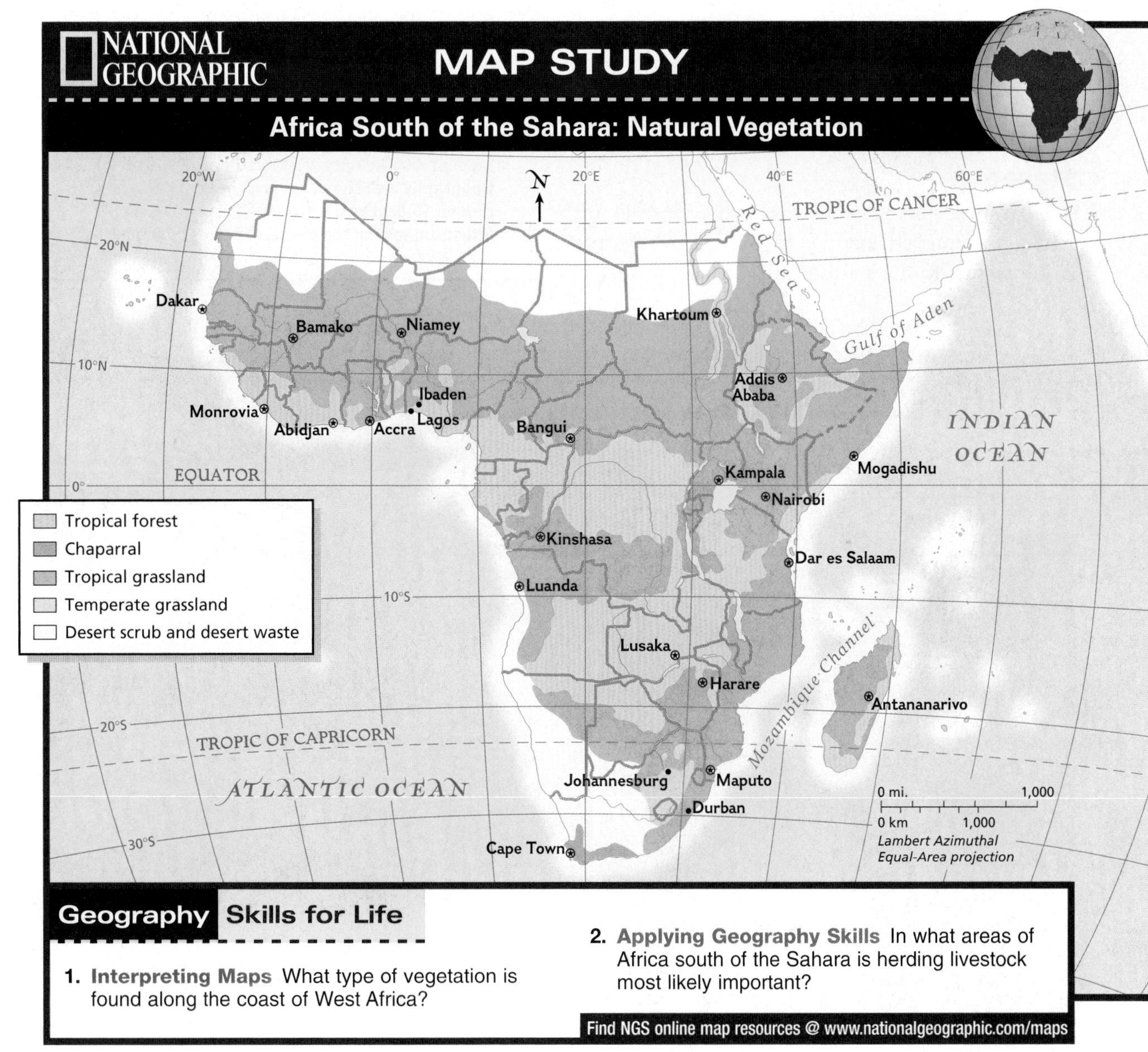

Geography Skills for Life

1. **Interpreting Maps** What type of vegetation is found along the coast of West Africa?

2. **Applying Geography Skills** In what areas of Africa south of the Sahara is herding livestock most likely important?

Find NGS online map resources @ www.nationalgeographic.com/maps

Tornadoes sometimes form when a harmattan and the southwest winds collide.

Trees are the main feature of the landscape in some parts of the savanna, while tall grasses cover other areas. Animals of many species graze in this zone. On the **Serengeti Plain**, one of the world's largest savanna plains, more than 1 million gnu, 60,000 zebras, and 150,000 gazelles roam, as well as hyenas, lions, giraffes, cheetahs, and other animals. Many of these animals live in the protected Serengeti National Park in Tanzania.

Dry Climates

Away from the Equator, tropical climates fade into semiarid steppe areas, which finally give way to the driest climate region of all—desert. Declining rainfall and growing populations have contributed to the expansion of the region's deserts.

Steppe

Separating the savanna from the deserts of Africa is semiarid steppe. In the south, steppe land extends to the southern tip of the continent. The northern steppe is called the **Sahel**—literally "shore" or "edge" in Arabic. This band of dry land, which extends from Senegal to Sudan, represents the southern "coast" of the Sahara. The Sahel has natural pastures of low-growing grasses, shrubs, and acacia trees. On average, 4 to 8 inches (10 to 20 cm) of rain falls annually, but this rainfall is concentrated in June, July, and August. The remaining months are generally very dry.

Economics

Desertification

Over the past 50 years, the Sahel has undergone much desertification—a process by which productive land turns into desert following the destruction

Student Web Activity Visit the **Glencoe World Geography** Web site at geography.glencoe.com and click on Student Web Activities—Chapter 20 for an activity on touring the physical features in Tanzania.

NATIONAL GEOGRAPHIC **World Explorer**

Geography Skills for Life **Tropical Climates**

Eastern Africa includes both tropical savanna (left) and tropical rain forest (right) climates.

Place What animals live on the Serengeti Plain?

▲ Desertification in the Sahel of West Africa

of vegetation. Some scientists claim that the Sahel's desertification is caused mainly by climate change that brings long periods of extreme dryness and water shortages. Lands managed well during drought periods can usually recover once rains return. Other scientists, however, believe that human and animal activities also contribute to desertification. People strip trees for firewood and clear too much land for farming, while livestock overgraze the short grasses. As a result, the land is depleted and topsoil is further eroded, reducing both the land's productivity and its ability to recover from drought.

Desert

Isolated parts of southern Africa swelter in a desert climate. In the east, hot, dry weather prevails in much of Kenya and Somalia. Along the Atlantic coast of Namibia, rocks, dunes, and scattered desert plants cover the **Namib Desert**. Joining the Namib, the **Kalahari Desert** occupies eastern Namibia, most of Botswana, and part of South Africa. A sand-swept expanse with few other features, most of the Kalahari is true desert, but parts of it do support some animals and a variety of plants, including grass and trees. In general, little rain falls in the desert, and average monthly temperatures are extremely high. Daily temperatures in the Kalahari vary greatly, however, ranging from 120°F (49°C) during the day to 50°F (10°C) at night.

Moderate Climates

Although less extensive than the main climate zones, moderate climate zones also exist in Africa south of the Sahara. As the map on page 506 shows, coastal areas of South Africa and highlands regions in East Africa enjoy moderate climates with comfortable temperatures and enough rainfall for farming. In the highlands, temperatures are somewhat lower, snow is not uncommon at high elevations, and vegetation abounds. The highlands areas can seem almost lush, as Curt Stager observed on his journey through East Africa:

> *"The Ethiopian Highlands are far cooler and [more moist] than the surrounding lowlands. Although plagued in recent years by drought, this area is, in normal times, an agricultural island in a desert sea."*
>
> Curt Stager, "Africa's Great Rift," *National Geographic*, May 1990

SECTION 2 ASSESSMENT

Checking for Understanding

1. **Define** leach, savanna, harmattan.
2. **Main Ideas** Use a table like the one below to fill in characteristics of Africa south of the Sahara. Then write a short description of one of the region's climate zones.

Climate Zone	Climate	Vegetation

Critical Thinking

3. **Making Predictions** Do you think desertification will continue in Africa south of the Sahara? Explain your answer.
4. **Identifying Cause and Effect** In what ways are people affecting Africa's tropical rain forests?
5. **Making Generalizations** How does physical geography affect the climate and vegetation in this region?

Analyzing Maps

6. **Region** Study the maps on pages 506 and 507. Which climate regions lie on the Equator? What kind of vegetation thrives there?

Applying Geography

7. **Rainfall's Impact** As a geographer studying rainfall in Africa south of the Sahara, write a report explaining how precipitation defines climate and vegetation there.

Understanding Time Zones

As the earth rotates on its axis, half of the planet experiences day and the other half experiences night. By international agreement there are 24 time zones around the world.

Learning the Skill

Each of the 24 time zones represents 15° longitude, or the distance that the earth rotates in one hour. The base time zone, called Greenwich Mean Time (GMT) or Universal Time, is set at the Prime Meridian (0°). As one travels west from Greenwich, the time becomes earlier; as one travels east, the time becomes later. The international date line generally follows the 180° meridian. Traveling west across this imaginary line, you add a day. Traveling east, you subtract a day.

The imaginary lines that divide time zones sometimes curve or form angles. The lines are drawn to allow for geographic or political needs. For example, certain lines curve around Pacific island groups so that island countries that cover relatively small areas will not have multiple time zones.

To determine the time and day of the week in different time zones, follow these steps:

- **Locate on the map a place for which you already know the time and day of the week.**
- **Locate the place for which you wish to know the time and day of the week.**
- **Count the time zones between the two places.**
- **Calculate the time by either adding or subtracting an hour for each time zone, depending on whether you are moving east or west.**
- **If you have crossed the International Date Line, identify the day.**

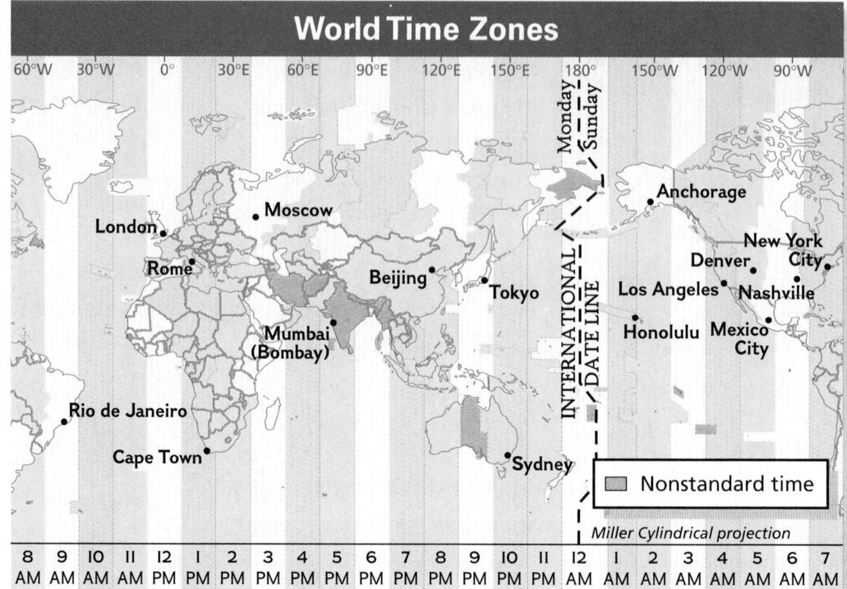

Practicing the Skill

Study the map and answer the questions.

1. How many time zones does continental Africa have?
2. Does Africa have more, fewer, or the same number of time zones as the United States?
3. If it is 4:00 P.M. Saturday in Cape Town, South Africa, what time and what day is it in Rio de Janeiro, Brazil?
4. If it is 10:00 A.M. Tuesday in Sydney, Australia, what time and what day is it in Honolulu, Hawaii?
5. Notice that some time zones have crooked boundaries. Why do you think that is?

Applying the Skill

Use a reference book or Internet sources to find a more detailed map of Africa's time zones. Notice how the lines are drawn in relation to cities, political divisions, or physical features. Then make a list of locations where adjusted lines occur. Write the reasons you think the adjustments were made.

Go To The **Glencoe Skillbuilder Interactive Workbook, Level 2** provides instruction and practice in key social studies skills.

20 SUMMARY & STUDY GUIDE

SECTION 1 The Land (pp. 499–504)

Terms to Know
- escarpment
- cataract
- rift valley
- fault
- delta
- estuary

Key Points
- Africa south of the Sahara is a series of step-like plateaus, rising in a few places to mountains and slashed in the east by a rift valley.
- High elevations and narrow coastal plains characterized by escarpments have made traveling to Africa's interior very difficult.
- The region's water systems include numerous long, large, or deep lakes; spectacular waterfalls; and great rivers that drain expansive basins.
- Minerals and water are the region's most abundant natural resources.

Organizing Your Notes
Use a table like the one below to help you organize important details about the physical features of Africa south of the Sahara.

Physical Feature	Location

SECTION 2 Climate and Vegetation (pp. 505–509)

Terms to Know
- leach
- savanna
- harmattan

Key Points
- Rainfall, tropical latitudes, nearness to the Equator, ocean air masses, and elevation are the main factors influencing climate variations in Africa south of the Sahara.
- The region can be divided into four main climate zones: tropical rain forest, savanna, steppe, and desert.
- Moderate climates such as humid subtropical and marine west coast are also found in Africa south of the Sahara.

Organizing Your Notes
Use a graphic organizer like the one below to organize your notes about each of the climate zones described in this section.

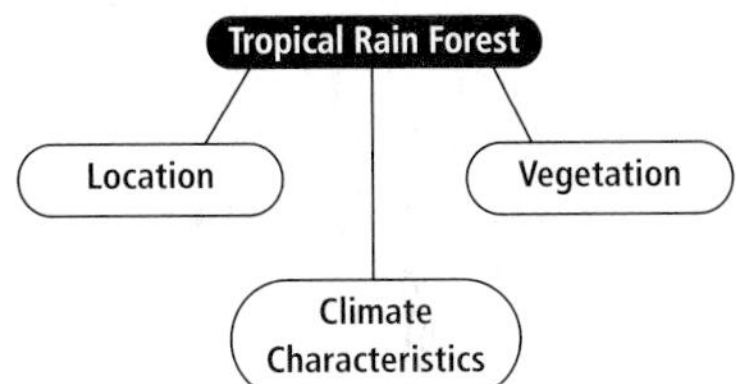

African wild dogs hunt in the Okavango Delta, Botswana

ASSESSMENT & ACTIVITIES

Reviewing Key Terms

On a sheet of paper, write the term that matches each definition. Refer to the Terms to Know in the Summary & Study Guide on page 511.

1. a crack in the earth's surface created by shifting of the earth's tectonic plates
2. tropical grassland with scattered trees
3. a towering waterfall
4. a triangular section of land formed by sand and silt carried downriver to a river's mouth
5. a steep, often jagged slope or cliff
6. a northeast trade wind crossing the Sahara
7. to dissolve and carry away
8. a passage where freshwater meets seawater
9. a long valley between faults in the earth, with volcanic mountains and deep lakes

Reviewing Facts

SECTION 1

1. Where are the main highlands areas and mountains in Africa south of the Sahara?
2. What three great river basins are located in Africa south of the Sahara?
3. What natural resources are especially plentiful in Africa south of the Sahara? Describe the locations of these resources.

SECTION 2

4. Describe vegetation changes in the Sahel and the causes that contribute to these changes.
5. What kind of vegetation grows in the savannas of this region?
6. What is the wettest climate zone in Africa south of the Sahara, and what types of vegetation grow there?

Critical Thinking

1. **Drawing Conclusions** What resources make Africa important to the world economy? Why?
2. **Making Generalizations** What general observations can you make about the areas of the region that have moderate climates?
3. **Drawing Conclusions** Create a Venn diagram to compare causes of rain forest deforestation and of desertification in the Sahel. Then propose steps to solve the problem.

Deforestation | Both | Desertification

NATIONAL GEOGRAPHIC

Locating Places

Africa South of the Sahara: Physical Geography

Match the letters on the map with the physical features of Africa south of the Sahara. Write your answers on a sheet of paper.

1. Lake Chad	5. Great Rift Valley	9. Lake Tanganyika
2. Kilimanjaro	6. Okavango Delta	10. Congo River
3. Kalahari Desert	7. Lake Victoria	11. Niger River
4. Lake Malawi	8. Zambezi River	12. Namib Desert

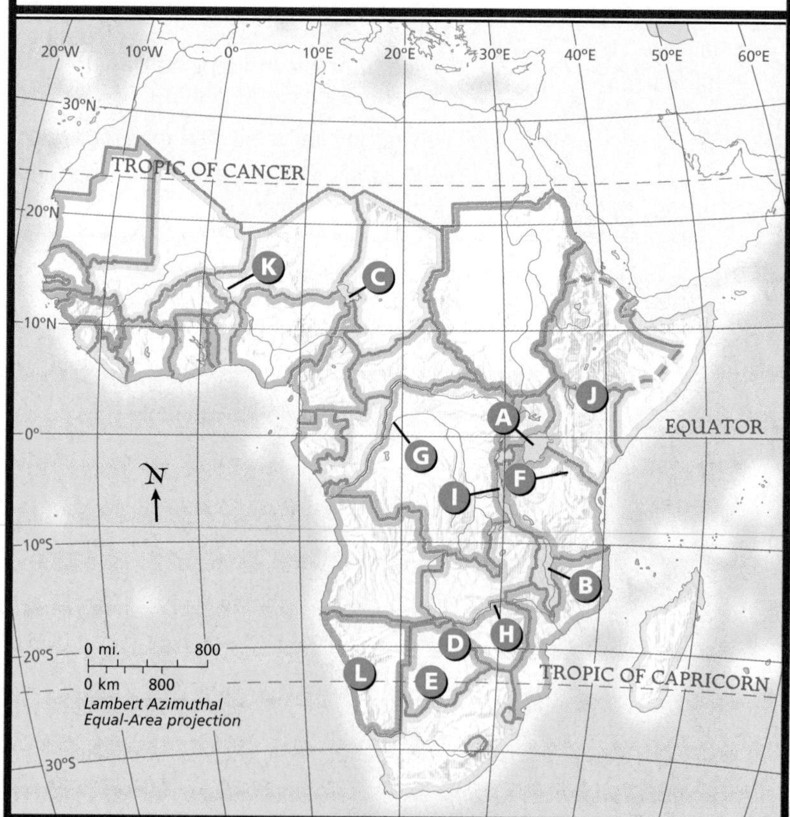

Self-Check Quiz Visit the **Glencoe World Geography** Web site at geography.glencoe.com and click on Self-Check Quizzes—Chapter 20 to prepare for the Chapter Test.

Using the Regional Atlas

Refer to the Regional Atlas on pages 486–489.

1. **Region** What rivers drain much of southern Africa?
2. **Location** What natural resources are found in the Ethiopian Highlands? The Katanga Plateau?

Thinking Like a Geographer

What challenges does the physical geography of Africa south of the Sahara pose to the development and distribution of the area's natural resources?

Problem-Solving Activity

Group Research Project Africa south of the Sahara has enormous potential for producing hydroelectric power. Work with a group to learn more about hydroelectricity in the region. Find out where water power has already been harnessed, and identify other sites that might be good for hydroelectric power plants. What problems might the physical geography pose to generating and distributing hydroelectricity? Suggest solutions to one or more problems, and share your findings with the class.

GeoJournal

Comparison-Contrast Essay Using the information you logged in your GeoJournal, write a descriptive paragraph about one of the significant physical features of Africa south of the Sahara. Then write a second paragraph comparing this feature with a similar physical feature of Latin America.

Technology Activity

Developing Multimedia Presentations Select a land or water feature of Africa south of the Sahara, and develop a multimedia presentation about it. Use Internet and library resources to gather information. Then design and draw maps and other visual aids to illustrate your work, and make your presentation to the class.

Standardized Test Practice

Study the time zone map below. Then choose the best answer for the following multiple-choice questions. If you have trouble answering the questions, use the process of elimination to narrow your choices.

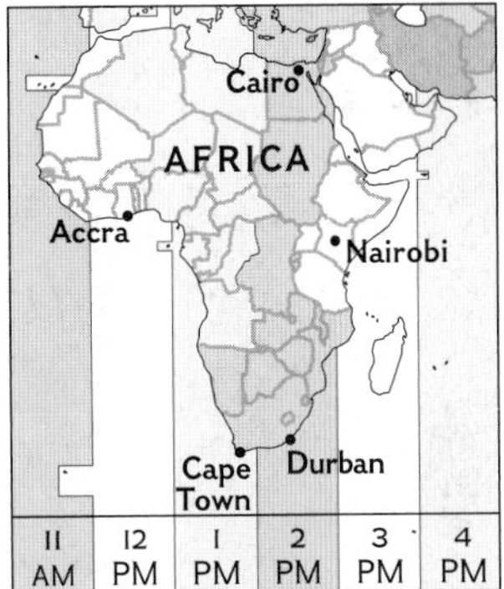

1. If it is noon in Accra, Ghana, what time is it in Cape Town, South Africa?

A 2 P.M.
B 2 A.M.
C 10 P.M.
D 11 P.M.

2. If you were standing in Nairobi, Kenya, at 2:00 in the afternoon, what time would it be in Durban, South Africa?

F 2 P.M.
G 11 A.M.
H 1 P.M.
J 4 P.M.

Be sure to pay close attention to the locations mentioned in the question. By studying the map, you can see that the time zone in which Nairobi lies is located next to the time zone in which Durban is located. Therefore, the difference between the times should be one hour. Notice that the sample times shown on the map are different from those in the question, however, so you will need to subtract to find the correct answer.

CHAPTER 21

The Cultural Geography of Africa South of the Sahara

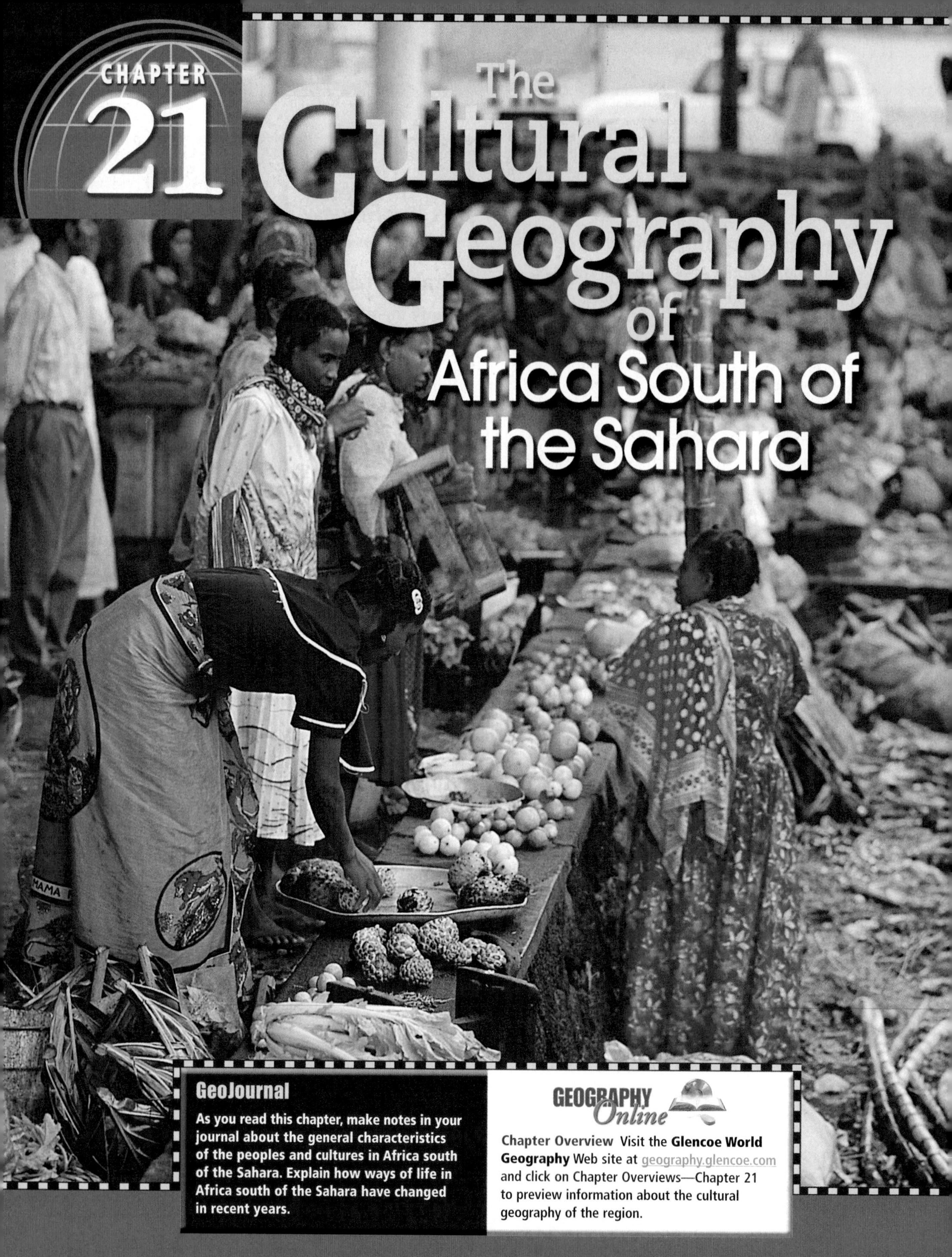

GeoJournal

As you read this chapter, make notes in your journal about the general characteristics of the peoples and cultures in Africa south of the Sahara. Explain how ways of life in Africa south of the Sahara have changed in recent years.

GEOGRAPHY Online

Chapter Overview Visit the **Glencoe World Geography** Web site at geography.glencoe.com and click on Chapter Overviews—Chapter 21 to preview information about the cultural geography of the region.

Population Patterns

Guide to Reading

Consider What You Know

As in other world regions you have studied, people in Africa south of the Sahara are moving from rural areas to cities. What changes might a new city resident face?

Read to Find Out

- Why are parts of Africa south of the Sahara densely populated?
- What are the obstacles to economic growth in the region?
- Who are the diverse peoples of Africa south of the Sahara?
- Why are the region's cities growing so rapidly?

Terms to Know

- sanitation
- urbanization
- service center

Places to Locate

- Nigeria
- Rwanda
- Namibia
- Zimbabwe
- Lagos
- Accra
- Kinshasa
- Nairobi
- Johannesburg

◀ *A market in the Comoros Islands*

NATIONAL GEOGRAPHIC

A Geographic View

To Market

The [riverside market] in Lukulu [Zambia] bustled with swarms of people picking through baskets of fish and used clothes lying in enormous piles or draped on racks or else flapping like pennants on long lines.

Julius Nkwita was selling small piles of dried fish, about 60 cents for a handful. But sometimes he swapped his fish for cups of flour or an item of clothing. . . . His wife and four of his children were in his home village, . . . while he stayed in his seasonal fishing camp—just a reed hut—with his son, fishing intensively.

—*Paul Theroux, "Down the Zambezi,"* National Geographic, *October 1997*

Seasonal fishing camp, central Africa

A small market town, a village, a temporary fishing camp—this is Julius Nkwita's world. On a trip through central Africa on the Zambezi River, writer Paul Theroux met Julius Nkwita and many others whose lives revolve around a small community. Like these people of the Zambezi River basin, the majority of Africans south of the Sahara live in rural areas. In this section you will trace population patterns in Africa south of the Sahara—the fastest-growing and third most populous region in the world.

Rapid Population Growth

Home to more than 673 million people, Africa south of the Sahara has about 10 percent of the world's population. It has both the highest birthrate and the highest death rate in the world. It also has the world's highest infant mortality rate and shortest life expectancy.

Despite a high death rate, births outnumber deaths in this region. In fact, population growth in Africa south of the Sahara surpasses that of every other region in the world, increasing at an average rate of 2.5 percent a year. At this rate, the total population of Africa south of the Sahara will more than double in just 35 years.

Nigeria, the most populous African country south of the Sahara, is one example of the region's rapid population growth. In 2001 about 126.6 million people lived in Nigeria. With an expected growth rate of 2.8 percent a year, it is likely that in 50 years Nigeria's population will reach more than 300 million.

One factor, however, may drastically limit population growth in the region over the next 10 years. The disease AIDS (acquired immunodeficiency syndrome) has spread rapidly. About 70 percent of the estimated 36 million people in the world currently carrying HIV, the virus that causes AIDS, live in Africa south of the Sahara. At the end of 2000, about 17 million Africans had already died of AIDS-related diseases.

Population Density and Distribution

Despite rapid population growth, Africa south of the Sahara has few people in relation to its vast land area. If the population were evenly spread across the region, there would be about 72 people per square mile (28 people per sq. km). However, the population of the region is not evenly distributed. **Rwanda**, one of the region's most densely populated countries, has 719 people per square mile (278 people per sq. km), whereas **Namibia** has only 6 people per square mile (3 people per sq. km).

Land and climate help explain this uneven distribution of people. Desert or steppe covers large areas of Africa south of the Sahara. Because living conditions there are difficult, few, if any, people live there. The land is generally too dry to support agriculture or the raising of livestock. As the map on page 488 shows, most of the region's people are crowded in the coastal belt of West Africa along the Gulf of Guinea and along the eastern coast of southern Africa. They are drawn to these areas because of easy access to water, fertile soil, and mild climates. As a result, agriculture, industry, and commerce are concentrated in these areas.

Population and Food Production

Soaring population growth combined with economic challenges have made it difficult for Africa south of the Sahara to feed its people. Agriculture—both subsistence and cash-crop farming—ranks as the region's main economic activity. About 70 percent of people in the region work as farmers. Yet they are producing less and eating less, while the population has almost tripled.

Factors such as the actions of governments and some farmers and the effects of climate have contributed to this critical situation. In recent years governments have geared their economies for exporting in order to boost national incomes. However, not enough food has been produced for domestic needs, making it necessary to import food. In addition, huge expanses of farmland in the region have been exhausted through intensive cultivation, loss of soil fertility, and devastating droughts.

Population and Health Care

In recent years, Africa south of the Sahara has made many advances in health care. However, famine and poor nutrition claim many lives, especially among infants and young children. Impure water is another cause of death. Only a third of rural Africans have clean water to drink, and only a fourth live where there is adequate **sanitation**, or disposal of waste products. Diseases such as malaria are widespread. Insects such as the mosquito and tsetse (SEHT•see) fly transmit viruses to people and animals.

AIDS, a worldwide disease caused by a virus that is spread from person to person, has reached epidemic proportions in the region. A child born in **Zimbabwe**, for instance, is more likely to die of AIDS than of any other cause. Treatments with drugs that help control the disease are available to patients in developed countries, but these treatments cost too much for most Africans or their governments to purchase. As a result of the lack of treatment, this deadly disease has drastically cut the average life expectancy throughout Africa south of the Sahara. In Zimbabwe, average life expectancy has fallen from 65 years to 39 years because of AIDS.

The disease is expected to reduce the populations of many of the region's countries significantly, with disastrous consequences. Workers will be in short supply, and industries may be forced to close. Families and communities will suffer as adults in the prime of life are lost to the disease. Children will lack caregivers. The United Nations estimates that by the year 2010, 10.7 million children in Africa under the age of 15 will have lost at least one parent to AIDS.

A Diverse Population

In both urban and rural areas, Africa south of the Sahara has a very diverse population. In fact, Africa is home to more ethnic groups than any other continent. Some 3,000 African ethnic groups make up the population. Other groups living in Africa include Europeans, South Asians, Arabs, and people of mixed backgrounds.

Culture

People Without Borders

A people known as the Sena live in a wide area in the marshes near the Zambezi River, which divides Zambia and Zimbabwe. The Sena often travel up the Zambezi in dugout canoes to sell fish and to buy needed items, such as nets, at markets in Malawi. They float downriver to Mozambique to trade fish for sugar. Writing about the Sena, noted travel author Paul Theroux observes:

> *"They come and go, from country to country, without passports—without even saying where they are going."*
>
> Paul Theroux, *"Down the Zambezi,"* National Geographic, October 1997

As in other parts of Africa, country borders separate people politically, but they do not usually disturb daily patterns of life. Throughout Africa south of the Sahara, members of individual ethnic groups speak the same language and share other cultural features, such as religion. They also have common ways of organizing community and family activities.

Growing Cities

Africa south of the Sahara is one of the least urbanized regions in the world, with only 30 percent of the population living in cities. The region's urban areas, however, are growing so rapidly that Africa has the world's fastest rate of **urbanization**, or movement of people from rural areas to cities. In 1950 only about 35 million Africans lived in cities. Today it is estimated that about 270 million Africans are urban dwellers.

Africans leave their rural villages for urban areas in order to find better job opportunities, health care, and public services. At the same time, population growth has caused cities to spread out into the countryside. Areas once made up of villages and towns

NATIONAL GEOGRAPHIC **World Explorer**

Geography Skills for Life

Medical Services In Botswana, one of the wealthiest countries in the region, government funds help support a growing health-care system.

Region How does Botswana's health-care system compare with other health-care systems in this region?

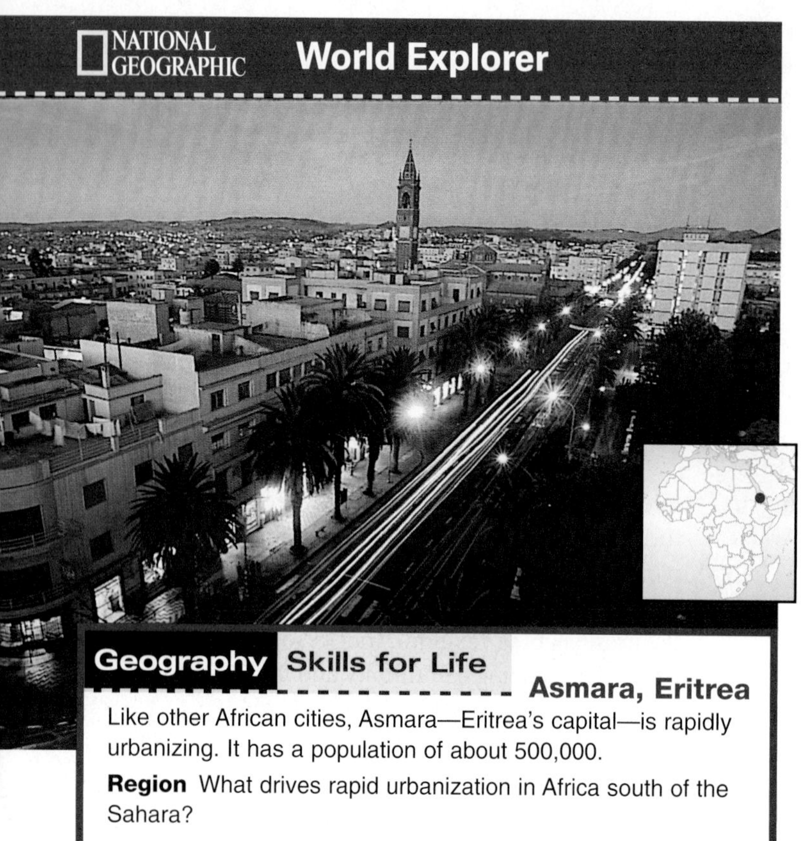

Geography Skills for Life — **Asmara, Eritrea**

Like other African cities, Asmara—Eritrea's capital—is rapidly urbanizing. It has a population of about 500,000.

Region What drives rapid urbanization in Africa south of the Sahara?

have mushroomed into service centers, convenient business locations for rural dwellers, who travel there by foot, bus, or boat.

Most cities in the region lie on the coast, along major rivers, or near areas rich in valuable resources. They developed largely as trading centers. The largest city in the region is the bustling seaport of **Lagos** (LAY•GAHS) in Nigeria, which has a population of more than 10 million. Other important cities include Cape Town, South Africa; Abidjan, Côte d'Ivoire; **Accra**, Ghana; and Dar es Salaam, Tanzania. **Kinshasa**, on the southern bank of the Congo River, is the political, cultural, and economic hub of the Democratic Republic of the Congo. In East Africa, inland cities—such as **Nairobi**, Kenya, and Addis Ababa, Ethiopia—have prospered from trade. **Johannesburg**, South Africa, also an inland city, owes its origins and growth to the mining of gold.

As in other regions of the world, Africa south of the Sahara faces many challenges because of rapid urbanization. Many African cities have towering skyscrapers and trendy shopping areas, but city residents often must endure traffic congestion, inadequate public services, overcrowded neighborhoods, and slums that lack water or sanitary facilities. In Chapter 22 you will learn how Africans south of the Sahara are meeting the challenges of their surroundings.

SECTION 1 ASSESSMENT

Checking for Understanding

1. **Define** sanitation, urbanization, service center.
2. **Main Ideas** In a diagram like the one below, make notes about population patterns in Africa south of the Sahara. Then write a brief paragraph about the region's population.

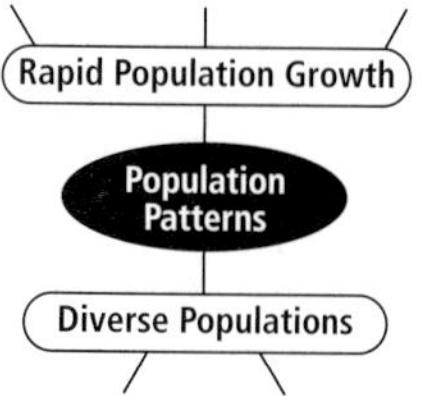

Critical Thinking

3. **Drawing Conclusions** How can Africa south of the Sahara have rapid population growth and yet have relatively few people?
4. **Identifying Cause and Effect** How do you think declining agricultural production in Africa south of the Sahara contributes to urbanization? Explain your answer.
5. **Predicting Consequences** How will inadequate health care ultimately affect economies in this region? What steps would you recommend to help solve this problem?

Analyzing Maps

6. **Region** Study the population density map on page 488. Which countries have low population densities? What physical features account for this fact? Might these countries' population patterns change in the future? Explain.

Applying Geography

7. **Uneven Population Density** Think about the population patterns in Africa south of the Sahara. Brainstorm the reasons most people in the region settle in coastal areas. Then write a paragraph that explains this pattern.

History and Government

Guide to Reading

Consider What You Know

Global regions influence one another through trade, migration, and the exchange of ideas. What effect do you think new contacts, products, and ideas have on cultural traditions in each region?

Read to Find Out

- What were the main achievements of the ancient civilizations of Africa south of the Sahara?
- How did European colonization disrupt African patterns of life?
- What challenges did countries of the region face after independence?

Terms to Know

- domesticate
- apartheid
- universal suffrage

Places to Locate

- Kush
- Axum
- Ghana
- Kumbi
- Mali
- Songhai
- Timbuktu
- Ethiopia
- Liberia

NATIONAL GEOGRAPHIC

A Geographic View

Home of the Zulu

The [South African] district of Msinga, . . . as deep into deep Zululand as you can go, is the strongest bastion [place of survival] of inherited Zulu culture. . . . As you drive from Greytown . . . through the Mpanza Valley, . . . the avocado, pecan, and macadamia plantations give way to aloes and thorn trees. You wind down the escarpment, and there below is a sweeping view over the green folds and steep valleys of Zululand, dotted with thatch huts and small patches of corn.

—*Peter Godwin, "Zulu: People of Heaven, Heirs to Violence,"* National Geographic, *August 2000*

Rural community, South Africa

Kraals—the traditional homesteads of the Zulu people—have long been a familiar sight in what is now South Africa. Like many other ethnic groups in Africa south of the Sahara, the Zulu are descendants of the Bantu peoples. Massive Bantu migrations and movements of other peoples shaped the region's early history and are still influential today.

African Roots

Tens of thousands of years ago, people were already moving from place to place across Africa to hunt and gather food. No written records exist of these people, but early paintings in places as widespread as Niger in the north and Namibia in the south offer clues to their ways of life. Scenes painted in caves and on rocks are filled with people hunting, fishing, and celebrating. Later paintings show new peoples involved in new activities—farming and herding.

First Civilizations

Around 2000 B.C. migrants fleeing a dramatic shift in climate joined other settlers in Africa south of the Sahara. For thousands of years, the climate to the north had been mild and wet. People who once hunted and scavenged for food learned to plant seeds and **domesticate**, or tame, animals. They developed agriculture in the Sahara area. Around 3000 to 2500 B.C., however, the climate became hotter and drier. Plants shriveled, forests perished, and rivers evaporated. Forced to move in order to survive, many people migrated south. They took with them their knowledge of raising crops and animals.

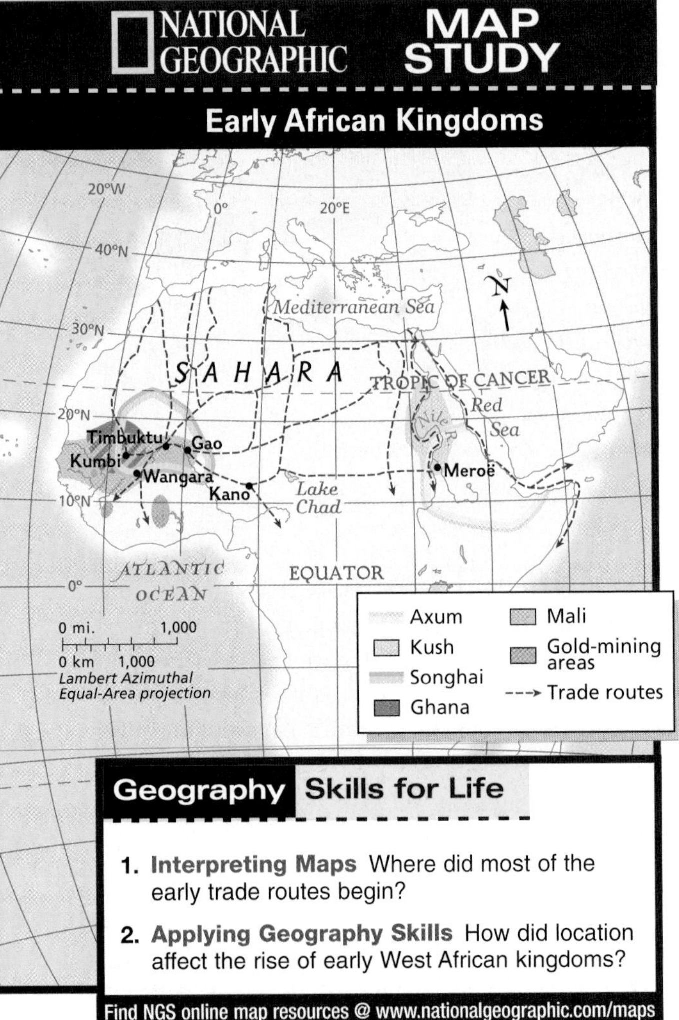

Geography Skills for Life

1. **Interpreting Maps** Where did most of the early trade routes begin?
2. **Applying Geography Skills** How did location affect the rise of early West African kingdoms?

Find NGS online map resources @ www.nationalgeographic.com/maps

In northeast Africa, the Nile Valley remained fertile and gave life to the great Egyptian civilization. Between 2000 and 1000 B.C., the Egyptians pressed south, bringing various cultures along the Nile under their control. When Egyptian civilization began to fade, the cultures under its sway rose to power. The kingdom of **Kush**, in what is now Sudan, extended its rule north into Egyptian territory. The Kushites then pushed south along the Nile, building a civilization around a new capital called Meroë (MEHR•oh•WEE). Kush flourished until the A.D. 300s, when its trade routes were attacked by **Axum**, a powerful trading empire in northern Ethiopia.

Empires in the West

Several centuries later, trading empires began to gain strength in West Africa. Today the West African countries of Ghana and Mali are named after two of these ancient empires. **Ghana**, one of the earliest of these trading kingdoms, emerged around A.D. 700. Its empire grew rich by trading gold for salt brought by camel caravans across the Sahara. Peoples south of the Sahara highly valued salt for use as a food preservative.

Gold was plentiful in Ghana. The Spanish-Arab geographer al-Bakri, who traveled to West Africa in the 1000s, reported, for example, that even the king's dogs wore collars of gold and silver. Ghana's wealth was reflected in its large capital, **Kumbi**. This prosperous empire, which created a tax collection system and charged tariffs on imports, flourished for almost 500 years.

The trading empires of **Mali** and **Songhai** (SAWNG•HY) succeeded Ghana and also grew rich from the gold-for-salt trade. Mali, which extended west to the Atlantic and was larger than Egypt, had as its center the wealthy city of **Timbuktu**. Songhai eventually took over Mali and then stretched east, prospering until about 1600, when it was overrun by Moroccans, a people from the north.

Bantu Migrations

In central and southern Africa, Bantu-speaking peoples had established settlements by A.D. 800. Although the origins of the Bantu and their routes of migration are debated, many historians believe that they spread across one-third of the continent. In addition to founding the central African kingdoms of Kongo (Congo), Luba, and Luanda, the Bantu established states to the southeast in what are today Tanzania, Malawi, Zambia, and Zimbabwe. The influence of the Bantu migration continues, with about 150 million Bantu speakers living in Africa today.

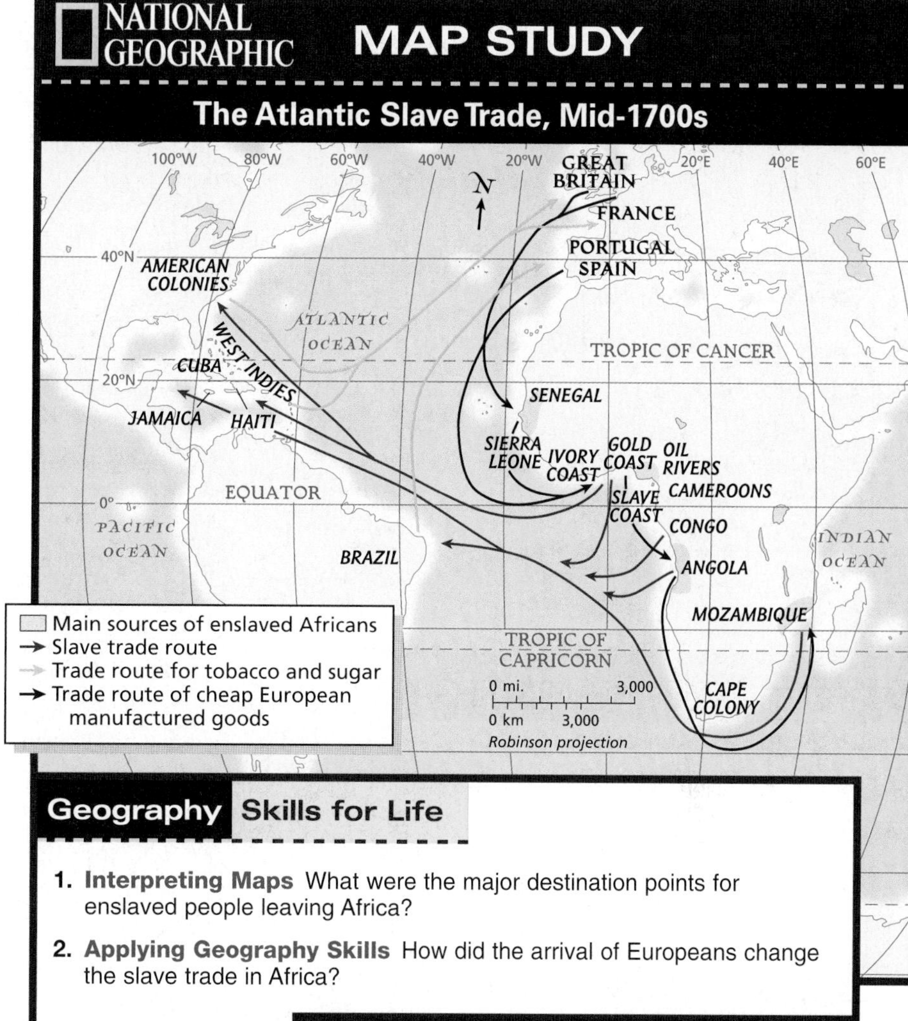

Geography Skills for Life

1. **Interpreting Maps** What were the major destination points for enslaved people leaving Africa?
2. **Applying Geography Skills** How did the arrival of Europeans change the slave trade in Africa?

Find NGS online map resources @ www.nationalgeographic.com/maps

European Colonization

Slowly, word of the wealth of Africa's kingdoms reached Europe. Europeans began trading with Africans as early as the 1200s, bringing gold and other African goods to Europe. By the time Columbus set sail, Portuguese explorers were sailing along the African coast. They set up trading posts and way stations along coastal areas, where enslaved Africans were held for transport. Foreign travelers who reached the trading centers of Timbuktu, Kano, Gao, and Wangara in the west and Kilwa, Mombasa, and Sofala in the east were impressed with the bustling, abundant markets and cultural life.

The Slave Trade in Africa

By the 1600s and 1700s, Europeans were trading extensively with Africans. They sought African gold, ivory, textiles, and enslaved workers. African chiefs and kings had enslaved and traded prisoners of war for centuries. Arab traders had brought enslaved Africans to the Islamic world since the A.D. 800s. The slave trade greatly increased when Europeans began shipping Africans to the Americas to work on large plantations where sugar, tobacco, rice, and cotton were cultivated.

Huge numbers of people from the African interior were sold into slavery. As early as 1526, Nzinga Mbemba, the king of Kongo, deplored the actions of some African rulers. He also complained to the king of Portugal about Portuguese slave merchants:

> *"[They] seize upon our subjects . . . and cause them to be sold; and so great, Sir, is their corruption . . . that our country is being utterly depopulated."*
>
> Nzinga Mbemba, quoted by Basil Davidson, in *African Kingdoms*, 1966

Once captured and sold, enslaved Africans faced a terrible trip across the Atlantic Ocean as human cargo in a ship's hold. This passage from Africa claimed millions of African lives. The loss of so

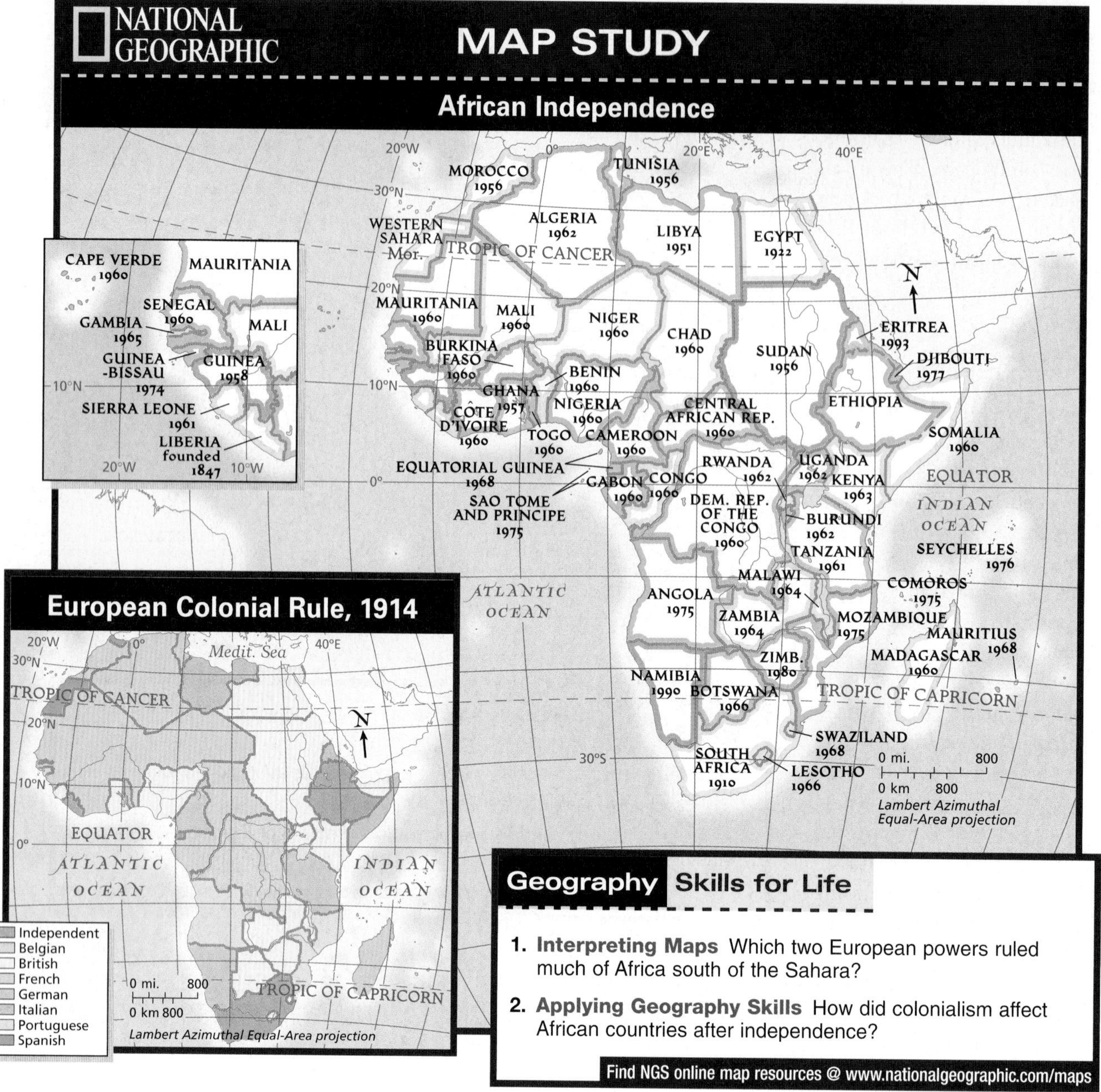

many young people to the slave trade was a major setback to the societies they left behind.

Government

Europe Divides and Rules

In the 1800s European powers regarded the region as a source of raw materials for their growing industries and a potential market for European manufactured goods. European countries quickly laid claim to African territory, and by 1914, all of Africa except **Ethiopia** and **Liberia** was under European control. In setting up their colonies, Europeans ignored African objections and created boundaries that often cut across ethnic homelands. By doing so they set African groups against one another and strengthened European rule in the region.

Among the earliest foreigners to explore Africa's interior, European missionaries often opposed the harsh treatment of Africans by the colonial traders and officials. Yet they, too, promoted European culture and weakened traditional African ways. European businessmen also disrupted African village

NATIONAL GEOGRAPHIC

MAP STUDY

Ethnic Groups, Colonial Rule, and Conflict

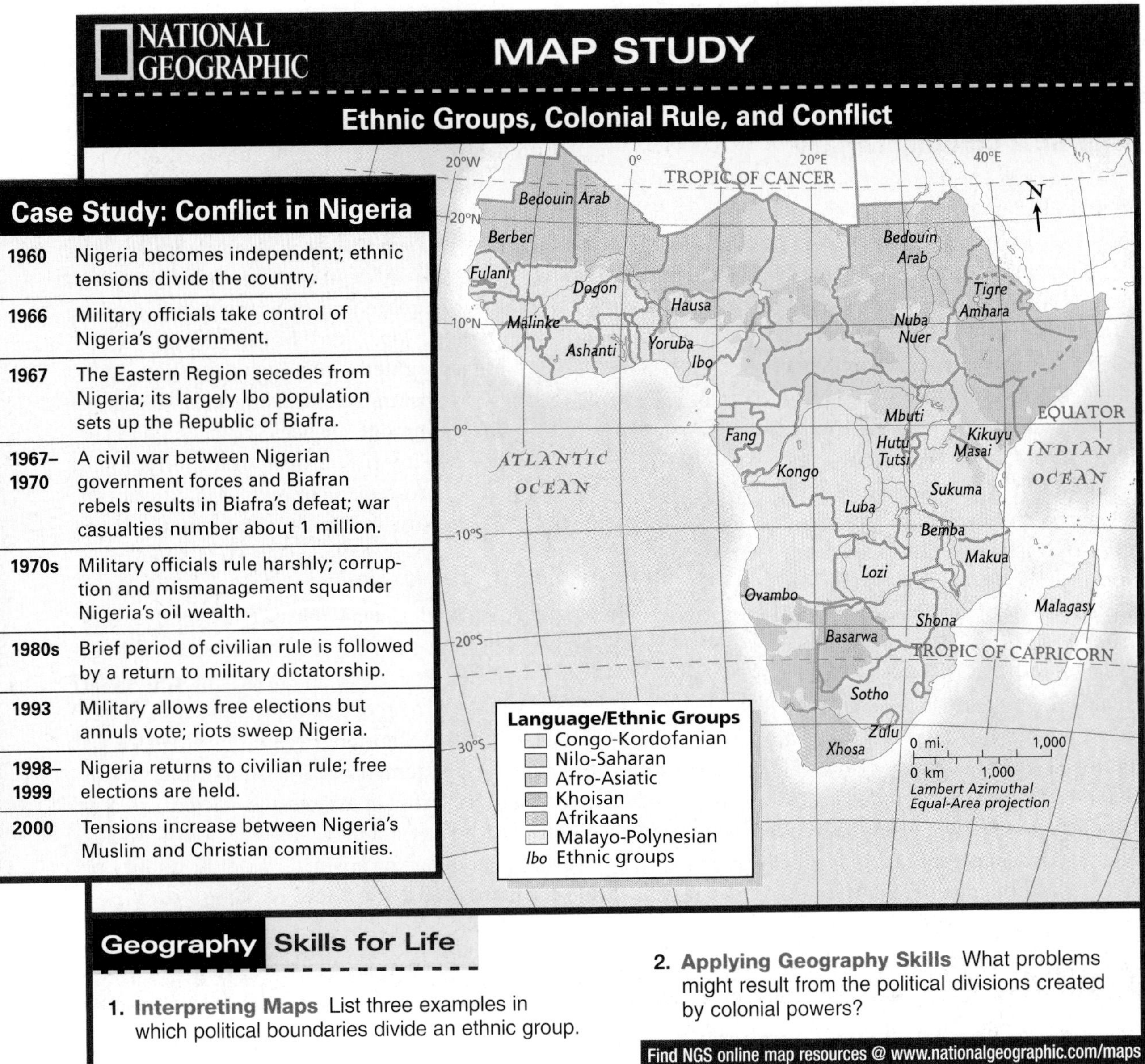

Case Study: Conflict in Nigeria

1960	Nigeria becomes independent; ethnic tensions divide the country.
1966	Military officials take control of Nigeria's government.
1967	The Eastern Region secedes from Nigeria; its largely Ibo population sets up the Republic of Biafra.
1967–1970	A civil war between Nigerian government forces and Biafran rebels results in Biafra's defeat; war casualties number about 1 million.
1970s	Military officials rule harshly; corruption and mismanagement squander Nigeria's oil wealth.
1980s	Brief period of civilian rule is followed by a return to military dictatorship.
1993	Military allows free elections but annuls vote; riots sweep Nigeria.
1998–1999	Nigeria returns to civilian rule; free elections are held.
2000	Tensions increase between Nigeria's Muslim and Christian communities.

Geography Skills for Life

1. **Interpreting Maps** List three examples in which political boundaries divide an ethnic group.
2. **Applying Geography Skills** What problems might result from the political divisions created by colonial powers?

Find NGS online map resources @ www.nationalgeographic.com/maps

life by replacing locally centered agriculture with huge plantation economies. These economies focused on the production of cash crops like coffee and tea for world markets.

From Colonies to Countries

Although European rule dealt serious blows to African life, many Africans benefited from new educational opportunities and city development. Soon some Africans demanded a share in government. By the mid-1900s, educated Africans had launched independence movements, and in the second half of the century, the colonies became independent. (See map on page 522.) These new countries faced difficult challenges, often the result of their colonial legacy. European powers, for example, had used African colonies as a source of raw materials for their industries. They also set up colonial economies that met European, rather than African, needs. Colonial governments did not involve Africans much in government, nor did they give Africans models for democracy. At independence, many of the new African countries adopted the political boundaries set earlier by the colonial powers. As the map above shows, these boundaries divided people of similar language and ethnic background. Within the new countries, rival ethnic groups struggled for power, and civil wars erupted.

Nigeria: A Colonial Legacy

An example of ethnic conflict in the region is Nigeria. The time line next to the map on page 523 shows key events in Nigeria's ongoing ethnic struggles since independence. Nigeria's problems stem from its colonial past.

In 1914 the British had formed the colony of Nigeria from several smaller ethnic territories. As a result, many different ethnic and religious groups lived within Nigeria's boundaries. In the north, various peoples had developed cultures based on centuries-old Islamic influences from North Africa. Those in the south had created ways of life based on traditional African religions or on Christianity. Despite these differences, Nigerians united to resist British rule. In 1960 the colony of Nigeria finally became an independent country. The ethnic and religious differences inherited from the past soon erupted in civil war, however. Although the civil war eventually ended, ethnic and religious divisions continue to plague Nigeria today as it moves from harsh military rule to democracy.

Nelson Mandela

South Africa: Road to Freedom

During the early 1900s, South Africa became independent of British rule. For most of the century, however, the country's white minority population ran the government. It imposed a policy known as **apartheid** (uh•PAHR•TAYT), or separation of the races, on South Africa's black majority and racially mixed peoples. Under apartheid, nonwhite South Africans were denied political rights and equality with whites in education, jobs, and housing. They were segregated into communities with substandard housing and few government services.

Internal unrest and international pressures finally forced South Africa to end apartheid in the early 1990s. Nelson Mandela, the country's most popular anti-apartheid leader, was released after 27 years in prison. In 1994 South Africa held its first election based on **universal suffrage**, or voting rights for all adult citizens. Nelson Mandela became South Africa's first black president. Within a short time, South Africa moved from a repressive society to one committed to democracy. Today, South Africa faces the challenge of ensuring a better quality of life for many nonwhite South Africans.

SECTION 2 ASSESSMENT

Checking for Understanding

1. **Define** domesticate, apartheid, universal suffrage.
2. **Main Ideas** Complete a time line of Africa's history based on the model below. Then use your time line to write a summary of the region's achievements and experiences of the past and present.

Important Dates in Africa South of the Sahara

2000 B.C. — A.D. 2000

Critical Thinking

3. **Making Generalizations** How did contact with other empires influence the West African empires?
4. **Drawing Conclusions** Why do you think the West African trading kingdoms were willing to trade gold for salt?
5. **Identifying Cause and Effect** In what ways did colonialism affect the region's development and set the stage for current conflicts in Africa south of the Sahara?

Analyzing Maps

6. **Movement** Study the map of early African kingdoms on page 520. To what seas and oceans did the early trade routes lead? What river was a major trade route?

Applying Geography

7. **Movement of People** List some of the major human migrations in Africa south of the Sahara. Then choose one migration, and write a paragraph about the motivation for and the effects of the migration.

SECTION 3

Guide to Reading

Consider What You Know

Think about ways in which Africans influence global culture. How do you think aspects of African culture have spread to other parts of the world?

Read to Find Out

- What languages do people in Africa south of the Sahara speak?
- What are the major religions in Africa south of the Sahara?
- What art forms have peoples of the region developed?
- How do lifestyles among peoples of the region differ? How are they similar?

Terms to Know

- mass culture
- lingua franca
- oral tradition
- extended family
- clan
- nuclear family

Places to Locate

- Eritrea
- Madagascar
- Tanzania
- Dar es Salaam

Cultures and Lifestyles

NATIONAL GEOGRAPHIC

A Geographic View

Wedding Traditions

Guests begin to arrive . . . [for] a wedding week of camel racing, dancing, and feasting on goat meat, wheat porridge, and sweet tea.

Bekitta, the bride, stays in seclusion, veiled behind an elaborate mask called a burga, which she has painstakingly decorated. . . .

When the day cools at sunset, a woman breaks into a dance. Clapping out the rhythm, the men sing: "The sun is setting, so we sing before the dark!" As she swirls in perfumed skirts, they punctuate their song with shouts of tribal pride: "Rashaida! Rashaida!"

—*Carol Beckwith and Angela Fisher, "African Marriage Rituals,"* National Geographic, *November 1999*

Rashaida woman of Eritrea

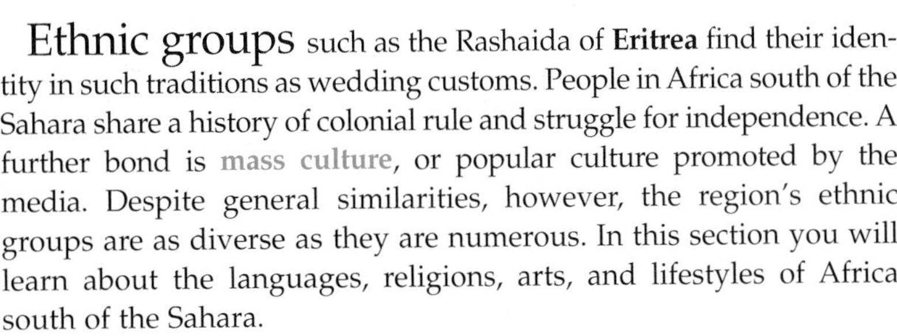

Ethnic groups such as the Rashaida of **Eritrea** find their identity in such traditions as wedding customs. People in Africa south of the Sahara share a history of colonial rule and struggle for independence. A further bond is **mass culture**, or popular culture promoted by the media. Despite general similarities, however, the region's ethnic groups are as diverse as they are numerous. In this section you will learn about the languages, religions, arts, and lifestyles of Africa south of the Sahara.

Languages

At least 2,000 different languages are spoken in Africa today. As the map on page 523 shows, language experts put the many ethnic groups and languages of Africa south of the Sahara into six major categories: Congo-Kordofanian, Nilo-Saharan, Afro-Asiatic, Khoisan, Malayo-Polynesian, and Afrikaans, an Indo-European language.

Languages in the African groups—Congo-Kordofanian, Nilo-Saharan, Afro-Asiatic, and Khoisan—are the most widely spoken and the most diverse. Originating on the African continent, these languages include some 800 Bantu-based Congo-Kordofanian languages spoken by peoples in central, eastern, and southern Africa. Among these are Swahili, Zulu, and Kongo. The Bantu-related languages of Guinea coast peoples also belong to the Congo-Kordofanian group.

The Sudanic peoples of the northwest and northeast corners of the region speak Afro-Asiatic languages. The Afro-Asiatic group includes African languages, such as Hausa and Fulani, as well as languages of North Africa and Southwest Asia, such as Berber and Arabic.

Some languages spoken in the region are non-African. People on the island of **Madagascar** speak the Malagasy language in the Malayo-Polynesian language group. Indo-European languages spoken in Africa include English, French, and Afrikaans. Derived from the dialect of early Dutch settlers in South Africa, Afrikaans also contains words adapted from English, French, German, and African languages. Africa's Indo-European languages were introduced by European traders, administrators, and missionaries. Some have become the official languages of today's African countries. French or English often serves as a **lingua franca**, or common language, throughout the region.

Religions

A variety of religions claim followers in Africa south of the Sahara. Most people are Christian or Muslim. Christians make up the largest religious group. Missionaries and traders from Egypt and the Mediterranean area introduced Christianity to Ethiopia in the A.D. 300s. The Ethiopian Coptic

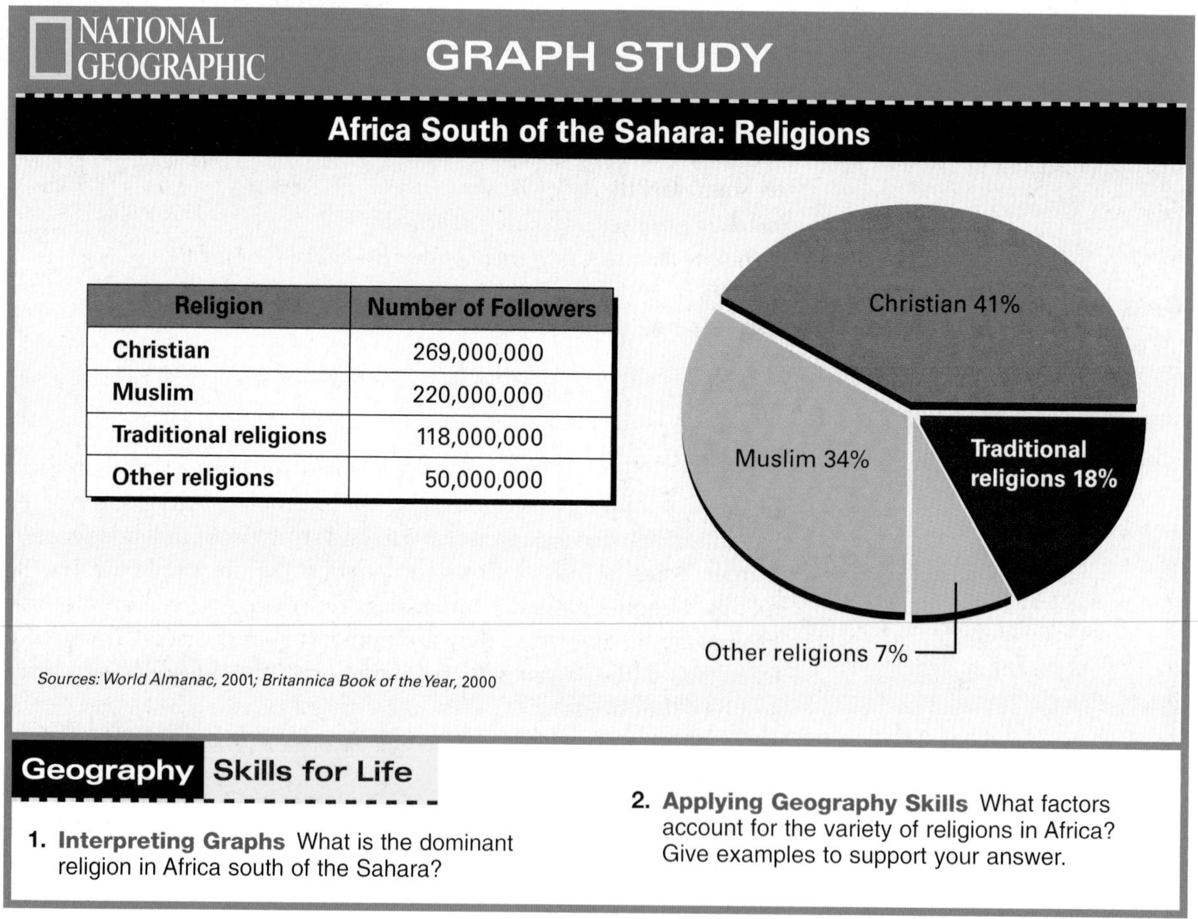

Religion	Number of Followers
Christian	269,000,000
Muslim	220,000,000
Traditional religions	118,000,000
Other religions	50,000,000

Sources: *World Almanac, 2001; Britannica Book of the Year, 2000*

Geography Skills for Life

1. **Interpreting Graphs** What is the dominant religion in Africa south of the Sahara?
2. **Applying Geography Skills** What factors account for the variety of religions in Africa? Give examples to support your answer.

Church has played an important role in Ethiopian life ever since. Christian beliefs did not spread among other African peoples until the colonial period, however. Since then, many Africans have adopted Christianity, especially along the coasts, where Africans had greater contact with foreigners. Most Muslims in the region live in West Africa, where Muslims ruled the kingdoms of Mali and Songhai along the Niger River during the 1400s and 1500s. Today Nigeria has the largest Islamic population of any African country south of the Sahara.

Traditional religions in Africa south of the Sahara are numerous and diverse, but they have many common elements. For example, most traditional religions profess a belief in the existence of a supreme being and a ranked order of lesser deities. In the late 1700s, Olaudah Equiano, an African known for his vivid account of slavery, stated that the supreme god of his people, the Igbo, was "one Creator of all things, and he lives in the sun, . . . and governs all events, especially . . . deaths." These same characteristics describe the supreme beings of other African groups. Most followers of traditional African religions also believe in the existence of nature spirits and honor distant ancestors and family members who have recently died.

Religion plays an integral role in everyday life in Africa. Although many followers of different religions live together peacefully, conflict sometimes occurs between competing religious groups. In recent years Nigeria and Sudan have been scenes of conflict among Christians, Muslims, and followers of traditional African religions.

NATIONAL GEOGRAPHIC **World Explorer**

Geography Skills for Life

Traditional Dancers Dance is an important religious and social tradition of the Masai of Kenya.

Region What are some common characteristics of traditional African religions?

Education

Africans have always valued education, but it has taken many different forms. In the past, African children did not attend school but apprenticed to trades such as wood carving and metalworking. Large-scale, formal schooling became widespread in the early 1900s, as European powers sought to fill civil service and industrial jobs with African workers.

Educational Advances

Since independence, higher education has expanded. In 1960 only 120,000 students in the region enrolled in universities, but by the late 1990s more than 2 million had. Public school attendance and literacy have also increased, but only about 60 percent of people aged 15 and older can read and write. Rural areas, which often are short of schools, materials, and qualified teachers, generally have lower literacy rates than do urban areas. In some places few children receive even an elementary education; parents there are too poor to send their children to school.

Culture

New Ways of Learning

Television has become an efficient teaching tool, but exposure to newer technology is limited. Fewer than 10 personal computers per 1,000 people exist in the region, and Internet service is not yet widely available. In some countries, such as South Africa and Zambia, however, use of the Internet is becoming more widespread.

The Arts

African art, often expressing traditional religious beliefs, comes in many forms, from ritual masks to

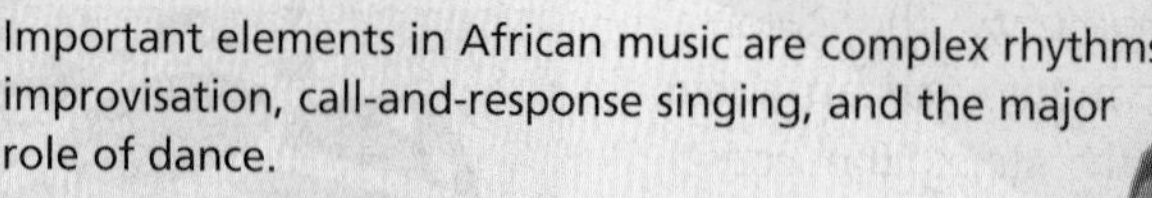

AFRICA SOUTH OF THE SAHARA

Important elements in African music are complex rhythms, improvisation, call-and-response singing, and the major role of dance.

Instrument Spotlight

The **talking drum** has an hourglass shape and a body carved from wood. On the open ends of the body, skins are held together by leather thongs or strings that stretch from one rim to the other. The drum is held under the arm and against the upper torso, and the drum head is struck with a curved wooden mallet. As the strings are squeezed down with the arm, the heads are stretched tighter and the pitch of the drum becomes higher. Many different pitches can be achieved by loosening and tightening the strings, and this gives the drum its characteristic "talking" sound.

Go To **World Music: A Cultural Legacy** Hear music of this region on Disc 2, Tracks 1–6.

rhythmic drum music to folktales. Strong examples of African visual arts include 2000-year-old terra-cotta heads produced by the Nok culture and bronze plaques that appeared in the palace courtyard of the Benin kingdom. Another art medium flourishing today is textiles, with patterns reflecting distinct ethnic groups—Ghana's *kente* cloth or East Africa's *khanga* cloth, for example.

Music and dance are art forms that are part of everyday African life. Entire communities participate, while dancers wearing masks honor specific deities, spirits of their ancestors, or a special occasion, such as a birth. Today African music is popular around the world and has influenced contemporary music. Paul Simon, Sting, and Peter Gabriel are only a few of the popular Western musicians who have borrowed from African music. In fact, the entire blues and jazz tradition of North America has its roots in the music enslaved Africans brought with them.

Oral literature, which is chanted, sung, or recited, has a strong tradition in Africa south of the Sahara. **Oral tradition**, the practice of passing down stories from generation to generation by word of mouth, is evident in folktales, myths, and proverbs and has helped preserve African history. Storytellers command great respect with tales of how the world began. In Mali, the Fulani people have this version:

> "At the beginning there was a huge drop of milk.
> Then Doondari came and created the stone.
> Then the stone created iron;
> And iron created fire;
> And fire created water;
> And water created air."
>
> Ulli Beier, trans. in *The Origins of Life and Death*, 1966

GEOGRAPHY Online

Student Web Activity Visit the **Glencoe World Geography** Web site at geography.glencoe.com and click on Student Web Activities—Chapter 21 for an activity about West African textiles.

Written literature developed mainly in northeast Africa, where societies came in contact with early Mediterranean systems of writing. In recent times written literature has become prominent in Africa south of the Sahara as well. Two Africans from the region have won the Nobel Prize in literature—Wole Soyinka of Nigeria in 1986 and Nadine Gordimer of South Africa in 1991.

Varied Lifestyles

Lifestyles in the region are as varied as the ethnic groups who live there. **Tanzania**, with some 120 ethnic groups, is a perfect example. The Sukuma farm the land south of Lake Victoria, and the Chaggas grow coffee in the plains around Kilimanjaro and conduct business in small cities like Moshi and Arusha. In the north live the nomadic cattleherders, the Masai, and in major cities, such as **Dar es Salaam**, Western urban lifestyles and dress prevail.

No matter how different their lifestyles, most Africans value strong family ties. In rural areas, most people still live in **extended families**, or households made up of several generations. Families also are organized into **clans**, large groups of people descended from an early common ancestor. Individuals often marry within their clan. In the cities, however, the **nuclear family**—made up of husband, wife, and children—is rapidly replacing the extended family.

NATIONAL GEOGRAPHIC **World Explorer**

Geography Skills for Life

Urban Shopping A supermarket in Gabarone, Botswana, provides shoppers with a variety of food products.

Place In which areas of Africa south of the Sahara do Western lifestyles prevail?

SECTION 3 ASSESSMENT

Checking for Understanding

1. **Define** mass culture, lingua franca, oral tradition, extended family, clan, nuclear family.
2. **Main Ideas** On a web diagram like the one below, fill in information about the cultural features of this region. Then write a paragraph describing one cultural feature.

Critical Thinking

3. **Comparing and Contrasting** How are urban families in the region different from rural families? How are they similar?
4. **Making Inferences** Why do you think storytellers are respected figures in African communities?
5. **Making Generalizations** Write three general statements to summarize religion in Africa south of the Sahara. Support your generalizations with specific examples from your reading.

Analyzing Maps

6. **Place** Study the map on page 523. What language/ethnic groups are common in countries bordering the Sahara?

Applying Geography

7. **Lifestyles** Think about the varied lifestyles in Africa south of the Sahara, and study the population density map on page 488. Would people living in extended families be more common in Namibia or South Africa? Explain.

GEOGRAPHY AND HISTORY

CONFLICT IN CENTRAL AFRICA: HUTU VERSUS TUTSI

THE STORY OF AFRICA in the last two centuries includes the tale of European conquest and its aftermath. Beginning in the late 1800s, European nations set up colonies in most of Africa. Although African nations have since gained independence, the lingering effects of European rule have caused instability and conflict in many regions.

Central Africa, home to the peoples of Rwanda and Burundi, is one of Africa's most unstable regions. Hutu and Tutsi peoples lived there peacefully for centuries. Since the late 1950s, however, the two groups have been at war. Their conflict has its roots in a tortured history in which Europe played a key role.

Exploration and Colonization

In the early 1800s, Africa was a mystery to Europeans. Few who ventured into Africa's interior came out alive. But with improved steam-powered transportation and new medicines to treat disease, mid-century explorers successfully made their way inland. One of the first explorers was British doctor and missionary David Livingstone. Besides finding unspoiled beauty and rich

Soldiers patrol in the Democratic Republic of the Congo, where Rwandans have come seeking refuge. ▶

◀ **Foreign ministers of 14 European nations decide the future of Africa at the Berlin Conference in 1884.**

NATIONAL GEOGRAPHIC SOCIETY

ethnic cultures, he discovered a flourishing slave trade. Livingstone called for Europeans to spread commerce, Christianity, and their civilization throughout Africa to stop the evil trade.

The European powers—mainly Britain, France, Portugal, and Germany—were competing to expand their empires and to protect trade routes. With the outcry against slavery, Europeans moved swiftly into Africa. In less than 40 years, they carved the continent into more than 40 colonies.

Europe's Legacy

The Europeans introduced new crops, legal systems, basic schooling, roads, and medicine—all of which brought benefits to Africans. For many Africans, however, especially those in Central Africa, European commercial and labor practices caused hardships and great loss of life.

At the Berlin Conference of 1884-1885, where European powers divided Africa, Rwanda and Burundi were given to Germany. In 1919 the lands were awarded to Belgium. Both the Germans and Belgians viewed the Tutsi as a superior people. Hence the Tutsi were favored and were promoted in society. Most Hutu were exploited farmers. They were denied positions of authority. The stage was set for conflict.

The killing began in 1959 with a Hutu uprising in Rwanda. When Burundi and Rwanda achieved independence in 1962, Hutu and Tutsi political groups began a struggle for control of the two countries. Violence led to more violence, spilling into the neighboring countries of Tanzania, Uganda, and the Democratic Republic of the Congo. Since the 1960s, more than a million Hutu and Tutsi men, women, and children have lost their lives. Millions more have been driven from their homes in a cycle of violence that seems to have no end.

Looking Ahead

Much of Africa today bears the scars of colonial rule. Why are peace and prosperity in Central Africa difficult to achieve?

1850s Dr. Livingstone (photo above) explores Africa

1884–1885 Leaders at Berlin Conference give Rwanda and Burundi to Germany

1881–1912 Europe occupies Africa, claiming more than 40 colonies

1919 Belgium controls Rwanda and Burundi

1959 Hutu rebel in Rwanda

1962 Rwanda and Burundi gain independence

1970s–1980s Violence escalates; refugees (background photo) flee to neighboring countries

1990s Rwandan and Burundian governments work toward democracy; fighting continues

1997 War crimes trials begin in Rwanda and Burundi

Making Generalizations

Suppose a friend says, "Both times I've been to Philadelphia, I heard someone on the street speaking Spanish. Philadelphia must have had a recent wave of immigrants from Spain." Your friend formed a generalization about Philadelphia. Whether the generalization is valid, or true, depends largely on the supporting details.

Learning the Skill

Generalizations are conclusions or judgments that people form based on the facts at hand. Using knowledge and experience, people make generalizations to help understand and explain the world. Geographers, like scientists, develop generalizations based on observation. Then they test these theories against further evidence they collect. This process is key to the scientific method.

Generalizations help us understand the world, but they can sometimes be misleading. For example, the generalization about Philadelphia was based on only two visits. But what are the real reasons the friend heard people speaking Spanish while visiting that city? Were the speakers longtime citizens who grew up speaking Spanish in a traditionally Hispanic neighborhood? To say that Philadelphia has had a recent wave of immigrants from Spain may be an *overgeneralization*, or a statement that is too broad.

Use the following steps to make useful generalizations:

- **Gather facts, examples, or statements related to the topic.**
- **Identify similarities or patterns among these facts.**
- **Use these similarities or patterns to form generalizations about the topic.**
- **Test your generalizations against other facts and examples.**

▲ *South Africans line up to vote in the country's historic election of 1994.*

Practicing the Skill

Complete the following activities about making generalizations.

1. Identify a generalization you have recently heard.
2. Describe ways you can avoid making an overgeneralization.
3. Write a generalization based on the following statements:

   In 1994 Nelson Mandela succeeded F.W. de Klerk as president of South Africa. Mandela's party, the African National Congress, received 63 percent of the vote; de Klerk's party, the National Party, received only 20 percent.

Applying the Skill

Read an article about Africa south of the Sahara in a newspaper or on an Internet news site. Write a generalization based on what you read. Provide details from the article to support your generalization.

The **Glencoe Skillbuilder Interactive Workbook, Level 2** provides instruction and practice in key social studies skills.

CHAPTER 21

SUMMARY & STUDY GUIDE

SECTION 1 Population Patterns (pp. 515–518)

Terms to Know

- sanitation
- urbanization
- service center

Key Points

- The uneven distribution of the 629 million people in Africa south of the Sahara is linked to the region's physical geography.
- The spread of AIDS has significantly impacted health and economic development in the region.
- Africa south of the Sahara is urbanizing faster than any other region in the world.
- Thousands of ethnic groups make up the population of Africa south of the Sahara.

Organizing Your Notes

Use a diagram like the one below to help you organize your notes for this section.

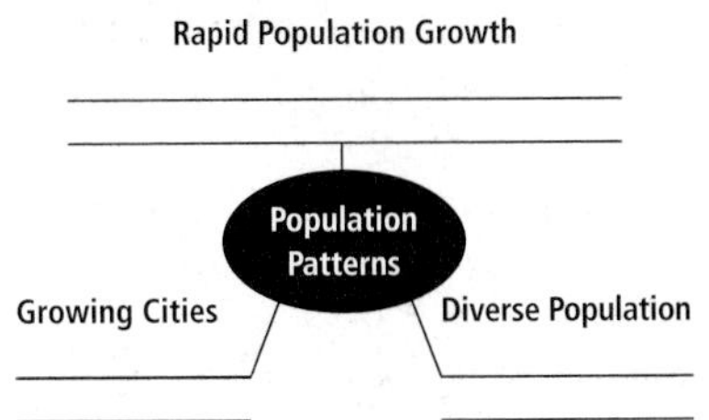

SECTION 2 History and Government (pp. 519–524)

Terms to Know

- domesticate
- apartheid
- universal suffrage

Key Points

- The movement of different groups, including the migrations of Bantu peoples, helped shape the history of Africa south of the Sahara.
- From the A.D. 700s to the 1600s, powerful trading empires arose and prospered in West Africa.
- European colonization cut across traditional ethnic territories.
- Most of the countries in Africa south of the Sahara won independence in the second half of the 1900s.

Organizing Your Notes

Use a table like the one below to organize your notes about each major stage in the history of Africa south of the Sahara.

Early Period	Colonial	Independent

SECTION 3 Cultures and Lifestyles (pp. 525–529)

Terms to Know

- mass culture
- lingua franca
- oral tradition
- extended family
- clan
- nuclear family

Key Points

- The many languages of Africans south of the Sahara contribute to the diversity of the region.
- The peoples of the region are followers of Christianity, Islam, or traditional African religions.
- The various art forms created by Africans south of the Sahara have influenced cultures around the world.
- Although they have diverse lifestyles, most peoples in the region value family ties, and many live in extended families.

Organizing Your Notes

Create an outline using the format below to help you organize your notes for this section.

Cultures and Lifestyles

I. Languages
  A. African
    1.
    2.
  B. Indo-European
    1.
    2.
II. Religions

ASSESSMENT & ACTIVITIES

Reviewing Key Terms

Write the key term that best completes each of the following sentences. Refer to the Terms to Know in the Summary & Study Guide on page 533.

1. A strong _____ has helped preserve many literary forms in Africa south of the Sahara.
2. A(n) _____ includes several generations.
3. The strict separation of races in South Africa was called _____.
4. Under _____, or equal voting rights, all adults of voting age may cast a vote.
5. _____ unites African people of different ethnic backgrounds.
6. Rapid _____ has caused problems in the region's cities.
7. A(n) _____ includes a husband and wife and their children.
8. _____ and disease are important health issues in Africa south of the Sahara.
9. People began to _____ animals many years ago.

Reviewing Facts

SECTION 1

1. What physical features influence population density in Africa south of the Sahara?
2. Why is food production in the region inadequate?

SECTION 2

3. What ancient kingdoms and empires developed in East Africa? In West Africa?
4. What was the major development in Africa after World War II?

SECTION 3

5. What religions are practiced in Africa south of the Sahara?
6. What is the origin of many of the art forms of Africa south of the Sahara?

Critical Thinking

1. **Predicting Consequences** In what ways has urbanization affected traditional ways of life in Africa south of the Sahara? Explain.
2. **Drawing Conclusions** How did trade play a major role in early African societies?
3. **Comparing and Contrasting** Use a Venn diagram like the one below to compare and contrast the role of music in African cultures and in American culture. Then write a paragraph summarizing your conclusions.

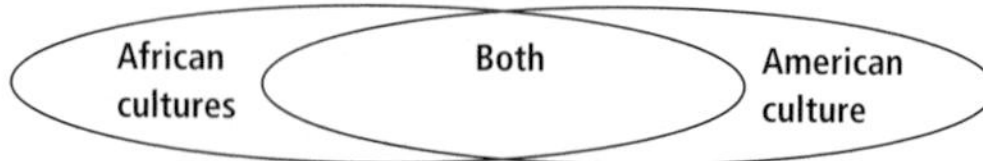

NATIONAL GEOGRAPHIC

Locating Places

Africa South of the Sahara: Political Geography

Match the letters on the map with the places in Africa south of the Sahara. Write your answers on a sheet of paper.

1. Angola	5. Namibia	9. Chad
2. Johannesburg	6. Botswana	10. Cameroon
3. Mali	7. Lagos	11. Ethiopia
4. Liberia	8. Rwanda	12. Kinshasa

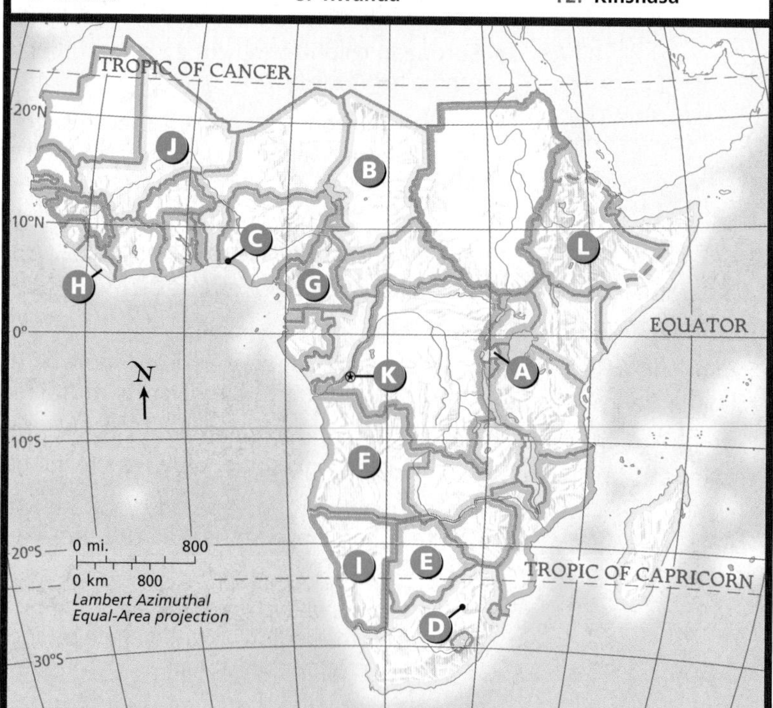

Self-Check Quiz Visit the **Glencoe World Geography** Web site at geography.glencoe.com and click on Self-Check Quizzes—Chapter 21 to prepare for the Chapter Test.

Using the Regional Atlas

Refer to the Regional Atlas on pages 486–489.

1. **Human-Environment Interaction** What is the relationship between areas of high population density and bodies of water?
2. **Location** Where are many of the capitals of West African countries located? Why do you think this pattern emerged?

Thinking Like a Geographer

How did past climate changes in the Sahara affect population patterns? What predictions can you make about future climate changes in the region?

Problem-Solving Activity

Contemporary Issues Case Study Many cities in Africa south of the Sahara have rapidly growing populations. Research and write a case study about an African city. Identify the processes causing the city's growth, such as location, resources, and transportation. Then describe the challenges the city faces and outline solutions. Design and draw maps and graphics for presentation.

GeoJournal

Newspaper Article Use your GeoJournal to write a newspaper article about an aspect of the cultural geography of Africa south of the Sahara. Remember to answer the 5-W questions (Who? What? Where? When? Why?). Your article should be impartial and factual, but be sure to include details that will hold your readers' interest.

Technology Activity

Using an Electronic Spreadsheet Using a world almanac, select 10 countries in Africa south of the Sahara, and find information about their birthrates, death rates, life expectancies, and rates of population change. Create a spreadsheet to compare and contrast these figures. If possible, also create graphs to illustrate your findings. Then write a summary of your analysis and possible explanations for your statistical findings.

Standardized Test Practice

Choose the best answer for the following multiple-choice question. If you have trouble answering the question, use the process of elimination to narrow your choices.

Facts About Tanzania

	1996	2001
Population	29,058,470	36,200,000
Percent urban	21	22
Percent rural	79	78
Population density of entire country	80 per sq. mi. (31 per sq. km)	99 per sq. mi. (38 per sq. km)
Population of current capital, Dar es Salaam	1,400,000	2,347,000

Sources: World Almanac, 1997, 2001; *2001 World Population Data Sheet*

1. Which of the following statements regarding Tanzania's population is NOT implied by the chart?

A Dar es Salaam's population increased by about 950,000 people.

B Between 1996 and 2000, many of Tanzania's residents moved out of cities and into traditional rural settings.

C The percentage of Tanzanians living in urban areas increased from 21 percent to 22 percent.

D Tanzania's population density was lower in 1996 than it was in 2000.

Some standardized test questions ask you to answer a question using a chart. Do not try to answer these types of questions from memory. First, read the question. Then skim the chart. Next, read each answer choice, deciding whether it is correct or incorrect by referring to the chart. Finally, choose the answer choice that is correct according to the chart.

CHAPTER 22

Africa South of the Sahara Today

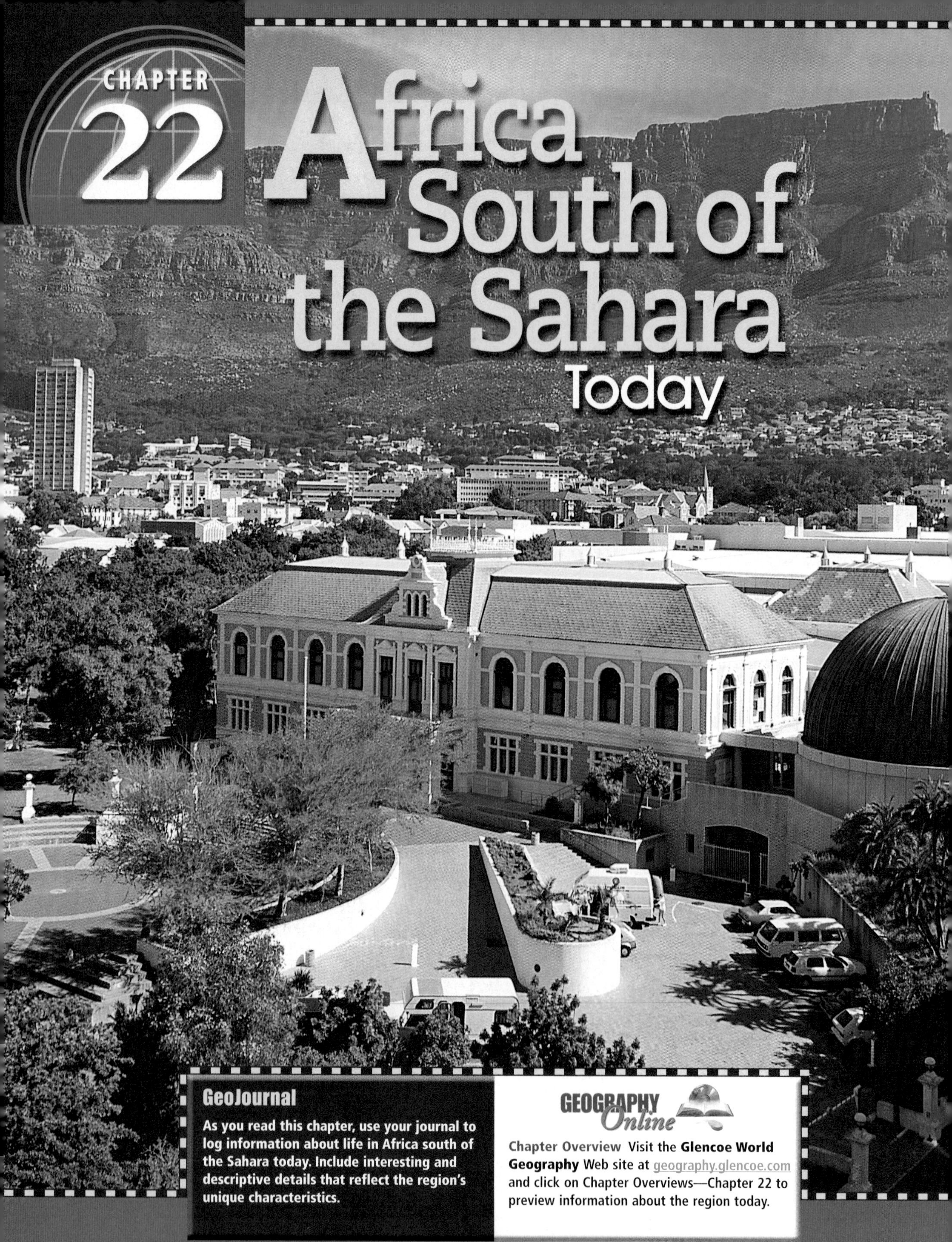

GeoJournal

As you read this chapter, use your journal to log information about life in Africa south of the Sahara today. Include interesting and descriptive details that reflect the region's unique characteristics.

GEOGRAPHY Online

Chapter Overview Visit the **Glencoe World Geography** Web site at geography.glencoe.com and click on Chapter Overviews—Chapter 22 to preview information about the region today.

SECTION 1

Living in Africa South of the Sahara

Guide to Reading

Consider What You Know

Think about the resources and products that come from Africa south of the Sahara and how people make their livings from them. If you wanted to create a mural showing what life in the region is like today, what images would you include?

Read to Find Out

- What are the most common farming methods in Africa south of the Sahara?
- How do mineral resources benefit the peoples of the region?
- Why has industrial development been slow in Africa south of the Sahara?
- How are transportation and communications in the region changing?

Terms to Know

- subsistence farming
- shifting farming
- sedentary farming
- commercial farming
- cash crop
- conservation farming
- infrastructure
- e-commerce

Places to Locate

- Zimbabwe
- Zambia
- South Africa
- Guinea
- Nigeria
- Kampala

◀ *View of South African Museum, Cape Town, South Africa*

NATIONAL GEOGRAPHIC

A Geographic View

Leap into the Future

With seven children, [Kawab Bulyar Lago] sold his animals and saddled himself with debt to send his son Paul to. . . school outside Marsabit [Kenya]. Now awaiting the results of the national exams that will determine his fate, Paul hopes to attend university and become either a doctor, a civil engineer, a teacher, or even a tour guide. "I'd prefer to be a doctor," he tells me one morning, "but anything would be all right.". . .

In the old days [Paul] would have inherited his father's herd. Today he inherits his hopes and dreams.

—*Wade Davis, "Vanishing Cultures,"* National Geographic, *August 1999*

Ariaal people, Kenya

Whichever career he pursues, Paul Lago's life will be different from his father's life. Like many other rural Africans, Paul's father herded camels and goats for a living. Today the lives of people throughout Africa south of the Sahara are changing as the region becomes more closely involved in the global economy. In this section you will learn about the region's changing economic activities— changes that offer new opportunities and challenges for the region's people.

Agriculture

Farming is the main economic activity in Africa south of the Sahara. More than two-thirds of the working population is involved in some form of agriculture. Some countries in the region still depend

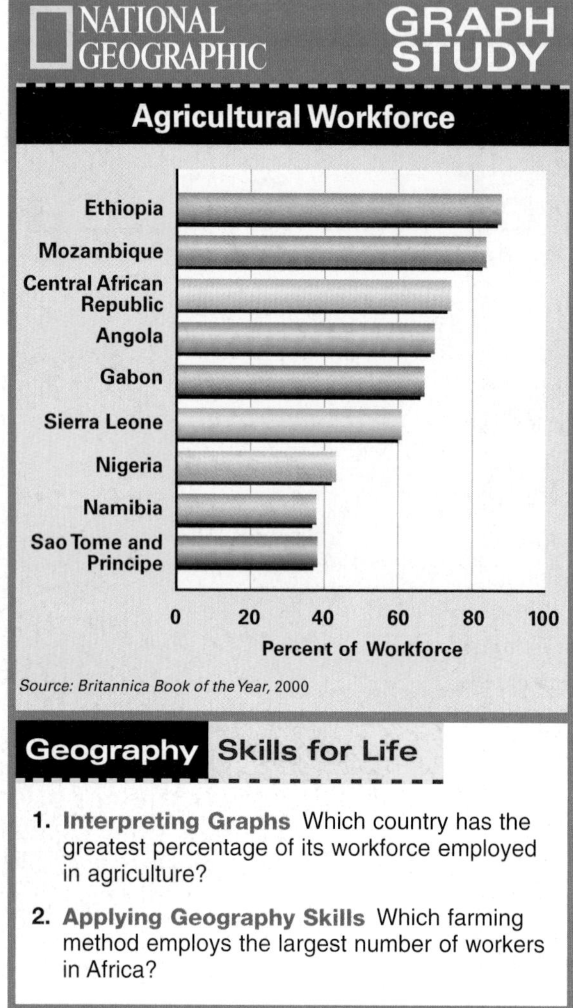

Geography Skills for Life

1. **Interpreting Graphs** Which country has the greatest percentage of its workforce employed in agriculture?
2. **Applying Geography Skills** Which farming method employs the largest number of workers in Africa?

on single-crop economies created during the colonial era. Others, however, produce a variety of agricultural goods.

Farming Methods and Export Crops

Most Africans south of the Sahara engage in **subsistence farming**, or small-scale agriculture that provides primarily for the needs of just a family or village. After they have met their own needs, farmers often sell any extra harvest or animals at a local market for cash or trade for other items they need or want.

African farmers use various methods to work the land. The Masai in Kenya and Tanzania and the Fulani in Nigeria and other parts of West Africa are nomadic herders. In the forest areas, farmers support themselves by **shifting farming**, a method in which farmers move every one to three years to find better soil. People practicing this method—also known as slash-and-burn farming—use basic tools, often just an ax and a hoe, to clear and cultivate land. They burn the trees and brush they have cut and then plant seeds in the ash-enriched soil. When the soil is no longer fertile, they move on, sometimes returning to a location after the soil has had time to renew itself. This method has been used to clear land for a variety of plantation crops, such as rubber in Liberia, cacao in Ghana, and coffee in Burundi.

Other farmers depend on **sedentary farming**, or agriculture conducted at permanent settlements. Sedentary farming is most common in areas with good soil. The Kikuyu in Kenya and the Hausa in Nigeria, for instance, farm permanent plots. Many people of European descent who have made their homes in South Africa, Kenya, and Zimbabwe also practice sedentary farming.

A small percentage of the population works at **commercial farming**, in which farms produce crops on a large scale. These **cash crops** are grown to be sold for profit instead of used by the farmer. Most commercial farms, such as those in Zimbabwe and South Africa, are large, foreign-owned plantations. They supply much of the world's palm oil, peanuts, cacao, and sisal, a vegetable fiber used for making rope.

The colonial economic systems played an important role in the growth of commercial farms in the region. Today the same commercial crops are the region's main agricultural exports. Côte d'Ivoire (KOHT dee•VWAHR), Nigeria, Ghana, and Cameroon, for instance, depend heavily on the sale of cacao, which is used to make cocoa and chocolate. Kenya, Tanzania, and Madagascar are large producers of coffee. Most of today's cash crops leave Africa to be processed elsewhere, just as they did during the colonial period.

The continued practice of cash-crop production has created problems for African economies. Reliance on one or two export crops is extremely risky. An unfavorable growing season or a drop in prices on the world market can have a disastrous effect on a country's entire economy.

Zimbabwe: Conflict Over Land

Cash-crop production also creates problems for farmers trying to meet their own food needs, because plantations and other large-scale farms take all the best land. For example, in **Zimbabwe**, a country with more than 11 million people, 40 percent of the farmland is controlled by only 4,000 commercial farmers and ranchers, descendants of Europeans who controlled the land in colonial times. Although the government has proposed land reform to distribute land more evenly, violence has broken out as small-scale farmers have tried to take over large-scale farms. The resulting conflict has slowed or completely halted production on commercial farms. These developments threaten Zimbabwe's economy, which currently depends heavily on commercial agriculture.

Meeting Challenges

Whether involved in large-scale or small-scale agriculture, farmers in the region face many challenges. Overgrazing, overworked soils, and a lack of technology have made farming difficult in many places. The use of heavy farm machinery, frequent tilling, and the clearing of forests for timber have caused soil erosion and desertification. Most subsistence farming in the region depends largely on human labor alone. Although men work primarily in cash-crop production, women often work at traditional subsistence farming, using basic tools and techniques. Food production has fallen far short of the needs of the region's booming population.

Gradually, however, farmers are beginning to employ new methods and tools. Farmers in **Zambia** have started to practice **conservation farming**, a land-management technique that helps protect farmland. By planting different crops where they will grow best, Zambian farmers actually conserve, or save, land for farming. In addition, better fertilizers and seeds have increased yields of maize and other crops. In Nigeria and other countries, farmers who depended solely on rain to water their fields now use irrigation to increase production.

Logging and Fishing

Although forests cover almost 25 percent of Africa south of the Sahara, human activities are destroying the region's forests at an alarming rate, upsetting unique ecosystems. The demand for farmland has led agricultural settlements to open up some of this land, using the slash-and-burn method of shifting farming. People in the region also cut wood from the rain forest and savanna woodlands to use as fuel. Logging companies also harvest and export valuable hardwoods, such as Rhodesian teak, ebony, African walnut, and rosewood.

Although logging creates serious consequences for Africa's forests, the lumber industry has a relatively small output. Logging in the region accounts for less than 10 percent of the world's lumber supply. Coastal countries with rain forests, such as Gabon, Equatorial Guinea, and South Africa, do export significant amounts of lumber and pulp.

Commercial fishing also represents only a small portion of the region's economic activity. Few countries build and support fleets of commercial fishing vessels. Africa also has a very narrow continental shelf, the shallow ocean area near a

NATIONAL GEOGRAPHIC **World Explorer**

Geography Skills for Life

Fresh Sardines Fishing is a major activity along Zimbabwe's Lake Kariba, one of the world's largest human-made lakes.

Human-Environment Interaction What inland countries in Africa south of the Sahara profit from fisheries on lakes and rivers?

continent's coast that usually contains abundant fish. Along the southwestern coast, commercial fishing vessels do catch large quantities of herring, sardines, and tuna for export. The richest fishing grounds in the region lie off the region's west coast. Countries bordering oceans—South Africa, Namibia, Angola, Nigeria, Ghana, and Senegal—haul in the largest catches. The economies of island countries in the region depend on the export of fish and fish products. In addition, the inland countries of Malawi, Uganda, Chad, and Mali profit from fisheries on lakes and rivers.

Mining Resources

Difficult and risky, mining is an important economic activity in the region. Gold mining is particularly dangerous. The extremely narrow seams of the valuable mineral are located deep in the earth. Such depths greatly increase the risk of rock bursts, or breaks in the earth's crust, under the stress of explosives and power tools, but mine workers need the wages to help support their families.

Mineral Wealth

The Witwatersrand, a gold deposit 300 miles (483 km) long, makes **South Africa** the world's largest producer of gold. The country also is a world leader in the production of gems and industrial diamonds mined from beneath the grassy plateau of Gauteng Province. South African miners also extract large quantities of coal, platinum, chromium, vanadium, and manganese for export.

Although South Africa's mineral wealth makes it one of the region's richest countries, foreign investors or companies owned by white South Africans reap the most benefits. Little money reaches black South African mine workers. Since the steady decline in gold prices that began in the 1980s, however, gold has also contributed less to South Africa's economy.

An Imbalance of Riches

The uneven distribution of mineral resources causes economic imbalances in Africa south of the Sahara. Most known mineral deposits lie along the Atlantic coast and south of the Equator. For example, **Guinea** (GIH•nee) has about one-third of the world's known reserves of bauxite, the main ore used in making aluminum. Immense oil reserves make **Nigeria** the region's only member of the Organization of Petroleum Exporting Countries (OPEC). In spite of rich mineral resources, many people in these two countries do not benefit directly from local resources, and they remain poor. Governments have badly managed the income from mineral wealth, and foreign mine owners often send their profits abroad.

Industrialization

Despite its large reserves of bauxite, Guinea cannot manufacture aluminum because it lacks the cheap energy, capital, and infrastructure—resources such as trained workers, facilities, and equipment—to build a refinery. Guinea is not unique. Most of the region's countries never developed manufacturing industries to process their natural resources. Today many countries in Africa south of the Sahara receive foreign loans to industrialize, but progress has been slow. Few countries have industrial centers for processing raw materials. As a result, most countries in the region continue to act as suppliers of raw materials for the industrialized countries of the world.

Development of Manufacturing

Since the 1960s the governments of newly independent African countries have encouraged industrial expansion. Demand for manufactured goods has increased, and locally produced goods have replaced some imported items. Today the region's industrial workers process food or produce textiles, paper goods, leather products, and cement. Some assemble electric motors, tractors, airplanes, and automobiles. Yet compared to manufacturing in other developing areas, such as Latin America, the economic role of manufacturing is small. By the late 1990s, only 15 percent of the region's GDP came from manufacturing.

Overcoming Obstacles

Africa south of the Sahara faces many obstacles to industrialization, including the lack of skilled workers. Educational systems are relatively new, and training programs are limited. Hydroelectric resources are plentiful but untapped, and power shortages often occur. Political conflicts interrupt economic planning and divert resources from development projects. In addition, countries must import food to feed their growing populations.

Although African products still do not reach many parts of the world, exports have been growing since World War II. Some countries in the region trade with Japan and the United States, but most rely on trading ties established with Western European countries before independence. Some countries are breaking old trading patterns to trade within the region. Various countries, for example, have formed regional trading associations, such as the Economic Community of West African States (ECOWAS), to exchange ideas and to protect their interests.

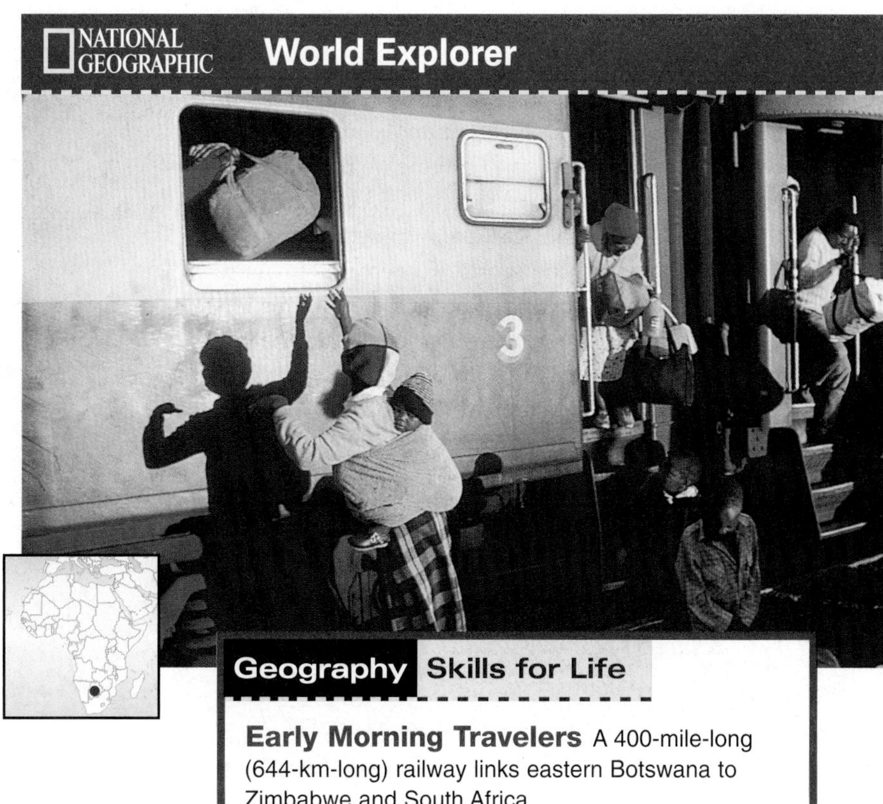

Geography Skills for Life

Early Morning Travelers A 400-mile-long (644-km-long) railway links eastern Botswana to Zimbabwe and South Africa.

Region What factors challenge the development of transportation in the region?

Transportation and Communications

Good transportation and communications systems are essential to industry and trade as well as to everyday life in the region. New transportation networks and technology are beginning to change lifestyles, but much remains to be done.

Creating and maintaining transportation systems in the region is difficult. Roads and railways must cross vast distances and changing terrain. Water transportation is limited because most rivers cannot be navigated from source to mouth, and the region has few natural harbors. In addition, there are few experts and skilled managers to plan and supervise transportation systems. In recent years wars and lack of money have kept many roads and rail lines from being repaired.

Roads and Railroads

Several countries, however, consider roads and railroads a top priority. Nigeria plans to link all parts of its railroad system, and Uganda is scheduling repairs on the heavily traveled Trans-African Highway, which runs from Mombasa, Kenya, to Lagos, Nigeria. Mauritania, Senegal, and the North African country of Morocco are discussing plans for a highway between Tangiers in Morocco and Dakar in Senegal that would eventually reach Lagos. This important project would link people and ideas in different parts of Africa.

Communications

In the area of communications, the region has long relied on radio, with state-run stations providing global programming. Television reaches fewer people because the land-relay systems for transmitting TV signals become very costly outside urban areas. Satellite technology should improve television's reach, however. Low literacy rates limit traditional media like newspapers and magazines, and in many countries, governments restrict the number of issues that can be published.

Telephone service is also limited, especially in rural areas. Across the region, only 14 main telephone lines serve each 1,000 people. However, satellite and wireless technology is expected to improve access to phone service and the Internet in Africa south of the Sahara.

Economics

Internet Commerce

Helen Mutono runs a small business selling baskets made by Ugandan women. Using e-commerce, or selling and buying on the Internet, Helen Mutono set up a Web site at a cybercafe in the Ugandan capital, **Kampala**, to sell baskets to customers around the world. Cybercafes provide Internet access for people who lack their own computers. For a fee, cybercafes allow customers to use Internet technology. The Internet broadens the market for locally made products, allowing customers from around the world to purchase unique products. As Helen Mutono notes:

> *"You can imagine trying to sell a basket that everybody can make locally. [The weavers] probably wouldn't be able to sell very many baskets, but to be able to market [the baskets] worldwide is . . . the greatest thing that could have happened for them."*
>
> quoted in "E-commerce: Uganda's Entrepreneurs Go Global," *BBC World Service* (online)

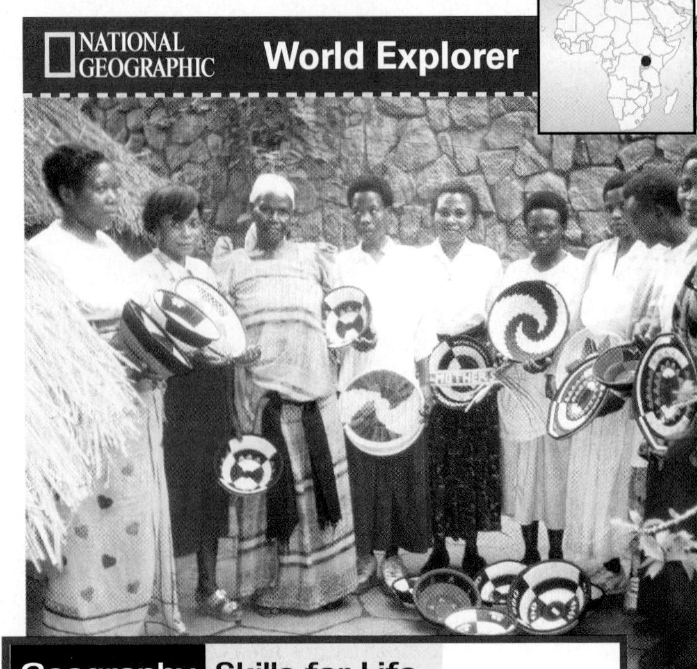

Geography Skills for Life

E-Commerce By selling baskets worldwide via the Internet, Helen Mutono and others have been able to buy needed clothes and send their children to school.

Region How does the Internet broaden the market for locally made products?

SECTION 1 ASSESSMENT

Checking for Understanding

1. **Define** subsistence farming, shifting farming, sedentary farming, commercial farming, cash crop, conservation farming, infrastructure, e-commerce.
2. **Main Ideas** Copy the web diagram below. List details about each of the five major economic activities in the region. Then choose one type, and write a paragraph about it.

Africa South of the Sahara: Logging and Fishing; Transportation and Communications; Agriculture; Mining; Industry

Critical Thinking

3. **Making Generalizations** What economic features do many countries in Africa south of the Sahara share? How have these shared features affected their economies?
4. **Making Predictions** What role might the Internet play in the region's economic development? Explain.
5. **Comparing and Contrasting** Compare and contrast the roles agriculture and industry each play in the region. Consider changes in the region's economies before and after independence as part of your analysis.

Analyzing Graphs

6. **Human-Environment Interaction** Study the graph on page 538. What can you conclude about the importance of agriculture in the economies of the countries represented in the graph?

Applying Geography

7. **Obstacles to Development** Think about the challenges to economic development in Africa south of the Sahara. Write a speech in which you define one of the most critical challenges, propose a way to overcome it, and explain why your proposal is a good idea.

Guide to Reading

Consider What You Know

Imagine that you are going to create a series of TV specials about Africa south of the Sahara. What issues would you include in the segment about the environment?

Read to Find Out

- Why have food shortages occurred in parts of Africa south of the Sahara?
- What steps are the African countries south of the Sahara taking to protect their environment?
- What is the outlook for the region's future development?

Terms to Know

- habitat
- extinction
- poaching
- ecotourism

Places to Locate

- Somalia
- Ethiopia
- Djibouti
- Sahel
- Sudan
- Eritrea
- Côte d'Ivoire
- Madagascar

People and Their Environment

NATIONAL GEOGRAPHIC

A Geographic View

Saving Forestlands

Traditional farmers use[d] slash-and-burn methods [in Madagascar], and a growing population . . . led to the clearing of more land. In the worst cases nearly a hundred tons of topsoil an acre were being lost each year. And while that flow has yet to be fully [stopped,] some progress has been made. Still, . . . unless these farming methods change, virtually all the island's forests will be gone within 25 years.

—*Virginia Morell, "Restoring Madagascar,"* National Geographic, *February 1999*

Village meeting in Madagascar

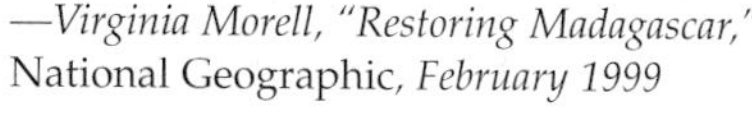

Africans south of the Sahara, like their neighbors around the globe, look to the future with hope. Yet the people in this region face tremendous difficulties in achieving a better life. Many environmental challenges threaten the region's supply of food, its health care, and its plant and animal life. In this section you will learn about these problems and the solutions proposed to deal with them.

Shadow of Hunger

Today millions of people in the region must focus on survival. Hunger is one of their bitterest enemies. In the 1990s, for example, many thousands of people died of starvation in the Horn of Africa—the bulge of land that juts into the Indian Ocean and includes the countries of **Somalia**, **Ethiopia**, and **Djibouti** (jih•BOO•tee). Drought and human activities, such as wars, contributed to the famine, or extreme scarcity of food. Today famine threatens many parts of Africa, which must look to the international community for food. Food donations often can help relieve famine if there are no barriers to distribution. However, they cannot end hunger caused by years of conflicts and natural disasters.

Desertification

Although never as fertile as land to the south and east, the **Sahel** region of West Africa once supported life. The Sahel is a band of semiarid land extending across the northern part of the region and bordering the Sahara. Not so long ago, nomadic peoples grazed livestock in the Sahel. Their animals helped fertilize the soil, and farming was possible. Today, however, a wide area of the Sahel has become desert. As the climate has become drier and as people and animals have stripped the Sahel of its vegetation, the desert has crept farther south, spreading in the countries of Mauritania, Mali, Niger, Chad, and Sudan.

Droughts, which have always occurred in the semiarid Sahel, have recently become severe there and in other parts of Africa south of the Sahara. Beginning in the 1960s, severe droughts in these areas helped turn farmland into wasteland. For example, in the early 1990s, drought in the Horn of Africa caused widespread famine. Since 1998 drought has killed crops and livestock across East Africa, threatening the lives of hundreds of thousands of people.

In 2000 the United Nations Food and Agriculture Organization (FAO) warned that famine could become a problem in central Africa because of unpredictable weather patterns and large numbers of refugees. In West Africa good harvests have boosted food supplies in most countries. Civil war, however, threatens to disrupt the distribution of food in Sierra Leone, Liberia, and Guinea.

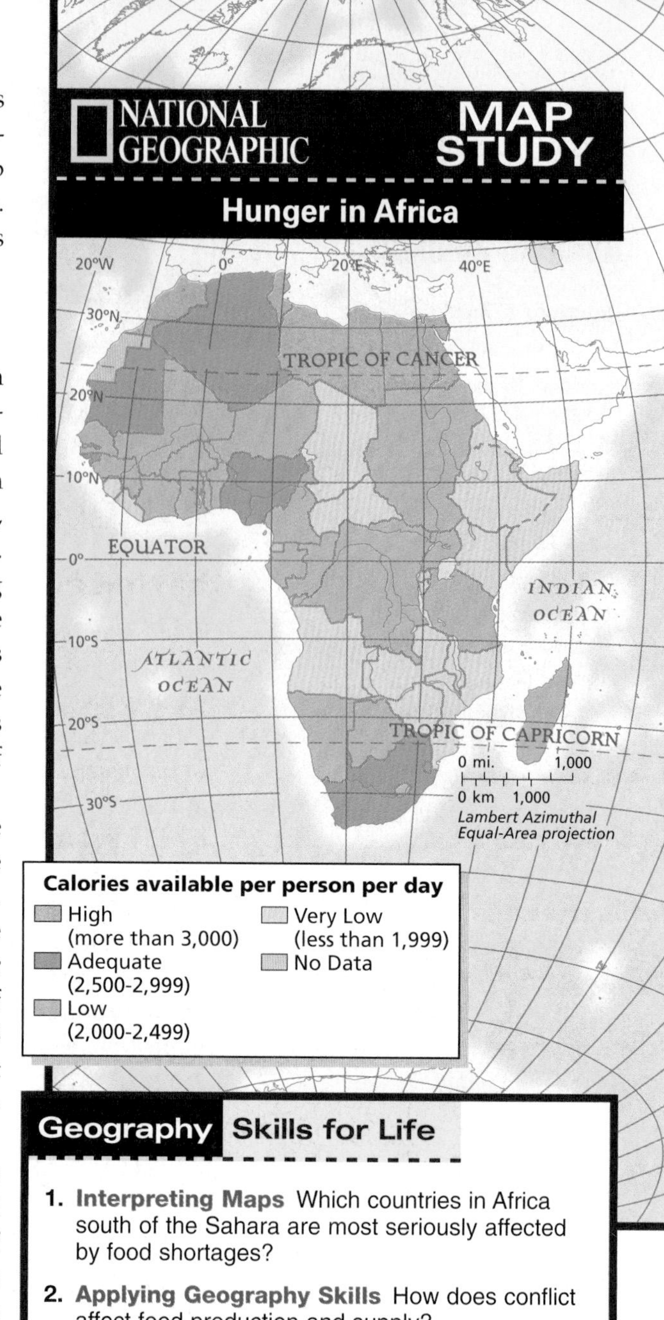

Geography Skills for Life

1. **Interpreting Maps** Which countries in Africa south of the Sahara are most seriously affected by food shortages?
2. **Applying Geography Skills** How does conflict affect food production and supply?

Find NGS online map resources @ www.nationalgeographic.com/maps

Conflict and Hunger

War continues to be a major cause of hunger and malnutrition in Africa south of the Sahara. Since 1990, conflicts in countries such as Liberia, Sudan, Somalia, and Rwanda have halted economic growth, caused widespread starvation, and cost the lives of countless Africans. Huge refugee populations fleeing war-torn areas have strained already meager food resources. Today civil conflict in Somalia endangers more than one million people, including relief workers. Looting and fighting severely hamper food distribution.

In **Sudan** about 2 million people are on the verge of starvation, according to UN estimates. Most of Sudan's people depend on subsistence farming, making them vulnerable to the country's periodic droughts. In addition, more than a decade of civil war between the Muslim Arab government and non-Muslim rebels in the south has torn Sudan

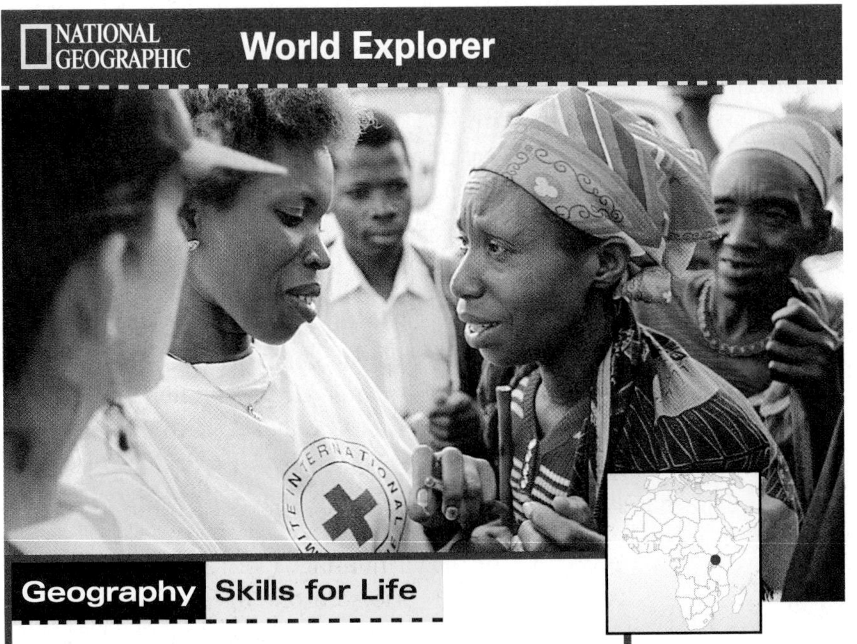

Geography Skills for Life

International Aid A Red Cross worker helps a Hutu mother in Rwanda find her lost children.

Region What factors lead to food shortages in Africa south of the Sahara?

apart and created the world's largest refugee population. Although international aid workers have tried to meet the enormous food needs of the refugees, warring factions continue to raise obstacles. In 2000, for example, rebel groups began to impose taxes on relief work, forcing many aid agencies to leave the country.

If the problem of hunger is to be solved, peace within the region is critical. Some countries and groups are moving toward peace. Ethiopia and **Eritrea**, for instance, finally signed a peace agreement in 2000 after two years of conflict. Tensions between the two countries remain high, and maintaining peace will be a great challenge. However, the Eritreans and Ethiopians have been working to undo the damage caused by drought and civil war.

History

Farming in Peace

After Eritrea gained its independence from Ethiopia in 1993, farmers in both countries worked to improve the land. Farmers in the northern Ethiopian province of Tigray terraced more than 250,000 acres (about 101,172 ha) of land and planted 42 million young trees to hold soil in place. They also built earthen dams to store precious rainwater. Grain crops thrived in their fields. In Eritrea, crops were so abundant that the government was able to reduce its request for relief from other countries by 50 percent.

When Ethiopia and Eritrea went to war over their shared border, however, many people lost their homes or lives. Then one of the worst droughts in years struck the region. Although drought continues, a shaky peace is allowing farmers to restore the land, bringing hope to the area's people.

Medical teams and relief workers with humanitarian organizations like Doctors Without Borders (Médecins Sans Frontières) and the International Red Cross have helped. Feeding centers, for example, have nursed many malnourished children and adults back to health in war-torn countries.

Land Use

People in the region are also struggling with problems of land use. At the start of the 2000s, tropical rain forests in the region were disappearing at a rate of more than 12 million acres (4.8 million ha) per year. The environmental impact of hunting and tourism has also raised difficult questions about the region's land use.

Destruction of the Rain Forest

In 1990 rain forests covered almost 1.5 billion acres (607 million ha) in this region. By 2000 126 million acres (51 million ha) of that forest had disappeared. **Côte d'Ivoire** has lost more than

Student Web Activity Visit the **Glencoe World Geography** Web site at geography.glencoe.com and click on Student Web Activities—Chapter 22 for an activity about life today in Africa south of the Sahara.

90 percent of its rain forest, and **Madagascar**, at one time a heavily wooded island, has also seen more than 90 percent of its rain forest disappear. On the continent as a whole, about half of the original rain forests are gone. Nearly 15,000 square miles (38,850 sq. km), an area almost the size of Switzerland, is being cleared every year.

Such statistics have alerted people in the region to the severity of the problem. Today various countries have created forest preserves to help save the rain forests. Many logging companies are also getting involved, using scientific tree farming and replanting projects to protect and renew forests.

Endangered Animals

As the rain forests disappear, many plant and animal species are put at risk. Deforestation destroys animal **habitats**, or living areas. Today hundreds of animal species in Madagascar that exist nowhere else are in danger of **extinction**, or disappearance from the earth.

the arts of AFRICA SOUTH OF THE SAHARA

WORLD CULTURE

Shona Sculpture

Since the 1950s, artists of the Shona ethnic group in Zimbabwe have carved attractive stone figures that are highly valued throughout the world. For the most part self-taught, Shona sculptors use simple hand tools. They rarely begin carving with a specific subject in mind; instead, the artists allow the qualities of the stone to determine the form they will create. Many artists believe that the spirit in the stone speaks to them, and they must work to set the spirit free. An example of Shona work is this stylized sculpture of an African bird.

The threat to wildlife exists elsewhere too. As the region's population grows, farmers have moved into some forested areas to find land for planting and grazing. Some grassy savannas, home to huge herds of animals such as elephants, giraffes, antelopes, and lions, are being plowed for farming. As a result, many species have greatly decreased in number.

Hunting also threatens the region's wildlife. During the colonial period, European hunters reduced animal populations significantly. For instance, during the 1900s the numbers of Zambian black lechwe, a kind of antelope, dwindled from 1 million to fewer than 8,000. In recent years hunters have continued to pursue African game for sport and profit. Two million elephants roamed the region in the early 1970s. Today fewer than 600,000 remain, largely because of **poaching**, or illegal hunting. Ivory from elephant tusks brings high prices despite international bans on its trade. Other animals at risk include the Cape Mountain zebra and the mountain gorilla.

Economics

Conservation and Tourism

To save endangered species, some countries have created huge game preserves. These preserves—which include Tanzania's Serengeti National Park, Kenya's Masai Mara, and Rwanda's Parc National des Volcans—have helped some animals make a comeback. The parks also attract millions of tourists each year. **Ecotourism**, or tourism based on concern for the environment, has become a big business in parts of the region, bringing millions of dollars into African economies. Despite the profits earned from such preserves, many people in the region, like V.N. Mthembu, object to them:

> *"My family lived in Ndumu [Game Reserve] until around 1960. They were moved outside when the park brought in rhinos. There is not enough land for everyone outside now and not enough water, especially in droughts. But the park has plenty of water and game. Why can't we come back inside to build homes and live?"*
>
> quoted by Douglas H. Chadwick, "A Place for Parks in the New South Africa," *National Geographic*, July 1996

Governments have tried to respond to such concerns by giving rural peoples an economic stake in the preserves. Some train to work in the preserves as trail guides or become involved in development planning.

Toward the Future

In Africa south of the Sahara, people are working to overcome some of the region's serious challenges, many of them inherited from the colonial period. The region has already taken important steps, however, toward preserving the environment and its precious natural resources. Efforts to encourage private enterprise have also had positive results. New ranching laws, for example, have allowed people to engage in crocodile farming, a highly profitable business that has also brought this species back from near extinction. Rhinoceroses and elephants are also beginning to thrive again as their habitats are protected and poaching is discouraged by stricter laws.

Increasingly, the protection of rain forests is a priority in the region. In 1999 leaders from six central African countries signed an agreement to preserve the forests. The effects of this and similar efforts have yet to be seen, but they are a strong signal that Africans today are moving toward a more positive future.

NATIONAL GEOGRAPHIC **World Explorer**

Geography Skills for Life

Protecting Wildlife A conservationist measures the growth of a young cheetah in Namibia.

Human-Environment Interaction How have African countries addressed the problem of endangered species?

SECTION 2 ASSESSMENT

Checking for Understanding

1. **Define** habitat, extinction, poaching, ecotourism.
2. **Main Ideas** Use a diagram like the one below for each challenge facing the people of Africa south of the Sahara. Then write a paragraph about one of the challenges, and write a proposal for meeting the challenge.

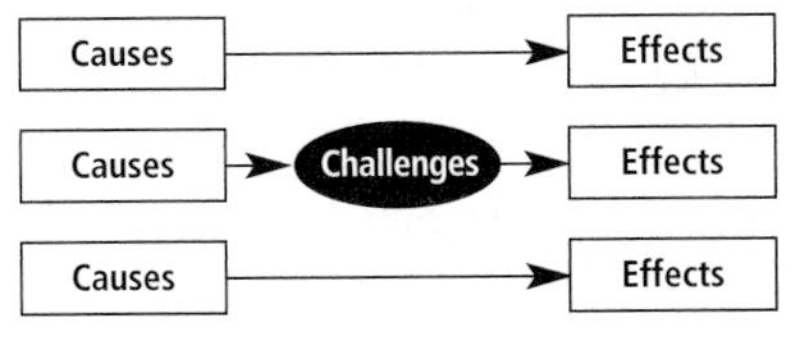

Critical Thinking

3. **Decision Making** List actions that governments in Africa south of the Sahara take to help eliminate hunger. Then arrange the items by order of importance and explain your reasoning.
4. **Finding and Summarizing the Main Idea** What is the central issue in the debate over the creation of game preserves?
5. **Problem Solving** How would you encourage logging companies and governments to help prevent deforestation? Consider the advantages and disadvantages.

Analyzing Maps

6. **Region** Study the map of hunger in Africa on page 544. What countries in Africa south of the Sahara generally have the least problem with hunger?

Applying Geography

7. **Global Issues** Think about the challenges Africa south of the Sahara faces today. Choose one problem in the region that might have an impact on the rest of the world. Make a chart or diagram that explains why the problem is a global issue.

Viewpoint

CASE STUDY on the Environment

Southern Africa's Dilemma: Renew the Ivory Trade?

African elephants—the biggest and strongest of all living land animals—once roamed in great numbers across the continent. During the last century, however, elephants were slaughtered by the tens of thousands for meat, for sport, and especially for their ivory tusks. In 1989 African elephants were placed on the endangered species list. Trade in elephant ivory was banned worldwide. Recently, however, three southern African nations were given approval to sell their stockpiles of ivory. Critics worry that these legalized sales will renew the demand for ivory and increase the killing of elephants for their tusks.

Standing more than 10 feet (3 m) tall and weighing nearly 10,000 pounds (4,500 kg), an African elephant (left) wades into the waters of the Okavango Delta in Botswana. Both male and female African elephants grow tusks—the world's main source of ivory. A lustrous, creamy-white material, ivory was once carved into everything from figurines and jewelry to billiard balls and piano keys.

Biologists estimate that in 1930, Africa was home to 5 to 10 million elephants. When the price of ivory soared in the 1970s, elephants became very valuable. Gun-carrying ivory poachers began illegally killing elephants for their tusks. By 1979 the elephant population had dropped to 1.3 million. As many as 80,000 elephants a year were shot for their ivory. By the late 1980s, only about 600,000 elephants were left in Africa.

In 1989 the nations that make up CITES—the Convention on International Trade in Endangered Species—placed the African elephant on the endangered species list. The sale of ivory was banned worldwide. During the 1990s, the ban was successful in protecting elephants. Demand for ivory dwindled, prices fell, and poaching declined. Many elephant populations began to increase, particularly in southern Africa.

In fact, elephants grew so plentiful in Botswana, Namibia, and Zimbabwe that in 1997, CITES changed the elephant's status in these countries from endangered to threatened. This change allowed the three nations to sell government stockpiles of ivory to Japan, as long as each country adopted strict anti-poaching measures. The sales went forward in 1999, but not without controversy.

◀ Watch out for elephants! Officials in southern Africa want to use money from ivory sales to protect elephants.

Supporters of the sale of ivory stress that only government stockpiles are being sold and that no elephants will be killed for ivory. Supporters argue that money from the sales can be spent on elephant conservation and on national parks. Further, a strict monitoring program will track poaching. If poaching increases, ivory sales will end.

Opponents of the ivory sales fear that even a partial lifting of the ban will lead to more poaching of elephants. Opponents argue that there are better ways to raise money, such as increasing park entrance fees. Moreover, opponents claim that programs to monitor and to report increased poaching of elephant populations could take years—too long to save the elephants.

What's Your Point of View?
Do you agree with the decision to allow limited ivory trade in southern Africa? Will resumption of trade give the green light to poachers?

Tons of tusks (below) are being sold to Japan. Will their sale lead carvers (right) and collectors to renew their interest in ivory? ▼

Reading Tables and Interpreting Statistics

Reading lists of facts and figures can be confusing. For this reason, statistics are often organized in tables, which display numerical information in rows and columns.

Learning the Skill

In a table, similar kinds of information are organized into columns and rows. Labels across the top and left-hand side give information about the figures in the table. Identifying patterns and relationships among the figures can reveal a great deal about a topic. In this table, several kinds of statistical data for different countries are compared. The left row of labels shows what specific information is included in the comparison, such as population density or infant mortality rate. The top row identifies the countries being compared.

To read tables and interpret statistics, follow these steps:

- **Read headings and labels to determine the kinds of information included in the table.**
- **Look up any unfamiliar terms in the table.**
- **Identify similarities, differences, and other relationships among the data.**
- **Use the data to draw conclusions.**

Population Information for Selected African Countries

	South Africa	Chad	Senegal
Total population	43,600,000	8,700,000	9,700,000
Population density	92/sq. mi.	18/sq. mi.	127/sq. mi.
Annual population growth	1.2%	3.3%	2.8%
Percent urban	54%	21%	43%
GDP (US dollars)	$290.6 billion	$7.5 billion	$15.6 billion
GDP per capita	$6,800	$1,000	$1,600
Life expectancy	53 years	50 years	52 years
Infant mortality rate (per 1,000 births)	57	103	59
Population per physician	1,529	27,765	14,825
Literacy rate	82%	48%	33%
Number of automobiles	4,350,000	9,630	110,000

Sources: 2001 World Population Data Sheet; World Almanac, 2001

Practicing the Skill

Study the table, and answer the following questions.

1. What countries are being compared in the table?
2. How do the countries rank according to total population? According to population density?
3. Which country has the highest GDP per capita?
4. Which country has the lowest annual population growth?
5. What is the relationship between infant mortality rate and life expectancy? Explain.
6. What is the relationship between infant mortality rate and GDP per capita? Explain.
7. What is the relationship between urbanization and the number of automobiles?

Applying the Skill

Choose three countries in Africa south of the Sahara. Using an almanac, identify three statistics about the countries. Then use these statistics to make a table that allows you to compare and contrast the countries. Discuss with a partner the relationships among data that your table reveals.

The **Glencoe Skillbuilder Interactive Workbook, Level 2** provides instruction and practice in key social studies skills.

CHAPTER 22

SUMMARY & STUDY GUIDE

SECTION 1 Living in Africa South of the Sahara (pp. 537–542)

Terms to Know

- subsistence farming
- shifting farming
- sedentary farming
- commercial farming
- cash crop
- conservation farming
- infrastructure
- e-commerce

Key Points

- Most people in Africa south of the Sahara engage in subsistence farming, and most countries in the region depend on the export of one or two cash crops.
- Mineral resources are not evenly distributed across Africa south of the Sahara, causing economic imbalances among the region's countries.
- Africa south of the Sahara has taken actions to break its dependence on old trading patterns, and manufacturing is gaining strength in the economies of some countries in the region.
- New transportation networks and new forms of communication are changing the lives of Africans south of the Sahara.

Organizing Your Notes

Use an outline like the one below to organize the notes you took as you read about living in Africa south of the Sahara.

Living in Africa South of the Sahara

I. Agriculture
   A. Farming Methods and Export Crops
      1.
      2.

SECTION 2 People and Their Environment (pp. 543–547)

Terms to Know

- habitat
- extinction
- poaching
- ecotourism

Key Points

- Desertification, drought, and conflict have contributed to hunger in Africa south of the Sahara.
- Deforestation, hunting, tourism, and meeting the basic needs of people are all issues in the debate over land use in the region.
- Africans south of the Sahara are working toward political stability and economic independence in the twenty-first century.

Organizing Your Notes

Use a diagram like the one below to organize your notes about each of the issues described in this section.

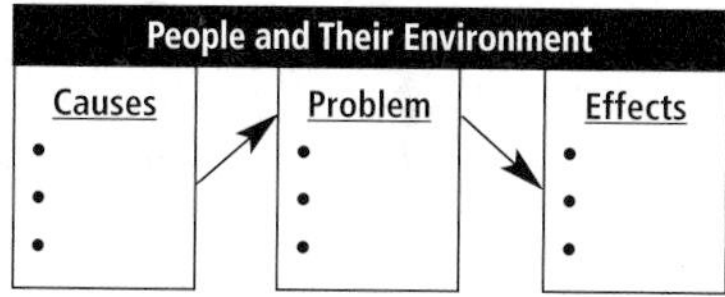

Children in a refugee camp, Democratic Republic of the Congo

ASSESSMENT & ACTIVITIES

Reviewing Key Terms

Write the key term that best matches each description. Refer to the Terms to Know in the Summary & Study Guide on page 551.

1. agriculture that provides for the needs of only a family or village
2. tourism based on concern for the environment
3. a method in which farmers move every one to three years to find better soil
4. crops grown for sale, not for use by the farmer
5. resources such as facilities and equipment
6. illegal hunting of animals
7. areas with conditions suitable for certain animals or plants
8. agriculture conducted at permanent settlements
9. the disappearance of a species from the earth
10. doing business on the Internet
11. a land-management technique that helps protect farmland

Reviewing Facts

SECTION 1

1. Why do Africans south of the Sahara share unequally in mineral wealth?
2. Why has Africa south of the Sahara been slow to industrialize?
3. Why have many people in the region relied primarily on radio for news and information?

SECTION 2

4. What factors have contributed to desertification in the Sahel?
5. What is the relationship between deforestation and endangered animals?
6. How does ecotourism affect the region's wildlife?

Critical Thinking

1. **Categorizing Information** What are the key elements shared by the region's economies? How important is each one?
2. **Making Inferences** How does conflict affect the region's overall quality of life?
3. **Predicting Consequences** Create a web diagram that lists the region's main challenges. Then list consequences for each one if no action is taken.

NATIONAL GEOGRAPHIC

Locating Places

Africa South of the Sahara: Physical-Political Geography

Match the letters on the map with the places and physical features of Africa south of the Sahara. Write your answers on a sheet of paper.

1. Mauritania	5. Côte d'Ivoire	9. Dar es Salaam
2. Nigeria	6. Sudan	10. Somalia
3. Madagascar	7. Zambia	11. Congo Basin
4. Eritrea	8. Nairobi	12. Orange River

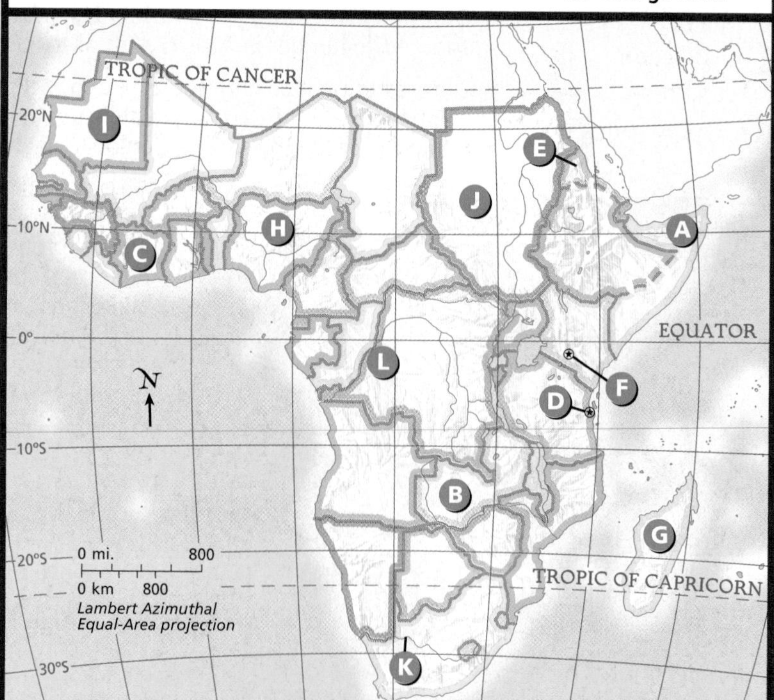

Self-Check Quiz Visit the **Glencoe World Geography** Web site at geography.glencoe.com and click on Self-Check Quizzes—Chapter 22 to prepare for the Chapter Test.

Using the Regional Atlas

Refer to the Regional Atlas on pages 486–489.

1. **Human-Environment Interaction** Describe the relationship between nomadic herding and the location of deserts.
2. **Region** Which two resources are concentrated in western and southern Africa?

Thinking Like a Geographer

As a geographer, would you favor setting aside more or less land for game preserves in Africa? Consider both the human and environmental concerns associated with the game preserves. What arguments would you present in support of your position?

Problem-Solving Activity

Problem-Solution Proposal The coastal zones of Africa south of the Sahara have great potential as sources of food, energy, and minerals. However, rapid, uncontrolled development of resources has led to environmental damage. Research to find out the specific causes of this damage and the ways in which it threatens the future of the region's ocean resources. Then write a proposal suggesting possible steps to prevent future harm to this environment.

GeoJournal

Public Service Announcement Use the information you logged in your GeoJournal to write a public service announcement explaining one of the efforts to address an environmental challenge in Africa south of the Sahara.

Technology Activity

Developing Multimedia Presentations Identify and research an endangered African animal. Collect information about the animal's habitat. Use mapping software, photographs, and other visual aids you download from Web sites to create a multimedia presentation. Show how the animal's location and habitat have been influenced by humans.

Standardized Test Practice

Use the table below to choose the best answer for the following multiple-choice question. If you have trouble answering the question, use the process of elimination to narrow your choices.

Population Change in Selected African Countries

	Deaths per 1,000 people		Life Expectancy	
	1996	2000	1996	2000
Botswana	17	22	46	40
Namibia	8	19	65	41
South Africa	10	15	60	54
Uganda	21	18	41	44
Zimbabwe	18	22	42	39

Sources: World Almanac, 1997; *World Almanac,* 2001

1. **What is the relationship between the change in the death rate (number of deaths per 1,000 people) and the change in life expectancy in the countries listed in the table?**

   A An increase in the death rate causes a decrease in life expectancy.

   B There is not a consistent relationship between changes in the death rate and changes in life expectancy.

   C In countries where the life expectancy decreased, the death rate increased.

   D Both the death rate and life expectancy are decreasing in these countries.

Before you read the answer choices, study the relationship between the death rate and the life expectancy of the African countries. Once you have drawn a conclusion about the relationship, read through the answer choices. Choose the one that most accurately supports the data in the chart.

South Asia

NGS ONLINE

WHY IT'S IMPORTANT—

Many of the countries of South Asia have earned their independence relatively recently, but they have their roots in very ancient civilizations. The rich culture, minerals, and spices of the area have attracted foreign invaders for hundreds of years. Since the subcontinent shook off the cloak of British colonial rule in the 20th century, political and religious rivalries within the region have threatened its peace and stability. The governments of South Asia are struggling to overcome their differences and increase the region's role in trade and technological development.

World Regions Video

To learn more about South Asia and its impact on your world, view the World Regions video "South Asia."

Monk in front of dome of Buddhist shrine, Nepal

8

What Makes South Asia a Region?

Like a giant pointed tooth, South Asia juts out of the Asian continent and into the salty waters of the Arabian Sea, the Indian Ocean, and the Bay of Bengal. Towering mountains separate this region from the rest of Asia. The greatest of these are the mountains of the Himalaya, which include Mount Everest—the tallest peak on Earth.

South of the Himalaya, the land descends to fertile lowlands that are watered by the Indus, Brahmaputra, and Ganges River systems. South Asia's southern tip is outlined by the Eastern and Western Ghats, ranges of low mountains that frame an arid tableland called the Deccan Plateau.

From snowy highlands to sun-scorched deserts, South Asia has a variety of climate zones. The climate is greatly affected by monsoons—seasonal winds that bring cycles of wet and dry weather to the region.

3

1 Bright bridles and nose rings adorn a camel in the Thar, or Great Indian, Desert. Straddling northwestern India and eastern Pakistan, the desert lies beyond the reach of heavy monsoon rains. Camels are a traditional means of transportation in this arid part of South Asia.

2 Water swirls down a street in Delhi during India's wet monsoon season. Each year as summer approaches, wind patterns shift and moist air from the Indian Ocean sweeps over the subcontinent. Once the rains begin, they may continue for 60 days or more.

3 Whitewashed walls echo the brilliance of snow-covered peaks in Namche Bazaar, a Sherpa village in Nepal. Sherpas are a people who live mainly among the mountains of the Himalaya, where they have won fame as guides on climbing expeditions.

4 Up to their knees in green shoots, a farmer and his cow pause in a paddy in Bangladesh. The rich soil of the Ganges River delta spreads across much of Bangladesh, helping to make this tiny country one of the world's leading producers of rice.

Population Giant

Over the centuries, the fertile floodplains of the Indus and Ganges Rivers have attracted many immigrants and invaders to South Asia, giving the region great diversity in peoples, languages, customs, and religious beliefs. Hinduism and Buddhism both originated in South Asia, whereas Islam arrived from the west. The British brought colonial rule, which lasted for nearly two centuries. The region won its independence in the mid-1900s, but not without political, religious, and economic upheaval.

South Asia remains culturally rich, but its burgeoning population—over one billion in India alone—struggles with a low standard of living. Subsistence farming and labor-intensive traditional industries form the basis of the region's economy.

1 Red powder coats the face of an Indian boy during the festival of Ganesh Chaturthi. The festival celebrates the birth of Ganesh, an elephant-headed Hindu god. Hinduism is the most widespread religion in South Asia today.

2 Filled to overflowing, a gaily painted city bus takes on passengers in a crowded street in Dhaka, the capital of Bangladesh. With a population of about 134 million, Bangladesh is one of the most densely populated countries in the world—and also one of the poorest and least developed.

3 Mirrored in still water, the Taj Mahal stands serenely outside the city of Agra, in northern India. The Taj Mahal was built in the 1600s by a Muslim ruler as a tomb for his favorite wife. Constructed of white marble, the building is decorated with verses from the Quran, the holy book of Islam.

4 Tender tea leaves are plucked by hand on a plantation in Sri Lanka, formerly called Ceylon. A legacy of British colonial rule, plantations produce much of the famous Ceylon tea that is a major product of this island nation. Sri Lanka gained its independence from Britain in 1948.

8

South Asia

PHYSICAL

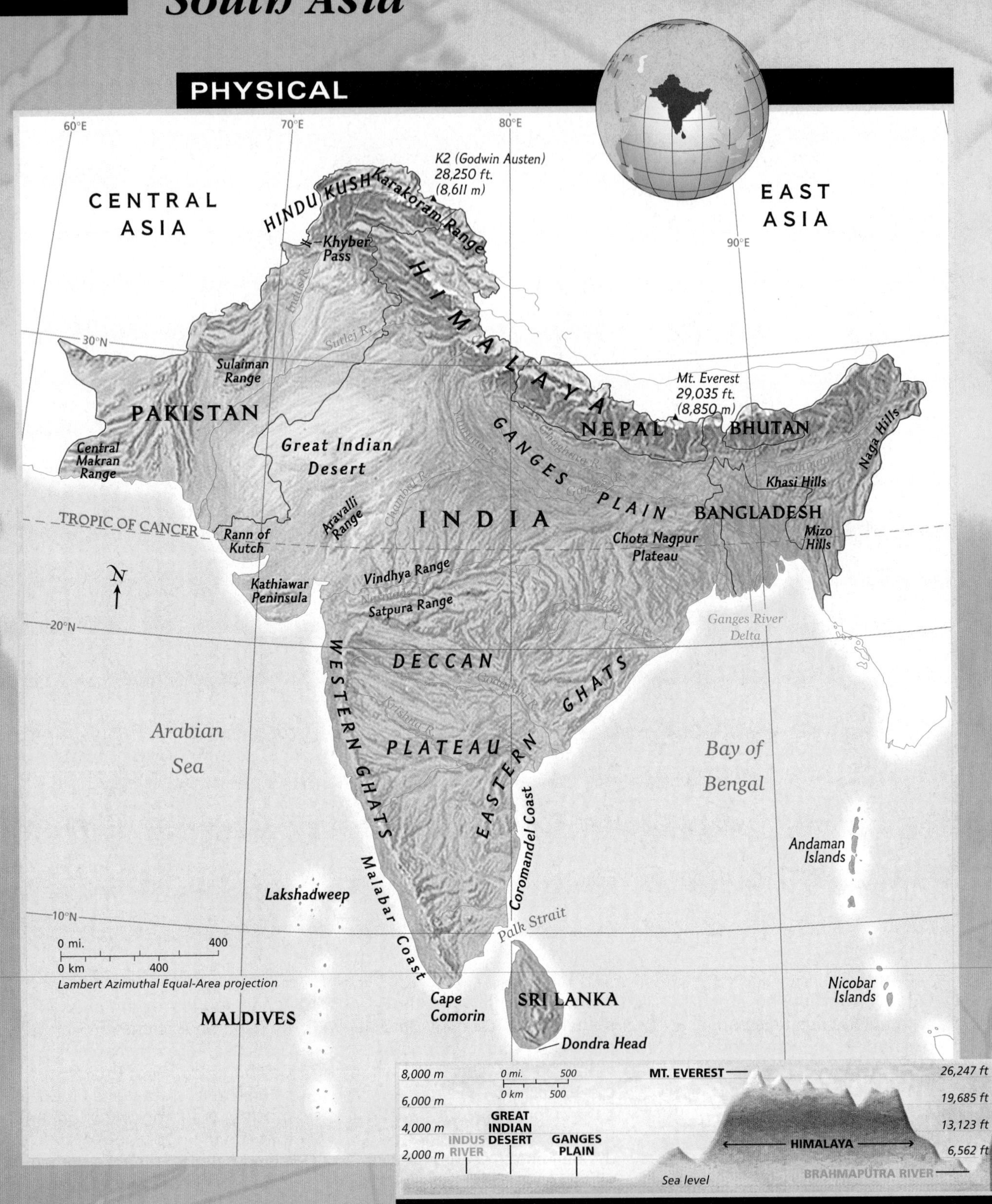

Elevation Profile

POLITICAL

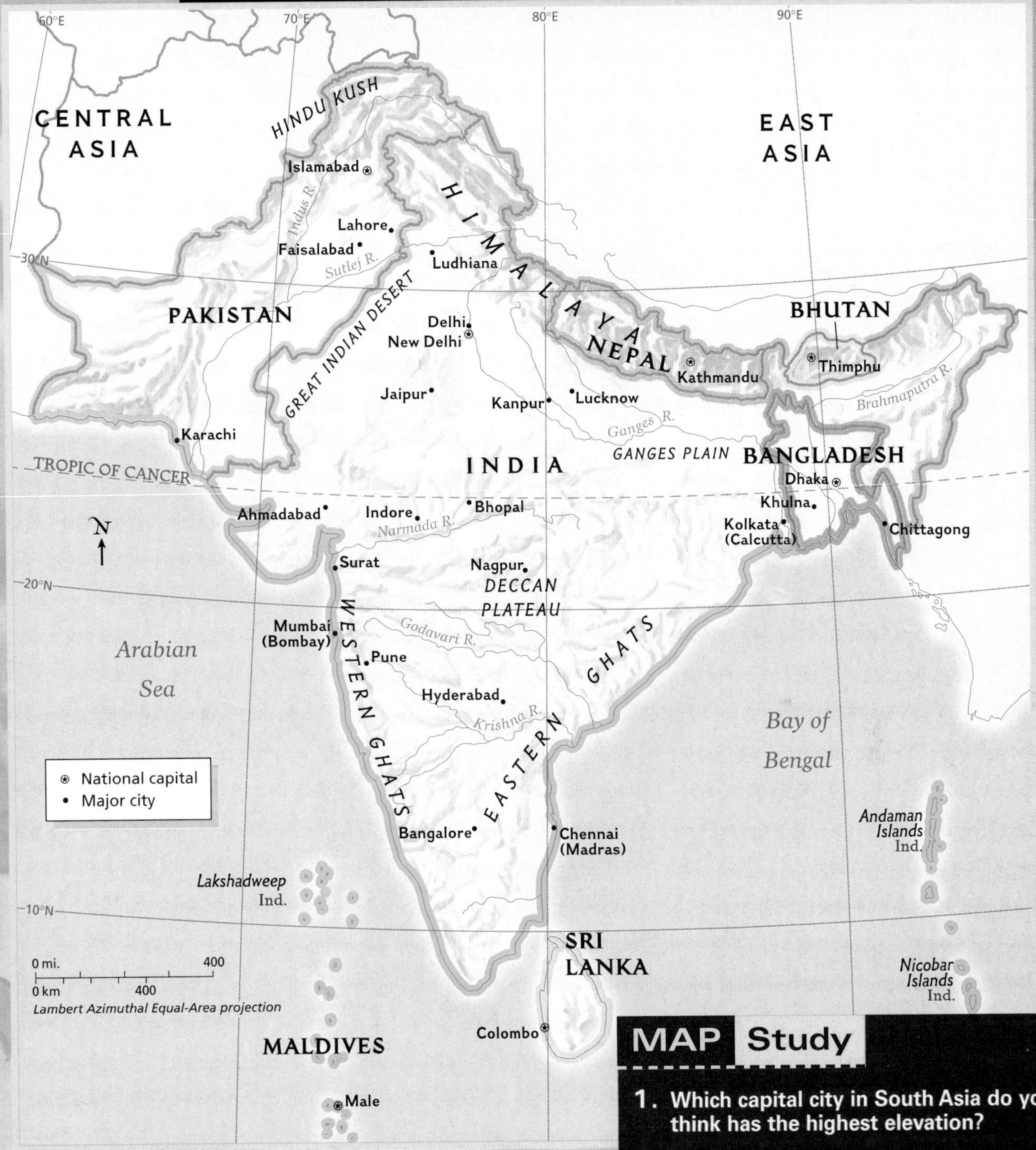

60°E
70°E
80°E
90°E
30°N
20°N
10°N
CENTRAL ASIA
HINDU KUSH
EAST ASIA
Islamabad
Indus R.
Lahore
Faisalabad
Sutlej R.
Ludhiana
HIMALAYA
PAKISTAN
GREAT INDIAN DESERT
Delhi
New Delhi
NEPAL
Kathmandu
BHUTAN
Thimphu
Brahmaputra R.
Jaipur
Kanpur
Lucknow
Ganges R.
Karachi
GANGES PLAIN
BANGLADESH
TROPIC OF CANCER
INDIA
Dhaka
Ahmadabad
Indore
Bhopal
Khulna
Narmada R.
Kolkata (Calcutta)
Chittagong
N
Surat
Nagpur
DECCAN PLATEAU
Mumbai (Bombay)
Godavari R.
WESTERN GHATS
EASTERN GHATS
Arabian Sea
Pune
Hyderabad
Krishna R.
Bay of Bengal

- ⊛ National capital
- • Major city

Bangalore
Chennai (Madras)
Andaman Islands Ind.
Lakshadweep Ind.
SRI LANKA
Nicobar Islands Ind.

0 mi. 400
0 km 400
Lambert Azimuthal Equal-Area projection

Colombo
MALDIVES
Male

MAP Study

1. Which capital city in South Asia do you think has the highest elevation?
2. What rivers join to form the Ganges River delta?

8

South Asia

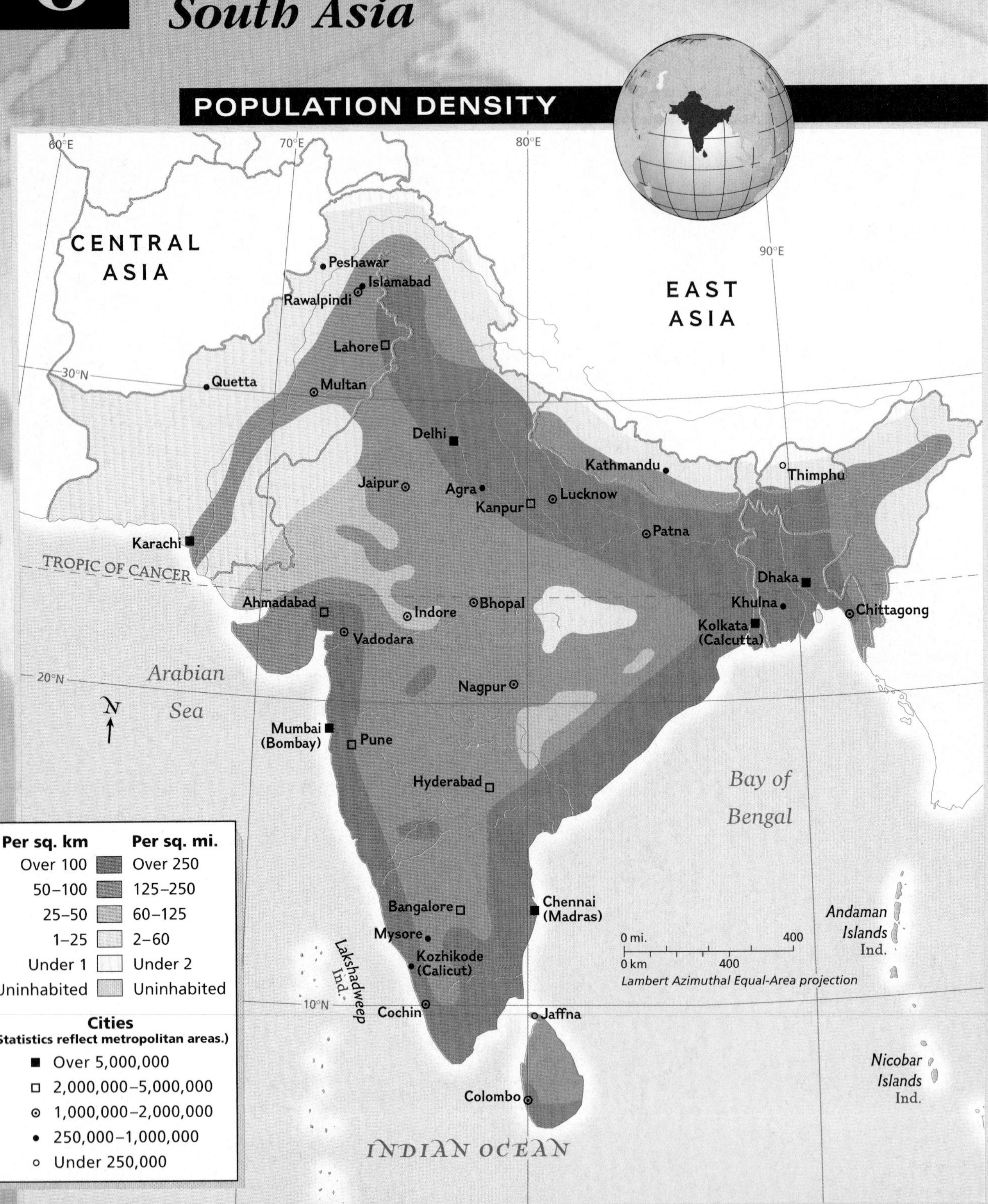

POPULATION DENSITY
60°E
70°E
80°E
90°E
30°N
20°N
10°N
TROPIC OF CANCER
CENTRAL ASIA
EAST ASIA
Peshawar
Islamabad
Rawalpindi
Lahore
Quetta
Multan
Delhi
Kathmandu
Thimphu
Jaipur
Agra
Kanpur
Lucknow
Patna
Karachi
Dhaka
Ahmadabad
Bhopal
Indore
Khulna
Chittagong
Vadodara
Kolkata (Calcutta)
Arabian Sea
Nagpur
Mumbai (Bombay)
Pune
Hyderabad
Bay of Bengal
Bangalore
Chennai (Madras)
Mysore
Kozhikode (Calicut)
Lakshadweep Ind.
Andaman Islands Ind.
Cochin
Jaffna
Colombo
Nicobar Islands Ind.
INDIAN OCEAN
0 mi. 400
0 km 400
Lambert Azimuthal Equal-Area projection
N
Per sq. km
Per sq. mi.
Over 100
Over 250
50–100
125–250
25–50
60–125
1–25
2–60
Under 1
Under 2
Uninhabited
Uninhabited
Cities
(Statistics reflect metropolitan areas.)
Over 5,000,000
2,000,000–5,000,000
1,000,000–2,000,000
250,000–1,000,000
Under 250,000

NATIONAL GEOGRAPHIC SOCIETY

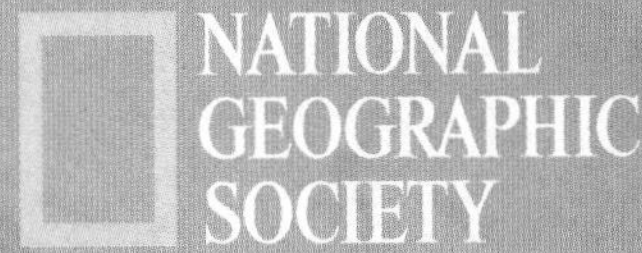

ECONOMIC ACTIVITY

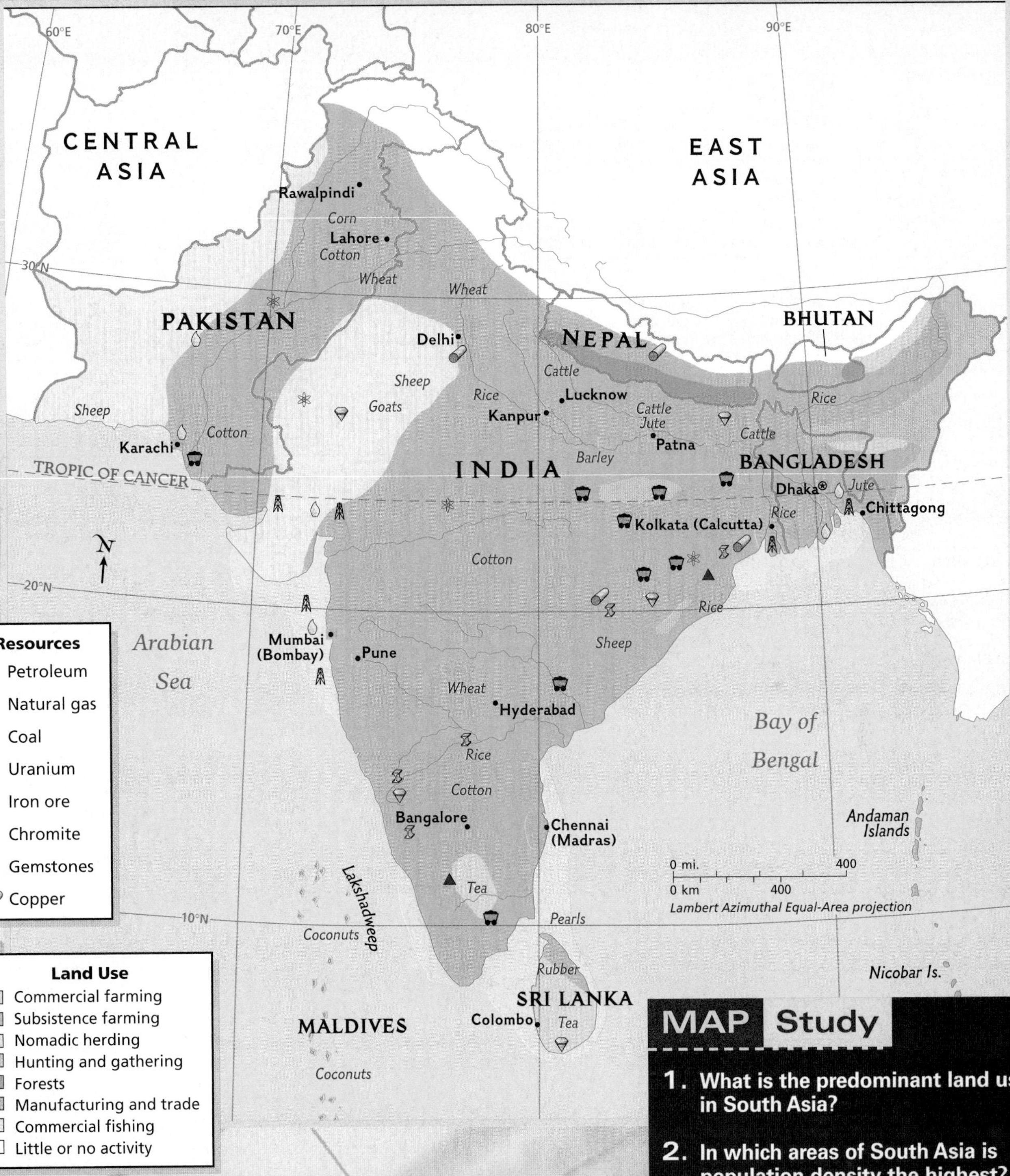

MAP Study

1. What is the predominant land use in South Asia?
2. In which areas of South Asia is population density the highest?

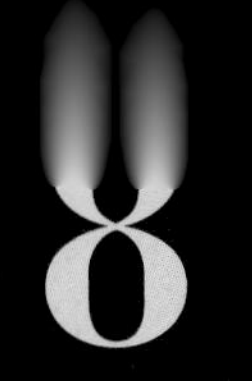

South Asia

COUNTRY PROFILES

COUNTRY * AND CAPITAL	FLAG AND LANGUAGE	POPULATION AND DENSITY	LANDMASS	MAJOR EXPORT	MAJOR IMPORT	CURRENCY	GOVERNMENT
BANGLADESH Dhaka	Bengali	133,500,000 2,401 per sq. mi. 927 per sq. km	55,598 sq. mi. 143,998 sq. km	Clothing	Machinery	Taka	Republic
BHUTAN Thimphu	Dzonkha, Local Languages	900,000 50 per sq. mi. 19 per sq. km	18,147 sq. mi. 47,001 sq. km	Cardamom	Fuels	Ngultrum	Constitutional Monarchy
INDIA New Delhi	Hindi, English, Local Languages	1,033,000,000 814 per sq. mi. 314 per sq. km	1,269,340 sq. mi. 3,287,606 sq. km	Gems & Jewelry	Crude Oil	Rupee	Federal Republic
MALDIVES Male	Maldivian Divehi, English	300,000 2,495 per sq. mi. 963 per sq. km	116 sq. mi. 300 sq. km	Fish	Machinery	Maldivian Rufiyaa	Republic
NEPAL Kathmandu	Nepali	23,500,000 413 per sq. mi. 159 per sq. km	56,826 sq. mi. 147,179 sq. km	Clothing	Petroleum Products	Nepalese Rupee	Constitutional Monarchy
PAKISTAN Islamabad	Urdu, English, Punjabi, Sindhi	145,000,000 472 per sq. mi. 182 per sq. km	307,375 sq. mi. 796,101 sq. km	Cotton	Petroleum	Pakistan Rupee	Federal Republic
SRI LANKA Colombo	Sinhalese, Tamil, English	19,500,000 771 per sq. mi. 298 per sq. km	25,332 sq. mi. 65,610 sq. km	Textiles	Machinery	Sri Lanka Rupee	Republic

* COUNTRIES AND FLAGS NOT DRAWN TO SCALE

FOR AN ONLINE UPDATE OF THIS INFORMATION, VISIT GEOGRAPHY.GLENCOE.COM AND CLICK ON "TEXTBOOK UPDATES."

▶ Boarding school students in Nepalganj, Nepal, study a computer.

NATIONAL
GEOGRAPHIC
SOCIETY

GLOBAL CONNECTION

SOUTH ASIA AND THE UNITED STATES

TEXTILES

▲ Freshly printed fabric drying in Jodhpur, India

There are few things more comfortable than a pair of well-worn blue jeans. Denim—that soft, strong cotton fabric with the rich blue color—has become an integral part of modern life. But "indigo-dyed" textiles are nothing new—they were being produced in India many centuries ago. In fact, the word *indigo* comes from the name "India"!

South Asia has been a world center of textile production for thousands of years. As long ago as 2700 B.C., people in the Indus River valley were cultivating cotton plants and weaving cotton fibers into cloth. Ancient Indian artisans elevated spinning and weaving to art forms. They were among the first in the world to master techniques for dyeing cotton and other types of fabric. Using extracts from more than 300 different native plants, along with other natural substances, the artisans created beautiful, brilliant fabric dyes. The dark blue dye known as indigo, for example, came from the indigo plant.

Indian textile makers pioneered another important technique—making dyes

▲ Colorful dyes for sale on a street stand in Bangalore, India

permanent, or "colorfast," so they would not wash out. The colorfastness of Indian fabrics, combined with their vivid colors and intricate woven and printed patterns, made these fabrics highly prized in Europe, Asia, and other regions. By the 1700s, India was the greatest exporter of textiles the world had ever known.

In England, printed Indian fabrics known as calico and chintz became wildly popular for both fashions and furnishings. Eventually, such fabrics made their way to the American colonies. So precious were these imported textiles that scraps of calico and chintz were saved and made into patchwork quilts—a thrifty gesture that would eventually become an American craft tradition.

Fabrics and patterns that originated in South Asia are now made in other places. However, India is still one of the world's leading producers of cotton, and the textile industry remains India's most important industry. India's eastern neighbor, Bangladesh, also has a thriving garment industry. Check the labels in your cotton clothes—chances are some were made in South Asian countries.

◀ Indian woman spinning cotton thread

CHAPTER 23

The Physical Geography of South Asia

GeoJournal

As you read this chapter, use your journal to record the geographic features of the countries of South Asia. Use descriptive terms to contrast the mountains, deserts, plains, and rivers of South Asia.

GEOGRAPHY Online

Chapter Overview Visit the **Glencoe World Geography** Web site at geography.glencoe.com and click on Chapter Overviews—Chapter 23 to preview information about the physical geography of the region.

The Land

Guide to Reading

Consider What You Know

Mount Everest in the Himalaya is the world's highest peak. You may often read or hear reports of climbers on Mount Everest and the difficulties they face. What features of high mountain ranges do you think contribute to the climbers' hardships?

Read to Find Out

- What landforms exist in South Asia?
- What are the three great river systems on which life in South Asia depends?
- How do the peoples of South Asia use the region's natural resources?

Terms to Know

- subcontinent
- alluvial plain
- mica

Places to Locate

- Himalaya
- Ganges Plain
- Vindhya Range
- Deccan Plateau
- Indus River
- Brahmaputra River
- Ganges River

◀ *Buddhist monastery in Bhutan*

NATIONAL GEOGRAPHIC

A Geographic View

India by Train

The valleys and these hillsides [in the north of India] are open to the distant plains, and so the traveler on the toy train has a view that seems almost unnatural, it is so dramatic. At Sonada it is like standing at the heights of a gigantic outdoor amphitheater and looking down and seeing the plains and the rivers, roads and crops printed upon it and flattened by the yellow heat.

—*Paul Theroux, "By Rail Across the Indian Subcontinent,"* National Geographic, *June 1984*

Train passing through Himalayan foothills

Novelist Paul Theroux described the varied and dramatic landscapes he saw while traveling South Asia by train. In this section you will explore the physical geography of South Asia—its majestic mountains, mighty rivers, and fertile plains.

A Separate Land

The seven countries that make up South Asia are separated from the rest of Asia by mountains. As a result, South Asia is called a **subcontinent**, a large, distinct landmass that is joined to a continent. In geologic terms South Asia contains some of the oldest and some of the youngest landforms on Earth.

Most of South Asia forms a peninsula of about 1.7 million square miles (4.4 million sq. km) touched by three bodies of water—the Arabian Sea to the west, the Indian Ocean to the south, and the Bay of Bengal to the east. The region also includes many small islands and the large island country of Sri Lanka, which lies off India's southern tip.

A Land of Great Variety

South Asia reveals a varied landscape. In the far north, some of the world's highest mountain ranges raise sharp, icy peaks above terraced foothills, high desert plateaus, and rich valleys. The older southern lands include eroded mountains and flat plateaus.

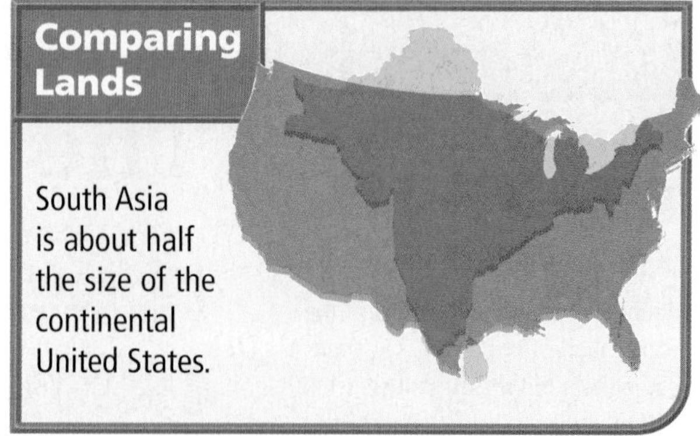

The Himalaya

According to the theory of continental drift, about 60 million years ago the Indian subcontinent was part of the same large landmass as Africa. After the subcontinent broke away, it collided with the southern edge of Asia. The force of this collision thrust up new mountain ranges, the **Himalaya**. These ranges spread more than 1,000 miles (1,609 km) across the northern edge of the peninsula and are hundreds of miles wide. Mount Everest, the world's highest peak, rises to 29,035 feet (8,850 m) above sea level in the Himalaya. A teenager describes climbing in the Himalaya:

> *"... I'm standing alone on a portion of the summit. ... On very clear days like this, some from Everest have claimed to see the curvature of the earth; others say they can see the Indian Ocean, hundreds of miles away. ... [I]t makes me feel very small. ..."*
>
> Mark Phetzer, *Within Reach: My Everest Story*, 1998

◀ *Fierce winds blow Buddhist prayer flags in Nepal.*

Other Northern Landforms

The Himalaya meet the Karakoram Mountains in the northernmost part of South Asia. Farther west, the Hindu Kush range completes the chain. Together, they create a high wall of mountains between the subcontinent and the rest of Asia. In the past, invaders from the north could only enter the region through a few narrow crossing places, such as the famous **Khyber Pass** between Pakistan and Afghanistan. The Himalaya also protected Nepal and Bhutan from outside influence until the 1900s.

At the foot of the Himalaya ranges, wide fertile plains are watered by the region's great rivers—the Indus, the Ganges (GAN•JEEZ), and the Brahmaputra. One-tenth of the world's people live in this crowded northern area referred to as the **Ganges Plain** (or Indo-Gangetic Plain). In the northeast of India lies the Chota Nogpur Plateau, a high tableland of forests.

Culture

Central Landforms

The collision between the Indian subcontinent and Asia also pushed up a mountain range in central India. Not as tall as the Himalaya, the **Vindhya Range** divides India into northern and southern regions. This physical division separates the two distinct cultures that have developed in India. The cuisine, architecture, and religious practices of the peoples of northern and southern India differ markedly, as you will read in the next chapter.

Southern Landforms

The southern regions of South Asia contrast with those of the north. At the base of the subcontinent, two chains of eroded mountains—the Eastern

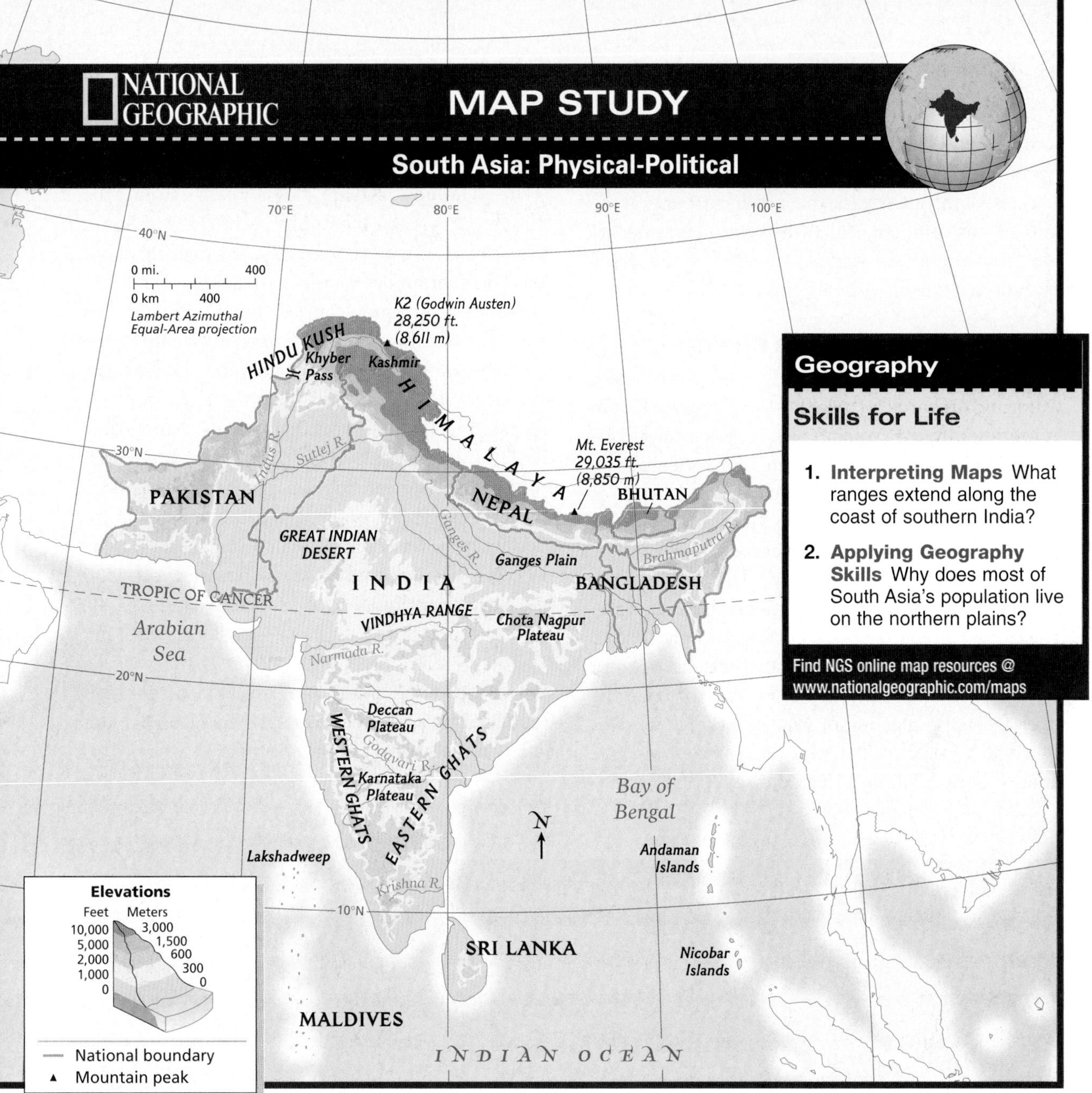

Geography Skills for Life

1. **Interpreting Maps** What ranges extend along the coast of southern India?
2. **Applying Geography Skills** Why does most of South Asia's population live on the northern plains?

Find NGS online map resources @ www.nationalgeographic.com/maps

Ghats and Western Ghats—form a triangle of rugged hills. Between them lies the **Deccan Plateau**. This plateau was part of the landmass from which the subcontinent broke away and is hundreds of millions of years old. Once covered with lava, the Deccan Plateau today has rich, black soil. The Western Ghats, however, prevent yearly rainy winds from reaching the plateau, leaving it arid, or extremely dry. The Karnataka Plateau south of the Deccan Plateau receives these rains instead, so hills there are lush and green. Spices growing on plantations in this area scent the air, and wild elephants move through the foliage of the plateau's dense rain forests.

Sri Lanka (SREE LAHN•kuh) is a teardrop-shaped island that broke away from the original Indian landmass. Maldives (MAWL•DEEVZ), the southernmost country in South Asia, is a chain of tiny coral atolls and volcanic outcroppings. Although Maldives covers 38,000 square miles (98,420 sq. km) of ocean, its land area totals only 115 square miles (298 sq. km).

Major River Systems

Rivers are the key to life in South Asia. From sources high in the Himalaya, three major river systems—the Indus, the Brahmaputra, and the Ganges—fan out across the northern part of the Indian subcontinent. All three rivers carry fertile soil from mountain slopes onto their floodplains as the rivers swell with seasonal rains.

Indus and Brahmaputra Rivers

The **Indus River** flows mainly through Pakistan, watering orchards of peaches and apples before emptying into the Arabian Sea. It also serves as an important transportation route. Historically, the Indus River valley is known as the cradle of ancient India, which, with Mesopotamia and Egypt, was one of the world's earliest civilizations.

The **Brahmaputra River** flows east through the Himalaya and then west into India and Bangladesh. There it joins the Ganges—to form a delta before emptying into the Bay of Bengal. The Brahmaputra is a major inland waterway. Ships can navigate the river from the the Bay of Bengal as far inland as Dibrugarh in the Indian state of Assam, about 800 miles (about 1,290 km) from the sea. The Brahmaputra also provides Bangladesh with 50 percent of its power through hydroelectricity.

Ganges River

The **Ganges River** flows east from the Himalaya. It is the most important river of South Asia, drawing waters from a basin covering about 400,000 square miles (about 1 million sq. km). Fed by water from snowcapped peaks, the Ganges retains its size throughout the year, even during the hot, dry season from April to June. During the summer monsoon period, heavy rains can cause devastating floods along the Ganges. Named for the Hindu goddess Ganga, the Ganges is revered by Hindus, who consider its waters to be sacred.

The land area through which the Ganges flows is known as the Ganges Plain. Almost all of the plain has been cleared of grasslands and forests to make way for crops, such as rice, sesame, sugarcane, jute, and beans. As India's most agriculturally productive area, the Ganges Plain is the world's longest **alluvial plain**, an area of fertile soil deposited by river flood waters. The Ganges Plain also is India's most densely populated area.

Geography Skills for Life

The Holy River The Ganges River is important to fishing, commerce, and agriculture. Millions of Hindus visit the river for ritual bathing (inset).

Human-Environment Interaction What crops are grown on the Ganges Plain?

NATIONAL GEOGRAPHIC **World Explorer**

Geography Skills for Life

Waterfalls in Nepal Twin waterfalls in the Annapurna region of Nepal cut through the sheer rock face of a Himalayan peak.

Human-Environment Interaction How might waterfalls benefit Nepal's economy?

Natural Resources

South Asia has a variety of natural resources. Dependent on these resources for their livelihood, South Asia's large populations and the fragile nature of some of their environments are ongoing challenges.

Water

The rivers of South Asia provide alluvial soil, drinking water, transportation, and hydroelectric power to the region's large, growing population. They also provide fish for local use and export.

Water resource management challenges South Asia because rivers cross national boundaries. Still, countries in the region sometimes work together on various projects. For example, India funded the Chhukha Hydel hydroelectric project in Bhutan. In return, India receives some of the energy generated there. Countries in the region also build dams to provide hydroelectric power and to open up new farmlands by ensuring consistent levels of water for irrigation. Mountainous Nepal, with its many waterfalls, has the potential for creating large amounts of hydroelectricity.

Such massive projects, however, often have drawbacks as well as benefits. Dam projects in India meet with resistance when they threaten to flood existing settlements. In Pakistan one of the largest dams in the world, the Tarbela Dam, will soon be unusable, choked with built-up silt from the Indus River.

Energy Resources

Petroleum reserves are known to lie along India's northwest coast, near the Ganges Delta, and in northern Pakistan. Offshore exploration in the Arabian Sea may eventually yield oil. Overall, though, South Asia depends on imported oil.

Student Web Activity Visit the **Glencoe World Geography** Web site at geography.glencoe.com and click on Student Web Activities—Chapter 23 for an activity on the formation of the Himalaya and attempts to reach the summit of Mount Everest.

Natural gas fields are found in southern Pakistan, in India's Ganges Delta, and in Bangladesh. India has a major uranium deposit north of the Eastern Ghats. Most South Asians, however, rely on energy from hydroelectricity, fuel wood, and coal.

Minerals

South Asia's mineral resources are rich, diverse, and widespread. India is a leading exporter of iron ore, and supplies 90 percent of the world's **mica**, a layered rock used in making electrical equipment. Deposits of manganese, chromite, and gypsum still await development. Nepal produces mica and small amounts of copper. Sri Lanka is one of the world's largest producers of graphite, the material used for the "lead" in pencils. Sri Lanka's other major mineral resources include sapphires, rubies, and about 40 other varieties of precious and semiprecious stones.

Geography Skills for Life

Timber Plantation In Bhutan, 90 percent of the workforce makes its living in agriculture and forestry.

Human-Environment Interaction How does overcutting impact the environment?

Timber

Timber is important to South Asia. The forests of Nepal and Bhutan contain conifers, including silver fir, and hardwoods such as oak, magnolia, beech, and birch. Severe overcutting threatens Nepal's timber, however, and could result in massive soil erosion. To preserve the fragile Himalayan environment, the government of Nepal is implementing conservation plans.

Timber resources also include India's prized sandalwood. Rain forests in southwest India yield sal and teak woods for export. To protect its rain forests, Sri Lanka since 1977 has banned timber exports.

SECTION 1 ASSESSMENT

Checking for Understanding

1. **Define** subcontinent, alluvial plain, mica.
2. **Main Ideas** On a table like the one below show examples of the physical features and natural resources of South Asia.

Country	Physical Features	Natural Resources

Critical Thinking

3. **Making Comparisons** How does the landscape of the Himalaya differ from that of the Deccan Plateau? How do these differences affect people's lives?
4. **Identifying Cause and Effect** Why are population densities so high on the Ganges Plain?
5. **Problem Solving** How would you address the problem of overcutting trees in Nepal? How does your solution affect the timber industry?

Analyzing Maps

6. **Region** Study the physical-political map on page 571. What areas of South Asia would you expect to be most agriculturally productive? Why?

Applying Geography

7. **Managing Resources** Think about the physical geography of South Asia. Create a sketch map highlighting potential sites of conflict over water management among the countries of South Asia.

Climate and Vegetation

Guide to Reading

Consider What You Know

The countries of South Asia often are affected by natural disasters of some kind. For example, in 1998 Bangladesh suffered the effects of a terrible flood. What problems can flooding cause?

Read to Find Out

- What are the five major climate regions of South Asia?
- How do seasonal weather patterns present challenges to the region's economy?
- How do elevation and rainfall affect South Asia's vegetation?

Terms to Know

- monsoon
- cyclone

Places to Locate

- Bay of Bengal
- Great Indian Desert

NATIONAL GEOGRAPHIC

A Geographic View

The Breath of Life

The eagle soared even higher in the updraft as I picked my way along the dark rocks beside the Arabian Sea. The winds shifted with promise, deepening the resonance of the surf, muffling even the crows that cackled and lurched along the seawalls. The water grew choppy, and the black thorns of fishermen's sails scratched the horizon. Surely the time [of the monsoon] was at hand.

—*Priit J. Vesilind, "Monsoons: Life Breath of Half the World,"* National Geographic, *December 1984*

Rain-swollen Mahandi River, India

Journalist Priit Vesilind captures in words the tension of waiting for South Asia's seasonal rains. The region, with its hot climates, comes alive when the rain-bearing winds sweep in.

South Asia's Climates

South Asia's climate and vegetation regions are a study in contrasts. Much of the subcontinent lies south of the Tropic of Cancer and has tropical climates with diverse vegetation. In the north and the west, however, the climate varies widely, from the highlands of the Himalaya to the deserts around the Indus River, where little vegetation grows.

Tropical and Subtropical Climates

Tropical rain forest climates, with a variety of vegetation, are located along the western coast of India, near the Ganges Delta in Bangladesh,

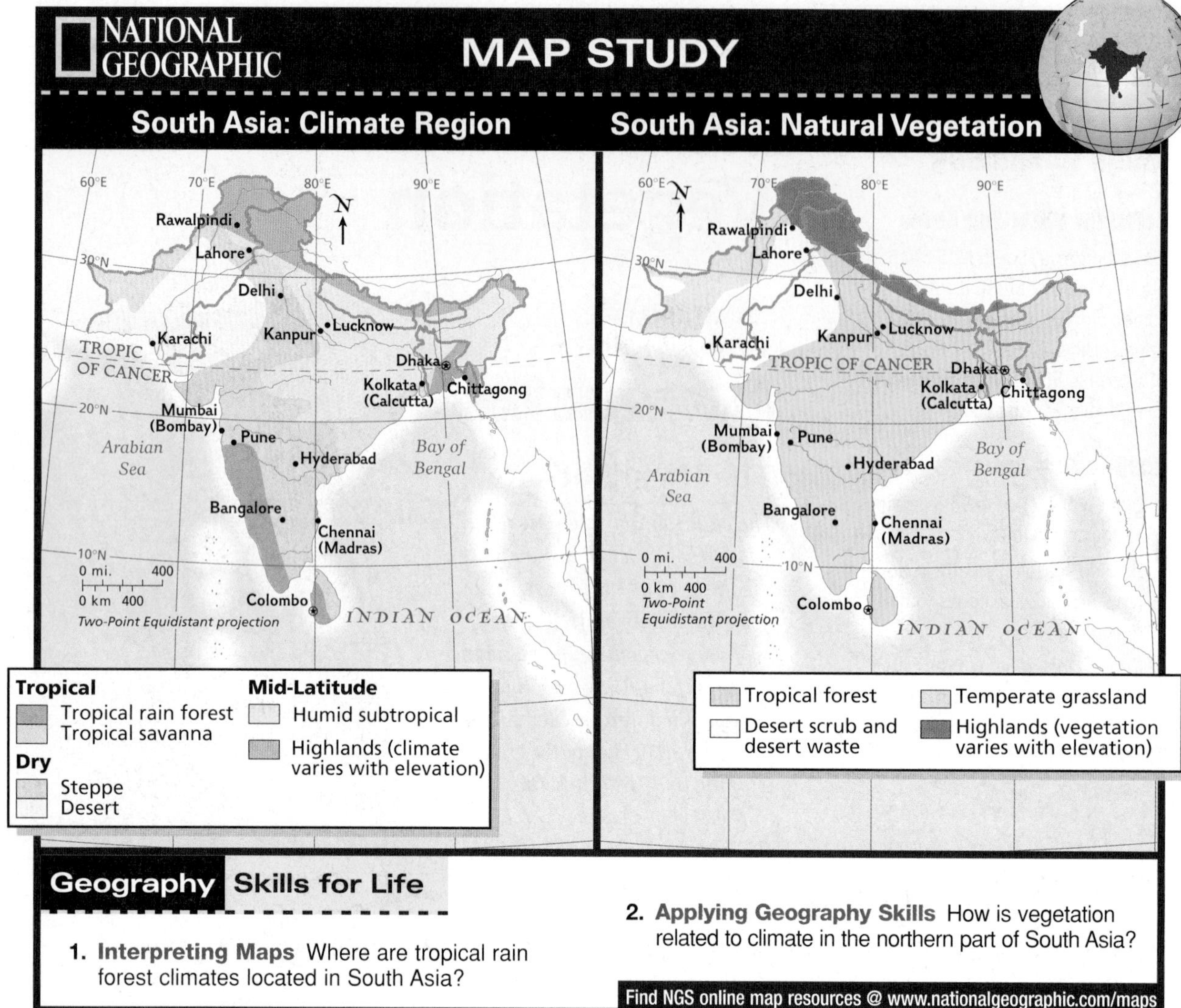

and in southern Sri Lanka. In the path of seasonal rains from the southwest, South Asia's rain forests absorb great quantities of moisture. The rain forests in western Sri Lanka, in southwest India, and in areas north of the **Bay of Bengal** have ebony trees, lush vines, and orchids. Tropical coniferous and deciduous trees surround the rain forests near the Western Ghats. In hot, damp Bangladesh, tropical forests of bamboo, mango, and palm trees thrive. The Sundarbans, a swampy area in southwestern Bangladesh, has the world's largest protected mangrove forest.

A tropical savanna climate surrounds the central Indian steppe and also is found in eastern Sri Lanka. The grasslands and tropical-moist deciduous forests of the savanna experience wet and dry seasons. In Sri Lanka dry evergreen forests and moist deciduous forests give way to drier grasslands at higher elevations.

A band of humid subtropical climate extends across Nepal, Bhutan, Bangladesh, and the northeastern part of India. Temperate mixed forests stretch across the borders of these countries in this area.

Highlands Climates

The coldest climate region of South Asia lies along its northern edge. In the Himalayan highlands and Karakoram peaks, snow never disappears. At the highest elevations, little vegetation can survive. Farther down these slopes, however, the climate turns milder and more temperate. In the upper area of this more temperate zone, coniferous and hardwood trees flourish. Grasslands and stands of bamboo cover the lower Himalayan foothills.

Dry Climates

Along the lower Indus River, a desert climate keeps the land arid and windswept. The **Great Indian Desert** (Thar Desert) lies to the east of the Indus. The vegetation here is desert scrub, low, thorny trees, and grasses. Livestock graze in some areas, and irrigation makes it possible to grow wheat near the Indus River. Much of this area, however, remains wasteland.

Surrounding this desert, except on the coast, is a steppe. Few trees grow in this semiarid grassland. In northwestern India annual rainfall averages less than 20 inches (51 cm). Another steppe area runs through the center of the Deccan Plateau between the Eastern and Western Ghats. The Ghats block rainfall here, making the area relatively arid. Dry, deciduous forests cover vast stretches of India's interior.

Monsoons

Much of South Asia experiences three distinct seasons—hot (from late February to June), wet (from June or July until September), and cool (from

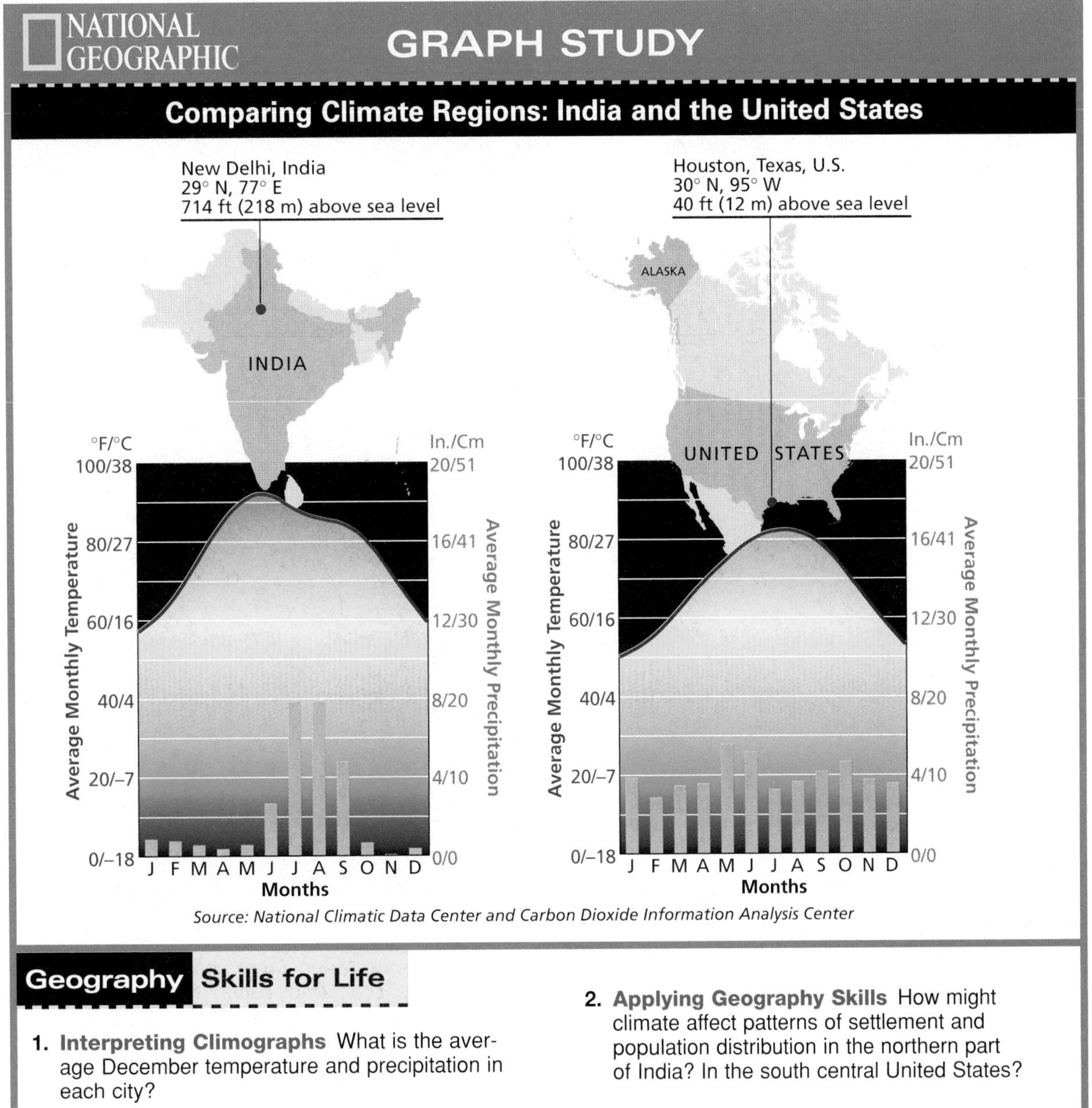

Geography Skills for Life

1. **Interpreting Climographs** What is the average December temperature and precipitation in each city?

2. **Applying Geography Skills** How might climate affect patterns of settlement and population distribution in the northern part of India? In the south central United States?

NATIONAL GEOGRAPHIC

MAP STUDY

South Asia: Monsoons

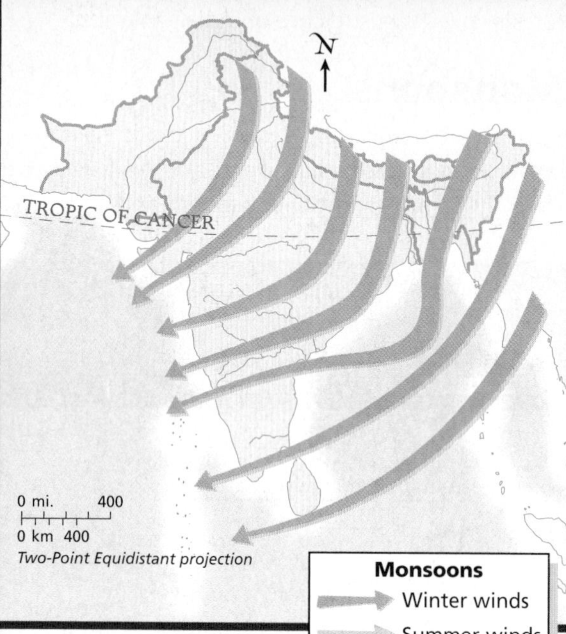

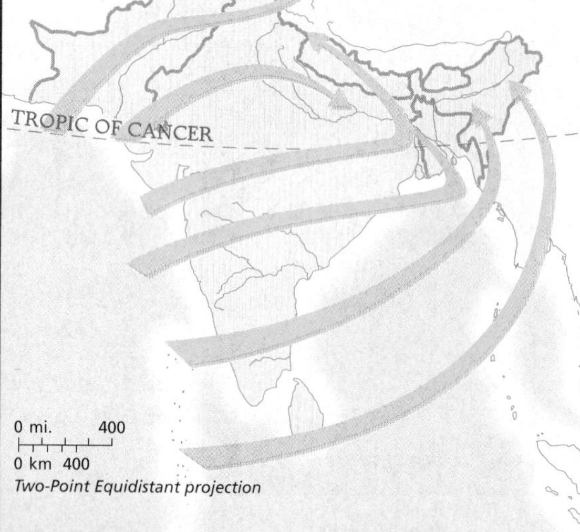

Geography Skills for Life

The same fields in west central India before (top right) and after (bottom right) the arrival of the monsoon rains reveal a stunning contrast between the dry and wet seasons.

1. **Interpreting Maps** From what direction do the winter monsoon winds come? What kind of weather do they bring?
2. **Applying Geography Skills** Describe the impact of the summer monsoon winds on South Asia.

Find NGS online map resources @ www.nationalgeographic.com/maps

October to late February). These periods depend on seasonal winds called **monsoons**. During the cool season, dry monsoon winds blow from the north and northeast. In the hot season, warm temperatures heat the air, which rises and triggers a change in wind direction. Moist ocean air then moves in from the south and southwest, bringing monsoon rains.

Monsoon Rains

The monsoon rains are heaviest in eastern South Asia. When the rains sweep over the Ganges-Brahmaputra delta, the Himalaya block them from moving north. As a result, the rains move west to the Ganges Plain, bringing rainfall needed for crops. It is no wonder, then, that people celebrate the monsoon rains, as an Indian writer describes:

> *"Kulfi [a woman shopping] watched with unbelieving elation as the approaching smell of rain spiked the air like a flower, as the clouds shifted in from the east. . . . Outside, she could hear the sound of cheering from the bazaar. 'Rain, rain, rain, rain.' And in the streets, she watched the children leap like frogs, unable to keep still in their excitement."*
>
> Kiran Desai, *Hullabaloo in the Guava Orchard*, 1997

Economics

Natural Disasters

Both the high temperatures of the hot season and the heavy rains of the wet season are mixed blessings in South Asia. High temperatures allow farmers to produce crops, including the rice that many in Bangladesh and India depend on, year-round as long as water supplies are good. The extreme heat can result in evaporation and dried-out, nutrient-poor soils, however.

The monsoon winds also have benefits and drawbacks. Rainfall waters crops, but areas outside the path of the monsoon, such as the Deccan Plateau and western Pakistan, may receive little or no rainfall during the year. When the people of Bangladesh are planting rice, and those on the Ganges Plain are planting their winter crops, other areas are scorched by drought.

Too much rain also can be a problem. In the low-lying delta country of Bangladesh, monsoons may cause flooding that kills people and livestock, leaves thousands homeless, and ruins crops.

Another kind of weather catastrophe sometimes strikes South Asia. A **cyclone** is a storm with high winds and heavy rains. A 1999 cyclone struck Orissa, India, with winds of more than 160 miles per hour (257 km per hour) and waves over 20 feet (6 m) high. The storm killed nearly 10,000 people and caused more than $20 million in damages.

SECTION 2 ASSESSMENT

Checking for Understanding

1. **Define** monsoon, cyclone.
2. **Main Ideas** On a chart like the one below, fill in the names of different areas of South Asia, and then write in the type of climate and vegetation found in each area.

Location	Climate	Vegetation

Critical Thinking

3. **Analyzing Information** Analyze the reaction of South Asia's environment to the monsoons.
4. **Decision Making** Suppose that you wanted to establish a lumber business in South Asia. Where would you locate it? Why?
5. **Comparing and Contrasting** Are the effects of the very hot temperatures in much of South Asia more positive or more negative? Explain.

Analyzing Maps

6. **Region** Compare the maps of South Asia's climate and vegetation on page 576. Explain how climate and vegetation are related in the region.

Applying Geography

7. **Visiting Sri Lanka** Think about the attractions of Sri Lanka's climate and vegetation. Write a descriptive paragraph urging people to visit and enjoy Sri Lanka's natural features.

Reading an Elevation Profile

If you were planning a long-distance cycling expedition, you might want to check elevations of places along your route. Elevation, the vertical distance above sea level of a place or landform, can be shown in a number of ways. An elevation profile gives you elevation information in a visual form.

Learning the Skill

An elevation profile presents visual information about the elevation of a particular area, route, or landform in a two-dimensional way. The base of an elevation profile is sea level, the point from which land elevation is measured. A vertical scale measures elevation above sea level.

Reading an elevation profile is similar to reading a line graph. The vertical scale corresponds to the *y*-axis. In some elevation profiles, a horizontal scale, corresponding to the *x*-axis, measures the length of the route, area, or landform in miles or kilometers. The profile, or top edge of the landscape shown, corresponds to the line in a line graph. This line shows elevation at specific points. Some elevation profiles provide information on more than one route, area, or landform, using different colors or patterns to distinguish each profile.

Follow these steps to read an elevation profile:

- **Look at the landscape profile as a whole.** This will give you a general sense of the variations in elevation shown.
- **Find the highest and lowest points.** Use the vertical scale to find their elevations. Calculate the approximate difference in elevation between the highest and lowest points.

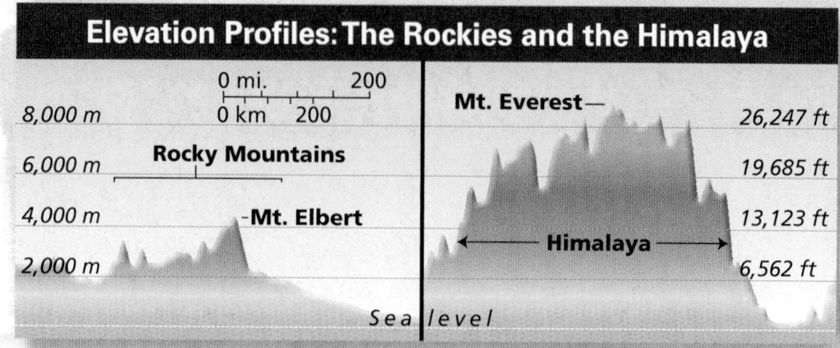

- **Use your finger to trace the profile.** If your finger must jump up and down to follow the profile, the area has dramatic differences in elevation.
- **If more than one area or landform is profiled, follow the procedure for each profile.** Then use the information to compare the profiled areas.

Practicing the Skill

Study the elevation profile contrasting the Rocky Mountains with the Himalaya. Then answer these questions.

1. What is the highest peak in the Rocky Mountains? About how many feet above sea level does it rise?
2. What is the approximate elevation of the highest peak in the Himalaya?
3. What is the approximate difference in elevation between the highest point in the Rocky Mountains and the highest point in the Himalaya?
4. Which range contains the greater variation in elevation?
5. Which range stretches over a greater distance?
6. What does the elevation profile reveal about the relative elevations of these mountain ranges?

Applying the Skill

Use a map to choose a bicycle route across your state. Identify several key points along your route, and check the elevations for each one. Then create an elevation profile of your route. Note the highest and lowest points on your profile. Where would your bike ride be easiest? Most difficult?

The **Glencoe Skillbuilder Interactive Workbook, Level 2** provides instruction and practice in key social studies skills.

SUMMARY & STUDY GUIDE

SECTION 1 The Land (pp. 569–574)

Terms to Know
- subcontinent
- alluvial plain
- mica

Key Points
- The landforms of South Asia include mountains, plateaus, plains, and islands.
- South Asia has three great river systems—the Indus, Brahmaputra, and Ganges—and the world's longest alluvial plain.
- South Asia has few significant oil reserves, but has substantial mineral deposits, including iron ore and mica.

Organizing Your Notes
Create an outline using the format below to help you organize your notes for this section.

South Asia's Land

I. A Separate Land
II. A Land of Great Variety
  A. The Himalaya
  B. Other Northern Landforms

SECTION 2 Climate and Vegetation (pp. 575–579)

Terms to Know
- monsoon
- cyclone

Key Points
- South Asia has highlands, tropical, and desert climates.
- The monsoon is a seasonal change in wind direction that brings heavy rainfall to much of South Asia from June to September.
- South Asia's vegetation is affected by elevation, rainfall, and human activity.

Organizing Your Notes
Use a table like the one below to help you organize the notes you took as you read this section.

Climate Region	Vegetation	Country or Area
tropical rain forest	ebony trees, lush vines, orchids	

Along the Ganges, Varanasi, India

ASSESSMENT & ACTIVITIES

Reviewing Key Terms

Write the letter of the key term that best matches each definition below.

a. subcontinent
b. alluvial plain
c. mica
d. monsoons
e. cyclone

1. seasonal winds
2. a storm with high winds and heavy rains
3. a layered mineral used to make electrical components
4. a very large, distinct landmass that is part of a continent
5. an area of rich, fertile soil found along a river

Reviewing Facts

SECTION 1

1. Why might the region of South Asia be referred to as "a land of great variety"?
2. How have the mountains of the Vindhya Range affected the people of India?
3. Why is the management of water resources important in South Asia?

SECTION 2

4. Where can you find a steppe climate region in South Asia?
5. When do the three seasons found in much of South Asia occur, and how would you describe each?
6. What factors enable South Asia's rain forests to thrive?

Critical Thinking

1. **Identifying Cause and Effect** In what way are the Himalaya responsible for the richness of the soil in the northern plains of the Indian subcontinent?
2. **Comparing and Contrasting** What are the advantages and disadvantages of the monsoons to South Asia?
3. **Predicting Consequences** Using a web diagram like the one below, show the consequences to the people of South Asia of possible weather conditions. Then choose one consequence and describe it in detail.

NATIONAL GEOGRAPHIC

Locating Places

South Asia: Physical-Political Geography

Match the letters on the map with the places and physical features of South Asia. Write your answers on a sheet of paper.

1. Arabian Sea
2. Bay of Bengal
3. Ganges River
4. Deccan Plateau
5. Sri Lanka
6. Himalaya
7. Brahmaputra River
8. Great Indian Desert
9. Pakistan
10. Indus River

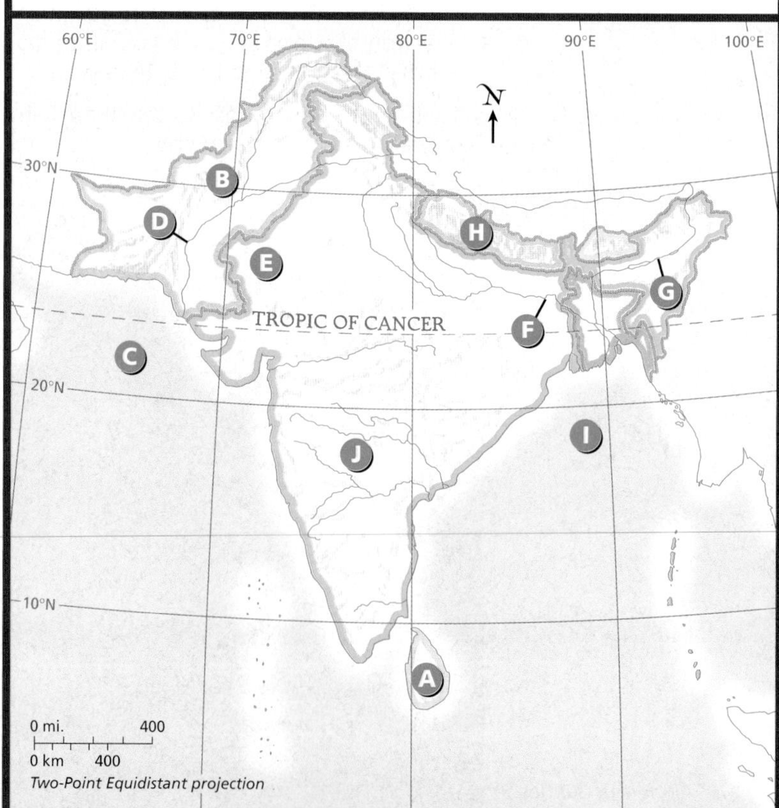

Self-Check Quiz Visit the **Glencoe World Geography** Web site at geography.glencoe.com and click on Self-Check Quizzes—Chapter 23 to prepare for the Chapter Test.

Using the Regional Atlas

Refer to the Regional Atlas on pages 560–563.

1. **Location** What mountains form the border between East Asia and South Asia?
2. **Place** Compare the political map with the population density map. Where is the area of lowest average population density along the India-Pakistan border?

Thinking Like a Geographer

Analyze the effects of physical geographic patterns on population in South Asia. What patterns favor high population density? Low population density?

Problem-Solving Activity

Contemporary Issues Case Study When natural disasters strike populated areas, their impact is worse in areas of high population density. In a group, research a recent natural disaster in South Asia, such as the 1999 cyclone in Orissa or the 1998 flood in Bangladesh. Find out the causes of the disaster and how it affected the area's population and natural resources. What efforts were taken following the disaster? Then, focusing on one of these efforts, present your group's findings to the class.

GeoJournal

Descriptive Writing Using your GeoJournal, write a description about the ways South Asians have adapted to or modified their environment. Then compare human-environment interaction in South Asia with that in your state and local community.

Technology Activity

Using an Electronic Spreadsheet

Use a spreadsheet program to organize information about elevations in South Asia. List at least six South Asian countries in the left column of a spreadsheet. Use a world atlas to find the highest point in each country. Then list the heights in the second column of the spreadsheet. Use the graphics feature of the program to make a bar graph to compare heights.

Standardized Test Practice

Study the elevation profile. Then choose the best answer for the following multiple-choice questions. If you have trouble answering the questions, use the process of elimination to narrow your choices.

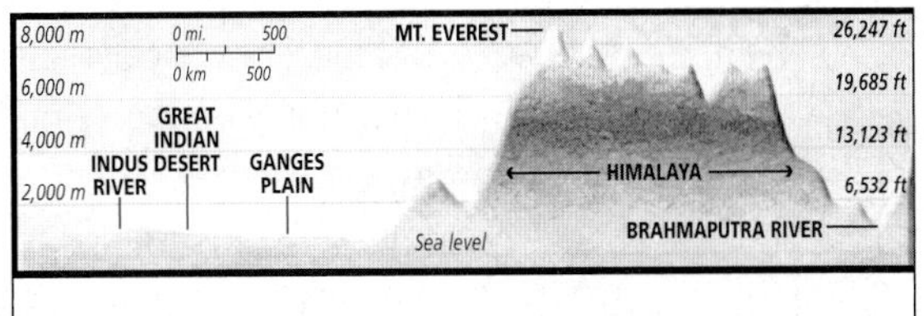

1. About how much higher is Mt. Everest than the Indus River?

A 2,500 feet
B 19,000 feet
C 25,000 feet
D 1,900 feet

Note that the question asks for the difference in height between the two locations. You can arrive at the answer by subtracting.

2. Based on elevation, which locations on the profile would be unsuited for farming?

F Ganges Plain
G Great Indian Desert
H Mt. Everest and the Great Indian Desert
J Mt. Everest

Read the question carefully. The phrase *based on elevation* is important. The Great Indian Desert is unsuited to farming, but not because of its elevation. Once you apply the standard asked for in the question, it is easy to eliminate wrong answers.

Geography Lab Activity

River and Stream Speed

▲*Flow calculations help people make effective use of river systems.*

Measuring the speed of a river or stream is the first step in determining the river's flow—the volume of water discharged over a period of time. Flow is calculated by multiplying the velocity, or speed, of a river (measured in feet or meters per second) by the area of the cross section of the river (a cutting made across, measured in square feet or square meters). Further calculations will yield the average flow in cubic feet per second or in gallons per day. Why measure the flow of a river or stream? This information can help water management engineers plan for emergencies such as drought or flooding.

Natural flow varies throughout the year, especially in South Asia, where seasonal weather patterns and human interaction with the environment affect the great river systems.

1 Materials

- Tape measure
- Ball of string or twine
- 4 wooden dowels or sticks
- Several medium-sized oranges or large craft sticks painted in bright colors
- Stopwatch
- Writing materials

2 Procedures

In this activity, you will use a simple method to measure the approximate speed (velocity) at which water moves in a stream.

1. Take all materials to a local stream that is no more than a few yards wide and is relatively free from vegetation and rocks.
2. Set up the measuring marks. Fix a dowel or stick in the ground, and tie one end of the string to it. Then toss the ball of string to a student on the opposite bank of the stream. Have the second student pull the string taut, tie it to another dowel fixed in the ground, and cut the end of the string. This will be Mark 1.
3. Use the tape measure to determine a point 10 feet (3 m) downstream from Mark 1. Insert another dowel at this point, and repeat the process of stringing a line across the stream, parallel to Mark 1. This will be Mark 2. If the stream you are measuring is very shallow or slow-moving, set the two lines only 5 (1.5 m) feet apart. If the stream is very fast, set the two lines 15 (4.6 m) feet apart.

4. Position an observer at Mark 1. Position another observer with the stopwatch at Mark 2. Have a third student go to a point several feet upstream and toss an orange or a painted craft stick into the water.
5. When the object crosses under the string at Mark 1, the first observer stationed there yells "Go!" and the second observer starts the stopwatch. When the object crosses under the string at Mark 2, the second observer stops the stopwatch.
6. Record the time, in seconds, that it took the object to pass from Mark 1 to Mark 2.
7. Repeat the process several times, recording the elapsed seconds. Calculate the average elapsed time in seconds. Then divide the distance in feet or meters between Mark 1 and Mark 2 by the average elapsed time. The result is the average stream speed, measured in feet or meters per second.
8. Note the weather for the days preceding your measurements. Did it rain, or were the days sunny? Why might this information be important?

3 Lab Report

1. Did you expect to find a faster or slower average stream speed, or were your findings consistent with what you expected?
2. How much variation in elapsed time did you observe when taking repeated measurements at the same site?
3. If you were asked to measure the speed of a river 100 yards (91 m) wide, how would you adapt this activity?
4. **Predicting Consequences** How might your measurement change if you dammed off the right half of the stream in the 10-foot (3-m) span you measured?

4 Find Out More

Contact a public works department for information on how flow calculations are used in your area (possibilities include environmental management, flood control, recreational use, agricultural irrigation, and urban water resource management). Choose one of these uses to research. Share your findings with the class.

Did You Know? Scientists have used the Acoustic Doppler Current Profiler (ADCP) to measure the flow of the Brahmaputra River in Bangladesh. The Brahmaputra often has severe floods. The ADCP allows scientists to measure river flow safely and accurately during flood conditions. The ADCP attaches to a boat and is connected to a computer that computes the river's flow, using data about depth, current, and direction.

Flow calculations are used to help design the irrigation systems that make South Asian desert lands able to be farmed.

CHAPTER 24

The Cultural Geography of South Asia

GeoJournal

As you read this chapter, note the diversity found in South Asia. Write a journal entry describing the culture of one particularly diverse area of South Asia. Be sure to make your descriptions as vivid and as accurate as possible.

GEOGRAPHY Online

Chapter Overview Visit the **Glencoe World Geography** Web site at geography.glencoe.com and click on Chapter Overviews—Chapter 24 to preview information about the cultural geography of the region.

SECTION 1

Population Patterns

Guide to Reading

Consider What You Know

India is South Asia's most populous country—with more than one billion people—and the second most populous country in the world. Do you know which country has more people than India?

Read to Find Out

- How do the peoples of South Asia reflect diversity?
- How is South Asia's large population distributed?
- How does life in the region's cities compare with life in traditional rural villages?

Terms to Know

- *jati*
- megalopolis

Places to Locate

- Islamabad
- Mumbai (Bombay)
- Kolkata (Calcutta)
- Delhi
- Dhaka
- Karachi

◀ *Schoolchildren in Pokhara, Nepal*

NATIONAL GEOGRAPHIC

A Geographic View

Scenes Along the Brahmaputra

[On a side stream of the Brahmaputra] there were men in long skirts, the descendants of ancient Aryans and Arab, Turkish, and Burmese traders. . . . Nearby, children bathed, men walked down planks with wicker baskets of coconuts, melons, and squash. . . . Three women fixed dinner in metal pots over a fire. The smell of mango, diesel fumes, and spices filled the air.

—*Jere Van Dyk, "Long Journey of the Brahmaputra,"* National Geographic, *November 1988*

Hindu temple near the Brahmaputra, India

Imagine taking a boat down the Brahmaputra River. You see great cities and small villages. You meet travelers on steep mountains and talk with families tending green rice fields. Life along the Brahmaputra reflects the color and diversity of all South Asia. In this section you will get a sense of that color and diversity as you learn about the peoples of this region.

Human Characteristics

One of the most significant characteristics of South Asia's population is its size. Over 1.3 billion people—more than one-fifth of the world's population—live in the region. Size is not the only distinguishing factor of South Asia's population, however. Diversity—the complex mix of religious, social, and cultural influences—is reflected in this region as in almost no other area on Earth. The peoples of the region speak hundreds of languages and practice several major religions. The region's diversity has fostered both tolerance and conflict.

India

India's population includes people from diverse groups. The largest number of Indians are descended from the Dravidians, who have lived in the south of India for 8,000 years, and the Aryans, who invaded from Central Asia more than 3,000 years ago. Also contributing to India's population mix are the descendants of British and Portuguese colonists as well as recent refugees from Tibet and Sri Lanka. Many Indians traditionally identify themselves by their religion—as Hindus, Muslims, Buddhists, Sikhs, Jains, or Christians. Hindus also identify themselves by a *jati*, a group that defines one's occupation and social position.

Pakistan and Bangladesh

Two South Asian countries—Pakistan and Bangladesh—were once part of British India. Pakistan and, later, Bangladesh became separate countries because of their distinct Muslim and ethnic heritages. More than 90 percent of the people of Pakistan and Bangladesh practice Islam. **Islamabad**, Pakistan's capital, is even named for the faith. This religious uniformity overshadows other cultural differences in Pakistan, which has at least five main ethnic groups. In Bangladesh most people are Bengali, an ethnic background they share with some of their Hindu neighbors in the Indian state of Bengal.

History

Sri Lanka's Sinhalese and Tamils

Sri Lanka has two main groups, which are fiercely divided along ethnic and religious lines. They speak different languages and live in different parts of the island country. The Buddhist Sinhalese are the majority and control the government. The other group—Hindu Tamils—have been fighting for an independent Tamil state in northern Sri Lanka since the early 1980s. Clashes between government forces and violent separatist groups like the Tamil Tigers have made this once peaceful, green island a war zone. Since 1984, more than 100,000 Sri Lankans have been killed or have disappeared. Almost a million people have been driven from their homes by ethnic violence—one of the largest such numbers ever recorded. The violence has disrupted the area's economy and demoralized its people.

Bhutan and Nepal

The peoples of Bhutan and Nepal differ in appearance from other South Asians, because their ancestors came from Mongolia. Bhutan's population is fairly evenly divided between the Bhote (BO•tay) people and those of Tibetan ancestry. Nepal, once a federation of tiny kingdoms, is home to a complex mix of ethnic groups. The group most familiar to people outside Nepal are the Sherpas, who are known for their mountaineering skills. One Sherpa, Tenzing Norgay, made the first successful ascent of Mount Everest with Sir Edmund Hillary in 1953.

Population Density and Distribution

With 780 people per square mile (301 people per sq. km), South Asia's population density is almost seven times the world average. Population growth rates in South Asia have traditionally been high, although educational and economic assistance efforts have slowed population growth

NATIONAL GEOGRAPHIC **World Explorer**

Geography Skills for Life

Himalayan Trekkers The mountain-dwelling Sherpas, such as the woman at right, are famed as guides to foreign expeditions in the Himalaya.

Human-Environment Interaction Which Sherpa made the first successful ascent of Mount Everest?

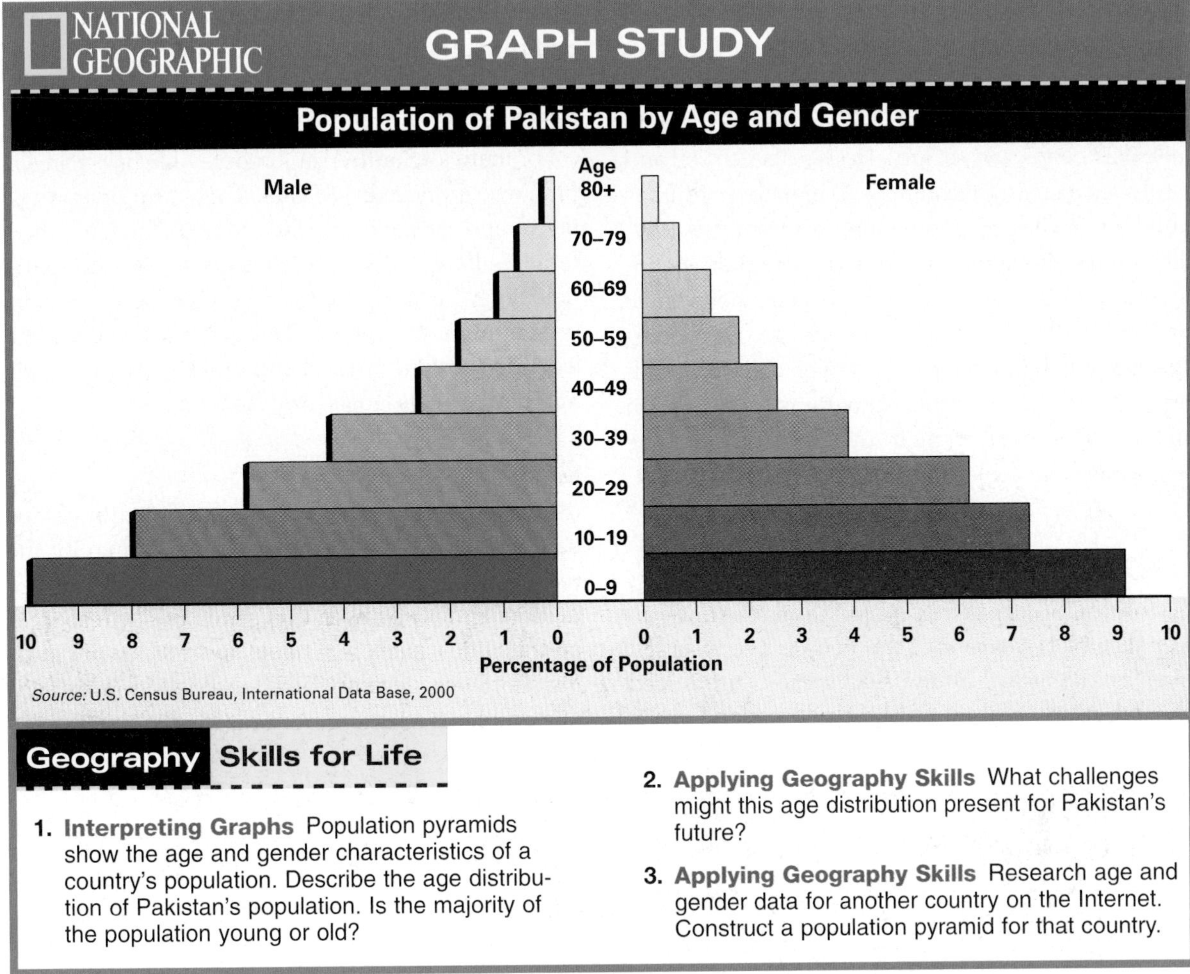

Geography Skills for Life

1. **Interpreting Graphs** Population pyramids show the age and gender characteristics of a country's population. Describe the age distribution of Pakistan's population. Is the majority of the population young or old?
2. **Applying Geography Skills** What challenges might this age distribution present for Pakistan's future?
3. **Applying Geography Skills** Research age and gender data for another country on the Internet. Construct a population pyramid for that country.

in some countries. Still, at present rates, South Asia will nearly double its current population by the year 2050.

Regional Variation

Although population densities are generally high throughout South Asia, the distribution of population varies from region to region. Factors such as climate, vegetation, and physical features have an impact on the number of people the land can support. The Great Indian Desert (Thar Desert) is sparsely populated, as are the mountainous highlands of western Pakistan. In southern Bhutan and Nepal, average population densities vary between 44 and 447 people per square mile (17 and 173 people per sq. km). To the north, however, population decreases as the elevation increases. An average of only 25 people per square mile (10 people per sq. km) make their homes in the Himalayan highlands because of unfavorable living conditions there.

The highest concentrations of population in South Asia are found on the fertile Ganges Plain (Indo-Gangetic Plain) and along the monsoon-watered coasts of the Indian peninsula. Because rice is an abundant and important food source, it is only natural that most South Asians live where rice is grown. Within parts of these agriculturally productive areas, densities exceed more than 2,000 people per square mile (772 people per sq. km). India's Deccan Plateau—not as populous as the Ganges Plain—supports up to 250 people per square mile (97 people per sq. km).

The large tea and rubber plantations of Sri Lanka require numerous workers. They come from the many villages that cluster around the plantations. The tiny coral islands of the Maldives are packed with 2,495 people per square mile (963 people per sq. km)!

Economics

Bangladesh Slows Its Growth

Bangladesh is the second most densely populated country in South Asia, with 2,401 people per square mile (927 people per sq. km). Despite its rich soil and improved farming techniques, Bangladesh still has difficulty feeding its population. As recently as 1991, the average Bengali woman had more than 4 children during her lifetime. A decade later, the average had lowered to 2.8 children per woman. To encourage Bengali women to have fewer children, both private and governmental programs give women small loans to start their own businesses. The programs have achieved some success.

Urban and Rural Life

Most of South Asia's population is rural. In Nepal only 11 percent of the people live in cities. Even in Pakistan, South Asia's most urbanized country, nearly two-thirds of the population lives in rural areas. The sharp differences between urban and rural life add to the region's many contrasts.

Rural Life

For many of South Asia's peoples, life has changed little over hundreds of years. They farm, live in villages, and struggle to grow enough food for their families. Part of their crop often goes to owners of the fields they farm. South Asia is also home to nomadic or seminomadic groups. These clans, usually large extended families, travel the desert and highlands and herd camels, goats, or yaks for a living.

Growing Urbanization

In recent years growing numbers of South Asians have been migrating to urban areas, drawn by the hope of better jobs and higher wages. As urban populations grow, however, they strain public resources and facilities, such as schools and hospitals. Housing shortages, overcrowding, and pollution are serious problems resulting from rapid urbanization.

NATIONAL GEOGRAPHIC **World Explorer**

Geography Skills for Life

Dhaka, Bangladesh

Although Bangladesh is one of the least urbanized countries in the region, its capital of Dhaka has a high population density.

Human-Environment Interaction What factors influence population density?

South Asia's Cities

South Asian cities are among the world's most densely populated urban areas. **Mumbai (Bombay)** is India's main port on the Arabian Sea as well as its largest city, with a population of more than 18.1 million. The city is also a leading industrial, financial, and filmmaking center. During the day, millions more people from outlying areas enter Mumbai to work. An American visiting Mumbai noted:

> *". . . [M]ost of [Mumbai's] newest citizens are from rural villages. Many of them are refugees from natural disasters such as floods and droughts. Others are refugees from the exacting demands of their own local societies."*
>
> John McCarry, "Bombay," *National Geographic*, March 1995

Kolkata (Calcutta), a thriving port city on a branch of the Ganges River, is the center of India's iron and steel industries. Here crumbling public buildings and high-rise slums contrast sharply with modern office towers and a modern subway system. Millions of people use the subway to travel to jobs in the city.

Delhi (DEH•lee), India's third largest city, is part of a **megalopolis**, or chain of closely linked metropolitan areas. Its sprawling land area encompasses the Old City, dating from the mid-1600s, and New Delhi, the modern capital built by British colonial rulers in the early 1900s. More than a million Delhi newcomers from rural areas have become "pavement dwellers"—people living on the streets in temporary settlements called *jhaggi bastis*.

The cities of Bangladesh and Pakistan are also crowded. **Dhaka**, the capital of Bangladesh, is the world's second most densely populated urban area after Lagos, Nigeria. Rural Pakistanis are drawn to the modern capital, Islamabad, where new housing projects struggle to keep up with a growing population, and to the booming port city of **Karachi**.

NATIONAL GEOGRAPHIC **World Explorer**

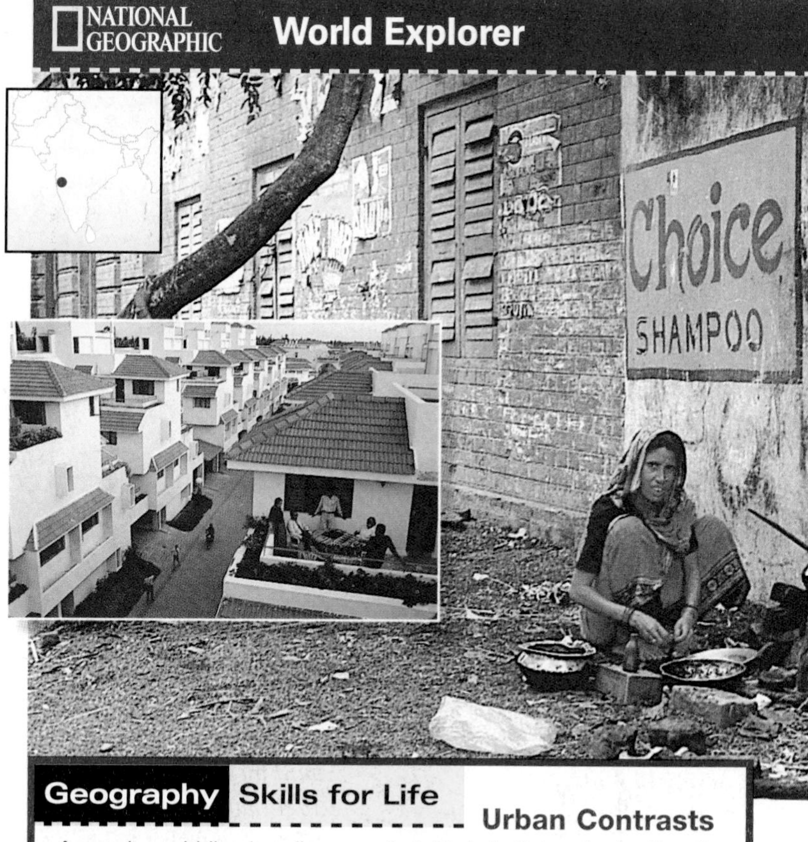

Geography Skills for Life

Urban Contrasts

A growing middle class lives comfortably in India's suburbs (inset), but the poor in India's cities must struggle to survive.

Movement Why do people in South Asia leave their families and rural villages to move to urban areas?

SECTION  ASSESSMENT

Checking for Understanding

1. **Define** *jati*, megalopolis.
2. **Main Ideas** Re-create the table below on a sheet of paper, and fill in the characteristics of the population in each South Asian country.

South Asian Countries	Population Characteristics

Critical Thinking

3. **Making Generalizations** Would you say that diversity has been more of a problem or a benefit for countries in South Asia? Why?
4. **Predicting Consequences** How might life in South Asia be affected in the next 50 years if present population growth rates and urbanization trends continue?
5. **Identifying Cause and Effect** What factors have contributed to the growth of South Asia's cities?

Analyzing Graphs

6. **Place** Study the graph on page 589. What percentage of people in Pakistan age 9 or under are female?

Applying Geography

7. **Effects of Geography** Study the population density map on page 562. Write a paragraph explaining how climate, physical features, and resources contribute to differences in population density.

History and Government

Guide to Reading

Consider What You Know

Many people who are not familiar with the Buddhist religion are nonetheless able to recognize the image of the Buddha. How have you seen the Buddha pictured?

Read to Find Out

- Where did South Asia's first civilization develop?
- What two major world religions originated in South Asia?
- How did invasions and conquests shape South Asia?
- What types of challenges are South Asian countries facing today?

Terms to Know

- dharma
- reincarnation
- karma
- nirvana
- raj

Places to Locate

- Indus River valley
- Mohenjo Daro
- Harappa
- Khyber Pass
- Mauryan Empire
- Gupta Empire

NATIONAL GEOGRAPHIC

A Geographic View

History as Architecture

Lahore [in Pakistan] . . . is an architectural accumulation of all of those who have conquered it. Mogul mosques stand next to Sikh temples, which stand next to British administration buildings. . . . Sitting near the Alamgiri Gate, a once private entrance to the royal quarters built by the emperor Aurangzeb in 1674 that is big enough for elephants to pass through, I fell into conversation with three college students. . . . Despite all the history around them . . . the young men were more interested in discussing the future than the past.

—*John McCarry, "The Promise of Pakistan,"* National Geographic, *October 1997*

Badshahi Mosque, Lahore

Modern life in Lahore unfolds amidst the architectural reminders of the city's fabled past. In fact, throughout all of South Asia, the past and present meet in many different and surprising ways. In this section you will explore South Asia's fascinating history—the story of a series of groups drawn to the region by its wealth of natural resources. Each successive group left its own permanent mark, making South Asia a region of great political and cultural diversity.

Early History

The earliest South Asians left few written records, but evidence of their great achievements in building and trade has been discovered in modern times. As the centuries passed, invaders from the northwest succeeded these early peoples. The influence of all these groups is still felt in South Asia today.

The Indus Valley Civilization

Around 2500 B.C. one of the world's great civilizations arose in the **Indus River valley**. This culture developed a writing system, a strong central government, and a thriving overseas trade. People built what may have been the world's first cities, **Mohenjo Daro** and **Harappa**. Made of bricks hardened by fire in kilns, these cities boasted sophisticated plumbing, sanitation systems, and other technology that would not be matched again for centuries.

Environmental changes may have led to the decline of this civilization between 1700 and 1500 B.C. The cities were most likely lost to flooding or drought as the Indus River changed its course.

The Aryans

As the Indus Valley civilization crumbled, a group of hunters and herders entered the region from the northwest. These people, the Aryans, settled down and began to farm. They left behind sacred writings called the Vedas.

The Vedas reveal Aryan ideas about religion and social structure. Society was organized into four groups—priests, warriors (or nobles), artisans and farmers, and enslaved people. At first the boundaries between groups were somewhat flexible; people of different classes could intermarry and change professions. Gradually, the social structure developed into a complex system of ranks that dictated from birth one's social status. This "caste" system prevailed in India for centuries and only now is gradually weakening.

Two Great Religions

Understanding the basic beliefs of Hinduism and Buddhism is a key to understanding South Asia's history and culture. These two religions, as well as other faiths, have had profound influence in the region.

Hinduism

Growing out of Aryan culture and religion, Hinduism is both a religion and a way of life. Hindu belief requires every person to carry out his or her **dharma** (DUHR•muh), or moral duty. Hindus also believe that after death people undergo **reincarnation**, or rebirth as another living being. This process occurs repeatedly until the individual overcomes personal weaknesses and earthly desires. At that point, a person leaves the cycle of rebirth and becomes reunited with the eternal being. In the law of **karma**, good deeds—actions in accord with one's dharma—move one toward this point, while bad deeds chain a person to the cycle of rebirth.

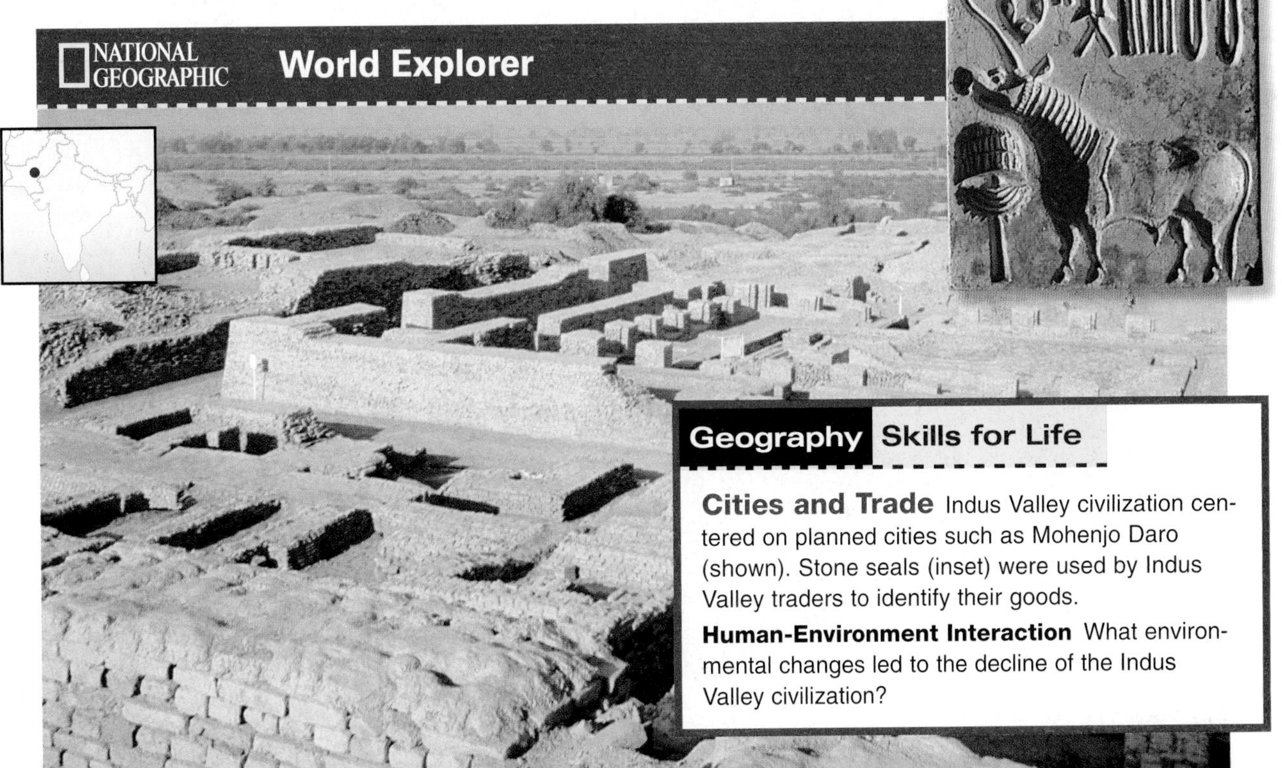

Geography Skills for Life

Cities and Trade Indus Valley civilization centered on planned cities such as Mohenjo Daro (shown). Stone seals (inset) were used by Indus Valley traders to identify their goods.

Human-Environment Interaction What environmental changes led to the decline of the Indus Valley civilization?

Hindus honor many gods and goddesses, which are often seen as different forms of the one eternal being. Many Hindus are tolerant of other religions, viewing them as different paths to the same goal.

Buddhism

Siddhartha Gautama (sih•DAHR•tuh GOW•tuh•muh) was born around 563 B.C. in what is today Nepal. Belonging to a noble Hindu family, Siddhartha lived a life of luxury. As he grew to manhood, however, he became aware of human suffering. Leaving his wealth and power behind, he went on a pilgrimage. Years of meditation and spiritual seeking led to the moment when Siddhartha perceived what he understood to be the true nature of human existence. He then became known as the Buddha, or the Awakened One.

The Buddha spent the rest of his life sharing his insights with others. He taught that people suffer because they are too attached to material things, which are temporary. The Buddha also taught people to think clearly, work diligently, and show compassion for all living things in order to escape desire and suffering and to be liberated from endless rebirth.

Like Hinduism, Buddhism developed a system of religious rituals, but it was primarily a practical way to achieve human happiness. By following Buddhist teachings, people could become enlightened, entering a state of insight, calm, and joy called **nirvana** (nir•VAHN•uh).

Culture

A Marriage of Influences

Because the Buddha rejected the rigid social system of his day, women and people of lower social classes embraced his teachings. Eventually, Buddhism spread from India to other countries. Sri Lanka became a Buddhist kingdom. In Nepal and Bhutan, new forms of Buddhism emerged that blended Hindu rituals with local practices. In India, Hinduism absorbed Buddhism but retained a tradition of honoring the Buddha.

Invasions and Empires

After the Aryans, other groups with new cultures invaded South Asia through the **Khyber Pass** in the northwestern Hindu Kush mountains. The **Mauryan Empire**, established by the first of these

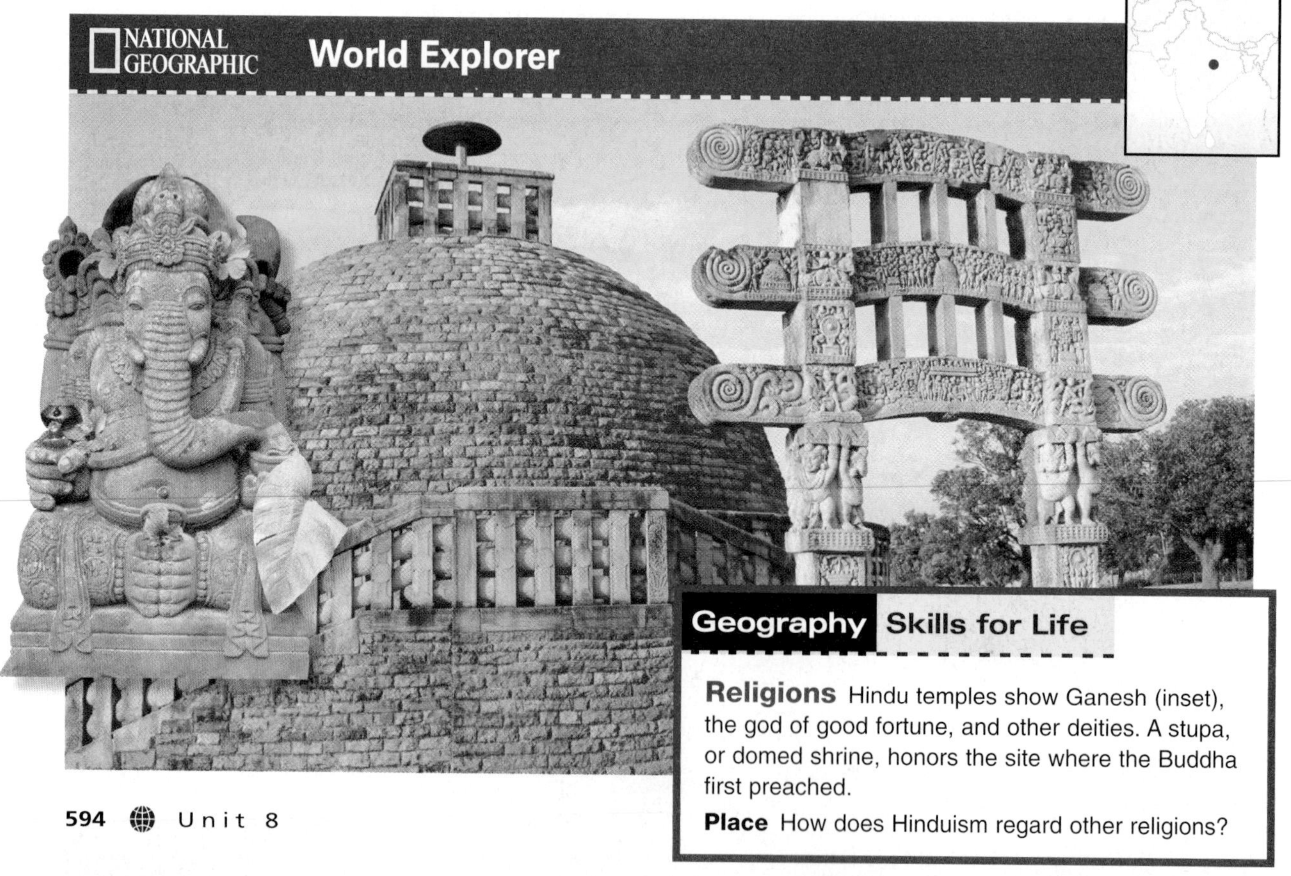

Geography Skills for Life

Religions Hindu temples show Ganesh (inset), the god of good fortune, and other deities. A stupa, or domed shrine, honors the site where the Buddha first preached.

Place How does Hinduism regard other religions?

groups, maintained control from about 320 to 180 B.C. and ruled all but the southernmost parts of the Indian peninsula. Asoka, the last and greatest Mauryan emperor, promoted Buddhism and nonviolence.

About 500 years later, the **Gupta Empire** came into power. From about A.D. 320 to 550, this Hindu civilization was one of the most advanced in the world. Science, technology, and the arts flourished. The numerals we call Arabic today were most likely developed in India during the Gupta period and introduced to Europe by Arab traders.

Lion sculpture from the palace of Asoka

Muslim missionaries and traders first entered India in the 700s. By the 1100s Muslim armies from Mongolia, Turkey, and Persia had conquered northern India. The Muslim-led Mogul Empire dominated the Indian subcontinent for several centuries. During this era, many South Asians converted to Islam.

The final invaders, Europeans, came by sea. Portuguese traders arrived first in about 1500. They were followed by the French and the British. In the late 1700s, the British expelled the French. Though Portuguese strongholds remained, the British were the major European power in South Asia at this time. The British called their Indian empire the British raj, the Hindi word for empire. The British introduced the English language to South Asia, restructured the educational system, built railroads, and developed a civil service. British influence is still seen in some elements of Indian culture.

Modern South Asia

Today South Asia is free of European control. Independence did not come easily, however, and these growing countries still struggle with the aftereffects of colonialism.

Independence

In the early and mid-1900s, India's fight for independence was led by Mohandas K. Gandhi. Using nonviolent methods, such as boycotting British products and staging peaceful demonstrations, Gandhi inspired the peoples of India to seek self-rule. A Hindu, Gandhi worked to end the rigid social system and promote local industry, such as spinning and weaving. Enduring prison and hunger strikes in the struggle for independence, Gandhi earned the name Mahatma, or "Great Soul." According to Gandhi,

> "*Nonviolence and truth* (Satya) *are inseparable and presuppose one another. There is no god higher than truth.*"
>
> S. Hobhouse, ed., *True Patriotism: Some Sayings of Mahatma Gandhi*, 1939

NATIONAL GEOGRAPHIC **MAP STUDY**

South Asian Empires, 300s B.C.–A.D. 400s

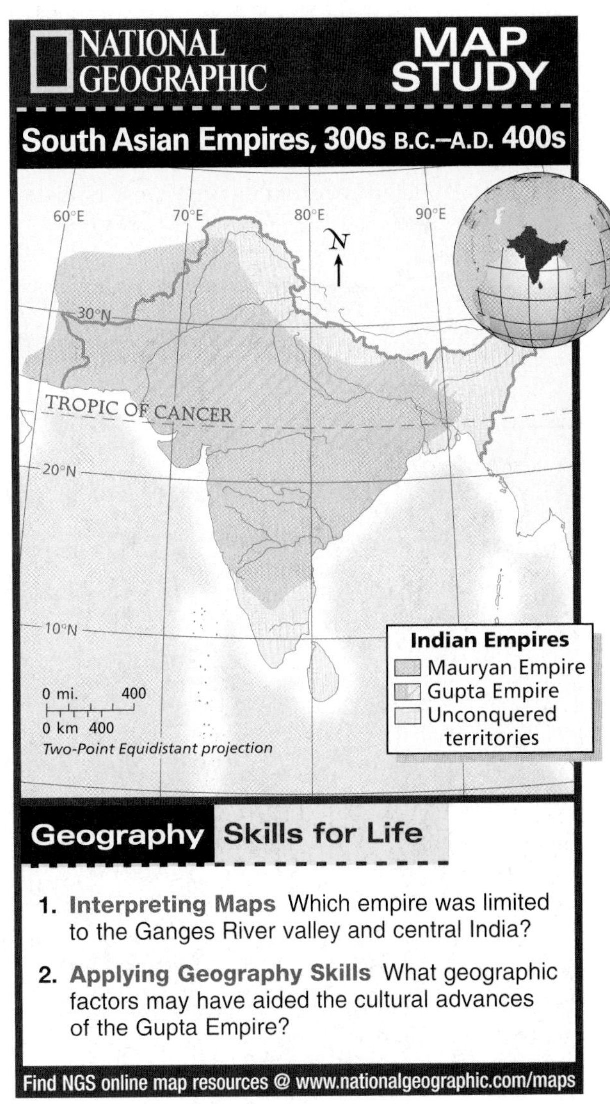

Geography Skills for Life

1. **Interpreting Maps** Which empire was limited to the Ganges River valley and central India?
2. **Applying Geography Skills** What geographic factors may have aided the cultural advances of the Gupta Empire?

Find NGS online map resources @ www.nationalgeographic.com/maps

In 1947 Britain finally granted independence to British India, and the land became two new countries. Areas with a Hindu majority became India, and those with a Muslim majority became Pakistan. Pakistan actually consisted of two isolated sections—East Pakistan and West Pakistan—separated by about 1,000 miles (1,609 km) of land belonging to India.

Dividing British India split many families. Hundreds of thousands of Hindus in Pakistan moved to India, and a similar number of Muslims in India moved to East or West Pakistan. Violence often marked the movements. Religious violence also claimed the life of Gandhi, who was assassinated in 1948 by a Hindu nationalist opposed to the division of India.

One year after granting India self-rule, Britain gave independence to Ceylon. In 1972 the island took back its ancient name, Sri Lanka. Nepal and Bhutan had always been independent of European rule. The Maldives, a group of islands in the Indian Ocean, won independence from Britain in 1965. In 1971 East Pakistan revolted against West Pakistan and became the new country of Bangladesh. The western part retained the name Pakistan.

Regional Conflicts

Tensions between India and Pakistan continued after independence. Some border areas, especially the former Indian provinces of Jammu and Kashmir, are still hotly disputed. Today both India and Pakistan have nuclear weapons, adding to the complexity of the conflict. Ethnic and religious tensions also trouble other parts of South Asia. Hindu and Muslim groups within India have clashed. Since the 1980s the Sri Lankan government has been troubled by ethnic Tamil rebel groups seeking a separate Tamil state.

Student Web Activity Visit the **Glencoe World Geography** Web site at geography.glencoe.com and click on Student Web Activities—Chapter 24 for an activity on Kashmir.

NATIONAL GEOGRAPHIC **World Explorer**

Geography Skills for Life

Migrations, 1947 Mohandas Gandhi (inset) mourned the violence between Hindus and Muslims that came with British India's division and the mass migration of people.

Place What are the causes of conflict in South Asia today?

Today's Governments

Today's South Asian governments are diverse. India, often called the world's largest democracy, is a federal parliamentary republic. For 40 years following India's independence, members of the Nehru (NAY•roo) family headed India's government. Jawaharlal (jah•wah•HAR•lahl) Nehru was India's prime minister from 1947 until his death in 1964. His daughter, Indira Gandhi, and later his grandson, Rajiv Gandhi, also led the country, but growing ethnic and religious conflict led to assassinations of Indira in 1984 and Rajiv in 1991. Since then, India's prime ministers have had less influence than the Nehru "dynasty." Workable parliamentary institutions have made India's democracy more secure.

Like India, Pakistan is a parliamentary republic, but instability and military rule have prevailed since 1971. A more stable democracy seemed likely in the 1990s under Benazir Ali Bhutto, the country's first female prime minister, and later under her successor, Nawaz Sharif. In 1999, however, charges of official corruption led to a military coup. Pakistan's new leader, General Pervez Musharaff, pledged to have a more democratic government.

Sri Lanka and Bangladesh also are parliamentary republics. Intense political or ethnic rivalries, however, have made stable political rule difficult. After independence, political assassinations or military takeovers marked both countries. In recent years democratic rule has been strengthened. In 1994 Chandrika Bandaranaike Kumartungah (chahn•DREE•kah BAHN•dah•rahn•EYE•keh KOO•mahr•TOON•gah) was elected Sri Lanka's first female president. Two years later, Sheikh Hasina Wazid was elected as Prime Minister in Bangladesh.

▲ *Sheikh Hasina Wazid (right), prime minister of Bangladesh, shown here with Megawati Sukarnoputri, president of Indonesia*

A few countries in the region today have traditional forms of government. For example, Bhutan and Nepal have monarchies that are trying to modernize and still keep some power. Once ruled by a sultan, the Maldives became a republic in 1968.

SECTION 2 ASSESSMENT

Checking for Understanding

1. **Define** dharma, reincarnation, karma, nirvana, raj.
2. **Main Ideas** Re-create the graphic organizer below on a sheet of paper, and complete it by filling in information about the successive groups that influenced South Asia.

Critical Thinking

3. **Making Inferences** Describe characteristics of South Asia during ancient, colonial, and modern eras.
4. **Comparing and Contrasting** In what ways are Hinduism and Buddhism similar? Different?
5. **Identifying Cause and Effect** How do present-day political borders in South Asia reflect ethnic and religious conflicts?

Analyzing Maps

6. **Location** Study the map of South Asian empires on page 595. Which empire extended beyond the borders of present-day Pakistan?

Applying Geography

7. **Geography and Religion** Think about the influences of religion on the history and culture of South Asia. How has the geography of the region impacted South Asia's religions?

GEOGRAPHY AND HISTORY

MOUNTAIN MADNESS: STRUGGLE FOR KASHMIR

A CASHMERE SWEATER, made of soft wool from the undercoat of the Kashmir goat, is a prized possession. So, too, is the Kashmir region, where the goat got its name. The problem: two countries claim Kashmir. No wonder, for Kashmir, situated high in the Himalaya on the northern tips of India and Pakistan, is renowned for its beauty and climate. Ancient mountain villages are reflected in the waters of its crystalline lakes. Fields of crocuses are harvested for the world's most expensive spice—saffron. However, decades of fighting have shattered this idyllic realm.

Tale of Two Religions

For centuries, Kashmir was part of the Indian kingdoms, ruled by maharajas, or princes. In 1846 Kashmir became the British Indian state of Jammu and Kashmir. When predominantly Hindu India won independence in 1947, Britain partitioned the western part of India to create Pakistan as a homeland for South Asia's Muslims. As the leaders of the existing Indian states decided which country to join, widespread rioting broke out between

Kashmiri Muslims struggle violently against Hindu India's rule. ▶

◀ **A flower merchant rows down a river in Kashmir, where years of warfare have shattered the calm.**

NATIONAL GEOGRAPHIC SOCIETY

Hindus and Muslims.

In the face of the chaos, the prince of Jammu and Kashmir sought to remain autonomous. As a Hindu, his loyalties were with India, but the majority of the population was Muslim. A Muslim uprising, perhaps supported by Pakistan, sent the prince fleeing to Delhi, where he signed his state over to India. India then sent troops to put down the uprising. The Pakistani army responded, and the first India-Pakistan war began.

India claimed a legal and historical right to Kashmir, but Pakistan insisted it would be a better homeland for the Muslim enclave. Each side also has strategic needs: India wants Kashmir as a buffer between itself and China, while Pakistan relies on river waters flowing from Kashmir for irrigation and electricity.

Demand for Independence

In 1949 the United Nations arranged a truce, which established a cease-fire line that split Kashmir unequally between India and Pakistan. But peace did not last. War broke out again in 1965. In 1972 an accord reaffirmed the original cease-fire line, now called the line of control. Yet troops on both sides regularly fire across it, killing civilians and wrecking villages.

In the late 1980s, a new crisis engulfed Kashmir. Muslim groups within Kashmir, demanding independence, began killing Indian soldiers and Kashmiri Hindus. India responded with force. Then the stakes were raised. In 1998, first India, then Pakistan confirmed the world's worst fears by conducting underground nuclear weapons tests. Today tensions continue between India and Pakistan, the world's newest nuclear powers, as they vie for Kashmir.

Looking Ahead

With nuclear weapons in the mix, other nations must pay attention. Why is this dispute so difficult for India and Pakistan to resolve? Why is it important to the world that they resolve it?

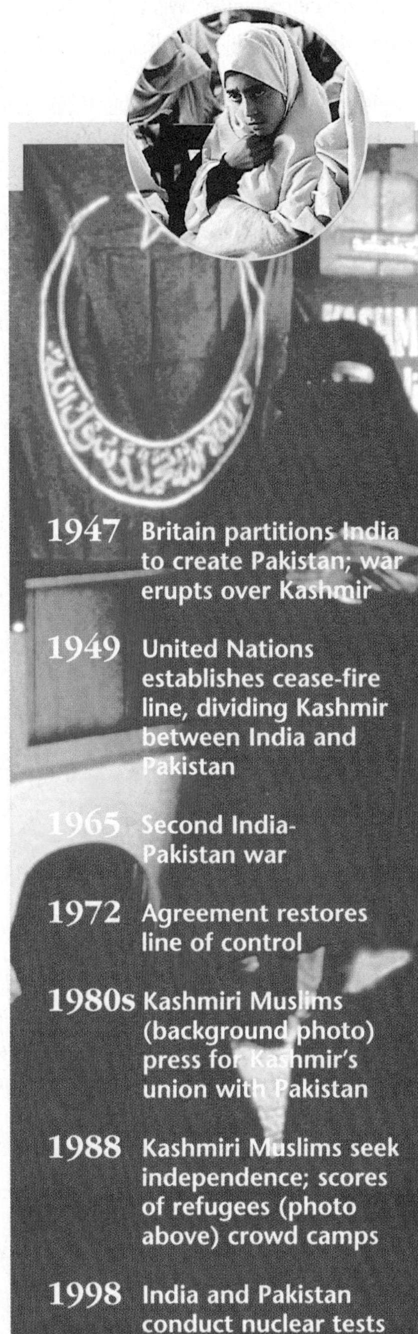

1947 Britain partitions India to create Pakistan; war erupts over Kashmir

1949 United Nations establishes cease-fire line, dividing Kashmir between India and Pakistan

1965 Second India-Pakistan war

1972 Agreement restores line of control

1980s Kashmiri Muslims (background photo) press for Kashmir's union with Pakistan

1988 Kashmiri Muslims seek independence; scores of refugees (photo above) crowd camps

1998 India and Pakistan conduct nuclear tests

Cultures and Lifestyles

Guide to Reading

Consider What You Know

Indians speak many variations of the same basic language. What are such language variations called?

Read to Find Out

- How do the lives of South Asia's peoples reflect the region's linguistic and religious diversity?
- What contributions to the arts has the region made?
- How are South Asian countries meeting challenges to improve the quality of life in the region?
- How is the rich cultural diversity of South Asia reflected in distinctive celebrations?

Terms to Know

- guru
- mantra
- sadhu
- stupa
- *dzong*

Places to Locate

- Taj Mahal
- New Delhi

NATIONAL GEOGRAPHIC

A Geographic View

A South Asian Celebration

Diwali, the five-day Festival of Lights, was my family's favorite among all the religious observances that crowd the Indian calendar. . . . It commemorates to the Hindus of the north the return to India of Lord Ram and his wife, Sita, after their victory over the . . . king of Sri Lanka. Tiny earthen oil lamps are lit to outline every house and hut to guide them on the journey home. Sikhs also celebrate on this night of Diwali. Even Muslim families sometimes join in.

—*Jeffrey C. Ward, "India: Fifty Years of Independence,"* National Geographic, *May 1997*

Diwali celebration

South Asia's ethnic diversity has produced a rich cultural blend, a mix of contrasting elements much like the spicy Indian stew called *masala* or the mixture of pungent spices that make up curry. As you read this section, note the gifts of art, music, architecture, and dance that South Asia shares with the world.

Languages

The peoples of South Asia speak 19 major languages and hundreds of local dialects. In India alone the government officially recognizes 14 languages, although Hindi is chief among them. English, the common language of international business and tourism, is also widely spoken in the parts of South Asia that were once under British rule.

Indo-European Languages

Most languages spoken in Pakistan, Bangladesh, and northern India fall into the Indo-European family of languages. These languages—Hindi, Urdu, and Bengali—trace their roots to the Aryan invaders of 3,000 years ago and are related to most of the major languages of Europe.

About half of India's people, especially those in the northern and central states, speak Hindi as their primary language. Urdu is Pakistan's official language, and Bengali is the official language of Bangladesh. In many northern areas, Indians speak Hindustani, a mixture of Hindi and Urdu. Nepali, Sinhalese, and Divehi, the official languages of Nepal, Sri Lanka, and the Maldives, respectively, also have Indo-European roots. Sanskrit, the classical Aryan language of the Vedas, is still used for religious, literary, and musical purposes.

Other Languages

Most of the population in southern India and Sri Lanka speak languages of the Dravidian family, whose roots go back to the earliest inhabitants of southern South Asia. Dravidian languages include Tamil, Telugu, Kannada, and Malayalam. In the north the languages of Bhutan and parts of Nepal reflect these countries' close ethnic and historical ties to East Asia.

Religions

Hinduism, Islam, and Buddhism are the major religions of South Asia. Most people in India and Nepal are Hindus. Hinduism is also practiced, to a lesser extent, in Bhutan, Sri Lanka, Pakistan, and Bangladesh. Pakistan, Bangladesh, and the Maldives were all founded as Islamic states, and the majority of the people in these countries are Muslims. India's 120 million Muslims form the country's second-largest religious group. Buddhism, although no longer a significant religion in India, remains strong in Sri Lanka, Bhutan, and Nepal.

Other religions practiced in South Asia include Jainism, Sikhism, Christianity, and Zoroastrianism. Jainism was founded in the 500s B.C. by Mahariva, a Hindu teacher. India's more than 3 million Jains practice strict nonviolence, believing that every living thing has a soul. Sikhism, founded in the early A.D. 1500s by a **guru**, or teacher, named Nanak, teaches that there is one God and that good deeds and meditation bring release from the cycle of reincarnation. Most of South Asia's 20 million Sikhs live in northwestern India, and many want an independent Sikh state there.

About 17 million Christians also live in South Asia, concentrated in urban areas in southern and northeastern India. The Indian city of Mumbai is home to some of the last living Zoroastrian followers, known as the Parsis, whose religious and cultural heritage comes from ancient Persia.

Culture

Religion and Daily Life

The influence of religion is ever present in South Asia. In Bhutan and Nepal, for example, colorful prayer flags wave in the wind, and prayer wheels twirl on many corners, sending out invocations. Monks chant **mantras**, or repetitive prayers. In India, Hindu holy men called

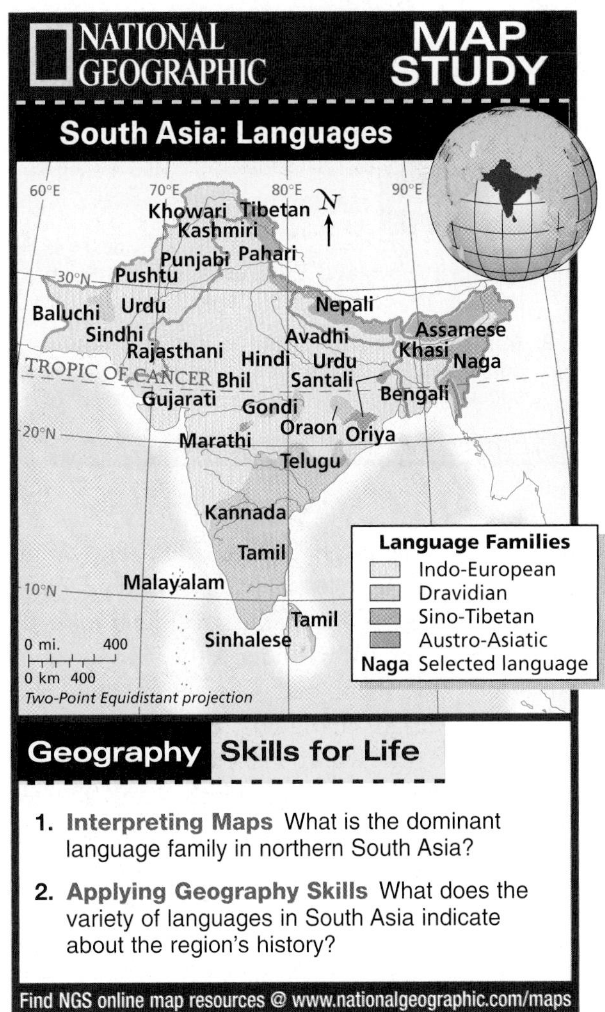

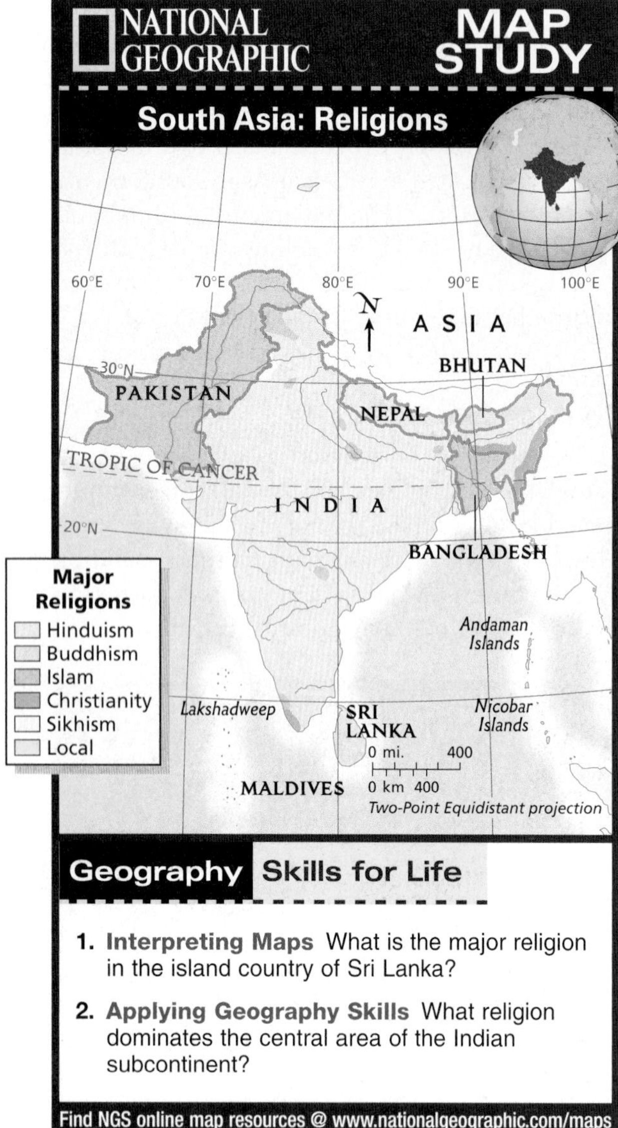

sadhus dress in bright yellow robes and roam from temple to temple, carrying only their blankets and begging bowls. In the streets and roads of India, where Hindus consider cattle sacred, thousands of cows roam freely, sometimes wearing garlands of bright marigolds. Buddhist pilgrims from around the world visit the shrines of Sri Lanka. In Pakistan and Bangladesh, many Muslim women wear the *chador*, the enveloping robe and veil that Islamic tradition requires for modesty.

Local communities of all these religions maintain places of worship, schools, clubs, and charitable foundations. Many religious groups, such as Hindus, have formed their own political parties. Through such organizations they try to influence the government to pass laws that deal with religious or social issues.

The Arts

Artistic expression is as much a part of South Asian life as religious practice. The South Asian environment, with its rich appeal to all the senses, nurtures a variety of distinctive, artistic expressions.

Literature

The South Asian literary tradition has its roots in religion. India's two great epic poems, the *Mahabharata* (muh•hah•BAH•ruh•tuh) and the *Ramayana* (rah•MAH•yah•nuh), combine Hindu social and religious beliefs with intricate plots and richly detailed characters. These two works, composed between 1500 and 500 B.C., endure today in public readings, mask and puppet theater, and even television series. An especially treasured portion of the *Mahabharata* is the *Bhagavad Gita* (BAH•guh•vahd GEE•tuh), or "song of the lord." In this dialogue between a warrior and his chariot driver, the Indian god-hero Krishna, the reader finds a message of devotion to duty and courage in the face of death.

Among writers in the 1900s, South Asia boasts the Muslim poet and philosopher Muhammad Iqbal, who wrote in the early part of the century. He was the first to propose the idea of an Islamic state in South Asia. The 1913 Nobel laureate Rabindranath Tagore was an Indian who wrote poetry, fiction, and drama in both English and Bengali. Tagore wrote India's national anthem, whose third verse proclaims:

> *"Eternal charioteer, thou drivest man's history along the road rugged with rises and falls of Nations. Amidst all tribulations and terror thy trumpet sounds to hearten those that despair and droop, and guide all people in their paths of peril and pilgrimage. Thou dispenser of India's destiny, victory, victory, victory to thee."*
>
> Rabindranath Tagore,
> "Jana Gana Mana," 1911

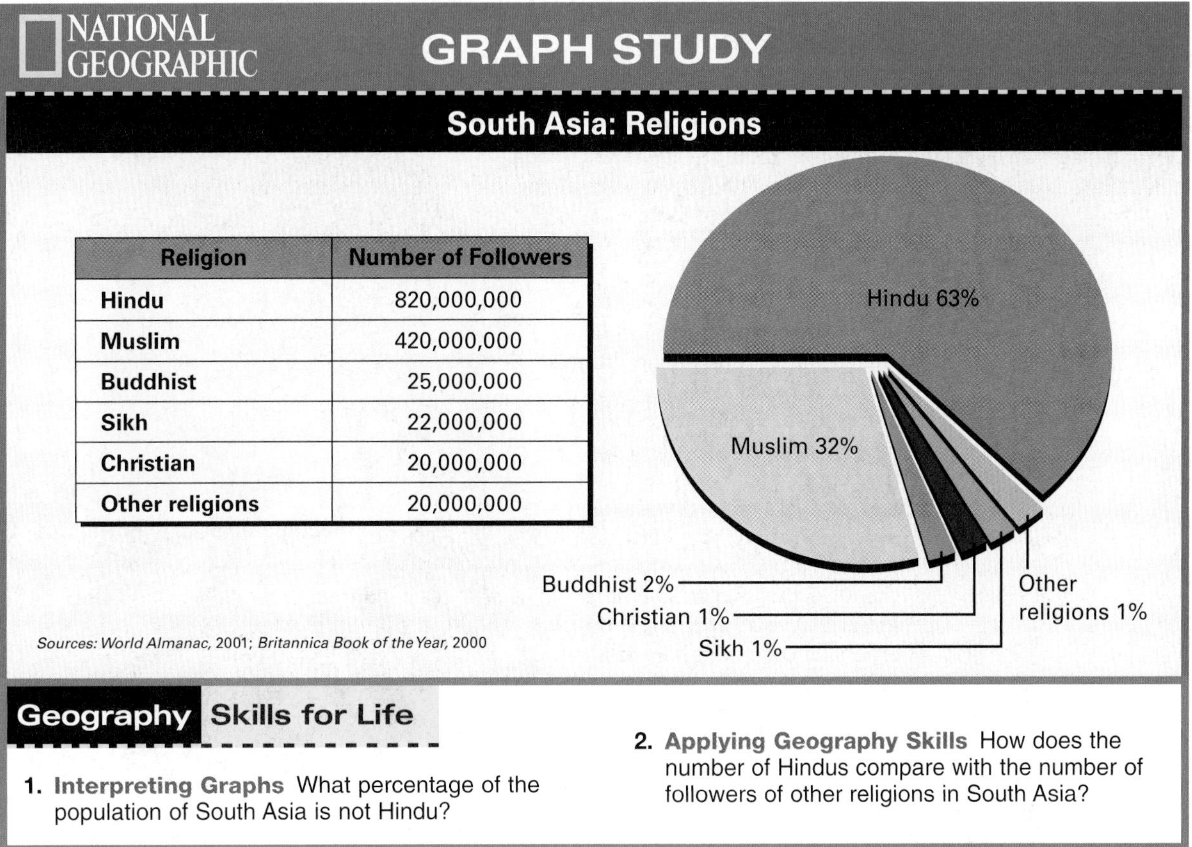

Religion	Number of Followers
Hindu	820,000,000
Muslim	420,000,000
Buddhist	25,000,000
Sikh	22,000,000
Christian	20,000,000
Other religions	20,000,000

Geography Skills for Life

1. **Interpreting Graphs** What percentage of the population of South Asia is not Hindu?
2. **Applying Geography Skills** How does the number of Hindus compare with the number of followers of other religions in South Asia?

Some contemporary South Asian novelists live in other countries but continue to write from a uniquely South Asian perspective. Salman Rushdie, born in Mumbai, has written controversial novels on Indian history and Islamic politics. Michael Ondaatje, born in Sri Lanka but now living in Canada, won England's prestigious Booker Prize for his novel *The English Patient*, which became an Academy Award-winning film.

Dance and Music

India has numerous classical dance styles, most of which are based on themes from Hindu mythology. The style known as *Bharata Natyam* (bah•RAH•tah NAHT•yam) is practiced mainly in the south. Based on the devotional postures of sacred temple dancers, these dances involve rapid whirling, stamping feet, and an elaborate language of hand gestures called mudras. The dancers, usually women, wear bright silk saris and jingling gold jewelry.

On India's west coast, an ancient style of dance called *Kathakali* (kah•tha•KAHL•lee) is now being revived. The male dancers wear huge, colorful masks, and their violent movements are rooted in martial arts postures.

Indian classical music is divided into two basic types: Hindustani, in the north, and Karnatic, in the south. The melody of each is called the raga, and the rhythm is called the tala.

Movies

Since the first Indian motion picture was made in 1896, movies have been a popular form of entertainment in India and Bangladesh. India's film industry, centered in Mumbai (nicknamed "Bollywood," a combination of Bombay and Hollywood), is the world's largest, producing more than 800 full-length feature films a year. When Satyajit Ray, India's most renowned director, died in 1992, more than half a million people joined his funeral procession.

Visual Arts and Architecture

Traditionally South Asians have used the visual arts to express religious beliefs and to document daily life. Stone carving and sculpture exist from as far back as the Indus Valley civilization, and some

music of SOUTH ASIA

The Indian subcontinent is the birthplace of some of the world's oldest and most complex musical forms. The traditional systems of raga (melody) and tala (rhythm) are at the root of the music of India.

Instrument Spotlight:
The **tabla** is the most popular percussion instrument of India. The cylindrical right-hand drum is carved of wood and is tuned by hammering wooden pegs underneath leather straps that hold the skin to the drum's top. The left-hand drum has a wider skin surface attached to a kettle-shaped metal bowl. Although the tabla first evolved in India about 500 years ago and is traditionally used in Indian classical music, these drums are now used in modern pop, jazz, and fusion music all over the world.

Go To **World Music: A Cultural Legacy** Hear music of this region on Disc 2, Tracks 7–13.

Mauryan Empire techniques for polishing marble have never been duplicated. Under Mogul emperors, traditional Muslim restrictions against depicting the human form loosened, and portraits and decorative miniature paintings flourished.

The elaborate Hindu temples of India, the Buddhist stupas, or domed shrines, of Nepal and Sri Lanka, and the fortified monasteries, or *dzong*, of Bhutan illustrate South Asia's artistic spirit. The **Taj Mahal** in Agra, India, and the Golden Temple of the Sikhs in Amritsar (UHM•RIHT•suhr), India, are world famous. A Muslim emperor built the Taj Mahal (shown on page 559) in the 1600s as a tomb for his beloved wife. Made of white marble, with towers and domes in the Islamic style, the structure has delicate screens, carved in the Hindu style.

Modern South Asian arts and architecture blend traditional and Western styles. By the mid-1900s South Asian painting and sculpture had an international flavor, and South Asian artists worked in a variety of different media. The mixture of traditional and modern forms is especially apparent in architecture. For example, the modern city of **New Delhi**, with its well-laid-out streets and Western-style government buildings, sits next to the historic city of Delhi, known for its mosques, ancient forts, and busy bazaars.

Quality of Life

The governments and economies of South Asia are still developing. Lifestyles there are a complicated mixture of the traditional and the modern, challenging South Asia's quality of life.

Health

Life expectancies in South Asia are generally lower than those in industrialized countries. Only Sri Lanka's life expectancy of 72 years comes close to that of the United States. Nepal's life expectancy, about 57 years, is the region's lowest, and in India, life expectancy is only about 61 years. In most countries in the region, figures for males and females are fairly close.

Tropical diseases, such as malaria, were once widespread but have been brought under control in much of South Asia. Other health problems continue, however. For example, South Asia and Southeast Asia together have the second-highest rate of HIV infection and AIDS in the world.

The scarcity of clean water in South Asia makes waterborne diseases such as cholera and dysentery common. About one-third of Nepal's infants die from dysentery before their first birthdays. Infant mortality rates are also high in Pakistan.

Food

Although improved farming techniques and government policies now make it theoretically possible for most of South Asia to feed its people, poor nutrition is still a problem. Almost one-third of South Asia's people are too poor to buy high-quality protein foods. To obtain needed protein, some South Asians eat soy-based tofu or beans.

Religious dietary restrictions prohibit Muslims from eating pork. Hindus cannot eat beef, and Jains and many Buddhists are vegetarian. Nevertheless, many South Asians enjoy cuisines of great variety.

Education

South Asia's standard of living is likely to rise with improved education. The region's governments are committed to raising literacy rates and extending educational opportunities to women and members of lower social classes.

▲ *A middle school in Hyderabad, Pakistan*

Celebrations

South Asia's cultural mix is underscored by its many celebrations. Muslims mark the end of the month-long daily Ramadan fast with feasting and family visits. Buddhists celebrate the birthday of Siddhartha. Hindus, Christians, Jains, and Sikhs all celebrate their traditional holidays. South Asians commemorate national holidays as well. For example, Indians mark the anniversary of the adoption of their constitution on January 26, known as Republic Day.

SECTION 3 ASSESSMENT

Checking for Understanding

1. **Define** guru, mantra, sadhu, stupa, *dzong*.
2. **Main Ideas** Create a graphic organizer like the one below, and fill in information about South Asian arts.

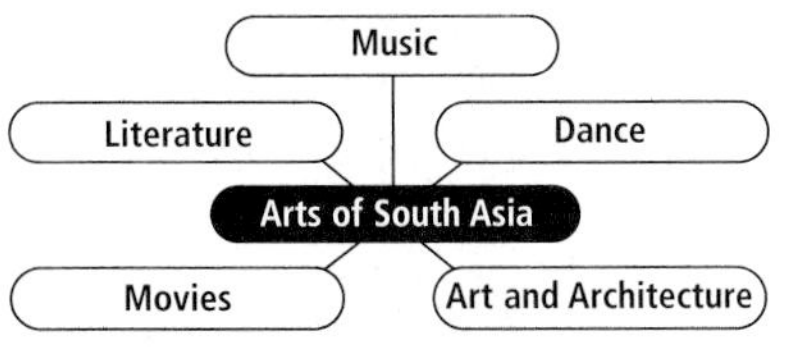

Critical Thinking

3. **Making Generalizations** What are the major languages of South Asia, and what do they have in common?
4. **Comparing and Contrasting** How might job opportunities for religious minorities in South Asia compare with those for religious minorities in other parts of the world?
5. **Problem Solving** What improvements in health and education might enrich the quality of life in South Asia?

Analyzing Maps

6. **Place** Look at the map of South Asian languages on page 601. What is the dominant language family in Pakistan? In Bhutan?

Applying Geography

7. **Identifying Relationships** Write a short essay describing the relationship among language families, religious groups, and national identities in South Asia.

Evaluating Information and Sources

You live in a world saturated with information and opinions. Finding information on any topic is not a problem. But how can you decide which information is useful and accurate?

Learning the Skill

Information that you find while researching can come from a variety of sources. However, not all of the information that you find may be useful or even accurate. It is important to evaluate the information you find in order to determine whether it is valid information. To evaluate information and sources:

- **Identify the reliability of the source.** Consider whether the source may be biased. For example, a statement published by an environmental group and one published by a large energy company may have different biases.
- **Summarize the key points of the information in a few sentences.**
- **Distinguish fact from opinion.** Look for ways that facts are chosen or left out to support the stated opinions.
- **Verify facts by cross-checking them in other sources.** Check encyclopedias, almanacs, and other references to be sure the information is accurate. Make sure you are getting complete information from your sources.
- **Follow up with additional research.** Look for additional information about your source and about the issue.

"British rule in India was not malign [or] needlessly cruel. . . . [T]he purpose of British rule was to educate Indians to be able to rule themselves and for the British to retire. . . . When freedom came, the British left us valuable legacies, which have come in very useful to us in ruling ourselves to some purpose."

—M.R. Masani, former opposition leader of the Indian Parliament

"[The British] tried to educate a certain middle class and allowed it all the facilities; but the basic reforms they did not carry out. Our literacy rates were so poor, and our technology has taken years to catch up with modern developments. . . . They needn't have left us to chaos, as they did, and divided our country. That was the worst—the partition of India. That was criminal: all the poisonous weeds have grown on that. . . ."

—Aruna Asaf Ali, Indian nationalist leader

Practicing the Skill

Read the passages about the effect of British rule in India. Then answer the questions.

1. How reliable are the speakers as sources of information?
2. Summarize each speaker's position.
3. Is the information in this source primarily fact or opinion? Explain.
4. What evidence does each speaker present?
5. What other information might you need to gain a deeper understanding of the topic?

Applying the Skill

Research a current South Asian issue in multiple sources—magazines, newspapers, or Internet sites. Analyze the usefulness of the articles as sources of information. Determine whether the articles present primarily facts or opinions. Also, evaluate the articles' validity, and identify any biases.

The **Glencoe Skillbuilder Interactive Workbook, Level 2** provides instruction and practice in key social studies skills.

CHAPTER 24 SUMMARY & STUDY GUIDE

SECTION 1 Population Patterns (pp. 587–591)

Terms to Know

- *jati*
- megalopolis

Key Points

- The population of South Asia reflects a rich and complex mix of religions, languages, and social groupings.
- South Asia has a high overall population density, but population distribution varies from region to region according to climate and terrain.
- There is a sharp contrast between urban and rural life in South Asia.

Organizing Your Notes

Create an outline using the format below to help you organize your notes for this section.

Population Patterns

I. Human characteristics
  A. India
    1. Descended from diverse groups
    2.
    3.
  B. Pakistan

SECTION 2 History and Government (pp. 592–597)

Terms to Know

- dharma
- reincarnation
- karma
- nirvana
- raj

Key Points

- One of the world's first civilizations developed in the Indus River valley.
- South Asia gave birth to two of the world's major religions, Hinduism and Buddhism.
- South Asia was shaped by a series of invasions and conquests, including the expansion of the British Empire into the region.
- South Asian countries today face the challenges of independence and establishing new governments.
- Several South Asian countries have had female leaders after becoming independent.

Organizing Your Notes

Use a table like the one below to help you organize important details from this section.

Country or Area	Early History	Government	Religions

SECTION 3 Cultures and Lifestyles (pp. 600–605)

Terms to Know

- guru
- mantra
- sadhu
- stupa
- *dzong*

Key Points

- South Asia is a land of many languages and religions.
- The diverse cultures of South Asia have made rich contributions to the arts.
- South Asia faces the challenge of improving the quality of life for much of its population.
- Even with the challenges it faces, South Asia benefits from its cultural diversity.

Organizing Your Notes

Use a web diagram like the one below to help you organize your notes for this section.

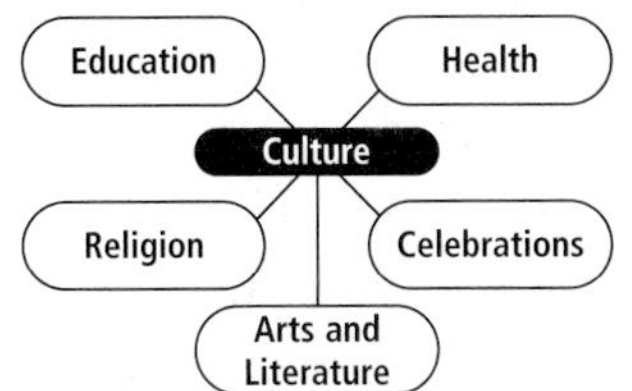

ASSESSMENT & ACTIVITIES

Reviewing Key Terms

Write the letter of the key term that best matches each definition below.

a. *jati*	**d.** reincarnation
b. megalopolis	**e.** guru
c. karma	**f.** *dzong*

1. a social group that defines a person's occupation and standing in the community
2. a teacher
3. rebirth
4. fortified monasteries
5. a large metropolitan area
6. good or bad deeds and their effects

Reviewing Facts

SECTION 1

1. How do many Indians traditionally identify themselves?
2. Where are the highest population densities in South Asia found?
3. What economic activities are important to Mumbai? What economic activities are important to Kolkata?

SECTION 2

4. Where did the first South Asian civilization develop?
5. Which ancient South Asian empire was one of the most advanced civilizations in the world?
6. Which European power ruled much of South Asia until the mid-1900s?

SECTION 3

7. Name the primary languages of Pakistan, Bangladesh, and northern India.
8. What religion teaches nonviolence and holds that every living being has a soul?
9. What art form is a major industry in India?

Critical Thinking

1. **Making Generalizations** How has the physical geography of South Asia contributed to the development of diverse cultures?
2. **Drawing Conclusions** What are some religious influences on South Asia's peoples?
3. **Categorizing Information** Create a time line showing important dates and events in South Asian history.

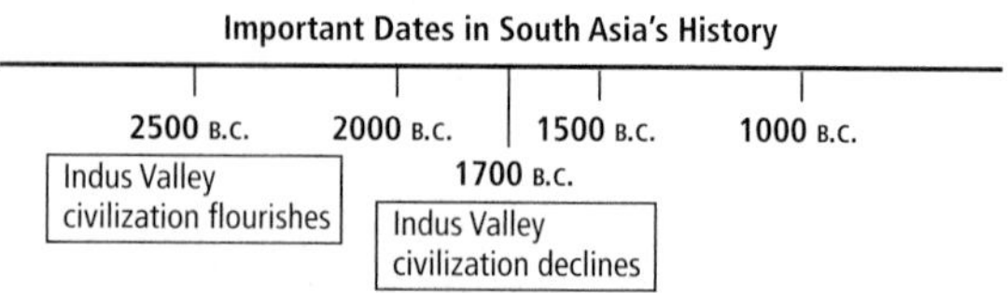

NATIONAL GEOGRAPHIC

Locating Places

South Asia: Physical-Political Geography

Match the letters on the map with the places and physical features of South Asia. Write your answers on a sheet of paper.

1. **Sri Lanka**	4. **New Delhi**	8. **Karachi**
2. **Mumbai (Bombay)**	5. **Nepal**	9. **Bhutan**
3. **Kolkata (Calcutta)**	6. **Hindu Kush**	10. **Dhaka**
	7. **Islamabad**	

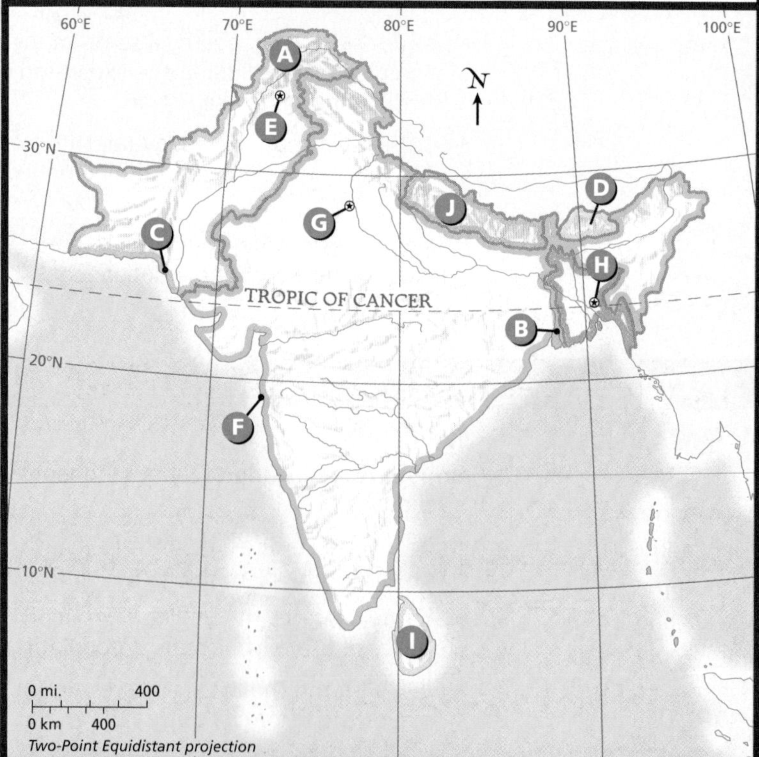

Self-Check Quiz Visit the **Glencoe World Geography** Web site at geography.glencoe.com and click on Self-Check Quizzes—Chapter 24 to prepare for the Chapter Test.

Using the Regional Atlas

Refer to the Regional Atlas on pp. 560–563.

1. **Location** Where are most of South Asia's largest cities located?
2. **Human-Environment Interaction** What physical features might account for the areas of low population density in the northern part of the region?

Thinking Like a Geographer

What factors, including physical geography, helped the process of diffusion of South Asian cultural influences to other parts of the world?

Problem-Solving Activity

Problem-Solution Proposal Using information from your text, the school library, or the Internet, write a report that proposes a solution to one of the following problems of South Asia: urban population density, conflicts in Sri Lanka or the Kashmir region, or nutrition and health. Your report should include an outline of the problem, recommendations for a solution to the problem, and a course of action. Design and draw graphic elements, such as charts, as needed.

GeoJournal

Persuasive Writing Using your GeoJournal data, write a short speech urging American high school students to become familiar with South Asian history and culture. Use descriptive language, appropriate vocabulary terms, and powerful verbs to convince your audience of the value of learning about South Asia.

Technology Activity

Developing Multimedia Presentations Work with a team to develop a multimedia presentation on one aspect of South Asia's history or culture. (Examples include the Gupta Empire, Indian dance, or Buddhism in Bhutan.) Use reference works and the Internet to develop your presentation. Present your work to the class.

Standardized Test Practice

Read the passage and choose the best answer for the following multiple-choice question. If you have trouble answering the question, use the process of elimination to narrow your choices.

"Raindrops keep falling on the head of anyone who takes a summer trip to Mawsynram, a hill town in northeast India. And falling. And falling. Two Indian meteorologists claim, and many U.S. specialists agree, that Mawsynram has ousted Hawaii's Waialeale as [the] earth's wettest spot, measured by average annual rainfall. Mawsynram gets an average 467.44 inches of rain a year, compared with 459.99 for Waialeale."

"Geographica," *National Geographic*, May 1993

1. **According to the reading, which of the following statements is NOT a fact?**

   A The average annual rainfall in Waialeale, Hawaii, is less than that of Mawsynram, India.

   B Mawsynram and Waialeale are wetter than other places on Earth.

   C All Indian and U.S. meteorologists agree that the average annual rainfall in Mawsynram, India, is greater than that in Waialeale, Hawaii.

   D Many U.S. specialists believe that with 467.44 inches of rain annually, Mawsynram, India, is Earth's wettest spot.

Learn to distinguish facts from nonfacts. Sometimes nonfacts contain phrases such as *I believe* or *in my view* or broad generalizations such as *every*, *all*, or *never*.

CHAPTER 25

South Asia Today

GeoJournal

As you read this chapter, use your journal to record information about economic activities and environmental issues in South Asia. Be sure to include details that illustrate each activity or issue.

GEOGRAPHY *Online*

Chapter Overview Visit the **Glencoe World Geography** Web site at geography.glencoe.com and click on Chapter Overviews—Chapter 25 to preview information about South Asia today.

SECTION 1

Living in South Asia

Guide to Reading

Consider What You Know

South Asia is world-renowned for its many fine fabrics—soft pashminas and cashmeres, bright cottons, and finely spun silks. What other items from South Asia might you find in stores in your community?

Read to Find Out

- How does agriculture provide a living for most of South Asia's people?
- What role do fisheries and mines have in South Asian economies?
- Where in South Asia is rapid industrial development taking place?
- What issues are raised by tourism in South Asia?

Terms to Know

- cash crop
- jute
- green revolution
- biomass
- cottage industry
- ecotourism

Places to Locate

- Bangalore
- Chittagong
- Hyderabad

◀ *Hindu temple, Delhi, India*

NATIONAL GEOGRAPHIC

A Geographic View

Ancient Rhythms

Despite . . . signs of change, much of Bhutan remains as it has always been, an unspoiled land of farmers and herders of yaks and cattle. Some 90 percent of Bhutanese live . . . as their [ancestors] did, following livestock through the high summer meadows, planting plots of rice and chiles in the valleys. People like . . . a woman I met in the northern village of Soe . . . still follow the ancient rhythms. . . . Together we watched pine smoke curl from her kitchen fire, sipped warm bowls of yak-butter tea, and talked about the sorts of things that concern farmers everywhere—the price of meat, the cost of clothing, [and] the health of the herd.

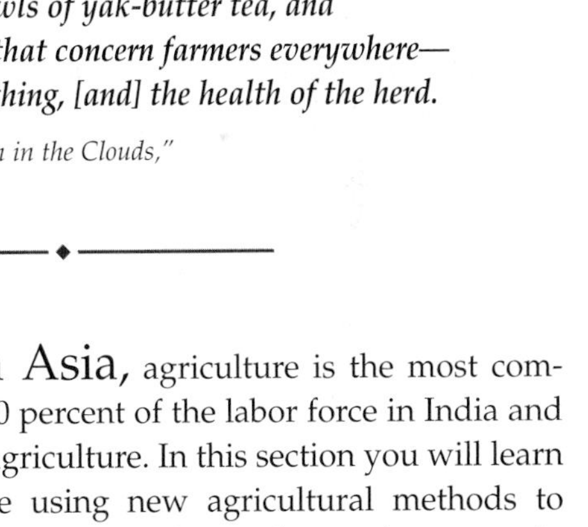

Bhutanese farmer drying chiles

—*Bruce W. Bunting, "Bhutan: Kingdom in the Clouds,"* National Geographic, *May 1991*

Throughout South Asia, agriculture is the most common occupation. More than 60 percent of the labor force in India and Bangladesh are employed in agriculture. In this section you will learn how South Asians today are using new agricultural methods to increase food production. You will also learn about other ways in which the peoples of South Asia earn a living.

Living From the Land

Most people in South Asia practice subsistence farming. Subsistence farmers often rely on labor-intensive farming methods. They may use digging sticks or hand plows to break up the soil, and they often sow seed by hand. To water their crops, farmers may hand-carry water for miles from a well or river, although some areas have irrigation systems.

Subsistence farmers also use animal power. Oxen and water buffalo pull wooden plows, carry heavy loads, and turn simple waterwheels for irrigation and mills for grinding grain. South Asians also use yaks, the long-haired cattle that flourish at high elevations; camels in desert areas; and elephants, which can do the heavy work of a tractor.

Farming depends on many changeable factors, such as rainfall, that are beyond the farmers' control. A family can lose its entire food supply in one season of drought, or crops might be eaten by wild animals. Even with the risks, however, subsistence farming allows many South Asians to be economically independent.

Agricultural Conditions

Farms in South Asia vary widely in size and appearance, based on geographic, historic, and cultural factors. In the Himalayan highlands of Nepal and Bhutan, farmers practice terracing, making use of every available inch of arable land on the steep slopes. Fruit orchards line the fertile highland valleys of Pakistan. In most of Bangladesh's delta region and along many of South Asia's great rivers, farmers work in water above their knees to grow rice. Farms in India are generally very small, with over one-third of them covering less than an acre.

History

Sri Lanka's Plantations

India's tiny farm plots stand in sharp contrast to the huge tea, rubber, and coconut plantations where many Sri Lankans work. British and Dutch colonizers established these large, technically sophisticated agricultural operations. The British moved their tea plantations from India to Sri Lanka (then called Ceylon) when Indian workers demanded better working conditions.

Although the European planters left Sri Lanka when the country gained its independence from the United Kingdom in 1948, plantations continue to employ about three-fourths of Sri Lanka's workers. The profitable plantations leave little land for growing crops to feed the country's own people, however, so Sri Lanka must import large quantities of basic foods, such as rice.

South Asian Crops

Cash crops bring much-needed income to South Asia. The tea, rubber, and coconuts of Sri Lanka are **cash crops**, farm products grown for sale or export. India also grows large quantities of cashews, coffee, and tea for export. Tea plants grow well in northeastern India's temperate highlands. However, balancing the physical needs of hungry people with the economic needs of growing countries is a challenge to the region.

Cotton is a key cash crop in South Asia. India and Pakistan are among the world leaders in cotton production. **Jute**, a fiber used to make string, rope, and cloth, is the major cash crop of Bangladesh and is grown mainly in the western lowlands bordering India. Sales of this fiber, called the "golden crop" for its color and value, account for a large part of Bangladesh's export income, although demand for jute is decreasing.

NATIONAL GEOGRAPHIC **World Explorer**

Geography Skills for Life

Cotton Production Workers pile cotton by hand to be stored in outdoor warehouses in India.

Human-Environment Interaction Why is it important for countries to raise other crops in addition to cash crops?

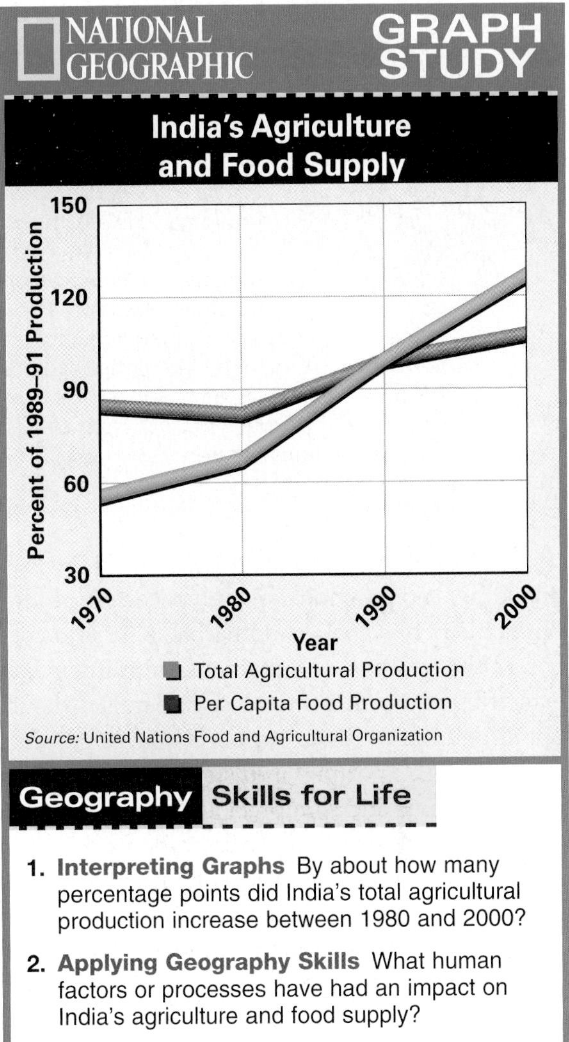

1. **Interpreting Graphs** By about how many percentage points did India's total agricultural production increase between 1980 and 2000?
2. **Applying Geography Skills** What human factors or processes have had an impact on India's agriculture and food supply?

India is one of the world's largest producers of bananas. Citrus fruits, chiles, and spices are grown for export in the steppe areas of India, Pakistan, and Bangladesh.

Grains provide South Asia with important food sources as well as profitable exports. Rice, the major food crop of South Asia, grows in the tropical rain forest climate of the Ganges Delta and along the peninsula's western Malabar Coast. India is second only to China in rice production, and Bangladesh ranks fourth in the world. Wheat is the main crop in the western Ganges Plain (Indo-Gangetic Plain) and in Pakistan's Indus River valley, but millet, corn, and sorghum also grow there. Peanuts grow along the Malabar Coast and the southern Deccan Plateau, and farmers grow sugarcane in most of India's lowlands.

Agricultural Improvements

Even with some success in slowing the population growth, feeding South Asia's people is an enormous challenge. Farmers are being trained to use modern technology and methods for irrigation, pest control, and fertilization to increase productivity. More planting cycles, for example, have been successful in Bangladesh, where farmers usually can harvest three rice crops per year. In Nepal's Kathmandu Valley, farmers are planting and harvesting winter wheat following the rice harvest.

Educational and governmental efforts have increased agricultural productivity. Research stations in Bhutan, for example, have helped farmers establish fruit orchards, and government-funded irrigation systems and higher rice prices encourage Sri Lankan farmers to grow more food crops.

The Green Revolution

Since the 1960s, an effort known as the **green revolution** has sought to increase and diversify crop yields in the world's developing countries. In India, as elsewhere, the green revolution has involved using carefully managed irrigation, fertilizers, and high-yielding varieties of crops. As a result, India's wheat and rice production has greatly increased. India is now able to store—and even export—grain. Not all the new methods work everywhere in South Asia, however. In parts of the region, monsoon rains allow only one planting cycle per year. Modernization also has costs. Irrigation and mechanization require expensive fuel, and in a region where not enough petroleum is available and many people burn **biomass**—plant materials and animal dung—as their only energy source, the costs are often too high.

Mining and Fishing

In addition to farming the soil, South Asians reap benefits from other natural resources in the region. Mining and fishing are profitable industries with the potential for growth in years to come.

Mineral Wealth

The Ganges Plain and parts of eastern India yield some of South Asia's richest mineral deposits. Iron ore, low-grade coal, bauxite, and copper are all

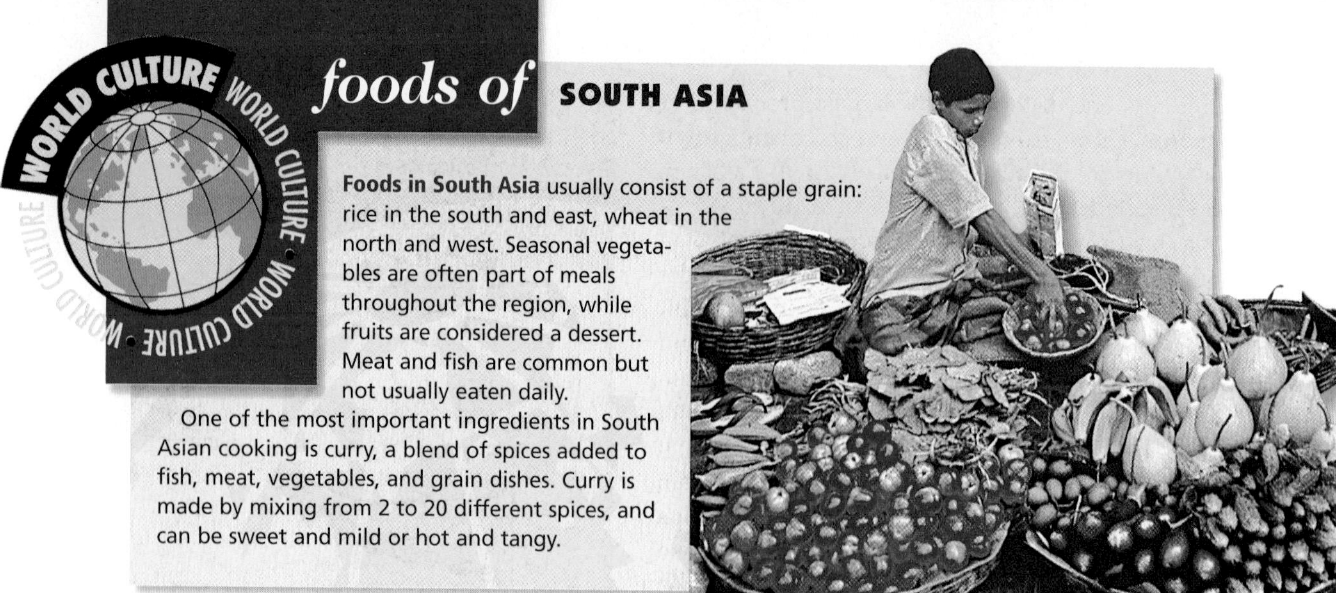

foods of SOUTH ASIA

Foods in South Asia usually consist of a staple grain: rice in the south and east, wheat in the north and west. Seasonal vegetables are often part of meals throughout the region, while fruits are considered a dessert. Meat and fish are common but not usually eaten daily.

One of the most important ingredients in South Asian cooking is curry, a blend of spices added to fish, meat, vegetables, and grain dishes. Curry is made by mixing from 2 to 20 different spices, and can be sweet and mild or hot and tangy.

mined in mountainous areas of eastern India. Bhutan is rich in coal, lead, marble, zinc, and copper, although its mountainous terrain makes extracting and processing these minerals difficult. The island of Sri Lanka supports a large graphite mining and exporting industry. Sri Lanka also mines precious and semiprecious stones.

Petroleum and natural gas reserves are found in several South Asian countries. India's oil fields are concentrated in the northeastern and northwestern areas of the country. Oil fields located in northeastern and southern Pakistan supply much of the country's energy needs. Pakistan also has significant natural gas reserves, especially in the western state of Baluchistan. Bangladesh, too, is rich in natural gas, a resource that offers an as-yet-untapped potential for export income to supplement the declining market for jute.

Fishing

Bordered partly by oceans and watered by great rivers, South Asia has rich fishing resources that provide needed income. Pakistan and Sri Lanka export shrimp, lobsters, and fresh and dried fish. Many people in India and Bangladesh fish for food. Bangladesh also has growing commercial fisheries, producing shrimp and frogs' legs for export.

In recent years the Indian government has encouraged deep-sea fishing by building processing plants and giving aid to oceangoing ships and fleets. More traditional local fishers see these developments as a threat to their livelihood.

South Asian Industries

Industrialization has proceeded along very different time lines in various South Asian countries. In India, industrialization began under British rule and was funded by European companies. In contrast, Bhutan, closed off from the outside world until 1975, still remains relatively isolated. Bhutan's government is moving ahead with industrial development slowly in order to preserve the country's natural and cultural resources.

Economics

India's Evolving Economy

After gaining independence in 1947, India introduced socialism, an economic policy that emphasized central planning. The government set goals for and closely regulated private industry. Many large industries were placed under direct government control, while others were partnerships between private owners and the government.

Wary of outside influences, India turned its back on foreign investment. It expanded home industries and reduced dependence on foreign trade to promote self-sufficiency. At first growth was steady, but by the 1960s the economy slowed, and India began to see the limitations of its policies in an increasingly global economy. Still, change came slowly. In the late 1980s, India's government still regulated or operated mining, banking and insurance, transportation, manufacturing, and construction industries. Then, in 1991, a financial crisis pushed India toward major economic reforms. It began moving toward a market economy.

> *"In 1991 India began opening its economy to wider trade, and the United States quickly became its primary trading and investment partner. . . . Foreign companies were thrilled by sheer numbers—an estimated 150 million potential middle-income consumers. . . . [F]oreign companies have also brought better job opportunities. . . ."*
>
> Erla Zwingle, "A World Together," *National Geographic*, August 1999

The government also deregulated many industries and turned over government-run companies to private ownership. These changes sparked economic growth that helped expand the middle class, which was believed to make up 20 to 25 percent of India's population by the late 1990s. As a result, the demand for consumer goods from shoes to luxury cars has expanded rapidly. Today India, along with the rest of South Asia, struggles to balance national interest and global interdependence.

Light Industry

Many South Asians work in light industry, producing consumer goods. Textiles are a major part of South Asia's manufacturing base, as they have been in India for hundreds of years. India's 25 million textile workers manufacture cotton, silk, and wool fabrics in a dazzling variety of patterns, colors, and styles. India's textile industry, centered in Mumbai, Nagpur, and Sholapur, also produces garments for export. Bangladesh entered the textile industry in 1979, and sales of finished garments provide the country with export income.

Some of the world's most prized wools—cashmere and pashmina—come from a rare breed of goat found only in the Jammu and Kashmir region. Used in high-quality, high-fashion garments, these wools are in great demand.

Other light industries throughout South Asia manufacture shoes, carpets, bicycles, and bicycle parts. These small industries are generally housed in factories employing fewer than 100 people, and they use traditional production techniques.

South Asia's broad involvement in light industry grows out of its history of **cottage industries**, businesses that employ workers in their homes. Indian villagers weave textiles and make shoes, jewelry, woodcarvings, furniture, and bowls. Cottage industries in India, Nepal, and Bhutan provide jobs, encourage traditional crafts, and supply needed export income.

Mohandas Gandhi, the leader of India's independence movement, chose the spinning wheel as a symbol of the strength India could draw from its cottage industries. In his later years, Gandhi dressed only in simple robes of *khadi* (KAW•dee), traditional homespun cotton fabric,

Geography Skills for Life

Selling Shawls These women selling shawls in Bangladesh are a few of the millions of textile workers and merchants in South Asia.

Place How do cottage industries contribute to the export income of countries in the region?

and was often pictured sitting at the spinning wheel where he spun *khadi* thread. Gandhi urged the Indian people to maintain their traditional, family-centered industries even as the country developed.

Heavy Industry

South Asia's industrial base includes heavy industries geared toward mass production. India manufactures iron, steel, cement, and heavy machinery in Bhadravati and **Bangalore**. Bangladesh also produces iron, steel, and cement.

India, Pakistan, and Bangladesh also recycle iron and steel in a unique industry called "ship breaking." In Bangladesh's port of **Chittagong**, thousands of workers use sledgehammers and blowtorches to dismantle aging or damaged ships from around the world. Melted-down parts are reforged into new iron and steel. In Bangladesh alone, ship breaking and related industries employ more than 1.3 million people.

NATIONAL GEOGRAPHIC **World Explorer**

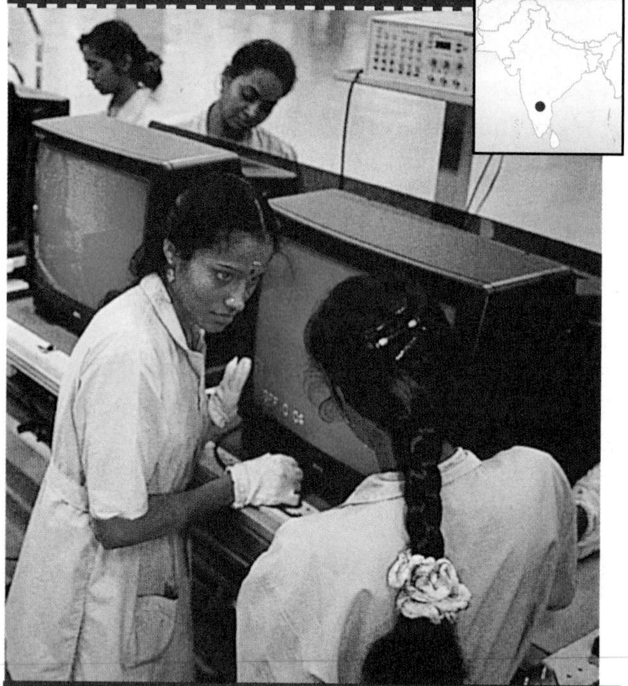

Geography Skills for Life

Technology Factories in India today produce a variety of electronic products such as televisions.

Place By how much is India's software trade with the United States expected to grow each year?

Service Industries

Since the late 1990s, service industries—transport, real estate, banking and insurance, and public administration—have become increasingly important in South Asia. India and, to a lesser extent, Pakistan have benefited the most. In India wholesale and retail trade and government services rank as the leading contributors to the country's service economy. The Indian government provides a variety of social services to its people, especially in health, education, and public administration.

The High-Technology Sector

High technology—including the manufacture of equipment for the computer, communications, and aerospace industries and the creation of computer software—is a growing industry in South Asia. Millions of Indians use the Internet, and Indian computer professionals are in high demand around the world. The southern Indian cities of Bangalore and **Hyderabad** (HY•duh•ruh•BAD) are called "India's Silicon Valley," a reference to the part of California where many computer industries flourish. Software manufacturing in these cities has helped make India the world's second-largest exporter of software. In 2000 the software trade between India and the United States alone yielded $5 billion in income for India, with a projected growth rate of about 60 percent per year.

India also has strong potential to be a developer of computer hardware. The increasing use of copper rather than aluminum in microchip manufacturing benefits India because of its abundant copper deposits. India already has a strong and growing industry in the manufacture of televisions and other communications equipment.

Tourism

Tourism income is important in several South Asian countries. Nepal draws tourists to hike and climb the Himalayan slopes and to hunt or photograph wild animals. India's temples and festivals attract more than 2 million visitors each year.

Continuing conflicts may discourage tourists, however. Sri Lanka's lush rain forests and tropical beaches once drew many tourists, but since the 1980s, violence between Hindu Tamils and Buddhist Sinhalese has emptied luxury hotels. Ongoing border disputes between India and Pakistan have all but eliminated tourism in Kashmir. Sporadic violence among religious groups in India also has discouraged foreign visitors.

In some South Asian countries, governments regulate tourism to protect threatened natural and cultural resources. For example, to preserve the Himalayan environment and its traditional culture, Bhutan issues fewer than 5,000 tourist visas each year. The Maldives restricts tourists to certain islands so that tourists do not interact with Maldivians who follow strict Islamic customs. **Ecotourism**, a form of tourism that encourages responsible interaction with the environment and endangered species, can support preservation efforts while contributing to South Asian economies.

Student Web Activity Visit the **Glencoe World Geography** Web site at geography.glencoe.com and click on Student Web Activities—Chapter 25 for an activity about business and tourism in India.

NATIONAL GEOGRAPHIC **World Explorer**

Geography Skills for Life

Beachside Paradise Among the most beautiful islands in the world, the Maldives also supports a diverse marine life.

Place Why does the Maldives restrict tourist access to some islands?

SECTION 1 ASSESSMENT

Checking for Understanding

1. **Define** cash crop, jute, green revolution, biomass, cottage industry, ecotourism.
2. **Main Ideas** Create a table like the one below, and fill in economic activities in South Asia and the challenges each represents.

South Asian Economic Activity	
Activity	Challenges

Critical Thinking

3. **Comparing and Contrasting** Compare and contrast cottage industries and commercial industries in the ways of operation, especially in regard to India's economy.
4. **Categorizing Information** Which of the region's industries focus on domestic needs, and which focus on exporting?
5. **Predicting Consequences** How might increased tourism affect life in the region?

Analyzing Maps

6. **Movement** Using the economic activity map on page 563, identify areas where nomadic herding is common. Explain why herding is the dominant economic activity in each of these areas.

Applying Geography

7. **Effects of Physical Geography** Think about farming methods in South Asia. Explain how farming methods are influenced by the region's physical geography.

SECTION 2

People and Their Environment

Guide to Reading

Consider What You Know

Tourism's growth rate in India is higher than the world average, and the Indian government considers tourism a high-priority industry. What effects do you think increased tourism might have on India's environment?

Read to Find Out

- How is South Asia handling the complex task of managing its rich natural resources?
- What environmental challenges does South Asia face in the years ahead?
- How do geographic factors impact the political and economic challenges of South Asia's future?

Terms to Know

- sustainable development
- poaching
- *Chipko*
- nuclear proliferation
- *Dalits*

Places to Locate

- Narmada River
- Bay of Bengal

NATIONAL GEOGRAPHIC

A Geographic View

A Threatened Treasure

I have been climbing since well before dawn, and now I am alone at 17,000 feet. . . . Around me in a vast arc stand the snowy crests of the majestic Annapurna Range. The day is cloudless, not a breath of wind. The solitary splendor is dazzling—until I glance down at my feet. There, frozen into the ice cap of Tharpu Chuli, lies a miniature garbage dump: discarded candy wrappers, film cartons, plastic bags, wads of tissue, and half-empty food cans, all of it left by foreign climbing groups. It is a familiar and sickening sight to old Himalaya hands—the growing pollution of a priceless heritage.

—Galen Rowell, "Annapurna: Sanctuary for the Himalayas," National Geographic, *September 1989*

Annapurna Range, Nepal

The tourism generated by trekking the Himalayan trails brings needed income to the kingdom of Nepal, but it also endangers the Himalayan ecosystem on which the entire Indian subcontinent depends. In this section you will learn about environmental and other challenges faced today by the countries of South Asia.

Managing Natural Resources

As you have learned, people and the environment interact and affect each other throughout the world. This interaction is especially significant in South Asia, where high population densities meet fragile ecosystems. As a result, South Asian countries seek to manage their resources wisely rather than just using them. A key to successful resource management is sustainable development, or using resources at a rate that does not deplete them for future generations.

Wildlife

South Asia is home to an astonishing variety of wildlife. Elephants, water buffalo, and monkeys flourish in the rain forests of India and Sri Lanka. Crocodiles and Bengal tigers roam in Bangladesh. In the high mountain passes of the Himalaya, the elusive snow leopard hunts alone above fields crowded with blue sheep, exotic birds, and rare butterflies.

The Hindu, Buddhist, and Jain traditions of South Asia promote respect for all living things. However, many of South Asia's animals have become endangered through contact with the region's growing human population. Deforestation and irrigation have reduced animals' natural habitats, driving them into areas where people live. Some animals have been overhunted by tourists or by farmers and herders seeking to protect their crops and flocks.

Governments in the region, assisted by international conservation organizations, are working to reverse some of South Asia's wildlife losses. The creation of wildlife reserves—protected habitats—and the passage of laws controlling hunting and logging have begun to make a difference. Providing South Asians with economic incentives to cooperate in conservation efforts may also be effective. However, challenges still remain. Farmers' crops often are threatened by foraging elephants, and poachers can realize huge profits by selling the hides of Bengal tigers. To eliminate poaching, or the illegal killing of protected animals, governments in the region need to find ways to encourage people to respect wildlife.

Water

Water is one of the most precious resources on the planet. Lack of access to clean water is a persistent problem in South Asia. Even in India, the most developed country in the region, 80 percent of the population has no access to sanitation facilities and must rely on water that is polluted by human waste and chemical runoff. The situation is worse in parts of Pakistan and in Bangladesh.

Because South Asia's climate varies greatly, the villagers of Rajasthan in northwestern India may be watching their crops and livestock perish from drought at the same time that farmers in Bangladesh are losing their homes to flooding. Building dams is one way to balance these extremes. Dams can change the course of rivers, reroute water for irrigation, and control flooding by holding water in reserve for times of drought.

Like many uses of technology, however, the building of dams has drawbacks as well as benefits. Dams trap silt that would otherwise flow downriver to enrich the soil. Reservoirs can trap bacteria, too, and become a source of disease. Also, building a dam usually results in the flooding of surrounding areas, displacing whole villages and disturbing the balance of wildlife and vegetation.

NATIONAL GEOGRAPHIC **World Explorer**

Geography Skills for Life

Bengal Tiger Some South Asian animals, such as this Bengal tiger, face extinction because of human activities such as poaching and clearing forests.

Human-Environment Interaction What steps are being taken to protect South Asia's wildlife?

NATIONAL GEOGRAPHIC **World Explorer**

Geography Skills for Life

Protesting the Dam
Activists concerned about environmental problems protest the building of a dam near Bhopal, India.

Human-Environment Interaction What are advantages and disadvantages of building dams?

Government

The Narmada River Dilemma

The 25-year effort to build a dam in India's **Narmada River** basin is a good example of the challenges of water management. Supporters of the project point out the benefits, including the irrigation of millions of acres of land currently subject to severe drought and the creation of hydroelectric power.

At the heart of the opposition to the project are environmentalists and the thousands of local peoples whose ancestral villages will be flooded as a result of the project. They point to other such projects in which farmers were uprooted and forced to resettle in cities or temporary camps. As work continues on the project, people on both sides of the controversy have begun talks to resolve their differences.

Forests

Centuries ago, much of South Asia was covered with forests. Today the region is in a state of environmental crisis because of deforestation. The problem has accelerated in recent years, driven by South Asia's growing population and the increasing interaction of humans with their environment. Commercial timber operations, an industry that began under British rule, have destroyed many of South Asia's old-growth forests. Other forest areas have been cleared to make way for human settlements.

Some deforestation is a result of traditional practices in South Asia. Slash-and-burn agriculture, no longer permitted in many places, is an ancient technique used by many hill peoples. In drought-stricken regions, villagers allow livestock to feed on leaves, slowly killing the trees. Most damaging of all is the widespread reliance on burning biomass, including the wood from trees, for fuel.

The effects of deforestation are devastating. The mangrove forests of Bangladesh's Sundarbans region, the area of swamp land near the Ganges River Delta, have over the years provided a barrier against erosion caused by cyclones. As the mangrove trees are cut, however, much of this protection against storms vanishes. Losing tropical rain forests also has other damaging effects. Rain forests usually grow in poor soil, where the trees' complex root systems efficiently absorb available nutrients and hold the topsoil in place. As rainfall filters slowly through layers of leafy branches, the surrounding air is cooled. When rain forests disappear, soil erodes, rains produce floods, and temperatures rise.

Culture

Protecting the Forests

Reforestation efforts, under way throughout South Asia, build on the region's traditional respect for trees. India's *Chipko*, or "tree-hugger," movement was founded by Sunderlal Bahaguna, a follower of Gandhi. Bahaguna has succeeded by reminding villagers of the importance of trees. *Chipko* nurseries provide seedlings for reforestation. The government ban on timber production in the Himalayan forests of Uttar Pradesh, advocated by Bahaguna, was the first of many such government efforts in the region.

Protecting forests is at the heart of the region's culture. As the poet Rabindranath Tagore wrote,

> *"India's civilization has been distinctive in locating its source of regeneration, material and intellectual, in the forest, not the city. India's best ideas have come when man is in communion with trees."*
>
> Rabindranath Tagore, quoted by Gita Mehta, *Snakes and Ladders: Glimpses of Modern India,* 1998

Seeking Solutions

As industrialization increases in South Asia, so does air pollution. Delhi, India, is now the world's fourth most polluted city. Scientists are studying the region to try to solve this and other problems.

Meteorologists are studying monsoon patterns in the **Bay of Bengal** in the hopes of reducing the devastation caused by these storms. The ability to predict with some accuracy the coming of the monsoon rains and their intensity could make enormous differences to South Asia's people.

Geographers are using satellite imaging to study the erosion in coastal deltas in Bangladesh. Millions of Bengali people live on thin, crusted islands formed from silt, which float on the surface of coastal waters. When the rains come, the rivers move silt—2 billion tons (1.8 billion metric tons) a year—into the Bay of Bengal. As a result, the average Bengali is displaced from his or her home seven times in a lifetime. If studies of silt erosion can identify solutions, scientists will be improving people's lives.

Finally, South Asia has the potential to help study global environmental issues. For example, an experimental station in the Maldives is measuring the possible effects of global warming on ocean levels.

NATIONAL GEOGRAPHIC **World Explorer**

Geography Skills for Life

Erosion

Severe flooding has eroded the bank near this woman's home in Bangladesh.

Human-Environment Interaction What factor makes erosion such a problem in Bangladesh?

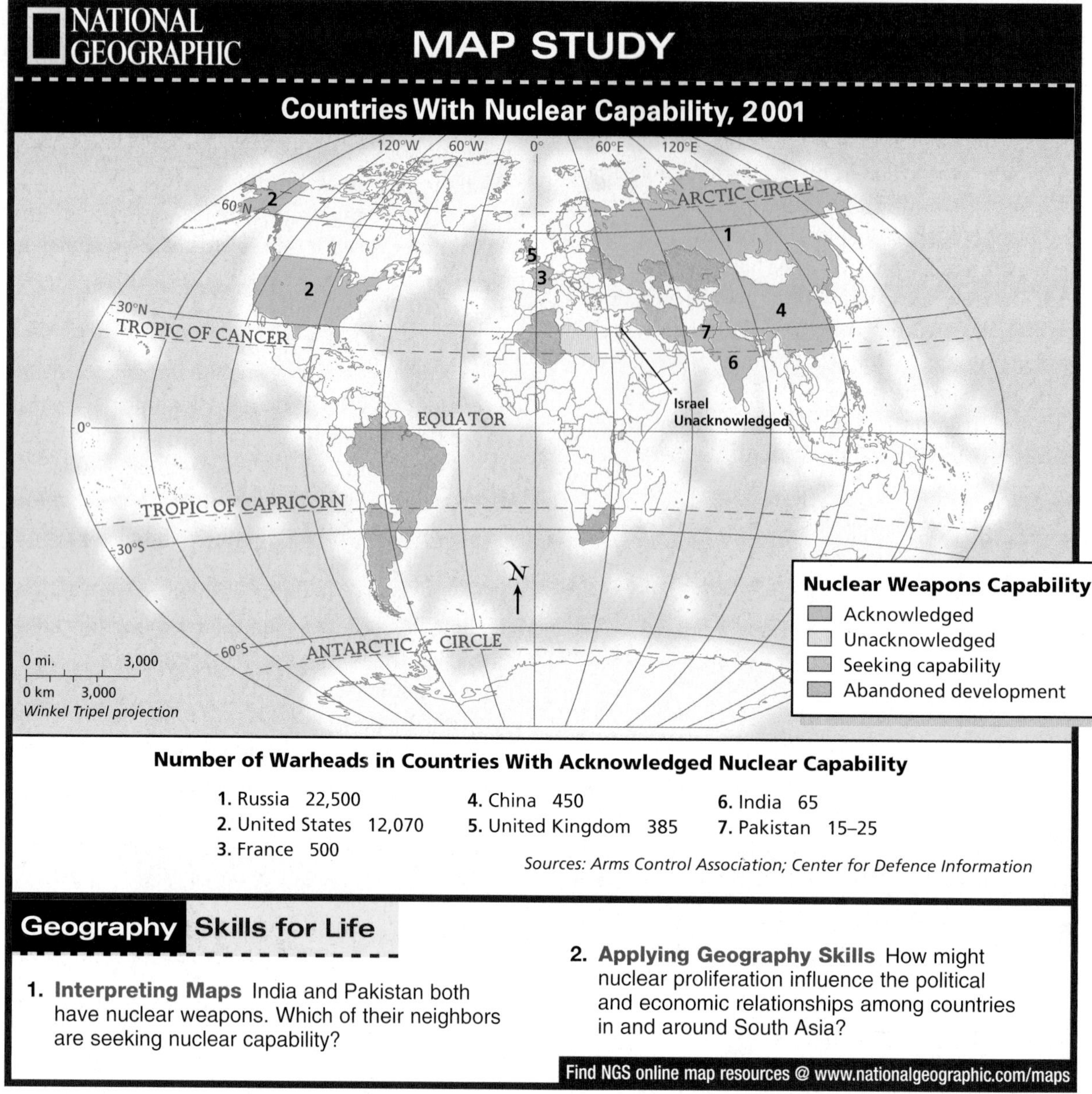

Number of Warheads in Countries With Acknowledged Nuclear Capability

1. Russia 22,500
2. United States 12,070
3. France 500
4. China 450
5. United Kingdom 385
6. India 65
7. Pakistan 15–25

Sources: Arms Control Association; Center for Defence Information

Geography Skills for Life

1. **Interpreting Maps** India and Pakistan both have nuclear weapons. Which of their neighbors are seeking nuclear capability?
2. **Applying Geography Skills** How might nuclear proliferation influence the political and economic relationships among countries in and around South Asia?

Find NGS online map resources @ www.nationalgeographic.com/maps

South Asia's Challenges

Geography holds a key to other challenges now facing South Asia. Conflict in the region has deep roots in issues of national autonomy and religious and ethnic concerns.

Conflict in Kashmir

Since 1947, India and Pakistan have disputed ownership of the largely Muslim territory of Kashmir. During the past 50 years, two of the three wars fought between India and Pakistan have focused on this territory. Today, Pakistan controls one-third of Kashmir; the remainder is held by India.

Indian and Pakistani troops patrol the Line of Control, the border between the two parts of Kashmir. Despite occasional peace talks, India and Pakistan accuse each other of violating this border. India also claims that Pakistan supports armed groups in Indian-ruled Kashmir that want an independent Kashmir.

The potential danger from this enduring conflict has escalated since 1998, when both India and Pakistan tested nuclear warheads. The map above shows the world's countries that have nuclear capability, including India and Pakistan. Both countries have spent huge sums to develop nuclear missiles. This new example of **nuclear proliferation**—the spreading development of nuclear arms—aroused international alarm. The costs of these nuclear programs hurt the people of both countries through the loss of much-needed funding for food and other human needs. In addition, economic sanctions leveled by the world's economic powers against India and Pakistan intensified the hardships of South Asia's people.

Internal Conflicts

Some South Asian conflicts occur within countries. The majority of people in Sri Lanka are Buddhist Sinhalese, who control the government. Tamils, who are Hindu, represent only about 20 percent of the population. Tamils accuse the government of discrimination, and some have taken up arms to create a separate Tamil state. In India differences between Hindu, Muslim, and Sikh militants often erupt into violence.

India also suffers from the legacy of its ancient system of social classes. Those traditionally assigned to the lowest social status—called the ***Dalits***, or "oppressed"—continue to experience discrimination and even, in some areas, violent assault. *Dalits* are denied housing, educational opportunities, and jobs, even though India's constitution outlaws such discrimination.

Promise and Possibility

South Asia's history of conflict rests side-by-side with its long tradition of tolerance for diversity. On the fiftieth anniversary of his country's independence, one Indian writer posed this challenge for the future. His comments could also apply to the rest of South Asia:

> *"In a country as diverse as India, the interests of various groups of Indians will tend to diverge, and political contention is inevitable. The major challenge for Indian democracy is therefore to absorb and resolve the clashes that may arise from contending interests, while ensuring the freedom, safety, and prosperity of all Indians."*
>
> Shashi Tharoor, *India: From Midnight to the Millennium*, 1997

SECTION 2 ASSESSMENT

Checking for Understanding

1. **Define** sustainable development, poaching, *Chipko*, nuclear proliferation, *Dalits*.
2. **Main Ideas** Create a web like the one below on a sheet of paper. Use it to fill in the information about how natural resources and conflicts present challenges in South Asia.

Critical Thinking

3. **Making Decisions** Which of the region's resource issues do you think should receive the most funding and attention? Explain your choice.
4. **Comparing and Contrasting** List examples from your reading to contrast the region's tolerance for diversity with its ongoing religious and ethnic conflicts.
5. **Making Inferences** In what ways does nuclear proliferation further complicate the already intense conflicts in South Asia? Give examples to support your answer.

Analyzing Maps

6. **Location** Study the map of countries with nuclear capability on page 622. On which continent are the most nuclear warheads located? The most countries with nuclear capability?

Applying Geography

7. **Writing a Letter** Imagine you are a local official writing to a South Asian government about a village hard hit by floods or drought. Analyze the environmental impact and suggest ways to resolve the problem.

Viewpoint

CASE STUDY on the Environment

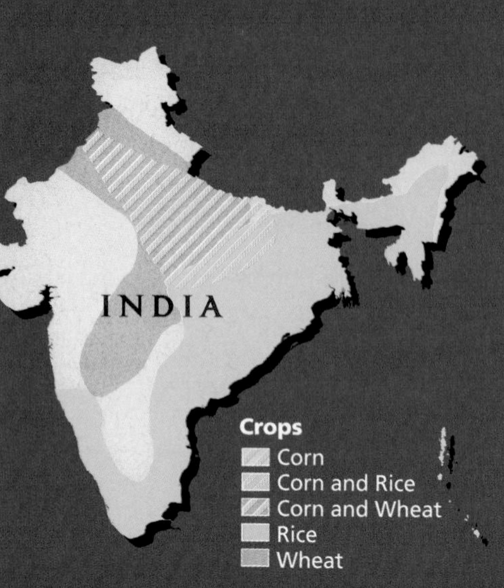

India's Green Revolution: Success or Failure?

Several decades ago, India's people were dying of starvation. Then came the green revolution. Hailed as the solution to India's chronic food shortages, the green revolution was an international effort to increase food production in less-developed countries. Starting in the 1960s, Indian farmers planted high-yield varieties of crops and used large amounts of fertilizers and pesticides to help the plants grow. By the 1970s, India was producing record harvests. Yet India's green revolution also has caused environmental damage and dependence on costly chemicals. Is India's green revolution a success or a failure?

At India's Rice Research Institute, scientists experiment with various methods of growing rice. The Institute's research aids Indian farmers (left) by introducing them to green revolution agricultural techniques.

The green revolution was designed to increase agricultural production and end hunger. In India, green revolution techniques encouraged farmers to turn more fields into cropland, raise more than one crop per year, and plant new high-yield variety (HYV) seeds—mainly wheat and rice.

India's green revolution has worked as people hoped it would. Grain harvests have soared, and India no longer imports grain. Yet there are problems. Compared with older strains of wheat and rice, new high-yield varieties need far more water, fertilizer, and pesticides to flourish. Huge irrigation projects deliver water to thirsty HYV plants. In some places, this has led to a buildup of salt in the soil, damaging once-fertile fields. Poor Indian farmers often go into debt to pay for expensive chemicals. Overuse of pesticides has gradually poisoned the soil and water in some areas. The chemicals also have led to pesticide-resistant crop pests.

India's growing population of one billion is the second-largest in the world. The country faces a critical decision: Should it continue to rely on green revolution technology?

Supporters of the green revolution point out that its techniques dramatically increased food production and alleviated hunger in India. They claim that new genetically engineered seeds will produce even higher yields. These new varieties will be resistant to pests, reducing the need for pesticides. Supporters also say that despite some exceptions, the green revolution has helped most farmers earn more money and raise their standards of living.

◀ Thanks to the green revolution, fewer children in India starve today.

Opponents of the green revolution argue that the new methods caused much environmental damage and widened the gap between rich and poor. Excessive use of chemicals pollutes water, poses health hazards, and leads to pest resistance. Opponents point out that farmers get caught in a cycle of using more and more chemicals to achieve healthy crops. Furthermore, some scientists warn that genetically modified seeds carry unknown risks and may create new environmental problems.

What's Your Point of View?
Should India continue to practice and improve upon green revolution techniques? Or should the country seek a new approach?

Indian farmers (below) harvest an abundant crop of rice. But abundance comes with a price—reliance on spraying crops with costly pesticides (right). ▼

Predicting Consequences

Making educated guesses about the outcome or consequences of an event or situation is useful in almost every area of life. Making good educated guesses is essential for successful decision making, problem solving, and planning.

Learning the Skill

Consequences are the results of actions or choices we make. For example, scoring high on an exam is one likely consequence of studying the night before. Often, predicting consequences is not so straightforward. An action or decision can have far-reaching or unintended consequences. One country's decision to provide aid to farmers, for instance, could lower the price of produce across an entire region. This would make it more difficult for farmers in other regions to compete.

Follow these steps to help you analyze information in order to predict consequences:

- **Gather information about the decision or action.**
- **Use your knowledge of history and human behavior to identify what consequences could result.**
- **Analyze each of the consequences by asking:** How likely is it that this will occur?
- **Determine whether this consequence will have other important consequences.**
- **Make a prediction using the information you have gathered.**

Practicing the Skill

Read and study the passage, and then answer the questions that follow.

"India is counting on information technology to create millions of new jobs and add billions of dollars to export earnings in the coming years. Yet it appears that unless more training and investment in education is made available, there may not be enough skilled workers to meet these ambitious goals. . . .

Most of the top schools producing computer and software workers send their graduates abroad to the United States and Europe, where Indian high technology professionals are in high demand. . . . Indian companies say they cannot compete with their counterparts abroad in salaries or benefits. . . . This is all happening while India's unemployment rate among unskilled workers is soaring. Economists say the only answer is a massive investment in primary and secondary education. But, with the Indian budget deep in deficit for years to come, it is hard to see where the money is going to come from."

—*Daniel Lak, "India at Risk of Tech Worker Shortage,"*
BBC News *(online), April 15, 2000*

1. What trend does the passage describe?
2. Do you think the trend the writer describes is likely to continue?
3. On what do you base this prediction?
4. What occurrences might have an effect on changing the trend?
5. What are three possible consequences or outcomes of this trend?
6. What are the possible benefits and drawbacks of the solution proposed by economists who study the issue?

Applying the Skill

Find a newspaper or news magazine article that describes a political, economic, or social problem in South Asia. Analyze the article, and describe how the people of South Asia are trying to solve the problem. Predict three consequences of the actions described. On what do you base your prediction?

Go To The **Glencoe Skillbuilder Interactive Workbook, Level 2** provides instruction and practice in key social studies skills.

25 SUMMARY & STUDY GUIDE

SECTION 1 Living in South Asia (pp. 611–617)

Terms to Know

- cash crop
- jute
- green revolution
- biomass
- cottage industry
- ecotourism

Reviewing Key Points

- Agriculture provides a living for most of South Asia's people, and it also provides cash crops for export.
- South Asia's mines and fisheries contribute to its exports.
- South Asia is experiencing rapid growth in the high-tech sector and continues to develop light and heavy industries.
- Tourism offers both benefits and challenges to the South Asian economy.

Organizing Your Notes

Create an outline, using the format below, to help you organize your notes for this section.

Living in South Asia

I. Living From the Land
    A. Agricultural Conditions
        1.
        2.
    B.
        1.
        2.
II. South Asian Crops

SECTION 2 People and Their Environment (pp. 618–623)

Terms to Know

- sustainable development
- poaching
- *Chipko*
- nuclear proliferation
- *Dalits*

Reviewing Key Points

- South Asia faces the complex task of managing its rich and varied natural resources.
- South Asia is seeking scientific solutions to its environmental challenges.
- Conflict in South Asia stems from issues of nationalism, religion, and ethnicity.

Organizing Your Notes

Use a table like the one below to help you organize important details from this section.

Natural Resources	Solutions	Conflicts

Shopping district, New Delhi, India

ASSESSMENT & ACTIVITIES

Reviewing Key Terms

Write the key term that best completes each of the following sentences. Refer to the Terms to Know in the Summary & Study Guide on page 627.

1. The spread of nuclear weapons is called _____.
2. _____ is a type of tourism that encourages responsible interaction with the environment.
3. Using resources at a rate that does not deplete them is called _____.
4. _____ are India's lowest social class.
5. People making products such as jewelry or textiles at home are working in a(n) _____.
6. The movement to increase food productivity through the use of experimental high-yield crops is called the _____.

Reviewing Facts

SECTION 1

1. What kinds of agricultural methods are used in South Asia?
2. How do mining and fishing contribute to the region's economy?
3. What are the benefits and challenges of tourism to the region today?

SECTION 2

4. What are South Asia's key natural resources, and where are they located?
5. What are the conflicting issues over India's Narmada River dam?
6. What are the causes and effects of the Kashmir conflict?

Critical Thinking

1. **Making Generalizations** What would you say is the greatest challenge facing South Asia today?
2. **Predicting Consequences** What might be the results of ongoing nuclear proliferation in South Asia?
3. **Identifying Cause and Effect** Create a cause-and-effect diagram like the one below for each of South Asia's current environmental challenges. Then fill in the details about the causes and effects of each.

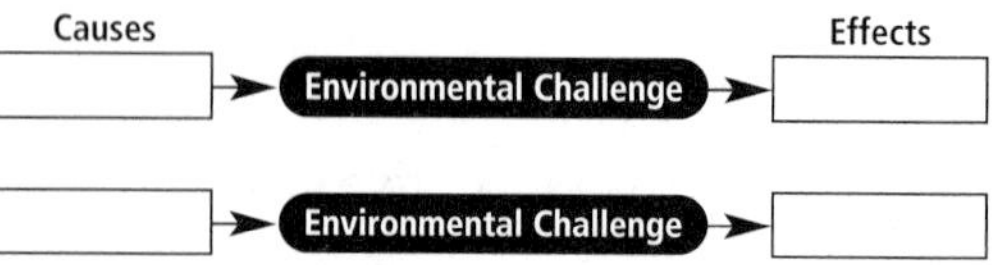

NATIONAL GEOGRAPHIC

Locating Places

South Asia: Physical-Political Geography

Match the letters on the map with the places and physical features of South Asia. Write your answers on a sheet of paper.

1. **Colombo**
2. **Lahore**
3. **Narmada River**
4. **Indus Valley**
5. **Western Ghats**
6. **Malabar Coast**
7. **Vindhya Range**
8. **Khyber Pass**
9. **Ganges Plain**
10. **Great Indian Desert**

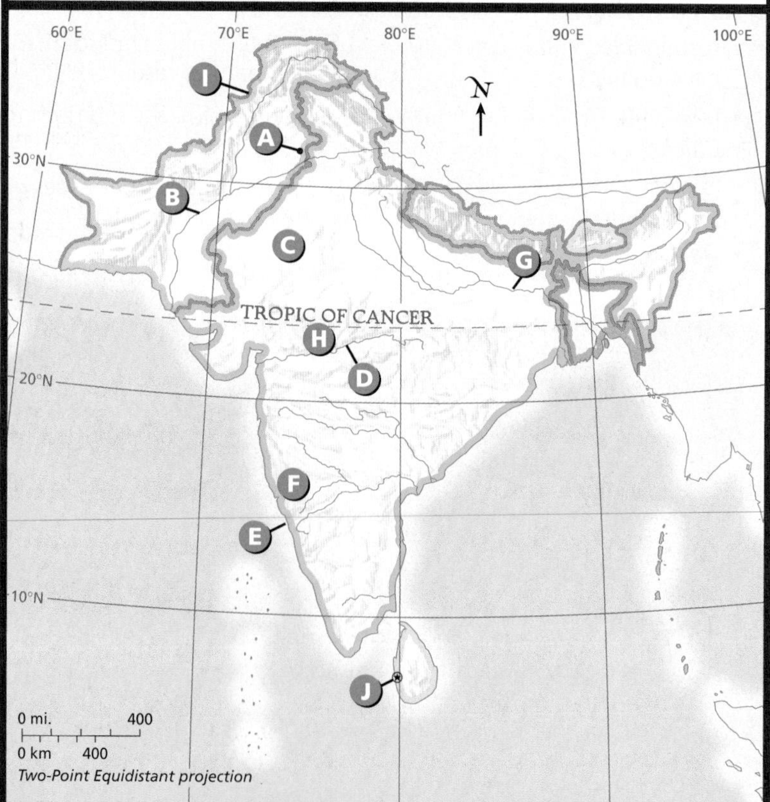

Self-Check Quiz Visit the **Glencoe World Geography** Web site at geography.glencoe.com and click on Self-Check Quizzes—Chapter 25 to prepare for the Chapter Test.

Using the Regional Atlas

Refer to the Regional Atlas on pages 560–563.

1. **Place** Which major physical feature allows subsistence farming west of the Great Indian Desert?
2. **Human-Environment Interaction** What natural resource do Pakistan and Bangladesh have in common?

Thinking Like a Geographer

Think about the diverse cultures and physical geography of South Asia. What do you think contributes to some of the problems or conflicts in South Asia today? Based on your understanding of this region's physical and human geography, what solutions might you propose?

Problem-Solving Activity

Contemporary Issues Case Study Prepare a case study on the use of the English language in South Asia. Gather data from print and electronic resources, and summarize the history and current status of English in the region. Also, consider why English today serves as a major international medium of communication.

GeoJournal

Creative Writing Referring to the notes you made in your journal, choose one economic activity. Imagine that you are employed in this activity, and write a description of a typical work day. If necessary, conduct additional research to add details to your account. Make your description as detailed as possible to capture the sense of what your job is like.

Technology Activity

Using E-Mail Choose an environmental issue in South Asia that you have read about in your text or in other news sources. List the important points about the issue, and then compose an e-mail letter to the editor of your local newspaper to bring the issue to the attention of others.

Standardized Test Practice

Read the passage and answer the question that follows. If you have trouble answering the question, use the process of elimination to narrow your choices.

South Asia is taking several steps to increase food production in the region's agricultural areas. Steps include increased planting cycles in Bangladesh and the use of technology such as modern irrigation techniques, pest control, soil fertilization, and new varieties of grain that increase crop yields. Farmers in Nepal now plant winter wheat in fields that used to lie fallow after the rice harvest. Research stations in Bhutan have helped farmers establish fruit orchards, and government-funded programs in Sri Lanka have encouraged farmers to grow more food crops.

1. **Which of the following reasons explains why South Asian countries are changing agricultural methods and using modern technology?**

   F South Asia is taking steps to increase manufacturing production.

   G Countries want to raise more food for their people.

   H Nepal does allow its agricultural fields to lie fallow.

   J Modern irrigation techniques will eliminate the threat of floods.

Never rely on your memory to answer questions derived from a passage. If you refer to the passage before you choose the correct answer, you will be less likely to make careless errors.

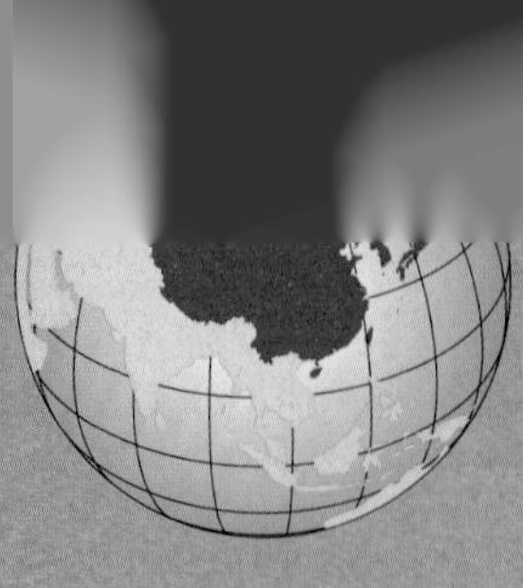

East Asia

WHY IT'S IMPORTANT—

East Asia and the United States are important trading partners. Many American companies manufacture goods in East Asia, and East Asia exports a variety of its own products to the United States. When you go shopping, notice the many items, ranging from cars and computers to clothing and furniture, that have been produced in East Asia or that are made of products exported from the region.

World Regions Video

To learn more about East Asia and its impact on your world, view the World Regions video "East Asia."

NGS ONLINE

www.nationalgeographic.com/education

Dancer in Hong Kong

What Makes East Asia a Region?

East Asia occupies much of the Asian mainland south of Russia. China takes up four-fifths of this region. With the exception of Mongolia, the other East Asian countries—Japan, North Korea, South Korea, and Taiwan—all lie on peninsulas and islands.

Towering mountains, such as the Himalaya and the Kunlun Shan, dominate the region's western landscape. Between these two ranges lies the Plateau of Xizang, the world's highest plateau. Two major rivers—the Yellow and the Yangtze—begin on the plateau and flow down onto fertile plains in eastern China.

Vast East Asia encompasses great variety in climate and vegetation, from the subarctic forests of northern Mongolia to the tropical rain forests on China's southernmost tip. Monsoons bring rain to coastal areas each summer, but the moist winds rarely reach the region's deep interior. In this arid heartland lie the parched and windswept Gobi and Taklimakan deserts.

1 A snug coat warms a boy in the chill, high-elevation air of Tibet, an area in southwestern China. Often called the "roof of the world," Tibet is perched on the lofty Plateau of Xizang, also known as the Plateau of Tibet. Valleys in Tibet are higher than the mountains of most countries.

2 Ancient limestone hills rise behind a rafter on the Li River, in southeastern China. On the raft are two large birds called cormorants, which are trained to dive for fish. Rivers and seas are important sources of food throughout East Asia.

3 Neatly terraced paddies follow the contours of steep hillsides in China. China is a huge country, but only about 10 percent of its land can be used for growing crops. Terraces allow farmers to grow rice in places that are fertile but sloping.

4 With frost on its fur, a Japanese macaque snoozes in a hot spring. Also called snow monkeys, Japanese macaques are adapted to the chilly climate of northern Japan, where cold winds scoop up moisture from the Sea of Japan and fling it back to Earth as snow.

Descended from Dynasties

East Asia can trace many of its cultural features to an ancient civilization that arose in China around 2000 B.C. In the centuries that followed, powerful dynasties ruled China, creating an enormous empire that influenced the cultural development of the entire region.

Today, East Asia is home to about one-fourth of the world's people. Most live crowded together in the region's fertile river valleys and coastal plains. Within each country, people tend to be ethnically similar.

During the twentieth century, the political and economic paths of East Asian countries diverged. China and North Korea adopted communist forms of government, while Japan, South Korea, and Taiwan developed capitalist, free-market economies.

1 **A huge portrait** of Communist leader Mao Zedong hangs above the Gate of Heavenly Peace, overlooking Tiananmen Square, in Beijing, China. In 1949 Mao stood at this site and established the People's Republic of China under Communist rule.

2 Built as a barrier to stop invaders from the north, China's Great Wall was started around 221 B.C., during the Qin dynasty. The wall winds for thousands of miles over plains and mountains and along desert borders. Erected entirely by hand, it is the longest structure ever built.

3 Neon lights glow as the sun sets over Tokyo, the capital of Japan. One of the largest, busiest, and most crowded cities in the world, Tokyo is Japan's center of commerce and culture. About one-fourth of Japan's population lives in the Tokyo area.

4 Standing serenely, an offshore torii, or gate, marks the entrance to one of Japan's most famous Shinto shrines. Shinto is an ancient religion that originated in Japan. Its followers worship *kami*—deities found in rivers, rocks, trees, and other elements of nature.

East Asia

PHYSICAL

Elevation Profile

POLITICAL

60°E
80°E
ARCTIC CIRCLE
100°E
120°E
140°E
160°E
60°N
50°N
40°N
30°N
20°N
10°N

0 mi. 1,000
0 km 1,000
Two-Point Equidistant projection

⊛ National capital
• Major city

RUSSIA
CENTRAL ASIA
Sea of Okhotsk
Amur R.
Hokkaido
Sapporo
ALTAY MOUNTAINS
Tuul R.
Ulaanbaatar
Songhua R.
Harbin
MONGOLIA
Liao R.
Honshu
Sea of Japan
TIAN SHAN
Tarim R.
GOBI
Shenyang
NORTH KOREA
Pyongyang
JAPAN
Tokyo
Yokohama
TAKLIMAKAN DESERT
Beijing
Seoul
Kyoto
Kobe
Osaka
Tianjin
SOUTH KOREA
Taegu
Pusan
Taiyuan
Zibo
KUNLUN SHAN
Yellow R.
North China Plain
Yellow Sea
Fukuoka
Shikoku
Zhengzhou
Kyushu
HIMALAYA
CHINA
Salween R.
Nanjing
Shanghai
East China Sea
Chengdu
Wuhan
Hangzhou
Ryukyu Is.
Chongqing
Yangtze R.
Okinawa
SOUTH ASIA
TROPIC OF CANCER
Taipei
TAIWAN
Philippine Sea
Red R.
Mekong R.
Xi R.
Guangzhou
Macau
Hong Kong
The People's Republic of China claims Taiwan as its 23rd province.
PACIFIC OCEAN
N
Bay of Bengal
SOUTHEAST ASIA
Hainan
South China Sea

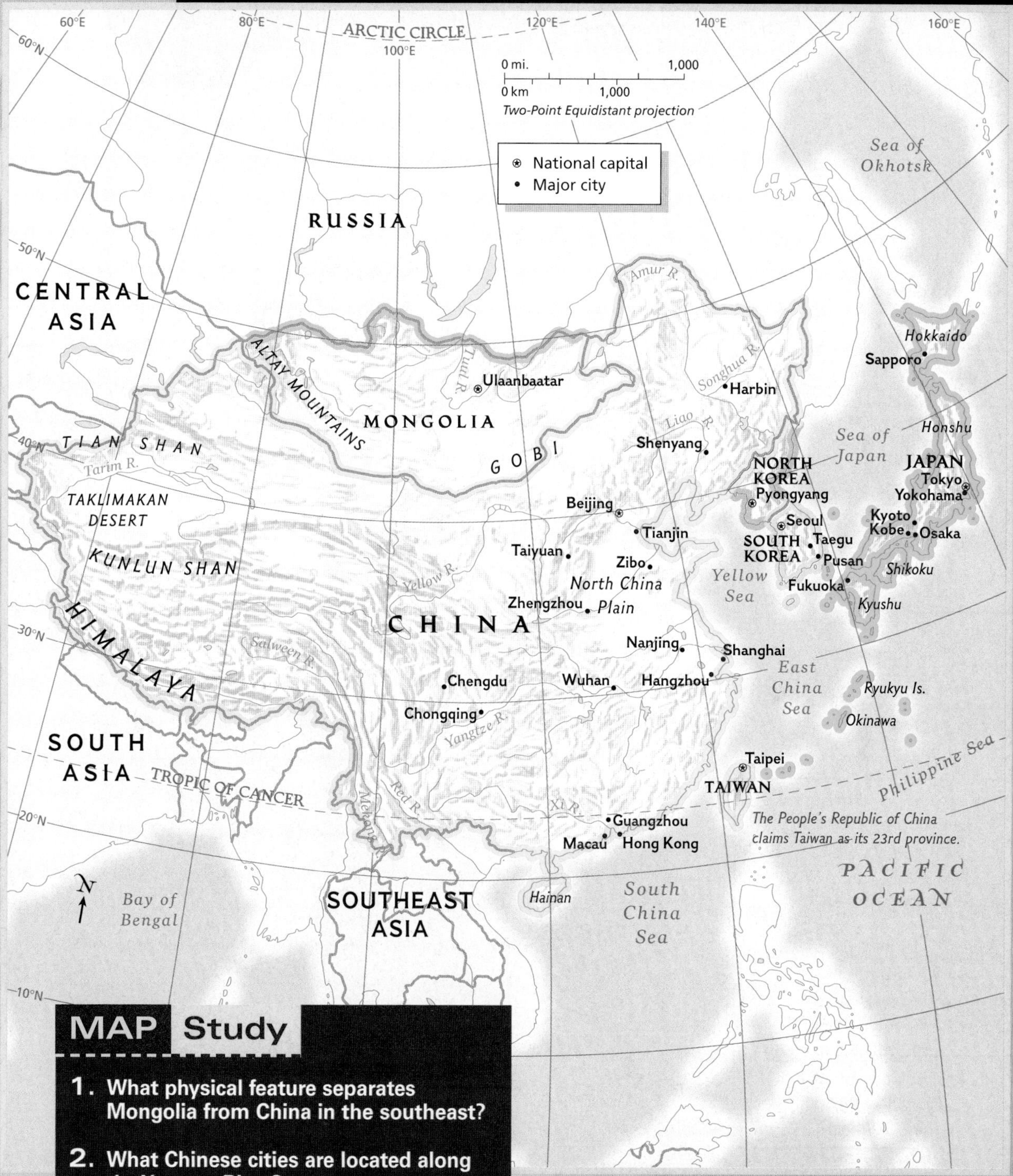

MAP Study

1. What physical feature separates Mongolia from China in the southeast?
2. What Chinese cities are located along the Yangtze River?

UNIT 9

REGIONAL ATLAS

East Asia

POPULATION DENSITY

NATIONAL GEOGRAPHIC SOCIETY

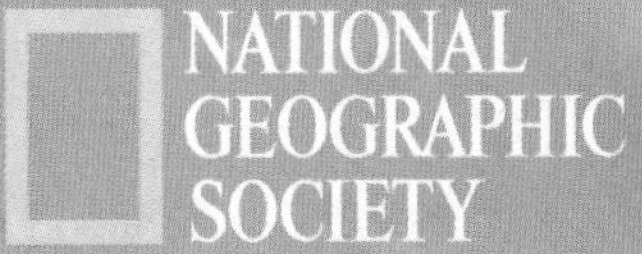

ECONOMIC ACTIVITY

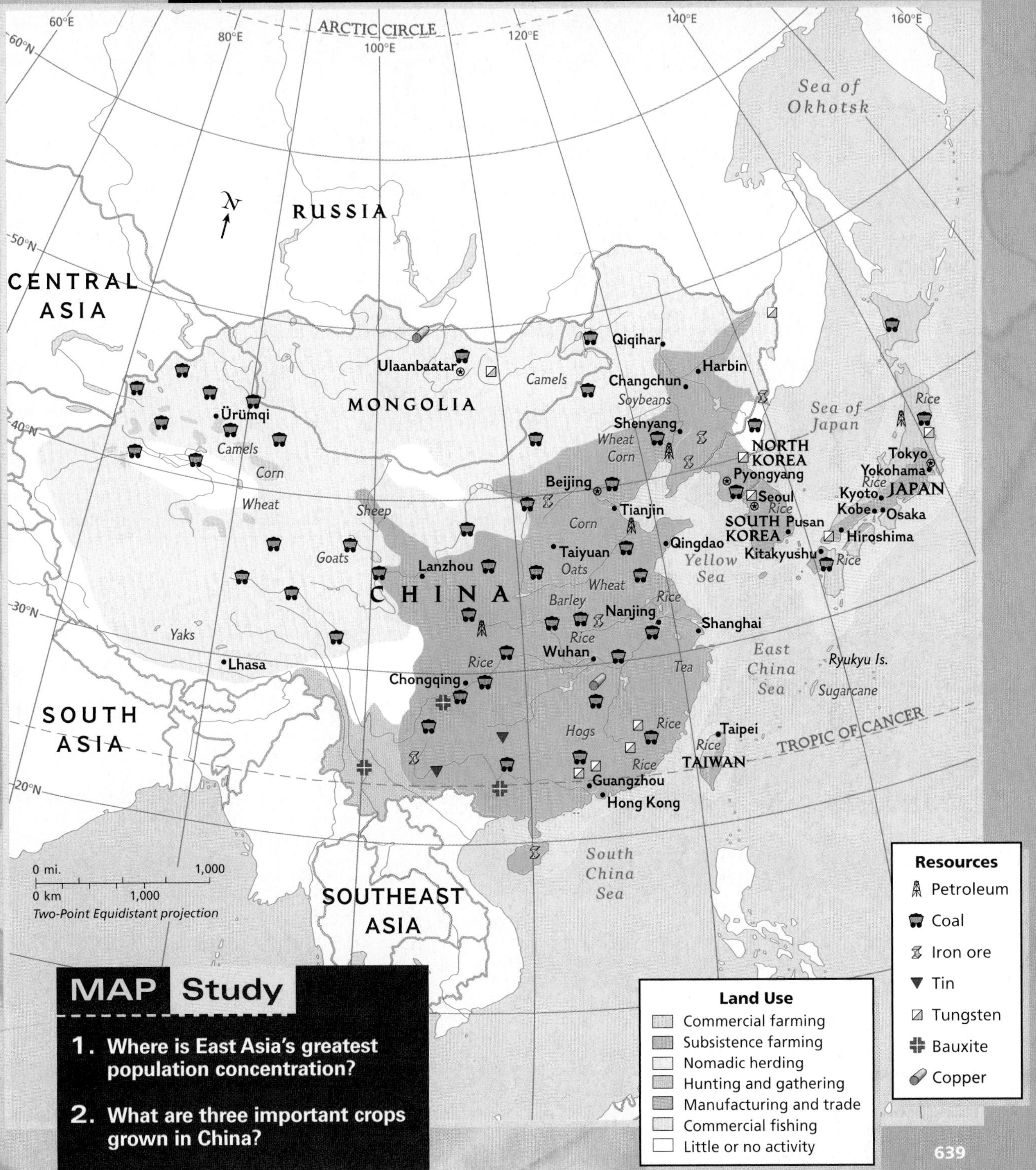

ARCTIC CIRCLE
60°E
80°E
100°E
120°E
140°E
160°E
60°N
50°N
40°N
30°N
20°N
N
Sea of Okhotsk
RUSSIA
CENTRAL ASIA
MONGOLIA
CHINA
NORTH KOREA
SOUTH KOREA
JAPAN
TAIWAN
SOUTH ASIA
SOUTHEAST ASIA
Ulaanbaatar
Qiqihar
Harbin
Changchun
Shenyang
Ürümqi
Beijing
Tianjin
Taiyuan
Qingdao
Lanzhou
Nanjing
Shanghai
Wuhan
Chongqing
Lhasa
Taipei
Guangzhou
Hong Kong
Pyongyang
Seoul
Pusan
Tokyo
Yokohama
Kyoto
Kobe
Osaka
Hiroshima
Kitakyushu
Sea of Japan
Yellow Sea
East China Sea
South China Sea
Ryukyu Is.
TROPIC OF CANCER
Camels
Soybeans
Wheat
Corn
Sheep
Goats
Oats
Barley
Rice
Yaks
Tea
Hogs
Sugarcane
0 mi. 1,000
0 km 1,000
Two-Point Equidistant projection
Resources
Petroleum
Coal
Iron ore
Tin
Tungsten
Bauxite
Copper
Land Use
Commercial farming
Subsistence farming
Nomadic herding
Hunting and gathering
Manufacturing and trade
Commercial fishing
Little or no activity

MAP Study

1. Where is East Asia's greatest population concentration?
2. What are three important crops grown in China?

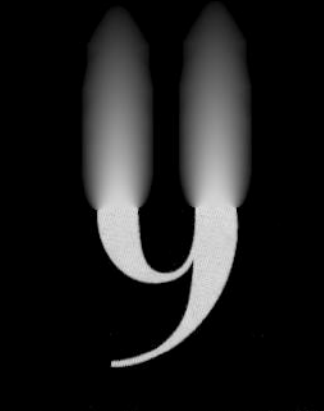

East Asia

COUNTRY PROFILES

COUNTRY * AND CAPITAL	FLAG AND LANGUAGE	POPULATION AND DENSITY	LANDMASS	MAJOR EXPORT	MAJOR IMPORT	CURRENCY	GOVERNMENT
CHINA Beijing	Mandarin Chinese	1,273,300,000 344 per sq. mi. 133 per sq. km	3,696,100 sq. mi. 9,572,899 sq. km	Machinery	Machinery	Yuan	Communist State
JAPAN Tokyo	Japanese	127,100,000 872 per sq. mi. 337 per sq. km	145,869 sq. mi. 377,801 sq. km	Machinery	Manufactured Goods	Yen	Constitutional Monarchy
MONGOLIA Ulaanbaatar	Khalkha Mongol	2,400,000 4 per sq. mi. 2 per sq. km	604,826 sq. mi. 1,566,499 sq. km	Copper	Fuels	Tugrik	Republic
NORTH KOREA Pyongyang	Korean	22,000,000 472 per sq. mi. 182 per sq. km	46,541 sq. mi. 120,538 sq. km	Minerals	Petroleum	Won	Communist State
SOUTH KOREA Seoul	Korean	48,800,000 1,274 per sq. mi. 492 per sq. km	38,324 sq. mi. 99,259 sq. km	Electronic Equipment	Machinery	Won	Republic
TAIWAN** Taipei	Mandarin Chinese	22,500,000 1,608 per sq. mi. 621 per sq. km	13,969 sq. mi. 36,179 sq. km	Textiles	Machinery	New Taiwan Dollar	Republic

* COUNTRIES AND FLAGS NOT DRAWN TO SCALE

**The People's Republic of China claims Taiwan as its 23rd province.

FOR AN ONLINE UPDATE OF THIS INFORMATION, VISIT GEOGRAPHY.GLENCOE.COM AND CLICK ON "TEXTBOOK UPDATES."

▶ Women inspecting cloth in textile mill, Kyoto, Japan

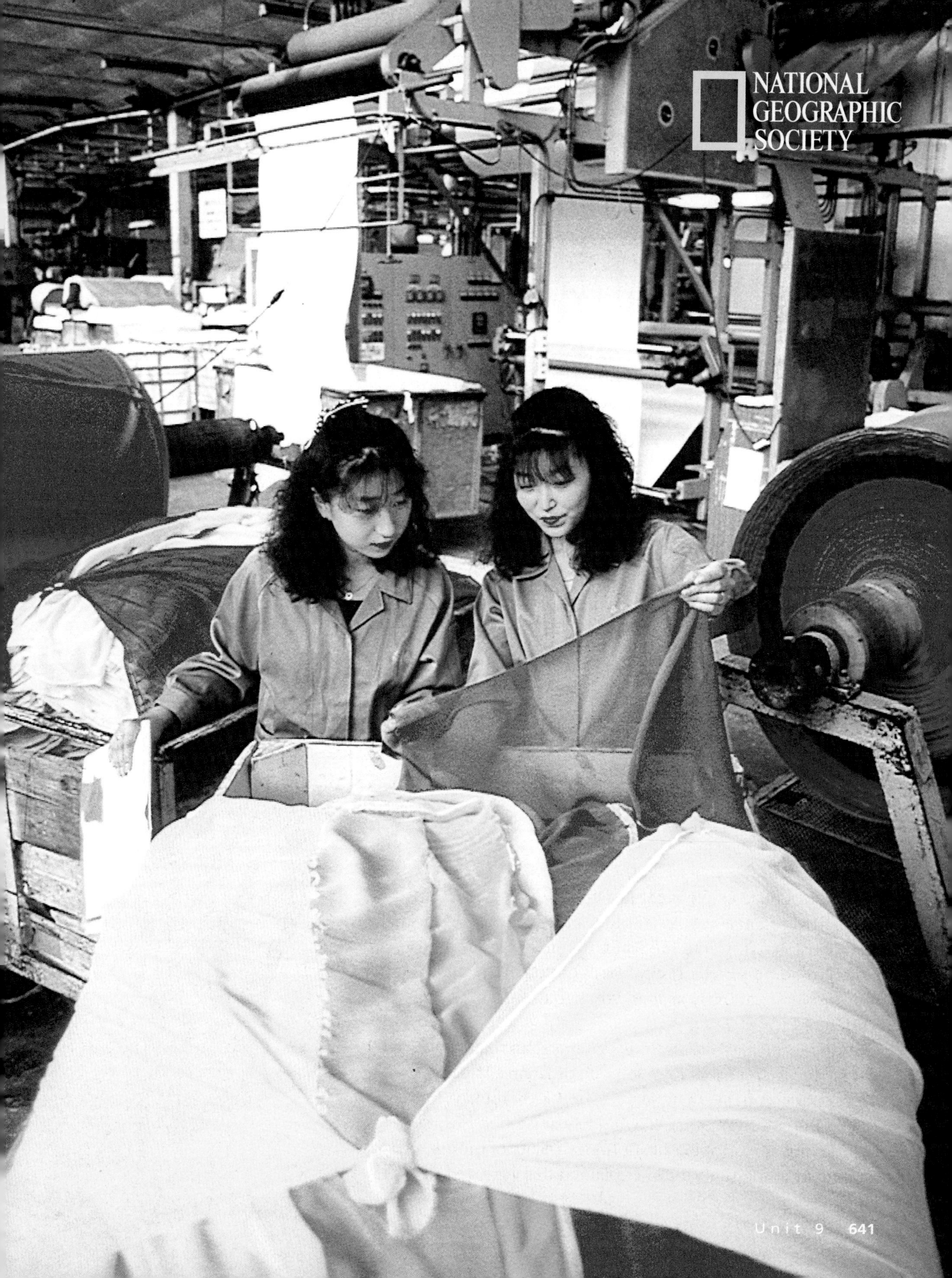
NATIONAL
GEOGRAPHIC
SOCIETY

9

GLOBAL CONNECTION

EAST ASIA AND THE UNITED STATES

ELECTRONICS

Turn on your TV, pop a tape into your VCR, play a CD on your stereo—chances are you're using a product that was made in Japan. Japan is one of the world's leading manufacturers of electronic goods.

Not only are many of our electronic gadgets made in Japan—quite a few were also invented there. That's the case with the portable personal stereo, the small tape or CD player with the lightweight headphones that people wear just about everywhere.

It's hard to imagine that a little over 20 years ago, personal stereos didn't exist. If you wanted to listen to music, your options were limited to home stereos, bulky boom boxes, or car audio systems. And, of course, all those around you had to listen, too, whether they wanted to or not.

Akio Morita and Masaru Ibuka changed all that. These two Japanese engineers founded one of Japan's largest electronics companies. One day in the 1970s, Ibuka walked into Morita's office lugging a heavy tape recorder and a pair of big headphones—state-of-the-art equipment at the time. Ibuka explained that he loved to

▲ Personal stereos for sale in a kiosk in Japan

listen to music, but he didn't want to disturb other people. It was the clunky tape recorder and earmuff headphones, or nothing at all.

Ibuka's dilemma made Morita think more seriously about an idea he'd been considering. Why not create a small, lightweight tape player with tiny headphones, so that people could conveniently take their music everywhere without bothering others? Morita instructed his engineers to remove the recording unit and speaker from a small cassette tape player, replace them with a tiny stereo amplifier, and then develop a very compact set of headphones to go with the device. Others at the company shook their heads, doubting that anyone would buy a tape machine that couldn't record. Still, in 1979, the first portable personal stereo hit the market. Within months, it was a runaway success. Morita's company could hardly keep pace with the demand.

Now many companies make personal stereos, which are among the most popular electronic devices in the world. This Japanese invention has changed the way people everywhere listen to music.

California surfers listening to personal stereos ▶

CHAPTER 26

The Physical Geography of East Asia

GeoJournal

As you read this chapter, use your journal to note the landforms and climate regions of East Asia. Write a series of descriptive paragraphs about these geographic features. Be sure to create a vivid, detailed description of each feature.

GEOGRAPHY Online

Chapter Overview Visit the **Glencoe World Geography** Web site at geography.glencoe.com and click on Chapter Overviews—Chapter 26 to preview information about the physical geography of the region.

The Land

Guide to Reading

Consider What You Know

Earthquakes, volcanic eruptions, and ocean flooding occur frequently in East Asia. What geographic factors are the most likely causes of these natural hazards?

Read to Find Out

- How are East Asia's landforms affected by the region's location on the Ring of Fire?
- How do the landforms of China differ from the rest of East Asia?
- What important natural resources are present in the region?

Terms to Know

- archipelago
- tsunami
- loess

Places to Locate

- Mongolia
- Hong Kong
- Macau
- South China Sea
- Korean Peninsula
- Japan
- Pamirs
- Himalaya
- Plateau of Tibet (Plateau of Xizang)
- Tarim Basin
- Taklimakan Desert
- Gobi
- Yellow River (Huang He)
- Yangtze River (Chang Jiang)
- Xi (West) River

◀ *Mt. Fuji, Japan*

NATIONAL GEOGRAPHIC

A Geographic View

China's Wild West

The curving road went on for another two miles across barren, rocky ground and ended at a meadow dotted with grazing yaks. We had entered a vast valley edged by a massive mountain, more shoulder than peak, its flank half-buried in sand. The meadow had been touched by spring, and at that seam of whitish sand and faint new green was a village. The low houses, strung along the base of the mountain, looked as if they had been there since the beginning of time.

—*Thomas B. Allen, "Xinjiang,"* National Geographic, *March 1996*

Outdoor classroom in Xinjiang, China

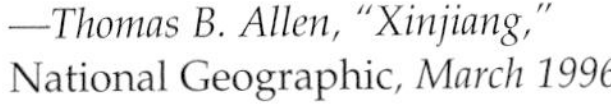

The wild and varied landscape of the western Chinese province of Xinjiang (SHIHN•JYAHNG) reflects the many contrasts and paradoxes of East Asia's physical geography. East Asia encompasses high mountains, rugged highlands, long and mighty rivers, barren deserts, fertile deltas and floodplains, miles of coastline, and countless islands dotting many seas. In this section you will read about East Asia's landforms and rich natural resources.

Land and Sea

The People's Republic of China makes up about 80 percent of the land area of East Asia and has the world's largest population—about 1.3 billion people. Of the world's countries, only Russia and Canada cover more land area than China. **Mongolia**, China's northern neighbor, occupies about 13 percent of East Asia's land. Mongolia's

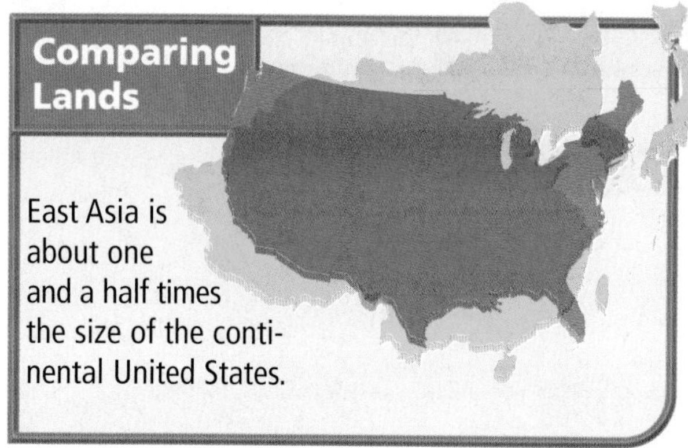

population is less than 1 percent of China's, making the country one of the world's most sparsely populated. The rest of East Asia is shared by the countries of Japan, Taiwan, North Korea, and South Korea. **Hong Kong** and **Macau**, two bustling ports on China's southern coast, were once European colonies and are now part of China.

Peninsulas, Islands, and Seas

Along the coast of East Asia, many peninsulas and islands dot the westernmost Pacific Ocean. These landforms divide the ocean into smaller bodies of water, including the Yellow Sea, the Sea of Japan, and the East China Sea. The **South China Sea**, stretching south from the island of Taiwan to the Philippines and the peninsula of Southeast Asia, carries one-third of the world's shipping traffic.

The **Korean Peninsula** juts southeast from China's Manchurian Plain, separating the Sea of Japan from the Yellow Sea. The peninsula, home to North Korea and South Korea, consists mainly of mountains surrounded by coastal plains.

Four large, mountainous islands and thousands of smaller ones form the **archipelago** (AHR•kuh•PEH•luh•GO), or island chain, of **Japan**. Honshu is the central and largest island, with Hokkaido to the north and Kyushu and Shikoku to the south. Most of Japan's major cities are on Honshu. Surrounding Japan are the Sea of Okhotsk on the north, the Sea of Japan and the East China Sea on the west, and the Philippine Sea on the south. On the east and southeast is the Pacific Ocean.

The Ring of Fire

An arc of islands east of China marks where the Pacific, Philippine, and Eurasian tectonic plates meet. These islands are part of the Ring of Fire, a circle of volcanoes bordering the Pacific Ocean. Most of these mountainous islands, including Japan and Taiwan, were formed by volcanic activity. Plate movements there cause frequent and often violent earthquakes and volcanic eruptions. Japan has about 50 active volcanoes and numerous hot springs formed through volcanic activity.

More than 1,000 small earthquakes shake Japan every year. Major quakes occur less often, but they may cause disastrous damage and loss of life in Japan's crowded cities. When an undersea earthquake generates a **tsunami** (soo•NAH•mee)—a huge tidal wave that gets higher and higher as it

NATIONAL GEOGRAPHIC **World Explorer**

Geography Skills for Life

Tsunami Damage Waves as high as an eight-story building caused devastation to Okushiri, Japan, in 1993.

Region What factors make the Ring of Fire susceptible to earthquakes and volcanic eruptions?

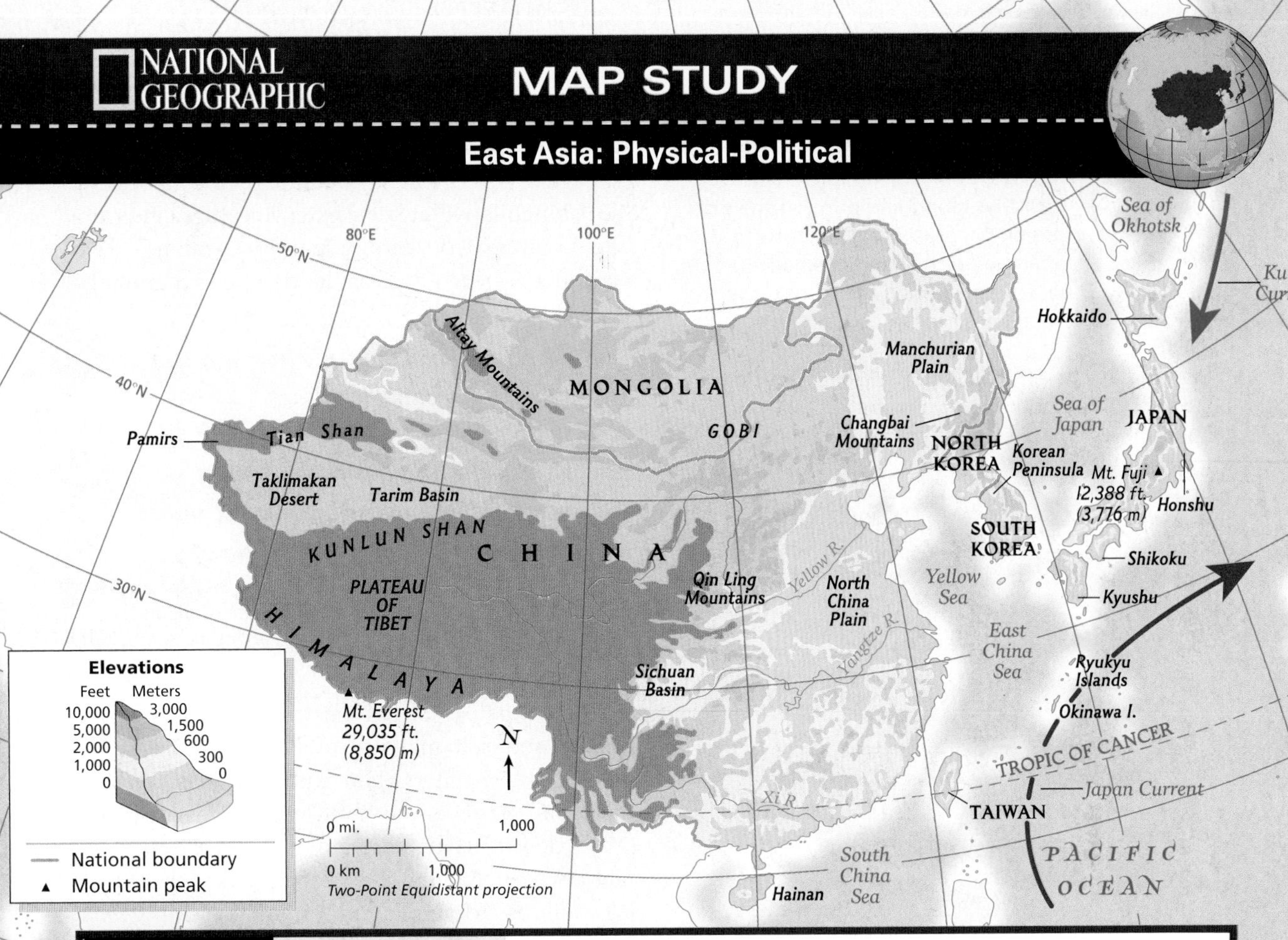

Geography Skills for Life

1. **Interpreting Maps** What rivers flow across the North China Plain?

2. **Applying Geography Skills** How have the Yellow and Yangtze Rivers impacted the lives of the Chinese?

Find NGS online map resources @ www.nationalgeographic.com/maps

approaches the coast—many lives may be lost. A tsunami that struck the Japanese island of Hokkaido in 1993 took 200 lives. Because earthquakes and tsunamis are difficult to predict, people along the Ring of Fire rely on special building methods and emergency preparedness to help reduce casualties.

Mountains, Highlands, and Lowlands

Mountain ranges and highlands mark the inland regions of East Asia. Most extremely rugged highlands areas are sparsely populated and have formed barriers to the movement of people and ideas. The region's only extensive lowland areas are China's Manchurian Plain and North China Plain. Narrow lowland plains also line many coastal areas.

East Asian Mountains

Numerous mountain ranges fan out from an area of high peaks and deep valleys called the **Pamirs** in western China. The ranges that begin in this remote interior region include the Kunlun Shan and Tian Shan. (*Shan* is Chinese for "mountains.") Farther north, the Altay Mountains form a natural barrier between Mongolia and China. To the south and west, the world's highest mountains, the **Himalaya**, separate China from South Asia. They include many peaks higher than 25,000 feet

the arts of EAST ASIA

Japanese Woodcuts Woodcut printing involves cutting a picture into a block of wood, applying ink to the surface of the cut block, and printing the picture on paper or some other surface. Japanese printmaking enjoyed a golden age around the turn of the nineteenth century.

Evening Snow, Mt. Fuji (about 1830) by Toyokuni II depicts a natural harmony within the environment that is typical of Japanese art. Compare this view with the photograph on page 644.

(7,620 m) above sea level. Mount Everest, the world's tallest peak at 29,035 feet (8,850 m), spans the border between China and Nepal.

The Kunlun Shan bends to become the Qin (CHIHN) Ling Mountains, crossing central China from west to east. To the east, the lower Changbai Mountains of Manchuria extend into the Korean Peninsula, where they are called the Northern Mountains. Coastal plains surround the high mountain interiors of Japan and Taiwan. Mount Fuji, at 12,388 feet (3,776 m), is a dramatic, cone-shaped, dormant volcano rising above the central plains of Japan's Honshu Island. Also called Fujiyama, Mount Fuji is an important spiritual symbol to Japan's people.

China's Plateaus, Basins, and Deserts

China contains the region's most diverse landforms. The **Plateau of Tibet**, in China's southwest quarter, is East Asia's highest plateau region. Because the Chinese name for Tibet is Xizang (SHEE•ZAHNG), the plateau is also known as the Plateau of Xizang. Its average elevation is about 15,000 feet (4,600 m). Other rugged highlands stretch north and eastward at lower elevations, averaging 4,900 feet (1,494 m). In the far north, the Mongolian Plateau's extensive highlands are mostly grassy pasture, ideal for grazing. Two visitors described the vast landscape and animals:

> *"They appeared suddenly from a ravine, two nomad horsemen driving a herd of sheep across the path of our truck. On and on the animals came, a sea of brown, black, and white against the golden grasses of the broad plain."*
>
> Cynthia Beall and Melvyn Goldstein, "Past Becomes Future for Mongolian Nomads," *National Geographic*, May 1993

Broad expanses of flat wastelands, including the deserts and salt marshes of the **Tarim Basin**, lie between the Kunlun Shan and Tian Shan. West of the Tarim Basin is the **Taklimakan Desert**, a dry, sandy desert. To the northeast is another desert, the **Gobi**, whose frequent dust storms make life difficult in southern Mongolia and north central China. China's high, interior deserts are dry and cold. By contrast, the huge, fertile Sichuan Basin between the Plateau of Tibet and the North China Plain has a mild climate and long growing season, making it an important agricultural area.

River Systems

East Asia's rivers serve densely populated urban centers as transport routes. They provide hydroelectric power for energy, and the fertile soil in their basins is used for farming.

China's Rivers

China's major rivers begin in the Plateau of Tibet and flow eastward to the Pacific Ocean. The **Yellow River**, known in Chinese as Huang He (HWAHNG HUH), is northern China's major river system. This river is called "yellow" because it carries tons of fine, yellowish-brown topsoil called loess (LEHS), blown by winds from the western deserts into the air and water. When deposited, the rich soil—along with water from the river—makes the

North China Plain a major wheat-farming area. Also called "China's sorrow," the Yellow River often floods its basin. Throughout history, it has flooded large areas, killing hundreds of thousands of people.

Central China's **Yangtze** (YANG•SEE) **River**, known in Chinese as the Chang Jiang, is Asia's longest river at 3,965 miles (6,380 km). It flows through spectacular gorges and broad plains and empties into the ocean at Shanghai. The Yangtze, a major transport route, provides water for a large agricultural area where more than half of China's rice and other grains grow. When completed in 2009, the river's Three Gorges Dam will be the world's largest dam (see the feature on pages 698–699).

The **Xi** (SHEE), or West, **River** is southern China's most important river system. Near the ports of Guangzhou and Macau, the soil deposits of the Xi form a huge, fertile delta, one of China's fast-developing areas.

The world's longest artificial waterway, China's Grand Canal, was begun in the 400s B.C. Over the centuries, the canal has been expanded and rebuilt. Today, the Grand Canal moves people and goods along a 1,085-mile (1,746-km) course from Beijing in the north to Hangzhou in the south.

Rivers in Japan and Korea

In contrast to China's long rivers, the rivers of Japan and Korea are short and swift. They flow through mountainous terrain, often forming spectacular waterfalls. During the wet season, they provide hydroelectric power. South Korea's chief rivers flow from inland mountains westward toward the Yellow Sea. The Han River flows through South Korea's capital, Seoul. In North Korea the Amnok (or Yalu) River flows west, forming the border with China.

Culture

The Power of Wind and Water

For centuries East Asians have chosen building sites and designed homes using feng shui (FUHNG SHWAY), from the Chinese words for "wind" and "water." By combining observations of the natural landscape with traditional spiritual teachings, the Chinese harmonize their buildings with the surrounding landforms, especially mountains and rivers. First used to locate favorable gravesites, feng shui is now used by architects, real estate agents, building contractors, and interior decorators worldwide.

Natural Resources

East Asia's rich mineral resources are unevenly distributed. China's huge land area contains the greatest share and widest range of minerals, including sizable reserves of iron ore, tin, tungsten, and

Student Web Activity Visit the **Glencoe World Geography** Web site at geography.glencoe.com and click on Student Web Activities—Chapter 26 for an activity about the physical geography of East Asia.

NATIONAL GEOGRAPHIC **World Explorer**

Geography Skills for Life

Yellow River Tributary A tributary of the Yellow River winds through central China, carrying the sediment that gives the Yellow River its name.

Place How do China's rivers compare with those in Japan and Korea?

gold. Large oil deposits lie in the South China Sea and in the Taklimakan Desert in the west. Abundant coal deposits also lie in northeastern China. Coal is mined in the Korean Peninsula and Mongolia. North Korea's rich deposits of economically useful minerals include iron ore and tungsten. South Korea has relatively few mineral reserves, though large deposits of graphite are found there. Taiwan's mineral reserves are small, and its coal reserves are almost exhausted.

Productive farmlands and forests are unevenly distributed in East Asia. For example, only 10 percent of China's land is suitable for agriculture. The southern "rice bowl" yields two harvests per year, making China the world's leading producer of rice. With nearly 25 percent of its land suitable for farming, South Korea produces two crops per year, one of rice and one of barley, in the prime farmland of the coastal south. By contrast, Mongolia can use less than 1 percent of its land for crops. Japan also has very limited farmland and poor soil. Only one-fourth of Taiwan's land is suitable for farming, but every available space is planted, chiefly with rice. Taiwan has valuable forests of cedar, hemlock, and oak.

Geography Skills for Life

Squid Harvest Coastal fishers in Japan dry their squid catch in the sun.

Region Which countries have the biggest deep-sea fishing industries?

East Asia's island countries and coastal areas depend on the sea for food. Japan, South Korea, Taiwan, and China have the world's biggest deep-sea fishing industries. China alone harvests about 18 million tons (16.3 million metric tons) of fish each year. Seafood farming has become a major industry in East Asia.

SECTION 1 ASSESSMENT

Checking for Understanding

1. **Define** archipelago, tsunami, loess.
2. **Main Ideas** On a table, fill in details about major features of the physical geography for each country in East Asia.

Area	Physical Geography

Critical Thinking

3. **Comparing and Contrasting** How are East Asia's coastal, island, and peninsula areas similar? How do they differ from inland areas?
4. **Drawing Conclusions** How does the technique of feng shui reflect East Asian beliefs about humans and their environment?
5. **Predicting Consequences** What consequences do you think will result from East Asia's use of its ocean resources?

Analyzing Maps

6. **Place** Study the physical-political map on page 647. How does the elevation of the North China Plain compare with that of the Plateau of Tibet?

Applying Geography

7. **Soil Building** Describe the soil-building process that takes place in northern China's Yellow River basin. How does this process influence the natural environment, the people, and economy of the area?

Climate and Vegetation

Guide to Reading

Consider What You Know

As you read in the last section, large areas of East Asia border the sea. How do you think climate and vegetation in coastal areas differ from those in inland areas?

Read to Find Out

- What accounts for East Asia's wide variety of climates?
- How do winds, ocean currents, and mountains influence the climates of East Asia?
- What conditions cause the extreme climates in much of China?
- What kinds of natural vegetation are found in East Asia's varied climate regions?

Terms to Know

- monsoon
- Japan Current
- typhoon

Places to Locate

- Taiwan
- Hainan
- Qin Ling Mountains

NATIONAL GEOGRAPHIC

A Geographic View

Weathering Uncertainty

Of course, living with uncertain circumstances is nothing new for Mongolia's nomads. For centuries they have weathered one of the earth's harshest and least predictable environments. Winter winds at camps high in mountain valleys can howl at minus 20°F to minus 50°F, and sudden blizzards can bury pastures and starve herds.

—*Cynthia Beall and Melvyn Goldstein, "Past Becomes Future for Mongolian Nomads,"* National Geographic, *May 1993*

Mongolian nomad on a camel

The nomads of Mongolia are among the few peoples who have adapted to living in East Asia's harshest climate regions. Following their herds across the high grasslands, these nomadic peoples take shelter in tentlike structures called yurts, built to be portable yet withstand the howling winter winds. Wind is a powerful force throughout East Asia, a region that depends on seasonal wind patterns for life-giving rains. In this section you will learn how physical features shape the climate and vegetation of this vast region.

Climate Regions

Latitude and physical features—such as mountain barriers, highlands, and coastal regions—shape East Asia's climates. Each climate region has distinct characteristics and unique vegetation. Dry highlands and grasslands dominate the north and west, with humid and temperate forests to the south and east.

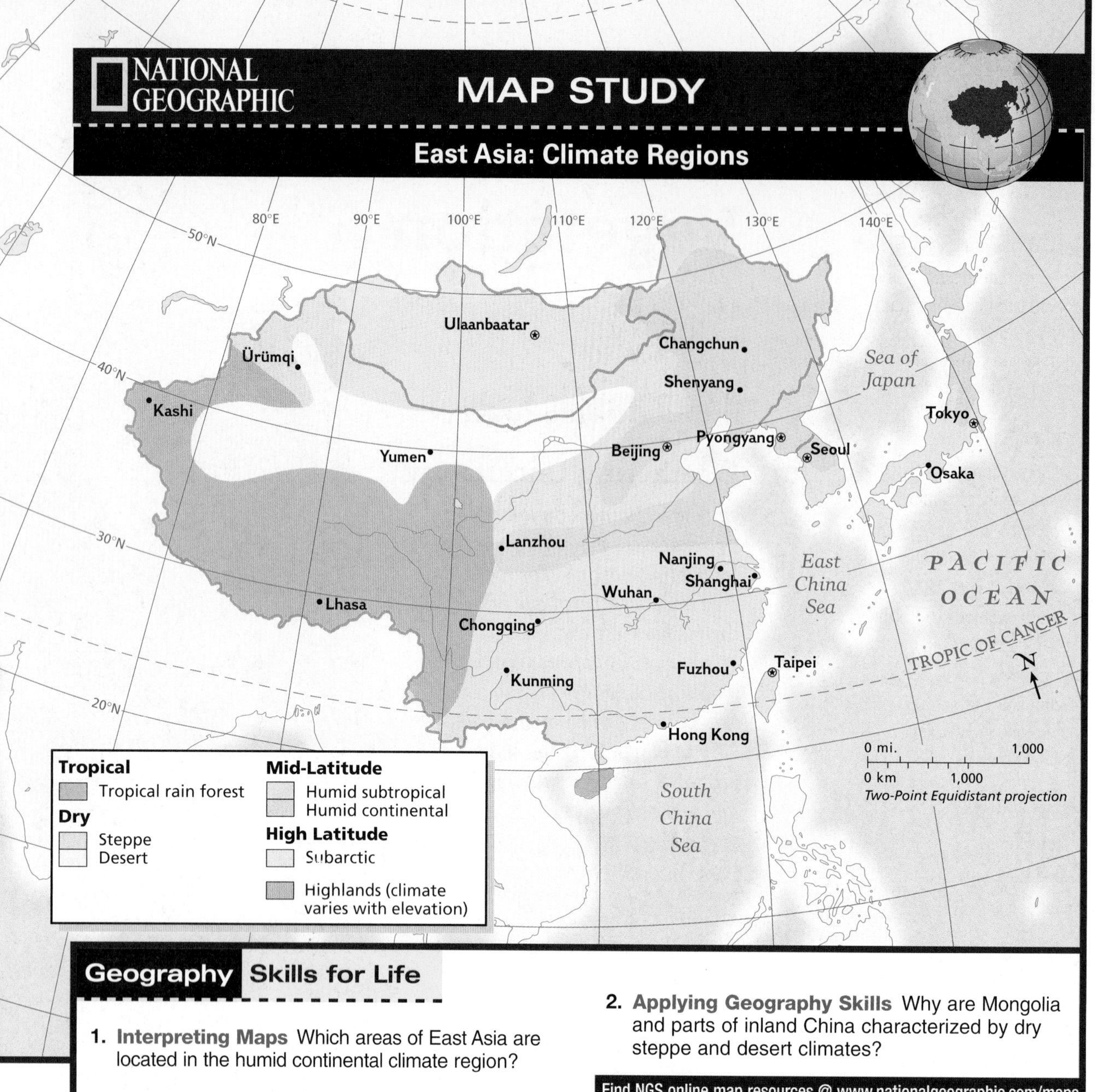

Geography **Skills for Life**

1. **Interpreting Maps** Which areas of East Asia are located in the humid continental climate region?

2. **Applying Geography Skills** Why are Mongolia and parts of inland China characterized by dry steppe and desert climates?

Find NGS online map resources @ www.nationalgeographic.com/maps

Mid-Latitude Climates

The southeastern quarter of East Asia, including **Taiwan** and parts of China, the Koreas, and Japan, has a humid subtropical climate, with warm or hot summers and heavy rains from the Pacific monsoon. In contrast, the northeastern quarter, including the northern parts of the Koreas and Japan, has a cooler, humid continental climate. Summers may be warm, but winters are cold and snowy.

Natural forests in mid-latitude climates consist of needle-leaved and broad-leaved evergreens and broad-leaved deciduous trees. Deciduous trees and broad-leaved evergreens also flourish in the humid subtropical regions. Bamboo, a treelike grass, grows abundantly in many of the warmer areas. This tough, versatile plant has more than a thousand uses, from herbal medicine, food, and decoration to construction of homes, skyscrapers, and bridges. Bamboo also provides the only food source for two of East Asia's rare mammals, the giant panda and the smaller, raccoon-like red panda. Other economically important native plants are the

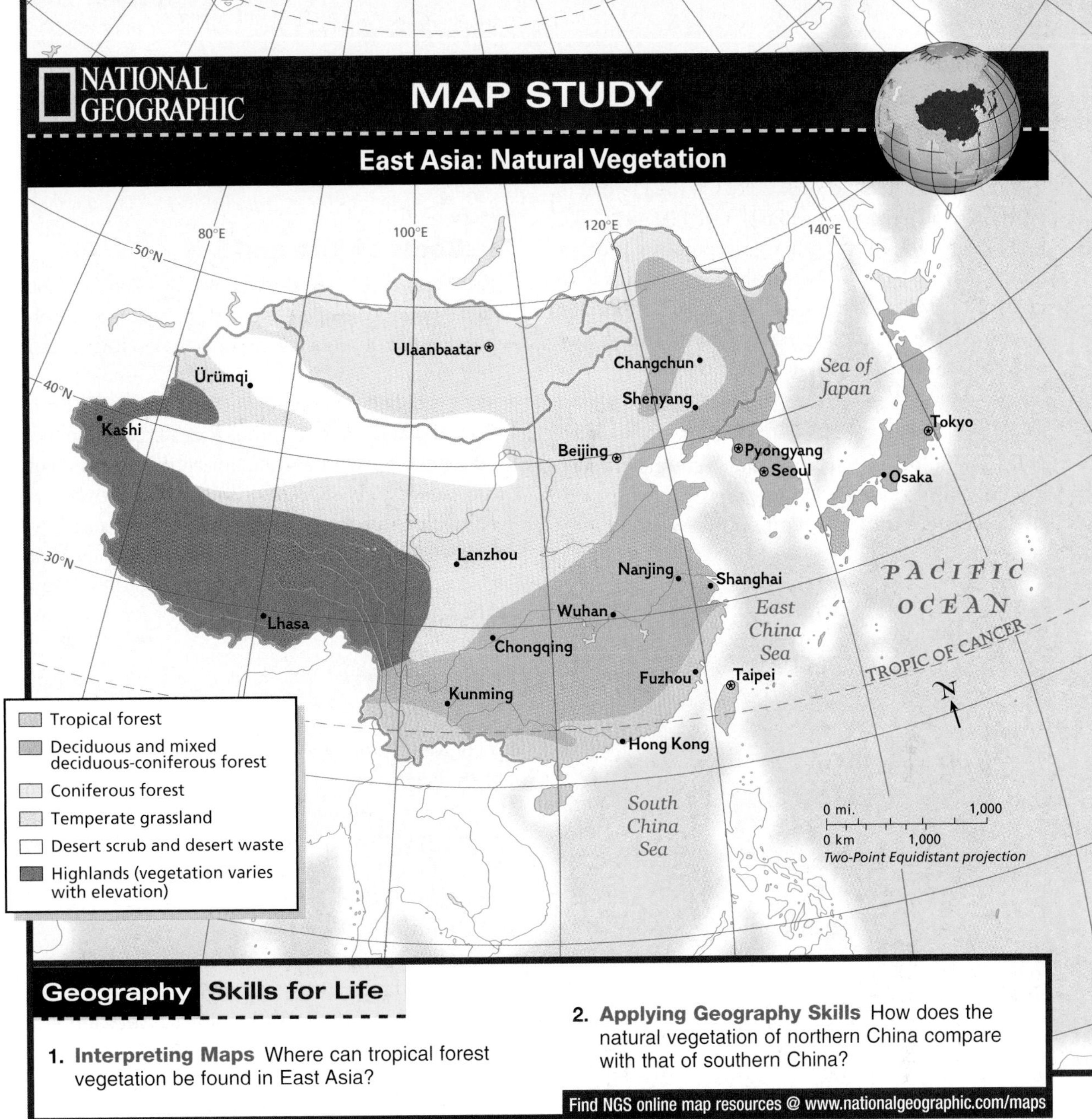

Geography Skills for Life

1. **Interpreting Maps** Where can tropical forest vegetation be found in East Asia?

2. **Applying Geography Skills** How does the natural vegetation of northern China compare with that of southern China?

Find NGS online map resources @ www.nationalgeographic.com/maps

mulberry tree, whose leaves provide food for silkworms, and the tea bush. Bamboo, tea, and silk are significant to East Asia's culture and economy and have become identified worldwide with the region.

Desert and Steppe Climates

Far away from the moist winds of the coast, deserts spread across Mongolia and inland northern China. Moisture that might reach these areas is blocked by the rain shadow effect caused by the surrounding mountains. Deserts are not always hot—the Gobi and Taklimakan are often cold and windy. In the northern and northwestern desert and steppe climates, temperature variation can be extreme, falling as much as 55°F (31°C) from daytime to nighttime. In the Gobi, temperatures average 73°F (23°C) in summer and 0°F (–18°C) in winter, but they may range from 100°F to –30°F (38°C to –35°C). Grasses and sparse trees are the natural vegetation of the large steppe climate east of the deserts and in most of Mongolia.

Highlands Climates

The climate in mountainous areas changes with elevation. Generally, the higher the elevation of an area, the cooler it is. East Asia's highlands climates, therefore, are usually cool or cold. On the Plateau of Tibet, with an elevation of 13,000 to 15,000 feet (3,962 to 4,572 m), the average high temperature reaches only about 58°F (14°C). Small alpine meadows with grass, flowers, and trees dot the lower mountain slopes. Above the timberline, where no trees grow, only mosses and colorful lichens thrive.

Tropical Rain Forest Climate

The island of **Hainan**, off China's southern coast, has a tropical rain forest climate. This area experiences year-round high temperatures and a very rainy summer monsoon. In tropical areas palms and tropical hardwoods thrive alongside broad-leaved evergreens and tropical fruit trees. Lush rain forest covers much of Hainan.

History

The Roots of Rice and Soy

Rice and soybeans—two of East Asia's most important food crops—were first cultivated from wild vegetation. Archaeologists have found evidence of rice cultivation in southern China as early as 5000 B.C. It then spread north to Japan, south to Indonesia, and west to India. Rice became a major food source for China's ancient civilization as well as for others in East and Southeast Asia. Soybeans, a valuable source of protein for people and livestock, were also first cultivated in East Asia around 5000 B.C. China's ancient peoples considered soybeans one of the five foods necessary for long life.

NATIONAL GEOGRAPHIC **World Explorer**

Geography Skills for Life

Rice Seedlings A farmer tends to rice plants in a flooded paddy.

Region How is rice important to East Asian culture?

Monsoons

In East Asia the air mass above the world's largest continent and the air mass above the world's largest ocean meet. The movement of these air masses causes prevailing winds, called **monsoons**, that bring seasonal weather patterns to East Asia. Along with inland highlands, mountains, and ocean currents in coastal areas, monsoons greatly influence East Asia's climate.

Monsoons blow in a steady direction for approximately half the year and then switch directions. The summer monsoon in East Asia blows from southeast to northwest, bearing heat and humidity from the Pacific Ocean. From April through October, especially near the coast and occasionally as far northwest as Mongolia, the winds cause intense downpours that provide more than 80 percent of the region's annual rainfall. From November to March, the winter monsoon brings cold, arctic air that usually blows from northwest to southeast. Inland, the winds tend to be dry, carrying clouds of dust from the Gobi. Along the coast, these winds pick up moisture in the Sea of Japan and bring heavy snow to Japan and the Korean Peninsula, especially in the north.

The East Asian economy depends on summer monsoons to bring the rains for crops. If the summer monsoons are late or do not bring enough rain, serious crop failures may occur. Too much rain brings disastrous flooding, as occurred in 1998 in the Chinese city of Harbin:

> "*In north east China, Harbin . . . faced another [overflow] as the third flood crest in recent weeks swept down the Songhua River.*"
>
> "China Floods 'Worst Ever'," *BBC News*, August 22, 1998

▲ The Gobi

In some parts of East Asia, mountains weaken the effects of the monsoons. The **Qin Ling Mountains** of central China, for example, act as a clear dividing line. South of the Qin Ling, the climate is warm and humid, and rice is the chief crop. To the north the mountains block the summer monsoons, so the climate north of the Qin Ling is cooler and drier, and wheat is the chief crop. The high mountains of the eastern Korean Peninsula act as a similar barrier against the winter monsoons, giving Korea's east coast warmer winters and lighter snowfalls.

Ocean Currents

Ocean currents, too, influence climate. Two such currents shape Japan's climate. The warm-water **Japan Current**, or Kuroshio, flows northward along the southern and southeastern coasts of the Japanese islands and adds moisture to the winter monsoon as it warms the land. The cold Kuril Current, or Oyashio, flows southwest from the Bering Sea along the Pacific coasts of Japan's northernmost islands. It brings harsh, cold winters to Hokkaido's east coast. In summer, when the cold ocean current meets the warm one near Hokkaido, a dense sea fog develops.

The interaction of ocean currents and winds frequently gives rise to violent storms called **typhoons**, which form in the Pacific and blow across coastal East Asia. Like hurricanes in the western Atlantic and Caribbean, typhoons tend to be most severe between late August and October. High winds, storm surges, and torrential rains during typhoons may cause heavy damage. Occasionally, though, a winter typhoon brings welcome rains during the normally dry part of the year.

SECTION 2 ASSESSMENT

Checking for Understanding

1. **Define** monsoon, Japan Current, typhoon.
2. **Main Ideas** Draw a Venn diagram like the one shown below. Use it to describe the similarities and differences in climate for western and eastern parts of East Asia.

East Asia (western part) | Both | East Asia (eastern part)

Critical Thinking

3. **Analyzing Information** Why is Mongolia more suitable for herding than for farming?
4. **Predicting Consequences** What economic effects would occur if the summer monsoon arrived months late in China?
5. **Identifying Cause and Effect** How do ocean currents affect East Asia's climate?

Analyzing Maps

6. **Region** Compare the maps on pages 652 and 653. What kinds of vegetation characterize high latitude climates?

Applying Geography

7. **Effects of Elevation** Write a paragraph analyzing how mountains, plateaus, and lowlands affect East Asia's climate and vegetation.

Developing Multimedia Presentations

You can take advantage of all available technologies and media forms to create classroom presentations. A multimedia presentation can engage the senses and capture the attention of your audience.

Learning the Skill

A multimedia presentation uses several types of media to present information. These media may include audio, text, and graphics, such as slides, transparencies, animation, or videos.

Any multimedia presentation should have a definite purpose. Before you begin to develop a presentation, state the purpose briefly in one or two sentences. Identifying the purpose will guide your research and help you select the media to use.

Choosing the appropriate media from those available will help you communicate information most effectively. Showing a videotape of a graph during a presentation will probably not capture your audience's attention. An overhead transparency might be a better tool for displaying a graph. To prepare a presentation on the recent migration patterns of East Asians, for example, you might display a combination of maps and photos showing past and present migrations.

Use these questions to develop multimedia presentations:

- **What is my purpose?**
- **Which forms of media will best show the kind of information I want to present?**
- **Which media are available?**
- **What computer software programs do I need, if any?**
- **Does my computer support these software programs?**

Practicing the Skill

Answer the following questions about developing multimedia presentations.

1. What media tools would be most effective for a presentation about an important leader in East Asia?
2. What media tools would be most effective for explaining population changes in East Asia?
3. What are some possible advantages and disadvantages of showing a Web site during a multimedia presentation?

Applying the Skill

Work with a group to plan and produce a multimedia presentation on a political, economic, or social issue in an East Asian country. Use the information in this chapter, and research print and Web sources to prepare your presentation. Share each presentation with the class.

26

SECTION 1 The Land (pp. 645–650)

Terms to Know

- archipelago
- tsunami
- loess

Key Points

- East Asia's location at the meeting point of tectonic plates leaves the region vulnerable to earthquakes, volcanic eruptions, and tsunamis.
- The region of East Asia consists of China, Mongolia, and North and South Korea on the Asian continent, plus the island countries of Japan and Taiwan.
- East Asia's rivers provide important transportation systems and support fertile farmlands.
- East Asia is rich in minerals, but they are unevenly distributed.
- Limited farmlands, long coastlines, and large populations have made the region dependent on the sea for food.

Organizing Your Notes

Create an outline using the format below to help you organize your notes for this section.

The Land

I. Land and Sea
   A. Peninsulas, Islands, and Seas
   B.
II. Mountains, Highlands, and Lowlands
   A.
   B.
III. River Systems
   A.
   B.
   C.
IV. Natural Resources

SECTION 2 Climate and Vegetation (pp. 651–655)

Terms to Know

- monsoon
- Japan Current
- typhoon

Key Points

- East Asia's natural vegetation tends to parallel the region's climate zones.
- East Asian countries rely on seasonal winds known as monsoons. The summer monsoons bring more than 80 percent of the region's rainfall.
- Ocean currents affect the climates of coastal and island regions. Powerful typhoons form in the Pacific and blow across coastal East Asia in later summer and early fall.
- East Asia's varied vegetation includes needle-leaved and broad-leaved evergreen trees, tropical plants, bamboo, tea, mulberry trees, and grasses as well as tropical rain forest vegetation.

Organizing Your Notes

Use a table like the one below to help you organize important details from this section.

Climate Zone	Location	Type of Vegetation
Humid Subtropical		
Humid Continental		
Desert		
Steppe		
Highlands		
Tropical Rain Forest		

◀ *Li River, China*

ASSESSMENT & ACTIVITIES

Reviewing Key Terms

Write the letter of the key term that best matches each description.

a. archipelago	**d.** monsoon
b. tsunami	**e.** Japan Current
c. loess	**f.** typhoon

1. large, fast-moving wave caused by an undersea earthquake
2. chain or group of islands
3. seasonal wind
4. powerful, hurricane-like storm generated in the western Pacific
5. warm-water stream that affects the climate in Japan
6. fine, windblown topsoil

Reviewing Facts

SECTION 1

1. On what landform are North and South Korea located?
2. What is the largest and most densely populated of the Japanese islands?
3. Describe the natural hazards that result from East Asia's location at the meeting point of three tectonic plates?
4. From which part of China do most of the region's great mountain ranges extend?
5. What are China's four major river or waterway systems?

SECTION 2

6. Which economically important plants thrive in East Asia's midlatitudes?
7. What climate factor influences East Asia in seasonal cycles?
8. How are economic activities affected by climate in East Asia?
9. Which ocean current brings cold winters to Hokkaido?

Critical Thinking

1. **Making Generalizations** How has the uneven distribution of natural resources most likely affected the economies of countries in the region?
2. **Analyzing Information** Why might the countries surrounding the South China Sea compete for control of its waters?
3. **Identifying Cause and Effect** Use a graphic organizer like the one below to fill in the effects that mountains have on the climate of East Asia.

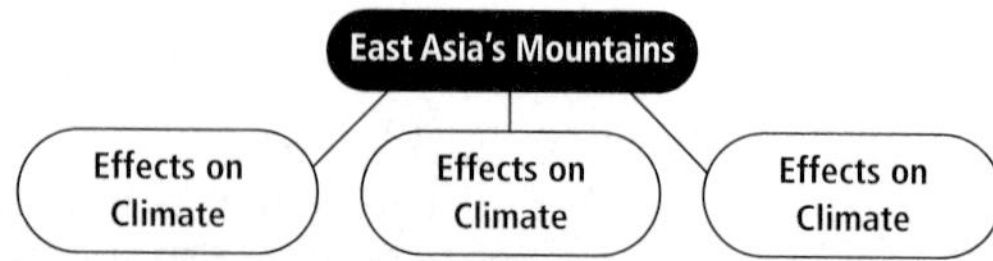

NATIONAL GEOGRAPHIC

Locating Places

East Asia: Physical-Political Geography

Match the letters on the map with the places and physical features of East Asia. Write your answers on a sheet of paper.

1. **Yellow River**	5. **Yellow Sea**	9. **Taiwan**
2. **Yangtze River**	6. **Plateau of Tibet**	10. **North China Plain**
3. **Mongolia**	7. **Himalaya**	
4. **Honshu**	8. **South Korea**	

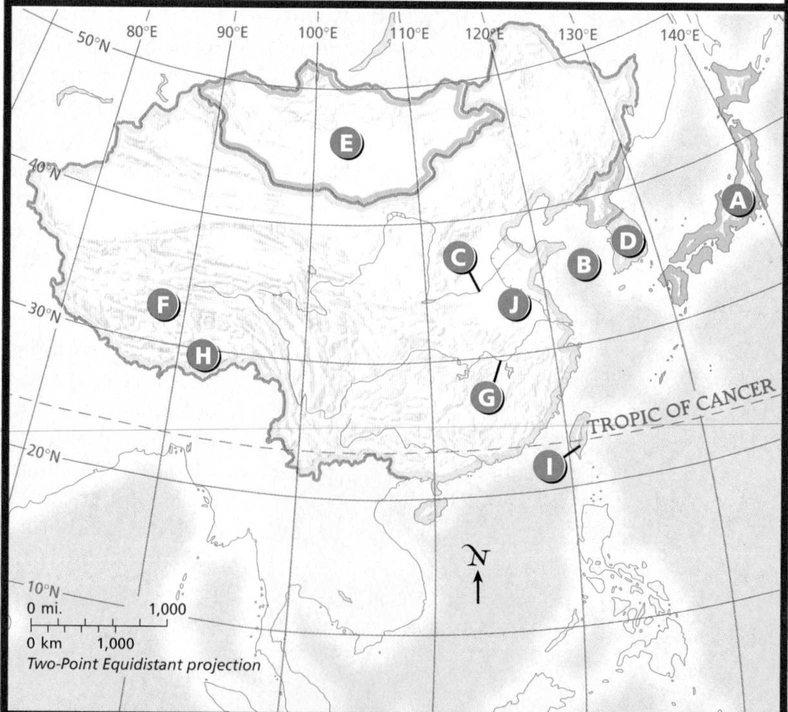

Self-Check Quiz Visit the **Glencoe World Geography** Web site at geography.glencoe.com and click on Self-Check Quizzes—Chapter 26 to prepare for the Chapter Test.

Using the Regional Atlas

Refer to the Regional Atlas on pages 636–639.

1. **Region** What rivers drain the Manchurian Plain?
2. **Location** Compare the physical and economic activity maps. What fossil fuels are found on the North China Plain?

Thinking Like a Geographer

Flooding on China's Yellow River periodically causes damage and loss of life. Use what you know about the physical geography of the region to write a paragraph explaining the causes of the flooding and suggesting possible solutions.

Problem-Solving Activity

Problem-Solution Proposal Conduct research on the growth of urbanization and manufacturing in East Asia. Analyze the effects of these processes on the climate of the region. Determine to what extent climate changes in East Asia can be related to global warming. Then prepare a proposal that suggests ways to avoid or reverse the causes or harmful consequences of climatic changes.

GeoJournal

Comparison-Contrast Essay Use GeoJournal data from this and previous units to write a descriptive essay that compares and contrasts cultural patterns of East Asia to those of two other global regions you have already studied.

Technology Activity

Creating an Electronic Database Use Internet and library resources to research recent significant earthquakes, tsunamis, volcanic eruptions, and typhoons in East Asia. Use a database program to organize your data into a table with headings for location, type, and severity of each event. Then write a paragraph describing the effects of physical processes, such as the wave action of tsunamis, on the specific locations.

Standardized Test Practice

Choose the best answer for the following multiple-choice question. If you have trouble answering the question, use the process of elimination to narrow your choices.

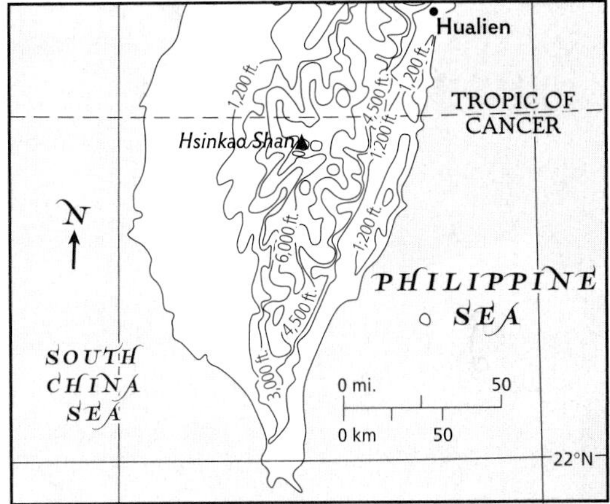

1. **If you were to hike straight up Hsinkao Shan, what would be the most gradual route to take?**

   A From the west
   B From the south
   C From the north
   D From the east

On a contour map, the closer the lines that show elevation (isolines), the faster the terrain rises and the steeper the topography. Where lines are far apart, the change in elevation is more gradual. To answer the question correctly, first find Hsinkao Shan on the map. Determine the side of the mountain where the lines seem farthest apart. Then choose the answer that best describes your observation. It also may be helpful to notice where the map lines are closest. You can then eliminate these choices from the answers.

CHAPTER 27

The Cultural Geography of East Asia

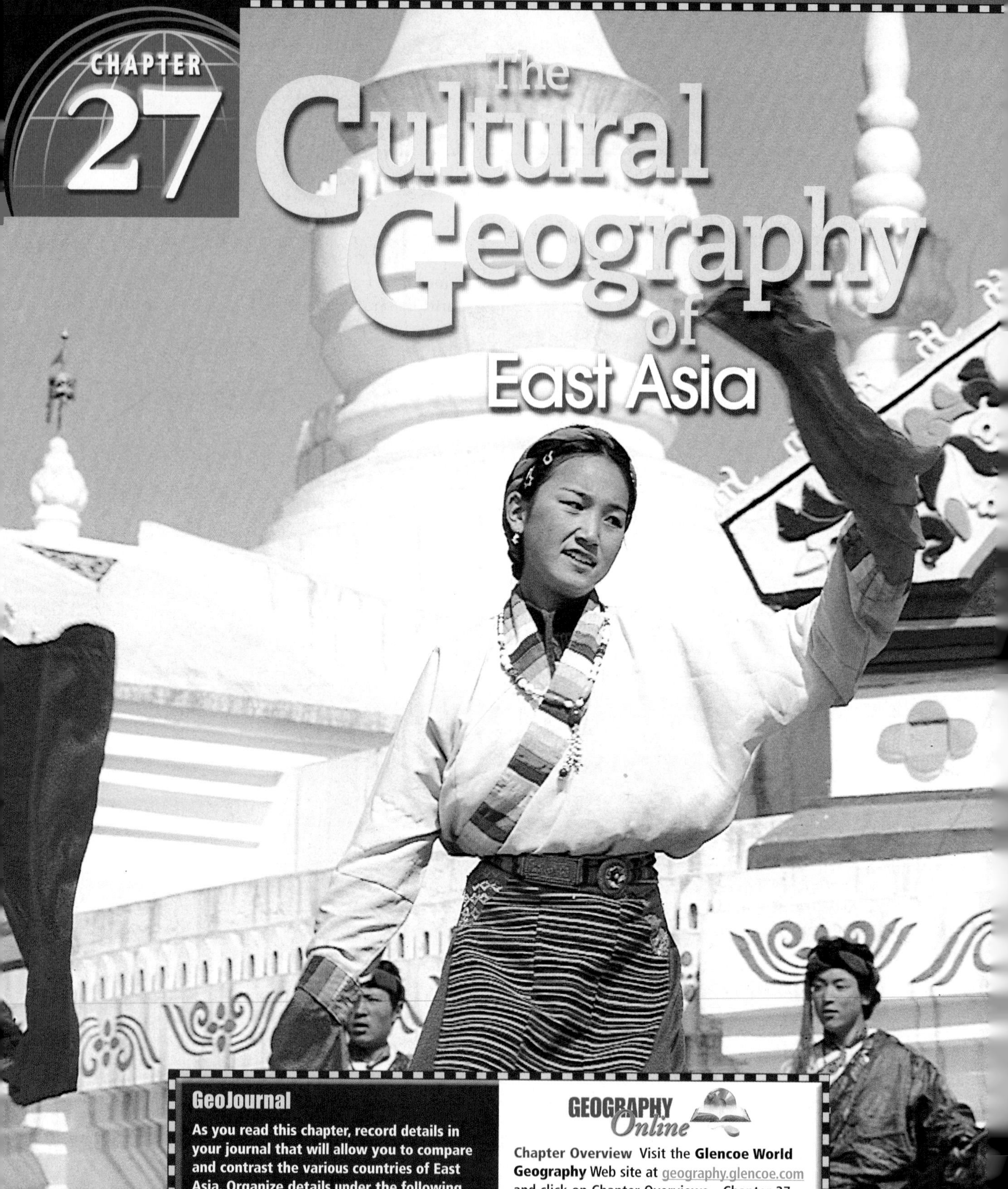

GeoJournal

As you read this chapter, record details in your journal that will allow you to compare and contrast the various countries of East Asia. Organize details under the following heads: population patterns, history and government, and cultures and lifestyles.

GEOGRAPHY Online

Chapter Overview Visit the **Glencoe World Geography** Web site at geography.glencoe.com and click on Chapter Overviews—Chapter 27 to preview information about the cultural geography of the region.

SECTION 

Population Patterns

Guide to Reading

Consider What You Know

In many parts of the world, people are migrating from rural areas to cities. What advantages and disadvantages do you think this trend brings?

Read to Find Out

- What ethnic groups make up East Asia's population?
- In what country do the majority of East Asians live?
- How is population in East Asia distributed?

Terms to Know

- aborigine
- homogeneous

Places to Locate

- Taipei
- Seoul
- Pyongyang
- Tokaido corridor
- Tokyo

NATIONAL GEOGRAPHIC

A Geographic View

Torrent of Commuters

There seems to be no end to Tokyo's congestion, no time of day when the city slackens pace to catch its breath. By 8 a.m., three million commuters are coursing through train and subway stations, joining 12 million residents of Tokyo proper on their purposeful way to work. . . . One rush hour morning . . . I got swept away in a pedestrian torrent . . . flowing in the opposite direction, and I was carried the distance of a city block. . . .

—*Arthur Zich, "Japan's Sun Rises Over the Pacific,"* National Geographic, *November 1991*

Tokyo street at night

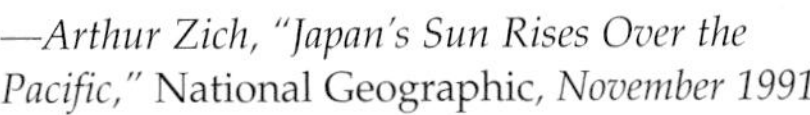

In Japan, as in other parts of East Asia, people are crowded onto relatively small lowland areas along rivers or on seacoasts. There, the largest cities are located. In this section you will learn what peoples make up East Asia's population, where East Asians live, and why many of them are migrating from rural areas to cities.

Human Characteristics

East Asia has more than 1.5 billion people—about 25 percent of the world's population. East Asians form many different ethnic groups, each with its own language and cultural traditions. Among the region's major ethnic groups are the Chinese, Tibetan, Japanese, Korean, and Mongolian.

China

When people in China say someone is Chinese, they use the Chinese word that means "a person of the Middle Kingdom." About 92 percent

◀ *Tibetan dancers, Beijing, China*

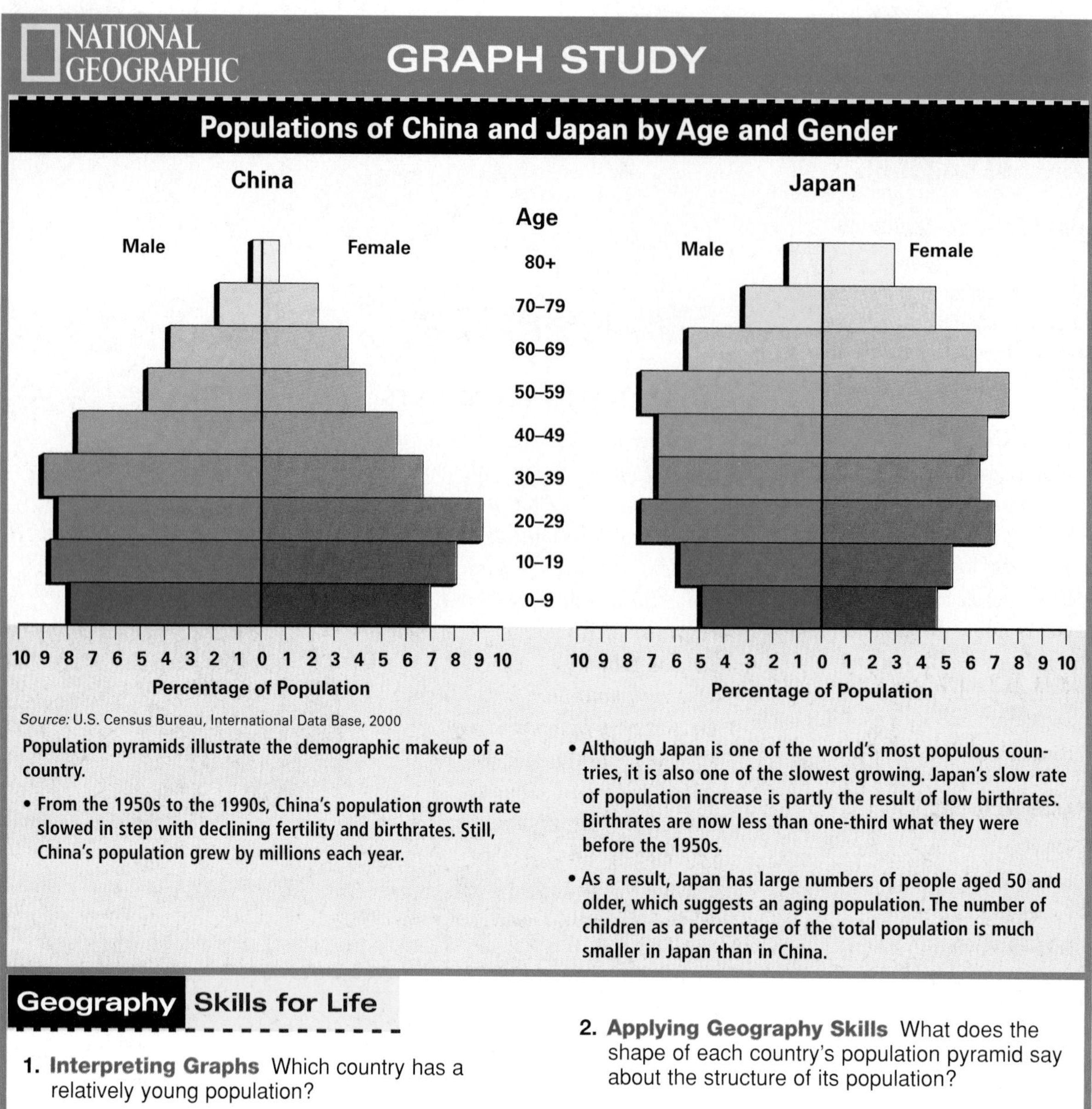

of China's 1.3 billion people belong to the Han, an ethnic group named for a powerful ancient Chinese ruling family. From 206 B.C. to A.D. 220, Han rulers developed a culture whose influence has lasted to the present.

The remaining 8 percent of China's population belong to about 55 different ethnic groups, most of whom live mainly in western and northern China. Although ruled by China, non-Chinese peoples such as the Tibetans have their own separate histories and cultures. For example, the Tibetan homeland of Tibet, located on a high Himalayan plateau, was once a Buddhist kingdom. Since China's takeover of Tibet in 1950, the Tibetans have resisted Chinese efforts to destroy their culture.

Off China's southeastern coast lies the island of Taiwan. Taiwan and China share a long history. Most of Taiwan's people are descended from Chinese who migrated to the island several hundred years ago. Another 15 percent of the Taiwanese population descend from Chinese who fled from China to Taiwan in 1949, after the Communists in China defeated the Nationalist government in a civil war. Taiwan's original inhabitants, or **aborigines**, are related to peoples in Southeast Asia and the Pacific area. They make up only about 2 percent of Taiwan's population.

Japan, Korea, and Mongolia

The populations of other East Asian countries have distinct ethnic groups. Japan is ethnically **homogeneous** (HOH•muh•JEE•nee•uhs)—having a population belonging to the same ethnic group. About 99 percent of Japan's population is ethnic Japanese, descendants of Asian migrants who crossed the Korean Peninsula to reach Japan centuries ago. The migrants forced Japan's earliest-known aboriginal people, the Ainu (EYE•noo), to move gradually north. Small numbers of Ainu still live on the island of Hokkaido (hoh•KY•doh).

Like Japan, Korea has long been ethnically homogeneous. Koreans trace their origins to early peoples from northern China and Central Asia. They have maintained their common identity despite long periods of foreign rule and today's division of the Korean Peninsula into communist North Korea and democratic South Korea.

The people of Mongolia are mostly ethnic Mongolians. Centuries ago their Mongol ancestors ruled the world's largest land empire, which stretched from China to eastern Europe. Today the Mongolians are divided into separate linguistic groups, but about 90 percent of them speak the Khalkha Mongolian language.

Where East Asians Live

Physical geography influences where East Asians live. Because much of East Asia is barren and mountainous, the region's population is distributed unevenly. Most East Asians settle in coastal areas or in fertile areas along rivers. In these places, among the most densely populated on Earth, the land and climate are favorable for agriculture, industry, and urban growth.

Population Distribution and Density

Despite China's large land area, more than 90 percent of Chinese live on only one-sixth of the land. Most inhabit the fertile valleys and plains of China's three great rivers: the Yellow (Huang He), Yangtze (Chang Jiang), and Xi. Large urban centers, such as Shanghai, Beijing, Tianjin, and Guangzhou, lie in river valleys or coastal plains. They have populations ranging from 6 million to more than 13.5 million. By contrast, the rugged western province of Xinjiang has a sparse population of farmers and herders living on scattered oases. About 2.5 million people live in Mongolia's vast interior steppes, a population density of only 4 people per square mile (2 people per sq. km).

Space is limited on Taiwan, where most of the island's 22 million people live in cities such as **Taipei** (TY•PAY) that lie on or close to the coast. In North and South Korea, most people inhabit coastal plains that wrap around the Korean Peninsula's mountainous interior. About two-thirds of the Korean population lives in rapidly growing cities, such as **Seoul** (SOHL) and **Pyongyang**.

Japan has limited land area for its large population. Forested mountains cover the central part of the country, leaving only valleys and coastal plains for settlement. About 78 percent of Japan's 126.7 million people live in coastal urban areas, such as the **Tokaido corridor**—a series of cities crowded together on the main island of Honshu. One of these cities, **Tokyo**, is the world's most populous urban area, with more than 28 million people. By contrast, Japan's northernmost large island, Hokkaido, remains rural with few people.

Culture

Japan's Urban Lifestyle

Urbanization shapes the physical surroundings and lifestyles of the Japanese people. Hundreds of skyscrapers tower over the busy streets of Japan's modern cities. Glaring neon signs advertise cars, electronics, and watches. As in most of East Asia's crowded cities, a childless couple might live in a tiny one- or two-bedroom apartment. Because of Japan's high population density and costly land, suburban homes are small compared to those in other developed countries.

The Japanese have adapted to their crowded conditions with an efficient transportation system. Commuters board the Shinkansen express, or bullet train, to get to their destinations. As the electric train pulls out of the station, its movement gently presses passengers back into their seats. In a few moments, the train reaches speeds of up to 160 miles per hour (257 km per hour) along the Tokaido corridor. The westbound train cruises from Tokyo through the urban,

Geography Skills for Life

Traveling in Japan Japan's high standard of living enables vacationers to go abroad or to seaside and mountain resorts.

Place What form of public transportation is popular in Japan today?

industrial areas of Nagoya, Osaka, and Kobe, to Fukuoka on the island of Kyushu, 604 miles (972 km) away. A trip that takes more than 11 hours of hectic driving and delays by car takes only about 5 hours by high-speed train.

Migration

In recent decades many people in China and South Korea have moved from rural, desert, or mountainous areas to cities. Although most Chinese still live and work on farms, millions of people continue to migrate to high-growth urban areas. Many are especially drawn to southeastern China, where China's communist government allows privately owned businesses in Hong Kong and in special economic zones. For factories in these special zones, the arrival of migrants means plenty of available labor, as one observer notes:

> *"These* wailai gongren*—literally, external coming workers—outnumber the Dongguan population, with more arriving all the time. 'When I need workers,' a sweater factory manager said, 'I just put a sign outside the gate.'"*
>
> Mike Edwards, "Boom Times on the Gold Coast of China," *National Geographic*, March 1997

In South Korea many people also have moved from rural areas, seeking industrial jobs in coastal cities. Politics, however, has affected migration on the Korean Peninsula. To escape communism, many people in the mid-1900s fled from North Korea to South Korea or to other countries, especially the United States and Canada, seeking political and economic freedom. Today South Korea has 48.8 million people, more than twice as many as North Korea, where the standard of living is much lower.

Challenges of Growth

Population changes and increasing urbanization have brought challenges to East Asia. In China and South Korea, for example, the steady migration from rural villages to cities has led to urban overcrowding. This population shift has contributed to farm labor shortages in the countryside. To stem migration from rural areas to already overcrowded urban areas, China, for example, has built dozens of new agricultural towns in remote areas. These towns are designed to provide more social services and a better quality of life for rural people. The Chinese government hopes that the benefits of the new towns will encourage people to stay on their farms.

▲ *A family in Xi'an, China*

Ever-growing populations in East Asia have put a strain on limited resources and services. Some of East Asia's governments see population control as another way to meet the challenges of population growth. In 1979 China began a policy that allowed each family to have no more than one child. Although not followed by all Chinese, the "one-child" policy until recently had been a factor in slowing China's population growth rate. Now that the policy is no longer strictly enforced, China's population growth rate is increasing once again. Statistics presented in the population pyramid of China on page 662 suggest that a higher birthrate is largely responsible for the increased population growth rate.

Population changes will continue to play an important part in East Asia's future. In the next section, you will learn about the values and traditions that sustain East Asians as they face the many challenges of the future.

SECTION 1 ASSESSMENT

Checking for Understanding

1. **Define** aborigine, homogeneous.
2. **Main Ideas** Create a graphic organizer like the one below, and fill in key points about population distribution and density, ethnic groups, and migration. Summarize one of the three topics in terms of population patterns in East Asia.

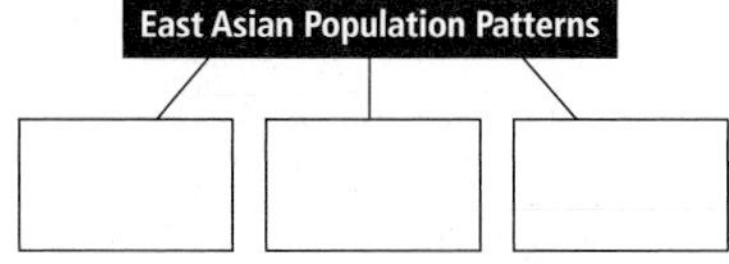

Critical Thinking

3. **Drawing Conclusions** How does high-speed transportation in Japan affect daily life and the economy?
4. **Identifying Cause and Effect** Why has migration to urban areas increased in East Asia in recent years?
5. **Making Inferences** How might population growth and the continued migration of people from rural to urban areas affect East Asia's agricultural future?

Analyzing Graphs

6. **Place** Study the graph on page 662. How might China's demographic makeup affect the rural/urban distribution of its population?

Applying Geography

7. **Geography and Cities** Study the physical and political maps on pages 636–637. Describe the type of physical feature East Asia's major cities have in common.

GEOGRAPHY AND HISTORY

A TALE OF TWO CHINAS

WOULD YOU VOTE FOR A PRESIDENTIAL CANDIDATE who occasionally donned a Superman costume? The citizens of Taiwan did when they elected Chen Shui-bian as their president. Neighboring China was infuriated, though not just because of the costume.

An Island Republic of China

Taiwan is a mountainous island located 90 miles (145 km) off China's coast. For most of its history, Taiwan has belonged to China. In 1949 the Chinese Nationalist Party, led by Chiang Kai-shek, lost its civil war against Mao Zedong's Communists. The battered Nationalist army fled to Taiwan with two million refugees. From Taipei, its capital-in-exile, the Nationalist regime maintained that it was the legitimate government of one China and vowed to recover control of the mainland. Taiwan called itself the Republic of China, while the Communist mainland took the name of the People's Republic of China.

United States intervention in the 1950s kept the more powerful Communists from conquering Taiwan. In step with the mainland, Taiwan pursued a goal of "one China"—two parts of one nation moving toward reunification. Taiwan wanted China's Communist government to change and to negotiate with Taiwan as an equal. Communist leaders, however, said no.

Taiwan's capital of Taipei bustles with economic activity. ▶

◀ **President Chen Shui-bian and Taiwan's first female vice president, Annette Lu, celebrate their win.**

NATIONAL GEOGRAPHIC SOCIETY

From Rice Fields to Computer Chips

As the two sides haggled, many nations shifted their allegiance from Taiwan to China. The United States improved its relations with China in the 1970s and ended diplomatic relations with Taiwan. Meanwhile, life on the island was changing as dramatically as its alliances.

When the Nationalists first arrived in Taiwan, they found farmers cultivating rice fields in fertile valleys and a small population of native people living in the mountains. Taipei was an overgrown shantytown. The Nationalists quickly and brutally seized power. They allowed no local representation, and freedoms were limited.

In 1975 President Chiang Kai-shek died. When his son Chiang Ching-kuo was elected president in 1978, he began to institute democratic reforms. He ended martial law and legalized opposition political parties. By the 1990s Taiwan was a shining example of democracy in Asia. Prosperity transformed the island into an economic powerhouse. By comparison, China's communist economy was stagnant.

In 2000 Chen Shui-bian of the Democratic Progressive Party became the first ethnic Taiwanese and the first non-Nationalist to be elected president. Chen supports Taiwanese independence from the mainland—a stance that evokes angry reactions and military threats from China. Ever mindful of China's threats of war, Chen is working to improve ties with the mainland.

Today the economies of China and Taiwan are intertwined. Taiwan has invested billions of dollars in factories on the mainland. China, and the rest of the world, relies on Taiwan for key computer parts. With its strategic location and hardworking population, Taiwan is an important player in the global economy.

Looking Ahead

China advocates a "one country, two systems" approach to reunification with Taiwan. Most Taiwanese, however, would prefer to remain separate from China. How might China and Taiwan reconcile their differences for a better future for both countries?

1949 Chinese Nationalists, defeated by Chinese Communists, flee to Taiwan and establish government

1975 Nationalist leader Chiang Kai-shek (photo above) dies

1978 Chiang Ching-kuo elected president of Taiwan

1980s Taiwan institutes democratic reforms

1990s Taiwanese demonstrators (background photo) call for independence from China

1996 Taiwan holds first presidential elections; tensions with China escalate

2000 Chen Shui-bian wins presidential election; China threatens war unless Taiwan resumes reunification talks

History and Government

Guide to Reading

Consider What You Know

Many of the inventions we take for granted, such as printing, gunpowder, paper money, the compass, and the wheelbarrow, originated in ancient East Asia. Why do you think these ideas did not spread to the West until many centuries later?

Read to Find Out

- Where did East Asia's ideas and traditions originate?
- How did East Asia first react to contact with the West?
- What major wars and revolutions occurred in East Asia?

Terms to Know

- culture hearth
- dynasty
- clan
- shogun
- samurai

Places to Locate

- Great Wall of China
- Guangzhou

NATIONAL GEOGRAPHIC

A Geographic View

China's Buried Army

"A creation of awesome scale and accomplishment—an unforgettable symbol of the power of China's first emperor . . . Qin Shi Huang [Di] wanted an army with him after he died," says museum director Yuan. "His underground empire was a miniature of his real one." More than 700,000 laborers toiled 36 years building his monument.

—O. Louis Mazzatenta, "China's Warriors Rise From the Earth," National Geographic, October 1996

Army of clay soldiers, China

In the Chinese city of Xi'an, archaeologists have unearthed thousands of life-size clay statues of soldiers and horses positioned as an army ready for battle. These burial statues were to protect the ancient Chinese ruler Qin Shi Huang Di (CHIHN SHIHR HWAHNG DEE) from threats in the afterlife. During the 200s B.C., Qin Huang Di ordered the building of the **Great Wall of China** to protect his empire. Archaeological finds, such as that of Qin Huang Di's tomb, reveal much about East Asia's long history and political heritage.

Ancient East Asia

East Asia is home to some of the world's oldest continuous civilizations. China, where the earliest East Asian civilization emerged, became the region's **culture hearth**, or a center from which ideas and practices spread to surrounding areas. Throughout history, China's influence helped shape East Asia's cultures. The Koreans and the Japanese, for example, blended Chinese ways with their own to form distinct cultural traditions.

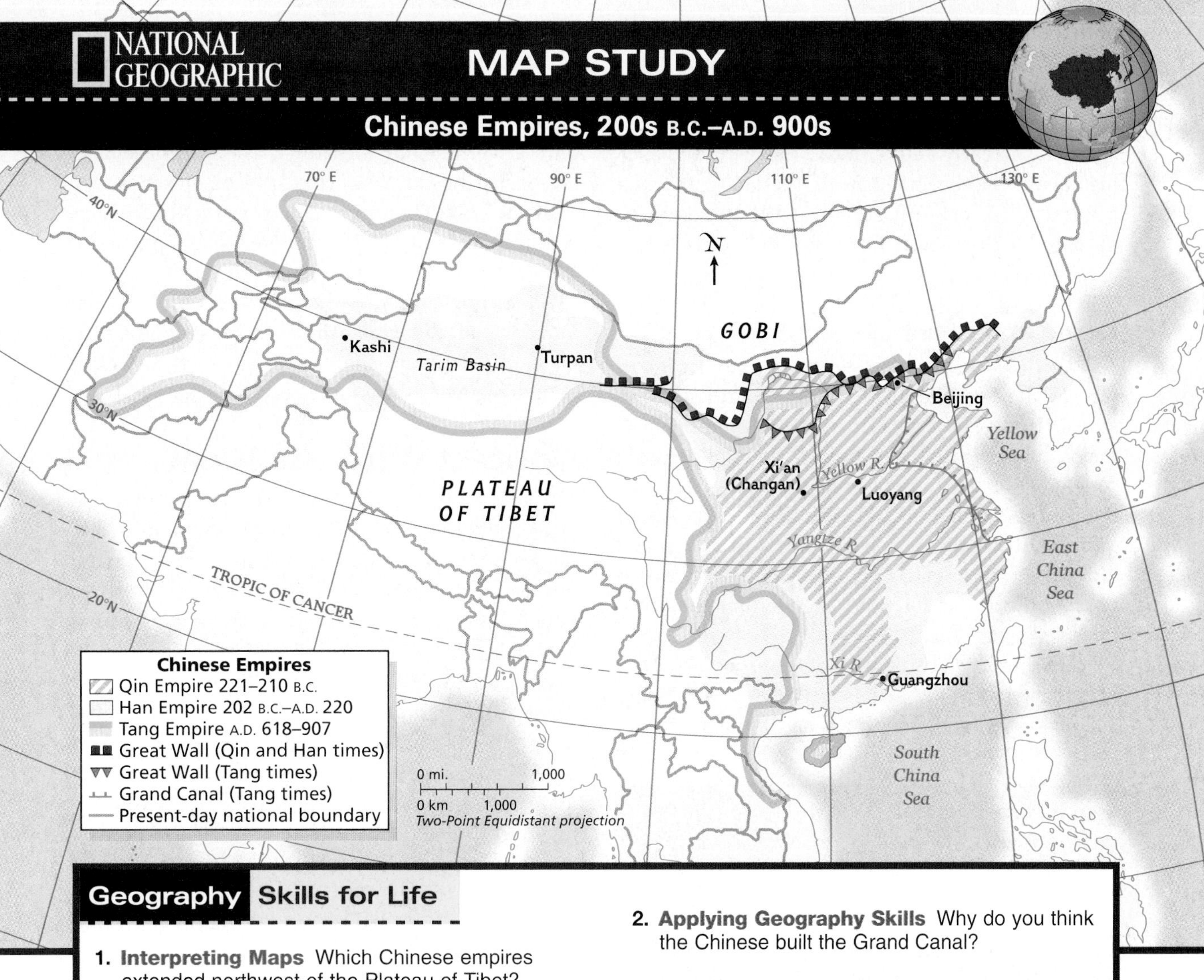

Geography **Skills for Life**

1. **Interpreting Maps** Which Chinese empires extended northwest of the Plateau of Tibet?

2. **Applying Geography Skills** Why do you think the Chinese built the Grand Canal?

Find NGS online map resources @ www.nationalgeographic.com/maps

Government

China's Dynasties

Although China's culture began more than 5,000 years ago in the valley of the Wei River, a tributary of the Yellow River, historical records were first kept under the Shang **dynasty**. The dynasty, or ruling family, took power about 1600 B.C. in the North China Plain. Like all succeeding dynasties, the Shang faced rebellions by local lords, attacks by Central Asian nomads, and natural disasters such as floods. When the government was stable, it could defend its people against some of these problems. Eventually, however, the dynasty weakened and fell. According to the Chinese, a fallen dynasty had lost "the mandate of heaven," the approval of the gods and goddesses.

After the Shang, the Zhou (JOH) dynasty ruled for 900 years, beginning about 1122 B.C. During the era of the Zhou dynasty, Chinese culture spread, trade grew, and the Chinese began making iron tools. China's best-known philosopher, Confucius (or Kongfuzi), lived during this time. He founded a system of thought based on discipline and moral conduct that for centuries influenced East Asian life. Another thinker, Laozi (or Lao-tzu), helped found Daoism, a philosophy of living in simplicity and harmony with nature.

After the Zhou, powerful dynasties expanded China's territory. In the 200s B.C., Qin Shi Huang Di united all of China and built the first section of the Great Wall to ward off attacks from Central Asia. Under the Han and Tang dynasties, traders and missionaries took Chinese culture to all of East Asia. In the early 1400s, under the Ming dynasty, the naval explorer Zheng He (JUNG HUH) reached as far as the coast of East Africa. The last dynasty, the Qing, ruled China from the mid-1600s to the early 1900s.

Korea and Japan

About 1200 B.C. Chinese settlers brought their culture to the neighboring Koreans. Buddhism later spread from China to Korea and became Korea's major religion. In the centuries that followed, a series of Korean dynasties, including the Silla and the Koryo, united the Korean Peninsula. About A.D. 1300 the Chinese seized control of Korea and introduced the philosophy of Confucius, which became the model for Korea's government, education, and family life.

The Korean Peninsula was for centuries a cultural bridge between the Asian mainland and Japan. As a result, China and Korea had a major impact on Japan's civilization. In the A.D. 400s Japan, once ruled by many **clans**, or family groups, united under the Yamato dynasty. Yamato rulers adopted China's philosophy, writing system, art, sciences, and governmental structure. The Japanese also were influenced by the works of Korean scholars.

By the 1100s the armies of local nobles had begun fighting for control of Japan. Yoritomo Minamoto became Japan's first **shogun**, or military ruler, in 1192. Supporting the shogun were professional warriors, or **samurai**. Although an emperor officially ruled Japan, the samurai helped powerful shoguns govern the country until the late 1800s.

NATIONAL GEOGRAPHIC World Explorer

Geography Skills for Life

Japanese History A Japanese man performs as a samurai in a historical reenactment.

Region Why were samurai important to Japanese history?

Contact With the West

By the 1600s Western countries had set up shipping routes to East Asia, hoping to share in the region's rich trade in silk and tea. China, Japan, and Korea, however, all rejected foreign efforts to penetrate their markets. Under Western pressure, China finally opened the port of **Guangzhou** to limited trade in 1834. Dissatisfied, Europeans used powerful warships to force China to open more ports. By the 1890s, European governments and Japan had claimed large areas of China as *spheres of influence*—areas in which they had exclusive trading rights. Deadlocked by rivalries, these powers reluctantly agreed in 1899 to a U.S. proposal to open China to all countries for trade.

During the 1800s the United States also worked to open Japan for trade. In 1854 U.S. naval officer Matthew C. Perry pressured the Japanese to change their policy. He and Japanese officials negotiated a treaty that ended centuries of Japanese isolation and opened Japan to trade with the United States. Not long afterward, rebel samurai forced shoguns to return full authority to the emperor. Japan's new government rapidly modernized the country's economy, government, and military forces.

Modern East Asia

During the 1900s East Asia as a whole was involved in two world wars. Meanwhile, each East Asian country faced its own internal upheavals.

Revolutionary China

In 1911 a revolution led by Sun Yat-sen ended the rule of emperors in China. By 1927 a military

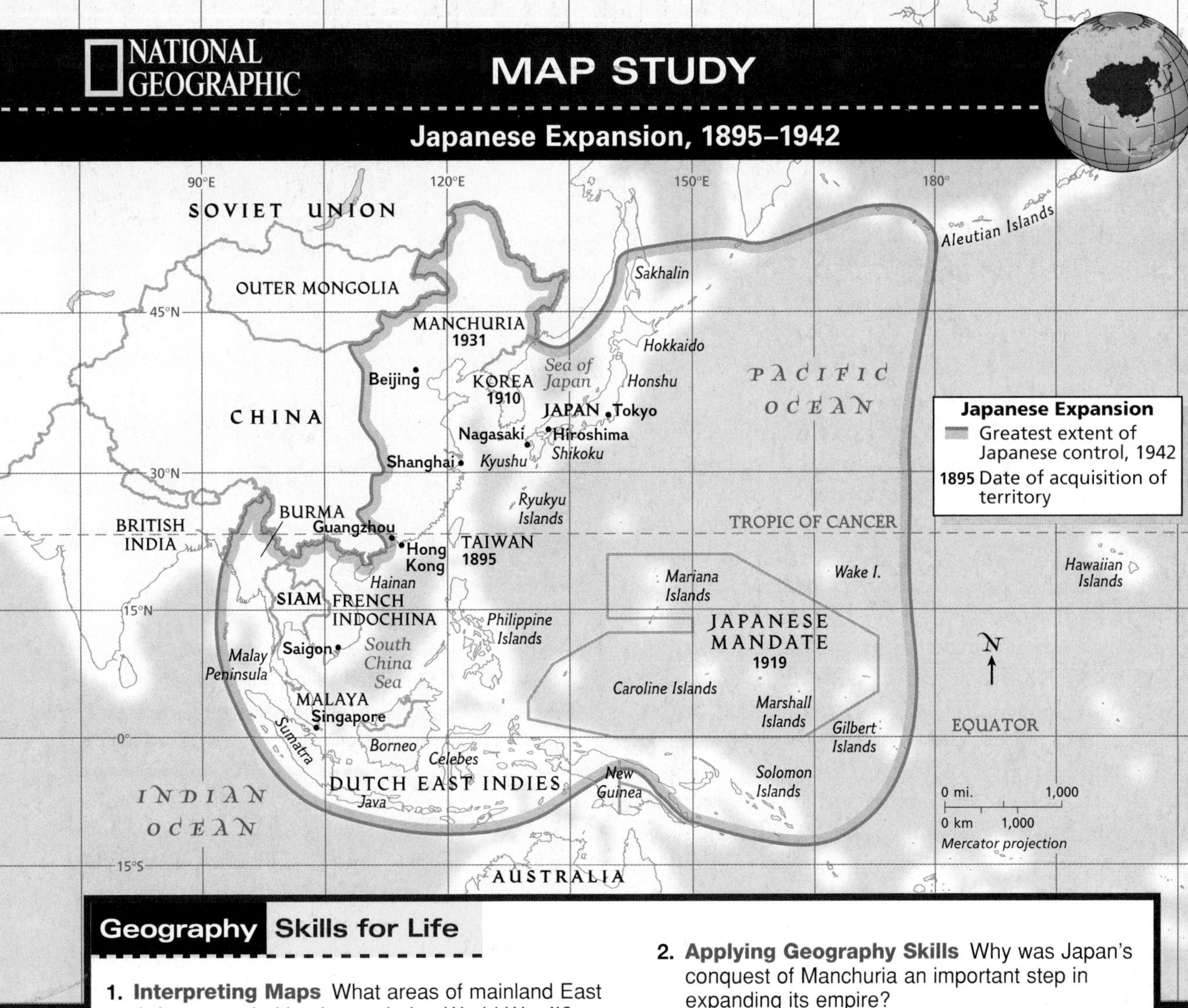

Geography Skills for Life

1. **Interpreting Maps** What areas of mainland East Asia were ruled by Japan during World War II?

2. **Applying Geography Skills** Why was Japan's conquest of Manchuria an important step in expanding its empire?

Find NGS online map resources @ www.nationalgeographic.com/maps

leader, Chiang Kai-shek, had formed the Nationalist government. Meanwhile, Chiang's communist rival, Mao Zedong, gained support from China's farmers. After years of civil war, the Communists won power in 1949 and set up the People's Republic of China on the Chinese mainland. The Nationalists fled to Taiwan and set up a government called the Republic of China.

In the late 1900s, the People's Republic of China maintained strict communist political rule. However, pressures to modernize gradually opened China's economy to free-market influences. Meanwhile, Taiwan built a powerful, export-based economy and carried out democratic reforms.

Japan's Transformation

From the 1890s to the 1940s, Japan used diplomacy and military force to build an empire that included Taiwan (then called Formosa), Korea, other parts of mainland Asia, and numerous Pacific islands. This expansion was one factor that led Japan to fight the United States and other Allied countries in World War II. After its defeat in 1945, Japan became a democracy. Stripped of its overseas territories and military might, Japan rebuilt its shattered economy and society. By the late 1900s, it had emerged as a global economic power with worldwide trading and business links. One retired official described the change this way:

> *"After the war, Japan was in chaos. There was regret, suffering. There were no rich then, only poor. We pulled together, worked hard, geared our economy for export. . . . Now we're prosperous, and we're bringing that prosperity to others."*
>
> Arthur Zich, "Japan's Sun Rises Over the Pacific," *National Geographic,* November 1991

Geography Skills for Life

A Historic Meeting A monk reads about the historic meeting of North Korea's Kim Jong Il and South Korea's Kim Dae Jung in June 2000.

Place Why was the meeting between the two leaders considered historic?

A Divided Korea

After World War II ended, Korea was divided into American-backed South Korea and communist-ruled North Korea. Wanting to unite Korea, North Korea invaded South Korea in 1950. During the Korean War, United Nations forces, led by the United States, rushed to South Korea's defense. By June 1951, each army had dug in along the thirty-eighth parallel. The stalemate ended with a truce in 1953. Millions of Koreans had died and both countries were devastated. Today, North Korea and South Korea are still separated by the cease-fire line along the thirty-eighth parallel.

North Korea's communist society often cannot meet the basic needs of its people. South Korea has become a democracy with a prosperous market economy. In 2000, relations between North Korea and South Korea began to improve after the leaders of the two countries held talks. That same year, South Korea's president, Kim Dae Jung, won the Nobel Peace Prize for his peacemaking efforts.

A Free Mongolia

Under the Soviet Union's influence, Mongolia was a communist state from 1924 to 1991. After the collapse of Soviet communism, the Mongolians adopted a democratic constitution that opened the way for free elections and a market economy, reflecting a growing openness to new ideas in East Asia.

SECTION 2 ASSESSMENT

Checking for Understanding

1. **Define** culture hearth, dynasty, clan, shogun, samurai.
2. **Main Ideas** On tables like the one below, summarize each East Asian country's history and government during each time period.

Country: ______________

Ancient Times: (3000 B.C. to A.D. 1600)	
Contact With the West (1600s–1900s)	
Modern Times (1900s to present)	

Critical Thinking

3. **Drawing Conclusions** Why were European powers dissatisfied with China's opening of the port of Guangzhou?
4. **Comparing and Contrasting** How were economic developments in Taiwan and South Korea during the 1900s similar and different?
5. **Making Inferences** Why do you think the Chinese Nationalists who fled to Taiwan called their government the Republic of China?

Analyzing Maps

6. **Human-Environment Interaction** Study the map of Chinese empires on page 669. Which rivers are linked by the Grand Canal?

Applying Geography

7. **Spread of Culture** Draw a map of East Asia to show the spread of Chinese culture in the region. Use arrows to show key movements. Then write an explanation of your map.

SECTION 3

Guide to Reading

Consider What You Know

East Asian food, art, pottery, and sports have become popular around the world. What foods and other products from East Asia are found in your community?

Read to Find Out

- What languages do the peoples of East Asia speak?
- What religions and philosophies do many people of East Asia follow?
- How do the standards of living of East Asians compare with one another?
- How does education in East Asia compare with education in North America?
- What traditional arts make East Asia unique?

Terms to Know

- ideogram
- shamanism
- lama
- acupuncture
- haiku
- calligraphy
- pagoda

Places to Locate

- Mongolia
- Tibet

Cultures and Lifestyles

NATIONAL GEOGRAPHIC

A Geographic View

A Spiritual Journey

As early as the [A.D. 400s], caves were carved into the sandstone cliffs of the Tian Shan range as shrines and places of worship for [Buddhists]. . . . Worshipers built these shrines in hopes of . . . personal well-being, a safe and prosperous journey, advancement in the next life, or perhaps the birth of many healthy sons. . . .

—*Reza, "Pilgrimage to China's Buddhist Caves,"* National Geographic, *April 1996*

Worshipers at Buddhist shrine, China

The peoples of East Asia have a long and rich cultural heritage. Since ancient times the ideas and practices of three religious traditions—Confucianism, Buddhism, and Shintoism—have profoundly influenced the region. In the modern era, communism also has had a major impact on the peoples and cultures of China, North Korea, and Mongolia. East Asians also have adopted many aspects of Western culture. In this section you will learn about the variety of cultures and lifestyles found in East Asia today.

East Asia's Languages

Because of their diverse backgrounds, people in East Asia speak languages from several different language families. The largest, Sino-Tibetan, which includes Chinese and Tibetan, comprises languages spoken by more than 1.2 billion people. Other principal languages of East Asia include Japanese, Korean, Khalkha Mongolian, and Uygur—spoken in western China.

China's Languages

Han Chinese, the most widely spoken language of China, has many dialects. Mandarin, the northern dialect, has become China's official language. It is taught in schools and used in business and government. Cantonese, another major dialect, is widely spoken in southeastern China. Other languages of China include Tibetan, Manchu, Uygur, and Mongolian dialects.

Unlike Western languages that use letters to stand for sounds in spoken language, Chinese languages use **ideograms**, pictures or symbols that stand for ideas. Chinese has thousands of ideograms. Each ideogram has one meaning, but combining it with other ideograms gives it a new meaning. For example, the ideogram for "man" next to the ideogram for "word" means standing by one's word, or "sincerity." Spoken Chinese languages also depend on tone, or pitch. Similar syllables, pronounced with different tones or inflections, take on different meanings.

Japanese and Korean Languages

Although the Japanese language developed in isolation, experts believe it may be distantly related to Korean and Mongolian. Over centuries, both Japanese and Korean languages borrowed words from Chinese. Japanese had no written form until the A.D. 400s, when Chinese writing and literature were introduced into Japan. Japan's first writing system was based on Chinese characters. Western languages, especially English, have also influenced Japanese and Korean languages.

Religion and Philosophy

East Asians hold a variety of philosophical and religious beliefs, including Confucianism, Buddhism, and Daoism. They also may follow more than one religion. Many Japanese, for example, practice both Buddhism and Shintoism, an ancient Japanese religion that stresses reverence for nature. Other religions of East Asia include Christianity,

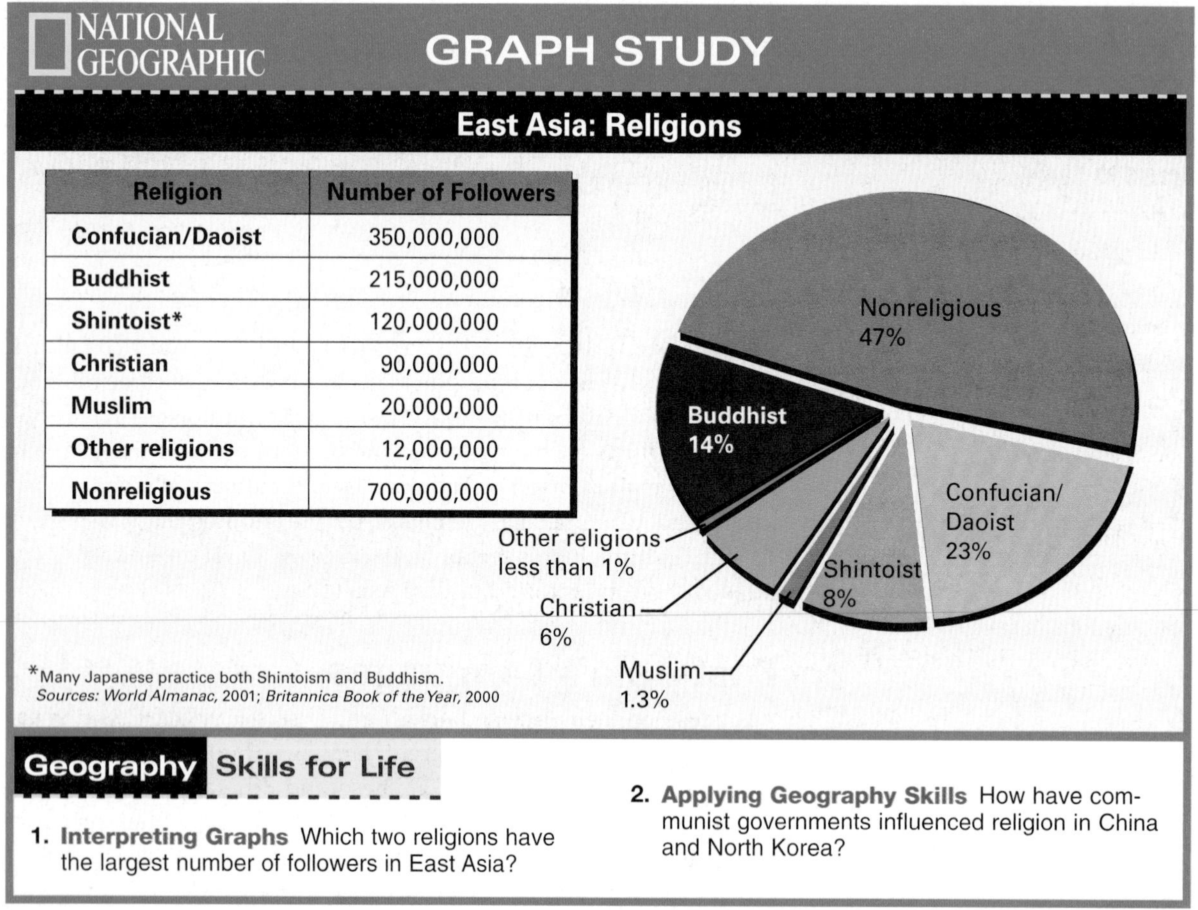

Religion	Number of Followers
Confucian/Daoist	350,000,000
Buddhist	215,000,000
Shintoist*	120,000,000
Christian	90,000,000
Muslim	20,000,000
Other religions	12,000,000
Nonreligious	700,000,000

*Many Japanese practice both Shintoism and Buddhism.
Sources: World Almanac, 2001; Britannica Book of the Year, 2000

Geography Skills for Life

1. **Interpreting Graphs** Which two religions have the largest number of followers in East Asia?
2. **Applying Geography Skills** How have communist governments influenced religion in China and North Korea?

widely practiced in Korea, and Islam, which has many followers among the Uygur people of western China. Some East Asians also practice shamanism, faith in leaders believed to have powers to heal the sick and to communicate with spirits.

Communist governments in China and North Korea strongly discourage all religious practices, but many people still hold to their traditional faiths. Before communism, Buddhist religious leaders called lamas ruled in Mongolia and Tibet. When communist governments came to power, they began to persecute Buddhists. **Mongolia** is now a democracy, and Mongolian citizens are again free to engage in religious practices.

In **Tibet**, however, the Chinese government continues to place harsh restrictions on the Buddhist population. For example, Tibetans risk arrest just for owning photographs of the Dalai Lama, Tibet's exiled spiritual leader. The Dalai Lama currently leads a worldwide movement in support of Tibetan rights from his place of exile in neighboring India.

The Dalai Lama

Standard of Living

During the late 1900s, booming economies in East Asia improved standards of living for many of the region's peoples. In wealthy countries, such as Japan, Taiwan, and South Korea, lifestyles have improved dramatically in the last few decades. Yet large gaps remain between the rich and the poor. In economically developing countries, such as China, glittering steel-and-glass skyscrapers in the cities stand in sharp contrast to the mud houses in the surrounding rural areas.

One indicator of a country's standard of living is gross domestic product (GDP) per capita, or the value of goods and services each person produces. By 2000, for example, Japan's GDP per capita was $32,350, the highest in the region. By comparison, China's GDP, at $750 per capita, was one of the lowest.

Economy

Japan's Downturn and Recovery

The Japanese traditionally have valued individual loyalty to society in return for society's protection and support. Japanese businesses often ran on the principles of teamwork and cooperation. White-collar workers had secure lifetime jobs with benefits, such as insurance programs, leave-of-absence policies, and opportunities to buy company stock.

In 1997 many of these traditions were pushed aside when Japan, along with other Asian countries, suffered a severe economic downturn. Thousands of companies went bankrupt, and financial pressures forced the companies that survived to operate more efficiently. For the first time, companies had to lay off large numbers of workers.

Since 2000, Japan's economy has still been faltering. Declining sales continue to force companies to lay off workers, and huge debts threaten to cripple the banking industry. Hesitant to spend, many Japanese consumers face housing shortages because of higher costs and lack of confidence in the economy. Meanwhile, Japan's business practices have changed. As companies focus more on profits and less on workers' job security, workers, in turn, have less loyalty to their companies.

China's New Direction

During the 1970s a new communist leadership came to power in China after the death of Mao Zedong. The most prominent leader, Deng Xiaoping (DUHNG SHOW•PIHNG), took China in a new economic direction, summed up in his phrase, "To get rich is glorious." After years of strict control

Student Web Activity Visit the **Glencoe World Geography** Web site at geography.glencoe.com and click on Student Web Activities—Chapter 27 for an activity about cultures and traditions of China, Japan, and the Koreas.

over China's economy, China's communist leaders began allowing some free enterprise as a result of economic and political setbacks during the 1950s and 1960s.

During the "Great Leap Forward" campaign of the 1950s, large government-owned farms had replaced the small-scale farm cooperatives. The new farms, however, failed to produce enough food for the country. About 20 million Chinese died of starvation, and the economy crumbled.

To move China forward, Deng Xiaoping allowed private ownership of businesses and farms. Chinese officials welcomed foreign businesses and technology to China. Foreign investment flowed into special economic zones where foreigners could own and operate businesses with little government interference. The resulting economic growth raised the standards of living of some Chinese. Despite progress, China's economy is still agricultural, and the majority of Chinese have a lower standard of living than do other East Asians.

Education and Health

Most East Asians highly value learning. Today elementary education is free throughout the region, and opportunities for higher education have expanded greatly. Better education and higher standards of living have also improved the region's health care.

Literacy and Learning

In the several East Asian countries that spend the most money for education, the literacy rate is high. Nearly all Japanese can read and write, and South Korea has a literacy rate of 98 percent. The literacy rate for Taiwanese and North Koreans is 95 percent. China and Mongolia, however, have a lower literacy rate of about 82 percent.

In the past only the wealthiest Chinese learned to read and write, but China's communist government has pushed to increase literacy. During the Cultural Revolution, a period of upheaval in the late 1960s,

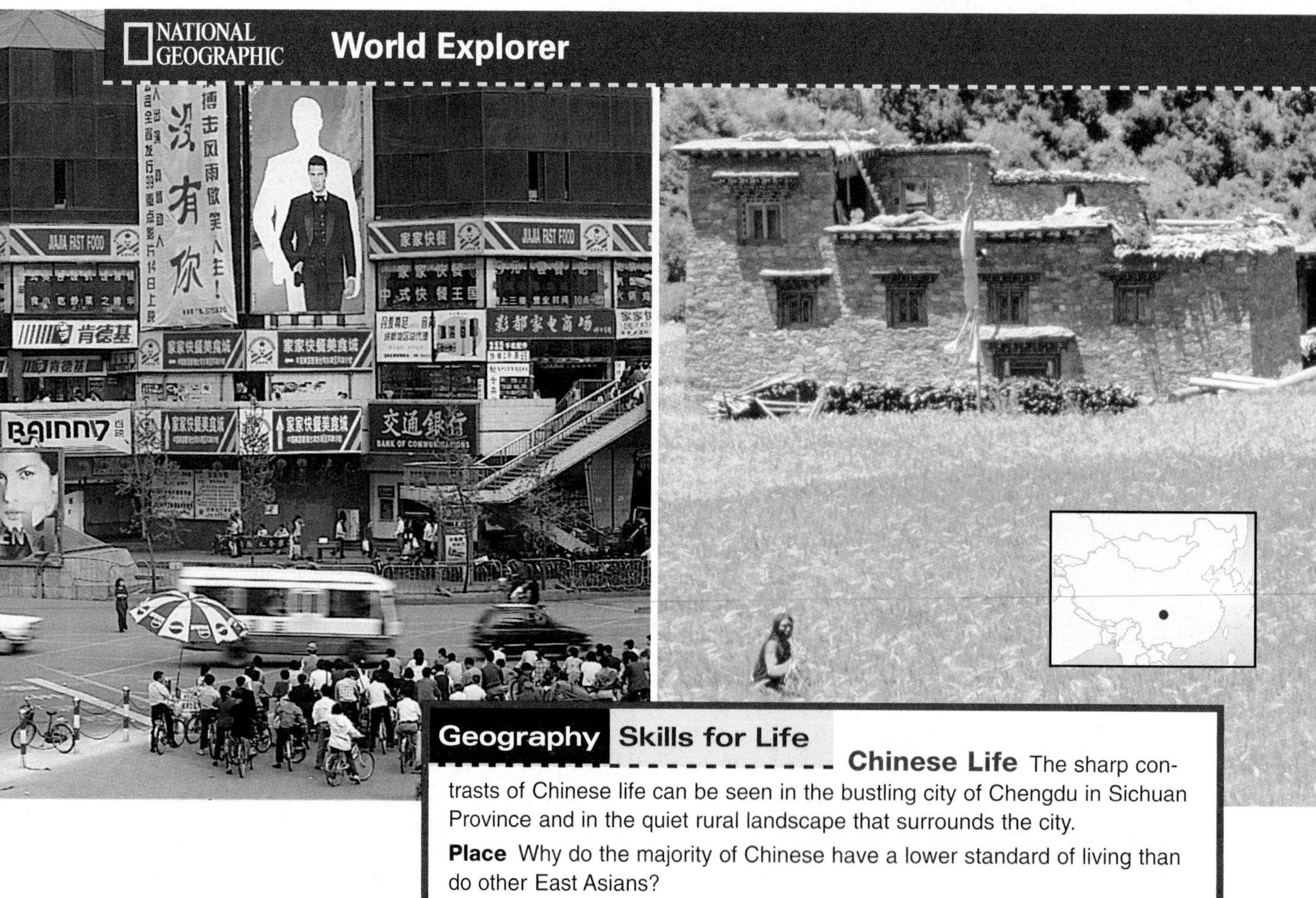

Geography Skills for Life

Chinese Life The sharp contrasts of Chinese life can be seen in the bustling city of Chengdu in Sichuan Province and in the quiet rural landscape that surrounds the city.

Place Why do the majority of Chinese have a lower standard of living than do other East Asians?

the growth of literacy, however, suffered a brief setback. During this time, schools and factories closed and people believed to be enemies of Mao Zedong's form of communism were persecuted. After Mao's death the Chinese government again emphasized education, and literacy has steadily risen.

Young South Koreans spend an average of 14 years in school and are among East Asia's best-educated students. South Korea and Taiwan believe that educational excellence supports the high performance of their economies.

▲ Chinese New Year celebrations (above) and Japanese professional baseball (right)

Health Care

Better health care has raised life expectancies, and infant mortality has declined in East Asia. The average life expectancy is about 74 years for women and 70 years for men. Communist governments generally pay for medical treatment. As China moves toward a market economy, however, its government no longer guarantees equal access to health care. As a result, the gap in the quality of health care between urban areas and rural areas is widening.

Many East Asians rely on both Western and traditional medical care, including herbal medicines. **Acupuncture**, an ancient practice that involves inserting fine needles into the body at specific points in order to cure disease or relieve pain, is popular in China. Both herbal medicine and acupuncture are widely accepted around the world.

Leisure Activities

East Asians engage in a variety of leisure activities, ranging from music to sports. Recreation frequently involves family activities. Because of small living quarters in many urban areas, people often socialize in public parks and restaurants.

Food

Although cooking styles vary throughout the region, East Asians prefer staple foods, such as rice, wheat, and millet. Many East Asians maintain vegetarian diets or get protein from fish. Western foods, such as beef, chicken, and dairy products, recently have become popular. As a result, more East Asians now have health problems associated with a Western diet.

Sports and Festivals

East Asians of all ages practice traditional exercises and martial arts, such as tai chi, tae kwon do, *gongfu* (kung fu), or karate. Japan's ancient sport of sumo wrestling draws thousands of fans to several tournaments each year. East Asians also enjoy many Western sports, such as baseball, soccer, and volleyball. Olympic champions in skiing, swimming, gymnastics, table tennis, and other sports have come from East Asia.

Colorful celebrations mark the seasons of the year, national holidays, and religious ideas or events in East Asia. Many people participate in parades and ceremonies related to the Confucian, Daoist, Buddhist, and Shintoist religions. People in East Asia also commemorate the Lunar New Year, which begins in late January or early February. The Lunar New Year reflects the lunar calendar, which is based on the phases of the moon instead of Earth's movements around the sun.

music of EAST ASIA

Under Chinese influence, music in East Asia has evolved over thousands of years. Used for both spiritual and entertainment purposes, music of the region is characterized by the use of strings, wind instruments, and percussion.

Instrument Spotlight
The **shakuhachi** is a bamboo flute from Japan with five finger holes. This unique wind instrument arrived in Japan through China during the A.D. 600s and 700s. Unlike other instruments, the shakuhachi developed a strong collection of solo pieces and was used by Zen Buddhist monks as a form of spiritual practice and for meditation. This music usually has a peaceful effect on both the player and the listener.

Go To **World Music: A Cultural Legacy** Hear music of this region on Disc 2, Tracks 14–18.

The Arts

Over the centuries East Asians have excelled in the arts. Their artistic and literary achievements are deeply rooted in the long history of the region. During ancient times Chinese styles in art and architecture influenced all of East Asia. Religions such as Confucianism, Daoism, Buddhism, and Shintoism also have inspired great art in the region. In modern times Mao Zedong's Cultural Revolution tried to wipe out the traditional arts of China in favor of communist-inspired art forms. After Mao's death, however, Chinese artists returned to their traditions.

Literature

In ancient China and Japan, poetry flourished among the educated members of society. Li Bo and Du Fu, for example, created some of China's best poetry. In their works these poets described human relationships and the beauty of nature. In A.D. 1010 a Japanese noblewoman, Lady Murasaki Shikibu, wrote one of the world's first novels, *The Tale of Genji*, about the life and loves of a prince at the emperor's court. The Japanese also developed a form of poetry called **haiku** that originally had only 3 lines and 17 syllables but now is written in many line and syllable combinations. A major theme is the fragile beauty of nature, as this example of haiku from the 1600s reveals:

> *"The red maple leaves shine so bright*
> *The wings of flying birds are scorched."*
>
> "In a Japanese Garden," *National Geographic*, November 1989

East Asia still produces notable writers. In 1994 Japanese writer Kenzaburo Oe won the Nobel Prize in literature for his works that connect the myths of traditional Japanese village life with life in the modern era. Exiled Chinese writer Gao Xingjian also won the Nobel Prize in literature in 2000.

Music and Theater

East Asian music is based on a five-tone scale with a melody line but no harmony. Over the centuries, instruments such as flutes, drums, and gongs accompanied dancers in temple rituals. Stringed instruments included the lute, the guitar, and the koto, a type of zither.

East Asians have many forms of drama. Chinese traditional opera uses elaborate costumes, music, and acrobatics or martial arts displays. Japan's lively Kabuki theater uses costumes, song, and dance. By contrast, the Japanese Noh drama has actors who tell stories only through precise movements. Traditional art in Korea may involve group folk dances. Most East Asian countries produce movies.

▲ *The ancient art of making porcelain is still practiced in China today.*

Visual Arts

Throughout history, East Asians have developed their own unique art forms. In China, Korea, and Japan, artists have painted the rugged landscapes of their countries. These paintings often include a verse made in elegant brush-stroke **calligraphy**, the art of beautiful writing. The Japanese also created vivid prints using carved wood blocks. Influential print artists include Hiroshige and Hokusai. Other Japanese art forms include origami, in which paper is folded into the shapes of animals and birds; the tea ceremony; formal landscaping; and ikebana, or flower arranging. In East Asia elegant Chinese pottery developed into a fine art over thousands of years. During the Tang dynasty, Chinese potters created the fine, thin porcelain known today as china. In Korea, during the Koryo dynasty, artists made graceful vases with a pale green glaze called celadon still highly valued all over the world. Buddhist temples in China, Korea, and Japan contain many statues and sculptures in stone, bronze, or jade.

Architecture

Except for skyscrapers, most East Asian architecture uses wood, brick, and stone. Bamboo is important in the architecture of Japan and southern China. Traditional East Asian buildings often have gracefully curved tile roofs in the **pagoda**, or tower, style.

Despite the changes and pressures brought by modernization, East Asians have kept alive their ancient art forms. These traditions help unite East Asia's diverse peoples into a cultural region.

SECTION 3 ASSESSMENT

Checking for Understanding

1. **Define** ideogram, shamanism, lama, acupuncture, haiku, calligraphy, pagoda.
2. **Main Ideas** On a table like the one below, fill in details about each country's languages, religions, education, health, standard of living, leisure, and arts.

China	Japan	N. Korea	S. Korea

Critical Thinking

3. **Comparing and Contrasting** Describe health care in East Asia. How is it different from health care in the United States?
4. **Making Generalizations** How have rising standards of living changed the lives of people in East Asia?
5. **Drawing Conclusions** How do East Asia's religions influence its art forms?

Analyzing Graphs

6. **Region** Study the graph on East Asia's religions on page 674. Christianity accounts for about what percentage of religious followers in East Asia?

Applying Geography

7. **Chinese Culture** Write a paragraph explaining the impact of the "Great Leap Forward" and the Cultural Revolution on Chinese culture.

Reading an Economic Activity Map

Geographers and researchers use economic activity maps as well as other specialized maps to help them understand a region. An economic activity map gives a quick overview of economic resources and activities.

Learning the Skill

By comparing activities on an economic activity map with information on other types of maps, such as political, climate, or population density maps, geographers can quickly see the distribution of economic resources. Geographers can also get an idea about a country's economic potential and the people's standard of living.

Economic activity maps use colors to represent dominant economic activities. Other maps may use patterns or symbols instead of colors. In all economic activity maps, the key or legend defines the colors and symbols.

To read an economic activity map, follow these steps:

- **Identify the geographic region shown on the map.**
- **Study the map key to understand all colors, symbols, and patterns used on the map.**
- **Study the map to determine what resources and economic activities are predominant in each area.**
- **Compare the map with other maps showing landforms, climate, and natural vegetation of the region.** Draw conclusions about the interaction of humans with the environment.

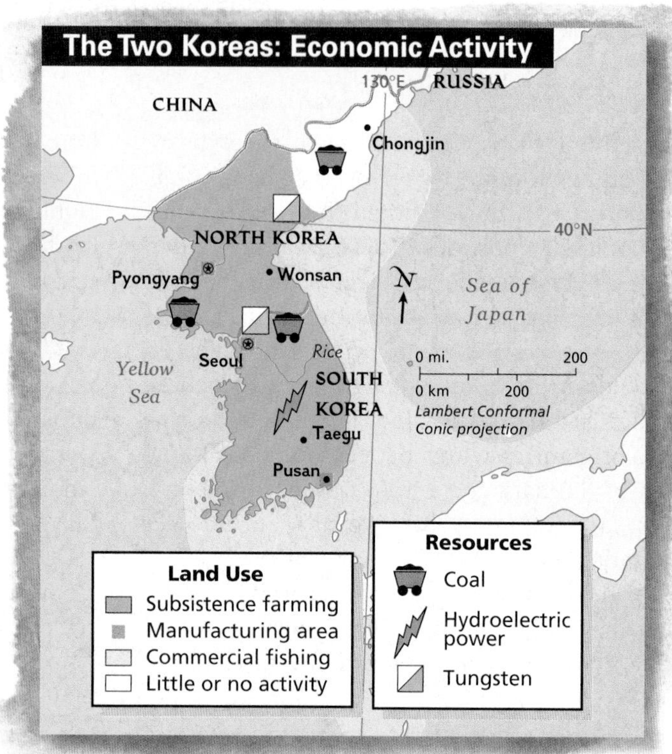

Practicing the Skill

Use the economic activity map above to answer the following questions.

1. Which color on the map represents subsistence farming?
2. Which country has more coal deposits?
3. Which area has little or no economic activity?
4. Which part of the region probably has the lowest standard of living? The highest? Explain your answer.

Use a reference book or Internet sources to create an economic activity map for your city or county. Draw an outline map of your region, and create symbols and colors to represent economic activities in your area. Be sure to include a map key.

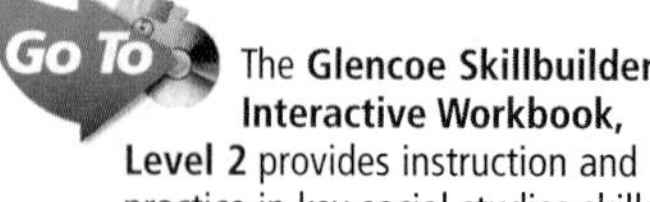

The **Glencoe Skillbuilder Interactive Workbook, Level 2** provides instruction and practice in key social studies skills.

CHAPTER 27 SUMMARY & STUDY GUIDE

SECTION 1 Population Patterns (pp. 661–665)

Terms to Know
- aborigine
- homogeneous

Key Points
- East Asia's 1.5 billion people are made up of many different ethnic groups with a variety of religions, languages, and cultures.
- Population in East Asia is unevenly distributed. It is concentrated in urban areas, in river valleys, and on coastal plains.
- Japan, Taiwan, and South Korea are highly urbanized countries. Mongolia is predominantly rural. In China most people live in rural areas.
- Massive migration from rural to urban areas has caused farm labor shortages in parts of East Asia.

Organizing Your Notes

Create a chart like the one below to help you organize your notes for this section. Fill in details for ethnic groups found in each country.

	Ethnic Group	Chief Population Distribution	Largest Area of Concentration
China			

SECTION 2 History and Government (pp. 668–672)

Terms to Know
- culture hearth
- dynasty
- clan
- shogun
- samurai

Key Points
- Confucianism and Daoism developed in China about 500 B.C. Buddhism spread from India throughout East Asia.
- China was ruled by a succession of dynasties until the early 1900s.
- Contact with the West forced East Asians to modernize.
- Revolutions and wars transformed East Asia in the 1900s.
- By the end of the 1900s, East Asian countries had important roles in the global economy.

Organizing Your Notes

On a web diagram like the one below, fill in important events in East Asia's history, including its various forms of government systems.

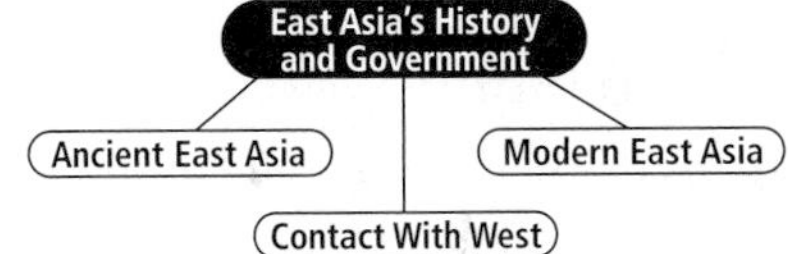

SECTION 3 Cultures and Lifestyles (pp. 673–679)

Terms to Know
- ideogram
- shamanism
- lama
- acupuncture
- haiku
- calligraphy
- pagoda

Key Points
- Sino-Tibetan languages and Korean and Japanese are the region's main languages.
- East Asians often adopt practices from more than one religious tradition.
- Rising standards of living since 1945 have brought dramatic improvements in education and health care for some countries.
- East Asians have a long history of traditional arts and activities.

Organizing Your Notes

Create an outline using the format below to help you organize your notes for this section.

East Asia's Languages and Religions

I. Languages
  A. China
    1. Mandarin
    2. Cantonese

ASSESSMENT & ACTIVITIES

Reviewing Key Terms

Write the key term that best completes each of the following sentences. Refer to the Terms to Know in the Summary & Study Guide on page 681.

1. A _____ was a professional soldier in early Japan.
2. Soldiers of ancient Japan were loyal to a military ruler known as a(n) _____.
3. _____ is the art of beautiful writing.
4. China was East Asia's _____, the center from which ideas spread.
5. A ruling family known as a(n) _____ formed China's early government.
6. Before ruling families, tribal groups, or _____, ruled in China.

Reviewing Facts

SECTION 1

1. Which countries in the region of East Asia are the most ethnically homogeneous?
2. What portions of East Asia are relatively unpopulated? Why?
3. What is Japan's most populous region? Why?

SECTION 2

4. During which dynasty did the philosophies of Confucius and Laozi emerge?
5. How did Japan build an empire in the early 1900s, and how did the empire come to an end?
6. How did the Communists in China come to power?

SECTION 3

7. Name four religious or philosophical traditions of East Asia.
8. Why is education a high priority in Taiwan and South Korea?
9. Name five art forms important in East Asia.

Critical Thinking

1. **Comparing and Contrasting** How do the standards of living vary among East Asian countries and between rural and urban areas?
2. **Making Inferences** Why are farmlands and the food supply of critical importance to China?
3. **Analyzing Consequences** Create a web diagram like the one below to show the effects of migration to urban areas in East Asian countries. Then write a paragraph explaining those effects.

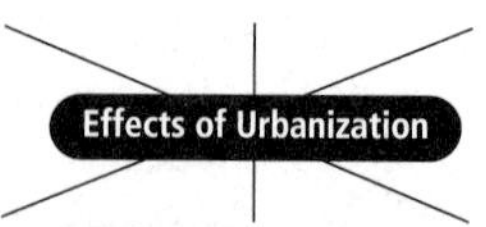

NATIONAL GEOGRAPHIC

Locating Places

East Asia: Physical-Political Geography

Match the letters on the map with the places and physical features of East Asia. Write your answers on a sheet of paper.

1. **Wuhan**
2. **Shanghai**
3. **Taipei**
4. **Yangtze**
5. **Tokyo**
6. **Beijing**
7. **Guangzhou**
8. **Ulaanbaatar**
9. **Seoul**
10. **Kyoto**

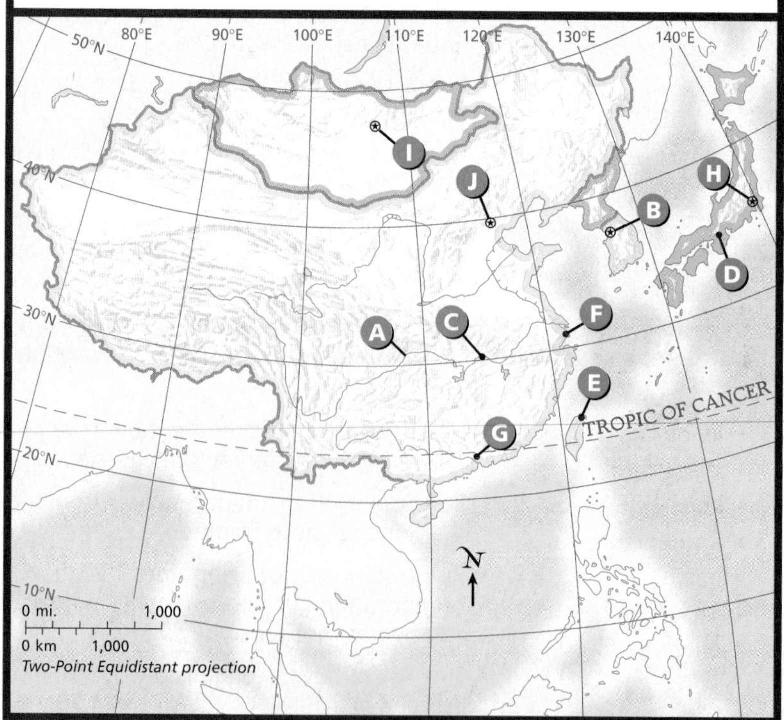

Self-Check Quiz Visit the **Glencoe World Geography** Web site at geography.glencoe.com and click on Self-Check Quizzes—Chapter 27 to prepare for the Chapter Test.

Using the Regional Atlas

Refer to the Regional Atlas on pages 636–639.

1. **Place** Which East Asian cities have populations over 5,000,000?
2. **Human-Environment Interaction** What natural resource may account for the areas of high population density in western China?

Thinking Like a Geographer

Think about the thousands of migrant workers settling in China's special economic zones. What are some of the problems created by this influx of people? As a geographer, what strategies would you suggest to help solve these problems? Explain your answer.

Problem-Solving Activity

Group Research Project Work with a group to research and evaluate the Chinese government's international reputation on human rights. Issues include the treatment of religious dissidents, political prisoners, ethnic Tibetans and exiled leaders, and students during the 1989 pro-democracy demonstrations at Tiananmen Square. Consider the following: What is China's current status on human rights? How will the Internet and communications technology affect this issue? Prepare a report stating your opinions and predicting future developments.

GeoJournal

Compare and Contrast Use the information you noted in your GeoJournal to write an essay comparing cultural aspects of two East Asian countries. Include specific examples.

Technology Activity

Creating an Electronic Database Create a database of the population densities of several East Asian countries, regions, or cities. Then use the database to help you draw an outline map to show population distribution using color codes, and include a map key.

Standardized Test Practice

Study the table. Then choose the best answer for the following multiple-choice question. If you have trouble answering the question, use the process of elimination to narrow your choices.

Economic Activities in East Asia

	Taiwan	China	Japan	South Korea
Economic Activity	%	%	%	%
Agriculture	2.9	18.4	1.7	4.9
Industry	34.0	48.7	36.0	43.5
Services	63.1	32.9	62.3	51.6
Labor Force				
Agriculture	8.0	50.0	5.0	12.0
Mining & Manufacturing	37.0	23.0	32.0	27.0
Services & Other	55.0	27.0	63.0	61.0

Source: The Economist Pocket World in Figures, 2001

1. **Based on the chart, which two countries have economic activities that are the most similar in all areas?**

   A South Korea and Taiwan
   B Taiwan and Japan
   C Taiwan and China
   D China and Japan

Charts and tables may reveal patterns or trends. Look for similarities in groups of numbers before you draw conclusions. In addition, numbers need not be exactly alike in a question such as this one. For example, although both China and South Korea have the same percentage for Services under Economic Activity, the other figures vary widely.

CHAPTER 28

East Asia Today

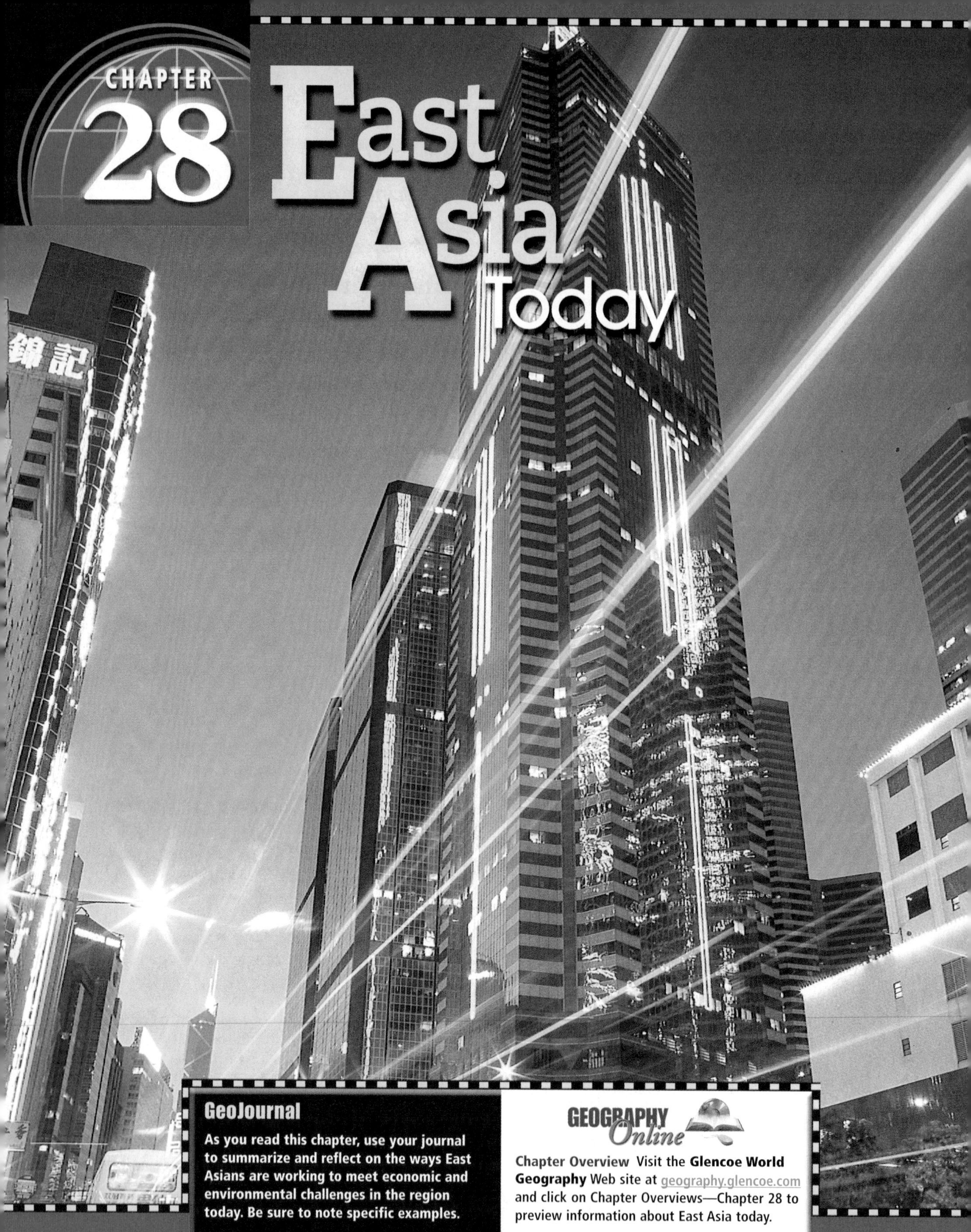

GeoJournal

As you read this chapter, use your journal to summarize and reflect on the ways East Asians are working to meet economic and environmental challenges in the region today. Be sure to note specific examples.

GEOGRAPHY Online

Chapter Overview Visit the **Glencoe World Geography** Web site at geography.glencoe.com and click on Chapter Overviews—Chapter 28 to preview information about East Asia today.

SECTION 1

Living in East Asia

Guide to Reading

Consider What You Know

People in East Asia, like those in other world regions, are experiencing rapid economic changes. What impact do you think rapid economic change has on the lives of people today?

Read to Find Out

- What types of governments and economies do East Asian countries have?
- What economic activities play an important role in East Asia?
- How are other countries in the region challenging Japan's economic dominance?
- How are the countries of East Asia economically interdependent?

Terms to Know

- command system
- commune
- cooperative
- Asia-Pacific Economic Cooperation Group (APEC)
- trade surplus
- trade deficit
- dissident
- economic sanctions
- World Trade Organization (WTO)
- merchant marine

Places to Locate

- Wuhan
- Tianjin
- Guangzhou

◀ *Hong Kong at night*

NATIONAL GEOGRAPHIC

A Geographic View

Japan's Economic Boom

. . . Out of the ashes [of Hiroshima's atomic bombing in 1945] has arisen a fully modern city. . . . The new Hiroshima is a self-proclaimed City of Peace, with a towering skyline, cosmopolitan shopping arcades, and more than 700 manicured parks. Its port sends out to New York, Shanghai, and London . . . the latest in consumer and industrial products.

—*Ted Gup, "Up From Ground Zero: Hiroshima,"* National Geographic, *August 1995*

Bullet train in Hiroshima, Japan

Beginning in the 1960s, East Asian countries such as Japan, South Korea, and Taiwan experienced tremendous economic growth. Then, in the 1990s, a severe economic downturn jolted much of East Asia, shaking public confidence and causing widespread hardships. By 2001 financial aid from Western countries and economic reforms at home had brought a slow recovery to the region. In this section you will learn how East Asians are adjusting to the challenges of living and working in the global economy.

Political and Economic Systems

As in other world regions, governments and economies are closely related in East Asia. East Asian economies include market systems based on private ownership; **command systems**, which are controlled by governments; and a mix of both systems. During the mid- to late 1900s, Japan, South Korea, and Taiwan developed democratic governments, prospered under market systems, and became global economic powers. Meanwhile, communist-ruled China and democratic Mongolia shifted from strict command systems to mixed economies with both command

▲ *Women making terraces to grow rice, China*

and market features. By 2001 North Korea had one of the world's few remaining command economies. Less economically developed than South Korea, North Korea is slowly reentering the global markets.

Agriculture

Since the mid-1900s most East Asian countries have shifted dramatically from rural-based agricultural economies to urban-based industrial ones. Agriculture, however, is still important in the region.

China

China has East Asia's most rural economy. About 50 percent of China's workers are farmers. Large numbers of people are needed to work the land because many farmers still use traditional tools. Still, China is a leading producer of rice, wheat, and tea. Chinese farmers also produce soybeans, cotton, jute, and silk and raise livestock.

Since 1949 China's communist government has made many changes to the country's agriculture. The Great Leap Forward campaign of the 1950s organized farmers into huge **communes**, large farming communities whose members shared work and products equally, but the government decided which farming methods to use. The results were disastrous. When crop production dropped, famines swept the country.

Then in the 1980s, Chinese leaders reversed their agricultural policies. They encouraged smaller farms, jointly run by households but with private garden plots. Farmers could sell and profit from any extra crops or animals. Chinese farmers now can grow enough food to feed the country.

Despite these agricultural reforms, large numbers of rural workers have begun moving to cities such as Shanghai, Hong Kong, and Beijing to take jobs in industry and commerce, where earnings are better.

Mongolia

Most of Mongolia is used for grazing herds of sheep, goats, camels, and cattle. Until the early 1990s, Mongolia modeled its command economy on the Soviet Union's. Large goverment-owned farms set targets for producing milk and wool, and grew food for people and fodder for animals. Although Mongolia's government still owns much of the country's limited farmland, Mongolian farmers and herders are slowly adapting to a market economy.

South and North Korea

Largely urbanized, with many industries, South Korea's agricultural workforce makes up only 12 percent of its population. Most South Korean farmers work on small family farms, but a farm labor shortage has developed as people continue to move to urban areas. To make up for this loss, South Korean agriculture increasingly uses modern machinery and more efficient farming practices.

In North Korea agriculture makes up 25 percent of the economy and employs about 40 percent of all workers. Farms in North Korea are organized into **cooperatives**, farms jointly operated by households. The communist government, however, controls crop

Student Web Activity Visit the **Glencoe World Geography** Web site at geography.glencoe.com and click on Student Web Activities—Chapter 28 for an activity about China's modern history.

production and distribution and rations agricultural products. Corn, wheat, and milk are in short supply, and North Korea cannot fully meet its own demand for rice, the country's major crop.

In the 1990s severe flooding destroyed North Korea's rice crop, causing food shortages. This disaster was heightened by government mismanagement of the economy and resulted in widespread famine. For the first time, North Korea accepted food aid from countries with market economies, such as the United States, Japan, and South Korea.

Japan and Taiwan

Japan and Taiwan are largely industrialized, but agriculture does play a role in their economies. Physical geography challenges farmers in both countries. Four-fifths of Japan is mountainous, so farmers have used terracing, modern machinery, fertilizers, and irrigation to dramatically raise Japan's crop yields since World War II. Japan's government also provides farmers with financial support to equalize rural and urban incomes. Despite such improvements, Japan still must import about 35 percent of its food.

Like Japan, Taiwan is mountainous. Rice, sugarcane, tea, bananas, and pineapples grow on limited, often terraced, farmland. Once an exporter of these crops, Taiwan now focuses on industrial exports and imports some food products.

Industry

Since the 1960s, East Asian countries such as Japan, South Korea, and Taiwan have become important industrial and trading countries. Although rich in minerals, North Korea lags far behind these three countries in industrial development. China is primarily agricultural but has a rapidly developing industrial economy. Mongolia also is building its industries, but its factories mostly process livestock and farm products.

Japan

With the aid of the United States, Japan's economy recovered quickly after World War II. A highly skilled workforce and the latest technology helped Japan dramatically develop its industries. Within a few decades, the Japanese were leading producers of ships, cars, cameras, computers, telecommunications products, and consumer goods. By the 1990s world demand for Japan's high-quality goods had made it a global economic power.

Japan, like other countries, faces challenges from the fast-paced, constantly changing global economy. Despite its overall success record, Japan and other parts of Asia suffered from a global economic slump in the 1990s. Japanese banks had invested in risky businesses. When they could not collect on their loans, they failed. Meanwhile, industrial production at home dropped, and unemployment soared. By 2001, Japan was considering banking and other economic reforms to get its sluggish economy moving again.

South and North Korea

After the Korean War, South Korea rapidly moved from an agricultural to an industrial economy. By the 1980s South Korean industries had begun exporting

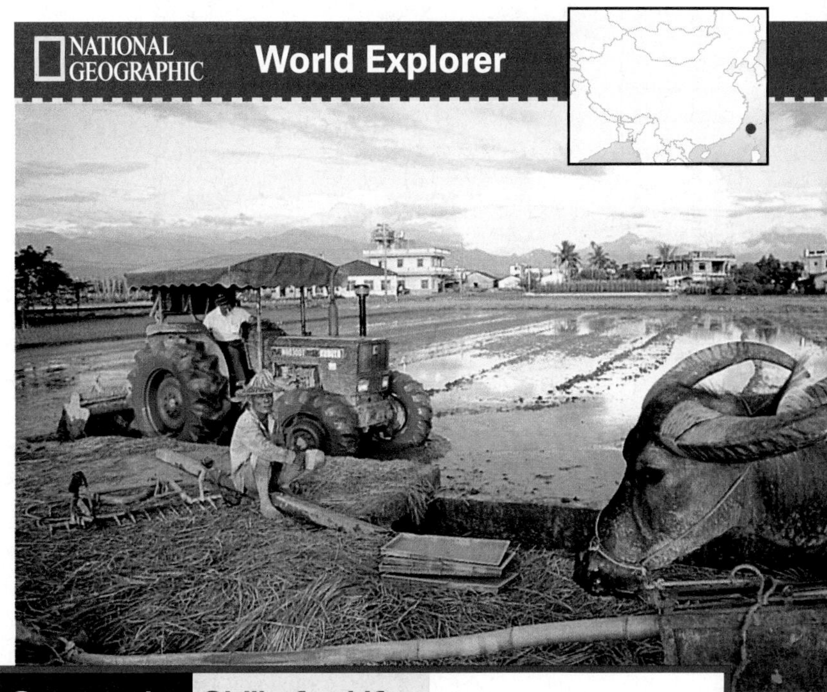

Geography Skills for Life

Old and New

The economic boom in Taiwan has brought many changes, including the modern farm equipment now used on traditional farms.

Place What trend characterizes Taiwan's agriculture?

ships, steel, electronic equipment, and motor vehicles. Like other Asian countries, South Korea suffered an economic downturn in the 1990s. With international financial aid, however, South Korea soon began rebuilding its economy.

In North Korea, government-owned heavy industries produce machinery, chemicals, and military equipment. Because so many resources go to North Korea's military forces, production of consumer goods suffers. Before the early 1990s, North Korea had depended on the Soviet Union for economic aid. When the Soviet Union broke up, North Korea's industrial output fell by half, forcing its communist leaders to trade with countries such as Japan and South Korea.

In 2000, relations improved between North Korea and South Korea. The leaders of the two countries agreed to trade with each other, and South Korea pledged economic aid to North Korea. Their governments also allowed a limited number of family visits across the border for the first time since the Korean War.

NATIONAL GEOGRAPHIC GRAPH STUDY

Selected Countries: Balance of Trade

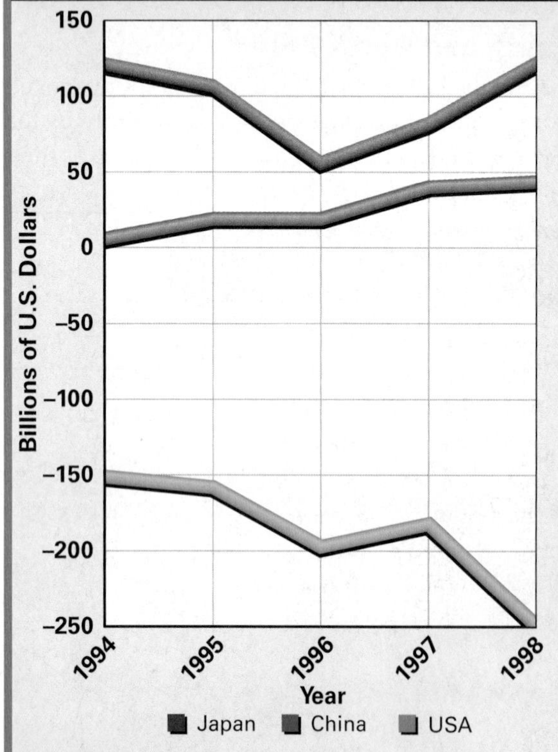

Billions of U.S. Dollars: 150, 100, 50, 0, -50, -100, -150, -200, -250

Year: 1994, 1995, 1996, 1997, 1998

Japan, China, USA

Sources: The World Almanac, 1995–2001

Geography Skills for Life

1. **Interpreting Graphs** How did China's balance of trade change between 1994 and 1998?
2. **Applying Geography Skills** Why do you think the Japanese balance of trade decreased and then increased between 1994 and 1998?

Taiwan

Taiwan has one of the world's most successful export-based economies. Until the 1960s, the island had exported a surplus of agricultural products, and invested the profits, as well as American and Japanese financial aid, in manufacturing. Taiwan's new industries specialized in textiles, plastics, and electronic goods for export. Their economic boom transformed Taiwan into a major trading country. Today about 60 percent of Taiwan's people work in service industries, such as finance and communications. Technology-based products, such as computers and precision instruments, are replacing traditional manufactured goods as Taiwan's major source of income.

China

When Chinese communist leaders came to power in 1949, they used government controls to boost industrial output. Today the Chinese government still controls major industries, such as textiles, clothing, footwear, toys, and plastics manufacturing. Many state-run factories, however, lack updated technology and incentives for improved performance.

To stimulate the economy, Chinese leaders since the 1970s have adopted some features of a market economy. For example, small, privately owned businesses are permitted to operate, and foreign companies and investments are welcomed. Many of China's most prosperous industries now lie in special economic zones along the southeastern coast, where they can operate without government controls on prices, production, or distribution.

With market reforms, China's economy is growing at a remarkably high annual rate of 8 percent. Standards of living have risen, especially in urban areas, but the country still faces economic challenges.

A large gap separates wealthier industrial areas on the coast from poorer agricultural regions in the interior. As unprofitable state-run industries close, unemployment increases. Industrial growth and few environmental safeguards have contributed to rising pollution.

▲A student protester faces down tanks in Beijing's Tiananmen Square.

Government

Hong Kong and Macau

The Chinese territories of Hong Kong and Macau are major industrial and trading centers. After more than 150 years of British rule, control of Hong Kong returned to China in 1997. Despite restricted political freedoms, Hong Kong is maintaining a market economy. Hong Kong's economic success provides great wealth to China. In 1999 Macau became part of China after centuries of Portuguese rule. Like Hong Kong, Macau's prosperous market economy benefits China.

Trade

In recent decades East Asian countries have become more interdependent with one another and the rest of the world. As in other world regions, East Asia has formed trading partnerships. For example, China, Japan, South Korea, and Taiwan are members of the **Asia-Pacific Economic Cooperation Group (APEC)**, which ensures that trade among member countries is efficient and fair. Still, trade disputes and deeply rooted political differences continue to affect the region's international relations.

Japan: Trade Surpluses

With few mineral resources, Japan depends on international trade for its economic well-being. To produce its vast range of products for foreign buyers, Japanese industries import raw materials such as iron ore and fuels. The Japanese government, however, places high taxes on many imported finished goods. These taxes protect many Japanese industries from foreign competition, but they restrict what other countries can sell to Japan.

The term *balance of trade* refers to the difference in value over time between a country's imports and exports. High import taxes, along with the high global demand for Japanese goods, cause Japan to have trade surpluses with other countries. A **trade surplus** occurs when exports exceed imports.

Trade surpluses bring increased wealth to Japan, but the resulting trade imbalances mean lower profits for Japan's trading partners. A **trade deficit** occurs when a country imports more goods from other countries than it exports to them. In recent years the United States and other countries have tried to persuade Japan to open its market. Results are mixed. Trade policy, therefore, continues to complicate Japan's relations with other countries.

China: Trade and Human Rights

In an effort to modernize its economy, China has sought increased trade with the United States and other countries with market economies. The United States also favors increased trade because of China's growing economy and more than a billion potential customers. A major stumbling block, however, is China's harsh treatment of **dissidents**, or citizens who speak out against government policies. For example, in 1989, Chinese students wanting democratic reform held a massive demonstration in Beijing's Tiananmen Square. The government sent in troops to brutally end the protest. This action brought China severe criticism.

The United States, Japan, and other important trading countries have tried to influence China to respect human rights. In response to the Tiananmen crackdown, these countries placed **economic sanctions**, or trade restrictions, on China. In response to economic losses, China released several dissidents from prison, and the United States lifted sanctions.

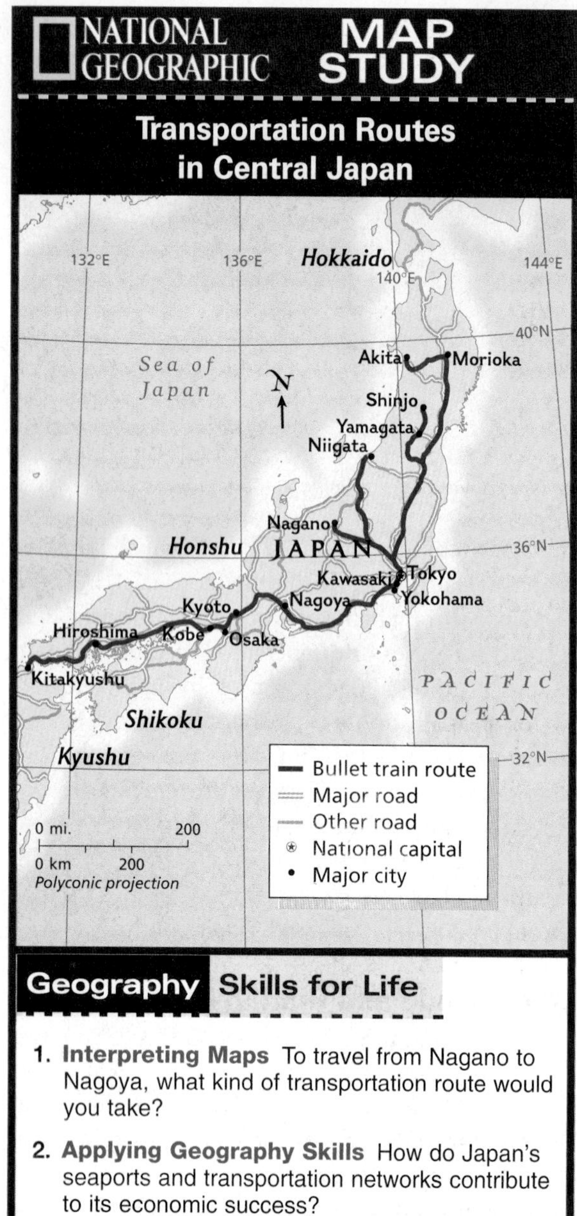

Geography Skills for Life

1. **Interpreting Maps** To travel from Nagano to Nagoya, what kind of transportation route would you take?
2. **Applying Geography Skills** How do Japan's seaports and transportation networks contribute to its economic success?

Find NGS online map resources @ www.nationalgeographic.com/maps

Many, however, remain dissatisfied with China's human rights record. The United States government hopes that trade might open China to democratic change. In 2000 the United States Congress granted full trading privileges to China. In the near future, China expects to be admitted to the **World Trade Organization (WTO)**, an international body that oversees trade agreements and settles trade disputes among countries. Western countries believe that China's human rights record will improve as it has more frequent contact with other countries.

Transportation and Communications

Before air travel became common, rugged mountains isolated most of East Asia. Today every country in the region has modern air services. Overland travel in mainland East Asia, however, involves long journeys by railroad or highway. Transportation and communications networks are concentrated in heavily populated areas, and rural areas often have little access to communications networks.

Land Travel

Land transportation varies throughout East Asia. Japan, South Korea, and Taiwan have nationwide highway and railroad networks. Japan's rail system includes high-speed trains, commuter trains, and subways. Elsewhere in the region, transportation links, especially highways, are not as developed. For example, Mongolia's roads are mostly unpaved, and people rely on the Trans-Mongolian Railway. Lack of inland transport has slowed the development of western China, which remains poor despite large oil and coal deposits.

The Chinese, however, have made progress in building roads and rail lines. In 2001, work began on what is hailed as the world's highest railway, linking Tibet to China's national rail system. The Chinese claim the railway will lessen Tibet's isolation and boost its economy. The railway's critics, however, say that it will draw more Chinese to Tibet, further diluting Tibet's culture.

Water Travel

China's rivers provide important routes from inland areas to seaports. The Yangtze River is China's most important waterway. The major port of Shanghai lies at its mouth:

> *"Today the port [of Shanghai] handles more than 160 million tons of cargo a year through loadings and unloadings along the 40 miles of wharves on the [Yangtze River]. . . ."*
>
> William Ellis, "Shanghai," *National Geographic*, March 1994

Large oceangoing ships travel 680 miles (1,094 km) inland on the Yangtze to the transportation center of **Wuhan** in central China. Other major ports are at the mouths of rivers—**Tianjin** (TYEHN•JIHN) on a tributary of the Yellow River and **Guangzhou** (GWAHNG•JOH) on the Xi River. The Grand Canal, the world's longest and oldest human-made waterway, runs from north to south, linking the Yangtze and Yellow Rivers.

Major seaports and merchant marine fleets used for commercial transport are vital to East Asia's export trade. Japan has more than 7,000 merchant vessels, and China's merchant marine fleet numbers about 2,400.

Communications

In North Korea and China, communist governments control communications, the news media, and citizens' access to the Internet. By contrast, people in democratic Japan, South Korea, and Taiwan enjoy a free press, and most own radios, televisions, and telephones (including pagers and cellular phones). In these countries, a wide variety of books, magazines, and newspapers is available, as is access to the Internet.

East Asian countries have overcome many obstacles in order to develop their economies. In the next section, you will learn of the environmental challenges that economic growth has brought to East Asia.

NATIONAL GEOGRAPHIC **World Explorer**

Geography Skills for Life

The Grand Canal The oldest part of the Grand Canal was built in the 400s B.C. Between 1958 and 1964, the Chinese government rebuilt or repaired vast stretches of the canal.

Movement How far inland can oceangoing ships travel on the Yangtze River?

SECTION 1 ASSESSMENT

Checking for Understanding

1. **Define** command system, commune, cooperative, Asia-Pacific Economic Cooperation Group (APEC), trade surplus, trade deficit, dissident, economic sanctions, World Trade Organization (WTO), merchant marine.
2. **Main Ideas** On a chart like the one below, describe the economy of each East Asian country.

Country	Type of Economy

Critical Thinking

3. **Identifying Cause and Effect** How did the Asian financial crisis of the 1990s affect the economies of China, Japan, and South Korea?
4. **Problem Solving** How might various East Asian countries provide food for their populations, despite their limited farmlands?
5. **Drawing Conclusions** What conclusions can you draw about the economic differences between North and South Korea? Explain your answer.

Analyzing Graphs

6. **Place** Study the graph on page 688. How does the U.S. balance of trade compare to that of Japan? China?

Applying Geography

7. **Economic Reform** Write a paragraph that compares China's economy before and after the economic reforms of the 1980s. Explain how the reforms have affected the lives of China's people.

People and Their Environment

Guide to Reading

Consider What You Know

Given East Asia's crowded cities and rapid industrial growth, what environmental challenges do you think the region might face?

Read to Find Out

- How have industrialization and urbanization in East Asia affected the environment?
- What steps are East Asians taking to solve environmental problems?
- What naturally occurring destructive forces does East Asia regularly face?

Terms to Know

- desertification
- chlorofluorocarbons
- aquaculture

Places to Locate

- Three Gorges Dam
- Inland Sea
- Kobe

NATIONAL GEOGRAPHIC

A Geographic View

The Price of Modernization

The most tangible cost of modernization is environmental. . . . Today greater Taipei's population has swollen to almost six million—nearly 30 percent of the island's total. . . . The city chokes on the fumes of 460,000 cars, 7,300 buses, 38,000 taxis, and 869,000 motorcycles, whose drivers park all over the sidewalks and often drive down them too.

— *Arthur Zich, "Taiwan: The Other China Changes Course,"* National Geographic, *November 1993*

Family on motor scooter, Taiwan

Modernization has brought higher standards of living and increasing global influence to the peoples of East Asia. Yet the benefits have come with serious costs, especially to the region's environment. For example, industrial expansion and urban development have heightened pollution of the air, land, and water throughout the region. In addition to environmental hazards, East Asians also regularly face both the dangers and the challenges of devastating natural disasters, such as floods, earthquakes, and typhoons.

The Power Dilemma

Throughout East Asia, economic growth has increased the demand for electric power to operate businesses and industries. In addition, rising standards of living mean that people tend to buy and use more appliances and electronic devices. Thus, finding adequate sources of electric power has become a vital issue for the countries of the region.

Fossil Fuels

Some of East Asia's power comes from hydroelectric plants, but most is produced from the burning of fossil fuels such as coal, oil, or natural gas. China, North Korea, and Mongolia produce most of their power using coal from large reserves. Japan, South Korea, and Taiwan, however, have few coal, natural gas, or oil deposits, so they must import these resources to produce energy. About 65 percent of Japan's electricity comes from plants burning coal, natural gas, or petroleum, as does roughly 60 percent of South Korea's power. Coal reserves have dwindled in Taiwan, so the country relies on imported petroleum as its primary source of energy.

Geography Skills for Life

Nuclear Energy

A cargo ship delivers plutonium, a radioactive element used to produce nuclear power, to a nuclear power plant in central Japan.

Human-Environment Interaction What are the positive and negative consequences of using nuclear power?

Burning fossil fuels, however, leads to acid rain, air pollution, and possibly global warming. East Asian governments have begun to search for cleaner power sources. China's massive **Three Gorges Dam** project on the Yangtze River, for example, aims to supply a huge amount of hydroelectric power to China's interior regions. The project should be completed by 2009.

Nuclear Energy

Japan, South Korea, and Taiwan rely on nuclear energy for 30 to 40 percent of their electrical power. Japan has more than 50 nuclear power generators, South Korea has 12, and Taiwan has 8. North Korea is not known to have any nuclear power facilities. China's few reactors currently produce just 1 percent of the country's electricity, but China's plans for the future include 100 more nuclear plants.

During the late 1990s, a series of accidents in Japan and South Korea exposed hundreds of people to radiation and raised public fears about the safety of nuclear power. People also worried that the earthquakes and volcanic activity common in the region could cause reactors to crack and release radiation.

After a 1999 nuclear accident, Japan began searching for alternatives to both nuclear and hydroelectric power. Since then, Japan has opened several plants that generate electricity from wind and solar energy.

Environmental Concerns

In many parts of East Asia, industrial and economic growth have been given more consideration than other issues. The effects of this growth on the region's environment have been largely ignored. Environmental challenges range from air pollution to the depletion of natural resources. East Asians are just beginning to take seriously issues dealing with health, environment, and quality of life.

China

In China's urban areas, the use of outdated technology in transportation and industry has caused major air pollution. In fact, 9 of the 10 cities worldwide with the worst air pollution are located in China. One major cause is China's heavy reliance on its huge reserves of relatively inexpensive coal. In northern industrial areas, windblown dust adds to the air pollution. As a result, large numbers of people living in the area suffer from lung disease.

Acid rain from burning coal also is a serious problem in the industrial region of southeast China.

Neighboring countries are affected by China's acid rain as well. Forests in Japan, for example, suffer not only from Japan's own coal-burning power plants but also from China's coal-generated pollution.

China and other rapidly urbanizing countries like Taiwan and South Korea also have trouble disposing of waste products. For example, 80 percent of the cities in China have no sewage treatment facilities.

Industrial waste from factories poses health risks to urban populations in parts of China. For many years an important metal company in the Chinese city of Shenyang spewed huge amounts of sulfur dioxide and other harmful chemicals into the atmosphere. In 2000 China responded to pleas from city residents who had long complained of health problems caused by the plant's toxic emissions. For the first time, China closed down a state-run factory for environmental reasons.

China has begun responding to other serious environmental issues, including deforestation. Each year China clears thousands of acres of forests and vegetation to meet the country's high demand for lumber. Deforestation caused by clear-cutting timber leads to soil erosion.

Without trees to slow runoff from rain, large-scale soil erosion and flooding occur. As heavy rains wash away large amounts of unprotected soil in deforested areas, soil deposits build up in rivers. The buildup causes waters to rise even higher than they otherwise might have, and flooding becomes more severe. In the late 1990s, a series of unusually heavy rains caused the Yangtze and Yellow Rivers to flood. Floodwaters destroyed property, altered the landscape of vast areas, and killed thousands of people.

In response to these disasters, China has begun planting trees on millions of acres along the deforested riverbanks. To help control flooding, the government ordered a major dam construction project along the Yellow River. Other steps in China's conservation and restoration plan include the creation of nature and wetland reserves and wildlife protection zones.

A related concern is **desertification**, the process in which grasslands became drier and desert areas expand. Along China's eastern borders with Mongolia, grasslands and desert meet. The grasses there are drought-resistant, but overgrazing and soil erosion have depleted much of the vegetation. The resulting desertification has contributed to dust and sand storms in northern and western areas of China.

North Korea, South Korea, and Taiwan

Urban areas of North Korea, South Korea, and Taiwan are plagued with air and water pollution due to lax industrial controls. Untreated sewage contaminates water supplies and threatens the health of humans and wildlife. For example, North Korea's safe drinking water supplies are inadequate.

Although nuclear energy provides inexpensive power for South Korea, the waste from such plants remains radioactive for tens of thousands of years and is difficult to dispose of safely. Nonetheless, South Korea continues to build nuclear power plants to meet its energy needs. Because North Korea has no nuclear power facilities, it is spared the task of managing nuclear waste, but the country still must confront the hazardous effects of using fossil fuels.

Geography Skills for Life

Endangered Forests Bamboo forests such as this one in southern China are threatened by air pollution.

Human-Environment Interaction How have East Asian countries tried to control air pollution?

Mongolia

Environmental challenges in Mongolia resemble those of western China. Deforestation caused by logging and desertification caused by overgrazing have contributed to soil erosion. Burning coal pollutes the air, and safe drinking water supplies are limited.

Government

Japan Leads the Cleanup

Japan's highly industrialized, crowded society was criticized for years for ignoring the environmental problems created by rapid technological growth. Since the 1970s, however, the Japanese government has encouraged industries to curb pollution. Japan today has emerged as a world leader in addressing environmental issues. By 2000 two Japanese automakers had introduced a hybrid car powered by a battery and a supplemental gasoline engine. The car uses less fuel, so it emits fewer pollutants. It may be years, however, before use of this new technology becomes widespread.

Pollution control has taken on a national urgency in Japan, where environmental laws are among the world's strictest. The government has urged other countries to reduce emissions of carbon dioxide and **chlorofluorocarbons (CFCs)**, gaseous substances found in liquid coolants. Once CFCs enter the atmosphere, they significantly contribute to the destruction of the earth's protective ozone layer.

Japan also offered "clean" technology and financial help for environmental projects to neighboring East Asian countries and other developing countries. Japan's 1990 Action Program to Arrest Global Warming required that 10 percent of its country's cars be replaced with less-polluting vehicles and that overall waste be reduced by 25 percent by the year 2000. Despite this resolve, Japan's total carbon dioxide emissions had increased in 1994 by more than 7 percent from 1990 levels.

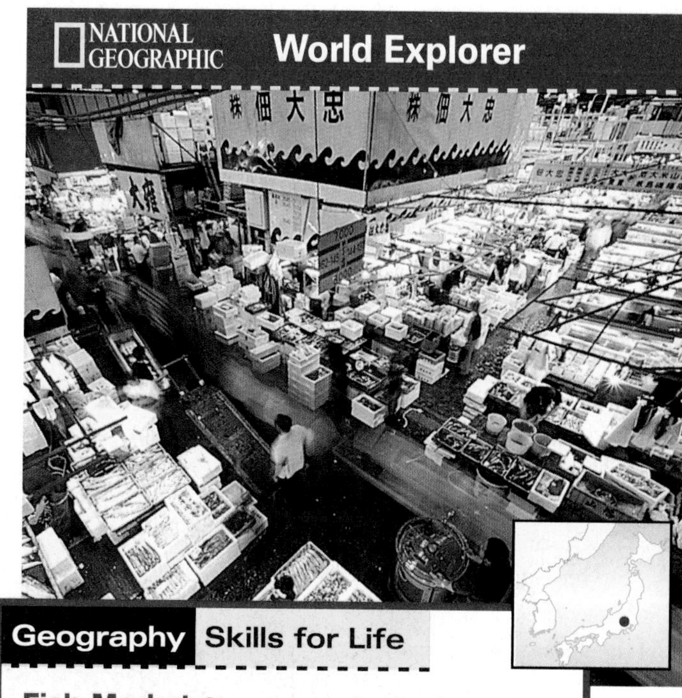

NATIONAL GEOGRAPHIC World Explorer

Geography Skills for Life

Fish Market Shoppers can find herring, salmon, trout, squid, and other seafoods at this Tokyo market.

Human-Environment Interaction How can countries in overfished sea areas obtain seafood?

Managing Ocean Resources

Another environmental issue in East Asia is the management of ocean resources. In most East Asian countries, oceans and seas provide an important source of food for local consumption and for export. Commercial fishing is a major industry in China, Japan, and South Korea. Fleets from these countries catch tons of snapper, tuna, squid, shrimp, and other seafood. Now, however, Japan imports large quantities of seafood because of decreasing quantities of fish in the region. The Japanese must find other ways to obtain seafood because they consume more seafood than any other people in the world.

In recent years East Asia's coastal waters, such as Japan's **Inland Sea**, have become overfished or polluted. As a result, commercial fishing companies from several countries have begun fishing farther from shore, in international waters. Many of these companies have giant factory ships that follow fishing fleets to quickly clean and freeze large catches of fish. The use of factory ships is discouraged internationally because the practice allows fishing fleets to harvest huge catches, leading to overfishing. One solution to overfishing is **aquaculture**, or the cultivation of fish and other seafood. Several countries in the region raise seafood, such as prawns, in ponds for export.

Japan, East Asia's largest consumer of whale meat, remains the target of global criticism for its whaling practices. According to international conservation groups, overhunting has caused a serious decline in the whale population. Despite a 1986 international treaty limiting whaling, Japanese fleets continue to hunt whales, including endangered species, in large numbers to satisfy the high demand for the expensive delicacy.

Natural Disasters

Because of its location and physical geography, East Asia has faced catastrophic natural disasters. China's Yellow and Yangtze Rivers can produce disastrous flooding. Attempts to control flooding have included building networks of drainage channels and irrigation canals to transport or redirect water quickly. Dams, dikes, and levees also have been built. Despite these measures, severe floods continue. More than 30,000 of China's dams, hastily built during the 1950s and 1960s, are now defective and at risk of failing.

To address these problems, China is dredging rivers and creating more flood-control projects. The world's largest public works project, and one of the most controversial, is the Three Gorges Dam, under construction upriver from Wuhan on the Yangtze River in central China. When finished, it will create a huge reservoir nearly 400 miles (644 km) long. The dam's critics argue that the project will force the relocation of almost 2 million people, put countless farms, villages, scenic canyons, and ancient temples underwater, and destroy the natural habitats for Siberian cranes and snub-nosed dolphins.

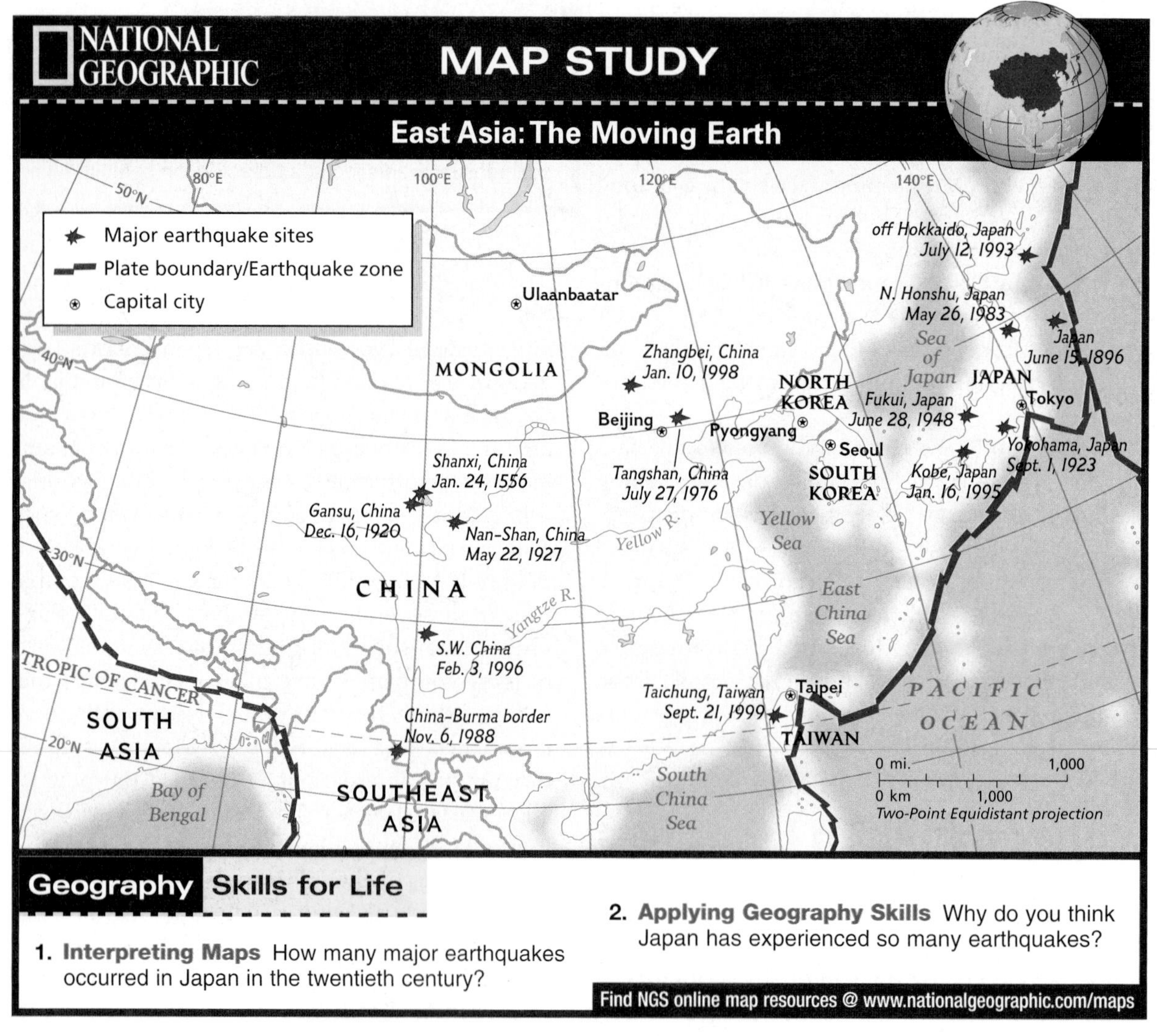

Geography Skills for Life

1. **Interpreting Maps** How many major earthquakes occurred in Japan in the twentieth century?

2. **Applying Geography Skills** Why do you think Japan has experienced so many earthquakes?

Find NGS online map resources @ www.nationalgeographic.com/maps

Most East Asian countries experience destructive earthquakes. A series of major earthquakes struck Taiwan in late 1999 and China's Yunnan Province in early 2000. Each year, about 1,500 small earthquakes shake Japan, which is located at plate boundaries (see the map on page 696) and is part of the Pacific Ocean's Ring of Fire. In 1995 a severe earthquake caused widespread damage around **Kobe** (koh•bay), a major Japanese port:

> *"In the aftermath of Kobe, the government has tried to improve prediction of quakes, but scientists still cannot provide crucial details of imminent jolts, such as when and where [they] will occur."*
>
> "Japan Recalls Quake Disaster," *BBC News* (online), January 17, 2000

Geography Skills for Life

Earthquake in Kobe The Kobe earthquake of 1995 destroyed about 100,000 buildings in the metropolitan area.

Human-Environment Interaction How many earthquakes shake Japan each year?

Japan also has about 50 active volcanoes. Undersea volcanoes or earthquakes can trigger huge tsunamis—waves that grow larger as they approach land and often cause massive destruction and loss of life when they hit land. Typhoons, violent tropical storms with circular winds of at least 74 miles per hour (119 km per hour), cause periodic devastation from high winds and flooding along East Asia's coasts.

East Asia has begun to address environmental issues. However, other challenges, such as flooding, erosion, desertification, and famine, continue to loom on East Asia's horizon.

SECTION 2 ASSESSMENT

Checking for Understanding

1. **Define** desertification, chlorofluorocarbons, aquaculture.
2. **Main Ideas** Use a graphic organizer like the one below to fill in examples of environmental concerns and natural disasters common in East Asia.

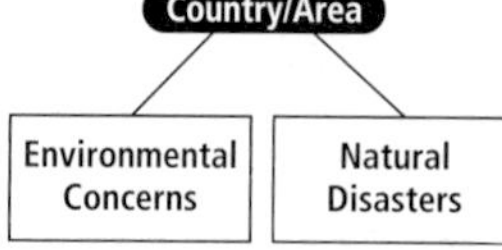

Critical Thinking

3. **Identifying Cause and Effect** What are some possible positive and negative aspects of using nuclear power in East Asia?
4. **Classifying Information** What are the leading sources of power in each East Asian country? Which of the countries use nuclear power?
5. **Drawing Conclusions** How are environmental challenges in other developing countries similar to those in East Asia?

Analyzing Maps

6. **Location** Study the map on page 696. When did the most recent earthquake in China occur? Where did it occur?

Applying Geography

7. **Environmental Solutions** Choose one of East Asia's environmental challenges, such as pollution, deforestation, or desertification. Write a paragraph outlining possible solutions.

Viewpoint

CASE STUDY on the Environment

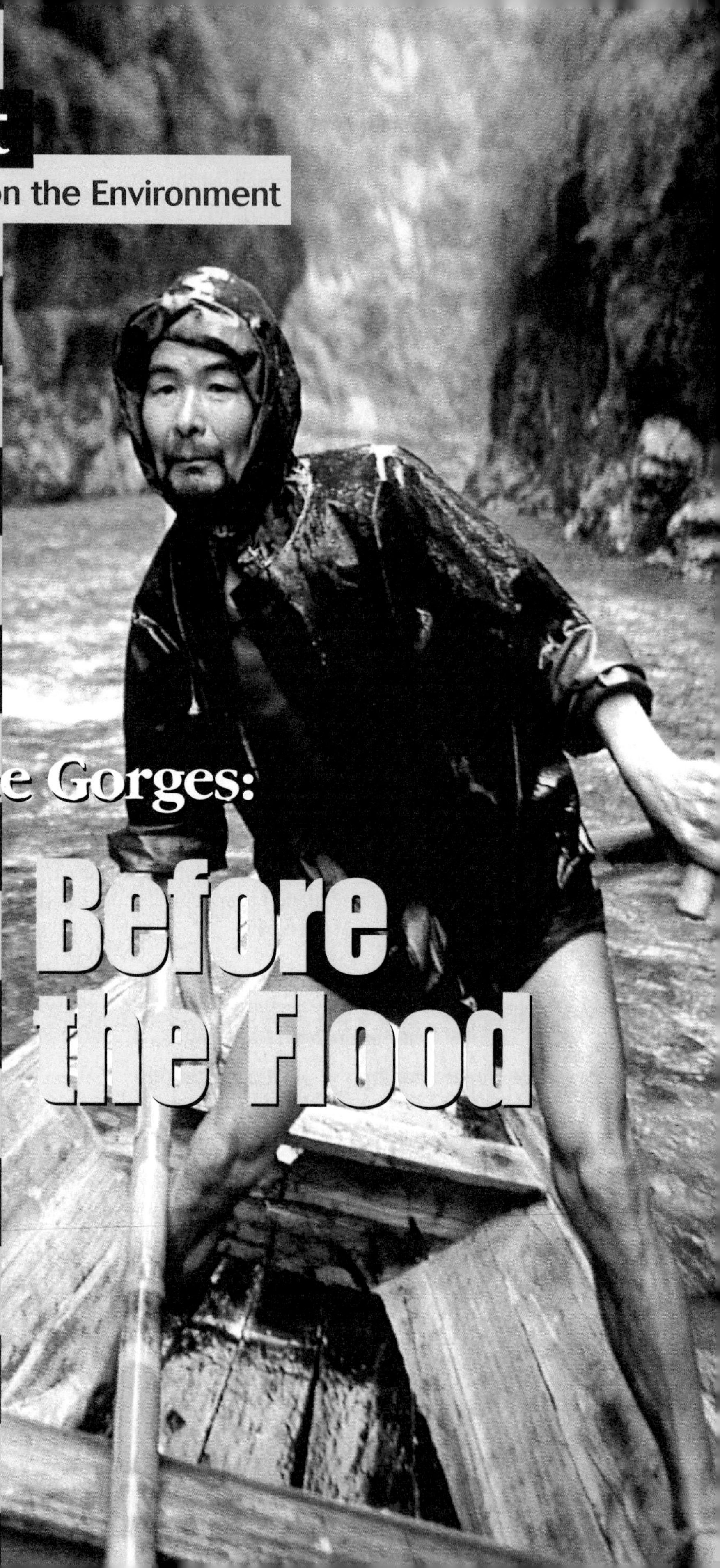

China's Three Gorges: Before the Flood

The Chinese call it Chang Jiang—"Long River." Elsewhere, most people know it as the Yangtze River. Nearly 4,000 miles (6,400 km) long, the Yangtze is the longest river in China. For thousands of years, the Yangtze has been both a positive and a negative force in the lives of many Chinese. The river provides water for 380 million people, and half of China's food is grown along its banks. Yet when the Yangtze overflows, its floodwaters can kill thousands of people and leave millions homeless. Now the Chinese government is trying to tame the Yangtze with a huge—and hugely controversial—dam.

Boatman Ma Linyou (left) deftly steers his wooden craft along the Yangtze, as it winds among the limestone cliffs of the scenic Three Gorges region. In less than a decade, these canyons will vanish forever under a huge volume of water when the Three Gorges Dam blocks the Yangtze's vigorous flow. The dam will create a deep reservoir nearly 400 miles (640 km) long.

Construction of the Three Gorges Dam began in 1994. Scheduled for completion in 2009, it will be the largest dam in the world. The dam is China's most ambitious construction project since building the Great Wall. The Chinese government claims that the dam's positive impact on the region will justify its staggering $25 billion price tag. The Three Gorges Dam could put an end to the devastating floods along the lower portions of the river. For centuries the floods have claimed lives, destroyed settlements, and damaged agricultural lands.

The dam will also be the world's largest hydroelectric plant, designed to generate more than 18 million kilowatts of electricity—an output equal to 18 nuclear power plants. China needs clean, renewable energy to replace the enormous amounts of coal the country burns, a practice that has led to severe air pollution and acid rain.

Supporters of the Three Gorges Dam emphasize its energy and commercial benefits. As China moves toward the future, it faces crippling power shortages. The dam will help solve the problem, generating 20 percent of China's electrical power. In addition, say supporters, the reservoir behind the dam will make more water available for irrigation. The reservoir also will allow large ships and tourist vessels to reach cities far upstream, thus sparking economic growth.

Opponents of the Three Gorges Dam object to its huge environmental and human costs. Environmentalists fear that as the water rises, pollutants in soil and chemicals in abandoned factories will leach into the river, threatening aquatic life and drinking water. Critics point out that the reservoir will submerge hundreds of farms, villages, and some 13 major cities. Nearly two million people will have to abandon their homes. In addition, archaeologists estimate that 8,000 unexcavated sites will be lost forever in a tomb of water.

Despite the opposition, China's government seems determined to finish the Three Gorges Dam. As the controversy rages, work continues on the great concrete barrier that will tame the Yangtze's flow.

◀ A displaced villager lugs his belongings downriver to a new home.

What's Your Point of View?
Do you think the Three Gorges Dam is a good idea? Why or why not?

The city of Fengdu today (below, left) and as it will look when the dam is complete (below, right). ▼

Before

After

Decision Making

From deciding what to eat for lunch to choosing a career goal, young people must make decisions every day. Some decisions are easier to make than others because they are less complex and have minor consequences. Thinking logically and carefully about more important decisions will help you choose wisely.

Learning the Skill

Decisions involve making a choice between alternatives. Each alternative has a likely consequence, or result. To make good decisions, consider as many of the likely consequences as possible before you take action. You can learn to improve your decision-making skills by following these basic steps:

- **State the situation or define the problem.** Ask: Why do I have to make a decision in this matter?
- **Gather all of the facts.** Ask: What information should influence my decision?
- **Identify and evaluate alternatives.** Ask: What are all of my options?
- **Predict future consequences.** Weigh the likely outcomes of each alternative.
- **Consider your personal values.** Use your values as guidelines for making the right decision.
- **Make your decision and act on it.** You should now feel confident that you have thought about the issue carefully.
- **Evaluate your decision.** Analyze whether you made the right choice. Ask: Would I make the same decision again?

▲ *Chinese president Jiang Zemin (left) confers with Premier Li Peng.*

Practicing the Skill

Answer the following questions about decision making.

1. Why is it important to consider more than one alternative when making a decision?
2. What are two reasons for predicting the possible consequences of each alternative you consider?
3. What might be the result of making a decision that conflicts with your values?
4. What can you learn from evaluating a decision you have made?

Applying the Skill

Identify a problem or issue in an East Asian country that requires a decision, and gather information about it. Write down the options available to the country, and assess the consequences of each option. Then write a paragraph explaining how you think the country should decide on the problem or issue.

The **Glencoe Skillbuilder Interactive Workbook, Level 2** provides instruction and practice in key social studies skills.

CHAPTER 28 SUMMARY & STUDY GUIDE

SECTION 1 Living in East Asia (pp. 685–691)

Terms to Know

- command system
- commune
- cooperative
- Asia-Pacific Economic Cooperation Group (APEC)
- trade surplus
- trade deficit
- dissident
- economic sanctions
- World Trade Organization (WTO)
- merchant marine

Key Points

- East Asian economies include market and command systems, as well as a mix of both.
- East Asia was once mainly agricultural, but trade and industry have brought prosperity and economic growth to most of its countries.
- Most Chinese work in agriculture, although industry and commerce are thriving in certain areas as a result of government-sponsored economic reforms.
- Japan is East Asia's leading industrial country, followed by Taiwan and South Korea.
- Trade and business investments bring together capitalist and communist countries in East Asia.

Organizing Your Notes

Create an outline using the format below to help you organize your notes for this section.

Governments and Economies

I. Political and Economic Systems
   A.
   B.
   C.
II. Agriculture
   A. China
   B.
   C.

SECTION 2 People and Their Environment (pp. 692–697)

Terms to Know

- desertification
- chlorofluorocarbons
- aquaculture

Key Points

- Rapid industrial growth in East Asia has caused environmental challenges that were ignored for decades.
- Japan, with its strict anti-pollution laws, has become a leader in protecting and cleaning up the environment.
- China's economic development and the needs of its large population have a decisive impact on the environment.
- East Asia is subject to natural disasters such as flooding, earthquakes, tsunamis, and typhoons.
- Human activities in East Asia—such as clear-cutting forests, farming, and mining—have caused environmental disasters such as erosion, desertification, and flooding.

Organizing Your Notes

Create a chart like the one below to help you organize important details from this section.

Locatio\ns	Problems/Causes	Effects/Solutions
	air pollution	
	deforestation	
	desertification	
	nuclear hazards	
	earthquakes	
	floods	

ASSESSMENT & ACTIVITIES

Reviewing Key Terms

Examine the pairs of words below. Then explain what each of the pairs has in common.

1. commune/cooperative
2. trade surplus/trade deficit
3. dissident/economic sanctions
4. desertification/chlorofluorocarbons
5. APEC/WTO

Reviewing Facts

SECTION 1

1. Explain how economies operate in each of the following East Asian countries: Japan, China, Taiwan, and North Korea.
2. What economic reforms have Chinese leaders introduced?
3. What contributes to Japan's trade surplus with other countries?
4. Which East Asian countries are members of APEC?
5. How have the United States and other countries tried to influence China's stance on human rights?

SECTION 2

6. Describe six serious environmental challenges facing East Asia.
7. Japan is looking for alternative electric power sources. What event helped cause Japanese interest in these alternatives?
8. Which three East Asian countries rely on nuclear power to meet at least 30 percent of their electricity needs?
9. How does China's heavy reliance on coal contribute to air pollution in the region?
10. How have commercial fishing companies and factory ships intensified overfishing in the region?

Critical Thinking

1. **Predicting Consequences** Study the map on page 639. What impact might a railroad between Tibet and the rest of China have on Tibet's economy?
2. **Comparing and Contrasting** Examine Japan's environmental record since 1990. In what ways has this record improved? What areas still need improvement?
3. **Identifying Cause and Effect** Complete the diagram below to show how China's state-run industries affect economic growth. Which effect do you think is most important?

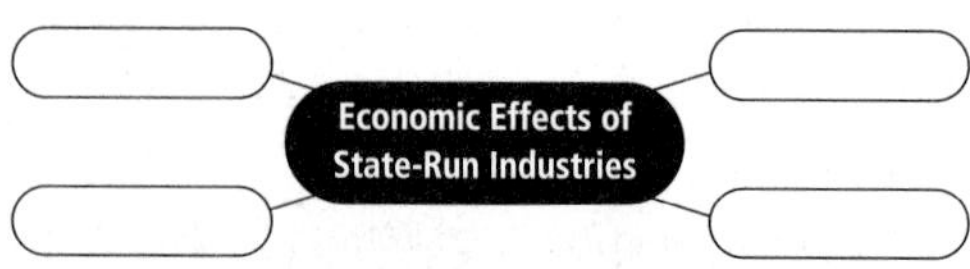

NATIONAL GEOGRAPHIC Locating Places

East Asia: Physical-Political Geography

Match the letters on the map with the places and physical features of East Asia listed below. Write your answers on a sheet of paper.

1. Mongolia
2. Yangtze River
3. Macau
4. Inland Sea
5. Kobe
6. Wuhan
7. Tianjin
8. Taiwan
9. South Korea
10. Yellow River

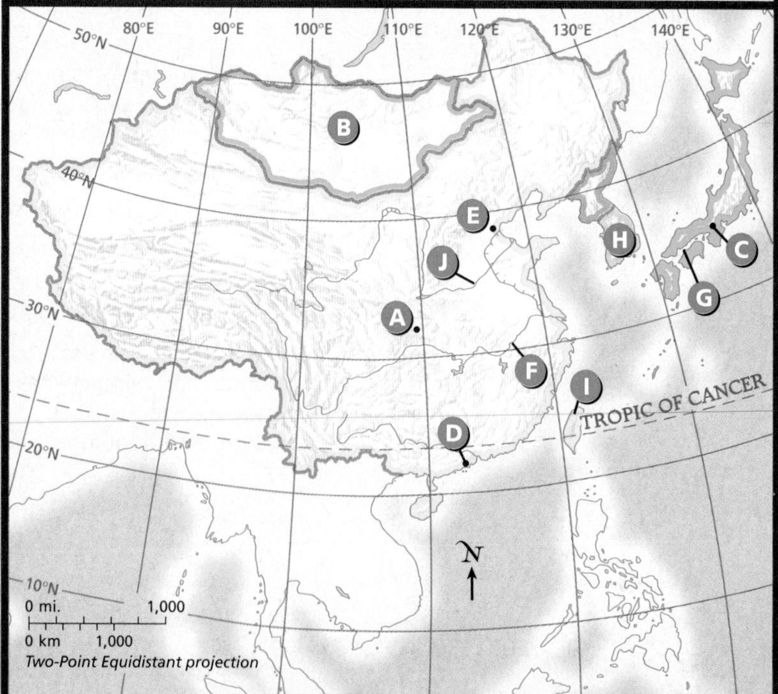

Self-Check Quiz Visit the **Glencoe World Geography** Web site at geography.glencoe.com and click on Self-Check Quizzes—Chapter 28 to prepare for the Chapter Test.

Using the Regional Atlas

Refer to the Regional Atlas on pages 636–639.

1. **Place** By giving the names of areas or countries, identify where each of the following natural resources in East Asia is found: tungsten, iron ore, copper, petroleum, tin, bauxite.
2. **Location** List the capitals of the East Asian countries and their absolute locations.

Thinking Like a Geographer

How might more and improved roads and railroads affect patterns of settlement, population distribution, and resource use in China's interior areas?

Problem-Solving Activity

Group Research Project In June 2000, North and South Korea entered into peaceful negotiations for the first time in 50 years. Work with a group to research the agreement the two countries reached. Evaluate whether unifying Korea is a realistic prospect for the future. Then write an editorial explaining your position.

GeoJournal

Evaluating Information Review the information you logged in your GeoJournal as you read this chapter. Choose one of the economic or environmental challenges. Then write an essay evaluating East Asia's success in using technology to meet and solve the challenge.

Technology Activity

Creating a Web Site Collect information, graphics, and photos for each country in East Asia, including examples of modern and traditional architecture, art, flags, clothing, and scenes from both rural and urban areas. Compose a fact sheet about each country that includes statistics on population, major cities, economic GNP or GDP, literacy rate, and other interesting information. Place your images and facts on your own Web site. Be sure to cite all your sources.

Standardized Test Practice

Read the excerpt below about Chinese writer Gao Xingjian, who won the Nobel Prize for literature in 2000. Then choose the best answer for the following multiple-choice question. If you have trouble answering the question, use the process of elimination to narrow your choices.

> *"Born in 1940 in Jiangxi province in eastern China, Mr. Gao earned a degree in French in Beijing and embarked on a life of letters [writing literature]. During the Cultural Revolution [1966-1976], he was sent to a re-education camp, where he spent six years at hard labor in the fields. He also burned a suitcase full of his early manuscripts. . . . [In] 1979 . . . he was first able to publish his work and to travel abroad. . . ."*
>
> —*New York Times On the Web* (online), October 12, 2000

1. Xingjian's relationship with the Chinese government can best be described as

A subservient.
B manipulative.
C passive.
D turbulent.

Return to the passage, and underline the parts that represent decisions Xingjian had to make or actions Xingjian had to take. Based on what you underline, try to summarize the government's role in Xingjian's life. Then read the answer choices, eliminating those that you know are incorrect. Last, determine an answer from the answer choices that remain.

Southeast Asia

WHY IT'S IMPORTANT—

Southeast Asia is a vital crossroads of trade and commerce. The region is rich in natural resources such as tin, petroleum, rubber, tea, spices, and valuable woods. In recent years, many Southeast Asians have migrated to the United States, bringing their own religions and cultures with them. You are probably familiar with the flavors of many Southeast Asian dishes, available now in restaurants in the United States.

World Regions Video To learn more about Southeast Asia and its impact on your world, view the World Regions video "Southeast Asia."

Terraced rice fields on the island of Bali, Indonesia

10

What Makes Southeast Asia a Region?

Lying east of India and south of China, Southeast Asia juts out from the rest of the Asian continent and then fragments into a jumble of islands that straddle the Equator. Two peninsulas form the mainland—the bulbous Indochina Peninsula and the narrow Malay Peninsula, which extends southward from the other like a long, gnarled finger. Millions of years ago, tectonic plates collided to form parallel mountain ranges that span the mainland from north to south. Great rivers, such as the Irrawaddy, Mekong, Chao Phraya, and Red, course through the valleys between these ranges and create fertile deltas where they meet the sea.

Most of the region's islands are mountainous, too, but their peaks were spawned by ancient volcanic eruptions. Active volcanoes remain a threat on these islands, which lie along the Pacific Ring of Fire.

Rain forests cover parts of Southeast Asia. Watered by monsoon rains, these forests—valued for their timber and wildlife—are decreasing because of extensive logging.

3

1

1 Ankle-deep in muddy water, a Philippine farmer plants rice seedlings in a flooded paddy. Southeast Asia's fertile soils and warm, wet climate are ideal for growing rice. Most farmers in Southeast Asia plant and harvest their crops by hand.

2

2 Using trunk and tusks, an Asian elephant piles up teak logs along a river in Myanmar. The lush forests of this country supply more than three-fourths of the world's teak, a beautiful, durable wood often used for furniture. Some Southeast Asian loggers use tractors, but elephants are cheaper and don't require roads.

3 Steering with slender paddles, a Vietnamese woman guides her boat through a shallow waterway on the delta of the Mekong River. From its source in China, the Mekong flows 2,600 miles (4,180 km) to the South China Sea. Like rivers throughout this region, the Mekong is a vital transportation route for people and goods.

4 Like a fountain of fire, Krakatau hurls lava into the night sky. One of Indonesia's many active volcanoes, Krakatau lies between the islands of Sumatra and Java. In 1883, 36,000 people died when Krakatau erupted violently, generating huge tidal waves that swept over the nearby islands.

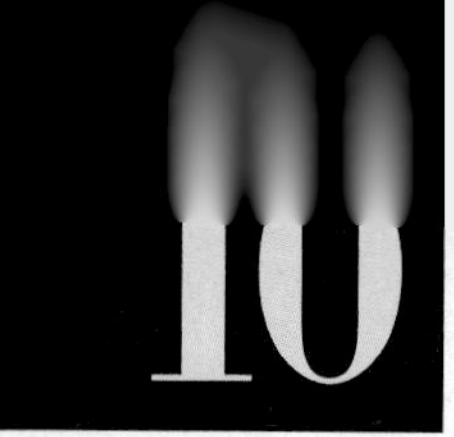

Ethnic Mosaic

Rugged mountains and rolling seas could not hold back the outsiders that have been drawn to Southeast Asia throughout its history. Some came to trade, some to settle, and others to forge empires. Beginning in the 1500s, Europeans laid claim to various parts of the region. Eventually, every Southeast Asian country except Thailand was a European colony.

Colonial rule ended in the mid-1900s, but the region was left fragmented and in turmoil. Struggles among ethnic groups and between Communist and non-Communist powers claimed thousands of lives.

Today, more than 500 million people live in this culturally diverse region. They speak hundreds of languages and dialects and practice several major religions. Despite rapid urbanization and industrialization in some places, most Southeast Asians still make their living traditionally, as farmers.

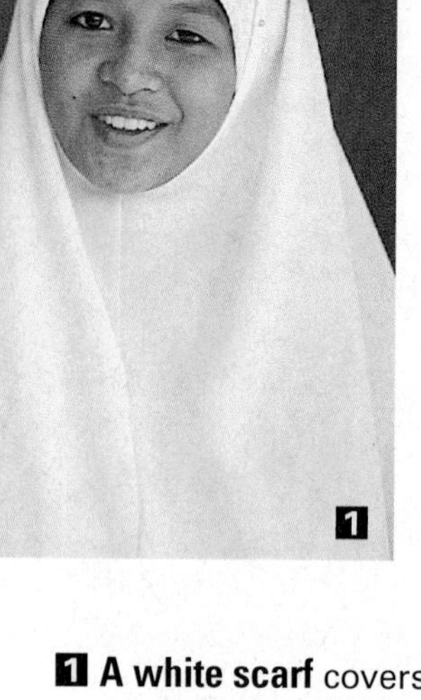

1 A white scarf covers the head of a Muslim girl in Malaysia. Arab and Indian traders brought the faith of Islam to Southeast Asia in the 1300s and 1400s. Today, Islam is the dominant religion on the Malay Peninsula and in Indonesia. In fact, Indonesia has more Muslims than any other country in the world.

2 Fruit vendors on bicycles offer bounty from the fields to buyers on the streets of Hanoi, Vietnam. Most Vietnamese, like other Southeast Asians, are farmers who raise rice, fruit, and other crops on small plots of land. Only a small percentage of Vietnamese people work in industry.

3 Beneath gilded towers, Buddhist monks descend the steps of a temple in Vientiane, the capital and largest city of Laos. After Islam, Buddhism is the second most widespread religion in Southeast Asia. It is the primary religion in Laos and on the rest of the Indochina Peninsula.

4 Southeast Asia's busiest port, Singapore lies at the tip of the Malay Peninsula, along the Strait of Malacca, the main shipping route between the Indian Ocean and the South China Sea. From this strategic location, the city handles much of the flow of goods into and out of Southeast Asia.

10

Southeast Asia

PHYSICAL

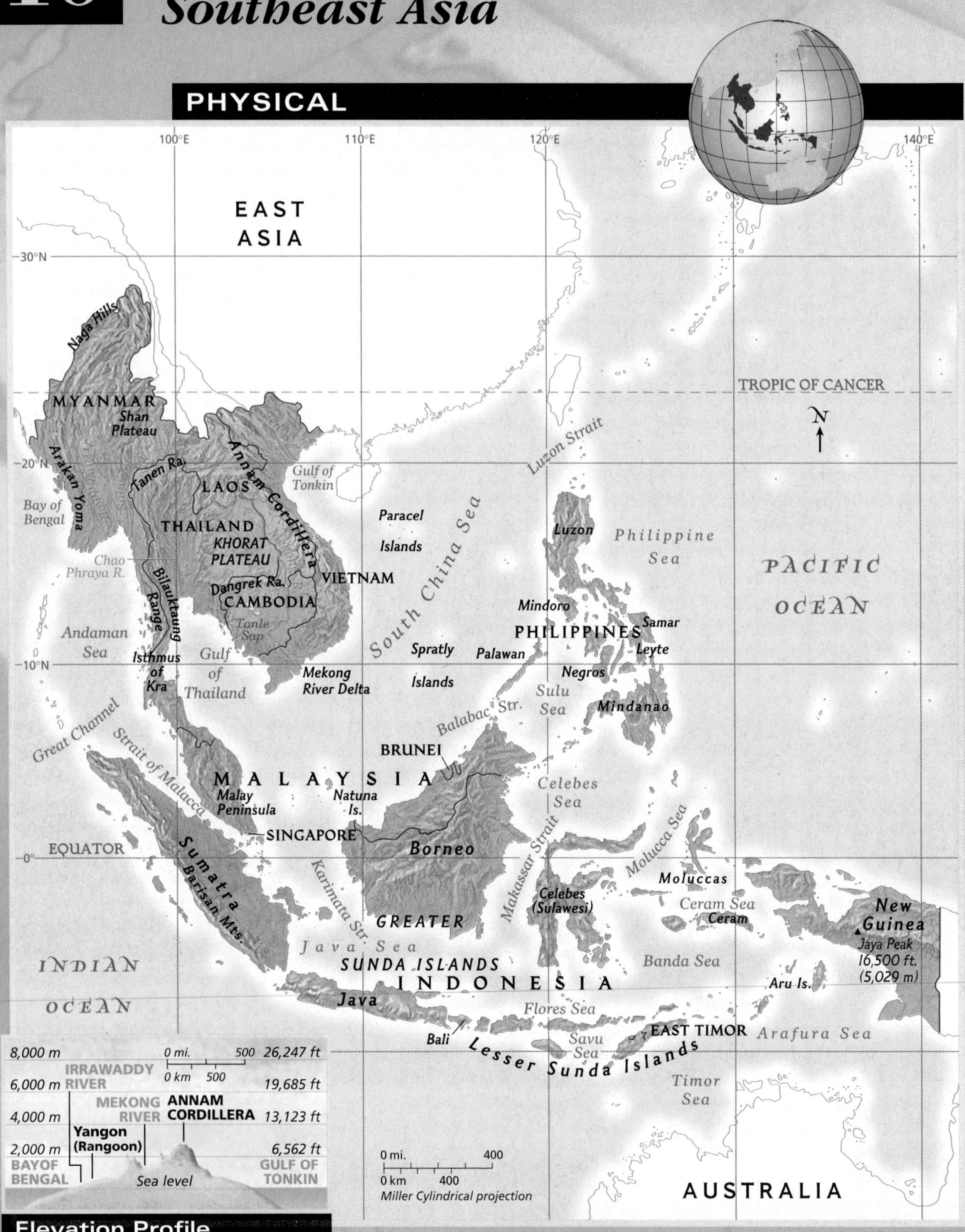

NATIONAL GEOGRAPHIC SOCIETY

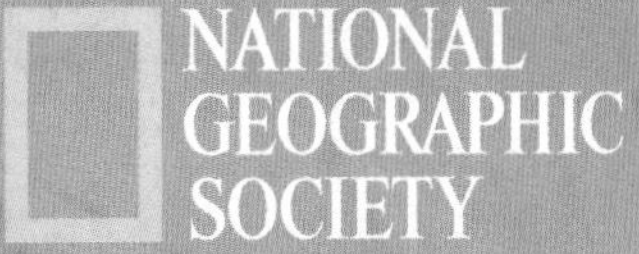

POLITICAL

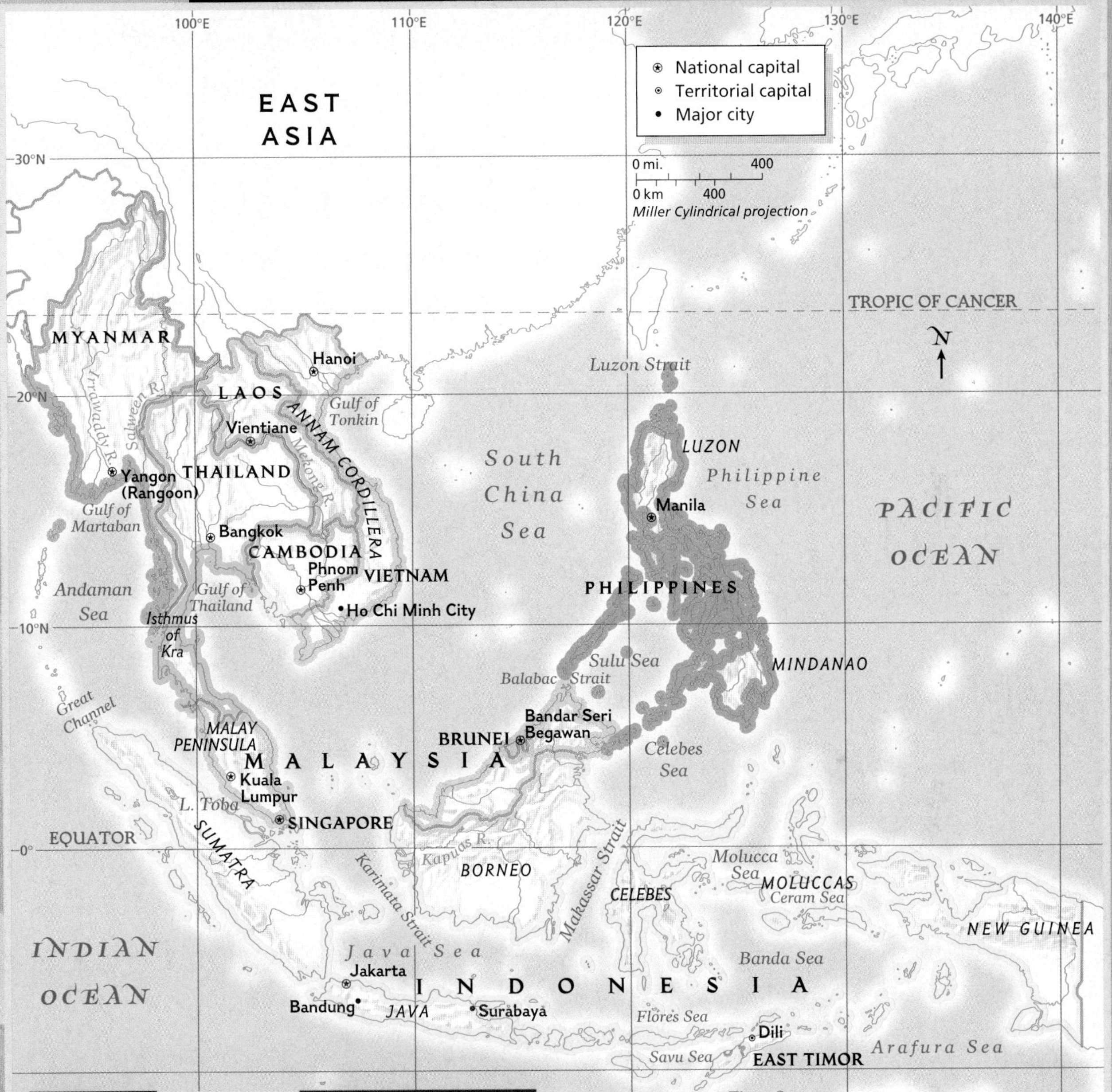

100°E
110°E
120°E
130°E
140°E
30°N
20°N
10°N
0°
National capital
Territorial capital
Major city
0 mi. 400
0 km 400
Miller Cylindrical projection
EAST ASIA
TROPIC OF CANCER
N
MYANMAR
Hanoi
LAOS
Gulf of Tonkin
Vientiane
ANNAM CORDILLERA
Irrawaddy R.
Salween R.
Mekong R.
Yangon (Rangoon)
THAILAND
Gulf of Martaban
Bangkok
CAMBODIA
Phnom Penh
VIETNAM
Ho Chi Minh City
Andaman Sea
Gulf of Thailand
Isthmus of Kra
South China Sea
Luzon Strait
LUZON
Philippine Sea
Manila
PACIFIC OCEAN
PHILIPPINES
Sulu Sea
Balabac Strait
MINDANAO
Great Channel
MALAY PENINSULA
Bandar Seri Begawan
BRUNEI
MALAYSIA
Celebes Sea
Kuala Lumpur
L. Toba
SINGAPORE
SUMATRA
EQUATOR
Kapuas R.
BORNEO
Karimata Strait
Makassar Strait
CELEBES
Molucca Sea
MOLUCCAS
Ceram Sea
NEW GUINEA
INDIAN OCEAN
Java Sea
Jakarta
Banda Sea
INDONESIA
Bandung
JAVA
Surabaya
Flores Sea
Dili
Savu Sea
EAST TIMOR
Arafura Sea
Timor Sea
AUSTRALIA

MAP Study

1. What are the capitals of the continental countries in Southeast Asia?
2. To what country does the island of Mindanao belong?

UNIT 10

REGIONAL ATLAS

Southeast Asia

POPULATION DENSITY

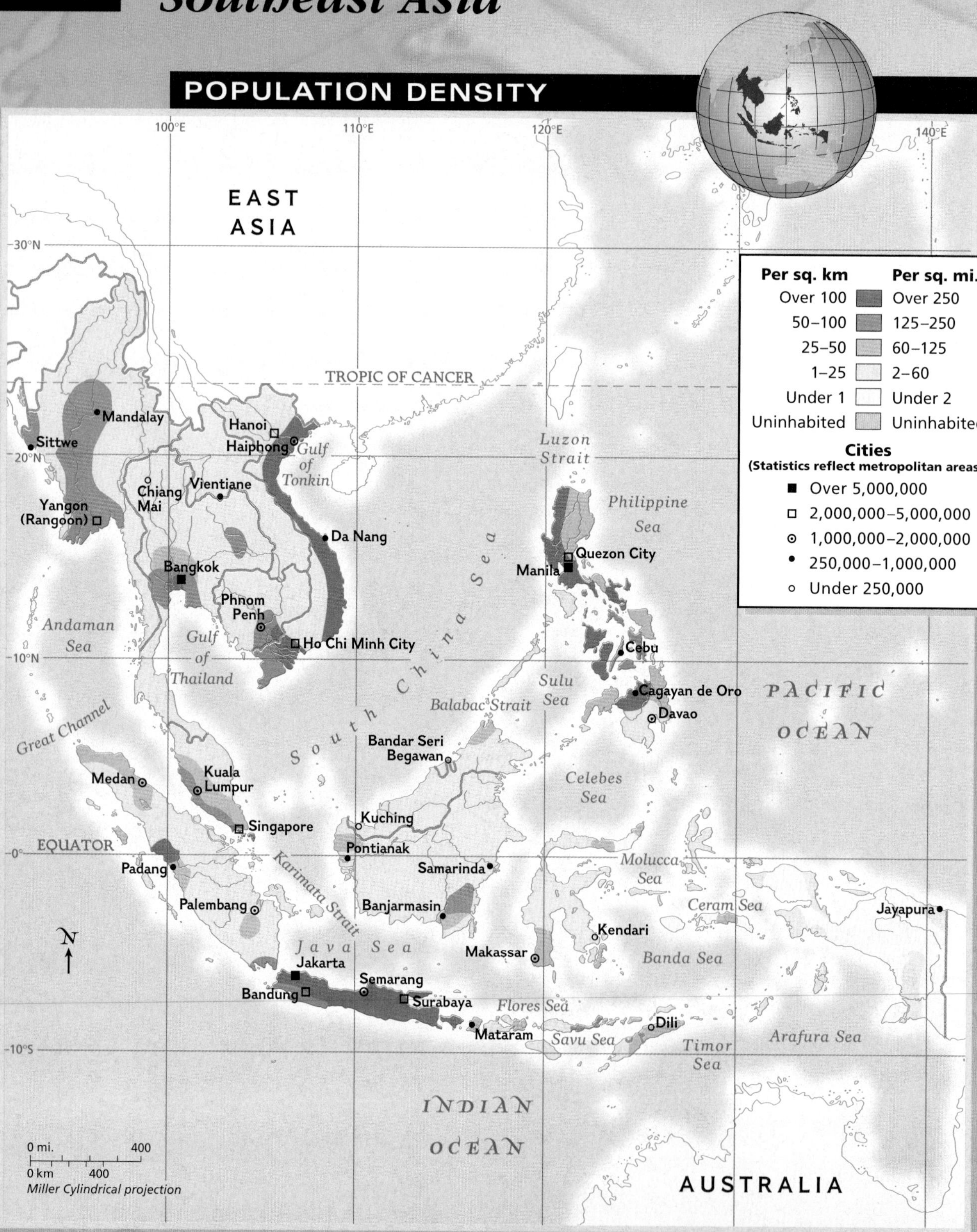

ECONOMIC ACTIVITY

Resources
- Petroleum
- Natural gas
- Coal
- Nickel
- Tungsten
- Copper
- Tin
- Gemstones
- Gold

Land Use
- Commercial farming
- Subsistence farming
- Hunting and gathering
- Manufacturing and trade
- Commercial fishing

100°E 110°E 120°E 130°E 140°E
30°N 20°N 10°N 0° 10°S

EAST ASIA
TROPIC OF CANCER
Luzon Strait
MYANMAR
Teak
Teak
Yangon (Rangoon)
Rice
LAOS
Hanoi
Rice
Gulf of Tonkin
THAILAND
Rice
Rice
Bangkok
Rubber
CAMBODIA
VIETNAM
Rice
Phnom Penh
Ho Chi Minh City
Rice
Andaman Sea
Gulf of Thailand
South China Sea
Philippine Sea
Sugarcane
Manila
PHILIPPINES
Sulu Sea
Coconuts
Abaca
Corn
PACIFIC OCEAN
Celebes Sea
BRUNEI
Tea
Medan
Kuala Lumpur
MALAYSIA
Rubber
SINGAPORE
Rice
Rubber
Coconuts
EQUATOR
SUMATRA
Coffee
Karimata Strait
BORNEO
Pearls
Spices
CELEBES
Rubber
Spices
Spices
Coconuts
Palembang
Spices
Java Sea
Jakarta
Bandung
JAVA
Surabaya
INDONESIA
Banda Sea
NEW GUINEA
Coconuts
INDIAN OCEAN
Cassava
Coconuts
Pearls
EAST TIMOR
Arafura Sea
AUSTRALIA

0 mi. 400
0 km 400
Miller Cylindrical projection

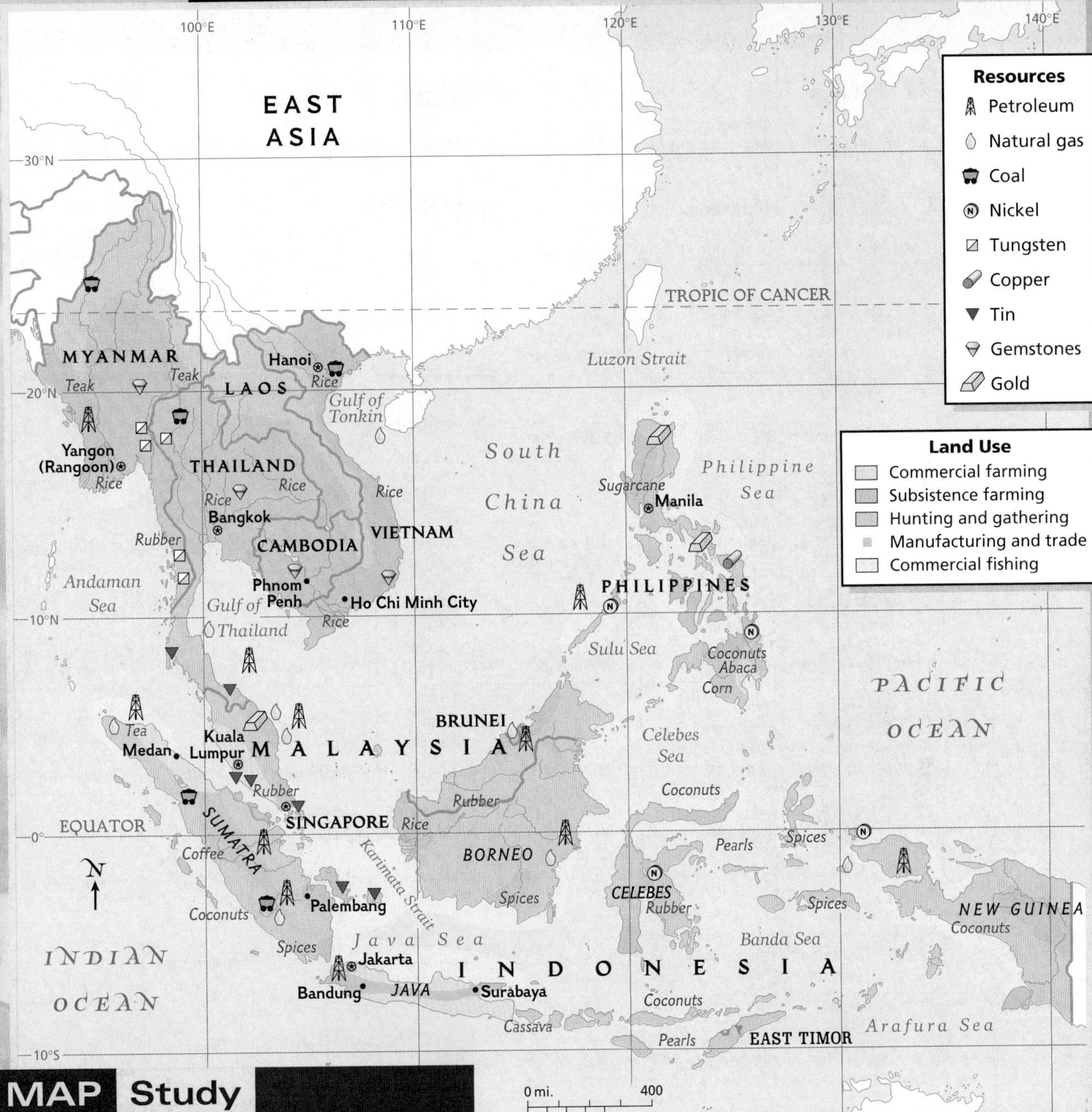

MAP Study

1. Which Southeast Asian countries have deposits of gold? Gemstones?
2. Which Southeast Asian island countries are the most densely populated?

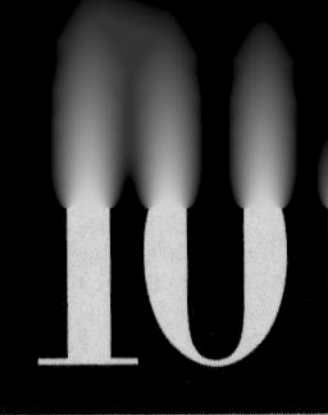

Southeast Asia

FOR AN ONLINE UPDATE OF THIS INFORMATION, VISIT GEOGRAPHY.GLENCOE.COM AND CLICK ON "TEXTBOOK UPDATES."

COUNTRY PROFILES

COUNTRY * AND CAPITAL	FLAG AND LANGUAGE	POPULATION AND DENSITY	LANDMASS	MAJOR EXPORT	MAJOR IMPORT	CURRENCY	GOVERNMENT
BRUNEI Bandar Seri Begawan	Malay, English, Chinese	300,000 156 per sq. mi. 60 per sq. km	2,228 sq. mi. 5,771 sq. km	Crude Oil	Machinery	Brunei Dollar	Constitutional Monarchy
CAMBODIA Phnom Penh	Khmer, French	13,100,000 187 per sq. mi. 72 per sq. km	69,900 sq. mi. 181,041 sq. km	Timber	Construction Materials	Riel	Constitutional Monarchy
EAST TIMOR Dili	Tetun, Javanese, Portuguese	800,000 134 per sq. mi. 52 per sq. km	5,741 sq. mi. 14,869 sq. km	Coconut Products	Manufactured Goods	Indonesian Rupiah	Republic
INDONESIA Jakarta	Bahasa Indonesia, Javanese	206,100,000 280 per sq. mi. 108 per sq. km	735,355 sq. mi. 1,904,569 sq. km	Crude Oil	Manufactured Goods	Rupiah	Republic
LAOS Vientiane	Lao, French	5,400,000 59 per sq. mi. 23 per sq. km	91,429 sq. mi. 236,800 sq. km	Wood Products	Machinery	Kip	Communist State
MALAYSIA Kuala Lumpur	Malay, English, Chinese	22,700,000 178 per sq. mi. 69 per sq. km	127,317 sq. mi. 329,749 sq. km	Electronic Equipment	Machinery	Ringgit	Constitutional Monarchy
MYANMAR Yangon (Rangoon)	Burmese, Local Languages	47,800,000 183 per sq. mi. 71 per sq. km	261,228 sq. mi. 676,581 sq. km	Beans	Machinery	Kyat	Military Dictatorship
PHILIPPINES Manila	Tagalog, English	77,200,000 666 per sq. mi. 257 per sq. km	115,830 sq. mi. 300,000 sq. km	Electronic Equipment	Raw Materials	Philippine Peso	Republic
SINGAPORE Singapore	Chinese, Malay, Tamil, English	4,100,000 17,320 per sq. mi. 6,687 per sq. km	239 sq. mi. 619 sq. km	Computer Equipment	Aircraft	Singapore Dollar	Republic
THAILAND Bangkok	Thai, Local Languages	62,400,000 315 per sq. mi. 122 per sq. km	198,116 sq. mi. 513,120 sq. km	Manufactured Goods	Machinery	Baht	Constitutional Monarchy
VIETNAM Hanoi	Vietnamese, Local Languages	78,700,000 623 per sq. mi. 241 per sq. km	128,066 sq. mi. 331,691 sq. km	Crude Oil	Machinery	Dong	Communist State

* COUNTRIES AND FLAGS NOT DRAWN TO SCALE

◀ Ruins of Buddhist temple c. A.D. 1000–1200, Pagan, Myanmar

GLOBAL CONNECTION

SOUTHEAST ASIA AND THE UNITED STATES

CUISINE

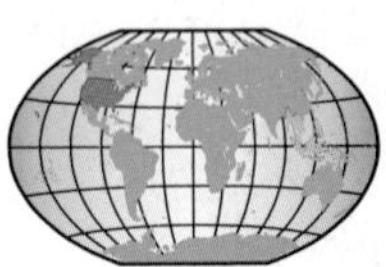

What's for dinner? Twenty or thirty years ago, the answer probably would have been "steak," "meatloaf," or "hamburgers." But now, you might hear "lemon grass chicken" or "laab moo"! Americans have developed a taste for foods from other lands. And the cuisines of two Southeast Asian countries—Thailand and Vietnam—have become especially popular in the United States.

Thai and Vietnamese cooks themselves have borrowed foods, flavors, and preparation methods from several of their neighbors, especially China and India. For example, many Thai and Vietnamese dishes are stir-fried, as is much Chinese food. Coconut milk is an ingredient picked up from India. Nevertheless, Thai and Vietnamese cuisines have their own distinctive flavors and characteristics.

Some of the common ingredients in Thai food are lemon grass, shrimp paste, Siamese ginger, and chilies—very hot chilies! These and other ingredients are combined to create complex and tantalizing tastes. Laab moo, for example, is a dish of minced pork seasoned with lemon juice, fish sauce, fresh mint, and green chilies. In a single forkful of a Thai dish, you might taste sweet, sour, salty, and hot flavors all at once.

Vietnamese food is often described as being similar to Thai food, but less intense,

▼ Enjoying a meal in Hanoi, Vietnam

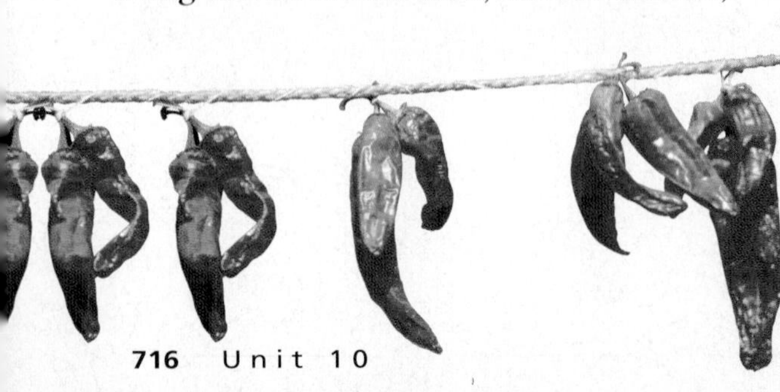

▲ Floating produce market in Thailand

with more subtle flavors. Some Vietnamese dishes might seem more like salads than main dishes to most Americans. Bits of cooked meat or fish are typically served with a platter of fresh lettuce, herbs, and vegetables. One ingredient found in almost all Vietnamese dishes is nuoc mam, a salty fish sauce. What salt is to American food and soy sauce is to Chinese dishes, nuoc mam is to Vietnamese cuisine.

How did Thai and Vietnamese foods get to the United States? In the 1970s, after the Vietnam War, Vietnamese refugees flocked to America. Many opened small restaurants. Thais had been coming to the United States as students since the 1960s. Many of the Thais settled in Los Angeles, where the climate may have reminded them of home. By 1990, there were more Thais living in greater Los Angeles than any other place outside Thailand—and the city had at least 200 Thai restaurants.

Now Thai and Vietnamese restaurants can be found in cities all across the United States. From the fiery flavors of Thailand to the more delicate tastes of Vietnam, Southeast Asian cuisines have found a home in America.

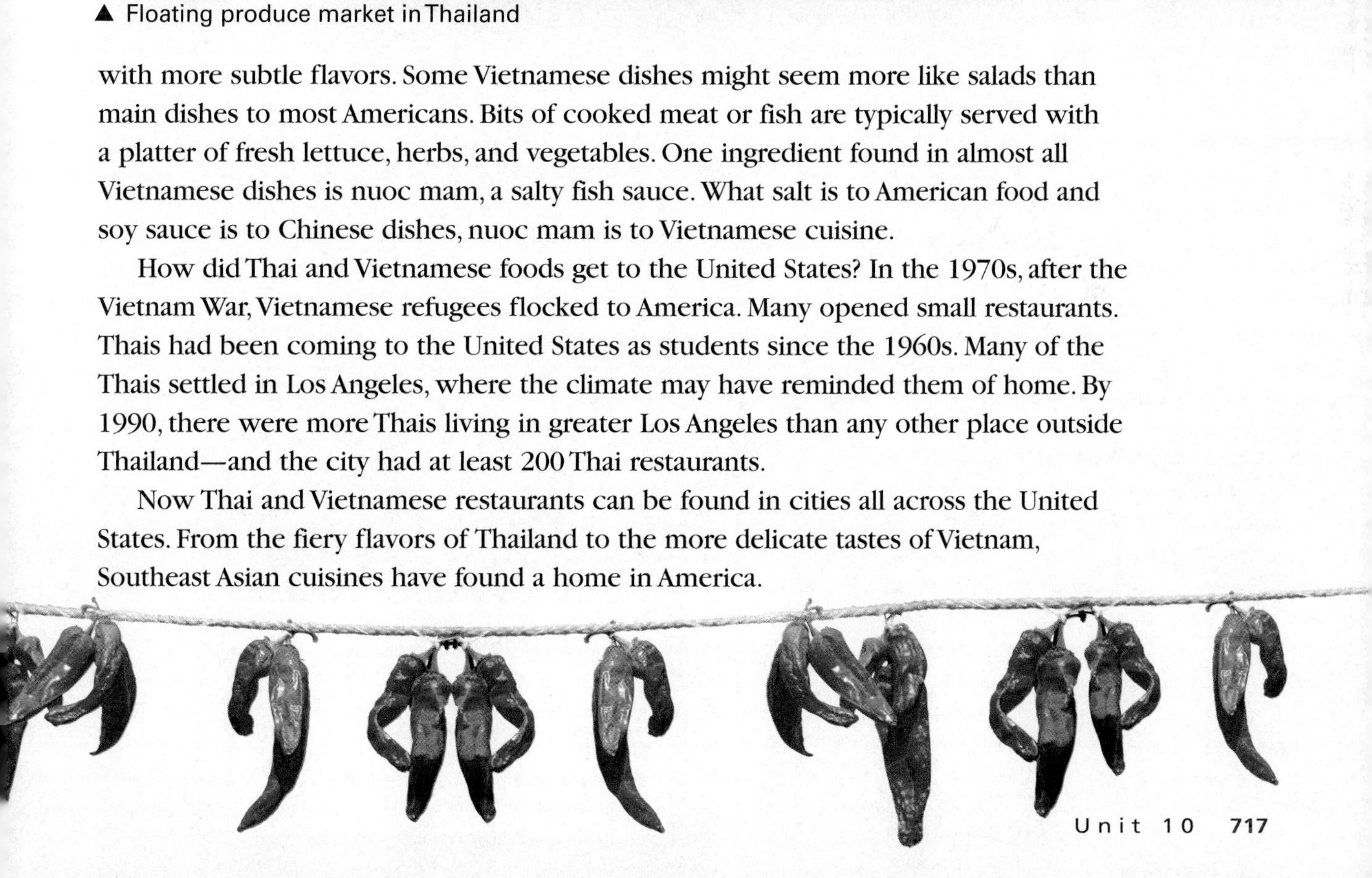

CHAPTER 29

The Physical Geography of Southeast Asia

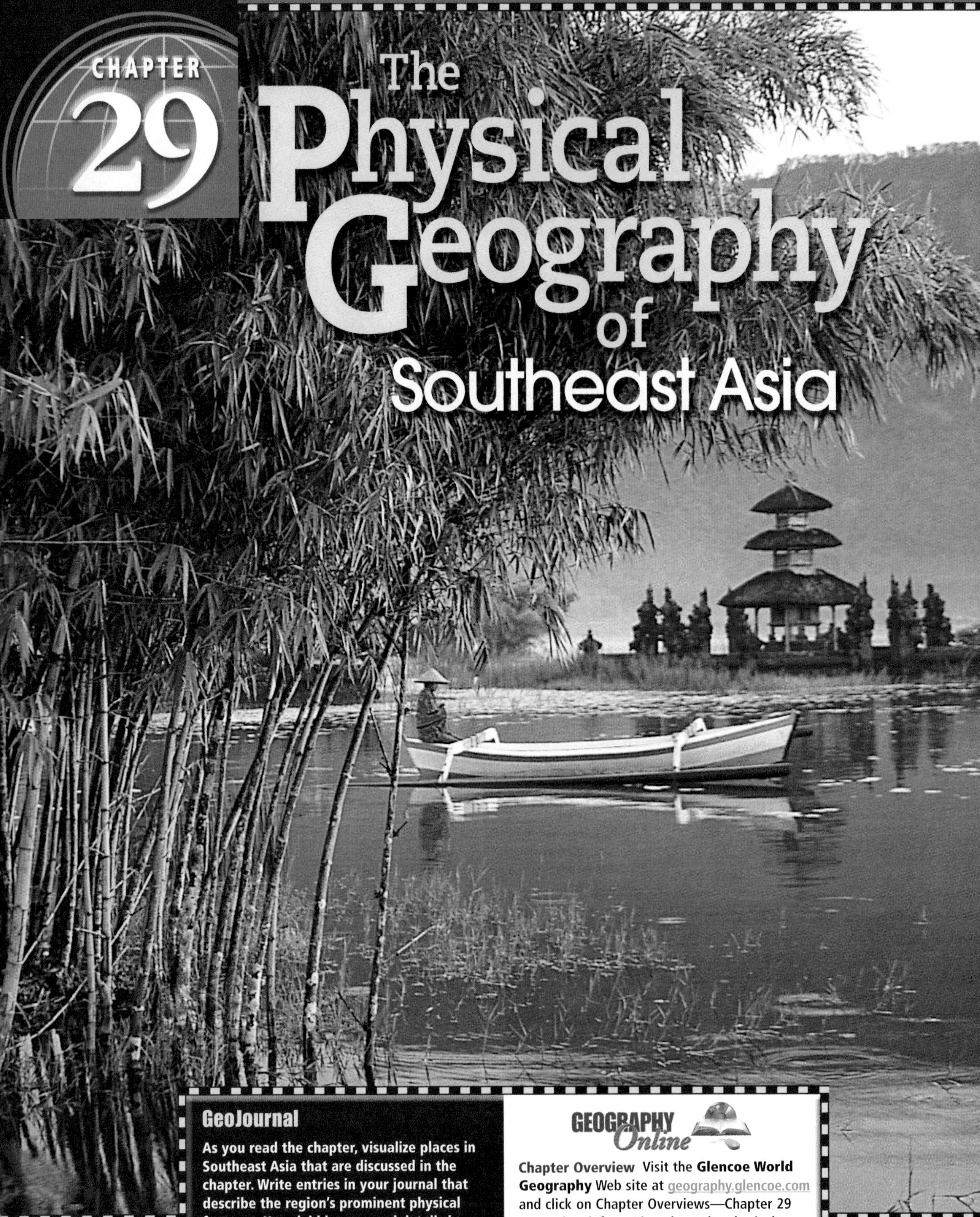

GeoJournal

As you read the chapter, visualize places in Southeast Asia that are discussed in the chapter. Write entries in your journal that describe the region's prominent physical features. Use vivid images and details in your entries.

GEOGRAPHY Online

Chapter Overview Visit the **Glencoe World Geography** Web site at geography.glencoe.com and click on Chapter Overviews—Chapter 29 to preview information about the physical geography of the region.

SECTION 1

The Land

Guide to Reading

Consider What You Know

You have learned how the physical geography of a region affects its economy. Southeast Asia is rich in tropical rain forests and water resources. What products do you know about that come from this region?

Read to Find Out

- How did tectonic plate movement, volcanic activity, and earthquakes form Southeast Asia?
- Why are the region's waterways important to its peoples?
- How do rich natural resources affect Southeast Asia's economy?

Terms to Know

- cordillera
- archipelago
- insular
- flora
- fauna

Places to Locate

- Indochina Peninsula
- Malay Peninsula
- Annam Cordillera
- Irrawaddy River
- Chao Phraya River
- Red River
- Mekong River

◀ *Pura (temple) Ulun Danu, Bali, Indonesia*

NATIONAL GEOGRAPHIC

A Geographic View

Journey to the Interior

At dawn the next day I set off upriver in a hollowed-out tree trunk with my guide. . . . [He] poles the dugout through the tea-colored water while I watch birds—kingfishers darting from the riverbanks, flocks of hornbills skimming above the treetops, their wings sounding like runners panting for breath.

The banks sprout wild breadfruits, bananas, and a host of palm trees, all tangled up with hanging vines. As the heat of the day intensifies, the river's green walls vibrate with the ringing of cicadas. Then the river grows shallower, forcing us to push the dugout over rocks. It is the dry season, something hard to fathom in a place drenched with more than 200 inches of rain a year.

—*Thomas O'Neill, "Irian Jaya, Indonesia's Wild Side,"* National Geographic, *February 1996*

Stilt houses in Irian Jaya

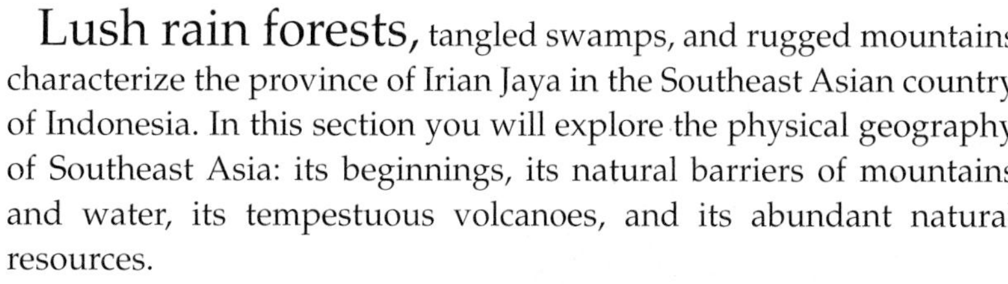

Lush rain forests, tangled swamps, and rugged mountains characterize the province of Irian Jaya in the Southeast Asian country of Indonesia. In this section you will explore the physical geography of Southeast Asia: its beginnings, its natural barriers of mountains and water, its tempestuous volcanoes, and its abundant natural resources.

Peninsulas and Islands

When the Eurasian, Philippine, and Indo-Australian tectonic plates collided millions of years ago, they formed the landmasses that are known today as Southeast Asia. The upheaval formed **cordilleras**,

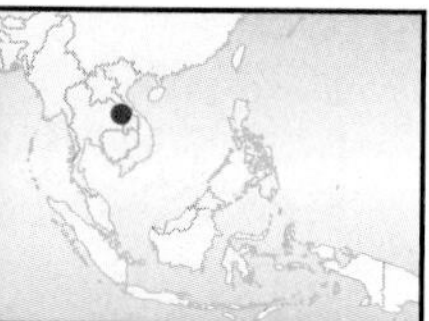

Geography Skills for Life

Hue, Vietnam Farmers tend their fields near Hue, a historic city in central Vietnam.

Place What countries lie on the Indochina Peninsula?

or parallel mountain ranges and plateaus, that extend into the Indochina Peninsula. Activity from related volcanoes and earthquakes created a series of archipelagos in the South Pacific. An archipelago is a group of islands.

Straddling the Equator, the peninsulas and islands of Southeast Asia combine mountainous terrain with a predominantly tropical climate. The region stretches from the Asian mainland almost to Australia and covers 1,570,000 square miles (4,066,300 sq. km). Two large land areas, the **Indochina Peninsula** and the **Malay Peninsula,** make up mainland Southeast Asia. South and east of this area lies the vast Malay Archipelago, sometimes called the East Indies. The Malay Archipelago, containing 20,000 islands, stretches from the Indian Ocean to the Pacific Ocean.

Comparing Lands

Southeast Asia is about half the size of the continental United States.

Mainland Southeast Asia

About half of Southeast Asia's 11 countries are located on the mainland. The rest are island countries, except for Malaysia, which is both a mainland and an island country. Laos is the region's only country without a coastline. The four mainland countries of Vietnam, Laos, Cambodia, and Myanmar (formerly called Burma) lie entirely on the Indochina Peninsula. Most of Thailand also is located there, but part of that country trails southward to the Malay Peninsula. Malaysia shares the Malay Peninsula with Thailand, while the rest of Malaysia is located on Borneo, an island east of the Malay Peninsula.

Island Southeast Asia

The insular, or island, countries of Southeast Asia include Brunei, East Timor, Indonesia, Singapore, and the Philippines. Brunei, almost surrounded by Malaysia, is a small country on the northern coast of Borneo. Indonesia is the largest island country

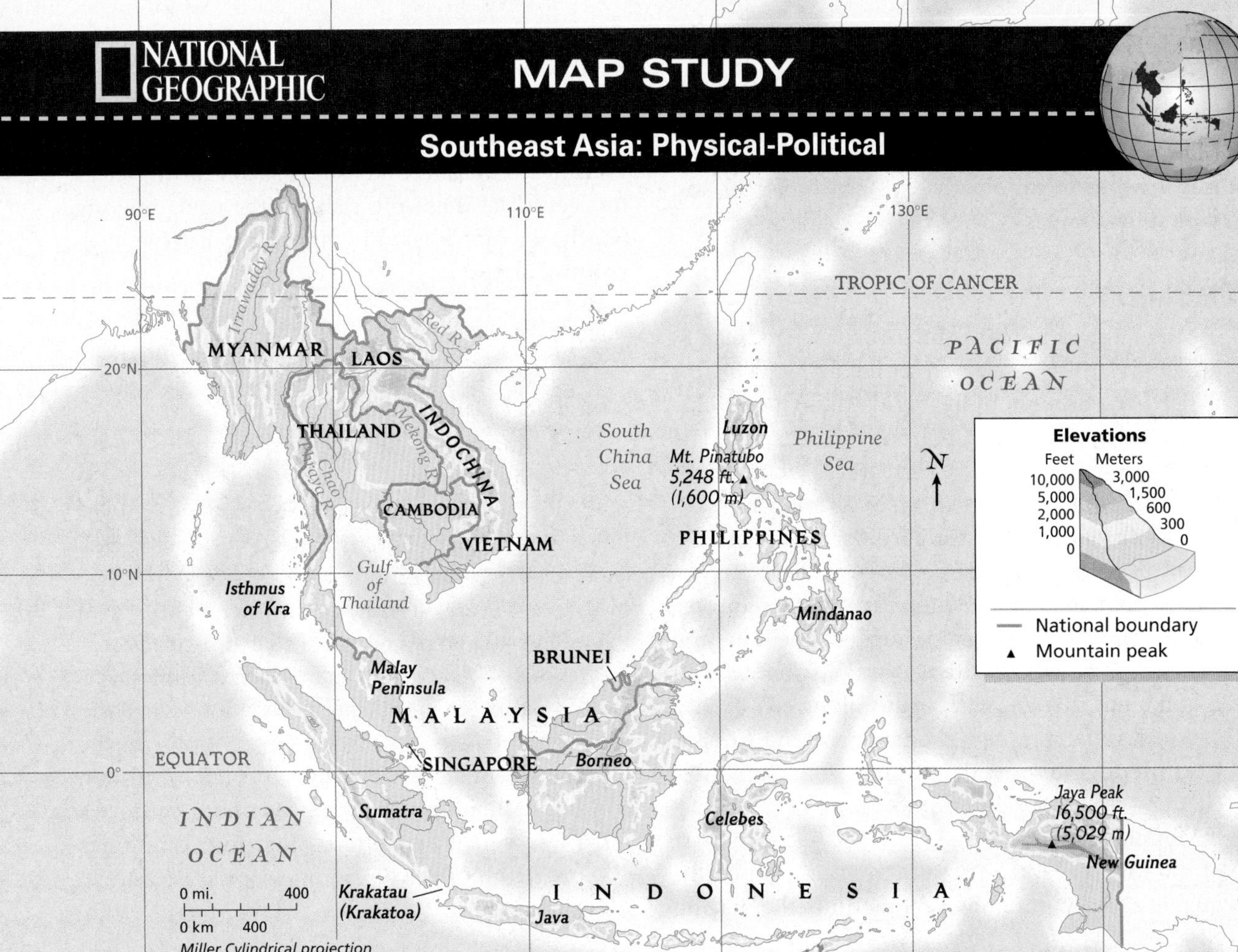

Geography Skills for Life

1. **Interpreting Maps** What is the only land-locked country in Southeast Asia?

2. **Applying Geography Skills** What challenge might Malaysia, Indonesia, and the Philippines face as island countries?

Find NGS online map resources @ www.nationalgeographic.com/maps

in the region. Its 13,677 islands span 3,000 miles (4,827 km) and two oceans, the Indian and the Pacific. Only about 6,000 islands are named, and fewer than 1,000 are permanently settled. East Timor, formerly a part of Indonesia, voted to become independent in 1999. Two years later, East Timorese held elections for a new national assembly.

The country of Singapore, a collection of one large island and more than 50 smaller ones, sits just off the southern tip of the Malay Peninsula. The country's capital is on the large island, and both the island and the capital city are called Singapore. The size of the islands varies greatly. The total area of the island of Singapore is 221 square miles (572 sq. km), and the total area of all the other islands is about 18 square miles (47 sq. km). Half of those islands are uninhabited.

Although more than 7,000 islands make up the Philippines, only around 900 are settled, and 11 islands account for over 95 percent of the country's area. As in Indonesia and Singapore, many of the Philippine islands have not been named.

Physical Features

Cordilleras loom above fertile fields. Rivers create transportation routes through lush vegetation. Majestic volcanoes add to the scenery. These physical features create Southeast Asia's colorful and varied landscapes.

Mountains

Mountains dominate Southeast Asian landscapes, although most peaks crest below 10,000 feet (3,048 m). Throughout the region these mountains create geographic and political boundaries. The Indochina Peninsula's western and northern highlands separate the region from India and China. To the south and east, three cordilleras run mainly north to south, forming natural barriers between and within mainland Southeast Asian countries. These parallel mountain ranges include the Arakan Yoma Range in western Myanmar; the Bilauktaung Range, which runs along the border between Myanmar and Thailand; and the **Annam Cordillera**, the mountain range that separates Vietnam from Laos and Cambodia.

Mountains on Southeast Asia's islands form part of the Ring of Fire, an area of volcanic and earthquake activity roughly surrounding the Pacific Ocean. These mountains are actually volcanoes, many of which are still active. Some islands of Indonesia and the Philippines are marked with craters formed by these volcanoes. Mineral-rich volcanic material that has broken down over the centuries has left rich, fertile soil, making Southeast Asia's islands highly productive agricultural areas.

History

Volcanoes of Indonesia and the Philippines

Three hundred twenty-seven volcanoes stretch across Indonesia. Java, an Indonesian island, is one of the Ring of Fire's most active areas. This geologic hot spot is home to 17 of Indonesia's 100 active volcanoes. In 1883, the eruption of Krakatau (Krakatoa) in Indonesia caused massive destruction and great loss of life. To avoid a repeat of such disastrous consequences, observers in Java monitor volcanic activity, prepared to alert the population when an eruption threatens.

Some scientists believe that the 1991 eruption of Mount Pinatubo was the twentieth century's most powerful eruption. Located 55 miles (89 km) north of the Philippine capital of Manila, Mount Pinatubo churned out lava that severely

NATIONAL GEOGRAPHIC **World Explorer**

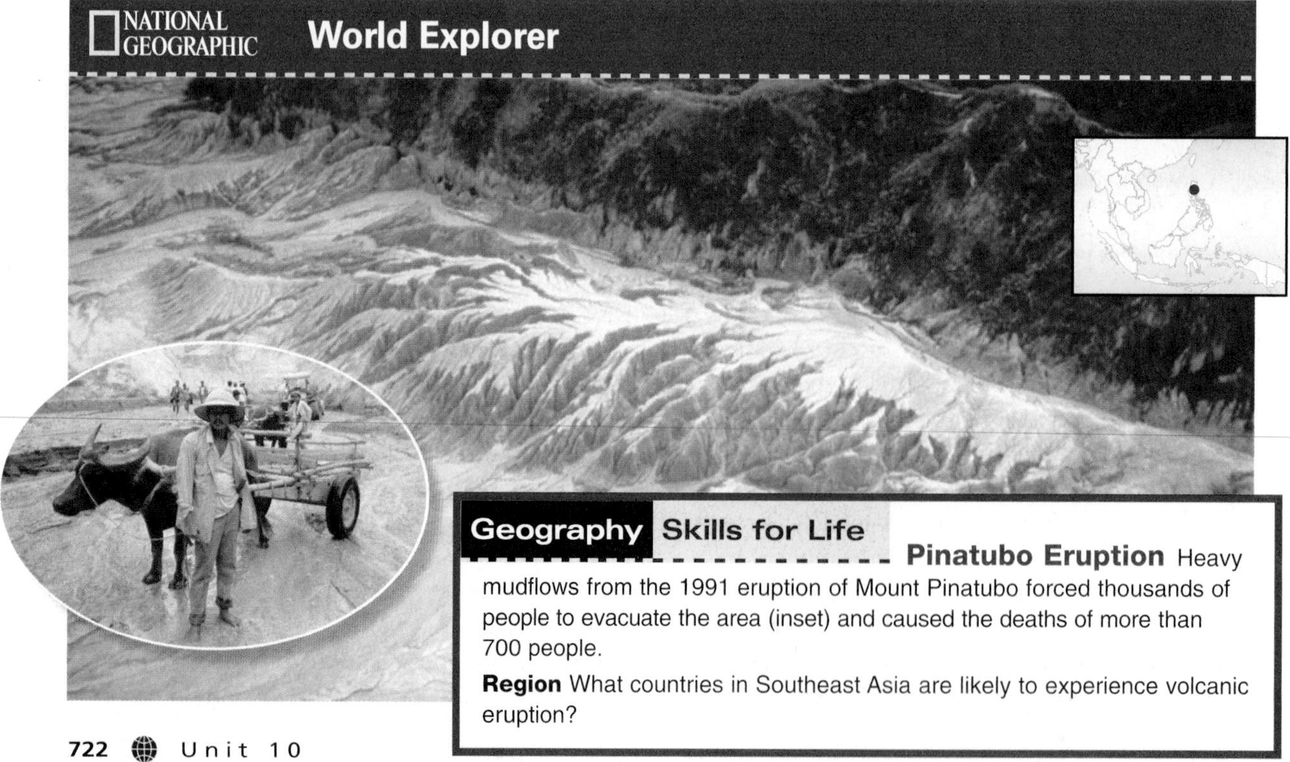

Geography Skills for Life

Pinatubo Eruption Heavy mudflows from the 1991 eruption of Mount Pinatubo forced thousands of people to evacuate the area (inset) and caused the deaths of more than 700 people.

Region What countries in Southeast Asia are likely to experience volcanic eruption?

damaged the town of Angeles. The volcano also blanketed the United States's Clark Air Force Base with volcanic ash nearly a foot deep.

Rivers

Southeast Asia's people rely on waterways for transportation, communication, and food. The rivers' silt and deposits of sediment also create fertile agricultural regions. Mainland rivers originate in the northern highlands of Southeast Asia and in southern China. Most of these rivers flow southward toward the Gulf of Thailand.

Major mainland rivers include the **Irrawaddy** in Myanmar, the **Chao Phraya** (chow PRY•uh) in Thailand, and the **Red (Hong)** in Vietnam. The **Mekong**, which begins its 2,600-mile (4,184-km) journey in China, forms the border between Thailand and Laos and then meanders through Cambodia and southern Vietnam before emptying into the South China Sea. Sediment deposited by the Mekong increases the shoreline around the delta by as much as 50 feet (15 m) per year.

Generally shorter than their mainland counterparts, rivers on Southeast Asia's islands flow in various directions. Most rivers in Indonesia run south to north, cutting vertically across the narrow islands. Borneo's rivers tend to start near the island's center, running outward toward the sea like spokes on a wheel. As one writer notes, traveling on Borneo's rivers reveals a dense, vibrant ecosystem:

> *"Poling our way along the inky green waterway, we glided upstream through quiet still-water bends in the river, where mats of fragrant white flowers had gathered, closing behind the stern of our 24-inch-wide dugout and concealing any sign of our passage."*
>
> Eric Hansen, *Stranger in the Forest: On Foot Across Borneo*, 1988

Geography Skills for Life

River Highway Two tugboats pull a chain of logs to a sawmill on the island of Borneo.

Place What are some important rivers in the region?

Natural Resources

In addition to the remarkable features found in the landscape, Southeast Asia also has rich natural resources. Fossil fuels, natural steam, minerals, and gems can be found in the region. The flora and fauna, or plants and animals, of Southeast Asia are among the most diverse on the earth and also a valuable natural resource of the region.

Energy Sources

The region has a plentiful supply of fossil fuels—coal, oil, and natural gas. Malaysia's second major export is petroleum, and the country's production of oil and natural gas has increased since the 1970s. Indonesia, Vietnam, and the Philippines mine coal, and Vietnam also has rich oil reserves offshore. Oil and natural gas deposits off Borneo's northern coast have made the sultan, or ruler, of Brunei one of the world's richest people. Indonesia also has large petroleum reserves. The island of Sumatra supplies two-thirds of Indonesia's oil, and oil and gas are the country's main exports. One of the leading producers of oil in the Far East, Indonesia is a member of OPEC (Organization of Petroleum Exporting Countries).

Minerals and Gems

Southeast Asia has an abundance of minerals. Indonesia mines nickel and iron, and the

Philippines mines copper. Thailand, Laos, Indonesia, and Malaysia mine tin. Indonesia and Malaysia are among the world's leading producers of tin.

Gems also are plentiful in the region. Sapphires and rubies can be found in Myanmar, Thailand, Cambodia, and Vietnam. In the Philippines pearls are harvested in the province of Sulu and on the island of Palawan. A giant pearl found off Palawan in 1934 weighed about 14 pounds (6.4 kg), making it the largest natural pearl ever harvested. Although most countries take advantage of the wealth provided by nature, some countries' resources remain underdeveloped. Myanmar, for example, has substantial deposits of tin, zinc, and other minerals, as well as jade, rubies, and sapphires, but mining employs less than 1 percent of Myanmar's workers.

Flora and Fauna

Southeast Asia's plant life is exotic and diverse. The region boasts the world's largest flower, the *Rafflesia arnoldii*, a spectacular plant with a blossom three feet wide. Southeast Asian flora, however, is more than just beautiful—it also contributes to the region's economy. For example, Thailand cultivates over 1,000 species of orchids, a valuable trade commodity. Workers tap rubber trees from Malaysia and process woods for export such as mahogany from the Phillippines and teak from Myanmar. Indonesia is the world's largest supplier of plywood.

Like the region's flora, Southeast Asian fauna is varied and distinctive. Elephants, tigers, rhinoceroses, and orangutans roam the region's wildlife sanctuaries and national parks. Southeast Asia is home to animals found nowhere else in the world, including Borneo's bearded pig, the Malaysian lacewing butterfly, and the Komodo dragon, an Indonesian native and the world's largest lizard.

Economics

Fishing

More than 2,500 species of fish swim the tropical waters of Southeast Asia. Fish thrive in the mainland rivers and in seas near the Philippines, Indonesia, and Myanmar. Fish farming is an important part of the region's economy. Southeast Asians consume seafood at almost twice the world's average rate. The region's fishers, who have traditionally maintained small operations, now compete with large fleets of trawlers. This competition has produced an increased fish yield that helps meet demand, and so overfishing is a concern. Luckily, demand for exported seafood has started to level off, which may ease the pressure to fish excessively.

As in other parts of the world, Southeast Asia's diverse landforms shape the climate and vegetation of the region. The next section will examine these features—the lush tropical vegetation of the region's rain forest, the seasonal grasslands of its savannas, and its highlands.

SECTION 1 ASSESSMENT

Checking for Understanding

1. **Define** cordillera, archipelago, insular, flora, fauna.
2. **Main Ideas** Re-create the table below, and fill in five Southeast Asian countries and examples of their physical features and natural resources.

Country	Physical Features	Natural Resources
Malaysia		

Critical Thinking

3. **Identifying Cause and Effect** Rich soil makes Southeast Asia a productive agricultural region. What makes this soil so fertile?
4. **Drawing Conclusions** Southeast Asia has a diversity of peoples and cultures. How might physical geography have shaped this diversity?
5. **Making Generalizations** What special challenges does the location of Laos, the only country in the region without a coastline, present?

Analyzing Maps

6. **Location** Review the text and analyze the physical-political map on page 721. Note the geographic features found on Southeast Asia islands. What geographic features do the islands of Borneo, Celebes, and New Guinea share?

Applying Geography

7. **Effects of Water** Write a paragraph explaining why the abundance of water in Southeast Asia can be both an asset and a challenge for the region's population.

Climate and Vegetation

Guide to Reading

Consider What You Know

You have learned that much of Southeast Asia lies near the Equator. Based on this knowledge, what types of climate and vegetation do you suppose dominate the region?

Read to Find Out

- What weather pattern influences the region's climate?
- What are the region's main climate types?
- What is the main type of natural vegetation found in the region?

Terms to Know

- endemic
- deciduous

Places to Locate

- Shan Plateau
- Myanmar
- New Guinea
- Borneo

NATIONAL GEOGRAPHIC

A Geographic View

World in Balance

This is the forest primeval. . . . The dappled splotches of sun and shade filtering through the leafy canopy 200 feet above wash over a rain forest that has been here since before humans appeared on earth. . . . It is a world in such careful balance that the mix of vegetation in these undisturbed jungle tracts has been essentially the same . . . for millions and millions of years.

—*T. R. Reid, "Malaysia: Rising Star,"* National Geographic, *August 1997*

Malaysian rain forest

The rain forests of Southeast Asia owe much of their ancient beauty to an equally ancient climate pattern—monsoons, or seasonal winds that blow over the northern part of the Indian Ocean and the land nearby. In summer, moist monsoons blow in from the cooler sea in the south and west toward the warmer land and bring abundant rain, enough to support the region's tropical rain forests. The ample rain falls on lush tropical plants whose exotic flowers perfume the air. In winter, air over the land is cooler than that over the sea, so the wind blows out to sea from the northeast as a dry monsoon. The rain forests themselves are aptly named. They are generally wet all year long.

Tropical Climate Regions

Tropical rain forest climate dominates Southeast Asia. Parts of the mainland and some of the islands have a tropical savanna or humid subtropical climate. These climate regions are characterized by grasslands and tropical forests that support a diverse ecosystem.

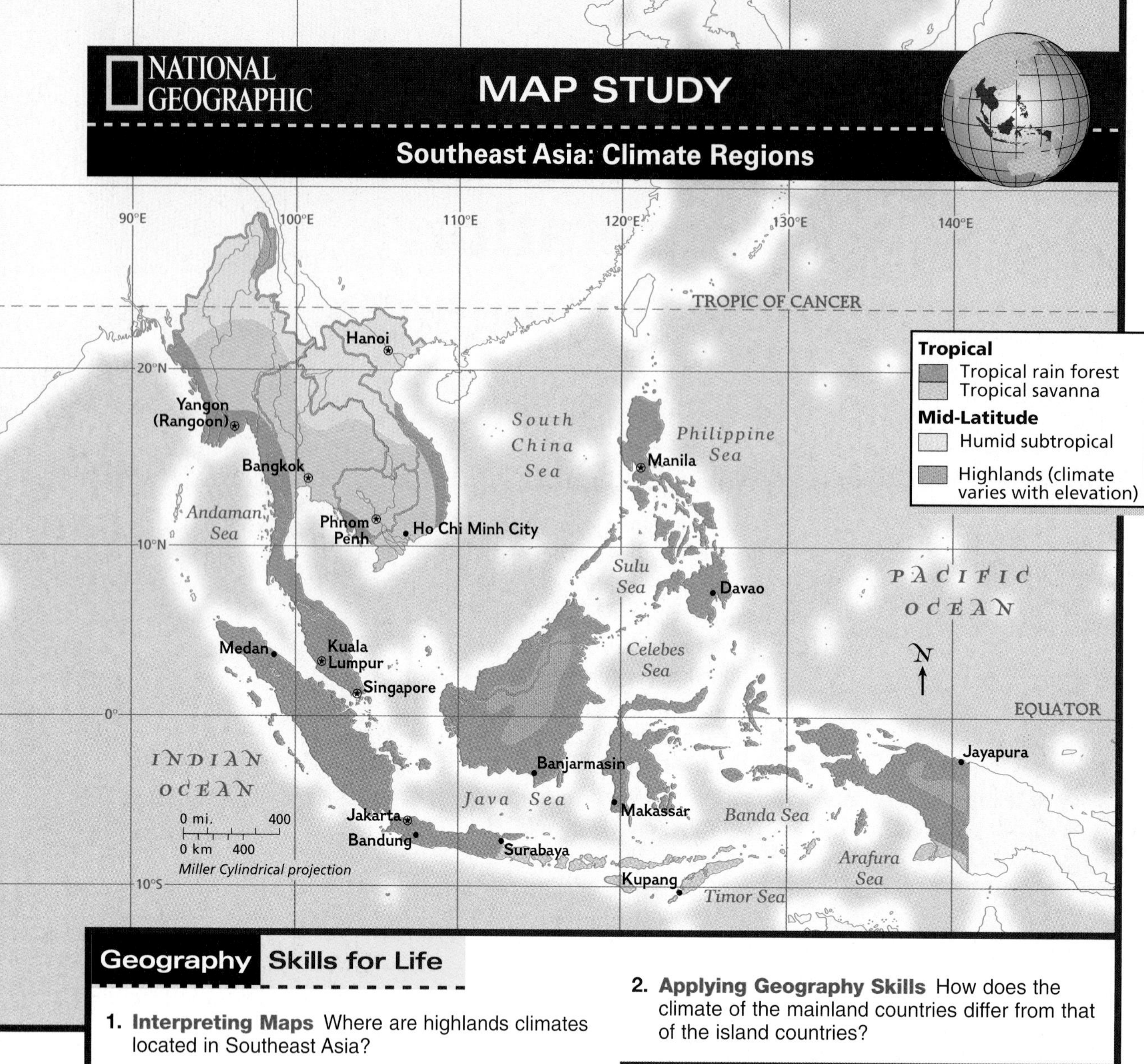

Geography Skills for Life

1. **Interpreting Maps** Where are highlands climates located in Southeast Asia?
2. **Applying Geography Skills** How does the climate of the mainland countries differ from that of the island countries?

Find NGS online map resources @ www.nationalgeographic.com/maps

Tropical Rain Forest Climate

Most of the region, including the islands and coastal areas, has a tropical rain forest climate. This climate is characterized by little variation in temperature and mostly wet conditions almost year-round. The 79°F (26°C) average daily temperature creates hot, humid, and rainy conditions. Rainfall averages between 79 and 188 inches (201 and 478 cm) per year, and the humidity hovers between 80 and 90 percent. Even more rain falls near the summit of Mount Isarog in the Philippines, described here by journalist Virginia Morell:

> *"Thick mats of spongy mosses cover every rock, tree trunk, and branch, forming an emerald carpet for the orchids and ferns that drape the limbs overhead—a lush testament to the 35 feet [420 inches, or 1,067 cm] of rain that can annually drench this mountain's summit."*
>
> Virginia Morell, "In Search of Solutions," *National Geographic*, February 1999

NATIONAL GEOGRAPHIC

MAP STUDY

Southeast Asia: Natural Vegetation

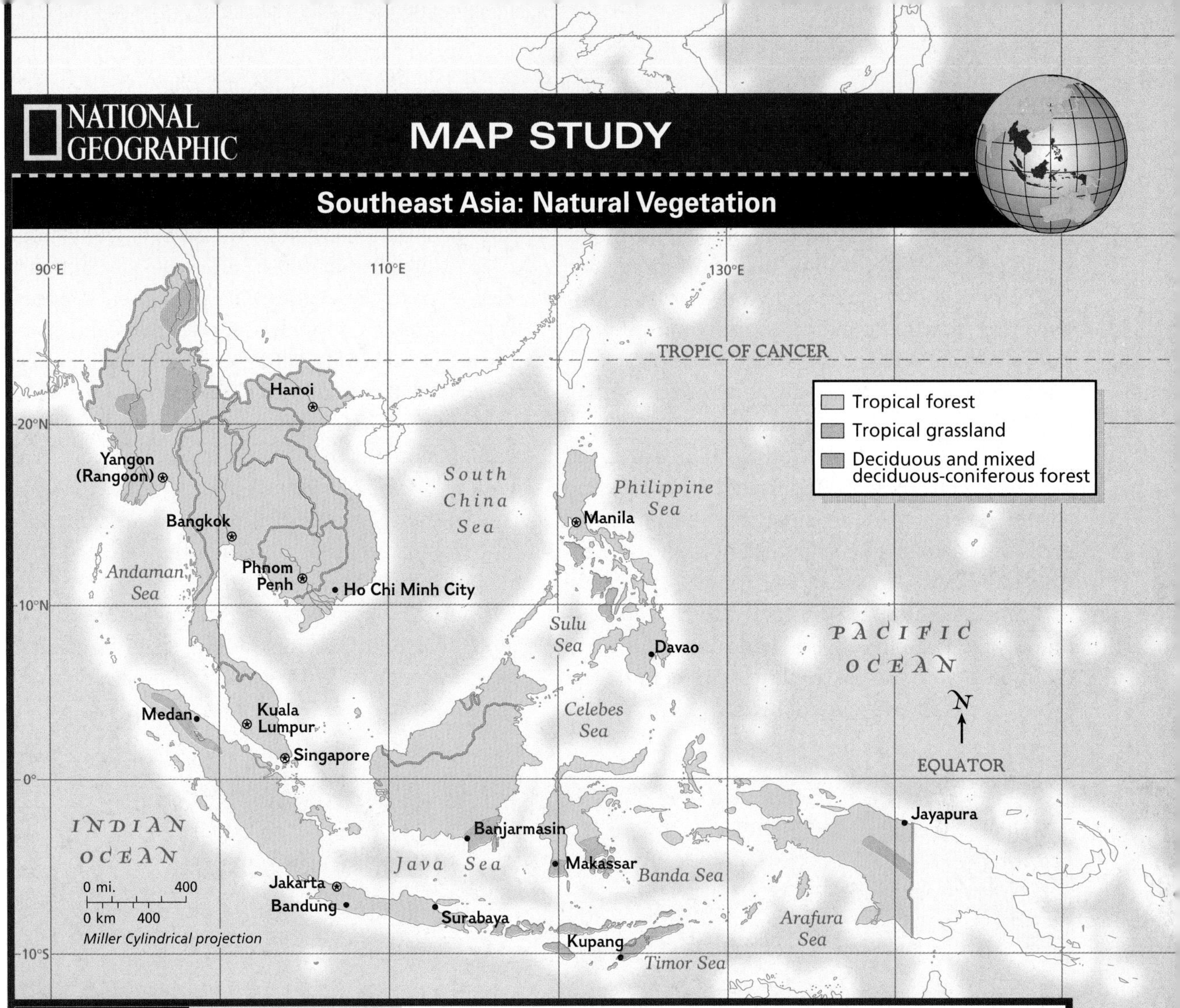

Tropical forest
Tropical grassland
Deciduous and mixed deciduous-coniferous forest

0 mi. 400
0 km 400
Miller Cylindrical projection

Geography Skills for Life

1. **Interpreting Maps** What is the dominant natural vegetation in Southeast Asia?
2. **Applying Geography Skills** What type of climate produces tropical grasslands on Indonesia's eastern islands?

Find NGS online map resources @ www.nationalgeographic.com/maps

The tropical rain forest climate supports a diverse ecosystem. More than 145,000 species of flowering plants blossom in Malaysia alone. The Malaysian rain forest, where vegetation types mix, may be the region's oldest forest, dating back many millions of years. Here there are several layers of vegetation between river valleys and higher elevations. Peat swamp forests thrive in the river valleys. Sandy coastal soil supports various shrubs, and mangrove swamp forests cover the tidal mud flats. Lowland areas with poor or shallow soil support forests of tall trees with leathery, evergreen leaves. Some of these trees produce aromatic resins, or organic compounds. Such resins are used to make medicines and varnishes, or chemicals that protect wood from water damage.

Student Web Activity Visit the **Glencoe World Geography** Web site at geography.glencoe.com and click on Student Web Activities—Chapter 29 for an activity about writing a visitor's journal about the physical features of Southeast Asia.

History

Singapore

Once an island covered by dense rain forest and surrounded by mangrove trees, **Singapore** developed into an urban area containing one of the world's highest population densities, more than 16,732 people per square mile (6,471 per sq. km). Towering apartment buildings now house Singapore's population of nearly 4,000,000 people.

As Singapore grew, an urbanized setting replaced much of its natural habitat. Many **endemic** species—those native to a particular area—are gone. Nearly 80 percent of the trees and shrubs now growing in Singapore are imported, some originating from such distant places as Central and South America. Singapore's vegetation makes it exceptional in another way. Singapore is one of only two cities in the world that have areas of tropical rain forests within their boundaries. (The other is Rio de Janeiro, Brazil.)

Tropical Savanna Climate

The second most prominent climate zone, the tropical savanna, sweeps southeastward across the Indochina Peninsula and along the southeastern parts of Indonesia. Unlike the steady, wet climate of the tropical rain forest, alternate wet and dry seasons characterize this climate, which supports tropical grasslands with scattered trees and some forests. On the Indochina Peninsula, the dry season may last from four to eight months each year.

On the mainland from around May through September, summer monsoon winds bring rain. The winter dry season extends from October to April. The first few months of this period are generally cooler, but the last few months become hot. In southern Indonesia, south of the Equator, the wet and dry cycles are reversed. From May to September, South Pacific tradewinds bring the hot, dry season. From October to April, the monsoons bring rain.

Geography Skills for Life

Monsoon Rains Commuters in Thailand travel by boat, using a plastic sheet to protect themselves against the torrential rains of the summer monsoons.

Region How do the summer monsoons differ from winter monsoons?

Humid Subtropical Climate

Parts of Southeast Asia's mainland, including most of Laos, a small part of Thailand, and northern Myanmar and Vietnam, have a humid subtropical climate. The northern reaches of Laos, Thailand, and Vietnam provide relief from the hot, humid temperatures. From November to April, the cool, dry temperatures there average around 61°F (16°C). In Myanmar the elevated **Shan Plateau** has lower temperatures than the rest of the country. The climate there resembles cooler climates elsewhere, and the plateau is sometimes called "tropical Scotland."

Geography Skills for Life

Highlands Forest Tropical deciduous forests are found in some highlands areas of Borneo (shown), Myanmar, and New Guinea.

Place What other kind of vegetation is found in the highlands areas of Myanmar?

Highlands Climate

In mountainous areas of **Myanmar**, **New Guinea**, and **Borneo**, highlands climates predominate. The much cooler temperatures of these areas set them apart from surrounding climate regions. Deciduous forests with moss-covered tree trunks are found on lower slopes. **Deciduous** trees are broad-leafed and lose their leaves in autumn. Evergreen forests appear at higher elevations. In Myanmar's highlands climate, forests of rhododendrons grow.

SECTION  2 ASSESSMENT

Checking for Understanding

1. **Define** endemic, deciduous.
2. **Main Ideas** Create a web like the one below. In the boxes, list each Southeast Asian climate region, its location, the kinds of vegetation found there and any identifying traits.

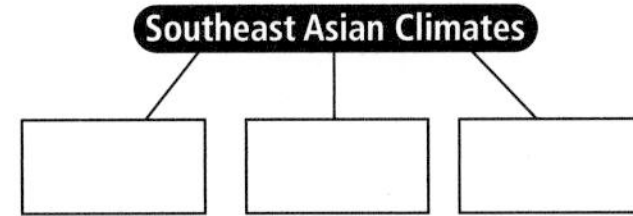

Critical Thinking

3. **Analyzing Information** How do monsoon winds impact climates and seasonal changes in Southeast Asia?
4. **Identifying Cause and Effect** Why are wet and dry seasons in Southeast Asia reversed on different sides of the Equator?
5. **Analyzing Information** How does a humid subtropical climate differ from a tropical rain forest climate?

Analyzing Maps

6. **Climate Regions** Study the map of climate regions on page 726. Which of the countries in Southeast Asia has the most varied climate?

Applying Geography

7. **Climate and Vegetation** Review Southeast Asia's climate regions and natural vegetation. How might these features influence the region's economic activities?

Writing About Geography

Writing well is an essential skill. In school you write research papers and answers to essay questions. Beyond the classroom you may have to write business letters or reports. The writing process can help you put your thoughts on paper.

Learning the Skill

The writing process has several steps: prewriting, writing, revising, proofreading, and publishing. Following this process allows you to organize your ideas and complete the writing task in a logical manner. Use the following steps to help you write about geography:

- **Prewriting is the research, writing, and organization you do before you begin your essay or report.** Select a topic, and define your purpose for writing about it. Identify the audience who will read your final product. Then do research to gather information. Organize your ideas using a graphic organizer such as a chart, a web diagram, or an outline.
- **Write your first draft.** As you write, follow the plan you created during the prewriting step. Do not worry about grammatically correct sentences in this stage. Focus on getting your main ideas and supporting details down on paper.
- **Revise your draft.** Look for places where you can add transitions between ideas, combine or rearrange paragraphs or sentences, or cut repetitive or unnecessary sections.

▲ *The newsroom of the* Chicago Defender

- **Proofread your draft.** Check your draft for grammar, spelling, and punctuation errors.
- **Publish your draft.** Create a clean draft, and present it to your audience.

Practicing the Skill

You have been assigned to write a travel brochure for a Southeast Asian country for an advertising agency. Answer the following questions about the writing process.

1. What will you need to do before you begin writing the first draft of your brochure?
2. Why might it be a good idea to let a day pass between writing and revising your brochure?
3. What are two resources you might use while proofreading your brochure?

Prewrite and then create a draft of a brief guide for someone who has just moved to the United States from Southeast Asia. Include information about the geography and life in the United States. Read your first draft carefully, and mark places that need to be revised. After revising, proofread and publish your writing.

The **Glencoe Skillbuilder Interactive Workbook, Level 2** provides instruction and practice in key social studies skills.

CHAPTER 29 SUMMARY & STUDY GUIDE

SECTION 1 The Land (pp. 719–724)

Terms To Know

- cordillera
- archipelago
- insular
- flora
- fauna

Key Points

- Southeast Asia's mountains were formed when the Indo-Australian, Philippine, and Eurasian tectonic plates collided.
- Straddling the Equator, Southeast Asia includes the Indochina and Malay Peninsulas as well as the 20,000 islands of the Malay Archipelago.
- About half of Southeast Asia's 11 countries are located on the mainland. The rest are island countries, except for Malaysia, which is both a mainland and an island country.
- Mountains and rivers dominate the region's landscape. The island mountains are part of the Pacific Ring of Fire.
- Rivers on the mainland of Southeast Asia are important for agriculture, communication, and transportation.
- Southeast Asia contains abundant natural resources, including fossil fuels, natural steam, minerals, and gems.

Organizing Your Notes

Use a web diagram like the one below to organize your notes about the islands and peninsulas, physical features, and natural resources of Southeast Asia.

Physical Features of Southeast Asia

SECTION 2 The Climate and Vegetation (pp. 725–729)

Terms To Know

- endemic
- deciduous

Key Points

- Monsoons cause two main seasons in Southeast Asia, one wet and one dry.
- Southeast Asia's major climate is tropical rain forest, although parts of the mainland and some of the islands have other types of climate.
- Humid subtropical climates predominate in Laos and in northern areas of Myanmar, Thailand, and Vietnam.
- Highlands climates are found in the mountains of Myanmar, Borneo, and New Guinea.
- Southeast Asia's lush vegetation is characteristic of tropical rain forest and tropical savanna climate regions.

Organizing Your Notes

Create an outline like the one below to help you organize your notes for this section. Copy the boldface headings and subheadings that appear in Section 2, and then list important points under each head.

Climate and Vegetation

I. Tropical Climate Regions
  A. Tropical rain forest climate
    1. steady rain and humidity
    2. average daily temperature of 79°F (26°C)
  B.

ASSESSMENT & ACTIVITIES

Reviewing Key Terms

Write the letter of the key term that best matches each definition below.

a. cordillera
b. archipelago
c. flora and fauna
d. insular
e. deciduous
f. endemic

1. group of islands
2. system of parallel mountain ranges
3. island
4. native to a particular area
5. trees that lose leaves in autumn
6. plants and animals

Reviewing Facts

SECTION 1

1. Which Southeast Asian countries lie partially or entirely on the Indochina Peninsula?
2. Name the five insular countries.
3. What geologic activities created Southeast Asia?
4. Explain why waterways are important to Southeast Asia's people.
5. Name a Southeast Asian resource found underground, another resource found underwater, and a third resource found in a tropical rain forest.

SECTION 2

6. What are the four main climate regions of Southeast Asia?
7. Where are the region's oldest forests found?
8. Where can highlands climates be found?
9. How is weather north of the Equator different from weather south of the Equator?
10. What is unusual about most of Singapore's vegetation?

Critical Thinking

1. Drawing Conclusions What geographic factors explain the large number of islands in Southeast Asia?

2. Making Inferences How might volcanoes affect the region's economy?

3. Identifying Cause-and-Effect Copy the web diagram below onto a sheet of paper. Complete the diagram to show how the tropical climate affects human activities in Southeast Asia. Then choose one effect, and write a paragraph explaining its impact on the people of Southeast Asia.

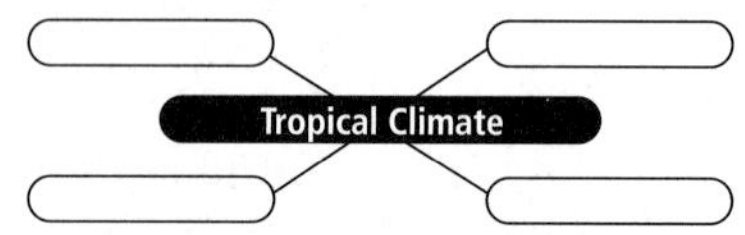

NATIONAL GEOGRAPHIC

Locating Places

Southeast Asia: Physical-Political Geography

Match the letters on the map with the places and physical features of Southeast Asia. Write your answers on a sheet of paper.

1. **Singapore**
2. **Irrawaddy River**
3. **Thailand**
4. **Malay Peninsula**
5. **Mekong River**
6. **Vietnam**
7. **Philippines**
8. **Java**
9. **Sumatra**
10. **Borneo**

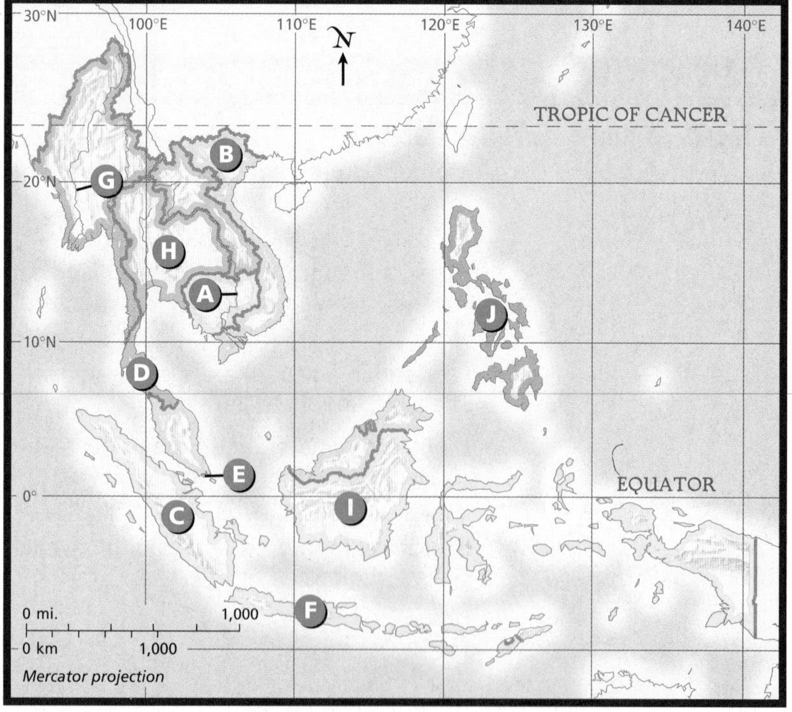

30°N
100°E
110°E
120°E
130°E
140°E
N
TROPIC OF CANCER
20°N
10°N
0°
EQUATOR
A B C D E F G H I J
0 mi. 1,000
0 km 1,000
Mercator projection

Self-Check Quiz Visit the **Glencoe World Geography** Web site at geography.glencoe.com and click on Self-Check Quizzes—Chapter 29 to prepare for the Chapter Test.

Using the Regional Atlas

Refer to the Regional Atlas on pages 710–713.

1. **Movement** What river provides Laos with its chief means of transportation?
2. **Place** Study the physical, economic activity, and population maps of the region. What three generalizations could you make about Borneo, given the information on these maps?

Thinking Like a Geographer

Think about the physical geography of Southeast Asia. Why do you suppose the capital cities of the region are located on or near water? As a geographer, would you encourage people to relocate to other areas in order to avoid overcrowding these cities? Explain.

Problem-Solving Activity

Group Research Project Working in a group of four, plan a trip through Southeast Asia. Decide what areas to visit, noting the kinds of landforms you would see in each place. Determine how to get from one place to another, and work together to create a map that shows your travel routes. Prepare a written itinerary, and present your travel plans to the class.

GeoJournal

Descriptive Writing Using the information you wrote in your GeoJournal as you read this chapter, write a newspaper story about the landscape of the region. You may wish to focus on a recent event, such as a volcanic eruption, flood, or other natural disaster.

Technology Activity

Using E-mail Use library or Internet resources to locate a postal or e-mail address for the United States Embassy in Manila, the Philippines. Compose and send a letter requesting information about the February 2000 eruption of the Mayon Volcano. Use the information you receive to create a bulletin board about the eruption.

Standardized Test Practice

Choose the best answer for the following multiple-choice question. If you have trouble answering the question, use the process of elimination to narrow your choices.

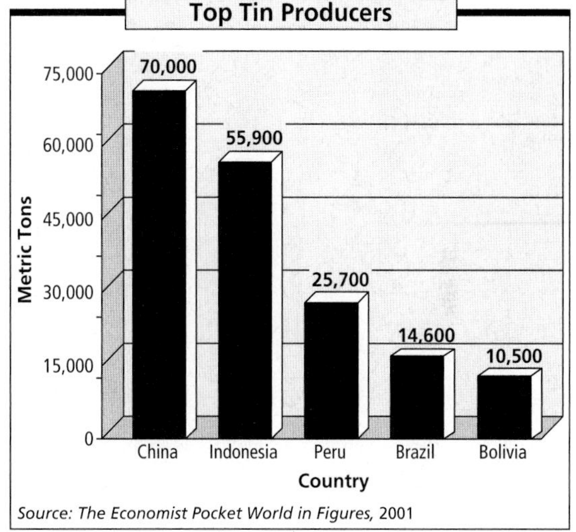

1. **About how much tin does Indonesia produce each year?**

   A 55,900 metric tons
   B 55,000,900 metric tons
   C 55.9 million metric tons
   D 55.9 billion metric tons

In order to understand any type of graph, look carefully around the graph for keys that show how it is organized. On this bar graph, the numbers along the left side represent the exact number shown. You do not have to multiply by millions or billions to find the number of metric tons.

CHAPTER 30

The Cultural Geography of Southeast Asia

GeoJournal

As you read this chapter, use your journal to describe the many ways of life in Southeast Asia. Use vivid details to depict homes, jobs, governments, and cultural activities.

GEOGRAPHY Online

Chapter Overview Visit the **Glencoe World Geography** Web site at geography.glencoe.com and click on Chapter Overviews—Chapter 30 to preview information about the cultural geography of the region.

Population Patterns

Guide to Reading

Consider What You Know

Many Southeast Asians have migrated to other countries, including the United States. What impact do you think migrants have on the cultures of their adopted countries?

Read to Find Out

- What are the various ethnic roots of Southeast Asia's peoples?
- Why do the majority of Southeast Asians live in river valley lowlands or on coastal plains?
- How have population movements and settlement patterns affected Southeast Asia?

Terms to Know

- urbanization
- primate city

Places to Locate

- Cambodia
- Vietnam
- Myanmar
- Indonesia
- Java
- Singapore
- Thailand
- Philippines
- Bangkok
- Jakarta

◀ *Master kite maker at work in Malaysia*

NATIONAL GEOGRAPHIC

A Geographic View

Traces of History

Home to nearly five million people, making it one of the world's most populated urban areas, Ho Chi Minh City [formerly Saigon, Vietnam] still bears traces of past foreign occupants. France, which made Saigon its first foothold in Indochina, left boulevards and a cathedral. The U.S., which based its military here during the Vietnam War, built an embassy complex and greatly expanded the airport. Now the Vietnamese take a turn, erecting hotels and factories.

—*Tracy Dahlby, "The New Saigon,"* National Geographic, *April 1995*

Hotel in Ho Chi Minh City, Vietnam

Vietnam's Western-style buildings are recent examples of a series of cultural influences—Chinese, Indian, Islamic, European, and American—that have shaped Southeast Asia over thousands of years. Each culture has added its own unique flavor to Southeast Asia's cultural mix. In this section you will learn about the diverse peoples of Southeast Asia, how physical geography affects where they live, and what challenges population changes are bringing to the region.

Human Characteristics

Southeast Asia's cultural geography is as varied as its physical geography. About 520 million people live on the many peninsulas and islands of Southeast Asia. Southeast Asia's population today includes descendants of indigenous peoples, Indians, Chinese, Arabs, and European colonists.

Geography Skills for Life **Indigenous People**

The young woman wearing a headdress (left) and the elderly people (right) are Iban, an indigenous group in East Malaysia.

Movement From which areas did people migrate to Southeast Asia 2,500 years ago?

Indigenous Peoples

Humans have lived in Southeast Asia for tens of thousands of years. About 2,500 years ago, groups of migrants from western China and eastern Tibet arrived in the region. Many of today's Southeast Asians are descendants of these early peoples. On the Southeast Asian mainland, the Khmers settled **Cambodia** and **Vietnam**, the Mons moved into **Myanmar**, and the Malays settled the Malay Peninsula. Some Malay groups also undertook sea voyages, settling the many islands that today form **Indonesia**. These indigenous peoples developed agricultural civilizations and borrowed from other peoples without losing their own identities.

Indian and Chinese Influences

Valuable spices grown in Southeast Asia drew outside traders to the region. While exchanging goods with Southeast Asians, these foreigners passed on new ideas and practices that blended with Southeast Asian traditions. Beginning in the A.D. 100s, merchants from India introduced the Hindu and Buddhist religions, art forms inspired by these religions, and a concept of government that glorified kings as both political and spiritual leaders. Meanwhile, Chinese traders and soldiers brought Chinese cultural influences to the region through Vietnam. During a thousand years of Chinese rule, the Vietnamese adopted China's writing system, Confucian traditions, and system of government. Today Indian and Chinese ethnic communities are scattered throughout Southeast Asia, particularly in Brunei, Malaysia, Thailand, and Vietnam. In Singapore today, people of ethnic Chinese ancestry make up 76 percent of the country's total population.

Islamic Influence

In search of spices, Arab and Indian traders brought cottons and silks to Southeast Asia beginning in the A.D. 800s. They and locally based Malay traders set up trade routes that linked Southeast Asia with other parts of Asia. During the 1200s, Southeast Asians—especially those in port towns—began to convert to Islam, the religion of these traders. Over the centuries, Islam spread from coastal areas to interior areas of the Indonesian islands and the Malay Peninsula. Today Muslims form the majority of the population in Brunei, Malaysia, and Indonesia.

Western Colonization

During the 1400s and 1500s, European explorers, like others before them, sought new sea routes to acquire Southeast Asia's spices and other rich natural resources. Their voyages eventually brought nearly all of the region, with the exception of Thailand, under European control. While exporting nutmeg, cloves, and pearls to Europe, European traders brought new products to Southeast Asia. For example, from Latin America the Spaniards introduced various chili peppers that added new flavor to Southeast Asian cooking. When drinking coffee became popular in Europe during the 1600s, the Dutch began cultivating coffee trees—originally from the Arabian Peninsula—on various Southeast Asian islands.

Population Growth

Many of Southeast Asia's 520 million people live in fertile river valleys or on the coastal plains. A ready supply of water, fertile land, adequate transportation, and available jobs have all contributed to these concentrations of people. In general, highlands areas have fewer people than lowlands, and rural areas have fewer people than the cities.

Population Density

Population density varies widely throughout Southeast Asia. Indonesia, the world's fourth most populous country, has more than 206 million people living on 13,600 scattered islands. The Indonesian island of **Java** is one of the most densely populated islands in the world. The overall population density of Indonesia is 280 people per square mile (108 people per sq. km). **Singapore**, the region's smallest country in land area, has the greatest population density—17,320 people per square mile (6,687 people per sq. km).

Population Growth Rates

The population of Southeast Asia is growing at a rate of 1.6 percent per year compared with the 1.3 percent average growth rate for the world. Some estimates indicate that more than 775 million people will live in the region by 2050, representing about a 50 percent increase over the number of people living there today. Some countries, such as **Thailand,** Indonesia, and Singapore, are working to slow their population growth rates. Singapore, in fact, has succeeded so well in reducing its population growth rate that there is concern the country may not have enough young workers to replace and support its aging population. As a result, married couples are now encouraged to have more children.

History

Cambodia: Population Decline

Since the 1970s Cambodia's population growth rate has been below the region's average. Between 1975 and 1979, Cambodia lost 38 percent of its population as a result of harsh rule by the Khmer Rouge communist government. Many people died as a result of starvation, torture, and executions. People considered to be intellectuals were often the first targets of the violence as described below:

> "*Even someone who as much [as] wore glasses was considered an intellect, [and] was killed. Thus began a vast extermination of all the wealthy and educated people in Cambodia. . . . The Khmer Rouge watched over the people constantly, making sure everything they did was right, and if they showed any signs of an education, they would be first tortured to confess, and then executed.*"
>
> Jerry Adler, "Pol Pot's Last Days," *Newsweek*, April 27, 1998

Movement to the Cities

For centuries, the majority of Southeast Asians lived in rural villages and farmed the land. Today increasing numbers of the region's people are moving from rural areas to urban centers. This population shift has resulted from political conflicts and government policies, but greater economic and educational opportunities available in cities have also been factors. The **Philippines**, for example, reflects this trend toward **urbanization**, or the shift from rural to urban lifestyles, in Southeast Asia. At the beginning of the 1900s, more than 80 percent of Filipinos lived in rural areas. Today about 53 percent of the Philippines' population lives in the countryside.

NATIONAL GEOGRAPHIC

CHART STUDY

Southeast Asia: Urban and Rural Growth (Selected Countries)

Country	Percent Urban	Percent Rural	Annual Urban Growth %	Annual Rural Growth %
Indonesia	39	61	3.4	0.3
Malaysia	57	43	2.9	0.1
Thailand	30	70	2.5	0.2
Vietnam	24	76	2.4	1.3
Philippines	47	53	3.1	0.1
Myanmar	27	73	3.4	0.9
Cambodia	16	84	4.4	1.0
Laos	17	83	5.2	2.0
Singapore	100	0	1.0	0.0

Sources: 2001 World Population Data Sheet; United Nations Population Division, 2000

Geography Skills for Life

1. **Interpreting Charts** Which country is the most urbanized? The least urbanized?

2. **Applying Geography Skills** How might migration and other human processes affect patterns of settlement in the region?

At least 11 Southeast Asian cities now have populations of more than 1 million. In some countries in the region, a single major city leads all other cities in attracting people, resources, and commerce. Such a magnet is called a **primate city**, an urban area that serves as a country's major port, economic center, and often its capital.

Bangkok, Thailand, and Jakarta, Indonesia, are examples of primate cities. Rapid growth in these and other urban areas has brought challenges as well as benefits. Thailand's capital, **Bangkok**, grew by 650 percent between 1950 and 1998, but the city's roads, housing, water and electric systems, and other public services could not adequately support all of the new migrants. About 1 million residents of Bangkok live in densely populated areas characterized by poor housing and poverty. Thailand is trying to solve these urban challenges by encouraging people to return to rural areas. The Thai government has offered incentives for industries to locate outside of cities. In spite of these efforts, however, the lure of urban jobs and lifestyles continues to drain small villages.

Indonesia also faces a movement of people from rural to urban areas. The major attraction for migrants in Indonesia is its capital, **Jakarta**, a city of more than 10 million on the densely populated island of Java. Some of these migrants are temporary residents seeking seasonal employment in the cities.

In an attempt to reduce urban overcrowding, Indonesia's government during the past 40 years has relocated 3 million people to the country's less densely populated outer islands. Although relocation has increased the rural population in some parts of Indonesia, it has done little to lessen overcrowding on Java. In addition, the mixing of peoples of different ethnic backgrounds has sparked conflict as groups compete for jobs, housing, and social services.

Outward Migrations

Since the 1970s, a number of Southeast Asians have left their homelands to settle in other parts of the world. Between 1975 and 1990, thousands of people left Vietnam to escape the widespread economic distress and political oppression that gripped the country. Since the mid-1970s, many people have left their homeland in Laos for similar reasons. Many of these Southeast Asian migrants came to settle in the United States. By 2000, for example, the United States population included 955,264 Vietnamese, 176,148 Cambodians, and 331,340 people of the Hmong and Lao ethnic groups. One effect of these outward migrations is that the countries of Southeast Asia lose skilled and educated workers who could contribute some of the valuable skills that their home countries need for sustained economic growth. Outward migration is only one factor that shapes the region's population patterns, however.

▲ *Yawaraj Road, in the heart of Bangkok's busy Chinese district*

Southeast Asia's physical features—the many islands and peninsulas—as well as its growing cities have also shaped the region's population patterns. In the next section, you will learn how historical events, such as migration and colonization, and contemporary politics have left their marks on Southeast Asia.

SECTION 1 ASSESSMENT

Checking for Understanding

1. **Define** urbanization, primate city.
2. **Main Ideas** On a web like the one below, list the factors that have influenced rural and urban settlement patterns for each country in Southeast Asia.

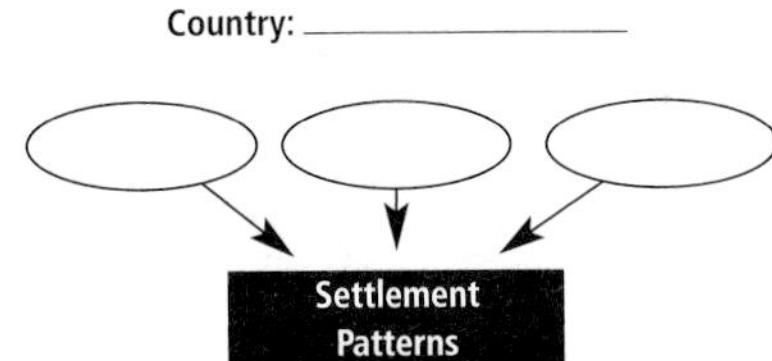

Critical Thinking

3. **Identifying Cause and Effect** Why have so many different peoples migrated to Southeast Asia over the centuries? How has this migration influenced the region's culture?
4. **Making Inferences** What do countries such as Indonesia hope to gain by slowing population growth?
5. **Comparing and Contrasting** How do migration patterns affect Indonesia's urban and rural populations?

Analyzing Maps

6. **Location** Study the political map on page 711 and the urban/rural growth chart on page 738. Are the countries with the lowest urban populations located on the mainland or on the islands?

Applying Geography

7. **Economic Effects** Think about population distribution in Southeast Asia. Write a paragraph explaining how environmental and economic factors have shaped settlement in the region.

SECTION 2

History and Government

Guide to Reading

Consider What You Know

Your history class may have taught you about the period of colonial rule that existed in Southeast Asia before the region's countries gained independence. What effects might foreign rule have on the people living in a colony?

Read to Find Out

- How did location influence the development of empires in Southeast Asia?
- What cultural influences have affected the region's peoples?
- What events led to the independence of Southeast Asian countries?

Terms to Know

- maritime
- sphere of influence
- buffer state

Places to Locate

- Mekong River
- Gulf of Thailand
- Indochina Peninsula
- Malay Peninsula
- Angkor Wat
- Strait of Malacca
- Sunda Strait
- East Timor

NATIONAL GEOGRAPHIC

A Geographic View

A Mighty Waterway

From its source, the Mekong [River] travels about half its length in China; then it borders or moves through Myanmar (formerly Burma), Laos, Thailand, Cambodia, and Vietnam. I would find it called by many names: River of Stone, Dragon Running River, Turbulent River, Mother River Khong, Big Water, the Nine Dragons. Along it empires, kingdoms, and colonial realms have risen and fallen. . . .

—*Thomas O'Neill, "The Mekong,"* National Geographic, *February 1993*

Boat traffic, Mekong River, Vietnam

The many names given to the **Mekong River** serve as reminders of Southeast Asia's rich and colorful history. Over the centuries the Mekong River has been a major waterway for the different civilizations that have flourished along its banks. In this section you will learn about Southeast Asia's ancient kingdoms, its era of European colonial rule, and its struggles for independence and democracy. You will also learn about the regional conflicts that have taken place in Southeast Asia during the past 50 years.

Early Civilizations

Early peoples in Southeast Asia were highly skilled farmers. Rice was the staple grain of these agricultural societies, as it is in Southeast Asia today. During this early period, farmers in the region grew vegetables and domesticated cattle and pigs. Early Southeast Asians also were advanced metalworkers. Bronze was first cast in Thailand in 3000 B.C., nearly one thousand years before the Chinese developed the same skill.

architecture of SOUTHEAST ASIA

WORLD CULTURE

Angkor Wat The temple complex at Angkor Wat forms the largest single religious building in the world. The complex covers nearly one square mile (2.6 sq. km) and is surrounded by an extensive moat. To ensure order and harmony in the universe, carvings depicting the Hindu gods and the Buddha cover the walls. At the center of the complex, the sanctuary stands 130 feet (40 m) high. The distinct style of Khmer architecture shows in the roof towers. Each pyramid-shaped tower consists of a series of tiers stacked one on top of the other, each smaller than the one beneath.

Many Southeast Asian cultural traditions arose during this period. Early Southeast Asians worshiped their ancestors as well as animal and nature spirits. In society, power and wealth were passed down through the mother's family.

Kingdoms and Empires

Many civilizations in early Southeast Asia developed on waterways or around strategic ports. Maritime, or seafaring, empires gained power by controlling shipping and trade. Land-based empires gained wealth from crops grown in fertile soil.

Funan

During the A.D. 100s, traders from India set up trading posts along what is today the **Gulf of Thailand** (Siam). Southeast Asians living in the area blended Indian traditions with their own. By the A.D. 200s, they had established the kingdom of Funan. The people of Funan adopted Hinduism and the Indian model of a centralized government under one powerful ruler. They became skillful goldsmiths and jewelers and developed an impressive irrigation system. As a maritime power, Funan traded with regions as far away as India, China, and Persia.

Khmer

An abundance of crops grown in fertile river valleys and deltas brought wealth to mainland Southeast Asia. During the A.D. 1100s and 1200s, the Khmer Empire flourished along the Mekong River and covered most of the **Indochina Peninsula** and the northern part of the **Malay Peninsula**. Technologically advanced in irrigation and agriculture, the Khmer used a complex system of lakes, canals, and irrigation channels to grow three or four rice crops annually.

Although agriculturally advanced, the Khmer are best known for their magnificent architecture. Located in present-day Cambodia, **Angkor Wat**, a Khmer temple more than 800 years old, was designed to resemble the home of the Hindu gods and goddesses. A mixture of Indian and local styles, Angkor Wat is both a Hindu temple and a tomb for Suryavarman II—the Khmer ruler who built it.

Srivijaya Empire

Based on the island of Sumatra, the Srivijaya Empire controlled the seas bordering Southeast Asia from A.D. 600 to 1300. Ancient trade routes from Africa and Southwest Asia to East Asia went through the **Strait of Malacca** and the **Sunda Strait** and linked the Indian Ocean, the Java Sea, and the South China Sea. The Srivijaya Empire used its navy to control these straits. Once its power was established, the empire gained wealth by taxing traders whose ships passed through these waters.

By the 1300s, the Srivijaya Empire had declined, but its legacy shaped later maritime territories in Southeast Asia. Today Singapore owes its economic prosperity to these same trade routes.

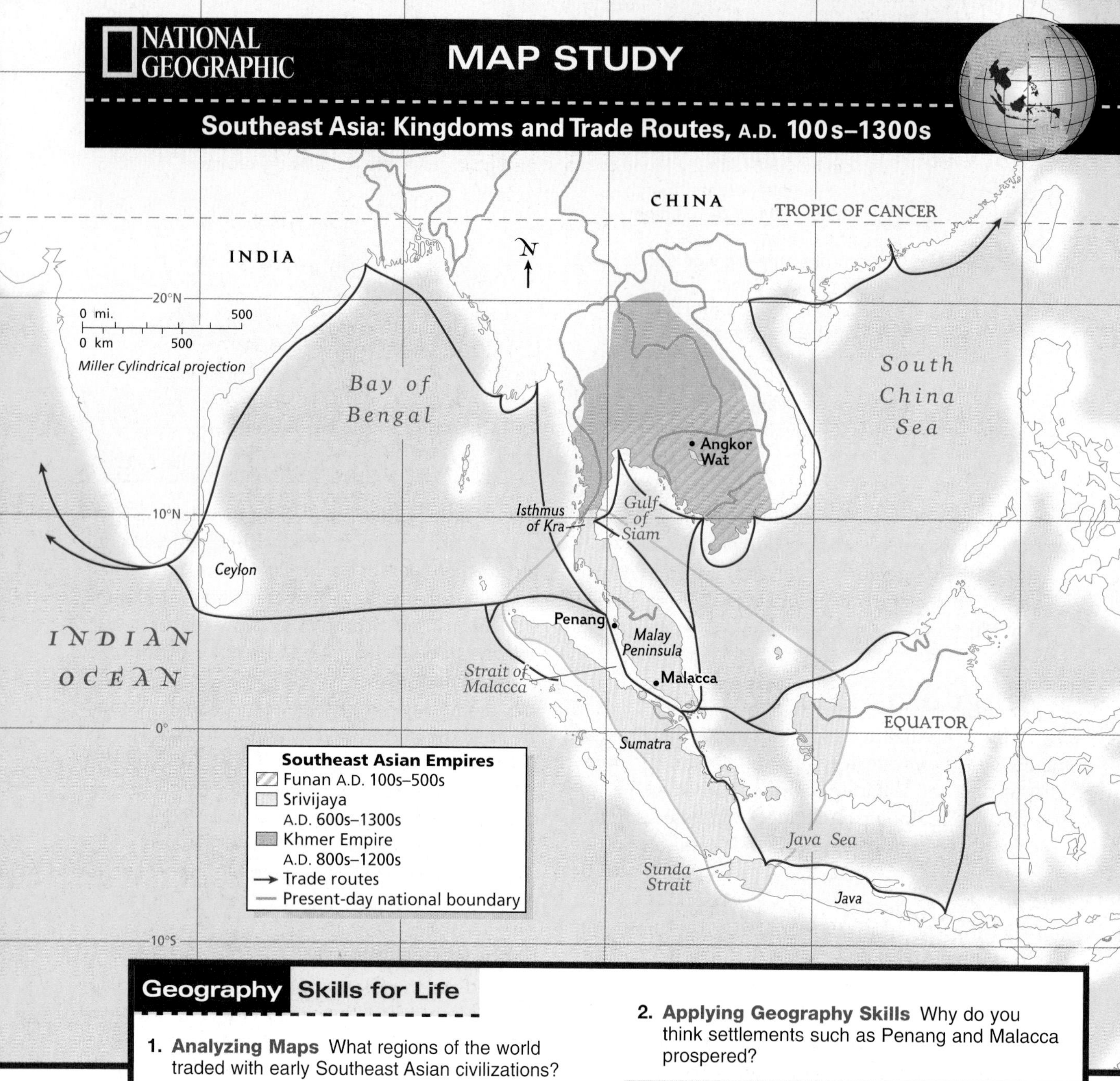

Vietnam

The Vietnamese people controlled the Indochina Peninsula from the Red (Hong) River delta in the north to coastal lands in the center. Throughout their history, the Vietnamese struggled against Chinese invaders. Finally, in 111 B.C. the Chinese emperor Wudi conquered the territory. The Chinese introduced their writing system and ideas about religion, philosophy, and government. Their control of the Vietnamese ended during the early A.D. 900s.

Islam

Muslim Arab merchants and missionaries from Southwest Asia traded and settled in Southeast Asian coastal areas during the A.D. 800s and 900s. Because of this influence, many coastal Southeast

Asians adopted Islamic ways and converted to the religion of Islam. After 1400, Islam quickly spread from coastal to interior areas in the Malay Peninsula and neighboring islands. During the 1400s, Malacca, on the Malay Peninsula, was an important seaport and Islamic cultural center.

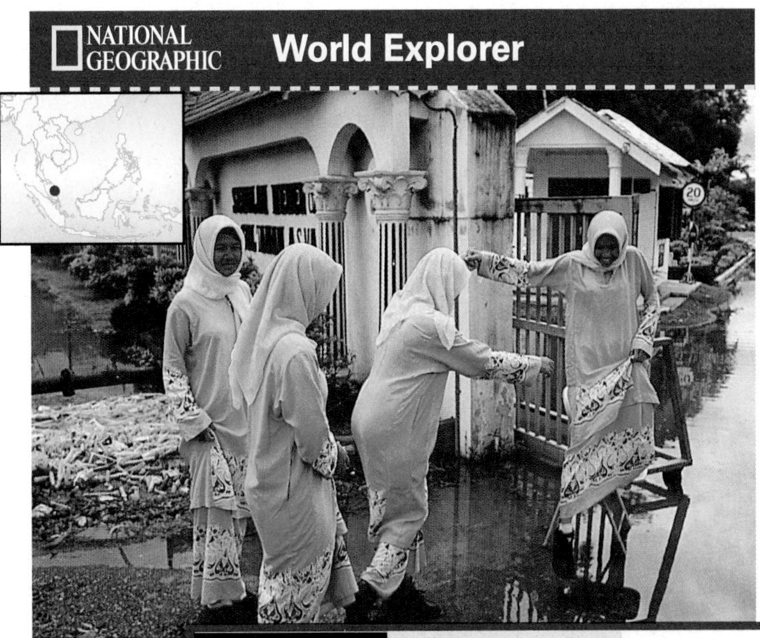

NATIONAL GEOGRAPHIC World Explorer

Geography Skills for Life

Islamic Influences Over half of Malaysians practice Islam, including most ethnic Malays.
Movement When did Islam spread to the Malay Peninsula?

Western Colonization

By the 1500s Europeans had arrived in Southeast Asia to trade, spread Christianity, and claim territory. The European powers at first set up **spheres of influence**—agreed-upon areas of control. They later acquired Southeast Asian lands as colonies. Dividing British- from French-ruled territories, the kingdom of Siam (present-day Thailand) served as a **buffer state**, or neutral territory between rival powers. Because of its position, Siam was the only Southeast Asian territory that remained free of European rule.

Western Holdings

During the early 1900s, the Netherlands, the United Kingdom, France, and the United States dominated Southeast Asia. The Netherlands claimed most of the islands that today make up Indonesia. The United Kingdom controlled what is now Myanmar, Malaysia, Singapore, and Brunei. France governed territories in Indochina that later became Cambodia, Vietnam, and Laos. The United States gained control of the Philippines in a war with Spain in 1898.

Economics

Effects of Western Rule

Europeans and Americans brought widespread changes to Southeast Asia. They built railroads, paved roads, and improved harbors to speed the movement of people and goods throughout the region. Westerners expanded tin mining and oil drilling, and they replaced small farms with large commercial plantations. The production of rice, rubber, coffee, and other products soared, and Westerners received enormous profits.

Western rivalries for control of resources and territory, however, increased military conflict in Southeast Asia. Western influences also altered traditional lifestyles. Colonial landowners and trading companies forced Southeast Asians—who received low, if any, wages—to grow cash crops, work in the mines, and cut trees for timber.

Southeast Asian agricultural workers alone could not meet the growing Western demand for labor. Plantation owners imported machinery, but they also hired Indian and Chinese immigrants to work in the mines and fields. Many of the migrant laborers and their families settled permanently in Southeast Asia, contributing to the ethnic diversity of the region.

Struggle for Freedom

During World War II, Japan forced Western countries out of Southeast Asia. After Japan's defeat in 1945, the Western countries tried to regain control. They met opposition, however, from Southeast Asians determined to gain their freedom. By 1965, after two decades of struggle, all of the countries of Southeast Asia had gained independence.

Culture

Regional Conflicts

After independence, political conflicts and wars raged throughout Southeast Asia. Local Communists fought other political groups in Indochina. In 1954 communist forces defeated the French in Vietnam, which was then divided into two independent parts: communist North Vietnam and non-communist South Vietnam. Vietnamese Communists used force to unite all of Vietnam under their rule by the mid-1970s. In Laos and Cambodia, Communists also fought newly independent governments for control. During the 1960s and early 1970s, the United States intervened in these Southeast Asian conflicts to block the spread of communism. The feature on pages 746–747 describes the Vietnam War and the United States's involvement in Southeast Asia.

Other Southeast Asian countries have faced ethnic conflict. In Malaysia, for example, ethnic Malays controlling the government have clashed with the Chinese and Indian communities that dominate the economy. The government has tried to boost Malay participation in business, an action that non-Malays regard as favoritism. Sometimes

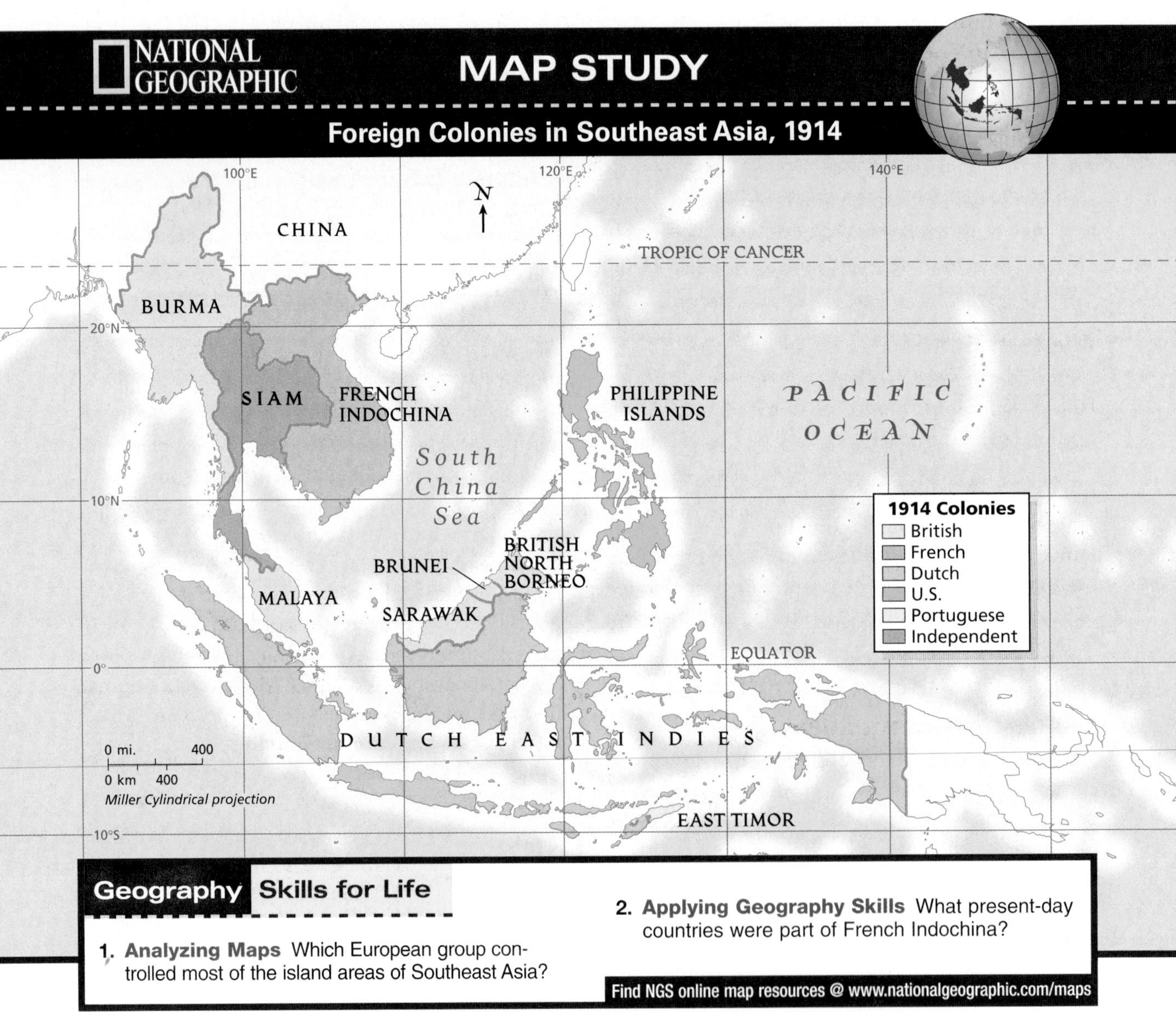

ethnic and religious groups within a country have waged struggles for independence. **East Timor**, a Portuguese-ruled territory seized by Indonesia in 1975, is an example. For 20 years East Timor's mostly Roman Catholic population resisted being absorbed into largely Muslim Indonesia. About 200,000 of East Timor's people died during a fierce conflict. A journalist visiting East Timor in the late 1980s described the fighting at that time:

> *"The consequences [of fighting] have been devastating. . . . 'Practically speaking,' [stated a local official], 'every family in East Timor has lost someone in this civil war.'"*
>
> Arthur Zich, "Indonesia: Two Worlds, Time Apart," *National Geographic*, January 1989

NATIONAL GEOGRAPHIC **World Explorer**

Geography Skills for Life

Voting for Freedom East Timorese greet a UN official sent to supervise a 1999 election in which East Timor's voters decided to separate their territory from Indonesia.

Place How did religion affect East Timor's relationship with Indonesia?

When Indonesia's corrupt dictatorship fell in 1999, East Timor finally broke away. Pro-Indonesian forces, however, spread unrest there. UN international peacekeeping forces arrived to keep order.

Forms of Government

Forms of government vary in Southeast Asia. Indonesia, the Philippines, and Singapore are democratic republics. In 1998 Indonesia moved toward democracy after years of dictatorship. Myanmar's military government has tried to crush the efforts of opposition leader Aung San Suu Kyi (AWNG SAHN SOO CHEE) to bring democracy peacefully to the country. Brunei, Cambodia, Malaysia, and Thailand are constitutional monarchies. Communist governments rule in Laos and Vietnam.

SECTION 2 ASSESSMENT

Checking for Understanding

1. **Define** maritime, sphere of influence, buffer state.
2. **Main Ideas** On a table like the one below, fill in and describe influences of outside cultures on the development of Southeast Asia.

Culture	Influences on Southeast Asia

Critical Thinking

3. **Comparing and Contrasting** How were the region's three early empires alike? Different?
4. **Making Inferences** What physical and human factors have shaped Southeast Asia's current political borders?
5. **Predicting Consequences** How might East Timor's independence influence the region?

Analyzing Maps

6. **Region** Study the map on page 742. Why were the Strait of Malacca and the Sunda Strait vital to maritime development?

Applying Geography

7. **Interpreting Historical Maps** Write a paragraph explaining how the map of foreign colonies on page 744 helps us understand the region's cultural diversity.

GEOGRAPHY AND HISTORY

THE LONG WAR: AMERICA IN VIETNAM

WITHIN THE JUNGLES OF VIETNAM, slim shafts of light penetrate the dense vegetation. In the 1960s and 1970s, American soldiers fought a war in these jungles, while their nation's leaders struggled over how to end it.

The United States became concerned about Vietnam after World War II, when the Cold War split the world's nations into two groups—those favoring the democratic United States, and those aligned with the Soviet Union and communism. Each side feared the other's dominance, and U.S. President Truman vowed to help any country threatened by communism. This policy, expanded by Presidents Eisenhower, Kennedy, and Johnson, led America to war in Vietnam.

America Intervenes

France ruled Vietnam from 1877 to the early 1940s. Japan occupied Vietnam during most of World War II. After the war, the United States supported France as it tried to resume rule. But Ho Chi Minh, a Communist and a Soviet ally, organized a revolt in northern Vietnam. In 1954 the Vietnamese won control, ending French rule. All parties signed a peace agreement, and Vietnam was divided into Communist North Vietnam, led by Ho Chi Minh, and non-Communist South Vietnam, eventually led by Ngo Dinh Diem.

Soldiers in Vietnam battled harsh terrain, as well as the enemy. ▶

◀ A former soldier demonstrates a trapdoor in the network of tunnels near Saigon (now Ho Chi Minh City). Today the tunnels are a tourist attraction.

NATIONAL GEOGRAPHIC SOCIETY

Diem proved unpopular, and rebel groups formed. Having North Vietnam's support, the rebels were called Viet Cong, or "Vietnamese Communists." In 1963 a military coup overthrew Diem. After U.S. President Lyndon Johnson announced that North Vietnam had attacked American ships in the Gulf of Tonkin, the United States took action. Soon American planes were bombing North Vietnam. In 1965 the first American troops landed to support South Vietnam.

A Losing Battle

American pilots flew B-52 bombers in air strikes against North Vietnam. In the south, Americans used helicopters, tanks, and well-armed ground troops to seek out Viet Cong. Chemicals, such as Agent Orange, were sprayed to kill the thick jungle vegetation. Modern weaponry, however, did not deter the Communist forces. Viet Cong and North Vietnamese fighters relied on guerrilla tactics, on knowledge of the terrain, and on weapons from the Soviet Union and China. Viet Cong hid out and attacked from 200 miles (320 km) of underground tunnels. Many American soldiers lost their lives trying to infiltrate the jungles and tunnels of Vietnam.

As the war dragged on, antiwar protests erupted in the United States. Under pressure to end the war, U.S. President Nixon began withdrawing troops. The last American forces left Vietnam in 1973. By war's end in 1975, more than 50,000 Americans and as many as 2 million Vietnamese were dead. By 1976 Vietnam was reunited, and Hanoi imposed harsh reforms on Saigon, which was renamed Ho Chi Minh City. In the following decade, more than a million refugees fled Vietnam's shores by boat. Tragically, half of these "boat people" died.

Today Americans still study lessons of the war. In Vietnam, north and south remain vastly different, with little economic development in the north and foreign investment pouring into the south.

Looking Ahead

The United States established diplomatic ties with Vietnam in 1995. What role might the United States play in Vietnam's economic recovery?

1954 Ho Chi Minh (photo above) and Communist fighters defeat France

1964 Gulf of Tonkin incident; U.S. bombs North Vietnam

1965 U.S. sends ground forces to aid South Vietnam

1967–1972 War continues; Americans protest (background photo)

1973 Paris Accords establish cease-fire; U.S. troops withdraw

1975 North Vietnam conquers South Vietnam

1978–1980s Boat people flee

1995 U.S. normalizes relations with Vietnam

SECTION 3

Cultures and Lifestyles

Guide to Reading

Consider What You Know

Southeast Asia is a culturally diverse region. Increasingly, Southeast Asian cultural influences are present in the Western world. What foods, clothing, or religions do you know of that are from Southeast Asian cultures?

Read to Find Out

- What makes Southeast Asia such an ethnically diverse region?
- How have outside influences affected the arts in Southeast Asia?
- How do people's lifestyles reflect Southeast Asia's diversity?

Terms to Know

- *wat*
- batik
- longhouse

Places to Locate

- Irrawaddy River
- Kuala Lumpur

NATIONAL GEOGRAPHIC

A Geographic View

Rural Progress

Thailand's economic success is most obvious in the cities, but it filters into the countryside as well. Where families once tended small [rice] paddies just outside Bangkok, large tractors now groom sweeping fields of commercial farms. . . . On the quiet side roads where I once slowed for water buffalo, I now dodged motorcycles piloted by young Thai men in love with speed.

—*Noel Grove, "The Many Faces of Thailand,"* National Geographic, *February 1996*

Shrimp farm outside Bangkok, Thailand

Throughout their history, Southeast Asians have successfully adapted new ideas and practices to indigenous cultural traditions. Today the peoples of Southeast Asia are learning to blend their cultural heritage with the fast-paced changes brought by the region's participation in a global economy. In this section you will learn about Southeast Asia's many cultures and lifestyles.

Cultural Diversity

Cultures in Southeast Asia reflect the region's ethnic diversity. In Vietnam, for example, a number of cultural traditions—Chinese, Hmong, Tai, Khmer, Man, and Cham—exist alongside the predominant Vietnamese culture. Indonesia has the region's largest number of ethnic and cultural groups. About 300 ethnic groups with more than 250 distinct languages live on Indonesia's many islands. Since independence, the Indonesian government has struggled to hold the country together. The collapse of its dictatorship and the breaking away of East Timor have encouraged independence movements in other parts of Indonesia to increase their demands.

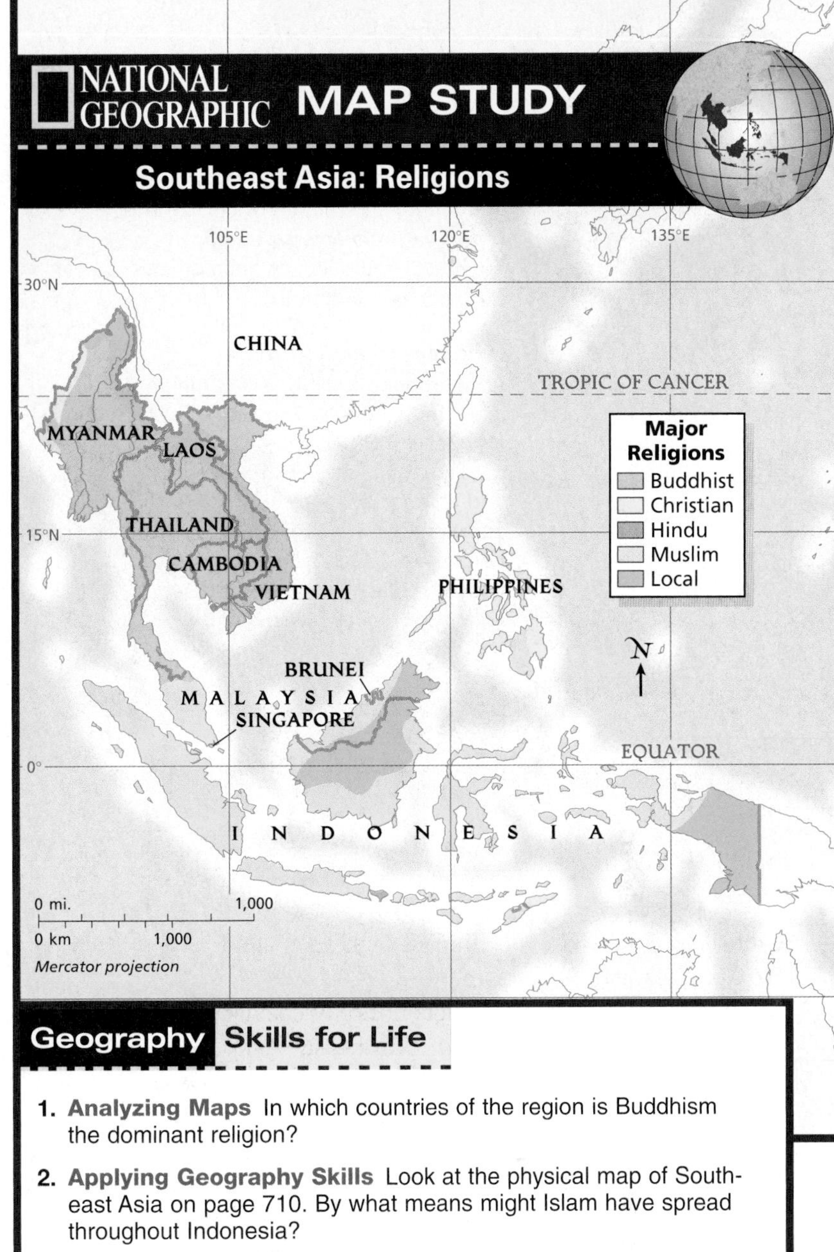

Geography Skills for Life

1. **Analyzing Maps** In which countries of the region is Buddhism the dominant religion?
2. **Applying Geography Skills** Look at the physical map of Southeast Asia on page 710. By what means might Islam have spread throughout Indonesia?

Find NGS online map resources @ www.nationalgeographic.com/maps

History

Languages

Hundreds of languages and dialects are spoken in Southeast Asia. Most of the region's languages stem from three major language families—Malayo-Polynesian, Sino-Tibetan, and Mon-Khmer.

Many of the languages spoken in Southeast Asia are the result of migration or colonization. In the Philippines, for example, Pilipino, English, and Spanish are the major languages. Pilipino, an official language of the Philippines, belongs to the Malayo-Polynesian language family and evolved from the speech of early migrants. Spanish was brought to the Philippines during the years of rule by Spain. English, the second official language, came later with rule by the United States.

Chinese, Malay, Tamil, and English are the official languages of Singapore, reflecting the importance of global trade to this tiny island country. In Malaysia, where British influence was strong during the 1800s and early 1900s, English is the language most often used in business and daily life. Affirming the country's traditional culture, however, the Malaysian government has made Malay the country's official language, especially in schools and universities. In Vietnam urban residents speak Vietnamese, Chinese dialects, French, or English. The presence of the three non-Vietnamese languages is a daily reminder of the influence that China, France, and the United States have had on Vietnam's history and culture.

Religions

Because of the many cultures that exist in Southeast Asia, nearly all of the world's major religions are represented in the region. Buddhism is the major religion of Myanmar, Thailand, Cambodia, Laos, and Vietnam. Many people living in Malaysia and Indonesia practice Islam. The majority of people in the Philippines are Roman Catholic. This Christian influence began when the Philippines came under the control of Spain during the 1500s. A great number of Southeast

music of SOUTHEAST ASIA

A variety of instruments, chants, vocal styles, and dances are found throughout Southeast Asia. The use of bronze and bamboo instruments is common in Thailand, Cambodia, the Philippines, Malaysia, and Indonesia.

Instrument Spotlight
A metal gong known as a **gamelan** is the most popular instrument of Indonesia. These bronze gongs are made in many shapes and sizes and are usually featured in ensembles along with drums, percussion, flutes, singers, and dancers. Gamelans originated in Java prior to the 1400s, and by the 1700s they were an important part of the royal courts. From Java this music tradition spread to Bali and other neighboring islands.

Go To **World Music: A Cultural Legacy** Hear music of this region on Disc 2, Tracks 19–24.

Asians—mainly those of Chinese ancestry—follow Confucianism or Daoism.

These different religious beliefs not only coexist but also mingle throughout Southeast Asia. In Vietnam people blend Buddhism, Confucianism, and, in some cases, Catholicism. A housewarming ceremony in Thailand might include blessings by a Buddhist monk and a Hindu priest, followed by offerings to ancestors and nature spirits. Hinduism, influenced by Buddhism and indigenous religions, is the basis for elaborate ceremonies on the Indonesian island of Bali.

The Arts

The civilizations of early India and China significantly shaped Southeast Asia's cultural development. Over the centuries, local artists and writers creatively adapted Indian and Chinese styles to their own needs. Hinduism and Buddhism also inspired literature, art, and architecture in Southeast Asia. During the era of Western colonization, European artistic and literary styles began to influence Southeast Asian arts and literature.

Architecture

Many beautiful examples of religious architecture exist throughout Southeast Asia. Elaborate Chinese-style pagodas and Indian-style *wats*, or temples, dot the landscape. Thousands of these religious buildings are located on the Indochina Peninsula alone.

Borobudur, a Buddhist shrine in Indonesia, is a stunning example of Southeast Asian religious art and architecture. Built of gray volcanic stone around A.D. 800 on the island of Java, this temple is larger than Europe's great cathedrals. A large tower shaped like a bell tops the pyramid-shaped monument. The shrine's three levels, connected by stairs, represent the three stages of the Buddha's journey to enlightenment.

Student Web Activity Visit the **Glencoe World Geography** Web site at geography.glencoe.com and click on Student Web Activities—Chapter 30 for an activity about Malaysia.

The royal city of Pagan (puh•GAHN) in Myanmar was the ancient capital of an early Burmese empire. From A.D. 1044 to about 1300, kings and commoners honored the Buddha by building more than 5,000 pagodas. More than 2,000 pagodas still stand along 8 miles (13 km) of the **Irrawaddy River**. Many of these ancient structures remain in excellent condition.

Christianity and Islam also have influenced Southeast Asian architecture. In the Philippines you can find Roman Catholic churches built in the Spanish colonial style. In Malaysia, Brunei, and Indonesia, where Islam is the major religion, the minarets of many beautiful mosques are prominent.

Modern architecture dominates the skyline of major Southeast Asian cities, such as Bangkok, Jakarta, Singapore, and Kuala Lumpur. **Kuala Lumpur**, Malaysia's capital, has an area called the Golden Triangle that includes luxury hotels, multistoried office buildings, and a development project known as the Kuala Lumpur City Center (KLCC). The KLCC has one of the world's tallest office buildings, the Petronas Twin Towers.

Crafts

The rich cultures of Southeast Asia have produced many fine crafts. Artisans in Myanmar and Vietnam produce glossy lacquerware. Boxes, trays, dishes, and furniture are covered with many layers of resin from the Asian sumac tree. Colored powders are used to paint designs on the pieces.

Creating lacquerware is time-consuming. Several weeks must pass between applications of layers of lacquer, and a piece may take up to a year to complete. An observer explains the state of mind an artisan requires to create this traditional craft:

> *"Good lacquer requires a mood of timelessness that even the visitor senses. Workers' time clocks, if such existed, would be marked in months, not hours."*
>
> W. E. Garrett, "Pagan, on the Road to Mandalay," *National Geographic*, March 1971

Using a method known as **batik** (buh•TEEK), Indonesians and Malaysians produce beautiful designs and patterns on cloth. First, they use wax or rice paste to create designs on the cloth. Then, they dye the fabric. The dyes form a pattern, coloring only the untreated parts of the cloth. Finally, the cloth is boiled to remove the wax. A colorful pattern or picture remains.

Literature

Early literature in Southeast Asia consisted of folktales, legends, and love stories passed orally from generation to generation. Indian, Chinese, and Islamic literature later had a great influence on local writers, whose works still showed their own distinct character. For example, in *Arjunavivaha*, a story about the life of a king in Java, the court poet Mpu Kanwa modified the Indian epic *Mahabharata* to fit Southeast Asian circumstances.

In recent times Southeast Asian authors have used Western styles and themes in their works. Many

NATIONAL GEOGRAPHIC **World Explorer**

Geography Skills for Life

An Emperor's Tomb With French backing Emperor Khai Dinh ruled Vietnam from 1916 to 1925. His tomb blends French and Vietnamese styles.

Place What important crafts are produced in Southeast Asia?

of the region's writers, however, have translated classic Southeast Asian literature into modern forms of language that can be read and understood by people today.

Culture

Dance and Drama

Performance arts remain immensely popular in Southeast Asia. Dance and drama are combined to retell legends or re-create historical events.

Traditional dances often make use of religious themes. On the island of Bali, in Indonesia, young women perform a dance called the *Legong*. Making graceful gestures, the dancers reenact episodes from the *Ramayana*, an ancient Indian story. Dances can also serve as reminders of the region's agricultural roots. In Cambodia, when the monsoon rains are late, dancers perform a type of rain dance called the *Leng Trot*.

Puppet plays are popular in many parts of Southeast Asia. These plays use historical and religious characters to perform tales. Sometimes a human dancer who imitates a puppet's movement performs the play.

Lifestyles

Southeast Asia's ethnic diversity leads to a wide variety of lifestyles in the region. Yet as global contacts have increased, similarities have also developed among the ways people in Southeast Asia live.

Health and Education

Since achieving independence, many Southeast Asian countries have enjoyed an improved quality of life. Industry has spread throughout the region, and per capita incomes have risen. Singapore's per capita gross domestic product (GDP) of $26,300 is comparable to that of the United States. The per capita GDPs of Laos, Cambodia, and Vietnam, however, are all lower than $2,000.

Life expectancy and infant mortality rates also have improved. The general levels of health in Southeast Asia still vary widely, with Singapore having the best overall health conditions. For example, average life expectancy is 78 years in Singapore, compared with only 52 years in Laos.

Since 1945, literacy has increased dramatically in the region, although educational opportunities are still limited in many areas. Governments continue efforts to make education available to everyone. Thailand has the highest literacy rate in the region (95 percent), and Laos has the lowest (57 percent).

Housing

Housing in Southeast Asia varies throughout the region, depending on physical geography. In cities, people often live in traditional brick or wooden houses. Some urban residents make their homes in high-rise apartments. Although many Southeast Asians still live in poor conditions, government-funded housing projects have improved the situations in some places.

Despite rapid urban growth, many Southeast Asians still live in small farming villages. A typical village consists of about 25 to 30 homes made of bamboo or wood. These houses are built to suit the environment. Most have roofs made of tiles, corrugated iron, or tin to keep out heavy rains. Most of these dwellings lack running water and electricity.

In some rural areas of Indonesia and Malaysia, people live in **longhouses**—elevated one-story buildings that house up to 100 people. Elevating the houses on poles helps ventilate and cool the structures and offers protection from insects, animals, and

NATIONAL GEOGRAPHIC **World Explorer**

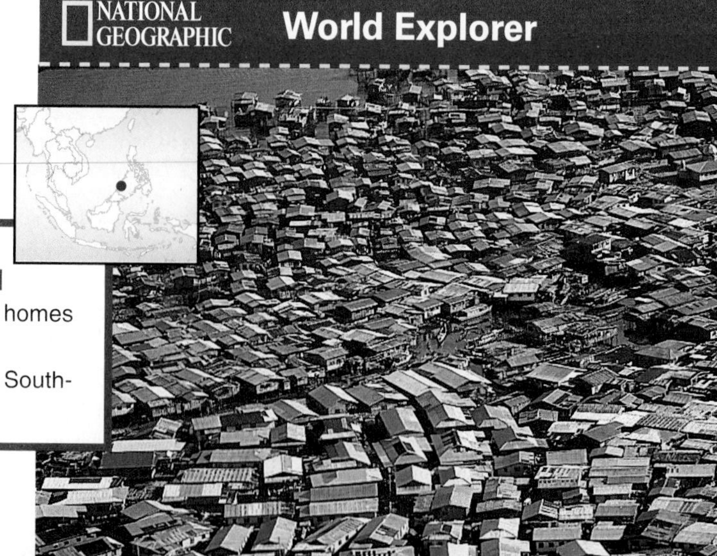

Geography Skills for Life

Cramped Housing Thousands of people live aboard floating homes in Sabah, a region of Malaysia.

Region What are typical farming communities like in Southeast Asia?

flooding. The residents of longhouses are usually members of several extended or related families.

Food, Recreation, and Celebrations

Most rural Southeast Asians live on the food they raise themselves. Throughout the region rice is the staple food and is usually served with spicy fish, chicken, vegetables, and sauces. Various countries have their own specialties. Some use curry and other spices; some make use of coconut milk.

Southeast Asians enjoy a variety of leisure activities. In large cities, such as Bangkok, Jakarta, and Singapore, people visit museums, theaters, parks, restaurants, and nightclubs. In rural areas people enjoy visiting their neighbors and celebrating family occasions such as weddings and birthdays.

People throughout the region enjoy sports such as soccer, basketball, and badminton. Traditional sports and pastimes are also popular. In Myanmar, people play a game called *chinlon*, in which players form a circle and try to keep a rattan ball in the air without using their hands. Indonesians practice a combination of dancing and self-defense known as *silat*. Thais enjoy a form of "kick" boxing that uses the feet as well as the hands.

Many Southeast Asian holidays are tied to religious observances. For example, Thailand celebrates *Songkran*, or the Water Festival, during the Buddhist New Year. People bathe statues of the Buddha and bless one another with a sprinkling of water. In January or February, Vietnam celebrates its New Year, called Tet. The celebration begins at the start of the lunar year and lasts three days.

NATIONAL GEOGRAPHIC **World Explorer**

Geography Skills for Life

Thailand's Monkey Feast Thais regard monkeys as symbols of good fortune. At a yearly festival in their honor, monkeys are provided with an abundance of food.

Region What kinds of celebrations do Southeast Asians enjoy?

SECTION 3 ASSESSMENT

Checking for Understanding

1. **Define** *wat*, batik, longhouse.
2. **Main Ideas** In a graphic organizer like the one below, list cultural groups that migrated to the region in one column and their contributions in the other.

Cultural Group	Cultural Contribution

Critical Thinking

3. **Making Generalizations** What cultural features reflect Southeast Asia's ethnic diversity?
4. **Problem Solving** How might a new art museum ensure that it reflects Southeast Asia's culture?
5. **Comparing and Contrasting** How might standards of living differ between rural and urban Southeast Asia?

Analyzing Maps

6. **Location** Look at the map of Southeast Asia's religions on page 749. In which country is Christianity the predominant religion?

Applying Geography

7. **Diversity** Trace the spread of foreign influences in Southeast Asia. How have these influences shaped Southeast Asian life and culture?

Understanding Cartograms

On most maps, land areas are drawn in proportion to their actual surface areas on the earth. A cartogram is a map in which size is based on some characteristic other than land area, such as population or economic factors.

Learning the Skill

A cartogram provides clear visual comparisons of the characteristic it measures. To read a cartogram, apply the following steps:

- **Read the map title and key to identify the kind of information presented in the cartogram.**
- **Look for relationships among the countries.** Determine which countries are largest and smallest.
- **Compare the cartogram with a standard land-area map.** Determine the degree of distortion of particular countries.
- **Study these relationships and comparisons.** Identify the most important information presented in the cartogram.

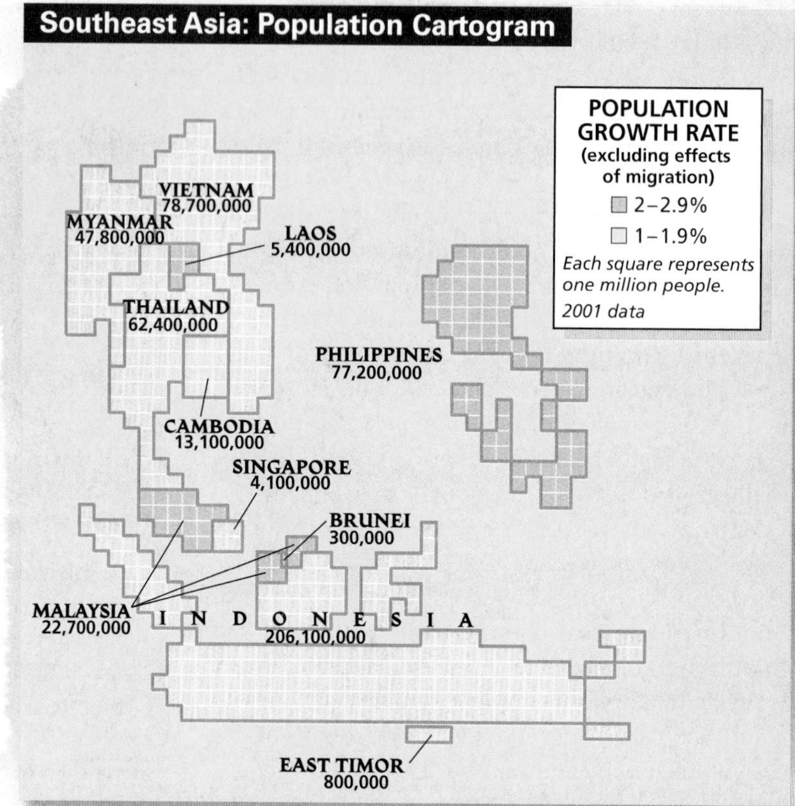

Practicing the Skill

Use the cartogram on this page to answer the following questions.

1. What data determine the relative sizes of countries on this cartogram?
2. What characteristics determine the color of the squares on this cartogram?
3. Compare the cartogram with the standard land-area map on page 721. How has the relative size of Singapore been changed on the cartogram? How would you explain this change?
4. From the information in this cartogram, would you expect Laos to have more squares than Vietnam in a cartogram based on 2010 data? Explain.
5. Suppose you want to compare the population densities of two countries in this region. Can this cartogram help you make this comparison? Explain.

Applying the Skill

Research the gross domestic product (GDP) of each country in Southeast Asia. Then create a cartogram that compares the GDP of these countries. Include a key for the symbols you use.

The **Glencoe Skillbuilder Interactive Workbook, Level 2** provides instruction and practice in key social studies skills.

30

SECTION 1 Population Patterns (pp. 735–739)

Terms to Know

- urbanization
- primate city

Key Points

- Southeast Asia has a diversity of ethnic and cultural groups.
- Most Southeast Asians live either in river valley lowlands or on coastal plains.
- Southeast Asian cities are growing rapidly as a result of migration from rural to urban areas.
- Since the 1970s, large numbers of Southeast Asians have migrated to escape political oppression and economic distress.

Organizing Your Notes

Use a web like the one below to help you organize the notes you took as you read this section. Fill in information about the population patterns of Southeast Asia.

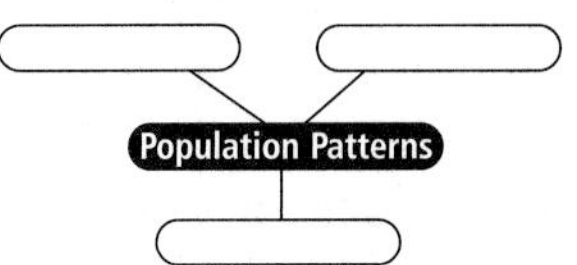

SECTION 2 History and Government (pp. 740–745)

Terms to Know

- maritime
- sphere of influence
- buffer state

Key Points

- Southeast Asia's early empires and kingdoms controlled shipping and trade that linked East Asia, South Asia, and Southwest Asia.
- European countries colonized all of Southeast Asia except Thailand (Siam). All of the region's countries are now independent.
- During the late 1900s, political conflict between communist and noncommunist forces divided much of Southeast Asia.

Organizing Your Notes

Use a cause-effect chart like the one below to help you organize the information you read in this section.

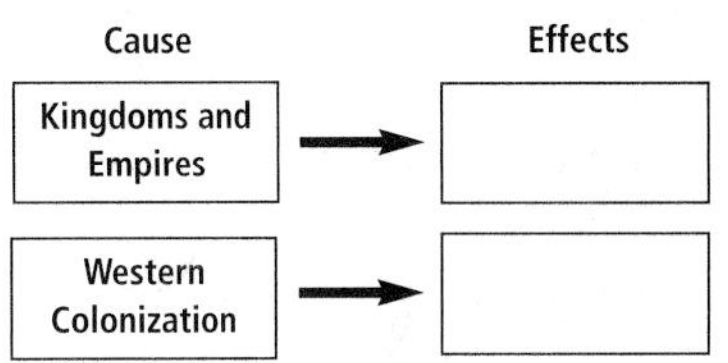

SECTION 3 Cultures and Lifestyles (pp. 748–753)

Terms to Know

- *wat*
- batik
- longhouse

Key Points

- Southeast Asian culture reflects the ways of life of peoples who migrated from other regions as well as those of indigenous peoples.
- Buddhism, Hinduism, and Islam greatly influenced Southeast Asian art, architecture, drama, and celebrations.
- In spite of rapid population growth, Southeast Asia's economic development has led to many improvements in the region's quality of life.

Organizing Your Notes

Use an outline like the one below to help you organize information in this section about cultures and lifestyles.

I. Cultural Diversity
  A. Languages
    1. Hundreds of languages
    2.

CHAPTER 30 ASSESSMENT & ACTIVITIES

Reviewing Key Terms

Write the key term that best completes each of the following sentences. Refer to the Terms to Know in the Summary & Study Guide on page 755.

1. Kuala Lumpur is Malaysia's _________.
2. A(n) _________ often houses a large, extended family.
3. Western countries set up _________ in Southeast Asia.
4. Southeast Asian _________, or seafaring, empires controlled shipping and trade.
5. Southeast Asian architecture includes _________, or temples inspired by India.
6. A neutral territory called a(n) _________ can prevent conflict between rival powers.

Reviewing Facts

SECTION 1

1. What geographic factors influence where Southeast Asians live?
2. Describe the characteristics of the region's urban and rural populations.

SECTION 2

3. Why did the early Southeast Asian kingdoms prosper?
4. How did colonization by Western countries affect the region?

SECTION 3

5. What foreign influences can be seen in Southeast Asia's arts?
6. How has the quality of life in Southeast Asia improved?

Critical Thinking

1. **Making Generalizations** Why are small farms unable to compete with plantations?
2. **Problem Solving** Identify Southeast Asia's greatest challenge, and propose a solution.
3. **Identifying Cause and Effect** Complete a flowchart like the one below to show the history of Southeast Asia from colonization to independence.

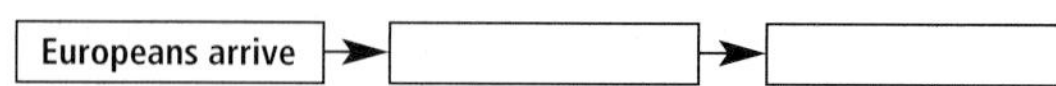

Using the Regional Atlas

Refer to the Regional Atlas on pages 710–713.

1. **Region** What areas of Southeast Asia are the most densely populated?
2. **Location** What Southeast Asian cities have populations of more than 2 million? What geographic factors do most of these cities have in common?

NATIONAL GEOGRAPHIC Locating Places

Southeast Asia: Physical-Political Geography

Match the letters on the map with the places and physical features of Southeast Asia. Write your answers on a sheet of paper.

1. Cambodia
2. Bangkok
3. Hanoi
4. Strait of Malacca
5. Gulf of Thailand
6. Manila
7. Sumatra
8. Kuala Lumpur
9. Indian Ocean
10. South China Sea

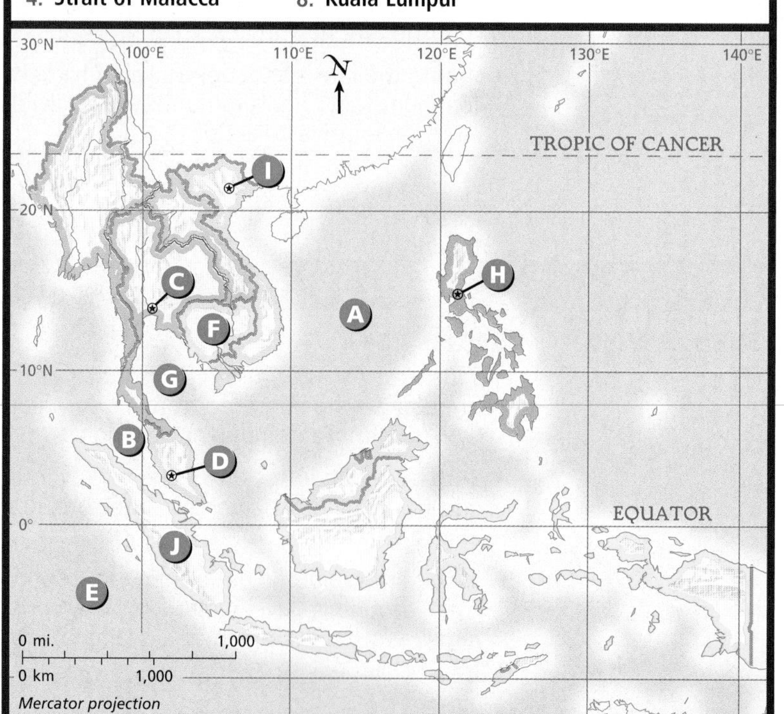

Self-Check Quiz Visit the **Glencoe World Geography** Web site at geography.glencoe.com and click on Self-Check Quizzes—Chapter 30 to prepare for the Chapter Test.

Thinking Like a Geographer

Use your textbook, library sources, and the Internet to answer the following questions about Southeast Asia: What geographic factors might have drawn foreigners to the region? How might foreign influences have shaped forms of government in the region?

Problem-Solving Activity

Contemporary Issues Case Study *Push factors*, such as unemployment or famine, are the unsatisfactory features of a place that cause people to emigrate. *Pull factors*, such as fertile soil or better job opportunities, are a place's attractive features that draw migrants from other areas. Research Southeast Asia's national and international migration patterns, and write a report explaining how push, pull, or both push-and-pull factors shape Southeast Asian migration today.

GeoJournal

Descriptive Writing Using the details you logged in your GeoJournal as you read this chapter, write a letter to a friend or relative about one cultural element in the region. Imagine that you are visiting the region and you want your friend or relative to have a vivid picture of the places you describe. Include word pictures that appeal to as many of the five senses as possible. Explain how this cultural element differs from that found in different parts of the United States.

Technology Activity

Developing Multimedia Presentations Use the Internet or the library to conduct research about one typical example of Southeast Asian religious architecture. Create a multimedia presentation about your temple or *wat* that uses narration, music, and images. Be sure to cite all the sources that you used to prepare your presentation, including print and Internet sources for text and photographs.

Standardized Test Practice

Choose the best answer for each of the following multiple-choice questions. If you have trouble answering the questions, use the process of elimination to narrow your choices.

1. Which countries' cultures most influenced Southeast Asia's religions?

A Japan and Korea
B China and the United States
C India and China
D Arabia and India

Think about what major religions are practiced in Southeast Asia. The answer that includes the cultures where those religions originated probably is the correct answer.

2. Southeast Asia is a region of highly diverse cultures for all of the following reasons EXCEPT:

F Trade and colonization from many regions spread new ideas.
G Diverse physical geography separated peoples who created their own traditions.
H High population density in many areas causes a variety of traditions.
J War and conquest by countries from outside the region forced changes in culture.

Consider how cultures from other world regions established a spice trade in Southeast Asia. Watch for key words such as *except, not,* and *only* that greatly affect what the question is asking. The question is asking for the *exception*. Therefore, answer choices that do reflect this important point of history can be eliminated.

CHAPTER 31

Southeast Asia Today

GeoJournal

As you read this chapter, use your journal to log information about the economies and the environmental challenges in Southeast Asia. Note interesting details that show similarities and differences among the region's countries.

GEOGRAPHY *Online*

Chapter Overview Visit the **Glencoe World Geography** Web site at geography.glencoe.com and click on Chapter Overviews—Chapter 31 to preview information about the region today.

Living in Southeast Asia

Guide to Reading

Consider What You Know

Southeast Asia is a region of peninsulas and islands. How do you think people living in such areas earn their incomes?

Read to Find Out

- Why is rice farming the most important agricultural activity in Southeast Asia?
- Why are the countries in the region industrializing at different rates?
- How are the economies of Southeast Asia becoming more interdependent?

Terms to Know

- paddy
- sickle
- subsistence crop
- cash crop
- lode
- interdependent
- Association of Southeast Asian Nations (ASEAN)
- free port

Places to Locate

- Brunei
- Manila

◀ *Buildings at night in Kuala Lumpur, Malaysia*

NATIONAL GEOGRAPHIC

A Geographic View

Open-Air Market

. . . I visited the . . . colorful open-air market in Kota Baharu, way up north near [Malaysia's] border with Thailand. The merchants were nearly all women. Wearing full-length batik sarongs of bright red, orange, pink, and purple, with coordinated scarves of emerald green or royal blue around their heads, they sat beside huge piles of fruit and vegetables, truckloads of fish and chicken, mountains of rice, and tall wicker baskets filled with eggs—turtle eggs, stork eggs, even chicken eggs.

—*T. R. Reid, "Malaysia: Rising Star,"* National Geographic, *August 1997*

Market in Malaysia

This market scene takes place in Malaysia, one of Southeast Asia's most rapidly developing countries. Like some other countries in the region, Malaysia is setting up new industries, yet it continues to rely on agriculture for its economic well-being. In this section you will learn about how people live and work in Southeast Asia today. You will also see how the region's countries face many of the same challenges and have come to depend on one another for increased economic growth.

Agriculture

Southeast Asia's fertile river valleys and plains are a major source of livelihood for its people. Southeast Asians depend on the rich variety of crops grown in these areas to supply their own food needs as well as to sell for income. Although Southeast Asia is industrializing rapidly, most of its workforce is still involved in agriculture. More than two-thirds of all workers in Cambodia and Laos are farmers.

Rice Cultivation

For 2,000 years, the Ifugao people of the Philippines have worked in terraced fields that follow the contours of the mountains. They plant their fields with rice, the most important crop in Southeast Asia. Southeast Asian farmers use more than half of the region's farmable land to grow this crop. A major food source for the region, rice is also a leading export product of Thailand, Cambodia, Vietnam, and Myanmar. A journalist traveling in Southeast Asia describes Thailand's rice trade:

> *"Since most rice is eaten in the countries where it's grown, the amount in world trade is small, only about 4 percent. The biggest exporter is Thailand, with 4.5 million tons a year. . . . [A]t Bangkok . . . [m]illed rice arrives by truck from the north—I see 100-kilo bags stacked 27 high—to be . . . packed for shipment to the Middle East, Europe, Africa."*
>
> Peter T. White, "Rice, the Essential Harvest," *National Geographic*, May 1994

Geography Skills for Life

Thai Rice Field A farmer uses the simple technology of a water buffalo and a wooden plow to prepare his field for planting.

Region Why does rice grow well in the region?

Rice grows well in Southeast Asia because most of the region has fertile soil, an abundant water supply, and a warm, wet climate. Some kinds of rice plants need a continuous supply of water from the time they are planted until just before harvest. Flooded rivers and abundant rainfall provide this water to Southeast Asia. In parts of Thailand, Cambodia, and Vietnam, seasonal flooding of the Chao Phraya and Mekong Rivers irrigates **paddies**, or flooded fields in which rice is grown. Rain also provides enough water to grow rice in the Irrawaddy River delta in Myanmar and in parts of the Philippines.

Farmers plant rice at the start of the rainy season, usually in May, and the crop is ready to harvest in October. They can then grow a second rice crop during the dry season by irrigating rice fields with water stored from rains and flooding rivers. Rice farming can be difficult work because many farmers do not use modern machinery. They plant and harvest their crops by hand, using simple tools such as **sickles**—long, sharp, curved knives. Water buffalo or oxen are often used to pull plows.

Other Crops

Southeast Asian farmers grow cassava, yams, corn, bananas, and other food crops in areas too dry for a second planting of rice. Some Indonesian farmers have begun to grow cassava, an edible root, as an alternative subsistence crop because it is easier to grow than rice. A **subsistence crop** is a crop grown mainly to feed the farmer's family. Many families in Southeast Asia have small subsistence garden plots that produce a variety of vegetables, and some people also raise pigs and poultry for food.

Plantations in Southeast Asia's coastal lowlands provide many of the region's **cash crops**—crops raised to be sold for profit. Rubber is an important cash crop, and Thailand, Indonesia, and Malaysia lead the world in natural rubber production. Sugarcane grows in the Philippines and on the Indonesian island of

Java. The Philippines is also one of the largest producers and exporters of coconuts. Other regional exports include coffee, palm oil, and spices.

Forests and Mines

Forestry, which includes jobs in logging, transporting logs, and manufacturing finished goods, is important to many Southeast Asian countries. It is a major industry in Vietnam, where factories produce plywood and lumber, pulp and paper, and furniture products. Myanmar leads the world in teakwood exports. Teakwood, ebony, mahogany, and bamboo, in the form of lumber and finished products, are vital to the economies of Malaysia, the Philippines, Indonesia, and Thailand. Although excessive logging has contributed to deforestation in the region, several Southeast Asian countries are working to make their economic goals compatible with environmental goals.

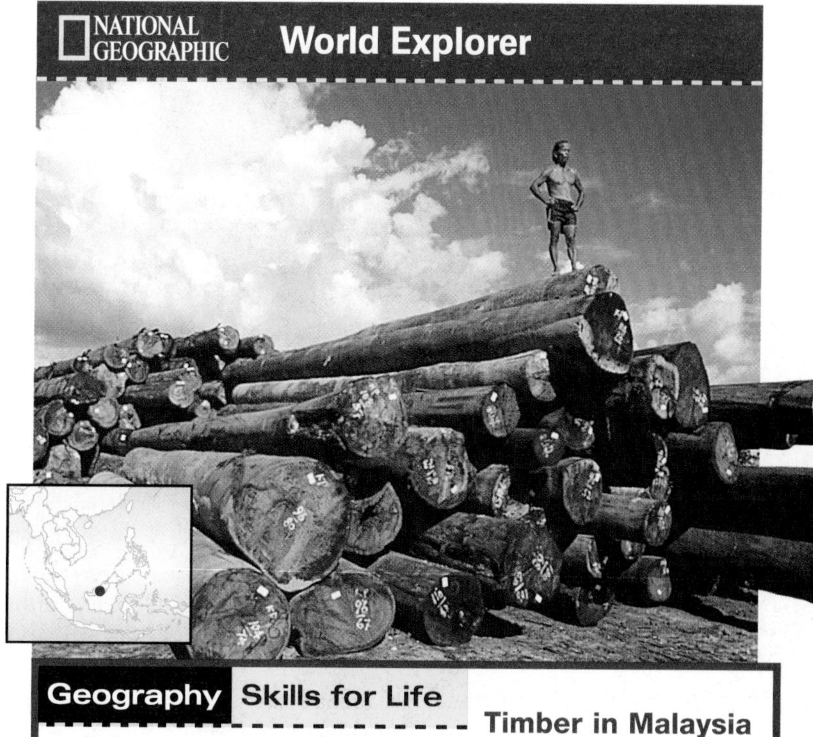

Geography Skills for Life

Timber in Malaysia

Logging is a major economic activity in East Malaysia.

Place Which economies are dependent on lumber and finished products?

Mineral Wealth

Rich mineral deposits lie within Southeast Asia's numerous mountains. Workers in several countries drill and blast their way to deposits of tin, iron ore, manganese, and tungsten. Malaysia, Thailand, and Indonesia are three of the world's leading producers of tin. Iron ore is excavated in Malaysia and the Philippines. Manganese, used to strengthen steel, is mined in the Philippines and Indonesia. Tungsten, used for electrical materials and in steel alloys, is found in Myanmar and Thailand.

The economies of Southeast Asia also benefit from oil extraction. Malaysia is rich in petroleum and natural gas reserves. Crude oil, natural gas, and petroleum products account for 95 percent of the export income of **Brunei** (bru•NY). This small country also has one of the world's largest natural gas plants. Indonesia, the largest producer of petroleum in the region, is one of the top ten producers in the Organization of Petroleum Exporting Countries (OPEC). Economic development in the Indonesian-owned western part of New Guinea and on the Indonesian islands of Sumatra, Java, and Borneo has been spurred by the building of pipelines. These pipelines carry oil from drilling sites to the coasts for shipping.

Economics

Irian Jaya's Resources

Indonesia's government has set aside large areas of Irian Jaya for resource development. Located on the western half of New Guinea, Irian Jaya has timber resources and rich lodes, or deposits of minerals. Many international companies are logging in mangrove swamps in Irian Jaya and surveying for gold, natural gas, oil, and uranium elsewhere.

Although Irian Jaya is rich in minerals, many of its people are poor. Groups favoring independence claim that the Indonesian government has allowed foreigners to extract resources but has invested little in improving health, education, and public services.

Industry

Industry is growing rapidly in Southeast Asia. In many places, workers are leaving farms to work

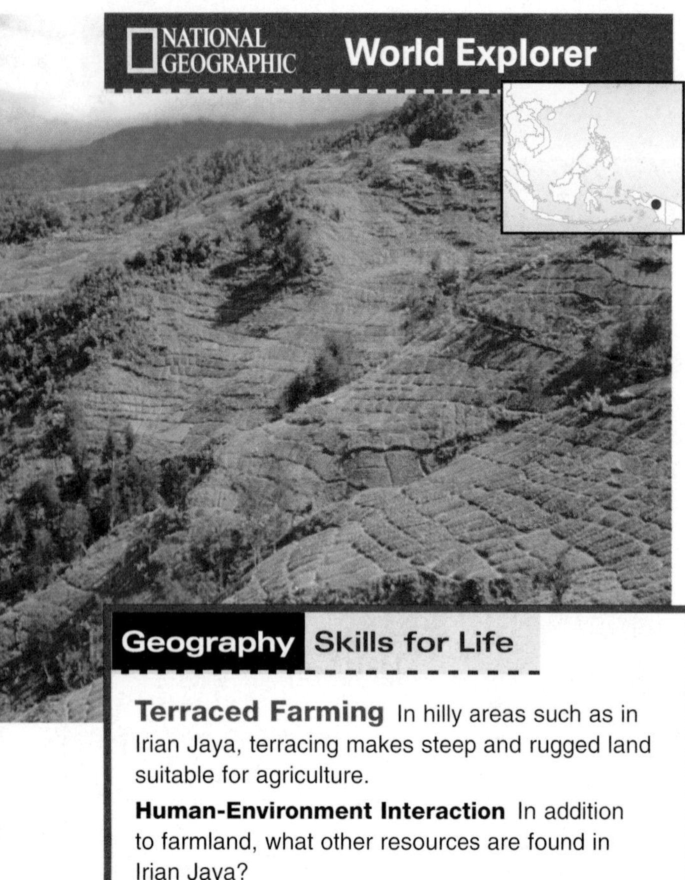

Geography Skills for Life

Terraced Farming In hilly areas such as in Irian Jaya, terracing makes steep and rugged land suitable for agriculture.

Human-Environment Interaction In addition to farmland, what other resources are found in Irian Jaya?

in urban manufacturing and service industries. Still, the industrial growth rate varies widely throughout the region. While Laos and Cambodia are mainly agricultural, Singapore, Malaysia, Thailand, and the Philippines are Southeast Asia's major industrializing countries. A center of world trade, Singapore focuses on producing goods for export. Factories in Malaysia, the Philippines, Indonesia, and Thailand manufacture textiles, clothing, and automobiles.

During the 1980s and early 1990s, the industrializing countries of Southeast Asia enjoyed an economic boom. This prosperity was based on plentiful natural resources, an abundant supply of inexpensive labor, and increased foreign investment. Massive debts, political corruption, and financial mismanagement, however, led to an economic crisis in the region in late 1997. Economic reforms allowed Thailand to emerge fairly quickly from the crisis. Since the crisis, both Thailand and the Philippines have had to balance industrial growth with investment in traditional economic activities such as agriculture and fish farming.

Economics

Singapore and Malaysia

Singapore has Southeast Asia's most developed economy. Its location and harbors make it a major port and manufacturing center. In addition, Singapore's government and businesses have carried out several policies that have led to strong economic growth. After independence in 1965, Singapore set up free-trade zones that attracted foreign investment. More recently, businesses have focused on developing communications, information, and financial services—activities less dependent on foreign investment. Singapore's economy also has moved away from labor-intensive industries, such as textiles, into electronics and oil refining. To ensure a supply of skilled workers for these industries, the government has made a strong commitment to education.

Singapore's neighbor, Malaysia, also has diversified, or increased its economic activities. Although Malaysia remains a major producer of natural rubber and palm oil, it now also manufactures a variety of goods, such as electronic and electrical products, cement, chemicals, and processed foods. The country also has developed heavy industries, such as steelmaking and automobile assembly. These manufactured products—along with natural rubber and palm oil—account for most of Malaysia's export earnings. Malaysia is also the world's largest exporter of microchips, making it an important center for information technology.

Less Industrialized Countries

Since the late 1990s, political instability and a rapidly growing population have slowed economic growth in Indonesia. The country supplies raw materials for world markets and is a major exporter of textiles and garments. Its labor force, however, currently lacks the technical skills and knowledge required for industrialization. Therefore, Indonesia depends heavily on foreign aid and investment to develop its industries.

Industrialization in other Southeast Asian countries, such as Laos, Vietnam, Cambodia, and Myanmar, is developing even more slowly than in Indonesia. Wars and political changes slowed economic growth in Laos, Vietnam, and Cambodia for many years. Landlocked and without ocean harbors, Laos remains largely agricultural. However, the attempt by its communist leaders to collectivize farming reduced incentives for farmers to produce. The country is rich in mineral resources but lacks up-to-date mining technology. Laos's future economic growth may depend on its rivers, which could provide hydroelectric power for the region.

Rapid population growth and inadequate transportation have hurt Vietnam's economic development. The country, however, has a large potential workforce in its literate population. Another possible boost to Vietnam's economy is its beautiful coastline, which is well suited to the development of tourism. Cambodia's economy suffers from outdated factories and the lack of a trained, experienced workforce. Myanmar's self-imposed isolation from world markets has long slowed its economic growth. Myanmar's gross national product per person is one of the lowest in the world, and manufacturing accounts for just one-tenth of the country's gross domestic product.

Interdependence

In recent years Southeast Asian countries have become more **interdependent**, or reliant on one another. As they draw closer together, economic and political developments in one country can affect other countries in the region. Two organizations formed to promote regional development, trade, and greater economic stability reflect this increasing interdependence. They are the Asian Development Bank (ADB) and the **Association of Southeast Asian Nations (ASEAN)**.

The ADB, based in the Philippines's capital of **Manila**, provides international loans to aid the economies of Asian member countries. In Southeast Asia these ADB loans support agricultural, transportation, and industrial development projects.

Indonesia, Malaysia, the Philippines, Singapore, and Thailand formed ASEAN in 1967 as an economic and political alliance. Brunei joined in 1984, and Vietnam, Cambodia, Laos, and Myanmar all became members by the late 1990s. ASEAN's main goals are to promote economic growth and to encourage cultural exchanges among member countries. ASEAN also provides an outlet for cooperation in a region that has long known economic and political conflict, although full political or economic unity is not its main focus.

ASEAN's founding members generally have had greater economic success than have other countries in Southeast Asia. Development has been slow or nonexistent in countries that joined ASEAN later and also in East Timor, which has not yet joined. ASEAN member countries try to balance diverse national goals while struggling

Geography Skills for Life

Making Silk A worker in Vietnam feeds mulberry leaves to silkworms. The fiber that silkworms produce is valuable in making finished silk (see inset above right).

Place What factors have hindered economic development in Vietnam?

for regional unity. In 1992 they agreed to establish a free-trade area and to reduce tariffs on nonagricultural products by 2008.

Transportation

Southeast Asia's peninsulas, islands, long coastlines, and many rivers make water transportation the most common way to move people and goods in the region. However, rain-swollen rivers in the tropical forests sometimes make travel slow and difficult. In some remote areas, such as Indonesia's territory of Irian Jaya, people receive supplies by air as well as by water.

Southeast Asia has long been the crossroads of major ocean trade routes. Today most shipping between Europe and East Asia passes through the Strait of Malacca, near Singapore. This transportation "choke point," or strategic location, enables Singapore to prosper as a **free port**, a place where goods can be unloaded, stored, and reshipped free of import duties. Other regional ports include the Indonesian cities of Palembang, on Sumatra, and the national capital, Jakarta, on Java. Manila, in the Philippines, is a major center for maritime trade in Asia. Vietnam's major international shipping port is Ho Chi Minh City.

Throughout Southeast Asia the quality of land transportation varies widely, partly because of differences in economic development. For example, Cambodia's original highway network was designed by French planners to link agricultural areas to the port of Saigon (now Ho Chi Minh City, Vietnam). Although the network no longer serves Cambodia's economic needs, the country lacks the resources to dramatically redesign the system. In contrast, the industrializing countries of Malaysia, Singapore, and Thailand, with their more successful economies, are able to fund improvements to roads.

Highways and railroads on Southeast Asia's peninsulas and larger islands generally link only major cities. Many people travel on bicycles, motor scooters, and oxcarts. In urban centers such as Jakarta, Indonesia, and Bangkok, Thailand, paved roads are also choked with trucks, automobiles, motorcycles, and buses.

Student Web Activity Visit the **Glencoe World Geography** Web site at geography.glencoe.com and click on Student Web Activities—Chapter 31 for an activity about life in Vietnam.

NATIONAL GEOGRAPHIC **World Explorer**

Geography Skills for Life

A Quiet Street Buddhist monks in Myanmar file past a family eating breakfast.

Region What physical features influence travel in most of rural Southeast Asia?

In most parts of rural Southeast Asia, travel is difficult because of dense forests, rugged terrain, and the seas that separate the region's islands. Outside major urban areas, unpaved roads are often impassable during heavy rains.

Communications

As with transportation, communications in Southeast Asia depend on a country's level of industrialization. Singapore's largely prosperous and urbanized population has a well-developed communications system. Rural dwellers in parts of Cambodia and Laos, however, have little access to newspapers, television, or the Internet.

In Southeast Asia's cities, good communications services help advance economic growth. The Internet and wireless communication have also benefited Southeast Asian commerce. Partly because of the region's rugged terrain, telecommunications service remains poor in rural areas. Satellite communication, however, is improving television and telephone transmissions.

Post offices, newspapers, books, and magazines are located in major urban centers such as Bangkok, Jakarta, Kuala Lumpur, and Singapore. Governments typically own and control radio stations and television networks. Most people own or have access to a radio, but television sets are less common. Singapore, Brunei, and the Philippines have the greatest number of television sets per person. Although most of Southeast Asia's countries are developing modern communications systems, it will take time for effective communications to reach every part of the region.

NATIONAL GEOGRAPHIC **World Explorer**

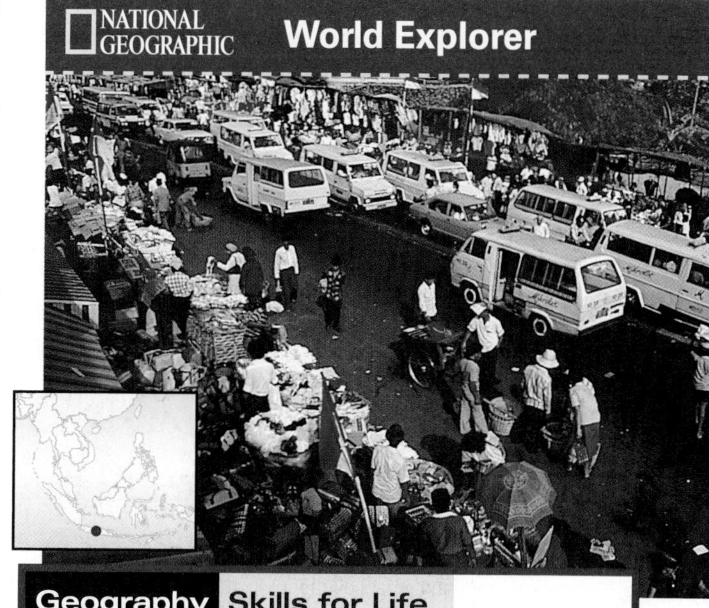

Geography Skills for Life

Urban Transportation City streets in Jakarta, Indonesia, are crowded with automobiles, buses, and pedestrians.

Region How is the distribution of communications similar to that of transportation in the region?

SECTION 1 ASSESSMENT

Checking for Understanding

1. **Define** paddy, sickle, subsistence crop, cash crop, lode, interdependent, Association of Southeast Asian Nations (ASEAN), free port.
2. **Main Ideas** On a table like the one below, list products and industries that support the economies of each of the region's countries.

Country	Products/Industries

Critical Thinking

3. **Making Comparisons** How is a country's workforce affected by growing cash crops? Subsistence crops?
4. **Making Generalizations** What might other Southeast Asian countries learn from Singapore's economic success?
5. **Predicting Consequences** How might rural ways of life change in Southeast Asia as communications services are developed? Provide examples to support your answer.

Analyzing Maps

6. **Place** Study the economic activity map on page 713. What geographic feature is common among most manufacturing and trade centers? Explain this relationship.

Applying Geography

7. **Water Transportation** Think about different kinds of transportation in Southeast Asia. Write a paragraph in which you discuss the day-to-day impacts of relying on water transportation.

People and Their Environment

Guide to Reading

Consider What You Know

Much of Southeast Asia has experienced rapid industrialization. What effect might industrial growth have on a region's natural surroundings?

Read to Find Out

- What dangers are posed by volcanoes, floods, and typhoons in Southeast Asia?
- How has economic progress increased environmental pollution in the region?
- What efforts are under way to protect the environment in Southeast Asia?

Terms to Know

- cyclone
- typhoon
- shifting cultivation

Places to Locate

- Ring of Fire
- Bali

NATIONAL GEOGRAPHIC

A Geographic View

Traffic Ballet

Nikorn Phasuk, a Bangkok policeman who is also known as Plastic Man, steps onto a stage of asphalt under the glare of a blazing sun. He crouches, then retreats with mincing footwork as he coaxes vehicles toward him with fluid arm gestures, part of an artful ballet he uses to keep traffic rolling, no small feat in the city that may have the most congested streets in the world.

—*Noel Grove, "The Many Faces of Thailand,"* National Geographic, *February 1996*

Rush hour in Bangkok, Thailand

In the heart of Bangkok, Thailand's capital, you can experience noisy, crowded, traffic-choked streets and intense heat rising from sunbaked pavement. Like other places in Southeast Asia, Bangkok faces a variety of environmental challenges. In this section you will learn about the natural and human factors that affect Southeast Asia's environment. You will also learn about the efforts of governments and citizens' groups to protect it.

Nature's Might

As you learned in Chapter 29, much of Southeast Asia is part of the **Ring of Fire**, the area of earthquake and volcanic activity that rims the Pacific Ocean. Residents of places along the Ring of Fire periodically face volcanic eruptions, flash floods, and typhoons. These natural disasters take their toll on human lives and on economic development. People's efforts to cope with the effects of disasters are part of everyday life in many parts of Southeast Asia.

Volcanoes

Volcanic mountains rise on most of the larger islands in the Philippines. Several of the volcanoes are active, and many Filipinos must cope with the constant threat of volcanic activity. In February 2000, thousands fled their homes as the Mayon Volcano, which had last erupted in 1993, spewed ash and lava over the landscape.

Another Philippine volcano, the 5,770-foot (1,759-m) Mount Pinatubo, erupted in June 1991. Scientists in the Philippines predicted the eruption, and government authorities ordered the evacuation of nearby towns. Still, the eruption killed more than 900 people and destroyed about 100,000 homes. Clouds of ash and dust blown into the atmosphere affected weather patterns worldwide.

> *"... Mount Pinatubo ... spewed 15 million tons of ash, rock, and sulfuric acid 22 miles into the stratosphere. Within three weeks, the debris had veiled the globe, reflecting sunlight back into space and chilling that year's winter by at least a full degree. ..."*
>
> Jack McClintock, "Under the Volcano," *Discover*, November 1999

Volcanoes also figure prominently in the culture of some Southeast Asian countries. For example, the Indonesian island of **Bali** (BAH•lee) is famous for a volcano that reaches 10,308 feet (3,142 m) high—Gunung Agung. The Balinese people regard the volcano as the sacred centerpiece of their Hindu faith, and they leave offerings of food and flowers on the crater's rim. Despite a 1993 eruption that took more than 1,500 lives, many Balinese still live near Gunung Agung, risking their lives and property.

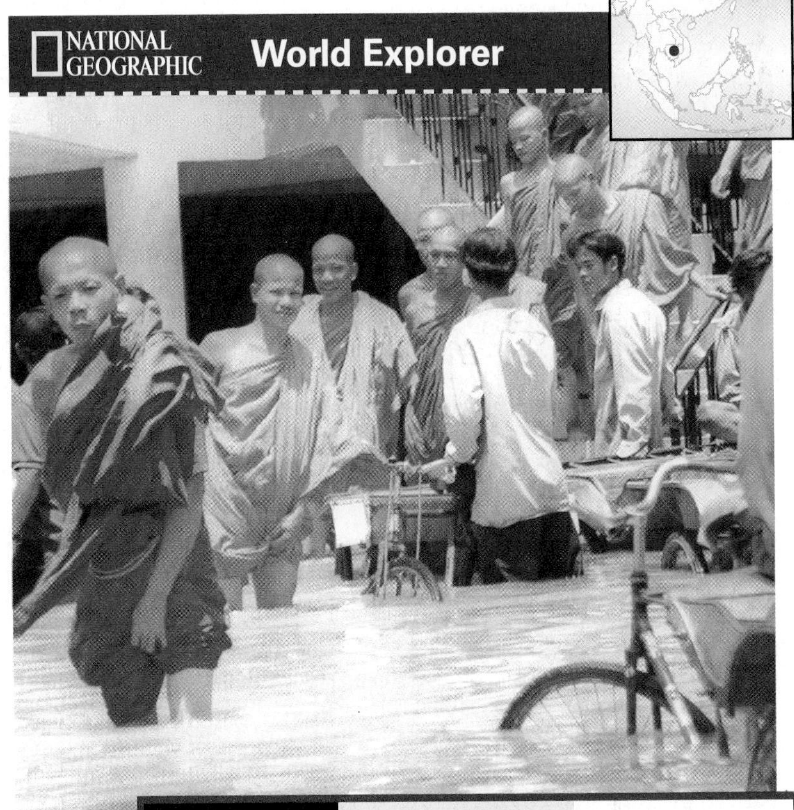

NATIONAL GEOGRAPHIC **World Explorer**

Geography Skills for Life **Flooding**

Southeast Asians, such as these Buddhist monks in Cambodia, cope with flooded streets during the rainy season.

Human-Environment Interaction What human activity contributes to the problem of flooding?

Floods and Typhoons

Flash floods in Southeast Asia kill hundreds of people a year and ruin about 10 million acres (4 million ha) of crops. Human activity often magnifies the effects of these floods. In 1991 and 1995, for example, major storms struck the Philippines. Because so much land had been cleared of forest, the storms caused widespread runoff and mudslides.

The rivers of mainland Southeast Asia undergo seasonal flooding every year. Flooding poses a particular threat to Bangkok, which is built on unstable land. Some sections of the city sink as much as 25 inches (64 cm) each year. The city's most recent serious flooding occurred in 1983, when one-fourth of its area was under water.

Tropical storms also often strike various parts of Southeast Asia. A **cyclone** is an area of low

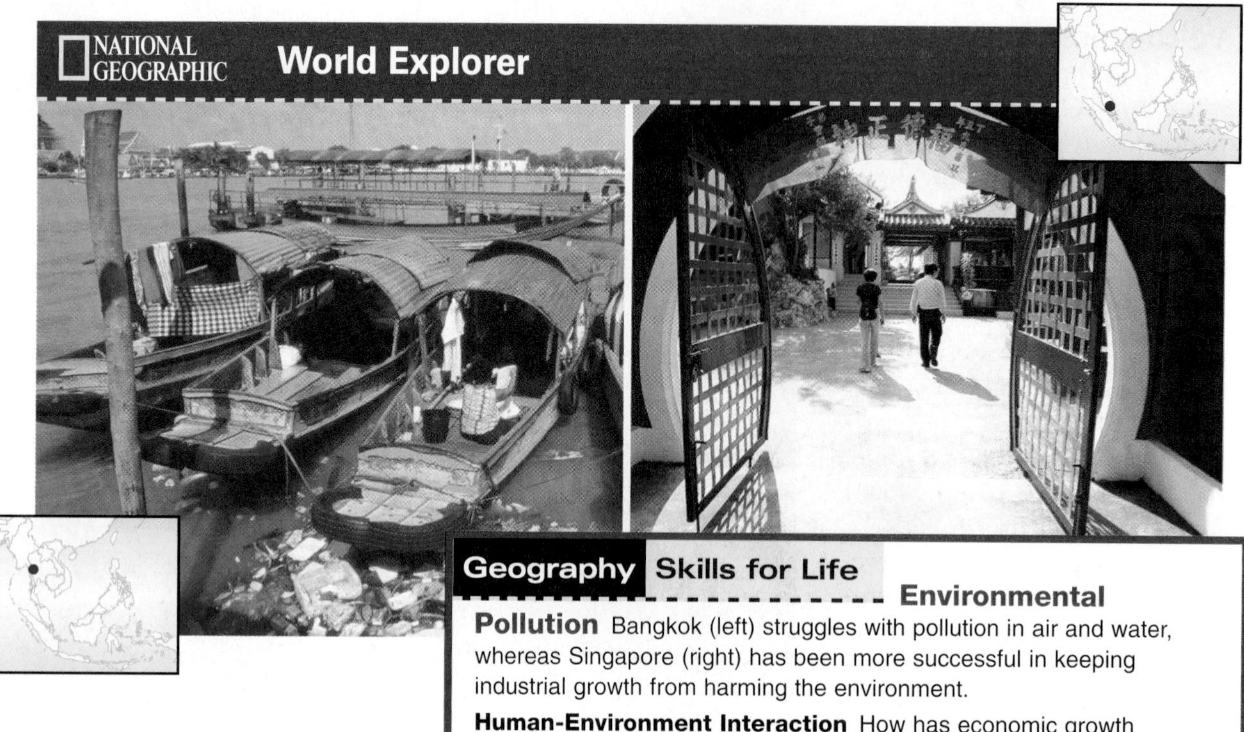

Geography Skills for Life

Environmental Pollution Bangkok (left) struggles with pollution in air and water, whereas Singapore (right) has been more successful in keeping industrial growth from harming the environment.

Human-Environment Interaction How has economic growth affected the environment and levels of pollution in Southeast Asia?

atmospheric pressure surrounded by circulating winds extending out from 100 to 1,000 miles (161 to 1,609 km). Tropical cyclones are particularly deadly storms. A **typhoon** is a tropical cyclone that forms in the Pacific Ocean 8° to 15° N of the Equator, often between July and November. Typhoon winds circulate in a counterclockwise direction.

Southeast Asia's typhoons form in the western Pacific Ocean, north of the island of New Guinea. Some travel north to Japan, while others move through the northern islands of the Philippines and then on to the Chinese mainland. Still others pass through the Philippines and reach Vietnam. Typhoons may have winds from 150 to 180 miles per hour (241 to 290 km per hour) and may be accompanied by rain, thunder, lightning, and high ocean waves that disrupt shipping.

Southeast Asians are taking steps to control the damage from typhoons. In Thailand, for example, planners in Bangkok are building dams to prevent typhoon-related flooding.

Environmental Pollution

Whether facing the commotion of a busy city, coaxing a modest crop from a small family plot, or taming a river's floodwaters to protect a cash crop, Southeast Asians, like people everywhere, affect their environments. In the face of technological advances and widespread air, water, and noise pollutants, Southeast Asia's people today try to balance environmental concerns with economic needs.

Cities

Increased prosperity in Southeast Asia has raised people's expectations about their quality of life. Economic growth, however, also depletes limited environmental resources. Increased manufacturing, for example, raises standards of living but also creates industrial waste. As societies become wealthier and more people buy automobiles, exhaust systems send toxic fumes into the air.

Growing populations and crowded conditions in cities such as Bangkok, Manila, and Jakarta raise concerns about adequate housing, water supplies, sanitation, and traffic control. Bangkok, for example, is a busy city of skyscrapers, factories, noisy expressways, and traffic jams. Dramatic population increases and industrialization even appear to be overheating Bangkok. In recent years, the city's heat, humidity, and pollution levels have increased at a rate higher than the global average. Higher temperatures affect both humans and their environment, causing health problems and trapping pollutants in the air that contribute to acid rain.

Because of strict law enforcement, Singapore is an exception in a region of polluted cities. One observer describes Singapore as a "world of almost surrealistic cleanliness and good behavior, prompted on every public wall by slogans of a watchful state." In Singapore, littering the sidewalk can bring a $250 fine.

Rural Areas

In some parts of Southeast Asia, pollution extends into the countryside, including the region's national parks. In Thailand's Tai Phi National Park, for example, 80 percent of the freshwater wells are contaminated as a result of poor waste disposal. The dumping of toxic wastes has created problems in other countries of Southeast Asia. In 1998, thousands of people in rural Cambodia fled their homes after finding out that tons of toxic materials, mislabeled as cement, had been dumped nearby. The waste, containing poisonous mercury, threatened their water supplies.

Volcanic eruptions and forest fires also pollute in rural areas, sometimes affecting cities as well. Forest fires on several islands of Indonesia in the 1990s, for example, created pollution and respiratory problems for people as far away as mainland Malaysia (see chart below). The smoke disrupted air traffic and shipping across the region.

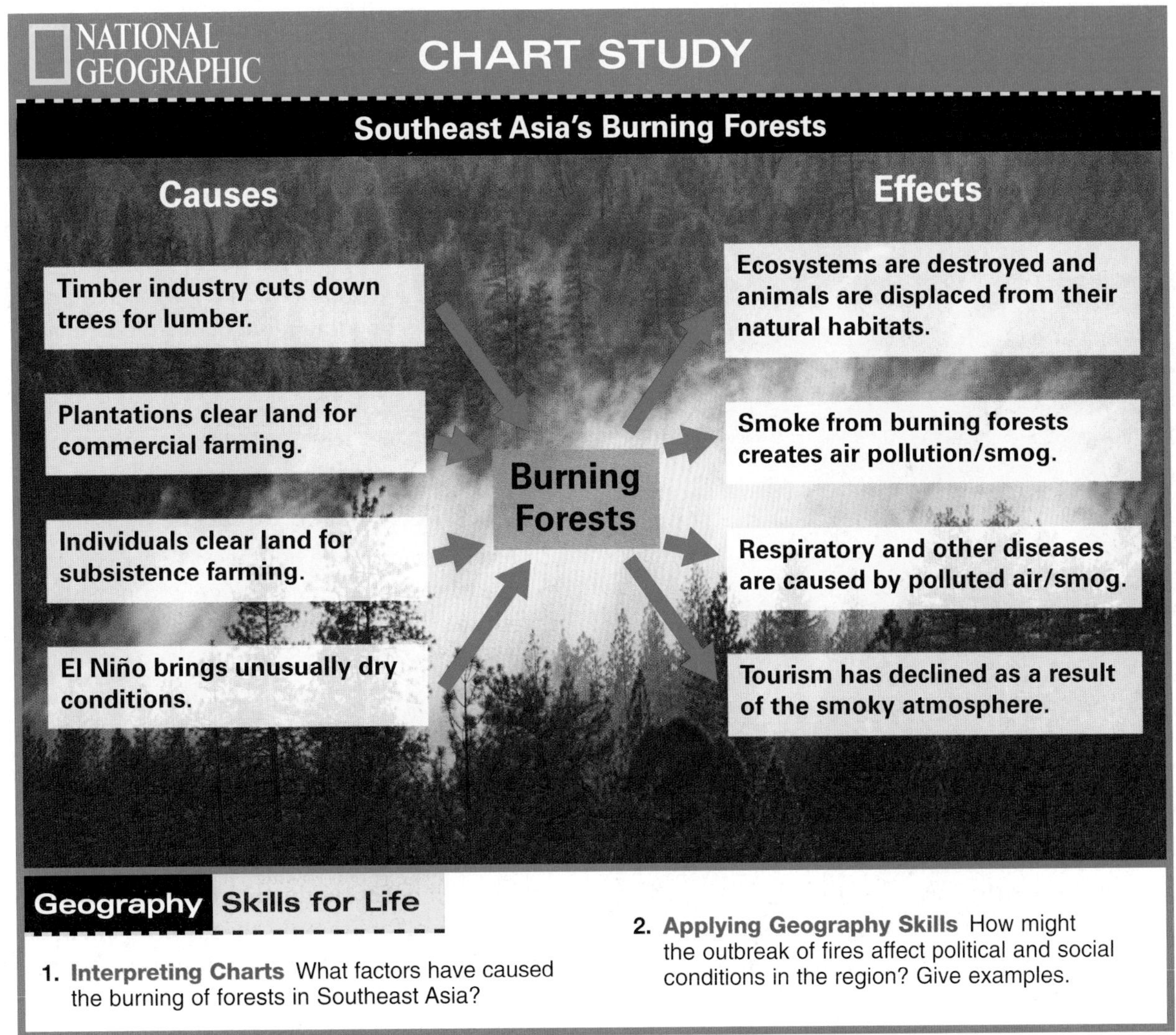

Geography Skills for Life

1. **Interpreting Charts** What factors have caused the burning of forests in Southeast Asia?
2. **Applying Geography Skills** How might the outbreak of fires affect political and social conditions in the region? Give examples.

Logging, Farming, and Mining

In Southeast Asia, some logging, farming, and mining practices have harmed the environment. The destruction of habitats and the demand for certain food items also endanger the region's wildlife.

Economies

Deforestation

Deforestation is a major concern throughout Southeast Asia. In Laos, Thailand, and Myanmar, teak and other timber provides important sources of income. Since the 1960s, commercial logging companies have set up modern logging processes and provided training and jobs for many Southeast Asians. The region's economies have benefited, but the widespread cutting of trees has steadily diminished the region's forests. Until recently, companies made few efforts to replant as they harvested. Without the trees' root systems, topsoil is no longer held in place. Heavy tropical rains easily erode topsoil, washing it into streams that crisscross the region. The topsoil clogs rivers and reduces the amount of water available for irrigation.

NATIONAL GEOGRAPHIC **World Explorer**

Geography Skills for Life

Avoiding Pollution A vendor sells air masks to protect against air pollution in Indonesia.

Place What are some of the negative results of burning forests for agriculture?

Excessive logging also has caused major flooding. Without forests to absorb downpours, flash floods on bare, muddy slopes have swept into valleys, killing hundreds of people and leaving thousands more injured and homeless.

Some farming methods contribute to deforestation and soil erosion. Throughout Southeast Asia, farmers carry out slash-and-burn agriculture—cutting down vegetation, burning it, and using the ashes for fertilizer. In highlands areas, farmers grow food crops by a method known as **shifting cultivation**, clearing forests to plant fields, cultivating the land for a few years, and then abandoning it. They then repeat the process in a new area.

Fires also have destroyed forested areas. Plantation owners in Southeast Asia often burn large areas of land in order to plant profitable cash crops. These fires are becoming more frequent, and they often destroy large areas of forests when, during periods of drought, they blaze out of control.

Mining

The mining of valuable minerals and metals has also led to environmental abuses. At Indonesia's largest gold mine, workers dump large amounts of rock waste into the Ajkwa River in Irian Jaya. This dumping will eventually divert the river from its original course, flooding more than 50 square miles (130 sq. km) of forest and displacing many people. Pollution from the rock waste has already begun to kill vegetation in the surrounding rain forest.

Environmental Protection

In recent years, some Southeast Asian countries have taken steps to protect their environments. To prevent further loss of rain forests, Thailand, Indonesia, the Philippines, and Malaysia have limited certain timber exports and have introduced reforestation programs. Such efforts, however, have proved difficult to enforce or carry out, and illegal logging is still taking its toll on the region's forests. Scientists predict that many unique environments—with their variety of plant and animal species—will be lost within a few years.

In Indonesia, for example, the government in the early 1980s introduced a plan to set aside large parts of the country as conservation areas. In recent years, however, this plan has been largely abandoned because of the government's grant of logging rights to timber companies and the outbreak of political turmoil.

In addition, illegal logging operations on the Indonesian islands of Sumatra and Kalimantan have destroyed much of the forests bordering national parks. Scientists visiting these areas state that the Indonesian government must enforce its own environmental laws, and the army may have to be used to stop illegal logging.

In other Southeast Asian countries, planned migration or resettlement has balanced environmental protection and economic development. Laos, for example, has tried to limit shifting cultivation by resettling highlands peoples on more fertile and arable plains.

Southeast Asian governments also are starting to deal with the impact of urban growth on the environment. Bangkok, Thailand, for example, is a major example of urban warming, caused by industrialization, crowded living and working areas, and the increased use of automobiles and other vehicles. To handle this problem, scientists have proposed several solutions. One includes the creation of "green zones," or areas within a city that are granted special environmental protection. Another suggests banning the construction of tall buildings near the sea, allowing winds to blow farther into the city and provide more ventilation. Despite enormous challenges, these and other proposals are helping Southeast Asians realize that they must work hard to protect the environment while developing their economies.

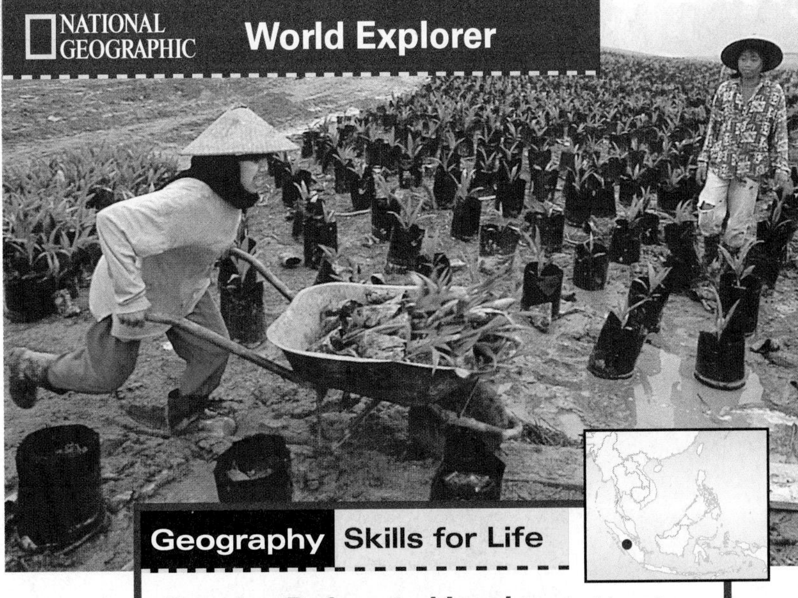

Farming Deforested Land Tropical forests continue to be cleared for settlement and agriculture, creating major ecological problems in Indonesia.

Human-Environment Interaction What steps have Southeast Asian countries taken to protect the forests?

SECTION 2 ASSESSMENT

Checking for Understanding

1. **Define** cyclone, typhoon, shifting cultivation.
2. **Main Ideas** On a cause-effect chart, fill in events, their causes, and their effects on Southeast Asia's environment.

Human Impact on the Environment		
Event	Cause	Effect

Critical Thinking

3. **Making Inferences** How do governments in Southeast Asia prepare for natural disasters?
4. **Predicting Consequences** What might be the consequences of clearing rain forests for housing in Indonesia?
5. **Drawing Conclusions** What might happen as Southeast Asia's less-industrialized countries develop manufacturing industries? Explain.

Analyzing Charts

6. **Human-Environment Interaction** Study the chart on page 769. What types of economic activities cause the burning of forests in Southeast Asia?

Applying Geography

7. **Using Resources** Evaluate the geographic impact of Southeast Asian government policies related to the use of resources.

Viewpoint

CASE STUDY on the Environment

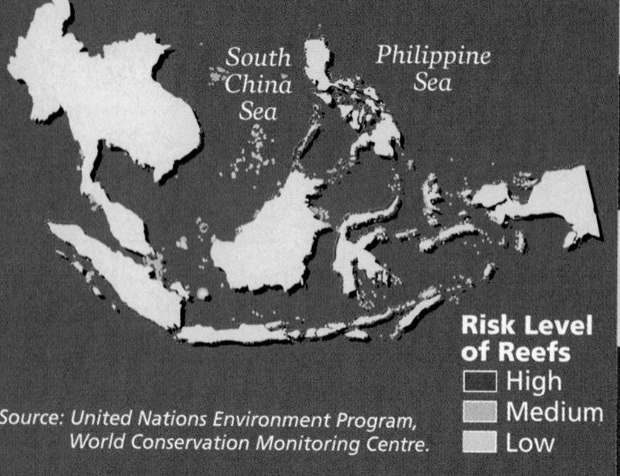

Source: United Nations Environment Program, World Conservation Monitoring Centre.

Southeast Asia's Reefs: Coral in Peril

Like underwater cities, coral reefs swarm with life in the warm, shallow oceans banding the Equator. Home to a fourth of all known marine species, coral reefs rival tropical rain forests in biodiversity. Like rain forests, too, coral reefs are at risk worldwide. Nowhere are these fragile habitats in more danger than in Southeast Asia, where local people use poisons and explosives to capture certain kinds of fish. These fish supply Asian restaurants and a world-wide aquarium industry. The trade generates huge profits for many people in the region, but their fishing methods are destroying the reefs.

Suspended in crystal clear waters, an Indonesian fisherman (left) drifts over a bamboo fish trap nestled among colorful corals. Corals look like rocks, but they are actually colonies of tiny animals called coral polyps. Each polyp secretes a limestone cup around itself, forming a limestone skeleton. The polyps attach to the skeletons of dead polyps, and the limestone formations—called coral reefs—grow larger and larger. Despite their rock-hard appearance, however, coral reefs are easily damaged. Worldwide, reefs are dying because of pollution, overfishing, coastal development, diseases, and rising ocean temperatures.

The reefs of Southeast Asia are among the world's richest—and some of the most endangered. The thriving trade in live reef fishes is largely to blame. Colorful tropical fish captured on coral reefs end up in collectors' aquariums. Fashionable restaurants in Hong Kong, Taiwan, and China are also to blame. Diners there pay big money to select live reef fish from a tank, and then have them killed, cooked, and served on the spot.

To capture fish alive, many Southeast Asian fishers use dynamite to stun fish—a technique called blast fishing. In addition, fishers squirt fish with cyanide. The poison temporarily immobilizes a fish, making it easy to catch. Explosives and cyanide, however, kill countless other fish and ocean creatures, including corals. Scientists estimate that 50 percent of Indonesian reefs and 80 percent of Philippine reefs are turning into aquatic graveyards.

◀ Philippine divers squirt deadly cyanide to stun and net fish—a practice that also kills corals.

Environmentalists claim that the use of explosives and cyanide to capture live reef fish is destroying entire coral reef ecosystems. Scientists warn that the loss of coral reefs will harm other ocean food chains, which could affect oceans worldwide. Concerned officials point out that once the reefs are gone, local communities will lose a primary source of food and a lucrative tourist industry.

Fishers working in the live reef-fish trade see it as an opportunity to raise their standard of living. Cyanide fishers often earn three times the salary of college-educated workers in the region. Some fishers claim that if they don't catch the reef fish now in demand, others will. Local people point out that Southeast Asia's coral reefs belong to Southeast Asian countries and that decisions about how to use their reef resources should be made by people living in the region.

What's Your Point of View?
Should people around the world be concerned about Southeast Asia's coral reefs? What might be done to restrict trade in live reef fish?

A coral grouper (below) lives and feeds on coral reefs in the Philippines. Many such fish end up in skillets, as Asian diners (right) pay top dollar for live reef fish. ▼

Drawing Conclusions

Drawing conclusions involves studying facts and details to understand how they are related and what they mean. By putting this information together, you can better understand an action or event.

Learning the Skill

When you draw a conclusion, you use facts, observation, and experience to form a judgment about an event. Drawing conclusions allows you to understand indirectly stated ideas and events, so you can apply your knowledge to similar situations.

Sometimes, however, people draw incorrect conclusions based on the information they have. Often the facts and details of a situation could logically lead to more than one conclusion. For example, if you see someone sweating, you might conclude that the person has been exercising. You might also conclude that this person may be sick with a fever. To determine which conclusion is correct, you would need to obtain and evaluate further information. To determine the accuracy of any conclusion, it is important to gather information that will prove or disprove it.

Follow these steps when drawing conclusions:

- **Review the facts that are stated directly.**
- **Use your own knowledge, experience, and insight to form conclusions about the facts.**
- **Find information that would help prove or disprove your conclusions.**

Asked to form an expedition overnight, Tu'o is ready by dawn. With six young Penan as our companions and Tu'o as guide, we leave Long Iman [in Malaysia] traveling up the Tutok by longboat to reach a trail that climbs steeply through gingers and wild durian [fruit]. . . .

For two long days we walk farther into the forest, following a route that rises and falls with each successive ridge. Delighted to be away from the settlement, the Penan watch the forest for signs, hunting hornbills at dusk, tracking deer and sun bears, gathering ripe fruits of mango trees. On the third morning our party crests a steep hill; we have reached the nomads. It is just after dawn, and the sound of the gibbons howling runs across the canopy. Smoke from cooking fires mingles with the cool mist. A hunting party returns. Tu'o bows his head in morning prayer. "Thank you for the sun rising, for the trees and the forest of abundance, the trees that were not made by man, but by you."

—*Wade Davis, "Vanishing Cultures,"*
***National Geographic**, August 1999*

Practicing the Skill

Read the excerpt above about traveling with the Penan people of Malaysia, and answer the following questions.

1. What important facts does the author include?
2. What information does Tu'o's morning prayer provide about his attitude toward the forest?
3. What conclusion can you draw about the Penan people based on this author's description?
4. What evidence do you have to support this conclusion?

Bring to class an article from a magazine, newspaper, or Internet source describing a current conflict in Southeast Asia. Using the steps on this page, draw conclusions about the causes of the conflict and its likely outcome. Summarize your conclusions in a paragraph.

Go To The **Glencoe Skillbuilder Interactive Workbook, Level 2** provides instruction and practice in key social studies skills.

CHAPTER 31

SUMMARY & STUDY GUIDE

SECTION 1 Living in Southeast Asia (pp. 759–765)

Terms to Know

- paddy
- sickle
- subsistence crop
- cash crop
- lode
- interdependent
- Association of Southeast Asian Nations (ASEAN)
- free port

Key Points

- Agriculture is the leading economic activity in Southeast Asia.
- The countries of the region are industrializing at different rates, which causes great variation in economies, occupations, transportation, and communications.
- Through ASEAN and other organizations that were formed to promote regional development and trade, the countries of Southeast Asia are becoming more interdependent.

Organizing Your Notes

Create an outline using the format below to help you organize your notes for this section.

Living in Southeast Asia

I. Agriculture
  A. Rice cultivation
    1. Most important crop

SECTION 2 People and Their Environment (pp. 766–771)

Terms to Know

- cyclone
- typhoon
- shifting cultivation

Key Points

- Volcanic eruptions, flash floods, and typhoons have serious effects on Southeast Asians' lives.
- Industrialization and economic development in Southeast Asia often result in the pollution of air, land, and water.
- The region's countries are taking steps to protect the environment.

Organizing Your Notes

Use a web diagram like the one below to help you organize important details from this section.

A Thai family stands by their house, built on a canal in Bangkok.

CHAPTER 31

ASSESSMENT & ACTIVITIES

Reviewing Key Terms

Write the key term that best completes each of the following sentences. Refer to the Terms to Know in the Summary & Study Guide on page 775.

1. Irian Jaya has rich ________, or deposits of minerals.
2. ____________ in the Philippines include coconuts and sugarcane.
3. A flooded field in which rice is grown is a ________.
4. Countries involved in oceangoing trade are attracted by a(n) _______ such as Singapore.
5. By becoming ________, the countries of Southeast Asia can build the region's economy.
6. Indonesia, Malaysia, the Philippines, Singapore, and Thailand were the first countries to join the ________.
7. Southeast Asia experiences a specific kind of ________ known as a typhoon.
8. A hand tool called a(n) _______ is still used to harvest crops.
9. In ________ farmers abandon their fields after a few years.

Reviewing Facts

SECTION 1

1. What is the occupation of most people in Southeast Asia?
2. What types of industries are the countries of Southeast Asia developing?

SECTION 2

3. List three negative impacts of industrialization in Southeast Asia. Give examples.
4. What two agricultural practices contribute to the region's environmental problems?

Critical Thinking

1. **Identifying Cause and Effect** What effects might continued mining and logging have on Irian Jaya's people?
2. **Making Inferences** How does interdependence help Southeast Asia's economic development and trade?
3. **Comparing and Constrasting** Using a Venn diagram like the one below, compare the environmental challenges that occur in Southeast Asian cities and rural areas.

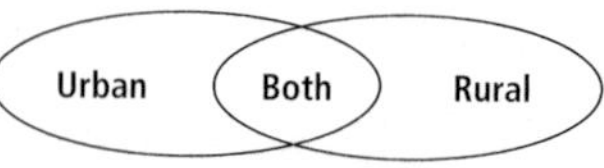

NATIONAL GEOGRAPHIC

Locating Places

Southeast Asia: Physical-Political Geography

Match the letters on the map with the places and physical features of Southeast Asia. Write your answers on a sheet of paper.

1. Chao Phraya River
2. Brunei
3. Myanmar
4. Phnom Penh
5. Bali
6. East Timor
7. Hanoi
8. Pacific Ocean
9. Jakarta
10. New Guinea

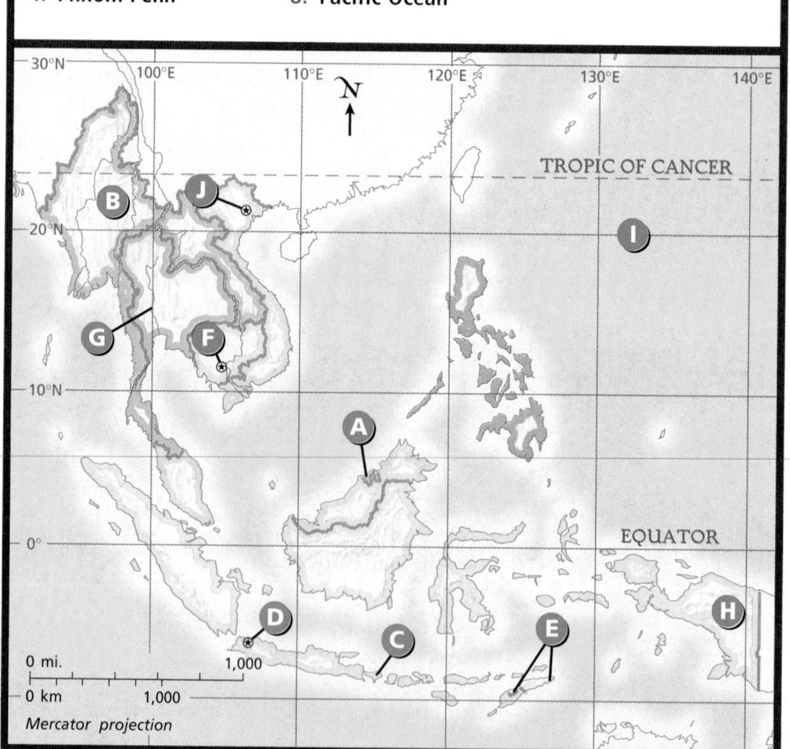

Self-Check Quiz Visit the **Glencoe World Geography** Web site at geography.glencoe.com and click on **Self-Check Quizzes—Chapter 31** to prepare for the Chapter Test.

Using the Regional Atlas

Refer to the Regional Atlas on pages 710–713.

1. **Location** What body of water separates northern Vietnam from the South China Sea?
2. **Human-Environment Interaction** What are the two main types of land use in mountainous areas in Southeast Asia?

Thinking Like a Geographer

Think about the physical geography of Southeast Asia. What is one of the region's important physical assets? As a geographer, how might you suggest that people utilize this asset to improve their lives?

Problem-Solving Activity

Group Research Project Work with a group to research how people in various world regions adapt to or modify their environment in order to control flooding. Compare methods of flood control in other regions with those used in Southeast Asia. Evaluate whether a method used in another region is workable in Southeast Asia.

GeoJournal

Expository Writing Using the information you logged in your GeoJournal as you read this chapter, choose one economic or environmental characteristic of Southeast Asia. Then write a short essay that compares and contrasts the characteristic among the region's countries. Use your textbook and the Internet as resources.

Technology Activity

Creating an Electronic Database Make a fact sheet for each Southeast Asian country. Include data about agriculture, industries, transportation, communications, and environmental challenges. Put this information into a database, and then write a paragraph comparing and contrasting two Southeast Asian countries. If possible, create graphic elements such as bar graphs or circle graphs to support your conclusions.

Standardized Test Practice

Study the bar graph below. Then choose the best answer for the multiple-choice question. If you have trouble answering the question, use the process of elimination to narrow your choices.

Media in Selected Southeast Asian Countries

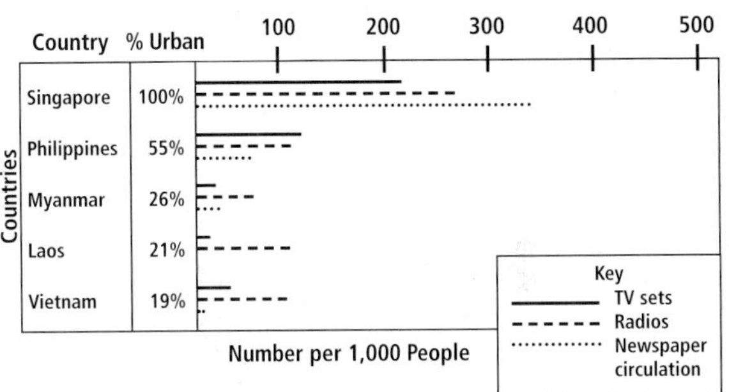

Source: The World Almanac and Book of Facts, 2000

1. **What conclusions can you draw from the graph?**

   A Newspapers are censored in Laos.
   B People in rural areas have less access to literacy programs.
   C Urban populations have more access to news sources.
   D Televisions are less expensive in Singapore than in other countries in Southeast Asia.

Study the title and labels on the graph to see what information is being presented. Note the important facts, and look at the relationships among the countries. Remember, an answer choice that may be true cannot be the correct answer if there is no information to support it in the graph.

UNIT 11

Australia, Oceania, and Antarctica

WHY IT'S IMPORTANT—

Vast and sparsely populated, the region of Australia, Oceania, and Antarctica is perhaps the most diverse of the world's regions. Parts of the region—Australia and Oceania—are developing close economic ties to other countries in the Pacific Rim, the area bordering the Pacific Ocean. Such ties to prosperous Pacific Rim nations will influence global trade and trading networks for decades to come. Cold, icy Antarctica lacks a permanent human population, but the data gathered there by scientists will broaden your understanding of the world's climates and resources in the years ahead.

World Regions Video

To learn more about Australia, Oceania, and Antarctica and their impact on your world, view the World Regions video "Australia, Oceania, and Antarctica."

Penguins in Antarctica

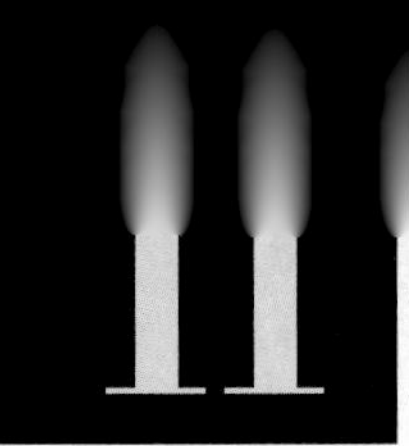

What Makes Australia, Oceania, and Antarctica a Region?

Both palm trees and polar ice lie within this diverse region that includes two continents—Australia and Antarctica—and some 25,000 islands scattered across vast expanses of the Pacific Ocean.

Australia is ancient and arid. Low mountains curve down its eastern coast, blocking rainfall to the flat interior where scrubland and deserts form what Australians call the "outback." Across the Tasman Sea lies New Zealand—lush, green, and mountainous. North of New Zealand's rugged shores lies the rest of Oceania, where groups of tropical islands dot the blue ocean waters like tiny jewels.

A different sort of jewel lies far to the south of New Zealand—Antarctica, a glittering kingdom of ice and snow that sits astride the bottom of the world.

1 Shouldering the day's catch, a spear fisher in the Cook Islands watches his companion take aim. The Cook Islands spread across 850,000 square miles (2.2 million sq. km) of ocean. Like people throughout Oceania, Cook Islanders depend on the sea for food.

2 A loaf-shaped mass of sandstone, Uluru (also known as Ayers Rock) looms over the flat landscape of central Australia. Uluru is sacred to Aborigines, the country's native inhabitants. Aborigines share their homeland with Australia's other natives—kangaroos and other animals found nowhere else on Earth.

3 Fringed with coral reefs, the thickly forested islands of Palau seem to float on the surface of blue Pacific waters. Palau is a chain of about 200 islands a few hundred miles east of the Philippines. Its coral reefs are among the world's most biologically diverse.

4 A sea of sheep parts for two bicyclists on New Zealand's South Island. Sheep greatly outnumber people in New Zealand, where pastures thrive in a climate that is mild year-round. The nation ranks as one of the world's leading producers of lamb, mutton, and wool.

11

Lands Down Under

Europeans were latecomers to this region, much of which lies "down under" the Equator. Australia's original settlers were Aborigines; the first settlers of New Zealand were the Maori. During the 1800s, the British colonized both lands. Today, Australia and the islands of Oceania are a blend of European, traditional Pacific, and Asian cultures. Antarctica has no permanent human inhabitants.

Although huge livestock ranches spread across Australia and New Zealand, life in these two countries is largely urban, with most people living in coastal cities. For many Pacific Islanders, life is more traditional, and people support themselves mainly through fishing and subsistence farming.

1 **A blend of cultures** is reflected by an Aborigine wearing western-style clothing. Aborigines are thought to have arrived in Australia from Asia at least 40,000 years ago. Today, many Aborigines are striving to preserve their ancient traditions while living in a modern world.

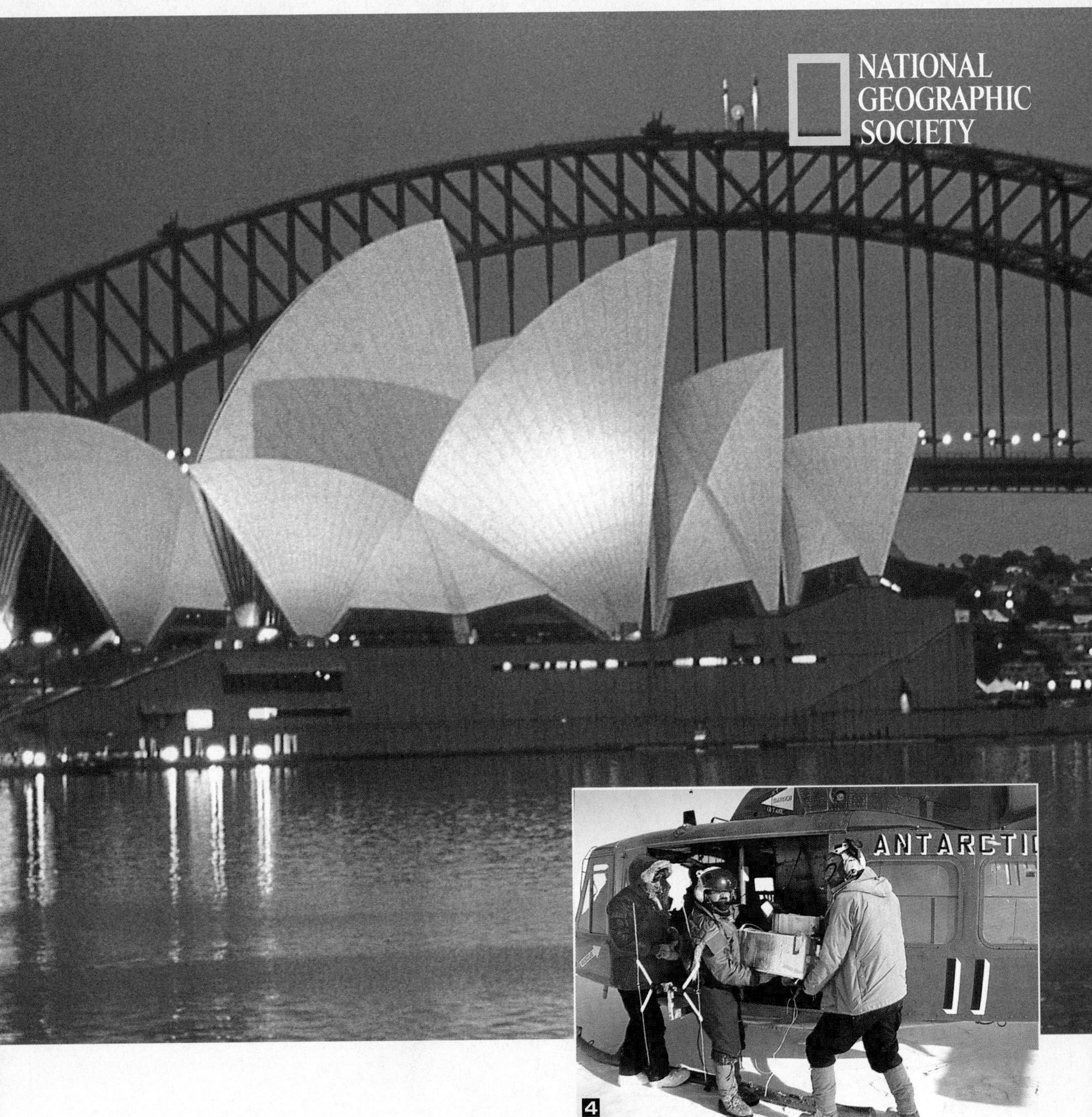

2 Thatched houses called *fale* sit beneath palm trees on an island in Samoa. Open sides allow cool ocean breezes to blow through the houses. Blinds made of palm leaves can be let down to keep out rain or glaring sun.

3 Mirrored in the waters of Sydney Harbor, the Sydney Opera House glows as evening falls. The white shells that form the building's roof resemble billowing sails—a fitting tribute to the city that is Australia's busiest seaport.

4 Bundled against the cold, scientists in Antarctica load equipment into a waiting helicopter. Antarctica is a continent reserved almost entirely for scientific research and exploration. The United States is one of many countries operating research stations here.

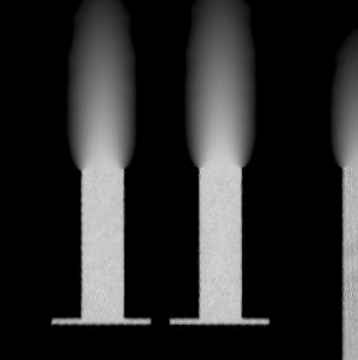

Australia, Oceania, and Antarctica

PHYSICAL

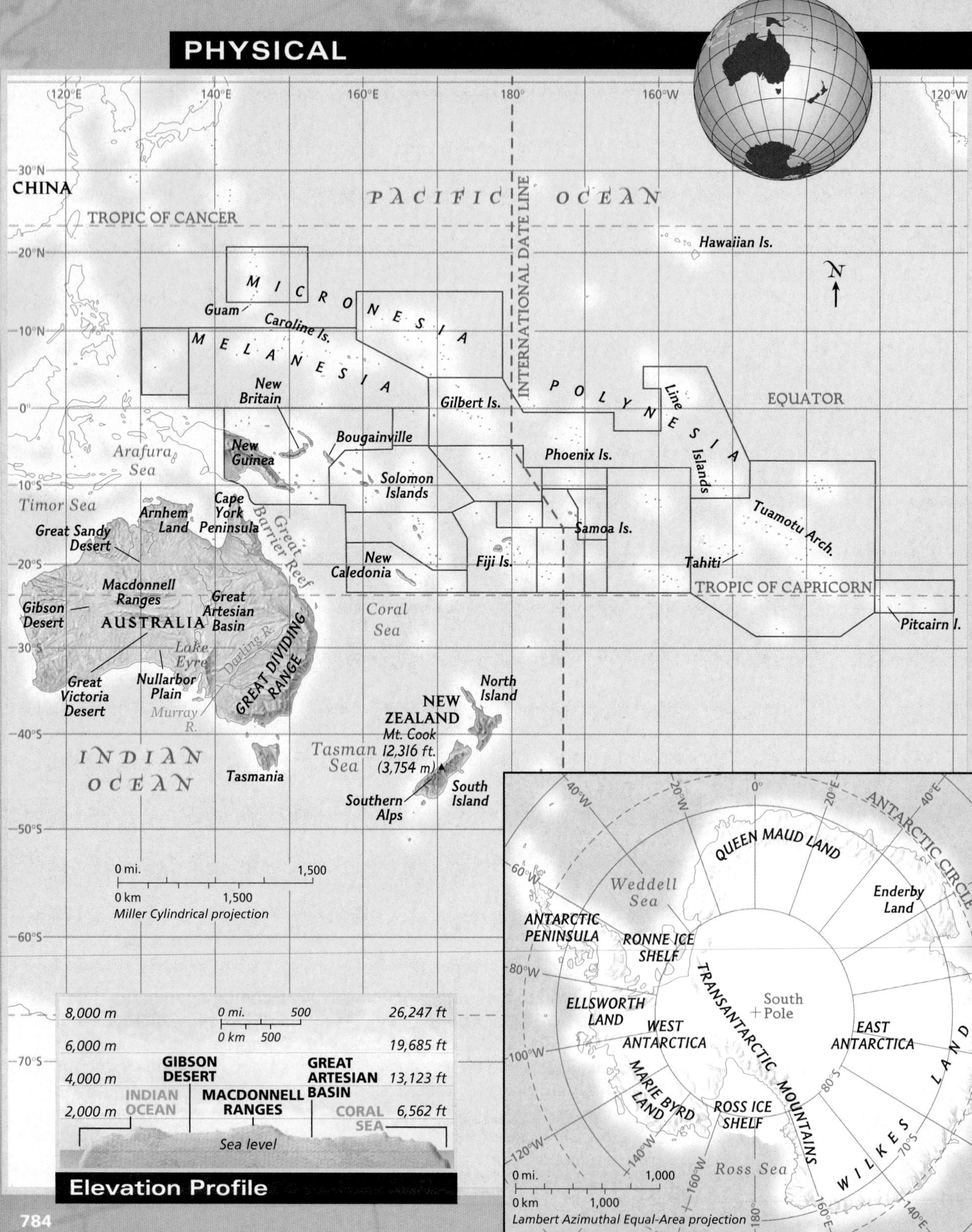

120°E 140°E 160°E 180° 160°W 120°W
30°N 20°N 10°N 0° 10°S 20°S 30°S 40°S 50°S 60°S 70°S

CHINA
TROPIC OF CANCER
PACIFIC OCEAN
INTERNATIONAL DATE LINE
Hawaiian Is.
N
MICRONESIA
Guam
Caroline Is.
MELANESIA
New Britain
Gilbert Is.
POLYNESIA
Line Islands
EQUATOR
Bougainville
New Guinea
Arafura Sea
Phoenix Is.
Solomon Islands
Timor Sea
Arnhem Land
Cape York Peninsula
Great Barrier Reef
Great Sandy Desert
Samoa Is.
Tuamotu Arch.
Tahiti
New Caledonia
Fiji Is.
TROPIC OF CAPRICORN
Macdonnell Ranges
Great Artesian Basin
Gibson Desert
AUSTRALIA
Coral Sea
Pitcairn I.
Lake Eyre
Darling R.
GREAT DIVIDING RANGE
Great Victoria Desert
Nullarbor Plain
Murray R.
North Island
NEW ZEALAND
Mt. Cook 12,316 ft. (3,754 m)
INDIAN OCEAN
Tasmania
Tasman Sea
South Island
Southern Alps

0 mi. 1,500
0 km 1,500
Miller Cylindrical projection

8,000 m — 26,247 ft
6,000 m — 19,685 ft
4,000 m — 13,123 ft
2,000 m — 6,562 ft
0 mi. 500
0 km 500
INDIAN OCEAN
GIBSON DESERT
MACDONNELL RANGES
GREAT ARTESIAN BASIN
CORAL SEA
Sea level

Elevation Profile

40°W 20°W 0° 20°E 40°E 60°E 80°E 100°E 120°E 140°E 160°E 180° 160°W 140°W 120°W 100°W 80°W 60°W 70°S 80°S 60°S
ANTARCTIC CIRCLE
QUEEN MAUD LAND
Weddell Sea
Enderby Land
ANTARCTIC PENINSULA
RONNE ICE SHELF
TRANSANTARCTIC MOUNTAINS
South Pole
ELLSWORTH LAND
WEST ANTARCTICA
EAST ANTARCTICA
MARIE BYRD LAND
ROSS ICE SHELF
WILKES LAND
Ross Sea
0 mi. 1,000
0 km 1,000
Lambert Azimuthal Equal-Area projection

NATIONAL GEOGRAPHIC SOCIETY

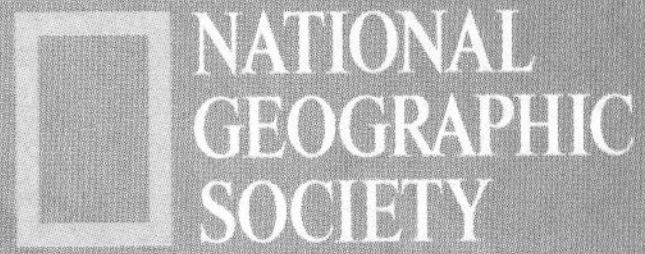

POLITICAL

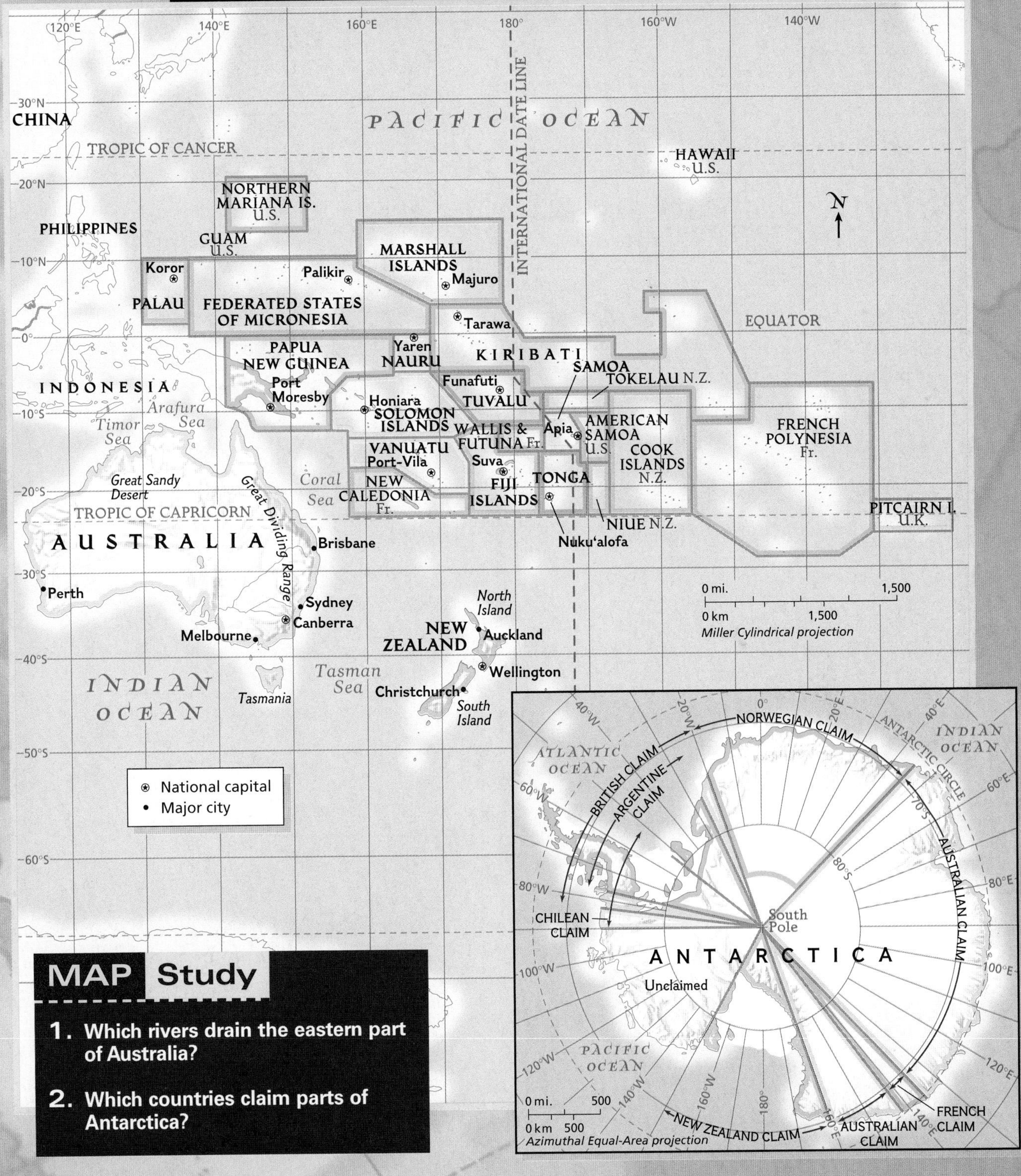

120°E
140°E
160°E
180°
160°W
140°W
30°N
20°N
10°N
0°
10°S
20°S
30°S
40°S
50°S
60°S
CHINA
PACIFIC OCEAN
INTERNATIONAL DATE LINE
TROPIC OF CANCER
HAWAII U.S.
NORTHERN MARIANA IS. U.S.
PHILIPPINES
GUAM U.S.
MARSHALL ISLANDS
Koror
Palikir
Majuro
PALAU
FEDERATED STATES OF MICRONESIA
Tarawa
EQUATOR
Yaren
NAURU
PAPUA NEW GUINEA
KIRIBATI
SAMOA
TOKELAU N.Z.
INDONESIA
Port Moresby
Funafuti
TUVALU
Arafura Sea
Timor Sea
Honiara
SOLOMON ISLANDS
WALLIS & FUTUNA Fr.
Apia
AMERICAN SAMOA U.S.
FRENCH POLYNESIA Fr.
VANUATU
Port-Vila
Suva
COOK ISLANDS N.Z.
Great Sandy Desert
Coral Sea
NEW CALEDONIA Fr.
FIJI ISLANDS
TONGA
Great Dividing Range
TROPIC OF CAPRICORN
PITCAIRN I. U.K.
NIUE N.Z.
AUSTRALIA
Brisbane
Nuku'alofa
Perth
Sydney
Canberra
North Island
NEW ZEALAND
Auckland
Melbourne
Wellington
INDIAN OCEAN
Tasman Sea
Christchurch
South Island
Tasmania
0 mi. 1,500
0 km 1,500
Miller Cylindrical projection
National capital
Major city
0°
20°W
20°E
40°W
40°E
60°W
60°E
80°W
80°E
100°W
100°E
120°W
120°E
140°W
140°E
160°W
160°E
180°
70°S
80°S
NORWEGIAN CLAIM
ANTARCTIC CIRCLE
INDIAN OCEAN
ATLANTIC OCEAN
BRITISH CLAIM
ARGENTINE CLAIM
AUSTRALIAN CLAIM
CHILEAN CLAIM
South Pole
ANTARCTICA
Unclaimed
PACIFIC OCEAN
0 mi. 500
0 km 500
Azimuthal Equal-Area projection
NEW ZEALAND CLAIM
AUSTRALIAN CLAIM
FRENCH CLAIM

MAP Study

1. Which rivers drain the eastern part of Australia?
2. Which countries claim parts of Antarctica?

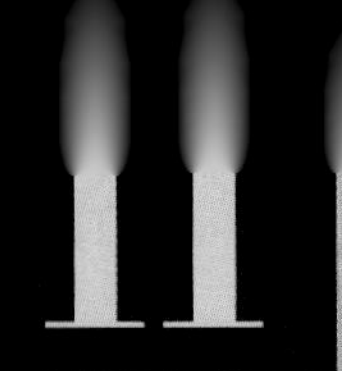

Australia, Oceania, and Antarctica

POPULATION DENSITY

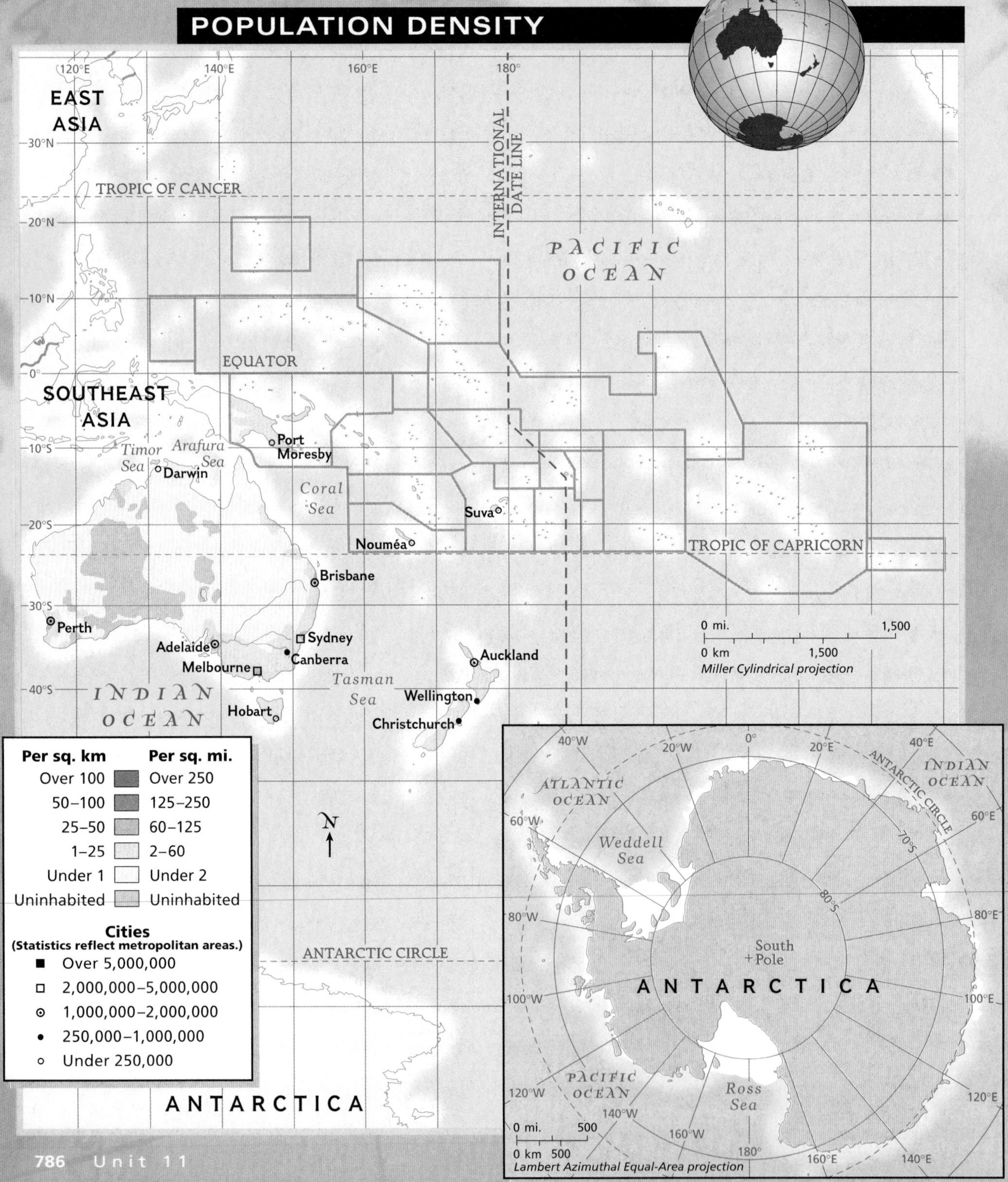

EAST ASIA

SOUTHEAST ASIA

PACIFIC OCEAN

INDIAN OCEAN

TROPIC OF CANCER

EQUATOR

TROPIC OF CAPRICORN

ANTARCTIC CIRCLE

INTERNATIONAL DATE LINE

120°E 140°E 160°E 180°

30°N 20°N 10°N 0° 10°S 20°S 30°S 40°S

Timor Sea

Arafura Sea

Coral Sea

Tasman Sea

Port Moresby

Darwin

Suva

Nouméa

Brisbane

Perth

Adelaide

Sydney

Canberra

Melbourne

Auckland

Wellington

Hobart

Christchurch

ANTARCTICA

N

0 mi. 1,500

0 km 1,500

Miller Cylindrical projection

Per sq. km	Per sq. mi.
Over 100	Over 250
50–100	125–250
25–50	60–125
1–25	2–60
Under 1	Under 2
Uninhabited	Uninhabited

Cities
(Statistics reflect metropolitan areas.)

- ■ Over 5,000,000
- □ 2,000,000–5,000,000
- ⊙ 1,000,000–2,000,000
- • 250,000–1,000,000
- ○ Under 250,000

ATLANTIC OCEAN

INDIAN OCEAN

PACIFIC OCEAN

Weddell Sea

Ross Sea

ANTARCTIC CIRCLE

South Pole

ANTARCTICA

40°W 20°W 0° 20°E 40°E 60°E 80°E 100°E 120°E 140°E 160°E 180° 160°W 140°W 120°W 100°W 80°W 60°W

70°S 80°S

0 mi. 500

0 km 500

Lambert Azimuthal Equal-Area projection

ECONOMIC ACTIVITY

Resources

- Petroleum
- Uranium
- Coal
- Iron ore
- Lead
- Manganese
- Nickel
- Zinc
- Gold
- Silver

0 mi. 1,500
0 km 1,500
Miller Cylindrical projection

Land Use

- Commercial farming
- Subsistence farming
- Livestock raising
- Hunting and gathering
- Forests
- Manufacturing and trade
- Commercial fishing
- Little or no activity

0 mi. 500
0 km 500
Lambert Azimuthal Equal-Area projection

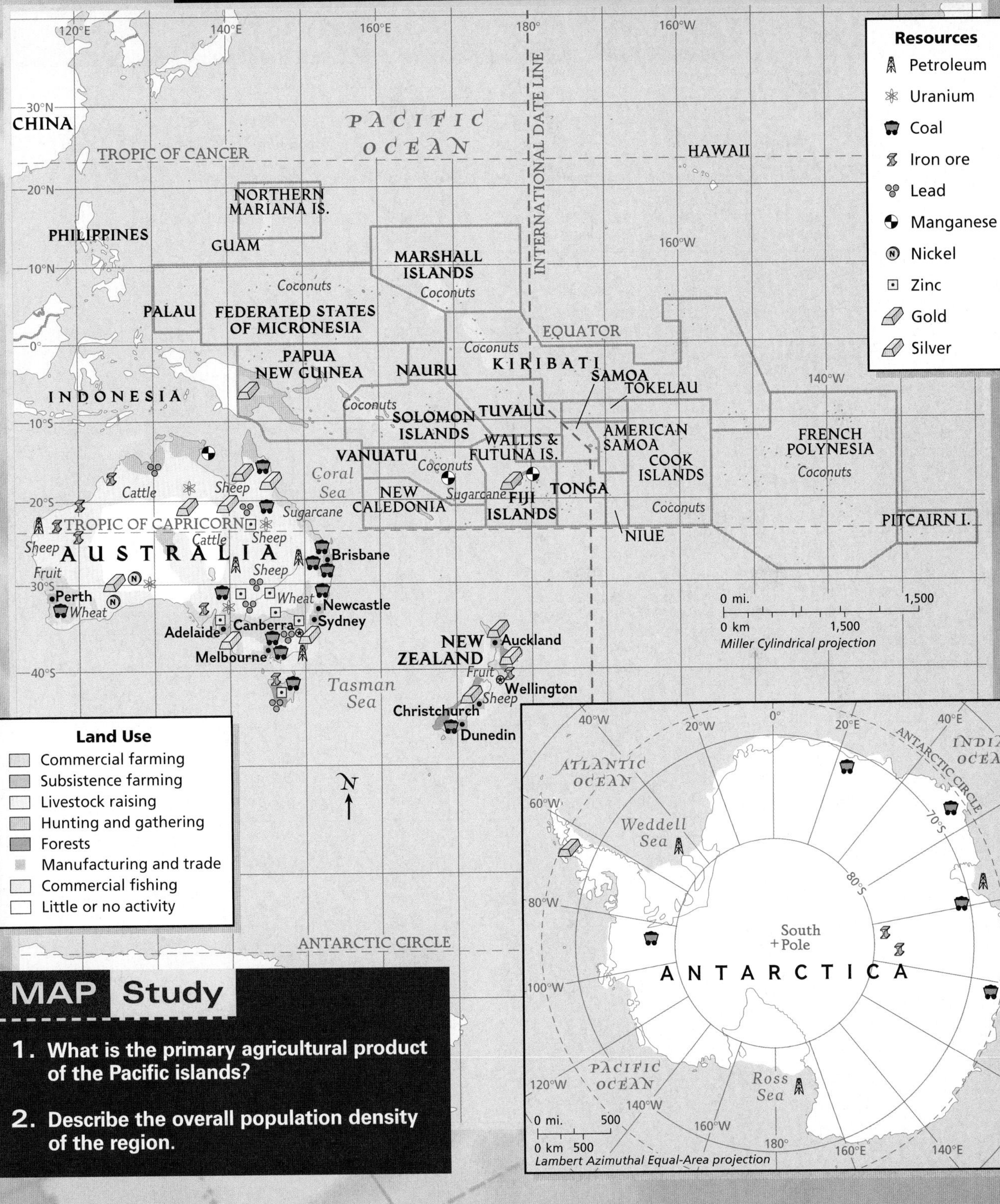

MAP Study

1. What is the primary agricultural product of the Pacific islands?
2. Describe the overall population density of the region.

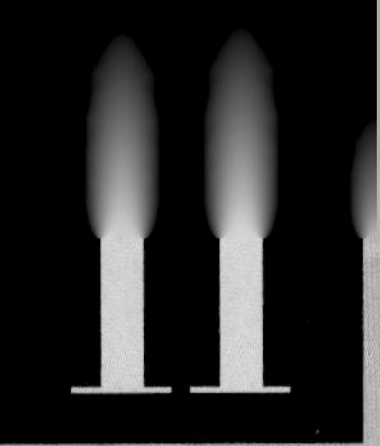

Australia, Oceania, and Antarctica

COUNTRY PROFILES

COUNTRY * AND CAPITAL	FLAG AND LANGUAGE	POPULATION AND DENSITY	LANDMASS	MAJOR EXPORT	MAJOR IMPORT	CURRENCY	GOVERNMENT
AUSTRALIA Canberra	English	19,400,000 6 per sq. mi. 2 per sq. km	2,988,888 sq. mi. 7,741,220 sq. km	Coal	Machinery	Australian Dollar	Parliamentary Democracy
FEDERATED STATES OF MICRONESIA Palikir	English, Local Languages	100,000 444 per sq. mi. 111 per sq. km	270 sq. mi. 699 sq. km	Fish	Foods	U.S. Dollar	Republic
FIJI Suva	English, Fijian, Hindi	800,000 119 per sq. mi. 46 per sq. km	7,054 sq. mi. 18,270 sq. km	Sugar	Machinery	Fiji Dollar	Republic
KIRIBATI Tarawa	English, Gilbertese	100,000 337 per sq. mi. 130 per sq. km	282 sq. mi. 730 sq. km	Coconut Products	Foods	Australian Dollar	Republic
MARSHALL ISLANDS Majuro	English, Local Languages	100,000 1,007 per sq. mi. 389 per sq. km	69 sq. mi. 179 sq. km	Coconut Products	Foods	U.S. Dollar	Republic
NAURU Yaren	Nauruan, English	10,000 1,412 per sq. mi. 545 per sq. km	9 sq. mi. 23 sq. km	Phosphates	Foods	Australian Dollar	Republic
NEW ZEALAND Wellington	English	3,900,000 37 per sq. mi. 14 per sq. km	104,452 sq. mi. 270,531 sq. km	Wool	Machinery	New Zealand Dollar	Parliamentary Democracy
PALAU Koror	English, Palauan	20,000 107 per sq. mi. 41 per sq. km	178 sq. mi. 461 sq. km	Fish	Machinery	U.S. Dollar	Republic
PAPUA NEW GUINEA Port Moresby	English, Local Languages	5,000,000 28 per sq. mi. 11 per sq. km	178,703 sq. mi. 462,841 sq. km	Gold	Machinery	Kina	Parliamentary Democracy
SAMOA Apia	Samoan, English	200,000 155 per sq. mi. 60 per sq. km	1,097 sq. mi. 2,841 sq. km	Coconut Products	Foods	Tala	Constitutional Monarchy

* COUNTRIES AND FLAGS NOT DRAWN TO SCALE

FOR AN ONLINE UPDATE OF THIS INFORMATION, VISIT GEOGRAPHY.GLENCOE.COM AND CLICK ON "TEXTBOOK UPDATES."

COUNTRY * AND CAPITAL	FLAG AND LANGUAGE	POPULATION AND DENSITY	LANDMASS	MAJOR EXPORT	MAJOR IMPORT	CURRENCY	GOVERNMENT
SOLOMON ISLANDS Honiara	English, Local Languages	500,000 41 per sq.mi. 16 per sq.km	11,158 sq.mi. 28,899 sq.km	Cocoa	Machinery	Solomon Islands Dollar	Parliamentary Democracy
TONGA Nuku'alofa	Tongan, English	100,000 349 per sq.mi. 156 per sq.km	290 sq.mi. 699 sq.km	Squash	Foods	Pa'anga	Constitutional Monarchy
TUVALU Funafuti	Tuvalu, English	10,000 1,100 per sq.mi. 435 per sq.km	10 sq.mi. 26 sq.km	Coconut Products	Foods	Tuvaluan Dollar	Parliamentary Democracy
VANUATU Port-Vila	Bislama, English, French	200,000 44 per sq.mi. 17 per sq.km	4,707 sq.mi. 12,191 sq.km	Coconut Products	Machinery	Vatu	Republic

* COUNTRIES AND FLAGS NOT DRAWN TO SCALE

▲ Aerial view of harbor and city, Papeete, Tahiti

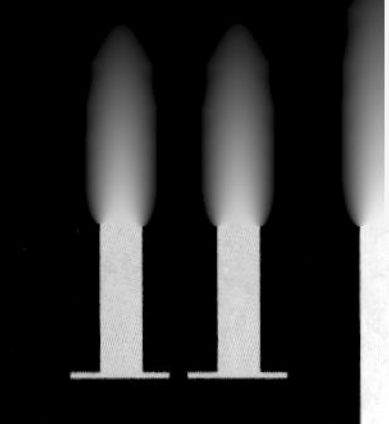

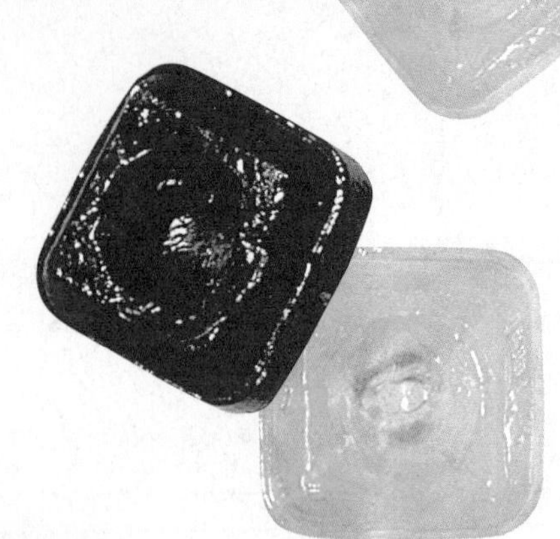

GLOBAL CONNECTION

AUSTRALIA AND THE UNITED STATES

EUCALYPTUS

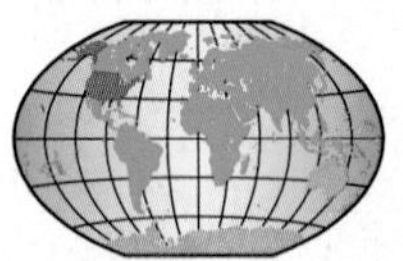

With their big noses, round faces, and cuddly teddy-bear appearance, koalas are among Australia's most famous native animals. If you wanted to see a koala in the wild, the best place to look would be in a eucalyptus tree. Koalas live in eucalyptus trees and eat almost nothing but eucalyptus leaves—about three pounds (1.4 kg) of leaves per day!

Like the koala, eucalyptus trees are native to Australia. More than 500 different kinds grow there. And for hundreds of years, they've been important not only to koalas, but to Australia's people as well.

Long before Europeans arrived in Australia, Aborigines used many different native plants to make medicines. They discovered that eucalyptus leaves contain a strong-smelling oil—eucalyptus oil—that has powerful antiseptic, or germ-fighting, properties. The Aborigines used the leaves to treat several common ailments, including infections, fevers, coughs, colds, and flu. They prepared the leaves in various ways so that the oil could be rubbed onto the skin, inhaled, or mixed with a liquid and swallowed.

When European colonists arrived in Australia, some were quick to recognize the value of Aboriginal remedies, especially those made from parts of eucalyptus trees. Eucalyptus preparations

▲ Koala nibbling eucalyptus leaf

▲ Eucalyptus tree in the Australian outback

▲ American cold sufferer who might benefit from eucalyptus oil

became so popular that colonists tried them for just about every imaginable ailment, from headaches and rheumatism to hair loss and stomach ailments.

The popularity of eucalyptus quickly spread beyond Australia's shores. Eucalyptus oil was one of the first products to be exported from the young colony. By the nineteenth century, millions of the trees themselves were being shipped to the far corners of the world. You can find eucalyptus trees growing in southern California, Florida, and other parts of the southern United States.

Today, eucalyptus oil is an important ingredient in many common medicines, especially cough and cold remedies. The next time you pop a cough drop into your mouth or use a "vapor rub" to clear a stuffy nose, check the label. Chances are good the product contains eucalyptus oil.

CHAPTER 32

The Physical Geography of Australia, Oceania, and Antarctica

GeoJournal

As you read this chapter, imagine that you are visiting interesting and beautiful locations in Australia, Oceania, and Antarctica. Write journal entries, using vivid details to explain why these places are appealing.

GEOGRAPHY Online

Chapter Overview Visit the **Glencoe World Geography** Web site at geography.glencoe.com and click on Chapter Overviews—Chapter 32 to preview information about the physical geography of the region.

SECTION 1

The Land

Guide to Reading

Consider What You Know

You may have seen photographs or movies showing the dry Australian outback or Antarctica's mountainous ice cap. What animals do you associate with these areas?

Read to Find Out

- How do mountains, plateaus, and lowlands differ in Australia and New Zealand?
- How have volcanoes and continental shelves formed the islands of Oceania?
- Why does the physical geography of Antarctica attract scientists?

Terms to Know

- artesian well
- coral
- atoll
- lagoon
- krill

Places to Locate

- Australia
- Great Dividing Range
- Nullarbor Plain
- Murray River
- Darling River
- Oceania
- Melanesia
- Micronesia
- Polynesia
- New Zealand
- North Island
- South Island
- Antarctica

◀ *Evening in Antarctica*

NATIONAL GEOGRAPHIC

A Geographic View

Australian Landscape

The land breathes magic. Not . . . colored scarves and playing cards, but real magic. Weepy eucalyptus trees with [curving sword]-shaped leaves. Dazzling-white ghost gums. Termite mounds: some red and bulbous as a Henry Moore sculpture; others, black and delicate as the spires of a Gothic cathedral. A glory of birds—sulfur-crested cockatoos that lift from trees in clouds, tiny bee-eaters, iridescent as opals.

—*Cathy Newman, "The Uneasy Magic of Australia's Cape York Peninsula,"* National Geographic, *June 1996*

Termite mound, Australia

On Australia's northeastern coast, the Cape York Peninsula displays a landscape of great contrasts. Rain forests, savannas, and wetlands form an exotic patchwork in this area. Australia, Oceania, and Antarctica together form the equally diverse South Pacific region, one that covers a huge portion of the globe. In this section you will explore the region's varied physical geography, including coastal lowlands bordering mountains and plateaus, islands rising from the sea, and a vast ice cap spanning a continent.

Australia: A Continent and a Country

As the only place on the earth that is both a continent and a country, **Australia** is unique. Although water surrounds Australia in the same way as an island, geographers classify it as a continent because of its tremendous size. Located in the Southern Hemisphere, its name comes from the Latin word *australis*, meaning "southern."

Mountains and Plateaus

A chain of hills and mountains known as the **Great Dividing Range** interrupts Australia's otherwise level landscape. The peaks stretch along Australia's eastern coast from the Cape York Peninsula to the island of Tasmania, separated from the mainland long ago by the sea. Most of Australia's rivers begin in the range, and they water the most fertile land in the country.

The Western Plateau, a low expanse of flat land in central and western Australia, covers almost two-thirds of the continent. Australians call this area where few people live the "outback." Across the plateau spread the hot sands of the Great Sandy, Great Victoria, and Gibson Deserts. Near the edges of the deserts, a few low mountain ranges and huge rock formations thrust up from the earth. When explorer Jean-Michel Cousteau visited the arid Western Plateau, he spoke of the land's effect on those few who inhabit it:

> *"For human or nonhuman, life in the vast dry sea, as we were soon to witness, demands extraordinary survival strategies, and those who endure do so with earthy ingenuity and tenacity."*
>
> Jean-Michel Cousteau, *Cousteau's Australia Journey*, 1993

South of the Great Victoria Desert lies the **Nullarbor Plain**. The name comes from the Latin *nullus arbor*, meaning "no tree." This dry, virtually treeless land ends abruptly in giant cliffs. Hundreds of feet below the cliffs lies the churning Great Australian Bight, a part of the Indian Ocean.

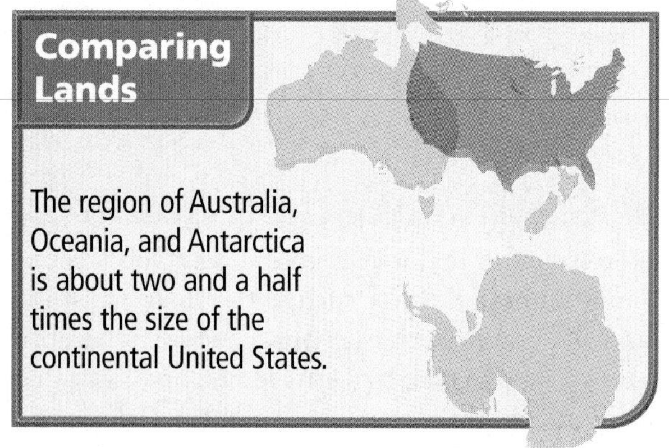

Comparing Lands

The region of Australia, Oceania, and Antarctica is about two and a half times the size of the continental United States.

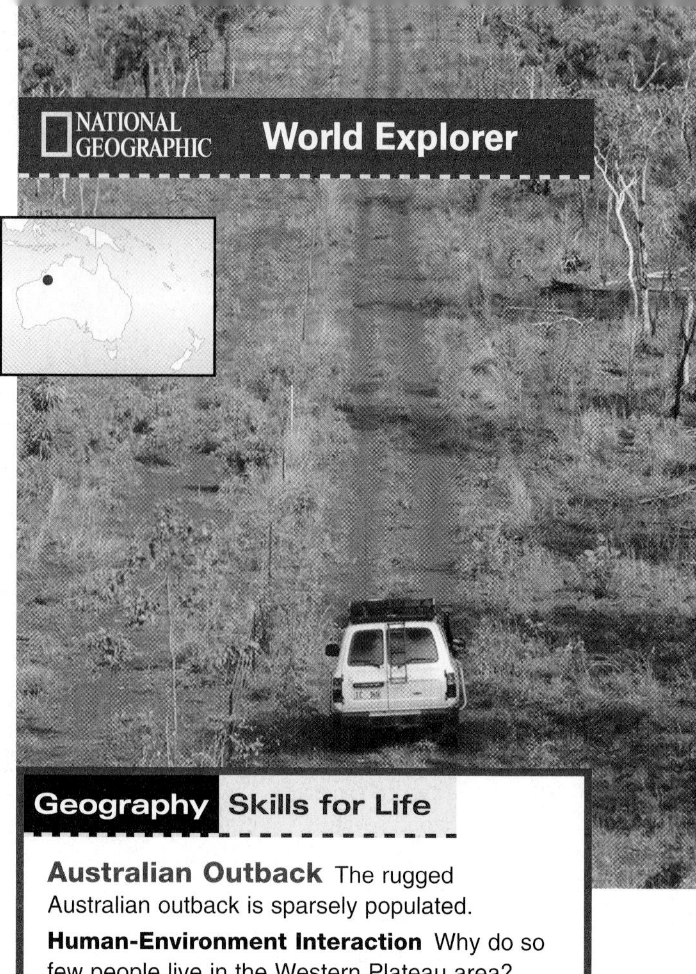

Geography Skills for Life

Australian Outback The rugged Australian outback is sparsely populated.

Human-Environment Interaction Why do so few people live in the Western Plateau area?

Central Lowlands

The Great Dividing Range and Western Plateau are separated by the Central Lowlands. This arid expanse of grassland and desert stretches across the east central part of Australia. After heavy rainfall, rivers and lakes throughout the area fill with water, but because rains are infrequent, most rivers and lakes remain dry much of the year. In the southeast, however, the **Murray River** and the **Darling River** supply water that supports farming. A vast treasure of pressurized underground water, known as the Great Artesian Basin, lies underneath the lowlands. Although the water that gushes from artesian wells, or wells from which pressurized water flows to the surface, is too salty for humans or crops, ranchers use it to water livestock.

Great Barrier Reef

Along Australia's northeastern coast lies the Great Barrier Reef. This famous natural wonder is the world's largest coral reef, home to brilliantly colored tropical fish and underwater creatures. Because of its unique beauty and the habitat it provides for multitudes of creatures, Australia has designated the reef a national park, and the United Nations has named it a World Heritage Site. Although its name suggests a single reef, the Great Barrier Reef is actually a string of more than 2,500 small reefs. Formed from coral, the limestone skeletons of a tiny sea animal, it extends 1,250 miles (2,012 km). This span equals the length of the coastline from New York City to Miami, Florida.

Economics

Natural Resources

Although only 10 percent of Australia's land can be farmed, agriculture is important to the country. Australian farmers make effective use of their land and water to grow wheat, barley, fruit,

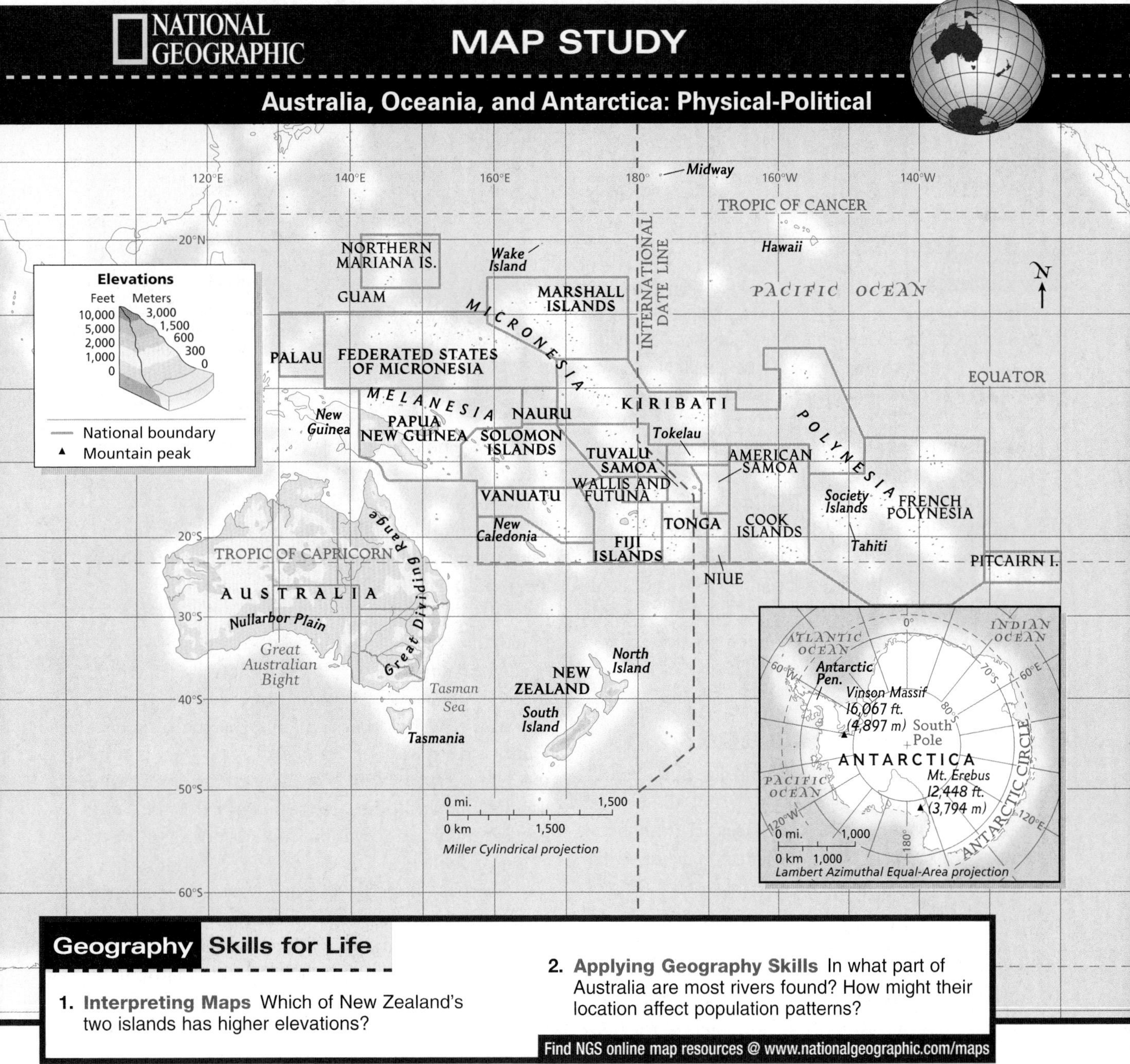

Geography Skills for Life

1. **Interpreting Maps** Which of New Zealand's two islands has higher elevations?

2. **Applying Geography Skills** In what part of Australia are most rivers found? How might their location affect population patterns?

Find NGS online map resources @ www.nationalgeographic.com/maps

Geography Skills for Life

Pacific Islands Volcanic peaks are found on high islands such as Tahiti (left), while low islands are known as atolls (right).

Region What is the third type of island found in the region?

and sugarcane. In arid areas, ranchers raise cattle, sheep, and chickens.

Australia also yields rich mineral resources, including one-fourth of the world's bauxite—the raw material for aluminum production—and most of the world's high-quality opals. Deposits of coal, iron ore, lead, zinc, gold, nickel, and petroleum also make the country one of the world's major mining areas.

Oceania: Island Lands

Thousands of islands, differing in size and extending across millions of square miles of the Pacific Ocean, form the region called **Oceania** (OH•shee•A•nee•uh). Created by colliding tectonic plates millions of years ago, the islands are part of the Ring of Fire, named for its volcanic and earthquake activity.

Island Clusters

Oceania's islands are classified into three clusters, based on location, how the islands formed, and the inhabitants' cultures. **Melanesia**, meaning "black islands," lies north and east of Australia. The "little islands" of **Micronesia** extend north of Melanesia. **Polynesia**, or "many islands," spans an area larger than either Melanesia or Micronesia, ranging from Midway Island in the north to **New Zealand** in the south.

Island Types

Earthquakes and volcanic eruptions still occur on many high islands, one of three island types in Oceania. The landscapes of high islands, such as Tahiti and many of the islands of Fiji, feature mountain ranges split by valleys that fan out into coastal plains. Bodies of freshwater dot the land, and the volcanic soil on high islands supports some agriculture.

Volcanoes shaped Oceania's many low islands differently than they shaped the high islands. Low islands, such as many of the Marshall Islands in Micronesia, are ring-shaped islands, known as **atolls**, formed by the buildup of coral reefs on the rim of submerged volcanoes. Atolls encircle **lagoons**, shallow pools of clear water, and usually rise only a few feet above sea level. Low islands have little soil and few natural resources.

Continental islands are the third type, formed by the rising and folding of ancient rock from the ocean floor. Most of Oceania's large islands, such as New Guinea and New Caledonia, fall into this category. Although volcanoes did not create these islands, many do have active volcanoes. Coastal areas consist of plains, swamps, and rivers. Beyond the coastal areas, the land rises into rugged interior mountains, plateaus, and steep valleys. Because of the variety of their rocks and soil, continental islands have most of Oceania's mineral deposits. Their mining industries produce oil, gold, nickel, and copper. Some larger forested islands support timber processing.

New Zealand: A Rugged Landscape

Located 1,200 miles (1,931 km) southeast of Australia, New Zealand's two largest islands make up 99 percent of the country's landmass. Both **North Island** and **South Island** display sandy beaches, emerald hillsides, and snow-tipped mountains.

North Island's northern region includes golden beaches, ancient forests, and rich soil that supports citrus orchards. A broad central plateau of volcanic stone features hot springs and several active volcanoes. Chief among them is Mount Ruapehu (ROO•uh•PAY•hoo), North Island's highest point. Mount Ruapehu often spews molten rock. Shining freshwater lakes—including Lake Taupo, New Zealand's largest lake—appear throughout the plateau. East of the plateau, a band of hills runs north and south. Here ranchers graze sheep and dairy cattle.

The towering, snowy peaks of the Southern Alps run along South Island's western edge. New Zealand's earliest inhabitants, the Maori, named the highest peak on South Island *Aorangi* (ow•RAHNG•ee), which means "cloud piercer." Today, Aorangi is known as Mount Cook and rises to 12,349 feet (3,764 m). This high country also features sparkling lakes, carved by glaciers, and tumbling rivers. Lowlands called the Canterbury Plains lie on the eastern coast. This land is New Zealand's flattest and most fertile area. Along the western coast, pounding surf meets rugged cliffs, deep fjords, and coastal caves.

Geography Skills for Life

Canterbury Plains In addition to producing grain, New Zealand's Canterbury Plains are economically important for the livestock they support.

Place How does the eastern coast of South Island differ from the western coast?

Natural Resources

New Zealand's fertile soil, perhaps its most important resource, greatly benefits the country's economy. About 55 percent of the land supports crops and livestock. New Zealand's sheep and wool products dominate exports, and its forests yield valuable timber. The country's rivers and dams produce abundant hydroelectric power, fulfilling about 75 percent of the country's needs. New Zealand also uses less typical means to generate power: geothermal energy is provided by water heated underground by volcanoes.

Warm and cold ocean currents meet in the waters off the New Zealand coasts, providing the country with a wide variety of fish. Tuna, marlin, and sharks are abundant in the warmer tropical currents, while cod and hake, a cod-like fish, thrive in the cold Antarctic currents.

Antarctica: A White Plateau

Antarctica, almost twice the size of Australia, lies at the southern extreme of the earth, beneath a massive ice cap. Antarctica's ice cap covers about 98 percent of the continent's landmass. The ice is as much as 2 miles (3.2 km) thick in places and holds 70 percent of the world's freshwater.

Like a jagged backbone, the Transantarctic Mountains extend northward across Antarctica and the Antarctic Peninsula to within 600 miles (966 km) of South America's Cape Horn. The mountains and the peninsula divide the continent into two areas. East of these mountains lies a high, ice-covered plateau. Coastal mountains and valleys near the plateau's edge form pathways for glaciers. To the west the landmass is largely below sea level, including underwater volcanoes.

Research Stations

Although Antarctica contains mineral resources, international agreements limit activity on Antarctica to scientific research. In year-round research stations, scientists from many countries gather fascinating information in this cold and barren land. They investigate weather patterns, measure environmental changes, and observe the sun and stars through an unpolluted atmosphere. The coastal sea also holds valuable resources. Fishing boats from several countries harvest krill, a shrimplike animal eaten by some whales. This plentiful, protein-rich food may someday help lessen world hunger.

Student Web Activity Visit the **Glencoe World Geography** Web site at geography.glencoe.com and click on Student Web Activities—Chapter 32 for an activity about the Ring of Fire.

SECTION 1 ASSESSMENT

Checking for Understanding

1. **Define** artesian well, coral, atoll, lagoon, krill.
2. **Main Ideas** Create a graphic organizer like the one below. List the features and resources for each region. Then choose two of the regions, and write a paragraph explaining how they differ.

Australia	Oceania	Antarctica
• • •	• • •	• • •

Critical Thinking

3. **Predicting Consequences** What group of people would be most affected if Australia's artesian wells dried up? Why?
4. **Comparing and Contrasting** Identify similarities and differences between New Zealand's two main islands and a high island such as Tahiti.
5. **Decision Making** Of the three types of islands found in Oceania, which type would you choose for a home? Explain why.

Analyzing Maps

6. **Location** Study the physical-political map on page 795. To which island region does Papua New Guinea belong?

Applying Geography

7. **Effects of Location** Consider the location of Oceania's islands in relation to other parts of the world. Write a paragraph explaining how this location might affect the development of natural resources.

SECTION 2

Climate and Vegetation

Guide to Reading

Consider What You Know

Scientists who live and work at research stations in Antarctica's harsh climate make exciting discoveries about the earth. Why might this bleak and icy land be a good location for studying ecology, biology, climatology, or astronomy?

Read to Find Out

- How do variations in rainfall affect Australia's climate and vegetation?
- How does elevation affect climate patterns in New Zealand?
- What vegetation survives in the cold, dry Antarctic climate?

Terms to Know

- wattle
- doldrums
- typhoon
- manuka
- lichen
- crevasse

Places to Locate

- Papua New Guinea
- Antarctic Peninsula

NATIONAL GEOGRAPHIC

A Geographic View

A Frozen Frontier

I have grown to love this cold, strange place. . . . Such a reaction may seem odd to those who have never heard the sigh of ice floes jostling on the swells. . . . Alighting here briefly, like a bird of passage, I have come to see this transient frontier not as a harsh place but as a living creature that nurtures a multitude of other lives. . . . We can't conquer it, settle it, even own it. The winter ice belongs only to itself.

—*Jane Ellen Stevens, "Exploring Antarctic Ice,"* National Geographic, *May 1996*

Ice shelf, Antarctica

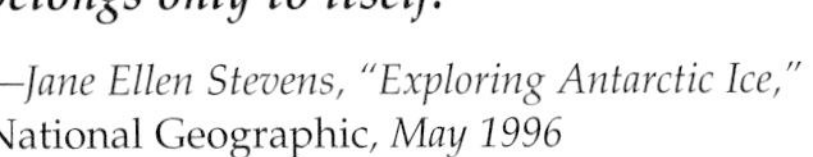

Just as there is a surprising variety of life in an area that appears to be a frozen desert, there are other startling geographic contrasts throughout Australia, Oceania, and Antarctica. In this section, you will learn about the climates and vegetation of one of the world's most geographically diverse regions.

Australia

In Australia, climate and vegetation vary greatly from area to area. The country's climate and vegetation regions include tropical rain forests in the northeast, dry desert expanses in the interior, and temperate areas of grasslands, scrub, and mixed forests along the eastern, southern, and southwestern coasts. Differences in rainfall cause these significant changes in climate and vegetation throughout Australia.

Subtropical high-pressure air masses block moisture-laden Pacific Ocean winds from reaching the Western Plateau, Australia's large

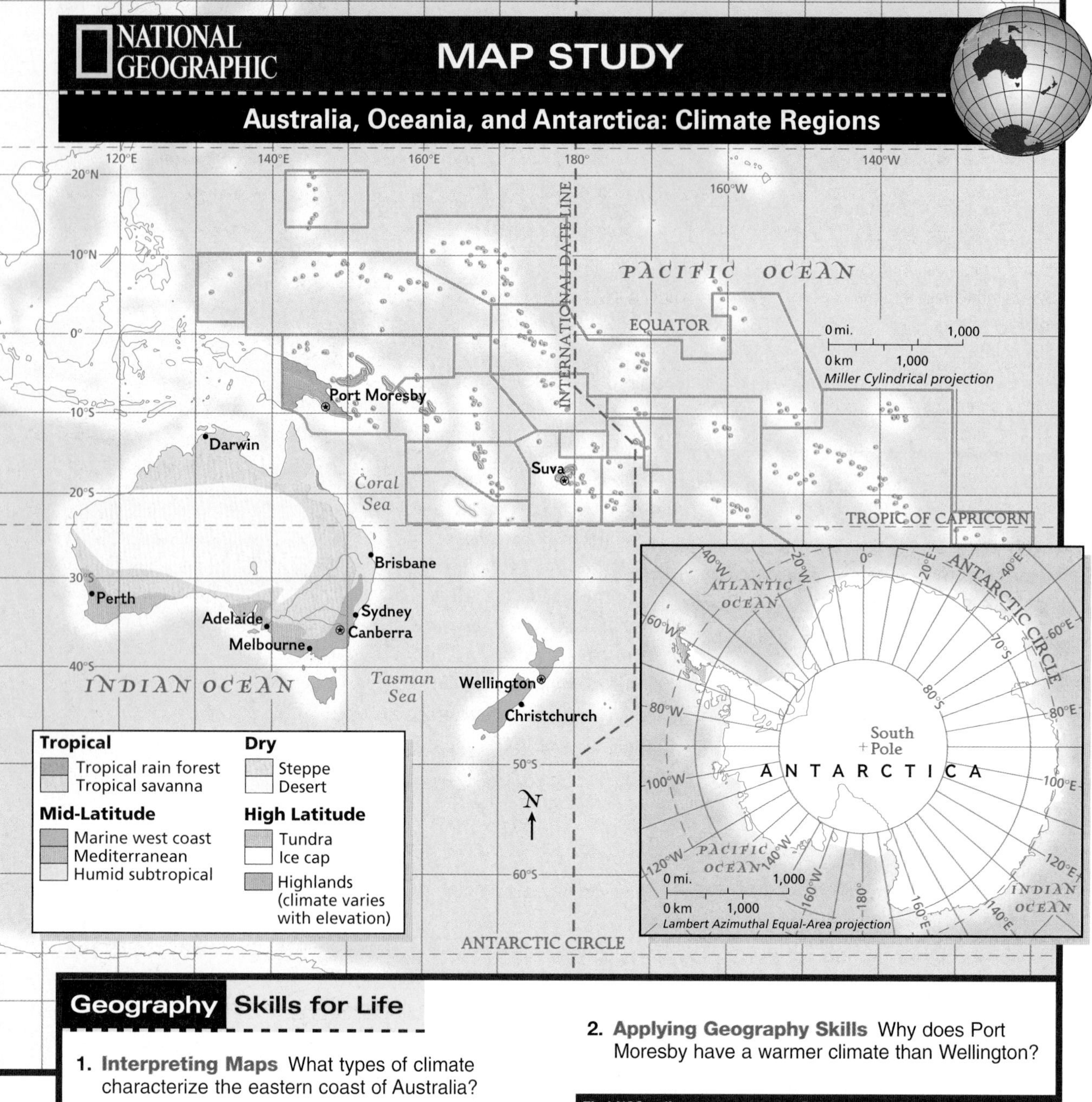

Geography Skills for Life

1. **Interpreting Maps** What types of climate characterize the eastern coast of Australia?

2. **Applying Geography Skills** Why does Port Moresby have a warmer climate than Wellington?

Find NGS online map resources @ www.nationalgeographic.com/maps

interior desert area. The sun scorches the land, but night temperatures drop dramatically. One traveler writes of the arid Western Plateau, as seen from a railroad car:

> *"... At twilight, the shrieking diesel horn scatters flights of long-beaked birds nesting in a sparse underbrush of burrs and thistles. ... Dawn purples a line of mesa-type, flat-topped hills outlined against a cloudless blue sky."*
>
> Hugh A. Mulligan, "The Ghan: Australia's Notoriously Lethargic Train to the Outback," *The Columbian*, November 11, 1999

With less than 10 inches (25 cm) of rain annually, there is not even enough vegetation for grazing.

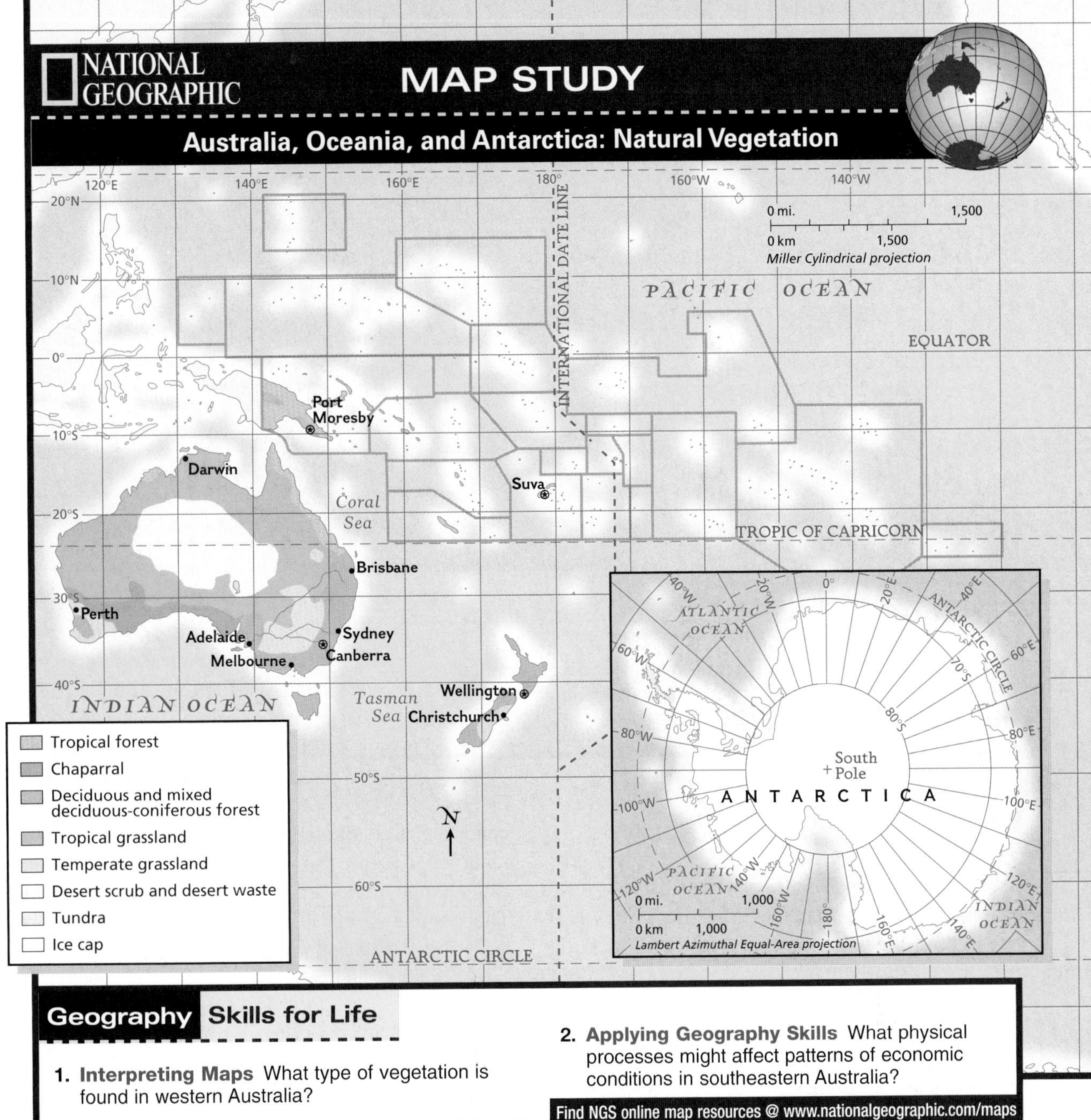

Geography Skills for Life

1. **Interpreting Maps** What type of vegetation is found in western Australia?

2. **Applying Geography Skills** What physical processes might affect patterns of economic conditions in southeastern Australia?

Find NGS online map resources @ www.nationalgeographic.com/maps

An area of milder steppe climate encircles Australia's desert region. Here more regular rainfall brings vegetation such as eucalyptus and acacia trees and small shrubs to life. Saplings of the acacia tree were used by early settlers to make **wattle**, a strong, interwoven wooden framework used for building homes. Rains fall only during the wet season, however, and the amount can vary greatly from year to year. Annual rainfall ranges from 10 to 20 inches (25 to 51 cm). Short grasses, ideal for grazing, also grow here, as do irrigated crops.

Australia's coastal areas have a variety of moister climates. The humid subtropical northeastern coast averages more than 20 inches (51 cm) of rain yearly. Less rain falls in the Mediterranean climate of the southern coasts and in the marine west coast climate along the southeastern coast. Coastal areas support most of Australia's agriculture.

Geography Skills for Life

A Diverse Region A ranching station in the Australian steppe (left), a tropical forest and island beach in Polynesia (middle), and a farm in the marine west coast climate of New Zealand (right) are just a sampling of the diverse landscapes of the region.
Place What kind of climate is found in most of New Zealand?

Oceania

Because much of Oceania lies between the Equator and the Tropic of Capricorn, most islands have a tropical rain forest climate. Most days are warm throughout the year, ranging from 70°F (21°C) to 80°F (27°C), though Pacific ocean winds cool atolls and the windward sides of higher islands. Some mountainous areas of **Papua New Guinea** even remain snow-covered year-round.

Seasons throughout most of Oceania alternate between wet and dry. The dry season features the cloudless blue skies often seen in travel advertisements, but the wet season brings constant rain and high humidity. The amount of rainfall varies from island to island. Low islands get little rainfall, but the larger landmasses of high islands give off warm, moisture-laden air. When this air rises and mixes with cool ocean breezes, heavy rains fall. Some high islands receive as much as 150 inches (381 cm) annually.

Only shrubs and grasses grow on dry, low islands, but coconut palms and other trees appear on islands with more rainfall. Hot, steamy rain forests thrive where heavy rains drench island interiors. A generally windless area called the **doldrums** occupies a narrow band near the Equator where opposing ocean currents meet. The eerie calm within the doldrums can change to violent storms called **typhoons**. Their forceful winds and heavy rain devastate land and vegetation and threaten lives.

New Zealand

A marine west coast climate is found in most of New Zealand. Ocean winds warm the land in winter and cool it in summer, preventing temperature extremes. Temperatures hover between 65°F (18°C) and 85°F (29°C) in summer and between 35°F (2°C) and 55°F (13°C) in winter. Abundant sunshine graces New Zealand's beaches and inland landscape, but clashing air masses may bring sudden clouds and rain.

Geographic differences also cause climatic variations. North Island's central plateau is warm and sunny during summer, but mountaintops may have snow year-round. Fierce winds or blizzards may strike these mountains at any time of year. Mountainous areas exposed to western winds generally have more rainfall than do other areas. Although the country as a whole averages 25 to 60 inches (64 to 152 cm) of rain annually, the Southern Alps on South Island have an average annual rainfall of 315 inches (800 cm). Humidity levels in inland areas are about 10 percent lower than coastal areas. Maurice Shadbolt, a popular travel writer, describes New Zealand as a "long, lean land fated to fickle weather." In fact, he says, "At its most temperamental, New Zealand can offer the traveller all four seasons in one day."

New Zealand's geographic isolation gives rise to unique plant life. Almost 90 percent of the country's indigenous plants are native only to New Zealand. Manuka, a small shrub, carpets land where prehistoric volcanic eruptions destroyed ancient forests. Early settlers from Great Britain cut down almost all of the pinelike kauri trees, but some still grow among thriving evergreen forests. In an effort to repair severe erosion damage in deforested areas, New Zealand's forest service has imported several tree species from Europe and North America. A species of pine tree native to California in the United States, for example, now grows in large areas of the volcanic plateau of North Island. Willows and poplars from Europe also help keep soil on hillsides from eroding.

Antarctica

Antarctica is the earth's highest, driest, windiest, and coldest continent. Though very cold year-round, Antarctica's climate exhibits some variation. Air loses moisture as it rises over Antarctica's plateau, making the plateau drier than Australia's deserts, but much colder. Temperatures may plunge as low as –129°F (–89°C) in winter. The Antarctic Plateau descends to coastal areas that have a milder, moister climate. Annual snowfall averages no more than 2 inches (5 cm) inland, but along the coast it often measures 24 inches (61 cm).

Despite the severe climate, some species of mosses and algae have adapted well to life on Antarctica. In rocky areas along the coasts, tiny sturdy plants called lichens thrive. Of the approximately 800 plant species in Antarctica, about 350 are lichens. These plants survive by remaining dormant for long periods and almost instantly beginning to photosynthesize during brief periods of milder weather. The continent's only two flowering plants grow in a small area on the **Antarctic Peninsula** that lies in a tundra climate zone. Summer temperatures there may reach almost 60°F (16°C).

Although frozen, Antarctica's ice is not motionless. The cap's tremendous weight causes the frozen mass to spread toward the coasts. As it moves, the ice breaks into pieces, causing huge crevasses, or cracks, as much as 100 feet (30 m) wide.

SECTION 2 ASSESSMENT

Checking for Understanding

1. **Define** wattle, doldrums, typhoon, manuka, lichen, crevasse.
2. **Main Ideas** Create a Venn diagram like the one below. Fill in the climate factors for each location, putting those factors that both places have in common in the area where the circles overlap.

Comparing Climates

Australia | Both | New Zealand

Critical Thinking

3. **Making Inferences** What type of climate does most of New Zealand have? Why might it appeal to settlers?
4. **Problem Solving** What problems might researchers encounter in Antarctica, and how could these conditions be overcome?
5. **Identifying Cause and Effect** How do Pacific ocean currents and winds affect the climate of Oceania? How do they affect vegetation?

Analyzing Maps

6. **Human-Environment Interaction** Study the vegetation map on page 801. What type of vegetation is suitable for raising livestock, and where in Australia is it found?

Applying Geography

7. **Understanding Climate Maps** Note the climate regions on the map on page 800. Write a paragraph explaining how climate relates to the way farmers operate in New Zealand and Australia.

Making Inferences

You see a police car stopped behind another car by the roadside. The emergency lights are flashing on the police car. You infer, or conclude, that the driver was speeding based on the information you already have from similar circumstances. Making an inference means using information to draw a conclusion.

Learning the Skill

When you hear about a country or an event in news reports or read about it in magazines and books, you may still have questions afterward. Most sources do not contain all the information on a subject, but they may offer enough information for you to infer, or figure out, the answers to your questions.

Different sources present information in different forms. Statistical charts, for example, often compare information from which you might infer differences, similarities, or trends over time. These steps will help you make inferences from a chart:

- **Read the title and other labels to know what information the chart presents.**
- **Determine whether the chart provides detailed information about one topic, compares two or more topics, or shows changes over time.** Some charts may give several different types of information.
- **Make a list of the information that is not given in the chart, or the questions that arise from it.**
- **Infer answers to your questions.** Make logical inferences based on the facts given.

Solomon Islands	
Population	500,000
Government	Under British rule from 1893 to 1978, when it gained independence; member of the Commonwealth of Nations
Capital	Honiara
Languages	120 indigenous languages; English
Land	About 900 islands scattered over approximately 11,000 square miles (28,900 sq. km) of ocean
Geography	Mountainous volcanic islands and small atolls; coral reefs
Climate	Tropical
Rainfall	Ranges from 120 to 140 inches (305 to 356 cm) annually
Temperature	Ranges from 70° to 90°F (21° to 32°C)
Vegetation	Tropical forests on main islands
Exports	Fish, timber, cocoa
Life Expectancy	67 years

Sources: National Geographic Atlas of the World, 7th edition; 2001 World Population Data Sheet

Practicing the Skill

A tourist traveling to the Solomon Islands might use a chart to learn about the country. Answer the following questions by making inferences about the Solomon Islands from the information in the chart above.

1. Given the information in this chart, is it more likely that the form of government in the Solomon Islands is a parliamentary democracy or a communist state? Explain.
2. What can you infer about the health of the people in the Solomon Islands? Explain.
3. What can you infer about the animal life on the Solomon Islands? Explain.
4. What sorts of activities might a tourist enjoy in the Solomon Islands? Explain.

Applying the Skill

Locate a chart with information about another country in this region from a newspaper, magazine, or Web site. Use the steps to make inferences from the facts presented. Draft several questions based on your inferences. Exchange charts with another student, and complete each other's questions.

The **Glencoe Skillbuilder Interactive Workbook, Level 2** provides instruction and practice in key social studies skills.

32 SUMMARY & STUDY GUIDE

SECTION 1 The Land (pp. 793–798)

Terms to Know

- artesian well
- coral
- atoll
- lagoon
- krill

Key Points

- Australia, both a country and a continent, encompasses mountains, central lowlands, and expansive deserts. Rich mineral deposits and productive farms and ranches contribute to the Australian economy.
- Oceania's thousands of islands extend across the southern Pacific Ocean. The islands of Oceania were formed either directly or indirectly by volcanic activity.
- New Zealand's main features are two large islands with mountain ranges, rivers, and lakes. The country boasts rich soil and timberland.
- Antarctica is an ice-covered continent. While Antarctica may have important mineral resources, its key resource is the information it offers to scientists.

Organizing Your Notes

Use a chart like the one below to help you organize information about the physical features and resources of Australia, Oceania, and Antarctica.

	Geographic Features	Natural Resources
Australia		
Oceania		
New Zealand		
Antarctica		

SECTION 2 Climate and Vegetation (pp. 799–803)

Terms to Know

- wattle
- doldrums
- typhoon
- manuka
- lichen
- crevasse

Key Points

- Australia generally has a hot, dry climate. Along the edges of the vast interior desert, the steppe receives sufficient rainfall for raising livestock. Only the coastal climates provide enough rainfall for growing crops without irrigation.
- Oceania enjoys a warm, moist tropical climate. Most islands have wet and dry seasons. The amount of rain during the wet season determines whether shrubs and grasses or dense rain forests will grow.
- New Zealand's marine west coast climate provides year-round rainfall, with temperatures that vary without being extreme.
- Antarctica's extremely cold and windy climate supports primarily lichens and mosses.

Organizing Your Notes

Use an outline like the one below to help you organize the information in this section about climate and vegetation.

Climate and Vegetation

I. Australia
   A. Mountains and Plateaus
   B. Central Lowlands
II. Oceania

ASSESSMENT & ACTIVITIES

Reviewing Key Terms

Write the letter of the key term that best matches each definition below.

a. artesian well	**f.** doldrums
b. coral	**g.** typhoon
c. atoll	**h.** manuka
d. lagoon	**i.** lichen
e. krill	**j.** crevasse

1. limestone skeletons of a tiny sea animal
2. windless area near the Equator
3. shrimplike animal
4. small, sturdy plants
5. well from which pressurized water flows to the surface
6. huge crack in an ice cap
7. small shrub that grows in New Zealand
8. pool of water inside an atoll
9. violent Pacific Ocean storm
10. low, ring-shaped island

Reviewing Facts

SECTION 1

1. What formation lies just off Australia's northeastern coast?
2. Name the three types of islands that are found in Oceania.
3. What is New Zealand's main natural resource?
4. List the local resources that help to meet New Zealand's energy needs.

SECTION 2

5. What climate supports most of Australia's agricultural lands?
6. Describe the factor that prevents temperature extremes in New Zealand.
7. What causes the motion of the Antarctic ice cap?

Critical Thinking

1. **Making Inferences** Based on the information in Section 1, would you infer that Australia does or does not have an even distribution of population across the continent?
2. **Comparing and Contrasting** How are Oceania's islands similar? Different?
3. **Identifying Cause and Effect** Create a chart like the one below, and fill in the effects of different climates on vegetation. Then choose one effect, and write a paragraph describing its possible economic impact.

	Australia	Oceania	New Zealand	Antarctica
Climate				
Vegetation				

NATIONAL GEOGRAPHIC

Locating Places

Australia and New Zealand: Physical-Political Geography

Match the letters on the map with the places and physical features of Australia and New Zealand. Write your answers on a sheet of paper.

1. Great Barrier Reef
2. Great Victoria Desert
3. Great Dividing Range
4. Tasmania
5. Cape York Peninsula
6. Great Australian Bight
7. Coral Sea
8. Lake Eyre
9. North Island
10. South Island

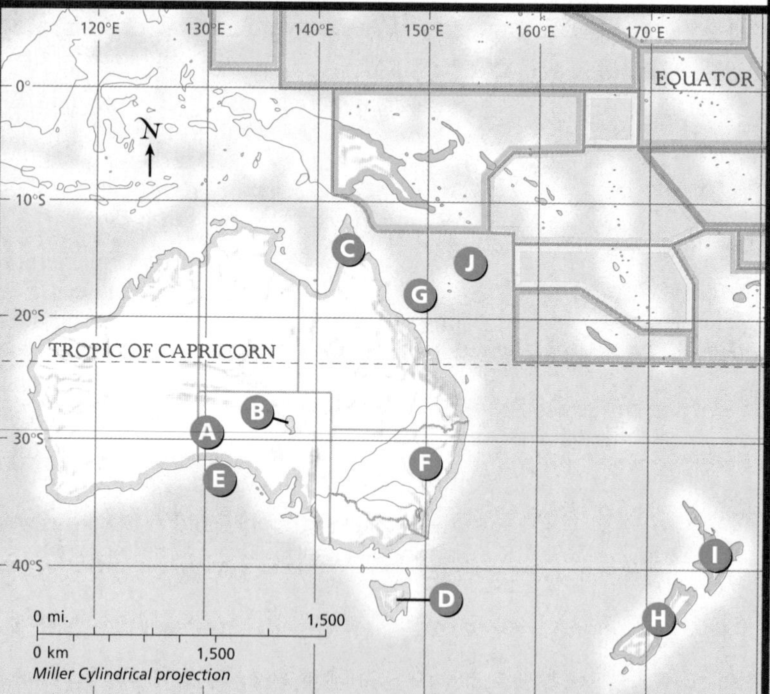

Self-Check Quiz Visit the **Glencoe World Geography** Web site at geography.glencoe.com and click on Self-Check Quizzes—Chapter 32 to prepare for the Chapter Test.

Using the Regional Atlas

Refer to the Regional Atlas on pages 784–787.

1. **Location** Which Australian city is located on the coast, just south of the Great Barrier Reef?
2. **Location** In which part of Australia are most of the coal deposits found?

Thinking Like a Geographer

Think about the activities of explorers, scientists, and tourists in Antarctica. What changes to Antarctica's physical geography might happen as a result? As a geographer, what safeguards would you suggest to preserve this unspoiled environment?

Problem-Solving Activity

Contemporary Issues Case Study Use print and nonprint resources to learn more about krill. Investigate how these tiny crustaceans fit into the food chain in the waters surrounding Antarctica. Find out about issues related to harvesting krill commercially as well as its potential for reducing world hunger. Write a brief report of your findings, and give recommendations for using krill responsibly.

GeoJournal

Travel Brochure Imagine that you are a travel writer, and draft a brochure about one location you wrote about in your GeoJournal. Include vivid details and information about the landforms, climate, and vegetation of the location you choose. Use your textbook and the Internet to make the brochure lively and interesting.

Technology Activity

Using the Internet for Research Search the Internet for photographs and information about plants mentioned in Section 2, such as acacia and manuka. Look for details about their habitats and their uses. Create a display of the information and images, and share the finished product with your class.

Standardized Test Practice

Use the chart below to choose the best answer for each of the following multiple-choice questions. If you have trouble answering the questions, use the process of elimination to narrow your choices.

Australian City	Average Yearly Precipitation (inches)	Average Temperature Range January (°F)	July (°F)
Alice Springs	10–20	75°–85°	45°–55°
Brisbane	over 30	75°–85°	55°–65°
Darwin	over 30	75°–85°	over 75°
Melbourne	20–30	65°–75°	45°–55°
Perth	over 30	75°–85°	45°–55°
Sydney	over 30	65°–75°	45°–55°

1. **If tourists were traveling to Australia in January and wanted to avoid both excessive heat and heavy rainfall, to which city should they travel?**

   A Melbourne
   B Brisbane
   C Darwin
   D Sydney

2. **What information in the chart shows that the Australian cities are in the Southern Hemisphere?**

   F July's temperatures are higher than January's.
   G January's temperatures are higher than July's.
   H The cities have abundant rain.
   J The cities have a dry season.

Read the chart and become familiar with the information it contains before you answer the questions. Do not, however, study the chart in depth. The quickest way to answer both question 1 and question 2 is to read through each answer choice and use the process of elimination to get rid of those that you think are wrong.

Geography Lab Activity

Air in Motion

▲Large tropical storms can be seen from space.

Winds are horizontal air movements caused by temperature differences among air masses. Surface winds are usually strongest during the day, when the sun heats the ground. The increased ground temperature causes the air to spread out, become lighter, and rise. As thin air rises, cold air moves down to take its place. This movement of air is the wind blowing. Winds usually, but not always, become gentler at night. Wind patterns have a significant impact on an area's climate, and they are often themselves affected by local weather patterns and conditions. People generally identify winds based on the direction from which they blow.

Tropical storms are created when an area of low atmospheric pressure is surrounded by circulating winds. Twenty to twenty-five typhoons blast across the Pacific Ocean each year. The word *typhoon* comes from the Chinese word *tai-fung*, which means "great wind." These storms, which are called tropical cyclones or hurricanes in other parts of the world, have spiraling winds that reach 100 to 150 miles per hour (161 to 241 km per hour).

1 Materials

- Drinking straw
- Scissors
- Thin, stiff plastic for the arrowhead and tail, 5 7/8 in × 5 7/8 in (15 cm × 15 cm)
- Clear tape
- Straight pin
- Wood block, 2 in × 2 in × 17 3/4 in (5 cm × 5 cm × 45 cm)
- Hammer
- Metal washer
- Photocopy of Figure 1—Compass
- Photocopy of Figure 2—Data chart

2 Procedures

In this activity, you will build and use a wind vane to see how local changes in wind direction are related to local weather changes.

1. To construct the arrow, make two small slits in each end of the drinking straw. The slits at the arrow end should be 1 1/8 inches (3 cm) long. The slits at the tail end should be 2 inches (5 cm) long. Make sure the slits align with each other.
2. Cut a small arrowhead and a large tail out of the plastic. Insert the arrowhead and the tail into the straw's slits, and secure them with a small amount of tape.
3. Balance the straw on your finger. NOTE: The balancing point may not be in the center of the straw. When you find this point, poke the straight pin through the straw. Enlarge the hole slightly.

4. Photocopy Figure 1 (the compass), and tape it to the top of the wood block. Using the hammer, gently drive the pin through the metal washer and into the center of the compass.
5. Take the wind vane outside to an open area.
6. Hold the wind vane so that the *N* on the block points north. The wind vane's arrow will point into the wind. Use the compass to determine the direction from which the wind is blowing. This is the wind direction.
7. Photocopy or draw Figure 2 (the data chart), and record the wind direction three times a day, for five days.

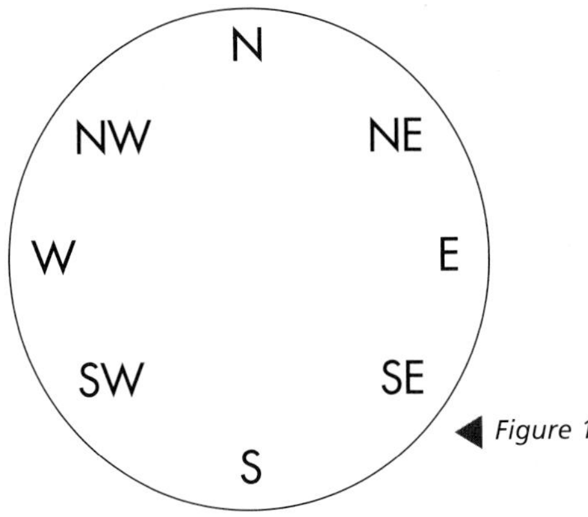

Figure 1

Date	Time	Wind Direction
	1. 2. 3.	
	1. 2. 3.	
	1. 2. 3.	
	1. 2. 3.	
	1. 2. 3.	

Figure 2

3 Lab Report

1. Why do you think the arrow of a wind vane points into the wind?
2. In which direction did your arrow point most often?
3. According to your results, how often does the wind direction change in your area?
4. **Drawing Conclusions** Weather stations take wind direction readings from the tops of tall buildings or high poles. Why do you think this is so?

4 Find Out More

In addition to measuring wind direction, you can measure wind speed. Use nylon thread to attach a table tennis ball to the center of the straight edge on a protractor. In the windiest area of the school grounds, hold the protractor with the straight edge up and level. Now face the wind. The angle made by the nylon line on the protractor will be the wind speed in degrees. The degree of wind speed converts to the following wind speeds:

10 degrees = 8 mph (13 km/h)
20 degrees = 12 mph (19.2 km/h)
30 degrees = 15 mph (24 km/h)
40 degrees = 17.9 mph (28.8 km/h)
50 degrees = 20.9 mph (33.6 km/h)
60 degrees = 25.8 mph (41.6 km/h)
70 degrees = 32.8 mph (52.8 km/h)

Did You Know? Meteorologists use technology to monitor tropical storms and issue warnings that can save lives and property. Specialists scan satellite photographs for thunderstorm clusters. They reexamine cluster images hourly for signs of rotating winds. If these conditions develop, tropical storm warnings go out to people on ships, on aircraft, and along coastlines.

CHAPTER 33

The Cultural Geography of Australia, Oceania, and Antarctica

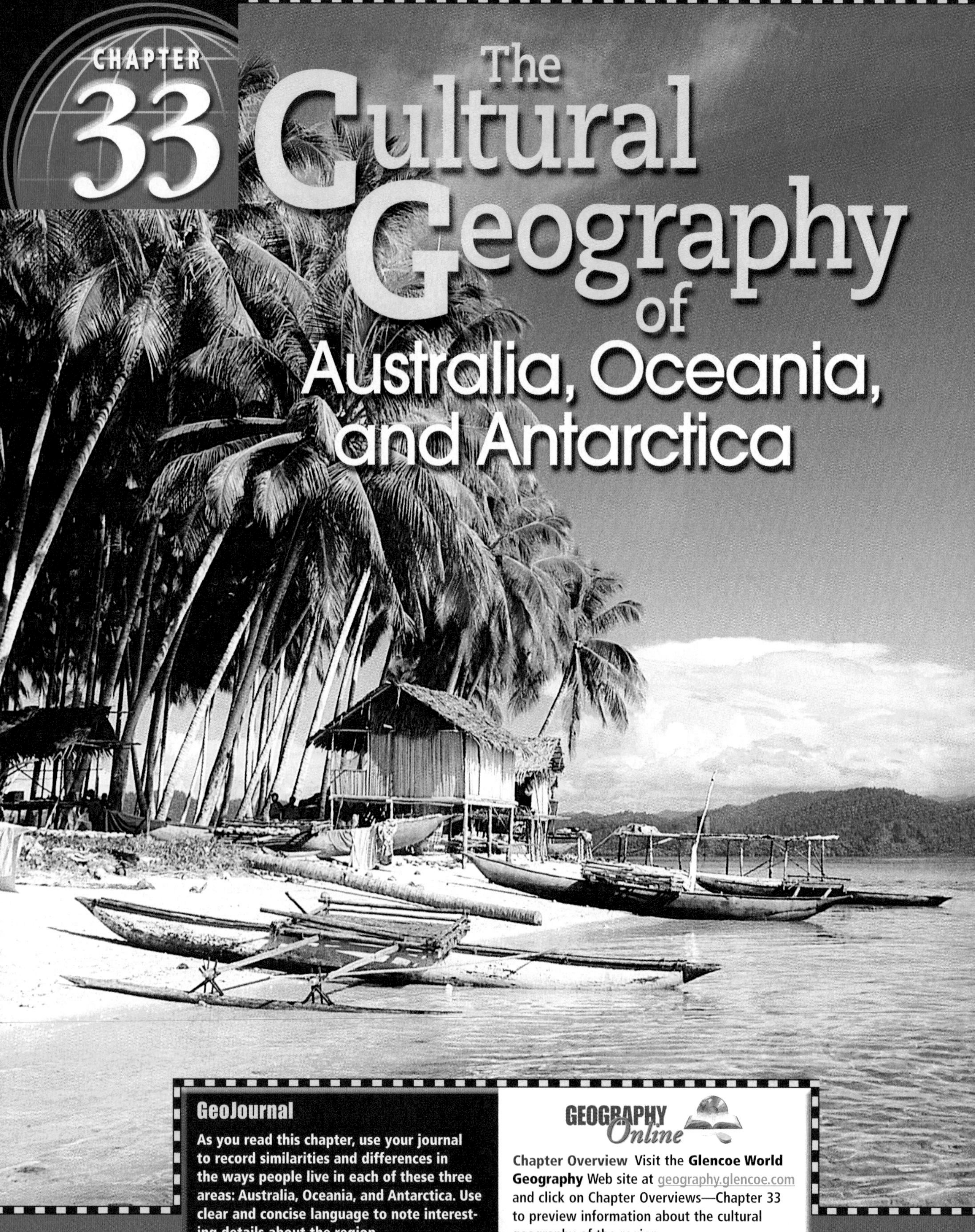

GeoJournal

As you read this chapter, use your journal to record similarities and differences in the ways people live in each of these three areas: Australia, Oceania, and Antarctica. Use clear and concise language to note interesting details about the region.

GEOGRAPHY Online

Chapter Overview Visit the **Glencoe World Geography** Web site at geography.glencoe.com and click on Chapter Overviews—Chapter 33 to preview information about the cultural geography of the region.

Population Patterns

Guide to Reading

Consider What You Know

What images have you seen in the news or in films of the various peoples living in Australia, Oceania, and Antarctica? What geographic factors might account for the ways people live in each of these areas?

Read to Find Out

- What peoples settled in Australia and Oceania?
- How does the region's geography affect population density, distribution, and growth?
- What factors account for settlement in urban and rural areas?

Terms to Know

- Strine
- pidgin English

Places to Locate

- Kiribati
- Sydney
- Melbourne

NATIONAL GEOGRAPHIC

A Geographic View

Dream Journey

Lying back and looking at the night sky, I felt pulled upward into that shimmering immensity. . . . Laserlike, a shooting star cuts the sky . . . and you suddenly understand how the Aborigines, who slept out here beneath these same stars for 50,000 years before the [Europeans] came, could devise their wonderful mythologies of the Sky Heroes who came down from the stars in that mystic Dreamtime and shaped the landscape.

—*Harvey Arden, "Journey Into Dreamtime,"* National Geographic, *January 1991*

Rock formations near Lake Argyle, Australia

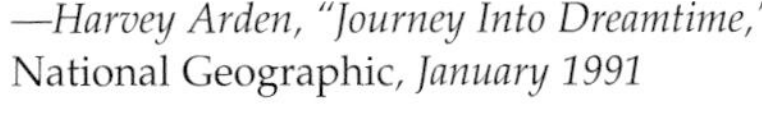

The Aborigines, Australia's earliest people, feel a direct relationship to the landscape that has shaped their movements throughout the island continent. Physical geography also has influenced migration and settlement patterns of other peoples in Australia and Oceania. In this section you will learn why Australia and Oceania have diverse cultures and what geographic factors influence where their populations live. You will also visit Antarctica, the cold, icy continent at the bottom of the world.

Human Characteristics

Australia and Oceania have populations with diverse ancestries—indigenous, European, and Asian. Both physical geography and the migration patterns of peoples have shaped the region's cultures.

◀ *Coastal scene, Papua New Guinea*

the arts of AUSTRALIA

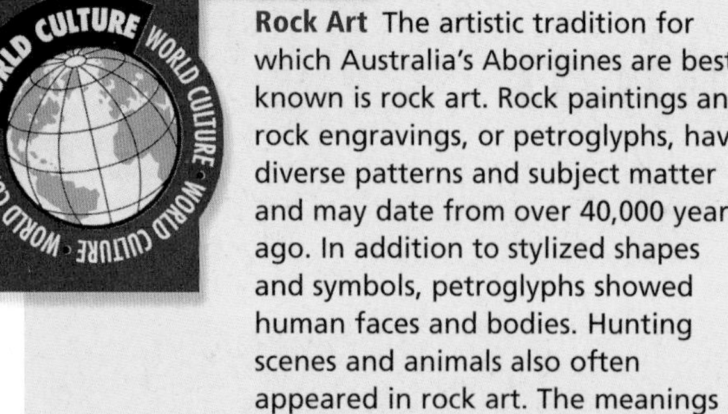

Rock Art The artistic tradition for which Australia's Aborigines are best known is rock art. Rock paintings and rock engravings, or petroglyphs, have diverse patterns and subject matter and may date from over 40,000 years ago. In addition to stylized shapes and symbols, petroglyphs showed human faces and bodies. Hunting scenes and animals also often appeared in rock art. The meanings of most of these paintings and petroglyphs, however, remain unknown.

Aborigines and Maori

Australia's Aborigines may have the oldest surviving culture in the world. The name given to them by European settlers is from the Latin *ab origine*, meaning "from the beginning." The first Aborigines probably arrived in Australia 40,000 to 60,000 years ago from Southeast Asia. They lived as nomadic hunters and gatherers in small kinship groups along the temperate coasts, in the northern rain forests, and across the vast interior deserts. Over the centuries, the Aborigines successfully learned to deal with the challenges posed by these environments. Today, Aborigines number about 315,000, making up about 2 percent of Australia's population.

New Zealand's indigenous peoples, known as the Maori (MOWR•ee), came from the Pacific islands of Polynesia. In New Zealand they hunted, fished, established villages, and raised crops. Many ancient Maori traditions still remain a part of Maori life. For example, Maori communities hold festive gatherings called *hui* in which important local events such as weddings, funerals, and the dedication of new buildings are celebrated. As a result of intermarriage with European settlers over the years, most Maori people today have at least some European ancestry.

Pacific Islanders

The islands of Oceania were probably first settled by peoples from Asia more than 30,000 years ago. Waves of migrants from Asia continued to arrive over many centuries, while groups already living in the Pacific area moved from island to island. Today many different peoples speaking hundreds of languages live on Oceania's scattered islands. However, there are three major indigenous groups—Melanesians, Micronesians, and Polynesians—based on the island cluster on which each group lives. People on all three island clusters generally support themselves by fishing or farming.

The first cluster is Melanesia, located in the southwestern Pacific Ocean. It includes independent island countries, such as Papua New Guinea, Fiji, and the Solomon Islands, as well as French-ruled New Caledonia. Melanesian cultures differ greatly, even among groups living in different parts of the same island.

Next is Micronesia, situated in the western Pacific east of the Philippines. Among the independent countries of Micronesia are the Federated States of Micronesia, Nauru, and **Kiribati** (KIHR•uh•BAH•tee). The area also includes the United States island territories of Guam and the Marianas. Micronesians also have several different languages and cultures.

The last cluster is Polynesia, located in the central Pacific area. Three independent countries—Samoa, Tonga, and Tuvalu—are found in Polynesia. Other island groups, known as French Polynesia, are under French rule and include Tahiti, Polynesia's largest island. Most Polynesians share similar languages and cultures.

Europeans

From the 1500s to the late 1700s, Europeans sailed the waters around Australia, New Zealand, and Oceania. They set up trading settlements and eventually colonized the region. Europeans, mainly of British descent, still make up most of the populations of both Australia and New Zealand. Smaller numbers of European groups live on various Pacific islands. For example, the French-ruled islands of Tahiti and New Caledonia are home to many people of European descent.

Asians

Asian communities also exist in the South Pacific area. Chinese traders and South Asian workers settled parts of Oceania during the 1800s, and today their descendants are included in the populations of places such as French Polynesia and Fiji. From the early 1900s to 1945, Japan ruled a number of Pacific islands, although few people of Japanese descent live there today. Australia and New Zealand once blocked non-European immigration, but the need for more workers finally led to more open immigration policies after the 1970s. Since then, increasing numbers of East Asians and Southeast Asians have migrated to Australia and New Zealand in search of economic opportunity.

Languages

Before the era of modern transportation and advanced communications, mountains, deserts, and ocean separated the peoples of the South Pacific area. As a result, isolated groups developed many different languages. Of the world's 3,000 languages, 1,200 are spoken today in Oceania alone, some by only a few hundred people.

European colonization brought European languages to the region. Today French is widely spoken in areas of Oceania that remain under French control. English is the major language of Australia and New Zealand. Australian English, called **Strine**, has a unique vocabulary made up of Aboriginal words, terms used by early settlers, and slang created by modern Australians. For example, Australians today call a barbecue a "barbie," and greet each other with the phrase "G'day." In many areas of Oceania, varieties of **pidgin English**, a blend of English and an indigenous language, developed to allow better communication among different groups.

Where People Live

Australia, Oceania, and Antarctica span a vast area; Australia and Oceania together cover about 5.7 percent of the earth's surface. However, a high percentage of the region's land is unsuited for human habitation. As a result, the region has only one-half of one percent of the world's population.

Population Distribution

Because of uninhabitable land and vast differences in physical features and climates, population in Australia and Oceania is unevenly distributed. Australia is the region's most heavily populated country. About two-thirds of the South Pacific area's 31 million people live in Australia, which has almost 90 percent of the region's habitable land. Very few people, however, live in Australia's dry central plateaus and deserts. Most live along the southeastern, eastern, and southwestern coasts, which have a mild climate, fertile soil, and access to sea transportation. Most of New Zealand's people also live in coastal areas.

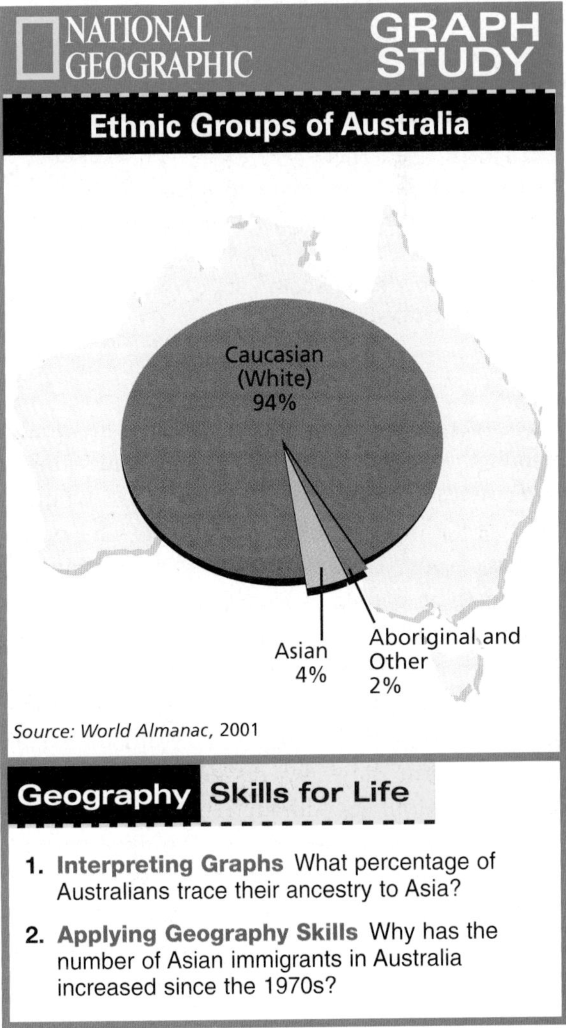

Source: *World Almanac*, 2001

Geography Skills for Life

1. **Interpreting Graphs** What percentage of Australians trace their ancestry to Asia?
2. **Applying Geography Skills** Why has the number of Asian immigrants in Australia increased since the 1970s?

NATIONAL GEOGRAPHIC **World Explorer**

Geography Skills for Life

Diverse Peoples The South Pacific region is home both to this indigenous man from Papua New Guinea (inset) and these children of European descent from Australia.

Place Where are the South Pacific's most urbanized areas located?

In Oceania, the population is divided unequally among the island countries. Papua New Guinea leads with about 5 million people, whereas Nauru—the world's smallest republic—has a population of only 11,000. Many more Pacific islanders live on their countries' coasts than in the often rugged interiors.

Antarctica's forbidding, icy terrain and merciless polar climate have never supported permanent human settlement. Conditions are difficult for all but short-term stays by research scientists and adventurous tourists. Although Antarctica measures about 5.5 million square miles (14.2 million sq. km), most research stations cluster along the Antarctic Peninsula, where summer temperatures may rise to a relatively mild 32°F (0°C). During this season the population of research stations reaches about 10,000, but only about 1,000 people remain during winter.

Population Density

Like population distribution, population density varies throughout the region. In Australia, for example, the population density averages only 6 people per square mile (2 per sq. km). In some interior rural areas, a person can travel 100 miles (161 km) without seeing another human being. In coastal urban areas, however, population density is much higher. Australia's urban areas are home to 85 percent of the country's total population. Like many developed countries, Australia has an aging population and a declining birthrate. Yet Australia's population probably will continue to increase because of immigration.

Oceania's population is growing at an average rate of 2.3 percent per year because it has a relatively young population. The land area of Oceania's 25,000 islands totals only 551,059 square miles (1,427,246 sq. km), and the population density varies greatly. Because Papua New Guinea has a large area, its population density is only 28 people per square mile (11 per sq. km). Tiny Nauru, measuring just 8 square miles (21 sq. km), has the highest population density in Oceania—about 1,412 people per square mile (545 per sq. km). In spite of its small area, mining of the island's rich phosphate deposits provides jobs and funding for economic development.

Urbanization

Most people in Australia and New Zealand live in cities or towns along the temperate coasts. The largest Australian cities are **Sydney** and **Melbourne**—each with more than 3 million residents. Sydney, located on the eastern coast, and Melbourne, on the southern coast, are port cities and commercial centers. Other coastal urban areas in Australia are Brisbane, Adelaide, and Perth. Few people, on the other hand, live in the hot, dry climate of Australia's interior.

New Zealand's ports of Auckland, Christchurch, and Wellington are Oceania's largest cities. These and other cities in the region offer newcomers opportunities for a high standard of living, quality health care, and excellent education.

Urban areas in Australia and Oceania draw people from within their own countries as well as from other countries. In Australia internal migration has led to shifts in population distribution. During the 1990s the population in rural areas declined while that of large cities and their suburbs grew rapidly. A similar pattern can be seen in Oceania, where 70 percent of the population lives in urban areas.

Government

Immigration

Rapid expansion of industry after World War II drew many immigrants to Australia. At first most immigrants came from European countries, such as the United Kingdom, Greece, Yugoslavia, and the Netherlands. In the 1980s Australia's industries still needed more workers, so the Australian government created programs to attract people from other regions. Today immigrants come from South Africa and various parts of Asia and Latin America. A number of them also come from Oceania. Population growth and uneven economic development in the various Pacific islands cause many young people and skilled workers to seek work elsewhere.

Student Web Activity Visit the **Glencoe World Geography** Web site at geography.glencoe.com and click on Student Web Activities—Chapter 33 for an activity about immigration and cultural diversity in Australia.

Publicly funded programs provide travel assistance to immigrants and help them adjust to Australian society. Most immigrant workers settle in major industrialized cities because of high-paying jobs. Today about 26 percent of Australia's population is foreign born. One worker from Lebanon describes his experience to a journalist:

> "*In this one factory you had people from maybe ten, twelve different countries, all speaking different languages. That's what Sydney was like. . . . It's a beautiful . . . country—beautiful. Great weather. Lovely lifestyle. Plenty of opportunity if you want to work hard.*"
>
> Bill Bryson, "Sydney," *National Geographic*, August 2000

Throughout Australia and Oceania, meeting the needs of a growing multiethnic population is a major concern. Diversity enriches the region's languages, arts, music, and lifestyles. At times, however, this same diversity may cause disagreements over issues such as immigration, health benefits, employment, and the effects of colonial rule. The next section highlights the legacy of the past and how it shapes life in Australia and Oceania today. Antarctica, with no permanent population, has a history that is unique to that icy continent.

SECTION 1 ASSESSMENT

Checking for Understanding

1. **Define** Strine, pidgin English.
2. **Main Ideas** On a chart like the one below, fill in three main ideas from the section and then list important supporting details for each idea.

Main Ideas	Supporting Details

Critical Thinking

3. **Categorizing Information** From what areas have peoples migrated to Australia and Oceania?
4. **Identifying Cause and Effect** What geographic factors cause most of Australia's population to cluster in coastal urban areas?
5. **Predicting Consequences** What are possible positive and negative effects as modern technology and transportation attract more people to the South Pacific region?

Analyzing Maps

6. **Region** Study the population density map on page 786. What are the most sparsely populated areas of the South Pacific region?

Applying Geography

7. **Movement and Population** Create two maps, one of migration patterns during the last 100 years, and the other showing population distribution in the region today.

SECTION 2

History and Government

Guide to Reading

Consider What You Know

Various indigenous peoples lived in Australia and Oceania before the Europeans established colonies there. What indigenous groups in various parts of the world have you read about or seen in movies or on television?

Read to Find Out

- What were the lifestyles of the region's indigenous peoples before colonization?
- How did colonial rule affect social, economic, and political structures?
- How do today's governments reflect the region's history?

Terms to Know

- clan
- boomerang
- trust territory
- dominion

Places to Locate

- Vanuatu
- Tonga

NATIONAL GEOGRAPHIC

A Geographic View

Pacific Origins

Samoa itself is said to mean "sacred center". . . . [T]his is where the world began as the creator, Tagaloalagi, first called forth earth, sea, and sky from rock. . . . Language links and artifacts suggest that the first distinctly Polynesian culture may have developed here some 3,000 years ago. Over the centuries that followed, seafarers in double-hulled sailing vessels stocked with pigs, dogs, and fruits spread that culture across much of the Pacific.

—*Douglas Chadwick, "The Samoan Way,"* National Geographic, *July 2000*

Samoan diver in Pacific waters

European and American influences in the past three centuries have profoundly changed the indigenous peoples and cultures of the South Pacific area. In this section you will learn about the early inhabitants of Australia and Oceania, the effects of Western settlement and rule in these areas, and the emergence of independent countries and new governments during the past 100 years.

Indigenous Peoples

Historians, anthropologists, archaeologists, and other scientists are continually uncovering new information about the history of early South Pacific peoples. At the same time, after years of Western dominance, indigenous peoples throughout the region are rediscovering their historical roots and are renewing their traditional cultures. All of these developments have heightened global interest in and appreciation of the South Pacific's pre-European past.

Go To **World Music: A Cultural Legacy** Hear music of this region on Disc 2, Tracks 25–29.

Early Migrations

Various peoples from Asia settled the region of Australia and Oceania more than 40,000 years ago. Some may have migrated to Australia over land bridges during the Ice Age, when ocean levels were much lower than they are today. Others probably used canoes and rafts to reach the South Pacific region. The reason they came to these areas is a mystery. Because of their connection to the sea, some of these peoples, especially those who came to the South Pacific region, may have regarded exploration as a natural part of daily life. Author Peter Crawford, impressed with the daring of these early explorers, described the early Polynesians:

> *"A tenacious, seafaring people had abandoned the shores of [S]outheast Asia and sailed into the Pacific. As their culture developed, they acquired new skills of survival, and new knowledge of the ocean world which became their home. . . . The vibrant Polynesian culture that grew and flourished . . . is testament to the invention and adaptability of its people."*
>
> Peter Crawford, *Nomads of the Wind: A Natural History of Polynesia*, 1993

Economics

Indigenous Lifestyles

In the hot, dry Australian interior, the early Aborigines led a nomadic life. They used well-traveled routes to reach water and seasonal food sources. These same routes made trading and social exchanges possible. **Clans**, or family groups, traveled together within their ancestral territories, carrying only baskets, bowls, spears, and sticks for digging. To hunt animals, Aboriginal men used a heavy throwing stick, called a **boomerang**, that soars or curves in flight, and the women and children gathered plants and seeds.

In Oceania people settled in family groups along island coasts. For food they relied on fish, turtles,

MAP STUDY

Australia and Oceania: Colonies, 1900

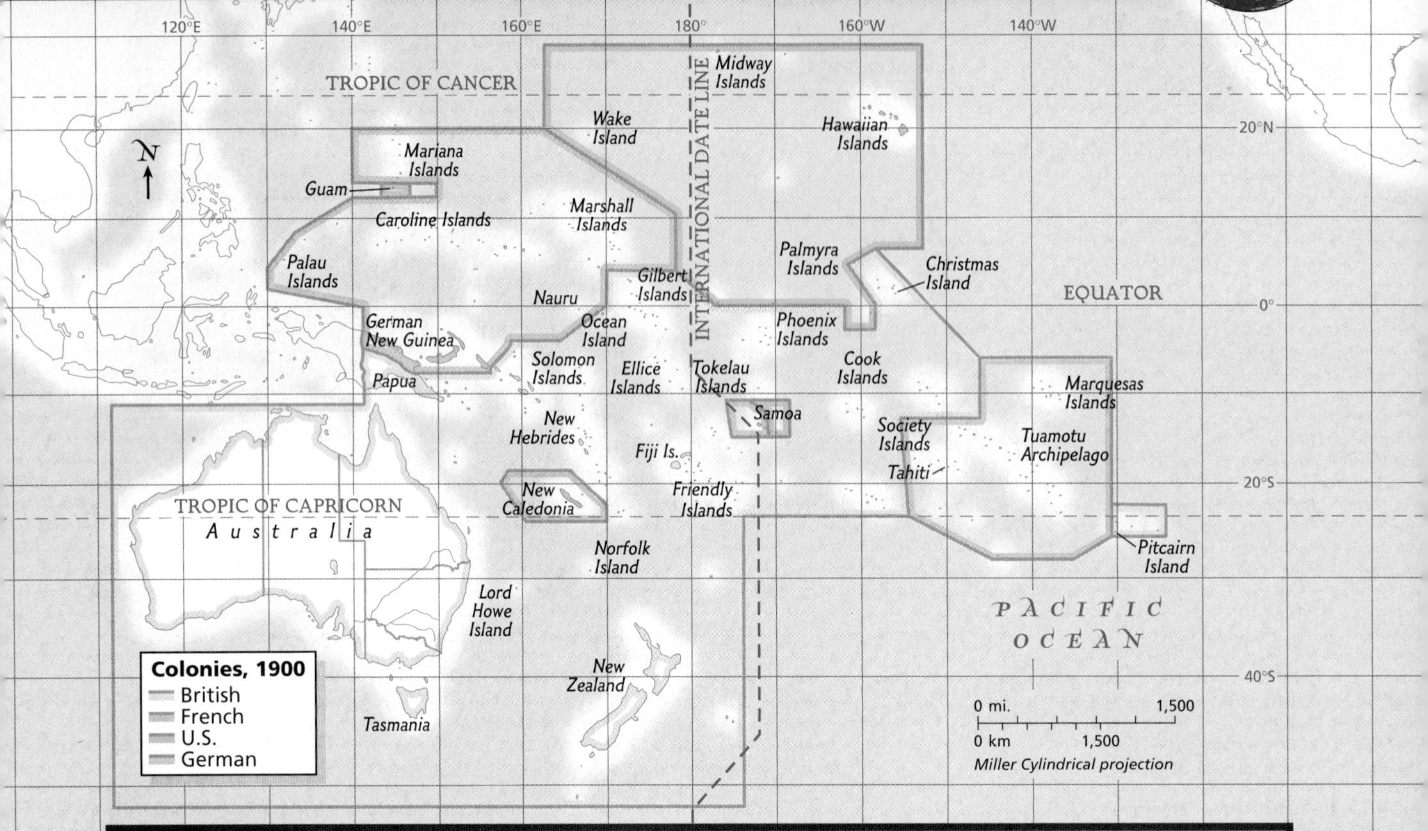

Geography Skills for Life

1. **Interpreting Maps** What two countries ruled the area of Samoa?

2. **Applying Geography Skills** Why do you think the United States took over Guam?

Find NGS online map resources @ www.nationalgeographic.com/maps

and shrimp as well as breadfruit and coconuts. Pacific islanders also cultivated root crops, such as taro and yams, and raised smaller animals, such as chickens and pigs. Well-built canoes made lengthy voyages possible, and trade gradually developed among the islands. To make trading easier, people on some islands used long strings of shell pieces as money. Today in New Britain, an island off the northeast coast of Papua New Guinea, shell money still is exchanged for canned goods or vegetables at markets.

Increased trade was accompanied by migrations among the islands. Between the A.D. 900s and 1300s, the Maori people left eastern Polynesia and settled the islands of New Zealand. On New Zealand's North Island and South Island, Maori groups hunted, fished, established villages, and farmed the land. Maori farmers, like the Pacific islanders, grew root crops, such as taro and yams, which they had brought from their Polynesian homeland.

European Colonization

From the 1500s to the 1700s, Europeans of various nationalities explored vast stretches of the South Pacific region. Perhaps the most well-known explorer was the British sailor James Cook, who undertook three voyages to the region between 1768 and

1779. Cook claimed eastern Australia for Great Britain, visited various South Pacific islands, circled Antarctica, and produced remarkably accurate records and maps of these places.

European Settlement

Great Britain at first used Australia as a colony for convicts sent out from overcrowded British prisons. The first shipload of prisoners arrived at Botany Bay, in what is today Sydney, in 1788. By the early 1850s, the imprisonment of British convicts in Australia had ended, and growing numbers of free British settlers were establishing coastal farms and settlements. Livestock, especially sheep, were introduced to the continent. As British textile manufacturers increased their demand for wool, Australian settlers profited greatly from exporting wool to the parent country. Another source of wealth for Australia was gold, which was discovered there in the early 1850s. The resulting gold rush nearly tripled Australia's population in 10 years and also promoted the mining of other mineral resources in the continent's interior.

Meanwhile, the British and other Europeans were also establishing settlements in Oceania. Attracted by excellent fishing waters and rich soil, British settlers arrived in New Zealand in the early 1800s. They brought with them sheep, cattle, and horses. By the end of the century, raising livestock had become a major part of New Zealand's economy. On some South Pacific islands, European businesspeople set up commercial plantations for growing sugarcane, pineapples, and other tropical products.

Indigenous Peoples

The arrival of Europeans in Australia and Oceania had a disastrous impact on indigenous peoples. As British migrants spread across Australia, they forcibly removed the Aborigines from the land and denied them basic rights. Many Aborigines resisted the European advance, but European diseases and weapons steadily reduced the Aboriginal population. In the mid-1800s, British-Australian authorities placed many Aborigines in reserves, or separate areas.

British settlement in New Zealand brought hardships to the Maori, who died from diseases carried by the newcomers. The Maori social structure also was weakened when the British colonists introduced new ways of farming and other aspects of European culture. As the number of European settlers increased, the British and some Maori chiefs signed a treaty in 1840 that guaranteed the Maori full rights under the British monarchy. Disagreements about the treaty, however, led to armed Maori resistance to British rule over the next 15 years. During these conflicts, many Maori were killed, and the Maori gradually lost most of their land to the British.

The Europeans also brought far-reaching changes to the other peoples of Oceania. Because European diseases had reduced indigenous island populations, the Europeans brought in workers from other Pacific islands and from more distant areas, such as South Asia. The resulting mix of cultures weakened indigenous societies and eventually led to ethnic conflicts. Meanwhile, Europeans sought to replace traditional ways of life with European beliefs and practices.

Struggle for Power

During the late 1800s and early 1900s, Britain, France, Germany, Spain, and the United States struggled for control of various Pacific islands. Many of these countries already had commercial interests in the area. The Europeans hoped to expand their influence and gain new sources of raw materials.

The two World Wars changed the course of Oceania's history. After World War I, many of Germany's Pacific colonies came under Japanese rule. Then in December 1941, Japanese airplanes bombed the United States Naval Base at Pearl Harbor in Hawaii. This attack brought the United States into World War II. During the conflict the United States and Japan fought a number of fierce battles on Pacific islands such as Guadalcanal and Iwo Jima.

Following Japan's defeat in World War II, Japan's South Pacific possessions, such as the islands of Micronesia, were turned over to the United States as trust territories. **Trust territories** were dependent areas that the United Nations placed under the temporary control of a foreign country. Since the 1970s most of these islands, including Palau, the Marshall Islands, and the Federated States of Micronesia, have become independent countries.

Independent Governments

Independence came to most of the South Pacific region during the 1900s. Australia and New Zealand became the region's first independent countries in the early 1900s. By the century's end, most of the Pacific islands had gained independence.

Australia and New Zealand

Australia and New Zealand both peacefully won their independence from British rule. In 1901 Britain's Australian colonies became states, united to form the Commonwealth of Australia. The new Australian country was a **dominion**, a largely self-governing country within the British Empire. Australia's form of government blended a United States-style federal system with a British-style parliamentary democracy. The British monarch—represented by a governor general—served as Australia's head of state, but a prime minister actually headed the national government.

In 1907 New Zealand became a self-governing dominion with a British parliamentary system. New Zealanders, however, contributed some political "firsts" of their own. In 1893 New Zealand became the first country in the world to legally recognize women's right to vote. New Zealand also was among the first countries to provide government assistance to the elderly, the sick, and the unemployed.

Until World War II, New Zealand and Australia maintained close economic, military, and political ties to Great Britain, now known as the United Kingdom. After 1945 British global influence declined, and the two Pacific countries looked increasingly to the United States for trade and military protection. In addition, Australia and New Zealand developed their own national characters based on increasingly diverse populations. The Aborigines and the Maori won greater recognition of their unique cultural identities, and many non-British immigrants settled in both countries. In the 1990s, Australians and New Zealanders took the first steps toward cutting ties to the British monarchy as a result of these changes. Many Australians today want their country to become a republic, with an Australian-born president as head of state.

NATIONAL GEOGRAPHIC **World Explorer**

Geography Skills for Life

Promoting Culture An indigenous group in Papua New Guinea perform traditional ceremonial dances.

Place When did most countries in Oceania gain their independence?

South Pacific Islands

Beginning in the 1960s, a number of the small islands in Oceania moved toward independence. Samoa—formerly Western Samoa—had been ruled by Germany until New Zealand assumed control after World War I. In 1962 Samoa became the first Pacific island territory to win its freedom. Today most of the South Pacific islands enjoy some form of independent government. For example, **Vanuatu**, once jointly governed by the United Kingdom and France, is a republic, and **Tonga**, formerly under British protection, is a constitutional monarchy. Some island countries, such as Fiji and the Solomon Islands, have been torn by ethnic conflict since independence. Many conflicts have roots in colonial times, when European rulers brought in foreign workers from other cultures, ignoring traditional ethnic and cultural patterns.

Antarctica

Europeans first sighted Antarctica during the early 1800s, but they believed that the icy continent had little, if any, commercial value. As a result, expeditions did not venture into Antarctica until much later. In the early 1900s, Norwegian explorer Roald Amundsen and British explorer Robert Scott, each with a team of four people, engaged in a dramatic race to be the first to reach the South Pole. Amundsen's team reached it on December 14, 1911; Scott's team arrived about a month later. Unfortunately, Scott and his team died on the return trip.

The race for the South Pole opened the rest of Antarctica for exploration. On their quests, Antarctic explorers looked for economic resources as well as adventure in the frozen landscape. The countries they represented hoped for new trading routes and seal-hunting areas as well as Antarctic mineral resources. Nonetheless, much of Antarctica remained unexplored until advances in radio communication and air travel made exploration easier and safer.

By the 1960s, scientists from 12 countries had established research centers in Antarctica. To preserve Antarctica as a peaceful scientific research site, the 12 countries signed the Antarctic Treaty in 1959. Since then, a number of other countries have agreed to abide by the treaty. In 1991 the treaty countries made an additional agreement to prohibit mining and to protect the environment of this unique continent.

NATIONAL GEOGRAPHIC World Explorer

Geography Skills for Life

Cold as Ice The Amundsen-Scott Station, managed by the United States, lies close to the South Pole on ice nearly 2 miles (3.2 km) deep.

Place What is the purpose of the Antarctic Treaty?

SECTION 2 ASSESSMENT

Checking for Understanding

1. **Define** clan, boomerang, trust territory, dominion.
2. **Main Ideas** Use a chart like the one below to organize factors that contributed to the region's cultural diversity and forms of government.

Indigenous Peoples	European Colonization	Power Struggles

Critical Thinking

3. **Identifying Cause and Effect** What effects, both positive and negative, resulted from European colonization in this region?
4. **Comparing and Contrasting** Compare and contrast the views of South Pacific indigenous peoples and European settlers about the land—its value, ownership, and use.
5. **Making Generalizations** How has Antarctica benefited from international cooperation?

Analyzing Maps

6. **Place** Study the map on page 785. Identify an island or a group of islands that is today under the rule of the United States.

Applying Geography

7. **Physical Geography and Migration** Think about why and how people and goods moved throughout Oceania. Write a paragraph describing the reasons for this migration and how significant physical features influenced it.

GEOGRAPHY AND HISTORY

JOURNEY TO THE BOTTOM OF THE WORLD

AN EXPEDITION CARRYING TONS OF CANDY, 500 cases of eggs, and 60,000 sheets of writing paper? Where might such an expedition be headed? What conditions would warrant such provisions? These were a small portion of the supplies on a ship that left New York City in 1928, headed for Antarctica. Also aboard were United States Navy officer Richard E. Byrd and a crew of 53 scientists and other professionals.

Lure of the Unknown

The Byrd Antarctic Expedition set out to establish a foothold in one of the most ferocious climates on Earth. Antarctica is the world's coldest place, where winter temperatures can drop to -129°F (-89°C). Thick ice buries most of the continent. Violent winds lash the Polar Plateau, where the South Pole lies. Glaciers spill out between mountain peaks that rim the coast, creating vast ice shelves that limit access by sea.

In 1928 little was known about Antarctica. Whalers and sealers had hunted its coastal waters in the 1800s. In 1911 Antarctica was the site of the tragic race to the South Pole—Roald Amundsen of Norway made it back, while British explorer Robert Falcon Scott and his team perished. Other than these brave souls, few people had ventured inland. Admiral Byrd was determined to change that.

Admiral Byrd (at left) and companions visit Little America. An American stamp commemorates one of Byrd's expeditions to Antarctica. ▶

◀ **Crunching through ice, the U.S.S. *Glacier* brings Admiral Byrd and crew back to Antarctica in 1955.**

NATIONAL GEOGRAPHIC SOCIETY

Little America

Before leaving New York, Byrd spent three years preparing for the inhuman conditions in Antarctica. He worked with numerous experts to determine vital supplies. Clothing was especially important. Reindeer fur proved the warmest and was used for parkas, pants, and boots. Other animal skins, such as sealskin, also were used.

Stopping in New Zealand, the last outpost of civilization, the expedition still had to negotiate iceberg-filled waters to reach the Ross Ice Shelf, the thick expanse of Antarctic ice that would be home for 14 months. Arriving in late 1928, the crew and 80 sled dogs moved more than 650 tons (590 t) of material from ship to shore. The crew built the first scientific station on the frozen continent. A village complete with multiple weather-tight buildings, bunkhouses, and storerooms, the station was named Little America.

Once Little America was established, Byrd launched his assault on Antarctica. Using an airplane he had brought by ship, Byrd and his crew made numerous flights over vast areas never seen by humans. Byrd's expedition accomplished many firsts: a flight over the South Pole, the mapping of 150,000 square miles (388,000 sq. km) of new territory, the invention of specialized instruments, and more.

Byrd returned to Antarctica four more times to supervise the completion of Little America II through V. His expeditions laid the groundwork for future research and international cooperation. Today the United States and many other countries maintain scientific stations in Antarctica. Scientists work on a variety of projects there, from studying animal behavior to monitoring ozone depletion and global warming.

1821 American seal hunters make first known landing on Antarctica

1901 British explorer Robert F. Scott begins first inland exploration

1911 Norwegian Roald Amundsen is first to reach South Pole

1928 Admiral Byrd (on medal, above) establishes Little America I (background photo)

1929 Byrd makes first flight over South Pole

1933–1955 Byrd establishes Little America II through V

1959 Twelve countries sign Antarctic Treaty, preserving Antarctica for peaceful endeavors

Looking Ahead

How did the hardships and dangers Byrd and his comrades endured benefit humankind? How might Antarctic research be important to the future of life on Earth?

Cultures and Lifestyles

Guide to Reading

Consider What You Know

World music, which includes musical expressions from many cultures, has become very popular in the United States. What instruments or types of music have you heard that come from other parts of the world?

Read to Find Out

- What role does religion play in the region's cultures?
- How have the peoples of Australia and Oceania expressed their heritages through the arts?
- How does everyday life in the region reflect cultural diversity?

Terms to Know

- subsistence farming
- *fale*

Places to Locate

- Papua New Guinea
- Samoa

NATIONAL GEOGRAPHIC

A Geographic View

Living in Australia

We're connected to Europe and North America culturally, but we're in an Asian time zone, which gives us an advantage. We have a highly educated workforce, . . . a first-rate international airport, good communications, and a stable and sophisticated financial system. We have a wonderful climate and attractive lifestyle—good restaurants, nice beaches, an optimistic way of looking at the world that I think outsiders find attractive. Once you develop a critical mass of those things, you find that more and more people want to come and be part of it.

—Sydney mayor Frank Sartor, quoted by Bill Bryson, "Sydney," National Geographic, *August 2000*

A girl visits Sydney, Australia

Australia, like other South Pacific countries, blends both European and indigenous elements in its culture. In recent years Asian influences also have increased in the region. In this section you will learn about the religions, arts, and lifestyles of the peoples of Australia and Oceania.

A Blend of Cultures

The movement of different peoples into the South Pacific region has contributed to the shaping of cultures there. Indigenous peoples developed lifestyles in harmony with their natural environment. Later, European immigrants brought their ways of life, using the environment to build Western-oriented societies.

NATIONAL GEOGRAPHIC

GRAPH STUDY

Australia and Oceania: Religions

Religion	Number of Followers
Roman Catholic	8,097,000
Protestant	7,279,000
Anglican	5,386,000
Eastern Orthodox	691,000
Hindu	349,000
Other religions	1,232,000
Nonreligious	3,628,000

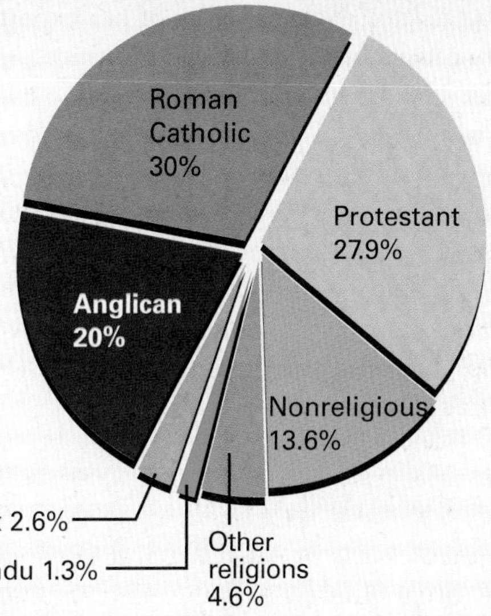

Roman Catholic 30%
Protestant 27.9%
Anglican 20%
Nonreligious 13.6%
Eastern Orthodox 2.6%
Hindu 1.3%
Other religions 4.6%

Sources: Britannica Book of the Year, 2000; *World Almanac,* 2001

Geography Skills for Life

1. **Interpreting Graphs** Which two religions have the largest number of followers in the region?

2. **Applying Geography Skills** Why do you think Christian religions are dominant in the region?

Religion

The religious traditions of the region's indigenous peoples focus on the relationship of humans to nature. Australia's Aborigines, for example, believe in the idea of Dreamtime, the early time when they say wandering spirits created land features, plants, animals, and humans. They believe that all natural things—rocks, trees, plants, animals, and humans—have a spirit and are interrelated. Europeans later brought Christianity, which attracted many followers among the indigenous peoples. Christianity is the most widely practiced religion in Australia and Oceania today.

The Arts

South Pacific peoples traditionally used art, music, dance, and storytelling to pass on knowledge from generation to generation. Australian Aborigines, for example, recorded their past in rock paintings and developed songs to pass on information about routes and landmarks. In New Zealand, Maori artisans developed skills in canoe making, basketry, tattooing, and woodcarving. Today Maori meeting houses are decorated with elaborate wood carvings.

After a time of copying European themes and styles, European artists in the region began looking to the South Pacific environment for inspiration.

> *"Strong emotional ties with the land . . . are not the sole preserve of the Aborigines. . . . Australian writers and poets, . . . composers and painters [have] come to realise that a tangle of eucalyptus trees, red gums in a dried-up steam bed, red rocks and dripping rain forest can have their own powerful visual appeal."*
>
> Roger Fenby, "Walkabout Oz," *BBC World Service* (online), August 4, 2000

In recent decades the South Pacific region has produced a number of outstanding musicians, writers, and artists. Australia's Joan Sutherland and New Zealand's Kiri Te Kanawa became famous opera performers. New Zealand author Sylvia Ashton Warner wrote of her experiences as a schoolteacher in Maori communities. Australian writer Thomas Keneally wrote the novel *Schindler's List*, which was later made into an award-winning motion picture.

Australia and New Zealand also have contributed well-known movie stars such as Mel Gibson, Nicole Kidman, and Russell Crowe. Filmmakers in both countries have made popular motion pictures, such as *Gallipoli*, *Crocodile Dundee*, *Muriel's Wedding*, and *The Piano*.

Everyday Life

In many parts of Australia and Oceania, people have urban lifestyles that reflect modern influences. In other places in the region, people live in a more traditional way.

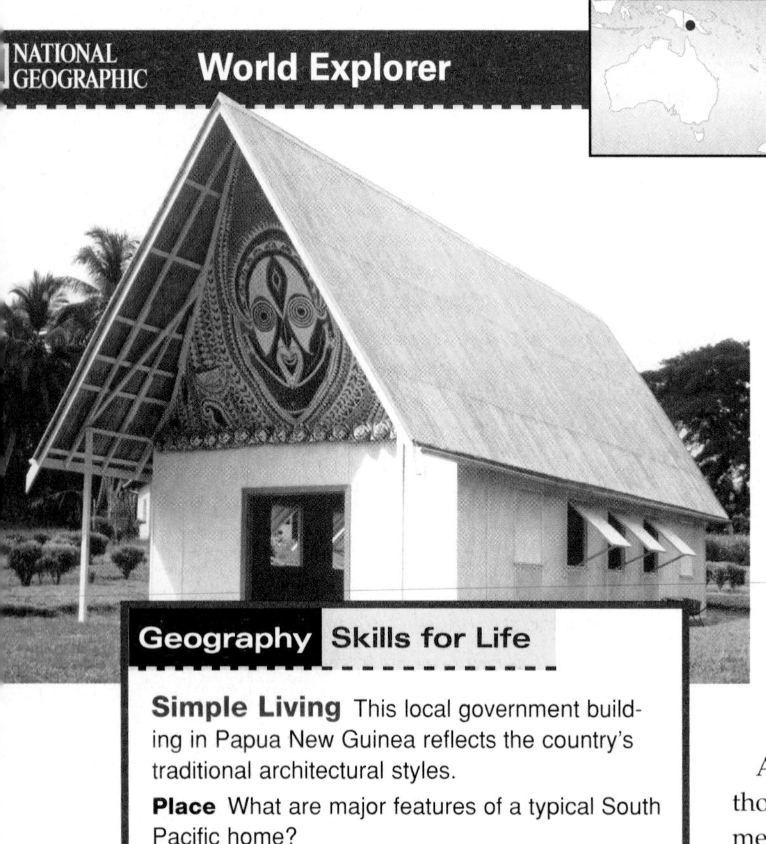

Geography Skills for Life

Simple Living This local government building in Papua New Guinea reflects the country's traditional architectural styles.

Place What are major features of a typical South Pacific home?

Economics

Traditional and Modern Lifestyles

Some Pacific island countries, such as **Papua New Guinea**, produce cash crops, including sugarcane, coffee, ginger, and copra—dried coconut meat. Others, such as Kiribati, have soil that is too poor for plantation agriculture. Many Pacific islanders work at **subsistence farming**, growing only enough for their own needs. These farmers grow bananas, coconuts, or sweet potatoes; raise chickens and pigs; or fish in ocean waters. Other islanders work in government offices, in the tourist trade, or in other service industries.

Kinship ties are the basis of traditional life throughout the region, but these bonds have weakened as young people find better job opportunities elsewhere. Even so, important events draw distant family members back home and help maintain the culture.

A typical traditional South Pacific home is very simple in design. On **Samoa**, this type of simple dwelling is called a *fale* and has a thatched roof for shelter and open sides that allow cooling ocean breezes to circulate. Blinds of coconut palm leaves can be lowered for privacy.

The simplicity of South Pacific island life contrasts greatly with the fast-paced, urbanized lifestyle in parts of Australia and New Zealand, where people are linked to the cities by roads and modern communications technology. A mild climate and nearness to the sea enable many people in the South Pacific region to enjoy outdoor activities.

Education and Health Care

The quality of education varies throughout the region. Both Australia and New Zealand provide free, compulsory education until age 15. Literacy rates are high in these two countries, and many students attend universities. Many students in Australia's remote outback receive and turn in assignments by mail or communicate with teachers by two-way radios.

Australians and New Zealanders, especially those in cities, generally have access to quality medical care and other social services. In some parts of Australia, rugged terrain and long distances make access to health care difficult. Modern technology, however, allows doctors to consult with patients through the use of two-way radios and through mobile clinics of the Flying Doctor Service.

Indigenous peoples, however, often do not receive these and other benefits. For example, many Aborigines suffer from poverty, malnutrition, and unemployment. In recent years the Australian government and private organizations have been trying to make up for past injustices, and the courts have recognized the claims of Aborigines to government assistance, land, and natural resources.

Many Pacific islanders also lack an adequate standard of living. On remote islands, fresh food, electricity, schools, and hospitals often are limited. Recently island countries, with international assistance, have begun to improve their quality of life.

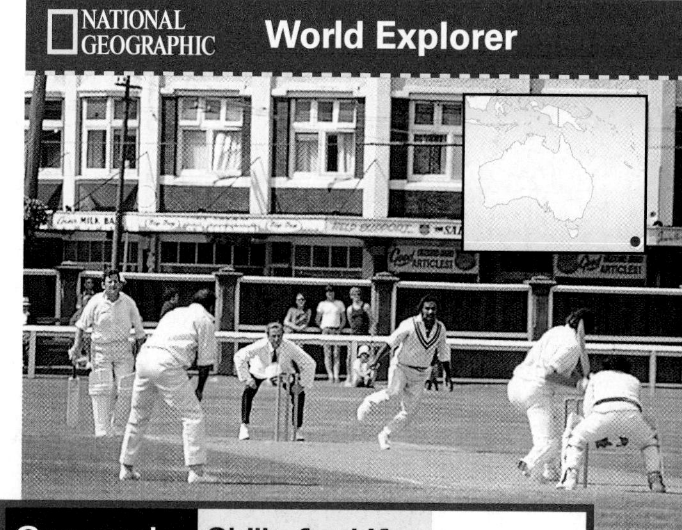

Geography Skills for Life

Cricket Cricket, first played in England during the late 1500s, today is a popular sport in New Zealand. **Place** What other sports are played in Australia and Oceania?

Sports and Leisure

Sports and leisure activities reflect the region's diversity. Western-style resorts attract tourists to the beaches, where they and the local people enjoy the traditional Pacific island sport of surfing. Traditional sports, such as outrigger canoe racing or spearfishing, are popular, as are Western sports. For example, British settlers brought cricket and rugby to Australia and New Zealand.

In former American territories, islanders play baseball. The French introduced cycling and archery to islands they controlled. Even small communities often have facilities for these and other sports, such as soccer, volleyball, and tennis. In urban areas of Australia and New Zealand, where Western influence dominates, leisure activities include boating, fishing, waterskiing, and other water sports along the metropolitan beaches.

In the next chapter, you will learn how people in Australia and Oceania are meeting the challenges of their environment.

SECTION 3 ASSESSMENT

Checking for Understanding

1. **Define** subsistence farming, *fale*.
2. **Main Ideas** On a web like the one below, fill in important ideas and supporting details from each section to describe the culture and lifestyles of the region.

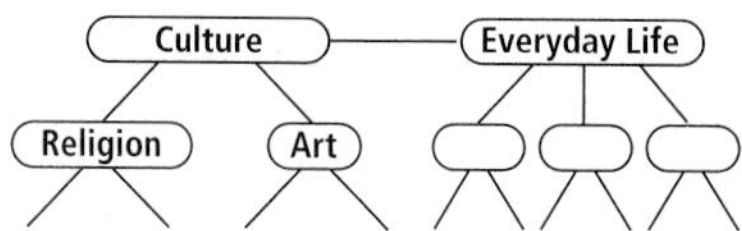

Critical Thinking

3. **Drawing Conclusions** How does the art of the South Pacific reflect the artists' physical environment?
4. **Comparing and Contrasting** How does education in the South Pacific differ from education in your community?
5. **Predicting Consequences** How might the Internet and e-mail change education in Oceania?

Analyzing Maps

6. **Location** Examine the political map on page 785. Which Pacific islands are administered by New Zealand?

Applying Geography

7. **Forms of Government** Compare the political-physical maps on pages 117 and 795. How might Australia, Canada, and the United States be similar in the way they distribute governmental powers?

Analyzing LANDSAT Images

Scientists and researchers who study the earth use satellites to help them gather data. Photographs taken by these satellites orbiting the earth provide a detailed record of conditions and changes on the earth's surface.

Learning the Skill

Scientists use LANDSAT images to receive a broad view of the surface of the earth. LANDSAT refers to a series of observation satellites that have been launched by the United States since 1972. The most recent satellite, LANDSAT 7, was launched on April 15, 1999. Orbiting at an altitude of 500 miles (805 km), LANDSAT spacecraft have recorded millions of images of the earth.

The main purpose of LANDSAT is to map and monitor natural resources and changes to the environment. Farmers, government officials, environmentalists, and the military use LANDSAT data, which can be helpful in making decisions that affect the health of the planet. For example, these satellites can identify the locations of tropical forests and provide information about the rates and effects of deforestation.

One of LANDSAT's main benefits is its ability to capture images of every place on Earth. LANDSAT 7 completes a full orbit of the earth every 99 minutes, allowing over 14 orbits a day. LANDSAT 7 is able to provide photographic coverage of the entire earth in only 16 days.

Follow these steps to analyze a LANDSAT image:

- **Read the title.** This feature explains the data being collected, the location, and the time period.

Deforestation in Rondônia, Brazil

1975

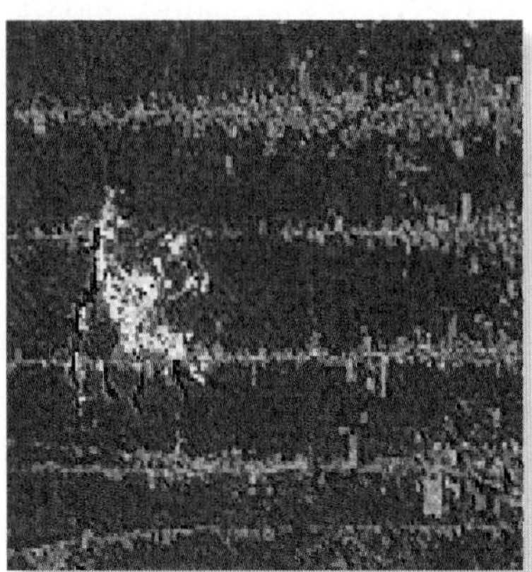

1992

- **Study the image carefully.** In the images on this page, red indicates healthy vegetation, light blue areas indicate deforested land, and light blue lines indicate roads.
- **Compare different images of the same place.** Notice changes that occur over time.
- **Think about what has caused the changes and how they may affect the area's physical and human geography.**

Practicing the Skill

The LANDSAT images on this page show an area of tropical forest in Brazil as it appeared in 1975 and in 1992. Use the images to answer the following questions.

1. Which image shows large areas of undisturbed tropical forest?
2. Compare the area in 1975 to the area in 1992.
3. How is the pattern of deforestation in the 1992 image connected to roadways?
4. How do you think these areas have changed in today's LANDSAT images? Explain your reasoning.

Locate LANDSAT images of Australia, Oceania, or Antarctica on the Internet. For each image, list its location and the kinds of data it includes. Choose one image, and write a paragraph describing two possible uses for the data.

33

SECTION 1 Population Patterns (pp. 811–815)

Terms to Know

- Strine
- pidgin English

Key Points

- Many different peoples settled in the South Pacific, resulting in diverse cultures and lifestyles.
- The population of the South Pacific is unevenly distributed because both the physical geography and the climate differ dramatically from place to place and because many areas cannot support life.
- Migration between and within South Pacific countries has influenced population patterns and caused a blending of cultures.

Organizing Your Notes

Use a graphic organizer like the one below to help you organize your notes about the population patterns of the South Pacific.

Populations	Migration

SECTION 2 History and Government (pp. 816–821)

Terms to Know

- clan
- boomerang
- trust territory
- dominion

Key Points

- Many of the area's earliest inhabitants came from Southeast Asia and survived by hunting, gathering, and, in some cases, farming.
- European countries were attracted to the area by its raw materials, rich fishing areas, and fertile coastal land.
- During the late 1800s and early 1900s, European countries, Japan, and the United States sought possessions in the region.
- Australia, New Zealand, and a number of Pacific islands are independent; a few island groups are still under foreign rule.

Organizing Your Notes

Create an outline using the format below to help you organize your notes for this section.

History and Government

I. Indigenous Peoples
   A. Early Migrations
   B. Indigenous Lifestyles
II.

SECTION 3 Cultures and Lifestyles (pp. 824–827)

Terms to Know

- subsistence farming
- *fale*

Key Points

- The culture of the South Pacific is a mixture of Western and indigenous lifestyles.
- Some people in the area still live in traditional villages; others live in modern urban areas.
- Modern technology helps provide services to people in some remote areas.

Organizing Your Notes

Use a web like the one below to help you organize your notes for this section.

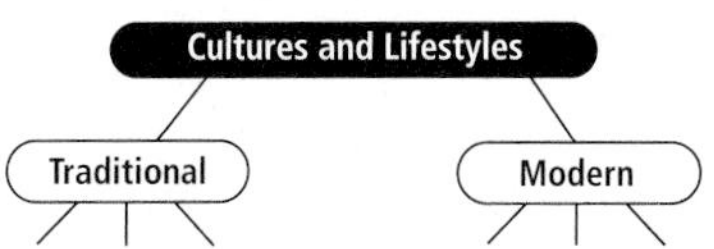

CHAPTER 33

ASSESSMENT & ACTIVITIES

Reviewing Key Terms

Write the key term that best completes each of the following sentences. Refer to the Terms to Know in the Summary & Study Guide on page 829.

1. The Micronesian islands became a(n) _______ after World War II.
2. In some parts of Oceania, _______ is spoken.
3. Australians speak _______, a dialect of English.
4. A(n) _______ provides simple shelter on tropical islands.
5. In 1901 Australia became a(n) _______ of Great Britain.
6. Some islanders still make their livings by _______.
7. The _______ was originally a hunting tool.
8. Each aboriginal family group traveled as a(n) _______.

Reviewing Facts

SECTION 1

1. Who were the original settlers of Australia, New Zealand, and Oceania?
2. How has geography influenced settlement patterns in the region?

SECTION 2

3. What ways of life did Pacific indigenous peoples practice?
4. In what ways did European settlement influence the region?
5. Why was the Antarctic treaty established in 1959?

SECTION 3

6. How have the arts enriched life in the South Pacific region?
7. What are some characteristics of modern lifestyles in Australia, New Zealand, and Oceania?

Critical Thinking

1. **Identifying Cause and Effect** How did the South Pacific's physical geography contribute to its cultural diversity?
2. **Comparing and Contrasting** In what ways were European influences similar in Australia and in New Zealand? Different?
3. **Problem Solving** Use a Venn diagram to compare the lifestyles and living standards of indigenous and European peoples in the region.

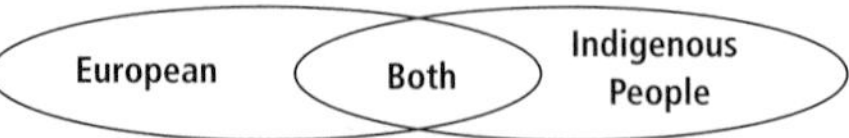

NATIONAL GEOGRAPHIC

Locating Places

Australia and Oceania: Political Geography

Match the letters on the map with the places in Australia and Western Oceania. Write your answers on a sheet of paper.

1. Papua New Guinea	4. Melbourne	7. Brisbane
2. Sydney	5. Canberra	8. Adelaide
3. Auckland	6. Perth	9. Wellington

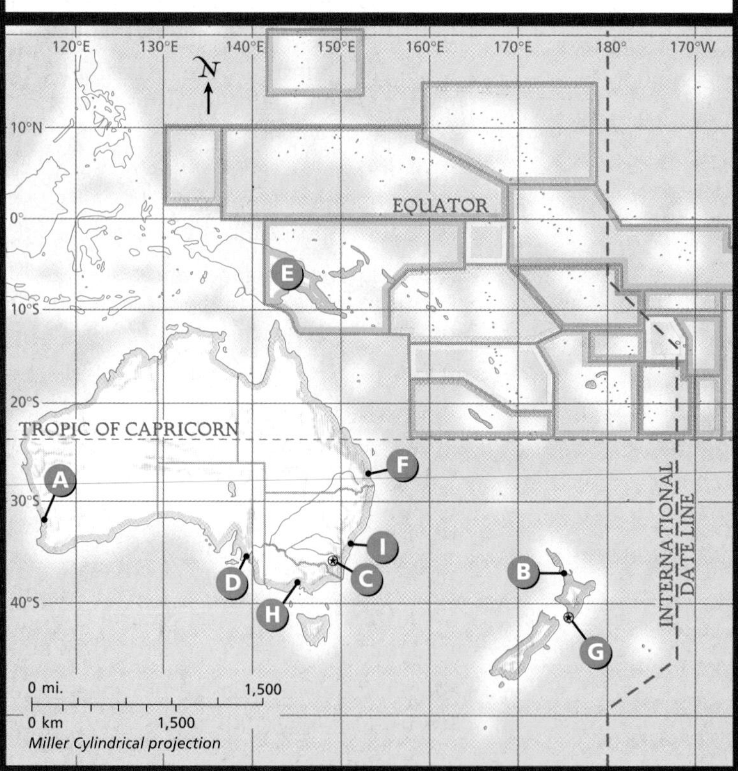

Self-Check Quiz Visit the **Glencoe World Geography** Web site at geography.glencoe.com and click on Self-Check Quizzes—Chapter 33 to prepare for the Chapter Test.

Using the Regional Atlas

Refer to the Regional Atlas on pages 784–787.

1. **Region** What part of Australia has most of the country's coal deposits?
2. **Human-Environment Interaction** Which physical features limit economic activity in central Australia?

Thinking Like a Geographer

Analyze the effects of processes, such as migration and colonization, on the traditional cultures of the South Pacific's indigenous peoples.

Problem-Solving Activity

Contemporary Issues Case Study The issue of land rights in Australia and New Zealand involves cultural divergence, or separation, between indigenous peoples and those currently using the land. Use print resources and the Internet to research the opposing viewpoints on this issue. Then, write a paragraph stating a possible solution.

GeoJournal

Expository Writing Using the information you logged in your GeoJournal as you read this chapter, write a paragraph comparing and contrasting two cultures in the region. Use your textbook and the Internet as resources to make your information as clear and accurate as possible. Provide visuals to illustrate your ideas.

Technology Activity

Using the Internet for Research Use the Internet to research a specific cultural group in the region. Identify at least three Web sites you used in your research. After you have completed your research, create a poster to illustrate one aspect of the group's culture, such as homes, clothing, or the arts.

Standardized Test Practice

Choose the best answer for each of the following multiple-choice questions. If you have trouble answering the questions, use the process of elimination to narrow your choices.

1. **Which of the following has NOT influenced population distribution in the South Pacific?**
   - A European colonization
   - B Geographic features
   - C Animal domestication
   - D Climate

Read the question carefully to determine what is being asked. For each answer choice, consider what factors may have the capacity to influence population distribution. Eliminate answer choices in which a direct correlation may be made. Do not forget to incorporate your knowledge of the region and cultures into your decision.

2. **When a group of people is described as *indigenous*, it means that they are**
   - F highly dependent on the agriculture of a region.
   - G the governing party of a region.
   - H the earliest inhabitants of a land.
   - J a culturally isolated group.

Consider all of the times you may have heard this word used and in what context you heard it being used. Try to find elements with the structure of the word, such as suffixes, prefixes, and roots, that may reveal something about its meaning.

CHAPTER 34

Australia, Oceania, and Antarctica Today

GeoJournal

As you read this chapter, use your journal to log the key economic activities of Australia, Oceania, and Antarctica. Note interesting details that illustrate the ways in which human activities and the region's environment are interrelated.

GEOGRAPHY Online

Chapter Overview Visit the **Glencoe World Geography** Web site at geography.glencoe.com and click on Chapter Overviews—Chapter 34 to preview information about the region today.

Guide to Reading

Consider What You Know

Environments in Australia, Oceania, and Antarctica range from tropical rain forests to icy wastelands. What attractions or activities might draw people to visit or live in a region with such extreme differences in the physical environment?

Read to Find Out

- How do people in Australia, New Zealand, and Oceania make their livings?
- What role does trade play in the economies of South Pacific countries?
- What means of transportation and communications are used in the region?

Terms to Know

- station
- grazier
- copra

Places to Locate

- Fiji
- Papua New Guinea
- Nauru

◀ *Skyline of Melbourne, Australia, at night*

Living in Australia, Oceania, and Antarctica

NATIONAL GEOGRAPHIC

A Geographic View

Antarctic Diving

There's something special about peering beneath the bottom of the world. When Antarctica's summer diving season begins in September the sun has been largely absent for six months, and the water . . . has become as clear as any in the world. Visibility is measured not in feet but in football fields. . . . Only here can you orbit an electric-blue iceberg while being serenaded by the eerie trills of Weddell seals.

—*Norbert Wu, "Under Antarctic Ice,"* National Geographic, *February 1999*

View from under Antarctic ice

The wonders hidden under Antarctic ice are among the many attractions of Australia, Oceania, and Antarctica. Tourism is a growing part of the region's economies. In this section you will learn how people in Australia and Oceania earn their livings despite remote geographic locations and challenging environments.

Agriculture

Agriculture is by far the most important economic activity in the South Pacific area. Australia and New Zealand—the region's major developed countries—export large quantities of farm products. Australia is the world's leading producer of wool, and New Zealand is known for the quality of its dairy products, lamb, beef, and wool.

Geography Skills for Life

Cattle Station

A rancher rounds up cattle on a station in southern Australia.

Place Why are Australian ranches so large?

Although only 5 percent of Australians work in agriculture, much of their country's vast land area is devoted to raising livestock—primarily sheep and cattle. Because of the generally dry climate, ranchers must roam over large areas to find enough vegetation to feed their herds. As a result, some Australian ranches, called **stations**, are gigantic—as large as 6,000 square miles (15,540 sq. km), about the size of Connecticut or Hawaii.

In addition, because of Australia's dry climate, only about 10 percent of its land is suitable for growing crops. Irrigation, fertilizers, and modern technology help Australian farmers make the best use of their limited croplands. Wheat, for example, is grown in the dry Central Lowlands. By contrast, sugarcane thrives in the wetter climate and fertile soil of Australia's northeastern coast.

About half of New Zealand's land is used for agriculture. New Zealand ranchers, known as **graziers**, raise sheep, beef, dairy cattle, and red deer. Surprisingly, the country has 25 times more farm animals than people! New Zealand's soil, more fertile than that of Australia, allows farmers to grow wheat, barley, potatoes, and fruits. One of New Zealand's most distinctive fruits is the kiwifruit, a small, green-fleshed fruit named for its resemblance to the kiwi, the flightless bird that is the country's national symbol.

Throughout Oceania, the lack of arable soil limits commercial agriculture. As a result, most island farmers practice subsistence farming. They grow starchy roots and tubers—taro, cassava, and sweet potatoes—and raise pigs and chickens. Fishing adds to the diet of many South Pacific peoples.

Some South Pacific islands, however, have areas of rich soil—often volcanic—and ample rainfall. These islands produce a variety of crops, such as tropical fruits, sugarcane, coffee, and coconut products, for export. The major South Pacific cash crop, produced widely in the region, is **copra** (KOH•pruh), or dried coconut meat. Among the island countries that export are **Fiji**, a producer of sugarcane, copra, and ginger, and **Papua New Guinea**, a supplier of coffee, copra, and cacao.

Mining and Manufacturing

A variety of mineral deposits exist in some parts of the South Pacific region. Australia is a leading exporter of diamonds, gold, bauxite, opals, and iron ore. Extracting these minerals, however, is hampered by high transportation costs inside and outside the country. In addition, public debate about Aboriginal land rights limits where mining can occur. For example, Australia has the world's largest undeveloped supply of uranium ore, but much of it lies within ancestral lands sacred to the Aborigines.

With some exceptions, few significant mineral resources are found in other areas of the South Pacific region. New Zealand has a large aluminum smelting industry, and Papua New Guinea's rich deposits of gold and copper have only recently been exploited. Kiribati and Nauru, once dependent on phosphate mining, now face dwindling supplies. They are now encouraging foreign investment and seeking aid to develop new economic activities.

Government

Mining in Antarctica

Antarctica holds enormous untapped mineral resources, including petroleum, gold, iron ore, and coal. Scientists have used core sampling—drilling cylindrical sections through the Antarctic ice cap—to identify the presence of these and other key minerals. Although seven countries have made territorial claims to Antarctica, the voluntary Protocol on Environmental Protection, signed by 44 nations in 1991, prohibits mining on the continent.

Manufacturing

Australia and New Zealand are the South Pacific region's major producers of manufactured goods. Because agriculture is important in these two countries, food processing is their most important manufacturing activity. Relatively isolated geographically, Australia and New Zealand must import costly machinery and raw materials in order to set up major manufacturing industries capable of producing exports. As a result, industries in the two countries generally manufacture products for home consumption. Goods that cannot be produced domestically are imported.

The rest of the South Pacific region is less industrially developed than Australia and New Zealand. Manufacturing in the islands of Oceania is limited to small-scale enterprises, such as textile production, clothing assembly, and mass production of craft items.

Service Industries

Throughout Australia and Oceania, service industries have emerged as major contributors to national economies. As in other developed countries, most people in Australia and New Zealand make their living in service industries. In Oceania few countries are large enough to support extensive service industries other than tourism. **Nauru**, however, has begun to attract international banking and investment companies as a way of ending its traditional dependence on phosphate mining.

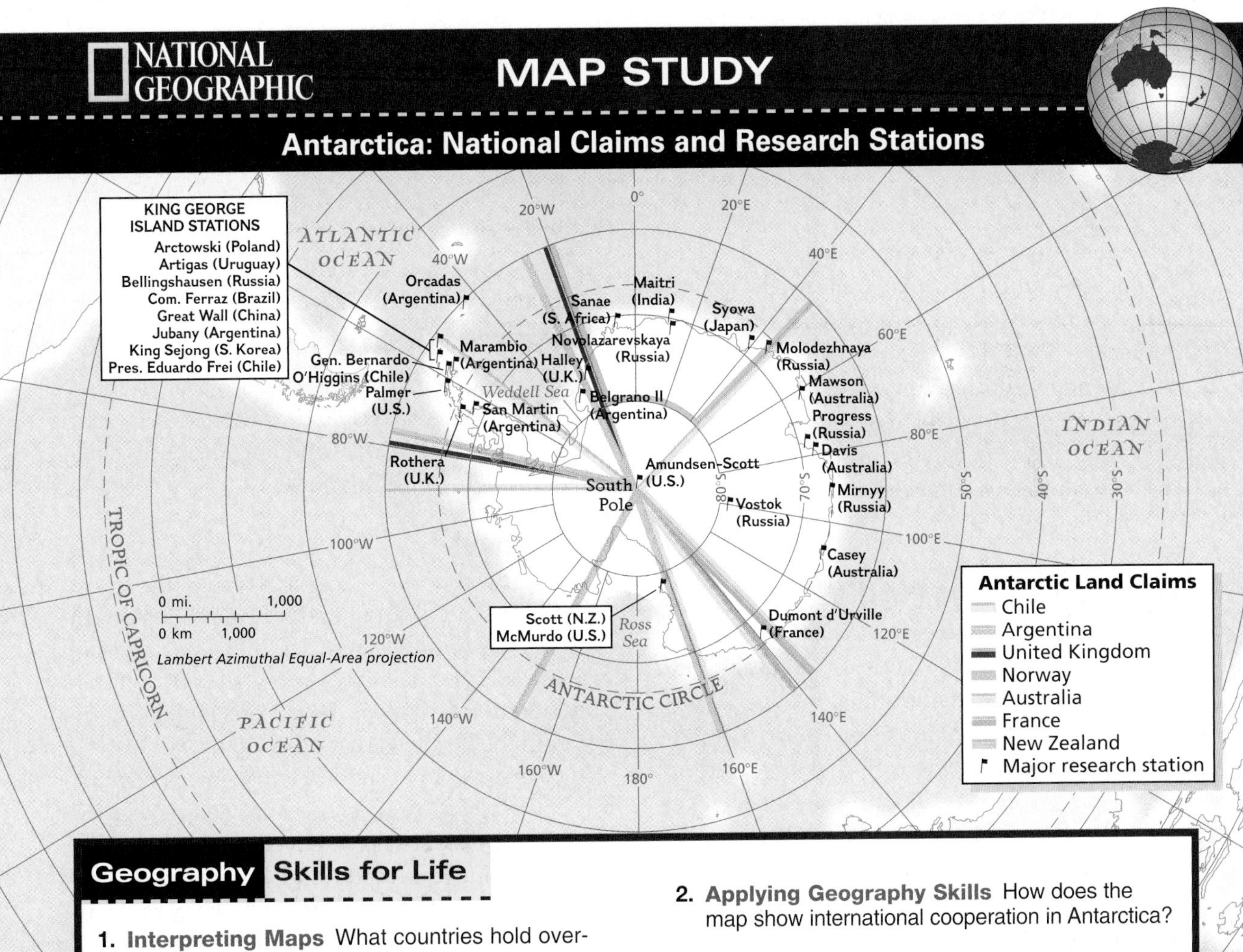

Geography Skills for Life

Battle Site Guadalcanal, one of the Solomon Islands, was the site of heavy fighting between the United States and Japan during World War II.

Place How do World War II battle sites benefit Pacific countries today?

Tourism

In recent decades the expansion of air travel has boosted tourism in Australia and Oceania. Each year thousands of tourists visit the region. Among the South Pacific region's attractions are its indigenous cultures, unique wildlife, and contrasting physical features—rock formations, tropical rain forests, geysers, mountain glaciers, sandy beaches, and coral reefs.

History

World War II in the Pacific

Today several countries in Oceania are promoting World War II battle sites on or near South Pacific islands as tourist destinations. Just as history buffs in the United States visit Civil War battlefields, people from the countries involved are now visiting World War II battle sites. For developing South Pacific countries, such as Vanuatu, the Marshall Islands, the Solomon Islands, and the Federated States of Micronesia, tourism provides a much-needed source of income.

Global Trade Links

In recent decades, improved transportation and communications links have increased trade between the once remote South Pacific region and other parts of the world. The South Pacific's agricultural and mining products are its greatest sources of export income. Countries in Oceania export copra, timber and wood products, fish, vegetables, and handicrafts. The spices of the vast South Pacific region are now found in kitchens around the world. For example, the islands of Micronesia are a major source of black pepper, and Tonga exports ginger and the costly vanilla beans used to flavor ice cream and baked goods. A number of South Pacific countries, however, must import food to supplement the subsistence crops.

During most of the 1900s, Australia and New Zealand traded exclusively with the United Kingdom and the United States. In recent years, however, these South Pacific countries have increased trade with their neighboring Asian countries of Japan, Taiwan, and China. In 1971 various island countries of Oceania set up the South Pacific Forum, an organization that promotes trade and economic growth. Because of few natural resources, some South Pacific islands are dependent to some degree on outside investment or foreign aid.

Transportation and Communications

Australia, Oceania, and Antarctica contain thousands of miles of coastland, barren desert, and solid ice. Physical barriers and long distances challenge travelers in the region.

Land Travel

Australia and New Zealand have the most developed road and rail systems in the region. In coastal areas of these countries, highways are well maintained, and subways provide public transportation in urban areas. Few roads, however, are found in the isolated Australian outback.

In Oceania many island countries are too small, too poor, or too rugged to have well-developed road or rail systems. Some governments, however, are improving the roads and bridges necessary for economic growth. Antarctica lacks permanent settlements and has no roads or rail systems.

Air and Water Travel

Long distances, harsh climates, or obstacles to land travel make air and water travel important to the region. Cargo ships and planes move imports and exports to and from far-flung Pacific territories. Commercial airlines and cruise ships bring travelers.

Water and air also provide important means of personal transportation. Pacific islanders began using outrigger canoes thousands of years ago, and many of Oceania's travelers continue to use boats today. Sailboats and motorized boats are common, and ferries link New Zealand's two major islands. Icebreakers—ships with reinforced bows—carry people and supplies to Antarctica as do small planes and helicopters, although winter blizzards often make transportation of any kind impossible. Severe winters isolate Antarctica:

> *"Along about February the annual exodus [from the research stations] begins in earnest. Once the cold season takes hold, planes stop making regular flights to inland stations, and the ice layer spreads out to sea, making access by ship nearly impossible. Only a few hundred residents stay through the winter."*
>
> Michael D. Lemonick, "McMurdo Station," *Time*, January 15, 1990

Planes also provide transportation between islands in the South Pacific. In Australia's outback almost every station or farm has at least one plane. Ranchers often use helicopters to herd cattle over thousands of acres of rough terrain.

Communications

In the South Pacific area, the same geographic obstacles that hinder land travel also make communications difficult. The development of modern technology, however, has helped increase contacts within Australia, Oceania, and Antarctica and with the rest of the world. In the Australian outback, some cattle stations are large enough to maintain their own post offices and telephone exchanges. Others use two-way radios to communicate. Emerging technologies, such as cellular, digital, and satellite communications and the Internet, are becoming common in developed areas. A continuing challenge is to provide developing Pacific countries with access to these technologies.

Student Web Activity Visit the **Glencoe World Geography** Web site at geography.glencoe.com and click on Student Web Activities—Chapter 34 for an activity about research in Antarctica.

SECTION 1 ASSESSMENT

Checking for Understanding

1. **Define** station, grazier, copra.
2. **Main Ideas** On a table like the one below, fill in details about the key agricultural and mining products of countries in this region. Then describe the role the region plays in world trade.

Country	Agricultural Products	Mining Products

Critical Thinking

3. **Identifying Cause and Effect** How does importing more manufactured goods than it exports affect a country's economy?
4. **Predicting Consequences** What might be the consequences of opening Antarctica to mining?
5. **Drawing Conclusions** Why are Australia and Oceania trading more with East Asia and Southeast Asia than with the West?

Analyzing Maps

6. **Place** Study the map on page 835. Which Latin American countries hold claims in Antarctica? Which have research stations there?

Applying Geography

7. **Economic Activities** Create a table that shows major economic activities for six countries in the region. Then explain why the economies of some countries focus on one major product.

SECTION 2

People and Their Environment

Guide to Reading

Consider What You Know

Australia's remarkable wildlife is recognizable around the world. What animals and plants unique to Australia can you name?

Read to Find Out

- Why do Australia, Oceania, and Antarctica face many environmental challenges?
- What effects did nuclear testing have on the region?
- Why are the thinning of the ozone layer and global warming special challenges for this region?

Terms to Know

- marsupial
- introduced species
- food web
- ozone layer
- El Niño-Southern Oscillation (ENSO)
- diatom

Places to Locate

- Tasmania
- Murray-Darling River Basin
- Great Barrier Reef

NATIONAL GEOGRAPHIC

A Geographic View

From Leafy Grove to Salty Swamp

Behind us a forest of dead eucalyptus trees stood in a salty swamp, a graveyard of skeletons with gray arms raised in good-bye. . . . Once a leafy grove in Western Australia, this salt lake rose from the ground when nearby woodlands were cleared for farms. Thirsty trees had absorbed rainwater and kept the water table from rising, but when they were cut, the water surfaced and brought salt with it. The result: saline ponds and dead fields.

—*Michael Parfit, "Australia: A Harsh Awakening,"* National Geographic, *July 2000*

Salty swamp, Australia

Beneath much of Australia's land surface there is a layer of salty subsoil or salty groundwater. Salts are carried to the surface as the water slowly evaporates. Scientists believe that 40 percent of Western Australia's productive wheat belt could be lost to salty swamps in the next two decades. Today Australia, like other countries, is experiencing the environmental consequences of human activity. In this section you will learn about environmental challenges in Australia, Oceania, and Antarctica as well as the efforts under way to remedy environmental damage.

Managing Resources

Australia, Oceania, and Antarctica hold some of the planet's richest and most diverse natural resources. Unfortunately, these resources have not always been well managed, and today the region faces many environmental issues. Conservation efforts, however, are gaining recognition in the region. Environmental issues concern voters and government leaders alike in Australia, New Zealand, and other South Pacific islands.

Australia's Unusual Animals

The continent of Australia, separated for so long from other landmasses, is home to many unique animal species. Kangaroos, koalas, and wallabies are just some of Australia's 144 species of **marsupials**—mammals whose young must mature in a pouch after they are born. The Australian island of **Tasmania** gave its name to the Tasmanian devil, a powerful meat-eating marsupial about the size of a badger. Australia's strangest wildlife may be the duck-billed platypus and the echidna, a spiny anteater—the only mammals in the world that lay eggs.

Australia's unusual wildlife species, however, have been seriously threatened by the human introduction of various nonnative animals. These **introduced species** include the hunting dogs called dingoes brought from Asia by migrating Aborigines. Sheep, cattle, foxes, cats, and rabbits were also brought by European settlers. In the absence of natural predators, these animals have multiplied and taken over the habitats of Australia's native species. Some of Australia's native species have become extinct, and at least 16 kinds of marsupials are now endangered. Efforts to restore Australia's ecological balance include the use of electric fencing to keep out nonnative animals, hunting and trapping programs, the introduction of natural predators, and the creation of native wildlife reserves.

NATIONAL GEOGRAPHIC **World Explorer**

Geography Skills for Life

Wildlife The kangaroo and the Tasmanian devil (inset) are uniquely Australian mammals.

Place Why does Australia have such a variety of unusual animal species?

Forest, Soil, and Water

The protection of forest, soil, and freshwater resources is a major concern throughout the South Pacific region. In Australia many sparse woodlands have been cleared for farms and grazing lands, leaving little protection against wind erosion. As in other parts of the world, soil conservation in the region is closely linked to reducing deforestation. Countries with valuable timber resources, such as New Zealand, Papua New Guinea, and Vanuatu, are developing plans to use forest resources without damaging the environment.

Drought, salt, irrigation, and agricultural runoff threaten Australia's freshwater sources. In the fertile **Murray-Darling River Basin**, one of the world's largest drainage basins, the use of water for agriculture and growing city populations has dramatically reduced the rivers' flow.

Oceania also faces challenges in managing its freshwater resources. Many small coral atolls and volcanic islands hold only limited supplies of freshwater. Agricultural runoff and inadequate sanitation cause pollution that further threatens these supplies. The lack of clean drinking water keeps the standard of living low and poses barriers to economic growth in some countries of Oceania.

Geography Skills for Life

Great Barrier Reef

The Great Barrier Reef (left) in Australia is home to hundreds of species of coral-forming organisms (right).

Human-Environment Interaction What human activities threaten the Great Barrier Reef?

Improvement will come with better management of runoff, construction of additional sanitation facilities, and development of less expensive ways of removing salt from ocean water.

Agricultural runoff, chemical fertilizers, and organic waste also threaten oceans in the South Pacific region. Toxic waste in particular endangers Australia's **Great Barrier Reef** and other Pacific coral reefs. Coral environments are increasingly stressed by tourists, boaters, and divers as well as oil-shale mining.

Pollution also affects all kinds of marine life, including the tiny organisms that make up coral reefs. Algae—on which these organisms thrive—and plankton are key parts of the ocean's **food web**, the interlinking chains of predators and their food sources in an ecosystem. As these tiny living things are destroyed, the larger plants and animals that rely on them for food also die off.

History

The Nuclear Legacy

The testing of nuclear weapons has had major effects on the region's environment. In the late 1940s and 1950s, the United States and other countries with nuclear capability carried out aboveground testing of nuclear weapons in the South Pacific. The dangers of such testing were gravely underestimated at the time. In 1954 the United States exploded a nuclear device on Bikini Atoll, in the Marshall Islands. The people of Bikini Atoll had been moved to safety, but those living on Rongelop Atoll, downwind of the explosion, were exposed to massive doses of radiation that resulted in deaths, illnesses, and genetic abnormalities.

Although the American testing was stopped, the effects of radiation exposure and environmental damage have continued through several generations. Today the atolls affected by the testing remain off-limits to human settlement. Recent studies, however, offer hopeful signs of eventual environmental recovery. In the 1990s the United States government provided $90 million to help decontaminate Bikini Atoll and set up a $45 million trust fund for blast survivors and their offspring from Rongelop Atoll.

The nuclear legacy also has had political effects. Antinuclear activism is a major factor in regional politics. In 1986 New Zealand banned nuclear-powered ships and those with nuclear weapons from entering its waters. Because of this ban, the United States withdrew from a defense agreement with New Zealand. In the mid-1990s, French plans to conduct nuclear tests on an atoll in French

Polynesia aroused antinuclear demonstrations. The international outcry led to an early halt to the tests.

Atmosphere and Climate

Like other world regions, Australia, Oceania, and Antarctica are threatened by global atmospheric and climate changes. In the 1970s scientists found a hole in the ozone layer over Antarctica:

> *"The mysterious stuff called ozone, which until then was known to the public chiefly as an . . . element of smog in overcrowded cities, was being destroyed in the stratosphere by chemicals made and released in the 20th century by humans. . . . The hole was real; the ozone had dropped by 50 percent. . . ."*
>
> Samuel W. Matthews, "Is Our World Warming?" *National Geographic*, October 1990

The ozone layer's protective gases prevent harmful solar rays from reaching the earth's surface. The ozone hole over Antarctica grew dramatically between 1975 and 1993, when it covered more than 9 million square miles (23 million sq. km). In 1989 a similar ozone hole developed over the Arctic.

The loss of protective ozone may be behind the global rise in the rates of skin cancer and cataracts, conditions caused by overexposure to the sun's ultraviolet rays. Increased solar radiation that reaches the earth through ozone holes may also contribute to global warming, the gradual rise in Earth's temperatures over the last century.

Climate and weather in the South Pacific region are highly sensitive to changes in the El Niño weather pattern called El Niño-Southern Oscillation (ENSO). This seasonal weather event can cause droughts in Australia and powerful cyclonic storms in the South Pacific. These ENSO-related weather patterns are believed to be increasing in frequency and severity and may also be linked to global warming.

Some scientists claim that continued rises in Earth's temperatures could be devastating. If polar ice caps were to melt and thermal expansion of ocean waters occurred, many of Oceania's islands would be flooded by rising ocean levels. Rising ocean temperatures also affect certain types of plankton and algae that grow in warm waters, causing overgrowth and the choking out of other life-forms. Diatoms—plankton that flourish in cold ocean waters—would die if temperatures rose, affecting life-forms that feed on them. Scientists in the region, especially in Antarctica, are studying global warming and are hoping to discover causes, predict consequences, and provide solutions.

SECTION 2 ASSESSMENT

Checking for Understanding

1. **Define** marsupial, introduced species, food web, ozone layer, El Niño-Southern Oscillation (ENSO), diatom.
2. **Main Ideas** On a chart like the one below, list resources and examples of their mismanagement in the region. Also list possible solutions.

Resource	Example of Mismanagement	Possible Solution

Critical Thinking

3. **Comparing and Contrasting** How are countries of the region similar and different in the challenges they face concerning water resources?
4. **Decision Making** Do you agree or disagree with New Zealand's nuclear ban? Explain your reasons.
5. **Problem Solving** What steps would you take to increase awareness about the risks of global warming? Explain.

Analyzing Maps

6. **Location** Study the physical-political map on page 796. Which countries are at the greatest risk from rising ocean levels as a result of continued global warming?

Applying Geography

7. **Effects of Mining** Study the map on page 787. Compare a mineral-rich area shown on the map to a mineral-rich area in another region. Explain the effects of mining on both environments.

Viewpoint

CASE STUDY on the Environment

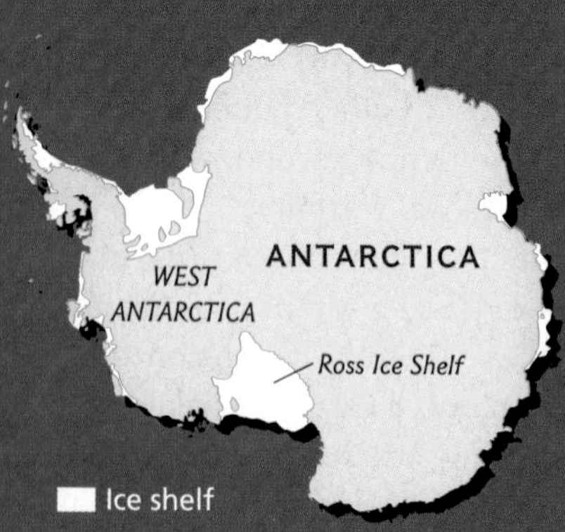

Antarctica's Melting Ice: Is Global Warming at Fault?

During the last century, Earth's average surface temperature crept steadily higher—a phenomenon called global warming. In the past few decades, vast expanses of Antarctic ice have started breaking up and large chunks have floated out to sea. Researchers speculate that if the huge West Antarctic ice sheet collapses and melts, sea levels could rise dramatically, causing flooding in coastal regions around the world. Is global warming responsible for Antarctica's melting ice?

In March 2000 an iceberg twice the size of Delaware broke free from Antarctica's Ross Ice Shelf, part of the West Antarctic ice sheet. On the other side of the continent, an entire ice shelf disintegrated in 1995. Why is this happening? Antarctica is the coldest place on Earth. Nevertheless, the continent is a little warmer than it used to be. The average temperature in parts of West Antarctica has increased by almost 5°F (3°C) in the last 50 years. During the 1900s, the average temperature worldwide rose by 1°F (.5°C).

Most scientists believe that rising global temperatures are partly due to an increased amount of carbon dioxide (CO_2) in the atmosphere. Much of the carbon dioxide is caused by human activities such as burning gasoline, coal, and other fossil fuels. In the atmosphere, carbon dioxide is a powerful heat absorber, trapping heat that radiates from the sun-warmed ground. The trapped heat leads to global warming.

Tavaerua Island (below) could disappear if sea level rises. Scientists (right) study Antarctic ice cores for clues to a changing climate. ▼

As global temperatures rise, ocean waters warm and then expand, and ice in places such as Antarctica begins to melt. The seas start creeping higher onto the edges of the continents. Sea levels in some parts of the world are already almost a foot (30 cm) higher than they were a century ago. However, this increase is trivial compared with the rise that could occur if the vast West Antarctic ice sheet melts. If this happens, sea levels could rise by 13 to 20 feet (4 to 6 m). Coastal communities worldwide would be flooded. Low-lying islands, such as Tuvalu and Kiribati in the Pacific, would disappear underwater.

Recent studies, however, indicate that the West Antarctic ice sheet has been receding for almost 8,000 years. Scientists have also uncovered evidence that the ice sheet may have collapsed about 400,000 years ago, before the last ice age. These findings have sparked a controversy.

◄ Cars spew carbon dioxide into the air, contributing to global warming.

Some scientists think that changes taking place in Antarctic ice are part of a natural cycle that has nothing to do with recent global warming. They point out that the West Antarctic ice sheet began shrinking before people started burning large amounts of fossil fuels and adding carbon dioxide to the atmosphere.

Other scientists think that recent changes in Antarctic ice sheets are a direct result of human-caused global warming. While these scientists admit there might be a natural cycle at work in Antarctica, they argue that global warming is speeding up that cycle.

What's Your Point of View?
Experts predict it will take 500 to 700 years for the West Antarctic ice sheet to melt completely, no matter what the cause. Should people today care about this issue? Why or why not?

Problem Solving

Individuals and groups often face problems that require critical thinking to solve. Identifying problems and evaluating possible solutions are important skills used by individual citizens, local and national governments, and world organizations.

Learning the Skill

Whether a problem is simple or complex, local or global, the same problem-solving steps can be applied. You can practice these steps in your everyday life, just as governments and organizations do when addressing major conflicts.

Here are the steps involved in problem solving:

- **Identify the problem.** State clearly the issue at hand and the reasons the problem must be solved.
- **Brainstorm possible solutions to the problem.** Be open-minded and creative. Take notes on all the possibilities suggested.
- **Evaluate the proposed solutions.** Evaluate each proposed solution by listing its advantages and disadvantages and anticipating its possible consequences.
- **Choose and implement the best solution.** Choose the best possibility, understanding that it may have some drawbacks. Put your solution into practice.
- **At a later time, review the success of the solution.** If implementing your solution has not improved the situation or has resulted in further problems, begin the process again.

Environmentalists say the Great Barrier Reef will be under threat if the Australian government allows oil explorations in the area. After years of controversy, the government has started testing ways of tapping oil reserves around one of the world's most spectacular sites. Experts say there is more oil to be tapped in the reef's coastal rock next to the coral than has ever been found on the entire American continent.

Environmentalists say the processes involved could destroy the delicate coral. . . . "To do that [extract the oil] requires a lot of energy and the oil you get is very carbon intensive, making the whole process a very dirty kind of mining."

More than one million people visit the reef each year but oil pollution has the potential to ruin the tourist industry. . . . [The government] says the country cannot afford to ignore the reef's precious resources. . . . [S]uch is the sensitivity of the issue, the authorities have only given the go-ahead for one pilot area to be exploited for oil.

—"World: Asia-Pacific Oil Threat to Great Barrier Reef,"
BBC News *(online), September 25, 1998*

Practicing the Skill

Read the excerpt above. Then use what you know about problem solving to answer these questions.

1. What is the problem?
2. What are the positions of environmental groups and the Australian government regarding the problem?
3. What are some possible solutions to the problem?
4. How has Australia tried to solve the problem?
5. How can the success of the solution be evaluated?

Applying the Skill

Work in a small group to find an environmental issue facing your community. As a group, apply the steps for problem solving to the issue you have chosen. Prepare a written report of your results. If possible, share your proposed solution with community authorities.

The **Glencoe Skillbuilder Interactive Workbook, Level 2** provides instruction and practice in key social studies skills.

34

SECTION 1 Living in Australia, Oceania, and Antarctica (pp. 833–837)

Terms to Know

- station
- grazier
- copra

Key Points

- Agriculture is the most important economic activity in the region, although mining is done in Australia and some island countries.
- Manufacturing in Australia and New Zealand centers on food processing, and the rest of the region engages in small-scale production of clothing and crafts.
- The importance of service industries, particularly tourism, is increasing in the economies of the region.
- Transportation and communications technologies, such as air travel, satellite communications, and the Internet, are helping people in the region to overcome geographic obstacles.

Organizing Your Notes

Create an outline using the format below to help you organize your notes for this section.

Living in the South Pacific

I. Agriculture
II. Mining and Manufacturing
  A. Mining in Antarctica
    1.
    2.

SECTION 2 People and Their Environment (pp. 838–841)

Terms to Know

- marsupial
- introduced species
- food web
- ozone layer
- El Niño-Southern Oscillation (ENSO)
- diatom

Key Points

- Australia, Oceania, and Antarctica have many natural resources, but the region's environment is threatened by human activity.
- Governments and individuals in the region are focusing on balanced management of water resources, forest, land, and wildlife.
- Nuclear testing conducted in Oceania during the 1940s and 1950s has had a lasting impact on people and the environment.
- Scientists are studying global warming and the thinning ozone layer to prevent potential risks.

Organizing Your Notes

Create a web diagram like the one below to help organize the notes you took for this section. Add other key ideas to the web, and draw lines to show connections between ideas.

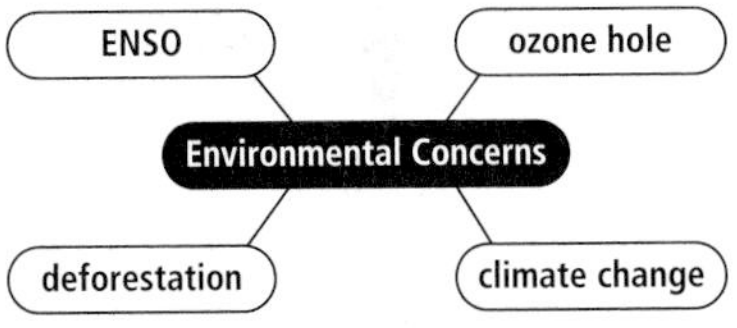

Thermal spring, Rotorua, New Zealand

CHAPTER 34

ASSESSMENT & ACTIVITIES

Reviewing Key Terms

Write the key term that best completes each of the following sentences. Refer to the Terms to Know in the Summary & Study Guide on page 845.

1. Ranchers on an Australian ________ will sometimes build fences to keep out ________.
2. The kangaroo, one type of ________, is native to Australia.
3. ________ are part of the ________ of larger life-forms.
4. Disruptions to weather patterns in the South Pacific caused by ________ may be increasing.
5. A New Zealand ________ makes a living by raising sheep, beef cattle, and dairy cattle.
6. Many countries in Oceania export ________.
7. Scientists discovered a reduction in the ________ in the 1970s.

Reviewing Facts

SECTION 1

1. How does the importance of agriculture, mining, and manufacturing vary among South Pacific countries?
2. What service industries are developing in Australia and Oceania?
3. How have changes in transportation and communications affected the location and patterns of economic activities in the South Pacific region?

SECTION 2

4. What are the major threats to the region's wildlife, forests, soil, and water?
5. What have been the effects of nuclear testing in Oceania?
6. What effects in the South Pacific have occurred because of atmospheric and climatic changes?

Critical Thinking

1. **Finding and Summarizing the Main Idea** What are three critical challenges to agriculture in Australia?
2. **Identifying Cause and Effect** In what ways could mining operations in Antarctica interfere with scientific research programs there?
3. **Problem Solving** Use a graphic organizer like the one below to describe three steps that countries in Oceania might take to reduce the impact of tourism on coral reefs.

Reduce tourist impact → ________ / ________ / ________

NATIONAL GEOGRAPHIC

Locating Places

Antarctica: Physical Geography

Match the letters on the map with the physical features of Antarctica. Write your answers on a sheet of paper.

1. Weddell Sea
2. Antarctic Circle
3. South Pole
4. Ross Sea
5. Antarctic Peninsula
6. Transantarctic Mountains

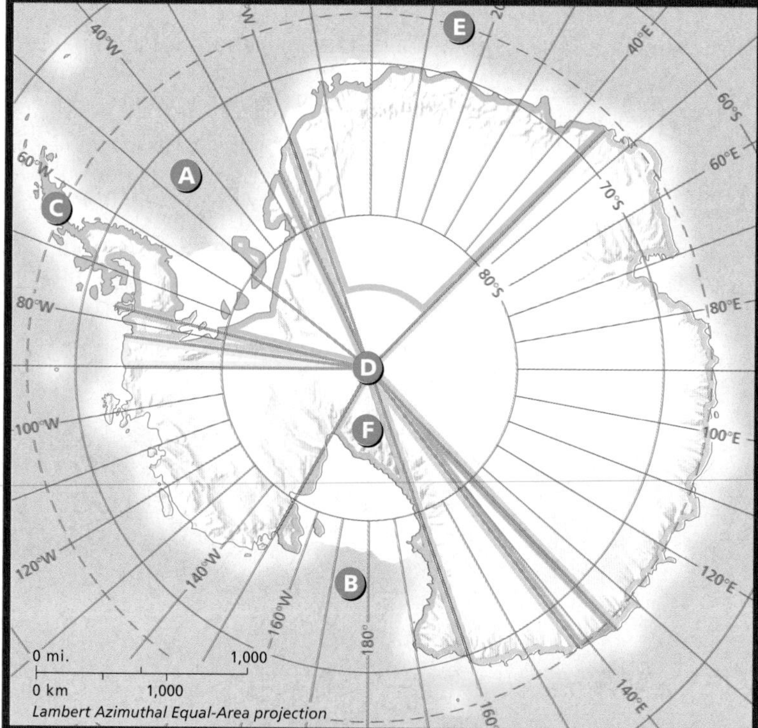

Self-Check Quiz Visit the **Glencoe World Geography** Web site at geography.glencoe.com and click on Self-Check Quizzes—Chapter 34 to prepare for the Chapter Test.

Using the Regional Atlas

Refer to the Regional Atlas on pages 784–787.

1. **Location** Where are most of the zinc deposits in the region?
2. **Human-Environment Interaction** Which physical features in the region are vulnerable to environmental damage from mining activities?

Thinking Like a Geographer

Using what you know about the physical geography of Oceania's islands, write a paragraph suggesting three ways these islands might address their lack of clean freshwater.

Problem-Solving Activity

Group Research Project With a small group of classmates, research one of the introduced species in Australia. Investigate the origins of the problem it has created, its effects on the environment, and suggested solutions. Brainstorm additional solutions, and evaluate each proposal. Prepare a report to the class on the solution you think is best.

GeoJournal

Descriptive Writing Using your GeoJournal data, select a human activity from each of the following areas: Australia, Oceania, and Antarctica. Then write a descriptive paragraph that compares how each of these activities has modified the physical environment.

Technology Activity

Using the Internet for Research Use the Internet to find information about global warming. List the sources you find on the Internet, and compare the different viewpoints on the issue of global warming. Then choose one solution that you support, and write an argument for adopting that solution.

Standardized Test Practice

Read the selection below. Then choose the best answer for the following multiple-choice question. If you have trouble answering the question, use the process of elimination to narrow your choices.

Rabbits are one of the more destructive wild animals that have been introduced into Australia. They damage the environment and reduce agricultural production. They compete with native wildlife for food and shelter, which reduces the populations of many native plants and animals. Because rabbits eat seedlings, there are fewer young plants to replace those that die naturally. Rabbits also compete with livestock for the same plants, eating them to below ground level. This loss of plant cover results in soil erosion.

1. **Based on the information in the paragraph, how do rabbits reduce agricultural production?**

   A They live in wheat-growing regions and eat the wheat seedlings.
   B They compete with native wildlife for food.
   C Dead plants are not replaced by enough new plants to prevent soil erosion.
   D They eat the plants that provide food for livestock and cause soil erosion by eliminating plant cover.

Look for the best answer choice for the question. The best answer choice is the one that offers *the most* correct information in response to the question.

Appendix

CONTENTS

HONORING AMERICA

Flag Etiquette

Over the years, Americans have developed rules and customs concerning the use and display of the flag. One of the most important things every American should remember is to treat the flag with respect.

- The flag should be raised and lowered by hand and displayed only from sunrise to sunset. On special occasions, it may be displayed at night, but it should be illuminated.
- The flag may be displayed on all days, weather permitting, particularly on national and state holidays and on historic and special occasions.
- No flag may be flown above the American flag or to the right of it at the same height.
- The flag should never touch the ground or floor beneath it.
- The flag may be flown at half-staff by order of the president, usually to mourn the death of a public official.
- The flag may be flown upside down only to signal distress.
- When the flag becomes old and tattered, it should be destroyed by burning. According to an approved custom, the Union (stars on blue field) is first cut from the flag; then the two pieces, which no longer form a flag, are burned.

★ ★ ★ ★ ★ ★ ★ ★

The Star-Spangled Banner

**O! say can you see, by the dawn's early light,
What so proudly we hail'd at the twilight's last gleaming,
Whose broad stripes and bright stars through the perilous fight,
O'er the ramparts we watched, were so gallantly streaming?
And the Rockets' red glare, the Bombs bursting in air,
Gave proof through the night that our Flag was still there;
O! say, does that star-spangled banner yet wave
O'er the Land of the free and the home of the brave!**

The Pledge of Allegiance

I pledge allegiance to the Flag of the United States of America and to the Republic for which it stands, one Nation under God, indivisible, with liberty and justice for all.

A

aborigine an area's original inhabitants (p. 662)
absolute location the exact position of a place on the earth's surface (p. 20)
accretion a slow process in which a sea plate slides under a continental plate, creating debris that can cause continents to grow outward (p. 40)
acid deposition wet or dry airborne acids that fall to the earth (p. 321)
acid rain precipitation carrying large amounts of dissolved acids which damages buildings, forests, and crops, and kills wildlife (pp. 166, 321)
acupuncture an ancient practice that involves inserting fine needles into the body at specific points in order to cure disease and ease pain (p. 677)
alluvial plain floodplain, such as the Indo-Gangetic Plain in South Asia, on which flooding rivers have deposited rich soil (p. 572)
alluvial-soil deposits rich soil made up of sand and mud deposited by running water (p. 423)
altiplano Spanish for "high plain," a region in Peru and Bolivia encircled by the Andes peaks (p. 194)
amendment in U.S. history, official changes made to the Constitution (p. 144)
apartheid policy of strict separation of the races adopted in South Africa in the 1940s (p. 524)
aquaculture the cultivation of seafood (p. 695)
aquifer underground water-bearing layers of porous rock, sand, or gravel (pp. 49, 470)
arable suitable for growing crops (p. 464)
archipelago a group or chain of islands (pp. 646, 720)
artesian water underground water supply that is under enough pressure to rise into wells without being pumped (p. 794)
Asian Pacific Economic Cooperation Group (APEC) a trade group, whose members are Japan, China, South Korea, and Taiwan, that ensures that trade among the member countries is efficient and fair (p. 689)
Association of Southeast Asian Nations (ASEAN) organization formed in 1967 to promote regional development and trade in Southeast Asia (p. 763)
atheism the belief that there is no God (p. 376)
atmosphere a layer of gases that surrounds the earth (p. 35)
atoll ring-shaped island formed by coral building up along the rim of an underwater volcano (p. 796)
autocracy government in which a single individual possesses the power and authority to rule (p. 87)
avalanche mass of ice, snow, or rock that slides down a mountainside (p. 279)
axis an imaginary line that runs through the center of the earth between the North and South Poles (p. 56)

B

batik method of dyeing cloth to produce beautiful patterns, developed in Indonesia and Malaysia (p. 751)
bazaar a traditional marketplace ranging from a single street of stalls to an entire city district (p. 456)
bedouin member of the nomadic desert peoples of North Africa and Southwest Asia (p. 455)
bilingual speaking or using two languages (p. 148)
Bill of Rights the first 10 amendments to the United States Constitution (p. 144)
biologist scientist who studies plant and animal life (p. 324)
biomass plant and animal waste used especially as a source of fuel (p. 613)
biosphere the part of the earth where life exists (p. 35)
birthrate the number of births per year for every 1,000 people (p. 76)
black market any illegal market where scarce or illegal goods are sold, usually at high prices (p. 388)
blizzard a snowstorm with winds of more than 35 miles per hour, temperatures below freezing, and visibility of less than 500 feet for 3 hours or more (p. 125)
Bolsheviks a revolutionary group in Russia led by Vladimir Ilyich Lenin (p. 370)
boomerang curved throwing stick used by Aborigines for hunting in Australia (p. 817)
buffer state neutral territory between rival powers (p. 743)

C

cabinet heads of departments in the U.S. government who advise the president (p. 144)
calligraphy the art of beautiful handwriting (p. 679)
campesinos farm workers; generally, people who live and work in rural areas (p. 238)
canopy top layer of a rain forest, where the tops of tall trees form a continuous layer of leaves (p. 200)
cartography the science of mapmaking (p. 24)
cash crop farm crop grown to be sold or traded rather than used by the farm family (pp. 238, 538, 612, 760)
cataract a large waterfall (p. 500)

caudillo a Latin American political leader from the late 1800s on, often a military dictator (p. 224)
cereal any grain, such as barley, oats, or wheat, grown for food (p. 431)
chaparral type of vegetation made up of dense forests of shrubs and short trees, common in Mediterranean climates (pp. 68, 281)
chernozem (cher•nuh•ZYAWM) rich, black topsoil found in the North European Plain, especially in Russia and Ukraine (p. 346)
chinampas floating farming islands made by the Aztec (p. 221)
chinook seasonal warm wind that blows down the Rockies in late winter and early spring (p. 124)
Chipko India's "tree-hugger" movement that protects forests through reforestation and by supporting limited timber production (p. 621)
chlorofluorocarbon chemical substance, found mainly in liquid coolants, that damages the earth's protective ozone layer (p. 695)
city-state in ancient Greece, independent community consisting of a city and the surrounding lands (p. 295)
clan tribal community or large group of people related to one another (pp. 529, 670, 817)
clear-cutting cutting down whole forests when removing timber (p. 165)
climate weather patterns typical for an area over a long period of time (p. 55)
Cold War power struggle between the Soviet Union and the United States after World War II (pp. 298, 371)
collective farm under communism, a large, state-owned farm on which farmers received wages plus a share of products and profits; also called a kolkhoz (p. 317)
command economy economic system in which economic decisions about production and distribution are made by some central authority (pp. 89, 388, 685)
commercial farming farming organized as a business (p. 538)
commodity goods produced for sale (pp. 158, 465)
commune a collective farming community whose members share work and products (p. 686)
communism society based on equality in which workers would control industrial production (pp. 298, 370)
condensation the process of excess water vapor changing into liquid water when warm air cools (p. 47)
coniferous trees such as evergreens that have cones and needle-shaped leaves, and keep their foliage throughout the winter (p. 68)
conquistador Spanish term for "conqueror," referring to soldiers who conquered Native Americans in Latin America (p. 222)
conservation farming a land-management technique that helps protect farmland (p. 539)
Constitution plan of government made for the United States in 1787 (p. 144)
consumer goods goods that directly satisfy human wants (p. 388)
continental drift the theory that the continents were once joined and then slowly drifted apart (p. 38)
continental shelf the part of a continent that extends underwater (p. 36)
cooperative a voluntary organization whose members work together and share expenses and profits (p. 686)
copra dried meat of a coconut (p. 834)
coral limestone deposits formed from the skeletons of tiny sea creatures (p. 795)
cordillera parallel chains or ranges of mountains (pp. 194, 719)
Coriolis effect an effect that causes the prevailing winds to blow diagonally rather than along strict north-south or east-west lines (p. 62)
cottage industry a business that employs workers in their homes (p. 615)
crevasse huge, deep crack that forms in thick ice or snow (p. 803)
Crusades series of religious wars (A.D. 1100–1300) in which European Christians tried to retake Palestine from Muslim rule (p. 296)
cultural diffusion the spread of new knowledge and skills from one culture to another (p. 84)
culture way of life of a group of people who share beliefs and similar customs (p. 80)
culture hearth a center where cultures developed and from which ideas and traditions spread outward (pp. 84, 447, 668)
culture region division of the earth based on a variety of factors, including government, social groups, economic systems, language, or religion (p. 83)
cuneiform Sumerian writing system using wedge-shaped symbols pressed into clay tablets (p. 447)
current cold or warm stream of seawater that flows in the oceans, generally in a circular pattern (p. 62)
cyclone storm with heavy rains and high winds that blow in a circular pattern around an area of low atmospheric pressure (pp. 579, 767)
czar ruler of Russia until the 1917 revolution (p. 369)

D

Dalits the "oppressed"; in India, people assigned to the lowest social class (p. 623)

death rate the number of deaths per year for every 1,000 people (p. 76)

deciduous trees, usually broad leaved such as oak and maple, that lose their leaves in autumn (p. 68)

deforestation the loss or destruction of forests, mainly for logging or farming (pp. 242, 507)

delta alluvial deposit at a river's mouth that looks like the Greek letter delta (Δ) (p. 503)

democracy any system of government in which leaders rule with consent of the citizens (p. 88)

desalination the removal of salt from seawater to make it usable for drinking and farming (pp. 48, 470)

desertification process in which arable land is turned into desert (pp. 508, 694)

developed country country that has a great deal of technology and manufacturing (p. 93)

developing country country in the process of becoming industrialized (pp. 93, 238)

dharma in Hinduism, a person's moral duty, based on class distinctions, which guides his or her life (p. 593)

dialect local form of a language used in a particular place or by a certain group (pp. 213, 302)

diatoms plankton that live in cold ocean water (p. 841)

dikes large banks of earth and stone that hold back water (p. 272)

dissident a citizen who speaks out against government policies (p. 689)

divide a high point or ridge that determines the direction rivers flow (p. 118)

doldrums a frequently windless area near the Equator (pp. 62, 802)

domesticate to adapt plants and animals from the wild to make them useful to people (pp. 446, 520)

dominion a partially self-governing country with close ties to another country (pp. 144, 820)

doubling time the number of years it takes a population to double in size (p. 76)

dry farming farming method used in dry regions in which land is plowed and planted deeply to hold water in the soil (pp. 143, 321)

dynasty a ruling house or continuing family of rulers, especially in China (p. 669)

dzong a fortified monastery of Bhutan, South Asia (p. 604)

E

e-commerce selling and buying on the Internet (p. 542)

economic sanctions trade restrictions (p. 689)

ecosystem the complex community of interdependent living things in a given environment (p. 22)

ecotourism tourism based on concern for the environment (pp. 546, 617)

El Niño a periodic reversal of the pattern of ocean currents and water temperatures in the mid-Pacific region (p. 63)

El Niño-Southern Oscillation (ENSO) a seasonal weather event that can cause droughts in Australia and powerful cyclones in the South Pacific (p. 841)

embargo a ban on trade (pp. 452, 468)

endemic native plant or animal species (p. 728)

Enlightenment a movement during the 1700s, emphasizing the importance of reason and questioning traditions and values (p. 297)

environmentalist person actively concerned with the quality and protection of the environment (p. 322)

equinox one of two days (about March 21 and September 23) on which the sun is directly above the Equator, making day and night equal in length (p. 56)

erosion wearing away of the earth's surface by wind, flowing water, or glaciers (p. 42)

escarpment steep cliff or slope between a higher and lower land surface (pp. 196, 500)

estuary an area where the tide meets a river current (pp. 197, 503)

ethnic cleansing the expelling from a country or killing of rival ethnic groups (p. 288)

ethnic diversity differences among groups of people based on their origins, languages, customs, or beliefs (p. 439)

ethnic group group of people who share common ancestry, language, religion, customs, or combination of such characteristics (pp. 82, 288, 363)

European Union an organization whose goal is to unite Europe so that goods, services, and workers can move freely among member countries (pp. 300, 313)

eutrophication process by which a body of water becomes too rich in dissolved nutrients, leading to plant growth that depletes oxygen (p. 168)

evaporation process of converting into vapor (p. 47)

exports commodities sent from one country to another for purposes of trade (p. 237)

extended family household made up of several generations of family members (pp. 229, 529)

extinction the disappearance or end of a species of animal or plant (p. 546)

fale traditional type of home in Samoa with open sides and thatched roof (p. 826)

fall line a boundary in the eastern United States where the higher land of the Piedmont drops to the lower Atlantic coastal plain (p. 118)

farm cooperative organization in which farmers share in growing and selling farm products (p. 317)

fault a crack or break in the earth's crust (pp. 40, 500)

fauna the animal life of a region (p. 724)

federal system form of government in which powers are divided between the national government and the state or provincial government (p. 87)

feudalism in medieval Europe and Japan, system of government in which powerful lords gave land to nobles in return for pledges of loyalty (p. 296)

fishery areas (freshwater or saltwater) in which fish or sea animals are caught (p. 120)

fjord (fee•YORD) long, steep-sided glacial valley now filled by seawater (p. 272)

flora the plant life of a region (p. 724)

foehn (FUHN) dry wind that blows from the leeward sides of mountains, sometimes melting snow and causing avalanches; term used mainly in Europe (p. 279)

fold a bend in layers of rock, sometimes caused by plate movement (p. 40)

food web the interlinking chains of predators and their food sources in an ecosystem (p. 840)

formal region a region defined by a common characteristic, such as production of a product (p. 21)

free port port city, such as Singapore, where goods can be unloaded, stored, and reshipped without the payment of import duties (p. 764)

free trade the removal of trade barriers so that goods can flow freely between countries (p. 94)

functional region a central point and the surrounding territory linked to it (p. 21)

fútbol Spanish term for soccer (p. 230)

gauchos the cowhands of Argentina and Uruguay (p. 197)

genetically modified foods foods whose genes have been altered to make them grow bigger or faster or more resistant to pests (p. 317)

geographic information systems computer tools for processing and organizing details and satellite images with other pieces of information (p. 25)

glaciation the process whereby glaciers form and spread (p. 272)

glacier large body of ice that moves across the surface of the earth (p. 42)

glasnost Russian term for a new "openness," part of Mikhail Gorbachev's reform plans (p. 371)

global warming gradual warming of the earth and its atmosphere that may be caused in part by pollution and an increase in the greenhouse effect (pp. 58, 322)

glyph picture writing carved in stone (p. 221)

Good Friday Peace Agreement paved the way for Protestant and Roman Catholic communities to share political power in Northern Ireland (p. 304)

grazier person who raises sheep or cattle (p. 834)

greenhouse effect the capacity of certain gases in the atmosphere to trap heat, thereby warming the earth (pp. 58, 322)

green revolution program, begun in the 1960s, to produce higher-yielding, more productive strains of wheat, rice, and other food crops (p. 613)

grid system pattern formed as the lines of latitude and longitude cross one another (p. 20)

gross domestic product (GDP) the value of goods and services created within a country in a year (p. 465)

groundwater water within the earth that supplies wells and springs (pp. 49, 167)

guru a teacher or spiritual guide (p. 601)

H

habitat area with conditions suitable for certain plants or animals to live (p. 546)

haiku form of Japanese poetry originally consisting of 17 syllables and three lines, often about nature (p. 678)

hajj in Islam, the yearly pilgrimage to Makkah (p. 466)

headwaters the sources of river waters (p. 118)

heavy industry the manufacture of machinery and equipment needed for factories and mines (p. 317)

hemisphere half of a sphere or globe, as in the earth's Northern and Southern Hemispheres (p. 20)

hieroglyphics Egyptian writing system using pictures and symbols to represent words or sounds (p. 447)

Holocaust the mass killings of 6 million Jews by Germany's Nazi leaders during World War II (p. 298)

homogeneous of the same or similar kind or nature (p. 663)

human-environment interaction the study of the interrelationship between people and their physical environment (p. 22)

human geography also called cultural geography; the study of human activities and their relationship to the cultural and physical environments (p. 24)

hurricane a large, powerful windstorm that forms over warm ocean waters (p. 125)

hydroelectric power electrical energy generated by falling water (pp. 197, 348)

hydrosphere the watery areas of the earth, including oceans, lakes, rivers, and other bodies of water (p. 35)

hypothesis a scientific explanation for an event (p. 69)

I

icon religious image, usually including a picture of Jesus, Mary, or a saint, used mainly by Orthodox Christians (p. 377)

ideogram a pictorial character or symbol that represents a specific meaning or idea (p. 674)

immigration the movement of people into one country from another (p. 133)

impressionism artistic style that developed in Europe in the late 1800s and tried to show the natural appearance of objects with dabs or strokes of color (p. 305)

indigenous native to a place (p. 212)

industrial capitalism an economic system in which business leaders use profits to expand their companies (p. 297)

industrialization transition from an agricultural society to one based on industry (p. 93)

infrastructure the basic urban necessities like streets and utilities (pp. 443, 540)

insular constituting an island, as in Java (p. 720)

intelligentsia intellectual elite (p. 378)

interdependent relying on one another for goods, services, and ideas (p. 763)

introduced species plants and animals placed in areas other than their native habitat (p. 839)

J

jai alai (HY•LY) traditional handball-type game popular with Mexicans and Cubans (p. 231)

Japan Current a warm-water ocean current that adds moisture to the winter monsoons (p. 655)

jati in traditional Hindu society, a social group that defines a family's occupation and social standing (p. 588)

jazz musical form that developed in the United States in the early 1900s, blending African rhythms and European harmonies (p. 148)

jute plant fiber used to make string and cloth (p. 612)

K

karma in Hindu belief, the sum of good and bad actions in one's present and past lives (p. 593)

kolkhoz in the Soviet Union, a small farm worked by farmers who shared in the farm's production and profits (p. 390)

krill tiny shrimplike sea animals that live in cold Antarctic oceans (p. 798)

kums term for deserts in Central Asia (p. 425)

L

lagoon shallow pool of water at the center of an atoll (p. 796)

lama Buddhist religious leader (p. 675)

language family group of related languages that have all developed from one earlier language (pp. 81, 303)

latifundia in Latin America, large agricultural estates owned by families or corporations (p. 238)

leach to wash nutrients out of the soil (p. 506)

leeward facing away from the direction from which the wind is blowing (p. 64)

lichens sturdy small plants that grow like a crust on rocks or tree trunks (p. 803)

light industry manufacturing aimed at making consumer goods such as textiles or food processing rather than heavy machinery (p. 317)

lingua franca a common language used among people with different native languages (p. 526)

literacy rate the percentage of people in a given place who can read and write (p. 150)

lithosphere surface land areas of the earth's crust, including continents and ocean basins (p. 35)

llanos (LAH•nohs) fertile plains in inland areas of Colombia and Venezuela (p. 196)

location a specific place on the earth (p. 20)

lode deposit of minerals (p. 761)

loess (LEHS) fine, yellowish-brown topsoil made up of particles of silt and clay, usually carried by the wind (pp. 42, 275, 648)

longhouse in rural areas of Indonesia and Malaysia, a large, elevated building where people from several related families live (p. 752)

M

Maastricht Treaty a 1992 meeting of European governments in Maastricht, the Netherlands, that formed the European Union (p. 314)

magma molten rock that is pushed up from the earth's mantle (p. 38)

malnutrition faulty or inadequate nutrition (p. 230)

mantle thick middle layer of the earth's interior structure, consisting of dense, hot rock (p. 38)

mantra in Hinduism, a sacred word or phrase repeated in prayers and chants (p. 601)

manuka small shrubs that grow in plateau regions of New Zealand (p. 803)

maquiladoras in Mexico, manufacturing plants set up by foreign firms (p. 239)

maritime concerned with travel or shipping by sea (p. 741)

market economy an economic system based on free enterprise, in which businesses are privately owned, and production and prices are determined by supply and demand (pp. 89, 157, 388)

marsupial mammal whose offspring mature in a pouch on the mother's abdomen (p. 839)

mass culture popular culture spread by media such as radio and television (p. 525)

megacities cities with more than 10 million people (p. 217)

megalopolis a "super-city" that is made up of several large and small cities such as the area between Boston and Washington, D.C. (pp. 136, 591)

meltwater water formed by melting snow and ice (p. 321)

merchant marine a country's fleet of ships that engage in commerce or trade (p. 691)

meteorology the study of weather and weather forecasting (p. 24)

metropolitan area region that includes a central city and its surrounding suburbs (p. 135)

mica silicate mineral that readily splits into thin, shiny sheets (p. 574)

Middle Ages the period of European history from about A.D. 500 to about 1500 (p. 296)

migration the movement of people from place to place (p. 79)

minifundia in Latin America, small farms that produce food chiefly for family use (p. 238)

mistral strong northerly wind from the Alps that can bring cold air to southern France (p. 280)

mixed economy an economy in which the government supports and regulates enterprise through decisions that affect the marketplace (p. 89)

mixed farming raising several kinds of crops and livestock on the same farm (p. 317)

mixed forest forest with both coniferous and deciduous trees (p. 68)

mobility able to move from place to place (p. 137)

monopoly total control of a type of industry by one person or one company (p. 162)

monotheism belief in one God (p. 448)

monsoon in Asia, seasonal wind that brings warm, moist air from the oceans in summer and cold, dry air from inland in winter (pp. 579, 654)

moraine piles of rocky debris left by melting glaciers (p. 42)

mosaic picture or design made with small pieces of colored stone, glass, shell, or tile (p. 228)

mosque in Islam, a house of public worship (p. 448)

movement ongoing movement of people, goods, and ideas (p. 22)

mural wall painting (p. 228)

N

nationalism belief in the right of each people to be an independent nation (p. 449)

nationalities large, distinct ethnic groups within a country (p. 364)

nationalize to place a company or industry under government control (p. 450)

Native American North America's first immigrant, who probably moved into the region from Asia thousands of years ago (p. 134)

natural increase the growth rate of a population; the difference between birthrate and death rate (p. 76)

natural resource substance from the earth that is not made by people but can be used by them (p. 91)

natural vegetation plant life that grows in a certain area if people have not changed the natural environment (p. 66)

nirvana in Buddhism, ultimate state of peace and insight toward which people strive (p. 594)

North American Free Trade Agreement (NAFTA) trade agreement made in 1994 by Canada, the United States, and Mexico (p. 240)

nuclear family family group made up of husband, wife, and children (p. 529)

nuclear proliferation the spreading development of nuclear arms (p. 623)

nuclear waste the by-product of producing nuclear power (p. 398)

oasis small area in a desert where water and vegetation are found (pp. 67, 429)

oligarchy system of government in which a small group holds power (p. 88)

oral tradition stories passed down from generation to generation by word of mouth (p. 528)

organic farming the use of natural substances rather than chemical fertilizers and pesticides to enrich the soil and grow crops (p. 318)

ozone layer atmospheric layer with protective gases that prevents solar rays from reaching the earth's surface (p. 841)

P

paddy flooded field in which rice is grown (p. 760)

pagoda a style of architecture most often found in traditional East Asian buildings, marked by gracefully curved tile roofs in the tower style (p. 679)

Pampas grassy, treeless plains of southern South America (p. 196)

parliament in Canada, national legislature made up of the Senate and the House of Commons (p. 145)

pastoralism the raising of livestock (p. 430)

patois dialects that blend elements of indigenous, European, African, and Asian languages (p. 213)

patriarch the head of the Eastern Orthodox Church (p. 377)

patriotism love for or devotion to one's country (p. 151)

perceptual region a region defined by popular feelings and images rather than by objective data (p. 21)

perestroika (PEHR•uh•STROY•kuh) in Russian, "restructuring"; part of Gorbachev's plan for reforming Soviet government (p. 371)

permafrost permanently frozen layer of soil beneath the surface of the ground (pp. 68, 281, 349)

pesticide chemical used to kill insects, rodents, and other pests (p. 398)

petrochemical chemical product derived from petroleum or natural gas (p. 465)

phosphate natural mineral containing chemical compounds often used in fertilizers (p. 426)

physical geography the study of Earth's physical features (p. 24)

pidgin English a dialect mixing English and a local language (p. 813)

pipeline long network of underground or aboveground pipes (p. 162)

place a particular space with physical and human meaning (p. 21)

plate tectonics the term scientists use to describe the activities of continental drift and magma flow which create many of Earth's physical features (p. 38)

poaching illegal hunting of protected animals (pp. 546, 619)

pogrom in czarist Russia, an attack on Jews carried out by government troops or officials (p. 378)

polder low-lying area from which seawater has been drained to create new farmland (p. 272)

pollution the existence of impure, unclean, or poisonous substances in the air, water, or land (p. 94)

population density the average number of people in a square mile or square kilometer (p. 77)

population distribution the pattern of population in a country, a continent, or the world (p. 77)

post-industrial an economy with less emphasis on heavy industry and manufacturing and more emphasis on services and technology (p. 158)

prairie an inland grassland area (pp. 68, 124)

precipitation moisture that falls to the earth as rain, sleet, hail, or snow (p. 47)

prevailing wind wind in a region that blows in a fairly constant directional pattern (p. 62)

primate city a city that dominates a country's economy, culture, and government and in which population is concentrated; usually the capital (pp. 217, 738)

privatization a change to private ownership of state-owned companies and industries (p. 389)

prophet person believed to be a messenger from God (p. 448)

Q

qanat underground canal used in water systems of ancient Persians (p. 447)

quipu (KEE•poo) knotted cords of various lengths and colors used by the Inca to keep financial records (p. 222)

R

radioactive material material contaminated by residue from the generation of nuclear energy (p. 397)

rain shadow dry area found on the leeward side of a mountain range (p. 64)

raj Hindu word for empire (p. 595)

realism artistic style portraying everyday life that developed in Europe during the mid-1800s (p. 305)

reforestation replanting young trees or seeds on lands where trees have been cut or destroyed (p. 244)

Reformation religious movement that began in Germany in the 1400s, leading to the establishment of Protestant churches (p. 297)

refugee one who flees his or her home for safety (p. 288)

regions places united by specific characteristics (p. 21)

reincarnated in Hindu belief, being reborn repeatedly in different forms, until one has overcome earthly desires (p. 593)

relative location location in relation to other places (p. 20)

Renaissance in Europe, a 300-year period of renewed interest in classical learning and the arts, beginning in the 1300s (p. 296)

reparations a payment for damages (p. 298)

republic form of government without a monarch in which people elect their officials (p. 142)

retooling converting old factories for use in new industries (p. 161)

revolution in astronomy, the earth's yearly trip around the sun, taking 365¼ days (p. 56)

rift valley a crack in the earth's surface created by shifting (p. 500)

romanticism artistic style emphasizing individual emotions that developed in Europe in the late 1700s and early 1800s as a reaction to industrialization (p. 305)

Russification in nineteenth-century Russia, a government program that required everyone in the empire to speak Russian and to become a Christian (p. 370)

S

sadhu a Hindu hermit or holy man (p. 602)

samurai in medieval Japan, a class of professional soldiers who lived by a strict code of personal honor and loyalty to a noble (p. 670)

sanitation disposal of waste products (p. 516)

satellite a country controlled by another country, notably Eastern European countries controlled by the Soviet Union by the end of World War II (p. 371)

savanna a tropical grassland containing scattered trees (p. 507)

sedentary farming farming carried on at permanent settlements (p. 538)

serf laborer obliged to remain on the land where he or she works (p. 369)

service center convenient business location for rural dwellers (p. 518)

service industry business that provides a service instead of making goods (p. 238)

shamanism belief in a leader who can communicate with spirits (p. 675)

shantytowns makeshift communities on the edges of cities (p. 244)

shifting cultivation clearing forests to plant fields for a few years and then abandoning them (p. 770)

shifting farming method in which farmers move every few years to find better soil (p. 538)

shogun military ruler in medieval Japan (p. 670)

sickle large, curved knife with a handle, used to cut grass or tall grains (p. 760)

sirocco hot desert wind that can blow air and dust from North Africa to western Europe's Mediterranean coast (pp. 280–281)

slash-and-burn farming traditional farming method in which all trees and plants in an area are cut and burned to add nutrients to the soil (p. 244)

smog haze caused by the interaction of ultraviolet solar radiation with chemical fumes from automobile exhausts and other pollution sources (pp. 69, 167)

socialism political philosophy in which the government owns the means of production (p. 370)

socialist realism realistic style of art and literature that glorified Soviet ideals and goals (p. 380)

socioeconomic status level of income and education (p. 150)

solstice one of two days (about June 21 and December 22) on which the sun's rays strike directly on the Tropic of Cancer or Tropic of Capricorn, marking the beginning of summer or winter (p. 57)

sovereignty self-rule (p. 365)

sovkhoz in the Soviet Union, a large farm owned and run by the state (p. 390)

sphere of influence area of a country in which a foreign power has political or economic control (p. 743)

spreading a process by which new land is created when sea plates pull apart and magma wells up between the plates (p. 40)

state farm under communism, a state-owned farm managed by government officials (p. 317)

station Australian term for an outlying ranch or large farm (p. 834)

steppe wide, grassy plains of Eurasia; also, similar semiarid climate regions elsewhere (p. 355)

Strine colloquial English spoken in Australia (p. 813)

stupa a dome-shaped structure that serves as a Buddhist shrine (p. 604)

subcontinent large landmass that is part of a continent but still distinct from it, such as India (p. 569)

subduction a process by which mountains can form as sea plates dive beneath continental plates (p. 39)

subsistence crop a crop grown mainly to feed the farmer's family (p. 760)

subsistence farming producing just enough food for a family or a village to survive (pp. 538, 826)

suburbs outlying communities around a city (p. 135)

Sunbelt mild climate region, southern United States (p. 135)

supercells violent thunderstorms that can spawn tornadoes (p. 124)

sustainable development technological and economic growth that does not deplete the human and natural resources of a given area (pp. 242, 619)

syncretism a blending of beliefs and practices from different religions into one faith (p. 228)

taiga Russian term for the vast subarctic forest, mostly evergreens, that covers much of Russia and Siberia (p. 353)

tariff a tax on imports or exports (p. 163)

temperature degree of hotness or coldness measured on a set scale, such as Fahrenheit or Celsius (p. 56)

tierra caliente Spanish term for "hot land"; the lowest altitude zone of Latin American highlands climates (p. 203)

tierra fría Spanish term for "cold land"; the highest altitude zone of Latin American highlands climates (p. 203)

tierra templada Spanish term for "temperate land"; the middle altitude zone of Latin American highlands climates (p. 203)

timberline elevation above which it is too cold for trees to grow (pp. 124, 279)

trade deficit spending more money on imports than earning from exports (pp. 163, 689)

trade surplus earning more money from export sales than spending for imports (pp. 163, 689)

traditional economy a system in which tradition and custom control all economic activity; exists in only a few parts of the world today (p. 89)

tributary smaller river or stream that feeds into a larger river (p. 118)

trust territory region placed by United Nations under temporary political and economic control of another country after World War II (p. 819)

tsunami Japanese term used for a huge sea wave caused by an undersea earthquake (p. 646)

tundra vast, treeless plains in cold northern climates, characterized by permafrost and small, low plants, such as mosses and shrubs (p. 352)

typhoon a violent tropical storm that forms in the Pacific Ocean, usually in late summer (pp. 655, 768, 802)

Underground Railroad an informal network of safehouses, in the United States, that helped thousands of enslaved people escape to freedom (p. 143)

unitary system a government in which all key powers are given to the national or central government (p. 87)

universal suffrage equal voting rights for all adult citizens of a nation (p. 524)

urbanization the movement of people from rural areas into cities (pp. 135, 216, 290, 517, 737)

viceroy representative of the Spanish monarch appointed to enforce laws in colonial Latin America (p. 222)

wadi in the desert, a streambed that is dry except during a heavy rain (p. 424)

wat in Southeast Asia, a temple (p. 750)

water cycle regular movement of water from ocean to air to ground and back to the ocean (p. 47)

wattle woven framework made from acacia saplings by early Australian settlers to build homes (p. 801)

weather condition of the atmosphere in one place during a short period of time (p. 55)

weathering chemical or physical processes, such as freezing, that break down rocks (p. 42)

welfare state nation in which the government assumes major responsibility for people's welfare in areas such as health and education (p. 306)

windward facing toward the direction from which the wind is blowing (p. 64)

World Trade Organization (WTO) an international body that oversees trade agreements and settles trade disputes among countries (p. 690)

ziggurat large step-like temple of mud brick built in ancient Mesopotamia (p. 454)

Gazetteer

A Gazetteer (GA•zuh•TIHR) is a geographic index or dictionary. It shows latitude and longitude for cities and certain other places. This Gazetteer lists most of the world's largest independent countries, their capitals, and several important geographic features. The page numbers tell where each entry can be found on a map in this book. As an aid to pronunciation, many entries are spelled phonetically.

A

Abidjan (AH•bee•JAHN) Capital and port city of Côte d'Ivoire, Africa. 5°N 4°W (p. 487)
Abu Dhabi (AH•boo DAH•bee) Capital of the United Arab Emirates, on the Persian Gulf. 24°N 54°E (p. 411)
Abuja (ah•BOO•jah) Capital of Nigeria. 8°N 9°E (p. 487)
Accra (AH•kruh) Capital and port city of Ghana. 6°N 0° longitude (p. 487)
Aconcagua Highest peak of the Andes and of the Western Hemisphere, in western Argentina near the Chilean border. 32°S 76°W (p. 182)
Addis Ababa (AHD•dihs AH•bah•BAH) Capital of Ethiopia. 9°N 39°E (p. 487)
Adriatic (AY•dree•A•tihk) **Sea** Arm of the Mediterranean Sea between the Balkan Peninsula and Italy. (p. 260)
Aegean (ee•JEE•uhn) **Sea** Arm of the Mediterranean Sea between Greece and Turkey. (p. 260)
Afghanistan (af•GA•nuh•STAN) Country in Central Asia, west of Pakistan. (p. 411)
Ahaggar Mountains Highest plateau region in the central Sahara. (p. 410)
Albania (al•BAY•nee•uh) Country on the east coast of the Adriatic Sea, south of Yugoslavia. (p. 261)
Algeria Country in North Africa. (p. 411)
Algiers (al•JIHRZ) Capital of Algeria. 37°N 3°E (p. 411)
Alps Mountain system extending through central Europe. (p. 260)
Altay Mountains Mountain system between western Mongolia and China and between Kazakhstan and southern Russia. (p. 636)
Amazon River River flowing through Peru and Brazil in South America and into the Atlantic Ocean. (p. 182)
Amman Capital of Jordan. 32°N 36°E (p. 411)
Amsterdam (AM•stuhr•DAM) Capital of the Netherlands. 52°N 5°E (p. 261)
Amu Darya River in Turkmenistan in central and western Asia. (p. 410)
Amur River River in northeast Asia. (p. 338)
Andes (AN•deez) Mountain system along western South America. (p. 182)
Andorra (an•DAWR•uh) Country in southern Europe, between France and Spain. (p. 261)
Angola (ang•GOH•luh) Country in Africa, south of the Democratic Republic of the Congo. (p. 487)
Ankara Capital of Turkey. 40°N 33°E (p. 411)
Antananarivo (AHN•tah•NAH•nah•REE•voh) Capital of Madagascar. 19°S 48°E (p. 487)
Antigua Island in the West Indies, part of independent Antigua and Barbuda. 18°N 61°W (p. 183)
Apennines Mountain range in central Italy. (p. 260)
Appalachian Mountains Mountain system in eastern North America. (p. 117)
Arabian Sea Part of the Indian Ocean between India and the Arabian Peninsula. (p. 410)
Aral Sea Inland sea between Kazakhstan and Uzbekistan. (p. 410)
Argentina (AHR•juhn•TEE•nuh) Country in South America, east of Chile. (p. 183)
Arkansas River River in south-central United States, emptying into the Mississippi River. (p. 117)
Armenia (ahr•MEE•nee•uh) Southeastern European country between the Black and Caspian Seas. (p. 261)
Ashkhabad (ASH•kuh•BAD) Capital of Turkmenistan. 40°N 58°E (p. 411)
Asmara Capital of Eritrea. 16°N 39°E (p. 487)
Astana Capital of Kazakhstan. 52°N 72°E (p. 411)
Asunción (ah•SOON•SYOHN) Capital of Paraguay. 25°S 58°W (p. 183)
Athens Capital of Greece. 38°N 24°E (p. 261)
Atlas Mountains Mountain range on the northern edge of the Sahara. (p. 410)
Australia Country and continent southeast of Asia. (p. 785)
Austria (AWS•tree•uh) Country in central Europe, east of Switzerland. (p. 261)
Azerbaijan (A•zuhr•by•JAHN) European-Asian country on the Caspian Sea. (p. 411)

B

Baghdad (BAG•DAD) Capital of Iraq. 33°N 44°E (p. 411)
Bahamas Independent state comprising a chain of islands, cays, and reefs southeast of Florida and north of Cuba. 24°N 76°W (p. 183)
Bahrain (bah•RAYN) Independent state in the western Persian Gulf. (p. 411)
Baku Capital of Azerbaijan. 40°N 50°E (p. 411)
Balkan Mountains Mountain range extending across central Bulgaria to the Black Sea. (p. 260)
Balkan Peninsula Peninsula in southeastern Europe bordered on the west by the Adriatic Sea. (p. 261)
Baltic Sea Arm of the Atlantic Ocean in northern Europe that connects with the North Sea. (p. 260)
Bamako (BAH•mah•KOH) Capital of Mali. 13°N 8°W (p. 487)

Bangkok Capital of Thailand. 14°N 100°E (p. 711)
Bangladesh (BAHNG•gluh•DESH) Country in South Asia, bordered by India and Myanmar. (p. 561)
Bangui (bahng•GEE) Capital of the Central African Republic. 4°N 19°E (p. 487)
Banjul Capital of Gambia. 13°N 17°W (p. 487)
Barbados Island country between the Atlantic Ocean and the Caribbean Sea. 14°N 59°W (p. 183)
Barbuda Island in the West Indies, part of independent Antigua and Barbuda. 18°N 62°W (p. 183)
Barents Sea Part of the Arctic Ocean, north of Norway and Russia. (p. 338)
Bay of Bengal Part of the Indian Ocean between eastern India and Southeast Asia. (p. 560)
Beijing Capital of China. 40°N 116°E (p. 637)
Beirut (bay•ROOT) Capital of Lebanon. 34°N 36°E (p. 411)
Belarus (BEE•luh•ROOS) Eastern European country west of Russia. (p. 261)
Belgium (BEHL•juhm) Country in northwestern Europe, south of the Netherlands. (p. 261)
Belgrade Capital of Serbia. 45°N 21°E (p. 261)
Belize (buh•LEEZ) Country in Central America. (p. 183)
Belmopan (BEHL•moh•PAHN) Capital of Belize. 17°N 89°W (p. 183)
Benin (buh•NEEN) Country in western Africa. (p. 487)
Ben Nevis Peak in the highlands region of the Grampian Mountains in Scotland. 54°N 5°W (p. 273)
Bering Sea Part of the north Pacific Ocean, extending between the United States and Russia. (p. 117)
Berlin Capital of Germany. 53°N 13°E (p. 261)
Bern Capital of Switzerland. 47°N 7°E (p. 261)
Bhutan (boo•TAHN) Country in the eastern Himalaya, northeast of India. 27°N 91°E (p. 561)
Bishkek (bihsh•KEHK) Capital and largest city of Kyrgyzstan. 43°N 75°E (p. 411)
Bissau (bih•SOW) Capital of Guinea-Bissau. 12°N 16°W (p. 487)
Black Sea Sea between Europe and Asia. (p. 260)
Bloemfontein (BLOOM•FAHN•TAYN) Judicial capital of the Republic of South Africa. 29°S 26°E (p. 487)
Bogotá (BOH•goh•TAH) Capital of Colombia. 5°N 74°W (p. 183)
Bolivia Republic in west central South America. (p. 183)
Bosnia-Herzegovina (BAHZ•nee•uh HERT•suh•goh•VEE•nuh) Southeastern European country between Yugoslavia and Croatia. (p. 261)
Bosporus Strait between European and Asian Turkey, connecting the Sea of Marmara with the Black Sea. (p. 410)
Botswana (baht•SWAH•nuh) Country in Africa, north of the Republic of South Africa. (p. 487)
Brahmaputra River River that begins in Tibet, flows through northeast India and Bangladesh, and empties into the Bay of Bengal. (p. 560)
Brasília (bruh•ZIHL•yuh) Capital of Brazil. 16°S 48°W (p. 183)
Bratislava (BRAH•tuh•SLAH•vuh) Capital and largest city of Slovakia. 48°N 17°E (p. 261)
Brazil (bruh•ZIHL) Largest country in South America, in east-central South America. (p. 183)
Brazzaville (BRA•zuh•VIHL) Capital of Congo. 4°S 15°E (p. 487)
Brunei (bru•NY) Country on the northern coast of the island of Borneo. (p. 711)
Brussels Capital of Belgium. 51°N 4°E (p. 261)
Bucharest (BOO•kuh•REHST) Capital of Romania. 44°N 26°E (p. 261)
Budapest Capital of Hungary. 48°N 19°E (p. 261)
Buenos Aires (BWAY•nuhs AR•eez) Capital of Argentina. 34°S 58°W (p. 183)
Bujumbura (BOO•juhm•BUR•uh) Capital of Burundi. 3°S 29°E (p. 487)
Bulgaria (BUHL•GAR•ee•uh) Country in southeastern Europe, south of Romania. (p. 261)
Burkina Faso (bur•KEE•nuh FAH•soh) Country in western Africa, south of Mali. (p. 487)
Burundi (bu•ROON•dee) Country in central Africa at the northern end of Lake Tanganyika. (p. 487)

C

Cairo (KY•roh) Capital of Egypt. 31°N 32°E (p. 411)
Cambodia (kam•BOH•dee•uh) Country in Southeast Asia, south of Thailand. (p. 711)
Cameroon (KA•muh•ROON) Country in west Africa, on the northeast shore of the Gulf of Guinea. (p. 487)
Canada Country in northern North America. (p. 117)
Canberra Capital of Australia. 35°S 149°E (p. 785)
Cape Town Legislative capital of the Republic of South Africa. 34°S 18°E (p. 487)
Cape Verde Republic consisting of a group of volcanic islands in the Atlantic Ocean. 15°N 26°W (p. 487)
Caracas (kah•RAH•kahs) Capital of Venezuela. 11°N 67°W (p. 183)
Caribbean (KAR•uh•BEE•uhn) **Sea** Part of the Atlantic Ocean, bounded by the West Indies, South America, and Central America. (p. 182)
Carpathian Mountains Mountain range in eastern Europe in Slovakia and Romania. (p. 260)
Caspian (KAS•pee•uhn) **Sea** Salt lake between Europe and Asia. (p. 260)
Caucasus Mountains Mountain range in southwestern Russia. (p. 410)
Central African Republic Country in central Africa, south of Chad. (p. 487)
Central Siberian Plateau Tableland area in Siberia. (p. 338)
Chad Country in north central Africa. (p. 487)
Chao Phraya (chow PRY•uh) River in Thailand, flowing south into the Gulf of Thailand. (p. 710)

Chile (CHIH•lee) Western South American country, along the Pacific Ocean. (p. 183)
China (People's Republic of China) Country in eastern and central Asia. (p. 637)
Chisinau (KEE•shee•NOW) Capital and largest city of Moldova. 47°N 29°E (p. 261)
Colombia Republic in northern South America. (p. 183)
Colombo Capital of Sri Lanka. 7°N 80°E (p. 561)
Colorado Plateau Highlands region in the western United States. (p. 117)
Colorado River River in the western United States that flows through the Grand Canyon. (p. 117)
Columbia Plateau Flat plains area primarily in western Washington State in the United States. (p. 117)
Comoros (KAH•muh•ROHZ) **Islands** Island country in the Indian Ocean between the island of Madagascar and Africa. 13°S 43°E (p. 487)
Conakry (KAH•nuh•kree) Capital of Guinea. 10°N 14°W (p. 487)
Congo Country in equatorial Africa. (p. 487)
Congo, Democratic Republic of the African country on the Equator, north of Zambia and Angola. (p. 487)
Congo River River that runs through the Democratic Republic of the Congo. (p. 486)
Copenhagen (KOH•puhn•HAY•guhn) Capital of Denmark. 56°N 12°E (p. 261)
Costa Rica (KAWS•tah REE•kuh) Central American country, south of Nicaragua. (p. 183)
Côte d'Ivoire (KOHT dee•VWAHR) West African country, south of Mali. (p. 487)
Croatia (kroh•AY•shuh) Southeastern European country on the Adriatic Sea. (p. 261)
Cuba Island country southeast of Florida. 21°N 80°W (p. 183)
Cyprus Island country in the eastern Mediterranean Sea, south of Turkey. 35°N 31°E (p. 261)
Czech (CHEHK) **Republic** Central European country south of Germany and Poland. (p. 261)

D

Dakar Capital of Senegal. 15°N 17°W (p. 487)
Damascus Capital of Syria. 34°N 36°E (p. 411)
Danube River River in Europe that begins in Germany and flows into the Black Sea. (p. 260)
Dardanelles Strait between European and Asian Turkey, connecting the Sea of Marmara with the Aegean Sea. (p. 260)
Dar es Salaam (DAHR EHS suh•LAHM) Capital of Tanzania. 7°S 39°E (p. 487)
Darling River River in southeast Australia. (p. 710)
Deccan Plateau The peninsula of India south of the Narmada River. (p. 560)
Denmark Country in northwestern Europe, between the Baltic and North Seas. (p. 261)
Dhaka Capital of Bangladesh. 24°N 90°E (p. 561)
Djibouti (juh•BOO•tee) Country in East Africa, on the Gulf of Aden. (p. 487)
Dnieper (NEE•puhr) **River** River that begins in Russia, flows through Belarus and Ukraine, and then drains into the Black Sea. (p. 260)
Dniester (NEE•stuhr) **River** River in south-central Europe that begins in Ukraine and flows southeast to the Black Sea. (p. 260)
Dodoma (doh•DOH•mah) Future capital of Tanzania. 7°S 36°E (p. 487)
Doha (DOH•hah) Capital of Qatar. 25°N 51°E (p. 411)
Dominica Island republic in the West Indies, lying in the center of the Lesser Antilles. 15°N 61°W (p. 183)
Dominican Republic Republic occupying the eastern two-thirds of Hispaniola Island in the West Indies. 19°N 70°W (p. 183)
Don River River in southwestern Russia. (p. 260)
Drakensberg (DRAH•kuhnz•BUHRG) **Range** Mountain range in South Africa. (p. 486)
Dublin Capital of Ireland. 53°N 6°W (p. 261)
Dushanbe (doo•SHAM•buh) Capital and largest city of Tajikistan. 39°N 69°E (p. 411)

E

Eastern Ghats Mountain range in southern India. (p. 560)
Ecuador (EH•kwuh•DAWR) Country in South America, south of Colombia. (p. 183)
Egypt (EE•jihpt) Country in northern Africa on the Mediterranean Sea. (p. 411)
Elbe River River in the Czech Republic and Germany. (p. 260)
Elburz Mountains Mountain range in northern Iran parallel to the shore of the Caspian Sea. (p. 410)
El Salvador (ehl SAL•vuh•DAWR) Country in Central America, southwest of Honduras. (p. 183)
Equatorial Guinea (EE•kwuh•TOHR•ee•uhl GIH•nee) Country in western Africa, south of Cameroon. (p. 487)
Eritrea (EHR•uh•TREE•uh) Country in northeast Africa, north of Ethiopia. (p. 487)
Ertis River River in northeastern Kazakhstan and the western part of Russia. *See also* Irtysh River. (p. 410)
Estonia (eh•STOH•nee•uh) Northern European country on the Baltic Sea. (p. 261)
Ethiopia (EE•thee•OH•pee•uh) Country in eastern Africa, north of Somalia and Kenya. (p. 487)
Euphrates (yu•FRAY•teez) **River** River in southwestern Asia that flows through Syria and Iraq and joins the Tigris River. (p. 410)

F

Fiji (FEE•jee) Country comprising an island group in the southwest Pacific Ocean. 19°S 175°E (p. 785)

Finland Country in northern Europe, east of Sweden. (p. 261)
France Country in western Europe. (p. 261)
Freetown Capital and port city of Sierra Leone, in western Africa. 9°N 13°W (p. 487)
French Guiana Overseas department of France on the northeast coast of South America. (p. 183)

Gabon (ga•BOHN) Country in western Africa, on the Atlantic Ocean. (p. 487)
Gaborone (GAH•boh•ROH•nay) Capital of Botswana, in southern Africa. 24°S 26°E (p. 487)
Gambia Country in western Africa. (p. 487)
Ganges (GAN•JEEZ) **Plain** A fertile plains region in northern India traversed by the Ganges River. (p. 560)
Ganges River River in northern India and Bangladesh that flows into the Bay of Bengal. (p. 560)
Georgetown Capital of Guyana. 8°N 58°W (p. 487)
Georgia Asian/European country bordering the Black Sea, south of Russia. (p. 411)
Germany (Federal Republic of Germany) Country in north central Europe. (p. 261)
Ghana (GAH•nuh) Country in western Africa, on the Gulf of Guinea. (p. 487)
Gobi Desert in Central Asia. (p. 636)
Godavari River River in central India. (p. 560)
Gran Chaco Region in south-central South America located in Paraguay, Bolivia, and Argentina. (p. 182)
Great Britain Kingdom in western Europe comprising England, Scotland, and Wales. (p. 261)
Great Dividing Range Chain of hills and mountains, on Australia's eastern coast. (p. 784)
Great Indian Desert Region of sandy desert in northwest India and southeast Pakistan. (p. 560)
Great Plains Rolling treeless area of central North America. (p. 117)
Great Salt Lake Large saltwater lake in Utah in the United States that has no outlet. (p. 117)
Great Slave Lake A lake in the south-central mainland of the Northwest Territories in Canada. (p. 117)
Greece Country in southern Europe, on the Balkan Peninsula. (p. 261)
Greenland Island in the northwestern Atlantic Ocean. 74°N 40°W (p. 117)
Grenada Island in the self-governing West Indies. 17°N 61°W (p. 183)
Guam Island in the western Pacific. It is an unincorporated United States territory. 13°N 144°E (p. 785)
Guatemala (GWAH•tuh•MAH•luh) Country in Central America, south of Mexico. (p. 183)
Guatemala City Capital of Guatemala and the largest city in Central America. 15°N 91°W (p. 183)
Guinea (GIH•nee) West African country on the Atlantic coast. 11°N 12°W (p. 487)
Guinea-Bissau (GIH•nee bih•SOW) West African country on the Atlantic coast. 12°N 20°W (p. 487)
Gulf of Aden Arm of the Indian Ocean between the Arabian Peninsula and Africa. (p. 410)
Gulf of Mexico Gulf on the southern coast of North America. (p. 182)
Gulf of Thailand Inlet of the South China Sea. (p. 710)
Guyana Republic in northern South America. (p. 183)

H

Hainan (HY•NAHN) Island province of China in the South China Sea. 19°N 109°E (p. 637)
Haiti (HAY•tee) Republic occupying the western third of Hispaniola Island in the West Indies. 19°N 72.25°W (p. 183)
Hanoi Capital of Vietnam. 21°N 106°E (p. 711)
Harare (huh•RAH•ray) Capital of Zimbabwe. 18°S 23°E (p. 487)
Havana Capital of Cuba. 23°N 82°W (p. 183)
Helsinki Capital of Finland. 60°N 24°E (p. 261)
Himalaya (HIH•muh•LAY•uh) Mountain range in South Asia, bordering the Indian subcontinent on the north. (p. 560)
Hindu Kush Mountain range in Central Asia. (p. 560)
Honduras (hahn•DUR•uhs) Central American republic. (p. 183)
Hong Kong Administrative district and port in southern China. 22°N 115°E (p. 637)
Hudson Bay Inland sea in east-central Canada. (p. 117)
Hungary (HUHNG•guh•ree) Central European country, south of Slovakia. (p. 261)

I

Iberian (eye•BIHR•ee•uhn) **Peninsula** Peninsula in southwestern Europe. (p. 260)
Iceland Island country between the north Atlantic and Arctic Oceans. 65°N 20°W (p. 261)
India South Asian country south of China. (p. 561)
Indochina Southeast peninsula of Asia. (p. 710)
Indonesia (IHN•duh•NEE•zhuh) Group of islands that forms the Southeast Asian country of the Republic of Indonesia. 5°S 119°E (p. 711)
Indus River River in Asia that rises in Tibet and flows through Pakistan to the Arabian Sea. (p. 560)
Iran (ih•RAHN) Southwest Asian country, formerly called Persia. (p. 411)
Iraq (ih•RAHK) Southwest Asian country, south of Turkey. (p. 411)
Ireland (EYER•luhnd) Island west of England, occupied by the Republic of Ireland and by Northern Ireland. 54°N 8°W (p. 261)
Irrawaddy River River in central Myanmar formed by the confluence of the Mali and Nmai Rivers. (p. 710)

Irtysh River River in northeast Kazakhstan and the western part of Russia, in Asia. (p. 410)
Islamabad (ihs•LAH•muh•BAHD) Capital of Pakistan. 34°N 73°E (p. 561)
Israel (IHZ•ree•uhl) Country in Southwest Asia, south of Lebanon. (p. 561)
Isthmus of Panama Narrow strip of land that forms the link in Central America between North America and South America. (p. 182)
Italy (IH•tuhl•ee) Southern European country, south of Switzerland and east of France. (p. 261)

Jakarta Capital of Indonesia. 6°S 107°E (p. 711)
Jamaica (juh•MAY•kuh) Island country in the West Indies. 18°N 78°W (p. 183)
Japan Country in East Asia, consisting of four main islands of Hokkaido, Honshu, Shikoku, and Kyushu, plus thousands of small islands. 37°N 134°E (p. 637)
Jerusalem (juh•ROO•suh•luhm) Capital of Israel and a holy city for Christians, Jews, and Muslims. 32°N 35°E (p. 411)
Jordan Country in Southwest Asia. (p. 411)
Jutland Peninsula extending north from Germany. (p. 260)

K

K2 (Godwin Austen) Himalayan mountain in Jammu and Kashmir. 35°N 76°E (p. 560)
Kabul Capital of Afghanistan. 35°N 69°E (p. 411)
Kalahari Desert Plateau and part desert located in the southern part of Africa. (p. 486)
Kamchatka Peninsula Peninsula in northeast Russia, in Asia. (p. 338)
Kampala (kahm•PAH•lah) Capital of Uganda. 0° latitude 32°E (p. 487)
Kara Sea Arm of the Arctic Ocean north of Russia. (p. 338)
Kathmandu (KAT•MAN•DOO) Capital of Nepal. 28°N 85°E (p. 561)
Kazakhstan (KA•zak•STAN) Large Asian country south of Russia, bordering the Caspian Sea. (p. 411)
Kenya (KEH•nyuh) Country in eastern Africa, south of Ethiopia. (p. 487)
Khartoum Capital of Sudan. 16°N 33°E (p. 487)
Khyber Pass Mountain pass between Afghanistan and Pakistan. 34°N 71°E (p. 560)
Kiev (KEE•EHF) Capital of Ukraine. 50°N 31°E (p. 261)
Kigali (kee•GAH•lee) Capital of Rwanda, in central Africa. 2°S 30°E (p. 487)
Kilimanjaro Highest mountain in Africa, located in Tanzania. 3°S 37°E (p. 486)
Kingston Capital of Jamaica. 18°N 77°W (p. 183)
Kinshasa (kihn•SHAH•suh) Capital of the Democratic Republic of the Congo. 4°S 15°E (p. 487)
Kiribati (KIHR•uh•BAS) One of the two Federated States of Micronesia. 5°S 170°W (p. 785)
Korean Peninsula Peninsula on which both North and South Korea are located. (p. 636)
Krishna River River of the Deccan Plateau in south India. (p. 560)
Kuala Lumpur (KWAH•luh LUM•PUR) Capital of Malaysia. 3°N 102°E (p. 711)
Kunlun Shan Mountain ranges in western China on the north edge of the Plateau of Tibet. (p. 636)
Kuwait (ku•WAYT) Country between Saudi Arabia and Iraq, on the Persian Gulf. (p. 411)
Kyrgyzstan (KIHR•gih•STAN) Small Central Asian country on China's western border. (p. 411)

L

Labrador Sea Part of the Atlantic Ocean south of Baffin Bay off the coast of Newfoundland. (p. 117)
Lagos Port city of Nigeria. 6°N 3°E (p. 487)
Lake Baikal Lake in southern Siberia, Russia. It is the largest freshwater lake in Eurasia. (p. 338)
Lake Chad Reservoir located in Chad. (p. 486)
Lake Erie One of the Great Lakes of the United States and Canada. (p. 117)
Lake Huron One of the Great Lakes of the United States and Canada. (p. 117)
Lake Malawi Lake in southeast Africa. (p. 486)
Lake Michigan One of the Great Lakes of the United States and Canada. (p. 117)
Lake Ontario The easternmost and smallest of the Great Lakes of the United States and Canada. (p. 117)
Lake Superior One of the Great Lakes of the United States and Canada. (p. 117)
Lake Tanganyika Lake in east-central Africa. (p. 486)
Lake Titicaca Lake on the border between Peru and Bolivia. Highest navigable lake in the world. (p. 182)
Lake Victoria Freshwater lake in Tanzania and Uganda. (p. 486)
Lake Volta Reservoir located in Ghana. (p. 486)
Lake Winnipeg Lake in south-central Manitoba, Canada. (p. 117)
Laos (LOWS) Southeast Asian country, south of China and west of Vietnam. (p. 711)
La Paz (lah PAHZ) Administrative capital of Bolivia, and the highest capital in the world. 17°S 68°W (p. 183)
Latvia (LAT•vee•uh) Northeastern European country on the Baltic Sea, west of Russia. (p. 261)
Lebanon (LEH•buh•nuhn) Country on the Mediterranean Sea, south of Syria. (p. 411)
Lena River River in east-central Russia. (p. 338)
Lesotho (luh•SOH•toh) Country in southern Africa. (p. 487)
Liberia (ly•BIHR•ee•uh) West African country, south of Guinea. 7°N 10°W (p. 487)
Libreville (LEE•bruh•VIHL) Capital and port city of Gabon. 1°N 9°E (p. 487)

Libya (LIH•bee•uh) North African country on the Mediterranean Sea, west of Egypt. (p. 411)
Liechtenstein (LIHK•tuhn•STYN) Small country in central Europe. (p. 261)
Lilongwe (lih•LAWNG•gway) Capital of Malawi. 14°S 34°E (p. 487)
Lima (LEE•muh) Capital of Peru. 12°S 77°W (p. 183)
Lisbon Capital of Portugal. 39°N 9°W (p. 261)
Lithuania (LIH•thuh•WAY•nee•uh) European country on the Baltic Sea, west of Belarus. (p. 261)
Ljubljana (lee•OO•blee•AH•nuh) Capital of Slovenia. 46°N 14°E (p. 261)
Llanos Vast plains in northern South America. (p. 182)
Loire River River in Europe that rises in southeastern France and empties into the Bay of Biscay. (p. 260)
Lomé (loh•MAY) Capital and port city of Togo in Africa. 6°N 1°E (p. 487)
London Capital of the United Kingdom, on the Thames River. 52°N 0° longitude (p. 261)
Luanda Capital of Angola. 9°S 13°E (p. 487)
Lusaka Capital of Zambia. 15°S 28°E (p. 487)
Luxembourg (LUHK•suhm•BUHRG) European country between France, Germany, and Belgium. (p. 261)

M

Macau (muh•KOW) Administrative district and port in southern China. (p. 638)
Macedonia (MA•suh•DOH•nee•uh) Republic in southeastern Europe, north of Greece. Macedonia also refers to a geographic region in the Balkan Peninsula. (p. 261)
Mackenzie River River in the western portion of the Northwest Territories in Canada. (p. 117)
Madagascar (MA•duh•GAS•kuhr) Island in the Indian Ocean, southeast of Africa. (p. 487)
Madrid Capital of Spain. 40°N 4°W (p. 261)
Malabo (mah•LAH•boh) Capital of Equatorial Guinea. 4°N 9°E (p. 487)
Malawi (muh•LAH•wee) Southeastern African country, south of Tanzania and east of Zambia. (p. 487)
Malaysia (muh•LAY•zhuh) Federation of states in Southeast Asia on the Malay Peninsula and the island of Borneo. (p. 711)
Maldives (MAWL•DEEVZ) Island country in the Indian Ocean near South Asia. 5°N 42°E (p. 561)
Mali (MAH•lee) Country in western Africa, south of Algeria. (p. 487)
Malta An independent state consisting of three islands in the Mediterranean Sea. 36°N 15°E (p. 261)
Managua (mah•NAH•gwah) Capital of Nicaragua. 12°N 86°W (p. 183)
Manila (muh•NIH•luh) Capital and port city of the Republic of the Philippines. 15°N 121°E (p. 711)
Marshall Islands Independent group of atolls and reefs in the western Pacific Ocean. 11°N 108°E (p. 785)
Maseru (MA•suh•ROO) Capital of Lesotho, in southern Africa. 29°S 27°E (p. 487)
Mauritania (MAWR•uh•TAY•nee•uh) West African country, north of Senegal. (p. 487)
Mauritius (maw•RIH•shuhs) Island country in the Indian Ocean east of Madagascar. 21°S 58°E (p. 487)
Mato Grasso Plateau Highlands area in southwest Brazil. (p. 182)
Mbabane (EHM•bah•BAH•nay) Capital of Swaziland, in southeastern Africa. 26°S 31°E (p. 487)
Mediterranean Sea Inland sea enclosed by Europe, Asia, and Africa. (p. 260)
Mekong River River in Southeast Asia that flows south through Laos, Cambodia, and Vietnam. (p. 710)
Meseta The plains of central Spain. (p. 260)
Mexico Country in North America, south of the United States. (p. 183)
Mexico City Capital and most populous city of Mexico. 19°N 99°W (p. 183)
Minsk (MIHNTSK) Capital of Belarus. 54°N 28°E (p. 261)
Mississippi River River in the central United States that rises in Minnesota and flows southeast into the Gulf of Mexico. (p. 117)
Missouri River River in the central United States that joins the Mississippi River. (p. 117)
Mogadishu (MAH•guh•DIH•shoo) Capital and major seaport of Somalia, in eastern Africa. 2°N 45°E (p. 487)
Moldova (mahl•DOH•vuh) European country between Ukraine and Romania. (p. 261)
Monaco (MAH•nuh•KOH) Independent principality in southern Europe, on the Mediterranean. (p. 261)
Mongolia (mahn•GOHL•yuh) Country in Asia between Russia and China. (p. 637)
Monrovia (muhn•ROH•vee•uh) Capital and major seaport of Liberia, in western Africa. 6°N 11°W (p. 411)
Mont Blanc The highest mountain of the Alps, in southeastern France. 46°N 7°E (p. 260)
Monte Carlo Capital of Monaco. 44°N 8°E (p. 261)
Montevideo (MAHN•tuh•vuh•DAY•oh) Capital of Uruguay. 35°S 56°W (p. 183)
Morocco (muh•RAH•koh) Country in northwestern Africa on the Mediterranean Sea and the Atlantic Ocean. (p. 411)
Moscow Capital of Russia. 56°N 38°E (p. 339)
Mount Ararat Mountain in eastern Turkey. 39°N 44°3 (p. 410)
Mount Elbrus Highest point in the Caucasus Mountains. 43°N 42°E (p. 410)
Mount Everest (EHV•ruhst) Highest mountain in the world, in the Himalaya mountain ranges between Nepal and Tibet. 28°N 87°E (p. 560)
Mount Fuji Peak in south-central Honshu, Japan. It is the highest peak in Japan. 35°N 138°E (p. 636)
Mount Logan Peak in northwest Arizona in the United States. 60°N 140°W (p. 117)

Mount McKinley Highest peak in North America, located in Denali National Park in Alaska. 63°N 151°W (p. 117)
Mount Pinatubo Active volcanic mountain in the Philippines. 15°N 170°E (p. 710)
Mount Whitney Peak in the Sierra Nevada range in central California. 36°N 118°W (p. 117)
Mozambique (MOH•zuhm•BEEK) Country in southeastern Africa, south of Tanzania. (p. 487)
Murray River River in Australia. (p. 784)
Muscat Capital of Oman. 23°N 59°E (p. 411)
Myanmar (MYAHN•MAHR) Country in Southeast Asia, south of China, formerly called Burma. (p. 711)

N

Nairobi Capital of Kenya. 1°S 37°E (p. 487)
Namib Desert Arid region along the coast of Namibia in southwestern Africa. (p. 486)
Namibia (nuh•MIH•bee•uh) Country in southwestern Africa, on the Atlantic Ocean. (p. 487)
Narmada River River in central India that flows into the Gulf of Khambat in the Arabian Sea. (p. 560)
Nassau (NA•SAW) Capital of the Bahamas. 25°N 77°W (p. 183)
Nauru (nah•OO•roo) One of the two Federated States of Micronesia. 32°S 166°E (p. 785)
N'Djamena (uhn•jah•MAY•nah) Capital of Chad. 12°N 15°E (p. 487)
Nepal (nuh•PAWL) Mountain country between India and China. (p. 561)
Netherlands Western European country on the North Sea. (p. 261)
New Delhi Capital of India. 29°N 77°E (p. 561)
New Zealand Major island country in the south Pacific, southeast of Australia. 42°S 175°E (p. 785)
Niamey (nee•AH•may) Capital and commercial center of Niger, in western Africa. 14°N 2°E (p. 487)
Nicaragua (NIH•kuh•RAH•gwuh) Republic in Central America. (p. 183)
Nicosia (NIH•kuh•SEE•uh) Capital of Cyprus. 35°N 33°E (p. 261)
Niger (NY•juhr) Landlocked country in western Africa, north of Nigeria. (p. 487)
Nigeria (ny•JIHR•ee•uh) Country in western Africa, south of Niger. (p. 487)
Niger River River in western Africa. (p. 486)
Nile River Longest river in the world, flowing north and east through eastern Africa. (p. 410)
North European Plain Plain that sweeps across western and central Europe into Russia and includes most of European Russia. (p. 260)
North Korea (kuh•REE•uh) Asian country in the northernmost part of the Korean Peninsula. (p. 637)
North Sea Arm of the Atlantic Ocean extending between the European continent on the south and east and Great Britain on the west. (p. 260)
Norway Country on the Scandinavian Peninsula. (p. 261)
Nouakchott (nu•AHK•SHAHT) Capital of Mauritania. 18°N 16°W (p. 487)
Nullarbor Plain Dry, treeless area that lies south of the Great Victorian Desert in Australia. (p. 784)

O

Ob River A river in western Russia. (p. 338)
Ohio River Major river in the midwestern United States, emptying into the Mississippi River. (p. 117)
Oman (oh•MAHN) Country on the Arabian Sea and the Gulf of Oman. (p. 411)
Orinoco River River in Venezuela. (p. 182)
Oslo Capital of Norway. 60°N 11°E (p. 261)
Ottawa Capital of Canada. 45°N 76°W (p. 107)
Ouagadougou (WAH•gah•DOO•goo) Capital of Burkina Faso, in western Africa. 12°N 2°W (p. 487)

P

Pakistan South Asian country on the Arabian Sea, northwest of India. (p. 561)
Palau (puh•LOW) Island country in the western Pacific Ocean. 7°N 135°E (p. 785)
Pamirs Mountainous region of Central Asia. (p. 410)
Pampas Plains area of South America. (p. 182)
Panama (PA•nuh•MAH) Republic in south Central America, on the Isthmus of Panama. (p. 183)
Panama City Capital of Panama. 9°N 79°W (p. 183)
Papua New Guinea (PA•pyuh•wuh noo GIH•nee) Independent island country in the south Pacific Ocean. 7°S 142°E (p. 785)
Paraguay (PAR•uh•GWY) Country in South America, north of Argentina. (p. 183)
Paraguay River River in south central South America. (p. 182)
Paramaribo (PAR•uh•MAR•uh•BOH) Capital and port city of Suriname. 6°N 55°W (p. 183)
Paraná River River in southeast central South America. (p. 182)
Paris Capital and river port of France. 49°N 2°E (p. 261)
Patagonia Plateau region of South America primarily in Argentina. (p. 182)
Peace River River in western Alberta, Canada. (p. 117)
Persian Gulf Arm of the Arabian Sea between Iran and Saudi Arabia. (p. 410)
Peru (puh•ROO) Country in South America, south of Ecuador and Colombia. (p. 183)
Philippines (FIH•luh•PEENZ) Country in the Pacific Ocean, southeast of China. (p. 711)
Phnom Penh (NAHM PEHN) Capital of Cambodia. 12°N 106°E (p. 711)
Poland Country on the Baltic Sea in eastern Europe. 52°N 18°E (p. 261)

Po River River in northern Italy that flows to the Adriatic Sea. (p. 260)
Port-au-Prince (POHRT•oh•PRIHNTS) Capital of Haiti. 19°N 72°W (p. 183)
Port Moresby (MOHRZ•bee) Capital of Papua New Guinea. 10°S 147°E (p. 785)
Porto-Novo (POHR•toh•NOH•voh) Capital and port city of Benin, in western Africa. 7°N 3°E (p. 487)
Portugal (POHR•chih•guhl) Country on the Iberian Peninsula, south and west of Spain. (p. 261)
Prague (PRAHG) Capital of the Czech Republic. 50°N 15°E (p. 261)
Pretoria (prih•TOHR•ee•uh) Administrative capital of the Republic of South Africa. 26°S 28°E (p. 487)
Puerto Rico Island in the West Indies. It is a self-governing commonwealth in union with the United States. 18°N 66°W (p. 183)
Pyongyang (PYAWNG•YAHNG) Capital of North Korea. 39°N 126°E (p. 637)
Pyrenees Mountains Mountain range extending along the border of France and Spain. (p. 260)

Q

Qatar (KAH•tuhr) Country on the southwestern shore of the Persian Gulf. (p. 411)
Qin Ling Mountain range in northern China. (p. 636)
Quito Capital of Ecuador. 0° latitude 79°W (p. 183)

R

Rabat Capital of Morocco. 34°N 7°W (p. 411)
Red River River in the south-central United States, emptying into the Mississippi River. (p. 117)
Red (Hong) River River in Vietnam that empties into the South China Sea. (p. 710)
Red Sea Inland sea between the Arabian Peninsula and northeast Africa. (p. 410)
Reykjavík (RAY•kyuh•VIHK) Capital of Iceland. 64°N 22°W (p. 261)
Rhine River in western Europe that flows to the North Sea. (p. 260)
Rhône River in Switzerland and France. (p. 260)
Riga Capital of Latvia. 57°N 24°E (p. 261)
Río de la Plata Estuary of the Paraná and Uruguay Rivers between Uruguay and Argentina. (p. 182)
Rio Grande River in the United States forming part of the boundary between the United States and Mexico. (p. 117)
Riyadh (ree•YAHD) Capital of Saudi Arabia. 25°N 47°E (p. 411)
Rocky Mountains An extensive mountain system in western North America. (p. 117)
Romania (ru•MAY•nee•uh) Country in eastern Europe, south of Ukraine. (p. 261)
Rome Capital of Italy. 42°N 13°E (p. 261)
Rub' al Khali Desert region in the southern Arabian Peninsula, also called the Empty Quarter. (p. 410)
Russia Largest country in the world, covering parts of Europe and Asia. (p. 339)
Rwanda (roo•AHN•dah) Country in Africa, south of Uganda. (p. 487)

S

Sahara Vast region of deserts and oases in North Africa. (p. 410)
St. Lawrence River River in southern Quebec and southeast Ontario, Canada. (p. 117)
St. Lucia Independent island state in the Caribbean Sea. 13°N 60°W (p. 183)
St. Vincent Principal island of St. Vincent and the Grenadines, south of St. Lucia. 13°N 61°W (p. 183)
Samoa Group of independent islands in the southwest Pacific Ocean. 13°S 172°W (p. 785)
Sanaa (sa•NAH) Capital of Yemen. 15°N 44°E (p. 411)
San José Capital of Costa Rica. 10°N 84°W (p. 183)
San Marino (SAN muh•REE•noh) Small European country, located on the Italian peninsula. (p. 261)
San Salvador (san SAL•vuh•DAWR) Capital of El Salvador. 14°N 89°W (p. 183)
Santiago Capital of Chile. 33°S 71°W (p. 183)
Santo Domingo (SAN•tuh duh•MIHNG•goh) Capital of the Dominican Republic. 19°N 70°W (p. 183)
São Francisco River River in eastern Brazil flowing into the Atlantic Ocean. (p. 182)
Sao Tome and Principe (SOWN•tuh MAY PRIHN•sih•pee) Small island country in the Gulf of Guinea off the coast of central Africa. 1°N 7°E (p. 487)
Sarajevo (SAR•uh•YAY•voh) Capital of Bosnia and Herzegovina. 43°N 18°E (p. 261)
Saskatchewan River River in south-central Canada that flows into Lake Winnipeg. (p. 117)
Saudi Arabia (SOW•dee uh•RAY•bee•uh) Country on the Arabian Peninsula. (p. 411)
Scandinavia A peninsula in northern Europe. (p. 261)
Sea of Japan Branch of the Pacific Ocean lying between Japan and the Korean Peninsula. (p. 636)
Sea of Okhotsk An inlet of the Pacific Ocean on the eastern coast of Russia. (p. 338)
Seine (SAYN) **River** French river that flows through Paris and into the English Channel. (p. 260)
Senegal (SEH•nih•GAWL) Country on the coast of western Africa, on the Atlantic Ocean. (p. 487)
Seoul (SOHL) Capital of South Korea. 38°N 127°E (p. 637)
Serbia (SUHR•bee•uh) European country south of Hungary. (p. 261)
Seychelles (say•SHEHLZ) Small island country in the Indian Ocean near East Africa. 6°S 56°E (p. 487)
Siberia An area in the region of north-central Asia, primarily in Russia. (p. 338)

Sierra Leone (see•EHR•uh lee•OHN) Country in western Africa, south of Guinea. (p. 487)
Sierra Madre del Sur Mountain range along the coast of southern Mexico. (p. 182)
Sierra Madre Occidental Mountain range running parallel to the Pacific Ocean coast in Mexico. (p. 182)
Sierra Madre Oriental Mountain range running parallel to the Gulf of Mexico coast in Mexico. (p. 182)
Sierra Nevada Mountain range in eastern California in the United States. (p. 117)
Sinai Peninsula Peninsula in northeast Egypt between the Gulf of Suez and the Gulf of Aqaba. (p. 410)
Singapore Multi-island country in Southeast Asia near the tip of the Malay Peninsula. 2°N 104°E (p. 711)
Skopje (SKAW•pyeh) Capital of the Republic of Macedonia. 42°N 21°E (p. 261)
Slave River River in west-central Canada between Lake Athabaska and Great Slave Lake. (p. 117)
Slovakia (sloh•VAH•kee•uh) Central European country south of Poland. (p. 261)
Slovenia (sloh•VEE•nee•uh) Small central European country on the Adriatic Sea, south of Austria. (p. 261)
Sofia Capital of Bulgaria. 43°N 23°E (p. 261)
Solomon Islands Independent island group in the west Pacific Ocean. 8°S 159°E (p. 784)
Somalia (soh•MAH•lee•uh) Country in east Africa, on the Gulf of Aden and the Indian Ocean. (p. 487)
South Africa Country at the southern tip of Africa. (p. 487)
South China Sea Part of the Pacific Ocean extending from Japan to the tip of the Malay Peninsula. (p. 636)
South Korea Country in Asia on the Korean Peninsula between the Yellow Sea and the Sea of Japan. (p. 637)
Spain Country on the Iberian Peninsula. (p. 261)
Sri Lanka (sree LAHNG•kuh) Island country in the Indian Ocean south of India. 9°N 83°E (p. 561)
Stockholm Capital of Sweden. 59°N 18°E (p. 261)
Strait of Gibraltar Passage connecting Mediterranean Sea to the Atlantic Ocean. (p. 260)
Strait of Hormuz Strait between the northern tip of Oman, the southeastern Arabian Peninsula, and the southern coast of Iran. (p. 410)
Strait of Malacca Ocean trade route running between Indonesia and Malaysia, near Singapore. (p. 710)
Sucre (SOO•kray) Constitutional capital of Bolivia. 19°S 65°W (p. 183)
Sudan Northeast African country on the Red Sea. (p. 487)
Suriname Republic in South America. (p. 183)
Suva Capital of Fiji. 18°S 177°E (p. 785)
Swaziland (SWAH•zee•LAND) South African country west of Mozambique. (p. 487)
Sweden Northern European country on the eastern side of the Scandinavian Peninsula. (p. 261)
Switzerland (SWIHT•suhr•luhnd) European country in the Alps, south of Germany. (p. 261)
Syr Darya River in west-central Asia in Kyrgyzstan, Uzbekistan, and Kazakhstan. (p. 410)
Syria (SIHR•ee•uh) Country in Asia on the eastern side of the Mediterranean Sea. (p. 411)

T

Taipei (TY•PAY) Capital of Taiwan. 25°N 122°E (p. 637)
Taiwan (TY•WAHN) Island country off the southeast coast of China, claimed by China. 24°N 122°E (p. 637)
Tajikistan (tah•JIH•kih•STAN) Central Asian country north of Afghanistan. (p. 411)
Taklimakan Desert Desert in western China. (p. 636)
Tallinn (TA•luhn) Capital and largest city of Estonia. 59°N 25°E (p. 261)
Tanzania (TAN•zuh•NEE•uh) East African country on the coast of the Indian Ocean. (p. 487)
Tashkent Capital of Uzbekistan. 41°N 69°E (p. 411)
Tasman Sea Part of the south Pacific Ocean between Australia and New Zealand. (p. 784)
Taurus Mountains Mountain range in southern Turkey. (p. 410)
Tbilisi (tuh•BEE•luh•see) Capital of the Republic of Georgia. 42°N 45°E (p. 411)
Tegucigalpa (tuh•GOO•suh•GAL•puh) Capital of Honduras. 14°N 87°W (p. 183)
Tehran (TAY•RAN) Capital of Iran. 36°N 52°E (p. 411)
Thailand (TY•LAND) Southeast Asian country south of Myanmar. (p. 711)
Thames (TEHMZ) **River** River in southern England that flows into the North Sea. (p. 260)
Thimphu (thihm•POO) Capital of Bhutan. 28°N 90°E (p. 561)
Tian Shan Mountain range in western China. (p. 636)
Tierra del Fuego Archipelago off southern South America. 54°N 68°W (p. 182)
Tirana (tih•RAH•nuh) Capital of Albania. 42°N 20°E (p. 261)
Togo (TOH•goh) West African country between Benin and Ghana, on the Gulf of Guinea. (p. 487)
Tokyo Capital of Japan. 36°N 140°E (p. 637)
Tonga South Pacific island country. 20°S 175°W (p. 785)
Trinidad and Tobago (TRIH•nih•DAD tuh•BAY•goh) Independent republic comprising the islands of Trinidad and Tobago, located in the Atlantic Ocean off the northeast coast of Venezuela. 11°N 61°W (p. 183)
Tripoli Capital of Libya. 33°N 13°E (p. 411)
Tunis Capital of Tunisia. 37°N 10°E (p. 411)
Tunisia (too•NEE•zhuh) North African country on the Mediterranean Sea between Libya and Algeria. (p. 411)
Turkey Country in southeastern Europe and western Asia. (p. 411)
Turkmenistan (tuhrk•MEH•nuh•STAN) Central Asian country on the Caspian Sea. (p. 411)
Tuvalu Independent island group in the western Pacific Ocean. 8°S 178°E (p. 785)

Uganda (oo•GAHN•duh) East African country south of Sudan. (p. 487)
Ukraine (yoo•KRAYN) Large eastern European country west of Russia, on the Black Sea. (p. 261)
Ulaanbaatar (OO•LAHN•BAH•TAWR) Capital of Mongolia. 48°N 107°E (p. 637)
United Arab Emirates Country of seven states on the eastern side of the Arabian Peninsula. (p. 411)
United Kingdom Country in western Europe made up of England, Scotland, Wales, and Northern Ireland. (p. 261)
United States Country in North America located between Canada and Mexico. (p. 117)
Ural Mountains Mountain range in Russia which marks the traditional boundary between European Russia and Asian Russia. (p. 338)
Ural River River in eastern Europe and western Asia, originating in the Ural Mountains. (p. 338)
Uruguay (UR•uh•GWY) South American country, south of Brazil on the Atlantic Ocean. (p. 183)
Uzbekistan (uz•BEH•kih•STAN) Central Asian country south of Kazakhstan. (p. 411)

Vanuatu (vahn•wah•TOO) Country made up of islands in the Pacific Ocean, east of Australia. 17°S 170°W (p. 785)
Vatican (VA•tih•kuhn) **City** Headquarters of the Roman Catholic Church, located in the city of Rome, Italy. 42°N 13°E (p. 261)
Venezuela Republic in northern South America. (p. 183)
Verkhoyansk Range Mountain range in northeastern Russia, just east of the Lena River. (p. 338)
Vesuvius Volcano on the east side of the Bay of Naples in Italy. 41°N 14°E (p. 261)
Vienna Capital of Austria. 48°N 16°E (p. 261)
Vientiane (vyehn•TYAHN) Capital of Laos. 18°N 103°E (p. 711)
Vietnam (vee•ET•NAHM) Southeast Asian country, east of Laos and Cambodia. (p. 711)
Vindhya Range Mountain range in central India. (p. 560)
Vistula River River in southwestern Poland that flows north into the Baltic Sea. (p. 260)
Volga River River in western Russia that flows south into the Caspian Sea. (p. 338)

Warsaw Capital of Poland. 52°N 21°E (p. 261)
Washington, D.C. Capital of the United States, near the Atlantic coast. 39°N 77°W (p. 107)
Wellington Capital of New Zealand. 41°S 175°E (p. 785)
Western Ghats Mountain range in southern India. (p. 560)
Western Sahara Territory in Northwest Africa. (p. 410)
West Siberian Plain Area of flat land that stretches from the Arctic Ocean to the grasslands of Central Asia. (p. 338)
Windhoek (VIHNT•HUK) Capital of Namibia, in southwestern Africa. 22°S 17°E (p. 487)

Xi (SHEE) **River** River in southeast China, known in its upper course as the Hongshui. (p. 636)

Yablonovyy Range Mountain range in southern Russia. (p. 338)
Yamoussoukro (YAH•muh•SOO•kroh) Second capital of Côte d'Ivoire, in western Africa. 7°N 6°W (p. 487)
Yangon Capital of Myanmar. 17°N 96°E (p. 711)
Yangtze River Major river in central China. (p. 636)
Yaoundé (yown•DAY) Capital of Cameroon, in western Africa. 4°N 12°E (p. 487)
Yellow River River in north-central and eastern China, also known as the Huang He. (p. 636)
Yellow Sea Large inlet of the Pacific Ocean between northeast China and the Korean Peninsula. (p. 636)
Yemen (YEH•muhn) Country on the Arabian Peninsula, south of Saudi Arabia. (p. 411)
Yenisey River A river in western Russia that flows north into the Kara Sea. (p. 338)
Yerevan (YEHR•uh•VAHN) Capital and largest city of Armenia. 40°N 44°E (p. 411)
Yucatán Peninsula Peninsula including parts of southeastern Mexico, Belize, and Guatemala in Central America. (p. 182)
Yugoslavia Republic in southeast Europe. (p. 261)
Yukon River River in the Yukon Territory, Canada. (p. 117)

Z

Zagreb (ZAH•grehb) Capital and largest city of Croatia. 46°N 16°E (p. 261)
Zagros Mountains Mountain system in southern and southwestern Iran. (p. 410)
Zambezi River River in south-central Africa. (p. 486)
Zambia (ZAM•bee•uh) Country in south-central Africa, east of Angola. (p. 487)
Zimbabwe (zim•BAH•bwee) Country in south-central Africa, southeast of Zambia. (p. 487)

A

aborigine/aborigen Habitante originario de un área (pág. 662)

absolute location/ubicación absoluta La posición exacta de un lugar en la superficie de la tierra (pág. 20)

accretion/acrecentamiento Un proceso lento en el cuál una plataforma marina se desliza por debajo de una plataforma continental, creando restos que pueden causar que los continentes crezcan hacia fuera (pág. 40)

acid deposition/deposición ácida Ácidos secos o húmedos que lleva el viento y que caen a la tierra (pág. 321)

acid rain/lluvia ácida Precipitación que lleva grandes cantidades de ácidos disueltos, los cuales dañan los edificios, los bosques y las cosechas y matan la fauna (págs. 166, 321)

acupuncture/acupuntura Una práctica antigua que involucra la inserción de agujas finas dentro del cuerpo en puntos específicos para curar enfermedades y aliviar el dolor (pág 677)

alluvial plain/llanura aluvial Llanura en la que los ríos inundados han depositado tierra rica, como la Llanura Indogangéctica en el Asia del Sur (pág. 572)

alluvial-soil deposits/depósitos de tierra aluvial Tierra rica compuesta de arena y lodo depositados por aguas corrientes (pág. 423)

altiplano Una región en Perú y Bolivia rodeada por la cordillera de los andes (pág. 194)

amendment/enmienda En la historia de los Estados Unidos, cambios oficiales que se le hacen a la Constitución (pág. 144)

apartheid/segregación racial Política de estricta separación de las razas adoptada en Sudáfrica en los años 40 (pág. 524)

aquaculture/acuacultura El cultivo de los mariscos (pág. 695)

aquifer/acuífero/aguas freáticas Capas subterráneas de roca porosa, arena o grava que acumulan agua (págs. 49, 470)

arable /arable Tierra idónea para cultivo (pág. 464)

archipelago/archipiélago Un grupo o cadena de islas (págs. 646, 720)

artesian water/agua artesiana Abastecimiento de agua subterránea que está bajo suficiente presión para subir a los pozos sin tener que ser bombeada (pág. 794)

Asian Pacific Economic Cooperation Group (APEC)/(APEC) Grupo de cooperación económica del Pacífico Asiático Un grupo de comercio, cuyos miembros son Japón, China, Corea del Sur y Taiwan, que asegura que el comercio entre los países miembros es eficiente y justo (pág. 689)

Association of Southeast Asian Nations (ASEAN)/ Asociación de Naciones del Asia del Sudeste (ANSEA) Organización formada en 1967 para promover el desarrollo y el comercio regional en el Asia del Sudeste (pág. 763)

atheism/ateísmo La creencia de que no hay Dios (pág. 376)

atmosphere/atmósfera Una capa de gases que rodea la Tierra (pág. 35)

atoll/atolón Isla en forma de aro formada por coral que se acumula por todo el borde de un volcán submarino (pág. 796)

autocracy/autocracia Gobierno en el cual un solo individuo posee el poder y la autoridad para gobernar (pág. 87)

avalanche/avalancha Masa de hielo, nieve o roca que se desliza por el lado de una montaña (pág. 279)

axis/eje Referente a la Tierra, una línea imaginaria que le atraviesa por el centro entre el Polo Norte y el Polo Sur (pág. 56)

B

batik/batik Método de teñir tela para producir bellos estampados, desarrollado en Indonesia y Malasia (pág. 751)

bazaar/bazar Un mercado tradicional con puestos. Puede estar en una sola calle o extenderse hasta un distrito completo (pág. 456)

bedouin/beduino Miembro de los pueblos nómadas del desierto del África del Norte y Asia del Sudoeste (pág. 455)

bilingual/bilingüe Que habla o usa dos idiomas (pág. 148)

Bill of Rights Las diez primeras enmiendas de la constitución estadounidense (pág. 144)

biologist/biólogo Científico que estudia la vida vegetal y animal (pág. 324)

biomass/biomasa Desperdicio vegetal y animal usado especialmente como fuente de combustible (pág. 613)

biosphere/biosfera La parte de la Tierra donde existe la vida (pág. 35)

birthrate/índice de natalidad El número de nacimientos por año por cada 1000 personas (pág. 76)

black market/mercado negro Cualquier mercado ilegal donde se venden productos escasos o ilegales, por lo general a precios altos (pág. 388)

blizzard/ventisca Una tormenta de nieve con vientos de más de 35 millas por hora, temperaturas abajo del punto de congelación y visibilidad de menos de 500 pies que dura 3 horas o más (pág. 125)

Bolsheviks/Bolcheviques Un grupo revolucionario en Rusia dirigido por Vladimir Ilyich Lenin (pág. 370)

boomerang/bumerang Palo curvo que se tira utilizado por los aborígenes de Australia para cazar (pág. 817)

buffer state/estado neutral Territorio neutral entre poderes rivales (pág. 743)

C

cabinet/gabinete Directores de departamentos en la rama ejecutiva del gobierno de EE.UU. quienes aconsejan al Presidente (pág. 144)

calligraphy/caligrafía El arte de escribir a mano de forma bella (pág. 679)

campesinos Trabajadores agrícolas; generalmente gente que vive y trabaja en áreas rurales (pág. 238)

canopy/follaje o copa Capa superior del bosque de lluvia, donde se juntan las puntas de los árboles altos formando una capa continua de hojas (pág. 200)

cartography/cartografía La ciencia de hacer mapas (pág. 24)

cash crop/cosecha comercial Cosecha agrícola cultivada para venderse o trocarse en lugar de usarse para la familia del agricultor (págs. 238, 538, 612, 760)

cataract/catarata Un gran salto de agua (pág. 500)

caudillo Término para un líder político de Latinoamérica desde los últimos años de los 1800 en adelante, con frecuencia un dictador militar (pág. 224)

cereal/cereal Cualquier grano, como la cebada, la avena o el trigo, que se cultiva para alimento (pág. 431)

chaparral/chaparral Tipo de vegetación compuesta de densos bosques de matorrales y arbustos, común en los climas mediterráneos (págs. 68, 281)

chernozem Capa vegetal de tierra negra y rica que se halla en la planicie del norte de Europa, especialmente en Rusia y Ucrania (pág. 346)

chinampas Islas flotantes de agricultura hechas por los aztecas (pág. 221)

chinook/chinuco Viento cálido estacional que sopla por las Montañas Rocosas al final del invierno y al principio de la primavera (pág. 124)

***Chipko*/Chipko** Movimiento en India en que la gente abraza árboles para proteger los bosques, reforestándolos y limitando la tala (pág. 621)

cholorofluorocarbons/clorofluorocarbonos También llamados CFCs; sustancias químicas que se hallan principalmente en líquidos refrigerantes que dañan la capa de ozono que protege a la Tierra (pág. 695)

city-state/ciudad-estado En la antigua Grecia, comunidad independiente que consistía de una ciudad y las tierras circundantes (pág. 295)

clan/clan Comunidad tribal o grupo grande de gente relacionada entre sí (págs. 529, 670, 817)

clear-cutting/deforestación Cortar todos los árboles de un bosque para utilizar la madera (pág. 165)

climate/clima Patrones del tiempo típicos de un área durante un largo período de tiempo (pág. 55)

Cold War/Guerra Fría Se refiere a la lucha por el poder entre la Unión Soviética y los Estados Unidos después de la Segunda Guerra Mundial (págs. 298, 371)

collective farm/granja colectiva Bajo el comunismo, una granja enorme, propiedad del estado, en la cual los agricultores recibían sueldos más una parte de los productos y ganancias; también denominada kolkhoz (pág. 317)

command economy/economía controlada Sistema económico en el que las decisiones económicas acerca de la producción y la distribución las toma alguna autoridad central (págs. 89, 388, 685)

commercial farming/agricultura comercial Agricultura organizada como un negocio (pág. 538)

commodity/mercancía Bienes producidos para su venta (págs. 158, 465)

commune/comuna En la China, una comunidad agrícola colectiva cuyos miembros compartían el trabajo y los productos (pág. 686)

communism/comunismo Sociedad basada en la igualdad en la que los trabajadores controlarían la producción industrial (págs. 298, 370)

condensation/condensación El proceso del cambio de vapor de agua a agua líquida cuando el aire caliente se enfría (pág. 47)

coniferous/coníferas Árboles que tienen conos y hojas con forma de agujas y que mantienen su follaje durante el invierno (pág. 68)

conquistador Término español para referirse a los soldados que conquistaron a los indios de Latinoamérica (pág. 222)

conservation farming/agricultura de conservación Técnica de administración de la tierra que ayuda a proteger la tierra agrícola (pág. 539)

Constitution/Constitución Plan que hizo el gobierno para los Estados Unidos en 1787 (pág. 144)

consumer goods/bienes de consumo Bienes que satisfacen directamente los deseos humanos (pág. 388)

continental drift/movimiento continental La teoría que dice que los continentes estaban juntos y se fueron separando lentamente (pág. 38)

continental shelf/plataforma continental La parte de un continente que se extiende bajo el agua (pág. 36)

cooperative/cooperativa Una organización voluntaria cuyos miembros trabajan juntos y comparten los gastos y las ganancias (pág. 686)

copra/copra La carne seca del coco (pág. 834)

coral/coral Depósitos de piedra caliza formados pro los esqueletos de animales marinos (pág. 795)

cordillera/cordillera Cadenas paralelas de montañas (págs. 194, 719)

Coriolis effect/efecto Coriolis Un efecto que causa que los vientos soplen diagonalmente en vez de sus líneas normales norte/sur o este/oeste (pág. 62)

cottage industry/industria casera Un negocio que emplea trabajadores en sus casas (pág. 615)

crevasse/grieta Una brecha enorme y honda que se forma en la nieve o en el hielo grueso (pág. 803)

Crusades/Cruzadas Serie de guerras religiosas (1100–1300 D.C.) en las que los cristianos europeos trataron de recuperar Palestina del control musulmán (pág. 296)

cultural diffusion/difusión cultural La difusión de conocimientos y costumbres de una cultura a otra (pág. 84)

culture/cultura Modo de vida de un grupo de gente que comparte creencias y costumbres similares (pág. 80)

culture hearth/hogar de la cultura Un centro donde las culturas se desarrollan y desde el cual las ideas y las tradiciones se difunden (págs. 84, 447, 668)

culture region/región cultural División de la tierra basada en una variedad de factores que incluyen el gobierno, los grupos sociales, los sistemas económicos, el lenguaje o la religión (pág. 83)

cuneiform/cuneiforme Sistema de escritura Sumeria que usa símbolos en forma de cuñas hundidas en tabletas de arcilla (pág. 447)

current/corriente Corriente de agua de mar fría o cálida que fluye en los océanos, por lo general en forma circular (pág. 62)

cyclone/ciclón Tormenta con lluvias y vientos fuertes que sopla en círculo alrededor de un área de baja presión atmosférica (págs. 579, 767)

czar/zar El emperador de Rusia hasta la revolución de 1917 (pág. 369)

D

***Dalits*/Dalit** Los oprimidos; clase social más baja de la India (pág. 623)

death rate/tasa de mortalidad El número de muertes por año por cada 1000 personas (pág. 76)

deciduous/deciduo Describe árboles, usualmente de follaje ancho como los robles y arces, que pierden las hojas en el otoño (pág. 68)

deforestation/deforestación La pérdida o destrucción de los bosques, debido principalmente a la tala de árboles para explotación forestal o agricultura (págs. 242, 507)

delta/delta Sección triangular de tierra que se forma en la boca de un río y que se parece a la letra griega delta (Δ) (pág. 503)

democracy/democracia Un sistema de gobierno en el cual los líderes gobiernan con el consentimiento de los ciudadanos (pág. 88)

desalination/desalinización La eliminación de la sal del agua de mar para que se pueda usar para beber y en la agricultura (págs. 48, 470)

desertification/desertificación Proceso en el cual la tierra arable se vuelve desierto, (págs. 508, 694)

developed country/país desarrollado País que tiene un gran avance en tecnología y manufactura (pág. 93)

developing country/país en vías de desarrollo País en el proceso de industrialización (págs. 93, 238)

dharma En hinduismo, el deber moral que guía la vida de una persona, de acuerdo a las diferencias de clase (pág. 593)

dialect/dialecto Variedad local de un lenguaje usado en un lugar en particular o por cierto grupo (págs. 213, 302)

diatoms/diátomo Plancton que vive en agua fría en los océanos (pág. 841)

dikes/diques Bancos grandes de tierra y piedras que detienen el agua (pág. 272)

dissident/disidente Un ciudadano que habla en contra de las políticas del gobierno (pág. 689)

divide/bifurcación Un punto alto o cresta que determina la dirección en la que fluyen los ríos (pág. 118)

doldrums/zona de calmas ecuatoriales Un área cerca del Ecuador frecuentemente sin viento (págs. 62, 802)

domesticate/domesticar El tomar plantas y animales silvestres para adaptarlos y hacerlos útiles para la gente (págs. 446, 520)

dominion/dominio Un país con gobierno propio parcial y con ligas cercanas a otro país (pág. 820)

Spanish Glossary

doubling time/tiempo de duplicación El número de años que le toma a una población duplicar su tamaño (pág. 76)

dry farming/agricultura seca Método de agricultura que se usa en regiones secas en el que la tierra se ara y se siembra profundo para retener el agua en la tierra (págs. 143, 321)

dynasty/dinastía Una casa gobernante o la continuación de una familia de gobernantes, especialmente en China (pág. 669)

dzong Un monasterio fortificado de Bhutan, Asia del sur (pág. 604)

E

e-commerce/comercio electrónico Comprar y vender en el Internet (pág. 542)

economic sanctions/sanciones económicas Restricciones de comercio (pág. 689)

ecosystem/ecosistema La compleja comunidad de seres vivientes que, en cierto medio ambiente, dependen los unos de los otros (pág. 22)

ecotourism/ecoturismo Turismo basado en preocupación por el medio ambiente (págs. 546, 617)

El Niño Un cambio total periódico del patrón de las corrientes del océano y las temperaturas del agua en la región del Pacífico medio (pág. 63)

El Niño-Southern Oscillation (ENSO)/ El Niño oscilación del sur Un cambio en el clima de temporada que puede causar sequías en Australia y ciclones poderosos en el Pacífico del sur (pág. 841)

embargo/embargo Una prohibición en el comercio (págs. 452, 468)

endemic/endémico Que describe especies de plantas o animales nativos de un área particular (pág. 728)

Enlightenment/Iluminación Un movimiento durante principios de los 1700 que enfatizaba la importancia de la razón y cuestionaba las tradiciones y valores (pág. 297)

environmentalist/ambientalista Persona activamente preocupada con la calidad y la protección del medio ambiente (pág. 322)

equinox/equinoccio Uno de dos días (el 21 marzo y el 23 de septiembre) cuando el Sol está directamente encima del ecuador, haciendo que el día y la noche sean de igual duración (pág. 56)

erosion/erosión El desgaste de la superficie de la Tierra por el viento, el fluir del agua o los glaciares (pág. 42)

escarpment/escarpa Risco o escarpadura entre una superficie de tierra alta y otra más baja (págs. 196, 500)

estuary/estuario Área en donde la marea se encuentra con la corriente de un río (págs. 197, 503)

ethnic cleansing/purificación de raza Cuando se saca de un país a personas de cierta raza o una matanza entre grupos étnicos rivales (pág. 288)

ethnic diversity/diversidad étnica Diferencias entre grupos o gente basadas en sus orígenes, lenguajes, costumbres o creencias (pág. 439)

ethnic group/grupo étnico Grupo de personas quienes comparten los mismos antepasados, lenguaje, religión, costumbres o una combinación de estas características (págs. 82, 288, 363)

European Union/Unión Europea Una organización cuya meta es unir a Europa para que los bienes, servicios y trabajadores puedan moverse libremente en todos los países miembros (págs. 300, 313)

eutrophication/eutroficación Proceso por el cual un cuerpo de agua se hace demasiado rico en nutrientes disueltos, haciendo que la abundancia de plantas agote el oxígeno (pág. 168)

evaporation/evaporación El cambio a vapor (pág. 47)

exports/exportaciones Recursos o bienes que se envían desde un país a otro para comercio (pág. 237)

extended family/parentela Una familia compuesta de varias generaciones de parientes; (págs. 229, 529)

extinction/extinción La desaparición o fin de una especie de animal o planta (pág. 546)

F

fale Tipo de casa tradicional en Samoa, con los lados abiertos y el techo de paja (pág. 826)

fall line/línea de caída Una frontera en el este de Estados Unidos donde la tierra más alta de Piamonte cae a la planicie más baja de la costa atlántica (pág. 118)

farm cooperative/cooperativa agrícola Organización en la que los agricultores comparten en el cultivo y la venta de los productos agrícolas (pág. 317)

fault/falla Una grieta o hendidura en la corteza terrestre (págs. 40, 500)

fauna/fauna La vida animal de una región (pág. 724)

federal system/sistema federal Divide los poderes de gobierno entre el gobierno nacional y el estatal o local (pág. 87)

feudalism/feudalismo Sistema de gobierno durante la época medieval en Europa y Japón, en el cual los poderosos señores feudales daban tierras a los nobles a cambio de su lealtad (pág. 296)

fishery/pesquería Zonas (de agua fresca o salada) donde se pescan peces u otros animales marinos (pág. 120)

fjord/fiordo Valle glacial con largos acantilados llenos de agua de mar (pág. 272)

flora/flora La vida de las plantas de una región (pág. 724)

foehn/fohn Viento seco y cálido que sopla del lado sotavento de las montañas, a veces derrite la nieve y provoca avalanchas; el término se usa predominantemente en Europa (pág. 279)

fold/pliegue Un doblez en las capas de roca, a veces causado por el movimiento de las plataformas (pág. 40)

food web/red alimenticia Los enlaces entre los depredadores y sus fuentes de comida en un ecosistema (pág. 840)

formal region/región formal Una región definida por características comunes, tales como la fabricación de un producto (pág. 21)

free port/zona libre Ciudad portuaria, como Singapur, donde los bienes se desembarcan, se almacenan y se vuelven a embarcar sin pago de impuestos portuarios (pág. 764)

free trade/libre comercio La eliminación de barreras comerciales para que los bienes puedan circular libremente entre países (pág. 94)

functional region/región funcional Un punto central y el territorio que lo rodea que está relacionado con ese punto central (pág. 230)

fútbol Se conoce en Norteamérica como el "soccer" (pág. 230)

G

gauchos Los vaqueros de Argentina y Uruguay (pág. 197)

genetically modified foods/alimentos modificados genéticamente Alimentos cuyos genes han sido alterados para hacerlos más grandes, crecer más rápido o ser más resistentes a las plagas (pág. 317)

geographic information systems/sistemas de información geográfica Herramientas de computadora para procesar y organizar los detalles y las imágenes de satélite junto con otras piezas de información (pág. 25)

glaciation/formación de glaciares El proceso por el cual se forman y se extienden los glaciares (pág. 272)

glacier/glaciar Grandes cuerpos de hielo que se mueven a través de la superficie de la Tierra (pág. 42)

glasnost/glasnost Término ruso para una nueva "apertura"; parte de los planes de reforma de Gorbachev (pág. 371)

global warming/calentamiento global Calentamiento gradual de la Tierra y su atmósfera que puede ser causado en parte por la contaminación y un aumento del efecto de invernadero (págs. 58, 322)

glyph/glifo Escritura a base de dibujos esculpida en piedra (pág. 221)

Good Friday Peace Agreement/Acuerdo de paz de Viernes Santo Por medio de este acuerdo, las comunidades Protestantes y Católicas Romanas comparten poder político en Irlanda del Norte (pág. 304)

grazier/ganadero Persona que cría ganado ovejuno y vacuno (pág. 834)

greenhouse effect/efecto invernadero La capacidad de ciertos gases en la atmósfera de atrapar el calor y hacer que la Tierra se caliente (págs. 58, 322)

green revolution/revolución verde Programa comenzado en los años 60 para producir cepas de mayor rendimiento de trigo, arroz y otras cosechas de alimento (pág. 613)

grid system/sistema de cuadrícula Patrón formado a medida que las líneas de latitud y longitud se intersectan (pág. 20)

gross domestic product (GDP)/producto interno bruto (PIB) El valor de los bienes y los servicios creados dentro de un país en un año (pág. 465)

groundwater/agua subterránea Agua que yace debajo de la superficie de la Tierra, que abastece pozos y manantiales (págs. 49, 167)

guru/gurú Maestro o guía espiritual (pág. 601)

H

habitat/hábitat Area con condiciones idóneas para que vivan ciertas plantas o animales (pág. 546)

haiku Forma de poesía japonesa que consiste de 17 sílabas y tres líneas, con frecuencia trata de la naturaleza (pág. 678)

hajj En el Islam, el peregrinaje anual a la Meca (pag. 466)

headwaters/cabeceras Las fuentes de aguas de río (pág. 118)

heavy industry/industria pesada La manufactura de maquinaria y equipo necesario para fábricas y minas (pág. 317)

hemisphere/hemisferio Mitad de una esfera o globo, como en los Hemisferios Norte y Sur de la Tierra (pág. 20)

hieroglyphics/jeroglíficos Sistema de escritura egipcia que usa dibujos y símbolos para representar palabras o sonidos (pág. 447)

Holocaust/Holocausto El asesinato masivo de 6 millones de judíos por los líderes Nazi de Alemania durante la Segunda Guerra Mundial (pág. 298)

homogeneous/homogéneo De la misma clase o naturaleza (pág. 663)

human-environment interaction/interacción humana-ambiental El estudio de las interrelaciones de la gente con sus ambientes físicos (pág. 22)

human geography/geografía humana También llamada geografía cultural; es el estudio de las actividades humanas y sus relaciones con los ambientes culturales y físicos (pág. 24)

hurricane/huracán Una gran tormenta de vientos fuertes que se forma sobre las aguas cálidas del océano (pág. 125)

hydroelectric power/energía hidroeléctrica Energía eléctrica generada por la caída del agua (págs. 197, 348)

hydrosphere/hidrosfera Las zonas acuosas de la Tierra que incluyen los océanos, lagos, ríos y otros cuerpos de agua (pág. 35)

hypothesis/hipótesis Una explicación científica para un evento (pág. 69)

icon/icono Imagen religiosa, usualmente incluye un cuadro de Jesús, María o un santo, usado principalmente por los cristianos ortodoxos (pág. 377)

ideogram/ideograma Un carácter pictórico o un símbolo que representa un significado específico o una idea (pág. 674)

immigration/inmigración El movimiento de gente saliendo de un país y entrando a otro (pág. 133)

impressionism/impresionismo Estilo artístico que se desarrolló en Europa en los últimos años de los 1800 y que trató de mostrar la apariencia natural de los objetos con toques o pinceladas de color (pág. 305)

indigenous/indígena Nativo de un lugar (pág. 212)

industrial capitalism/capitalismo industrial Un sistema económico en el cual los líderes de negocios usan sus utilidades para expandir sus compañías (pág. 297)

industrialization/industrialización Transición de una sociedad agrícola a una industrial (pág. 93)

infrastructure/infraestructura Las necesidades urbanas básicas como calles y servicios (págs. 443, 540)

insular/insular Constituye una isla, como en Java (pág. 270)

intelligentsia/intelectualidad Círculo intelectual (pág. 378)

interdependent/interdependiente Depender los unos de los otros para bienes, servicios e ideas (pág. 763)

introduced species/especies introducidas Plantas y animales colocadas en áreas diferentes a su hábitat natural (pág. 839)

jai alai/jai alai Juego de mano tradicional popular en México y Cuba (pág. 231)

Japan Current/Corriente de Japón Corriente oceánica de agua templada que le proporciona humedad a los monzones de invierno (pág. 655)

jati En la sociedad tradicional Hindú, un grupo social que define la ocupación de la familia y su posición social (pág. 588)

jazz/jazz Forma musical que se desarrolló en los Estados Unidos a principios de los 1900, mezcla ritmos africanos con harmonías europeas (pág. 148)

jute/yute Fibra vegetal que se usa para hacer cordón y tela (pág. 612)

karma/karma En la creencia Hindú, la suma de las acciones buenas y malas en la vida presente y las vidas pasadas de una persona (pág. 593)

kolkhoz En la Unión Soviética, una granja pequeña trabajada por granjeros que comparten la producción y las utilidades (pág. 390)

krill Diminutos animales marinos parecidos a los camarones que viven en los frígidos océanos de la Antártica (pág. 798)

kums Término regional para los desiertos de Asia central (pág. 425)

lagoon/laguna Estanque de poca profundidad en el centro de un atolón (pág. 796)

lama/lama Líder religioso budista (pág. 675)

language family/familia de lenguajes Grupo de lenguajes relacionados que se desarrollaron de un lenguaje anterior (págs. 81, 303)

***latifundia*/latifundios** En la moderna Latinoamérica, extensas propiedades de cultivo que pertenecen a familias o corporaciones (pág. 238)

leach/lixiviar Lavar la tierra para sacarle los nutrientes (pág. 506)

leeward/sotavento Con cara en contra de la dirección de donde sopla el viento (pág. 64)

lichens/líquenes Plantas diminutas y fuertes que crecen como una corteza sobre las rocas y los troncos de los árboles (pág. 803)

light industry/industria ligera Manufactura con fines de producir bienes como textiles o alimentos para el consumidor en vez de maquinaria pesada (pág. 317)

lingua franca/lengua franca Un lenguaje común usado entre la gente con diferentes lenguas nativas (pág. 526)

literacy rate/índice de alfabetización El porcentaje de personas en un lugar dado que puede leer y escribir (pág. 150)

lithosphere/litosfera Áreas de superficie terrestre de la corteza de la Tierra, incluye los continentes y las cuencas de los océanos (pág. 35)

llanos Llanuras fértiles tierra adentro en áreas de Colombia y Venezuela (pág. 196)

location/ubicación Un lugar específico en la Tierra (pág. 20)

lode/veta Depósito de minerales (pág. 761)

loess/loes Capa superficial del suelo, fina, amarillenta y marrón compuesta de partículas de limo y arcilla, usualmente arrastrada por el viento (págs. 42, 275, 648)

longhouse/casa larga En zonas rurales de Indonesia y Malasia, un edificio alto donde viven miembros de varias familias que están relacionadas entre sí (pág. 752)

M

Maastricht Treaty/Tratado de Maastricht Una reunión en 1992 de los gobiernos europeos en Maastricht, Holanda, en la que se formó la Unión Europea (pág. 314)

magma/magma Roca fundida que es enviada hacia arriba desde el manto de la Tierra (pág. 38)

malnutrition/malnutrición Nutrición pobre o inadecuada (pág. 230)

mantle/manto Gruesa capa mediana de la estructura interior de la Tierra, consiste de densa piedra caliente (pág. 38)

mantra/mantra En Hinduismo, una palabra o frase sagrada que se repite en los rezos y cantos (pág. 601)

manuka Pequeños arbustos que crecen en las mesetas de Nueva Zelanda (pág. 803)

maquiladoras En México, plantas de manufactura montadas por firmas extranjeras (pág. 239)

maritime/marítimo Relacionado a los viajes o la transportación por mar (pág. 741)

market economy/economía de mercado Un sistema económico basado en la libre empresa, en el que los negocios son de propiedad privada y la producción y los precios se determinan por la oferta y la demanda (págs. 89, 157, 388)

marsupial/marsupial Tipo de mamífero que da a luz a crías que maduran en una bolsa en el abdomen de la madre (pág. 839)

mass culture/cultura popular Cultura de las masas extendida por los medios de comunicación como la radio y la televisión (pág. 525)

megacities/megaciudad Una ciudad con más de 10 millones de habitantes (pág. 217)

megalopolis/megalópolis Una "super-ciudad" compuesta por varias ciudades grandes y pequeñas, como el área entre Boston y Washington, D.C. (págs. 136, 591)

meltwater/aguanieve Agua formada al derretirse la nieve y el hielo (pág. 321)

merchant marine/marina mercante La flota de barcos de un país que participa en el intercambio comercial (pág. 691)

meteorology/meteorlogía El estudio del clima y el pronóstico del tiempo (pág. 24)

metropolitan area/área metropolitana Región que incluye una ciudad central y sus suburbios circundantes (pág. 135)

mica/mica Mineral de silicato que típicamente se parte en delgadas láminas brillantes (pág. 574)

Middle Ages/Edad Media El período de la historia europea que va aproximadamente desde 500 D.C. hasta 1500 D.C. (pág. 296)

migration/migración El movimiento de gente de un lugar a otro (pág. 79)

***minifundia*/minifundios** En Latinoamérica, pequeñas granjas que producen alimento principalmente para el uso de la familia (pág. 238)

mistral/mistral Viento fuerte del norte que sopla desde los Alpes y que trae aire frío al sur de Francia (pág. 280)

mixed economy/economía mixta Una economía en la cual el gobierno apoya y regula la empresa tomando decisiones que afectan el mercado (pág. 89)

mixed farming/agricultura mixta Cultivo de varias clases de cosechas y ganado en la misma granja (pág. 317)

mixed forest/bosque mixto Tierras forestales con árboles de coníferas y deciduos (pág. 68)

mobility/mobilidad Poder moverse de un lugar a otro (pág. 137)

monopoly/monopolio Control total de un tipo de industria por una persona o una compañía (pág. 162)

monotheism/monoteísmo Creencia en un solo Dios (pág. 448)

monsoon/monzón En Asia, viento veraniego que trae aires cálidos y húmedos desde los océanos, aire seco y frío desde las tierras interiores en el invierno (págs. 579, 654)

moraine/morena Montones de deshechos rocosos que dejaron los glaciares al fundirse (pág. 42)

mosaic/mosaico Dibujo o diseño hecho con pequeños pedazos de piedra, vidrio, concha o azulejo de colores (pág. 228)

mosque/mezquita En Islam, una casa de oración pública (pág. 448)

movement/movimiento Movimiento actual de gente, bienes e ideas (pág. 22)

mural/mural Pintura en las paredes (pág. 228)

N

nationalism/nacionalismo Creencia en el derecho de cada persona a ser una nación independiente (pág. 449)

nationalities/nacionalidades Grandes y diversos grupos étnicos distintos dentro de un país (pág. 364)

nationalize/nacionalizar Colocar una compañía o industria bajo el control del gobierno (pág. 450)

Native American/Nativo Americano Los primeros inmigrantes de Norte América, quienes vinieron a la región probablemente de Asia hace cientos de años (pág. 134)

natural increase/incremento natural La tasa de crecimiento de una población; la diferencia entre la tasa ce crecimiento y la de mortalidad (pág. 76)

natural resource/recurso natural Sustancia de la tierra que no está hecha por la gente pero que ellos la pueden usar (pág. 91)

natural vegetation/vegetación natural Vida vegetal que crece en ciertas zonas si la gente no cambia el medio ambiente (pág. 66)

nirvana En Budismo, el máximo estado de paz y entendimiento que la gente trata de alcanzar (pág. 594)

North American Free Trade Agreement (NAFTA)/Tratado de Libre Comercio de Norteamérica (TLCNA) Tratado comercial hecho en 1994 entre Canadá, los Estados Unidos y México (pág. 240)

nuclear family/familia Grupo familiar constituido de esposo, esposa e hijos (pág. 529)

nuclear proliferation/proliferación nuclear El desarrollo expansivo de armas nucleares (pág. 623)

nuclear waste/desperdicio nuclear El producto secundario de producir energía nuclear (pág. 398)

O

oasis Pequeña área en el desierto donde hay agua y vegetación (págs. 67, 429)

oligarchy/oligarquía Sistema de gobierno en el cual un grupo pequeño tiene el poder (pág. 88)

oral tradition/tradición hablada Historias pasadas de generación en generación por palabra de boca en boca (pág. 528)

organic farming/agricultura orgánica El uso de sustancias naturales en lugar de fertilizantes y pesticidas para enriquecer la tierra y cultivar las cosechas (pág. 318)

ozone layer/capa de ozono Capa atmosférica con gases protectores que evitan que los rayos solares alcancen la superficie de la tierra (pág. 841)

P

paddy/arrozal Campos inundados en los que se cultiva el arroz (pág. 760)

pagoda/pagoda Un estilo de arquitectura encontrado más a menudo en los edificios tradicionales de Asia del este, marcados por techos de azulejo graciosamente curvados en estilo de torres (pág. 679)

Pampas/Pampa Llanos con pasto, sin árboles en América del Sur (pág. 196)

Parliament/parlamento En Canadá, la legislatura nacional conformada por el Senado y la Casa de los Comunes (pág. 145)

pastoralism/pastoreo críar y pastorear ganado (pág. 430)

patois/dialecto Dialectos que mezclan elementos de los lenguajes indígenas, europeos, africanos y asiáticos (pág. 213)

patriarch/patriarca El dirigente de una Iglesia Ortodoxa del Este (pág. 377)

patriotism/patriotismo Amor o devoción de una persona por su país (pág. 151)

perceptual region/región perceptual Una región definida por sentimientos e imágenes populares en vez de datos objetivos (pág. 21)

perestroika En Rusia, "reestructuración"; parte del plan de Gorbachev para reformar el gobierno soviético (pág. 371)

permafrost Capa de tierra por debajo de la superficie del suelo que está permanentemente congelada (págs. 68, 281, 349)

pesticide/pesticida Sustancia química que se usa para matar insectos, roedores y otras plagas (pág. 398)

petrochemical/petroquímico Producto químico derivado del petróleo o del gas natural (pág. 465)

phosphate/fosfato Mineral natural que contiene compuestos químicos que a menudo se usan en fertilizantes (pág. 426)

physical geography/geografía física El estudio de las características físicas de la Tierra (pág. 24)

pidgin English/Inglés corrompido Un dialecto que mezcla el inglés con los idiomas locales (pág. 813)

pipeline/tubería Una red grande de tuberías que pueden estar bajo tierra o sobre la tierra (pág. 162)

place/lugar Un espacio particular con significado físico y humano (pág. 21)

plate tectonics/placa tectónica Es el término científico que se usa para describir las actividades del movimiento continental y del flujo del magma, los cuales crean muchas de las características físicas de la Tierra (pág. 38)

poaching/cazar o pescar en veda La caza o pesca prohibida de animales protegidos (págs. 546, 619)

pogrom En la Rusia zarista, un ataque contra los judíos llevado a cabo por las tropas o los oficiales del gobierno (pág. 378)

polder/pólder Zona en Holanda, bajo el nivel del mar, de la cual se ha drenado el agua del mar para crear nuevas tierras de agricultura (pág. 272)

pollution/polución La existencia de sustancias sucias, impuras o venenosas en el aire, el agua o la tierra (pág. 94)

population density/densidad de población Número promedio de personas en una milla o kilómetro cuadrado (pág. 77)

population distribution/distribución demográfica El patrón de población en un país, un continente o en el mundo (pág. 77)

post-industrial/postindustrial Una economía con menos énfasis en la industria pesada y de manufactura y más énfasis en servicios y tecnología (pág. 158)

prairie/pradera Pastizal tierra adentro (págs. 68, 124)

precipitation/precipitación Humedad que cae a la tierra, como lluvia, nevisca, granizo o nieve (pág. 47)

prevailing wind/viento dominante Viento en una región que sopla en una dirección casi constantemente (pág. 62)

primate city/ciudad prócer Ciudad que domina la economía, cultura y el gobierno de un país y en la que la población está concentrada; usualmente la capital (págs. 217, 738)

privatization/privatización El cambio de propiedad de las compañías e industrias, pasando de ser propiedad del estado a ser propiedad privada (pág. 389)

prophet/profeta Persona de quien se cree que es un mensajero de Dios (pág. 448)

qanat Canales subterráneos que se usaban en los sistemas de acueductos en la antigua Persia (pág. 447)

quipu Cuerda con cordones anudados de varios largos y colores que usaban los incas para llevar sus registros financieros (pág. 222)

R

radioactive material/material radioactivo Material contaminado por residuos que provienen de la generación de energía nuclear (pág. 397)

rain shadow/sombra de lluvia Zona seca que se halla del lado sotavento de una cordillera (pág. 64)

raj Palabra hindú para imperio (pág. 595)

realism/realismo Estilo artístico que capta la vida cotidiana que se desarrolló en Europa a mediados de los años 1800 (pág. 305)

reforestation/reforestación La siembra de árboles jóvenes o semillas en terrenos donde los árboles han sido talados o destruidos (pág. 244)

Reformation/Reforma Movimiento religioso que comenzó en Alemania en los 1400 y que condujo al establecimiento de Iglesias Protestantes (pág. 297)

refugee/refugiado Persona que tiene que dejar su hogar y huir a otro lugar para estar a salvo (pág. 288)

regions/regiones Lugares unidos por características específicas (pág. 21)

reincarnated/reencarnado En la creencia Hindú, el renacer repetidas veces en distintas formas, hasta que uno ha vencido todos los deseos terrenales (pág. 593)

relative location/ubicación relativa La posición de un lugar en relación a otros lugares (pág. 20)

Renaissance/Renacimiento En Europa, un período de renovado interés por las enseñanzas y artes clásicas que comenzó cerca del 1300 y duró 300 años (pág. 296)

reparations/reparaciones Un pago por daños (pág. 298)

republic/república Forma de gobierno sin monarca, en el cual la gente elige a sus oficiales (pág. 142)

retooling/modificación Cambiar las fábricas antiguas para usarlas en industrias nuevas (pág. 161)

revolution/revolución En astronomía, el viaje de la Tierra alrededor del Sol que toma 365$^1/4$ días (pág. 56)

rift valley/valle hundido Una grieta en la superficie terrestre creada por los movimientos de tierra (pág. 500)

romanticism/romanticismo Estilo artístico que enfati-zaba las emociones individuales. Se desarrolló en Europa en los últimos años de los 1700 y principios de los 1800 como una reacción a la industrialización (pág. 305)

Russification/Rusificación En la Rusia del siglo XIX, programa de gobierno que requería que todo el mundo en el imperio hablara ruso y se convirtiera al Cristianismo (pág. 370)

sadhu Un ermitaño hindú o un hombre santo (pág. 602)

samurai/samurai En el Japón medieval, una clase de soldados profesionales quienes se regían por un código estricto de honor personal y lealtad a un noble (pág. 670)

sanitation/sanidad Eliminación de productos de desperdicio (pág. 516)

satellite/satélite Un país controlado por otro, como los países de Europa del este controlados por la Unión Soviética a finales de la Segunda Guerra Mundial (pág. 371)

savanna/sabana Un pastizal tropical que contiene árboles dispersos (pág. 507)

sedentary farming/agricultura sedentaria Agricultura que toma lugar en asentamientos permanentes (pág. 538)

serf/siervo Obrero de la propiedad de un noble y obligado a permanecer en la tierra donde trabajaba (pág. 369)

service center/centro de servicio Ubicación de negocios conveniente para los habitantes del campo (pág. 518)

service industry/industria de servicios Negocios que proveen servicios, en vez de producir bienes (pág. 238)

shamanism/shamanismo Creencia en un líder que se puede comunicar con los espíritus (pág. 675)

shantytowns/barrios Comunidades improvisadas a las orillas de las ciudades (pág. 244)

shifting cultivation/rotación de cultivos Deforestar para plantar campos por unos cuantos años y abandonarlos (pág. 770)

shifting farming/cultivo migratorio Método en el cual los agricultores se mudan después de unos cuantos años para hallar mejor tierra (pág. 538)

shogun/shogún En el Japón medieval, un gobernante militar del país (pág. 670)

sickle/hoz Cuchillo grande y curvo con un mango, usado para cortar pasto o granos altos (pág. 760)

sirocco/siroco Viento caliente del desierto que sopla aire y polvo desde el África del Norte hasta la costa mediterránea de Europa (págs. 280-281)

slash-and-burn farming/agricultura de corte y quema Método tradicional de cultivo en el que todos los árboles y plantas en el área se cortan y se queman para añadir nutrientes al suelo (pág. 244)

smog/smog Neblina irritante causada por la interacción de la radiación solar ultravioleta con humos químicos del escape de los automóviles y otras fuentes de contaminación (págs. 69, 167)

socialism/socialismo Filosofía política en la que el gobierno era propietario de los medios de producción (págs. 370)

socialist realism/realismo socialista En la Unión Soviética, estilo realista de arte y literatura que glorificaba los ideales y las metas soviéticas (pág. 380)

socioeconomic status/posición socioeconómica Nivel de ingresos y educación (pág. 150)

solstice/solsticio Uno de dos días (el 21 de junio y el 22 de diciembre) en que los rayos del Sol dan directamente en el Trópico de Cáncer o el Trópico de Capricornio, para marcar el principio del verano o el invierno (pág. 57)

sovereignty/soberanía Autogobierno (pág. 365)

sovkhoz En la Unión Soviética, una enorme granja del estado y que el estado mismo opera (pág. 390)

sphere of influence/esfera de influencia Zona del país en la que una potencia extranjera tiene control político o económico (pág. 743)

spreading/extensión Un proceso por el cual se crea nueva tierra cuando las plataformas marinas se separan y el magma fluye hacia arriba entre las plataformas (pág. 40)

state farm/granja estatal Bajo el comunismo, una granja de propiedad del estado dirigida por oficiales del gobierno (pág. 317)

station/estación Término australiano para un rancho o granja grande en áreas remotas (pág. 834)

steppe/estepa Vastas llanuras de pastizales de Eurasia; también regiones en otros lugares con climas semiáridos similares (pág. 355)

Strine Inglés coloquial que se habla en Australia (pág. 813)

stupa/stupa Estructura en forma de cúpula que sirve de santuario budista (pág. 604)

subcontinent/subcontinente Gran área de tierra que forma parte de un continente pero es diferente, como la India (pág. 569)

subduction/subducción Proceso por el que la corteza oceánica se hunde bajo la continental (pág. 39)

subsistence crop/cultivo de subsistencia Un cultivo que sirve principalmente para alimentar a la familia del agricultor (pág. 760)

subsistence farming/agricultura de subsistencia Agricultura tradicional cuya meta es producir lo suficiente para que una familia o una aldea coma y sobreviva (págs. 538, 826)

suburbs/suburbios Comunidades en las afueras y que circundan una ciudad central (pág. 135)

Sunbelt/Franja del Sol Parte sur de los Estados Unidos, denominada así debido a su clima (pág. 135)

supercells/superceldas Tormenta de rayos y truenos violenta que puede producir tornados (pág. 124)

sustainable development/desarrollo sostenido Crecimiento tecnológico y económico que no agota los recursos humanos y naturales de un área dada (págs. 242, 619)

syncretism/sincretismo Una mezcla de creencias y prácticas de diferentes religiones en una fe (pág. 228)

T

taiga Término ruso para un vasto bosque que cubre gran parte de Rusia y Siberia (pág. 353)

tariff/tarifa Un impuesto sobre bienes importados o exportados (pág. 163)

temperature/temperatura Una medida de cuán caliente o cuán frío está algo, generalmente se mide en grados con una escala fija, como Fahrenheit o Centígrados (pág. 56)

tierra caliente La zona más baja de los climas de montaña de Latinoamérica (pág. 203)

tierra fría La zona de altitud más alta de los climas de montaña de Latinoamérica (pág. 203)

tierra templada La zona de altitud media de los climas de montaña de Latinoamérica (pág. 203)

timberline/límite de la vegetación selvática Elevación por encima de la cual hace demasiado frío para que crezcan los árboles (págs. 124, 279)

trade deficit/déficit comercial Gastar más dinero en importaciones que lo que se gana en exportaciones (págs. 163, 689)

trade surplus/superávit comercial Ganar más dinero por exportaciones que lo que se gasta en importaciones (págs. 163, 689)

traditional economy/economía tradicional Sistema en que las costumbres determinan las actividades económicas; existe en pocos lugares (pág. 89)

tributary/afluente Pequeños ríos y arroyos que se alimentan de los ríos más grandes (pág. 118)

trust territory/territorio en fideicomiso Región, que después de la Segunda Guerra Mundial fue colocada temporáneamente por las Naciones Unidas bajo el control político y económico de otra nación (pág. 819)

tsunami Término japonés usado para una ola de mar enorme causada por un maremoto (pág. 646)

tundra Vastas llanuras sin árboles en climas fríos del norte, caracterizadas por la capa del subsuelo conge-lada y pequeñas plantas como musgos y arbustos (pág. 352)

typhoon/tifón Un ciclón que se forma en el Océano Pacífico, usualmente al final del verano (págs. 655, 768, 802)

U

Underground Railroad/Ferrocarril subterráneo Una red informal de refugios, en los Estados Unidos, que ayudó a miles de esclavos a escapar hacia la libertad (pág. 143)

unitary system/sistema unitario Le da todos los poderes clave al gobierno nacional o central (pág. 87)

universal suffrage/sufragio universal Derechos de igualdad de votación para todos los adultos residentes de una nación (pág. 524)

urbanization/urbanización El movimiento de personas de las áreas rurales a las ciudades (págs. 135, 216, 290, 517, 737)

V

viceroy/virrey Representante del monarca español asignado para hacer cumplir las leyes en las colonias americanas (pág. 222)

W

wadi En el desierto, cauce seco excepto durante las lluvias fuertes (pág. 424)

wat En el Asia del Sudeste, un templo (pág. 750)

water cycle/ciclo de agua Movimiento regular del agua desde el océano hasta el aire y el suelo y de vuelta al océano (pág. 47)

wattle/zarzo Árboles de acacia; los primeros colonos de Australia entretejían los zarzos jóvenes para construir sus casas (pág. 801)

weather/tiempo Condición de la atmósfera en un lugar durante un período de tiempo corto (pág. 55)

weathering/erosión Procesos químicos o físicos, como congelación, que desbaratan las rocas (pág. 42)

welfare state/estado benefactor Nación en la que el gobierno asume la responsabilidad principal por el bienestar de la gente en cuanto a salud y educación (pág. 306)

windward/barlovento Con la cara hacia la dirección desde donde el viento está soplando (pág. 64)

World Trade Organization/Organización Mundial del Comercio Organismo internacional que supervisa acuerdos comerciales y resuelve disputas comerciales entre países (pág. 690)

Z

ziggurat/zigurat En la antigua Mesopotamia, gran templo escalonado hecho de ladrillos de arcilla (pág. 454)

Index

c = chart m = map
d = diagram p = photo
g = graph ptg = painting

A

N

Acknowledgments

Abbreviation Key: NGIC = National Geographic Image Collection

Cover Bill Brooks/Masterfile; **ix** VCG/FPG; **vi** Gianni Dagli Orti/CORBIS; **vii** Will & Deni McIntyre/Stone; **viii** Christopher Arnesen/Stone; **xxii** (l)Judy Griesedieck/CORBIS, (r)Jan Butchofsky-Houser/CORBIS; **xxiii** (t)Reuters NewMedia Inc./CORBIS, (cl)Morton Beebe, S.F./CORBIS;(cr)Mark Gibson/CORBIS, (b)Roger Ressmeyer/CORBIS; **1** Richard T. Nowitz/CORBIS; **5** Roger Ressmeyer/CORBIS; **16–17** Joseph McBride/Stone; **18** Joel Satore/NGIC; **19** Richard Olsenius/NGIC; **23** Peter Poole/NGIC; **24** James L. Stanfield/NGIC; **25** Michael K. Nichols/NGIC; **29** Roger Ressmeyer/CORBIS; **32** Courtesy European Space Agency; **33** CORBIS; **36** (l)NASA/Roger Ressmeyer/CORBIS, (c)David Paterson/CORBIS;(r)Darrell Gulin/CORBIS; **37** James L. Stanfield/NGIC; **42** Dr. John Reinhard/NGIC; **44** Jeff Lepore/Photo Researchers; **45** (t)Maurizio Lanini/CORBIS, (l)Wolfgang Kaehler/CORBIS;(r)Nik Wheeler/CORBIS; **46** Jay Dickman/NGIC; **48** Steven L. Raymond/NGIC; **49** PhotoDisc; **54** Annie Griffiths Belt/NGIC; **55** Richard Olsenius/NGIC; **56** Peter Essick/NGIC; **59** Rick Tomlinson/NGIC; **65** Thomas Nebia/NGIC; **68** Annie Griffiths Belt/NGIC; **74** Pablo Corral/NGIC; **75** Priit J. Vesilind/NGIC; **77** Michael Nichols/NGIC; **80** Stuart Franklin/NGIC; **86** Mark Segal/Stone; **87** Peter Turnley/CORBIS; **88** David Turnley/CORBIS; **89** Walter Hodges/Stone; **91** Joe McNally/NGIC; **93** Gail Mooney/CORBIS; **100–101** CORBIS; **102** (l)Annie Griffiths Belt;(r)Donald Nausbaum/Stone; **103** (t)Raymond Gehman/NGIC, (b)CORBIS **104** (l)Tom Bean; (r)William A. Allard; **105** (t)Lorraine C. Parow/First Light, (b)David Hiser/Photographers/Aspen/PNI; **112** Rick Stewart/Allsport; **112–113** Jose Azel/Aurora/PNI; **113** (t)Jose Azel, (c) David Burnett/Contact Press Images/PNI, (b)PhotoDisc; **114** Carr Clifton/Minden Pictures; **115** William A. Allard/NGIC; **118** Raymond Gehman/NGIC; **119** Charles E. Rotkin/CORBIS; **120** Richard Olsenius/NGIC; **121** Robert Clark/NGIC; **124** (l)Bettmann/CORBIS; (r)Arthur Rothstein/CORBIS; **125** Larry Ulrich Stock Photography, Inc.; **130** Raymond Gehman/CORBIS; 132 Lowell Georgia/CORBIS; 133 Bob Krist/CORBIS; **134** Catherine Karnow/CORBIS; **136** US Geological Survey/Science Photo Library/Photo Researchers; **137** Annie Griffiths Belt/CORBIS; **138** Jeff Greenberg/PhotoEdit/PictureQuest; **139** (t)Louis S. Glanzman, (b)Sarah Leen; **140** Joseph Sohm/Visions of America/CORBIS; **141** Mark Gibson/CORBIS; **143** (l)Layne Kennedy/CORBIS, (r)CORBIS; **144** Bettmann/CORBIS; **145** Paul A. Souders/CORBIS; **146** Christie's Images/CORBIS; **147** Mike Habermann, courtesy Music of the World; **148** AP/Wide World Photos; **149** Robbie Jack/CORBIS; **151** Phillip Gould/CORBIS; **152** Leif Skoogfers/CORBIS; **156** Terry Vine/Stone; **157** Randy Olson/NGIC; **158** Richard Olsenius/NGIC; **159** PhotoDisc; **160** (l)file photo; (r)Bettmann/CORBIS; **161** (l)file photo; (r)Ed Young/CORBIS; **162** PhotoDisc; **163** David S. Boyer/NGIC; **163** David Ball/The Stock Market; **164** Thomas E. Franklin/Bergen Record/SABA/CORBIS; **165** Peter Essick/NGIC; **167** Johnathan S. Blair/NGIC; **168** (t)Nashua River Watershed Association/NGIC, (b)George Steinmetz/NGIC; **170** James Zipp/Photo Researchers; **171** (t)J. McDonald/Bruce Coleman Inc., (l)Marc Epstein/Visuals Unlimited; (r)Raymond K. Gehman/NGIC; **173** Bob Sacha/NGIC; **176–177** Kenneth Garrett/NGIC; 178 (l)Stuart Westmorland/Stone; (r)Jacques Jangoux/Stone; **179** (t)Hans Strand/Stone, (b)O. Louis Mazzatenta; **180** Robert Frerck/Stone; **181** (t)Bill Bachmann/Stone, (b)Alex Webb; **190** Artville LLC; **191** (t)Peter Essick, (l)PhotoDisc, (r)Jeff Greenberg/Southern Stock/PictureQuest; **192** Pal Hermansen/Stone; **193** Stephen G. St.John/NGIC; **196** Ary Diesendruck/Stone; **197** (l)Stephanie Maze/NGIC, (r)Kit Houghton Photography/CORBIS; **198** Gianni Dagli Orti/CORBIS; **199** Gordon Wiltsie/NGIC; **202** (l)Tom Bean/Stone; (r)Stone; **205** George F. Mobley/NGIC; **208** (t)Albert Moldvay/NGIC, (b)courtesy of Environmental Systems Research Institute, Inc.; **209** Bob Jordan/AP/Wide World Photos; **210** Steve Winter/NGIC; **211** Dudley Brooks/NGIC; **214** Stephanie Maze/NGIC; **215** Wolfgang Kaehler/CORBIS; **218** Danny Lehman; **219** (t)The Granger Collection, New York, (c)Culver Pictures, Inc., (b)Panama Canal Co.; **220** Tomasz Tomaszewski/NGIC; **223** (l)Sisse Brimberg/NGIC, (r)Stephanie Maze/CORBIS; **224** (l)CORBIS, (lc)Bettmann/CORBIS, (rc)Martin Boneo/National History Museum of Buenos Aires, (r)Bettmann/CORBIS; **225** AFP/CORBIS; **226** Jeremy Horner/CORBIS; **228** Jack Vartoogian; **229** Charles Lenars/CORBIS; **230** Darren Modricker/CORBIS; **231** TempSport/CORBIS; **236** Richard T. Nowitz/NGIC; **237** Bruce Dale/NGIC; **238** William A. Allard/NGIC; **240** Nathan Benn/NGIC; **241** William A. Allard/NGIC; **242** Buddy Mays/CORBIS; **245** James P. Blair/NGIC; **246** Sarah Leen/NGIC; **247** Howard Davies/CORBIS; **248** Denise Tackett/Tom Stack & Associates; **249** (t)Joel Creed/ Ecoscene/CORBIS, (l)Carlos Humberto T.D.C/Contact Press Images/PNI, (r)Paul Edmonson/Stone; **251** Bettman/CORBIS; **254–255** Hans Wiesenhofer/West Stock, Inc.; **256** (t)Jonathan Blair, (b)Michael St. Maur Sheil **256–257** Stephen Studd/Stone, (b)Jodi Cobb; **258** (l)Mark A. Leman/Stone; (r)Anthony Suau; **258–259** Rick Morley/West Stock; **259** Joanna B. Pinneo; **268** (l)Wolfgang Kaehler/CORBIS, (r)Randy Wells/Stone; **269** (t)Paul Chesley/Stone, (b)Larry Brownstein/RAINBOW/PictureQuest; **270** Larry Ulrich Stock Photography; **271** Steve Winter/NGIC; **272** Paul Almasy/CORBIS; **275** Gerd Ludwig/NGIC; **276** Nicole Galeazzi/Omni-Photo Communications; **277** Philip Gould/CORBIS; **281** Chris Lisle/CORBIS; **283** Mattias Klum/NGIC; **286** Ric Ergenbright; **287** Joe McNally/NGIC; **28**8 The Purcell Team/CORBIS; **289** Dallas and John Heaton/CORBIS; **290** Craig Lovell/CORBIS; **291** Alexandra Boulat; **292** AFP/CORBIS; **293** (t)Francoise de Mulder/CORBIS, (b)David Turnley/CORBIS; **294** Dean Conger/CORBIS; **296** Mike Southern/CORBIS; **297** Martin Jones/CORBIS; **298** James L. Stanfield/NGIC; **300** Manfred Vollmer/CORBIS; **301** Dave Bartruff/CORBIS; **303** Jack Vartoogian; **305** Alexandra Avakian; **307** Winifield I. Parks Jr./NGIC; **312** Stephen Simpson/FPG; **313** Tomasz Tomaszewski; **314** AP/ Wide World Photos; **314–315** CORBIS; **315** (l)Lionel Cironneau/Wide World Photos, (r)Bob Edme/Stringer/Associated Press; **317** Steven L. Raymer/NGIC; **318** Todd Gipstein/NGIC; **319** Jodi Cobb/NGIC; **320–321** James Nachtwey; **322** James P. Blair/NGIC; **323** James P. Blair/NGIC; **324** Joel Sartore; **325** Cotton Coulson/NGIC; **326** Oliver Strewe/Stone; **327** (t)Werner H. Muller/Peter Arnold, Inc., (l)P. Frischmuth/Argus Fotoarchiv/Peter Arnold, Inc., (r)Thomas Hartrich/Sovfoto/Eastfoto/PictureQuest; **328** Art Wolfe/Stone; **329** Ben Osborne/Stone; **336–337** Steve Raymer; **338** (l)Fabian/Corbis Sygma; (r)Bob Krist/Stone; **339** (t)Jerry Kobalenko, (b)Nick Nichols/NGIC; **340** (t)Gerd Ludwig, (b)Steve Raymer; **341** (t)James P. Blair, (b)Macduff Everton; **344** Bryan & Cherry Alexander Photography; **345** James P. Blair/NGIC; **346** Dean Conger/CORBIS; **346** (l)Archivo Iconografico, S.A./CORBIS, (r)Peter Fownes/Stock South/PictureQuest; **347** (t)Bob Daemmrich/Stock Boston/PictureQuest, (b)Kevin R. Morris; **348** William R. Curtsinger/NGIC; **349** Wolfgang Kaehler/CORBIS; **350** Dean Conger/NGIC; **351** Sarah Leen/NGIC; **354** Bryan & Cherry Alexander Photography; **355** N/A/Sovfoto/Eastfoto/PictureQuest; **357** Keren Su/CORBIS; **360** AFP/CORBIS; **361** Matt Meadows; **362** Sarah Leen/Matrix; **363** Peter Turnley/CORBIS; **365** James P. Blair/NGIC; **366** Steven L. Raymer/NGIC; **367** Bettman/CORBIS; **369** Steve Raymer/NGIC; **370** Larry Stein/Black Star/PictureQuest; **371** State Central Archive of Film and Photographic Documents, Moscow; **372** Steven L. Raymer/NGIC; **373** David Turnley/CORBIS; **376** Gerd Ludwig; **377** Hulton Getty Collection/Stone; **378** Klaus Reisinger/Black Star; **380** Jack Vartoogian; **382** Peter Essick/NGIC; **386** Alain Le Garsmeur/Stone; **387** Gerd Ludwig/NGIC; **388** AP/Wide World Photos; **388–389** Archivo Iconografico, S. A./CORBIS; **389** Peter Turnley/CORBIS; **390** Steven L. Raymer/NGIC; **391** Thomas T. Hammond/NGIC; **392** Dennis Chamberlin/NGIC; **396** Gerd Ludwig/NGIC; **396** David Falconer; **397** Wolfgang Kaehler/CORBIS, (t)Michael Yamashita, (l)Wolfgang Kaehler, (r)N/A/Sovfoto/Eastfoto/PNI **398** Sarah Leen/NGIC; **399** Gerd Ludwig/NGIC; **401** Demitrio Carrasco/Stone; **404–405** Annie Griffiths Belt; **406** (l)George Chan/Stone, (r)Lonnie Duka/Stone; **407** (l)Robert Everts/Stone, (r)Charles Crowell/Balck Star/PictureQuest; **408** (l)Sylvain Grandadam/Stone, (r)Steve McCurry; **409** (t)Robert Frerck/Stone, (b)Stephen Studd/Stone; **417** James Strachan/Stone; **418** (t)CMCD/PhotoDisc, (b)Alon Reininger/Contact Press Images/PictureQuest; **419** (t)David Rubinger/CORBIS, (b)Rohan/Stone; **420** Victoria Pearson/Stone; **421** James L. Stanfield/NGIC; **422** (l)Richard T. Nowitz/CORBIS, (r)Bojan Brecelj/CORBIS; **424** K.M. Westermann/CORBIS; **425** Reuters NewMedia Inc./CORBIS; **426** David Turnley/CORBIS; **427** Tiziana and Gianni Baldizzone/

CORBIS; **430** Kim Hart/Black Star/PictureQuest; **431** Tomasz Tomaszewski/NGIC; **433** Robert Stahl/Stone; **436** Matt Meadows; **437** Ed Kashi/NGIC; **438** Will & Deni McIntyre/Stone; **439** Steven L. Raymer/NGIC; **442** Richard T. Nowitz/NGIC; **444** John Moore/Image Works; **445** (t)Wayne Eastep/Stone, (c)Owen Franken/CORBIS, (b)Noel Quido/ Liaison; **446** James L. Stanfield/NGIC; **450** AFP/CORBIS; **453** Stuart Franklin/NGIC; **455** Mrs. Lynn Abercrombie/NGIC; **456** AP/ Wide World Photos; **462** Annie Griffiths Belt/NGIC; **463** George F. Mobley/NGIC; **466** Jack Vartoogian; **467** Alain Le Garsmeur/CORBIS; **469** Digitized NASA Photography (c)1996 Corbis; **472** George F. Mobley/NGIC; **475** (t)Ed Kashi, (l)James L. Stanfield, (r)Zafer Kizilkaya; **477** Reza/NGIC; **480–481** Betty Press/Woodfin Camp & Associates; **482** (l)John Chard/Stone, (r)Dave Saunders/Stone; **483** (t)Patrick Ward/Corbis, (b)Manoj Shah/Stone; **484** (l)Ferorelli, (r)Nik Wheeler/CORBIS; **485** (t)Chris Harvey/Stone, (b)Paul Kenward/ Stone; **495** Bruno de Hogues/Stone; **496** (t)CORBIS, (b)Bettmann/ CORBIS; **497** (t)D. Boone/CORBIS, (b)Jak Kilby/Arena/StageImage; **498** Gerry Ellis/Minden Pictures; **499** Chris Johns/NGIC; **502** (l)Yann Arthus-Bertrand/CORBIS; (r)John Elk III; **505** Chris Johns/NGIC; **508** (l)Digital Stock; (r)Gerry Ellis/Minden Pictures; **509** Steve McCurry/Magnum Photos; **511** Chris Johns/NGIC; **514** Wolfgang Kaehler; **515** Robert Caputo; **517** Peter Essick/NGIC; **518** Robert Caputo; **519** Richard S. Durrance/NGIC; **524** David Turnley/CORBIS; **525** Dave Bartruff/CORBIS; **527** Wolfgang Kaehler; **528** Jack Vartoogian; **529** Peter Essick/NGIC; **530** Peter Turnley/CORBIS; **531** (t)Culver Pictures, (c)Hulton-Deutsch Collection/CORBIS, (b)Peter Turnley/CORBIS; **532** AP/Wide World Photos; **536** Wolfgang Kaehler; **537** Maria Stenzel/NGIC; **539** Chris Johns/NGIC; **541** Peter Essick/ NGIC; **542** Courtesy PEOPLink; **543** Frans Lanting/NGIC; **545** Michael K. Nichols/NGIC; **546** Jen K. DeVere; **547** Chris Johns/NGIC; **548** Kevin Schafer; Martha Hill/Allstock/PictureQuest; **549** (t)Paul A. Souders/CORBIS, (l)Beverly Joubert/NGIC; (r)Jodi Cobb/NGIC; **551** Michael K. Nichols/NGIC; **554–555** Paul Steel/The Stock Market; **556** (l)Stewart Cohen/Stone; (r)Steve McCurry; **557** (t)Adrian Masters/ Stone, (b)James P. Blair; **558** (l)Steve McCurry, (r)Dick Durrance II; **558–559** Hilarie Kavanagh/Stone; **559** Martin Puddy/Stone; **565** Maggie Steber; **566** (t)Robert Holmes/CORBIS, (b)Ric Ergenbright/ CORBIS; **567** (t)David Cumming/Eye Ubiquitous/CORBIS, (b)Jagdish Agarwal/Stock Connection/PictureQuest; **568** Erik Sampers/Liaison; **569** Steve McCurry/NGIC; **570** Hugh Sitton/Stone; **572** (l)Mark Downey/Lucid Images/PictureQuest, (r)NGIC; **573** Galen Rowell/ Mountain Light Photography; **574** Jagdish Agarwal/Dinodia Picture Agency; **575–578** Steve McCurry/NGIC; **581** David Cumming; Eye Ubiquitous/CORBIS; **584** Rich LaSalle/Stone; **585** Johnathan S. Blair/NGIC; **586** John Beatty/Stone; **587** Raghubir Singh/NGIC; **588** David Robbins/Stone; **590** Karen Kasmauski/NGIC; **591** Steve McCurry/NGIC; **592** James L. Stanfield/NGIC; **593** (l)Diego Lezama Orezzoli/CORBIS, (r)Randy Olson/NGIC; **594** (l)Paul Seheult; Eye Ubiquitous/CORBIS, (r)Chris Lisle/CORBIS; **595** May Cooper/Images of India; **596** (l)N/A/Archive Photos/PictureQuest, (r)Bettmann/ CORBIS; **597** AFP/CORBIS; **598** AFP/CORBIS; **599** (t)Steve McCurry, (c)Robb Kendrick/AURORA, (b)Steve McCurry; **600** George F. Mobley/ NGIC; **604** Courtesy of Music of the World; **605** Johnathan S. Blair/NGIC; **610** SuperStock; **611** James Stanfield/NGIC; **612** Dan Gair/Liaison; **614** Steve McCurry/NGIC; **615** Carolyn Schaefer/ Liaison; **616** Steve McCurry/NGIC; **617** Pete Seaward/Stone; **618** Galen Rowell; **619** Michael K. Nichols/NGIC; **620** Reuters NewMedia Inc./CORBIS; **621** AFP/CORBIS; **624** M. Amirtham, Dinodia Picture Agency; **625** (t)Steve McCurry, (l)Robb Kendrick/ AURORA; (r)Robb Kendrick; **627** John Elk III; **632** (l)Keren Su/Stone; (r)Edward J. Kazmirski/Image Techniques; **632–633** IFA/Bruce Coleman Inc.; **635** (t)Keren Su/Stone, (b)Renee Lynn/Stone; **636** (l)Bruce Coleman Inc./IFA, (r)Dallas and John Heaton/CORBIS; **637** (t)Chad Ehlers/Stone, (b)Paul Chesley/Stone; **643** CORBIS; **644** SuperStock, (t)Peter L. Chapman/Stock Boston Inc./PictureQuest, (b)Donna Day/Stone; **645** Reza/NGIC, (t)Paul Chesley/Stone, (b)Catherine Karnow/CORBIS; **646** Katsumi Kasahara/AP/Wide World Photos; **648** Christie's Images/CORBIS; **649** Julia Waterlow; Eye Ubiquitous/CORBIS; **650** Peter Essick/NGIC; **651** Cynthia Bell and Melvyn Goldstein; **654** Digital Stock; **655** Dean Conger/CORBIS; **656** Aaron Haupt; **657** Keren Su/China Span; **660** Wolfgang Kaehler; **661** Chad Ehlers/Stone; **664** Peter Essick/NGIC; **665** Christopher Arnesen/Stone; **668** O. Louis Mazzatenta/NGIC; **670** AP/Wide World Photos; **672** Reuters New Media Inc./CORBIS; **672** Dilip Mehta/ Contact Press Images/PictureQuest; **673** Tom Nebbia/CORBIS; (t)Reuters Newmedia Inc./Archive Photos, (c)Archive Photos/ PictureQuest, (b)Eddie Shih/AP/Wide World Photos; **675** AFP/ CORBIS; **676** (l)Keren Su/China Span, (r)Tiziana and Gianni Baldizzone/ CORBIS; **677** (l)Bob Krist/Stone, (r)Reuters NewMedia Inc./CORBIS; **678** Jack Vartoogian; **679** James L. Stanfield/NGIC; **684** Hugh Sitton/Stone; **685** David Samuel Robbins/CORBIS; **686** Yann Layma/Stone; **687** Jodi Cobb/NGIC; **689** Jeff Widener/AP/Wide World Photos; **691** Dean Conger/CORBIS; **692** Jodi Cobb/NGIC; **693** Reuters NewMedia Inc./CORBIS; **694** Keren Su/Stone; **695** James L. Stanfield/NGIC; **697** Katsumi Kasahara/AP/Wide World Photos; **698–699** Bob Sacha; **700** AP/Wide World Photos; **704–705** Denis Waugh/Stone; **706** (l)Penny Tweedie/Stone, (r)Steve McCurry; **707** (t)John William Banagan/The Image Bank/PictureQuest, (b)Bruno Barbey/Magnum/PictureQuest; **708** (l)Paul Chesley/NGIC; (r)Paul Chesley/Stone; **709** (t)Paul Chesley/Stone, (b)R. Ian Lloyd; **715** Hugh Sitton/Stone; **716** (t)Michael Freeman/CORBIS, (b)Bruno Barbey/ Magnum/PictureQuest; **716–717** Macduff Everton/CORBIS; **717** Herb Schmitz/Stone; **718** Keren Su/China Span; **719** George Steinmetz/ NGIC; **720** David Alan Harvey/NGIC; **722** Roger Ressmeyer/CORBIS; **723** James P. Blair/NGIC; **725** Mattia Klum/NGIC; **728** Paul Chesley/Stone; **729** Christophe Loviny/CORBIS; **730** AP/Wide World Photos; **734** Hugh Sitton/Stone; **735** John Elk III; **736** (l)Dave Bartruff/CORBIS, (r)Reinhard Eisele/CORBIS; **739** Paul Chesley/ NGIC; **740** Michael S. Yamashita/NGIC; **741** Kevin R. Morris/CORBIS; **743** Jodi Cobb/NGIC; **745** AP/Wide World Photos; **746** Wilbur E. Garrett; **747** (t)David Alan Harvey, (c)Bettmann/CORBIS, (b)Bettmann/ CORBIS; **748** Yann Arthus-Bertrand/CORBIS; **750** Bob Haddad; **751** David Alan Harvey/NGIC; **752–753** Jodi Cobb/NGIC; **758** Hugh Sitton/Stone; **759** Glen Allison/Stone; **760** Dean Conger/CORBIS; **761** Paul Chesley/NGIC; **762** Albrecht G. Schaefer/CORBIS; **763** Stephanie Maze/NGIC; **764** Steve McCurry/NGIC; **765** Michael K. Nichols/NGIC; **766** Jodi Cobb/NGIC; **767** AP/Wide World Photos; **768** (l)Kevin R. Morris/CORBIS, (r)John Elk III; **769** Digital Stock; **770–771** Michael Yamashita/NGIC; **772** David Doubilet; **773** (t)Howard Hall/HHP, (l)Robert Yin/CORBIS; (r)Michael Yamashita/CORBIS; **775** Brian Vikander/CORBIS; **778–779** K.B.Sandved/Photo Researchers; **780** (l)Nicholas DeVore/Stone, (r)Doug Armand/Stone; **781** (t)Paul Chesley/Stone, (b)Paul A. Souders/CORBIS; **782** (l)Paul Chesley/ Photographers/Aspen/PictureQuest, (r)David Hiser; **783** (t)Sam Abell/NGIC, (b)Peter Arnold, Inc.; **789** Paul Chesley/Stone; **790** (t)Brian Gordan Green, (b)C. & S. Pollitt/ANT Photo Library; **791** (t)Ed Collacott/Stone, (c)Brian Gordan Green, (b)Stuart Westmorland/ CORBIS; **792** Ben Osborne/Stone; **793** Sam Abell/NGIC; **794** David Samuel Robbins/CORBIS; **796** (l)Earl & Nazima Kowall/CORBIS, (r)Darryl Torckler/Stone; **797** Robert Holmes/CORBIS; **799** Maria Stenzel/NGIC; **802** (l)Michael S. Yamashita/CORBIS, (c)Wolfgang Kaehler, (r)Nik Wheeler/CORBIS; **808** U.S. Government Nuclear Regulatory Commission/NGIC; **810** Wolfgang Kaehler; **811** Sam Abell/NGIC; **812** Medford Taylor/NGIC; **814** (l)Digital Stock, (r)R. Ian Lloyd; **816** Randy Olson/NGIC; **817** Penny Tweedie/CORBIS; **820** Peter Skinner/Danita Delimont; **821** George F. Mobley/NGIC; **822** (l)Leonard de Selva/CORBIS, (r)Bettmann/CORBIS; **823** (t)Andrew H. Brown; NGIC; **824** Annie Griffiths Belt/NGIC; **826** Charles & Josette Lenars/CORBIS; **827** Bates Littlehales/NGIC; **828** USGS/EROS; **832** Glen Allison/Stone; **833** Norbert Wu/NGIC; **834** Medford Taylor/NGIC; **836** Wolfgang Kaehler/CORBIS; **838** Cary Wolinsky/ NGIC; **839** (t)NGIC, (b)Charles Philip Canglialosi/CORBIS; **840** (l)CORBIS, (r)Ralph A. Clevenger/CORBIS; **842** Michael Van Woert/ NOAA Corps Photo Collection; **843** (t)Robert W. Ginn/PhotoEdit/ PictureQuest, (l)Don King/The Image Bank/PictureQuest, (r)Maria Stenzel; **845** VCG/FPG.